STUDENT´S SOLUTIONS MANUAL

ELEMENTARY AND INTERMEDIATE ALGEBRA

CONCEPTS AND APPLICATIONS

SEVENTH EDITION

Marvin L. Bittinger

Indiana University Purdue University Indianapolis

David J. Ellenbogen

Community College of Vermont

Barbara L. Johnson

Ivy Tech Community College of Indiana

Reproduced by Pearson from electronic files supplied by the author.

Publishing as Pearson, 501 Boylston Street, Boston, MA 02116.

ISBN-13: 978-0-13-446290-5
ISBN-10: 0-13-446290-4

2 17

www.pearsonhighered.com

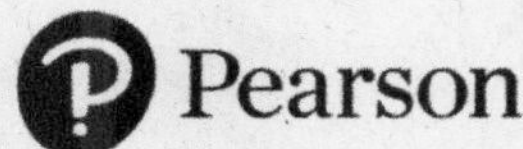

CONTENTS

Chapter 1

Introduction to Algebraic Expressions

Exercise Set 1.1

1. In the expression $4 + x$, the number 4 is a *constant*.

3. To *evaluate* an algebraic expression, we substitute a number for each variable and carry out the operations.

5. $10n - 1$ does not contain an equals sign, so it is an expression.

7. $2x - 5 = 9$ contains an equals sign, so it is an equation.

9. $45 = a - 1$ contains an equals sign, so it is an equation.

11. $2x - 3y = 8$ contains an equals sign, so it is an equation.

13. Substitute 9 for a and multiply.
$5a = 5 \cdot 9 = 45$

15. Substitute 4 for r and subtract.
$12 - 4 = 8$

17. $\frac{a}{b} = \frac{45}{9} = 5$

19. $\frac{x+y}{4} = \frac{2+14}{4} = \frac{16}{4} = 4$

21. $\frac{p-q}{7} = \frac{55-20}{7} = \frac{35}{7} = 5$

23. $\frac{5z}{y} = \frac{5 \cdot 9}{15} = \frac{45}{15} = 3$

25. $$\begin{aligned} bh &= (6 \text{ ft})(4 \text{ ft}) \\ &= (6)(4)(\text{ft})(\text{ft}) \\ &= 24 \text{ ft}^2, \text{ or } 24 \text{ square feet} \end{aligned}$$

27. $$\begin{aligned} A &= \frac{1}{2}bh \\ &= \frac{1}{2}(5 \text{ cm})(6 \text{ cm}) \\ &= \frac{1}{2}(5)(6)(\text{cm})(\text{cm}) \\ &= \frac{5}{2} \cdot 6 \text{ cm}^2 \\ &= 15 \text{ cm}^2, \text{ or } 15 \text{ square centimeters} \end{aligned}$$

29. $$\begin{aligned} A &= bh \\ &= (67 \text{ ft})(12 \text{ ft}) \\ &= (67)(12)(\text{ft})(\text{ft}) \\ &= 804 \text{ ft}^2, \text{ or } 804 \text{ square feet} \end{aligned}$$

31. Let r represent Ron's age. Then we have $r + 5$, or $5 + r$.

33. $6b$, or $b \cdot 6$

35. $c - 9$

37. $6 + q$, or $q + 6$

39. $p - t$

41. $y - x$

43. $x \div w$, or $\frac{x}{w}$

45. Let l and h represent the box's length and height, respectively. Then we have $l + h$, or $h + l$.

47. $9 \cdot 2m$, or $2m \cdot 9$

49. Let y represent "some number." Then we have $\frac{1}{4}y - 13$, or $\frac{y}{4} - 13$.

51. Let a and b represent the two numbers. Then we have $5(a - b)$.

53. Let w represent the number of women attending. Then we have 64% of w, or $0.64w$.

55. Let x represent the number.
Translating:

What number	added to	73	is	201?
↓	↓	↓	↓	↓
x	$+$	73	$=$	201

$x + 73 = 201$

57. Let x represent the number.

Rewording:	42	times	what number	is	2352?
	↓	↓	↓	↓	↓
Translating:	42	$\cdot$	x	$=$	2352

$42x = 2352$

59. Let s represent the number of unoccupied squares.

Rewording:	The number of unoccupied squares	added to	19	is	64.
	↓	↓	↓	↓	↓
Translating:	s	$+$	19	$=$	64

$s + 19 = 64$

61. Let x represent the total amount of waste generated, in millions of tons.

Rewording: $\underbrace{34.5\%}$ of $\underbrace{\text{the total amount of waste}}$ is $\underbrace{\text{87 million tons.}}$

$$\downarrow \quad \downarrow \quad \downarrow \quad \downarrow \quad \downarrow$$

Translating: $34.5\% \quad \cdot \quad x \quad = \quad 87$

$34.5\% \cdot x = 87$, or $0.345x = 87$

63. We look for a pattern in the data. We try subtracting.

$8-3=5$ $\quad 11-6=5$
$9-4=5$ $\quad 12-7=5$
$10-5=5$ $\quad 13-8=5$

The amount is the same, 5, for each pair of numbers. Let a represent the age of the child and f represent the number of grams of dietary fiber.
We reword and translate as follows:

Rewording: $\underbrace{\text{dietary fiber}}$ is $\underbrace{\text{child's age}}$ added to $\underbrace{5}$

$$\downarrow \quad \downarrow \quad \downarrow \quad \downarrow \quad \downarrow$$

Translating: $f \quad = \quad a \quad + \quad 5$

$f = a + 5$

65. We look for a pattern in the data. We try subtracting.

$6.59 - 4.17 = 2.42$
$7.18 - 4.76 = 2.42$
$8.76 - 6.34 = 2.42$

The amount is the same, 2.42, for each pair of numbers. Let n represent the nonmachinable cost and n represent the machinable cost.
We reword and translate as follows:

$\underbrace{\text{nonmachinable cost}}$ is $\underbrace{\text{machinable cost}}$ added to $\underbrace{2.42}$

$$\downarrow \quad \downarrow \quad \downarrow \quad \downarrow \quad \downarrow$$

$n \quad = \quad m \quad + \quad 2.42$

$n = m + 2.42$

67. We look for a pattern in the data. We try dividing.

$\frac{10,000}{1} = 10,000$ $\quad \frac{30,000}{3} = 10,000$
$\frac{20,000}{2} = 10,000$ $\quad \frac{40,000}{4} = 10,000$

The amount is the same, 10,000, for each pair of numbers. Let v represent the number of vehicle miles traveled and d represent the number drivers.
We reword and translate as follows:

Rewording: $\underbrace{\text{number of miles traveled}}$ is $\underbrace{10,000}$ times $\underbrace{\text{number of drivers}}$

$$\downarrow \quad \downarrow \quad \downarrow \quad \downarrow \quad \downarrow$$

Translating: $v \quad = 10,000 \quad \cdot \quad d$

$v = 10,000d$

69. The sum of two numbers m and n is $m+n$, and twice the sum is $2(m+n)$.
Choice (f) is the correct answer.

71. Twelve more than a number t is $t+12$. If this expression is equal to 5, we have the equation $t+12=5$. Choice (d) is the correct answer.

73. The sum of a number t and 5 is $t+5$, and 3 times the sum is $3(t+5)$. Choice (g) is the correct answer.

75. The product of two numbers a and b is ab, and 1 less than this product is $ab-1$. If this expression is equal to 48, we have the equation $ab-1=48$. Choice (e) is the correct answer.

77. *Writing Exercise.*

79. *Writing Exercise.*

81. Area of sign: $A = \frac{1}{2}(3\text{ ft})(2.5\text{ ft}) = 3.75\text{ ft}^2$
Cost of sign: $\$120(3.75) = \$450$

83. When x is twice y, then y is one-half x, so $y = \frac{12}{2} = 6$.

$$\frac{x-y}{3} = \frac{12-6}{3} = \frac{6}{3} = 2$$

85. When a is twice b, then b is one-half a, so $b = \frac{16}{2} = 8$.

$$\frac{a+b}{4} = \frac{16+8}{4} = \frac{24}{4} = 6$$

87. The next whole number is one more than $w+3$:
$w+3+1 = w+4$

89. $l+w+l+w$, or $2l+2w$

91. If t is Molly's race time, then Dion's race time is $t+3$ and Ellie's race time is
$t+3+5 = t+8$.

93. *Writing Exercise.*

Exercise Set 1.2

1. *Equivalent* expressions represent the same number.

3. The result of addition is called a *sum*.

5. Commutative

7. Distributive

9. Commutative

11. $t+11$ Changing the order

13. $8x+4$

15. $3y+9x$

17. $5(1+a)$

19. $x \cdot 7$ Changing the order

21. ts

23. $5+ba$

25. $(a+1)5$

27. $x+(8+y)$

29. $(u+v)+7$

31. $ab+(c+d)$

33. $10(xy)$

35. $(2a)b$

37. $(3\cdot 2)(a+b)$

39. $s+(t+6)=(s+t)+6$
$=(t+6)+s$

41. $(17a)b=17(ab)$ Using the associative law
$(17a)b=b(17a)$ Using the commutative law
$=b(a17)$ Using the commutative law again
Answers may vary.

43. $(1+x)+2=(x+1)+2$ Commutative law
$=x+(1+2)$ Associative law
$=x+3$ Simplifying

45. $(m\cdot 3)7=m(3\cdot 7)$ Associative law
$=m\cdot 21$ Simplifying
$=21m$ Commutative law

47. $x+xyz+1$
The term are separated by plus signs. They are x, xyz, and 1.

49. $2a+\frac{a}{3b}+5b$
The terms are separated by plus signs. They are $2a$, $\frac{a}{3b}$, and $5b$.

51. $4x, 4y$

53. $5n=5\cdot n$
The factors are 5 and n.

55. $3(x+y)=3\cdot(x+y)$
The factors are 3 and $(x+y)$.

57. The factors are 7, a and b.

59. $(a-b)(x-y)=(a-b)\cdot(x-y)$
The factors are $(a-b)$ and $(x-y)$.

61. $2(x+15)=2\cdot x+2\cdot 15=2x+30$

63. $4(1+a)=4\cdot 1+4\cdot a=4+4a$

65. $10(9x+6)=10\cdot 9x+10\cdot 6=90x+60$

67. $5(r+2+3t)=5\cdot r+5\cdot 2+5\cdot 3t=5r+10+15t$

69. $(a+b)2=a(2)+b(2)=2a+2b$

71. $(x+y+2)5=x(5)+y(5)+2(5)=5x+5y+10$

73. $2\cdot a+2\cdot b$ Using the distributive law
$=2(a+b)$ The common factor is 2.
Check: $2(a+b)=2\cdot a+2\cdot b=2a+2b$

75. $7+7y=7\cdot 1+7\cdot y$ The common factor is 7.
$=7(1+y)$ Using the distributive law
Check: $7(1+y)=7\cdot 1+7\cdot y=7+7y$

77. $32x+2=2\cdot 16x+2\cdot 1=2(16x+1)$
Check: $2(16x+1)=2\cdot 16x+2\cdot 1=32x+2$

79. $5x+10+15y=5\cdot x+5\cdot 2+5\cdot 3y=5(x+2+3y)$
Check: $5(x+2+3y)=5\cdot x+5\cdot 2+5\cdot 3y$
$=5x+10+15y$

81. $7a+35b=7\cdot a+7\cdot 5b=7(a+5b)$
Check: $7(a+5b)=7\cdot a+7\cdot 5b=7a+35b$

83. $44x+11y+22z=11\cdot 4x+11\cdot y+11\cdot 2z$
$=11(4x+y+2z)$
Check: $11(4x+y+2z)=11\cdot 4x+11\cdot y+11\cdot 2z$
$=44x+11y+22z$

85. $3(2+x)$
$=3(x+2)$ Commutative law of addition
$=3\cdot x+3\cdot 2$ Distributive law
$=3x+6$ Multiplying

87. $7(2x+3y)$
$=7(2x)+7(3y)$ Distributive law
$=(7\cdot 2)x+(7\cdot 3)y$ Associative law of multiplication
$=14x+21y$ Multiplying

89. *Writing Exercise.*

91. *Writing Exercise.*

93. $[2(x+1)]+3x$
$=[2\cdot x+2\cdot 1)]+3x$ Distributive law
$=[2x+2]+3x$ Multiplying
$=2x+[2+3x]$ Associative law of addition
$=2x+[3x+2]$ Commutative law of addition
$=[2x+3x]+2$ Associative law of addition
$=[(2+3)x]+2$ Distributive law
$=[5x]+2$ Adding
$=5x+2$

95. The expressions are equivalent by the distributive law.
$8+4(a+b)=8+4a+4b=4(2+a+b)$

97. The expressions are not equivalent.
Let $m=1$. Then we have:

$7\div 3\cdot 1=\frac{7}{3}\cdot 1=\frac{7}{3}$, but
$1\cdot 3\div 7=3\div 7=\frac{3}{7}$.

99. The expressions are not equivalent.
Let $x = 1$ and $y = 0$. Then we have:
$30 \cdot 0 + 1 \cdot 15 = 0 + 15 = 15$, but
$5[2(1 + 3 \cdot 0)] = 5[2(1)] = 5 \cdot 2 = 10$.

101. *Writing Exercise.*

103. Answers may vary.
a. Let x represent the number of overtime hours worked in one week.
Aidan: $10(1.5x + 40)$
Beth: $10 \cdot 40 + 10(1.5)x$
Cody: $15x + 400$
b. We simplify each expression.
Aiden: $10(1.5x + 40) = 10(1.5x) + 10(40)$
$= 15x + 400$
Beth: $10 \cdot 40 + 10(1.5)x = 400 + 15x$
Cody: $15x + 400$
All expressions are equivalent to $15x + 400$.

Exercise Set 1.3

1. The top number in a fraction is called the *numerator*.

3. To divide two fractions, multiply by the *reciprocal* of the divisor.

5. Since $35 = 5 \cdot 7$, choice (b) is correct.

7. Since 65 is an odd number and has more than two different factors, choice (d) is correct.

9. 9 is composite because it has more than two different factors. They are 1, 3, and 9.

11. 41 is prime because it has only two different factors, 41 and 1.

13. 77 is composite because it has more than two different factors. They are 1, 7, 11, and 77.

15. 2 is prime because it has only two different factors, 2 and 1.

17. The terms "prime" and "composite" apply only to natural numbers. Since 0 is not a natural number, it is neither prime nor composite.

19. Factorizations:
$1 \cdot 50, 2 \cdot 25, 5 \cdot 10$
List all of the factors of 50:
1, 2, 5, 10, 25, 50

21. Factorizations:
$1 \cdot 42$, $2 \cdot 21$, $3 \cdot 14$, $6 \cdot 7$
List all of the factors of 42:
1, 2, 3, 6, 7, 14, 21, 42

23. $39 = 3 \cdot 13$

25. We begin factoring 30 in any way that we can and continue factoring until each factor is prime.
$30 = 2 \cdot 15 = 2 \cdot 3 \cdot 5$

27. We begin by factoring 27 in any way that we can and continue factoring until each factor is prime.
$27 = 3 \cdot 9 = 3 \cdot 3 \cdot 3$

29. We begin by factoring 150 in any way that we can and continue factoring until each factor is prime.
$150 = 2 \cdot 75 = 2 \cdot 3 \cdot 25 = 2 \cdot 3 \cdot 5 \cdot 5$

31. 31 has exactly two different factors, 31 and 1. Thus, 31 is prime.

33. $210 = 2 \cdot 105 = 2 \cdot 3 \cdot 35 = 2 \cdot 3 \cdot 5 \cdot 7$

35. $115 = 5 \cdot 23$

37. $\frac{21}{35} = \frac{7 \cdot 3}{7 \cdot 5}$ Factoring numerator and denominator
$= \frac{7}{7} \cdot \frac{3}{5}$ Rewriting as a product of two fractions
$= 1 \cdot \frac{3}{5}$ $\quad \frac{7}{7} = 1$
$= \frac{3}{5}$ Using the identity property of 1

39. $\frac{16}{56} = \frac{2 \cdot 8}{7 \cdot 8} = \frac{2}{7} \cdot \frac{8}{8} = \frac{2}{7} \cdot 1 = \frac{2}{7}$

41. $\frac{12}{48} = \frac{1 \cdot 12}{4 \cdot 12}$ Factoring and using the identity property of 1 to to write 12 as $1 \cdot 12$
$= \frac{1}{4} \cdot \frac{12}{12}$
$= \frac{1}{4} \cdot 1 = \frac{1}{4}$

43. $\frac{52}{13} = \frac{13 \cdot 4}{13 \cdot 1} = 1 \cdot \frac{4}{1} = 4$

45. $\frac{19}{76} = \frac{1 \cdot 19}{4 \cdot 19}$ Factoring and using the identity property of 1 to write 19 as $1 \cdot 19$
$= \frac{1}{4} \cdot \frac{\cancel{19}}{\cancel{19}}$ Removing a factor equal to 1: $\frac{19}{19} = 1$
$= \frac{1}{4}$

47. $\frac{150}{25} = \frac{6 \cdot 25}{1 \cdot 25}$ Factoring and using the identity property of 1 to write 25 as $1 \cdot 25$
$= \frac{6}{1} \cdot \frac{\cancel{25}}{\cancel{25}}$ Removing a factor equal to 1: $\frac{25}{25} = 1$
$= \frac{6}{1}$
$= 6$ Simplifying

49. $\frac{42}{50} = \frac{2 \cdot 21}{2 \cdot 25}$ Factoring the numerator and the denominator
$= \frac{\cancel{2} \cdot 21}{\cancel{2} \cdot 25}$ Removing a factor equal to 1: $\frac{2}{2} = 1$
$= \frac{21}{25}$

51. $\frac{120}{82}=\frac{2\cdot 60}{2\cdot 41}$ Factoring

$=\frac{\cancel{2}\cdot 60}{\cancel{2}\cdot 41}$ Removing a factor equal to 1: $\frac{2}{2}=1$

$=\frac{60}{41}$

53. $\frac{210}{98}=\frac{2\cdot 7\cdot 15}{2\cdot 7\cdot 7}$ Factoring

$=\frac{\cancel{2}\cdot\cancel{7}\cdot 15}{\cancel{2}\cdot\cancel{7}\cdot 7}$ Removing a factor equal to 1: $\frac{2\cdot 7}{2\cdot 7}=1$

$=\frac{15}{7}$

55. $\frac{1}{2}\cdot\frac{3}{5}=\frac{1\cdot 3}{2\cdot 5}$ Multiplying numerators and denominators

$=\frac{3}{10}$

57. $\frac{9}{2}\cdot\frac{4}{3}=\frac{9\cdot 4}{2\cdot 3}=\frac{3\cdot\cancel{3}\cdot\cancel{2}\cdot 2}{\cancel{2}\cdot\cancel{3}}=6$

59. $\frac{1}{8}+\frac{3}{8}=\frac{1+3}{8}$ Adding numerators; keeping the common denominator

$=\frac{4}{8}$

$=\frac{1\cdot\cancel{4}}{2\cdot\cancel{4}}=\frac{1}{2}$ Simplifying

61. $\frac{4}{9}+\frac{13}{18}=\frac{4}{9}\cdot\frac{2}{2}+\frac{13}{18}$ Using 18 as the common denominator

$=\frac{8}{18}+\frac{13}{18}$

$=\frac{21}{18}$

$=\frac{7\cdot\cancel{3}}{6\cdot\cancel{3}}=\frac{7}{6}$ Simplifying

63. $\frac{3}{a}\cdot\frac{b}{7}=\frac{3b}{7a}$ Multiplying numerators and denominators

65. $\frac{4}{n}+\frac{6}{n}=\frac{10}{n}$ Adding numerators; keeping the common denominator

67. $\frac{3}{10}+\frac{8}{15}=\frac{3}{10}\cdot\frac{3}{3}+\frac{8}{15}\cdot\frac{2}{2}$ Using 30 as the common denominator

$=\frac{9}{30}+\frac{16}{30}$

$=\frac{25}{30}$

$=\frac{5\cdot\cancel{5}}{6\cdot\cancel{5}}=\frac{5}{6}$ Simplifying

69. $\frac{11}{7}-\frac{4}{7}=\frac{7}{7}=1$

71. $\frac{13}{18}-\frac{4}{9}=\frac{13}{18}-\frac{4}{9}\cdot\frac{2}{2}$ Using 18 as the common denominator

$=\frac{13}{18}-\frac{8}{18}$

$=\frac{5}{18}$

73. $\frac{11}{30}-\frac{2}{9}=\frac{11}{30}\cdot\frac{3}{3}-\frac{2}{9}\cdot\frac{10}{10}$ Using 90 as the common denominator

$=\frac{33}{90}-\frac{20}{90}$

$=\frac{13}{90}$

75. $\frac{7}{6}\div\frac{3}{5}=\frac{7}{6}\cdot\frac{5}{3}$ Multiplying by the reciprocal of the divisor

$=\frac{35}{18}$

77. $12\div\frac{4}{9}=\frac{12}{1}\cdot\frac{9}{4}=\frac{\cancel{4}\cdot 3\cdot 9}{1\cdot\cancel{4}}=27$

79. Note that we have a number divided by itself. Thus, the result is 1. We can also do this exercise as follows:

$\frac{7}{13}\div\frac{7}{13}=\frac{7}{13}\cdot\frac{13}{7}=\frac{7\cdot 13}{7\cdot 13}=1$

81. $\frac{\frac{2}{7}}{\frac{5}{3}}=\frac{2}{7}\div\frac{5}{3}=\frac{2}{7}\cdot\frac{3}{5}=\frac{2\cdot 3}{7\cdot 5}=\frac{6}{35}$

83. $\frac{9}{\frac{1}{2}}=9\div\frac{1}{2}=\frac{9}{1}\cdot\frac{2}{1}=\frac{9\cdot 2}{1\cdot 1}=18$

85. *Writing Exercise.*

87. *Writing Exercise*

89.

Product	56	63	36	72	140	96	168
Factor	7	7	2	36	14	8	8
Factor	8	9	18	2	10	12	21
Sum	15	16	20	38	24	20	29

91. $\frac{16\cdot 9\cdot 4}{15\cdot 8\cdot 12}=\frac{\cancel{4}\cdot\cancel{4}\cdot\cancel{3}\cdot\cancel{3}\cdot\cancel{2}\cdot 2}{\cancel{3}\cdot 5\cdot\cancel{2}\cdot\cancel{4}\cdot\cancel{3}\cdot\cancel{4}}=\frac{2}{5}$

93. $\frac{45pqrs}{9prst}=\frac{5\cdot\cancel{9}\cdot\cancel{p}\cdot q\cdot\cancel{r}\cdot\cancel{s}}{\cancel{9}\cdot\cancel{p}\cdot\cancel{r}\cdot\cancel{s}\cdot t}=\frac{5q}{t}$

95. $\frac{15\cdot 4xy\cdot 9}{6\cdot 25x\cdot 15y}=\frac{\cancel{15}\cdot\cancel{2}\cdot 2\cdot\cancel{x}\cdot\cancel{y}\cdot\cancel{3}\cdot 3}{\cancel{2}\cdot\cancel{3}\cdot 25\cdot\cancel{x}\cdot\cancel{15}\cdot\cancel{y}}=\frac{6}{25}$

97. $\frac{\frac{27ab}{15mn}}{\frac{18bc}{25np}}=\frac{27ab}{15mn}\div\frac{18bc}{25np}=\frac{27ab}{15mn}\cdot\frac{25np}{18bc}$

$=\frac{27ab\cdot 25np}{15mn\cdot 18bc}=\frac{\cancel{3}\cdot\cancel{9}\cdot a\cdot\cancel{b}\cdot\cancel{5}\cdot 5\cdot\cancel{n}\cdot p}{\cancel{3}\cdot\cancel{5}\cdot m\cdot\cancel{n}\cdot 2\cdot\cancel{9}\cdot\cancel{b}\cdot c}$

$=\frac{5ap}{2cm}$

99. $\dfrac{5\frac{3}{4}rs}{4\frac{1}{2}st} = \dfrac{\frac{23}{4}rs}{\frac{9}{2}st} = \dfrac{\frac{23rs}{4}}{\frac{9st}{2}} = \dfrac{23rs}{4} \div \dfrac{9st}{2}$

$= \dfrac{23rs}{4} \cdot \dfrac{2}{9st} = \dfrac{23rs \cdot 2}{4 \cdot 9st}$

$= \dfrac{23 \cdot r \cdot \cancel{s} \cdot \cancel{2}}{\cancel{2} \cdot 2 \cdot 9 \cdot \cancel{s} \cdot t} = \dfrac{23r}{18t}$

101. $A = lw = \left(\frac{4}{5}\text{ m}\right)\left(\frac{7}{9}\text{ m}\right)$

$= \left(\frac{4}{5}\right)\left(\frac{7}{9}\right)(\text{m})(\text{m})$

$= \frac{28}{45}\text{m}^2$, or $\frac{28}{45}$ square meters

103. $P = 4s = 4\left(3\frac{5}{9}\text{ m}\right) = 4 \cdot \frac{32}{9}\text{ m} = \frac{128}{9}\text{ m}$,

or $14\frac{2}{9}$m

105. There are 12 edges, each with length $2\frac{3}{10}$ cm. We multiply to find the total length of the edges.

$12 \cdot 2\frac{3}{10}\text{ cm} = 12 \cdot \frac{23}{10}\text{ cm}$

$= \frac{12 \cdot 23}{10}\text{ cm}$

$= \frac{\cancel{2} \cdot 6 \cdot 23}{\cancel{2} \cdot 5}\text{ cm}$

$= \frac{138}{5}\text{ cm, or } 27\frac{3}{5}\text{ cm}$

Exercise Set 1.4

1. Since $\frac{3}{20} = 0.15$, we can write $\frac{3}{20}$ as a *terminating* decimal.

3. 0 is the only *whole number* that is not a natural number.

5. The *opposite* of 1 is –1.

7. $-n$ is the opposite of n.

9. $-10 < x$

11. The real number –9500 corresponds to borrowing $9500. The real number 5000 corresponds to the award of $5000.

13. The real number 100 corresponds to 100°F. The real number –80 corresponds to 80°F below zero.

15. The real number –777.68 corresponds to a 777.68-point fall. The real number 936.42 corresponds to a 936.42-point gain.

17. The real number 8 corresponds to an 8-yd gain, and the real number –5 corresponds to a 5-yd loss.

19.

–5 –4 –3 –2 –1 0 1 2 3 4 5

21. The graph of –4.3 is $\frac{3}{10}$ of a unit to the left of –4.

–4.3

–5 –4 –3 –2 –1 0 1 2 3 4 5

23. Since $\frac{10}{3} = 3\frac{1}{3}$, its graph is $\frac{1}{3}$ of a unit to the right of 3.

$\frac{10}{3}$

–5 –4 –3 –2 –1 0 1 2 3 4 5

25. From lowest to highest: points-per-game

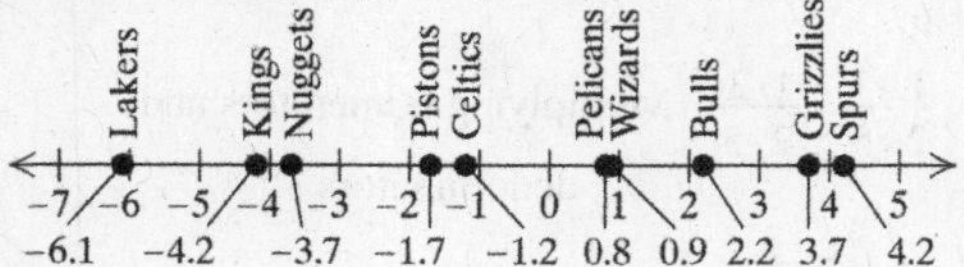

27. $\frac{7}{8}$ means $7 \div 8$, so we divide.

$$\begin{array}{r} 0.875 \\ 8\overline{)7.000} \\ \underline{6\,4} \\ 60 \\ \underline{56} \\ 40 \\ \underline{40} \\ 0 \end{array}$$

We have $\frac{7}{8} = 0.875$.

29. We first find decimal notation for $\frac{3}{4}$. Since $\frac{3}{4}$ means $3 \div 4$, we divide.

$$\begin{array}{r} 0.75 \\ 4\overline{)3.00} \\ \underline{2\,8} \\ 20 \\ \underline{20} \\ 0 \end{array}$$

Thus, $\frac{3}{4} = 0.75$, so $-\frac{3}{4} = -0.75$.

31. $\frac{7}{6}$ means $7 \div 6$, so we divide.

$$\begin{array}{r} 1.166\ldots \\ 6\overline{)7.000} \\ \underline{6} \\ 10 \\ \underline{6} \\ 40 \\ \underline{36} \\ 40 \end{array}$$

Thus $\frac{7}{6} = 1.1\overline{6}$, so $-\frac{7}{6} = -1.1\overline{6}$.

33. $\frac{2}{3}$ means $2 \div 3$, so we divide.

$$\begin{array}{r} 0.666\ldots \\ 3\overline{)2.000} \\ \underline{1\,8} \\ 20 \\ \underline{18} \\ 20 \\ \underline{18} \\ 2 \end{array}$$

We have $\frac{2}{3} = 0.\overline{6}$.

35. We first find decimal notation for $\frac{1}{2}$. Since $\frac{1}{2}$ means $1 \div 2$, we divide.

$$\begin{array}{r} 0.5 \\ 2\overline{)1.0} \\ \underline{1\,0} \\ 0 \end{array}$$

Thus, $\frac{1}{2} = 0.5$, so $-\frac{1}{2} = -0.5$.

37. Since the denominator is 100, we know that $\frac{13}{100} = 0.13$. We could also divide 13 by 100 to find this result.

39.

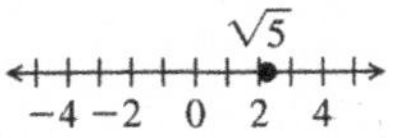

41.

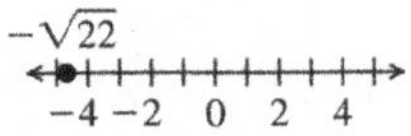

43. Since 5 is to the right of 0, we have $5 > 0$.

45. Since –9 is to the left of 9, we have $-9 < 9$.

47. Since –8 is to the left of –5, we have $-8 < -5$.

49. Since –5 is to the right of –11, we have $-5 > -11$.

51. Since -12.5 is to the left of –10.2, we have $-12.5 < -10.2$.

53. We convert to decimal notation. $\frac{5}{12} = 0.41\overline{6}$ and $\frac{11}{25} = 0.44$. Thus, $\frac{5}{12} < \frac{11}{25}$.

55. $-2 > x$ has the same meaning as $x < -2$.

57. $10 \le y$ has the same meaning as $y \ge 10$.

59. $-83,\ -4.7,\ 0,\ \frac{5}{9},\ 2.\overline{16},\ 62$

61. –83, 0, 62

63. All are real numbers.

65. $|-58| = 58$ since –58 is 58 units from 0.

67. $|-12.2| = 12.2$ since –12.2 is 12.2 units from 0.

69. $|\sqrt{2}| = \sqrt{2}$ since $\sqrt{2}$ is $\sqrt{2}$ units from 0.

71. $\left|-\frac{9}{7}\right| = \frac{9}{7}$ since $-\frac{9}{7}$ is $\frac{9}{7}$ units from 0.

73. $|0| = 0$ since 0 is 0 units from itself.

75. $|x| = |-8| = 8$

77. *Writing Exercise.*

79. *Writing Exercise.*

81. *Writing Exercise.*

83. List the numbers as they occur on the number line, from left to right: –23, –17, 0, 4

85. Converting to decimal notation, we can write $\frac{4}{5}, \frac{4}{3}, \frac{4}{8}, \frac{4}{6}, \frac{4}{9}, \frac{4}{2}, -\frac{4}{3}$ as 0.8, $1.3\overline{3}$, 0.5, $0.6\overline{6}$, $0.4\overline{4}$, 2, $-1.3\overline{3}$, respectively. List the numbers (in fractional form) as they occur on the number line, from left to right:

$$-\frac{4}{3}, \frac{4}{9}, \frac{4}{8}, \frac{4}{6}, \frac{4}{5}, \frac{4}{3}, \frac{4}{2}$$

87. $|4| = 4$ and $|-7| = 7$, so $|4| < |-7|$.

89. $|23| = 23$ and $|-23| = 23$, so $|23| = |-23|$.

91. x represents an integer whose distance from 0 is less than 3 units. Thus, $x = -2, -1, 0, 1, 2$.

93. $0.1\overline{1} = \frac{0.3\overline{3}}{3} = \frac{\frac{1}{3}}{3} = \frac{1}{3} \cdot \frac{1}{3} = \frac{1}{9}$

95. $5.5\overline{5} = 50(0.1\overline{1}) = 50 \cdot \frac{1}{9} = \frac{50}{9}$

(See Exercise 93.)

97. $a < 0$

99. $|x| \le 10$

101. *Writing Exercise.*

Mid-Chapter Review

1.
$$\begin{aligned} \frac{x-y}{3} &= \frac{22-10}{3} && \text{Substituting} \\ &= \frac{12}{3} && \text{Subtracting} \\ &= 4 && \text{Dividing} \end{aligned}$$

2.
$$\begin{aligned} 14x + 7 &= 7 \cdot 2x + 7 \cdot 1 \\ &= 7(2x+1) \end{aligned}$$

3.
$$\begin{aligned} x + y &= 3 + 12 \quad \text{Substituting} \\ &= 15 \end{aligned}$$

4. $\frac{2a}{5} = \frac{2 \cdot 10}{5} = 4$

5. $d - 10$

6. Let h represent the number of hours worked; $8h$

7. Let s represent the number of students originally enrolled.
$s - 5 = 27$

8. No; $13t = 13 \cdot 8 = 104 \neq 94$

9. $10x + 7$

10. $(3a)b$

11. $4(2x+8) = 8x + 32$

12. $3(2m+5n+10) = 6m + 15n + 30$

13. $18x + 15 = 3(6x) + 3(5) = 3(6x+5)$

14. $9c + 12d + 3 = 3(3c + 4d + 1)$

15. $84 = 2 \cdot 2 \cdot 3 \cdot 7$

16. $\frac{135}{315} = \frac{\cancel{3} \cdot \cancel{3} \cdot 3 \cdot \cancel{5}}{\cancel{3} \cdot \cancel{3} \cdot \cancel{5} \cdot 7} = \frac{3}{7}$

17. $\frac{11}{12} - \frac{3}{8} = \frac{22}{24} - \frac{9}{24} = \frac{13}{24}$

18. $\frac{8}{15} \div \frac{6}{11} = \frac{8}{15} \cdot \frac{11}{6} = \frac{\cancel{2} \cdot 2 \cdot 2 \cdot 11}{3 \cdot 5 \cdot \cancel{2} \cdot 3} = \frac{44}{45}$

19. Graph –2.5.

−2.5

−4 −2 0 2 4

20.
$$\begin{array}{r} 0.15 \\ 20\overline{)3.00} \\ \underline{20} \\ 100 \\ \underline{100} \\ 0 \end{array}$$

$\frac{3}{20} = 0.15$, so $-\frac{3}{20} = -0.15$.

21. Since –16 is to the right of –24, we have $-16 > -24$.

22. $-\frac{3}{22} \approx -0.13636$ and $-\frac{2}{15} \approx -0.13333$.

Thus, $-\frac{3}{22} < -\frac{2}{15}$

23. $9 \le x$ has the same meaning as $x \ge 9$.

24. $|-5.6| = 5.6$

25. 0

Exercise Set 1.5

1. To add –3 + (–6), *add* 3 and 6 and make the answer *negative*.

3. To add –11 + 5, *subtract* 5 from 11 and make the answer *negative*.

5. The addition –7 + 0 = –7 illustrates the *identity* property of 0.

7. Choice (f), –3*n*, has the same variable factor as 8*n*.

9. Choice (e), 9, is a constant as is 43.

11. Choice (b), 5*x*, has the same variable factor as –2*x*.

13. Start at 5. Move 8 units to the left.

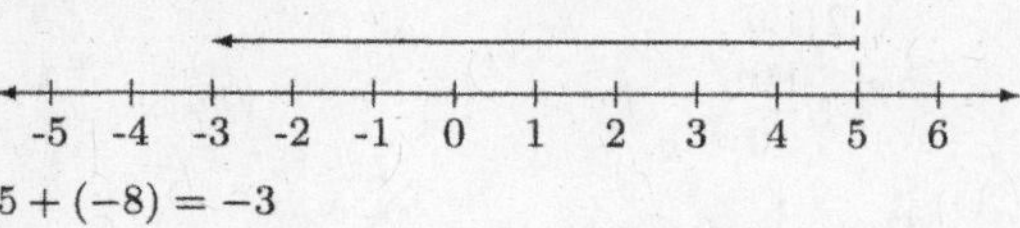

$5 + (-8) = -3$

15. Start at –6. Move 10 units to the right.

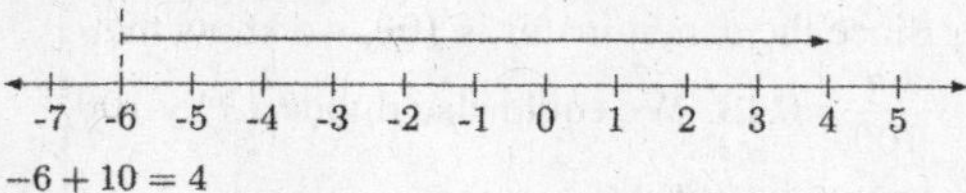

$-6 + 10 = 4$

17. Start at –7. Move 0 units.

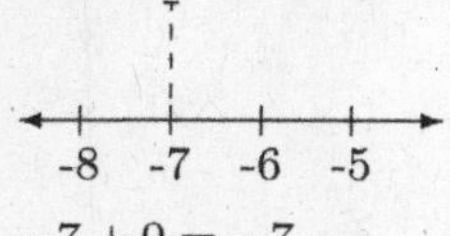

$-7 + 0 = -7$

19. Start at –3. Move 5 units to the left.

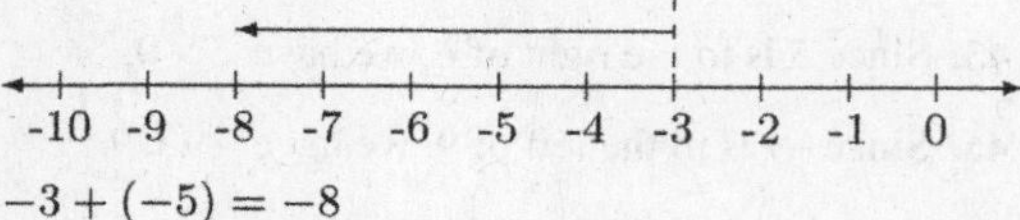

$-3 + (-5) = -8$

21. $-6 + (-5)$ Two negatives. Add the absolute values, getting 11. Make the answer negative.
$-6 + (-5) = -11$

23. $10 + (-15)$ The absolute values are 10 and 15. The difference is $15 - 10$, or 5. The negative number has the larger absolute value, so the answer is negative.
$10 + (-15) = -5$

25. $12 + (-12)$ The numbers have the same absolute value. The sum is 0.
$12 + (-12) = 0$

27. $-24 + (-17)$ Two negatives. Add the absolute values, getting 41. Make the answer negative.
$-24 + (-17) = -41$

29. $-13 + 13$ The numbers have the same absolute value. The sum is 0.
$-13 + 13 = 0$

31. $20+(-11)$ The absolute values are 20 and 11. The difference is $20-11$, or 9. The positive number has the larger absolute value, so the answer is positive.
$20+(-11)=9$

33. $-36+0$ One number is 0. The answer is the other number.
$-36+0=-36$

35. $-3+14$ The absolute values are 3 and 14. The difference is $14-3$, or 11. The positive number has the larger absolute value, so the answer is positive.
$-3+14=11$

37. $-24+(-19)$ Two negatives. Add the absolute values, getting 43. Make the answer negative.
$-24+(-19)=-43$

39. $19+(-19)$ The numbers have the same absolute value. The sum is 0.
$19+(-19)=0$

41. $23+(-5)$ The absolute values are 23 and 5. The difference is $23-5$ or 18. The positive number has the larger absolute value, so the answer is positive.
$23+(-5)=18$

43. $69+(-85)$ The absolute values are 69 and 85. The difference is $85-69$, or 16. The negative number has the larger absolute value, so the answer is negative.
$69+(-85)=-16$

45. $-3.6+2.8$ The absolute values are 3.6 and 2.8. The difference is $3.6-2.8$, or 0.8. The negative number has the larger absolute value, so the answer is negative.
$-3.6+2.8=-0.8$

47. $-5.4+(-3.7)$ Two negatives. Add the absolute values, getting 9.1. Make the answer negative.
$-5.4+(-3.7)=-9.1$

49. $\frac{4}{5}+\left(-\frac{1}{5}\right)$ The absolute values are $\frac{4}{5}$ and $\frac{1}{5}$. The positive number has the larger absolute value, so the answer is positive. $\frac{4}{5}+\left(-\frac{1}{5}\right)=\frac{3}{5}$

51. $\frac{-4}{7}+\frac{-2}{7}$ Two negatives. Add the absolute values, getting $\frac{6}{7}$. Make the answer negative.
$\frac{-4}{7}+\frac{-2}{7}=\frac{-6}{7}$

53. $-\frac{2}{5}+\frac{1}{3}$ The absolute values are $\frac{2}{5}$ and $\frac{1}{3}$. The difference is $\frac{6}{15}-\frac{5}{15}$, or $\frac{1}{15}$. The negative number has the larger absolute value, so the answer is negative. $-\frac{2}{5}+\frac{1}{3}=-\frac{1}{15}$

55. $\frac{-4}{9}+\frac{2}{3}$ The absolute values are $\frac{4}{9}$ and $\frac{2}{3}$. The difference is $\frac{6}{9}-\frac{4}{9}$, or $\frac{2}{9}$. The positive number has the larger absolute value, so the answer is positive.
$\frac{-4}{9}+\frac{2}{3}=\frac{2}{9}$

57. $35+(-14)+(-19)+(-5)$
$=35+[(-14)+(-19)+(-5)]$ Using the associative law of addition
$=35+(-38)$ Adding the negatives
$=-3$ Adding a positive and a negative

59. $-4.9+8.5+4.9+(-8.5)$
Note that we have two pairs of numbers with different signs and the same absolute value: -4.9 and 4.9, 8.5, and -8.5. The sum of each pair is 0, so the result is $0+0$, or 0.

61. Rewording: First decrease plus second decrease plus first increase is change in price.

Translating: $(-15¢) + (-3¢) + 17¢ =$ change in price

Since $-15+(-3)+17=-18+17$
$=-1$,
the price dropped 1¢ during the given period.

63. Rewording: July bill plus payment plus August charges is new balance.

Translating: $-82 + 50 + (-63) =$ new balance

Since $-82+50+(-63)=-32+(-63)$
$=-95$,
Chloe's new balance was \$95.

65. Rewording: First try yardage plus second try yardage plus third try yardage is total gain or loss.

Translating: $(-13) + 12 + 21 =$ total gain or loss

Since $(-13)+12+21$
$=-1+21$
$=20$,
the total gain was 20 yd.

67. Rewording: first drop plus rise plus

Translating: $\left(-\frac{2}{5}\right) \quad + \quad \frac{9}{10} \quad +$

drop is total level change.

$\left(-\frac{6}{5}\right) \quad = \quad$ total level change

Since $-\frac{2}{5}+\frac{9}{10}+\left(-\frac{6}{5}\right)=-\frac{4}{10}+\frac{9}{10}+\left(-\frac{12}{10}\right)$
$=\frac{5}{10}-\frac{12}{10},$
$=-\frac{7}{10},$

The lake dropped by $\frac{7}{10}$ ft.

69. Rewording: Base plus height is peak

Translating: $-19{,}864 \quad + \quad 33{,}480 \quad = \quad$ peak

Since $-19{,}684+33{,}480=13{,}796,$ the elevation of the peak is 13,796 ft above sea level.

71. $7a+10a=(7+10)a$ Using the distributive law
$=17a$

73. $-3x+12x=(-3+12)x$ Using the distributive law
$=9x$

75. $7m+(-9m)=[7+(-9)]m=-2m$

77. $-8y+(-2y)=[-8+(-2)]y=-10y$

79. $-3+8x+4+(-10x)$
$=-3+4+8x+(-10x)$ Using the commutative law of addition
$=(-3+4)+[8+(-10)]x$ Using the distributive law
$=-2x+1$ Adding

81. $6m+9n+(-9n)+(-10m)$
$=9n+(-9n)+6m+(-10m)$ Using the commutative law of addition
$=[9+(-9)]n+[6+(-10)]m$ Using the distributive law
$=-4m$ Adding

83. $-4x+6.3+(-x)+(-10.2)$
$=-4x+(-x)+6.3+(-10.2)$ Using the commutative law of addition
$=[-4+(-1)]x+6.3+(-10.2)$ Using the distributive law
$=-5x-3.9$ Adding

85. Perimeter $=8+5x+9+7x$
$=8+9+5x+7x$
$=(8+9)+(5+7)x$
$=17+12x$

87. Perimeter $=3t+3r+7+5t+9+4r$
$=3t+5t+3r+4r+7+9$
$=(3+5)t+(3+4)r+(7+9)$
$=8t+7r+16$

89. Perimeter $=9+6n+7+8n+4n$
$=9+7+6n+8n+4n$
$=(9+7)+(6+8+4)n$
$=16+18n$

91. *Writing Exercise.*

93. *Writing Exercise.*

95. Starting with the final value, we "undo" the deposit and original amount by adding their opposites. The result is the amount of the check.

Rewording: Final value plus opposite of deposit plus

Translating: $-\$42.37 \quad + \quad (-\$152) \quad +$

opposite of original amount is check amount.

$-\$257.33 \quad = \quad$ check amount.

Since $-42.37+(-152)+(-257.33)$
$=(-194.37)+(-257.33)$
$=-451.70,$

So the amount of the check was \$451.70.

97. $4x+\_\_+(-9x)+(-2y)$
$=4x+(-9x)+\_\_+(-2y)$
$=[4+(-9)]x+\_\_+(-2y)$
$=-5x+\_\_+(-2y)$

This expression is equivalent to $-5x-7y,$ so the missing term is the term which yields $-7y$ when added to $-2y$. Since $-5y+(-2y)=-7y,$ the missing term is $-5y$.

99. $3m+2n+\_\_+(-2m)$
$=2n+\_\_+(-2m)+3m$
$=2n+\_\_+(-2+3)m$
$=2n+\_\_+m$

This expression is equivalent to $2n+(-6m),$ so the missing term is the term which yields $-6m$ when added to m. Since $-7m+m=-6m,$ the missing term is $-7m$.

101. Note that, in order for the sum to be 0, the two missing terms must be the opposites of the given terms. Thus, the missing terms are $-7t$ and -23.

103. $-3+(-3)+2+(-2)+1=-5$

Since the total is 5 under par after the five rounds and $-5=-1+(-1)+(-1)+(-1)+(-1),$ the golfer was 1 under par on average.

Exercise Set 1.6

1. The numbers 5 and –5 are *opposites* of each other.

3. We subtract by adding the *opposite* of the number being subtracted.

5. $-x$ is read "the opposite of x," so choice (d) is correct.

7. $12-(-x)$ is read "twelve minus the opposite of x," so choice (f) is correct.

9. $x-(-12)$ is read "x minus negative twelve," so choice (a) is correct.

11. $-x-x$ is read "the opposite of x minus x," so choice (b) is correct.

13. $6-10$ is read "six minus ten."

15. $2-(-12)$ is read "two minus negative twelve."

17. $-x-y$ is read "the opposite of x minus y."

19. $-3-(-n)$ is read "negative three minus the opposite of n."

21. The opposite of 51 is –51 because $51+(-51)=0$.

23. The opposite of $-\frac{11}{3}$ is $\frac{11}{3}$ because $-\frac{11}{3}+\frac{11}{3}=0$.

25. The opposite of –3.14 is 3.14 because $-3.14+3.14=0$.

27. If $x=-45$, then $-x=-(-45)=45$. (The opposite of –45 is 45.)

29. If $x=-\frac{14}{3}$, then $-x=-\left(-\frac{14}{3}\right)=\frac{14}{3}$.
(The opposite of $-\frac{14}{3}$ is $\frac{14}{3}$.)

31. If $x=0.101$, then $-x=-(0.101)=-0.101$.
(The opposite of 0.101 is –0.101.)

33. If $x=37$, then $-(-x)=-(-37)=37$
(The opposite of the opposite of 37 is 37.)

35. If $x=-\frac{2}{5}$, then $-(-x)=-\left[-\left(-\frac{2}{5}\right)\right]=-\frac{2}{5}$.
(The opposite of the opposite of $-\frac{2}{5}$ is $-\frac{2}{5}$.)

37. When we change the sign of –1 we obtain 1.

39. When we change the sign we obtain –15.

41. $7-10=7+(-10)=-3$

43. $0-6=0+(-6)=-6$

45. $-4-3=-4+(-3)=-7$

47. $-9-(-3)=-9+3=-6$

49. Note that we are subtracting a number from itself. The result is 0. We could also do this exercise as follows:
$-8-(-8)=-8+8=0$

51. $14-19=14+(-19)=-5$

53. $30-40=30+(-40)=-10$

55. $-9-(-9)=-9+9=0$
(See Exercise 49.)

57. $5-5=5+(-5)=0$
(See Exercise 49.)

59. $4-(-4)=4+4=8$

61. $-7-4=-7+(-4)=-11$

63. $6-(-10)=6+10=16$

65. $-4-15=-4+(-15)=-19$

67. $-6-(-5)=-6+5=-1$

69. $5-(-12)=5+12=17$

71. $0-(-3)=0+3=3$

73. $-5-(-2)=-5+2=-3$

75. $-7-14=-7+(-14)=-21$

77. $0-11=0+(-11)=-11$

79. $-8-0=-8+0=-8$

81. $-52-8=-52+(-8)=-60$

83. $2-25=2+(-25)=-23$

85. $-4.2-3.1=-4.2+(-3.1)=-7.3$

87. $-1.3-(-2.4)=-1.3+2.4=1.1$

89. $3.2-8.7=3.2+(-8.7)=-5.5$

91. $0.072-1=0.072+(-1)=-0.928$

93. $\frac{2}{11}-\frac{9}{11}=\frac{2}{11}+\left(-\frac{9}{11}\right)=-\frac{7}{11}$

95. $\frac{-1}{5}-\frac{3}{5}=\frac{-1}{5}+\left(\frac{-3}{5}\right)=\frac{-4}{5}$, or $-\frac{4}{5}$

97. $-\frac{4}{17}-\left(-\frac{9}{17}\right)=-\frac{4}{17}+\frac{9}{17}=\frac{5}{17}$

99. $16-(-12)-1-(-2)+3=16+12+(-1)+2+3$
$=32$

101. $-31+(-28)-(-14)-17$
$=(-31)+(-28)+14+(-17)$
$=-62$

103. $-34-28+(-33)-44$
$=(-34)+(-28)+(-33)+(-44)$
$=-139$

105. $-93+(-84)-(-93)-(-84)$
Note that we are subtracting –93 from –93 and –84 from –84. Thus, the result will be 0. We could also do this exercise as follows:
$-93+(-84)-(-93)-(-84)$
$=-93+(-84)+93+84$
$=0$

107. $-3y-8x=-3y+(-8x)$, so the terms are $-3y$ and $-8x$.

109. $9-5t-3st=9+(-5t)+(-3st)$, so the terms are 9, $-5t$, and $-3st$.

111. $10x-13x$
$=10x+(-13x)$ Adding the opposite
$=(10+(-13))x$ Using the distributive law
$=-3x$

113. $7a-12a+4$
$=7a+(-12a)+4$ Adding the opposite
$=(7+(-12))a+4$ Using the distributive law
$=-5a+4$

115. $-8n-9+7n$
$=-8n+(-9)+7n$ Adding the opposite
$=-8n+7n+(-9)$ Using the commutative law of addition
$=-n-9$ Adding like terms

117. $5-3x-11=5+(-3x)+(-11)$
$=-3x+5+(-11)$
$=-3x-6$

119. $2-6t-9-2t=2+(-6t)+(-9)+(-2t)$
$=2+(-9)+(-6t)+(-2t)$
$=-7-8t$

121. $5y+(-3x)-9x+1-2y+8$
$=5y+(-3x)+(-9x)+1+(-2y)+8$
$=5y+(-2y)+(-3x)+(-9x)+1+8$
$=3y-12x+9$

123. $13x-(-2x)+45-(-21)-7x$
$=13x+2x+45+21+(-7x)$
$=13x+2x+(-7x)+45+21$
$=8x+66$

125. $-8-32=-8+(-32)=-40$

127. $18-(-25)=18+25=43$

129. We subtract –5.2 from 3.8.
Translate: $3.8-(-5.2)$
Simplify: $3.8-(-5.2)=3.8+5.2=9$

131. We subtract –79 from 114.
Translate: $114-(-79)$
Simplify: $114-(-79)=114+79=193$

133. We subtract the lower elevation from the higher elevation:
$$550-(-400)=550+400=950$$
It falls 950 m.

135. We subtract the lower temperature from the higher temperature:
$$134-(-79.8)=134+79.8=213.8$$
The temperature range is 213.8°F.

137. We subtract the lower differential from the higher differential:
$$7.3-(-1.6)=7.3+1.6=8.9$$
The differential improved by 8.9 points.

139. *Writing Exercise.*

141. Answers will vary. The symbol "–" can represent the opposite, or a negative, or subtraction.

143. If the clock reads 8:00 A.M. on the day following the blackout when the actual time is 3:00 P.M., then the clock is 7 hr behind the actual time. This indicates that the power outage lasted 7 hr, so power was restored 7 hr after 4:00 P.M., or at 11:00 P.M. on August 14.

145. False. For example, let $m=-3$ and $n=-5$. Then $-3>-5$, but $-3+(-5)=-8 \not> 0$.

147. True. For example, for $m=4$ and $n=-4$, $4=-(-4)$ and $4+(-4)=0$; for $m=-3$ and $n=3$, $-3=-3$ and $-3+3=0$.

149. [(-)] [9] [-] [(-)] [7] [ENTER]

Connecting the Concepts

1. $-8+(-2)=-10$
2. $-8\cdot(-2)=16$
3. $-8\div(-2)=\frac{-8}{-2}=4$
4. $-8-(-2)=-8+2=-6$
5. $\frac{3}{5}-\frac{8}{5}=-\frac{5}{5}=-1$
6. $\frac{12}{5}+\left(\frac{-7}{5}\right)=\frac{5}{5}=1$
7. $(1.3)(-2.9)=-3.77$
8. $-44.1\div 6.3=-7$
9. $-38-(-38)=0$
10. $-46-46=-92$

Exercise Set 1.7

1. The product of two negative numbers is *positive.*

3. Division by zero is *undefined.*

5. The *opposite* of a negative number is positive.

7. The product of two reciprocals is 1.

9. The sum of a pair of additive inverses is 0.

11. The number 0 has no reciprocal.

13. The number 1 is the multiplicative identity.

15. A nonzero number divided by itself is 1.

17. $-4\cdot10=-40$ Think: $4\cdot10=40$, make the answer negative.

19. $-8\cdot7=-56$ Think: $8\cdot7=56$, make the answer negative.

21. $4\cdot(-10)=-40$

23. $-9\cdot(-8)=72$ Multiplying absolute values; the answer is positive.

25. $-19\cdot(-10)=190$

27. $11\cdot(-12)=-132$

29. $4.5\cdot(-28)=-126$

31. $-5\cdot(-2.3)=11.5$

33. $-(25)\cdot0=0$ The product of 0 and any real number is 0.

35. $\frac{2}{5}\cdot\left(-\frac{5}{7}\right)=-\left(\frac{2\cdot5}{5\cdot7}\right)=-\left(\frac{2}{7}\cdot\frac{5}{5}\right)=\frac{-2}{7}$

37. $-\frac{3}{8}\cdot\left(-\frac{2}{9}\right)=\frac{\cancel{3}\cdot\cancel{2}\cdot1}{4\cdot\cancel{2}\cdot\cancel{3}\cdot3}=\frac{1}{12}$

39. $(-5.3)(2.1)=-11.13$

41. $-\frac{5}{9}\cdot\frac{3}{4}=-\frac{5\cdot\cancel{3}}{\cancel{3}\cdot3\cdot4}=-\frac{5}{12}$

43. $3\cdot(-7)\cdot(-2)\cdot6$
$=-21\cdot(-12)$ Multiplying the first two numbers and the last two numbers
$=252$

45. 0, The product of 0 and any real number is 0.

47. $-\frac{1}{3}\cdot\frac{1}{4}\cdot\left(-\frac{3}{7}\right)=-\frac{1}{12}\cdot\left(-\frac{3}{7}\right)=\frac{3}{12\cdot7}$
$=\frac{\cancel{3}\cdot1}{\cancel{3}\cdot4\cdot7}=\frac{1}{28}$

49. $-2\cdot(-5)\cdot(-3)\cdot(-5)=10\cdot15=150$

51. 0, The product of 0 and any real number is 0.

53. $(-8)(-9)(-10)=72(-10)=-720$

55. $(-6)(-7)(-8)(-9)(-10)=42\cdot72\cdot(-10)$
$=3024\cdot(-10)=-30,240$

57. $18\div(-2)=-9$ Check: $-9\cdot(-2)=18$

59. $\frac{36}{-9}=-4$ Check: $-4\cdot(-9)=36$

61. $\frac{-56}{8}=-7$ Check: $-7\cdot8=-56$

63. $\frac{-48}{-12}=4$ Check: $4(-12)=-48$

65. $-72\div8=-9$ Check: $-9\cdot8=-72$

67. $-10.2\div(-2)=5.1$ Check: $5.1(-2)=-10.2$

69. $-100\div(-11)=\frac{100}{11}$

71. $\frac{400}{-50}=-8$ Check: $-8\cdot(-50)=400$

73. Undefined

75. $-4.8\div1.2=-4$ Check: $-4(1.2)=-4.8$

77. $\frac{0}{-9}=0$

79. $\frac{9.7(-2.8)0}{4.3}$
Since the numerator has a factor of 0, the product in the numerator is 0. The denominator is nonzero, so the quotient is 0.

81. $\frac{-8}{3}=\frac{8}{-3}$ and $\frac{-8}{3}=-\frac{8}{3}$

83. $\frac{29}{-35}=\frac{-29}{35}$ and $\frac{29}{-35}=-\frac{29}{35}$

85. $-\frac{7}{3}=\frac{-7}{3}$ and $-\frac{7}{3}=\frac{7}{-3}$

87. $\frac{-x}{2}=\frac{x}{-2}$ and $\frac{-x}{2}=-\frac{x}{2}$

89. The reciprocal of $\frac{4}{-5}$ is $\frac{-5}{4}\left(\text{or equivalently, }-\frac{5}{4}\right)$ because $\frac{4}{-5}\cdot\frac{-5}{4}=1$.

91. The reciprocal of $\frac{51}{-10}$ is $-\frac{10}{51}$ because $\frac{51}{-10}\cdot\left(-\frac{10}{51}\right)=1$.

93. The reciprocal of -10 is $\frac{1}{-10}\left(\text{or equivalently, }-\frac{1}{10}\right)$ because $-10\left(\frac{1}{-10}\right)=1$.

95. The reciprocal of 4.3 is $\frac{1}{4.3}$ because $4.3\left(\frac{1}{4.3}\right)=1$.

Since $\frac{1}{4.3}=\frac{1}{4.3}\cdot\frac{10}{10}=\frac{10}{43}$, the reciprocal can also be expressed as $\frac{10}{43}$.

97. The reciprocal of $\frac{-1}{4}$ is $\frac{4}{-1}$ (or equivalently, -4) because $\frac{-1}{4}\cdot\frac{4}{-1}=1$.

99. The reciprocal of 0 does not exist. (There is no number n for which $0\cdot n=1$.)

101. $\left(\frac{-7}{4}\right)\left(-\frac{3}{5}\right)=\left(-\frac{7}{4}\right)\left(-\frac{3}{5}\right)$ Rewriting $\frac{-7}{4}$ as $-\frac{7}{4}$
$=\frac{21}{20}$

103. $\frac{-3}{8}+\frac{-5}{8}=\frac{-8}{8}=-1$

105. $\left(\frac{-9}{5}\right)\left(\frac{5}{-9}\right)$
Note that this is the product of reciprocals. Thus, the result is 1.

107. $\left(-\frac{3}{11}\right)-\left(-\frac{6}{11}\right)=\frac{-3}{11}+\frac{6}{11}=\frac{3}{11}$

109. $\frac{7}{8}\div\left(-\frac{1}{2}\right)=\frac{7}{8}\cdot\left(-\frac{2}{1}\right)=-\frac{14}{8}=-\frac{7\cdot\cancel{2}}{\cancel{2}\cdot4\cdot1}=-\frac{7}{4}$

111. $\frac{9}{5}\cdot\frac{-20}{3}=-\frac{9}{5}\cdot\frac{20}{3}=-\frac{\cancel{3}\cdot3\cdot4\cdot\cancel{5}}{\cancel{5}\cdot\cancel{3}}=-12$

113. $\left(-\frac{18}{7}\right)+\left(\frac{3}{-7}\right)=\frac{-18}{7}+\frac{-3}{7}=\frac{-21}{7}=-3$

115. $-\frac{5}{9}\div\left(-\frac{5}{9}\right)$
Note that we have a number divided by itself. Thus, the result is 1.

117. $\frac{-3}{10}+\frac{2}{5}=\frac{-3}{10}+\frac{2}{5}\cdot\frac{2}{2}=\frac{-3}{10}+\frac{4}{10}=\frac{1}{10}$

119. $\frac{7}{10}\div\left(\frac{-3}{5}\right)=\frac{7}{10}\div\left(-\frac{3}{5}\right)=\frac{7}{10}\cdot\left(-\frac{5}{3}\right)=-\frac{35}{30}$
$=-\frac{7\cdot\cancel{5}}{2\cdot\cancel{5}\cdot3}=-\frac{7}{6}$

121. $\frac{14}{-9}\div\frac{0}{3}=\frac{14}{-9}\cdot\frac{3}{0}$ Undefined

123. $\frac{-4}{15}+\frac{2}{-3}=\frac{-4}{15}+\frac{-2}{3}=\frac{-4}{15}+\frac{-2}{3}\cdot\frac{5}{5}=\frac{-4}{15}+\frac{-10}{15}$
$=\frac{-14}{15}$, or $-\frac{14}{15}$

125. *Writing Exercise.*

127. *Writing Exercise.*

129. Let a and b represent the numbers. The $a+b$ is the sum and $\frac{1}{a+b}$ is the reciprocal of the sum.

131. Let a and b represent the numbers. Then $a+b$ is the sum and $-(a+b)$ is the opposite of the sum.

133. Let x represent a real number. $x=-x$

135. For 2 and 3, the reciprocal of the sum is $\frac{1}{(2+3)}$ or $\frac{1}{5}$. But $\frac{1}{5}\neq\frac{1}{2}+\frac{1}{3}$.

137. The starting temperature is -3°F.

Rewording: starting temp. rise 2° for 3hr rise 3° for 6hr

Translating: $-3 \quad + \quad 2 \quad \cdot \quad 3 \quad + \quad 3 \quad \cdot \quad 6$

fall 2° for 3hr fall 5° for 2hr

$- \quad 2 \quad \cdot \quad 3 \quad - \quad 5 \quad \cdot \quad 2$

Since $-3+2\cdot3+3\cdot6-2\cdot3-5\cdot2$
$=-3+6+18-6-10=5$.

The temperature forecast at 8PM is 5°F.

139. $-n$ and $-m$ are both positive, so $\frac{-n}{-m}$ is positive.

141. $-m$ is positive, so $\frac{n}{-m}$ is negative and $-\left(\frac{n}{-m}\right)$ is positive.

143. $-n$ and $-m$ are both positive, so $-n-m$, or $-n+(-m)$ is positive; $\frac{n}{m}$ is also positive, so $(-n-m)\frac{n}{m}$ is positive.

145.

$a(-b)+ab=a[-b+b]$	Distributive law
$=a(0)$	Law of opposites
$=0$	Multiplicative property of 0

Therefore, $a(-b)$ is the opposite of ab by the law of opposites.

Exercise Set 1.8

1. Since 6 is the exponent in x^6, the correct choice is (c).

3. Since 2 is the exponent that indicates a square, as in 10^2, the correct choice is (a).

5. Since the terms $3y^4$ and $-y^4$ have the same base and exponent, they are like terms. The correct choice is (e).

7. $4+8\div 2\cdot 2$
There are no grouping symbols or exponential expressions, so we multiply and divide from left to right. This means that we divide first.

9. $5-2(3+4)$
We perform the operation in the parentheses first. This means that we add first.

11. $18-2[4+(3-2)]$
We perform the operation in the innermost grouping symbols first. This means that we perform the subtraction in the parentheses first.

13. $\underbrace{x\cdot x\cdot x\cdot x\cdot x\cdot x}_{\text{6 factors}}=x^6$

15. $\underbrace{(-5)(-5)(-5)}_{\text{3 factors}}=(-5)^3$

17. $3t\cdot 3t\cdot 3t\cdot 3t\cdot 3t=(3t)^5$

19. $2\cdot \underbrace{n\cdot n\cdot n\cdot n}_{\text{4 factors}}=2n^4$

21. $4^2=4\cdot 4=16$

23. $(-3)^2=(-3)(-3)=9$

25. $-3^2=-(3\cdot 3)=-9$

27. $4^3=4\cdot 4\cdot 4=16\cdot 4=64$

29. $(-5)^4=(-5)(-5)(-5)(-5)=25\cdot 25=625$

31. $7^1=7$ (1 factor)

33. $(-2)^5=(-2)(-2)(-2)(-2)(-2)=-32$

35. $(3t)^4=(3t)(3t)(3t)(3t)$
$=3\cdot 3\cdot 3\cdot 3\cdot t\cdot t\cdot t\cdot t=81t^4$

37. $(-7x)^3=(-7x)(-7x)(-7x)$
$=(-7)(-7)(-7)(x)(x)(x)=-343x^3$

39. $5+3\cdot 7=5+21$ Multiplying
$=26$ Adding

41. $10\cdot 5+1\cdot 1=50+1$ Multiplying
$=51$ Adding

43. $5-50\div 5\cdot 2=5-10\cdot 2$ Dividing
$=5-20$ Multiplying
$=-15$ Subtracting

45. $14\cdot 19\div(19\cdot 14)$
Since $14\cdot 19$ and $19\cdot 14$ are equivalent, we are dividing the product $14\cdot 19$ by itself. Thus the result is 1.

47. $3(-10)^2-8\div 2^2$
$=3(100)-8\div 4$ Simplifying the exponential expressions
$=300-8\div 4$ Multiplying and dividing
$=300-2$ from left to right
$=298$ Subtracting

49. $8-(2\cdot 3-9)$
$=8-(6-9)$ Multiplying inside the parentheses
$=8-(-3)$ Subtracting inside the parentheses
$=8+3$ Removing parentheses
$=11$ Adding

51. $(8-2)(3-9)$
$=6(-6)$ Subtracting inside the parentheses
$=-36$ Multiplying

53. $13(-10)^2+45\div(-5)$
$=13(100)+45\div(-5)$ Simplifying the exponential expression
$=1300+45\div(-5)$ Multiplying and dividing
$=1300-9$ from left to right
$=1291$ Subtracting

55. $5+3(2-9)^2=5+3(-7)^2$
$=5+3\cdot 49=5+147=152$

57. $[2\cdot(5-8)]^2-12=[2\cdot(-3)]^2-12$
$=(-6)^2-12$
$=36-12$
$=24$

59. $\dfrac{7+2}{5^2-4^2}=\dfrac{9}{25-16}=\dfrac{9}{9}=1$

61. $8(-7)+|3(-4)|=-56+|-12|=-56+12=-44$

63. $36\div(-2)^2+4[5-3(8-9)^5]$
$=36\div 4+4[5-3(-1)^5]$
$=9+4[5+3]$
$=9+32$
$=41$

65. $\dfrac{(7)^2-(-1)^7}{5\cdot 7-4\cdot 3^2-2^2}=\dfrac{49-(-1)}{35-4\cdot 9-4}$
$=\dfrac{49+1}{35-36-4}=\dfrac{50}{-5}=-10$

67. $\dfrac{-3^3-2\cdot 3^2}{8\div 2^2-(6-|2-15|)}=\dfrac{-27-2\cdot 9}{8\div 4-(6-|-13|)}$
$=\dfrac{-27-18}{2-(6-13)}=\dfrac{-45}{2-(-7)}$
$=\dfrac{-45}{9}=-5$

69. $9-4x=9-4\cdot 7$ Substituting 7 for x
$=9-28$ Multiplying
$=-19$

71. $24 \div t^3 = 24 \div (-2)^3$ Substituting -2 for t
$= 24 \div (-8)$ Simplifying the exponential expression
$= -3$ Dividing

73. $45 \div a \cdot 5 = 45 \div (-3) \cdot 5$ Substituting -3 for a
$= -15 \cdot 5$ Multiplying and dividing in order from left to right
$= -75$

75. $5x \div 15x^2$
$= 5 \cdot 3 \div 15(3)^2$ Substituting 3 for x
$= 5 \cdot 3 \div 15 \cdot 9$ Simplifying the exponential expression
$= 15 \div 15 \cdot 9$ Multiplying and dividing
$= 1 \cdot 9$ in order from
$= 9$ left to right

77. $45 \div 3^2 x(x-1)$
$= 45 \div 3^2 \cdot 3(3-1)$ Substituting 3 for x
$= 45 \div 3^2 \cdot 3(2)$ Subtracting inside the parentheses
$= 45 \div 9 \cdot 3(2)$ Evaluating the exponential expression
$= 5 \cdot 3(2)$ Dividing and
$= 15(2)$ multiplying from
$= 30$ left to right

79. $-x^2 - 5x = -(-3)^2 - 5(-3) = -9 - 5(-3)$
$= -9 + 15 = 6$

81. $\dfrac{3a - 4a^2}{a^2 - 20} = \dfrac{3 \cdot 5 - 4(5)^2}{(5)^2 - 20} = \dfrac{3 \cdot 5 - 4 \cdot 25}{25 - 20}$
$= \dfrac{15 - 100}{5} = \dfrac{-85}{5} = -17$

83. $3x - 2 + 5(2x + 7)$
$= 3x - 2 + 10x + 35$ Distributing the 5
$= 13x + 33$ Adding like terms

85. $2x^2 + 5(3x^2 - x) - 12x$
$= 2x^2 + 15x^2 - 5x - 12x$ Distributing the 5
$= 17x^2 - 17x$ Adding like terms

87. $9t - 7r + 2(3r + 6t)$
$= 9t - 7r + 6r + 12t$ Distributing the 2
$= 21t - r$ Adding like terms

89. $5t^3 + t + 3(t - 2t^3)$
$= 5t^3 + t + 3t - 6t^3$ Distributing the 3
$= -t^3 + 4t$ Adding like terms

91. $-(9x + 1) = -9x - 1$ Removing parentheses and changing the sign of each term

93. $-[7n + 8] = 7n - 8$ Removing grouping symbols and changing the sign of each term

95. $-(4a - 3b + 7c) = -4a + 3b - 7c$

97. $-(3x^2 + 5x - 1) = -3x^2 - 5x + 1$

99. $8x - (6x + 7)$
$= 8x - 6x - 7$ Removing parentheses and changing the sign of each term
$= 2x - 7$ Collecting like terms

101. $2x - 7x - (4x - 6) = 2x - 7x - 4x + 6 = -9x + 6$

103. $15x - y - 5(3x - 2y + 5z)$
$= 15x - y - 15x + 10y - 25z$ Multiplying each term in parentheses by -5
$= 9y - 25z$

105. $3x^2 + 11 - (2x^2 + 5) = 3x^2 + 11 - 2x^2 - 5$
$= x^2 + 6$

107. $12a^2 - 3ab + 5b^2 - 5(-5a^2 + 4ab - 6b^2)$
$= 12a^2 - 3ab + 5b^2 + 25a^2 - 20ab + 30b^2$
$= 37a^2 - 23ab + 35b^2$

109. $-7t^3 - t^2 - 3(5t^3 - 3t)$
$= -7t^3 - t^2 - 15t^3 + 9t$
$= -22t^3 - t^2 + 9t$

111. $5(2x - 7) - [4(2x - 3) + 2]$
$= 5(2x - 7) - [8x - 12 + 2]$
$= 5(2x - 7) - [8x - 10]$
$= 10x - 35 - 8x + 10$
$= 2x - 25$

113. *Writing Exercise.*

115. *Writing Exercise.*

117. $5t - \{7t - [4r - 3(t - 7)] + 6r\} - 4r$
$= 5t - \{7t - [4r - 3t + 21] + 6r\} - 4r$
$= 5t - \{7t - 4r + 3t - 21 + 6r\} - 4r$
$= 5t - \{10t + 2r - 21\} - 4r$
$= 5t - 10t - 2r + 21 - 4r$
$= -5t - 6r + 21$

119. $\{x - [f - (f - x)] + [x - f]\} - 3x$
$= \{x - [f - f + x] + [x - f]\} - 3x$
$= \{x - [x] + [x - f]\} - 3x$
$= \{x - x + x - f\} - 3x$
$= x - f - 3x$
$= -2x - f$

121. *Writing Exercise.*

123. True; $m - n = -n + m = -(n - m)$

125. False; let $m = 2$ and $n = 1$. Then $-2(1 - 2)$
$= -2(-1) = 2$, but $-(2 \cdot 1 + 2^2) = -(2 + 4) = -6$.

127. $[x + 3(2 - 5x) \div 7 + x](x - 3)$
When $x = 3$, the factor $x - 3$ is 0, so the product is 0.

129. $\frac{x^2+2^x}{x^2-2^x}=\frac{3^2+2^3}{3^2-2^3}=\frac{9+8}{9-8}=\frac{17}{1}=17$

131. $4\cdot 20^3+17\cdot 20^2+10\cdot 20+0\cdot 1$
$=4\cdot 8000+17\cdot 400+10\cdot 20+0\cdot 1$
$=32{,}000+6800+200+0$
$=39{,}000$

133. The tower is composed of cubes with sides of length x. The volume of each cube is $x\cdot x\cdot x$, or x^3. Now we count the number of cubes in the tower. The two lowest levels each contain 3×3, or 9 cubes. The next level contains one cube less than the two lowest levels, so it has $9-1$, or 8 cubes. The fourth level from the bottom contains one cube less than the level below it, so it has $8-1$, or 7 cubes. The fifth level from the bottom contains one cube less than the level below it, so it has $7-1$, or 6 cubes. Finally, the top level contains one cube less than the level below it, so it has $6-1$, or 5 cubes. All together there are $9+9+8+7+6+5$, or 44 cubes, each with volume x^3, so the volume of the tower is $44x^3$.

Chapter 1 Review

1. True
2. True
3. False
4. True
5. False
6. False
7. True
8. False
9. False
10. True

11. $8t=8\cdot 3=24$ Substitute 3 for t

12. $9-y^2=9-(-5)^2$ Substitute -5 for y
$=9-25$
$=-16$

13. $-10+a^2\div(b+1)$ Substitute 5 for a and -6 for b
$=-10+5^2\div(-6+1)$
$=-10+25\div(-5)$
$=-10+(-5)$
$=-15$

14. $y-7$

15. $xz+10$, or $10+xz$

16. Let b represent Brandt's speed and w represent the wind speed; $15(b-w)$

17. Let b represent the number of calories per hour that Katie burns while backpacking.

Translating:	b	is	two	times	237.
	↓	↓	↓	↓	↓
	b	=	2	·	237

or $b=2\cdot 237$

18. Let c represent the number of calories burned and t represent the number hours.
$c=200t$

19. $3t+5=t\cdot 3+5$

20. $(2x+y)+z=2x+(y+z)$

21. Answers may vary.
$4(xy)=(4x)y$
$4(xy)=4(yx)$
$4(xy)=4(yx)=(4y)x$

22. $6(3x+5y)=6\cdot 3x+6\cdot 5y$
$=18x+30y$

23. $8(5x+3y+2)=8\cdot 5x+8\cdot 3y+8\cdot 2$
$=40x+24y+16$

24. $21x+15y=3\cdot 7x+3\cdot 5y$
$=3(7x+5y)$

25. $22a+99b+11=11\cdot 2a+11\cdot 9b+11\cdot 1$
$=11(2a+9b+1)$

26. $56=7\cdot 8=7\cdot 2\cdot 2\cdot 2$ or $2\cdot 2\cdot 2\cdot 7$

27. $\frac{20}{48}=\frac{\cancel{2}\cdot\cancel{2}\cdot 5}{\cancel{2}\cdot\cancel{2}\cdot 2\cdot 2\cdot 3}=\frac{5}{12}$

28. $\frac{18}{8}=\frac{\cancel{2}\cdot 3\cdot 3}{\cancel{2}\cdot 2\cdot 2}=\frac{9}{4}$

29. $\frac{5}{12}+\frac{3}{8}=\frac{10}{24}+\frac{9}{24}=\frac{10+9}{24}=\frac{19}{24}$

30. $\frac{9}{16}\div 3=\frac{9}{16}\cdot\frac{1}{3}$ Multiply by the reciprocal of the divisor.
$=\frac{\cancel{3}\cdot 3\cdot 1}{16\cdot\cancel{3}}$
$=\frac{3}{16}$

31. $\frac{2}{3}-\frac{1}{15}=\frac{5}{5}\cdot\frac{2}{3}-\frac{1}{15}$ Use 15 as the common denominator.
$=\frac{5\cdot 2}{15}-\frac{1}{15}$
$=\frac{10-1}{15}=\frac{9}{15}$
$=\frac{\cancel{3}\cdot 3}{\cancel{3}\cdot 5}=\frac{3}{5}$

32. $\frac{9}{10}\cdot\frac{6}{5}=\frac{9\cdot 6}{10\cdot 5}=\frac{9\cdot \cancel{2}\cdot 3}{\cancel{2}\cdot 5\cdot 5}=\frac{27}{25}$

33. The real number –3600 corresponds to borrowing \$3600. The real number 1350 corresponds to the savings account value of \$1350.

34. 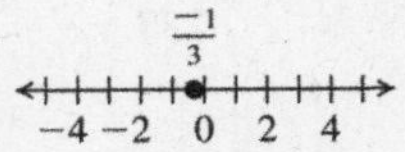

35. $-3 < x$ has the same meaning as $x > -3$.

36. $-10 < 0$

37. $-\frac{4}{9}=-\left(\frac{4}{9}\right)=-(4\div 9)$, so we divide.

$$\begin{array}{r} 0.444 \\ 9\overline{)4000} \\ \underline{36} \\ 40 \\ \underline{36} \\ 40 \\ \underline{36} \\ 4 \end{array}$$

$\frac{4}{9}=0.\overline{4}$, so $-\frac{4}{9}=-0.\overline{4}$

38. $|-1|=1$, since –1 is 1 unit from 0.

39. $-(-x)=-(-(-12))$ Substitute –12 for x
$=-(12)=-12$

40. $-3+(-7)=-10$

41. $-\frac{2}{3}+\frac{1}{12}=\frac{-8}{12}+\frac{1}{12}$ The absolute values are $\frac{8}{12}$ and $\frac{1}{12}$. The difference is $\frac{8}{12}-\frac{1}{12}$, or $\frac{8-1}{12}=\frac{7}{12}$. The negative number has the greater absolute value, so the answer is negative. $-\frac{2}{3}+\frac{1}{12}=-\frac{7}{12}$

42. $-3.8+5.1+(-12)+(-4.3)+10$
$=(5.1+10)+[(-3.8)+(-12)+(-4.3)]$
$=15.1+[-20.1]$ Adding positive numbers and adding negative numbers
$=-5$ Adding a positive and a negative number.

43. $-2-(-10)=-2+10=8$

44. $\frac{-9}{10}-\frac{1}{2}=\frac{-9}{10}+\left(\frac{-1}{2}\right)$
$=\frac{-9}{10}+\left(\frac{-5}{10}\right)=\frac{-14}{10}=\frac{-7}{5}$

45. $-2.7(3.4)=-(2.7\cdot 3.4)=-9.18$

46. $\frac{2}{3}\cdot\left(-\frac{3}{7}\right)=-\left(\frac{2}{3}\cdot\frac{3}{7}\right)=-\left(\frac{2\cdot 3}{3\cdot 7}\right)=-\frac{2}{7}$

47. $2\cdot(-7)\cdot(-2)\cdot(-5)=-14\cdot 10=-140$

48. $35\div(-5)=-7$

49. $-5.1\div 1.7=-3$

50. $-\frac{3}{5}\div\left(-\frac{4}{15}\right)=\frac{3}{5}\cdot\frac{15}{4}=\frac{3\cdot 3\cdot \cancel{5}}{\cancel{5}\cdot 4}=\frac{9}{4}$

51. $120-6^2\div 4\cdot 8=120-36\div 4\cdot 8$
$=120-9\cdot 8$
$=120-72$
$=48$

52. $(120-6^2)\div 4\cdot 8=(120-36)\div 4\cdot 8$
$=84\div 4\cdot 8$
$=21\cdot 8$
$=168$

53. $(120-6^2)\div(4\cdot 8)=(120-36)\div(4\cdot 8)$
$=84\div 32$
$=\frac{84}{32}$
$=\frac{\cancel{4}\cdot 21}{\cancel{4}\cdot 8}$
$=\frac{21}{8}$

54. $16\div(-2)^3-5[3-1+2(4-7)]$
$=16\div(-8)-5[3-1+2(4-7)]$
$=-2-5[3-1+2(-3)]$
$=-2-5[-4]$
$=-2+20$
$=18$

55. $|-3\cdot 5-4\cdot 8|-3(-2)=|-15-32|+6$
$=|-47|+6=53$

56. $\frac{4(18-8)+7\cdot 9}{9^2-8^2}=\frac{4(18-8)+7\cdot 9}{81-64}$
$=\frac{4(10)+7\cdot 9}{81-64}$
$=\frac{40+63}{81-64}$
$=\frac{103}{17}$

57. $11a+2b+(-4a)+(-3b)$
$=11a+(-4a)+2b+(-3b)$
$=[11+(-4)]a=[2+(-3)]b$
$=7a-b$

58. $7x-3y-11x+8y$
$=7x+(-11x)+(-3y)+8y$
$=[7+(-11)]x+(-3+8)y$
$=-4x+5y$

59. The opposite of –7 is 7, since $7+(-7)=0$.

60. The reciprocal of –7 is $-\frac{1}{7}$, since $-\frac{1}{7}\cdot -7=1$.

61. $\underbrace{2x \cdot 2x \cdot 2x \cdot 2x}_{\text{4 Factors}} = (2x)^4$

62. $(-5x)^3 = -5x \cdot (-5x) \cdot (-5x)$
$= -5 \cdot (-5) \cdot (-5) \cdot x \cdot x \cdot x$
$= -125x^3$

63. $2a - (5a - 9) = 2a + (-5a) + 9$
$= [2 + (-5)]a + 9 = -3a + 9$

64. $11x^4 + 2x + 8(x - x^4) = 11x^4 + 2x + 8 \cdot x - 8x^4$
$= 11x^4 - 8x^4 + 2x + 8x$
$= 3x^4 + 10x$

65. $2n^2 - 5(-3n^2 + m^2 - 4mn) + 6m^2$
$= 2n^2 + 15n^2 - 5m^2 + 20mn + 6m^2$
$= (2 + 15)n^2 + (-5 + 6)m^2 + 20mn$
$= 17n^2 + m^2 + 20mn$

66. $8(x + 4) - 6 - [3(x - 2) + 4]$
$= 8x + 32 - 6 - [3x - 6 + 4]$
$= 8x + 32 - 6 - [3x - 2]$
$= 8x + 26 - 3x + 2$
$= (8 - 3)x + 26 + 2$
$= 5x + 28$

67. *Writing Exercise.*

68. *Writing Exercise.*

69. *Writing Exercise.*

70. *Writing Exercise.*

71. Substitute 1 for a, 2 for b, and evaluate:
$a^{50} - 20a^{25}b^4 + 100b^8 = 1^{50} - 20 \cdot 1^{25}2^4 + 100 \cdot 2^8$
$= 1 - 20 \cdot 1 \cdot 16 + 100 \cdot 256$
$= 1 - 320 + 25{,}600$
$= 25{,}281$

72. a. Since $0.090909\ldots + 0.181818\ldots = 0.272727\ldots$
we have $\frac{1}{11} + \frac{2}{11} = \frac{3}{11}$; $0.272727\ldots = \frac{3}{11}$

b. Since $10 \cdot 0.090909\ldots = 0.909090\ldots$,
we have $10 \cdot \frac{1}{11} = \frac{10}{11}$; $0.909090\ldots = \frac{10}{11}$

73. $-\left|\frac{7}{8} - \left(-\frac{1}{2}\right) - \frac{3}{4}\right|$ Use 8: the common denominator
$= -\left|\frac{7}{8} - \left(-\frac{4}{8}\right) - \frac{6}{8}\right|$
$= -\left|\frac{7}{8} + \frac{4}{8} + \frac{-6}{8}\right|$
$= -\left|\frac{5}{8}\right| = -\frac{5}{8}$

74. $(|2.7 - 3| + 3^2 - |-3|) \div (-3)$
$= (|2.7 - 3| + 9 - |-3|) \div (-3)$
$= (|-0.3| + 9 - |-3|) \div (-3)$
$= (0.3 + 9 - 3) \div (-3)$
$= 6.3 \div (-3)$
$= -2.1$

75. i

76. j

77. a

78. h

79. k

80. b

81. c

82. e

83. d

84. f

85. g

Chapter 1 Test

1. Substitute 10 for x and 5 for y.
$\frac{2x}{y} = \frac{2 \cdot 10}{5} = \frac{2 \cdot 2 \cdot \cancel{5}}{\cancel{5}} = 4$

2. Let x and y represent the numbers; $xy - 9$

3. $A = \frac{1}{2} \cdot b \cdot h$
$= \frac{1}{2} \cdot (16 \text{ ft}) \cdot (30 \text{ ft})$
$= \frac{1}{2} \cdot 16 \cdot 30 \cdot \text{ft} \cdot \text{ft}$
$= 8 \cdot 30 \text{ ft} \cdot \text{ft}$
$= 240 \text{ ft}^2$, or 240 square feet

4. $3p + q = q + 3p$

5. $x \cdot (4 \cdot y) = (x \cdot 4) \cdot y$

6. Let t represent the number of golden lion tamarins living in zoos; $1500 = t + 1050$

7. $7(5 + x) = 7 \cdot 5 + 7 \cdot x = 35 + 7x$

8. $-5(y - 2) = -5 \cdot y - 5(-2) = -5y + 10$

9. $11 + 44x = 11 \cdot 1 + 11 \cdot 4x = 11(1 + 4x)$

10. $7x + 7 + 49y = 7 \cdot x + 7 \cdot 1 + 7 \cdot 2y$
$= 7(x + 1 + 7y)$

11. $300 = 2 \cdot 150 = 2 \cdot 2 \cdot 75$
$= 2 \cdot 2 \cdot 3 \cdot 25 = 2 \cdot 2 \cdot 3 \cdot 5 \cdot 5$

12. $\frac{24}{56} = \frac{3 \cdot \cancel{8}}{7 \cdot \cancel{8}} = \frac{3}{7}$

13. $-4 < 0$, since -4 is left of 0.

14. $-3 > -8$, since -3 is right of -8.

15. $\left|\frac{9}{4}\right| = \frac{9}{4}$, since $\frac{9}{4}$ is $\frac{9}{4}$ units from 0.

16. $|-3.8| = 3.8$, since -3.8 is 3.8 units from 0.

17. $\frac{2}{3}$, since $\frac{2}{3} + \frac{-2}{3} = 0$

18. $\frac{-7}{4}$, since $\frac{-7}{4} \cdot \frac{-4}{7} = 1$

19. Let $x = -10$, then $-x = -(-10) = 10$.

20. $-5 \ge x$ has the same meaning as $x \le -5$.

21. $3.1 - (-4.7) = 3.1 + 4.7 = 7.8$

22. $-8 + 4 + (-7) + 3 = [-8 + (-7)] + [4 + 3]$
$= -15 + 7 = -8$

23. $-\frac{1}{8} - \frac{3}{4} = -\frac{1}{8} - \frac{2}{2} \cdot \frac{3}{4}$
$= -\frac{1}{8} - \frac{6}{8} = \frac{-1-6}{8} = -\frac{7}{8}$

24. $4 \cdot (-12) = -(4 \cdot 12) = -48$

25. $-\frac{1}{2}\left(-\frac{4}{9}\right) = \frac{1}{2} \cdot \frac{4}{9} = \frac{1 \cdot \not{2} \cdot 2}{\not{2} \cdot 9} = \frac{2}{9}$

26. $-\frac{3}{5} \div \left(-\frac{4}{5}\right) = \frac{3}{5} \cdot \frac{5}{4} = \frac{3 \cdot \not{5}}{\not{5} \cdot 4} = \frac{3}{4}$

27. $4.864 \div (-0.5) = -9.728$

28. $10 - 2(-16) \div 4^2 + |2 - 10|$
$= 10 - 2(-16) \div 16 + |2 - 10|$
$= 10 + 32 \div 16 + |-8|$
$= 10 + 2 + 8$
$= 20$

29. $9 + 7 - 4 - (-3) = 9 + 7 + (-4) + 3$
$= 9 + 7 + 3 + (-4)$
$= 19 + (-4) = 15$

30. $256 \div (-16) \cdot 4 = -16 \cdot 4 = -64$

31. $2^3 - 10[4 - 3(-2 + 18)] = 8 - 10[4 - 3(-2 + 18)]$
$= 8 - 10[4 - 16 \cdot 3]$
$= 8 - 10[4 - 48]$
$= 8 - 10[4 + (-48)]$
$= 8 - 10[-44]$
$= 8 - (-440)$
$= 8 + 440 = 448$

32. $18y + 30a - 9a + 4y = 30a + (-9a) + 18y + 4y$
$= [30 + (-9)]a + (18 + 4)y$
$= 21a + 22y$

33. $(-2x)^4 = -2x \cdot (-2x) \cdot (-2x) \cdot (-2x)$
$= -2 \cdot (-2) \cdot (-2) \cdot (-2) \cdot x \cdot x \cdot x \cdot x = 16x^4$

34. $4x - (3x - 7) = 4x - 3x + 7$
$= (4 - 3)x + 7 = x + 7$

35. $4(2a - 3b) + a - 7 = 8a - 12b + a - 7$
$= (8a + a) - 12b - 7$
$= (8 + 1)a - 12b - 7$
$= 9a - 12b - 7$

36. $3[5(y - 3) + 9] - 2(8y - 1)$
$= 3[5y - 15 + 9] - 16y + 2$
$= 3[5y - 6] - 16y + 2$
$= 15y - 18 - 16y + 2$
$= -y - 16$

37. y is 4 less than half of x

Rewording:	half of x	less 4	is	y
	↓	↓	↓	↓
Translating:	$\frac{1}{2}x$	-4	$=$	y

Since $x = 20$, $\frac{1}{2}(20) - 4 = y$
$10 - 4 = y$
$y = 6$

Substitute 20 for x and 6 for y.

$\frac{5y - x}{2} = \frac{5 \cdot 6 - 20}{2}$
$= \frac{30 - 20}{2} = \frac{10}{2} = 5$

38. $9 - (3 - 4) + 5 = 15$

39. $n \ge 0$

40. $a - \{3a - [4a - (2a - 4a)]\}$
$= a - \{3a - [4a - (-2a)]\}$
$= a - \{3a - [4a + 2a]\}$
$= a - \{3a - 6a\}$
$= a - \{-3a\}$
$= a + 3a = 4a$

41. Let $a = 2, b = 3, c = 4$.
$a|b - c| = |ab| - |ac|$
$2|3 - 4| \stackrel{?}{=} |2 \cdot 3| - |2 \cdot 4|$
$2|-1| \stackrel{?}{=} |6| - |8|$
$2 \cdot 1 \stackrel{?}{=} 6 - 8$
$2 \ne -2$
False.

Chapter 2

Equations, Inequalities, and Problem Solving

Exercise Set 2.1

1. The equations $x+3=7$ and $6x=24$ are *equivalent equations*. Choice (c) is correct.

3. A *solution* is a replacement that makes an equation true. Choice (f) is correct.

5. *The multiplication principle* is used to solve $\frac{2}{3}\cdot x=-4$. Choice (d) is correct.

7. Substitute 4 for x.

$$\begin{array}{c|c} \multicolumn{2}{c}{6-x=-2} \\ \hline 6-4 & -2 \end{array}$$

$$2\stackrel{?}{=}-2 \quad \text{FALSE}$$

Since the left-hand and right-hand sides differ, 4 is not a solution.

9. Substitute 18 for t.

$$\begin{array}{c|c} \multicolumn{2}{c}{\frac{2}{3}t=12} \\ \hline \frac{2}{3}(18) & 12 \end{array}$$

$$12\stackrel{?}{=}12 \quad \text{TRUE}$$

Since the left-hand and right-hand sides are the same, 18 is a solution.

11. Substitute –2 for x.

$$\begin{array}{c|c} \multicolumn{2}{c}{x+7=3-x} \\ \hline -2+7 & 3-(-2) \\ & 3+2 \end{array}$$

$$5\stackrel{?}{=}5 \quad \text{TRUE}$$

Since the left-hand and right-hand sides are the same, –2 is a solution.

13. Substitute –20 for n.

$$\begin{array}{c|c} \multicolumn{2}{c}{4-\frac{1}{5}n=8} \\ \hline 4-\frac{1}{5}(-20) & 8 \\ 4+4 & \end{array}$$

$$8\stackrel{?}{=}8 \quad \text{TRUE}$$

Since the left-hand and right-hand sides are the same, –20 is a solution.

15.
$$\begin{aligned} x+10&=21 \\ x+10-10&=21-10 \\ x&=11 \end{aligned}$$

Check:
$$\begin{array}{c|c} \multicolumn{2}{c}{x+10=21} \\ \hline 11+10 & 21 \end{array}$$

$$21\stackrel{?}{=}21 \quad \text{TRUE}$$

The solution is 11.

17.
$$\begin{aligned} y+7&=-18 \\ y+7-7&=-18-7 \\ y&=-25 \end{aligned}$$

Check:
$$\begin{array}{c|c} \multicolumn{2}{c}{y+7=-18} \\ \hline -25+7 & -18 \end{array}$$

$$-18\stackrel{?}{=}-18 \quad \text{TRUE}$$

The solution is –25.

19.
$$\begin{aligned} -6&=y+25 \\ -6-25&=y+25-25 \\ -31&=y \end{aligned}$$

Check:
$$\begin{array}{c|c} \multicolumn{2}{c}{-6=y+25} \\ \hline -6 & -31+25 \end{array}$$

$$-6\stackrel{?}{=}-6 \quad \text{TRUE}$$

The solution is –31.

21.
$$\begin{aligned} x-18&=23 \\ x-18+18&=23+18 \\ x&=41 \end{aligned}$$

Check:
$$\begin{array}{c|c} \multicolumn{2}{c}{x-18=23} \\ \hline 41-18 & 23 \end{array}$$

$$23\stackrel{?}{=}23 \quad \text{TRUE}$$

The solution is 41.

23.
$$\begin{aligned} 12&=-7+y \\ 7+12&=7+(-7)+y \\ 19&=y \end{aligned}$$

Check:
$$\begin{array}{c|c} \multicolumn{2}{c}{12=-7+y} \\ \hline 12 & -7+19 \end{array}$$

$$12\stackrel{?}{=}12 \quad \text{TRUE}$$

The solution is 19.

25.
$$\begin{aligned} -5+t&=-11 \\ 5+(-5)+t&=5+(-11) \\ t&=-6 \end{aligned}$$

Check:
$$\begin{array}{c|c} \multicolumn{2}{c}{-5+t=-11} \\ \hline -5+(-6) & -11 \end{array}$$

$$-11\stackrel{?}{=}-11 \quad \text{TRUE}$$

The solution is –6.

27.
$$r+\frac{1}{3}=\frac{8}{3}$$
$$r+\frac{1}{3}-\frac{1}{3}=\frac{8}{3}-\frac{1}{3}$$
$$r=\frac{7}{3}$$

Check: $r+\frac{1}{3}=\frac{8}{3}$

$\frac{7}{3}+\frac{1}{3} \;\big|\; \frac{8}{3}$

$\frac{8}{3}\stackrel{?}{=}\frac{8}{3}$ TRUE

The solution is $\frac{7}{3}$.

29.
$$x-\frac{3}{5}=-\frac{7}{10}$$
$$x-\frac{3}{5}+\frac{3}{5}=-\frac{7}{10}+\frac{3}{5}$$
$$x=-\frac{7}{10}+\frac{3}{5}\cdot\frac{2}{2}$$
$$x=-\frac{7}{10}+\frac{6}{10}$$
$$x=-\frac{1}{10}$$

Check: $x-\frac{3}{5}=-\frac{7}{10}$

$-\frac{1}{10}-\frac{3}{5} \;\big|\; -\frac{7}{10}$

$-\frac{1}{10}-\frac{6}{10}$

$-\frac{7}{10}\stackrel{?}{=}-\frac{7}{10}$ TRUE

The solution is $-\frac{1}{10}$.

31.
$$x-\frac{5}{6}=\frac{7}{8}$$
$$x-\frac{5}{6}+\frac{5}{6}=\frac{7}{8}+\frac{5}{6}$$
$$x=\frac{7}{8}\cdot\frac{3}{3}+\frac{5}{6}\cdot\frac{4}{4}$$
$$x=\frac{21}{24}+\frac{20}{24}$$
$$x=\frac{41}{24}$$

Check: $x-\frac{5}{6}=\frac{7}{8}$

$\frac{41}{24}-\frac{5}{6} \;\big|\; \frac{7}{8}$

$\frac{41}{24}-\frac{20}{24} \;\big|\; \frac{21}{24}$

$\frac{21}{24}\stackrel{?}{=}\frac{21}{24}$ TRUE

The solution is $\frac{41}{24}$.

33.
$$-\frac{1}{5}+z=-\frac{1}{4}$$
$$\frac{1}{5}-\frac{1}{5}+z=\frac{1}{5}-\frac{1}{4}$$
$$z=\frac{1}{5}\cdot\frac{4}{4}-\frac{1}{4}\cdot\frac{5}{5}$$
$$z=\frac{4}{20}-\frac{5}{20}$$
$$z=-\frac{1}{20}$$

Check: $-\frac{1}{5}+z=-\frac{1}{4}$

$-\frac{1}{5}+\left(-\frac{1}{20}\right) \;\big|\; -\frac{1}{4}$

$-\frac{4}{20}+\left(-\frac{1}{20}\right) \;\big|\; -\frac{5}{20}$

$-\frac{5}{20}\stackrel{?}{=}-\frac{5}{20}$ TRUE

The solution is $-\frac{1}{20}$.

35.
$$m-2.8=6.3$$
$$m-2.8+2.8=6.3+2.8$$
$$m=9.1$$

Check: $m-2.8=6.3$

$9.1-2.8 \;\big|\; 6.3$

$6.3\stackrel{?}{=}6.3$ TRUE

The solution is 9.1.

37.
$$-9.7=-4.7+y$$
$$4.7+(-9.7)=4.7+(-4.7)+y$$
$$-5=y$$

Check: $-9.7=-4.7+y$

$-9.7 \;\big|\; -4.7+(-5)$

$-9.7\stackrel{?}{=}-9.7$ TRUE

The solution is –5.

39. $8a=56$

$\frac{8a}{8}=\frac{56}{8}$ Dividing both sides by 8

$1\cdot a=7$ Simplifying

$a=7$ Identity property of 1

Check: $8a=56$

$8\cdot 7 \;\big|\; 56$

$56\stackrel{?}{=}56$ TRUE

The solution is 7.

41. $84=7x$

$\frac{84}{7}=\frac{7x}{7}$ Dividing both sides by 7

$12=1\cdot x$

$12=x$

Check: $84=7x$

$84 \;\big|\; 7\cdot 12$

$84\stackrel{?}{=}84$ TRUE

The solution is 12.

43. $-x=38$
$-1\cdot x=38$
$-1\cdot(-1\cdot x)=-1\cdot 38$
$1\cdot x=-38$
$x=-38$

Check: $\dfrac{-x=38}{-(-38)\ \big|\ 38}$
$38\stackrel{?}{=}38$ TRUE

The solution is –38.

45. $-t=-8$
The equation states that the opposite of t is the opposite of 8. Thus, $t=8$. We could also do this exercise as follows.
$-t=-8$
$-1(-t)=-1(-8)$ Multiplying both sides by –1
$t=8$

Check: $\dfrac{-t=-8}{-(8)\ \big|\ -8}$
$-8\stackrel{?}{=}-8$ TRUE

The solution is 8.

47. $-7x=49$
$\dfrac{-7x}{-7}=\dfrac{49}{-7}$
$1\cdot x=-7$
$x=-7$

Check: $\dfrac{-7x=49}{-7(-7)\ \big|\ 49}$
$49\stackrel{?}{=}49$ TRUE

The solution is –7.

49. $0.2m=10$
$\dfrac{0.2m}{0.2}=\dfrac{10}{0.2}$
$m=50$

Check: $\dfrac{0.2m=10}{0.2(50)\ \big|\ 10}$
$10\stackrel{?}{=}10$ TRUE

The solution is 50.

51. $-1.2x=0.24$
$\dfrac{-1.2x}{-1.2}=\dfrac{0.24}{-1.2}$
$x=-0.2$

Check: $\dfrac{-1.2x=0.24}{-1.2(-0.2)\ \big|\ 0.24}$
$0.24\stackrel{?}{=}0.24$ TRUE

The solution is –0.2.

53. $-1.3a=-10.4$
$\dfrac{-1.3a}{-1.3}=\dfrac{-10.4}{-1.3}$
$a=8$

Check: $\dfrac{-1.3a=-10.4}{-1.3(8)\ \big|\ -10.4}$
$-10.4\stackrel{?}{=}-10.4$ TRUE

The solution is 8.

55. $\dfrac{y}{8}=11$
$\dfrac{1}{8}\cdot y=11$
$8\left(\dfrac{1}{8}\right)\cdot y=8\cdot 11$
$y=88$

Check: $\dfrac{\frac{y}{8}=11}{\frac{88}{8}\ \big|\ 11}$
$11\stackrel{?}{=}11$ TRUE

The solution is 88.

57. $\dfrac{4}{5}x=16$
$\dfrac{5}{4}\cdot\dfrac{4}{5}x=\dfrac{5}{4}\cdot 16$
$x=\dfrac{5\cdot \cancel{4}\cdot 4}{\cancel{4}\cdot 1}$
$x=20$

Check: $\dfrac{\frac{4}{5}x=16}{\frac{4}{5}\cdot 20\ \big|\ 16}$
$16\stackrel{?}{=}16$ TRUE

The solution is 20.

59. $\dfrac{-x}{6}=9$
$-\dfrac{1}{6}\cdot x=9$
$-6\left(-\dfrac{1}{6}\right)\cdot x=-6\cdot 9$
$x=-54$

Check: $\dfrac{\frac{-x}{6}=9}{\frac{-(-54)}{6}\ \big|\ 9}$
$\dfrac{54}{6}$
$9\stackrel{?}{=}9$ TRUE

The solution is –54.

61.
$$\frac{1}{9}=\frac{z}{-5}$$
$$\frac{1}{9}=-\frac{1}{5}\cdot z$$
$$-5\cdot\frac{1}{9}=-5\cdot\left(-\frac{1}{5}\cdot z\right)$$
$$-\frac{5}{9}=z$$

Check: $\dfrac{1}{9}=\dfrac{z}{-5}$
$$\begin{array}{c|c} \frac{1}{9} & \frac{-5/9}{-5} \\ & -\frac{5}{9}\cdot\frac{1}{-5} \end{array}$$
$$\frac{1}{9}\stackrel{?}{=}\frac{1}{9}\quad\text{TRUE}$$

The solution is $-\frac{5}{9}$.

63. $-\frac{3}{5}r=-\frac{3}{5}$

The solution of the equation is the number that is multiplied by $-\frac{3}{5}$ to get $-\frac{3}{5}$. That number is 1. We could also do this exercise as follows:
$$-\frac{3}{5}r=-\frac{3}{5}$$
$$-\frac{5}{3}\cdot\left(-\frac{3}{5}r\right)=-\frac{5}{3}\left(-\frac{3}{5}\right)$$
$$r=1$$

Check: $-\frac{3}{5}r=-\frac{3}{5}$
$$\begin{array}{c|c} -\frac{3}{5}\cdot 1 & -\frac{3}{5} \end{array}$$
$$-\frac{3}{5}\stackrel{?}{=}-\frac{3}{5}\quad\text{TRUE}$$

The solution is 1.

65.
$$\frac{-3r}{2}=-\frac{27}{4}$$
$$-\frac{3}{2}r=-\frac{27}{4}$$
$$-\frac{2}{3}\cdot\left(-\frac{3}{2}r\right)=-\frac{2}{3}\cdot\left(-\frac{27}{4}\right)$$
$$r=\frac{\not{2}\cdot\not{3}\cdot 3\cdot 3}{\not{3}\cdot\not{2}\cdot 2}$$
$$r=\frac{9}{2}$$

Check: $\dfrac{-3r}{2}=-\dfrac{27}{4}$
$$\begin{array}{c|c} -\frac{3}{2}\cdot\frac{9}{2} & -\frac{27}{4} \end{array}$$
$$-\frac{27}{4}\stackrel{?}{=}-\frac{27}{4}\quad\text{TRUE}$$

The solution is $\frac{9}{2}$.

67.
$$4.5+t=-3.1$$
$$4.5+t-4.5=-3.1-4.5$$
$$t=-7.6$$
The solution is –7.6.

69.
$$-8.2x=20.5$$
$$\frac{-8.2x}{-8.2}=\frac{20.5}{-8.2}$$
$$x=-2.5$$
The solution is –2.5.

71.
$$x-4=-19$$
$$x-4+4=-19+4$$
$$x=-15$$
The solution is –15.

73.
$$t-3=8$$
$$t-3+3=-8+3$$
$$t=-5$$
The solution is –5.

75.
$$-12x=14$$
$$\frac{-12x}{-12}=\frac{14}{-12}$$
$$1\cdot x=-\frac{7}{6}$$
$$x=-\frac{7}{6}$$
The solution is $-\frac{7}{6}$.

77.
$$48=-\frac{3}{8}y$$
$$-\frac{8}{3}\cdot 48=-\frac{8}{3}\left(-\frac{3}{8}y\right)$$
$$-\frac{8\cdot\not{3}\cdot 16}{\not{3}}=y$$
$$-128=y$$
The solution is –128.

79.
$$a-\frac{1}{6}=-\frac{2}{3}$$
$$a-\frac{1}{6}+\frac{1}{6}=-\frac{2}{3}+\frac{1}{6}$$
$$a=-\frac{4}{6}+\frac{1}{6}$$
$$a=-\frac{3}{6}$$
$$a=-\frac{1}{2}$$
The solution is $-\frac{1}{2}$.

81.
$$-24=\frac{8x}{5}$$
$$-24=\frac{8}{5}x$$
$$\frac{5}{8}(-24)=\frac{5}{8}\cdot\frac{8}{5}x$$
$$-\frac{5\cdot\not{8}\cdot 3}{\not{8}\cdot 1}=x$$
$$-15=x$$
The solution is –15.

83. $-\frac{4}{3}t = -12$
$-\frac{3}{4}\left(-\frac{4}{3}t\right) = -\frac{3}{4}(-12)$
$t = \frac{3 \cdot \not{4} \cdot 3}{\not{4}}$
$t = 9$
The solution is 9.

85. $-483.297 = -794.053 + t$
$-483.297 + 794.053 = -794.053 + t + 794.053$
$310.756 = t$ Using a calculator
The solution is 310.756.

87. *Writing Exercise.*

89. $\frac{1}{3}y - 7$

91. $35a + 55c + 5 = 5(7a + 11c + 1)$

93. *Writing Exercise.*

95. $mx = 11.6m$
$\frac{mx}{m} = \frac{11.6m}{m}$
$x = 11.6$
The solution is 11.6.

97. $cx + 5c = 7c$
$cx + 5c - 5c = 7c - 5c$
$cx = 2c$
$\frac{cx}{c} = \frac{2c}{c}$
$x = 2$
The solution is 2.

99. $7 + |x| = 30$
$-7 + 7 + |x| = -7 + 30$
$|x| = 23$
x represents a number whose distance from 0 is 23. Thus $x = -23$ or $x = 23$.

101. $t - 3590 = 1820$
$t - 3590 + 3590 = 1820 + 3590$
$t = 5410$
$t + 3590 = 5410 + 3590$
$t + 3590 = 9000$

103. To "undo" the last step, divide 225 by 0.3.
$225 \div 0.3 = 750$
Now divide 750 by 0.3.
$750 \div 0.3 = 2500$
The answer should be 2500 not 225.

Exercise Set 2.2

1. To isolate x in $x - 4 = 7$, we would use the *Addition principle*. Add 4 to both sides of the equation.

3. To clear fractions or decimals, we use the *Multiplication principle.*

5. To solve $3x - 1 = 8$, we use the *Addition principle* first. Add 1 to both sides of the equation.

7. $3x - 1 = 7$
$3x - 1 + 1 = 7 + 1$ Adding 1 to both sides
$3x = 7 + 1$
Choice (c) is correct.

9. $6(x - 1) = 2$
$6x - 6 = 2$ Using the distributive law
Choice (a) is correct.

11. $4x = 3 - 2x$
$4x + 2x = 3 - 2x + 2x$ Adding $2x$ to both sides
$4x + 2x = 3$
Choice (b) is correct.

13. $2x + 9 = 25$
$2x + 9 - 9 = 25 - 9$ Subtracting 9 from both sides
$2x = 16$ Simplifying
$\frac{2x}{2} = \frac{16}{2}$ Dividing both sides by 2
$x = 8$ Simplifying

Check:
$2x + 9 = 25$	
$2 \cdot 8 + 9$	25
$16 + 9$	

$25 \stackrel{?}{=} 25$ TRUE

The solution is 8.

15. $7t - 8 = 27$
$7t - 8 + 8 = 27 + 8$ Adding 8 to both sides
$7t = 35$
$\frac{7t}{7} = \frac{35}{7}$ Dividing both sides by 7
$t = 5$

Check:
$7t - 8 = 27$	
$7 \cdot 5 - 8$	27
$35 - 8$	

$27 \stackrel{?}{=} 27$ TRUE

The solution is 5.

17. $3x - 9 = 1$
$3x - 9 + 9 = 1 + 9$
$3x = 10$
$\frac{3x}{3} = \frac{10}{3}$
$x = \frac{10}{3}$

Check:
$3x - 9 = 1$	
$3 \cdot \frac{10}{3} - 9$	1
$10 - 9$	

$1 \stackrel{?}{=} 1$ TRUE

The solution is $\frac{10}{3}$.

19. $8z+2=-54$
$8z+2-2=-54-2$
$8z=-56$
$\frac{8z}{8}=\frac{-56}{8}$
$z=-7$

Check: $8z+2=-54$
$8(-7)+2 \mid -54$
$-56+2$
$-54\stackrel{?}{=}-54$ TRUE

The solution is –7.

21. $-37=9t+8$
$-37-8=9t+8-8$
$-45=9t$
$\frac{-45}{9}=\frac{9t}{9}$
$-5=t$

Check: $-37=9t+8$
$-37 \mid 9\cdot(-5)+8$
$-45+8$
$-37\stackrel{?}{=}-37$ TRUE

The solution is –5.

23. $12-t=16$
$-12+12-t=-12+16$
$-t=4$
$\frac{-t}{-1}=\frac{4}{-1}$
$t=-4$

Check: $12-t=16$
$12-(-4) \mid 16$
$12+4$
$16\stackrel{?}{=}16$ TRUE

The solution is –4.

25. $-6z-18=-132$
$-6z-18+18=-132+18$
$-6z=-114$
$\frac{-6z}{-6}=\frac{-114}{-6}$
$z=19$

Check: $-6z-18=-132$
$-6\cdot 19-18 \mid -132$
$-114-18$
$-132\stackrel{?}{=}-132$ TRUE

The solution is 19.

27. $5.3+1.2n=1.94$
$1.2n=-3.36$
$\frac{1.2n}{1.2}=\frac{-3.36}{1.2}$
$n=-2.8$

Check: $5.31+1.2n=1.94$
$5.3+1.2(-2.8) \mid 1.94$
$5.3+(-3.36)$
$1.94\stackrel{?}{=}1.94$ TRUE

The solution is –2.8.

29. $32-7x=11$
$-32+32-7x=-32+11$
$-7x=-21$
$\frac{-7x}{-7}=\frac{-21}{-7}$
$x=3$

Check: $32-7x=11$
$32-7\cdot 3 \mid 11$
$32-21$
$11\stackrel{?}{=}11$ TRUE

The solution is 3.

31. $\frac{3}{5}t-1=8$
$\frac{3}{5}t-1+1=8+1$
$\frac{3}{5}t=9$
$\frac{5}{3}\cdot\frac{3}{5}t=\frac{5}{3}\cdot 9$
$t=\frac{5\cdot \cancel{3}\cdot 3}{\cancel{3}\cdot 1}$
$t=15$

Check: $\frac{3}{5}t-1=8$
$\frac{3}{5}\cdot 15-1 \mid 8$
$9-1$
$8\stackrel{?}{=}8$ TRUE

The solution is 15.

33. $6+\frac{7}{2}x=-15$
$-6+6+\frac{7}{2}x=-6-15$
$\frac{7}{2}x=-21$
$\frac{2}{7}\cdot\frac{7}{2}x=\frac{2}{7}(-21)$
$x=-\frac{2\cdot 3\cdot \cancel{7}}{\cancel{7}\cdot 1}$
$x=-6$

Check: $6+\frac{7}{2}x=-15$
$6+\frac{7}{2}(-6) \mid -15$
$6+(-21)$
$-15\stackrel{?}{=}-15$ TRUE

The solution is –6.

35. $-\frac{4a}{5}-8=2$

$-\frac{4a}{5}-8+8=2+8$

$-\frac{4a}{5}=10$

$-\frac{5}{4}\left(-\frac{4a}{5}\right)=-\frac{5}{4}\cdot 10$

$a=-\frac{5\cdot 5\cdot \not{2}}{2\cdot \not{2}}$

$a=-\frac{25}{2}$

Check: $-\frac{4a}{5}-8=2$

$-\frac{4}{5}\left(-\frac{25}{2}\right)-8$	2
$10-8$	

$2\stackrel{?}{=}2$ TRUE

The solution is $-\frac{25}{2}$.

37. $6x+10x=18$

$16x=18$ Combining like terms

$\frac{16x}{16}=\frac{18}{16}$

$x=\frac{9}{8}$

Check: $6x+10x=18$

$6\left(\frac{9}{8}\right)+10\left(\frac{9}{8}\right)$	18
$\frac{27}{4}+\frac{45}{4}$	

$18\stackrel{?}{=}18$ TRUE

The solution is $\frac{9}{8}$.

39. $4x-6=6x$

$-6=6x-4x$ Subtracting $4x$ from both sides

$-6=2x$ Simplifying

$\frac{-6}{2}=\frac{2x}{2}$ Dividing both sides by 2

$-3=x$

Check: $4x-6=6x$

$4(-3)-6$	$6(-3)$
$-12-6$	-18

$-18\stackrel{?}{=}-18$ TRUE

The solution is -3.

41. $2-5y=26-y$

$2-5y+y=26-y+y$ Adding y to both sides

$2-4y=26$ Simplifying

$-2+2-4y=-2+26$ Adding -2 to both sides

$-4y=24$ Simplifying

$\frac{-4y}{-4}=\frac{24}{-4}$ Dividing both sides by -4

$y=-6$

Check: $2-5y=26-y$

$2-5(-6)$	$26-(-6)$
$2+30$	$26+6$

$32\stackrel{?}{=}32$ TRUE

The solution is -6.

43. $6x+3=2x+3$

$6x-2x=3-3$

$4x=0$

$\frac{4x}{4}=\frac{0}{4}$

$x=0$

Check: $6x+3=2x+3$

$6\cdot 0+3$	$2\cdot 0+3$
$0+3$	$0+3$

$3\stackrel{?}{=}3$ TRUE

The solution is 0.

45. $5-2x=3x-7x+25$

$5-2x=-4x+25$

$4x-2x=25-5$

$2x=20$

$\frac{2x}{2}=\frac{20}{2}$

$x=10$

Check: $5-2x=3x-7x+25$

$5-2\cdot 10$	$3\cdot 10-7\cdot 10+25$
$5-20$	$30-70+25$
-15	$-40+25$

$-15\stackrel{?}{=}-15$ TRUE

The solution is 10.

47. $7+3x-6=3x+5-x$

$3x+1=2x+5$ Combining like terms on each side

$3x-2x=5-1$

$x=4$

Check: $7+3x-6=3x+5-x$

$7+3\cdot 4-6$	$3\cdot 4+5-4$
$7+12-6$	$12+5-4$
$19-6$	$17-4$

$13\stackrel{?}{=}13$ TRUE

The solution is 4.

49. $\frac{2}{3}+\frac{1}{4}t=2$

The number 12 is the least common denominator, so we multiply by 12 on both sides.

$12\left(\frac{2}{3}+\frac{1}{4}t\right)=12\cdot 2$

$12\cdot\frac{2}{3}+12\cdot\frac{1}{4}t=24$

$8+3t=24$

$3t=24-8$

$3t=16$

$t=\frac{16}{3}$

Check: $\frac{2}{3}+\frac{1}{4}t=2$

$$\begin{array}{c|c} \frac{2}{3}+\frac{1}{4}\left(\frac{16}{3}\right) & 2 \\ \frac{2}{3}+\frac{4}{3} & \end{array}$$

$2 \stackrel{?}{=} 2$ TRUE

The solution is $\frac{16}{3}$.

51. $\frac{2}{3}+4t=6t-\frac{2}{15}$

The number 15 is the least common denominator, so we multiply by 15 on both sides.

$$\begin{aligned} 15\left(\frac{2}{3}+4t\right)&=15\left(6t-\frac{2}{15}\right)\\ 15\cdot\frac{2}{3}+15\cdot 4t&=15\cdot 6t-15\cdot\frac{2}{15}\\ 10+60t&=90t-2\\ 10+2&=90t-60t\\ 12&=30t\\ \frac{12}{30}&=t\\ \frac{2}{5}&=t \end{aligned}$$

Check: $\frac{2}{3}+4t=6t-\frac{2}{15}$

$$\begin{array}{c|c} \frac{2}{3}+4\cdot\frac{2}{5} & 6\cdot\frac{2}{5}-\frac{2}{15} \\ \frac{2}{3}+\frac{8}{5} & \frac{12}{5}-\frac{2}{15} \\ \frac{10}{15}+\frac{24}{15} & \frac{36}{15}-\frac{2}{15} \end{array}$$

$\frac{34}{15} \stackrel{?}{=} \frac{34}{15}$ TRUE

The solution is $\frac{2}{5}$.

53. $\frac{1}{3}x+\frac{2}{5}=\frac{4}{5}+\frac{3}{5}x-\frac{2}{3}$

The number 15 is the least common denominator, so we multiply by 15 on both sides.

$$\begin{aligned} 15\left(\frac{1}{3}x+\frac{2}{5}\right)&=15\left(\frac{4}{5}+\frac{3}{5}x-\frac{2}{3}\right)\\ 15\cdot\frac{1}{3}x+15\cdot\frac{2}{5}&=15\cdot\frac{4}{5}+15\cdot\frac{3}{5}x-15\cdot\frac{2}{3}\\ 5x+6&=12+9x-10\\ 5x+6&=2+9x\\ 5x-9x&=2-6\\ -4x&=-4\\ \frac{-4x}{-4}&=\frac{-4}{-4}\\ x&=1 \end{aligned}$$

Check: $\frac{1}{3}x+\frac{2}{5}=\frac{4}{5}+\frac{3}{5}x-\frac{2}{3}$

$$\begin{array}{c|c} \frac{1}{3}\cdot 1+\frac{2}{5} & \frac{4}{5}+\frac{3}{5}\cdot 1-\frac{2}{3} \\ \frac{1}{3}+\frac{2}{5} & \frac{4}{5}+\frac{3}{5}-\frac{2}{3} \\ \frac{5}{15}+\frac{6}{15} & \frac{12}{15}+\frac{9}{15}-\frac{10}{15} \\ \frac{11}{15} & \frac{11}{15} \end{array}$$

$\frac{11}{15} \stackrel{?}{=} \frac{11}{15}$ TRUE

The solution is 1.

55. $2.1x+45.2=3.2-8.4x$

Greatest number of decimal places is 1

$10(2.1x+45.2)=10(3.2-8.4x)$

Multiplying by 10 to clear decimals

$$\begin{aligned} 10(2.1x)+10(45.2)&=10(3.2)-10(8.4x)\\ 21x+452&=32-84x\\ 21x+84x&=32-452\\ 105x&=-420\\ x&=\frac{-420}{105}\\ x&=-4 \end{aligned}$$

Check: $2.1x+45.2=3.2-8.4x$

$$\begin{array}{c|c} 2.1(-4)+45.2 & 3.2-8.4(-4) \\ -8.4+45.2 & 3.2+33.6 \end{array}$$

$36.8 \stackrel{?}{=} 36.8$ TRUE

The solution is –4.

57. $0.76+0.21t=0.96t-0.49$

Greatest number of decimal places is 2

$100(0.76+0.21t)=100(0.96t-0.49)$

Multiplying by 100 to clear decimals

$$\begin{aligned} 100(0.76)+100(0.21t)&=100(0.96t)-100(0.49)\\ 76+21t&=96t-49\\ 76+49&=96t-21t\\ 125&=75t\\ \frac{125}{75}&=t\\ \frac{5}{3}&=t, \text{ or}\\ 1.\overline{6}&=t \end{aligned}$$

The answer checks. The solution is $\frac{5}{3}$, or $1.\overline{6}$.

59. $\frac{2}{5}x-\frac{3}{2}x=\frac{3}{4}x+3$

The least common denominator is 20.

$$\begin{aligned} 20\left(\frac{2}{5}x-\frac{3}{2}x\right)&=20\left(\frac{3}{4}x+3\right)\\ 20\cdot\frac{2}{5}x-20\cdot\frac{3}{2}x&=20\cdot\frac{3}{4}x+20\cdot 3\\ 8x-30x&=15x+60\\ -22x&=15x+60\\ -22x-15x&=60\\ -37x&=60\\ x&=-\frac{60}{37} \end{aligned}$$

Check: $\frac{2}{5}x-\frac{3}{2}x=\frac{3}{4}x+3$

$$\begin{array}{c|c} \frac{2}{5}\left(-\frac{60}{37}\right)-\frac{3}{2}\left(-\frac{60}{37}\right) & \frac{3}{4}\left(-\frac{60}{37}\right)+3 \\ -\frac{24}{37}+\frac{90}{37} & -\frac{45}{37}+\frac{111}{37} \\ \frac{66}{37} \stackrel{?}{=} \frac{66}{37} & \text{TRUE} \end{array}$$

The solution is $-\frac{60}{37}$.

61.
$$\begin{aligned} \frac{1}{3}(2x-1)&=7 \\ 3\cdot\frac{1}{3}(2x-1)&=3\cdot 7 \\ 2x-1&=21 \\ 2x&=22 \\ x&=11 \end{aligned}$$

Check: $\frac{1}{3}(2x-1)=7$

$$\begin{array}{c|c} \frac{1}{3}(2\cdot 11-1) & 7 \\ \frac{1}{3}\cdot 21 & \\ 7 \stackrel{?}{=} 7 & \text{TRUE} \end{array}$$

The solution is 11.

63.
$$\begin{aligned} 7(2a-1)&=21 \\ 14a-7&=21 && \text{Using the distributive law} \\ 14a&=21+7 && \text{Adding 7} \\ 14a&=28 \\ a&=2 && \text{Dividing by 14} \end{aligned}$$

Check: $7(2a-1)=21$

$$\begin{array}{c|c} 7(2\cdot 2-1) & 21 \\ 7(4-1) & \\ 7\cdot 3 & \\ 21 \stackrel{?}{=} 21 & \text{TRUE} \end{array}$$

The solution is 2.

65. We can write $11=11(x+1)$ as $11\cdot 1=11(x+1)$. Then $1=x+1$, or $x=0$. The solution is 0.

67.
$$\begin{aligned} 2(3+4m)-6&=48 \\ 6+8m-6&=48 \\ 8m&=48 && \text{Combining like terms} \\ m&=6 \end{aligned}$$

Check: $2(3+4m)-6=48$

$$\begin{array}{c|c} 2(3+4\cdot 6)-6 & 48 \\ 2(3+24)-6 & \\ 2\cdot 27-6 & \\ 54-6 & \\ 48 \stackrel{?}{=} 48 & \text{TRUE} \end{array}$$

The solution is 6.

69. $2x=x+x$
$2x=2x$ TRUE
Identity; the solution set is all real numbers.

71.
$$\begin{aligned} 2r+8&=6r+10 \\ 2r+8-10&=6r+10-10 \\ 2r-2&=6r && \text{Combining like terms} \\ -2r+2r-2&=-2r+6r \\ -2&=4r \\ \frac{-2}{4}&=\frac{4r}{4} \\ -\frac{1}{2}&=r \end{aligned}$$

Check: $2r+8=6r+10$

$$\begin{array}{c|c} 2\left(-\frac{1}{2}\right)+8 & 6\left(-\frac{1}{2}\right)+10 \\ -1+8 & -3+10 \\ 7 \stackrel{?}{=} 7 & \text{TRUE} \end{array}$$

The solution is $-\frac{1}{2}$.

73.
$$\begin{aligned} 4y-4+y+24&=6y+20-4y \\ 5y+20&=2y+20 \\ 5y-2y&=20-20 \\ 3y&=0 \\ y&=0 \end{aligned}$$

Check: $4y-4+y+24=6y+20-4y$

$$\begin{array}{c|c} 4\cdot 0-4+0+24 & 6\cdot 0+20-4\cdot 0 \\ 0-4+0+24 & 0+20-0 \\ 20 \stackrel{?}{=} 20 & \text{TRUE} \end{array}$$

The solution is 0.

75.
$$\begin{aligned} 3(x+4)&=3(x-1) \\ 3x+12&=3x-3 \\ 3x-3x+12&=3x-3x-3 \\ 12&=-3 \quad \text{FALSE} \end{aligned}$$

Contradiction; there is no solution.

77.
$$\begin{aligned} 19-3(2x-1)&=7 \\ 19-6x+3&=7 \\ 22-6x&=7 \\ -6x&=7-22 \\ -6x&=-15 \\ x&=\frac{15}{6} \\ x&=\frac{5}{2} \end{aligned}$$

Check: $19-3(2x-1)=7$

$$\begin{array}{c|c} 19-3\left(2\cdot\frac{5}{2}-1\right) & 7 \\ 19-3(5-1) & \\ 19-3(4) & \\ 19-12 & \\ 7 \stackrel{?}{=} 7 & \text{TRUE} \end{array}$$

The solution is $\frac{5}{2}$.

79. $2(3t+1)-5=t-(t+2)$
$6t+2-5=t-t-2$
$6t-3=-2$
$6t=-2+3$
$6t=1$
$t=\frac{1}{6}$

Check: $2(3t+1)-5=t-(t+2)$

$2\left(3\cdot\frac{1}{6}+1\right)-5$	$\frac{1}{6}-\left(\frac{1}{6}+2\right)$
$2\left(\frac{1}{2}+1\right)-5$	$\frac{1}{6}-\frac{1}{6}-2$
$2\cdot\frac{3}{2}-5$	-2

$-2\overset{?}{=}-2$ TRUE

The solution is $\frac{1}{6}$.

81. $19-(2x+3)=2(x+3)+x$
$19-2x-3=2x+6+x$
$16-2x=3x+6$
$16-6=3x+2x$
$10=5x$
$2=x$

Check: $19-(2x+3)=2(x+3)+x$

$19-(2\cdot2+3)$	$2(2+3)+2$
$19-(4+3)$	$2\cdot5+2$
$19-7$	$10+2$

$12\overset{?}{=}12$ TRUE

The solution is 2.

83. $4+7x=7(x+1)$
$4+7x=7x+7$
$4+7x-7x=7x-7x+7$
$4=7$ FALSE

Contradiction; there is no solution.

85. $\frac{3}{4}(3t-4)=15$
$\frac{4}{3}\cdot\frac{3}{4}(3t-4)=\frac{4}{3}\cdot15$
$3t-4=20$
$3t=24$
$t=8$

Check: $\frac{3}{4}(3t-4)=15$

$\frac{3}{4}(3\cdot8-4)$	15
$\frac{3}{4}\cdot(24-4)$	
$\frac{3}{4}\cdot20$	

$15\overset{?}{=}15$ TRUE

The solution is 8.

87. $\frac{1}{6}\left(\frac{3}{4}x-2\right)=-\frac{1}{5}$
$30\cdot\frac{1}{6}\left(\frac{3}{4}x-2\right)=30\left(-\frac{1}{5}\right)$
$5\left(\frac{3}{4}x-2\right)=-6$
$\frac{15}{4}x-10=-6$
$\frac{15}{4}x=4$
$4\cdot\frac{15}{4}x=4\cdot4$
$15x=16$
$x=\frac{16}{15}$

Check: $\frac{1}{6}\left(\frac{3}{4}x-2\right)=-\frac{1}{5}$

$\frac{1}{6}\left(\frac{3}{4}\cdot\frac{16}{15}-2\right)$	$-\frac{1}{5}$
$\frac{1}{6}\left(\frac{4}{5}-2\right)$	
$\frac{1}{6}\left(-\frac{6}{5}\right)$	

$-\frac{1}{5}\overset{?}{=}-\frac{1}{5}$ TRUE

The solution is $\frac{16}{15}$.

89. $0.7(3x+6)=1.1-(x-3)$
$2.1x+4.2=1.1-x+3$
$2.1x+4.2=-x+4.1$
$10(2.1x+4.2)=10(-x+4.1)$ Clearing decimals
$21x+42=-10x+41$
$21x=-10x+41-42$
$21x=-10x-1$
$31x=-1$
$x=-\frac{1}{31}$

The check is left to the student. The solution is $-\frac{1}{31}$.

91. $2(7-x)-20=7x-3(2+3x)$
$14-2x-20=7x-6-9x$
$-2x-6=-2x-6$ TRUE

Identity; the solution set is all real numbers.

93. $a+(a-3)=(a+2)-(a+1)$
$a+a-3=a+2-a-1$
$2a-3=1$
$2a=1+3$
$2a=4$
$a=2$

Check: $a+(a-3)=(a+2)-(a+1)$

$2+(2-3)$	$(2+2)-(2+1)$
$2-1$	$4-3$

$1\overset{?}{=}1$ TRUE

The solution is 2.

95. *Writing Exercise.*

97. $\frac{2}{9}+\frac{1}{6}=\frac{4}{18}+\frac{3}{18}=\frac{7}{18}$

99. $9\overline{)1.000}$ = 0.111...

$$\begin{array}{r} 0.111\ldots \\ 9\overline{)1.000} \\ \underline{9} \\ 10 \\ \underline{9} \\ 10 \\ \underline{9} \\ 1 \end{array}$$

$\frac{1}{9} = 0.\overline{1}$, so $-\frac{1}{9} = -0.\overline{1}$.

101. *Writing Exercise.*

103.
$$\begin{aligned} 8.43x - 2.5(3.2 - 0.7x) &= -3.455x + 9.04 \\ 8.43x - 8 + 1.75x &= -3.455x + 9.04 \\ 10.18x - 8 &= -3.455x + 9.04 \\ 10.18x + 3.455x &= 9.04 + 8 \\ 13.635x &= 17.04 \\ x &= 1.\overline{2497}, \text{ or } \frac{1136}{909} \end{aligned}$$

The solution is $1.\overline{2497}$, or $\frac{1136}{909}$.

105.
$$\begin{aligned} -2[3(x-2)+4] &= 4(5-x) - 2x \\ -2[3x - 6 + 4] &= 20 - 4x - 2x \\ -2[3x - 2] &= 20 - 6x \\ -6x + 4 &= 20 - 6x \\ 4 &= 20 \quad \text{Adding } 6x \text{ to both sides} \end{aligned}$$

This is a contradiction. No solution.

107.
$$\begin{aligned} 3(x+5) &= 3(5+x) \\ 3x + 15 &= 15 + 3x \\ 3x + 15 - 15 &= 15 - 15 + 3x \\ 3x &= 3x \end{aligned}$$

Identity; the solution set is all real numbers.

109.
$$\begin{aligned} 2x(x+5) - 3(x^2 + 2x - 1) &= 9 - 5x - x^2 \\ 2x^2 + 10x - 3x^2 - 6x + 3 &= 9 - 5x - x^2 \\ -x^2 + 4x + 3 &= 9 - 5x - x^2 \\ 4x + 3 &= 9 - 5x \quad \text{Adding } x^2 \\ 4x + 5x &= 9 - 3 \\ 9x &= 6 \\ x &= \frac{2}{3} \end{aligned}$$

The solution is $\frac{2}{3}$.

111. $[7 - 2(8 \div (-2))]x = 0$

Since $7 - 2(8 \div (-2)) \neq 0$ and the product on the left side of the equation is 0, then x must be 0.

113. Let x represent the number of miles.

Translating we have:

$$\frac{3}{200}x + \frac{1}{4}(2) = 8$$

Solve the equation for x.

$$\begin{aligned} \frac{3}{200}x + \frac{1}{4}(2) &= 8 \\ \frac{3}{200}x + \frac{1}{2} &= 8 \\ 200\left(\frac{3}{200}x + \frac{1}{2}\right) &= 200(8) \\ 3x + 100 &= 1600 \\ 3x &= 1500 \\ \frac{3x}{3} &= \frac{1500}{3} \\ x &= 500 \end{aligned}$$

He will drive 500 miles.

Exercise Set 2.3

1. False. For example, π represents a constant.

3. The distance around a circle is its *circumference*.

5. We substitute 0.9 for t and calculate d.

$$d = 344t = 344 \cdot 0.9 = 309.6$$

The fans were 309.6 m from the stage.

7. We substitute 21,345 for n and calculate f.

$$f = \frac{n}{15} = \frac{21{,}345}{15} = 1423$$

There are 1423 full-time equivalent students.

9. We substitute 0.025 for I and 0.044 for U and calculate f.

$$\begin{aligned} f &= 8.5 + 1.4(I - U) \\ &= 8.5 + 1.4(0.025 - 0.044) \\ &= 8.5 + 1.4(-0.019) \\ &= 8.5 - 0.0266 \\ &= 8.4734 \end{aligned}$$

The federal funds rate should be 8.4734.

11. Substitute 1 for t and calculate n.

$$\begin{aligned} n &= 0.5t^4 + 3.45t^3 - 96.65t^2 + 347.7t \\ &= 0.5(1)^4 + 3.45(1)^3 - 96.65(1)^2 + 347.7(1) \\ &= 0.5 + 3.45 - 96.65 + 347.7 \\ &= 255 \end{aligned}$$

255 mg of ibuprofen remain in the bloodstream.

13.
$$\begin{aligned} A &= bh \\ \frac{A}{h} &= \frac{bh}{h} \quad \text{Dividing both sides by } h \\ \frac{A}{h} &= b \end{aligned}$$

15.
$$\begin{aligned} I &= Prt \\ \frac{I}{rt} &= \frac{Prt}{rt} \quad \text{Dividing both sides by } rt \\ \frac{I}{rt} &= P \end{aligned}$$

17.
$$\begin{aligned} H &= 65 - m \\ H + m &= 65 \quad \text{Adding } m \text{ to both sides} \\ m &= 65 - H \quad \text{Subtracting } H \text{ from both sides} \end{aligned}$$

19. $P = 2l + 2w$

$P - 2w = 2l + 2w - 2w$ Subtracting $2w$ from both sides

$P - 2w = 2l$

$\frac{P-2w}{2} = \frac{2l}{2}$ Dividing both sides by 2

$\frac{P-2w}{2} = l$, or $\frac{P}{2} - w = l$

21. $A = \pi r^2$

$\frac{A}{r^2} = \frac{\pi r^2}{r^2}$

$\frac{A}{r^2} = \pi$

23. $A = \frac{1}{2}bh$

$2A = 2 \cdot \frac{1}{2}bh$ Multiplying both sides by 2

$2A = bh$

$\frac{2A}{b} = \frac{bh}{b}$ Dividing both sides by h

$\frac{2A}{b} = h$

25. $E = mc^2$

$\frac{E}{m} = \frac{mc^2}{m}$ Dividing both sides by m

$\frac{E}{m} = c^2$

27. $Q = \frac{c+d}{2}$

$2Q = 2 \cdot \frac{c+d}{2}$ Multiplying both sides by 2

$2Q = c + d$

$2Q - c = c + d - c$ Subtracting c from both sides

$2Q - c = d$

29. $p - q + r = 2$

$p + r = 2 + q$

$p + r - 2 = q$

31. $w = \frac{r}{f}$

$f \cdot w = f \cdot \frac{r}{f}$ Multiplying both sides by f

$fw = r$

33. $H = \frac{TV}{550}$

$\frac{550}{V} \cdot H = \frac{550}{V} \cdot \frac{TV}{550}$ Multiplying both sides by $\frac{550}{V}$

$\frac{550H}{V} = T$

35. $F = \frac{9}{5}C + 32$

$F - 32 = \frac{9}{5}C$

$\frac{5}{9}(F - 32) = \frac{5}{9} \cdot \frac{9}{5}C$

$\frac{5}{9}(F - 32) = C$

37. $2x - y = 1$

$2x - y + y - 1 = 1 + y - 1$ Adding $y - 1$ to both sides

$2x - 1 = y$

39. $2x + 5y = 10$

$5y = -2x + 10$

$y = \frac{-2x+10}{5}$

$y = -\frac{2}{5}x + 2$

41. $4x - 3y = 6$

$-3y = -4x + 6$

$y = \frac{-4x+6}{-3}$

$y = \frac{4}{3}x - 2$

43. $9x + 8y = 4$

$8y = -9x + 4$

$y = \frac{-9x+4}{8}$

$y = -\frac{9}{8}x + \frac{1}{2}$

45. $3x - 5y = 8$

$-5y = -3x + 8$

$y = \frac{-3x+8}{-5}$

$y = \frac{3}{5}x - \frac{8}{5}$

47. $z = 13 + 2(x + y)$

$z - 13 = 2(x + y)$

$z - 13 = 2x + 2y$

$z - 13 - 2y = 2x$

$\frac{z-13-2y}{2} = x$

$\frac{1}{2}z - \frac{13}{2} - y = x$

49. $t = 27 - \frac{1}{4}(w - l)$

$t - 27 = -\frac{1}{4}(w - l)$

$-4(t - 27) = w - l$ Multiplying by -4

$-4(t - 27) = w - l$

$-4(t - 27) - w = -l$

$4(t - 27) + w = l$ Multiplying by -1

51. $A = at + bt$

$A = t(a + b)$ Factoring

$\frac{A}{a+b} = t$ Dividing both sides by $a + b$

53.
$$A = \frac{1}{2}ah + \frac{1}{2}bh$$
$$2A = 2\left(\frac{1}{2}ah + \frac{1}{2}bh\right)$$
$$2A = ah + bh$$
$$2A = h(a+b)$$
$$\frac{2A}{a+b} = h$$

55.
$$R = r + \frac{400(W-L)}{N}$$
$$N \cdot R = N\left(r + \frac{400(W-L)}{N}\right) \quad \text{Multiplying both sides by } N$$
$$NR = Nr + 400(W-L)$$
$$NR = Nr + 400W - 400L$$
$$NR + 400L = Nr + 400W \quad \text{Adding } 400L \text{ to both sides}$$
$$400L = Nr + 400W - NR \quad \text{Adding } -NR \text{ to both sides}$$
$$L = \frac{Nr + 400W - NR}{400}, \text{ or } L = W - \frac{N(R-r)}{400}$$

57. *Writing Exercise.*

59. $-2+5-(-4)-17 = -2+5+4-17$
$= 3+4-17 = 7-17 = -10$

61. $4.2(-11.75)(0) = 0$

63. $20 \div (-4) \cdot 2 - 3$
$= -5 \cdot 2 - 3$ Dividing and multiplying from left to right
$= -10 - 3$
$= -13$ Subtracting

65. *Writing Exercise.*

67.
$$K = 21.235w + 7.75h - 10.54a + 102.3$$
$$2852 = 21.235(80) + 7.75(190) - 10.54a + 102.3$$
$$2852 = 1698.8 + 1472.5 - 10.54a + 102.3$$
$$2852 = 3273.6 - 10.54a$$
$$-421.6 = -10.54a$$
$$40 = a$$

The man is 40 years old.

69. First we substitute 54 for A and solve for s to find the length of a side of the cube.
$$A = 6s^2$$
$$54 = 6s^2$$
$$9 = s^2$$
$$3 = s \quad \text{Taking the positive square root}$$

Now we substitute 3 for s in the formula for the volume of a cube and compute the volume.
$$V = s^3 = 3^3 = 27$$

The volume of the cube is 27 in^3.

71.
$$c = \frac{w}{a} \cdot d$$
$$ac = a \cdot \frac{w}{a} \cdot d$$
$$ac = wd$$
$$a = \frac{wd}{c}$$

73.
$$ac = bc + d$$
$$ac - bc = d$$
$$c(a-b) = d$$
$$c = \frac{d}{a-b}$$

75.
$$3a = c - a(b+d)$$
$$3a = c - ab - ad$$
$$3a + ab + ad = c$$
$$a(3+b+d) = c$$
$$a = \frac{c}{3+b+d}$$

77. $K = 21.235w + 7.75h - 10.54a + 102.3$
$$K = 21.235\left(\frac{w}{2.2046}\right) + 7.75\left(\frac{h}{0.3937}\right) - 10.54a + 102.3$$
$$K = 9.632w + 19.685h - 10.54a + 102.3$$

Mid-Chapter Review

1. $2x + 3 - 3 = 10 - 3$
$$2x = 7$$
$$\frac{1}{2} \cdot 2x = \frac{1}{2} \cdot 7$$
$$x = \frac{7}{2}$$

The solution is $\frac{7}{2}$.

2. $6 \cdot \frac{1}{2}(x-3) = 6 \cdot \frac{1}{3}(x-4)$
$$3(x-3) = 2(x-4)$$
$$3x - 9 + 9 = 2x - 8 + 9$$
$$3x - 9 = 2x - 8$$
$$3x = 2x + 1$$
$$3x - 2x = 2x + 1 - 2x$$
$$x = 1$$

3.
$$x - 2 = -1$$
$$x - 2 + 2 = -1 + 2$$
$$x = 1$$

The solution is 1.

4.
$$2 - x = -1$$
$$-x = -3$$
$$-1(-x) = -1(-3)$$
$$x = 3$$

5. $3t = 5$
$$\frac{3t}{3} = \frac{5}{3}$$
$$t = \frac{5}{3}$$

The solution is $\frac{5}{3}$.

6.
$$-\frac{3}{2}x = 12$$
$$-\frac{2}{3}\left(-\frac{3}{2}x\right) = -\frac{2}{3} \cdot 12$$
$$x = -8$$

7. $\frac{y}{8}=6$
$8\cdot\frac{y}{8}=8\cdot 6$
$y=48$
The solution is 48.

8. $0.06x=0.03$
$\frac{0.06x}{0.06}=\frac{0.03}{0.06}$
$x=0.5$

9. $3x-7x=20$
$-4x=20$
$\frac{-4x}{-4}=\frac{20}{-4}$
$x=-5$
The solution is –5.

10. $9x-7=17$
$9x=24$
$x=\frac{8}{3}$

11. $4(t-3)-t=6$
$4t-12-t=6$
$3t-12=6$
$3t-12+12=6+12$
$3t=18$
$\frac{3t}{3}=\frac{18}{3}$
$t=6$
The solution is 6.

12. $8n-(3n-5)=5-n$
$8n-3n+5=5-n$
$5n+5=5-n$
$6n+5=5$
$6n=0$
$n=0$

13. $\frac{9}{10}y-\frac{7}{10}=\frac{21}{5}$
$10\left(\frac{9}{10}y-\frac{7}{10}\right)=10\left(\frac{21}{5}\right)$
$9y-7=42$
$9y-7+7=42+7$
$9y=49$
$\frac{9y}{9}=\frac{49}{9}$
$y=\frac{49}{9}$
The solution is $\frac{49}{9}$.

14. $2(t-5)-3(2t-7)=12-5(3t+1)$
$2t-10-6t+21=12-15t-5$
$-4t+11=7-15t$
$11t+11=7$
$11t=-4$
$t=-\frac{4}{11}$

15. $\frac{2}{3}(x-2)-1=-\frac{1}{2}(x-3)$
$6\cdot\left(\frac{2}{3}(x-2)-1\right)=6\cdot\left(-\frac{1}{2}(x-3)\right)$
$4(x-2)-6=-3(x-3)$
$4x-8-6=-3x+9$
$4x-14=-3x+9$
$4x-14+3x=-3x+9+3x$
$7x-14=9$
$7x=9+14$
$7x=23$
$\frac{7x}{7}=\frac{23}{7}$
$x=\frac{23}{7}$
The solution is $\frac{23}{7}$.

16. $E=wA$
$\frac{E}{w}=A$

17. $Ax+By=C$
$By=C-Ax$
$\frac{By}{B}=\frac{C-Ax}{B}$
$y=\frac{C-Ax}{B}$

18. $at+ap=m$
$a(t+p)=m$
$a=\frac{m}{t+p}$

19. $m=\frac{F}{a}$
$a\cdot m=a\cdot\frac{F}{a}$
$am=F$
$\frac{am}{m}=\frac{F}{m}$
$a=\frac{F}{m}$

20. $v=\frac{b-f}{t}$
$t\cdot v=t\cdot\frac{b-f}{t}$
$tv=b-f$
$tv+f=b$

Exercise Set 2.4

1. To convert from percent notation to decimal notation, move the decimal point two places to the *left* and drop the percent symbol.

3. The expression 1.3% is written in *percent* notation.

5. The *sale* price is the original price minus the discount.

7. “What percent of 57 is 23?” can be translated as $n\cdot 57=23$, so choice (d) is correct.

9. "23 is 57% of what number?" can be translated as $23 = 0.57y$, so choice (e) is correct.

11. "57 is what percent of 23?" can be translated as $n \cdot 23 = 57$, so choice (c) is correct.

13. "What is 23% of 57?" can be translated as $a = (0.23)57$, so choice (f) is correct.

15. "23% of what number is 57?" can be translated as $57 = 0.23y$, so choice (b) is correct.

17. $47\% = 47.0\%$

47% 0.47.0

Move the decimal point 2 places to the left.

$47\% = 0.47$

19. $5\% = 5.0\%$

5% 0.05.0

Move the decimal point 2 places to the left.

$5\% = 0.05$

21. $3.2\% = 3.20\%$

3.2% 0.03.20

Move the decimal point 2 places to the left.

$3.2\% = 0.032$

23. $10\% = 10.0\%$

10% 0.10.0

Move the decimal point 2 places to the left.

$10\% = 0.10$, or 0.1

25. 6.25% 0.06.25

Move the decimal point 2 places to the left.

$6.25\% = 0.0625$

27. 0.2% 0.00.2

Move the decimal point 2 places to the left.

$0.2\% = 0.002$

29. $175\% = 175.0\%$ 1.75.0

Move the decimal point 2 places to the left.

$175\% = 1.75$

31. 0.21

First move the decimal point two places to the right; then write a % symbol: 0.21. → 21%

33. 0.047

First move the decimal point two places to the right; then write a % symbol: 0.04.7 → 4.7%

35. 0.7

First move the decimal point two places to the right; then write a % symbol: 0.70. → 70%

37. 0.0009

First move the decimal point two places to the right; then write a % symbol: 0.00.09 → 0.09%

39. 1.06

First move the decimal point two places to the right; then write a % symbol: 1.06. → 106%

41. $\frac{3}{5}$ $\left(\text{Note: } \frac{3}{5} = 0.6\right)$

Move the decimal point two places to the right; then write a % symbol: 0.60. → 60%

43. $\frac{8}{25}$ $\left(\text{Note: } \frac{8}{25} = 0.32\right)$

First move the decimal point two places to the right; then write a % symbol: 0.32. → 32%

45. ***Translate.***

What percent of 76 is 19?

↓ ↓ ↓ ↓ ↓

$y \cdot 76 = 19$

We solve the equation and then convert to percent notation.

$$y \cdot 76 = 19$$
$$y = \frac{19}{76}$$
$$y = 0.25 = 25\%$$

The answer is 25%.

47. ***Translate.***

14 is 30% of what number?

↓ ↓ ↓ ↓ ↓

$14 = 30\% \cdot y$

We solve the equation.

$$14 = 0.3y \quad (30\% = 0.3)$$
$$\frac{14}{0.3} = y$$
$$46.\overline{6} = y$$

The answer is $46.\overline{6}$, or $46\frac{2}{3}$, or $\frac{140}{3}$.

49. ***Translate.***

0.3 is 12% of what number?

↓ ↓ ↓ ↓ ↓

$0.3 = 12\% \cdot y$

We solve the equation.

$$0.3 = 0.12y \quad (12\% = 0.12)$$
$$\frac{0.3}{0.12} = y$$
$$2.5 = y$$

The answer is 2.5.

51. ***Translate.***

What number is 1% of one million?

$y = 1\% \cdot 1{,}000{,}000$

We solve the equation.

$y = 0.01 \cdot 1{,}000{,}000$ $(1\% = 0.01)$

$y = 10{,}000$ Multiplying

The answer is 10,000.

53. ***Translate.***

What percent of 60 is 75?

$y \cdot 60 = 75$

We solve the equation and then convert to percent notation.

$y \cdot 60 = 75$

$y = \frac{75}{60}$

$y = 1.25 = 125\%$

The answer is 125%.

55. ***Translate.***

What is 2% of 40?

$x = 2\% \cdot 40$

We solve the equation.

$x = 0.02 \cdot 40$ $(2\% = 0.02)$

$x = 0.8$ Multiplying

The answer is 0.8.

57. Observe that 25 is half of 50. Thus, the answer is 0.5, or 50%. We could also do this exercise by translating to an equation.

Translate.

25 is what percent of 50?

$25 = y \cdot 50$

We solve the equation and convert to percent notation.

$25 = y \cdot 50$

$\frac{25}{50} = y$

$0.5 = y$, or $50\% = y$

The answer is 50%.

59. ***Translate.***

What percent of 69 is 23?

$y \cdot 69 = 23$

We solve the equation and convert to percent notation.

$y \cdot 69 = 23y = \frac{23}{69}y = 0.33\overline{3} = 33.\overline{3}\%$ or $33\frac{1}{3}\%$

The answer is $33.\overline{3}\%$ or 33 1/3%.

61. First we reword and translate, letting c represent pet cats from animal shelters, in millions.

What is 15% of 95.6?

$c = 0.15 \cdot 95.6$

$c = 0.15 \cdot 95.6 = 14.34$

There are 14.34 million cats from animal shelters.

63. First we reword and translate, letting c represent pet cats from local animal rescue groups, in millions.

What is 2% of 95.6?

$c = 0.02 \cdot 95.6$

$c = 0.02 \cdot 95.6 = 1.912$

There are 1.912 million cats from local animal rescue groups.

65. First we reword and translate, letting c represent the number of credits Cody has completed.

What is 60% of 125?

$c = 0.6 \cdot 125$

$c = 0.6 \cdot 125 = 75$

Cody has completed 75 credits.

67. First we reword and translate, letting b represent the number of at-bats.

172 is 31.4% of what number?

$172 = 0.314 \cdot b$

$\frac{172}{0.314} = b$

$548 \approx b$

Andrew McCutchen had 548 at-bats.

69. a) First we reword and translate, letting p represent the unknown percent.

What percent of \$25 is \$4?

$p \cdot 25 = 4$

$\frac{p \cdot 25}{25} = \frac{4}{25}$

$p = 0.16 = 16\%$

The tip was 16% of the cost of the meal.

b) We add to find the total cost of the meal, including tip:

\$25 + \$4 = \$29

71. To find the percent of teachers who worked at public and private schools, we first reword and translate, letting p represent the unknown percent.

3.1 million is what percent of 3.5 million?

$3.1 = p \cdot 3.5$

$\frac{3.1}{3.5} = p$

$0.886 \approx p$

$88.6\% \approx p$

About 88.6% of teachers worked in public schools. To find the percent of teacher who worked in private schools, we subtract:

$100\% - 88.6\% = 11.4\%$

About 11.4% of of teachers worked in private schools.

73. Let $I =$ the amount of interest Glenn will pay. Then we have:

$$\begin{array}{ccccc} I & \text{is} & 6.8\% & \text{of} & \$2400. \\ \downarrow & \downarrow & \downarrow & \downarrow & \downarrow \\ I & = & 0.068 & \cdot & \$2400 \\ I & = & \$163.20 & & \end{array}$$

Glenn will pay \$163.20 interest.

75. If $n =$ the number of women who had babies in good or excellent health, we have:

$$\begin{array}{ccccc} n & \text{is} & 95\% & \text{of} & 300. \\ \downarrow & \downarrow & \downarrow & \downarrow & \downarrow \\ n & = & 0.95 & \cdot & 300 \\ n & = & 285 & & \end{array}$$

285 women had babies in good or excellent health.

77. A self-employed person must earn 120% as much as a non-self-employed person. Let $a =$ the amount Tia would need to earn, in dollars per hour, on her own for a comparable income.

$$\begin{array}{ccccc} a & \text{is} & 120\% & \text{of} & \$16. \\ \downarrow & \downarrow & \downarrow & \downarrow & \downarrow \\ a & = & 1.2 & \cdot & 16 \\ a & = & 19.20 & & \end{array}$$

Tia would need to earn \$19.20 per hour on her own.

79. We reword and translate.

$$\begin{array}{ccccc} \underbrace{\text{What percent}} & \text{of} & 2.6 & \text{is} & 12? \\ \downarrow & \downarrow & \downarrow & \downarrow & \downarrow \\ p & \cdot & 2.6 & = & 12 \end{array}$$

$$p \cdot 2.6 = 12$$
$$p \approx 462 = 462\%$$

The actual cost exceeds the initial estimate by about 462%.

81. First we reword and translate.

$$\begin{array}{ccccc} \text{What} & \text{is} & 16.5\% & \text{of} & 191? \\ \downarrow & \downarrow & \downarrow & \downarrow & \downarrow \\ a & = & 0.165 & \cdot & 191 \end{array}$$

Solve. We convert 16.5% to decimal notation and multiply.

$$a = 0.165 \cdot 191$$
$$a = 31.515 \approx 31.5$$

About 31.5 lb of the author's body weight is fat.

83. Let $m =$ the number of e-mails that are spam and viruses. Then we have:

$$\begin{array}{ccccc} \underbrace{\text{What percent}} & \text{of} & 294 & \text{is} & 265? \\ \downarrow & \downarrow & \downarrow & \downarrow & \downarrow \\ p & \cdot & 294 & = & 265 \end{array}$$

$$p = \frac{265}{294}$$
$$p \approx 0.90 = 90\%$$

About 90% of e-mail is spam and viruses.

85. The number of calories in a serving of cranberry juice is 240% of the number of calories in a serving of cranberry juice drink. Let $c =$ the number of calories in a serving of cranberry juice. Then we have:

$$\begin{array}{ccccc} \underbrace{120 \text{ calories}} & \text{is} & 240\% & \text{of} & c. \\ \downarrow & \downarrow & \downarrow & \downarrow & \downarrow \\ 120 & = & 2.40 & \cdot & c \end{array}$$

$$\frac{120}{2.40} = c$$
$$50 = c$$

There are 50 calories in a serving cranberry juice.

87. (a) In the survey report, 40% of all sick days on Monday or Friday sounds excessive. However, for a traditional 5-day business week, 40% is the same as $\frac{2}{5}$. That is, just 2 days out of 5.

(b) In the FBI statistics, 26% of home burglaries occurring between Memorial Day and Labor Day sounds excessive. However, 26% of a 365-day year is 73 days, For the months of June, July, and August there are at least 90 days. So 26% is less than the rate for other times during the year, or less than expected for a 90-day period.

89. The opposite of $-\frac{1}{3}$ is $\frac{1}{3}$.

91. $-(-(-12)) = -12$

93. *Writing Exercise.*

95. Let $p =$ the population of Bardville. Then we have:

$$\begin{array}{ccccccc} 1332 & \text{is} & 15\% & \text{of} & 48\% & \text{of} & \underbrace{\text{the population.}} \\ \downarrow & \downarrow & \downarrow & \downarrow & \downarrow & \downarrow & \downarrow \\ 1332 & = & 0.15 & \cdot & 0.48 & \cdot & p \end{array}$$

$$\frac{1332}{0.15(0.48)} = p$$
$$18{,}500 = p$$

The population of Bardville is 18,500.

97. Since $6 \text{ ft} = 6 \times 1 \text{ ft} = 6 \times 12 \text{ in.} = 72 \text{ in.}$, we can express 6 ft 4 in. as 72 in. + 4 in., or 76 in. We reword and translate. Let $a =$ Jaraan's final adult height.

$$\begin{array}{ccccc} \underbrace{76 \text{ in.}} & \text{is} & 96.1\% & \text{of} & \underbrace{\text{adult height}} \\ \downarrow & \downarrow & \downarrow & \downarrow & \downarrow \\ 76 & = & 0.961 & \cdot & a \end{array}$$

$$\frac{76}{0.961} = a$$
$$79 \approx a$$

Note that 79 in. = 72 in. + 7 in. = 6 ft 7 in. Jaraan's final adult height will be about 6 ft 7 in.

99. Using the formula for the area A of a rectangle with length l and width w, $A = l \cdot w$, we first find the area of the photo.

$$A = 8 \text{ in.} \times 6 \text{ in.} = 48 \text{ in}^2$$

Next we find the area of the photo that will be visible using a mat intended for a 5-in. by 7-in. photo.

$A = 7 \text{ in.} \times 5 \text{ in.} = 35 \text{ in}^2$

Then the area of the photo that will be hidden by the mat is $48 \text{ in}^2 - 35 \text{ in}^2$, or 13 in^2.

We find what percentage of the area of the photo this represents.

$$\underbrace{\text{What percent}} \quad \text{of} \quad \underbrace{48 \text{ in}^2} \quad \text{is} \quad \underbrace{13 \text{ in}^2}?$$

$$p \cdot 48 = 13$$

$$\frac{p \cdot 48}{48} = \frac{13}{48}$$

$$p \approx 0.27$$

$$p \approx 27\%$$

The mat will hide about 27% of the photo.

101. *Writing Exercise.*

Exercise Set 2.5

1. In order, the steps are:
1) Familiarize.
2) Translate.
3) Carry out.
4) Check.
5) State.

3. To write the answer clearly, use the step *State*.

5. To reword the problem, use the step *Translate*.

7. To recall a formula, use the step *Familiarize*.

9. ***Familiarize***. Let n = the number. Then three less than two times the number is $2n - 3$.

Translate.

$$\underbrace{\text{Three less than twice a number}} \quad \text{is} \quad 19.$$

$$2n - 3 = 19$$

Carry out. We solve the equation.

$2n - 3 = 19$

$2n = 22$ Adding 3

$n = 11$ Dividing by 2

Check. Twice 11 is 22, and three fewer is 19. The answer checks.

State. The number is 11.

11. ***Familiarize***. Let a = the number. Then "five times the sum of 3 and twice some number" translates to $5(2a + 3)$.

Translate.

$$\underbrace{\text{Five times the sum of 3 and twice some number}} \quad \text{is} \quad 70.$$

$$5(2a + 3) = 70$$

Carry out. We solve the equation.

$5(2a + 3) = 70$

$10a + 15 = 70$ Using the distributive law

$10a = 55$ Subtracting 15

$a = \frac{11}{2}$ Dividing by 10

Check. The sum of $2 \cdot \frac{11}{2}$ and 3 is 14, and $5 \cdot 14 = 70$. The answer checks.

State. The number is $\frac{11}{2}$.

13. ***Familiarize***. Let d = the kayaker's distance, in miles, from the finish. Then the distance from the start line is $4d$.

Translate.

$$\underbrace{\text{Distance from finish}} \quad \text{plus} \quad \underbrace{\text{distance from start}} \quad \text{is} \quad 20.5 \text{ mi.}$$

$$d + 4d = 20.5$$

Carry out. We solve the equation.

$d + 4d = 20.5$

$5d = 20.5$

$d = 4.1$

Check. If the kayakers are 4.1 mi from the finish, then they are $4 \cdot (4.1)$, or 16.4 mi from the start. Since $4.1 + 16.4$ is 20.5, the total distance, the answer checks.

State. The kayakers had traveled approximately 16.4 mi.

15. ***Familiarize***. Let d = the distance, in miles, that Juan Pablo Montoya had traveled to the given point after the start. Then the distance from the finish line was $500 - d$ miles.

Translate.

$$\underbrace{\text{Distance to finish}} \quad \text{plus} \quad \underbrace{\text{20 mi more}} \quad \text{was} \quad \underbrace{\text{distance to start.}}$$

$$500 - d + 20 = d$$

Carry out. We solve the equation.

$500 - d + 20 = d$

$520 - d = d$

$520 = 2d$

$260 = d$

Check. If Juan Pablo Montoyawas 260 mi from the start, he was 500 – 260, or 240 mi from the finish. Since 240 is 20 more than 260, the answer checks.

State. Juan Pablo Montoyahad traveled 260 mi at the given point.

17. ***Familiarize***. Let n = the number of the smaller apartment number. Then $n+1=$ the number of the larger apartment number.

Translate.

$$\underbrace{\text{Smaller number}}_{\downarrow\; n} \quad \underset{\downarrow\; +}{\text{plus}} \quad \underbrace{\text{larger number}}_{\downarrow\; (n+1)} \quad \underset{\downarrow\; =}{\text{is}} \quad \underset{\downarrow\; 2409}{2409}$$

Carry out. We solve the equation.

$$\begin{aligned} n+(n+1)&=2409\\ 2n+1&=2409\\ 2n&=2408\\ n&=1204 \end{aligned}$$

If the smaller apartment number is 1204, then the other number is $1204+1$, or 1205.

Check. 1204 and 1205 are consecutive numbers whose sum is 2409. The answer checks.

State. The apartment numbers are 1204 and 1205.

19. ***Familiarize***. Let n = the smaller house number. Then $n+2=$ the larger number.

Translate.

$$\underbrace{\text{Smaller number}}_{\downarrow\; n} \quad \underset{\downarrow\; +}{\text{plus}} \quad \underbrace{\text{larger number}}_{\downarrow\; (n+2)} \quad \underset{\downarrow\; =}{\text{is}} \quad \underset{\downarrow\; 572}{572.}$$

Carry out. We solve the equation.

$$\begin{aligned} n+(n+2)&=572\\ 2n+2&=572\\ 2n&=570\\ n&=285 \end{aligned}$$

If the smaller number is 285, then the larger number is $285+2$, or 287.

Check. 285 and 287 are consecutive odd numbers and $285+287=572$. The answer checks.

State. The house numbers are 285 and 287.

21. ***Familiarize***. Let x = the first page number. Then $x+1=$ the second page number, and $x+2=$ the third page number.

Translate.

$$\underbrace{\text{The sum of three consecutive page numbers}}_{\downarrow\; x+(x+1)+(x+2)} \quad \underset{\downarrow\; =}{\text{is}} \quad \underset{\downarrow\; 99}{99.}$$

Carry out. We solve the equation.

$$\begin{aligned} x+(x+1)+(x+2)&=99\\ 3x+3&=99\\ 3x&=96\\ x&=32 \end{aligned}$$

If x is 32, then $x + 1$ is 33 and $x + 2 = 34$.

Check. 32, 33, and 34 are consecutive integers, and $32+33+34=99$. The result checks.

State. The page numbers are 32, 33, and 34.

23. ***Familiarize***. Let m = the man's age. Then $m-2=$ the woman's age.

Translate.

$$\underbrace{\text{Man's age}}_{\downarrow\; m} \quad \underset{\downarrow\; +}{\text{plus}} \quad \underbrace{\text{Woman's age}}_{\downarrow\; (m-2)} \quad \underset{\downarrow\; =}{\text{is}} \quad \underset{\downarrow\; 206}{206.}$$

Carry out. We solve the equation.

$$\begin{aligned} m+(m-2)&=206\\ 2m-2&=206\\ 2m&=208\\ m&=104 \end{aligned}$$

If m is 104, then $m-2$ is 102.

Check. 104 is 2 more than 102, and 104 + 102 = 206. The answer checks.

State. The man was 104 yr old, and the woman was 102 yr old.

25. ***Familiarize***. Familiarize. Let d = the number of dollars lost, in millions on *The 13th Warrior*. Then d + 12.7 is the number dollars lost, in millions on *Mars Needs Moms*.

Translate.

$$\underbrace{\text{The 13th Warrior}}_{\downarrow\; d} \quad \underset{\downarrow\; +}{\text{plus}} \quad \underbrace{\text{Mars Needs Moms}}_{\downarrow\; d+12.7} \quad \underset{\downarrow\; =}{\text{is}} \quad \underset{\downarrow\; 209.3}{209.3}$$

Carry out. We solve the equation.

$$\begin{aligned} d+d+12.7&=209.3\\ 2d&=196.6\\ d&=98.3 \end{aligned}$$

If d is 98.3, then d + 12.7 is 98.3 + 12.7 = 111.

Check. Their total is 98.3 + 111 = 209.3. The answer checks.

State. *The 13th Warrior* lost $98.3 million and *Mars Needs Moms* lost $111 million.

27. ***Familiarize***. The page numbers are consecutive integers. If we let x = the smaller number, then $x+1=$ the larger number.

Translate. We reword the problem.

$$\underbrace{\text{First integer}}_{\downarrow\; x} \quad \underset{\downarrow\; +}{+} \quad \underbrace{\text{Second integer}}_{\downarrow\; (x+1)} \quad \underset{\downarrow\; =}{=} \quad \underset{\downarrow\; 281}{281}$$

Carry out. We solve the equation.

$$\begin{aligned} x+(x+1)&=281\\ 2x+1&=281 && \text{Combining like terms}\\ 2x&=280 && \text{Adding }-1\text{ on both sides}\\ x&=140 && \text{Dividing on both sides by 2} \end{aligned}$$

Check. If $x=140$, then $x+1=141$. These are consecutive integers, and $140+141=281$. The answer checks.

State. The page numbers are 140 and 141.

29. ***Familiarize***. Let $w =$ the width, in meters. Then $w+4$ is the length. The perimeter is twice the length plus twice the width.

Translate.

Twice the width	plus	twice the length	is	92.
↓	↓	↓	↓	↓
$2w$	$+$	$2(w+4)$	$=$	92

Carry out. We solve the equation.

$$2w+2(w+4)=92$$
$$2w+2w+8=92$$
$$4w=84$$
$$w=21$$

Then $w+4=21+4=25$.

Check. The length, 25 m is 4 more than the width, 21 m. The perimeter is $2\cdot 21\text{ m}+2\cdot 25\text{ m}=42\text{ m}+50\text{ m}=92\text{ m}$. The answer checks.

State. The length of the garden is 25 m and the width is 21 m.

31. ***Familiarize***. Let $w =$ the width, in inches. Then $2w =$ the length. The perimeter is twice the length plus twice the width. We express $10\frac{1}{2}$ as 10.5.

Translate.

Twice the length	plus	twice the width	is	10.5 in.
↓	↓	↓	↓	↓
$2\cdot 2w$	$+$	$2w$	$=$	10.5

Carry out. We solve the equation.

$$2\cdot 2w+2w=10.5$$
$$4w+2w=10.5$$
$$6w=10.5$$
$$w=1.75, \text{ or } 1\frac{3}{4}$$

Then $2w=2(1.75)=3.5$, or $3\frac{1}{2}$.

Check. The length, $3\frac{1}{2}$ in., is twice the width, $1\frac{3}{4}$ in. The perimeter is $2\left(3\frac{1}{2}\text{ in.}\right)+2\left(1\frac{3}{4}\text{ in.}\right)=$ $7\text{ in.}+3\frac{1}{2}\text{ in.}=10\frac{1}{2}\text{ in.}$ The answer checks.

State. The actual dimensions are $3\frac{1}{2}$ in. by $1\frac{3}{4}$ in.

33. ***Familiarize***. We draw a picture. We let $x =$ the measure of the first angle. Then $3x =$ the measure of the second angle, and $x+30 =$ the measure of the third angle.

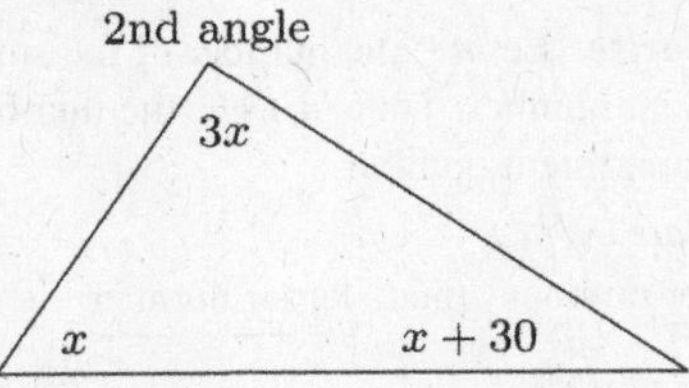

Recall that the measures of the angles of any triangle add up to 180°.

Translate.

Measure of first angle	+	measure of second angle	+	measure of third angle	is	180°.
↓	↓	↓	↓	↓	↓	↓
x	$+$	$3x$	$+$	$x+30$	$=$	180

Carry out. We solve the equation.

$$x+3x+(x+30)=180$$
$$5x+30=180$$
$$5x=150$$
$$x=30$$

Possible answers for the angle measures are as follows:

First angle: $x=30°$

Second angle: $3x=3(30)°=90°$

Third angle: $x+30°=30°+30°=60°$

Check. Consider 30°, 90° and 60°. The second angle is three times the first, and the third is 30° more than the first. The sum of the measures of the angles is 180°. These numbers check.

State. The measure of the first angle is 30°, the measure of the second angle is 90°, and the measure of the third angle is 60°.

35. ***Familiarize***. Let $x =$ the measure of the first angle. Then $4x =$ the measure of the second angle, and $x+4x+5=5x+5$ is the measure of the third angle.

Translate.

Measure of first angle	+	measure of second angle	+	measure of third angle	is	180°.
↓	↓	↓	↓	↓	↓	↓
x	$+$	$4x$	$+$	$(5x+5)$	$=$	180

Carry out. We solve the equation.

$$x+4x+(5x+5)=180$$
$$10x+5=180$$
$$10x=175$$
$$x=17.5$$

If $x=17.5$, then $4x=4(17.5)=70$, and $5x+5=5(17.5)+5=87.5+5=92.5$.

Check. Consider 17.5°, 70°, and 92.5°. The second is four times the first, and the third is 5° more than the sum of the other two. The sum of the measures of the angles is 180°. These numbers check.

State. The measure of the second angle is 70°.

37. ***Familiarize.*** Let b = the length of the bottom section of the rocket, in feet. Then $\frac{1}{6}b =$ the length of the top section, and $\frac{1}{2}b =$ the length of the middle section.

Translate.

Length of top section	+	length of middle section	+	length of bottom section	is	240 ft
↓	↓	↓	↓	↓		
$\frac{1}{6}b$	+	$\frac{1}{2}b$	+	b	=	240

Carry out. We solve the equation. First we multiply by 6 on both sides to clear the fractions.

$$\frac{1}{6}b+\frac{1}{2}b+b=240$$
$$6\left(\frac{1}{6}b+\frac{1}{2}b+b\right)=6\cdot 240$$
$$6\cdot\frac{1}{6}b+6\cdot\frac{1}{2}b+6\cdot b=1440$$
$$b+3b+6b=1440$$
$$10b=1440$$
$$b=144$$

Then $\frac{1}{6}b=\frac{1}{6}\cdot 144=24$ and $\frac{1}{2}b=\frac{1}{2}\cdot 144=72$.

Check. 24 ft is $\frac{1}{6}$ of 144 ft, and 72 ft is $\frac{1}{2}$ of 144 ft. The sum of the lengths of the sections is 24 ft + 72 ft + 144 ft = 240 ft. The answer checks.

State. The length of the top section is 24 ft, the length of the middle section is 72 ft, and the length of the bottom section is 144 ft.

39. ***Familiarize.*** Let r = the speed downstream. Then $r - 10$ = the speed upstream. Then, since $d = r\cdot t$, we multiply to find each distance.

Downstream distance $= r(2)$ mi;

Upstream distance $= (r-10)(3)$ mi.

Translate.

Distance downstream	plus	distance upstream	is	total distance.
↓	↓	↓	↓	↓
$r(2)$	+	$(r-10)(3)$	=	30

Carry out. We solve the equation.

$$2r+3(r-10)=30$$
$$2r+3r-30=30$$
$$5r-30=30$$
$$5r=60$$
$$r=12$$

$r-10=2$

Check. Distance = speed × time.
Distance downstream $=12(2)=24$ mi
Distance upstream $=(12-10)(3)=6$ mi
The total distance is 24 mi + 6 mi, or 30 mi. The answer checks.

State. The speed downstream was 12 mph.

41. ***Familiarize.*** Let d = the distance Phoebe ran. Then $17 - d$ = the distance Phoebe walked. Then, since $t=\frac{d}{r}$, we divide to find each time.

Time running $=\frac{d}{12}$ hr;

Time walking $=\frac{17-d}{5}$ hr;

Translate.

Time running	is	time walking.
↓	↓	↓
$\frac{d}{12}$	=	$\frac{17-d}{5}$

Carry out. We solve the equation.

$$60\cdot\frac{d}{12}=60\cdot\frac{17-d}{5}$$
$$5d=12(17-d)$$
$$5d=204-12d$$
$$17d=204$$
$$d=12$$

$17-d=5$

Check. Time = distance ÷ speed.

Time running $=\frac{12}{12}=1$ hr;

Time walking $=\frac{17-12}{5}=1$ hr;

The time running is the same as the time walking The answer checks.

State. Phoebe ran for 1 hour.

43. Let $p =$ the percent increase. The population increased by $660 - 570 = 90$.
Rewording and Translating:

Population increase	is	what percent	of	original population.
↓	↓	↓	↓	↓
90	=	p	·	570

$$\frac{90}{570}=p$$
$$0.158\approx p$$

The percent increase was about 15.8%.

45. Let $p =$ the percent increase. The budget increased by \$1,800,000 – \$1,600,000 = \$200,000.
Rewording and Translating:

Budget increase	is	what percent	of	original budget.
↓	↓	↓	↓	↓
200,000	=	p	·	1,600,000

$$\frac{200{,}000}{1{,}600{,}000}=p$$
$$0.125=p$$

The percent increase is 12.5%.

47. Let $b =$ the bill without tax.
Rewording and Translating:

The bill plus tax is $1310.75.

$b + 0.07b = 1310.75$

$$1.07b = 1310.75$$
$$b = \frac{1310.75}{1.07}$$
$$b = 1225$$

The bill without tax is $1225.

49. Let $s =$ the sales tax.
Rewording and Translating:

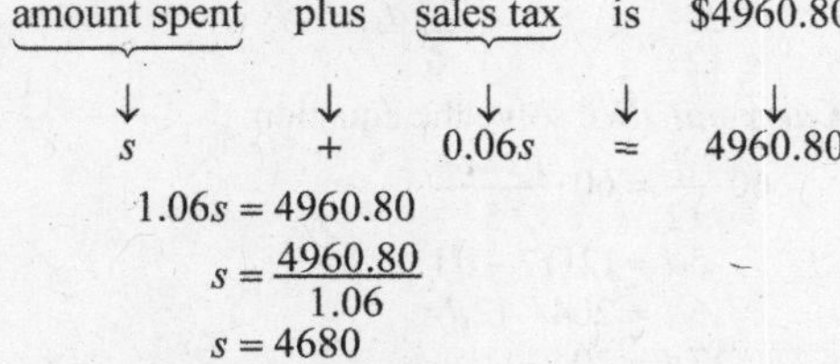

amount spent plus sales tax is $4960.80.
s + 0.06s = 4960.80

$$1.06s = 4960.80$$
$$s = \frac{4960.80}{1.06}$$
$$s = 4680$$

$4960.80 - 4680 = 280.80$
The sales tax is $280.80.

51. ***Familiarize***. Let $p =$ the regular price of the camera. At 30% off, Raena paid $(100 - 30)\%$, or 70% of the regular price.

Translate.

$224 is 70% of the regular price.

$224 = 0.70 \cdot p$

Carry out. We solve the equation.
$$224 = 0.70p$$
$$320 = p$$

Check. 70% of $320, or 0.70($320), is $224. The answer checks.

State. The regular price was $320.

53. ***Familiarize***. Let $s =$ the annual salary of Bradley's previous job. With a 15% pay cut, Bradley received $(100 - 15)\%$, or 85% of the salary of the previous job.

Translate.

$30,600 is 85% of the previous salary.

$30,600 = 0.85 \cdot s$

Carry out. We solve the equation.
$$30,600 = 0.85s$$
$$36,000 = s$$

Check. 85% of $36,000, or 0.85($36,000), is $30,600. The answer checks.

State. Bradley's previous salary was $36,000.

55. ***Familiarize***. Let $g =$ the original amount of the grocery bill. Saving 85% pay cut, Marie paid $(100 - 85)\%$, or 15% of the original amount.

Translate.

$15 is 15% of the original bill.

$15 = 0.15 \cdot g$

Carry out. We solve the equation.
$$15 = 0.15g$$
$$100 = g$$

Check. 15% of $100, or 0.15($100), is $15. The answer checks.

State. Marie's original grocery bill was $100.

57. ***Familiarize***. Let $c =$ the cost of a 30-sec slot in 2013, in millions of dollars. The increase was 20% of c, or $0.20c$.

Translate.

Cost in 2013 plus increase is $4.8.

$c + 0.20c = 4.8$

Carry out. We solve the equation.
$$c + 0.20c = 4.8$$
$$1.2c = 4.8$$
$$c = 4$$

Check. 20% of 4, or 0.20(4), is 0.8 and 4 + 0.8 is 4.8, the cost in 2016. The answer checks.

State. In 2013, a 30-sec slot cost $4 million.

59. ***Familiarize***. Let a the selling price of the house. Then the commission on the selling price is 6% times a, or $0.06a$.

Translate.

Selling price minus commission is $117,500.

$a - 0.06a = 117,500$

Carry out. We solve the equation.
$$a - 0.06a = 117,500$$
$$0.94a = 117,500$$
$$a = 125,000$$

Check. A selling price of $125,000 gives a commission of $7500. Since $\$125,000 - \$7500 = \$117,500$, the answer checks.

State. They must sell the house for $125,000.

61. ***Familiarize***. Let $m =$ the number of miles that can be traveled for $19. Then the total cost of the taxi ride, in dollars, is $3.25 + 1.80m$.

Translate.

Cost of taxi ride is $19.

$3.25 + 1.80m = 19$

Carry out. We solve the equation.

$$3.25+1.80m=19$$
$$1.80m=15.75$$
$$m=\frac{15.75}{1.80}=8.75\text{ mi}$$

Check. The mileage charge is \$1.80(8.75), or \$15.75, and the total cost of the ride is \$3.25 + \$15.75 = \$19. The answer checks.

State. Debbie can travel 8.75 mi, or $8\frac{3}{4}$ mi.

63. ***Familiarize***. The total cost is the daily rate plus the mileage charge. Let d = the distance that can be traveled, in miles, in one day for \$100. The mileage charge is the cost per mile times the number of miles traveled, or $0.55d$.

Translate.

Daily rate plus mileage charge is \$100.

$$39.95 + 0.55d = 100$$

Carry out. We solve the equation.

$$39.95+0.55d=100$$
$$0.55d=60.05$$
$$d\approx 109.2$$

Check. For a trip of 109.2 mi, the mileage charge is $\$0.55(109.2)$, or \$60.06, and $\$39.95+\$60.06\approx\$100$. The answer checks.

State. Concert Productions can travel 109.2 mi in one day and stay within their budget.

65. ***Familiarize***. Let x = the measure of one angle. Then $90-x=$ the measure of its complement.

Translate.

Measure of one angle is 15° more than twice the measure of its complement.

$$x = 15 + 2(90-x)$$

Carry out. We solve the equation.

$$x=15+2(90-x)$$
$$x=15+180-2x$$
$$x=195-2x$$
$$3x=195$$
$$x=65$$

If x is 65, then $90-x$ is 25.

Check. The sum of the angle measures is 90°. Also, 65° is 15° more than twice its complement, 25°. The answer checks.

State. The angle measures are 65° and 25°.

67. ***Familiarize***. Let x = the measure of one angle. Then $180-x=$ the measure of its supplement.

Translate.

Measure of one angle is $3\frac{1}{2}$ times measure of second angle.

$$x = 3\tfrac{1}{2}\cdot(180-x)$$

Carry out. We solve the equation.

$$x=3\tfrac{1}{2}(180-x)$$
$$x=630-3.5x$$
$$4.5x=630$$
$$x=140$$

If $x=140$, then $180-140=40°$.

Check. The sum of the angles is 180°. Also 140° is three and a half times 40°. The answer checks.

State. The angles are 40° and 140°.

69. ***Familiarize***. Let l = the length of the paper, in cm. Then $l-6.3=$ the width. The perimeter is twice the length plus twice the width.

Translate.

Twice the length plus twice the width is 99 cm.

$$2l + 2(l-6.3) = 99$$

Carry out. We solve the equation.

$$2l+2(l-6.3)=99$$
$$2l+2l-12.6=99$$
$$4l-12.6=99$$
$$4l=111.6$$
$$l=27.9$$

Then $l-6.3=27.9-6.3=21.6$.

Check. The width, 21.6 cm, is 6.3 cm less than the length, 27.9 cm. The perimeter is 2(27.9 cm) +2(21.6 cm) = 55.8 cm +43.2 cm = 99 cm. The answer checks.

State. The length of the paper is 27.9 cm, and the width is 21.6 cm.

71. ***Familiarize***. Let a = the amount Janeka invested. Then the simple interest for one year is $1\%\cdot a$, or $0.01a$.

Translate.

Amount invested plus interest is \$1555.40.

$$a + 0.01a = 1555.40$$

Carry out. We solve the equation.

$$a+0.01a=1555.40$$
$$1.01a=1555.40$$
$$a=1540$$

Check. An investment of \$1540 at 1% simple interest earns $0.01(\$1540)$, or \$15.40, in one year. Since $\$1540+\$15.40=\$1555.40$, the answer checks.

State. Janeka invested \$1540.

73. ***Familiarize.*** Let w = the winning score. Then $w-340 =$ the losing score.

Translate.

Winning score	plus	losing score	was	1320 points.
↓	↓	↓	↓	↓
w	$+$	$w-340$	$=$	1320

Carry out. We solve the equation.

$$w+(w-340)=1320$$
$$2w-340=1320$$
$$2w=1660$$
$$w=830$$

Then $w-340=830-340=490$.

Check. The winning score, 830, is 340 points more than the losing score, 490. The total of the two scores is $830+490=1320$ points. The answer checks.

State. The winning score was 830 points.

75. ***Familiarize.*** We will use the equation $c=1.2x+32.94$.

Translate. We substitute 50.94 for c.

$$50.94=1.2x+32.94.$$

Carry out. We solve the equation.

$$50.94=1.2x+32.94.$$
$$18=1.2x$$
$$15=x$$

Check. When $x=15$, we have $c=1.2(15)+32.94=$ $18+32.94=50.94$. The answer checks.

State. The cost of a dinner for 10 people will be $50.94 in 2015.

77. ***Familiarize.*** We will use the equation

$$T=\frac{1}{4}N+40.$$

Translate. We substitute 80 for T.

$$80=\frac{1}{4}N+40$$

Carry out. We solve the equation.

$$80=\frac{1}{4}N+40$$
$$40=\frac{1}{4}N$$
$$160=N \quad \text{Multiplying by 4 on both sides}$$

Check. When $N=160$, we have $T=\frac{1}{4}\cdot 160+40=$ $40+40=80$. The answer checks.

State. A cricket chirps 160 times per minute when the temperature is 80°F.

79. *Writing Exercise.*

81. $4(2n+8t+1)=8n+32t+4$

83. $x-3[2x-4(x-1)+2]$

$$=x-3[2x-4x+4+2]$$
$$=x-3[-2x+6]$$
$$=x+6x-18$$
$$=7x-18$$

85. *Writing Exercise.*

87. ***Familiarize.*** Let c = the amount the meal originally cost. The 15% tip is calculated on the original cost of the meal, so the tip is $0.15c$.

Translate.

Original cost	plus	tip	less	$10	is	$32.55.
↓	↓	↓	↓	↓	↓	↓
c	$+$	$0.15x$	$-$	10	$=$	32.55

Carry out. We solve the equation.

$$c+0.15c-10=32.55$$
$$1.15c-10=32.55$$
$$1.15c=42.55$$
$$c=37$$

Check. If the meal originally cost $37, the tip was 15% of $37, or 0.15($37), or $5.55. Since $37 + $5.55 − $10 = $32.55, the answer checks.

State. The meal originally cost $37.

89. ***Familiarize.*** Let s = one score. Then four score = $4s$ and four score and seven = $4s+7$.

Translate. We reword.

1776	plus	four score and seven	is	1863
↓	↓	↓	↓	↓
1776	$+$	$(4s+7)$	$=$	1863

Carry out. We solve the equation.

$$1776+(4s+7)=1863$$
$$4s+1783=1863$$
$$4s=80$$
$$s=20$$

Check. If a score is 20 years, then four score and seven represents 87 years. Adding 87 to 1776 we get 1863. This checks.

State. A score is 20.

91. ***Familiarize.*** Let n = the number of half dollars. Then the number of quarters is $2n$; the number of dimes is $2\cdot 2n$, or $4n$; and the number of nickels is $3\cdot 4n$, or $12n$. The total value of each type of coin, in dollars, is as follows.

Half dollars: $0.5n$

Quarters: $0.25(2n)$, or $0.5n$

Dimes: $0.1(4n)$, or $0.4n$

Nickels: $0.05(12n)$, or $0.6n$

Then the sum of these amounts is $0.5n+0.5n+0.4n+0.6n$, or $2n$.

Translate.

Total amount of change	is	\$10.
↓	↓	↓
$2n$	$=$	10

Carry out. We solve the equation.

$$2n = 10$$
$$n = 5$$

Then $2n = 2 \cdot 5 = 10$, $4n = 4 \cdot 5 = 20$, and $12n = 12 \cdot 5 = 60$.

Check. If there are 5 half dollars, 10 quarters, 20 dimes, and 60 nickels, then there are twice as many quarters as half dollars, twice as many dimes as quarters, and 3 times as many nickels as dimes. The total value of the coins is

$$\$0.5(5) + \$0.25(10) + \$0.1(20) + \$0.05(60)$$
$$= \$2.50 + \$2.50 + \$2 + \$3 = \$10.$$

The answer checks.

State. The shopkeeper got 5 half dollars, 10 quarters, 20 dimes, and 60 nickels.

93. ***Familiarize.*** Let p = the price before the two discounts. With the first 10% discount, the price becomes 90% of p, or $0.9p$. With the second 10% discount, the final price is 90% of $0.9p$, or $0.9(0.9p)$.

Translate.

10% discount	and	10% discount of price	is	\$77.75.
↓	↓	↓	↓	↓
0.9	$\cdot$	$0.9p$	$=$	77.75

Carry out. We solve the equation.

$$0.9(0.9p) = 77.75$$
$$0.81p = 77.75$$
$$p = 95.99$$

Check. Since 90% of \$95.99 is \$86.39, and 90% of \$86.39 is \$77.75, the answer checks.

State. The original price before discounts was \$95.99.

95. ***Familiarize.*** Let n = the number of DVDs purchased. Assume that at least two more DVDs were purchased. Then the first DVD costs \$9.99 and the total cost of the remaining $(n-1)$ DVDs is $\$6.99(n-1)$. The shipping and handling costs are \$3 for the first DVD, \$1.50 for the second (half of \$3), and a total of $\$1(n-2)$ for the remaining $n-2$ DVDs.

Translate.

1st DVD	plus	remaining DVDs	plus	1st S&H charges
↓	↓	↓	↓	↓
9.99	$+$	$6.99(n-1)$	$+$	3

… plus	2nd S&H charges	plus	remaining S&H charges	is	\$45.45.
↓	↓	↓	↓	↓	↓
$+$	1.50	$+$	$1(n-2)$	$=$	45.45

Carry out. We solve the equation.

$$9.99 + 6.99(n-1) + 3 + 1.5 + (n-2) = 45.45$$
$$9.99 + 6.99n - 6.99 + 4.5 + n - 2 = 45.45$$
$$7.99n + 5.5 = 45.45$$
$$7.99n = 39.95$$
$$n = 5$$

Check. If there are 5 DVDs, the cost of the DVDs is \$9.99 + \$6.99(5 – 1), or \$9.99 + \$27.96, or \$37.95. The cost for shipping and handling is \$3 + \$1.50 + \$1(5 – 2) = \$7.50. The total cost is \$37.95 + \$7.50, or \$45.45. The answer checks.

State. There were 5 DVDs in the shipment.

97. ***Familiarize.*** Let $d =$ the distance, in miles, that Mya traveled. At \$0.50 per $\frac{1}{5}$ mile, the mileage charge can also be given as 5(\$0.50), or \$2.5 per mile. Since it took 20 min to complete what is usually a 10-min drive, the taxi was stopped in traffic for $20-10$, or 10 min.

Translate.

Initial charge	plus	\$2.50 per mile	plus	stopped in traffic charge	is	\$23.80.
↓	↓	↓	↓	↓	↓	↓
2.80	$+$	$2.50d$	$+$	$0.60(10)$	$=$	23.80

Carry out. We solve the equation.

$$2.80 + 2.5d + 0.6(10) = 23.80$$
$$2.8 + 2.5d + 6 = 23.80$$
$$2.5d + 8.8 = 23.80$$
$$2.5d = 15$$
$$d = 6$$

Check. Since \$2.5(6) = \$15, and \$0.60(10) = \$6, and \$15 + \$6 + \$2.80 = \$23.80, the answer checks.

State. Mya traveled 6 mi.

99. *Writing Exercise.*

101. ***Familiarize.*** Let w = the width of the rectangle, in cm. Then $w + 4.25 =$ the length.

Translate.

The perimeter	is	101.74 cm.
↓	↓	↓
$2(w+4.25)+2w$	$=$	101.74

Carry out. We solve the equation.

$$2(w+4.25) + 2w = 101.74$$
$$2w + 8.5 + 2w = 101.74$$
$$4w + 8.5 = 101.74$$
$$4w = 93.24$$
$$w = 23.31$$

Then $w + 4.25 = 23.31 + 4.25 = 27.56$.

Check. The length, 27.56 cm, is 4.25 cm more than the width, 23.31 cm. The perimeter is 2(27.56) cm + 2(23.31 cm) = 55.12 cm + 46.62 cm = 101.74 cm. The answer checks.

State. The length of the rectangle is 27.56 cm, and the width is 23.31 cm.

Connecting the Concepts

1. $x-6=15$
$x=21$ Adding 6 to both sides
The solution is 21.

2. $x-6\le 15$
$x\le 21$ Adding 6 to both sides
The solution is $\{x|x\le 21\}$, or $(-\infty, 21]$.

3. $3x=-18$
$x=-6$ Dividing both sides by 3
The solution is –6.

4. $3x>-18$
$x>-6$ Dividing both sides by 3
The solution is $\{x|x>-6\}$, or $(-6, \infty)$.

5. $7-3x\ge 8$
$-3x\ge 1$ Subtracting 7 from both sides
$x\le -\frac{1}{3}$ Dividing both sides by –3 and reversing the direction of the inequality symbol

The solution is $\left\{x|x\le -\frac{1}{3}\right\}$, or $\left(-\infty, -\frac{1}{3}\right]$.

6. $7-3x=8$
$-3x=1$ Subtracting 7 from both sides
$x=-\frac{1}{3}$ Dividing both sides by –3
The solution is $-\frac{1}{3}$.

7. $\frac{n}{6}-6=5$
$\frac{n}{6}=11$ Adding 6 to both sides
$n=66$ Multiplying both sides by 6
The solution is 66.

8. $\frac{n}{6}-6<5$
$\frac{n}{6}<11$ Adding 6 to both sides
$n<66$ Multiplying both sides by 6
The solution is $\{n|n<66\}$, or $(-\infty, 66)$.

9. $10\ge -2(a-5)$
$10\ge -2a+10$ Using the distributive law
$0\ge -2a$ Subtracting 10 from both sides
$0\le a$ Dividing both sides by –2 and reversing the direction of the inequality symbol
The solution is $\{a|a\ge 0\}$

10. $10=-2(a-5)$
$10=-2a+10$ Using the distributive law
$0=-2a$ Subtracting 10 from both sides
$0=a$ Dividing both sides by –2
The solution is 0.

Exercise Set 2.6

1. The number –2 is one *solution* of the inequality $x<0$.

3. The interval [6, 10] is an example of a *closed* interval.

5. $y-10\ge 4$
$y\ge 14$ Add 10

7. $3n>90$
$n>30$ Dividing by 3

9. $-5x\le 30$
$x\ge -6$ Dividing by –5 and reversing the inequality symbol

11. $-2t>-14$
$t<7$ Dividing by –2 and reversing the inequality symbol

13. $x>-4$

a. Since $4>-4$ is true, 4 is a solution.
b. Since $-6>-4$ is false, –6 is not a solution.
c. Since $-4>-4$ is false, –4 is not a solution.

15. $y\le 19$

a. Since $18.99\le 19$ is true, 18.99 is a solution.
b. Since $19.01\le 19$ is false, 19.01 is not a solution.
c. Since $19\le 19$ is true, 19 is a solution.

17. $c\ge -7$

a. Since $0\ge -7$ is true, 0 is a solution.
b. Since $-5\frac{4}{5}\ge -7$ is true, $-5\frac{4}{5}$ is a solution.
c. Since $1\frac{1}{3}\ge -7$ is true, $1\frac{1}{3}$ is a solution.

19. The solutions of $y<2$ are those numbers less than 2. They are shown on the graph by shading all points to the left of 2. The parenthesis at 2 indicates that 2 is not part of the graph.

$y<2$

–4 –2 0 2 4

21. The solutions of $x\ge -1$ are those numbers greater than or equal to –1. They are shown on the graph by shading all points to the right of –1. The bracket at –1 indicates that the point –1 is part of the graph.

$x\ge -1$

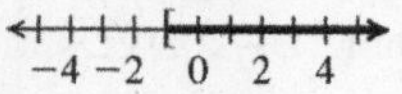

–4 –2 0 2 4

23. The solutions of $0\le t$, or $t\ge 0$, are those numbers greater than or equal to zero. They are shown on the graph by shading all points to the right of 0. The bracket at 0 indicates that 0 is part of the graph.

$0\le t$

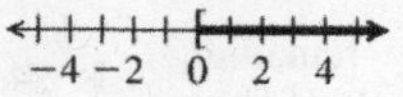

–4 –2 0 2 4

25. In order to be solution of the inequality $-5 \le x < 2$, a number must be a solution of both $-5 \le x$ and $x < 2$. The solution set is graphed as follows:

$$-5 \le x < 2$$

−6 −4 −2 0 2 4

The bracket at -5 means that -5 is part of the graph. The parenthesis at 2 means that 2 is not part of the graph.

27. In order to be a solution of the inequality $-4 < x < 0$, a number must be a solution of both $-4 < x$ and $x < 0$. The solution set is graphed as follows:

$$-4 < x < 0$$

−6 −4 −2 0 2

The parentheses at -4 and 0 mean that -4 and 0 are not part of the graph.

29. $y < 6$

Using set-builder notation, we write the solution set as $\{y|y < 6\}$. Using interval notation, we write $(-\infty, 6)$.

To graph the solution, we shade all numbers to the left of 6 and use a parenthesis to indicate that 6 is not a solution.

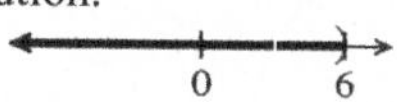

31. $x \ge -4$

Using set-builder notation, we write the solution set as $\{x|x \ge -4\}$. Using interval notation, we write $[-4, \infty)$.

To graph the solution, we shade all numbers to the right of −4 and use a bracket to indicate that −4 is a solution.

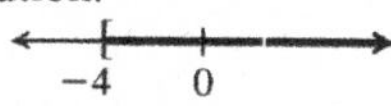

33. $t > -3$

Using set-builder notation, we write the solution set as $\{t|t > -3\}$. Using interval notation, we write $(-3, \infty)$.

To graph the solution, we shade all numbers to the right of −3 and use a parenthesis to indicate that −3 is not a solution.

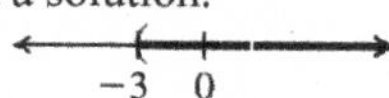

35. $x \le -7$

Using set-builder notation, we write the solution set as $\{x|x \le -7\}$. Using interval notation, we write $(-\infty, -7]$.

To graph the solution, we shade all numbers to the left of −7 and use a bracket to indicate that −7 is a solution.

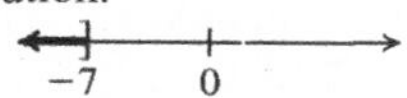

37. All points to the right of −4 are shaded. The parenthesis at −4 indicates that −4 is not part of the graph. Set-builder notation: $\{x|x > -4\}$. Interval notation: $(-4, \infty)$.

39. All points to the left of 2 are shaded. The bracket at 2 indicates that 2 is part of the graph. Set-builder notation: $\{x|x \le 2\}$. Interval notation: $(-\infty, 2]$.

41. All points to the left of −1 are shaded. The parenthesis at −1 indicates that −1 is not part of the graph. Set-builder notation: $\{x|x < -1\}$. Interval notation: $(-\infty, -1)$.

43. All points to the right of 0 are shaded. The bracket at 0 indicates that 0 is part of the graph. Set-builder notation: $\{x|x \ge 0\}$. Interval notation: $[0, \infty)$

45.
$$\begin{aligned} y+6 &> 9 \\ y+6-6 &> 9-6 \quad \text{Adding } -6 \text{ to both sides} \\ y &> 3 \quad \text{Simplifying} \end{aligned}$$

The solution set is $\{y|y > 3\}$, or $(3, \infty)$.

0 3

47.
$$\begin{aligned} n-6 &< 11 \\ n-6+6 &< 11+6 \quad \text{Adding 6 to both sides} \\ n &< 17 \quad \text{Simplifying} \end{aligned}$$

The solution set is $\{n|n < 17\}$, or $(17, \infty)$.

0 17

49.
$$\begin{aligned} 2x &\le x-9 \\ 2x-x &\le x-9-x \\ x &\le -9 \end{aligned}$$

The solution set is $\{x|x \le -9\}$, or $(-\infty, -9]$.

−9 0

51.
$$\begin{aligned} 5 &\ge t+8 \\ 5-8 &\ge t+8-8 \\ -3 &\ge t \text{ or } t \le -3 \end{aligned}$$

The solution set is $\{t|t \le -3\}$, or $(-\infty, -3]$.

−3 0

53.
$$\begin{aligned} t-\frac{1}{8} &> \frac{1}{2} \\ t-\frac{1}{8}+\frac{1}{8} &> \frac{1}{2}+\frac{1}{8} \\ t &> \frac{4}{8}+\frac{1}{8} \\ t &> \frac{5}{8} \end{aligned}$$

The solution set is $\left\{t \middle| t > \frac{5}{8}\right\}$, or $\left(\frac{5}{8}, \infty\right)$.

0 $\frac{5}{8}$

55. $-9x+17>17-8x$
$-9x+17-17>17-8x-17$ Adding -17
$-9x>-8x$
$-9x+9x>-8x+9x$ Adding $9x$
$0>x$
The solution set is $\{x|x<0\}$, or $(-\infty, 0)$.

0

57. $-23<-t$
The inequality states that the opposite of 23 is less than the opposite of t. Thus, t must be less than 23, so the solution set is $\{t|t<23\}$. To solve this inequality using the addition principle, we would proceed as follows:
$-23<-t$
$t-23<0$ Adding t to both sides
$t<23$ Adding 23 to both sides
The solution set is $\{t|t<23\}$, or $(-\infty, 23)$

0 23

59. $4x<28$
$\frac{1}{4}\cdot 4x<\frac{1}{4}\cdot 28$ Multiplying by $\frac{1}{4}$
$x<7$
The solution set is $\{x|x<7\}$, or $(-\infty, 7)$.

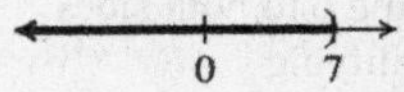

0 7

61. $-24>8t$
$-3>t$
The solution set is $\{t|t<-3\}$, or $(-\infty,-3)$.

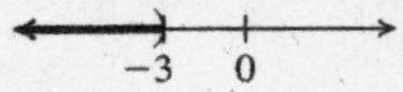

−3 0

63. $1.8\geq -1.2n$
$\frac{-1}{1.2}\cdot 1.8\leq\frac{-1}{1.2}(-1.2n)$ Multiplying by $\frac{-1}{1.2}$
Reversing the inequality
$-1.5\leq n$
The solution set is $\{n|n\geq -1.5\}$, or $[-1.5, \infty)$.

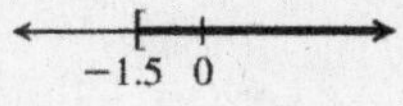

−1.5 0

65. $-2y\leq\frac{1}{5}$
$-\frac{1}{2}\cdot(-2y)\geq -\frac{1}{2}\cdot\frac{1}{5}$ Reversing the inequality
$y\geq -\frac{1}{10}$
The solution set is $\left\{y\middle|y\geq -\frac{1}{10}\right\}$, or $\left[-\frac{1}{10}, \infty\right)$.

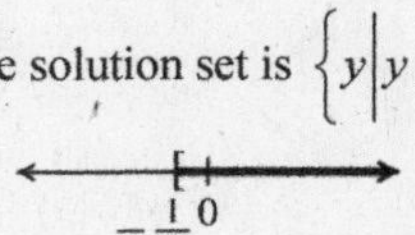

$-\frac{1}{10}$ 0

67. $-\frac{8}{5}>2x$
$\frac{1}{2}\cdot\left(-\frac{8}{5}\right)>\frac{1}{2}\cdot(2x)$
$-\frac{8}{10}>x$
$-\frac{4}{5}>x$, or $x<-\frac{4}{5}$
The solution set is $\left\{x\middle|x<-\frac{4}{5}\right\}$, or $\left(-\infty,-\frac{4}{5}\right)$.

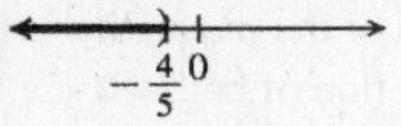

$-\frac{4}{5}$ 0

69. $2+3x<20$
$2+3x-2<20-2$ Adding -2 to both sides
$3x<18$ Simplifying
$x<6$ Multiplying both sides by $\frac{1}{3}$
The solution set is $\{x|x<6\}$, or $(-\infty, 6)$.

71. $4t-5\leq 23$
$4t-5+5\leq 23+5$ Adding 5 to both sides
$4t\leq 28$
$\frac{1}{4}\cdot 4t\leq\frac{1}{4}\cdot 28$ Multiplying both sides by $\frac{1}{4}$
$t\leq 7$
The solution set is $\{t|t\leq 7\}$, or $(-\infty, 7]$.

73. $39>3-9x$
$39-3>3-9x-3$ Adding -3
$36>-9x$
$-\frac{1}{9}\cdot 36<-\frac{1}{9}\cdot(-9x)$ Multiplying by $-\frac{1}{9}$
Reversing the inequality
$-4<x$
The solution set is $\{x|x>-4\}$, or $(-4, \infty)$.

75. $5-6y>25$
$-5+5-6y>-5+25$
$-6y>20$
$-\frac{1}{6}\cdot(-6y)<-\frac{1}{6}\cdot 20$ Reversing the inequality.
$y<-\frac{20}{6}$
$y<-\frac{10}{3}$
The solution set is $\left\{y\middle|y<-\frac{10}{3}\right\}$, or $\left(-\infty,-\frac{10}{3}\right)$.

77. $-3<8x+7-7x$
$-3<x+7$ Collecting like terms
$-3-7<x+7-7$
$-10<x$
The solution set is $\{x|x>-10\}$, or $(-10, \infty)$.

79. $6-4y>6-3y$
$6-4y+4y>6-3y+4y$ Adding $4y$
$6>6+y$
$-6+6>-6+6+y$ Adding − 4
$0>y$, or $y<0$
The solution set is $\{y|y<0\}$, or $(-\infty, 0)$.

81. $2.1x + 43.2 > 1.2 - 8.4x$
$10(2.1x + 43.2) > 10(1.2 - 8.4x)$ Multiplying by 10 to clear decimals
$21x + 432 > 12 - 84x$
$21x + 84x > 12 - 432$ Adding $84x$ and -432
$105x > -420$
$x > -4$ Multiplying by $\frac{1}{105}$

The solution set is $\{x|x > -4\}$, or $(-4, \infty)$.

83. $1.7t + 8 - 1.62t < 0.4t - 0.32 + 8$
$0.08t + 8 < 0.4t + 7.68$ Collecting like terms
$100(0.08t + 8) < 100(0.4t + 7.68)$ Multiplying by 100
$8t + 800 < 40t + 768$
$-8t - 768 + 8t + 800 < 40t + 768 - 8t - 768$
$32 < 32t$
$1 < t$

The solution set is $\{t|t > 1\}$, or $(1, \infty)$.

85. $\frac{x}{3} + 4 \le 1$
$3\left(\frac{x}{3} + 4\right) \le 3 \cdot 1$ Multiplying by 3 to clear the fraction
$x + 12 \le 3$
$x \le -9$

The solution set is $\{x|x \le -9\}$, or $(-\infty, -9]$.

87. $3 < 5 - \frac{t}{7}$
$-2 < -\frac{t}{7}$
$-7(-2) > -7\left(-\frac{t}{7}\right)$
$14 > t$

The solution set is $\{t|t < 14\}$, or $(-\infty, 14)$.

89. $4(2y - 3) \le -44$
$8y - 12 \le -44$ Removing parentheses
$8y \le -32$ Adding 12
$y \le -4$ Multiplying by $\frac{1}{8}$

The solution set is $\{y|y \le -4\}$, or $(-\infty, -4]$.

91. $8(2t + 1) > 4(7t + 7)$
$16t + 8 > 28t + 28$
$-12t + 8 > 28$
$-12t > 20$
$t < -\frac{5}{3}$ Multiplying by $-\frac{1}{12}$ and reversing the symbol

The solution set is $\left\{t|t < -\frac{5}{3}\right\}$, or $\left(-\infty, -\frac{5}{3}\right)$.

93. $3(r - 6) + 2 < 4(r + 2) - 21$
$3r - 18 + 2 < 4r + 8 - 21$
$3r - 16 < 4r - 13$
$-16 + 13 < 4r - 3r$
$-3 < r$, or $r > -3$

The solution set is $\{r|r > -3\}$, or $(-3, \infty)$.

95. $\frac{4}{5}(3x - 4) \le 20$
$\frac{5}{4} \cdot \frac{4}{5}(3x + 4) \le \frac{5}{4} \cdot 20$
$3x + 4 \le 25$
$3x \le 21$
$x \le 7$

The solution set is $\{x|x \le 7\}$, or $(-\infty, 7]$.

97. $\frac{2}{3}\left(\frac{7}{8} - 4x\right) - \frac{5}{8} < \frac{3}{8}$
$\frac{2}{3}\left(\frac{7}{8} - 4x\right) < 1$ Adding $\frac{5}{8}$
$\frac{7}{12} - \frac{8}{3}x < 1$ Removing parentheses
$12\left(\frac{7}{12} - \frac{8}{3}x\right) < 12 \cdot 1$ Clearing fractions
$7 - 32x < 12$
$-32x < 5$
$x > -\frac{5}{32}$

The solution is $\left\{x|x > -\frac{5}{32}\right\}$, or $\left(-\frac{5}{32}, \infty\right)$.

99. *Writing Exercise.*

101. $5x - 2(3 - 6x) = 5x - 6 + 12x = 17x - 6$

103. $x - 2[4y + 3(8 - x) - 1]$
$= x - 2[4y + 24 - 3x - 1]$
$= x - 2[4y - 3x + 23]$
$= x - 8y + 6x - 46$
$= 7x - 8y - 46$

105. *Writing Exercise.*

107. $x < x + 1$
When any real number is increased by 1, the result is greater than the original number. Thus the solution set is $\{x|x \text{ is a real number}\}$, or $(-\infty, \infty)$.

109. $27 - 4[2(4x - 3) + 7] \ge 2[4 - 2(3 - x)] - 3$
$27 - 4[8x - 6 + 7] \ge 2[4 - 6 + 2x] - 3$
$27 - 4[8x + 1] \ge 2[-2 + 2x] - 3$
$27 - 32x - 4 \ge -4 + 4x - 3$
$23 - 32x \ge -7 + 4x$
$23 + 7 = 4x + 32x$
$30 \ge 36x$
$\frac{5}{6} \ge x$

The solution set is $\left\{x|x \le \frac{5}{6}\right\}$, or $\left(-\infty, \frac{5}{6}\right]$.

111. $-(x + 5) \ge 4a - 5$
$-x - 5 \ge 4a - 5$
$-x \ge 4a - 5 + 5$
$-x \ge 4a$
$-1(-x) \le -1 \cdot 4a$
$x \le -4a$

The solution set is $\{x|x \le -4a\}$, or $(-\infty, -4a]$.

113. $y < ax + b$ Assume $a > 0$.

$y - b < ax$

$\frac{y-b}{a} < x$ Since $a > 0$, the inequality symbol stays the same.

The solution set is $\left\{x \middle| x > \frac{y-b}{a}\right\}$, or $\left(\frac{y-b}{a}, \infty\right)$.

115. $|x| > -3$

Since absolute value is always nonnegative, the absolute value of any real number will be greater than -3. Thus, the solution set is $\{x|x \text{ is a real number}\}$, or $(-\infty, \infty)$.

117. a. No. The percentage of calories from fat is $\frac{54}{150} = 0.36$, or 36%, which is greater than 30%.

b. There is more than 6 g of fat per serving.

Exercise Set 2.7

1. If Matt's income is always \$500 per week or more, then his income *is at least* \$500.

3. If Lori works out 30 min or more every day, then her exercise time is *no less than* 30 min.

5. a is at least b can be translated as $b \le a$.

7. a is at most b can be translated as $a \le b$.

9. b is no more than a can be translated as $b \le a$.

11. b is less than a can be translated as $b < a$.

13. Let n represent the number. Then we have $n < 10$.

15. Let t represent the temperature. Then we have $t \le -3$.

17. Let d represent the number of years of driving experience. Then we have $d \ge 5$.

19. Let a represent the age of the Mayan altar. Then we have $a > 1200$.

21. Let h represent Bianca's hourly wage. Then we have $h \ge 12$.

23. Let s represent the number of hours of sunshine. Then we have $1100 < s < 1600$.

25. ***Familiarize***. Let s = the length of the service call, in hours. The total charge is \$55 plus \$40 times the number of hours RJ's was there.

Translate.

\$55 charge	plus	hourly rate	times	number of hours	is greater than	\$150.
↓	↓	↓	↓	↓	↓	↓
55	+	40	·	s	>	150

Carry out. We solve the inequality.

$$55 + 40s > 150$$
$$40s > 95$$
$$s > 2.375$$

Check. As a partial check, we show that the cost of a 2.375 hour service call is \$150.

$$\$55 + \$30(2.375) = \$55 + \$95 = \$150$$

State. The length of the service call was more than 2.375 hr.

27. ***Familiarize***. Let q = Robbin's undergraduate grade point average. Unconditional acceptance is 500 plus 200 times the grade point average.

Translate.

GMAT score of 500	plus	200	times	grade point average	is at least	1020.
↓	↓	↓	↓	↓	↓	↓
500	+	200	·	q	≥	1020

Carry out. We solve the inequality.

$$500 + 200q \ge 1020$$
$$200q \ge 520$$
$$q \ge 2.6$$

Check. As a partial check we show that the acceptance score is 1020.

$$500 + 200(2.6) = 500 + 520 = 1020.$$

State. For unconditional acceptance, Robbin must have a GPA of at least 2.6.

29. ***Familiarize***. The average of the five scores is their sum divided by the number of tests, 5. We let s represent Rod's score on the last test.

Translate. The average of the five scores is given by

$$\frac{73 + 75 + 89 + 91 + s}{5}.$$

Since this average must be at least 85, this means that it must be greater than or equal to 85. Thus, we can translate the problem to the inequality

$$\frac{73 + 75 + 89 + 91 + s}{5} \ge 85.$$

Carry out. We first multiply by 5 to clear the fraction.

$$5\left(\frac{73 + 75 + 89 + 91 + s}{5}\right) \ge 5 \cdot 85$$
$$73 + 75 + 89 + 91 + s \ge 425$$
$$328 + s \ge 425$$
$$s \ge 97$$

Check. As a partial check, we show that Rod can get a score of 97 on the fifth test and have an average of at least 85:

$$\frac{73 + 75 + 89 + 91 + 97}{5} = \frac{425}{5} = 85.$$

State. Scores of 97 and higher will earn Rod an average quiz grade of at least 85.

31. ***Familiarize.*** Let c = the number of credits Millie must complete in the fourth quarter.

Translate.

$$\underbrace{\text{Average number of credits}}\ \underbrace{\text{is at least}}\ 7.$$

$$\frac{5+7+8+c}{4} \geq 7$$

Carry out. We solve the inequality.

$$\frac{5+7+8+c}{4} \geq 7$$
$$4\left(\frac{5+7+8+c}{4}\right) \geq 4\cdot 7$$
$$5+7+8+c \geq 28$$
$$20+c \geq 28$$
$$c \geq 8$$

Check. As a partial check, we show that Millie can complete 8 credits in the fourth quarter and average 7 credits per quarter.

$$\frac{5+7+8+8}{4} = \frac{28}{4} = 7$$

State. Millie must complete 8 credits or more in the fourth quarter.

33. ***Familiarize.*** The average number of plate appearances for 10 days is the sum of the number of appearance per day divided by the number of days, 10. We let p represent the number of plate appearances on the tenth day.

Translate. The average for 10 days is given by

$$\frac{5+1+4+2+3+4+4+3+2+p}{10}.$$

Since the average must be at least 3.1, this means that it must be greater than or equal to 3.1. Thus, we can translate the problem to the inequality

$$\frac{5+1+4+2+3+4+4+3+2+p}{10} \geq 3.1.$$

Carry out. We first multiply by 10 to clear the fraction.

$$10\left(\frac{5+1+4+2+3+4+4+3+2+p}{10}\right) \geq 10\cdot 3.1$$
$$5+1+4+2+3+4+4+3+2+p \geq 31$$
$$28+p \geq 31$$
$$p \geq 3$$

Check. As a partial check, we show that 3 plate appearances in the 10th game will average 3.1.

$$\frac{5+1+4+2+3+4+4+3+2+3}{10} = \frac{31}{10} = 3.1$$

State. On the tenth day, 3 or more plate appearances will give an average of at least 3.1.

35. ***Familiarize.*** We first make a drawing. We let b represent the length of the base. Then the lengths of the other sides are $b-2$ and $b+3$.

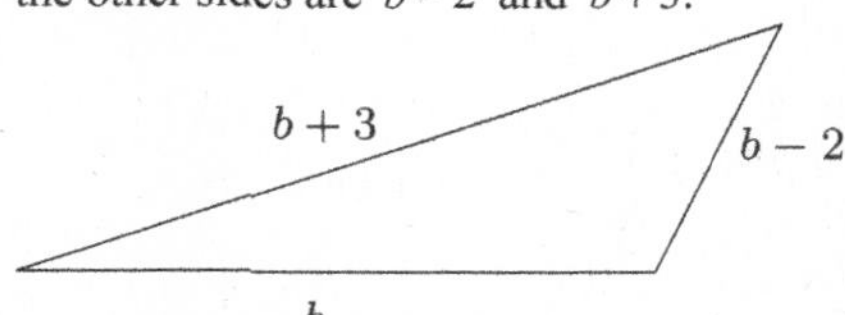

The perimeter is the sum of the lengths of the sides or $b+b-2+b+3$, or $3b+1$.

Translate.

$$\underbrace{\text{The perimeter}}\ \underbrace{\text{is greater than}}\ \underbrace{\text{19 cm.}}$$

$$3b+1 > 19$$

Carry out.

$$3b+1 > 19$$
$$3b > 18$$
$$b > 6$$

Check. We check to see if the solution seems reasonable.

When $b = 5$, the perimeter is $3\cdot 5+1$, or 16 cm.

When $b = 6$, the perimeter is $3\cdot 6+1$, or 19 cm.

When $b = 7$, the perimeter is $3\cdot 7+1$, or 22 cm.

From these calculations, it would appear that the solution is correct.

State. For lengths of the base greater than 6 cm the perimeter will be greater than 19 cm.

37. ***Familiarize.*** Let d = the depth of the well, in feet. Then the cost on the pay-as-you-go plan is $\$500 + \$8d$. The cost of the guaranteed-water plan is \$4000. We want to find the values of d for which the pay-as-you-go plan costs less than the guaranteed-water plan.

Translate.

$$\underbrace{\text{Cost of pay-as-you-go plan}}\ \underbrace{\text{is less than}}\ \underbrace{\text{cost of guaranteed-water plan}}$$

$$500+8d < 4000$$

Carry out.

$$500+8d < 4000$$
$$8d < 3500$$
$$d < 437.5$$

Check. We check to see that the solution is reasonable.

When $d = 437$, $\$500+\$8\cdot 437 = \$3996 < \$4000$

When $d = 437.5$, $\$500+\$8(437.5) = \$4000$

When $d = 438$, $\$500+\$8(438) = \$4004 > \$4000$

From these calculations, it appears that the solution is correct.

State. It would save a customer money to use the pay-as-you-go plan for a well of less than 437.5 ft.

39. ***Familiarize.*** Let v = the blue book value of the car. Since the car was repaired, we know that \$8500 does not exceed $0.8v$ or, in other words, $0.8v$ is at least \$8500.

Translate.

80% of the blue book value	is at least	\$8500.
↓	↓	↓
$0.8v$	$\geq$	8500

Carry out.

$$0.8v \geq 8500$$
$$v \geq \frac{8500}{0.8}$$
$$v \geq 10{,}625$$

Check. As a partial check, we show that 80% of \$10,625 is at least \$8500:

$0.8(\$10{,}625) = \$8500$

State. The blue book value of the car was at least \$10,625.

41. ***Familiarize.*** Let L = the length of the package.

Translate.

Length	and	girth	is less than	84 in
↓	↓	↓	↓	↓
L	$+$	29	$<$	84

Carry out.

$$L + 29 < 84$$
$$L < 55$$

Check. We check to see if the solution seems reasonable.

When $L = 60$, $60 + 29 = 89$ in.
When $L = 55$, $55 + 29 = 84$ in.
When $L = 50$, $50 + 29 = 79$ in.
From these calculations, it would appear that the solution is correct.

State. For lengths less than 55 in, the box is considered a "package."

43. ***Familiarize.*** We will use the formula $F = \frac{9}{5}C + 32$.

Translate.

Fahrenheit temperature	is above	98.6°.
↓	↓	↓
F	$>$	98.6

Substituting $\frac{9}{5}C + 32$ for F, we have

$$\frac{9}{5}C + 32 > 98.6.$$

Carry out. We solve the inequality.

$$\frac{9}{5}C + 32 > 98.6$$
$$\frac{9}{5}C > 66.6$$
$$C > \frac{333}{9}$$
$$C > 37$$

Check. We check to see if the solution seems reasonable.

When $C = 36$, $\frac{9}{5} \cdot 36 + 32 = 96.8$.

When $C = 37$, $\frac{9}{5} \cdot 37 + 32 = 98.6$.

When $C = 38$, $\frac{9}{5} \cdot 38 + 32 = 100.4$.

It would appear that the solution is correct, considering that rounding occurred.

State. The human body is feverish for Celsius temperatures greater than 37°.

45. ***Familiarize.*** Let h = the height of the triangle, in ft. Recall that the formula for the area of a triangle with base b and height h is $A = \frac{1}{2}bh$.

Translate.

Area	less than or equal to	12 ft^2.
↓	↓	↓
$\frac{1}{2}(8)h$	$\leq$	12

Carry out. We solve the inequality.

$$\frac{1}{2}(8)h \leq 12$$
$$4h \leq 12$$
$$h \leq 3$$

Check. As a partial check, we show that a length of 3 ft will result in an area of 12 ft^2.

$$\frac{1}{2}(8)(3) = 12$$

State. The height should be no more than 3 ft.

47. ***Familiarize.*** Let r = the amount of fat in a serving of the peanut butter, in grams. If reduced fat peanut butter has at least 25% less fat than regular peanut butter, then it has at most 75% as much fat as the regular peanut butter.

Translate.

12 g of fat	is at most	75%	of	the amount of fat in regular peanut butter.
↓	↓	↓	↓	↓
12	$\leq$	0.75	$\cdot$	r

Carry out.

$$12 \leq 0.75r$$
$$16 \leq r$$

Check. As a partial check, we show that 12 g of fat does not exceed 75% of 16 g of fat:

$0.75(16) = 12$

State. A serving of regular peanut butter contains at least 16 g of fat.

49. ***Familiarize***. Let t = the number of years after 2004. To simplify, the number of dogs is in millions.

Translate.

Number of dogs in 2004	plus	1.1 dogs per year	times
↓	↓	↓	↓
73.9	+	1.1	·

number of years	exceeds	90 dogs.
↓	↓	↓
t	>	90

Carry out. We solve the inequality.

$$73.9 + 1.1t > 90$$
$$1.1t > 16.1$$
$$t > 14.6$$
$$2004 + 15 = 2019$$

Check. As a partial check, we can show that the number of dogs is 90 million 15 years after 2004.

$$73.9 + 1.1 \cdot 15 = 73.9 + 16.5 = 90.4$$

State. The there will be more than 90 million dogs living as household pets in 2019 and after.

51. ***Familiarize***. Let n = the number of text messages. The total cost is the monthly fee of $1.99 each day for 22 days, or 1.99(22) = $43.78, plus 0.02 times the number of text messages, or $0.02n$.

Translate.

Day fee	plus	text messages	cannot exceed	$60
↓	↓	↓	↓	↓
$43.78	+	$0.02n$	≤	60

Carry out. We solve the inequality.

$$43.78 + 0.02n \le 60$$
$$0.02n \le 16.22$$
$$n \le 811$$

Check. As a partial check, if the number of text messages is 811, the budget of $60 will not be exceeded.

State. Liam can send or receive 811 text messages and stay within his budget.

53. ***Familiarize***. We will use the formula $R = -0.0065t + 4.3259$.

Translate.

The world record	is less than	3.6 minutes.
↓	↓	↓
$-0.0065t + 4.3259$	<	3.6

Carry out. We solve the inequality.

$$-0.0065t + 4.3259 < 3.6$$
$$-0.0065t < -0.7259$$
$$t > 111.68$$

Check. As a partial check, we can show that the record is more than 3.6 min 111 yr after 1900 and is less than 3.6 min 112 yr after 1900.

For $t = 111$, $R = -0.0065(111) + 4.3259 = 3.7709$.
For $t = 112$, $R = -0.0065(112) + 4.3259 = 3.5979$.

State. The world record in the mile run is less than 3.6 min more than 112 yr after 1900, or in years after 2012.

55. ***Familiarize***. We will use the equation $y = 0.122x + 0.912$.

Translate.

The cost	is at most	$14.
↓	↓	↓
$0.122x + 0.912$	≤	14

Carry out. We solve the inequality.

$$0.122x + 0.912 \le 14$$
$$0.122x \le 13.088$$
$$x \le 107$$

Check. As a partial check, we show that the cost for driving 107 mi is $14.

$$0.122(107) + 0.912 \approx 14$$

State. The cost will be at most $14 for mileages less than or equal to 107 mi.

57. *Writing Exercise.*

59. $7 + xy$

61. Changing the sign of 18 gives −18.

63. *Writing Exercise.*

65. ***Familiarize***. We use the formula $F = \frac{9}{5}C + 32$.

Translate. We are interested in temperatures such that $5° < F < 15°$. Substituting for F, we have:

$$5 < \frac{9}{5}C + 32 < 15$$

Carry out:

$$5 < \frac{9}{5}C + 32 < 15$$
$$5 \cdot 5 < 5\left(\frac{9}{5}C + 32\right) < 5 \cdot 15$$
$$25 < 9C + 160 < 75$$
$$-135 < 9C < -85$$
$$-15 < C < -9\frac{4}{9}$$

Check. The check is left to the student.

State. Green ski wax works best for temperatures between −15°C and $-9\frac{4}{9}$°C.

67. Since $8^2 = 64$, the length of a side must be less than or equal to 8 cm (and greater than 0 cm, of course). We can also use the five-step problem-solving procedure.

Familiarize. Let s represent the length of a side of the square. The area s is the square of the length of a side, or s^2.

Translate.

$\underbrace{\text{The area}}$ $\underbrace{\text{is no more than}}$ 64 cm^2.

$\downarrow$ $\downarrow$ $\downarrow$

s^2 $\le$ 64

Carry out.

$s^2 \le 64$
$|s| \le 8$

Then $-8 \le s \le 8$.

Check. Since the length of a side cannot be negative we only consider positive values of s, or $0 < s \le 8$. We check to see if this solution seems reasonable.

When $s = 7$, the area is 7^2, or 49 cm^2.

When $s = 8$, the area is 8^2, or 64 cm^2.

When $s = 9$, the area is 9^2, or 81 cm^2.

From these calculations, it appears that the solution is correct.

State. Sides of length 8 cm or less will allow an area of no more than 64 cm^2. (Of course, the length of a side must be greater than 0 also.)

69. ***Familiarize.*** Let p = the price of Neoma's tenth book. If the average price of each of the first 9 books is \$12, then the total price of the 9 books is $9 \cdot \$12$, or \$108. The average price of the first 10 books will be $\frac{\$108 + p}{10}$.

Translate.

$\underbrace{\text{The average price of 10 books}}$ $\underbrace{\text{is at least}}$ \$15.

$\downarrow$ $\downarrow$ $\downarrow$

$\frac{108 + p}{10}$ $\ge$ 15

Carry out. We solve the inequality.

$$\frac{108 + p}{10} \ge 15$$
$$108 + p \ge 150$$
$$p \ge 42$$

Check. As a partial check, we show that the average price of the 10 books is \$15 when the price of the tenth book is \$42.

$$\frac{\$108 + \$42}{10} = \frac{\$150}{10} = \$15$$

State. Neoma's tenth book should cost at least \$42 if she wants to select a \$15 book for her free book.

71. *Writing Exercise.*

Chapter 2 Review

1. True

2. False

3. True

4. True

5. True

6. False

7. True

8. True

9.
$x + 9 = -16$
$x + 9 - 9 = -16 - 9$ Adding -9
$x = -25$ Simplifying

The solution is -25.

10.
$-8x = -56$
$\left(-\frac{1}{8}\right)(-8x) = \left(-\frac{1}{8}\right)(-56)$ Multiplying by $-\frac{1}{8}$
$x = 7$ Simplifying

The solution is 7.

11.
$-\frac{x}{5} = 13$
$-5\left(-\frac{x}{5}\right) = -5(13)$ Multiplying by -5
$x = -65$ Simplifying

The solution is -65.

12.
$x - 0.1 = 1.01$
$x - 0.1 + 0.1 = 1.01 + 0.1$ Adding 0.1
$x = 1.11$ Simplifying

The solution is 1.11.

13.
$-\frac{2}{3} + x = -\frac{1}{6}$
$6\left(-\frac{2}{3} + x\right) = 6\left(-\frac{1}{6}\right)$ Multiplying by 6
$-4 + 6x = -1$ Simplifying
$-4 + 6x + 4 = -1 + 4$ Adding 4
$6x = 3$ Simplifying
$x = \frac{1}{2}$ Multiplying by $\frac{1}{6}$

The solution is $\frac{1}{2}$.

14.
$4y + 11 = 5$
$4y + 11 - 11 = 5 - 11$ Adding -11
$4y = -6$ Simplifying
$y = \frac{-6}{4} = -\frac{3}{2}$ Multiplying by $\frac{1}{4}$ and reducing

The solution is $-\frac{3}{2}$.

15.
$5 - x = 13$
$5 - x - 5 = 13 - 5$ Adding -5
$-x = 8$ Simplifying
$x = -8$ Multiplying by -1

The solution is -8.

16.
$$\begin{aligned} 3t+7&=t-1 \\ 3t+7-7&=t-1-7 && \text{Adding } -7 \\ 3t&=t-8 && \text{Simplifying} \\ 3t-t&=t-8-t && \text{Adding } -t \\ 2t&=-8 && \text{Simplifying} \\ t&=-4 && \text{Multiplying by } \tfrac{1}{2} \end{aligned}$$
The solution is –4.

17.
$$\begin{aligned} 7x-6&=25x \\ 7x-6-7x&=25x-7x && \text{Adding } -7x \\ -6&=18x && \text{Simplifying} \\ -\tfrac{1}{3}&=x && \text{Multiplying by } \tfrac{1}{18} \end{aligned}$$
The solution is $-\frac{1}{3}$.

18.
$$\begin{aligned} \tfrac{1}{4}x-\tfrac{5}{8}&=\tfrac{3}{8} \\ 8\left(\tfrac{1}{4}x-\tfrac{5}{8}\right)&=8\left(\tfrac{3}{8}\right) && \text{Multiplying by 8} \\ 2x-5&=3 && \text{Simplifying} \\ 2x-5+5&=3+5 && \text{Adding 5} \\ 2x&=8 && \text{Simplifying} \\ x&=4 && \text{Multiplying by } \tfrac{1}{2} \end{aligned}$$
The solution is 4.

19.
$$\begin{aligned} 14y&=23y-17-9y \\ 14y&=14y-17 \\ 14y-14y&=14y-14y-17 \\ 0&=-17 \quad \text{FALSE} \end{aligned}$$
Contradiction; there is no solution.

20.
$$\begin{aligned} 0.22y-0.6&=0.12y+3-0.8y \\ 0.22y-0.6&=-0.68y+3 && \text{Simplifying} \\ 0.22y-0.6+0.68y&=-0.68y+3+0.68y && \text{Adding } 0.68y \\ 0.9y-0.6&=3 && \text{Simplifying} \\ 0.9y-0.6+0.6&=3+0.6 && \text{Adding 0.6} \\ 0.9y&=3.6 && \text{Simplifying} \\ y&=4 && \text{Multiplying by } \tfrac{1}{0.9} \end{aligned}$$
The solution is 4.

21.
$$\begin{aligned} \tfrac{1}{4}x-\tfrac{1}{8}x&=3-\tfrac{1}{16}x \\ 16\left(\tfrac{1}{4}x-\tfrac{1}{8}x\right)&=16\left(3-\tfrac{1}{16}x\right) && \text{Multiplying by 16} \\ 4x-2x&=48-x && \text{Distributive Law} \\ 2x&=48-x && \text{Simplifying} \\ 2x+x&=48-x+x && \text{Adding } x \\ 3x&=48 && \text{Simplifying} \\ x&=16 && \text{Multiplying by } \tfrac{1}{3} \end{aligned}$$
The solution is 16.

22.
$$\begin{aligned} 6(4-n)&=18 \\ 24-6n&=18 && \text{Distributive Law} \\ 24-6n-24&=18-24 && \text{Adding } -24 \\ -6n&=-6 && \text{Simplifying} \\ n&=1 && \text{Multiplying by } -\tfrac{1}{6} \end{aligned}$$
The solution is 1.

23.
$$\begin{aligned} 4(5x-7)&=-56 \\ 20x-28&=-56 && \text{Distributive Law} \\ 20x-28+28&=-56+28 && \text{Adding 28} \\ 20x&=-28 && \text{Simplifying} \\ x&=-\tfrac{28}{20} && \text{Multiplying by } \tfrac{1}{20} \\ x&=-\tfrac{7}{5} && \text{Simplifying} \end{aligned}$$
The solution is $-\frac{7}{5}$.

24.
$$\begin{aligned} 8(x-2)&=4(x-4) \\ 8x-16&=4x-16 && \text{Distributive Law} \\ 8x-16+16&=4x-16+16 && \text{Adding 16} \\ 8x&=4x && \text{Simplifying} \\ 8x-4x&=4x-4x && \text{Adding } -4x \\ 4x&=0 && \text{Simplifying} \\ x&=0 && \text{Multiplying by } \tfrac{1}{4} \end{aligned}$$
The solution is 0.

25.
$$\begin{aligned} 3(x-4)+2&=x+2(x-5) \\ 3x-12+2&=x+2x-10 \\ 3x-10&=3x-10 \quad \text{TRUE} \end{aligned}$$
Identity; the solution is all real numbers.

26.
$$\begin{aligned} C&=\pi d \\ C\left(\tfrac{1}{\pi}\right)&=\pi d\left(\tfrac{1}{\pi}\right) && \text{Multiplying by } \tfrac{1}{\pi} \\ \tfrac{C}{\pi}&=d && \text{Simplifying} \end{aligned}$$

27.
$$\begin{aligned} V&=\tfrac{1}{3}Bh \\ 3\cdot V&=3\left(\tfrac{1}{3}Bh\right) && \text{Multiplying by 3} \\ 3V&=Bh && \text{Simplifying} \\ \tfrac{1}{h}(3V)&=\tfrac{1}{h}(Bh) && \text{Multiplying by } \tfrac{1}{h} \\ \tfrac{3V}{h}&=B && \text{Simplifying} \end{aligned}$$

28.
$$\begin{aligned} 5x-2y&=10 \\ -5x+5x-2y&=-5x+10 && \text{Adding } -5x \\ -2y&=-5x+10 && \text{Simplifying} \\ -\tfrac{1}{2}(-2y)&=-\tfrac{1}{2}(-5x+10) && \text{Multiplying by } -\tfrac{1}{2} \\ y&=\tfrac{5}{2}x-5 && \text{Simplifying} \end{aligned}$$

29.
$$\begin{aligned} tx&=ax+b \\ tx-ax&=ax+b-ax && \text{Adding } -ax \\ tx-ax&=b && \text{Simplifying} \\ x(t-a)&=b && \text{Factoring } x \\ x&=\frac{b}{t-a} && \text{Multiplying by } \frac{1}{t-a} \end{aligned}$$

30. 1.2% 0.01.2

Move the decimal 2 places to the left. 1.2% = 0.012

31. $\frac{11}{25}=\frac{4}{4}\cdot\frac{11}{25}=\frac{44}{100}=0.44$

First, move the decimal point two places to the right; then write a % symbol: The answer is 44%.

32. ***Translate.***

What percent of 60 is 42?

$y \cdot 60 = 42$

We solve the equation and then convert to percent notation.

$$y \cdot 60 = 42$$
$$y = \frac{42}{60}$$
$$y = 0.70 = 70\%$$

The answer is 70%.

33. ***Translate.***

49 is 35% of What number?

$49 = 0.35 \cdot y$

We solve the equation and then convert to percent notation.

$$49 = 0.35y$$
$$\frac{49}{0.35} = y$$
$$140 = y$$

The answer is 140.

34. $x \le -5$

We substitute -3 for x giving $-3 \le -5$, which is a false statement since -3 is not to the left of -5 on the number line, so -3 is not a solution of the inequality $x \le -5$.

35. $x \le -5$

We substitute -7 for x giving $-7 \le -5$, which is a true statement since -7 is to the left of -5 on the number line, so -7 is a solution of the inequality $x \le -5$.

36. $x \le -5$

We substitute 0 for x giving $0 \le -5$, which is a false statement since 0 is not to the left of -5 on the number line, so 0 is not a solution of the inequality $x \le -5$.

37.

$5x - 6 < 2x + 3$
$5x - 6 + 6 < 2x + 3 + 6$ Adding 6
$5x < 2x + 9$ Simplifying
$5x - 2x < 2x + 9 - 2x$ Adding $-2x$
$3x < 9$ Simplifying
$x < 3$ Multiplying by $\frac{1}{3}$

The solution set is $\{x | x < 3\}$. The graph is as follows:

$5x - 6 < 2x + 3$

−4 −2 0 2 4

38. $-2 < x \le 5$

The solution set is $\{x | -2 < x \le 5\}$. The graph is as follows:

$-2 < x \le 5$

−4 −2 0 2 4 6

39. $t > 0$

The solution set is $\{t | t > 0\}$. The graph is as follows:

$t > 0$

−4 −2 0 2 4

40.

$t + \frac{2}{3} \ge \frac{1}{6}$
$6\left(t + \frac{2}{3}\right) \ge 6\left(\frac{1}{6}\right)$ Multiplying by 6
$6t + 4 \ge 1$ Simplifying
$6t + 4 - 4 \ge 1 - 4$ Adding -4
$6t \ge -3$ Simplifying
$\frac{1}{6}(6t) \ge \frac{1}{6}(-3)$ Multiplying by $\frac{1}{6}$
$t \ge -\frac{1}{2}$ Simplifying

The solution set is $\left\{t \middle| t \ge -\frac{1}{2}\right\}$, or $\left[-\frac{1}{2}, \infty\right)$.

41.

$2 + 6y > 20$
$2 + 6y - 2 > 20 - 2$ Adding -2
$6y > 18$ Simplifying
$\frac{1}{6}(6y) > \frac{1}{6} \cdot 18$ Multiplying by $\frac{1}{6}$
$y > 3$ Simplifying

The solution set is $\{y | y > 3\}$, or $(3, \infty)$.

42.

$7 - 3y \ge 27 + 2y$
$7 - 3y - 7 \ge 27 + 2y - 7$ Adding -7
$-3y \ge 20 + 2y$ Simplifying
$-3y - 2y \ge 20 + 2y - 2y$ Adding $-2y$
$-5y \ge 20$ Simplifying
$y \le -4$ Multiplying by $-\frac{1}{5}$ and reversing the inequality symbol

The solution set is $\{y | y \le -4\}$, or $(-\infty, -4]$.

43.

$-4y < 28$
$-\frac{1}{4}(-4y) > -\frac{1}{4} \cdot 28$ Multiplying by $-\frac{1}{4}$ and reversing the inequality symbol
$y > -7$ Simplifying

The solution set is $\{y | y > -7\}$, or $(-7, \infty)$.

44.

$3 - 4x < 27$
$3 - 4x - 3 < 27 - 3$ Adding -3
$-4x < 24$ Simplifying
$-\frac{1}{4}(-4x) > -\frac{1}{4} \cdot 24$ Multiplying by $-\frac{1}{4}$ and reversing the inequality symbol
$x > -6$ Simplifying

The solution set is $\{x | x > -6\}$, or $(-6, \infty)$.

45.
$$\begin{aligned} 4-8x &< 13+3x \\ 4-8x-4 &< 13+3x-4 && \text{Adding } -4 \\ -8x &< 9+3x && \text{Simplifying} \\ -8x-3x &< 9+3x-3x && \text{Adding } -3x \\ -11x &< 9 && \text{Simplifying} \\ -\frac{1}{11}(-11x) &> -\frac{1}{11}\cdot 9 && \text{Multiplying by } -\frac{1}{11} \\ x &> -\frac{9}{11} && \text{Simplifying} \end{aligned}$$

The solution set is $\left\{x \middle| x > -\frac{9}{11}\right\}$, or $\left(-\frac{9}{11}, \infty\right)$.

46.
$$\begin{aligned} 13 &\le -\frac{2}{3}t+5 \\ 13-5 &\le -\frac{2}{3}t+5-5 && \text{Adding } -5 \\ 8 &\le -\frac{2}{3}t && \text{Simplifying} \\ -\frac{3}{2}(8) &\ge \left(-\frac{3}{2}\right)\left(-\frac{2}{3}t\right) && \text{Multiplying by } -\frac{3}{2} \\ -12t &\ge t && \text{Simplifying} \end{aligned}$$

The solution set is $\{t | t \le -12\}$, or $(-\infty, -12]$.

47.
$$\begin{aligned} 7 &\le 1-\frac{3}{4}x \\ 7-1 &\le 1-\frac{3}{4}x-1 && \text{Adding } -1 \\ 6 &\le -\frac{3}{4}x && \text{Simplifying} \\ -\frac{4}{3}\cdot 6 &\ge -\frac{4}{3}\left(-\frac{3}{4}x\right) && \text{Multiplying by } -\frac{4}{3} \\ -8 &\ge x && \text{Simplifying} \end{aligned}$$

The solution set is $\{x | x \le -8\}$, or $(-\infty, -8]$.

48. ***Familiarize***. Let $x =$ the total number of cats placed.

Translate.

30% of total cats was 280?

$0.30 \cdot x = 280$

Carry out. We solve the equation.

$$\begin{aligned} 0.30x &= 280 \\ x &= \frac{280}{0.3} \\ x &\approx 933 \end{aligned}$$

Check. If 933 was the total number of cats placed, then 30% of 933 is 0.30(933), about 280. This checks.

State. There were about 933 cats adopted through FACE in 2014.

49. ***Familiarize***. Let $x =$ the length of the first piece, in ft. Since the second piece is 2 ft longer than the first piece, it must be $x+2$ ft.

Translate.

The sum of the lengths of the two pieces is 32 ft.

$x+(x+2) = 32$

Carry out. We solve the equation.

$$\begin{aligned} x+(x+2) &= 32 \\ 2x+2 &= 32 \\ 2x &= 30 \\ x &= 15 \end{aligned}$$

Check. If the first piece is 15 ft long, then the second piece must be 15 + 2, or 17 ft long. The sum of the lengths of the two pieces is 15 ft + 17 ft, or 32 ft. The answer checks.

State. The lengths of the two pieces are 15 ft and 17 ft.

50. ***Familiarize***. Let $x =$ the number of Indian students. Then $2x-6000$ is the number of Chinese students.

Translate.

The number of Indian students plus The number of Chinese students is 294,000

$x + 2x-6000 = 294,000$

Carry out. We solve the equation.

$$\begin{aligned} x+2x-6000 &= 294,000 \\ 3x-6000 &= 294,000 \\ 3x &= 300,000 \\ x &= 100,000 \end{aligned}$$

$2x-000 = 2\cdot 100,000-6000 = 194,000$

Check. If the number of Indian students is 100,000 and the number of Chinese students is 194,000, then the total is 100,000 + 194,000, or 294,000. The answer checks.

State. There were 100,000 Indian students and 194,000 Chinese students.

51. ***Familiarize***. Let $x =$ the original number of new international students in 2014.

Translate.

Students in 2014 plus 14.18% increase is 1,130,000.

$x + 0.1418x = 1,130,000$

Carry out. We solve the equation.

$$\begin{aligned} x+0.1418x &= 1,130,000 \\ 1.1418x &= 1,130,000 \\ x &= \frac{1,130,000}{1.1418} \\ x &\approx 990,000 \end{aligned}$$

Check. If the number of new international students in 2014 was about 990,000, then that number plus an increase of 14.18% is $990,000+0.1418\cdot 990,000$, is about 1,130,000. The answer checks.

State. There were about 990,000 new international students in 2014.

52. ***Familiarize***. Let $x =$ the first odd integer and let $x+2 =$ the next consecutive odd integer.

Translate.

$$\underbrace{\text{The sum of the two consecutive odd integers}}_{x+(x+2)} \quad \underbrace{\text{is}}_{=} \quad \underbrace{116}_{116}$$

Carry out. We solve the equation.

$$\begin{aligned} x+(x+2) &= 116 \\ 2x+2 &= 116 \\ 2x &= 114 \\ x &= 57 \end{aligned}$$

Check. If the first odd integer is 57, then the next consecutive odd integer would be 57 + 2 or 59. The sum of these two integers is 57 + 59, or 116. This result checks.

State. The integers are 57 and 59.

53. ***Familiarize***. Let $x =$ the length of the rectangle, in cm. The width of the rectangle is $x-6$ cm. The perimeter of a rectangle is given by $P = 2l + 2w$, where l is the length and w is the width.

Translate.

$$\underbrace{\text{The perimeter of the rectangle}}_{2x+2(x-6)} \quad \underbrace{\text{is}}_{=} \quad \underbrace{56 \text{ cm}}_{56}$$

Carry out. We solve the equation.

$$\begin{aligned} 2x+2(x-6) &= 56 \\ 2x+2x-12 &= 56 \\ 4x-12 &= 56 \\ 4x &= 68 \\ x &= 17 \end{aligned}$$

Check. If the length is 17 cm, then the width is 17 cm − 6 cm, or 11 cm. The perimeter is $2 \cdot 17 \text{ cm} + 2 \cdot 11 \text{ cm}$, or 34 cm + 22 cm, or 56 cm. These results check.

State. The length is 17 cm and the width is 11 cm.

54. ***Familiarize***. Let $x =$ the regular price of the picnic table. Since the picnic table was reduced by 25%, it actually sold for 75% of its original price.

Translate.

$$\underbrace{75\%}_{0.75} \quad \underbrace{\text{of}}_{\cdot} \quad \underbrace{\text{the original price}}_{x} \quad \underbrace{\text{is}}_{=} \quad \underbrace{\$120?}_{120}$$

Carry out. We solve the equation.

$$\begin{aligned} 0.75x &= 120 \\ x &= \frac{120}{0.75} \\ x &= 160 \end{aligned}$$

Check. If the original price was \$160 with a 25% discount, then the purchaser would have paid 75% of \$160, or $0.75 \cdot \$160$, or \$120. This result checks.

State. The original price was \$160.

55. ***Familiarize***. Let $x =$ the measure of the first angle. The measure of the second angle is $x + 50°$, and the measure of the third angle is $2x - 10°$. The sum of the measures of the angles of a triangle is 180°.

Translate.

$$\underbrace{\text{The sum of the measures of the angles}}_{x+(x+50)+(2x-10)} \quad \underbrace{\text{is}}_{=} \quad \underbrace{180°}_{180}$$

Carry out. We solve the equation.

$$\begin{aligned} x+(x+50)+(2x-10) &= 180 \\ 4x+40 &= 180 \\ 4x &= 140 \\ x &= 35 \end{aligned}$$

Check. If the measure of the first angle is 35°, then the measure of the second angle is $35° + 50°$, or 85°, and the measure of the third angle is $2 \cdot 35° - 10°$, or 60°. The sum of the measures of the first, second, and third angles is $35° + 85° + 60°$, or 180°. These results check.

State. The measures of the angles are 35°, 85°, and 60°.

56. ***Familiarize***. Let $x =$ the amount spent in the sixth month.

Translate.

$$\underbrace{\text{Entertainment average}}_{\frac{98+89+110+85+83+x}{6}} \quad \underbrace{\text{does not exceed}}_{\leq} \quad \underbrace{\$95}_{95}$$

Carry out. We solve the inequality.

$$\begin{aligned} \frac{98+89+110+85+83+x}{6} &\leq 95 \\ 6\left(\frac{98+89+110+85+83+x}{6}\right) &\leq 6 \cdot 95 \\ 98+89+110+85+83+x &\leq 570 \\ 465+x &\leq 570 \\ x &\leq 105 \end{aligned}$$

Check. As a partial check we calculate the average spent if \$105 was spent on the sixth month. The average is $\frac{98+89+110+85+83+105}{6} = \frac{570}{6} = 95$. The results check.

State. Kathleen can spend \$105 or less in the sixth month without exceeding her budget.

57. ***Familiarize***. Let $n =$ the number of copies. The total cost is the setuup fee of \$6 plus \$4 per copy, or $4n$.

Translate.

$$\underbrace{\text{Set up fee}}_{6} \quad \underbrace{\text{plus}}_{+} \quad \underbrace{\text{cost per copy}}_{4n} \quad \underbrace{\text{cannot exceed}}_{\leq} \quad \underbrace{\$65}_{65}$$

Carry out. We solve the inequality.

$$6+4n \le 65$$
$$4n \le 59$$
$$n \le \frac{59}{4}$$
$$n \le 14.75$$

Check. As a partial check, if the number of copies is 14, the total cost $\$6+\$4\cdot 14$, or \$62 does not exceed the budget of \$65.

State. Myra can make 14 or fewer copies.

58. *Writing Exercise.*

59. *Writing Exercise.*

60. ***Familiarize***. Let x = the amount of time the average child spends reading or doing homework.

Translate.

Time spent reading or doing homework	plus	108% more	is	3 hr 20 min.
↓	↓	↓	↓	↓
x	$+$	$1.08x$	$=$	$3\frac{1}{3}$

Carry out. We solve the equation.

$$x+1.08x = 3\tfrac{1}{3}$$
$$2.08x = 3\tfrac{1}{3}$$
$$x \approx 1.6 \text{ hr} \approx 1 \text{ hr } 36 \text{ min}$$

Check. If the amount of time spent reading or doing homework is 1 hr 36 min, then that time plus an increase of 108% more is 1 hr 36 min + $1.08\cdot 1$ hr 36 min, is about 3 hrs 20 min.

The answer checks.

State. About 1hr 36 min is spent reading or doing homework.

61. ***Familiarize***. Let x = the length of the Nile River, in mi. Let $x+65$ represent the length of the Amazon River, in mi.

Translate.

The combined length of both rivers	is	8385 mi
↓	↓	↓
$x+(x+65)$	$=$	8385

Carry out. We solve the equation.

$$x+(x+65) = 8385$$
$$2x+65 = 8385$$
$$2x = 8320$$
$$x = 4160$$

Check. If the Nile River is 4160 mi long, then the Amazon River is 4160 mi + 65 mi, or 4225 mi. The combined length of both rivers is then 4160 mi + 4225 mi, or 8385 mi. These results check.

State. The Amazon River is 4225 mi long, and the Nile River is 4160 mi long.

62. $2|n|+4=50$

$$2|n| = 46$$
$$|n| = 23$$

The distance from some number n and the origin is 23 units. Thus, $n=-23$ or $n=23$.

63. $|3n|=60$

The distance from some number, $3n$, to the origin is 60 units. So we have:

$$3n=-60 \quad or \quad 3n=60$$
$$n=-20 \quad or \quad n=20$$

The solutions are –20 and 20.

64.

$$y = 2a-ab+3$$
$$y = a(2-b)+3$$
$$y-3 = a(2-b)$$
$$\frac{y-3}{2-b} = a$$

The solution is $a=\frac{y-3}{2-b}$.

65. **1.** w; $12w$
2. $12w$; $0.3(12w)$
3. $0.3(12w)$; $\frac{0.3(12w)}{9}$

So $F=\frac{0.3(12w)}{9}$ or $F=0.4w$.

Chapter 2 Test

1.

$$t+7=16$$
$$t+7-7=16-7 \quad \text{Adding } -7$$
$$t=9 \quad \text{Simplifying}$$

The solution is 9.

2.

$$6x=-18$$
$$\frac{1}{6}(6x)=\frac{1}{6}(-18) \quad \text{Multiplying by } \frac{1}{6}$$
$$x=-3 \quad \text{Simplifying}$$

The solution is –3.

3.

$$-\frac{4}{7}x=-28$$
$$-\frac{7}{4}\left(-\frac{4}{7}\right)=-\frac{7}{4}(-28) \quad \text{Multiplying by } -\frac{7}{4}$$
$$x=49 \quad \text{Simplifying}$$

The solution is 49.

4.

$$3t+7=2t-5$$
$$3t+7-7=2t-5-7 \quad \text{Adding } -7$$
$$3t=2t-12 \quad \text{Simplifying}$$
$$3t-2t=2t-12-2t \quad \text{Adding } -2t$$
$$t=-12 \quad \text{Simplifying}$$

The solution is –12.

5. $\frac{1}{2}x - \frac{3}{5} = \frac{2}{5}$

$\frac{1}{2}x - \frac{3}{5} + \frac{3}{5} = \frac{2}{5} + \frac{3}{5}$ Adding $\frac{3}{5}$

$\frac{1}{2}x = 1$ Simplifying

$x = 2$ Multiplying by 2

The solution is 2.

6. $8 - y = 16$

$8 - y - 8 = 16 - 8$ Adding -8

$-y = 8$ Simplifying

$y = -8$ Multiplying by -1

The solution is -8.

7. $4.2x + 3.5 = 1.2 - 2.5x$

$-3.5 + 4.2x + 3.5 = 1.2 - 2.5x - 3.5$ Adding -3.5

$4.2x = -2.5x - 2.3$ Simplifying

$4.2x + 2.5x = -2.5x - 2.3 + 2.5x$ Adding $2.5x$

$6.7x = -2.3$ Simplifying

$\frac{1}{6.7}(6.7x) = \frac{1}{6.7}(-2.3)$ Multiplying by $\frac{1}{6.7}$

$x = -\frac{23}{67}$ Simplifying

The solution is $-\frac{23}{67}$.

8. $4(x+2) = 36$

$4x + 8 = 36$ Distributive Law

$4x + 8 - 8 = 36 - 8$ Adding -8

$4x = 28$ Simplifying

$\frac{1}{4}(4x) = \frac{1}{4}(28)$ Multiplying by $\frac{1}{4}$

$x = 7$ Simplifying

The solution is 7.

9. $\frac{5}{6}(3x+1) = 20$

$\frac{6}{5}\left[\frac{5}{6}(3x+1)\right] = \frac{6}{5} \cdot 20$ Multiplying by $\frac{6}{5}$

$3x + 1 = 24$ Simplifying

$3x + 1 - 1 = 24 - 1$ Adding -1

$3x = 23$ Simplifying

$\frac{1}{3}(3x) = \frac{1}{3}(23)$ Multiplying by $\frac{1}{3}$

$x = \frac{23}{3}$ Simplifying

The solution is $\frac{23}{3}$.

10. $13t - (5 - 2t) = 5(3t - 1)$

$13t - 5 + 2t = 15t - 5$

$15t - 5 = 15t - 5$ TRUE

Identity; the solution is all real numbers.

11. $x + 6 > 1$

$x + 6 - 6 > 1 - 6$ Adding -6

$x > -5$ Simplifying

The solution set is $\{x|x > -5\}$, or $(-5, \infty)$.

12. $14x + 9 > 13x - 4$

$14x + 9 - 9 > 13x - 4 - 9$ Adding -9

$14x > 13x - 13$ Simplifying

$14x - 13x > 13x - 13 - 13x$ Adding $-13x$

$x > -13$ Simplifying

The solution set is $\{x|x > -13\}$, or $(-13, \infty)$.

13. $-5y \geq 65$

$y \leq -13$

The solution set is $\{y|y \leq -13\}$, or $(-\infty, -13]$.

14. $4n + 3 < -17$

$4n + 3 - 3 < -17 - 3$ Adding -3

$4n < -20$ Simplifying

$\frac{1}{4}(4n) < \frac{1}{4}(-20)$ Multiplying by $\frac{1}{4}$

$n < -5$ Simplifying

The solution set is $\{n|n < -5\}$, or $(-\infty, -5)$.

15. $3 - 5x > 38$

$3 - 5x - 3 > 38 - 3$ Adding -3

$-5x > 35$ Simplifying

$-\frac{1}{5}(-5x) < -\frac{1}{5}(35)$ Multiplying by $-\frac{1}{5}$ and reversing the inequality symbol

$x < -7$ Simplifying

The solution set is $\{x|x < -7\}$, or $(-\infty, -7)$.

16. $\frac{1}{2}t - \frac{1}{4} \leq \frac{3}{4}t$

$\frac{1}{2}t - \frac{1}{4} - \frac{1}{2}t \leq \frac{3}{4}t - \frac{1}{2}t$ Adding $-\frac{1}{2}t$

$-\frac{1}{4} \leq \frac{1}{4}t$ Simplifying

$4\left(-\frac{1}{4}\right) \leq 4\left(\frac{1}{4}t\right)$ Multiplying by 4

$-1 \leq t$ Simplifying

The solution set is $\{t|t \geq -1\}$, or $[-1, \infty)$.

17. $5 - 9x \geq 19 + 5x$

$5 - 9x - 5 \geq 19 + 5x - 5$ Adding -5

$-9x \geq 14 + 5x$ Simplifying

$-9x - 5x \geq 14 + 5x - 5x$ Adding $-5x$

$-14x \geq 14$ Simplifying

$-\frac{1}{14}(-14x) \leq -\frac{1}{14}(14)$ Multiplying by $-\frac{1}{14}$ and reversing the inequality symbol

$x \leq -1$ Simplifying

The solution set is $\{x|x \leq -1\}$, or $(-\infty, -1]$.

18. $A = 2\pi rh$

$\frac{1}{2\pi h} \cdot A = \frac{1}{2\pi h}(2\pi rh)$ Multiplying by $\frac{1}{2\pi h}$

$\frac{A}{2\pi h} = r$ Simplifying

The solution is $r = \frac{A}{2\pi h}$.

19.
$$
\begin{aligned}
w &= \frac{P+l}{2} \\
2\cdot w &= 2\left(\frac{P+l}{2}\right) && \text{Multiplying by 2} \\
2w &= P+l && \text{Simplifying} \\
2w-P &= P+l-P && \text{Adding } -P \\
2w-P &= l && \text{Simplifying}
\end{aligned}
$$
The solution is $l = 2w - P$.

20. $230\% = 230 \times 0.01$ Replacing % by $\times 0.01$
$= 2.3$

21. 0.003 First move the decimal point two places to the right; then write a % symbol. The answer is 0.3%.

22. ***Translate.***

What number	is	18.5%	of	80?
↓	↓	↓	↓	↓
x	=	0.185	·	80

We solve the equation.
$$x = 0.185 \cdot 80$$
$$x = 14.8$$
The solution is 14.8.

23. ***Translate.***

What percent	of	75	is	33?
↓	↓	↓	↓	↓
y	·	75	=	33

We solve the equation and then convert to percent notation.
$$y \cdot 75 = 33$$
$$y = \frac{33}{75}$$
$$y = 0.44 = 44\%$$
The solution is 44%.

24.
$$y < 4$$
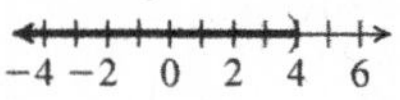

25.
$$-2 \le x \le 2$$
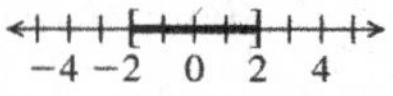

26. ***Familiarize.*** Let w = the width of the calculator, in cm. Then the length is $w+4$, in cm. The perimeter of a rectangle is given by $P = 2l + 2w$.

Translate.

The perimeter of the rectangle	is	36.
↓	↓	↓
$2(w+4)+2w$	=	36

Carry out. We solve the equation.
$$
\begin{aligned}
2(w+4)+2w &= 36 \\
2w+8+2w &= 36 \\
4w+8 &= 36 \\
4w &= 28 \\
w &= 7
\end{aligned}
$$

Check. If the width is 7 cm, then the length is $7+4$, or 11 cm. The perimeter is then $2\cdot 11 + 2\cdot 7$, or $22+14$, or 36 cm. These results check.

State. The width is 7 cm and the length is 11 cm.

27. ***Familiarize.*** Let x = the distance from start. Then $3x$ mi is the distance to the end.

Translate.

Distance from start	and	distance to end	is	whole trip.
↓	↓	↓	↓	↓
x	+	$3x$	=	240

Carry out. We solve the equation.
$$
\begin{aligned}
x+3x &= 240 \\
4x &= 240 \\
x &= 60 \\
3x &= 180
\end{aligned}
$$

Check.
$$60+180=240.$$

State. Kari has biked a distance of 60 miles so far.

28. ***Familiarize.*** Let x = the length of the first side, in mm. Then the length of the second side is $x+2$ mm, and the length of the third side is $x+4$ mm. The perimeter of a triangle is the sum of the lengths of the three sides.

Translate.

The perimeter of the triangle	is	249 mm.
↓	↓	↓
$x+(x+2)+(x+4)$	=	249

Carry out. We solve the equation.
$$
\begin{aligned}
x+(x+2)+(x+4) &= 249 \\
3x+6 &= 249 \\
3x &= 243 \\
x &= 81
\end{aligned}
$$

Check. If the length of the first side is 81 mm, then the length of the second side is $81+2$, or 83 mm, and the length of the third side is $81+4$, or 85 mm. The perimeter of the triangle is $81+83+85$, or 249 mm. These results check.

State. The lengths of the sides are 81 mm, 83 mm, and 85 mm.

29. ***Familiarize.*** Let $x =$ the electric bill before the temperature of the water heater was lowered. If the bill dropped by 7%, then the Kellys paid 93% of their original bill.

Translate.

93%	of	the original bill	is	$60.45.
↓	↓	↓	↓	↓
0.93	·	x	=	60.45

Carry out. We solve the equation.

$$0.93x = 60.45$$
$$x = \frac{60.45}{0.93}$$
$$x = 65$$

Check. If the original bill was $65, and the bill was reduced by 7%, or $0.07 \cdot \$65$, or $4.55, the new bill would be $65 – $4.55, or $60.45. This result checks.

State. The original bill was $65.

30. ***Familarize.*** Let $x =$ the number of trips.

Translate.

Monthly pass	must not exceed	cost of individual trips
↓	↓	↓
79	<	$2.25x$

Carry out. We solve the inequality.

$$79 < 2.25x$$
$$\frac{79}{2.25} < x$$
$$35.1 < x$$

Check. As a partial check, we let $x = 36$ trips and determine the cost. The cost would be $36(2.25) = \$81$. If the number of trips were less, the cost would be under $81, so the result checks.

State. Gail should make more than 35 one-way trips per month.

31.

$$c = \frac{2cd}{a-d}$$
$$(a-d)c = (a-d)\left(\frac{2cd}{a-d}\right) \quad \text{Multiplying by } a-d$$
$$ac - dc = 2cd \quad \text{Simplifying}$$
$$ac - dc + dc = 2cd + dc \quad \text{Adding } dc$$
$$ac = 3cd \quad \text{Simplifying}$$
$$\frac{1}{3c}(ac) = \frac{1}{3c}(3cd) \quad \text{Multiplying by } \frac{1}{3c}$$
$$\frac{a}{3} = d \quad \text{Simplifying}$$

The solution is $d = \frac{a}{3}$.

32.

$$3|w| - 8 = 37$$
$$3|w| - 8 + 8 = 37 + 8 \quad \text{Adding 8}$$
$$3|w| = 45 \quad \text{Simplifying}$$
$$\frac{1}{3}(3|w|) = \frac{1}{3} \cdot 45 \quad \text{Multiplying by } \frac{1}{3}$$
$$|w| = 15 \quad \text{Simplifying}$$

This tells us that the number w is 15 units from the origin. The solutions are $w = -15$ and $w = 15$.

33. Let $h =$ the number of hours of sun each day. Then we have $4 \le h \le 6$.

34. ***Familiarize.*** Let $x =$ the number of tickets given away. The following shows the distribution of the tickets:

First person received $\frac{1}{3}x$ tickets.

Second person received $\frac{1}{4}x$ tickets.

Third person received $\frac{1}{5}x$ tickets.

Fourth person received 8 tickets.
Fifth person received 5 tickets.

Translate.

The number of tickets the five people received	is	the total number of tickets.
↓	↓	↓
$\frac{1}{3}x + \frac{1}{4}x + \frac{1}{5}x + 8 + 5$	=	x

Carry out. We solve the equation.

$$\frac{1}{3}x + \frac{1}{4}x + \frac{1}{5}x + 8 + 5 = x$$
$$60\left(\frac{1}{3}x + \frac{1}{4}x + \frac{1}{5}x + 8 + 5\right) = 60x$$
$$60\left(\frac{1}{3}x + \frac{1}{4}x + \frac{1}{5}x + 13\right) = 60x$$
$$20x + 15x + 12x + 780 = 60x$$
$$47x + 780 = 60x$$
$$780 = 13x$$
$$60 = x$$

Check. If the total number of tickets given away was 60, then the first person received $\frac{1}{3}(60)$, or 20 tickets; the second person received $\frac{1}{4}(60)$, or 15 tickets; the third person received $\frac{1}{5}(60)$, or 12 tickets. We are told that the fourth person received 8 tickets, and the fifth person received 5 tickets. The sum of the tickets distributed is $20 + 15 + 12 + 8 + 5$, or 60 tickets. These results check.

State. There were 60 tickets given away.

Chapter 3

Introduction to Graphing

Exercise Set 3.1

1. The letter in the third quadrant is *E*.

3. The letter marking the second coordinate is *G*.

5. The letter marking the ordered pair is *H*.

7. The x-values extend from –9 to 4 and the y-values range from –1 to 5, so (a) is the best choice.

9. The x-values extend from –2 to 4 and the y-values range from –9 to 1, so (b) is the best choice.

11. We go to the top of the bar that is above the body weight 100 lb. Then we move horizontally from the top of the bar to the vertical scale listing numbers of drinks. It appears that consuming approximately 2 drinks in one hour will give a 100 lb person a blood-alcohol level of 0.08%.

13. For 3 drinks in one hour, we use the horizontal line at 3. For persons weighing 140 lb or less, their blood-alcohol level is 0.08% or more For persons weighing more than 140 lbs. their blood-alcohol level is under 0.08%. Thus, the person weighs more than 140 lbs.

15. We locate 1980 on the horizontal axis and then move up to the line. From there we move left to the vertical axis and read the percent. In 1980, we estimate 12% of bachelor's degrees were conferred in education.

17. We locate 9 on the vertical axis and move right to the line. From there we move down to the horizontal scale and read the year. We see that in 1985 and 1995, 9% of bachelor's degrees were conferred in education.

19. We locate the highest point on the graph. From there we move down to the horizontal scale and read the year. We see that approximately the highest percentage of bachelor's degrees conferred in education was in 1990.

21. Starting at the origin:
(1, 2) is 1 unit right and 2 units up;
(–2, 3) is 2 units left and 3 units up;
(4, –1) is 4 units right and 1 unit down;
(–5, –3) is 5 units left and 3 units down;
(4, 0) is 4 units right and 0 units up or down;
(0, –2) is 0 units right or left and 2 units down.

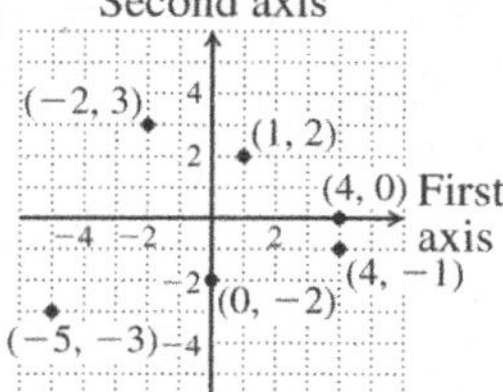

23. Starting at the origin:
(4, 4) is 4 units right and 4 units up;
(–2, 4) is 2 units left and 4 units up;
(5, –3) is 5 units right and 3 units down;
(–5, –5) is 5 units left and 5 units down;
(0, 4) is 0 units right or left and 4 units up;
(0, –4) is 0 units right or left and 4 units down;
(–4, 0) is 4 units left and 0 units up or down.
(0, 0) is 0 units right and 0 units up or down;

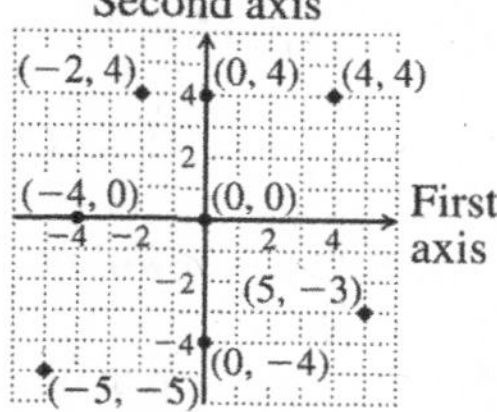

25. We plot the points (2009, 161), (2010, 247), (2011, 367), (2012, 423), (2013, 498), and (2014, 561), and connect adjacent points with line segments.

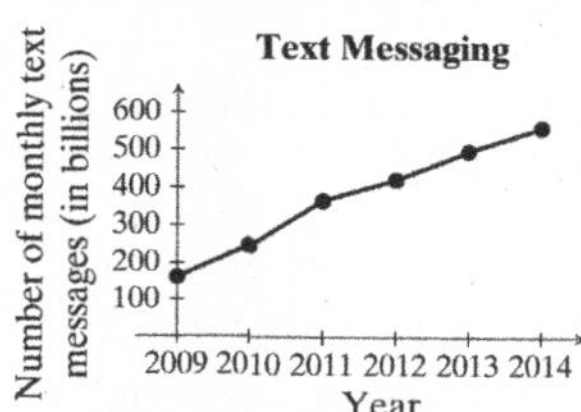

27.

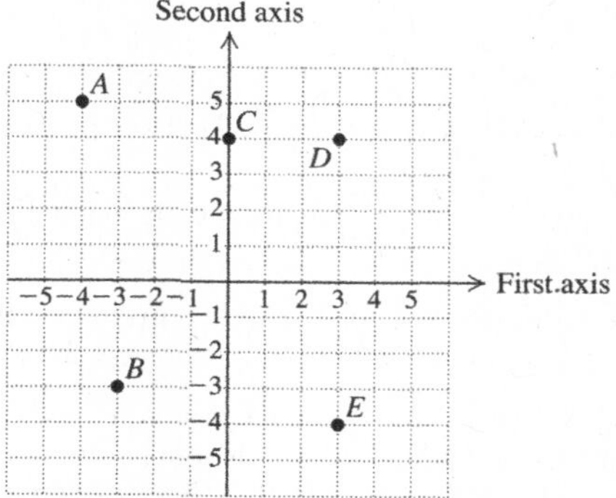

Point *A* is 4 units left and 5 units up. The coordinates of *A* are (–4, 5).

Point *B* is 3 units left and 3 units down. The coordinates of *B* are (–3,–3).

Point *C* is 0 units right or left and 4 units up. The coordinates of *C* are (0, 4).

Point *D* is 3 units right and 4 units up. The coordinates of *D* are (3, 4).

Point *E* is 3 units right and 4 units down. The coordinates of *E* are (3,–4).

29.

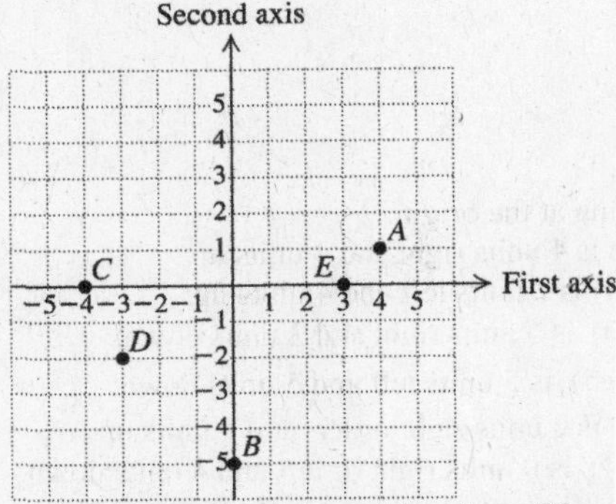

Point A is 4 units right and 1 unit up. The coordinates of A are (4, 1).

Point B is 0 units right or left and 5 units down. The coordinates of B are (0,–5).

Point C is 4 units left and 0 units up or down. The coordinates of C are (–4, 0).

Point D is 3 units left and 2 units down. The coordinates of D are (–3,–2).

Point E is 3 units right and 0 units up or down. The coordinates of E are (3, 0).

31. Since the first coordinate is positive and the second coordinate negative, the point $(7,-2)$ is located in quadrant IV.

33. Since both coordinates are negative, the point $(-4,-3)$ is in quadrant III.

35. Since the first coordinate is 0, and the second coordinate negative, the point (0,–3) is on the y-axis.

37. Since the first coordinate is negative and the second coordinate is positive, the point $(-4.9,8.3)$ is in quadrant II.

39. Since the first coordinate is negative, and the second coordinate is 0, the point $\left(-\frac{5}{2},\ 0\right)$ is on the x-axis.

41. Since both coordinates are positive, the point (160, 2) is in quadrant I.

43. First coordinates are positive in the quadrants that lie to the right of the origin, or in quadrants I and IV.

45. Points for which both coordinates are positive lie in quadrant I, and points for which both coordinates are negative lie in quadrant III. Thus, both coordinates have the same sign in quadrants I and III.

47. Since the x-values range from –75 to 9, the 10 horizontal squares must span $9-(-75)$, or 84 units. Since 84 is close to 100 and it is convenient to count by 10's, we can count backward from 0 eight squares to –80 and forward from 0 two squares to 20 for a total of $8+2$, or 10 squares.

Since the y-values range from –4 to 5, the 10 vertical squares must span $5-(-4)$, or 9 units. It will be convenient to count by 2's in this case. We count down from 0 five squares to –10 and up from 0 five squares to 10 for a total of $5+5$, or 10 squares. (Instead, we might have chosen to count by 1's from –5 to 5.)

Then we plot the points (–75, 5), (–18,–2), and (9,–4).

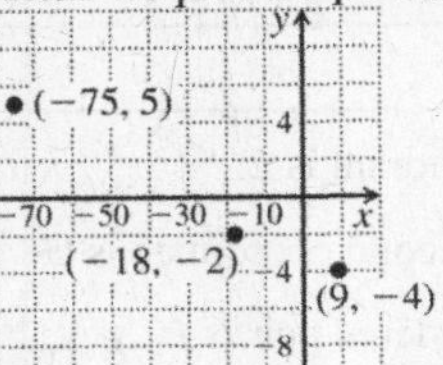

49. Since the x-values range from –5 to 5, the 10 horizontal squares must span $5-(-5)$, or 10 units. It will be convenient to count by 2's in this case. We count backward from 0 five squares to –10 and forward from 0 five squares to 10 for a total of $5+5$, or 10 squares.

Since the y-values range from –14 to 83, the 10 vertical squares must span $83-(-14)$, or 97 units. To include both –14 and 83, the squares should extend from about -20 to 90, or $90-(-20)$, or 110 units. We cannot do this counting by 10's, so we use 20's instead. We count down from 0 four units to –80 and up from 0 six units to 120 for a total of $4+6$, or 10 units. There are other ways to cover the values from –14 to 83 as well.

Then we plot the points (–1, 83), (–5,–14), and (5, 37).

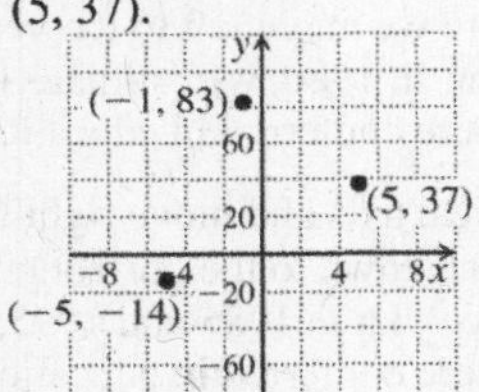

51. Since the x-values range from –16 to 3, the 10 horizontal squares must span $3-(-16)$, or 19 units. We could number by 2's or 3's. We number by 3's, going backward from 0 eight squares to -24 and forward from 0 two squares to 6 for a total of $8+2$, or 10 squares.

Since the y-values range from –4 to 15, the 10 vertical squares must span $15-(-4)$, or 19 units. We will number the vertical axis by 3's as we did the horizontal axis. We go down from 0 four squares to -12 and up from 0 six squares to 18 for a total of $4+6$, or 10 squares.

Then we plot the points (–10,–4), (–16, 7), and (3, 15).

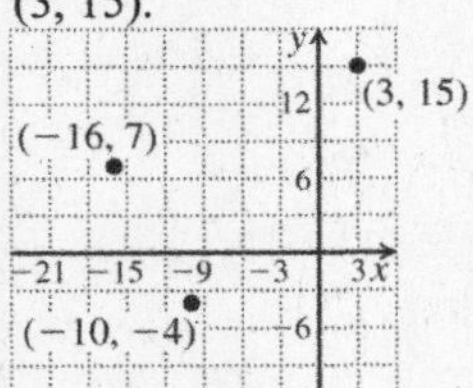

53. Since the x-values range from -100 and 800, the 10 horizontal squares must span $800-(-100)$, or 900 units. Since 900 is close to 1000 we can number by 100's. We go backward from 0 two squares to -200 and forward from 0 eight squares to 800 for a total of $2+8$, or 10 squares. (We could have numbered from -100 to 900 instead.)

Since the y-values range from -5 to 37, the 10 vertical squares must span $37-(-5)$, or 42 units. Since 42 is close to 50, we can count by 5's. We go down from 0 two squares to -10 and up from 0 eight squares to 40 for a total of $2+8$, or 10 squares.

Then we plot the points $(-100,-5)$, $(350, 20)$, and $(800, 37)$.

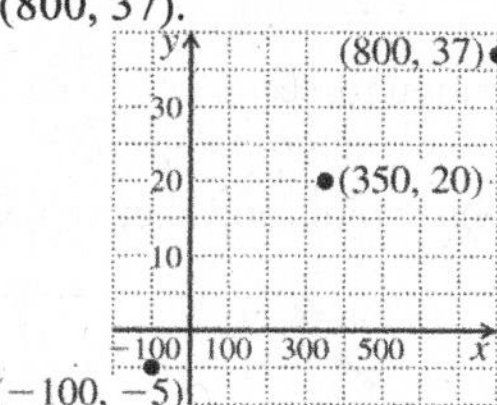

55. Since the x-values range from -124 to 54, the 10 horizontal squares must span $54-(-124)$, or 178 units. We can number by 25's. We go backward from 0 six squares to -150 and forward from 0 four squares to 100 for a total of $6+4$, or 10 squares.

Since the y-values range from -238 to 491, the 10 vertical squares must span $491-(-238)$, or 729 units. We can number by 100's. We go down from 0 four squares to -400 and up from 0 six squares to 600 for a total of $4+6$, or 10 squares.

Then we plot the points $(-83, 491)$, $(-124,-95)$, and $(54,-238)$.

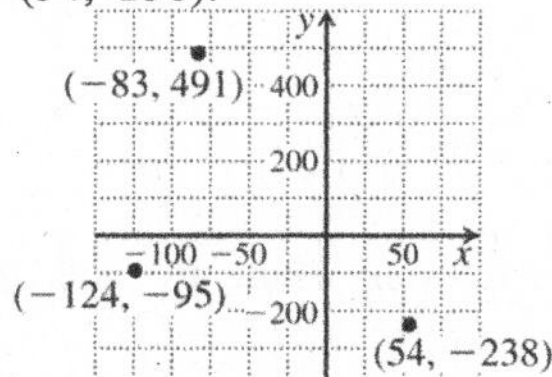

57. *Writing Exercise.*

59. $-\frac{1}{2}+\frac{1}{3}=-\frac{3}{6}+\frac{2}{6}=-\frac{1}{6}$

61. $3(-40)=-120$

63. $(-1)(-2)(-3)(-4)=24$

65. *Writing Exercise.*

67. The coordinates have opposite signs, so the point could be in quadrant II or quadrant IV.

69.

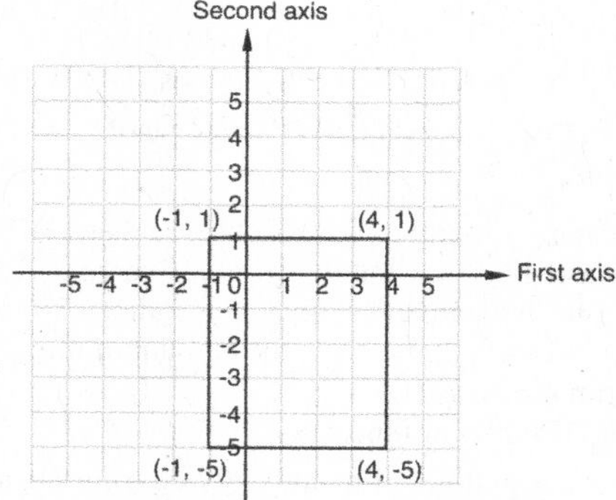

The coordinates of the fourth vertex are $(-1,-5)$.

71. Answers may vary.
We select eight points such that the sum of the coordinates for each point is 7.

$(0, 7)$ $0+7=7$
$(1, 6)$ $1+6=7$
$(2, 5)$ $2+5=7$
$(3, 4)$ $3+4=7$
$(4, 3)$ $4+3=7$
$(5, 2)$ $5+2=7$
$(6, 1)$ $6+1=7$
$(7, 0)$ $7+0=7$

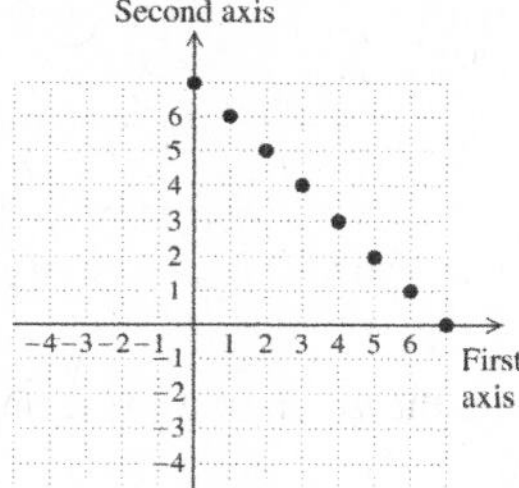

73.

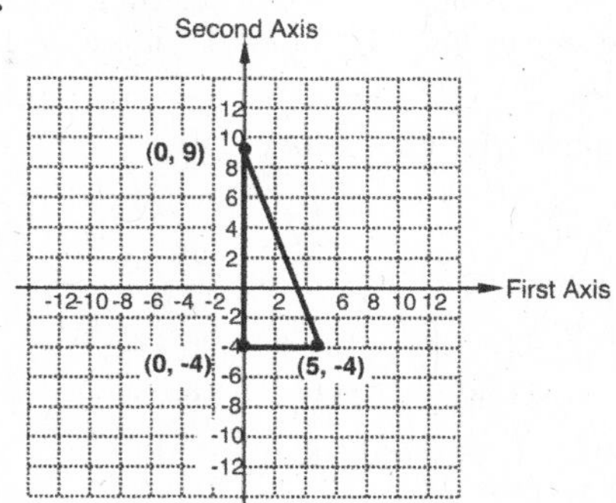

The base is 5 units and the height is 13 units.

$A=\frac{1}{2}bh=\frac{1}{2}\cdot 5\cdot 13=\frac{65}{2}$ sq units, or $32\frac{1}{2}$ sq units

75. ***Familiarize.*** From the pie chart we see that 13.7% of solid waste is yard trimmings. Let y = the amount of yard trimmings, in pounds, in the solid waste generated per day by the average American in 2012.

Translate. We reword the problem.

What	is	13.7%	of	4.38?
↓	↓	↓	↓	↓
y	=	13.7%	·	4.38

Carry out.

$y = 0.137(4.38) \approx 0.6$ lb

Check. We can repeat the calculations.

State. In 2012, about 0.6 lb of waste per day was yard trimmings.

77. ***Familiarize.*** From the pie chart we see that 13.7% of solid waste is yard trimmings. From Exercise 75 we know that an average American generated 4.38 lb of waste per day in 2012. Then the amount of this that is yard trimmings is

$0.137(4.38)$, or 0.60006 lb.

Let $y =$ the number of pounds of yard trimmings the average American recycled each day in 2012.

Translate. We reword the problem.

What	is	13.5%	of	0.60006 lb?
↓	↓	↓	↓	↓
y	$=$	13.5%	$\cdot$	0.60006

Carry out.

$y = 0.135(0.60006) \approx 0.08$

Check. The result checks.

State. The average American recycled about 0.08 lb of yard trimmings per day in 2012.

79. Latitude 27° North, Longitude 81° West

Exercise Set 3.2

1. False. A linear equation in two variables has infinitely many ordered pairs that are solutions.

3. True. All of the points on the graph of the line are solutions to the equation.

5. True. A solution may be found by selecting a value for x and solving for y. The ordered pair is a solution to the equation.

7. We substitute 2 for x and 1 for y.

$$\begin{array}{c|l} \multicolumn{2}{c}{y = 4x - 7} \\ \hline 1 & 4(2) - 7 \\ & 8 - 7 \end{array}$$

$$1 \stackrel{?}{=} 1 \text{ TRUE}$$

Since $1 = 1$ is true, the pair (2, 1) is a solution.

9. We substitute 5 for x and 1 for y.

$$\begin{array}{c|l} \multicolumn{2}{c}{3y + 4x = 19} \\ \hline 3(1) + 4(5) & 19 \\ 3 + 20 & \end{array}$$

$$23 \stackrel{?}{=} 19 \text{ FALSE}$$

Since $23 = 19$ is false, the pair (5, 1) is not a solution.

11. We substitute 3 for m and -1 for n.

$$\begin{array}{c|l} \multicolumn{2}{c}{4m - 5n = 7} \\ \hline 4(3) - 5(-1) & 7 \\ 12 + 5 & \end{array}$$

$$17 \stackrel{?}{=} 7 \text{ FALSE}$$

Since $17 = 7$ is false, the pair $(3, -1)$ is not a solution.

13. To show that a pair is a solution, we substitute, replacing x with the first coordinate and y with the second coordinate in each pair.

$$\begin{array}{c|l} \multicolumn{2}{c}{y = x + 3} \\ \hline 2 & -1 + 3 \end{array} \qquad \begin{array}{c|l} \multicolumn{2}{c}{y = x + 3} \\ \hline 7 & 4 + 3 \end{array}$$

$$2 \stackrel{?}{=} 2 \text{ TRUE} \qquad 7 \stackrel{?}{=} 7 \text{ TRUE}$$

In each case the substitution results in a true equation. Thus, $(-1, 2)$ and $(4, 7)$ are both solutions of $y = x + 3$. We graph these points and sketch the line passing through them.

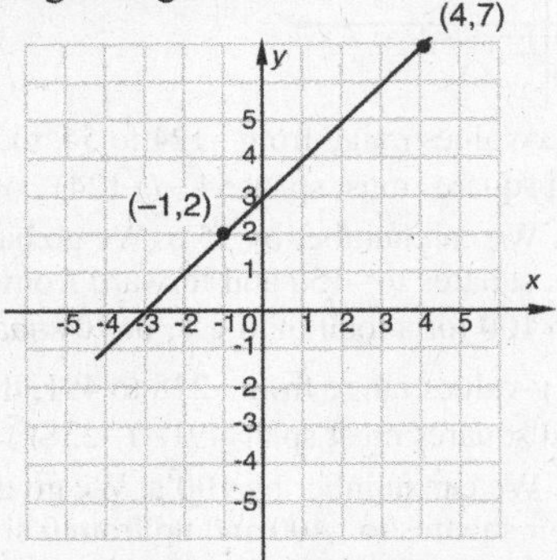

The line appears to pass through (0, 3) also. We check to determine if (0, 3) is a solution of $y = x + 3$.

$$\begin{array}{c|l} \multicolumn{2}{c}{y = x + 3} \\ \hline 3 & 0 + 3 \end{array}$$

$$3 \stackrel{?}{=} 3 \text{ TRUE}$$

Thus, (0, 3) is another solution. There are other correct answers, including $(-5, -2)$, $(-4, -1)$, $(-3, 0)$, $(-2, 1)$, $(1, 4)$, $(2, 5)$, and $(3, 6)$.

15. To show that a pair is a solution, we substitute, replacing x with the first coordinate and y with the second coordinate in each pair.

$$\begin{array}{c|l} \multicolumn{2}{c}{y = \frac{1}{2}x + 3} \\ \hline 5 & \frac{1}{2} \cdot 4 + 3 \\ & 2 + 3 \end{array} \qquad \begin{array}{c|l} \multicolumn{2}{c}{y = \frac{1}{2}x + 3} \\ \hline 2 & \frac{1}{2}(-2) + 3 \\ & -1 + 3 \end{array}$$

$$5 \stackrel{?}{=} 5 \text{ TRUE} \qquad 2 \stackrel{?}{=} 2 \text{ TRUE}$$

In each case the substitution results in a true equation. Thus, (4, 5) and (−2, 2) are both solutions of $y = \frac{1}{2}x + 3$. We graph these points and sketch the line passing through them.

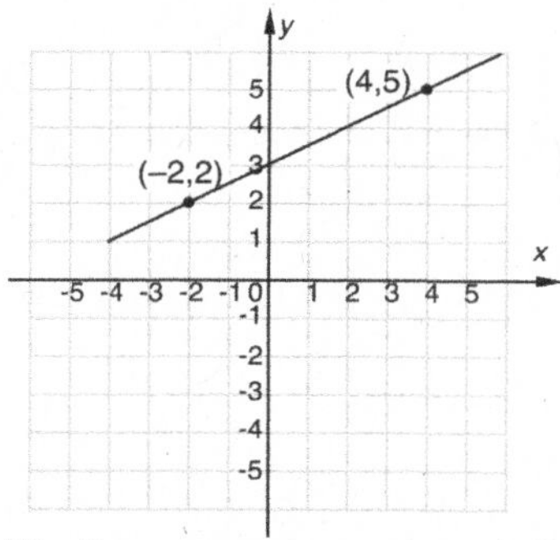

The line appears to pass through (0, 3) also. We check to determine if (0, 3) is a solution of $y = \frac{1}{2}x + 3$.

$$\begin{array}{c|c} y & \frac{1}{2}x+3 \\ \hline 3 & \frac{1}{2}\cdot 0+3 \end{array}$$

$3 \stackrel{?}{=} 3$ TRUE

Thus, (0, 3) is another solution. There are other correct answers, including $(-6,0)$, (−4, 1), (2, 4), and (6, 6).

17. To show that a pair is a solution, we substitute, replacing x with the first coordinate and y with the second coordinate in each pair.

$$\begin{array}{c|c} y+3x & 7 \\ \hline 1+3\cdot 2 & 7 \\ 1+6 & \end{array} \qquad \begin{array}{c|c} y+3x & 7 \\ \hline -5+3\cdot 4 & 7 \\ -5+12 & \end{array}$$

$7 \stackrel{?}{=} 7$ TRUE $\qquad$ $7 \stackrel{?}{=} 7$ TRUE

In each case the substitution results in a true equation. Thus, $(2,1)$ and $(4,-5)$ are both solutions of $y+3x=7$. We graph these points and sketch the line passing through them.

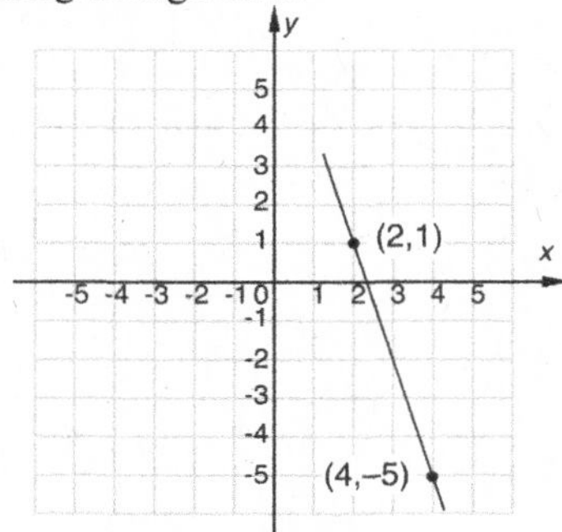

The line appears to pass through (1, 4) also. We check to determine if (1, 4) is a solution of $y+3x=7$.

$$\begin{array}{c|c} y+3x & 7 \\ \hline 4+3\cdot 1 & 7 \\ 4+3 & \end{array}$$

$7 \stackrel{?}{=} 7$ TRUE

Thus, (1, 4) is another solution. There are other correct answers, including $(3,-2)$.

19. To show that a pair is a solution, we substitute, replacing x with the first coordinate and y with the second coordinate in each pair.

$$\begin{array}{c|c} 4x-2y & 10 \\ \hline 4\cdot 0-2(-5) & 10 \end{array} \qquad \begin{array}{c|c} 4x-2y & 10 \\ \hline 4\cdot 4-2\cdot 3 & 10 \\ 16-6 & \end{array}$$

$10 \stackrel{?}{=} 10$ TRUE $\qquad$ $10 \stackrel{?}{=} 10$ TRUE

In each case the substitution results in a true equation. Thus, $(0,-5)$ and (4, 3) are both solutions of $4x-2y=10$. We graph these points and sketch the line passing through them.

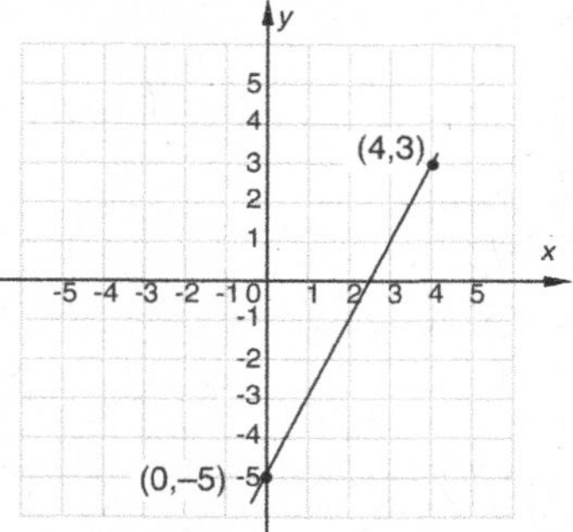

The line appears to pass through $(2,-1)$ also. We check to determine if $(2,-1)$ is a solution of $4x-2y=10$.

$$\begin{array}{c|c} 4x-2y & 10 \\ \hline 4\cdot 2-2(-1) & 10 \\ 8+2 & \end{array}$$

$10 \stackrel{?}{=} 10$ TRUE

Thus, $(2,-1)$ is another solution. There are other correct answers, including $(1,-3)$, $(2,-1)$, (3, 1), and (5, 5).

21. $y = x+1$

We select some x-values and compute y-values.

When $x=0$, $y=0+1=1$.

When $x=3$, $y=3+1=4$.

When $x=-5$, $y=-5+1=-4$.

x	y
0	1
3	4
−5	−4

Plot these points, draw the line they determine, and label the graph $y = x+1$.

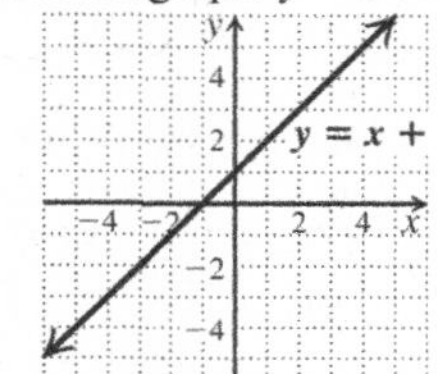

23. $y=-x$

We select some convenient x-values and compute y-values.

When $x=0,\ y=-0=0$.

When $x=-2,\ y=-(-2)=2$.

When $x=3,\ y=-3$.

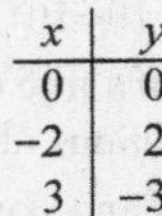

x	y
0	0
−2	2
3	−3

Plot these points, draw the line they determine, and label the graph $y=-x$.

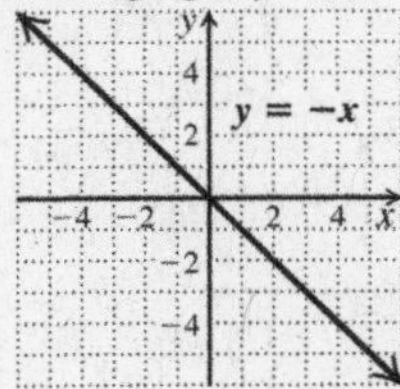

25. $y=2x$

We select some x-values and compute y-values.

When $x=0,\ y=2(0)=0$.

When $x=1,\ y=2(1)=2$.

When $x=-1,\ y=2(-1)=-2$.

x	y
0	0
1	2
−1	−2

Plot these points, draw the line they determine, and label the graph $y=2x$.

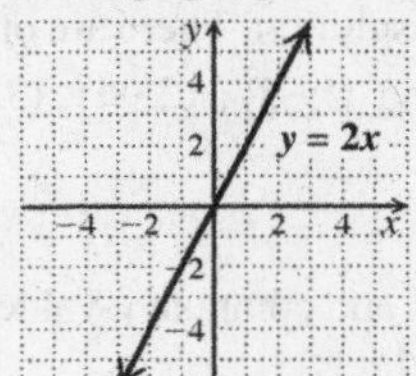

27. $y=2x+2$

We select some x-values and compute y-values.

When $x=0,\ \ y=2(0)+2=0+2=2$.

When $x=-3,\ \ y=2(-3)+2=-6+2=-4$.

When $x=1,\ \ y=2\cdot 1+2=2+2=4$.

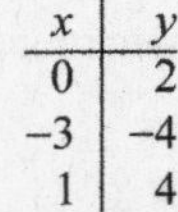

x	y
0	2
−3	−4
1	4

Plot these points, draw the line they determine, and label the graph $y=2x+2$.

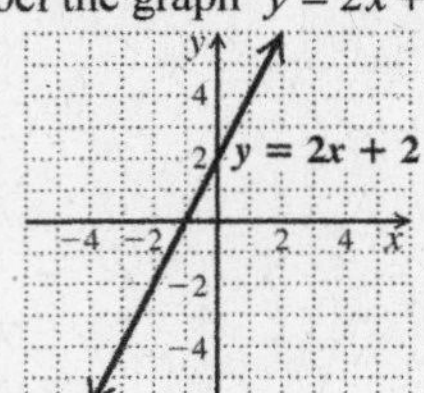

29. $y=-\frac{1}{2}x=-\frac{1}{2}x+0$

We select some x-values and compute y-values.

When $x=0,\ y=-\frac{1}{2}(0)=0$.

When $x=2,\ y=-\frac{1}{2}(2)=-1$.

When $x=-2,\ y=-\frac{1}{2}(-2)=1$.

x	y
0	0
2	−1
−2	1

Plot these points, draw the line they determine, and label the graph $y=-\frac{1}{2}x$.

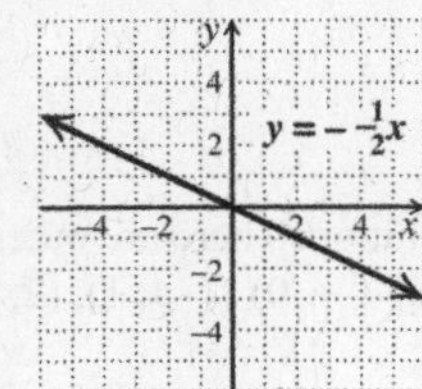

31. $y=\frac{1}{3}x-4=\frac{1}{3}x+(-4)$

When $x=0,\ y=\frac{1}{3}(0)-4=0-4=-4$.

We find two other points, using multiples of 3 for x to avoid fractions.

When $x=-3,\ y=\frac{1}{3}(-3)-4=-1-4=-5$.

When $x=3,\ y=\frac{1}{3}\cdot 3-4=1-4=-3$.

x	y
0	−4
−3	−5
3	−3

Plot these points, draw the line they determine, and label the graph $y=\frac{1}{3}x-4$.

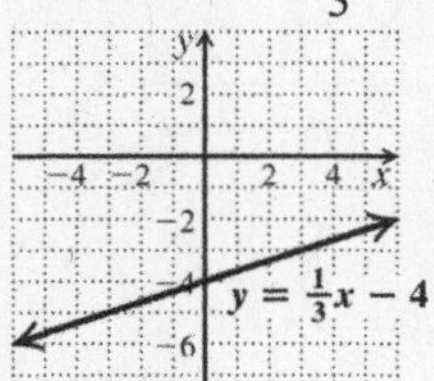

33. $x+y=4$

$y=-x+4$

We select some x-values and compute y-values.

When $x=0,\ \ y=0+4=4$.

When $x=-1,\ y=-(-1)+4=1+4=5$.

When $x=2,\ \ y=-2+4=2$.

x	y
0	4
−1	5
2	2

Plot these points, draw the line they determine, and label the graph $x+y=4$.

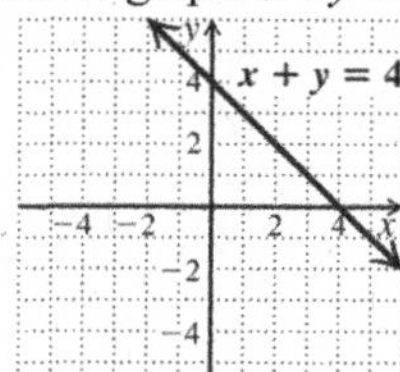

35. $x-y=-2$

$y=x+2$

We select some x-values and compute y-values.

When $x=0$, $y=0+2=2$.

When $x=1$, $y=1+2=3$.

When $x=-1$, $y=-1+2=1$.

x	y
0	2
1	3
−1	1

Plot these points, draw the line they determine, and label the graph $x-y=-2$.

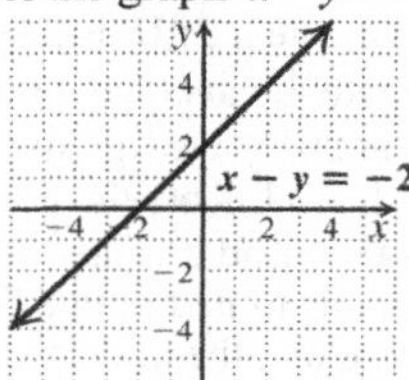

37. $x+2y=-6$

$2y=-x-6$

$y=-\frac{1}{2}x-3$

When $x=0$, $y=-\frac{1}{2}(0)-3=0-3=-3$.

We find two other points, using multiples of 2 for x to avoid fractions.

When $x=-4$, $y=-\frac{1}{2}(-4)-3=2-3=-1$.

When $x=2$, $y=-\frac{1}{2}\cdot 2-3=-1-3=-4$.

x	y
0	−3
−4	−1
2	−4

Plot these points, draw the line they determine, and label the graph $x+2y=-6$.

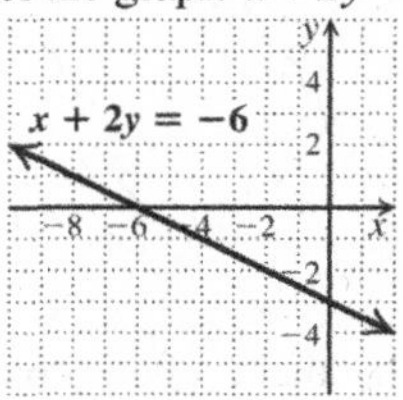

39. $y=-\frac{2}{3}x+4$

When $x=0$, $y=-\frac{2}{3}\cdot 0+4=0+4=4$.

We find two other points, using multiples of 3 for x to avoid fractions.

When $x=3$, $y=-\frac{2}{3}\cdot 3+4=-2+4=2$.

When $x=6$, $y=-\frac{2}{3}\cdot 6+4=-4+4=0$.

x	y
0	4
3	2
6	0

Plot these points, draw the line they determine, and label the graph $y=-\frac{2}{3}x+4$.

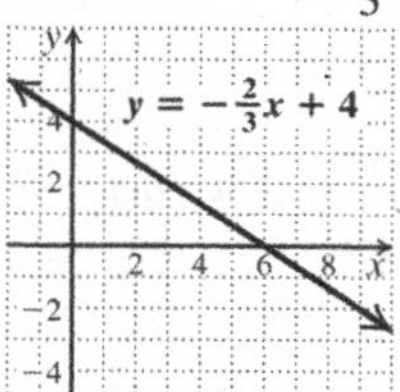

41. $4x=3y$

$y=\frac{4}{3}x$

We select some x-values and compute y-values.

When $x=0$, $y=\frac{4}{3}(0)=0$.

When $x=3$, $y=\frac{4}{3}(3)=4$.

When $x=-3$, $y=\frac{4}{3}(-3)=-4$.

x	y
0	0
3	4
−3	−4

Plot these points, draw the line they determine, and label the graph $4x=3y$.

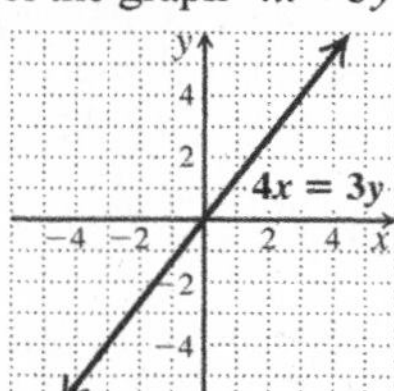

43. $5x-y=0$

$y=5x$

We select some x-values and compute y-values.

When $x=0$, $y=5(0)=0$.

When $x=1$, $y=5(1)=5$.

When $x=-1$, $y=5(-1)=-5$.

x	y
0	0
1	5
−1	−5

Plot these points, draw the line they determine, and

label the graph $5x - y = 0$.

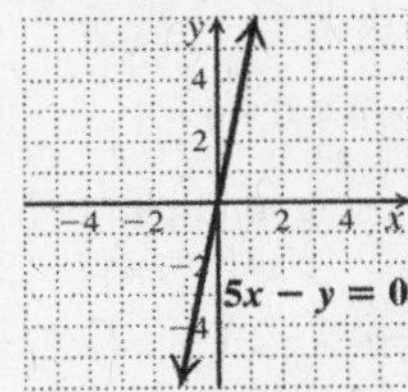

45. $6x - 3y = 9$

$-3y = -6x + 9$

$y = 2x - 3$

We select some x-values and compute y-values.

When $x = 0$, $y = 2(0) - 3 = 0 - 3 = -3$.

When $x = -1$, $y = 2(-1) - 3 = -2 - 3 = -5$.

When $x = 3$, $y = 2 \cdot 3 - 3 = 6 - 3 = 3$.

x	y
0	−3
−1	−5
3	3

Plot these points, draw the line they determine, and label the graph $6x - 3y = 9$.

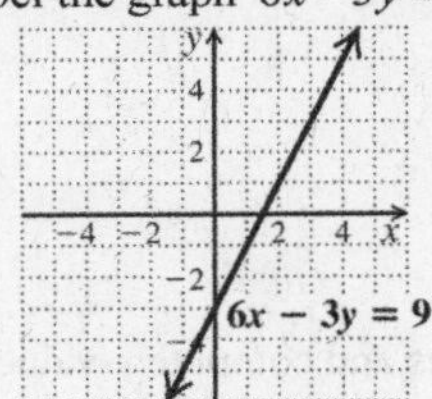

47. $6y + 2x = 8$

$6y = -2x + 8$

$y = -\frac{1}{3}x + \frac{4}{3}$

We select some x-values and compute y-values.

When $x = 0$, $y = -\frac{1}{3}(0) + \frac{4}{3} = 0 + \frac{4}{3} = \frac{4}{3}$.

When $x = -2$, $y = -\frac{1}{3}(-2) + \frac{4}{3} = \frac{2}{3} + \frac{4}{3} = 2$.

When $x = 1$, $y = -\frac{1}{3} \cdot 1 + \frac{4}{3} = -\frac{1}{3} + \frac{4}{3} = 1$.

x	y
0	$\frac{4}{3}$
−2	2
1	1

Plot these points, draw the line they determine, and label the graph $6y + 2x = 8$.

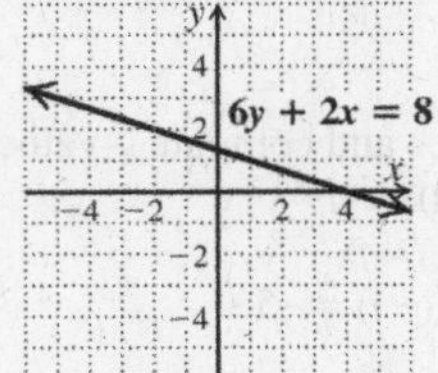

49. We graph $a = 0.08t + 2.5$ by selecting values for t and then calculating the associated values for a.

If $t = 0$, $a = 0.08(0) + 2.5 = 2.5$.

If $t = 10$, $a = 0.08(10) + 2.5 = 3.3$.

If $t = 20$, $a = 0.08(20) + 2.5 = 4.1$.

t	a
0	2.5
10	3.3
20	4.1

We plot the points and draw the graph.

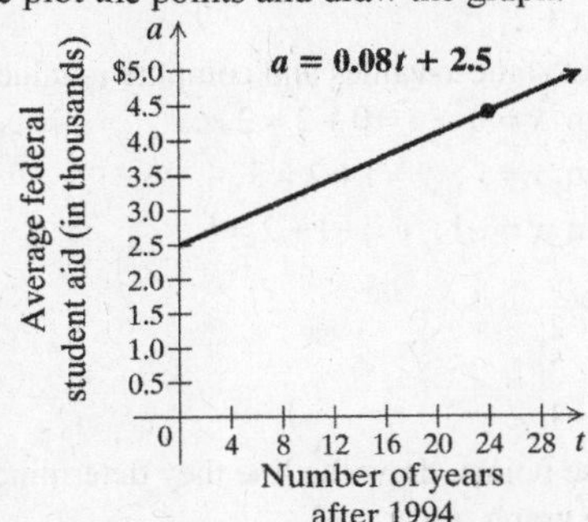

Since $2018 - 1994 = 24$, the year 2018 is 24 years after 1994. Thus, to estimate the average amount of financial aid per student in 2018, we find the second coordinate associated with 24. Locate the point on the line that is above 24 and then find a value on the vertical axes that corresponds to that point. That value is about 4.5, so we estimate that the average amount of financial aid per student in 2018 is \$4.5 thousand or \$4500.

51. We graph $t = -0.1s + 13.1$ by selecting values for t and then calculating the associated values for s.

If $t = 50$, $t = -0.1(50) + 13.1 = 8.1$.

If $t = 55$, $t = -0.1(55) + 13.1 = 7.6$.

If $t = 60$, $t = -0.1(60) + 13.1 = 7.1$.

s	t
50	8.1
55	7.6
60	7.1

We plot the points and draw the graph.

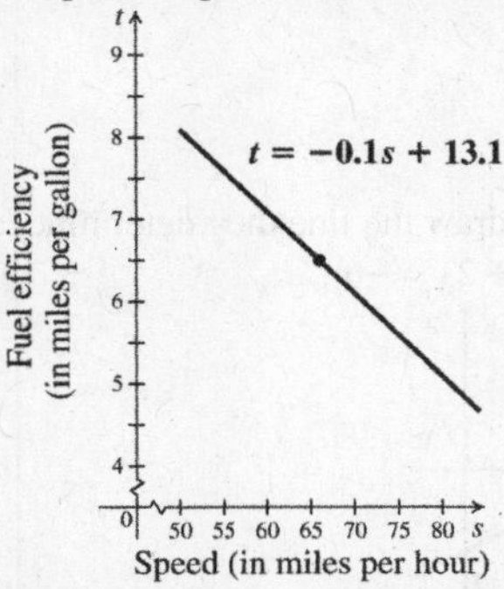

Locate the point on the line that is above 66 and then find the value on the vertical axes that corresponds to that point. That value is 6.5, so we estimate the fuel efficiency to be 6.5 mpg.

53. We graph $p = 0.7n + 16$ by selecting values for n and then calculating the associated values for p.

If $n = 10$, $p = 0.7(10) + 16 = 23$.

If $n = 20$, $p = 0.7(20) + 16 = 30$.

If $n = 30$, $p = 0.7(30) + 16 = 37$.

n	p
10	23
20	30
30	37

We plot the points and draw the graph.

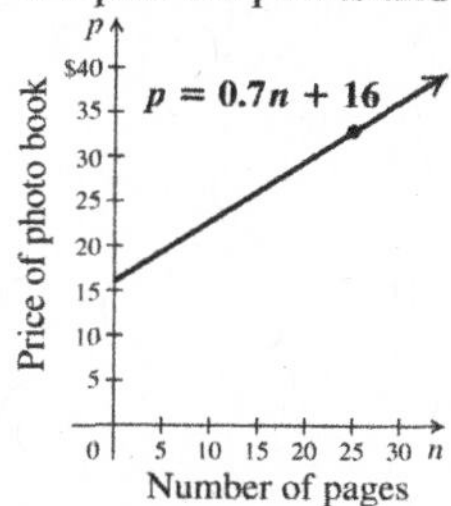

Locate the point on the line that is above 25 and then find the value on the vertical axes that corresponds to that point. That value is 34, so we estimate the price of a photobook containing 25 pages as $34.

55. We graph $a = 0.23d - 7$ by selecting values for d and then calculating the associated values for a.

If $d = 100$, $a = 0.23(100) - 7 = 16$.

If $d = 200$, $a = 0.23(200) - 7 = 39$.

If $d = 300$, $a = 0.23(300) - 7 = 62$.

We plot the points and draw the graph.

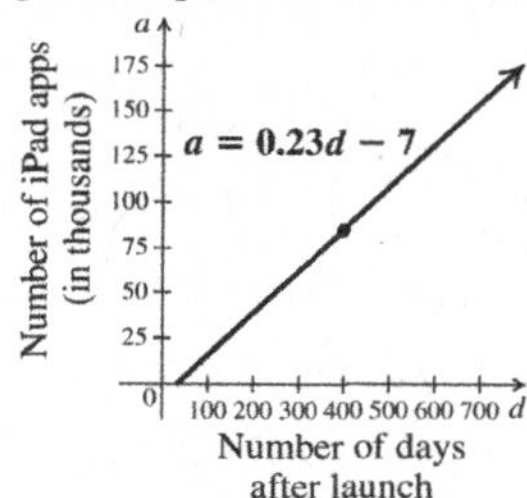

To predict the number of apps we find the second coordinate associated with 400. The value is 85, so we predict that about 85,000 apps will be created 400 days after launch.

57. We graph $T = \frac{5}{4}c + 2$. Since number of credits cannot be negative, we select only nonnegative values for c.

If $c = 4$, $T = \frac{5}{4}(4) + 2 = 7$.

If $c = 8$, $T = \frac{5}{4}(8) + 2 = 12$.

If $c = 12$, $T = \frac{5}{4}(12) + 2 = 17$.

We plot the points and draw the graph.

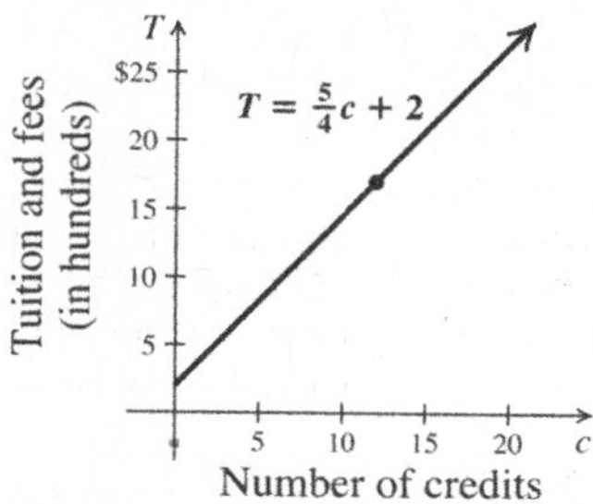

Four three-credit courses total $4 \cdot 3$ or 12, credits. Locate the point on the graph, the value is 17. So tuition and fees will cost $17 hundred, or $1700.

59. *Writing Exercise.*

61.
$$\begin{aligned} 5x + 3(2 - x) &= 12 \\ 5x + 6 - 3x &= 12 \\ 2x + 6 &= 12 \\ 2x &= 6 \\ x &= 3 \end{aligned}$$

Check: $5x + 3(2 - x) = 12$

$5(3) + 3(2 - 3)$	12
$15 + 3(-1)$	
$15 - 3$	
12	

$12 \stackrel{?}{=} 12$ TRUE

The solution is 3.

63.
$$\begin{aligned} A &= \frac{T + Q}{2} \\ 2A &= T + Q \\ 2A - T &= Q \end{aligned}$$

65. $Ax + By = C$

$By = C - Ax$ Subtracting Ax

$y = \frac{C - Ax}{B}$ Dividing by B

67. *Writing Exercise.*

69. Let s represent the gear that Laura uses on the southbound portion of her ride and n represent the gear she uses on the northbound portion. Then we have $s + n = 24$. We graph this equation, using only positive integer values for s and n.

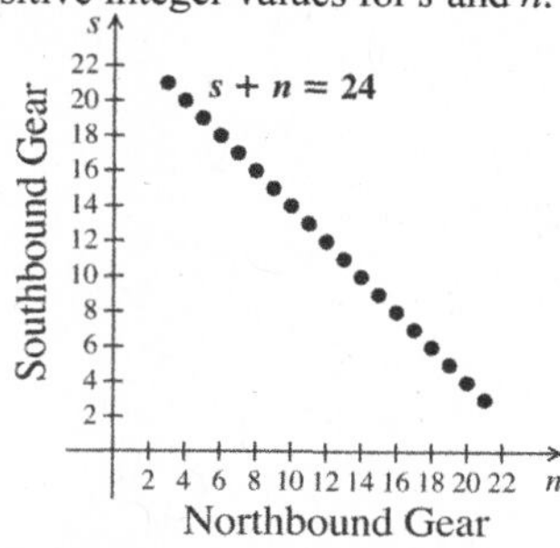

71. Note that the sum of the coordinates of each point on the graph is 5. Thus, we have $x + y = 5$, or $y = -x + 5$.

73. Note that each y-coordinate is 2 more than the corresponding x-coordinate. Thus, we have $y = x + 2$.

75. The equation is $25d + 5l = 225$.
Since the number of dinners cannot be negative, we choose only nonnegative values of d when graphing the equation. The graph stops at the horizontal axis since the number of lunches cannot be negative.

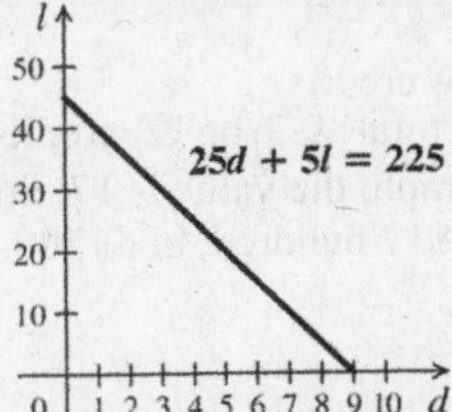

We see that three points on the graph are (1, 40), (5, 20), and (8, 5). Thus, three combinations of dinners and lunches that total $225 are
1 dinner, 40 lunches,
5 dinners, 20 lunches,
8 dinners, 5 lunches.

77. $y = -|x|$

x	y
−3	−3
−2	−2
−1	−1
0	0
1	−1
2	−2
3	−3

y = −|x|

79. $y = x^2$

x	y
−3	9
−2	4
−1	1
0	0
1	1
2	4
3	9

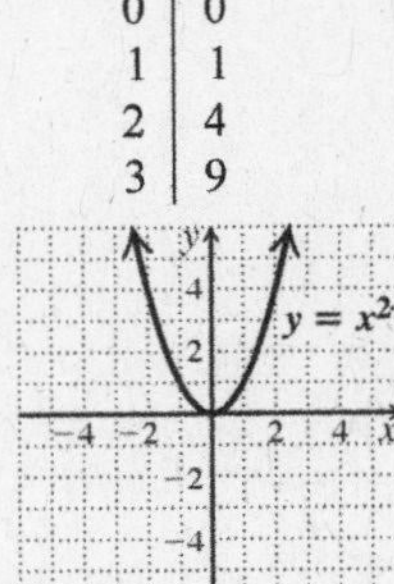

81.

y = −2.8x + 3.5
10
−10 10
−10

83.

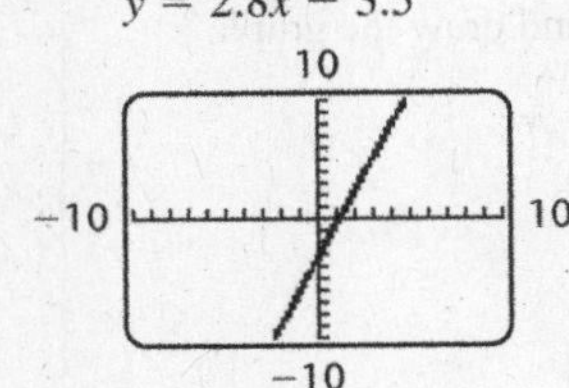

85.

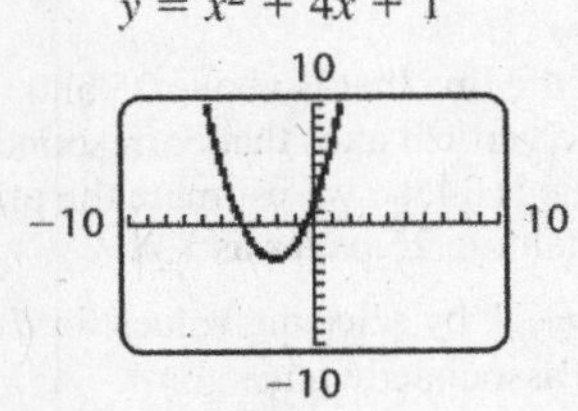

87. *Writing Exercise.*

89. At 55 mph, fuel efficiency is
$t = -0.1(55) + 13.1 = 7.6$ mpg
At \$3.50 per gallon for 500 miles, the cost is
$\frac{500 \cdot 3.50}{7.6} = \$230.26$
At 70 mph, fuel efficiency is
$t = -0.1(70) + 13.1 = 6.1$ mpg
At \$3.50 per gallon for 500 miles, the cost is
$\frac{500 \cdot 3.50}{6.1} = \$286.89$
The savings is \$286.89 − \$230.26 = \$56.63, which is equal to $\frac{\$56.63}{\$3.50} \approx 16.2$ gallons.

Exercise Set 3.3

1. A horizontal line has a *y-intercept*.

3. To find a *y-intercept*, replace x with 0 and solve for y.

5. The point (−3, 0) could be an *x-intercept* of a graph.

7. The graph of $x = -4$ is a vertical line, so (f) is the most appropriate choice.

9. The point (0, 2) lies on the y-axis, so (d) is the most appropriate choice.

11. The point (3, −2) does not lie on an axis, so it could be used as a check when we graph using intercepts. Thus (b) is the most appropriate choice.

13. a. The graph crosses the y-axis at (0, 5), so the y-intercept is (0, 5).

b. The graph crosses the x-axis at (2, 0), so the x-intercept is (2, 0).

15. a. The graph crosses the y-axis at $(0,-4)$, so the y-intercept is (0,–4).

b. The graph crosses the x-axis at (3, 0), so the x-intercept is (3, 0).

17. a. The graph crosses the y-axis at $(0,-2)$, so the y-intercept is $(0,-2)$.

b. The graph crosses the x-axis at (–3, 0) and (3, 0), so the x-intercepts are (–3, 0) and (3, 0).

19. a. The graph crosses the y-axis at (0, 0), so the y-intercept is (0, 0).

b. The graph crosses the x-axis at (–2, 0), (0, 0) and (5, 0), so the x-intercepts are (–2, 0), (0, 0) and (5, 0).

21. $3x+5y=15$

a. To find the y-intercept, let $x=0$. This is the same as temporarily ignoring the x-term and then solving.

$$5y=15$$
$$y=3$$

The y-intercept is (0, 3).

b. To find the x-intercept, let $y=0$. This is the same as temporarily ignoring the y-term and then solving.

$$3x=15$$
$$x=5$$

The x-intercept is (5, 0).

23. $9x-2y=36$

a. To find the y-intercept, let $x=0$. This is the same as temporarily ignoring the x-term and then solving.

$$-2y=36$$
$$y=-18$$

The y-intercept is (0,–18).

b. To find the x-intercept, let $y=0$. This is the same as temporarily ignoring the y-term and then solving.

$$9x=36$$
$$x=4$$

The x-intercept is (4, 0).

25. $-4x+5y=80$

a. To find the y-intercept, let $x=0$. This is the same as temporarily ignoring the x-term and then solving.

$$5y=80$$
$$y=16$$

The y-intercept is (0, 16).

b. To find the x-intercept, let $y=0$. This is the same as temporarily ignoring the y-term and then solving.

$$-4x=80$$
$$x=-20$$

The x-intercept is $(-20,0)$.

27. $x=12$

Observe that this is the equation of a vertical line 12 units to the right of the y-axis. Thus, (a) there is no y-intercept and (b) the x-intercept is (12, 0).

29. $y=-9$

Observe that this is the equation of a horizontal line 9 units below the x-axis. Thus, (a) the y-intercept is (0, –9) and (b) there is no x-intercept.

31. $3x+5y=15$

Find the y-intercept:

$$5y=15 \quad \text{Ignoring the } x\text{-term}$$
$$y=3$$

The y-intercept is (0, 3).

Find the x-intercept:

$$3x=15 \quad \text{Ignoring the } y\text{-term}$$
$$x=5$$

The x-intercept is (5, 0).

To find a third point we replace x with –5 and solve for y.

$$3(-5)+5y=15$$
$$-15+5y=15$$
$$5y=30$$
$$y=6$$

The point (–5, 6) appears to line up with the intercepts, so we draw the graph.

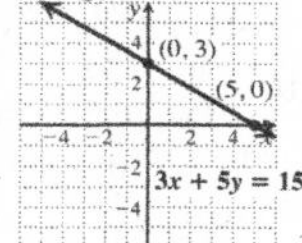

33. $x+2y=4$

Find the y-intercept:

$$2y=4 \quad \text{Ignoring the } x\text{-term}$$
$$y=2$$

The y-intercept is (0, 2).

Find the x-intercept:

$$x=4 \quad \text{Ignoring the } y\text{-term}$$

The x-intercept is (4, 0).

To find a third point we replace x with 2 and solve for y.

$$2+2y=4$$
$$2y=2$$
$$y=1$$

The point (2, 1) appears to line up with the intercepts, so we draw the graph.

(0, 2)
(4, 0)
x + 2y = 4

35. $-x+2y=8$

Find the *y*-intercept:

$2y=8$ Ignoring the *x*-term

$x=4$

The *y*-intercept is $(0,4)$.

Find the *x*-intercept:

$-x=8$ Ignoring the *y*-term

$x=-8$

The *x*-intercept is (–8, 0).

To find a third point we replace *x* with 4 and solve for *y*.

$-4+2y=8$

$2y=12$

$y=6$

The point (4, 6) appears to line up with the intercepts, so we draw the graph.

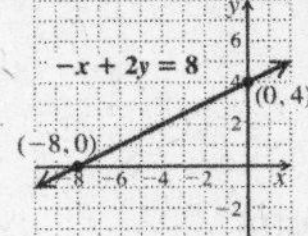

37. $3x+y=9$

Find the *y*-intercept:

$y=9$ Ignoring the *x*-term

The *y*-intercept is (0, 9).

Find the *x*-intercept:

$3x=9$ Ignoring the *y*-term

$x=3$

The *x*-intercept is (3, 0).

To find a third point we replace *x* with 2 and solve for *y*.

$3\cdot 2+y=9$

$6+y=9$

$y=3$

The point (2, 3) appears to line up with the intercepts, so we draw the graph.

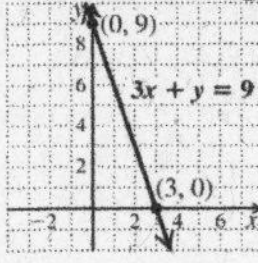

39. $y=2x-6$

Find the *y*-intercept:

$y=-6$ Ignoring the *x*-term

The *y*-intercept is $(0,-6)$.

Find the *x*-intercept:

$0=2x-6$ Replacing *y* with 0

$6=2x$

$3=x$

The *x*-intercept is (3, 0).

To find a third point we replace *x* with 2 and find *y*.

$y=2\cdot 2-6=4-6=-2$

The point $(2,-2)$ appears to line up with the intercepts, so we draw the graph.

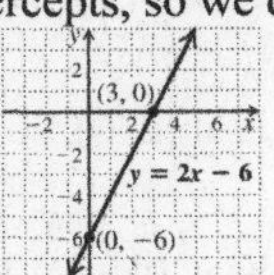

41. $5x-10=5y$

We can leave the equation in the given form or rewrite it in the form $Ax+By=C$. We will use the given form.

Find the *y*-intercept:

$-10=5y$ Ignoring the *x*-term

$-2=y$

The *y*-intercept is $(0,-2)$.

To find the *x*-intercept, let $y=0$.

$5x-10=5\cdot 0$

$5x-10=0$

$5x=10$

$x=2$

The *x*-intercept is (2, 0).

To find a third point we replace *x* with 5 and solve for *y*.

$5\cdot 5-10=5y$

$25-10=5y$

$15=5y$

$3=y$

The point (5, 3) appears to line up with the intercepts, so we draw the graph.

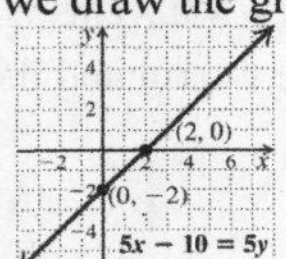

43. $2x-5y=10$

Find the *y*-intercept:

$-5y=10$ Ignoring the *x*-term

$y=-2$

The *y*-intercept is $(0,-2)$.

Find the *x*-intercept:

$2x=10$ Ignoring the *y*-term

$x=5$

The *x*-intercept is (5, 0).

To find a third point we replace *x* with –5 and solve for *y*.

$2(-5)-5y=10$

$-10-5y=10$

$-5y=20$

$y=-4$

The point $(-5,-4)$ appears to line up with the intercepts, so we draw the graph.

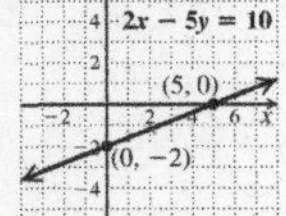

45. $6x + 2y = 12$

Find the y-intercept:

$2y = 12$ Ignoring the x-term

$y = 6$

The y-intercept is (0, 6).

Find the x-intercept:

$6x = 12$ Ignoring the y-term

$x = 2$

The x-intercept is (2, 0).

To find a third point we replace x with 3 and solve for y.

$$\begin{aligned} 6\cdot 3 + 2y &= 12 \\ 18 + 2y &= 12 \\ 2y &= -6 \\ y &= -3 \end{aligned}$$

The point $(3,-3)$ appears to line up with the intercepts, so we draw the graph.

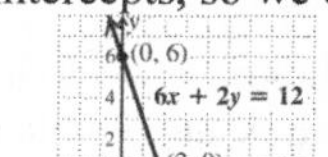

47. $4x + 3y = 16$

Find the y-intercept:

$3y = 16$ Ignoring the x-term

$y = \frac{16}{3}$

The y-intercept is $\left(0, \frac{16}{3}\right)$.

Find the x-intercept:

$4x = 16$ Ignoring the y-term

$x = 4$

The x-intercept is (4, 0).

To find a third point we replace x with –2 and solve for y.

$$\begin{aligned} 4(-2) + 3y &= 16 \\ -8 + 3y &= 16 \\ 3y &= 24 \\ y &= 8 \end{aligned}$$

The point (–2, 8) appears to line up with the intercepts, so we draw the graph.

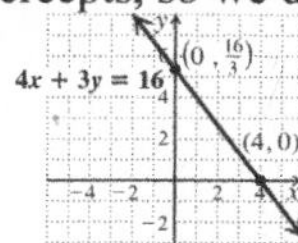

49. $2x + 4y = 1$

Find the y-intercept:

$4y = 1$ Ignoring the x-term

$y = \frac{1}{4}$

The y-intercept is $\left(0, \frac{1}{4}\right)$.

Find the x-intercept:

$2x = 1$ Ignoring the y-term

$x = \frac{1}{2}$

The x-intercept is $\left(\frac{1}{2}, 0\right)$.

To find a third point we replace x with $-\frac{3}{2}$ and solve for y.

$$\begin{aligned} 2\left(-\frac{3}{2}\right) + 4y &= 1 \\ -3 + 4y &= 1 \\ 4y &= 4 \\ y &= 1 \end{aligned}$$

The point $\left(-\frac{3}{2}, 1\right)$ appears to line up with the intercepts, so we draw the graph.

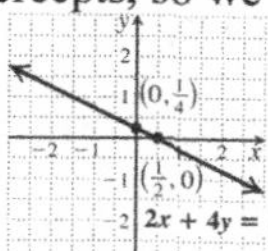

51. $5x - 3y = 180$

Find the y-intercept:

$-3y = 180$ Ignoring the x-term

$y = -60$

The y-intercept is $(0,-60)$.

Find the x-intercept:

$5x = 180$ Ignoring the y-term

$x = 36$

The x-intercept is (36, 0).

To find a third point we replace x with 6 and solve for y.

$$\begin{aligned} 5\cdot 6 - 3y &= 180 \\ 30 - 3y &= 180 \\ -3y &= 150 \\ y &= -50 \end{aligned}$$

This means that $(6,-50)$ is on the graph.

To graph all three points, the y-axis must go to at least 60 and the x-axis must go to at least 36. Using a scale of 10 units per square allows us to display both intercepts and $(6,-50)$ as well as the origin.

The point $(6,-50)$ appears to line up with the intercepts, so we draw the graph.

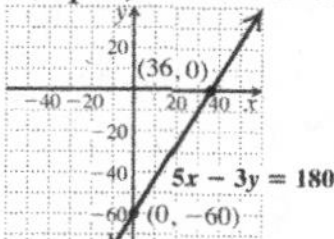

53. $y = -30 + 3x$

Find the y-intercept:

$y = -30$ Ignoring the x-term

The y-intercept is $(0,-30)$.

To find the x-intercept, let $y = 0$.

$$\begin{aligned} 0 &= -30 + 3x \\ 30 &= 3x \\ 10 &= x \end{aligned}$$

The x-intercept is (10, 0).

To find a third point we replace x with 5 and solve for y.

$y = -30 + 3 \cdot 5$
$y = -30 + 15$
$y = -15$

This means that $(5, -15)$ is on the graph.

To graph all three points, the y-axis must go to at least –30 and the x-axis must go to at least 10. Using a scale of 5 units per square allows us to display both intercepts and $(5, -15)$ as well as the origin.

The point $(5, -15)$ appears to line up with the intercepts, so we draw the graph.

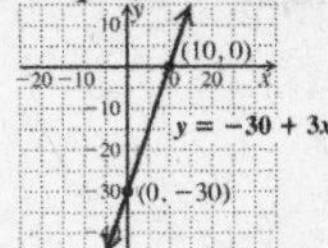

55. $-4x = 20y + 80$

To find the y-intercept, we let $x = 0$.

$-4 \cdot 0 = 20y + 80$
$0 = 20y + 80$
$-80 = 20y$
$-4 = y$

The y-intercept is $(0, -4)$.

Find the x-intercept:

$-4x = 80$ Ignoring the y-term
$x = -20$

The x-intercept is (–20, 0).

To find a third point we replace x with –40 and solve for y.

$-4(-40) = 20y + 80$
$160 = 20y + 80$
$80 = 20y$
$4 = y$

This means that (–40, 4) is on the graph.

To graph all three points, the y-axis must go at least from –4 to 4 and the x-axis must go at least from –40 to –20. Since we also want to include the origin we can use a scale of 10 units per square on the x-axis and 1 unit per square on the y-axis.

The point $(-40, 4)$ appears to line up with the intercepts, so we draw the graph.

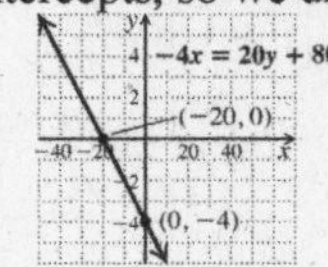

57. $y - 3x = 0$

Find the y-intercept:

$y = 0$ Ignoring the x-term

The y-intercept is (0, 0). Note that this is also the x-intercept.

In order to graph the line, we will find a second point.

When $x = 1$, $y - 3 \cdot 1 = 0$
$y - 3 = 0$
$y = 3$

Thus, a second point is (1, 3).

To find a third point we replace $x = -1$ and solve for y.

$y - 3(-1) = 0$
$y + 3 = 0$
$y = -3$

The point $(-1, -3)$ appears to line up with the other two points, so we draw the graph.

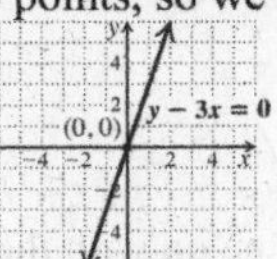

59. $y = 1$

Any ordered pair (x, 1) is a solution. The variable y must be 1, but the x variable can be any number we choose. A few solutions are listed below. Plot these points and draw the line.

x	y
–3	1
0	1
2	1

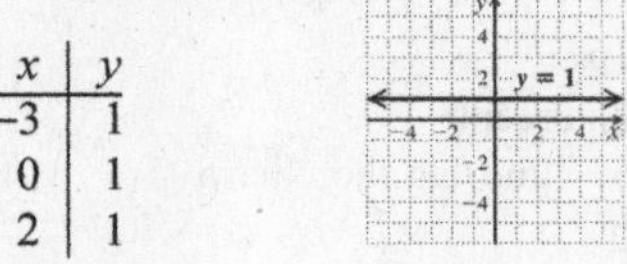

61. $x = 3$

Any ordered pair (3, y) is a solution. The variable x must be 3, but the y variable can be any number we choose. A few solutions are listed below. Plot these points and draw the line.

x	y
3	–2
3	0
3	4

63. $y = -2$

Any ordered pair (x,–2) is a solution. The variable y must be –2, but the x variable can be any number we choose. A few solutions are listed below. Plot these points and draw the line.

x	y
–3	–2
0	–2
4	–2

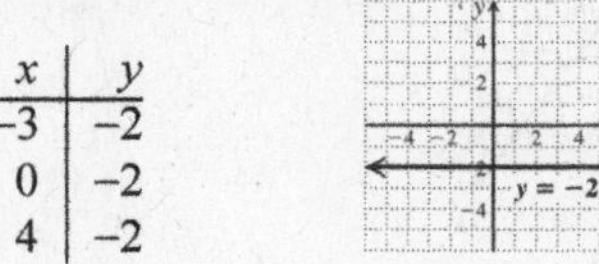

65. $x = -1$

Any ordered pair (–1, y) is a solution. The variable x must be –1, but the y variable can be any number we choose. A few solutions are listed below. Plot these points and draw the line.

x	y
–1	–3
–1	0
–1	2

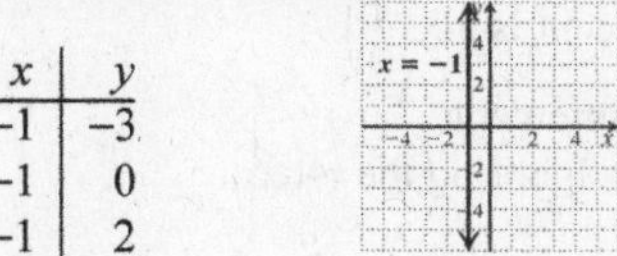

67. $y = -15$

Any ordered pair $(x, -15)$ is a solution. The variable y must be -15, but the x variable can be any number we choose. A few solutions are listed below. Plot these points and draw the line.

x	y
-1	-15
0	-15
3	-15

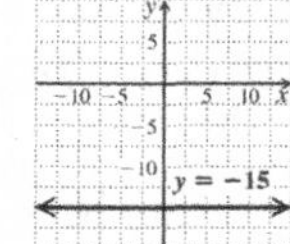

69. $y = 0$

Any ordered pair $(x, 0)$ is a solution. A few solutions are listed below. Plot these points and draw the line.

x	y
-4	0
0	0
2	0

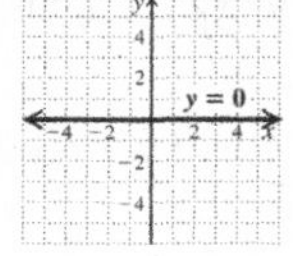

71. $x = -\frac{5}{2}$

Any ordered pair $\left(-\frac{5}{2}, y\right)$ is a solution. A few solutions are listed below. Plot these points and draw the line.

x	y
$-\frac{5}{2}$	-3
$-\frac{5}{2}$	0
$-\frac{5}{2}$	5

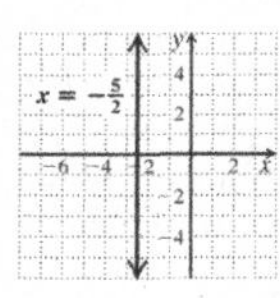

73. $-4x = -100$

$x = 25$ Dividing by -4

The graph is a vertical line 25 units to the right of the y-axis.

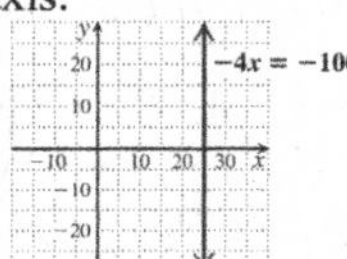

75. $35 + 7y = 0$

$$7y = -35$$
$$y = -5$$

The graph is a horizontal line 5 units below the x-axis.

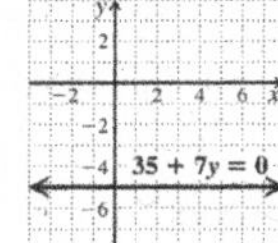

77. Note that every point on the horizontal line passing through $(0, -1)$ has -1 as the y-coordinate. Thus, the equation of the line is $y = -1$.

79. Note that every point on the vertical line passing through (4, 0) has 4 as the x-coordinate. Thus, the equation of the line is $x = 4$.

81. Note that every point on the vertical line passing through (0, 0) has 0 as the x-coordinate. Thus, the equation of the line is $x = 0$.

83. *Writing Exercise.*

85. $d - 7$

87. Let n represent the number. Then we have $7 + 4n$.

89. Let x and y represent the numbers. Then we have $2(x + y)$.

91. *Writing Exercise.*

93. The x-axis is a horizontal line, so it is of the form $y = b$. All points on the x-axis are of the form $(x, 0)$, so b must be 0 and the equation is $y = 0$.

95. A line parallel to the y-axis has an equation of the form $x = a$. Since the x-coordinate of one point on the line is -2, then $a = -2$ and the equation is $x = -2$.

97. Since the x-coordinate of the point of intersection must be -3 and y must equal 4, the point of intersection is $(-3, 4)$.

99. The y-intercept is (0, 5), so we have $y = mx + 5$. Another point on the line is $(-3, 0)$ so we have

$$0 = m(-3) + 5$$
$$-5 = -3m$$
$$\frac{5}{3} = m$$

The equation is $y = \frac{5}{3}x + 5$, or $5x - 3y = -15$, or $-5x + 3y = 15$.

101. Substitute 0 for x and -8 for y.

$$4 \cdot 0 = C - 3(-8)$$
$$0 = C + 24$$
$$-24 = C$$

103. $Ax + D = C$

$$Ax = C - D$$
$$x = \frac{C - D}{A}$$

The x-intercept is $\left(\frac{C - D}{A}, 0\right)$.

105. Find the y-intercept:

$-7y = 80$ Covering the x-term

$$y = -\frac{80}{7} = -11.\overline{428571}$$

The y-intercept is $\left(0, -\frac{80}{7}\right)$, or $(0, -11.\overline{428571})$.

Find the x-intercept:

$2x = 80$ Covering the y-term

$x = 40$

The x-intercept is $(40, 0)$.

107. From the equation we see that the y-intercept is $(0,-9)$.
To find the x-intercept, let $y=0$.

$$0=0.2x-9$$
$$9=0.2x$$
$$45=x$$

The x-intercept is $(45,0)$.
Find the x-intercept:

$25x=1$ Covering the y-term

$x=\frac{1}{25}$, or 0.04

The x-intercept is $\left(\frac{1}{25}, 0\right)$, or $(0.04, 0)$.

109. Find the y-intercept.

$25y=1$ Covering the x-term

$y=\frac{1}{25}$, or 0.04

The y-intercept is $\left(0, \frac{1}{25}\right)$, or $(0, 0.04)$.

Find the x-intercept:

$50x=1$ Covering the y-term

$x=\frac{1}{50}$, or 0.02

The x-intercept is $\left(\frac{1}{50}, 0\right)$, or $(0.02, 0)$.

Exercise Set 3.4

1. True

3. False. Speed equals distance traveled divided by time.

5. $\frac{100 \text{ miles}}{5 \text{ hours}}=20$ miles per hour, or miles/hour

7. $\frac{300 \text{ dollars}}{150 \text{ miles}}=2$ dollars per mile, or dollars/mile

9. a. We divide the number of miles traveled by the number of gallons of gas used for distance traveled.

Rate, in miles per gallon

$$=\frac{14{,}131 \text{ mi}-13{,}741 \text{ mi}}{13 \text{ gal}}$$
$$=\frac{390 \text{ mi}}{13 \text{ gal}}$$
$$=30 \text{ mi/gal}$$
$$=30 \text{ miles per gallon}$$

b. We divide the cost of the rental by the number of days. From June 5 to June 8 is 8 – 5, or 3 days.

Average cost, in dollars per day

$$=\frac{118 \text{ dollars}}{3 \text{ days}}$$
$$\approx 39.33 \text{ dollars/day}$$
$$\approx \$39.33 \text{ per day}$$

c. We divide the number of miles traveled by the number of days. The car was driven 390 miles, and was rented for 3 days.

Rate, in miles per day

$$=\frac{390 \text{ mi}}{3 \text{ days}}$$
$$=130 \text{ mi/day}$$
$$=130 \text{ mi per day}$$

d. Note that $118 = 11,800¢. The car was driven 390 miles.

$$\text{Rate, in cents per mile} =\frac{11{,}800¢}{390 \text{ mi}}$$
$$\approx 30¢ \text{ per mile}$$

11. a. From 9:00 to 11:00 is $11-9$, or 2 hr.

$$\text{Average speed, in miles per hour} =\frac{14 \text{ mi}}{2 \text{ hr}}$$
$$=7 \text{ mph}$$

b. From part (a) we know that the bike was rented for 2 hr.

$$\text{Rate, in dollars per hour} =\frac{\$15}{2 \text{ hr}}=\$7.50 \text{ per hr}$$

c. $\text{Rate, in dollars per mile} =\frac{\$15}{14 \text{ mi}}\approx \$1.07 \text{ per mi}$

13. a. It is 3 hr from 9:00 A.M. to noon and 2 more hours from noon to 2:00 P.M., so the proofreader worked 3 + 2, or 5 hr.

$$\text{Rate, in dollars per hour} =\frac{\$110}{5 \text{ hr}}$$
$$=\$22 \text{ per hr}$$

b. The number of pages proofread is 195 – 92, or 103.

$$\text{Rate, in pages per hour} =\frac{103 \text{ pages}}{5 \text{ hr}}$$
$$=20.6 \text{ pages per hr}$$

c. $$\text{Rate, in dollars per page} =\frac{\$110}{103 \text{ pages}}$$
$$\approx \$1.07 \text{ per page}$$

15. Increase in debt: 18,884 billion – 15,041 billion or 3843 billion.
Change in time $2016-2011=5$ years

$$\text{Rate of increase} =\frac{\text{Change in debt}}{\text{Change in time}}$$
$$=\frac{\$3843 \text{ billion}}{5 \text{ yr}}$$
$$=\$768.6 \text{ billion/yr}$$

17. a. The elevator traveled $34-5$, or 29 floors in 2:40 – 2:38, or 2 min.

$$\text{Average rate of travel} =\frac{29 \text{ floors}}{2 \text{ min}}$$
$$=14.5 \text{ floors per min}$$

b. In part (a) we found that the elevator traveled 29 floors in 2 min. Note that $2 \text{ min} = 2\times 1 \text{ min}$ $= 2\times 60 \text{ sec} = 120 \text{ sec}$.

$$\text{Average rate of travel} =\frac{120 \text{ sec}}{29 \text{ floors}}$$
$$\approx 4.14 \text{ sec per floor}$$

19. Ascended $29,029 \text{ ft} - 17,700 \text{ ft} = 11,329 \text{ ft}$. The time of ascent: 8 hr, 10 min, or $8 \text{ hr} + 10 \text{ min}$ $= 480 \text{ min} + 10 \text{ min} = 490 \text{ min}$.

a. $\text{Rate, in feet per minute} = \dfrac{11,329 \text{ ft}}{490 \text{ min}} \approx 23.12 \text{ ft/min}$

b. $\text{Rate, in minutes per foot} = \dfrac{490 \text{ min}}{11,329 \text{ ft}} \approx 0.04 \text{ min/ft}$

21. The rate of increase of the average hourly wage for instructional teacher aide is given in dollars per year, so we list an average hourly wage on the vertical axis and year on the horizontal axis. We can count by increments of $0.50 on the vertical axis and label the units in dollars. Plot the points (2011, $13.55) and (2011 + 1, $13.55 + $0.50), or (2012, $14.05) and draw a line through the two points.

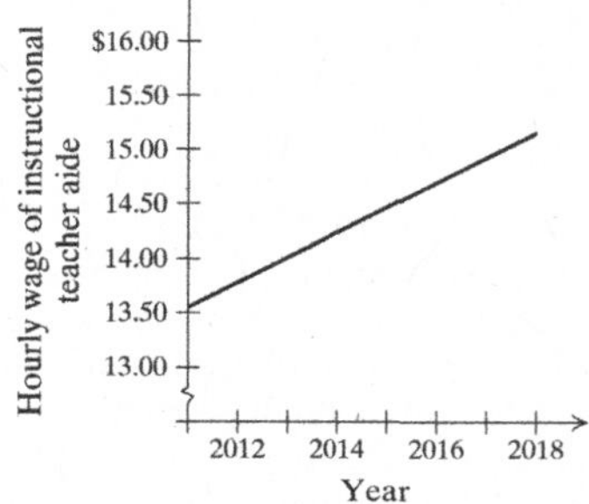

23. The rate is given in tons per year, so we list the amount in tons of paper on the vertical axis and year on the horizontal axis. We can count by increments of 5 million on the vertical axis. We plot the point (2008, 28 million). Then to display the rate decreasing, we move from that point to a point that represents 2 million less a year later. The coordinates of this point are $(2008+1,\ 28-2 \text{ million})$ or (2009, 26 million). Finally, we draw a line through the two points.

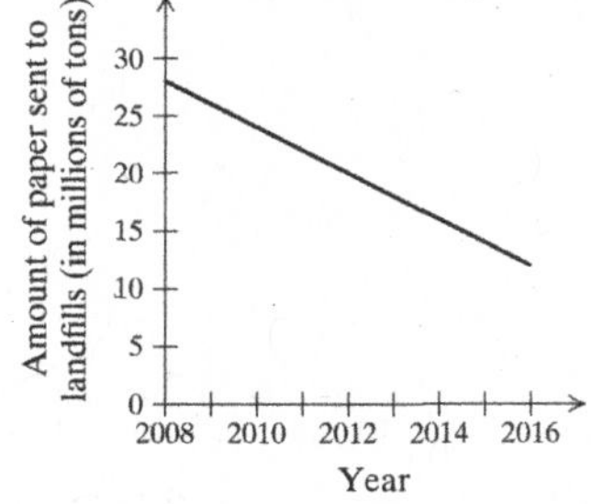

25. The rate is given in miles per hour, so we list the number of miles traveled on the vertical axis and the time of day on the horizontal axis. If we count by 100's of miles on the vertical axis we can easily reach 230 without needing a terribly large graph. We plot the point (3:00, 230). Then to display the rate of travel, we move from that point to a point that represents 90 more miles traveled 1 hour later. The coordinates of this point are (3:00 + 1 hr, 230 + 90), or (4:00, 320). Finally, we draw a line through the two points.

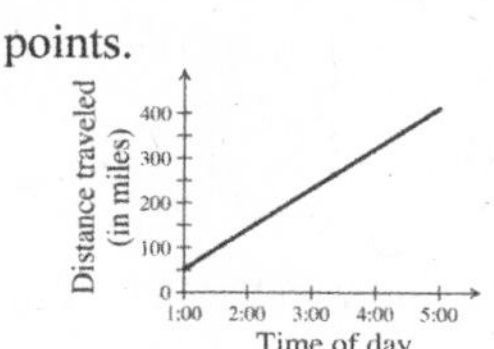

27. The rate is given in dollars per hour so we list money earned on the vertical axis and the time of day on the horizontal axis. We can count by $20 on the vertical axis and reach $50 without needing a terribly large graph. Next we plot the point (2:00 P.M., $50). To display the rate we move from that point to a point that represents $15 more 1 hour later. The coordinates of this point are $(2+1, \$50+\$15)$, or (3:00 P.M., $65). Finally, we draw a line through the two points.

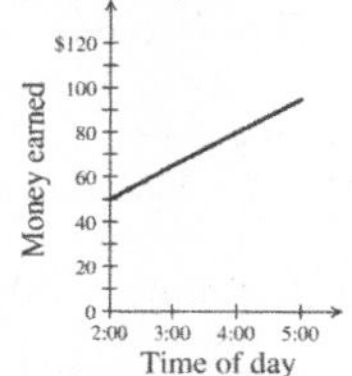

29. The rate is given in cost per minute so we list the amount of the telephone bill on the vertical axis and the number of additional minutes on the horizontal axis. We begin with $7.50 on the vertical axis and count by $0.25. A jagged line at the base of the axis indicates that we are not showing amounts smaller than $7.50. We begin with 0 additional minutes on the horizontal axis and plot the point (0, $7.50). We move from there to a point that represents $0.10 more 1 minute later. The coordinates of this point are $(0+1 \text{ min}, \$7.50+\$0.10)$, or (1 min, $7.60). Then we draw a line through the two points.

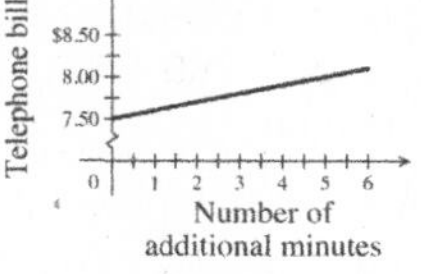

31. The points (10:00, 30 calls) and (1:00, 90 calls) are on the graph. This tells us that in the 3 hr between 10:00 and 1:00 there were 90 – 30 = 60 calls completed. The rate is

$$\frac{60 \text{ calls}}{3 \text{ hr}} = 20 \text{ calls/hour.}$$

33. The points (12:00, 100 mi) and (2:00, 250 mi) are on the graph. This tells us that in the 2 hr between 12:00 and 2:00 the train traveled 250 – 100 = 150 mi. The rate is

$$\frac{150 \text{ mi}}{2 \text{ hr}} = 75 \text{ mi per hour.}$$

35. The points (5 min, 15¢) and (10 min, 30¢) are on the graph. This tells us that in 10 – 5 = 5 min the cost of the call increased $30¢ - 15¢ = 15¢$. The rate is

$$\frac{15¢}{5 \text{ min}} = 3¢ \text{ per min.}$$

37. The points (2001, 3.4 million) and (2011, 3.0 million) are on the graph. This tells us that in $2011-2001=10$ yrs the population changes $3.0-3.4=-0.4$ million. The rate is
$$\frac{-0.4\text{ million}}{10\text{ yr}}=-40{,}000\text{ people/year.}$$

39. The points (80 mi, 2 gal) and (240 mi, 6 gal) are on the graph. This tells us that when driven $240-80=160$ mi the vehicle consumed $6-2=4$ gal of gas. The rate is
$$\frac{4\text{ gal}}{160\text{ mi}}=0.025\text{ gal/mi.}$$

41. Since swimming is the slowest of the three sports and biking is the fastest, the slope of the line representing swimming speed will be the least steep of the three and that representing biking speed will be the steepest. The second segment of graph (e) rises most steeply and the third segment is the least steep of the three segments. Thus this graph represents running followed by biking and then swimming.

43. Since swimming is the slowest of the three sports and biking is the fastest, the slope of the line representing swimming speed will be the least steep of the three and that representing biking speed will be the steepest. The first segment of graph (d) is the least steep and the second segment is the steepest of the three segments. Thus this graph represents swimming followed by biking and then running.

45. Since swimming is the slowest of the three sports and biking is the fastest, the slope of the line representing swimming speed will be the least steep of the three and that representing biking speed will be the steepest. The first segment of graph (b) is the steepest and the second segment is the least steep of the three segments. Thus this graph represents biking followed by swimming and then running.

47. *Writing Exercise.*

49. $150=2\cdot3\cdot5\cdot5$

51.
$$\begin{array}{r} 1.375 \\ 8\overline{)11.000} \\ \underline{8} \\ 3\,0 \\ \underline{2\,4} \\ 60 \\ \underline{56} \\ 40 \\ \underline{40} \\ 0 \end{array}$$

$\frac{11}{8}=1.375$, so $-\frac{11}{8}=-1.375$.

53. The opposite of $\frac{3}{2}$ is $-\frac{3}{2}$.

55. *Writing Exercise.*

57. Let t = flight time and a = altitude. While the plane is climbing at a rate of 6300 ft/min, the equation $a=6300t$ describes the situation. Solving $31{,}500=6300t$, we find that the cruising altitude of 31,500 ft is reached after about 5 min. Thus we graph $a=6300t$ for $0\le t\le5$.
The plane cruises at 31,500 ft for 3 min, so we graph $a=31{,}500$ for $5<t\le8$. After 8 min the plane descends at a rate of 3500 ft/min and lands. The equation $a=31{,}500-3500(t-8)$, or $a=-3500t+59{,}500$, describes this situation. Solving $0=-3500t+59{,}500$, we find that the plane lands after about 17 min. Thus we graph $a=-3500t+59{,}500$ for $8<t\le17$. The entire graph is show below.

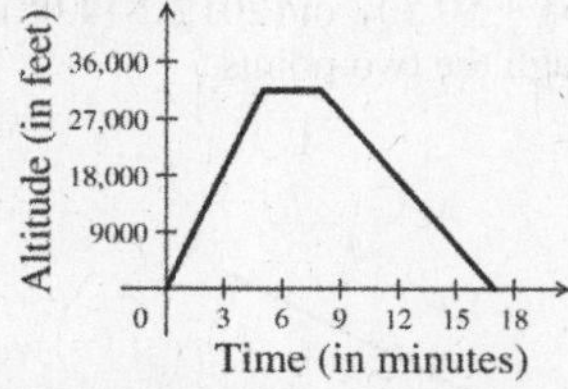

59. Let the horizontal axis represent the distance traveled, in miles, and let the vertical axis represent the fare, in dollars. Use increments of 1/5, or 0.2 mi, on the horizontal axis and of \$1 on the vertical axis. The fare for traveling 0.2 mi is \$2.50 + \$0.50 · 1, or \$3 and for 0.4 mi, or 0.2 mi×2, we have \$2.50 + \$0.50(2), or \$3.50. Plot the points (0.2 mi, \$3) and (0.4 mi, \$3.50) and draw the line through them.

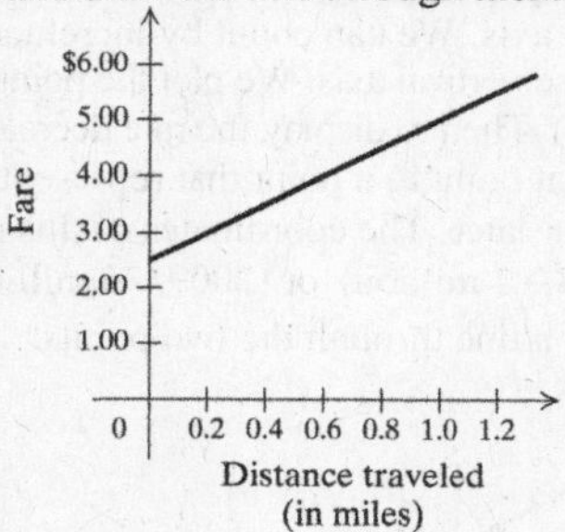

61. $95\text{ mph}+39\text{ mph}=134\text{ mph}$

$\frac{134\text{ mi}}{1\text{ hr}}$ gives us $\frac{1\text{ hr}}{134\text{ mi}}$.

$$\frac{1\text{ hr}}{134\text{ min}}=\frac{1\cancel{\text{ hr}}}{134\text{ mi}}\cdot\frac{60\text{ min}}{1\cancel{\text{ hr}}}\approx0.45\text{ min per mi}$$

63. First we find Anne's speed in minutes per kilometer.
$$\text{Speed}=\frac{15.5\text{ min}}{7\text{ km}-4\text{ km}}=\frac{15.5\text{ min}}{3\text{ km}}$$
Now we convert min/km to min/mi.
$$\frac{15.5\text{ min}}{3\text{ km}}\approx\frac{15.5\text{ min}}{3\cancel{\text{ km}}}\cdot\frac{1\cancel{\text{ km}}}{0.621\text{ mi}}\approx\frac{15.5\text{ min}}{1.863\text{ mi}}$$
At a rate of $\frac{15.5\text{ min}}{1.863\text{ mi}}$, to run a 5-mi race it would take $\frac{15.5\text{ min}}{1.863\cancel{\text{ mi}}}\cdot5\cancel{\text{ mi}}\approx41.6$ min.
(Answers may vary due to conversion factor used.)

65. First we find Doug's rate. Then we double it to find Trevor's rate. Note that 50 minutes = $\frac{50}{60}$ hr = $\frac{5}{6}$ hr.

Doug's rate

$= \frac{\text{change in number of bushels picked}}{\text{corresponding change in time}}$

$= \frac{5\frac{1}{2} - 4 \text{ bushels}}{\frac{5}{6} \text{ hr}}$

$= \frac{1\frac{1}{2} \text{ bushels}}{\frac{5}{6} \text{ hr}}$

$= \frac{3}{2} \cdot \frac{6}{5} \frac{\text{bushels}}{\text{hr}}$

$= \frac{9}{5}$ bushels per hour, or 1.8 bushels per hour

Then Trevor's rate is $2(1.8) = 3.6$ bushels per hour.

Exercise Set 3.5

1. Slope is the rate at which y is changing with respect to x.

3. The slope of a line can be expressed using rise and run as follows: $\frac{rise}{run}$.

5. If a line slants up from left to right, the sign of its slope is *positive*, and if a line slants down from left to right, the sign of its slope is *negative*.

7. A teenager's height increases over time, so the rate is positive.

9. The water level decreases during a drought, so the rate is negative.

11. A person's IQ does not change during sleep, so the rate is zero.

13. The number of people present decreases in the moments following the final buzzer, so the rate is negative.

15. The rate can be found using the coordinates of any two points on the line. We use (4, \$100) and (8, \$200).

Rate $= \frac{\text{change in compensation}}{\text{corresponding change in number of blogs}}$

$= \frac{\$200 - \$100}{8 - 4}$

$= \frac{\$100}{4 \text{ blogs}}$

$= \$25$ per blog

17. The rate can be found using the coordinates of any two points on the line. We use (2006, \$17 billion) and (2014, \$13 billion).

Rate $= \frac{\text{change in price}}{\text{corresponding change in time}}$

$= \frac{\$13 - \$17}{2014 - 2006}$

$= \frac{-\$4}{8 \text{ years}}$

$= -\$\frac{1}{2}$ billion per year

19. The rate can be found using the coordinates of any two points on the line. We use (50, 490) and (150, 540), where 50 and 150 are in \$1000's.

Rate $= \frac{\text{change in score}}{\text{corresponding change in income}}$

$= \frac{540 - 490 \text{ points}}{150 - 50}$

$= \frac{50 \text{ points}}{100}$

$= \frac{1}{2}$ point per \$1000 income

21. The rate can be found using the coordinates of any two points on the line. We use (0 min, 54°) and (27 min, –4°).

Rate $= \frac{\text{change in temperature}}{\text{corresponding change in time}}$

$= \frac{-4° - 54°}{27 \text{ min} - 0 \text{ min}}$

$= \frac{-58°}{27 \text{ min}}$

$\approx -2.1°$per min

23. We can use any two points on the line, such as (0, 1) and (3, 5).

$m = \frac{\text{change in } y}{\text{change in } x}$

$= \frac{5-1}{3-0} = \frac{4}{3}$

25. We can use any two points on the line, such as (0, 2) and (3, 3).

$m = \frac{\text{change in } y}{\text{change in } x}$

$= \frac{3-2}{3-0} = \frac{1}{3}$

27. We can use any two points on the line, such as (0, 2) and (2, 0).

$m = \frac{\text{change in } y}{\text{change in } x}$

$= \frac{2-0}{0-2} = \frac{2}{-2} = -1$

29. This is the graph of a horizontal line. Thus, the slope is 0.

31. We can use any two points on the line, such as (2, 4) and (4, 0).

$m = \frac{\text{change in } y}{\text{change in } x}$

$= \frac{0-4}{4-2} = \frac{-4}{2} = -2$

33. This is the graph of a vertical line. Thus, the slope is undefined.

35. We can use any two points on the line, such as (0, 2) and (3, 1).

$$m = \frac{\text{change in } y}{\text{change in } x} = \frac{1-2}{3-0} = -\frac{1}{3}$$

37. Two points on the line are (2,–4) and (3, 1).

$$m = \frac{1-(-4)}{3-2} = \frac{5}{1} = 5$$

39. (1, 3) and (5, 8)

$$m = \frac{8-3}{5-1} = \frac{5}{4}$$

41. (–2, 4) and (3, 0)

$$m = \frac{4-0}{-2-3} = \frac{4}{-5} = -\frac{4}{5}$$

43. (–4, 0) and (5, 6)

$$m = \frac{6-0}{5-(-4)} = \frac{6}{9} = \frac{2}{3}$$

45. (0, 7) and (–3, 10)

$$m = \frac{10-7}{-3-0} = \frac{3}{-3} = -1$$

47. (–2, 3) and (–6, 5)

$$m = \frac{5-3}{-6-(-2)} = \frac{2}{-6+2} = \frac{2}{-4} = -\frac{1}{2}$$

49. $\left(-2, \frac{1}{2}\right)$ and $\left(-5, \frac{1}{2}\right)$

Observe that the points have the same *y*-coordinate. Thus, they lie on a horizontal line and its slope is 0. We could also compute the slope.

$$m = \frac{\frac{1}{2}-\frac{1}{2}}{-2-(-5)} = \frac{\frac{1}{2}-\frac{1}{2}}{-2+5} = \frac{0}{3} = 0$$

51. (5,–4) and (2,–7)

$$m = \frac{-7-(-4)}{2-5} = \frac{-3}{-3} = 1$$

53. (6,–4) and (6, 5)

Observe that the points have the same *x*-coordinate. Thus, they lie on a vertical line and its slope is undefined. We could also compute the slope.

$$m = \frac{-4-5}{6-6} = \frac{-9}{0}, \text{ undefined}$$

55. The line $y = 5$ is a horizontal line. A horizontal line has slope 0.

57. The line $x = -8$ is a vertical line. Slope is undefined.

59. The line $x = 9$ is a vertical line. The slope is undefined.

61. The line $y = -10$ is a horizontal line. A horizontal line has slope 0.

63. The grade is expressed as a percent.

$$m = \frac{230}{1600} = 0.14375 = 14.375\%$$

65. The slope is expressed as a percent.

$$m = \frac{28}{80} = 0.35 = 35\%.$$

67. 2 ft 5 in. = 2·12 in. + 5 in. = 24 in. + 5 in. = 29 in.
8 ft 2 in. = 8·12 in. + 2 in. = 96 in. + 2 in. = 98 in.

$$m = \frac{29}{98}$$

69. Dooley Mountain rises $5400 - 3500 = 1900$ ft.

$$m = \frac{1900}{37000} \approx 0.051 \approx 5.1\%$$

Yes, it qualifies as part of the Tour de France.

71. *Writing Exercise.*

73. $3(4+a) = 12+3a$

75. $-(-15) = 15$

77. $5t \cdot 5t \cdot 5t = (5t)^3$

79. *Writing Exercise.*

81. From the dimensions on the drawing, we see that the ramps labeled *A* have a rise of 61 cm and a run of 167.6 cm.

$$m = \frac{61 \text{ cm}}{167.6 \text{ cm}} \approx 0.364, \text{ or } 36.4\%$$

83. If the line passes through (2, 5) and never enters the second quadrant, then it slants up from left to right or is vertical. This means that its slope is positive. The line slants least steeply if it passes through (0, 0). In this case, $m = \frac{5-0}{2-0} = \frac{5}{2}$. Thus, the numbers the line could have for it slope are $\left\{m \middle| m \geq \frac{5}{2}\right\}$.

85. Let t = the number of units each tick mark on the vertical axis represents. Note that the graph drops 4 units for every 3 units of horizontal change. Then we have:

$$\frac{-4t}{3} = -\frac{2}{3}$$

$-4t = -2$ Multiplying by 3

$t = \frac{1}{2}$ Dividing by – 4

Each tick mark on the vertical axis represents $\frac{1}{2}$ unit.

Mid-Chapter Review

1. y-intercept: $y - 3 \cdot 0 = 6$

$y = 6$

The y-intercept is (0, 6).

x-intercept: $0 - 3x = 6$

$-3x = 6$

$x = -2$

The x-intercept is (–2, 0).

2. $m = \frac{y_2 - y_1}{x_2 - x_1} = \frac{-1-5}{3-1}$

$= \frac{-6}{2}$

$= -3$

3. Graph (0,–3).

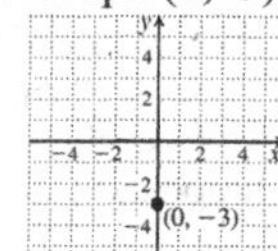

4. IV

5. To show that a pair is a solution, we substitute, replacing x with the first coordinate and y with the second coordinate in each pair.

$y = 5 - x$	
-3	$5-(-2)$
	$5+2$

$-3 \stackrel{?}{=} 7$ FALSE

No

6. Graph $y = x - 3$.

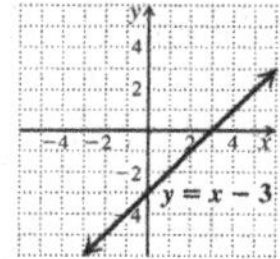

7. $y = -3x$

The equation is equivalent to $y = -3x + 0$. The y-intercept is (0, 0). We find two other points.

When $x = -1$, $y = -3(-1) = 3$.

When $x = 1$, $y = -3$.

x	y
0	0
–1	3
1	–3

Plot these points, draw the line they determine, and label the graph $y = -3x$.

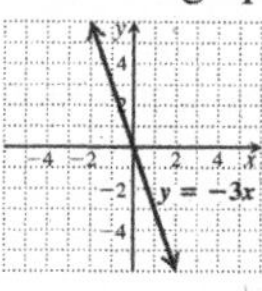

8. Graph $3x - y = 2$.

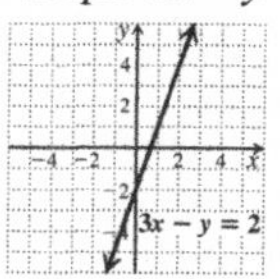

9. $4x - 5y = 20$

$-5y = -4x + 20$

$y = \frac{4}{5}x - 4$

$y = \frac{4}{5}x + (-4)$

The y-intercept is $(0,-4)$. We find two other points.

When $x = -1$, $y = \frac{4}{5}(-1) - 4 = -\frac{4}{5} - 4 = -4\frac{4}{5}$.

When $x = 5$, $y = \frac{4}{5} \cdot 5 - 4 = 4 - 4 = 0$.

x	y
0	–4
–1	$-4\frac{4}{5}$
5	0

Plot these points, draw the line they determine, and label the graph $4x - 5y = 20$.

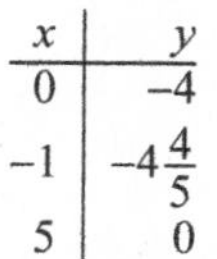

10. Graph $y = -2$.

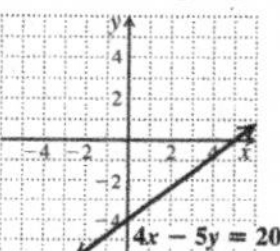

11. $x = 1$

Any ordered pair $(1, y)$ is a solution. The variable x must be 1, but the y variable can be any number we choose. A few solutions are listed below. Plot these points and draw the line.

x	y
1	–2
1	0
1	4

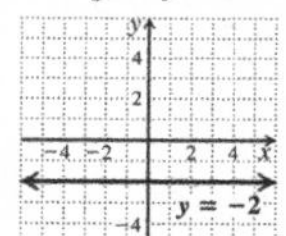

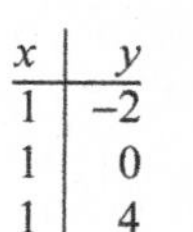

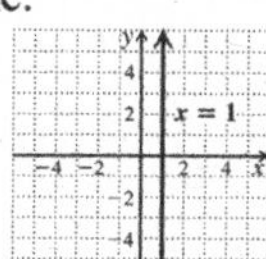

12. Rate, in homes per month

$= \frac{38 \text{ homes} - 10 \text{ homes}}{2 \text{ months}}$

$= \frac{28 \text{ homes}}{2 \text{ months}}$

$= 14$ homes/month

13. Ascended $14,255 \text{ ft} - 9600 \text{ ft} = 4655 \text{ ft}$. The distance of ascent: 15,840

Grade $= \frac{4655 \text{ ft}}{15,840 \text{ ft}} \approx 0.29 = 29\%$

14. $m=\frac{8-(-2)}{1-(-5)}=\frac{10}{6}=\frac{5}{3}$

15. (1, 2) and (4,–7)

$m=\frac{-7-2}{4-1}=\frac{-9}{3}=-3$

16. $m=\frac{-2-0}{0-0}=\frac{-2}{0}$ undefined

17. (6,–3) and (2,–3)

$m=\frac{-3-(-3)}{2-6}=\frac{-3+3}{-4}=0$

18. 0

19. The line $x=-7$ is a vertical line. The slope is undefined.

20. $2y-3x=12$

Find the x-intercept:

$-3x=12$ Ignoring the y-term

$x=-4$

The x-intercept is (–4, 0).

Find the y-intercept:

$2y=12$ Ignoring the x-term

$y=6$

The y-intercept is (0, 6).

Exercise Set 3.6

1. Slope

3. y-intercept

5. y-intercept

7. We can read the slope, 3, directly from the equation. Choice (f) is correct.

9. We can read the slope, $\frac{2}{3}$, directly from the equation. Choice (d) is correct.

11. $y=3x-2=3x+(-2)$

The y-intercept is $(0,-2)$, so choice (e) is correct.

13. Slope $\frac{2}{3}$; y-intercept (0, 1)

We plot (0, 1) and from there move up 2 units and right 3 units. This locates the point (3, 3). We plot (3, 3) and draw a line passing through (0, 1) and (3, 3).

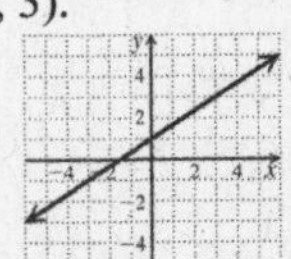

15. Slope $\frac{5}{3}$, y-intercept $(0,-2)$

We plot $(0,-2)$ and from there move up 5 units and right 3 units. This locates the point (3, 3). We plot (3, 3) and draw a line passing through $(0,-2)$ and (3, 3).

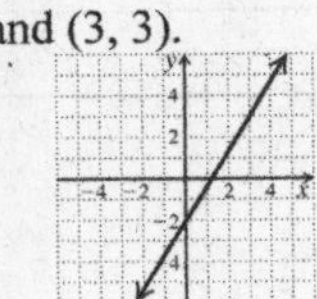

17. Slope $-\frac{1}{3}$; y-intercept (0, 5)

We plot (0, 5). We can think of the slope as $\frac{-1}{3}$, so from (0, 5) we move down 1 unit and right 3 units. This locates the point (3, 4). We plot (3, 4) and draw a line passing through (0, 5) and (3, 4).

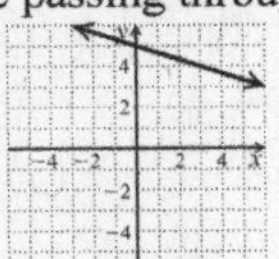

19. Slope 2; y-intercept (0, 0)

We plot (0, 0). We can think of the slope as $\frac{2}{1}$, so from (0, 0) we move up 2 units and right 1 unit. This locates the point (1, 2). We plot (1, 2) and draw a line passing through (0, 0) and (1, 2).

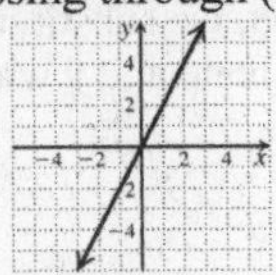

21. Slope –3; y-intercept (0, 2)

We plot (0, 2). We can think of the slope as $\frac{-3}{1}$, so from (0, 2) we move down 3 units and right 1 unit. This locates the point $(1,-1)$. We plot $(1,-1)$ and draw a line passing through (0, 2) and $(1,-1)$.

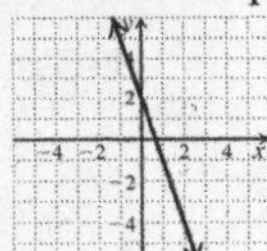

23. Slope 0; y-intercept $(0,-5)$

Since the slope is 0, we know the line is horizontal, so from $(0,-5)$ we move right 1 unit. This locates the point $(1,-5)$. We plot $(1,-5)$ and draw a line passing through $(0,-5)$ and $(1,-5)$.

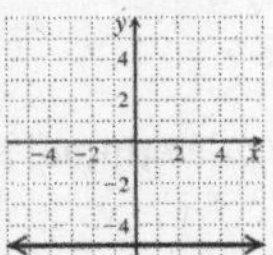

25. We read the slope and y-intercept from the equation.

$y=-\frac{2}{7}x+5$

The slope is $-\frac{2}{7}$. The y-intercept is (0, 5).

27. We read the slope and y-intercept from the equation.

$$y = \frac{1}{3}x + 7$$

The slope is $\frac{1}{3}$. The y-intercept is (0, 7).

29.
$$y = \frac{9}{5}x - 4$$
$$y = \frac{9}{5}x + (-4)$$

The slope is $\frac{9}{5}$, and the y-intercept is $(0, -4)$.

31. We solve for y to rewrite the equation in the form $y = mx + b$.

$$-3x + y = 7$$
$$y = 3x + 7$$

The slope is 3, and the y-intercept is (0, 7).

33.
$$4x + 2y = 8$$
$$2y = -4x + 8$$
$$y = \frac{1}{2}(-4x + 8)$$
$$y = -2x + 4$$

The slope is –2, and the y-intercept is (0, 4).

35. Observe that this is the equation of a horizontal line that lies 3 units above the x-axis. Thus, the slope is 0, and the y-intercept is (0, 3). We could also write the equation in slope-intercept form.

$$y = 3$$
$$y = 0x + 3$$

The slope is 0, and the y-intercept is (0, 3).

37.
$$2x - 5y = -8$$
$$-5y = -2x - 8$$
$$y = -\frac{1}{5}(-2x - 8)$$
$$y = \frac{2}{5}x + \frac{8}{5}$$

The slope is $\frac{2}{5}$, and the y-intercept is $\left(0, \frac{8}{5}\right)$.

39.
$$9x - 8y = 0$$
$$-8y = -9x$$
$$y = \frac{9}{8}x \text{ or } y = \frac{9}{8}x + 0$$

Slope: $\frac{9}{8}$, y-intercept: (0, 0)

41. We use the slope-intercept equation, substituting 5 for m and 7 for b:

$$y = mx + b$$
$$y = 5x + 7$$

43. We use the slope-intercept equation, substituting $\frac{7}{8}$ for m and –1 for b:

$$y = mx + b$$
$$y = \frac{7}{8}x - 1$$

45. We use the slope-intercept equation, substituting $-\frac{5}{3}$ for m and –8 for b:

$$y = mx + b$$
$$y = -\frac{5}{3}x - 8$$

47. We use the slope-intercept equation, substituting 0 for m and $\frac{1}{3}$ for b.

$$y = mx + b$$
$$y = 0x + \frac{1}{3}$$
$$y = \frac{1}{3}$$

49. From the graph we see that the y-intercept is (0, 85). We also see that the point (10, 110) is on the graph. We find the slope:

$$m = \frac{110 - 85}{10 - 0} = \frac{25}{10} = 2.5$$

Substituting 2.5 for m and 85 for b in the slope-intercept equation $y = mx + b$, we have

$$y = 2.5x + 85$$

where y is the number of jobs and x is the number of years after 2012.

51. From the graph we see that the y-intercept is (0, 30). We also see that the point (10, 45) is on the graph. We find the slope:

$$m = \frac{45 - 30}{10 - 0} = \frac{15}{10} = 1.5$$

Substituting 1.5 for m and 30 for b in the slope-intercept equation $y = mx + b$, we have

$$y = 1.5x + 30,$$

where y is the cost of a 1-month gym membership, in dollars, and x is the number of classes taken.

53. $y = \frac{2}{3}x + 2$

First we plot the y-intercept (0, 2). We can start at the y-intercept and use the slope, $\frac{2}{3}$, to find another point. We move up 2 units and right 3 units to get a new point (3, 4). Thinking of the slope as $\frac{-2}{-3}$ we can start at (0, 2) and move down 2 units and left 3 units to get another point (–3, 0).

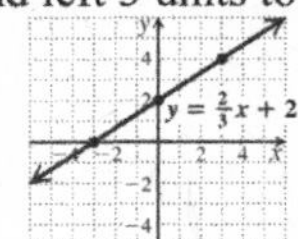

55. $y=-\frac{2}{3}x+3$

First we plot the y-intercept (0, 3). We can start at the y-intercept and, thinking of the slope as $\frac{-2}{3}$, find another point by moving down 2 units and right 3 units to the point (3, 1). Thinking of the slope as $\frac{2}{-3}$ we can start at (0, 3) and move up 2 units and left 3 units to get another point (–3, 5).

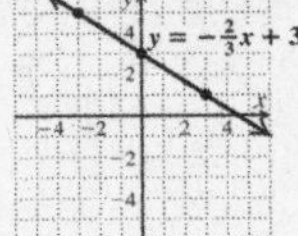

57. $y=\frac{3}{2}x+3$

First we plot the y-intercept (0, 3). We can start at the y-intercept and use the slope, $\frac{3}{2}$, to find another point. We move up 3 units and right 2 units to get a new point (2, 6). Thinking of the slope as $\frac{-3}{-2}$ we can start at (0, 3) and move down 3 units and left 2 units to get another point (–2, 0).

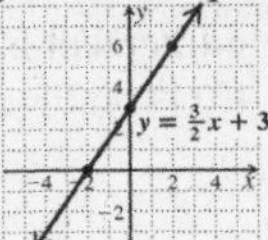

59. $y=-\frac{4}{3}x+3$

First we plot the y-intercept (0, 3). We can start at the y-intercept and, thinking of the slope as $\frac{-4}{3}$ find another point by moving down 4 units and right 3 units to the point $(3,-1)$. Thinking of the slope as $\frac{-4}{3}$ we can start at (3,–1) and move down 4 units and right 3 units to get another point (6,–5).

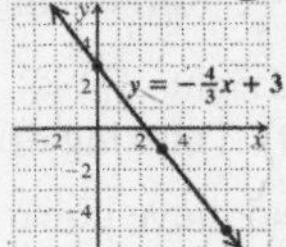

61. We first rewrite the equation in slope-intercept form.

$2x+y=1$

$y=-2x+1$

Now we plot the y-intercept (0, 1). We can start at the y-intercept and, thinking of the slope as $\frac{-2}{1}$, find another point by moving down 2 units and right 1 unit to the point $(1,-1)$. In a similar manner, we can move from the point $(1,-1)$ to find a third point $(2,-3)$.

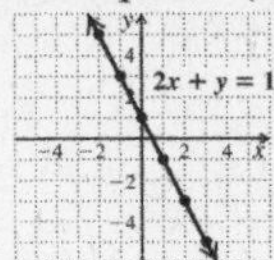

63. We first rewrite the equation in slope-intercept form.

$3x+y=0$

$y=-3x$, or $y=-3x+0$

Now we plot the y-intercept (0, 0). We can start at the y-intercept and, thinking of the slope as $\frac{-3}{1}$, find another point by moving down 3 units and right 1 unit to the point $(1,-3)$. Thinking of the slope as $\frac{3}{-1}$ we can start at (0, 0) and move up 3 units and left 1 unit to get another point (–1, 3).

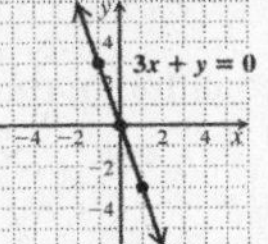

65. We first rewrite the equation in slope-intercept form.

$4x+5y=15$

$5y=-4x+15$

$y=\frac{1}{5}(-4x+15)$

$y=-\frac{4}{5}x+3$

Now we plot the y-intercept (0, 3). We can start at the y-intercept and, thinking of the slope as $\frac{-4}{5}$, find another point by moving down 4 units and right 5 units to the point $(5,-1)$. Thinking of the slope as $\frac{4}{-5}$, we can start at (0, 3) and move up 4 units and left 5 units to get another point (–5, 7).

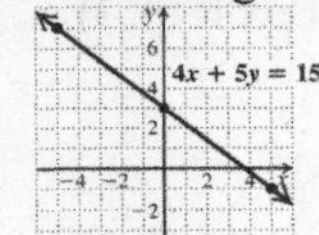

67. We first rewrite the equation in slope-intercept form.

$x-4y=12$

$-4y=-x+12$

$y=-\frac{1}{4}(-x+12)$

$y=\frac{1}{4}x-3$

Now we plot the y-intercept $(0,-3)$. We can start at the y-intercept and use the slope, $\frac{1}{4}$, to find another point. We move up 1 unit and right 4 units to the point $(4,-2)$. Thinking of the slope as $\frac{-1}{-4}$ we can start at $(0,-3)$ and move down 1 unit and left 4 units to get another point $(-4,-4)$.

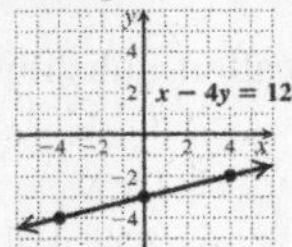

69. We first solve for y and determine the slope of each line.

$x+2=y$
$y=x+2$ Reversing the order

The slope of $y=x+2$ is 1.

$y-x=-2$
$y=x-2$

The slope of $y=x-2$ is 1.
The slopes are the same; the lines are parallel.

71. We first solve for y and determine the slope of each line.

$y+9=3x$
$y=3x-9$

The slope of $y=3x-9$ is 3.

$3x-y=-2$
$3x+2=y$
$y=3x+2$ Reversing the order

The slope of $y=3x+2$ is 3.
The slopes are the same; the lines are parallel.

73. We determine the slope of each line.
The slope of $f(x)=3x+9$ is 3.

$2y=8x-2$
$y=4x-1$

The slope of $y=4x-1$ is 4.
The slopes are not the same; the lines are not parallel.

75. We determine the slope of each line.

$x-2y=3$
$-2y=-x+3$
$y=\frac{1}{2}x-\frac{3}{2}$

The slope of $x-2y=3$ is $\frac{1}{2}$.

$4x+2y=1$
$2y=-4x+1$
$y=-2x+\frac{1}{2}$

The slope of $4x+2y=1$ is -2.
The product of their slopes is $\left(\frac{1}{2}\right)(-2)$, or -1; the lines are perpendicular.

77. We determine the slope of each line.
The slope of $f(x)=3x+1$ is 3.

$6x+2y=5$
$2y=-6x+5$
$y=-3x+\frac{5}{2}$

The slope of $6x+2y=5$ is -3.
The product of their slopes is $3(-3)$, or $-9\neq-1$, so the lines are not perpendicular.

79. *Writing Exercise.*

81. First we reword and translate, letting p represent the unknown percent.

What percent of \$25 is \$3?

$p \cdot 25 = 3$

$\frac{p\cdot 25}{25}=\frac{3}{25}$
$p=0.12=12\%$

The tip was 12% of the cost of the meal.

83. First we reword and translate letting b = the balance at the beginning of the month. The increase in the account was 15% of b, or $0.15b$.

Beginning balance plus increase is \$2760.

$b + 0.15b = 2760$

$b+0.15b=2760$
$1.15b=2760$
$b=2400$

The balance at the beginning of the month was \$2400.

85. *Writing Exercise.*

87. When $x=0$, $y=b$, so $(0, b)$ is on the line. When $x=1$, $y=m+b$, so $(1, m+b)$ is on the line. Then,

slope $=\frac{(m+b)-b}{1-0}=m.$

89. Rewrite each equation in slope-intercept form.

$-4x+8y=5$
$8y=4x+5$
$y=\frac{1}{8}(4x+5)$
$y=\frac{1}{2}x+\frac{5}{8}$

The slope is $\frac{1}{2}$.

$4x-3y=0$
$-3y=-4x$
$y=-\frac{1}{3}(-4x)$
$y=\frac{4}{3}x$, or $y=\frac{4}{3}x+0$

The y-intercept is $(0, 0)$.
The equation of the line is $y=\frac{1}{2}x+0$, or $y=\frac{1}{2}x$.

91. Rewrite each equation in slope-intercept form.

$4x+5y=9$
$5y=-4x+9$
$y=\frac{1}{5}(-4x+9)$
$y=-\frac{4}{5}x+\frac{9}{5}$

The slope of the line is $-\frac{4}{5}$.

$2x+3y=12$
$3y=-2x+12$
$y=\frac{1}{3}(-2x+12)$
$y=-\frac{2}{3}x+4$

The y-intercept of the line is $(0,4)$.
The equation of the line is $y=-\frac{4}{5}x+4$.

93. Rewrite the first equation in slope-intercept form.

$$3x-5y=8$$
$$-5y=-3x+8$$
$$y=-\frac{1}{5}(-3x+8)$$
$$y=\frac{3}{5}x-\frac{8}{5}$$

The slope is $\frac{3}{5}$.

The slope of a line perpendicular to this line is a number m such that

$\frac{3}{5}m=-1$, or $m=-\frac{5}{3}$.

Now rewrite the second equation in slope-intercept form.

$$2x+4y=12$$
$$4y=-2x+12$$
$$y=\frac{1}{4}(-2x+12)$$
$$y=-\frac{1}{2}x+3$$

The y-intercept of the line is (0, 3).

The equation of the line is $y=-\frac{5}{3}x+3$.

95. Rewrite $2x+5y=6$ in slope-intercept form.

$$2x+5y=6$$
$$5y=-2x+6$$
$$y=\frac{1}{5}(-2x+6)$$
$$y=-\frac{2}{5}x+\frac{6}{5}$$

The slope is $-\frac{2}{5}$.

The slope of a line perpendicular to this line is a number m such that

$-\frac{2}{5}m=-1$, or $m=\frac{5}{2}$.

We graph the line whose equation we want to find. First we plot the given point (2, 6). Now think of the slope as $\frac{-5}{-2}$. From the point (2, 6) go down 5 units and left 2 units to the point (0, 1). Plot this point and draw the graph.

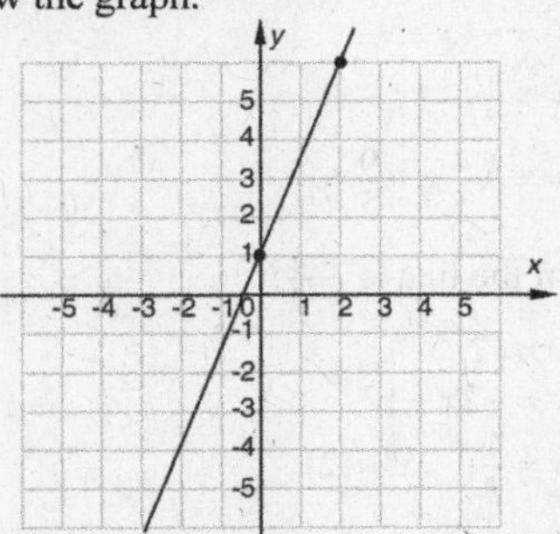

We see that the y-intercept is (0, 1), so the desired equation is $y=\frac{5}{2}x+1$.

Connecting the Concepts

1. Slope-intercept form
2. Standard form
3. None of these
4. Standard form
5. Point-slope form
6. None of these
7. $2x=5y+10$
   $2x-5y=10$
8. $y=2x+7$
   $-2x+y=7$, or $2x-y=-7$
9. $2x-7y=8$
   $-7y=-2x+8$
   $y=\frac{-2x+8}{-7}$
   $y=\frac{2}{7}x-\frac{8}{7}$
10. $y+5=-(x+3)$
    $y+5=-x-3$
    $y=-x-8$

Exercise Set 3.7

1. False
3. False; given just one point, there is an infinite number of lines that can be drawn through it.
5. True
7. $y-3=\frac{1}{4}(x-5)$
   $y-y_1=m(x-x_1)$
   $m=\frac{1}{4}$, $x_1=5$, $y_1=3$, so the slope is $\frac{1}{4}$ and a point (x_1, y_1) on the graph is (5, 3).
9. $y+1=-7(x-2)$
   $y-(-1)=-7(x-2)$
   $y-y_1=m(x-x_1)$
   $m=-7$, $x_1=2$, $y_1=-1$, so the slope is –7 and a point (x_1, y_1) on the graph is (2, –1).
11. $y-6=-\frac{10}{3}(x+4)$
    $y-6=-\frac{10}{3}(x-(-4))$
    $y-y_1=m(x-x_1)$
    $m=-\frac{10}{3}$, $x_1=-4$, $y_1=6$, so the slope is $-\frac{10}{3}$ and a point (x_1, y_1) on the graph is (–4, 6).

13.
$$y = 5x$$
$$y - 0 = 5(x - 0)$$
$$y - y_1 = m(x - x_1)$$
$m = 5$, $x_1 = 0$, $y_1 = 0$, so the slope is 5 and a point (x_1, y_1) on the graph is (0, 0).

15. $y - y_1 = m(x - x_1)$ Point-slope equation
$y - 2 = 3(x - 5)$ Substituting 3 for m, 5 for x_1, and 2 for y_1

To graph the equation, we count off a slope of $\frac{3}{1}$, starting at (5, 2), and draw the line.

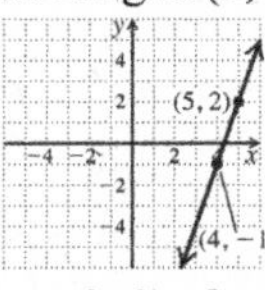

$y - 2 = 3(x - 5)$

17. $y - y_1 = m(x - x_1)$ Point-slope equation
$y - 2 = -4(x - 1)$ Substituting −4 for m, 1 for x_1, and 2 for y_1

To graph the equation, we count off a slope of $-\frac{4}{1}$, starting at (1, 2), and draw the line.

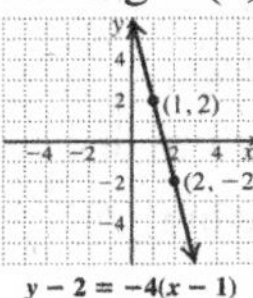

$y - 2 = -4(x - 1)$

19. $y - y_1 = m(x - x_1)$ Point-slope equation
$y - (-4) = \frac{1}{2}[x - (-2)]$ Substituting $\frac{1}{2}$ for m, −2 for x_1, and −4 for y_1

To graph the equation, we count off a slope of $\frac{1}{2}$, starting at $(-2, -4)$, and draw the line.

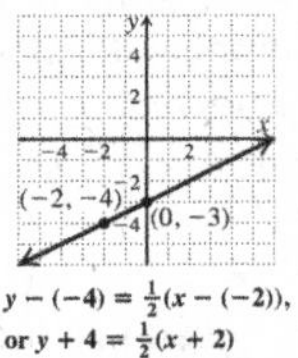

$y - (-4) = \frac{1}{2}(x - (-2))$, or $y + 4 = \frac{1}{2}(x + 2)$

21. $y - y_1 = m(x - x_1)$ Point-slope equation
$y - 0 = -1(x - 8)$ Substituting −1 for m, 8 for x_1, and 0 for y_1

To graph the equation, we count off a slope of $\frac{-1}{1}$, starting at (8, 0), and draw the line.

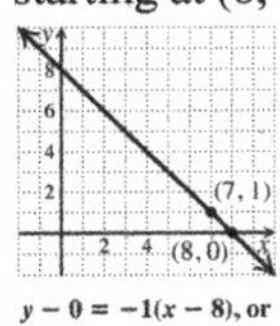

$y - 0 = -1(x - 8)$, or $y = -(x - 8)$

23. From $y = 3x - 7$, the parallel slope is 3. The y-intercept is (0, 4).
$$y = mx + b$$
$$y = 3x + 4$$

25. From $y = -\frac{3}{4}x + 1$, the perpendicular slope is $\frac{4}{3}$. The y-intercept is (0,–12).
$$y = mx + b$$
$$y = \frac{4}{3}x - 12$$

27.
$$2x - 3y = 4$$
$$-3y = -2x + 4$$
$$y = \frac{2}{3}x - \frac{4}{3}$$
The parallel slope is $\frac{2}{3}$. The y-intercept is $\left(0, \frac{1}{2}\right)$.
$$y = mx + b$$
$$y = \frac{2}{3}x + \frac{1}{2}$$

29.
$$x + y = 18$$
$$y = -x + 18$$
The perpendicular slope is 1. The y-intercept is (0,–32).
$$y = mx + b$$
$$y = x - 32$$

31. $m = 6$, (7, 1)
$y - y_1 = m(x - x_1)$ Point-slope equation
$y - 1 = 6(x - 7)$

33. $m = -5$, (3, 4)
$y - y_1 = m(x - x_1)$ Point-slope equation
$y - 4 = -5(x - 3)$

35. $m = \frac{1}{2}$, (–2,–5)
$y - y_1 = m(x - x_1)$ Point-slope equation
$y - (-5) = \frac{1}{2}(x - (-2))$

37. $m = -1$, (9, 0)
$y - y_1 = m(x - x_1)$ Point-slope equation
$y - 0 = -1(x - 9)$

39.
$y - y_1 = m(x - x_1)$ Point-slope equation
$y - (-4) = 2(x - 1)$ Substituting 2 for m, 1 for x_1, and −4 for y_1
$y + 4 = 2x - 2$ Simplifying
$y = 2x - 6$ Subtracting 4 from both sides
$f(x) = 2x - 6$ Using function notation

To graph the equation, we count off a slope of $\frac{2}{1}$, starting at (1, –4) or (0, –6), and draw the line.

$f(x) = 2x - 6$

41. $y-y_1=m(x-x_1)$ Point-slope equation

$y-8=-\frac{3}{5}[x-(-4)]$ Substituting $-\frac{3}{5}$ for m, -4 for x_1, and 8 for y_1

$y-8=-\frac{3}{5}(x+4)$

$y-8=-\frac{3}{5}x-\frac{12}{5}$ Simplifying

$y=-\frac{3}{5}x+\frac{28}{5}$ Adding 8 to both sides

$f(x)=-\frac{3}{5}x+\frac{28}{5}$ Using function notation

To graph the equation, we count off a slope of $\frac{-3}{5}$, starting at (–4, 8), and draw the line.

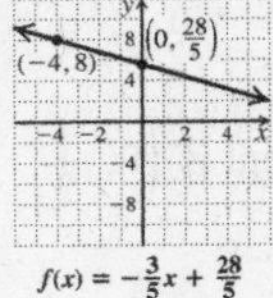

$f(x) = -\frac{3}{5}x + \frac{28}{5}$

43. $y-y_1=m(x-x_1)$ Point-slope equation

$y-(-4)=-0.6[x-(-3)]$ Substituting -0.6 for m, -3 for x_1, and -4 for y_1

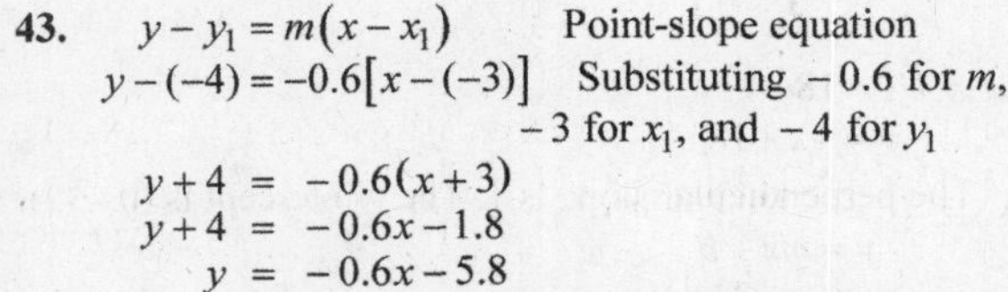

$y+4 = -0.6(x+3)$

$y+4 = -0.6x-1.8$

$y = -0.6x-5.8$

$f(x) = -0.6x-5.8$ Using function notation

To graph the equation, we count off a slope of -0.6, or $\frac{-3}{5}$, starting at (–3, –4), and draw the line.

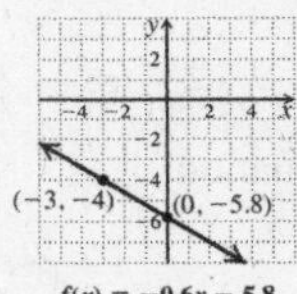

$f(x) = -0.6x - 5.8$

45. $m=\frac{2}{7}$; $(0,-6)$

Observe that the slope is $\frac{2}{7}$ and the y-intercept is $(0,-6)$. Thus, we have $f(x)=\frac{2}{7}x-6$.

To graph the equation, we count off a slope of $\frac{2}{7}$, starting at $(0,-6)$, and draw the line.

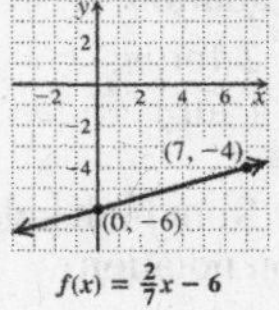

$f(x) = \frac{2}{7}x - 6$

47. $y-y_1=m(x-x_1)$ Point-slope equation

$y-6=\frac{3}{5}[x-(-4)]$ Substituting $\frac{3}{5}$ for m, -4 for x_1, and 6 for y_1

$y-6=\frac{3}{5}(x+4)$

$y-6=\frac{3}{5}x+\frac{12}{5}$

$y=\frac{3}{5}x+\frac{42}{5}$

$f(x)=\frac{3}{5}x+\frac{42}{5}$ Using function notation

To graph the equation, we count off a slope of $\frac{3}{5}$, starting at (–4, 6), and draw the line.

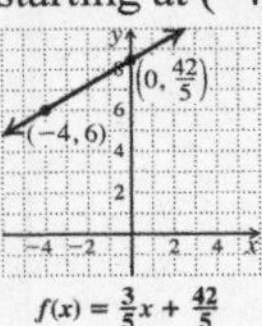

$f(x) = \frac{3}{5}x + \frac{42}{5}$

49. First solve the equation for y and determine the slope of the given line.

$x-2y=3$ Given line

$-2y=-x+3$

$y=\frac{1}{2}x-\frac{3}{2}$

The slope of the given line is $\frac{1}{2}$.

The slope of every line parallel to the given line must also be $\frac{1}{2}$. We find the equation of the line with slope $\frac{1}{2}$ and containing the point (2, 5).

$y-y_1=m(x-x_1)$ Point-slope equation

$y-5=\frac{1}{2}(x-2)$ Substituting

$y-5=\frac{1}{2}x-1$

$y=\frac{1}{2}x+4$

51. First solve the equation for y and determine the slope of the given line.

$x+y=7$ Given line

$y=-x+7$

The slope of the given line is –1.
The slope of every line parallel to the given line must also be –1. We find the equation of the line with slope –1 and containing the point (–3, 2).

$y-y_1=m(x-x_1)$ Point-slope equation

$y-2=-1(x-(-3))$ Substituting

$y-2=-1(x+3)$

$y-2=-x-3$

$y=-x-1$

53. First solve the equation for y and determine the slope of the given line.

$2x+3y=-7$ Given line

$3y=-2x-7$

$y=-\frac{2}{3}x-\frac{7}{3}$

The slope of the given line is $-\frac{2}{3}$.
The slope of every line parallel to the given line must also be $-\frac{2}{3}$. We find the equation of the line with slope $-\frac{2}{3}$ and containing the point $(-2,\ -3)$.

$$\begin{aligned} y - y_1 &= m(x - x_1) && \text{Point-slope equation} \\ y-(-3) &= -\tfrac{2}{3}[x-(-2)] && \text{Substituting} \\ y+3 &= -\tfrac{2}{3}(x+2) \\ y+3 &= -\tfrac{2}{3}x - \tfrac{4}{3} \\ y &= -\tfrac{2}{3}x - \tfrac{13}{3} \end{aligned}$$

55. $x = 2$ is a vertical line. A line parallel to it that passes through $(5,-4)$ is the vertical line 5 units to the right of the y-axis, or $x = 5$.

57. First solve the equation for y and determine the slope of the given line.

$$\begin{aligned} 2x - 3y &= 4 && \text{Given line} \\ -3y &= -2x + 4 \\ y &= \tfrac{2}{3}x - \tfrac{4}{3} \end{aligned}$$

The slope of the given line is $\frac{2}{3}$.
The slope of perpendicular line is given by the opposite of the reciprocal of $\frac{2}{3}$, $-\frac{3}{2}$. We find the equation of the line with slope $-\frac{3}{2}$ and containing the point (3, 1).

$$\begin{aligned} y - y_1 &= m(x - x_1) && \text{Point-slope equation} \\ y-1 &= -\tfrac{3}{2}(x-3) && \text{Substituting} \\ y-1 &= -\tfrac{3}{2}x + \tfrac{9}{2} \\ y &= -\tfrac{3}{2}x + \tfrac{11}{2} \end{aligned}$$

59. First solve the equation for y and determine the slope of the given line.

$$\begin{aligned} x + y &= 6 && \text{Given line} \\ y &= -x + 6 \end{aligned}$$

The slope of the given line is –1.
The slope of perpendicular line is given by the opposite of the reciprocal of –1, 1. We find the equation of the line with slope 1 and containing the point (–4, 2).

$$\begin{aligned} y - y_1 &= m(x - x_1) && \text{Point-slope equation} \\ y-2 &= 1(x-(-4)) && \text{Substituting} \\ y-2 &= x+4 \\ y &= x+6 \end{aligned}$$

61. First solve the equation for y and determine the slope of the given line.

$$\begin{aligned} 3x - y &= 2 && \text{Given line} \\ -y &= -3x + 2 \\ y &= 3x - 2 \end{aligned}$$

The slope of the given line is 3.
The slope of perpendicular line is given by the opposite of the reciprocal of 3, $-\frac{1}{3}$. We find the equation of the line with slope $-\frac{1}{3}$ and containing the point (1, –3).

$$\begin{aligned} y - y_1 &= m(x - x_1) && \text{Point-slope equation} \\ y-(-3) &= -\tfrac{1}{3}(x-1) && \text{Substituting} \\ y+3 &= -\tfrac{1}{3}x + \tfrac{1}{3} \\ y &= -\tfrac{1}{3}x - \tfrac{8}{3} \end{aligned}$$

63. First solve the equation for y and find the slope of the given line.

$$\begin{aligned} 3x - 5y &= 6 \\ -5y &= -3x + 6 \\ y &= \tfrac{3}{5}x - \tfrac{6}{5} \end{aligned}$$

The slope of the given line is $\frac{3}{5}$. The slope of a perpendicular line is given by the opposite of the reciprocal of $\frac{3}{5}$, $-\frac{5}{3}$.

We find the equation of the line with slope $-\frac{5}{3}$ and containing the point $(-4,-7)$.

$$\begin{aligned} y - y_1 &= m(x - x_1) && \text{Point-slope equation} \\ y-(-7) &= -\tfrac{5}{3}[x-(-4)] && \text{Substituting} \\ y+7 &= -\tfrac{5}{3}(x+4) \\ y+7 &= -\tfrac{5}{3}x - \tfrac{20}{3} \\ y &= -\tfrac{5}{3}x - \tfrac{41}{3} \end{aligned}$$

65. $y = 5$ is a horizontal line, so a line perpendicular to it must be vertical. The equation of the vertical line containing (–3, 7) is $x = -3$.

67. First find the slope of the line:

$$m = \frac{7-3}{3-2} = \frac{4}{1} = 4$$

Use the point-slope equation with $m = 4$ and $(2,\ 3) = (x_1,\ y_1)$. (We could let $(3,\ 7) = (x_1,\ y_1)$ instead to obtain an equivalent equation.)

$$\begin{aligned} y-3 &= 4(x-2) \\ y-3 &= 4x-8 \\ y &= 4x-5 \\ f(x) &= 4x-5 && \text{Using function notation} \end{aligned}$$

69. First find the slope of the line:

$$m = \frac{5-(-4)}{3.2-1.2} = \frac{5+4}{2} = \frac{9}{2} = 4.5$$

Use the point-slope equation with $m = 4.5$ and $(1.2,\ -4) = (x_1,\ y_1)$.

$$\begin{aligned} y-(-4) &= 4.5(x-1.2) \\ y+4 &= 4.5x-5.4 \\ y &= 4.5x-9.4 \\ f(x) &= 4.5x-9.4 && \text{Using function notation} \end{aligned}$$

71. First find the slope of the line:

$$m=\frac{-1-(-5)}{0-2}=\frac{-1+5}{0-2}=\frac{4}{-2}=-2$$

Use the slope-intercept equation with $m=-2$ and $b=-1$.

$$y=-2x-1$$
$$f(x)=-2x-1 \quad \text{Using function notation}$$

73. First find the slope of the line:

$$m=\frac{-5-(-10)}{-3-(-6)}=\frac{-5+10}{-3+6}=\frac{5}{3}$$

Use the point-slope equation with $m=\frac{5}{3}$ and $(-3,\ -5)=(x_1,\ y_1)$.

$$y-(-5)=\frac{5}{3}(x-(-3))$$
$$y+5=\frac{5}{3}(x+3)$$
$$y+5=\frac{5}{3}x+5$$
$$y=\frac{5}{3}x$$
$$f(x)=\frac{5}{3}x \quad \text{Using function notation}$$

75. $y=-6$

77. $x=-10$

79. a. First find the slope of the line passing through the points (2, 27) and (10, 20).

$$m=\frac{20-27}{10-2}=\frac{-7}{8}=-0.875$$

Now write an equation of the line. We use (10, 20) in the point-slope equation and then write an equivalent slope-intercept equation.

$$y-y_1=m(x-x_1)$$
$$y-20=-0.875(x-10)$$
$$y-20=-0.875x+8.75$$
$$y=-0.875x+28.75$$

b. Since 2008 is 8 yr after 2000, we substitute 8 for x to calculate the percentage of women in 2008.

$$y=-0.875(8)+28.75=21.75\approx 22$$

In 2008, there were about 22% of U.S. women ages 60 and older with high cholesterol.

c. Since 2015 is 15 yr after 2000, we substitute 15 for x to calculate the percentage in 2015.

$$y=-0.875(15)+28.75=15.625\approx 16$$

In 2015, there will be about 16% of U.S. women ages 60 and older with high cholesterol.

81. a. First find the slope of the line passing through (13, 42) and (23, 50). In each case, we let the first coordinate represent the number of years after 2007.

$$m=\frac{50-42}{23-13}=\frac{8}{10}=0.8$$

Now write an equation of the line. We use (13, 42) in the point-slope equation and then write an equivalent slope-intercept equation.

$$y-y_1=m(x-x_1)$$
$$y-42=0.8(x-13)$$
$$y-42=0.8x-10.4$$
$$y=0.8x+31.6$$

b. Since 2040 is 33 yr after 2007, we substitute 33 for x to calculate the SCC in 2007.

$$y=0.8(33)+31.6=\$58 \text{ per ton}$$

Since 2050 is 43 yr after 2007, we substitute 43 for x to calculate the SCC in 2015.

$$y=0.8(43)+31.6=\$66 \text{ per ton}$$

83. a. First find the slope of the line through (2, 10.05) and (6, 5.47). In each case, we let the first coordinate represent the number of years after 2008 and the second billions of dollars.

$$m=\frac{5.47-10.05}{6-2}=\frac{-4.58}{4}=-1.145$$

We use (2, 10.05) in the point-slope equation and then write an equivalent slope-intercept equation.

$$y-y_1=m(x-x_1)$$
$$y-10.05=-1.145(x-2)$$
$$y-10.05=-1.145x+2.29$$
$$y=-1.145x+12.34$$

b. Since 2016 is 8 yr after 2008, we substitute 8 for x to find the revenue in 2008.

$$y=-1.145(8)+12.34=\$3.18 \text{ billion}$$

c. To find zero revenue, substitute 0 for y and solve for x.

$$0=-1.145x+12.34$$
$$1.145x=12.34$$
$$x=\frac{12.34}{1.145}\approx 10.8$$

If the trend continues, about 11 years after 2008, or in 2019, there will be no sales of physical video and computer games.

85. *Writing Exercise.*

87.

$$\frac{3}{8}x=-24$$
$$\frac{8}{3}\cdot\frac{3}{8}x=\frac{8}{3}(-24)$$
$$x=-64$$

The solution is –64.

89.

$$\frac{t}{3}=6$$
$$3\cdot\frac{t}{3}=3\cdot 6$$
$$t=18$$

The solution is 18.

91.
$$2(x-7) > 5x+3$$
$$2x-14 > 5x+3$$
$$2x-5x-14 > 3$$
$$-3x-14 > 3$$
$$-3x > 3+14$$
$$-3x > 17$$
$$\frac{-3x}{-3} < \frac{17}{-3}$$
$$x < -\frac{17}{3}$$

The solution is $\left\{x \middle| x < -\frac{17}{3}\right\}$ or $\left(-\infty, -\frac{17}{3}\right)$.

93. *Writing Exercise.*

95. $y-3=0(x-52)$
Observe that the slope is 0. Then this is the equation of a horizontal line that passes through (52, 3). Thus, its graph is a horizontal line 3 units above the x-axis.

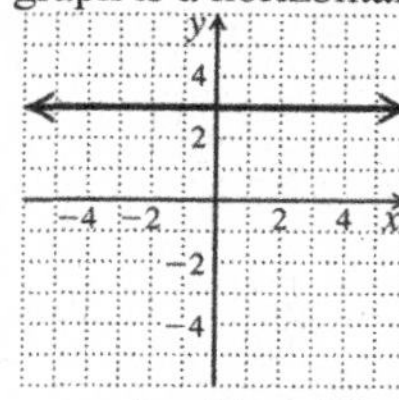

$y - 3 = 0(x - 52)$

97. First we find the slope of the line using any two points on the line. We will use $(3,-3)$ and $(4,-1)$.
$$m=\frac{-3-(-1)}{3-4}=\frac{-2}{-1}=2$$
Then we write an equation of the line in point-slope form using either of the points above.
$$y-(-3)=2(x-3)$$
Finally, we find an equivalent equation in slope-intercept form.
$$y-(-3)=2(x-3)$$
$$y+3=2x-6$$
$$y=2x-9$$

99. $x-3y=6$
$$-3y=-x+6$$
$$y=\frac{1}{3}x-2$$
The y-intercept is $(0,-2)$.

Then $m=\frac{-2-(-1)}{0-5}=\frac{-1}{-5}=\frac{1}{5}$.
$$y-(-1)=\frac{1}{5}(x-5)$$
$$y+1=\frac{1}{5}x-1$$
$$y=\frac{1}{5}x-2$$

101. First we find the slope of the given line:
$$4x-8y=12$$
$$-8y=-4x+12$$
$$y=\frac{1}{2}x-\frac{3}{2}$$
The slope is $\frac{1}{2}$.
Then we use the point-slope equation to find the equation of a line with slope $\frac{1}{2}$ containing the point (−2, 0):
$$y-y_1=m(x-x_1)$$
$$y-0=\frac{1}{2}(x-(-2))$$
$$y=\frac{1}{2}(x+2)$$
$$y=\frac{1}{2}x+1$$

103. Find the slope of $7y-kx=9$:
$$7y-kx=9$$
$$7y=kx+9$$
$$y=\frac{k}{7}x+\frac{9}{7}$$
The slope is $\frac{k}{7}$.

Find the slope of the line containing $(2,-1)$ and $(-4, 5)$.
$$m=\frac{5-(-1)}{-4-2}=\frac{6}{-6}=-1$$
If the lines are perpendicular, the product of their slopes must be −1:
$$\frac{k}{7}(-1)=-1$$
$$-\frac{k}{7}=-1$$
$$k=7 \text{ Multiplying by } -7$$

105. $\frac{x}{10}-\frac{y}{3}=1$ or $\frac{x}{10}+\frac{y}{-3}=1$

Using the form $\frac{x}{a}+\frac{y}{b}=1$
The x-intercept is (10, 0).
The y-intercept is $(0,-3)$.

107. $y=1.876712329x+11.28767123$

Chapter 3 Review

1. True
2. True
3. False, slope-intercept form is $y=mx+b$.
4. False, the only horizontal line that has an x-intercept is the line $y=0$, or the x-axis
5. True
6. True
7. True
8. False, to write the equation of the line, both the slope and a point on the line are needed.
9. True
10. True

11. We locate 35 to 44 on the horizontal axis and move up to the line. From there we move left to the vertical scale and read the number of volunteers, 13 (million people). Then, we locate 25 to 34 on the horizontal axis and move up to the line. From there we move left to the vertical scale and read the number of volunteers, 9 (million people).

The difference is 13 – 9, or 4 million people.

12. From the bar graph we see that 4 (million people) ages 16 to 19 volunteered. Then the percent who volunteered in environmental organizations is

$0.02(4) = 0.08$ million people, or 80,000 people.

13.-15. We plot the points $(5,-1)$, $(2, 3)$ and $(-4, 0)$.

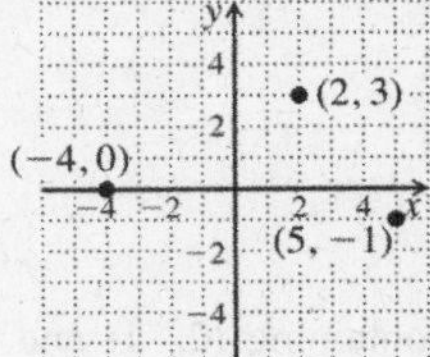

16. Since both coordinates are negative, the point $(-8,-7)$ is in quadrant III.

17. Since the first coordinate is positive and the second point is negative, the point $(15.3,-13.8)$ is in quadrant IV.

18. Since the first coordinate is negative and the second coordinate is positive, the point $\left(-\frac{1}{2}, \frac{1}{10}\right)$ is in quadrant II.

19. Point A is 5 units left and 1 unit down. The coordinates of A are (–5,–1).

20. Point B is 2 units left and 5 units up. The coordinates of B are (–2, 5).

21. Point C is 3 units right and 0 units up or down. The coordinates of C are (3, 0).

22. We plot the points (–65,–2), (–10, 6) and (25, 7).

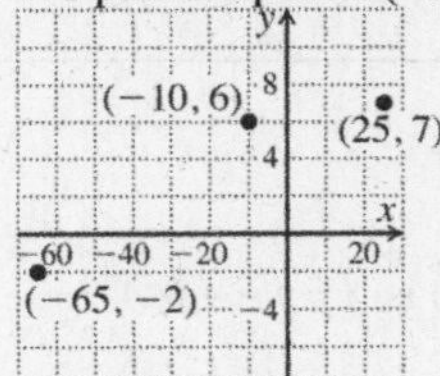

23. a.. We substitute 3 for x and 1 for y.

$$\begin{array}{c|c} y = 2x+7 & \\ \hline 1 & 2(3)+7 \\ & 6+7 \end{array}$$

$1 \stackrel{?}{=} 13$ FALSE

No, the pair (3,1) is not a solution.

b. We substitute –3 for x and 1 for y.

$$\begin{array}{c|c} y = 2x+7 & \\ \hline 1 & 2(-3)+7 \\ & -6+7 \end{array}$$

$1 \stackrel{?}{=} 1$ TRUE

Yes, the pair (–3, 1) is a solution.

24. To show that a pair is a solution, we substitute for x and y.

$$\begin{array}{c|c} 2x-y=3 & \\ \hline 2\cdot 0-(-3) & 3 \\ 0+3 & 3 \end{array} \qquad \begin{array}{c|c} 2x-y=3 & \\ \hline 2\cdot 2-1 & 3 \\ 4-1 & 3 \end{array}$$

$3 \stackrel{?}{=} 3$ TRUE $\qquad 3 \stackrel{?}{=} 3$ TRUE

In each case, the result is true.

Thus $(0,-3)$ and $(2,1)$ are solutions.

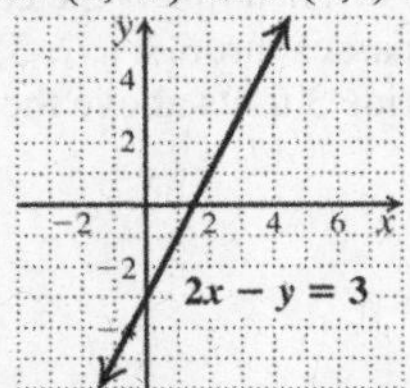

The line appears to pass through $(1,-1)$ also. We check.

$$\begin{array}{c|c} 2x-y=3 & \\ \hline 2(1)-(-1) & 3 \\ 2+1 & \end{array}$$

$3 \stackrel{?}{=} 3$ TRUE

There are other correct answers, including $(-3,-9)$, $(-2,-7)$, $(-1,-5)$ and $(3, 3)$.

25. $y = x - 5$

The y-intercept is (0, –5). We find two other points.

When $x = 5$, $y = 5 - 5 = 0$.

When $x = 3$, $y = 3 - 5 = -2$.

x	y
0	–5
5	0
3	–2

We plot these points, draw the line and label the graph $y = x - 5$.

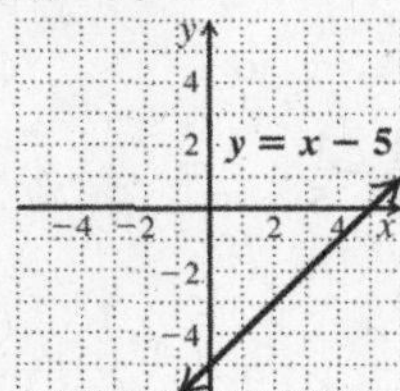

26. $y = -\frac{1}{4}x$

The y-intercept is (0, 0). We find two other points.

When $x = 4$, $y = -\frac{1}{4}\cdot 4 = -1$.

When $x = 8$, $y = -\frac{1}{4}\cdot 8 = -2$.

x	y
0	0
4	−1
8	−2

We plot these points, draw the line and label the graph $y=-\frac{1}{4}x$.

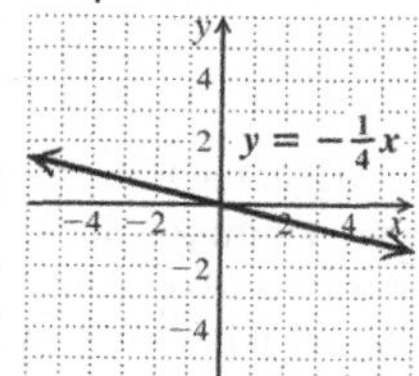

27. $y=-x+4$

The y-intercept is (0, 4). We find two other points.
When $x=4$, $y=-4+4=0$.
When $x=2$, $y=-2+4=2$.

x	y
0	4
4	0
2	2

We plot these points, draw the line and label the graph $y=-x+4$.

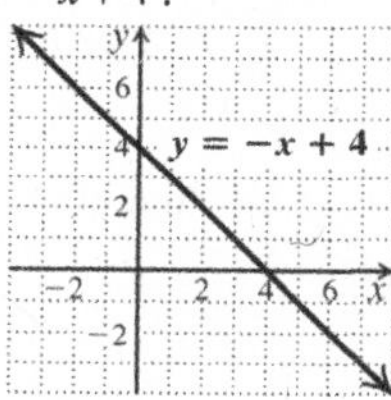

28. $4x+y=3$

The y-intercept is (0, 3). We find two other points.
When $x=1$, $4\cdot1+y=3$, $y=-1$.
When $x=-1$, $4(-1)+y=3$, $y=7$. .

x	y
0	3
1	−1
−1	7

We plot these points, draw the line and label the graph $4x+y=3$.

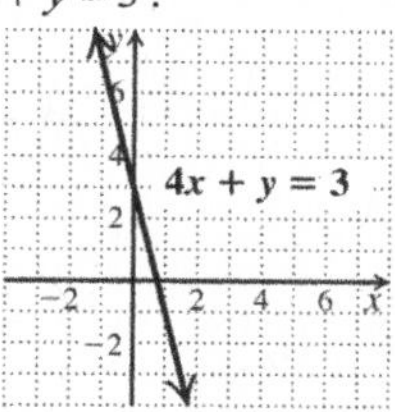

29. $4x+5=3$

$x=-\frac{1}{2}$

Any order pair $(-\frac{1}{2},y)$ is a solution. The variable x must be $-\frac{1}{2}$, but the y variable can be any number. A few are listed below.

When $y=2$, $x=-\frac{1}{2}$.

When $y=-2$, $x=-\frac{1}{2}$.

x	y
$-\frac{1}{2}$	0
$-\frac{1}{2}$	2
$-\frac{1}{2}$	−2

We plot these points, draw the line and label the graph $4x+5=3$.

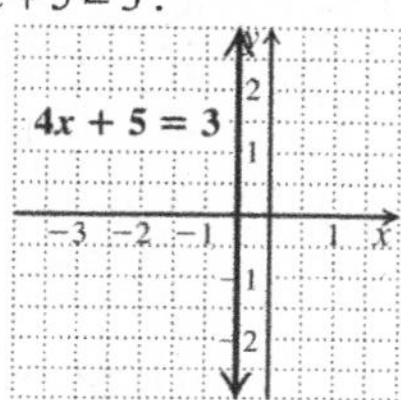

30. $5x-2y=10$

The y-intercept is $(0,-5)$. We find two other points.
When $x=2$, $5(2)-2y=10$, $y=0$.
When $x=-2$, $5(-2)-2y=10$, $y=-10$.

x	y
0	−5
2	0
−2	−10

We plot these points, draw the line and label the graph $5x-2y=10$.

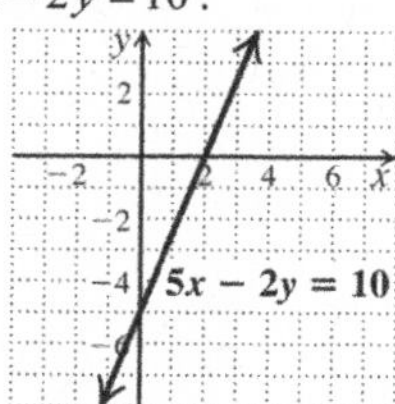

31. $y=6$

Any ordered pair (x, 6) is a solution. The variable y must be 6, but the x variable can be any number. A few are listed below. Plot these points and draw the graph.

x	y
0	6
1	6
2	6

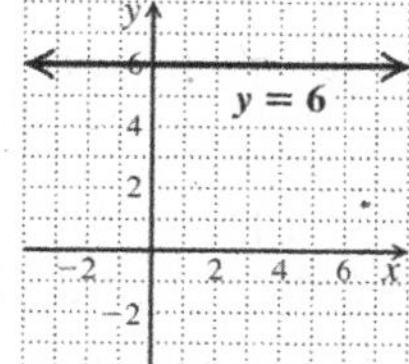

32. $y = \frac{2}{3}x - 5$

First we plot the y-intercept $(0,-5)$. We can start at the y-intercept and use the slope, $\frac{2}{3}$, to find another point. We move up 2 units and right 3 units to get a new point $(3,-3)$. Thinking of the slope as $\frac{2}{3}$, we move up 2 units and right 3 units to get another point $(6,-1)$.

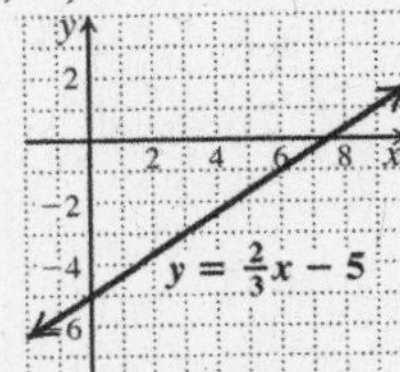

33. $2x + y = 4$

The y-intercept is (0, 4). We find two other points.
When $x = 1$, $2(1) + y = 4$, $y = 2$.
When $x = -2$, $2(-2) + y = 4$, $y = 8$.

x	y
0	4
1	2
−2	8

Plot these points, draw the line, and label the graph $2x + y = 4$.

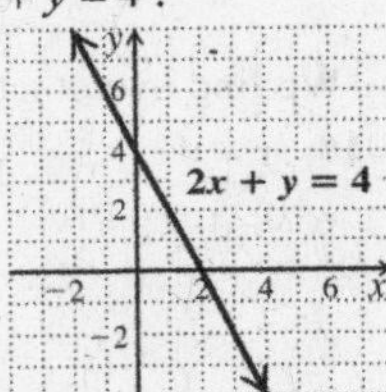

34. $y + 2 = -\frac{1}{2}(x - 3)$

We plot the (3,–2) move up 1 unit and left 2 units to the point $(1,-1)$ and draw the line.

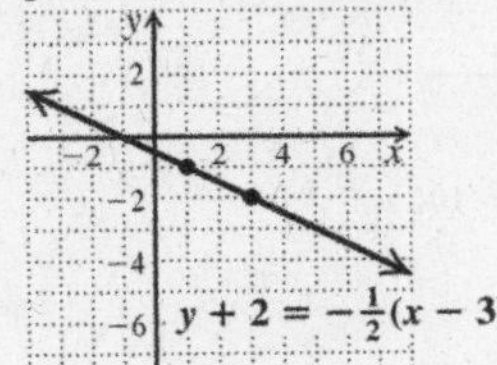

35. We graph $g = 1.75t + 5$ by selecting values of t and calculating the values for g.
If $t = 0$, $g = 1.75(0) + 5 = 5$.
If $t = 8$, $g = 1.75(8) + 5 = 19$.
If $t = 12$, $g = 1.75(12) + 5 = 26$.

t	g
0	5
8	19
12	26

We plot these points, draw the line and label the graph.

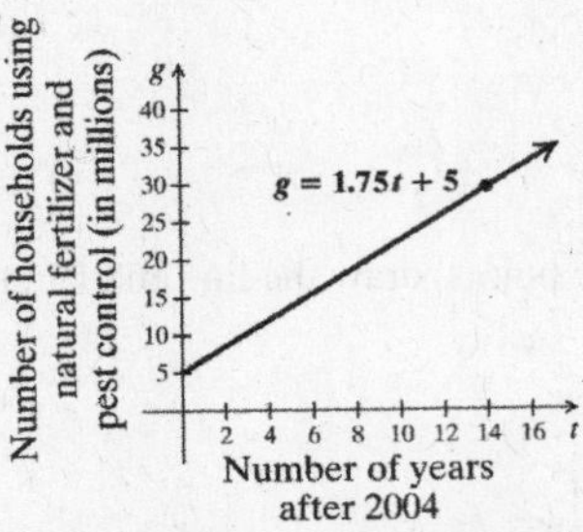

Since $2018 - 2004 = 14$. Locate the point on the line above 14 and find the corresponding value on the vertical axis. The value is 30, so we estimate about 30 million households use natural fertilizer pest control.

36. From 4:00pm to 4:45pm = 45 minutes

a. Average speed, in miles per minute

$$\frac{23-17 \text{ miles}}{45 \text{ min}} = \frac{6}{45} = \frac{2}{15} \text{ mi/min}$$

b. Rate in minutes per mile

$$\frac{45 \text{ min}}{6 \text{ miles}} = 7.5 \text{ min/mile}$$

37. The points (100 mi, 5 gal) and (200 mi, 10 gal) are on the graph. This tells $200 - 100 = 100$ mi and $10 - 5 = 5$ gal. The rate is

$$\frac{5 \text{ gal}}{100 \text{ mi}} = \frac{1}{20} \text{ gal/mi}$$

38. We can use any two points on the line, such as $(0,-2)$ and $(2,-2)$.

$$m = \frac{\text{change in } y}{\text{change in } x} = \frac{-2-(2)}{2-0} = \frac{0}{2} = 0$$

39. We can use any two points on the line, such as $(-1,-2)$ and (2, 5).

$$m = \frac{\text{change in } y}{\text{change in } x} = \frac{5-(-2)}{2-(-1)} = \frac{7}{3}$$

40. We can use any two points on the line, such as (–5, 1) and $(2,-2)$.

$$m = \frac{\text{change in } y}{\text{change in } x} = \frac{-2-1}{2-(-5)} = \frac{-3}{7}$$

41. (–2, 5) and (3,–1)

$$m = \frac{-1-5}{3-(-2)} = \frac{-6}{5}$$

42. (6, 5) and (–2, 5)

$$m = \frac{5-5}{-2-6} = \frac{0}{-8} = 0$$

43. (–3, 0) and (–3, 5)

$$m = \frac{5-0}{-3-(-3)} = \frac{5}{0}, \text{ undefined}$$

44. (–8.3, 4.6) and (–9.9, 1.4)

$$m = \frac{1.4-4.6}{-9.9-(-8.3)} = \frac{3.2}{1.6} = 2$$

45. $5x - 8y = 80$

Find the x-intercept:

$5x = 80$ Ignoring the y-term

$x = 16$

The x-intercept is (16, 0).

Find the y-intercept:

$-8y = 80$ Ignoring the x-term

$y = -10$

The y-intercept is $(0, -10)$.

46. Rewrite the equation in slope-intercept form.

$$3x + 5y = 45$$
$$5y = -3x + 45$$
$$y = -\frac{3}{5}x + 9$$

The slope is $-\frac{3}{5}$ and the y-intercept is (0, 9).

47. The grade is expressed as a percent

$$m = \frac{1}{12} \approx 0.08\overline{3} \approx 8.\overline{3}\%$$

48. First solve for y and determine the slope of each line.

$$y + 5 = -x$$
$$y = -x - 5$$

The slope of $y + 5 = -x$ is -1.

$$x - y = 2$$
$$y = x - 2$$

The slope of $x - y = 2$ is 1.

The product of their slopes is $(-1)(1)$, or -1; the lines are perpendicular.

49. First solve for y and determine the slope of each line.

$$3x - 5 = 7y$$
$$y = \frac{3}{7}x - \frac{5}{7}$$

The slope of $3x - 5 = 7y$ is $\frac{3}{7}$.

$$7y - 3x = 7$$
$$y = \frac{3}{7}x + 1$$

The slope of $7y - 3x = 7$ is $\frac{3}{7}$.

The slopes are the same, so the lines are parallel.

50. Use the slope-intercept equation, substituting $\frac{3}{8}$ for m and 7 for b:

$$y = mx + b$$
$$y = \frac{3}{8}x + 7$$

51. $y - y_1 = m(x - x_1)$

We substitute $-\frac{1}{3}$ for m, –2 for x_1, and 9 for y_1.

$$y - 9 = -\frac{1}{3}(x - (-2))$$

52. $y - y_1 = m(x - x_1)$

We substitute 5 for m, 3 for x_1, and –10 for y_1.

$$y - (-10) = 5(x - 3)$$
$$y + 10 = 5x - 15$$
$$y = 5x - 25$$

53. (–2, 5) and (3, 10)

First we find the slope.

$$m = \frac{10-5}{3-(-2)} = \frac{10-5}{3+2} = \frac{5}{5} = 1$$

Then we write an equation of the line in point-slope form using either of the points above.

$$y - 10 = 1(x - 3)$$

Finally, we find an equivalent equation in slope-intercept form.

$$y - 10 = 1(x - 3)$$
$$y - 10 = x - 3$$
$$y = x + 7$$

54. a. First find the slope of the line passing through the points (4, 27.7) and (14, 32.1). Let the x represent the number of years after 2000. Then y is the percentage of the U.S. population with a bachelor's degree or higher.

$$m = \frac{32.1-27.7}{14-4} = \frac{4.4}{10} = 0.44$$

We use the point (4, 27.7), in point-slope form to find the line.

$$y - 27.7 = 0.44(x - 4)$$
$$y = 0.44x + 25.94$$

b. In 2012, $x = 12$: $y = 0.44(12) + 25.94 \approx 31.2\%$

c. In 2020, $x = 20$: $y = 0.44(20) + 25.94 \approx 34.7\%$

55. *Writing Exercise.*

56. *Writing Exercise.*

57. $y = mx + 3$, we substitute (–2, 5) and solve.

$$5 = m(-2) + 3$$
$$5 = -2m + 3$$
$$2 = -2m$$
$$-1 = m$$

58. Plot the three given points and observe that the coordinates of the fourth vertex is $(-2, -3)$. The length of the rectangle is $7 - (-2) = 9$ units, and the width is $2 - (-3) = 5$ units.

$A = lw = 9(5) = 45$ sq units

$P = 2l + 2w = 2(9) + 2(5) = 28$ units

59. $y = 4 - |x|$

x	y
0	4
1	3
−1	3

Answers may vary.

Chapter 3 Test

1. We locate 2007 on the horizontal axis and move up to the line. From there we move left to the vertical scale and read the income, \$53,000. Then, we locate 2010 on the horizontal axis and move up to the line. From there we move left to the vertical scale and read the income, \$49,000.
The difference is \$53,000 − \$49,000 = \$4000.

2. We locate the highest point on the graph. From there we move down to the horizontal scale and read the year. We see that in the years 2000 and 2007, the median household income was highest.

3. The point having coordinates $(-2,-10)$ is located in quadrant III.

4. The point having coordinates (−1.6, 2.3) is located in quadrant II.

5. Point *A* has coordinates (3, 4).

6. Point *B* has coordinates $(0,-4)$.

7. Point *C* has coordinates (−5, 2).

8. $y = 2x - 1$
Slope: 2; *y*-intercept: $(0,-1)$. First we plot the *y*-intercept $(0,-1)$. We can start at the *y*-intercept and thinking of the slope as $\frac{2}{1}$ we find another point. We move up 2 units and right 1 unit to get a new point (1, 1). Thinking of the slope as $\frac{4}{2}$ we can start at $(0,-1)$ and move up 4 units and right 2 units to get another point (2, 3). To finish, we draw and label the line.

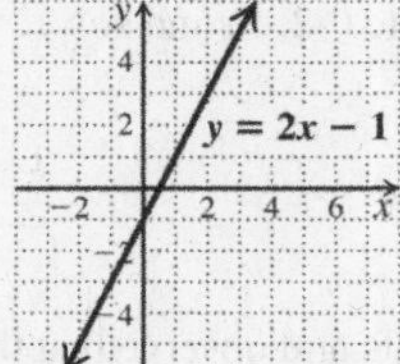

9. $2x - 4y = -8$
We rewrite the equation in slope-intercept form.

$$2x - 4y = -8$$
$$-4y = -2x - 8$$
$$4y = 2x + 8$$
$$y = \frac{1}{2}x + 2$$

Slope: $\frac{1}{2}$; *y*-intercept: (0, 2). First we plot the *y*-intercept (0, 2). We can start at the *y*-intercept and use the slope $\frac{1}{2}$ to find another point. We move up 1 unit and right 2 units to get a new point (2, 3). Thinking of the slope as $\frac{-1}{-2}$ we can start at (0, 2) and move down 1 unit and left 2 units to get another point (−2, 1). To finish, we draw and label the line.

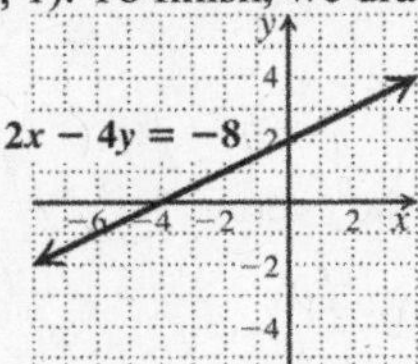

10. $y + 1 = 6$
$y = 5$ Solving for *y*.
This is a horizontal line with *y*-intercept (0, 5).

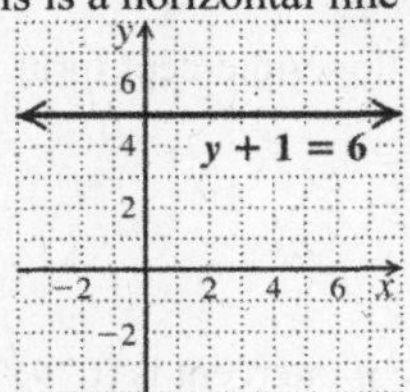

11. $y = \frac{3}{4}x$
We rewrite this equation in slope-intercept form.

$$y = \frac{3}{4}x + 0$$

Slope: $\frac{3}{4}$; *y*-intercept: (0, 0)
First we plot the *y*-intercept (0, 0). We can start at the *y*-intercept and using the slope as $\frac{3}{4}$ we find another point. We move up 3 units and right 4 units to get a new point (4, 3). Thinking of the slope as $\frac{-3}{-4}$ we can start at (0, 0) and move down 3 units and left 4 units to get another point (−4,−3). To finish, we draw and label the line.

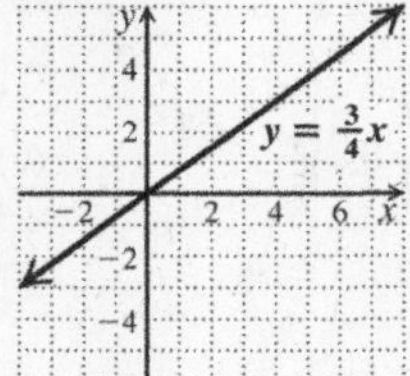

12. $2x - y = 3$
We rewrite this equation in slope-intercept form.

$$2x - y = 3$$
$$-y = -2x + 3$$
$$y = 2x - 3$$

Slope: 2; *y*-intercept: $(0,-3)$
First we plot the *y*-intercept $(0,-3)$. We can start at the *y*-intercept and thinking of the slope as $\frac{4}{2}$ we move up 4 units and right 2 units to get a new point

(2, 1). To finish, we draw and label the line.

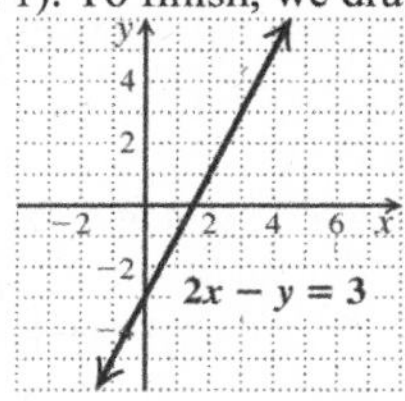

13. $x = -1$ This is a vertical line with x-intercept (–1, 0).

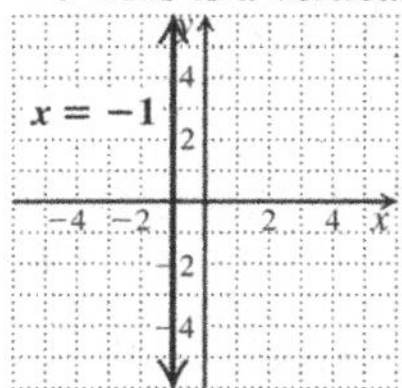

14. $y+4=-\frac{1}{2}(x-3)$

Rewriting this equation in point-slope form, we have:

$$y-(-4)=-\frac{1}{2}(x-3)$$

We have slope of $-\frac{1}{2}$ and a point having coordinates $(3,-4)$. Thinking of the slope as $\frac{-1}{2}$, we start at $(3,-4)$ and move down 1 unit and right 2 units to get a new point $(5,-5)$. Thinking of the slope as $\frac{1}{-2}$, we move up 1 unit and left 2 units to get the point $(1,-3)$. To finish, we draw and label the line.

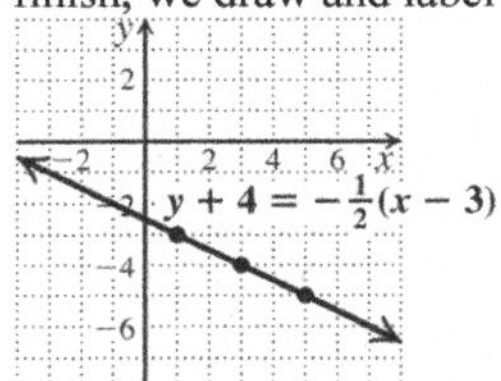

15. $(3,-2)$ and (4, 3)

$$m=\frac{3-(-2)}{4-3}=\frac{3+2}{4-3}=\frac{5}{1}=5$$

16. (–5, 6) and $(-1,-3)$

$$m=\frac{-3-6}{-1-(-5)}=\frac{-3-6}{-1+5}=\frac{-9}{4}$$

17. (4, 7) and $(4,-8)$

$$m=\frac{-8-7}{4-4}=\frac{-15}{0}$$

The slope is undefined.

18.
$$\begin{aligned}\text{rate}&=\frac{\text{change in distance}}{\text{change in time}}\\&=\frac{6\text{ km}-3\text{ km}}{2{:}24\text{ P.M.}-2{:}15\text{ P.M.}}\\&=\frac{3\text{ km}}{9\text{ min}}\\&=\frac{1}{3}\text{ km/min}\end{aligned}$$

19. $\text{grade}=\frac{\text{rise}}{\text{run}}=\frac{63\text{ ft}}{200\text{ ft}}=0.315=31.5\%$

20. $5x-y=30$

To find the x-intercept, we let $y=0$ and solve for x.

$$\begin{aligned}5x-0\cdot0&=30\\5x-0&=30\\5x&=30\\x&=6\end{aligned}$$

The x-intercept is (6, 0).

To find the y-intercept, we let $x=0$ and solve for y.

$$\begin{aligned}5\cdot0-y&=30\\-y&=30\\y&=-30\end{aligned}$$

The y-intercept is $(0,-30)$.

21. $y-8x=10$

Rewrite the equation in slope-intercept form.

$$\begin{aligned}y-8x&=10\\y&=8x+10\end{aligned}$$

The line has slope 8 and y-intercept (0, 10).

22. Slope: $-\frac{1}{3}$; y-intercept: $(0,-11)$

The slope-intercept equation is $y=-\frac{1}{3}x-11$.

23. Write both equations in slope-intercept form.

$$\begin{aligned}4y+2&=3x & -3x+4y&=-12\\y&=\frac{3}{4}x-\frac{1}{2} & y&=\frac{3}{4}x-3\\m&=\frac{3}{4} & m&=\frac{3}{4}\end{aligned}$$

The slopes are the same, so the lines are parallel.

24. Write both equations in slope-intercept form.

$$\begin{aligned}y&=-2x+5 & 2y-x&=6\\m&=-2 & y&=\frac{1}{2}x+3\\ & & m&=\frac{1}{2}\end{aligned}$$

The product of their slopes is $(-2)\left(\frac{1}{2}\right)$, or –1; the lines are perpendicular.

25. $2x-5y=8$

$$y=\frac{2}{5}x-\frac{8}{5}$$

The slope is $\frac{2}{5}$.

The slope of a perpendicular line is given by the opposite of the reciprocal of $\frac{2}{5}$, $-\frac{5}{2}$.

$$\begin{aligned}y-2&=-\frac{5}{2}[x-(-3)]\\y&=-\frac{5}{2}x-\frac{11}{2}\end{aligned}$$

26. a. Plot (20, 150) and (60, 120)
To finish we draw and label the line.

$$m=\frac{150-120}{20-60}=\frac{30}{-40}=-\frac{3}{4}$$

Using point-slope form:

$$r-120=\frac{-3}{4}(a-60)$$
$$r=-\frac{3}{4}a+165$$

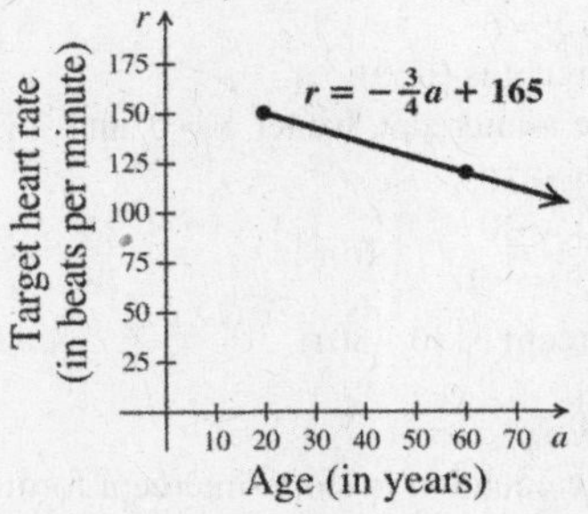

b. Using equation and let $a=36$

$$r=-\frac{3}{4}(36)+165$$
$$=-27+165$$
$$=138$$

The target heart rate for a 36-year-old is 138 beats per minute.

27. We first find the slope of the line having equation $2x-5y=6$ by writing it in slope-intercept form.

$$2x-5y=6$$
$$-5y=-2x+6$$
$$y=\frac{2}{5}x-\frac{6}{5}$$

This line has slope $\frac{2}{5}$ and any line parallel to this line would also have slope $\frac{2}{5}$.
Second, we must find the y-intercept of the line $3x+y=9$ by writing it in slope-intercept form.

$$3x+y=9$$
$$y=-3x+9$$

The y-intercept of this line is (0, 9).
We use the slope-intercept form of the line to determine the equation of a line having slope $\frac{2}{5}$ and y-intercept (0, 9).

$$y=\frac{2}{5}x+9$$

28. The height of the square is $4-(-1)=4+1=5$ and the width is $2-(-3)=2+3=5$. Therefore, the area of the square is $5\times5=25$ square units and the perimeter is $4\times5=20$ units.

29. (–2, 14) and (17,–5)
First determine the slope of the line containing these two points. $m=\frac{-5-14}{17-(-2)}=\frac{-19}{19}=-1$

Using the coordinates of either point and the point-slope formula, we can determine the equation of the line containing the two points. We choose the point (–2, 14).

$$y-14=-1(x-(-2))$$
$$y-14=-1(x+2)$$
$$y-14=-x-2$$
$$y=-x+12$$

Any points whose coordinates satisfy this equation lie on the same line. Answers will vary, but (0, 12), (–3, 15), and (5, 7) all satisfy this equation and are therefore points on the same line.

Chapter 4

Polynomials

Exercise Set 4.1

1. By the rule for raising a product to a power, choice (e) is correct.

3. By the power rule, choice (b) is correct.

5. By the definition of 0 as an exponent, choice (g) is correct.

7. By the rule for raising a quotient to a power, choice (c) is correct.

9. The base is $2x$. The exponent is 5.

11. The base is x. The exponent is 3.

13. The base is $\frac{4}{y}$. The exponent is 7.

15. $d^3 \cdot d^{10} = d^{3+10} = d^{13}$

17. $a^6 \cdot a = a^6 \cdot a^1 = a^{6+1} = a^7$

19. $6^5 \cdot 6^{10} = 6^{5+10} = 6^{15}$

21. $(3y)^4(3y)^8 = (3y)^{4+8} = (3y)^{12}$

23. $(5p)^0(5p)^1 = 1 \cdot (5p) = 5p$

25. $(x+3)^5(x+3)^8 = (x+3)^{5+8} = (x+3)^{13}$

27. $(a^2b^7)(a^3b^2) = a^2b^7a^3b^2$ Using an associative law
$= a^2a^3b^7b^2$ Using a commutative law
$= a^5b^9$ Adding exponents

29. $r^3 \cdot r^7 \cdot r^0 = r^{3+7+0} = r^{10}$

31. $(mn^5)(m^3n^4) = mn^5m^3n^4$
$= m^1m^3n^5n^4$
$= m^{1+3}n^{5+4}$
$= m^4n^9$

33. $\frac{7^5}{7^2} = 7^{5-2} = 7^3$, or 343 Subtracting exponents

35. $\frac{t^8}{t} = t^{8-1} = t^7$ Subtracting exponents

37. $\frac{(5a)^7}{(5a)^6} = (5a)^{7-6} = (5a)^1 = 5a$

39. $\frac{(x+y)^8}{(x+y)^8}$
Observe that we have an expression divided by itself. Thus, the result is 1.
We could also do this exercise as follows:
$\frac{(x+y)^8}{(x+y)^8} = (x+y)^{8-8} = (x+y)^0 = 1$

41. $\frac{(r+s)^{12}}{(r+s)^4} = (r+s)^{12-4} = (r+s)^8$

43. $\frac{12d^9}{15d^2} = \frac{12}{15}d^{9-2} = \frac{4}{5}d^7$

45. $\frac{8a^9b^7}{2a^2b} = \frac{8}{2} \cdot \frac{a^9}{a^2} \cdot \frac{b^7}{b^1} = 4a^{9-2}b^{7-1} = 4a^7b^6$

47. $\frac{x^{12}y^9}{x^0y^2} = x^{12-0}y^{9-2} = x^{12}y^7$

49. When $t = 15$, $t^0 = 15^0 = 1$. (Any nonzero number raised to the 0 power is 1.)

51. When $x = -22$, $5x^0 = 5(-22)^0 = 5 \cdot 1 = 5$.

53. $7^0 + 4^0 = 1 + 1 = 2$

55. $(-3)^1 - (-3)^0 = -3 - 1 = -4$

57. $(x^3)^{11} = x^{3 \cdot 11} = x^{33}$ Multiplying exponents

59. $(5^8)^4 = 5^{8 \cdot 4} = 5^{32}$ Multiplying exponents

61. $(t^{20})^4 = t^{20 \cdot 4} = t^{80}$

63. $(10x)^2 = 10^2x^2 = 100x^2$

65. $(-2a)^3 = (-2)^3a^3 = -8a^3$

67. $(-5n^7)^2 = (-5)^2(n^7)^2 = 25n^{7 \cdot 2} = 25n^{14}$

69. $(a^2b)^7 = (a^2)^7(b^7) = a^{14}b^7$

71. $(r^5t)^3(r^2t^8) = (r^5)^3(t)^3r^2t^8 = r^{15}t^3r^2t^8 = r^{17}t^{11}$

73. $(2x^5)^3(3x^4) = 2^3(x^5)^3(3x^4) = 8x^{15} \cdot 3x^4 = 24x^{19}$

75. $\left(\frac{x}{5}\right)^3 = \frac{x^3}{5^3} = \frac{x^3}{125}$

77. $\left(\frac{7}{6n}\right)^2 = \frac{7^2}{(6n)^2} = \frac{49}{6^2n^2} = \frac{49}{36n^2}$

79. $\left(\frac{a^3}{b^8}\right)^6 = \frac{(a^3)^6}{(b^8)^6} = \frac{a^{18}}{b^{48}}$

81. $\left(\frac{x^2y}{z^3}\right)^4 = \frac{(x^2y)^4}{(z^3)^4} = \frac{(x^2)^4(y^4)}{z^{12}} = \frac{x^8y^4}{z^{12}}$

83. $\left(\frac{a^3}{-2b^5}\right)^4 = \frac{(a^3)^4}{(-2b^5)^4} = \frac{a^{12}}{(-2)^4(b^5)^4} = \frac{a^{12}}{16b^{20}}$

85. $\left(\frac{5x^7y}{-2z^4}\right)^3 = \frac{(5x^7y)^3}{(-2z^4)^3} = \frac{5^3(x^7)^3y^3}{-2^3(z^4)^3} = \frac{125x^{21}y^3}{-8z^{12}}$

87. $\left(\frac{4x^3y^5}{3z^7}\right)^0$

Observe that for $x \neq 0$, $y \neq 0$, and $z \neq 0$, we have a nonzero number raised to the 0 power. Thus, the result is 1.

89. *Writing Exercise.*

91. $-\frac{x}{7} = 3$

$-x = 21$

$x = -21$

The solution is -21.

93. $\frac{1}{2}x + \frac{1}{3} = \frac{1}{6}x$

$x\left(\frac{1}{2} - \frac{1}{6}\right) = -\frac{1}{3}$

$x\left(\frac{3}{6} - \frac{1}{6}\right) = -\frac{1}{3}$

$\frac{1}{3}x = -\frac{1}{3}$ Since $\frac{2}{6} = \frac{1}{3}$

$x = -1$

The solution is -1.

95. $8 - 2(n-7) > 9 - (3-n)$

$8 - 2n + 14 > 9 - 3 + n$

$-2n + 22 > n + 6$

$-3n > 6 - 22$

$-3n > -16$

$n < \frac{16}{3}$

The solution is $\left\{n \mid n < \frac{16}{3}\right\}$, or $\left(-\infty, \frac{16}{3}\right)$.

97. *Writing Exercise.*

99. *Writing Exercise.*

101. Choose any number except 0. For example, let $x = 1$.

$3x^2 = 3 \cdot 1^2 = 3 \cdot 1 = 3$, but

$(3x)^2 = (3 \cdot 1)^2 = 3^2 = 9$.

103. Choose any number except 0 or 1. For example, let $t = -1$.

Then $\frac{t^6}{t^2} = \frac{(-1)^6}{(-1)^2} = \frac{1}{1} = 1$, but $t^3 = (-1)^3 = -1$

105. $y^{4x} \cdot y^{2x} = y^{4x+2x} = y^{6x}$

107. $\frac{x^{5t}(x^t)^2}{(x^{3t})^2} = \frac{x^{5t}x^{2t}}{x^{6t}} = \frac{x^{7t}}{x^{6t}} = x^t$

109. $\frac{t^{26}}{t^x} = t^x$

$t^{26-x} = t^x$

$26 - x = x$ Equating exponents

$26 = 2x$

$13 = x$

The solution is 13.

111. Since the bases are the same, the expression with the larger exponent is larger. Thus, $4^2 < 4^3$.

113. $4^3 = 64$, $3^4 = 81$, so $4^3 < 3^4$.

115. $25^8 = (5^2)^8 = 5^{16}$

$125^5 = (5^3)^5 = 5^{15}$

$5^{16} > 5^{15}$, or $25^8 > 125^5$.

117. $2^{22} = 2^{10} \cdot 2^{10} \cdot 2^2 \approx 10^3 \cdot 10^3 \cdot 4 \approx 1000 \cdot 1000 \cdot 4$

$\approx 4{,}000{,}000$

Using a calculator, we find that $2^{22} = 4{,}194{,}304$. The difference between the exact value and the approximation is $4{,}194{,}304 - 4{,}000{,}000$, or 194,304.

119. $2^{31} = 2^{10} \cdot 2^{10} \cdot 2^{10} \cdot 2 \approx 10^3 \cdot 10^3 \cdot 10^3 \cdot 2$

$\approx 1000 \cdot 1000 \cdot 1000 \cdot 2 = 2{,}000{,}000{,}000$

Using a calculator, we find that $2^{31} = 2{,}147{,}483{,}648$. The difference between the exact value and the approximation is

$2{,}147{,}483{,}648 - 2{,}000{,}000{,}000 = 147{,}483{,}648$.

121. $1.5 \text{ MB} = 1 \cdot 5 \times 1000 \text{ KB}$

$= 1 \cdot 5 \times 1000 \times 1 \times 2^{10}$ bytes

$= 1{,}536{,}000$ bytes

$\approx 1{,}500{,}000$ bytes

Connecting the Concepts

1. $x^4x^{10} = x^{4+10} = x^{14}$

2. $x^{-4}x^{-10} = x^{-4-10} = x^{-14} = \frac{1}{x^{14}}$

3. $\frac{x^{-4}}{x^{10}} = x^{-4-10} = x^{-14} = \frac{1}{x^{14}}$

4. $\frac{x^4}{x^{-10}} = x^{4+10} = x^{14}$

5. $\left(x^{-4}\right)^{-10} = x^{-4(-10)} = x^{40}$

6. $\left(x^4\right)^{10} = x^{4\cdot 10} = x^{40}$

7. $\frac{1}{c^{-8}} = c^8$

8. $c^{-8} = \frac{1}{c^8}$

9. $\left(\frac{a^3}{b^4}\right)^5 = \frac{a^{3\cdot 5}}{b^{4\cdot 5}} = \frac{a^{15}}{b^{20}}$

10. $\left(\frac{a^3}{b^4}\right)^{-5} = \frac{a^{3(-5)}}{b^{4(-5)}} = \frac{a^{-15}}{b^{-20}} = \frac{b^{20}}{a^{15}}$

Exercise Set 4.2

1. True

3. True

5. $\left(\frac{x^3}{y^2}\right)^{-2} = \left(\frac{y^2}{x^3}\right)^2 = \frac{(y^2)^2}{(x^3)^2} = \frac{y^4}{x^6} \Rightarrow$ (c)

7. $\left(\frac{y^{-2}}{x^{-3}}\right)^{-3} = \frac{(y^{-2})^{-3}}{(x^{-3})^{-3}} = \frac{y^6}{x^9} \Rightarrow$ (a)

9. Since the number of centimeters in the length of an Olympic marathon is much greater than 10, use a positive power of 10.

11. Since the mass of a hydrogen atom represented in grams is much less than 1, use a negative power of 10.

13. Since the time between leap years, represented in seconds, is much greater than 10, use a positive power of 10.

15. $2^{-3} = \frac{1}{2^3} = \frac{1}{8}$

17. $(-2)^{-6} = \frac{1}{(-2)^6} = \frac{1}{64}$

19. $t^{-9} = \frac{1}{t^9}$

21. $8x^{-3} = 8\cdot\frac{1}{x^3} = \frac{8}{x^3}$

23. $\frac{1}{a^{-8}} = a^8$

25. $7^{-1} = \frac{1}{7^1} = \frac{1}{7}$

27. $3a^8b^{-6} = 3a^8\cdot\frac{1}{b^6} = \frac{3a^8}{b^6}$

29. $\left(\frac{x}{2}\right)^{-5} = \left(\frac{2}{x}\right)^5 = \frac{2^5}{x^5} = \frac{32}{x^5}$

31. $\frac{z^{-4}}{3x^5} = \frac{1}{3x^5z^4}$

33. $\frac{1}{9^2} = 9^{-2}$

35. $\frac{1}{y^3} = y^{-3}$

37. $\frac{1}{t} = \frac{1}{t^1} = t^{-1}$

39. $2^{-5}\cdot 2^8 = 2^{-5+8} = 2^3$, or 8

41. $x^{-3}\cdot x^{-9} = x^{-12} = \frac{1}{x^{12}}$

43. $t^{-3}\cdot t = t^{-3}\cdot t^1 = t^{-3+1} = t^{-2} = \frac{1}{t^2}$

45. $(5a^{-2}b^{-3})(2a^{-4}b) = 10a^{-2-4}b^{-3+1} = 10a^{-6}b^{-2} = \frac{10}{a^6b^2}$

47. $(n^{-5})^3 = n^{-5\cdot 3} = n^{-15} = \frac{1}{n^{15}}$

49. $(t^{-3})^{-6} = t^{-3(-6)} = t^{18}$

51. $(mn)^{-7} = \frac{1}{(mn)^7} = \frac{1}{m^7n^7}$

53. $(3x^{-4})^2 = 3^2(x^{-4\cdot 2}) = 9x^{-8} = \frac{9}{x^8}$

55. $(5r^{-4}t^3)^2 = 5^2(r^{-4})^2(t^3)^2$
$= 25r^{-8}t^6 = \frac{25t^6}{r^8}$

57. $\frac{t^{12}}{t^{-2}} = t^{12-(-2)} = t^{14}$

59. $\frac{y^{-7}}{y^{-3}} = y^{-7-(-3)} = y^{-4} = \frac{1}{y^4}$

61. $\frac{15y^{-7}}{3y^{-10}} = 5y^{-7-(-10)} = 5y^3$

63. $\frac{2x^6}{x} = 2\frac{x^6}{x^1} = 2x^{6-1} = 2x^5$

65. $-\frac{15a^{-7}}{10b^{-9}} = -\frac{3b^9}{2a^7}$

67. $\dfrac{t^{-7}}{t^{-7}}$

Note that we have an expression divided by itself. Thus, the result is 1. We could also find this result as follows:

$\dfrac{t^{-7}}{t^{-7}} = t^{-7-(-7)} = t^0 = 1$

69. $\dfrac{3t^4}{s^{-2}u^{-4}} = 3s^2t^4u^4$

71. $(x^4y^5)^{-3} = (x^4)^{-3}(y^5)^{-3}$
$= x^{-12}y^{-15} = \dfrac{1}{x^{12}y^{15}}$

73. $(3m^{-5}n^{-3})^{-2} = 3^{-2}m^{-5(-2)}n^{-3(-2)} = 3^{-2}m^{10}n^6$
$= \dfrac{m^{10}n^6}{9}$

75. $(a^{-5}b^7c^{-2})(a^{-3}b^{-2}c^6) = a^{-5+(-3)}b^{7+(-2)}c^{-2+6}$
$= a^{-8}b^5c^4 = \dfrac{b^5c^4}{a^8}$

77. $\left(\dfrac{a^4}{3}\right)^{-2} = \left(\dfrac{3}{a^4}\right)^2 = \dfrac{3^2}{(a^4)^2} = \dfrac{9}{a^8}$

79. $\left(\dfrac{m^{-1}}{n^{-4}}\right)^3 = \dfrac{(m^{-1})^3}{(n^{-4})^3} = \dfrac{m^{-3}}{n^{-12}} = \dfrac{n^{12}}{m^3}$

81. $\left(\dfrac{2a^2}{3b^4}\right)^{-3} = \left(\dfrac{3b^4}{2a^2}\right)^3 = \dfrac{(3b^4)^3}{(2a^2)^3} = \dfrac{3^3(b^4)^3}{2^3(a^2)^3} = \dfrac{27b^{12}}{8a^6}$

83. $\left(\dfrac{5x^{-2}}{3y^{-2}z}\right)^0$

Any nonzero expression raised to the 0 power is equal to 1. Thus, the answer is 1.

85. $\dfrac{-6a^3b^{-5}}{-3a^7b^{-8}} = \dfrac{-6}{-3}a^{3-7}b^{-5+8} = 2a^{-4}b^3 = \dfrac{2b^3}{a^4}$

87. $\dfrac{10x^{-4}yz^7}{8x^7y^{-3}z^{-3}} = \dfrac{10}{8}x^{-4-7}y^{1+3}z^{7+3} = \dfrac{5}{4}x^{-11}y^4z^{10}$
$= \dfrac{5y^4z^{10}}{4x^{11}}$

89. $4.92\times10^3 = 4.920\times10^3$ The decimal point moves right three places.
$= 4920$

91. 8.92×10^{-3}

Since the exponent is negative, the decimal point will move to the left.

.008.92 The decimal point moves left 3 places.

$8.92\times10^{-3} = 0.00892$

93. 9.04×10^8

Since the exponent is positive, the decimal point will move to the right.

9.04000000.

8 places

$9.04\times10^8 = 904{,}000{,}000$

95. $3.497\times10^{-6} = 0000003.497\times10^{-6}$ The decimal moves left six places.
$= 0.000003497$

97. $36{,}000{,}000 = 3.6\times10^m$

We move the decimal point 7 places to the right. Thus, m is 7.

$36{,}000{,}000 = 3.6\times10^7$

99. $0.00583 = 5.83\times10^m$

To write 5.83 as 0.00583 we move the decimal point 3 places to the left. Thus, m is –3 and

$0.00583 = 5.83\times10^{-3}$.

101. $78{,}000{,}000{,}000 = 7.8\times10^m$

To write 7.8 as 78,000,000,000 we move the decimal point 10 places to the right. Thus, m is 10 and

$78{,}000{,}000{,}000 = 7.8\times10^{10}$.

103. $0.000001032 = 1.032\times10^m$

We move the decimal point 6 places to the left. Thus, $m = -6$.

$0.000001032 = 1.032\times10^{-6}$

105. $(3\times10^5)(2\times10^8) = (3\cdot2)\times(10^5\cdot10^8)$
$= 6\times10^{5+8}$
$= 6\times10^{13}$

107. $(3.8\times10^9)(6.5\times10^{-2}) = (3.8\cdot6.5)\times(10^9\cdot10^{-2})$
$= 24.7\times10^7$

The answer is not yet in scientific notation since 24.7 is not a number between 1 and 10. We convert to scientific notation.

$24.7\times10^7 = (2.47\times10)\times10^7 = 2.47\times10^8$

109. $(8.7\times10^{-12})(4.5\times10^{-5})$
$= (8.7\cdot4.5)\times(10^{-12}\cdot10^{-5})$
$= 39.15\times10^{-17}$

The answer is not yet in scientific notation since 39.15 is not a number between 1 and 10. We convert to scientific notation.

$39.15\times10^{-17} = (3.915\times10)\times10^{-17} = 3.915\times10^{-16}$

111. $\dfrac{8.5\times10^8}{3.4\times10^{-5}} = \dfrac{8.5}{3.4}\times\dfrac{10^8}{10^{-5}}$
$= 2.5\times10^{8-(-5)}$
$= 2.5\times10^{13}$

113. $(4.0\times10^3)\div(8.0\times10^8)=\frac{4.0}{8.0}\times\frac{10^3}{10^8}$
$=0.5\times10^{3-8}$
$=0.5\times10^{-5}$
$=5.0\times10^{-6}$

115. $\frac{7.5\times10^{-9}}{2.5\times10^{12}}=\frac{7.5}{2.5}\times\frac{10^{-9}}{10^{12}}$
$=3.0\times10^{-9-12}$
$=3.0\times10^{-21}$

117. *Writing Exercise.*

119. $\frac{1}{6}-\frac{1}{3}=\frac{1}{6}-\frac{2}{6}=\frac{1}{6}$

121. $(-2a)^5=(-2)^5a^5=-32a^5$

123. $24\div6\cdot2-3[4(3-1)-7]$
$=24\div6\cdot2-3[4(2)-7]$
$=24\div6\cdot2-3[8-7]$
$=24\div6\cdot2-3[1]$
$=4\cdot2-3$
$=8-3$
$=5$

125. *Writing Exercise.*

127. *Writing Exercise.*

129. $8^{-3}\cdot32\div16^2=(2^3)^{-3}\cdot2^5\div(2^4)^2$
$=2^{-9}\cdot2^5\div2^8$
$=2^{-4}\div2^8$
$=2^{-12}$

131. $\frac{125^{-4}(25^2)^4}{125}=\frac{(5^3)^{-4}((5^2)^2)^4}{5^3}$
$=\frac{5^{-12}(5^4)^4}{5^3}$
$=\frac{5^{-12}\cdot5^{16}}{5^3}$
$=\frac{5^4}{5^3}=5^1$
$=5$

133. $\left[(5^{-3})^2\right]^{-1}=5^{(-3)(2)(-1)}=5^6$

135. $3^{-1}+4^{-1}=\frac{1}{3}+\frac{1}{4}=\frac{4}{12}+\frac{3}{12}=\frac{7}{12}$

137. ***Familiarize***. Let n = the number of car miles.

Translate. We reword the problem.

Miles for one tree	times	the number of trees	is	n
↓	↓	↓	↓	↓
500	×	600,00	=	n

Carry out. We solve the equation.
$500\times600{,}000=n$
$300{,}000{,}000=n$
$n=3\times10^8$

Check. Review the computation. The answer is reasonable.

State. Trees can clean 3×10^8 miles of car traffic in a year.

139. Find the number of light-years in 5.88×10^{17} miles.

$5.88\times10^{17}\text{ mi}\times\frac{1\text{ light-year}}{5.88\times10^{12}\text{ mi}}=1\times10^5\text{ light-years}$

It is 1×10^5 light-years from one end of the galaxy to the other.

141. ***Familiarize***. Let s = the number of strands. Convert meters to nanometers.

$4\times10^{-5}\text{m}=4\times10^{-5}\text{m}\cdot\frac{1\times10^9\text{nm}}{1\text{ m}}=4\times10^4\text{nm}$

Translate.

Hair width	is	how many	times	DNA strands?
↓	↓	↓	↓	↓
4×10^4	=	s	×	2

Carry out.
$4\times10^4=2s$
$2\times10^4=s$

Check. We review calculations.

State. It takes 2×10^4 strands of DNA to equal the width of a human hair.

Exercise Set 4.3

1. The only expression with 4 terms is (b).

3. Expression (h) has three terms and they are written in descending order.

5. Expression (g) has two terms, and the degree of the leading term is 7.

7. Expression (a) has two terms, but it is not a binomial because $\frac{2}{x^2}$ is not a monomial.

9. Yes

11. No. The expression $\frac{x^2+x+1}{x^3-7}$ is not a polynomial because it represents a quotient, not a sum.

13. Yes

15. $8x^3-11x^2+6x+1$

The terms are $8x^3$, $-11x^2$, $6x$, and 1.

17. $-t^6-3t^3+9t-4$

The terms are $-t^6$, $-3t^3$, $9t$, and -4.

19. Three monomials are added, so $x^2-23x+17$ is a trinomial.

21. The polynomial $x^3-7x+2x^2-4$ is a polynomial with no special name because it is composed of four monomials.

23. Two monomials are added, so $y+8$ is a binomial.

25. The polynomial 17 is a monomial because it is the product of a constant and a variable raised to a whole number power. (In this case the variable is raised to the power 0.)

27. $7x^2+8x^5-4x^3+6-\frac{1}{2}x^4$

Term	Coefficient	Degree of the Term	Degree of the Polynomial
$8x^5$	8	5	
$-\frac{1}{2}x^4$	$-\frac{1}{2}$	4	
$-4x^3$	-4	3	5
$7x^2$	7	2	
6	6	0	

29. $8x^4+2x$

Term	Coefficient	Degree
$8x^4$	8	4
$2x$	2	1

31. $9t^2-3t+4$

Term	Coefficient	Degree
$9t^2$	9	2
$-3t$	-3	1
4	4	0

33. x^4-x^3+4x-3

Term	Coefficient	Degree
x^4	1	4
$-x^3$	-1	3
$4x$	4	1
-3	-3	0

35. $5t+t^3+8t^4$

a.

Term	$5t$	t^3	$8t^4$
Degree	1	3	4

b. The term of highest degree is $8t^4$. This is the leading term. Then the leading coefficient is 8.

c. Since the term of highest degree is $8t^4$, the degree of the polynomial is 4.

37. $3a^2-7+2a^4$

a.

Term	$3a^2$	-7	$2a^4$
Degree	2	0	4

b. The term of highest degree is $2a^4$. This is the leading term. Then the leading coefficient is 2.

c. Since the term of highest degree is $2a^4$, the degree of the polynomial is 4.

39. $8+6x^2-3x-x^5$

a.

Term	8	$6x^2$	$-3x$	$-x^5$
Degree	0	2	1	5

b. The term of highest degree is $-x^5$. This is the leading term. Then the leading coefficient is -1 since $-x^5=-1\cdot x^5$.

c. Since the term of highest degree is $-x^5$, the degree of the polynomial is 5.

41. $5n^2+n+6n^2=(5+6)n^2+n=11n^2+n$

43. $$3a^4-2a+2a+a^4=(3+1)a^4+(-2+2)a$$
$$=4a^4+0a=4a^4$$

45. $$4b^3+5b+7b^3+b^2-6b$$
$$=(4+7)b^3+b^2+(5-6)b$$
$$=11b^3+b^2-b$$

47. $$10x^2+2x^3-3x^3-4x^2-6x^2-x^4$$
$$=-x^4+(2-3)x^3+(10-4-6)x^2=-x^4-x^3$$

49. $$\frac{1}{5}x^4+7-2x^2+3-\frac{2}{15}x^4+2x^2$$
$$=\left(\frac{1}{5}-\frac{2}{15}\right)x^4+(-2+2)x^2+(7+3)$$
$$=\left(\frac{3}{15}-\frac{2}{15}\right)x^4+0x^2+10=\frac{1}{15}x^4+10$$

51. $$8.3a^2+3.7a-8-9.4a^2+1.6a+0.5$$
$$=(8.3-9.4)a^2+(3.7+1.6)a-8+0.5$$
$$=-1.1a^2+5.3a-7.5$$

53. For $x=3$: $$-4x+9=-4\cdot 3+9$$
$$=-12+9$$
$$=-3$$

For $x=-3$: $$-4x+9=-4(-3)+9$$
$$=12+9$$
$$=21$$

55. For $x=3$: $$2x^2-3x+7=2\cdot 3^2-3\cdot 3+7$$
$$=2\cdot 9-3\cdot 3+7$$
$$=18-9+7$$
$$=16$$

For $x=-3$: $$2x^2-3x+7=2(-3)^2-3(-3)+7$$
$$=2\cdot 9-3(-3)+7$$
$$=18+9+7=34$$

57. For $x = 3$:

$$-3x^3 + 7x^2 - 4x - 8 = -3 \cdot 3^3 + 7 \cdot 3^2 - 4 \cdot 3 - 8$$
$$= -3 \cdot 27 + 7 \cdot 9 - 12 - 8$$
$$= -81 + 63 - 12 - 8 = -38$$

For $x = -3$:

$$-3x^3 + 7x^2 - 4x - 8 = -3(-3)^3 + (-3)^2 - 4(-3) - 8$$
$$= -3(-27) + 7 \cdot 9 + 12 - 8 = 148$$

59. For $x = 3$: $2x^4 - \frac{1}{9}x^3 = 2 \cdot 3^4 - \frac{1}{9}3^3$

$$= 2 \cdot 81 - \frac{1}{9} \cdot 27$$
$$= 162 - 3 = 159$$

For $x = -3$: $2x^4 - \frac{1}{9}x^3 = 2(-3)^4 - \frac{1}{9}(-3)^3$

$$= 2 \cdot 81 - \frac{1}{9}(-27)$$
$$= 162 + 3$$
$$= 165$$

61. $11.12t^2 = 11.12(10)^2 = 11.12(100) = 1112$
The skydivers had fallen approximately 1112 ft 10 seconds after jumping from the plane.

63. $2\pi r = 2(3.14)(10)$ Substituting 3.14 for π and 10 for r
$= 62.8$
The circumference is 62.8 cm.

65. $\pi r^2 = 3.14(7)^2$ Substituting 3.14 for π and 7 for r
$= 3.14(49)$
$= 153.86$
The area is 153.86 m^2.

67. $s(2.9) = 16(2.9)^2$ Substituting 2.9 for t
$= 134.56$ ft
He dropped about 135 ft.

69. $N(3) = \frac{1}{3}(3)^3 + \frac{1}{2}(3)^2 + \frac{1}{6}(3)$ Substituting 3 for x
$= 9 + \frac{9}{2} + \frac{1}{2}$
$= 14$ oranges
For 3 layers, 14 oranges are required.

$N(5) = \frac{1}{3}(5)^3 + \frac{1}{2}(5)^2 + \frac{1}{6}(5)$ Substituting 5 for x
$= \frac{125}{3} + \frac{25}{2} + \frac{5}{6}$
$= 55$ oranges
For 5 layers, 55 oranges are required.

71. We first locate 2 on the horizontal axis. From there we move vertically to the graph and then horizontally to the C-axis. This locates a C-value of about 2.3. Thus, the amount of Gentamicin in the bloodstream after 2 hours is about 2.3 mcg/mL.

73. *Writing Exercise.*

75. Graph $y = 3x$.

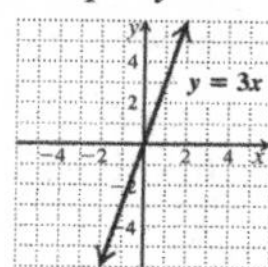

77. Graph $3x - y = 3$.

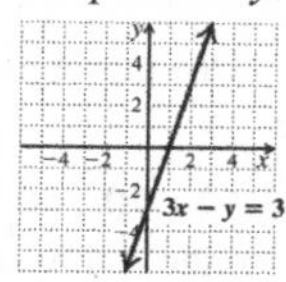

79. Graph $x = 2$.

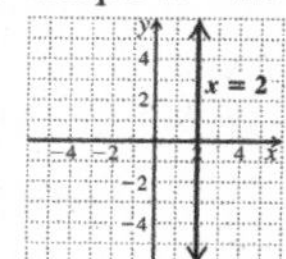

81. *Writing Exercise.*

83. Answers may vary. Choose an ax^5-term where a is an even integer. Then choose three other terms with different degrees, each less than degree 5, and coefficients $a + 2$, $a + 4$, and $a + 6$, respectively, when the polynomial is written in descending order. One such polynomial is $2x^5 + 4x^4 + 6x^3 + 8$.

85. $(3x^2)^3 + 4x^2 \cdot 4x^4 - x^4(2x)^2 + [(2x)^2]^3 - 100x^2(x^2)^2$
$= 27x^6 + 4x^2 \cdot 4x^4 - x^4 \cdot 4x^2 + (2x)^6 - 100x^2 \cdot x^4$
$= 27x^6 + 16x^6 - 4x^6 + 64x^6 - 100x^6$
$= 3x^6$

87. First locate 5 on the vertical axis. Then move horizontally to the graph. We meet the curve at 2 places. At each place move down vertically to the horizontal axis and read the corresponding x-value. We see that the times for a 5 mcg/mL concentration are about 3.4 hr and 8.5 hr.

91. We first find q, the quiz average, and t, the test average.

$$q = \frac{60 + 85 + 72 + 91}{4} = \frac{308}{4} = 77$$
$$t = \frac{89 + 93 + 90}{3} = \frac{272}{3} \approx 90.7$$

Now we substitute in the polynomial.

$$A = 0.3q + 0.4t + 0.2f + 0.1h$$
$$= 0.3(77) + 0.4(90.7) + 0.2(84) + 0.1(88)$$
$$= 23.1 + 36.28 + 16.8 + 8.8$$
$$= 84.98$$
$$\approx 85.0$$

t	$-t^2 + 6t - 4$
1	1
2	4
3	5
4	4
5	1

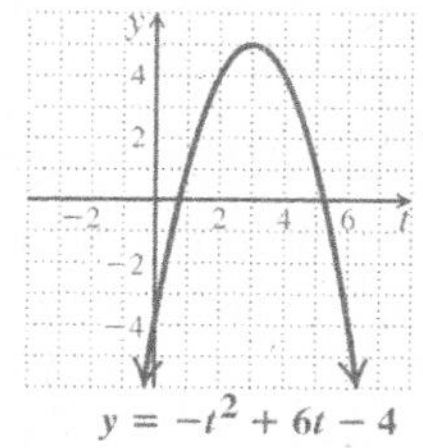

93. When $t = 3$,
$-t^2 + 10t - 18 = -3^2 + 10 \cdot 3 - 18 = -9 + 30 - 18 = 3$
When $t = 4$,
$-t^2 + 10t - 18 = -4^2 + 10 \cdot 4 - 18 = -16 + 40 - 18 = 6$
When $t = 5$,
$-t^2 + 10t - 18 = -5^2 + 10 \cdot 5 - 18 = -25 + 50 - 18 = 7$
When $t = 6$,
$-t^2 + 10t - 18 = -6^2 + 10 \cdot 6 - 18 = -36 + 60 - 18 = 6$
When $t = 7$,
$-t^2 + 10t - 18 = -7^2 + 10 \cdot 7 - 18 = -49 + 70 - 18 = 3$
We complete the table. Then we plot the points and connect them with a smooth curve.

t	$-t^2 + 10t - 18$
3	3
4	6
5	7
6	6
7	3

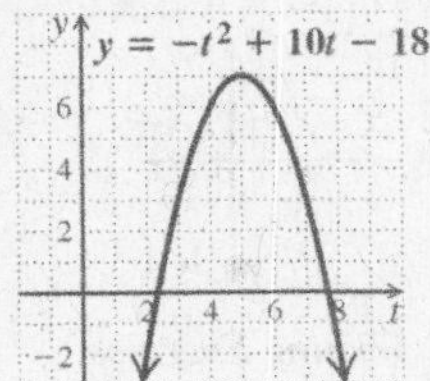

Exercise Set 4.4

1. To subtract a polynomial, add the *opposite* of the polynomial.

3. To write an equivalent expression for the *opposite* of a polynomial, we change the *sign* of every term.

5. Since the right-hand side has collected like terms, the correct expression is x^2 to make
$(3x^2 + 2) + (6x^2 + 7) = (3 + 6)x^2 + (2 + 7)$.

7. Since the right-hand side is the result of using subtraction (the distributive law), the correct operation is – to make
$(9x^3 - x^2) - (3x^3 + x^2) = 9x^3 - x^2 - 3x^3 - x^2$.

9. $(3x + 2) + (x + 7) = (3 + 1)x + (2 + 7) = 4x + 9$

11. $(2t + 7) + (-8t + 1) = (2 - 8)t + (7 + 1) = -6t + 8$

13. $(x^2 + 6x + 3) + (-4x^2 - 5)$
$= (1 - 4)x^2 + 6x + (3 - 5)$
$= -3x^2 + 6x - 2$

15. $(7t^2 - 3t - 6) + (2t^2 + 4t + 9)$
$= (7 + 2)t^2 + (-3 + 4)t + (-6 + 9) = 9t^2 + t + 3$

17. $(4m^3 - 7m^2 + m - 5) + (4m^3 + 7m^2 - 4m - 2)$
$= (4 + 4)m^3 + (-7 + 7)m^2 + (1 - 4)m + (-5 - 2)$
$= 8m^3 - 3m - 7$

19. $(3 + 6a + 7a^2 + a^3) + (4 + 7a - 8a^2 + 6a^3)$
$= (1 + 6)a^3 + (7 - 8)a^2 + (6 + 7)a + (3 + 4)$
$= 7a^3 - a^2 + 13a + 7$

21. $(3x^6 + 2x^4 - x^3 + 5x) + (-x^6 + 3x^3 - 4x^2 + 7x^4)$
$= (3 - 1)x^6 + (2 + 7)x^4 + (-1 + 3)x^3 - 4x^2 + 5x$
$= 2x^6 + 9x^4 + 2x^3 - 4x^2 + 5x$

23. $\left(\frac{3}{5}x^4 + \frac{1}{2}x^3 - \frac{2}{3}x + 3\right) + \left(\frac{2}{5}x^4 - \frac{1}{4}x^3 - \frac{3}{4}x^2 - \frac{1}{6}x\right)$
$= \left(\frac{3}{5} + \frac{2}{5}\right)x^4 + \left(\frac{1}{2} - \frac{1}{4}\right)x^3 - \frac{3}{4}x^2 + \left(-\frac{2}{3} - \frac{1}{6}\right)x + 3$
$= x^4 + \left(\frac{2}{4} - \frac{1}{4}\right)x^3 - \frac{3}{4}x^2 + \left(\frac{-4}{6} - \frac{1}{6}\right)x + 3$
$= x^4 + \frac{1}{4}x^3 - \frac{3}{4}x^2 - \frac{5}{6}x + 3$

25. $(5.3t^2 - 6.4t - 9.1) + (4.2t^3 - 1.8t^2 + 7.3)$
$= 4.2t^3 + (5.3 - 1.8)t^2 - 6.4t + (-9.1 + 7.3)$
$= 4.2t^3 + 3.5t^2 - 6.4t - 1.8$

27.
$$\begin{array}{r} -4x^3 + 8x^2 + 3x - 2 \\ -4x^2 + 3x + 2 \\ \hline -4x^3 + 4x^2 + 6x + 0 \end{array}$$
$-4x^3 + 4x^2 + 6x$

29.
$$\begin{array}{rrrrr} 0.05x^4 & +\ 0.12x^3 & -\ 0.5x^2 & & \\ & -\ 0.02x^3 & +\ 0.02x^2 & +\ 2x & \\ 1.5x^4 & & +\ 0.01x^2 & & +\ 0.15 \\ & 0.25x^3 & & & +\ 0.85 \\ -0.25x^4 & & +\ 10x^2 & & -\ 0.04 \\ \hline 1.3x^4 & +\ 0.35x^3 & +\ 9.53x^2 & +\ 2x & +\ 0.96 \end{array}$$

31. Two forms of the opposite of $-3t^3 + 4t - 7$ are
i) $-(-3t^3 + 4t - 7)$ and
ii) $3t^3 - 4t + 7$. (Changing the sign of every term)

33. Two forms for the opposite of $x^4 - 8x^3 + 6x$ are
i) $-(x^4 - 8x^3 + 6x)$ and
ii) $-x^4 + 8x^3 - 6x$ (Changing the sign of every term)

35. We change the sign of every term inside parentheses.
$-(3a^4 - 5a^2 + 1.2) = -3a^4 + 5a^2 - 1.2$

37. We change the sign of every term inside parentheses.
$-(-4x^4 + 6x^2 + \frac{3}{4}x - 8) = 4x^4 - 6x^2 - \frac{3}{4}x + 8$

39. Change the sign of every term inside parentheses.
$(3x + 1) - (5x + 8) = 3x + 1 - 5x - 8$
$= -2x - 7$

41. $(-9t + 12) - (t^2 + 3t - 1) = -9t + 12 - t^2 - 3t + 1$
$= -t^2 - 12t + 13$

43. $(4a^2 + a - 7) - (3 - 8a^3 - 4a^2)$
$= 4a^2 + a - 7 - 3 + 8a^3 + 4a^2$
$= 8a^3 + 8a^2 + a - 10$

45. $(7x^3-2x^2+6)-(6-2x^2+7x^3)$

Observe that we are subtracting the polynomial $7x^3-2x^2+6$ from itself. The result is 0.

47. $(3+5a+3a^2-a^3)-(2+4a-9a^2+2a^3)$

$=3+5a+3a^2-a^3-2-4a+9a^2-2a^3$

$=1+a+12a^2-3a^3$

49. $\left(\frac{5}{8}x^3-\frac{1}{4}x-\frac{1}{3}\right)-\left(-\frac{1}{2}x^3+\frac{1}{4}x-\frac{1}{3}\right)$

$=\frac{5}{8}x^3-\frac{1}{4}x-\frac{1}{3}+\frac{1}{2}x^3-\frac{1}{4}x+\frac{1}{3}$

$=\frac{9}{8}x^3-\frac{2}{4}x$

$=\frac{9}{8}x^3-\frac{1}{2}x$

51. $(0.07t^3-0.03t^2+0.01t)-(0.02t^3+0.04t^2-1)$

$=0.07t^3-0.03t^2+0.01t-0.02t^3-0.04t^2+1$

$=0.05t^3-0.07t^2+0.01t+1$

53.

$$\begin{array}{r} x^3+3x^2+1 \\ \underline{-(x^3+x^2-5)} \end{array}$$

$$\begin{array}{r} x^3+3x^2+1 \\ \underline{-x^3-x^2+5} \\ 2x^2+6 \end{array}$$

55.

$$\begin{array}{r} 4x^4-2x^3 \\ \underline{-(7x^4+6x^3+7x^2)} \end{array}$$

$$\begin{array}{r} 4x^2-2x^3 \\ \underline{-7x^4-6x^3-7x^2} \\ -3x^4-8x^3-7x^2 \end{array}$$

57. a.

A x 3x B x x

Familiarize. The area of a rectangle is the product of the length and the width.

Translate. The sum of the areas is found as follows:

Area of A	+	Area of B	+	Area of C	+	Area of D
$= 3x\cdot x$	+	$x\cdot x$	+	$4\cdot x$	+	$x\cdot x$

Carry out. We collect like terms.

$3x^2+x^2+4x+x^2=5x^2+4x$

Check. We can go over our calculations. We can also assign some value to x, say 2, and carry out the computation of the area in two ways.

Sum of areas: $3\cdot 2\cdot 2+2\cdot 2+4\cdot 2+2\cdot 2$

$=12+4+8+4=28$

Substituting in the polynomial:

$5(2)^2+4\cdot 2=20+8=28$

Since the results are the same, our solution is probably correct.

State. A polynomial for the sum of the areas is $5x^2+4x$.

b. For $x=5$: $5x^2+4x=5\cdot 5^2+4\cdot 5$

$=5\cdot 25+4\cdot 5=125+20$

$=145$

When $x=5$, the sum of the areas is 145 square units.

For $x=7$: $5x^2+4x=5\cdot 7^2+4\cdot 7$

$=5\cdot 49+4\cdot 7=245+28$

$=273$

When $x=7$, the sum of the areas is 273 square units.

59. The perimeter is the sum of the lengths of the sides.

$4y+4+7+2y+7+6+(3y+2)+7y$

$=(4+2+3+7)y+(4+7+7+6+2)$

$=16y+26$

61.

r 11 9 r $r+9$ $r+11$

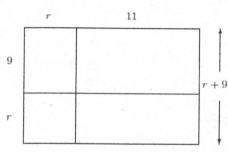

The length and width of the figure can be expressed as $r+11$ and $r+9$ respectively. The area of this figure (a rectangle) is the product of the length and width. An algebraic expression for the area is $(r+11)\cdot(r+9)$.

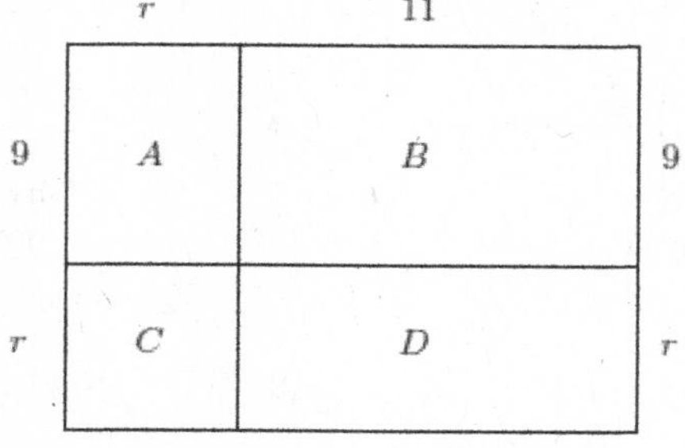

The area of the figure can be found by adding the areas of the four rectangles A, B, C, and D.

Area of A	+	Area of B	+	Area of C	+	Area of D
$= 9\cdot r$	+	$11\cdot 9$	+	$r\cdot r$	+	$11\cdot r$
$= 9r$	+	99	+	r^2	+	$11r$

The algebraic expressions $9r+99+r^2+11r$ and $(r+11)\cdot(r+9)$ represent the same area.

$(r+11)\cdot(r+9)=9r+99+r^2+11r$

63.

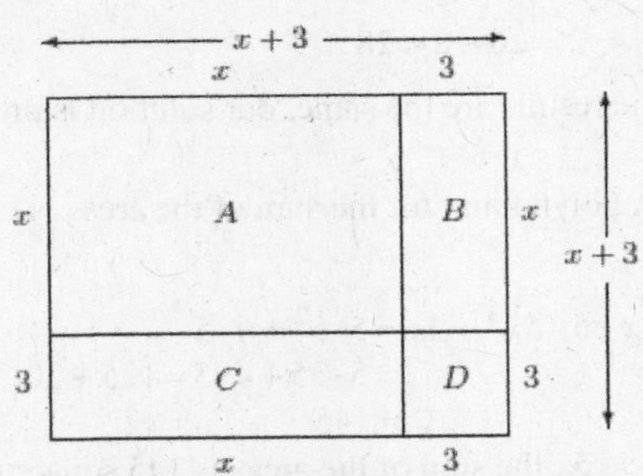

The length and width of the figure can each be expressed as $x+3$. The area can be expressed as $(x+3)\cdot(x+3)$, or $(x+3)^2$.

Another way to express the area is to find an expression for the sum of the areas of the four rectangles *A*, *B*, *C*, and *D*. The area of each rectangle is the product of its length and width.

Area of *A*	+	Area of *B*	+	Area of *C*	+	Area of *D*
$= x\cdot x$	+	$3\cdot x$	+	$3\cdot x$	+	$3\cdot 3$
$= x^2$	+	$3x$	+	$3x$	+	9

The algebraic expressions $(x+3)^2$ and $x^2+3x+3x+9$ represent the same area.

$$(x+3)^2 = x^2+3x+3x+9$$

65. Recall that the area of a rectangle is length times width.

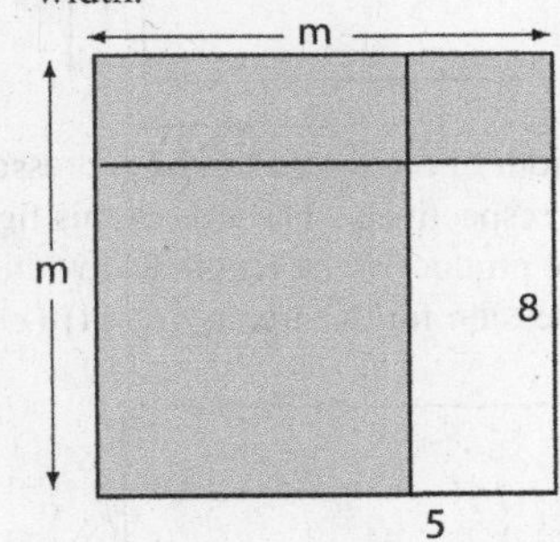

Area of entire square	−	Area not shaded	=	Shaded area
$m\cdot m$	−	$8\cdot 5$	=	Shaded area
		m^2-40	=	Shaded area

67. Recall that the area of a circle is the product of π to and the square of the radius, r^2.

$$A=\pi r^2$$

The area of a square is the length of on side, s, squared.

$$A=s^2$$

Area of circle	−	Area of square	=	Shaded area
πr^2	−	$7\cdot 7$	=	Shaded area
πr^2	−	49	=	Shaded area

69. ***Familiarize.*** Recall that the area of a rectangle is the product of the length and the width and that, consequently, the area of a square with side s is s^2. The remaining floor area is the area of the entire floor less the area of the bath enclosure, in square feet.

Translate.

Area of entire floor	−	Area of bath enclosure	=	Remaining floor area
x^2	−	$2\cdot 6$	=	Remaining floor area

Carry out. We simplify the expression.

$$x^2-2\cdot 6=x^2-12$$

Check out. We go over the calculations. The answer checks.

State. A polynomial for the remaining floor area is (x^2-12) ft^2.

71. ***Familiarize.*** Recall that the area of a square with side *z* is z^2. Recall that the area of a circle with radius 6 is πr^2 or $\pi\cdot 6^2$. The remaining area of the garden is the area of the garden less the area of the patio, in square feet.

Translate.

Area of garden	−	Area of patio	is	Remaining garden area
z^2	−	$\pi\cdot 6^2$	=	Remaining garden area

Carry out. We simplify the expression.

$$(z^2-36\pi)\text{ft}^2.$$

Check. We go over the calculations. The answer checks.

State. A polynomial for the remaining area of the garden is $(z^2-36\pi)$ ft^2.

73. ***Familiarize.*** Recall that the area of a square with side s is s^2 and the area of a circle with radius r is πr^2. The radius of the circle is half the diameter, or $\frac{d}{2}m$.

The area of the mat outside the circle is the area of the entire mat less the area of the circle, in square meters.

Translate.

Area of mat	−	Area of circle	is	Area outside the circle
12^2	−	$\pi\cdot\left(\frac{d}{2}\right)^2$	=	Area outside the circle

Carry out. We simplify the expression.

$$12^2-\pi\cdot\left(\frac{d}{2}\right)^2=144-\pi\cdot\frac{d^2}{4}=144-\frac{d^2}{4}\pi$$

Check out. We go over the calculations. The answer checks.

State. A polynomial for the area of the mat outside the wrestling circle is $\left(144-\frac{d^2}{4}\pi\right)$ m^2.

75. *Writing Exercise.*

77. We use the slope-intercept equation, substituting $\frac{1}{3}$ for m and 2 for b.

$$y=mx+b$$
$$y=\frac{1}{3}x+2$$

79. Since the slope is 0, this is a horizontal line.

$$y=10$$

81. We use the slope-intercept equation, substituting $-\frac{4}{7}$ for m and -4 for b.

$$y=mx+b$$
$$y=-\frac{4}{7}x-4$$

83. *Writing Exercise.*

85. $(6t^2-7t)+(3t^2-4t+5)-(9t-6)$
$=6t^2-7t+3t^2-4t+5-9t+6$
$=9t^2-20t+11$

87. $4(x^2-x+3)-2(2x^2+x-1)$
$=4x^2-4x+12-4x^2-2x+2$
$=(4-4)x^2+(-4-2)x+(12+2)$
$=-6x+14$

89. $(345.099x^3-6.178x)-(94.508x^3-8.99x)$
$=345.099x^3-6.178x-94.508x^3+8.99x$
$=250.591x^3+2.812x$

91. ***Familiarize***. The surface area is $2lw+2lh+\ 2wh$, where $l=$ length, $w=$ width and $h=$ height of the rectangular solid. Here we have $l=3, w=w$, and $h=7$.

Translate. We substitute in the formula above.

$2\cdot3\cdot w+2\cdot3\cdot7+2\cdot w\cdot7$

Carry out. We simplify the expression.

$2\cdot3\cdot w+2\cdot3\cdot7+2\cdot w\cdot7$
$=6w+42+14w$
$=20w+42$

Check. We can go over the calculations. We can also assign some value to w, say 6, and carry out the computation in two ways.
Using the formula:
$2\cdot3\cdot6+2\cdot3\cdot7+2\cdot6\cdot7=36+42+84=162$
Substituting in the polynomial:
$20\cdot6+42=120+42=162$
Since the results are the same, our solution is probably correct.

State. A polynomial for the surface area is $20w+42$.

93. ***Familiarize***. The surface area is $2lw+2lh+\ 2wh$, where $l=$ length, $w=$ width, and $h=$ height of the rectangular solid. Here we have $l=x, w=x$, and $h=5$.

Translate. We substitute in the formula above.

$2\cdot x\cdot x+2\cdot x\cdot5+2\cdot x\cdot5$

Carry out. We simplify the expression.

$2\cdot x\cdot x+2\cdot x\cdot5+2\cdot x\cdot5$
$=2x^2+10x+10x$
$=2x^2+20x$

Check. We can go over the calculations. We can also assign some value to x, say 3, and carry out the computation in two ways.
Using the formula:
$2\cdot3\cdot3+2\cdot3\cdot5+2\cdot3\cdot5=18+30+30=78$
Substituting in the polynomial:
$2\cdot3^2+20\cdot3=2\cdot9+60=18+60=78$
Since the results are the same, our solution is probably correct.

State. A polynomial for the surface area is $2x^2+20x$.

95. a. $P=R-C$
$=175x-0.4x^2-(5000+0.6x^2)$
$=175x-0.4x^2-5000-0.6x^2$
$=(-0.4-0.6)x^2+175x-5000$
$=-x^2+175x-5000$

b. For $x=75$:

$P=-x^2+175x-5000$
$P=-(75)^2+175(75)-5000$
$P=-5625+13{,}125-5000$
$P=2500$
The total profit is \$2500.

c. For $x=120$:

$P=-x^2+175x-5000$
$P=-120^2+175(120)-5000$
$P=-14{,}400+21{,}000-5000$
$P=1600$
The total profit is \$1600.

Mid-Chapter Review

1. $\dfrac{x^{-3}y}{x^{-4}y^7} = x^{-3-(-4)}y^{1-7} = x^1y^{-6} = \dfrac{x}{y^6}$

2. $\dfrac{x^{-3}y}{x^{-4}y^7} = x^{-3-(-4)}y^{1-7} = x^1y^{-6} = \dfrac{x}{y^6}$

3. $(x^2y^5)^8 = x^{2\cdot 8}y^{5\cdot 8} = x^{16}y^{40}$

4. $(4x)^0 = 1$

5. $\dfrac{3a^{11}}{12a} = \dfrac{3}{12}a^{11-1} = \dfrac{1}{4}a^{10}$

6. $\dfrac{-48ab^7}{18ab^6} = -\dfrac{48}{18}a^{1-1}b^{7-6} = -\dfrac{8}{3}b$

7. $5x^{-2} = \dfrac{5}{x^2}$

8. $(a^{-2}bc^3)^{-1} = a^{-2(-1)}b^{-1}c^{3(-1)} = a^2b^{-1}c^{-3} = \dfrac{a^2}{bc^3}$

9. $\left(\dfrac{2a^2}{3}\right)^{-3} = \left(\dfrac{3}{2a^2}\right)^3 = \dfrac{27}{8a^6}$

10. $\dfrac{8m^{-2}n^3}{12m^4n^7} = \dfrac{8}{12}m^{-2-4}n^{3-7} = \dfrac{2}{3}m^{-6}n^{-4} = \dfrac{2}{3m^6n^4}$

11. $1.89\times 10^{-6} = 0.00000189$

12. $27{,}000{,}000{,}000 = 2.7\times 10^{10}$

13. 3

14. $3a^2 - 6a - a^2 + 7 + a - 10 = 2a^2 - 5a - 3$

15. $(3x^2 - 2x + 6) + (5x - 3)$ Addition
$= 3x^2 - 2x + 5x + 6 - 3$
$= 3x^2 + 3x + 3$

16. $(9x + 6) - (2x - 1)$ Subtraction
$= 9x + 6 - 2x + 1$
$= 7x + 7$

17. $(4x^2 - x - 7) - (10x^2 - 3x + 5)$
$= 4x^2 - x - 7 - 10x^2 + 3x - 5$
$= -6x^2 + 2x - 12$

18. $(t^6 + 3t^6 - 8t^2) + (5t^7 - 3t^6 + 8t^2)$
$= t^9 + 3t^6 - 8t^2 + 5t^7 - 3t^6 + 8t^2$
$= t^9 + 5t^7$

19. $(3a^4 - 9a^3 - 7) - (4a^3 + 13a^2 - 3)$
$= 3a^4 - 9a^3 - 7 - 4a^3 - 13a^2 + 3$
$= 3a^4 - 13a^3 - 13a^2 - 4$

20. $(x^4 - 2x^2 - \frac{1}{2}x) - (x^5 - x^4 + \frac{1}{2}x)$
$= x^4 - 2x^2 - \frac{1}{2}x - x^5 + x^4 - \frac{1}{2}x$
$= -x^5 + 2x^4 - 2x^2 - x$

Exercise Set 4.5

1. $= (5x)(2x+3) - 4(2x+3)$ Using the commutative law for multiplication. Choice (c) is correct.

3. $= 10x^2 + 15x - 8x - 12$ Multiplying monomials. Choice (b) is correct.

5. $(3x^5)7 = (3\cdot 7)x^5 = 21x^5$

7. $(-x^3)(x^4) = (-1\cdot x^3)(x^4) = -1(x^3\cdot x^4) = -1\cdot x^7 = -x^7$

9. $(-x^6)(-x^2) = (-1\cdot x^6)(-1\cdot x^2)$
$= (-1)(-1)(x^6\cdot x^2)$
$= x^8$

11. $4t^2(9t^2) = (4\cdot 9)(t^2\cdot t^2) = 36t^4$

13. $(0.3x^3)(-0.4x^6) = 0.3(-0.4)(x^3\cdot x^6) = -0.12x^9$

15. $\left(-\frac{1}{4}x^4\right)\left(\frac{1}{5}x^8\right) = \left(-\frac{1}{4}\cdot\frac{1}{5}\right)(x^4\cdot x^8) = -\frac{1}{20}x^{12}$

17. $(-5n^3)(-1) = (-5)(-1)n^3 = 5n^3$

19. $(-4y^5)(6y^2)(-3y^3) = -4(6)(-3)(y^5\cdot y^2\cdot y^3) = 72y^{10}$

21. $5x(4x+1) = 5x(4x) + 5x(1) = 20x^2 + 5x$

23. $(a-9)3a = a\cdot 3a - 9\cdot 3a = 3a^2 - 27a$

25. $x^2(x^3+1) = x^2(x^3) + x^2(1) = x^5 + x^2$

27. $-3n(2n^2 - 8n + 1)$
$= (-3n)(2n^2) + (-3n)(-8n) + (-3n)(1)$
$= -6n^3 + 24n^2 - 3n$

29. $-5t^2(3t+6) = -5t^2(3t) - 5t^2(6) = -15t^3 - 30t^2$

31. $\frac{2}{3}a^4\left(6a^5 - 12a^3 - \frac{5}{8}\right)$
$= \frac{2}{3}a^4(6a^5) - \frac{2}{3}a^4(12a^3) - \frac{2}{3}a^4\left(\frac{5}{8}\right)$
$= \frac{12}{3}a^9 - \frac{24}{3}a^7 - \frac{10}{24}a^4$
$= 4a^9 - 8a^7 - \frac{5}{12}a^4$

33. $0.2x^3(1.2x^2 - 0.5x) = 0.2x^3(1.2x^2) - 0.2x^3(0.5x)$
$= 0.24x^5 - 0.1x^4$

35. $(x+3)(x+4) = (x+3)x + (x+3)4$
$= x\cdot x + 3\cdot x + x\cdot 4 + 3\cdot 4$
$= x^2 + 3x + 4x + 12$
$= x^2 + 7x + 12$

37. $(t+7)(t-3) = (t+7)t + (t+7)(-3)$
$= t \cdot t + 7 \cdot t + t(-3) + 7(-3)$
$= t^2 + 7t - 3t - 21$
$= t^2 + 4t - 21$

39. $(a-0.6)(a-0.7)$
$= (a-0.6)a + (a-6)(-0.7)$
$= a \cdot a - 0.6 \cdot a + a(-0.7) + (-0.6)(-0.7)$
$= a^2 - 0.6a - 0.7a + 0.42$
$= a^2 - 1.3a + 0.42$

41. $(x+3)(x-3) = (x+3)x + (x+3)(-3)$
$= x \cdot x + 3 \cdot x + x(-3) + 3(-3)$
$= x^2 + 3x - 3x - 9$
$= x^2 - 9$

43. $(4-x)(7-2x) = (4-x)7 + (4-x)(-2x)$
$= 4 \cdot 7 - x \cdot 7 + 4(-2x) - x(-2x)$
$= 28 - 7x - 8x + 2x^2$
$= 28 - 15x + 2x^2$

45. $\left(t+\frac{3}{2}\right)\left(t+\frac{4}{3}\right) = \left(t+\frac{3}{2}\right)t + \left(t+\frac{3}{2}\right)\left(\frac{4}{3}\right)$
$= t \cdot t + \frac{3}{2} \cdot t + t \cdot \frac{4}{3} + \frac{3}{2} \cdot \frac{4}{3}$
$= t^2 + \frac{3}{2}t + \frac{4}{3}t + 2$
$= t^2 + \frac{9}{6}t + \frac{8}{6}t + 2$
$= t^2 + \frac{17}{6}t + 2$

47. $\left(\frac{1}{4}a+2\right)\left(\frac{3}{4}a-1\right)$
$= \left(\frac{1}{4}a+2\right)\left(\frac{3}{4}a\right) + \left(\frac{1}{4}a+2\right)(-1)$
$= \frac{1}{4}a\left(\frac{3}{4}a\right) + 2 \cdot \frac{3}{4}a + \frac{1}{4}a(-1) + 2(-1)$
$= \frac{3}{16}a^2 + \frac{3}{2}a - \frac{1}{4}a - 2$
$= \frac{3}{16}a^2 + \frac{6}{4}a - \frac{1}{4}a - 2$
$= \frac{3}{16}a^2 + \frac{5}{4}a - 2$

49. Illustrate $x(x+5)$ as the area of a rectangle with width x and length $x+5$.

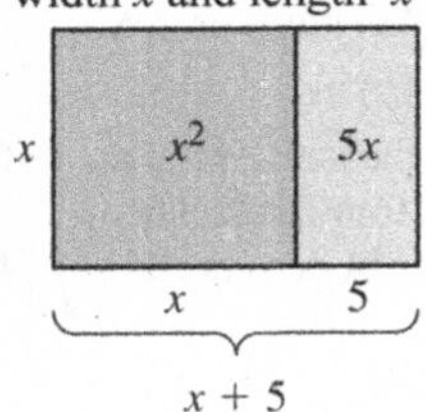

51. Illustrate $(x+1)(x+2)$ as the area of a rectangle with $x+1$ and length $x+2$.

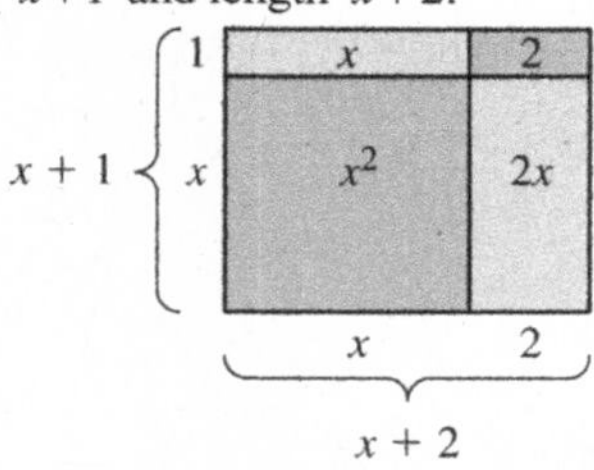

53. Illustrate $(x+5)(x+3)$ as the area of a rectangle with length $x+5$ and width $x+3$.

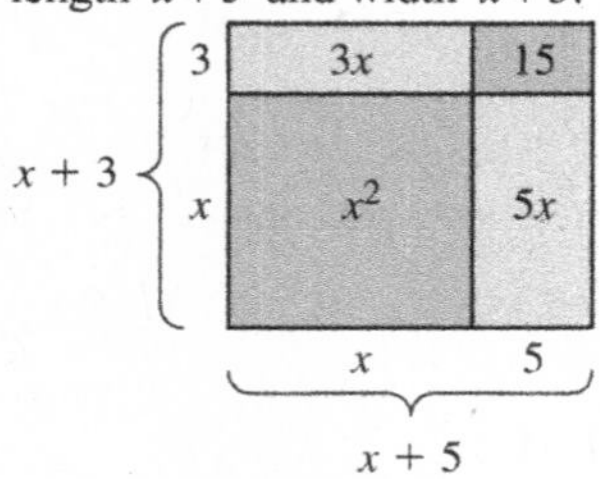

55. $(x^2 - x + 3)(x+1)$
$= (x^2 - x + 3)x + (x^2 - x + 3)1$
$= x^3 - x^2 + 3x + x^2 - x + 3$
$= x^3 + 2x + 3$

A partial check can be made by selecting a convenient replacement for x, say 1, and comparing the values of the original expression and the result.

$(1^2 - 1 + 3)(1+1)$	$1^3 + 2 \cdot 1 + 3$
$= (1 - 1 + 3)(1+1)$	$= 1 + 2 + 3$
$= 3 \cdot 2$	$= 6$

Since the value of both expressions is 6, the multiplication is very likely correct.

57. $(2a+5)(a^2 - 3a + 2)$
$= (2a+5)a^2 - (2a+5)(3a) + (2a+5)2$
$= 2a \cdot a^2 + 5 \cdot a^2 - 2a \cdot 3a - 5 \cdot 3a + 2a \cdot 2 + 5 \cdot 2$
$= 2a^3 + 5a^2 - 6a^2 - 15a + 4a + 10$
$= 2a^3 - a^2 - 11a + 10$

A partial check can be made as in Exercise 55.

59. $(y^2 - 7)(3y^4 + y + 2)$
$= (y^2 - 7)(3y^4) + (y^2 - 7)y + (y^2 - 7)(2)$
$= y^2 \cdot 3y^4 - 7 \cdot 3y^4 + y^2 \cdot y - 7 \cdot y + y^2 \cdot 2 - 7 \cdot 2$
$= 3y^6 - 21y^4 + y^3 - 7y + 2y^2 - 14$
$= 3y^6 - 21y^4 + y^3 + 2y^2 - 7y - 14$

A partial check can be made as in Exercise 55.

61. $(3x+2)(7x+4x+1) = (3x+2)(11x+1)$
$= (3x+2)(11x) + (3x+2)(1)$
$= 3x \cdot 11x + 2 \cdot 11x + 3x \cdot 1 + 2 \cdot 1$
$= 33x^2 + 22x + 3x + 2$
$= 33x^2 + 25x + 2$

63. $\left(3x+\frac{1}{2}\right)(x^2-4x-2)$
$=\left(3x+\frac{1}{2}\right)(x^2)-\left(3x+\frac{1}{2}\right)(4x)-\left(3x+\frac{1}{2}\right)(2)$
$=3x^3+\frac{1}{2}x^2-12x^2-2x-6x-1$
$=3x^3-\frac{23}{2}x^2-8x-1$

65. $(1.2a^3-0.2a^2+5a)(0.4a^2-10)$
$=(1.2a^3-0.2a^2+5a)(0.4a^2)$
$\quad-(1.2a^3-0.2a^2+5a)(10)$
$=0.48a^5-0.08a^4+2a^3-12a^3+2a^2-50a$
$=0.48a^5-0.08a^4-10a^3+2a^2-50a$

67.
x^2+5x-1 Line up like terms
x^2-x+3 in columns
$3x^2+15x-3$ Multiplying by 3
$-x^3-5x^2+x$ Multiplying by $-x$
$x^4+5x^3-x^2$ Multiplying by x^2
$x^4+4x^3-3x^2+16x-3$
A partial check can be made as in Exercise 55.

69.
$5t^2-t+\frac{1}{2}$
$2t^2+t-4$
$-20t^2+4t-2$ Multiplying by -4
$5t^3-t^2+\frac{1}{2}t$ Multiplying by t
$10t^4-2t^3+t^2$ Multiplying by $2t^2$
$10t^4+3t^3-20t^2+\frac{9}{2}t-2$
A partial check can be made as in Exercise 55.

71. We will multiply horizontally while still aligning like terms.
$(x+1)(x^3+7x^2+5x+4)$
$=x^4+7x^3+5x^2+4x$ Multiplying by x
$\quad+x^3+7x^2+5x+4$ Multiplying by 1
$=x^4+8x^3+12x^2+9x+4$
A partial check can be made as in Exercise 55.

73. *Writing Exercise.*

75.
$A=\frac{b+c}{2}$
$2\cdot A=2\cdot\frac{b+c}{2}$
$2A=b+c$
$2A-b=b+c-b$
$2A-b=c$

77. Graph $t\geq-3$.

-5 -4 -3 -2 -1 0 1 2 3 4 5

79. Two points on the line are (2, 3) and (5, –1).
$m=\frac{-1-3}{5-2}=\frac{-4}{3}=-\frac{4}{3}$

81. *Writing Exercise.*

83. The shaded area is the area of the large rectangle, $6y(14y-5)$ less the area of the unshaded rectangle, $3y(3y+5)$. We have:
$6y(14y-5)-3y(3y+5)$
$=84y^2-30y-9y^2-15y$
$=75y^2-45y$

85. Let n = the missing number.

n nx 3n
x x² 3x
x 3

The area of the figure is $x^2+3x+nx+3n$. This is equivalent to $x^2+8x+15$, so we have $3x+nx=8x$ and $3n=15$. Solving either equation for n, we find that the missing number is 5.

87.

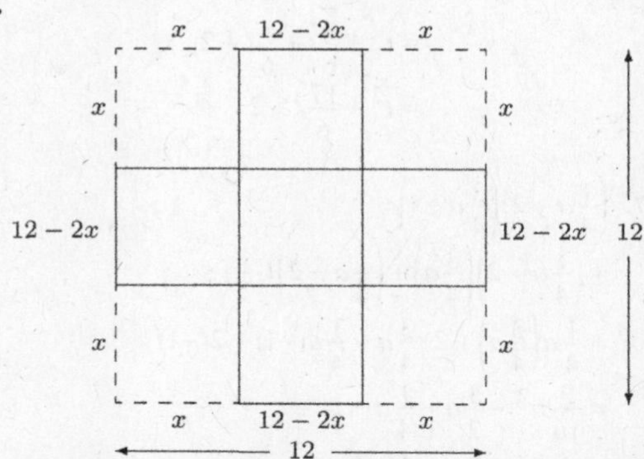

The dimensions, in inches, of the box are $12-2x$ by $12-2x$ by x. The volume is the product of the dimensions (volume = length × width × height):
Volume: $=(12-2x)(12-2x)x$
$=(144-48x+4x^2)x$
$=(144x-48x^2+4x^3)\text{ in}^3$, or
$(4x^3-48x^2+144x)\text{ in}^3$

The outside surface area is the sum of the area of the bottom and the areas of the four sides. The dimensions, in inches, of the bottom are $12-2x$ by $12-2x$, and the dimensions, in inches, of each side are x by $12-2x$.

Surface area = Area of bottom + 4 · Area of each side
$=(12-2x)(12-2x)+4\cdot x(12-2x)$
$=144-24x-24x+4x^2+48x-8x^2$
$=144-48x+4x^2+48x-8x^2$
$=(144-4x^2)\text{in}^2$, or $(-4x^2+144)\text{ in}^2$

89.

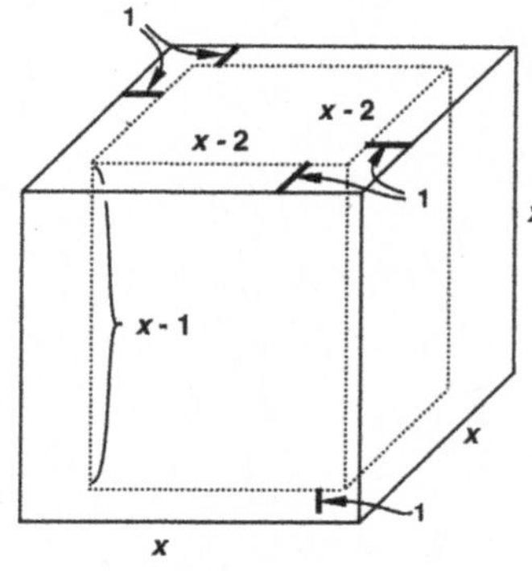

The interior dimensions of the open box are $x-2$ cm by $x-2$ cm by $x-1$ cm.

Interior volume: $=(x-2)(x-2)(x-1)$

$=(x^2-4x+4)(x-1)$

$=(x^3-5x^2+8x-4)\text{ cm}^3$

91. Let x = the radius.

The plastic cube has volume $V=(2x)^3=8x^3$.
The plastic sphere inside the cube has volume $V=\frac{4}{3}\pi x^3$.

a. The volume of air inside the cube outside the sphere is the difference between volumes.

$$V=8x^3-\frac{4}{3}\pi x^3=\left(8-\frac{4}{3}\pi\right)x^3$$

b. If the diameter is 2.9 in., then the radius is half that amount, or 1.45 in. Substitute 1.45 for x in the formula from part (a).

$$V=\left(8-\frac{4}{3}\pi\right)(1.45)^3\approx 11.6\text{ in}^3$$

93. $(x+3)(x+6)+(x+3)(x+6)$

$=(x+3)x+(x+3)6+(x+3)x+(x+3)6$

$=x^2+3x+6x+18+x^2+3x+6x+18$

$=2x^2+18x+36$

95. $(x+5)^2-(x-3)^2$

$=(x+5)(x+5)-(x-3)(x-3)$

$=(x+5)x+(x+5)5-[(x-3)x-(x-3)3]$

$=x^2+5x+5x+25-(x^2-3x-3x+9)$

$=x^2+10x+25-(x^2-6x+9)$

$=x^2+10x+25-x^2+6x-9$

$=16x+16$

97. $(x-3)(x-3)(x-3)$

$=(x^2-3x-3x+9)(x-3)$

$=(x^2-6x+9)(x-3)$

$=x^3-6x^2+9x-3x^2+18x-27$

$=x^3-9x^2+27x-27$

99.

Exercise Set 4.6

1. It is true that FOIL is simply a memory device for finding the product of two binomials.

3. This statement is false. See Example 2(d).

5. $(x^2+2)(x+3)$

F O I L

$=x^2\cdot x+x^2\cdot 3+2\cdot x+2\cdot 3$

$=x^3+2x+3x^2+6$, or x^3+3x^2+2x+6

7. $(t^4-2)(t+7)$

F O I L

$=t^4\cdot t+t^4\cdot 7-2\cdot t-2\cdot 7$

$=t^5+7t^4-2t-14$

9. $(y+2)(y-3)$

F O I L

$=y\cdot y+y\cdot(-3)+2\cdot y+2\cdot(-3)$

$=y^2-3y+2y-6$

$=y^2-y-6$

11. $(3x+2)(3x+5)$

F O I L

$=3x\cdot 3x+3x\cdot 5+2\cdot 3x+2\cdot 5$

$=9x^2+15x+6x+10$

$=9x^2+21x+10$

13. $(5x-3)(x+4)$

F O I L

$=5x\cdot x+5x\cdot 4-3\cdot x-3\cdot 4$

$=5x^2+20x-3x-12$

$=5x^2+17x-12$

15. $(3-2t)(5-t)$

F O I L

$=3\cdot 5+3\cdot t-2t\cdot 5+(-2t)(-t)$

$=15-3t-10t+2t^2$

$=15-13t+2t^2$

17. $(x^2+3)(x^2-7)$

F O I L

$=x^2\cdot x^2-x^2\cdot 7+3\cdot x^2-3\cdot 7$

$=x^4-7x^2+3x^2-21$

$=x^4-4x^2-21$

19. $\left(p-\frac{1}{4}\right)\left(p+\frac{1}{4}\right)$

F O I L

$=p\cdot p+p\cdot\frac{1}{4}+\left(-\frac{1}{4}\right)\cdot p+\left(-\frac{1}{4}\right)\cdot\frac{1}{4}$

$=p^2+\frac{1}{4}p-\frac{1}{4}p-\frac{1}{16}$

$=p^2-\frac{1}{16}$

21. $(x-0.3)(x-0.3)$

F O I L

$= x\cdot x - x\cdot 0.3 + -0.3\cdot x + (-0.3)(-0.3)$

$= x^2 - 0.3x - 0.3x + 0.09$

$= x^2 - 0.6x + 0.09$

23. $(10x^2+3)(10x^2-3)$

F O I L

$= 100x^4 - 30x^2 + 30x^2 - 9$

$= 100x^4 - 9$

25. $(1-5t^3)(1+5t^3)$

F O I L

$= 1 + 5t^3 - 5t^3 - 25t^6$

$= 1 - 25t^6$

27. $(t-2)^2$ $\quad (A-B)^2 = A^2 - 2\cdot A\cdot B + B^2$

$= t^2 - 2\cdot t\cdot 2 + 2^2$

$= t^2 + 4t + 4$

29. $(x+10)(x+10)$

$= (x+10)^2$ $\quad (A+B)^2 = A^2 + 2\cdot A\cdot B + B^2$

$= x^2 + 2\cdot x\cdot 10 + 10^2$

$= x^2 + 20x + 100$

31. $(3x+2)^2$ $\quad (A+B)^2 = A^2 + 2\cdot A\cdot B + B^2$

$= (3x)^2 - 2\cdot 3x\cdot 2 + 2^2$

$= 9x^2 - 12x + 4$

33. $(1-10a)^2$ $\quad (A-B)^2 = A^2 - 2\cdot A\cdot B + B^2$

$= 1^2 - 2\cdot 1\cdot (10a) + (10a)^2$

$= 1 - 20a + 100a^2$

35. $(x^3+12)^2$ $\quad (A+B)^2 = A^2 + 2\cdot A\cdot B + B^2$

$= (x^3)^2 - 2\cdot x^3\cdot 12 + 12^2$

$= x^6 - 24x^3 + 144$

37. $(x^2+3)(x^3-1)$ Product of two binomials; use FOIL

$= x^5 - x^2 + 3x^3 - 3$

39. $(x+8)(x-8)$ Product of sum and difference of the same two terms

$= x^2 - 8^2$

$= x^2 - 64$

41. $(-3n+2)(n+7)$ Product of two binomials; use FOIL

$= -3n^2 - 21n + 2n + 14$

$= -3n^2 - 19n + 14$

43. $(x+3)^2$

$= x^2 + 2\cdot x\cdot 3 + 3^2$ Square of a binomial

$= x^2 + 6x + 9$

45. $(7x^3-1)^2$ Square of a binomial

$= (7x^3)^2 - 2\cdot 7x^3\cdot 1 + (-1)^2$

$= 49x^6 - 14x^3 + 1$

47. $(9a^3+1)(9a^3-1)$

$= (9a^3)^2 - 1^2$

$= 81a^6 - 1$

49. $(x^4+0.1)(x^4-.01)$

$= (x^4)^2 - 0.1^2$

$= x^8 - 0.01$

51. $\left(t-\frac{3}{4}\right)\left(t+\frac{3}{4}\right)$

$= t^2 - \left(\frac{3}{4}\right)^2$

$= t^2 - \frac{9}{16}$

53. $(1-3t)(1+5t^2)$ Product of two binomials; use FOIL

$= 1 + 5t^2 - 3t - 15t^3$

$= 1 - 3t + 5t^2 - 15t^3$

55. $\left(a-\frac{2}{5}\right)^2$ Square of a binomial

$= a^2 - 2\cdot a\cdot \frac{2}{5} + \left(\frac{2}{5}\right)^2$

$= a^2 - \frac{4}{5}a + \frac{4}{25}$

57. $(t^4+3)^2$ Square of a binomial

$= (t^4)^2 + 2\cdot t^4\cdot 3 + 3^2$

$= t^8 + 6t^4 + 9$

59. $(5x-9)(9x+5)$ Product of two binomials; use FOIL

$= 45x^2 + 25x - 81x - 45$

$= 45x^2 - 56x - 45$

61. $7n^3(2n^2-1)$

$= 7n^3\cdot 2n^2 - 7n^3\cdot 1$ Multiplying each term of

$= 14n^5 - 7n^3$ the binomial by the monomial

63. $(a-3)(a^2+2a-4)$

$= a^3 + 2a^2 - 4a$ Multiplying horizontally

$\quad - 3a^2 - 6a + 12$ and aligning like terms

$= a^3 - a^2 - 10a + 12$

65. $(7-3x^4)(7-3x^4)$

$= 7^2 - 2\cdot 7\cdot 3x^4 + (-3x^4)^2$ Squaring a binomial

$= 49 - 42x^4 + 9x^8$

67. $(2-3x^4)^2 = 2^2 - 2\cdot 2\cdot 3x^4 + (3x^4)^2$

$= 4 - 12x^4 + 9x^8$

69. $(5t+6t^2)^2 = (5t)^2 + 2\cdot 5t\cdot 6t^2 + (6t^2)^2$

$= 25t^2 + 60t^3 + 36t^4$

71. $5x(x^2+6x-2)$

$= 5x\cdot x^2 + 5x\cdot 6x + 5x(-2)$ Multiplying each term of the trinomial by the monomial

$= 5x^3 + 30x^2 - 10x$

73. $(q^5+1)(q^5-1)$

$= (q^5)^2 - 1^2$

$= q^{10} - 1$

75. $3t^2(5t^3 - t^2 + t)$

$= 3t^2\cdot 5t^3 + 3t^2(-t^2) + 3t^2\cdot t$ Multiplying each term of the trinomial by the monomial

$= 15t^5 - 3t^4 + 3t^3$

77. $(6x^4 - 3x)^2$ Squaring a binomial

$= (6x^4)^2 - 2\cdot 6x^4\cdot 3x + (-3x^2)$

$= 36x^8 - 36x^5 + 9x^2$

79. $(9a+0.4)(2a^3+0.5)$ Product of two binomials; use FOIL

$= 9a\cdot 2a^3 + 9a\cdot 0.5 + 0.4\cdot 2a^3 + 0.4\cdot 0.5$

$= 18a^4 + 4.5a + 0.8a^3 + 0.2$, or

$18a^4 + 0.8a^3 + 4.5a + 0.2$

81. $\left(\frac{1}{5} - 6x^4\right)\left(\frac{1}{5} + 6x^4\right)$

$= \left(\frac{1}{5}\right)^2 - \left(6x^4\right)^2$

$= \frac{1}{25} - 36x^8$

83. $(a+1)(a^2-a+1)$

$= a^3 - a^2 + a$ Multiplying horizontally

$\quad a^2 - a + 1$ and aligning like terms

$= a^3 \qquad + 1$

85.

3 | A | B

x | C | D

x 3

We can find the shaded area in two ways.

Method 1: The figure is a square with side $x+3$, so the area is $(x+3)^2 = x^2 + 6x + 9$.

Method 2: We add the areas of *A*, *B*, *C*, and *D*.

$3\cdot x + 3\cdot 3 + x\cdot x + x\cdot 3 = 3x + 9 + x^2 + 3x$

$= x^2 + 6x + 9.$

Either way we find that the total shaded area is $x^2 + 6x + 9$.

87.

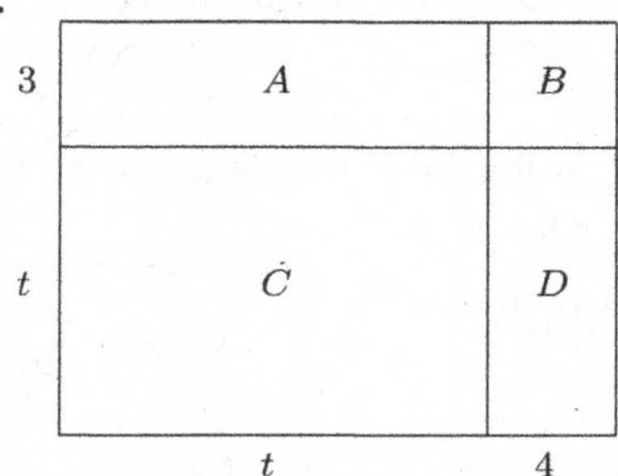

We can find the shaded area in two ways.

Method 1: The figure is a rectangle with dimensions $t+3$ by $t+4$, so the area is

$(t+3)(t+4) = t^2 + 4t + 3t + 12 = t^2 + 7t + 12.$

Method 2: We add the areas of *A*, *B*, *C*, and *D*.

$3\cdot t + 3\cdot 4 + t\cdot t + t\cdot 4 = 3t + 12 + t^2 + 4t = t^2 + 7t + 12.$

Either way, we find that the area is $t^2 + 7t + 12$.

89.

x 7

x | A | B

3 | C | D

We can find the shaded area in two ways.

Method 1: The figure is a rectangle with dimensions $x+7$ by $x+3$, so the area is

$(x+7)(x+3) = x^2 + 3x + 7x + 21$

$= x^2 + 10x + 21$

Method 2: We add the areas of *A*, *B*, *C*, and *D*.

$x\cdot x + x\cdot 7 + 3\cdot x + 3\cdot 7 = x^2 + 10x + 21$

Either way, we find that the area is $x^2 + 10x + 21$.

91.

2 | A | B

$5t$ | C | D

$5t$ 2

We can find the shaded area in two ways.

Method 1: The figure is a square with side $5t+2$, so the area is $(5t+2)^2 = 25t^2 + 20t + 4$.

Method 2: We add the areas of A, B, C, and D.

$$5t\cdot 5t+5t\cdot 2+2\cdot 5t+2\cdot 2=25t^2+10t+10t+4$$
$$=25t^2+20t+4$$

Either way, we find that the total shaded area is $25t^2+20t+4$.

93. We draw a square with side $x+5$.

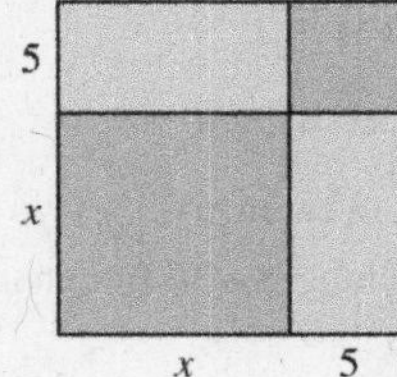

95. We draw a square with side $3+x$.

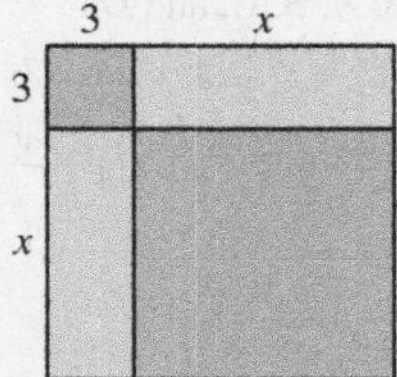

97. We draw a square with side $2t+1$.

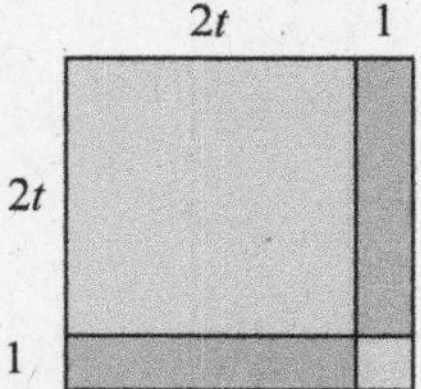

99. *Writing Exercise.*

101. ***Familiarize.*** Let w = the energy, in kilowatt-hours per month, used by the washing machine. Then $3w$ = the amount of energy used by the refrigerator, and $6w$ = the amount of energy used by the freezer.

Translate.

Washing Machine	and	refrigerator	and	freezer	is	Total energy
w	$+$	$3w$	$+$	$6w$	$=$	1200

Solve. We solve the equation.

$$w+3w+6w=1200$$
$$10w=1200$$
$$w=120$$

Then $3w=3\cdot 120=360$

and $6w=6\cdot 120=720$

Check. The energy used by the refrigerator, 360 kWh, is 3 times the energy used by the washing machine. The energy used by the freezer, 720 kWh is 6 times the energy used by the washing machine. Also, 120 + 360 + 720 = 1200, the total energy used.

State. The washing machine used 120 kWh/year, the refrigerator used 360 kWh/year, and the freezer used 720 kWh/year.

103. $3ab=c$

$a=\frac{c}{3b}$ Dividing both sides by $3b$

105. *Writing Exercise.*

107. $(4x^2+9)(2x+3)(2x-3)$

$=(4x^2+9)(4x^2-9)$

$=16x^4-81$

109. $(3t-2)^2(3t+2)^2$

$=[(3t-2)(3t+2)]^2$

$=(9t^2-4)^2$

$=81t^4-72t^2+16$

111. $(t^3-1)^4(t^3+1)^4$

$=[(t^3-1)(t^3+1)]^4$

$=(t^6-1)^4$

$=[(t^6-1)^2]^2$

$=(t^{12}-2t^6+1)^2$

$=(t^{12}-2t^6+1)(t^{12}-2t^6+1)$

$=t^{24}-2t^{18}+t^{12}-2t^{18}+4t^{12}-2t^6+t^{12}-2t^6+1$

$=t^{24}-4t^{18}+6t^{12}-4t^6+1$

113. $18\times 22=(20-2)(20+2)=20^2-2^2$

$=400-4=396$

115. $(x+2)(x-5)=(x+1)(x-3)$

$x^2-5x+2x-10=x^2-3x+x-3$

$x^2-3x-10=x^2-2x-3$

$-3x-10=-2x-3$ Subtracting x^2

$-3x+2x=10-3$ Adding $2x$ and 10

$-x=7$

$x=-7$

The solution is -7.

117.

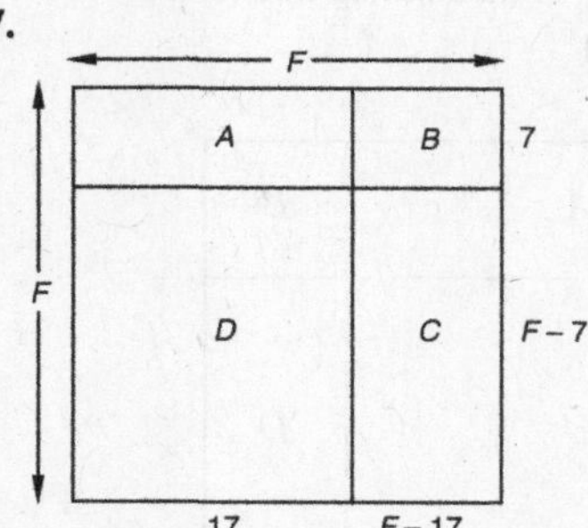

The area of the entire figure is F^2. The area of the unshaded region, C is $(F-7)(F-17)$. Then one expression for the area of the shaded region is

$$F^2-(F-7)(F-17).$$

To find a second expression we add the areas of regions $A, B,$ and D. We have:

$$17\cdot 7+7(F-17)+17(F-7)$$
$$=119+7F-119+17F-119$$
$$=24F-119$$

It is possible to find other equivalent expressions also.

119. The dimensions of the shaded area, regions A and D together, are $y+1$ by $y-1$ so the area is $(y+1)(y-1)$.

To find another expression we add the areas of regions A and D. The dimensions of region A are y by $y-1$, and the dimensions of region D are $y-1$ by 1, so the sum of the areas is $y(y-1)+(y-1)(1)$, or $y(y-1)+y-1$.

It is possible to find other equivalent expressions also.

121.

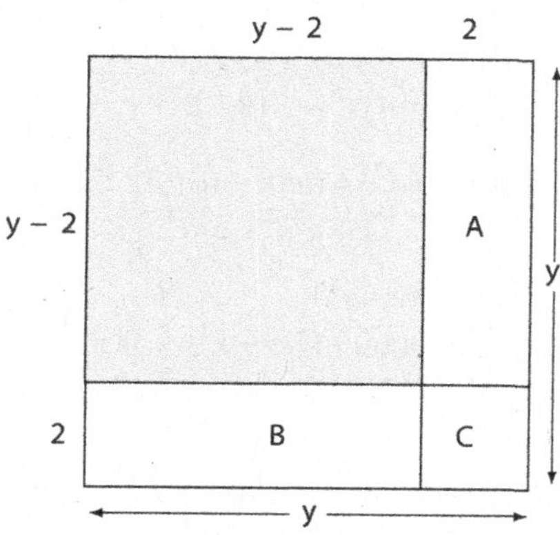

The shaded area is $(y-2)^2$. We find it as follows:

Shaded area	=	Area of square	−	Area of A	−	Area of B	−	Area of B
$(y-2)^2$	=	y^2		$-2(y-2)$		$-2(y-2)$	−	$2\cdot 2$

$$(y-2)^2 = y^2-2y+4-2y+4-4$$
$$(y-2)^2 = y^2-4y+4$$

123.

Exercise Set 4.7

1. Since $2x^2y$ and $5x^2y$ are like terms, choice (b).

3. Since 4 is the degree of the polynomial, choice (a).

5. $(3x+5y)^2$ is the square of a binomial, choice (a).

7. $(5a+6b)(-6b+5a)$, or $(5a+6b)(5a-6b)$ is the product of the sum and difference of the same two terms, choice (b).

9. $(r-3s)(5r+3s)$ is neither the square of a binomial nor the product of the sum and difference of the same two terms, so choice (c) is appropriate.

11. We replace x with 5 and y with -2.

$$\begin{aligned} x^2-2y^2+3xy &= 5^2-2(-2)^2+3\cdot 5(-2) \\ &= 25-8-30 \\ &= -13 \end{aligned}$$

13. We replace x with 2, y with -3, and z with -4.

$$xy^2z-z = 2(-3)^2(-4)-(-4) = -72+4 = -68$$

15. Evaluate the polynomial for $g=231$, $l=135$, and $c=86$.

$$\begin{aligned} &11.5g+7.55l+12.5c-4016 \\ &= 11.5(231)+7.55(135)+12.5(86)-4016 \\ &= 734.75 \end{aligned}$$

The elephant weighs about 735 kg.

17. Evaluate the polynomial for $h=7$, $r=1\frac{1}{2}=\frac{3}{2}$, and $\pi\approx 3.14$.

$$\begin{aligned} 2\pi rh+\pi r^2 &= 2(3.14)\left(\frac{3}{2}\right)(7)+3.14\left(\frac{3}{2}\right)^2 \\ &\approx 65.94+7.065 \\ &\approx 73.005 \end{aligned}$$

The surface area is about 73.005 in^2.

19. Evaluate the polynomial for $h=50$, $v=18$, and $t=2$.

$$\begin{aligned} h+vt-4.9t^2 &= 50+18\cdot 2-4.9(2)^2 \\ &= 50+36-19.6 \\ &= 66.4 \end{aligned}$$

The ball will be 66.4 m above the ground 2 seconds after it is thrown.

21. $3x^2y+5xy+2y^2-11$

Term	Degree	
$3x^2y$	3	
$-5y^2$	2	
$2y^2$	2	
-11	0	(Think: $-11=-11x^0$)

The degree of the polynomial is the degree of the term of highest degree. The term of highest degree is $3x^2y$. Its degree is 3, so the degree of the polynomial is 3.

23. $7-abc+a^2b+9ab^2$

Term	Degree
7	0
$-abc$	3
a^2b	3
$9ab^2$	3

The terms of highest degree are $-abc$, a^2b and $9ab^2$. Each has degree 3. The degree of the polynomial is 3.

25. $3r+s-r-7s=(3-1)r+(1-7)s=2r-6s$

27. $5xy^2-2x^2y+x+3x^2$

There are no like terms, so none of the terms can be combined.

29.
$$\begin{aligned} &6u^2v-9uv^2+3vu^2-2v^2u+11u^2 \\ &= (6+3)u^2v+(-9-2)uv^2+11u^2 \\ &= 9u^2v-11uv^2+11u^2 \end{aligned}$$

31. $5a^2c - 2ab^2 + a^2b - 3ab^2 + a^2c - 2ab^2$
$= (5+1)a^2c + (-2-3-2)ab^2 + a^2b$
$= 6a^2c - 7ab^2 + a^2b$

33. $(6x^2 - 2xy + y^2) + (5x^2 - 8xy - 2y^2)$
$= (6+5)x^2 + (-2-8)xy + (1-2)y^2$
$= 11x^2 - 10xy - y^2$

35. $(3a^4 - 5ab + 6ab^2) - (9a^4 + 3ab - ab^2)$
$= 3a^4 - 5ab + 6ab^2 - 9a^4 - 3ab + ab^2$
Adding the opposite
$= (3-9)a^4 + (-5-3)ab + (6+1)ab^2$
$= -6a^4 - 8ab + 7ab^2$

37. $(5r^2 - 4rt + t^2) + (-6r^2 - 5rt - t^2) + (-5r^2 + 4rt - t^2)$
Observe that the polynomials $5r^2 - 4rt + t^2$ and $-5r^2 + 4rt - t^2$ are opposites. Thus, their sum is 0 and the sum in the exercise is the remaining polynomial, $-6r^2 - 5rt - t^2$.

39. $(x^3 - y^3) - (-2x^3 + x^2y - xy^2 + 2y^3)$
$= x^3 - y^3 + 2x^3 - x^2y + xy^2 - 2y^3$
$= 3x^3 - 3y^3 - x^2y + xy^2$, or
$3x^3 - x^2y + xy^2 - 3y^3$

41. $(2y^4x^3 - 3y^3x) + (5y^4x^3 - y^3x) - (9y^4x^3 - y^3x)$
$= (2+5-9)y^4x^3 + (-3-1+1)y^3x$
$= -2y^4x^3 - 3y^3x$

43. $(4x+5y) + (-5x+6y) - (7x+3y)$
$= 4x + 5y - 5x + 6y - 7x - 3y$
$= (4-5-7)x + (5+6-3)y$
$= -8x + 8y$

45. $(4c-d)(3c+2d)$
F O I L
$= 12c^2 + 8cd - 3cd - 2d^2$
$= 12c^2 + 5cd - 2d^2$

47. $(xy-1)(xy+5)$
F O I L
$= x^2y^2 + 5xy - xy - 5$
$= x^2y^2 + 4xy - 5$

49. $(2a-b)(2a+b)$ $[(A+B)(A-B) = A^2 - B^2]$
$= 4a^2 - b^2$

51. $(5rt-2)(4rt-3)$
F O I L
$= 20r^2t^2 - 15rt - 8rt + 6$
$= 20r^2t^2 - 23rt + 6$

53. $(m^3n+8)(m^3n-6)$
F O I L
$= m^6n^2 - 6m^3n + 8m^3n - 48$
$= m^6n^2 + 2m^3n - 48$

55. $(6x-2y)(5x-3y)$
F O I L
$= 30x^2 - 18xy - 10xy + 6y^2$
$= 30x^2 - 28xy + 6y^2$

57. $(pq+0.1)(-pq+0.1)$
$= (0.1+pq)(0.1-pq)$ $[(A+B)(A-B) = A^2 - B^2]$
$= 0.01 - p^2q^2$

59. $(x+h)^2 = x^2 + 2xh + h^2$
$[(A+B)^2 = A^2 + 2AB + B^2]$

61. $(4a-5b)^2 = 16a^2 - 40ab + 25b^2$
$[(A-B)^2 = A^2 - 2AB + B^2]$

63. $(ab+cd^2)(ab-cd^2) = (ab)^2 - (cd^2)^2$
$= a^2b^2 - c^2d^4$

65. $(2xy + x^2y + 3)(xy + y^2)$
$= (2xy + x^2y + 3)(xy) + (2xy + x^2y + 3)(y^2)$
$= 2x^2y^2 + x^3y^2 + 3xy + 2xy^3 + x^2y^3 + 3y^2$
$= x^3y^2 + x^2y^3 + 2x^2y^2 + 2xy^3 + 3xy + 3y^2$

67. $(a+b-c)(a+b+c) = [(a+b)-c][(a+b)+c]$
$= (a+b)^2 - c^2$
$= a^2 + 2ab + b^2 - c^2$

69. $[a+b+c][a-(b+c)] = [a+(b+c)][a-(b+c)]$
$= a^2 - (b+c)^2$
$= a^2 - (b^2 + 2bc + c^2)$
$= a^2 - b^2 - 2bc - c^2$

71. The figure is a square with side $x+y$. Thus the area is $(x+y)^2 = x^2 + 2xy + y^2$.

73. The figure is a triangle with base $ab+2$ and height $ab-2$. Its area is
$\frac{1}{2}(ab+2)(ab-2) = \frac{1}{2}(a^2b^2 - 4) = \frac{1}{2}a^2b^2 - 2.$

75. The figure is a rectangle with dimensions $a+b+c$ by $a+d+c$. Its area is
$(a+b+c)(a+d+c)$
$= [(a+c)+b][(a+c)+d]$
$= (a+c)^2 + (a+c)d + b(a+c) + bd$
$= a^2 + 2ac + c^2 + ad + cd + ab + bc + bd$

77. The figure is a parallelogram with base $m-n$ and height $m+n$. Its area is $(m-n)(m+n) = m^2 - n^2$.

79. We draw a rectangle with dimensions $r+s$ by $u+v$.

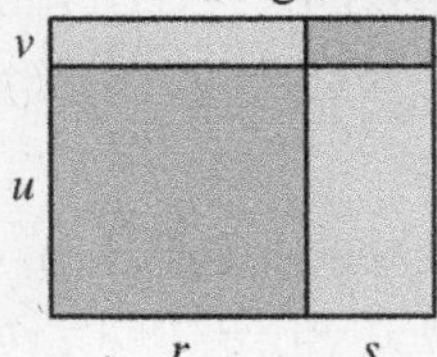

81. We draw a rectangle with dimensions $a+b+c$ by $a+d+f$.

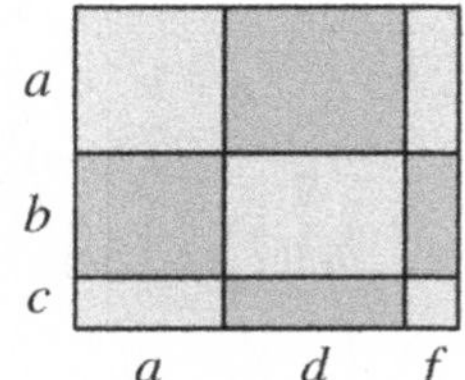

83. *Writing Exercise.*

85. $-16+20\div 2^2\cdot 5-3$
$=-16+20\div 4\cdot 5-3$
$=-16+5\cdot 5-3$
$=-16+25-3$
$=6$

87. $2[3-4(5-6)^2-1]+10$
$=2[3-4(-1)^2-1]+10$
$=2[3-4(1)-1]+10$
$=2[3-4-1]+10$
$=2[-2]+10$
$=-4+10$
$=6$

89. $2a-3(-x-5a+7)$
$=2a+3x+15a-21$
$=17a+3x-21$

91. *Writing Exercise.*

93. It is helpful to add additional labels to the figure.

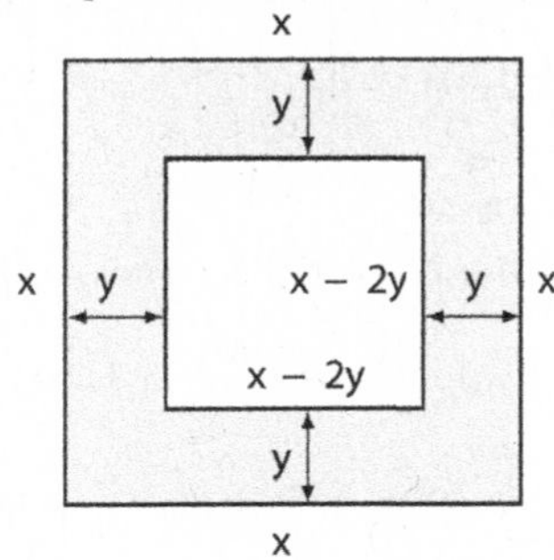

The area of the large square is $x\cdot x$, or x^2. The area of the small square is $(x-2y)(x-2y)$, or $(x-2y)^2$.

Area of shaded region	=	Area of large square	−	Area of small square
Area of shaded region	=	x^2	−	$(x-2y)^2$

$=x^2-(x^2-4xy+4y^2)$
$=x^2-x^2+4xy-4y^2$
$=4xy-4y^2$

95. The unshaded region is a circle with radius $a-b$. Then the shaded area is the area of a circle with radius a less the area of a circle with radius $a-b$. Thus, we have:

$$\begin{aligned}\text{Shaded area} &= \pi a^2-\pi(a-b)^2\\ &=\pi a^2-\pi(a^2-2ab+b^2)\\ &=\pi a^2-\pi a^2+2\pi ab-\pi b^2\\ &=2\pi ab-\pi b^2\end{aligned}$$

97. The figure can be thought of as a cube with side x, a rectangular solid with dimensions x by x by y, a rectangular solid with dimensions x by y by y, and a rectangular solid with dimensions y by y by $2y$. Thus the volume is

$x^3+x\cdot x\cdot y+x\cdot y\cdot y+y\cdot y\cdot 2y$, or
$x^3+x^2y+xy^2+2y^3$

99. The surface area of the solid consists of the surface area of a rectangular solid with dimensions x by x by h less the areas of 2 circles with radius r plus the lateral surface area of a right circular cylinder with radius r and height h. Thus, we have

$2x^2+2xh+2xh-2\pi r^2+2\pi rh$, or
$2x^2+4xh-2\pi r^2+2\pi rh$

101. *Writing Exercise.*

103. For the formula $=2*A4+3*B4$, we substitute 5 for $A4$ and 10 for $B4$.

$$\begin{aligned}=2*A4+3*B4&=2\cdot 5+3\cdot 10\\ &=10+30\\ &=40\end{aligned}$$

The value of $D4$ is 40.

105. Replace t with 2 and multiply.

$$\begin{aligned}P(1+r)^2&=P(1+2r+r^2)\\ &=P+2Pr+Pr^2\end{aligned}$$

107. Substitute \$10,400 for P, 4.5%, or 0.045 for r, and 5 for t.

$$\begin{aligned}P(1+r)^t&=\$10,400(1+0.045)^5\\ &\approx \$12,960.29\end{aligned}$$

Exercise Set 4.8

1. Here, $x-3$ represents the divisor.

3. Here, x^2-x+9 represents the dividend.

5. $\dfrac{40x^6-25x^3}{5}=\dfrac{40x^6}{5}-\dfrac{25x^3}{5}$
$=\dfrac{40}{5}x^6-\dfrac{25}{5}x^3$ Dividing coefficients
$=8x^6-5x^3$

To check, we multiply the quotient by 5:

$(8x^6-5x^3)5=40x^6-25x^3$

The answer checks.

7. $\dfrac{u-2u^2+u^7}{u}=\dfrac{u}{u}-\dfrac{2u^2}{u}+\dfrac{u^7}{u}$
$=1-2u+u^6$

Check: We multiply.

$u(1-2u+u^6)=u-2u^2+u^7$

9. $(18t^3-24t^2+6t)\div(3t)$
$=\dfrac{18t^3-24t^2+6t}{3t}$
$=\dfrac{18t^3}{3t}-\dfrac{24t^2}{3t}+\dfrac{6t}{3t}$
$=6t^2-8t+2$

Check: We multiply.

$3t(6t^2-8t+2)=18t^3-24t^2+6t$

11. $(42x^5-36x^3+9x^2)\div(6x^2)$
$=\dfrac{42x^5-36x^3+9x^2}{6x^2}$
$=\dfrac{42x^5}{6x^2}-\dfrac{36x^3}{6x^2}+\dfrac{9x^2}{6x^2}$
$=7x^3-6x+\dfrac{3}{2}$

Check: We multiply.

$6x^2\left(7x^3-6x+\dfrac{3}{2}\right)=42x^5-36x^3+9x^2$

13. $(32t^5+16t^4-8t^3)\div(-8t^3)$
$=\dfrac{32t^5+16t^4-8t^3}{-8t^3}$
$=\dfrac{32t^5}{-8t^3}+\dfrac{16t^4}{-8t^3}-\dfrac{8t^3}{-8t^3}$
$=-4t^2-2t+1$

Check: We multiply.

$-8t^3(-4t^2-2t+1)=32t^5+16t^4-8t^3$

15. $\dfrac{8x^2-10x+1}{2x}$
$=\dfrac{8x^2}{2x}-\dfrac{10x}{2x}+\dfrac{1}{2x}$
$=4x-5+\dfrac{1}{2x}$

Check: We multiply.

$2x\left(4x-5+\dfrac{1}{2x}\right)=8x^2-10x+1$

17. $\dfrac{5x^3y+10x^5y^2+15x^2y}{5x^2y}$
$=\dfrac{5x^3y}{5x^2y}+\dfrac{10x^5y^2}{5x^2y}+\dfrac{15x^2y}{5x^2y}$
$=x+2x^3y+3$

Check: We multiply.

$5x^2y(x+2x^3y+3)=5x^3y+10x^5y^2+15x^2y$

19. $\dfrac{9r^2s^2+3r^2s-6rs^2}{-3rs}$
$=\dfrac{9r^2s^2}{-3rs}+\dfrac{3r^2s}{-3rs}-\dfrac{6rs^2}{-3rs}$
$=-3rs-r+2s$

Check: We multiply.

$-3rs(-3rs-r+2s)=9r^2s^2+3r^2s-6rs^2$

21.
$$
\begin{array}{r l}
 & x-6 \\
x-2\,\big) & x^2+8x+12 \\
 & \underline{x^2-2x} \\
 & -6x+12 \leftarrow (x^2+8x)-(x^2-2x)=-6x \\
 & \underline{-6x+12} \\
 & 0 \leftarrow (-6x+12)-(-6x+12)=0
\end{array}
$$

The answer is $x-6$.

23.
$$
\begin{array}{r l}
 & t-5 \\
t-5\,\big) & t^2+10t-20 \\
 & \underline{t^2+5t} \\
 & -5t-20 \leftarrow (t^2-10t)-(t^2-5t)=-5t \\
 & \underline{-5t+25} \\
 & 45 \leftarrow (-5t-20t)-(-5t+25)=45
\end{array}
$$

The answer is $t-5-\dfrac{45}{t-5}$.

25.
$$
\begin{array}{r l}
 & 2x-1 \\
x+6\,\big) & 2x^2+11x-5 \\
 & \underline{2x^2+12x} \\
 & -x-5 \leftarrow (2x^2+11x)-(2x^2+12x)=-x \\
 & \underline{-x-6} \\
 & 1 \leftarrow (-x-5)-(-x-6)=1
\end{array}
$$

The answer is $2x-1+\dfrac{1}{x+6}$.

27.
$$
\begin{array}{r l}
 & t^2-3t+9 \\
t+3\,\big) & t^3+0t^2+0t+27 \leftarrow \text{Writing in missing terms} \\
 & \underline{t^3+3t^2} \\
 & -3t^2+0t \quad \leftarrow (t^3+0t^2)-(t^3+3t^2)=-3t^2 \\
 & \underline{-3t^2-9t} \\
 & 9t+27 \leftarrow (-3t^2)-(-3t^2-9t^2)=9t \\
 & \underline{9t+27} \\
 & 0 \leftarrow (9t+27)-(9t+27)=0
\end{array}
$$

The answer is t^2-3t+9.

29.
$$
\begin{array}{r l}
 & a+5 \\
a-5\,\big) & a^2+0a-21 \leftarrow \text{Writing in missing term} \\
 & \underline{a^2-5a} \\
 & 5a-21 \leftarrow (a^2+0a)-(a^2-5a)=5a \\
 & \underline{5a-25} \\
 & 4 \leftarrow (5a^2-21)-(5a-25)=4
\end{array}
$$

The answer is $a+5+\dfrac{4}{a-5}$.

31.
$$\begin{array}{r} 3x-5 \\ 2x+3\overline{)6x^2-x-15} \\ \underline{6x^2+9x} \\ -10x-15 \\ \underline{-10x-15} \\ 0 \end{array}$$

The answer is $3x-5$.

33.
$$\begin{array}{rl} x-3 & \\ 5x-1\overline{)5x^2-16x+0} & \leftarrow \text{Writing in missing term} \\ \underline{5x^2-x} & \\ -15x+0 & \leftarrow (5x^2-16x)-(5x^2-x)=-15x \\ \underline{-15x+3} & \\ -3 & \leftarrow (-15x+0)-(-15x+3)=-3 \end{array}$$

The answer is $x-3-\frac{3}{5x-1}$.

35.
$$\begin{array}{rl} 3a+1 & \\ 2a+5\overline{)6a^2+17a+8} & \\ \underline{6a^2+15a} & \\ 2a+8 & \leftarrow (6a^2+17a)-(6a^2+15a)=2a \\ \underline{2a+5} & \\ 3 & \leftarrow (2a+8)-(2a+5)=3 \end{array}$$

The answer is $3a+1+\frac{3}{2a+5}$.

37.
$$\begin{array}{rl} t^2-3t+1 & \\ 2t-3\overline{)2t^3-9t^2+11t-3} & \leftarrow \text{Writing in missing term} \\ \underline{2t^3-3t^2} & \\ -6t^2+11t & \leftarrow (2t^3-9t^2)-(2t^3-3t^2)=-6t^2 \\ \underline{-6t^2+9t} & \\ 2t-3 & \leftarrow (-6t^2+11t)-(-6t^2+9t^2)=2t \\ \underline{2t-3} & \\ 0 & \leftarrow (2t-3)-(2t-3)=0 \end{array}$$

The answer is t^2-3t+1.

39.
$$\begin{array}{rl} x^2+1 & \\ x-1\overline{)x^3-x^2+x-1} & \\ \underline{x^3-x^2} & \\ x-1 & \leftarrow (x^3-x^2)-(x^3-x^2)=0 \\ \underline{x-1} & \\ 0 & \leftarrow (x-1)-(-x-1)=0 \end{array}$$

The answer is x^2+1.

41.
$$\begin{array}{rl} t^2-1 & \\ t^2+5\overline{)t^4+4t^2+3t-6} & \\ \underline{t^4+5t^2} & \\ -t^2+3t-6 & \leftarrow (t^4-4t^2)-(t^4+5t^2)=-t^2 \\ \underline{-t^2\quad -5} & \\ 3t-1 & \leftarrow (-t^2+3t-6)-(-t^2-5)=3t-1 \end{array}$$

The answer is $t^2-1+\frac{3t-1}{t^2+5}$.

43.
$$\begin{array}{r} 2x^2+1 \\ 2x^2-3\overline{)4x^4-4x^2-x-3} \\ \underline{4x^4-6x^2} \\ 2x^2-x-3 \\ \underline{2x^2\quad -3} \\ -x \end{array}$$

The answer is $2x^2+1-\frac{x}{2x^2-3}$.

45. *Writing Exercise.*

47. $y=-\frac{2}{3}x+4$

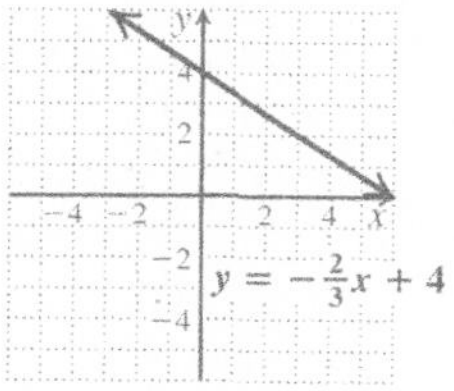

49. $y=4x+\frac{7}{2}$

$m=4$

y-intercept $\left(0,\ \frac{7}{2}\right)$

51. *Writing Exercise.*

53. $(10x^{9k}-32x^{6k}+28x^{3k})\div(2x^{3k})$

$=\frac{10x^{9k}-32x^{6k}+28x^{3k}}{2x^{3k}}$

$=\frac{10x^{9k}}{2x^{3k}}-\frac{32x^{6k}}{2x^{3k}}+\frac{28x^{3k}}{2x^{3k}}$

$=5x^{9k-3k}-16x^{6k-3k}+14x^{3k-3k}$

$=5x^{6k}-16x^{3k}+14$

55.
$$\begin{array}{r} 3t^{2h}+2t^h-5 \\ 2t^h+3\overline{)6t^{3h}+13t^{2h}-4t^h-15} \\ \underline{6t^3h+9t^2h} \\ 4t^{2h}-4t^h \\ \underline{4t^{2h}+6t^h} \\ -10t^h-15 \\ \underline{-10t^h-15} \\ 0 \end{array}$$

The answer is $3t^{2h}+2t^h-5$.

57.
$$\begin{array}{r} a+3 \\ 5a^2-7a-2\overline{)5a^3+8a^2-23a-1} \\ \underline{5a^3-7a^2-2a} \\ 15a^2-21a-1 \\ \underline{15a^2-21a-6} \\ 5 \end{array}$$

The answer is $a+3+\frac{5}{5a^2-7a-2}$.

59. $(4x^5 - 14x^3 - x^2 + 3) + (2x^5 + 3x^4 + x^3 - 3x^2 + 5x)$
$= 6x^5 + 3x^4 - 13x^3 - 4x^2 + 5x + 3$

$$\begin{array}{r}
2x^2 + x - 3 \\
3x^3 - 2x - 1\overline{)6x^5 + 3x^4 - 13x^3 - 4x^2 + 5x + 3} \\
\underline{6x^5 \qquad\quad - 4x^3 - 2x^2} \qquad\qquad\quad \\
3x^4 - 9x^3 - 2x^2 + 5x \qquad \\
\underline{3x^4 \qquad\quad - 2x^2 - x} \qquad \\
-9x^3 \qquad\quad + 6x + 3 \\
\underline{-9x^3 \qquad\quad + 6x + 3} \\
0
\end{array}$$

The answer is $2x^2 + x - 3$.

61.
$$\begin{array}{r}
x - 3 \\
x - 1\overline{)x^2 - 4x + c} \\
\underline{x^2 - x} \qquad \\
-3x + c \\
\underline{-3x + 3} \\
c - 3
\end{array}$$

We set the remainder equal to 0.
$c - 3 = 0$
$c = 3$
Thus, c must be 3.

63.
$$\begin{array}{r}
c^2x + (2c + c^2) \\
x - 1\overline{)c^2x^2 + 2cx + 1} \\
\underline{c^2x^2 - c^2x} \qquad\qquad \\
(2c + c^2)x + 1 \\
\underline{(2c + c^2)x - (2c + c^2)} \\
1 + (2c + c^2)
\end{array}$$

We set the remainder equal to 0.
$c^2 + 2c + 1 = 0$
$(c + 1)^2 = 0$
$c + 1 = 0$ or $c + 1 = 0$
$c = -1$ or $c = -1$
Thus, c must be -1.

65. a. Divide $a^3 + 3a^2 + 3a + 1$ by $a + 1$.

$$\begin{array}{r}
a^2 + 2a + 1 \\
a + 1\overline{)a^3 + 3a^2 + 3a + 1} \\
\underline{a^3 + a^2} \qquad\qquad \\
2a^2 + 3a \qquad \\
\underline{2a^2 + 2a} \qquad \\
a + 1 \\
\underline{a + 1} \\
0
\end{array}$$

The length of one side is $(a + 1)$ cm.

b. Area of a square is s^2.
$A = (a + 1)^2 = (a^2 + 2a + 1)\text{ cm}^2$

Chapter 4 Review

1. True; see Section 4.4 in the text.

2. True; see Section 4.6 in the text.

3. True; see Section 4.6 in the text.

4. False; FOIL can be used in the product of two binomials.

5. False; the degree of the polynomial is the degree of the leading term.

6. False; scientific notation is used for very large and very small numbers.

7. True; see Section 4.7 in the text.

8. True; see Section 4.2 in the text.

9. $n^3 \cdot n^8 \cdot n = n^{3+8+1} = n^{12}$

10. $(7x)^8 \cdot (7x)^2 = (7x)^{(8+2)} = (7x)^{10}$

11. $t^6 \cdot t^0 = t^{6+0} = t^6$

12. $\dfrac{4^5}{4^2} = 4^{5-2} = 4^3$ or 64

13. $\dfrac{(a+b)^4}{(a+b)^4} = 1$

14. $\dfrac{-18c^9d^3}{2c^5d} = \dfrac{-18}{2} \cdot c^{9-5}d^{3-1} = -9c^4d^2$

15. $\left(-2xy^2\right)^3 = (-2)^3 x^{1\cdot 3} y^{2\cdot 3} = -8x^3y^6$

16. $(2x^3)(-3x)^2 = 2x^3(-3)^2x^2 = 2x^3 \cdot 9x^2 = 18x^5$

17. $(a^2b)(ab)^5 = a^2b \cdot a^5b^5 = a^{2+5}b^{1+5} = a^7b^6$

18. $\left(\dfrac{2t^5}{3s^4}\right)^2 = \dfrac{2^2t^{5\cdot 2}}{3^2s^{4\cdot 2}} = \dfrac{4t^{10}}{9s^8}$

19. $8^{-6} = \dfrac{1}{8^6}$

20. $\dfrac{1}{a^9} = a^{-9}$

21. $4^5 \cdot 4^{-7} = 4^{5-7} = 4^{-2} = \dfrac{1}{4^2}$ or $\dfrac{1}{16}$

22. $\dfrac{6a^{-5}b}{3a^8b^{-8}} = 2a^{-5-8}b^{1+8} = 2a^{-13}b^9 = \dfrac{2b^9}{a^{13}}$

23. $(w^3)^{-5} = w^{3(-5)} = w^{-15} = \dfrac{1}{w^{15}}$

24. $(2x^{-3}y)^{-2} = (2)^{-2}x^{-3(-2)}y^{-2} = 4^{-1}x^6y^{-2}$
$= \dfrac{x^6}{4y^2}$

25. $\left(\dfrac{2x}{y}\right)^{-3} = \dfrac{2^{-3}x^{-3}}{y^{-3}} = \dfrac{y^3}{8x^3}$

26. 4.7×10^8 Since the exponent is positive, the decimal point will move to the right
4.70000000.
8 places
$4.7\times10^8 = 470{,}000{,}000$

27. $0.0000109 = 1.09\times10^m$
To write 1.09 as 0.0000109, we move the decimal point 5 places to the left. Thus, m is -5 and
$0.0000109 = 1.09\times10^{-5}$

28. $(3.8\times10^4)(5.5\times10^{-1}) = 20.9\times10^3$
$= 2.09\times10^4$

29. $\frac{1.28\times10^{-8}}{2.5\times10^{-4}} = \frac{1.28}{2.5}\times\frac{10^{-8}}{10^{-4}}$
$= \frac{1.28}{2.5}\times10^{-8+4}$
$= 0.512\times10^{-4}$
$= 5.12\times10^{-5}$

30. $4y^5+7y^2-3y-2$
The terms are $4y^5,\ 7y^2,\ -3y$, and -2.

31. $7n^4-\frac{5}{6}n^2-4n+10$
The coefficients are 7, $-\frac{5}{6}$, -4, and 10.

32. $4t^2+6+15t^5$

a.

Term	$4t^2$	6	$15t^5$
Degree	2	0	5

b. The term of highest degree is $15t^5$. This is the leading term. Then the leading coefficient is 15 since $15t^5 = 15\cdot t^5$.

c. Since the term of highest degree is $15t^5$, the degree of the polynomial is 5.

33. $-2x^5+7+3x^2+x$

a.

Term	$-2x^5$	7	$-3x^2$	x
Degree	5	0	2	1

b. The term of highest degree is $-2x^5$. This is the leading term. Then the leading coefficient is -2 since $-2x^2 = -2\cdot x^5$.

c. Since the term of highest degree is $-2x^5$, the degree of the polynomial is 5.

34. Three monomials are added, so $4x^3-5x+3$ is a trinomial.

35. The polynomial $4-9t^3-7t^4+10t^2$ has four monomials so it is a polynomial with no special name.

36. There is just one monomial, so $7y^2$ is a monomial.

37. $-4t^3+2t+4t^3+8-t-9$
$= (-4+4)t^3+(2-1)t+(8-9)$
$= t-1$

38. $-a+\frac{1}{3}+20a^5-1-6a^5-2a^2$
$= (20-6)a^5-2a^2-a+\left(\frac{1}{3}-1\right)$
$= 14a^5-2a^2-a-\frac{2}{3}$

39. For $x=-2$: $9x-6$
$= 9(-2)-6$
$= -18-6$
$= -24$

40. For $x=-2$: x^2-3x+6
$= (-2)^2-3(-2)+6$
$= 4+6+6$
$= 16$

41. $(8x^4-x^3+x-4)+(x^5+7x^3-3x-5)$
$= 8x^4-x^3+x-4+x^5+7x^3-3x-5$
$= x^5+8x^4+(-1+7)x^3+(1-3)x+(-4-5)$
$= x^5+8x^4+6x^3-2x-9$

42. $(5a^5-2a^3-9a^2)+(2a^5+a^3)+(-a^5-3a^2)$
$= 5a^5-2a^3-9a^2+2a^5+a^3-a^5-3a^2$
$= (5+2-1)a^5+(-2+1)a^3+(-9-3)a^2$
$= 6a^5-a^3-12a^2$

43. $(3x^5-4x^4+2x^2+3)-(2x^5-4x^4+3x^3+4x^2-5)$
$= 3x^5-4x^4+2x^2+3-2x^5+4x^4-3x^3-4x^2+5$
$= (3-2)x^5+(-4+4)x^4-3x^3+(2-4)x^2+(3+5)$
$= x^5-3x^3-2x^2+8$

44.
$$\begin{array}{rrrrr} -\frac{3}{4}x^4 & +\frac{1}{2}x^3 & & & +\frac{7}{8} \\ & -\frac{1}{4}x^3 & -x^2 & -\frac{7}{4}x & \\ +\frac{3}{2}x^4 & & +\frac{2}{3}x^2 & & -\frac{1}{2} \\ \hline +\frac{3}{4}x^4 & +\frac{1}{4}x^3 & -\frac{1}{3}x^2 & -\frac{7}{4}x & +\frac{3}{8} \end{array}$$

45.
$$\begin{array}{l} 2x^5 \quad -x^3 \quad +x+3 \\ -(3x^5-x^4+4x^3+2x^2-x+3) \\ \hline 2x^5 \quad -x^3 \quad +x+3 \\ -3x^5+x^4-4x^3-2x^2+x-3 \\ \hline -x^5+x^4-5x^3-2x^2+2x \end{array}$$

46. Let w = the width, then $w+3$ is the length, in meters.

a. Recall that the perimeter of a rectangle is the sum of all sides, so
perimeter $= 2w+2(w+3) = 2w+2w+6$
$= 4w+6.$

b. Recall that the area of a rectangle is the product of the length and the width. So,
area $= w(w+3) = w^2 + 3w$.

47. $5x^2(-6x^3) = 5(-6)x^{2+3} = -30x^5$

48. $(7x+1)^2 = (7x)^2 + 2\cdot 7x\cdot 1 + 1^2$ Squaring a binomial
$= 49x^2 + 14x + 1$

49. $(a-7)(a+4)$
F O I L
$= a^2 + 4a - 7a - 28$
$= a^2 - 3a - 28$

50. $(d-8)(d+8) = d^2 - 8^2 \quad (A+B)(A-B) = A^2 - B^2$
$= d^2 - 64$

51. $(4x^2 - 5x + 1)(3x - 2)$
$= (4x^2 - 5x + 1)3x + (4x^2 - 5x + 1)(-2)$
$= 12x^3 - 15x^2 + 3x - 8x^2 + 10x - 2$
$= 12x^3 + (-15-8)x^2 + (3+10)x - 2$
$= 12x^3 - 23x^2 + 13x - 2$

52. $3t^2(5t^3 - 2t^2 + 4t) = 15t^5 - 6t^4 + 12t^3$

53. $(2a+9)(2a-9) = (2a)^2 - 9^2$
$= 4a^2 - 81$

54. $(x-0.8)(x-0.5)$
F O I L
$= x^2 - 0.5x - 0.8x + 0.4$
$= x^2 - 1.3x + 0.4$

55. $(x^4 - 2x + 3)(x^3 + x - 1)$
$= x^4(x^3 + x - 1) - 2x(x^3 + x - 1) + 3(x^3 + x - 1)$
$= x^7 + x^5 - x^4 - 2x^4 - 2x^2 + 2x + 3x^3 + 3x - 3$
$= x^7 + x^5 - 3x^4 + 3x^3 - 2x^2 + 5x - 3$

56. $(4y^3 - 5)^2$
$= (4y^3)^2 - 2\cdot 4y^3\cdot 5 + 5^2 \quad (A-B)^2 = A - 2AB + B^2$
$= 16y^6 - 40y^3 + 25$

57. $(2t^2 + 3)(t^2 - 7)$
F O I L
$= 2t^4 - 14t^2 + 3t^2 - 21$
$= 2t^4 - 11t^2 - 21$

58. $\left(a - \frac{1}{2}\right)\left(a + \frac{2}{3}\right)$
F O I L
$= a^2 + \frac{2}{3}a - \frac{1}{2}a - \frac{1}{2}\cdot\frac{2}{3}$
$= a^2 + \frac{1}{6}a - \frac{1}{3}$

59. $(-7+2n)(7+2n) = (2n-7)(2n+7)$
$= (2n)^2 - 7^2 \quad (A+B)(A-B) = A^2 - B^2$
$= 4n^2 - 49$

60. For $x = -1$ and $y = 2$:
$2 - 5xy + y^2 - 4xy^3 + x^6$
$= 2 - 5(-1)(2) + 2^2 - 4(-1)(2)^3 + (-1)^6$
$= 2 + 10 + 4 + 32 + 1$
$= 49$

61. $x^5y - 7xy + 9x^2 - 8$

Term	Coefficient	Degree
x^5y	1	6
$-7xy$	-7	2
$9x^2$	9	2
-8	-8	0

The term of highest degree is x^5y. Its degree is 6, so the degree of the polynomial is 6.

62. $a^3b^8c^2 - c^{22} + a^5c^{10}$

Term	Coefficient	Degree
$a^3b^8c^2$	1	13
$-c^{22}$	-1	22
a^5c^{10}	1	15

The term of highest degree is $-c^{22}$. Its degree is 22, so the degree of the polynomial is 22.

63. $u + 3v - 5u + v - 7$
$= (1-5)u + (3+1)v - 7$
$= -4u + 4v - 7$

64. $6m^3 + 3m^2n + 4mn^2 + m^2n - 5mn^2$
$= 6m^3 + (3+1)m^2n + (4-5)mn^2$
$= 6m^3 + 4m^2n - mn^2$

65. $(4a^2 - 10ab - b^2) + (-2a^2 - 6ab + b^2)$
$= 4a^2 - 10ab - b^2 - 2a^2 - 6ab + b^2$
$= (4-2)a^2 + (-10-6)ab + (-1+1)b^2$
$= 2a^2 - 16ab$

66. $(6x^3y^2 - 4x^2y - 6x) - (-5x^3y^2 + 4x^2y + 6x^2 - 6)$
$= 6x^3y^2 - 4x^2y - 6x + 5x^3y^2 - 4x^2y - 6x^2 + 6$
$= (6+5)x^3y^2 + (-4-4)x^2y - 6x^2 - 6x + 6$
$= 11x^3y^2 - 8x^2y - 6x^2 - 6x + 6$

67. $(2x+5y)(x-3y)$
F O I L
$= 2x^2 - 6xy + 5x - 15y^2$
$= 2x^2 - xy - 15y^2$

68. $(5ab - cd^2)^2 = (5ab)^2 - 2\cdot 5ab\cdot cd^2 + (cd^2)^2$
$= 25a^2b^2 - 10abcd^2 + c^2d^4$

69. The figure is a triangle with base $x+y$ and height $x-y$. Its area is
$\frac{1}{2}(x+y)(x-y)=\frac{1}{2}(x^2-y^2)=\frac{1}{2}x^2-\frac{1}{2}y^2.$

70. $(3y^5-y^2+12y)\div(3y)$
$=\frac{3y^5-y^2+12y}{3y}$
$=\frac{3y^5}{3y}-\frac{y^2}{3y}+\frac{12y}{3y}$
$=y^4-\frac{1}{3}y+4$

71.
$$\begin{array}{r|l} & 3x^2-7x+4 \\ \hline 2x+3 & 6x^3-5x^2-13x+13 \\ & \underline{6x^3+9x^2} \\ & -14x^2-13x \\ & \underline{-14x^2-21x} \\ & 8x+13 \\ & \underline{8x+12} \\ & 1 \end{array}$$

The answer is $3x^2-7x+4+\frac{1}{2x+3}$.

72.
$$\begin{array}{r|l} & t^3\quad+2t-3 \\ \hline t+1 & t^4+t^3+2t^2-t-3 \\ & \underline{t^4+t^3} \\ & 2t^2-t \\ & \underline{2t^2+2t} \\ & -3t-3 \\ & \underline{-3t-3} \\ & 0 \end{array}$$

The answer is t^3+2t-3.

73. *Writing Exercise.*

74. *Writing Exercise.*

75. a. For $(x^5-6x^2+3)(x^4+3x^3+7)$, the highest terms of each factor are x^5 and x^4. Their product is x^9, which is degree 9. Thus, the degree of the product is 9.

b. For $(x^7-4)^4$, the term of highest degree is x^7, which is degree 7. Taking that term to the fourth power results in a term of degree 28.

76. $\left(-3x^5 \cdot 3x^3-x^6(2x)^2+(3x^4)^2+(2x^2)^4-20x^2(x^3)^2\right)^2$
$=\left(-9x^8-x^6\cdot 4x^2+9x^8+16x^8-20x^8\right)^2$
$=(-8x^8)^2$
$=64x^{16}$

77. Let $c=$ the coefficient of x^3.
Let $2c=$ the coefficient of x^4.
Let $2c-3=$ the coefficient of x.
Let $2c-3-7=$ the constant $\cdot$ (the remaining term)
The coefficient of x^2 is 0.
Solve:
$$\begin{aligned} 2c+c+0+2c-3+2c-3-7&=15 \\ 7c-13&=15 \\ 7c&=28 \\ c&=4 \end{aligned}$$
Coefficient of x^3, $c=4$
Coefficient of x^4, $2c=2.4=8$
Coefficient of x^2, 0
Coefficient of x, $2c-3=2.4-3=8-3=5$
Constant, $2c-3-7=2.4-10=8-10=-2$
The polynomial is $8x^4+4x^3+5x-2$.

78. $\left[(x-5)-4x^3\right]\left[(x-5)+4x^3\right]$
Let $(x-5)=A$, $4x^3=B$.
Then $(A-B)(A+B)=A^2-B^2$
$=(x-5)^2-(4x^3)^2$
$=x^2-10x+25-16x^6$
$=-16x^6+x^2-10x+25$

79.
$$\begin{aligned} (x-7)(x+10)&=(x-4)(x-6) \\ x^2+10x-7x-70&=x^2-6x-4x+24 \\ x^2+3x-70&=x^2-10x+24 \\ 3x-70&=-10x+24 && \text{Subtracting } x^2 \\ 3x&=-10x+94 && \text{Adding 70} \\ 13x&=94 && \text{Adding } 10x \\ x&=\frac{94}{13} \end{aligned}$$
The solution is $\frac{94}{13}$.

80. $(1.14\times10^6\text{ ml}^3)\left(2\times10^5\ \frac{\text{platelets}}{\text{ml}^3}\right)$
$=(1.14)(2)\cdot(10^6)(10^5)$ platelets
$=2.28\times10^{11}$ platelets

Chapter 4 Test

1. $x^7\cdot x\cdot x^5=x^{7+1+5}=x^{13}$

2. $\frac{3^8}{3^7}=3^{8-7}=3^1=3$

3. $\frac{(3m)^4}{(3m)^4}=1$

4. $(t^5)^9=t^{5\cdot9}=t^{45}$

5. $(-3y^2)^3=(-3)^3\cdot(y^2)^3=-27y^{2\cdot3}=-27y^6$

6. $(5x^4y)(-2x^5y)^3 = (5x^4y)(-2)^3(x^5y)^3$
$= 5x^4y\cdot(-8)x^{5\cdot3}y^{1\cdot3}$
$= 5x^4y\cdot(-8)\cdot x^{15}y^3$
$= 5(-8)x^{4+15}y^{1+3}$
$= -40x^{19}y^4$

7. $\dfrac{24a^7b^4}{20a^2b} = \dfrac{24}{20}a^{7-2}b^{4-1} = \dfrac{6}{5}a^5b^3$

8. $\left(\dfrac{4p}{5q^3}\right)^2 = \dfrac{4^2p^2}{5^2q^{3\cdot2}} = \dfrac{16p^2}{25q^6}$

9. $y^{-7} = \dfrac{1}{y^7}$

10. $\dfrac{1}{5^6} = 5^{-6}$

11. $t^{-4}\cdot t^{-5} = t^{-4+(-5)} = t^{-9} = \dfrac{1}{t^9}$

12. $\dfrac{9x^3y^2}{3x^8y^{-3}} = 3x^{3-8}y^{2+3} = 3x^{-5}y^5 = \dfrac{3y^5}{x^5}$

13. $(2a^3b^{-1})^{-4} = 2^{-4}(a^3)^{-4}(b^{-1})^{-4}$
$= 2^{-4}\cdot a^{-12}\cdot b^4$
$= \dfrac{1}{2^4}\cdot\dfrac{1}{a^{12}}\cdot b^4$
$= \dfrac{b^4}{16a^{12}}$

14. $\left(\dfrac{ab}{c}\right)^{-3} = \left(\dfrac{c}{ab}\right)^3 = \dfrac{c^3}{a^3b^3}$

15. $3{,}060{,}000{,}000 = 3.06\times10^9$

16. $5\times10^{-8} = 0.00000005$

17. $\dfrac{5.6\times10^6}{3.2\times10^{-11}} = \dfrac{5.6}{3.2}\times\dfrac{10^6}{10^{-11}}$
$= 1.75\times10^{6-(-11)}$
$= 1.75\times10^{17}$

18. $(2.4\times10^5)(5.4\times10^{16}) = (2.4\times5.4)(10^5\times10^{16})$
$= 12.96\times10^{5+16}$
$= 1.296\times10^1\times10^{21}$
$= 1.296\times10^{1+21}$
$= 1.296\times10^{22}$

19. Two monomials are added so $4x^2y-7y^3$ is a binomial.

20. $3x^5 - x + \dfrac{1}{9}$

The coefficients are $3, -1$ and $\dfrac{1}{9}$.

21. $2t^3 - t + 7t^5 + 4$
The degrees of the terms are 3, 1, 5, 0; the leading term is $7t^5$; the leading coefficient is 7; the degree of the polynomial is 5.

22. x^2+5x-1 for $x=-3$
$(-3)^2+5(-3)-1 = 9-15-1 = -7$

23. $y^2-3y-y+\dfrac{3}{4}y^2 = \dfrac{7}{4}y^2-4y$

24. $3-x^2+8x+5x^2-6x-2x+4x^3$
$= 4x^3+(-1+5)x^2+(8-6-2)x+3$
$= 4x^3+4x^2+3$

25. $(3x^5+5x^3-5x^2-3)+(x^5+x^4-3x^2+2x-4)$
$= (3+1)x^5+x^4+5x^3+(-5-3)x^2+2x+(-3-4)$
$= 4x^5+x^4+5x^3-8x^2+2x-7$

26. $\left(x^4+\dfrac{2}{3}x+5\right)+\left(4x^4+5x^2+\dfrac{1}{3}x\right)$
$= (1+4)x^4+5x^2+\left(\dfrac{2}{3}x+\dfrac{1}{3}\right)x+5$
$= 5x^4+5x^2+x+5$

27. $(5a^4+3a^3-a^2-2a-1)-(7a^4-a^2-a+6)$
$= 5a^4+3a^3-a^2-2a-1-7a^4+a^2+a-6$
$= (5-7)a^4+3a^3+(-1+1)a^2+(-2+1)a+(-1-6)$
$= -2a^4+3a^3-a-7$

28. $(t^3-0.3t^2-20)-(t^4-1.5t^3+0.3t^2-11)$
$= t^3-0.3t^2-20-t^4+1.5t^3-0.3t^2+11$
$= -t^4+(1+1.5)t^3+(-0.3-0.3)t^2+(-20+11)$
$= -t^4+2.5t^3-0.6t^2-9$

29. $-2x^2(3x^2-3x-5)$
$= -2x^2\cdot3x^2-2x^2\cdot(-3x)-2x^2(-5)$
$= -6x^4+6x^3+10x^2$

30. $\left(x-\dfrac{1}{3}\right)^2 = x^2-2\cdot x\cdot\dfrac{1}{3}+\dfrac{1}{3}^2$
$= x^2-\dfrac{2}{3}x+\dfrac{1}{9}$

31. $(5t-7)(5t+7) = (5t)^2-7^2$
$= 25t^2-49$

32. $(3b+5)(2b-1) = 6b^2-3b+10b-5$
$= 6b^2-7b-5$

33. $(x^6-4)(x^8+4) = x^{14}+4x^6-4x^8-16$
$= x^{14}-4x^8+4x^6-16$

34. $(8-y)(6+5y) = 48+40y-6y-5y^2$
$= 48+34y-5y^2$

35. $(2x+1)(3x^2-5x-3)$

$$\begin{array}{r} 3x^2-5x-3 \\ 2x+1 \\ \hline 3x^2-5x-3 \\ 6x^3-10x^2-6x \\ \hline 6x^3-7x^2-11x-3 \end{array}$$

36. $(8a^3+3)^2 = (8a^3)^2 + 2(8a^3)(3) + 3^2$
$= 64a^6 + 48a^3 + 9$

37. $2x^2y - 3y^2$ for $x=-3$, and $y=2$

$2(-3)^2(2) - 3(2)^2 = 36 - 12 = 24$

38. $2x^3y - y^3 + xy^3 + 8 - 6x^3y - x^2y^2 + 11$
$= (2-6)x^3y - x^2y^2 + xy^3 - y^3 + (8+11)$
$= -4x^3y - x^2y^2 + xy^3 - y^3 + 19$

39. $(8a^2b^2 - ab + b^3) - (-6ab^2 - 7ab - ab^3 + 5b^3)$
$= 8a^2b^2 - ab + b^3 + 6ab^2 + 7ab + ab^3 - 5b^3$
$= 8a^2b^2 + (-1+7)ab + 6ab^2 + ab^3 + (1-5)b^3$
$= 8a^2b^2 + 6ab + 6ab^2 + ab^3 - 4b^3$

40. $(3x^5 - y)(3x^5 + y) = (3x^5)^2 - y^2$
$= 9x^{10} - y^2$

41. $(12x^4 + 9x^3 - 15x^2) \div (3x^2)$
$= \frac{12x^4}{3x^2} + \frac{9x^3}{3x^2} - \frac{15x^2}{3x^2}$
$= 4x^2 + 3x - 5$

42.
$$\begin{array}{r} 6x^2-20x+26 \\ x+2\,\overline{)\,6x^3-8x^2-14x+13} \\ \underline{6x^3+12x^2} \\ -20x^2-14x \\ \underline{-20x^2-40x} \\ 26x+13 \\ \underline{26x+52} \\ -39 \end{array}$$

The answer is $6x^2 - 20x + 26 - \frac{39}{x+2}$.

43. ***Familiarize***. Let l = the length of the box. Express the height and width in terms of l. V=product of the dimensions.

Translate

height	is	length	less 1
↓	↓	↓	↓
h	=	l	-1
length	is	width	and 2 more
↓	↓	↓	↓
l	=	w	$+2$

or $l-2=w$
$V = l \cdot w \cdot h$
$V = l \cdot (l-2) \cdot (l-1)$

Carry out. We multiply.
$V = l \cdot [(l-2)(l-1)]$
$= l \cdot [l^2 - l - 2l + 2]$
$= l \cdot [l^2 - 3l + 2]$
$= l^3 - 3l^2 + 2l$

Check. We can do a partial check by choosing values of l, w, and h and calculating the volume in two ways.
Let $l = 4$
$h = l - 1 = 4 - 1 = 3$
$w = l - 2 = 4 - 2 = 2$

$V = l \cdot w \cdot h$	$V = l^3 - 3l^2 + 2l$
$= 4 \cdot 3 \cdot 2$	$= 4^3 - 3 \cdot 4^2 + 2 \cdot 4$
$= 24$	$= 64 - 48 + 8$
	$= 24$

Since the volumes are equal, our answer is probably correct.

State. A polynomial for the volume is $l^3 - 3l^2 + 2l$.

44. $2^{-1} - 4^{-1} = \frac{1}{2} - \frac{1}{4} = \frac{1}{4}$

45. ***Familiarize***. We need to multiply to determine the hours wasted. There are 60 sec in a minute and 60 min in an hr. 1 sec $= \frac{1}{60} \cdot \frac{1}{60} = \frac{1}{3600}$ hr.

So 4 sec $= \frac{4}{3600}$ hr

Translate. We multiply the number of spam emails by the time wasted on each spam email. 265 billion $= 2.65 \times 10^{11}$.

Carry out

$s = 2.65 \times 10^{11} \cdot \frac{4}{3600} \approx 2.9 \times 10^8$ hr

Check. We recalculate to check our solution. The answer checks.

State. About 2.9×10^8 hr each day are wasted due to spam.

Chapter 5

Polynomials and Factoring

Exercise Set 5.1

1. False. The largest common variable factor is the smallest power of x in the original polynomial.

3. True

5. Since $7a \cdot 5ab = 35a^2b$, choice (h) is most appropriate.

7. $5x+10=5(x+2)$ and $4x+8=4(x+2)$, so $x+2$ is a common factor of $5x+10$ and $4x+8$ and choice (b) is most appropriate.

9. $3x^2(3x^2-1)=9x^4-3x^2$, so choice (c) is most appropriate.

11. $3a+6a^2=3a(1+2a)$, so $1+2a$ is a factor of $3a+6a^2$ and choice (d) is most appropriate.

13. Answers may vary.
$14x^3=(14x)(x^2)=(7x^2)(2x)=(-2)(-7x^3)$

15. Answers may vary.
$-15a^4=(-15)(a^4)=(-5a)(3a^3)=(-3a^2)(5a^2)$

17. Answers may vary.
$25t^5=(5t^2)(5t^3)=(25t)(t^4)=(-5t)(-5t^4)$

19. $8x+24=8\cdot x+8\cdot 3$
$=8(x+3)$

21. $2x^2+2x-8=2\cdot x^2+2\cdot x-2\cdot 4$
$=2(x^2+x-4)$

23. $3t^2+t=t\cdot 3t+t\cdot 1$
$=t(3t+1)$

25. $-5y^2-10y=-5y\cdot y-5y\cdot 2$
$=-5y(y+2)$

27. $x^3+6x^2=x^2\cdot x+x^2\cdot 6$
$=x^2(x+6)$

29. $16a^4-24a^2=8a^2\cdot 2a^2-8a^2\cdot 3$
$=8a^2(2a^2-3)$

31. $-6t^6+9t^4-4t^2=-t^2\cdot 6t^4-t^2(-9t^2)-t^2\cdot 4$
$=-t^2(6t^4-9t^2+4)$

33. $6x^8+12x^6-24x^4+30x^2$
$=6x^2\cdot x^6+6x^2\cdot 2x^4-6x^2\cdot 4x^2+6x^2\cdot 5$
$=6x^2(x^6+2x^4-4x^2+5)$

35. $x^5y^5+x^4y^3+x^3y^3-x^2y^2$
$=x^2y^2\cdot x^3y^3+x^2y^2\cdot x^2y+x^2y^2\cdot xy-x^2y^2\cdot 1$
$=x^2y^2(x^3y^3+x^2y+xy-1)$

37. $-35a^3b^4+10a^2b^3-15a^3b^2$
$=-5a^2b^2\cdot 7ab^2-5a^2b^2(-2b)-5a^2b^2\cdot 3a$
$=-5a^2b^2(7ab^2-2b+3a)$

39. $n(n-6)+3(n-6)$
$=(n-6)(n+3)$ Factoring out the common binomial factor $n-6$

41. $x^2(x+3)-7(x+3)$
$=(x+3)(x^2-7)$ Factoring out the common binomial factor $x+3$

43. $y^2(2y-9)+(2y-9)$
$=y^2(2y-9)+1(2y-9)$
$=(2y-9)(y^2+1)$ Factoring out the common factor $2y-9$

45. $x^3+2x^2+5x+10$
$=(x^3+2x^2)+(5x+10)$
$=x^2(x+2)+5(x+2)$ Factoring each binomial
$=(x+2)(x^2+5)$ Factoring out the common factor $x+2$

47. $9n^3-6n^2+3n-2$
$=3n^2(3n-2)+1(3n-2)$
$=(3n-2)(3n^2+1)$

49. $4t^3-20t^2+3t-15$
$=4t^2(t-5)+3(t-5)$
$=(t-5)(4t^2+3)$

51. $7x^3+5x^2-21x-15=x^2(7x+5)-3(7x+5)$
$=(7x+5)(x^2-3)$

53. $6a^3+7a^2+6a+7=a^2(6a+7)+1(6a+7)$
$=(6a+7)(a^2+1)$

55. $2x^3+12x^2-x+6=2x^2(x+6)-1(x+6)$
$=(x+6)(2x^2-1)$

57. We try factoring by grouping.

$p^3 + p^2 - 3p + 10 = p^2(p+1) - (3p - 10)$, or

$p^3 - 3p + p^2 + 10 = p(p^2 - 3) + p^2 + 10$

Because we cannot find a common binomial factor, this polynomial cannot be factored by grouping.

59. $y^3 + 8y^2 - 2y - 16 = y^2(y+8) - 2(y+8)$
$= (y+8)(y^2 - 2)$

61. $2x^3 + 36 - 8x^2 - 9x = 2x^3 - 8x^2 - 9x + 36$
$= 2x^2(x-4) - 9(x-4)$
$= (x-4)(2x^2 - 9)$

63. *Writing Exercise.*

65. $\frac{2}{5} \div \frac{10}{3} = \frac{2}{5} \cdot \frac{3}{10} = \frac{3}{25}$

67. $(2xy^{-4})^{-1} = 2^{-1}x^{-1}y^4 = \frac{y^4}{2x}$

69. $(3x^2 - x - 3) - (-x^2 - 6x + 10)$
$= 3x^2 - x - 3 + x^2 + 6x - 10$
$= 4x^2 + 5x - 13$

71. *Writing Exercise.*

73. $4x^5 + 6x^2 + 6x^3 + 9 = 2x^2(2x^3 + 3) + 3(2x^3 + 3)$
$= (2x^3 + 3)(2x^2 + 3)$

75. $2x^4 + 2x^3 - 4x^2 - 4x = 2x(x^3 + x^2 - 2x - 2)$
$= 2x(x^2(x+1) - 2(x+1))$
$= 2x(x+1)(x^2 - 2)$

77. $5x^5 - 5x^4 + x^3 - x^2 + 3x - 3$
$= 5x^4(x-1) + x^2(x-1) + 3(x-1)$
$= (x-1)(5x^4 + x^2 + 3)$

We could also do this exercise as follows:

$5x^5 - 5x^4 + x^3 - x^2 + 3x - 3$
$= (5x^5 + x^3 + 3x) - (5x^4 + x^2 + 3)$
$= x(5x^4 + x^2 + 3) - 1(5x^4 + x^2 + 3)$
$= (5x^4 + x^2 + 3)(x-1)$

79. Answers may vary. $8x^4y^3 - 24x^3y^3 + 16x^2y^4$

Exercise Set 5.2

1. False. The product of factors that have the same sign is positive.

3. True

5. If b is positive and c is positive, then p will be *positive* and q will be *positive*.

7. If p is negative and q is negative, then b must be *negative* and c must be *positive*.

9. If b, c, and p are all negative, then q must be *positive*.

11. $x^2 + 8x + 16$

Since the constant term and the coefficient of the middle term are both positive, we look for a factorization of 16 in which both factors are positive. Their sum must be 8.

Pairs of factors	Sums of factors
1, 16	17
2, 8	10
4, 4	8

The numbers we want are 4 and 4.

$x^2 + 8x + 16 = (x+4)(x+4)$

13. $x^2 + 11x + 10$

Since the constant term and the coefficient of the middle term are both positive, we look for a factorization of 10 in which both factors are positive. Their sum must be 11.

Pairs of factors	Sums of factors
1, 10	11
2, 5	10

The numbers we want are 1 and 10.

$x^2 + 11x + 10 = (x+1)(x+10)$

15. $t^2 - 9t + 14$

Since the constant term is positive and the coefficient of the middle term is negative, we look for a factorization of 14 in which both factors are negative.

Their sum must be –9.

Pairs of factors	Sums of factors
−1, −14	−15
−2, −7	−9

The numbers we want are –2 and –7.

$t^2 - 9t + 14 = (t-2)(t-7)$

17. $b^2 - 5b + 4$

Since the constant term is positive and the coefficient of the middle term is negative, we look for a factorization of 4 in which both factors are negative. Their sum must be –5.

Pairs of factors	Sums of factors
−1, −4	−5
−2, −2	−4

The numbers we want are –1 and –4.

$b^2 - 5b + 4 = (b-1)(b-4)$

19. $d^2 - 7d + 10$

Since the constant term is positive and the coefficient of the middle term is negative, we look for a factorization of 10 in which both factors are negative. Their sum must be –7.

Pairs of factors	Sums of factors
−1, −10	−11
−2, −5	−7

The numbers we want are –2 and –5.

$d^2 - 7d + 10 = (d-2)(d-5)$.

21. $x^2 - 2x - 15$
The constant term, –15, must be expressed as the product of a negative number and a positive number. Since the sum of those two numbers must be negative, the negative number must have the greater absolute value.

Pairs of factors	Sums of factors
1, –15	–14
3, –5	–2

The numbers we need are 3 and –5.

$$x^2 - 2x - 15 = (x+3)(x-5).$$

23. $x^2 + 2x - 15$
The constant term, –15, must be expressed as the product of a negative number and a positive number. Since the sum of those two numbers must be positive, the positive number must have the greater absolute value.

Pairs of factors	Sums of factors
–1, 15	14
–3, 5	2

The numbers we need are –3 and 5.

$$x^2 + 2x - 15 = (x-3)(x+5).$$

25. $2x^2 - 14x - 36 = 2(x^2 - 7x - 18)$
After factoring out the common factor, 2, we consider $x^2 - 7x - 18$. The constant term, –18, must be expressed as the product of a negative number and a positive number. Since the sum of those two numbers must be negative, the negative number must have the greater absolute value.

Pairs of factors	Sums of factors
1, –18	–17
2, –9	–7
3, –6	–3

The numbers we need are 2 and–9. The factorization of $x^2 - 7x - 18$ is $(x-9)(x+2)$. We must not forget the common factor, 2. Thus,

$$2x^2 - 14x - 36 = 2(x^2 - 7x - 18) = 2(x-9)(x+2).$$

27. $-x^3 + 6x^2 + 16x = -x(x^2 - 6x - 16)$
After factoring out the common factor, –x, we consider $x^2 - 6x - 16$. The constant term, –16, must be expressed as the product of a negative number and a positive number.
Since the sum of those two numbers must be negative, the negative number must have the greater absolute value.

Pairs of factors	Sums of factors
1, –16	–15
2, –8	–6
4, –4	0

The numbers we need are 2 and -8. The factorization of $x^2 - 6x - 16$ is $(x+2)(x-8)$. We must not forget the common factor, $-x$. Thus,

$$\begin{aligned}-x^3 + 6x^2 + 16x &= -x(x^2 - 6x - 16)\\ &= -x(x+2)(x-8)\end{aligned}$$

29. $4y - 45 + y^2 = y^2 + 4y - 45$
The constant term, –45, must be expressed as the product of a negative number and a positive number. Since the sum of those two numbers must be positive, the positive number must have the greater absolute value.

Pairs of factors	Sums of factors
–1, 45	44
–3, 15	12
–5, 9	4

The numbers we need are –5 and 9.

$$4y - 45 + y^2 = (y-5)(y+9)$$

31. $x^2 - 72 + 6x = x^2 + 6x - 72$
The constant term, –72, must be expressed as the product of a negative number and a positive number. Since the sum of those two numbers must be positive, the positive number must have the greater absolute value.

Pairs of factors	Sums of factors
–1, 72	71
–2, 36	34
–3, 24	21
–4, 18	14
–6, 12	6

The numbers we need are –6 and 12.

$$x^2 - 72 + 6x = (x-6)(x+12)$$

33. $-5b^2 - 35b + 150 = -5(b^2 + 7b - 30)$
After factoring out the common factor, -5, we consider $b^2 + 7b - 30$. The constant term, -30, must be expressed as the product of a negative number and a positive number. Since the sum of those two numbers must be positive, the positive number must have the greater absolute value.

Pairs of factors	Sums of factors
–1, 30	29
–2, 15	13
–3, 10	7
–5, 6	1

The numbers we need are –3 and 10. The factorization of $b^2 + 7b - 30$ is $(b-3)(b+10)$. We must not forget the common factor. Thus,

$$\begin{aligned}-5b^2 - 35b + 150 &= -5(b^2 + 7b - 30)\\ &= -5(b-3)(b+10)\end{aligned}$$

35. After factoring out the common factor, x^3, we consider $x^2 - x - 2$. The constant term, –2, must be expressed as the product of a negative number and a positive number. Since the sum of those two numbers must be negative, the negative number must have the greater absolute value. The only possible factors that fill these requirements are 1 and –2. These are the numbers we need. The factorization of $x^2 - x - 2$ is $(x+1)(x-2)$. We must not forget the common factor, x^3. Thus,

$x^5 - x^4 - 2x^3 = x^3(x^2 - x - 2) = x^3(x+1)(x-2)$.

37. $x^2 + 5x + 10$
Since the constant term and the coefficient of the middle term are both positive, we look for a factorization of 10 in which both factors are positive. Their sum must be 5. The only possible pairs of positive factors are 1 and 10, and 2 and 5 but neither sum is 5. Thus, this polynomial is not factorable into polynomials with integer coefficients. It is prime.

39. $32 + 12t + t^2 = t^2 + 12t + 32$
Since the constant term is positive and the coefficient of the middle term is positive, we look for a factorization of 32 in which both terms are positive. Their sum must be 12.

Pairs of factors	Sums of factors
1, 32	33
2, 16	18
4, 8	12

The numbers we want are 4 and 8.

$32 + 12t + t^2 = (t+4)(t+8)$

41. $x^2 + 20x + 99$
We look for two factors, both positive, whose product is 99 and whose sum is 20.
They are 9 and $11: 9 \cdot 11 = 99$ and $9 + 11 = 20$.

$x^2 + 20 + 99 = (x+9)(x+11)$

43. $3x^3 - 63x^2 - 300x = 3x(x^2 - 21x - 100)$
After factoring out the common factor, $3x$, we consider $x^2 - 21x - 100$. We look for two factors, one positive, one negative, whose product is –100 and whose sum is –21.
They are –25 and $4: -25(4) = -100$ and $-25 + 4 = -21$.

$x^2 - 21x - 100 = (x-25)(x+4)$, so

$3x^3 - 63x^2 - 300x = 3x(x-25)(x+4)$.

45. $-4x^2 - 40x - 100 = -4(x^2 + 10x + 25)$
$= -4(x+5)(x+5)$

47. $y^2 - 20y + 96$
We look for two factors, both negative, whose product is 96 and whose sum is -20. They are -8 and -12.

$y^2 - 20y + 96 = (y-8)(y-12)$

49. $-a^6 - 9a^5 + 90a^4 = -a^4(a^2 + 9a - 90)$
After factoring out the common factor, $-a^4$, we consider $a^2 + 9a - 90$. We look for two factors, one positive and one negative, whose product is –90 and whose sum is 9. They are –6 and 15.

$a^2 + 9a - 90 = (a-6)(a+15)$, so

$-a^6 - 9a^5 + 90a^4 = -a^4(a-6)(a+15)$.

51. $t^2 + \frac{2}{3}t + \frac{1}{9}$
We look for two factors, both positive, whose product is $\frac{1}{9}$ and whose sum is $\frac{2}{3}$. They are $\frac{1}{3}$ and $\frac{1}{3}$.

$t^2 + \frac{2}{3}t + \frac{1}{9} = \left(t + \frac{1}{3}\right)\left(t + \frac{1}{3}\right)$, or $\left(t + \frac{1}{3}\right)^2$

53. $11 + w^2 - 4w = w^2 - 4w + 11$
Since the constant term is positive and the coefficient of the middle term is negative, we look for a factorization of 11 in which both factors are negative. Their sum must be –4. The only possible pair of factors is –1 and –11, but their sum is not –4. Thus, this polynomial is not factorable into polynomials with integer coefficients. It is prime.

55. $p^2 - 7pq + 10q^2$
Think of $-7q$ as a "coefficient" of p. Then we look for factors of $10q^2$ whose sum is $-7q$. They are $-5q$ and $-2q$.

$p^2 - 7pq + 10q^2 = (p-5q)(p-2q)$.

57. $m^2 + 5mn + 5n^2 = m^2 + 5nm + 5n^2$
We look for factors of $5n^2$ whose sum is $5n$. The only reasonable possibilities are shown below.

Pairs of factors	Sums of factors
$5n$, n	$6n$
$-5n$, $-n$	$-6n$

There are no factors whose sum is $5n$. Thus, the polynomial is not factorable into polynomials with integer coefficients. It is prime.

59. $s^2 - 4st - 12t^2 = s^2 - 4ts - 12t^2$
We look for factors of $-12t^2$ whose sum is $-4t$. They are $-6t$ and $2t$.

$s^2 - 4st - 12t^2 = (s-6t)(s+2t)$

61. $6a^{10} + 30a^9 - 84a^8 = 6a^8(a^2 + 5a - 14)$
After factoring out the common factor, $6a^8$, we consider $a^2 + 5a - 14$. We look for two factors, one positive and one negative, whose product is -14 and whose sum is 5. They are -2 and 7.

$a^2 + 5a - 14 = (a-2)(a+7)$, so

$6a^{10} + 30a^9 - 84a^8 = 6a^8(a-2)(a+7)$.

63. *Writing Exercise.*

65. $\frac{y}{2} = -5$
$y = -10$

67. $3x + 7 = 12$
$3x = 5$
$x = \frac{5}{3}$

69. $x - (2x - 5) = 16$
$x - 2x + 5 = 16$
$-x + 5 = 16$
$-x = 11$
$x = -11$

71. *Writing Exercise.*

73. $a^2 + ba - 50$
We look for all pairs of integer factors whose product is –50. The sum of each pair is represented by b.

Pairs of factors whose product is – 50	Sums of factors
–1, 50	49
1, – 50	–49
–2, 25	23
2, – 25	–23
–5, 10	5
5, – 10	–5

The polynomial $a^2 + ba - 50$ can be factored if b is 49, –49, 23, –23, 5, or – 5.

75. $y^2 - 0.2y - 0.08$
We look for two factors, one positive and one negative, whose product is –0.08 and whose sum is –0.2. They are –0.4 and 0.2.
$y^2 - 0.2y - 0.08 = (y - 0.4)(y + 0.2)$

77. $-\frac{1}{3}a^3 + \frac{1}{3}a^2 + 2a = -\frac{1}{3}a(a^2 - a - 6)$
After factoring out the common factor, $-\frac{1}{3}a$, we consider $a^2 - a - 6$. We look for two factors, one positive and one negative, whose product is –6 and whose sum is –1. They are 2 and –3.
$a^2 - a - 6 = (a + 2)(a - 3)$, so
$-\frac{1}{3}a^3 + \frac{1}{3}a^2 + 2a = -\frac{1}{3}a(a + 2)(a - 3).$

79. $x^{2m} + 11x^m + 28 = (x^m)^2 + 11x^m + 28$
We look for numbers p and q such that $x^{2m} + 11x^m + 28 = (x^m + p)(x^m + q)$. We find two factors, both positive, whose product is 28 and whose sum is 11. They are 4 and 7.

81. $(a+1)x^2 + (a+1)3x + (a+1)2$
$= (a+1)(x^2 + 3x + 2)$
After factoring out the common factor $a + 1$, we consider $x^2 + 3x + 2$. We look for two factors, whose product is 2 and whose sum is 3. They are 1 and 2.
$x^2 + 3x + 2 = (x+1)(x+2)$, so
$(a+1)x^2 + (a+1)3x + (a+1)2$
$= (a+1)(x+1)(x+2).$

83. $6x^2 + 36x + 54 = 6(x^2 + 6x + 9)$
$= 6(x+3)(x+3)$
$= 6(x+3)^2$
Since the surface area of a cube with sides is given by $6s^2$, we know that this cube has side $x + 3$. The volume of a cube with side s is given by s^3, so the volume of this cube is
$(x+3)^3$, or $x^3 + 9x^2 + 27x + 27$.

85. The shaded area consists of the area of a rectangle with sides x and $x + x$, or $2x$, and $\frac{3}{4}$ of the area of a circle with radius x. It can be expressed as follows:
$x \cdot 2x + \frac{3}{4}\pi x^2 = 2x^2 + \frac{3}{4}\pi x^2 = x^2\left(2 + \frac{3}{4}\pi\right)$, or
$\frac{1}{4}x^2(8 + 3\pi)$

87. The shaded area consists of the area of a square with side $x + x + x$, or $3x$, less the area of a semicircle with radius x. It can be expressed as follows:
$3x \cdot 3x - \frac{1}{2}\pi x^2 = 9x^2 - \frac{1}{2}\pi x^2 = x^2\left(9 - \frac{1}{2}\pi\right)$

89. $x^2 + 4x + 5x + 20 = x^2 + 9x + 20$
$= (x+4)(x+5)$

Exercise Set 5.3

1. Since $11x^2 + 3x + 1$ is a prime polynomial, choice (d) is correct.

3. Since the first term of $5x^2 - 14x - 3$ is $5x^2$, the only possibilities for the first terms of the binomial factors are $5x$ and x, so choice (b) is correct.

5. $2x^2 + 7x - 4$
(1) There is no common factor (other than 1 or –1).
(2) Because $2x^2$ can be factored as $2x \cdot x$, we have this possibility:
$(2x + \quad)(x + \quad)$
(3) There are 3 pairs of factors of –4 and they can be listed two ways:
–4, 1 4, –1 2, –2
and 1, –4 –1, 4 –2, 2
(4) Look for Outer and Inner products resulting from steps (2) and (3) for which the sum is $7x$. We can immediately reject all possibilities in which a factor has a common factor, such as $(2x - 4)$ or $(2x + 2)$, because we determined at the outset that there is no common factor other than 1 and –1.

We try some possibilities:

$(2x+1)(x-4) = 2x^2 - 7x - 4$

$(2x-1)(x+4) = 2x^2 + 7x - 4$

The factorization is $(2x - 1)(x + 4)$

7. $3x^2 - 17x - 6$

(1) There is no common factor (other than 1 or –1).

(2) Because $3x^2$ can be factored as $3x \cdot x$, we have this possibility:

$(3x+\ \ )(x+\ \ )$

(3) There are 4 pairs of factors of –6 and they can be listed two ways:

	–6, 1	6, –1	–3, 2	3, –2
and	1, –6	–1, 6	2, –3	–2, 3

(4) Look for Outer and Inner products resulting from steps (2) and (3) for which the sum is $-17x$. We can immediately reject all possibilities in which either factor has a common factor, such as $(3x - 6)$ or $(3x + 3)$, because at the outset we determined that there is no common factor other than 1 or –1. We try some possibilities:

$(3x+2)(x-3) = 3x^2 - 7x - 6$

$(3x+1)(x-6) = 3x^2 - 17x - 6$

The factorization is $(3x + 1)(x - 6)$.

9. $4t^2 + 12t + 5$

(1) There is no common factor (other than 1 or –1).

(2) Because $4t^2$ can be factored as $4t \cdot t$ or $2t \cdot 2t$, we have these possibilities:

$(4t+\ \ )(t+\ \ )$ and $(2t+\ \ )(2t+\ \ )$

(3) Since the middle term is positive, we need consider only the factorizations with positive factors:

5, 1 and –1, –5

(4) Look for Outer and Inner products resulting from steps (2) and (3) for which the sum is $12t$. We try some possibilities:

$(4t+1)(t+5) = 4t^2 + 21t + 5$

$(2t+1)(2t+5) = 4t^2 + 12t + 5$

The factorization is $(2t + 1)(2t + 5)$.

11. $15a^2 - 14a + 3$

(1) There is no common factor (other than 1 or –1).

(2) Because $15a^2$ can be factored as $15a \cdot a$ or $5a \cdot 3a$, we have these possibilities:

$(15a+\ \ )(a+\ \ )$ and $(5a+\ \ )(3a+\ \ )$

(3) Since the middle term is negative, we need consider only the factorizations with negative factors:

–3, –1 and –1, –3

(4) Look for Outer and Inner products resulting from steps (2) and (3) for which the sum is $-14a$. We can immediately reject all possibilities in which either factor has a common factor, such as $(15a+3)$ or $(3a-3)$, because at the outset we determined that there is no common factor other than 1 or –1.

We try some possibilities:

$(15a-1)(a-3) = 15a^2 - 46a + 3$

$(5a-3)(3a-1) = 15a^2 - 14a + 3$

The factorization is $(5a-3)(3a-1)$.

13. $6x^2 + 17x + 12$

(1) There is no common factor (other than 1 or –1).

(2) Because $6x^2$ can be factored as $6x \cdot x$ and $3x \cdot 2x$, we have these possibilities:

$(6x+\ \ )(x+\ \ )$ and $(3x+\ \ )(2x+\ \ )$

(3) Since all coefficients are positive, we need consider only positive pairs of factors of 12. There are 3 pairs and they can be listed two ways:

	1, 12	2, 6	3, 4
and	12, 1	6, 2	4, 3

(4) We can immediately reject all possibilities in which either factor has a common factor, such as $(6x + 12)$ or $(3x + 3)$, because at the outset we determined that there is no common factor other than 1 or –1 We try some possibilities:

$(6x+1)(x+12) = 6x^2 + 73x + 12$

$(3x+4)(2x+3) = 6x^2 + 17x + 12$

The factorization is $(3x + 4)(2x + 3)$.

15. $6x^2 - 10x - 4$

(1) We factor out the largest common factor, 2:

$2(3x^2 - 5x - 2)$.

Then we factor the trinomial $3x^2 - 5x - 2$.

(2) Because $3x^2$ can be factored as $3x \cdot x$, we have this possibility:

$(3x+\ \ )(x+\ \ )$

(3) There are 2 pairs of factors of –2 and they can be listed two ways:

	–2, 1	2, –1
and	1, –2	–1, 2

(4) Look for Outer and Inner products resulting from steps (2) and (3) for which the sum is $-5x$ We try some possibilities:

$(3x-2)(x+1) = 3x^2 + x - 2$

$(3x+2)(x-1) = 3x^2 - x - 2$

$(3x+1)(x-2) = 3x^2 - 5x - 2$

The factorization of $3x^2 - 5x - 2$ is $(3x+1)(x-2)$. We must include the common factor in order to get a factorization of the original trinomial.

$6x^2 - 10x - 4 = 2(3x+1)(x-2)$

17. $7t^3 + 15t^2 + 2t$

(1) We factor out the common factor, t:

$t(7t^2 + 15t + 2)$.

Then we factor the trinomial $7t^2 + 15t + 2$.

(2) Because $7t^2$ can be factored as $7t \cdot t$, we have this possibility:

$(7t+\quad)(t+\quad)$

(3) Since all coefficients are positive, we need consider only positive factors of 2. There is only 1 such pair and it can be listed two ways:

2, 1 1, 2

(4) Look for Outer and Inner products resulting from steps (2) and (3) for which the sum is $15t$. We try some possibilities:

$$(7t+2)(t+1) = 7t^2+9t+2$$
$$(7t+1)(t+2) = 7t^2+15t+2$$

The factorization of $7t^2+15t+2$ is $(7t+1)(t+2)$. We must include the common factor in order to get a factorization of the original trinomial.

$$7t^3+15t^2+2t = t(7t+1)(t+2)$$

19. $10-23x+12x^2 = 12x^2-23x+10$

(1) There is no common factor (other than 1 or –1).

(2) Because $12x^2$ can be factored as $12x \cdot x$, $6x \cdot 2x$, and $4x \cdot 3x$, we have these possibilities:

$(12x+\quad)(x+\quad)$, $(6x+\quad)(2x+\quad)$, and $(4x+\quad)(3x+\quad)$

(3) Since the sign of the middle term is negative but the sign of the last term is positive, we need consider only negative factors of 10.

–10, –1 –5, –2
and –1, –10 –2, –5

(4) We can immediately reject all possibilities in which either factor has a common factor, such as $(2x-10)$ or $(4x-2)$, because we determined at the outset that there is no common factor other than 1 or –1. We try some possibilities:

$$(12x-5)(x-2) = 12x^2-29x+10$$
$$(4x-1)(3x-10) = 12x^2-43x+10$$
$$(4x-5)(3x-2) = 12x^2-23x+10$$

The factorization is $(4x-5)(3x-2)$.

21. $-35x^2-34x-8$

(1) We factor out –1 in order to have a trinomial with a positive leading coefficient.

$$-35x^2-34x-8 = -1(35x^2+34x+8)$$

Now we factor $35x^2+34x+8$.

(2) Because $35x^2$ can be factored as $35x \cdot x$ or $7x \cdot 5x$, we have these possibilities:

$(35x+\quad)(x+\quad)$ and $(7x+\quad)(5x+\quad)$

(3) Since all coefficients are positive, we need consider only positive pairs of factors of 8. There are 2 such pairs and they can be listed two ways:

8, 1 4, 2
and 1, 8 2, 4

(4) We try some possibilities:

$$(35x+8)(x+1) = 35x^2+43x+8$$
$$(7x+8)(5x+1) = 35x^2+47x+8$$
$$(7x+4)(5x+2) = 35x^2+34x+8$$

The factorization of $35x^2+34x+8$ is $(7x+4)(5x+2)$.
We must include the factor of –1 in order to get a factorization of the original trinomial.

$-35x^2-34x-8 = -1(7x+4)(5x+2)$, or $-(7x+4)(5x+2)$.

23. $4+6t^2-13t = 6t^2-13t+4$

(1) There is no common factor (other than 1 or –1).

(2) Because $6t^2$ can be factored as $6t \cdot t$ or $3t \cdot 2t$, we have these possibilities:

$(6t+\quad)(t+\quad)$ and $(3t+\quad)(2t+\quad)$

(3) Since the sign of the middle term is negative but the sign of the last term is positive, we need to consider only negative factors of 4. There is only 1 such pair and it can be listed two ways:

–4, –1 and –1, –4

(4) We can immediately reject all possibilities in which either factor has a common factor, such as $(6t-4)$ or $(2t-4)$, because we determined at the outset that there is no common factor other than 1 or –1. We try some possibilities:

$$(6t-1)(t-4) = 6t^2-25t+4$$
$$(3t-4)(2t-1) = 6t^2-11t+4$$

These are the only possibilities that do not contain a common factor. Since neither is the desired factorization, we must conclude that $4+6t^2-13t$ is prime.

25. $25x^2+40x+16$

(1) There is no common factor (other than 1 or –1).

(2) Because $25x^2$ can be factored as $25x \cdot x$ or $5x \cdot 5x$, we have these possibilities:

$(25x+\quad)(x+\quad)$ and $(5x+\quad)(5x+\quad)$

(3) Since all coefficients are positive, we need consider only positive pairs of factors of 16. There are 3 such pairs and two of them can be listed two ways:

16, 1 8, 2 4, 4
and 1, 16 2, 8

(4) We try some possibilities:

$$(25x+16)(x+1) = 25x^2+41x+16$$
$$(5x+8)(5x+2) = 25x^2+50x+16$$
$$(5x+4)(5x+4) = 25x^2+40x+16$$

The factorization is $(5x+4)(5x+4)$, or $(5x+4)^2$.

27. $20y^2+59y-3$

(1) There is no common factor (other than 1 or -1).

(2) Because $20y^2$ can be factored as $20y\cdot y$, $10y\cdot 2y$, or $5y\cdot 4y$, we have these possibilities:

$(20y+\ \ )(y+\ \ )$ and $(10y+\ \ )(2y+\ \ )$ and $(5y+\ \ )(4y+\ \ )$

(3) There are 2 such pairs of factors of -3, which can be listed two ways:

$-3, 1$ $\quad 3, -1$
and $1, -3$ $\quad -1, 3$

(4) Look for Outer and Inner products resulting from steps (2) and (3) for which the sum is $59y$. We try some possibilities:

$$(20y-3)(y+1)=20y^2+17y-3$$
$$(10y-3)(2y+1)=20y^2+4y-3$$
$$(5y-1)(4y+3)=20y^2+11y-3$$
$$(20y-1)(y+3)=20y^2+59y-3$$

The factorization is $(20y-1)(y+3)$.

29. $14x^2+73x+45$

(1) There is no common factor (other than 1 or -1).

(2) Because $14x^2$ can be factored as $14x\cdot x$, and $7x\cdot 2x$, we have two possibilities:

$(14x+\ \ )(x+\ \ )$ and $(7x+\ \ )(2x+\ \ )$

(3) Since all coefficients are positive, we need consider only positive pairs of factors of 45. There are 3 such pairs and they can be listed two ways.

$45, 1$ $\quad 15, 3$ $\quad 9, 5$
and $1, 45$ $\quad 3, 15$ $\quad 5, 9$

(4) Look for Outer and Inner products from steps (2) and (3) for which the sum is $73x$. We try some possibilities:

$$(14x+45)(x+1)=14x^2+59x+45$$
$$(14x+15)(x+3)=14x^2+57x+45$$
$$(7x+1)(2x+45)=14x^2+317x+45$$
$$(7x+5)(2x+9)=14x^2+73x+45$$

The factorization is $(7x+5)(2x+9)$.

31. $-2x^2+15+x=-2x^2+x+15$

(1) We factor out -1 in order to have a trinomial with a positive leading coefficient.

$$-2x^2+x+15=-1(2x^2-x-15)$$

Now we factor $2x^2-x-15$.

(2) Because $2x^2$ can be factored as $2x\cdot x$ we have this possibility:

$(2x+\ \ )(x+\ \ )$

(3) There are 4 pairs of factors of -15 and they can be listed two ways:

$-15, 1$ $\quad 15, -1$ $\quad -5, 3$ $\quad 5, -3$
and $1, -15$ $\quad -1, 15$ $\quad 3, -5$ $\quad -3, 5$

(4) We try some possibilities:

$$(2x-15)(x+1)=2x^2-13x-15$$
$$(2x-5)(x+3)=2x^2+x-15$$
$$(2x+5)(x-3)=2x^2-x-15$$

The factorization of $2x^2-x-15$ is $(2x+5)(x-3)$. We must include the factor of -1 in order to get a factorization of the original trinomial.

$$-2x^2+15+x=-1(2x+5)(x-3) \text{ or } -(2x+5)(x-3)$$

33. $-6x^2-33x-15$

(1) Factor out -3. This not only removes the largest common factor, 3. It also produces a trinomial with a positive leading coefficient.

$$-3(2x^2+11x+5)$$

Then we factor the trinomial $2x^2+11x+5$.

(2) Because $2x^2$ can be factored as $2x\cdot x$ we have this possibility:

$(2x+\ \ )(x+\ \ )$

(3) Since all coefficients are positive, we need consider only positive pairs of factors of 5. There is one such pair and it can be listed two ways:

$5, 1$ and $1, 5$

(4) We try some possibilities:

$$(2x+5)(x+1)=2x^2+7x+5$$
$$(2x+1)(x+5)=2x^2+11x+5$$

The factorization of $2x^2+11x+5$ is $(2x+1)(x+5)$. We must include the common factor in order to get a factorization of the original trinomial.

$$-6x^2-33x-15=-3(2x+1)(x+5)$$

35. $10a^2-8a-18$

(1) Factor out the common factor, 2:

$$2(5a^2-4a-9)$$

Then we factor the trinomial $5a^2-4a-9$.

(2) Because $5a^2$ can be factored as $5a\cdot a$, we have this possibility:

$(5a+\ \ )(a+\ \ )$

(3) There are 3 pairs of factors of -9, and they can be listed two ways.

$-9, 1$ $\quad 9, -1$ $\quad 3, -3$
and $1, -9$ $\quad -1, 9$ $\quad -3, 3$

(4) Look for Outer and Inner products resulting from steps (2) and (3) for which the sum is $-4a$. We try some possibilities:

$$(5a-3)(a+3)=5a^2+12a-9$$
$$(5a+9)(a-1)=5a^2+4a-9$$
$$(a+1)(5a-9)=5a^2-4a-9$$

The factorization of $5a^2-4a-9$ is $(a+1)(5a-9)$. We must include the common factor in order to get a factorization of the original trinomial.

$$2(a+1)(5a-9)$$

37. $12x^2 + 68x - 24$

(1) Factor out the common factor, 4:

$4(3x^2 + 17x - 6)$

Then we factor the trinomial $3x^2 + 17x - 6$.

(2) Because $3x^2$ can be factored as $3x \cdot x$ we have this possibility:

$(3x + \quad)(x + \quad)$

(3) There are 4 pairs of factors of –6 and they can be listed two ways:

6, –1   –6, 1   3, –2   –3, 2
and –1, 6   1, –6   –2, 3   2, –3

(4) We can immediately reject all possibilities in which either factor has a common factor, such as $(3x + 6)$ or $(3x - 3)$, because we determined at the outset that there is no common factor other than 1 or –1. We try some possibilities:

$(3x - 1)(x + 6) = 3x^2 + 17x - 6$

The factorization of $3x^2 + 17x - 6$ is $(3x - 1)(x + 6)$. We must include the common factor in order to get a factorization of the original trinomial.

$12x^2 + 68x - 24 = 4(3x - 1)(x + 6)$

39. $4x + 1 + 3x^2 = 3x^2 + 4x + 1$

(1) There is no common factor (other than 1 or –1).

(2) Because $3x^2$ can be factored as $3x \cdot x$ we have this possibility:

$(3x + \quad)(x + \quad)$

(3) Since all coefficients are positive, we need consider only positive pairs of factors of 1. There is one such pair: 1, 1.

(4) We try the possible factorization:

$(3x + 1)(x + 1) = 3x^2 + 4x + 1$

The factorization is $(3x + 1)(x + 1)$.

41. $x^2 + 3x - 2x - 6 = x(x + 3) - 2(x + 3)$
$= (x + 3)(x - 2)$

43 . $8t^2 - 6t - 28t + 21 = 2t(4t - 3) - 7(4t - 3)$
$= (4t - 3)(2t - 7)$

45. $6x^2 + 4x + 15x + 10 = 2x(3x + 2) + 5(3x + 2)$
$= (3x + 2)(2x + 5)$

47. $2y^2 + 8y - y - 4 = 2y(y + 4) - 1(y + 4)$
$= (y + 4)(2y - 1)$

49. $6a^2 - 8a - 3a + 4 = 2a(3a - 4) - 1(3a - 4)$
$= (3a - 4)(2a - 1)$

51. $16t^2 + 23t + 7$

(1) First note that there is no common factor (other than 1 or –1).

(2) Multiply the leading coefficient, 16, and the constant, 7:

$16 \cdot 7 = 112$

(3) We look for factors of 112 that add to 23. Since all coefficients are positive, we need to consider only positive factors.

Pairs of factors	Sums of factors
1, 112	113
2, 56	58
4, 28	32
8, 14	22
16, 7	23

The numbers we need are 16 and 7.

(4) Rewrite the middle term:

$23t = 16t + 7t$

(5) Factor by grouping:

$$16t^2 + 23t + 7 = 16t^2 + 16t + 7t + 7$$
$$= 16t(t + 1) + 7(t + 1)$$
$$= (t + 1)(16t + 7)$$

53. $-9x^2 - 18x - 5$

(1) We factor out –1 in order to have a trinomial with a positive leading coefficient.

$-9x^2 - 18x - 5 = -1(9x^2 + 18x + 5)$

Now we factor $9x^2 + 18x + 5$.

(2) Multiply the leading coefficient, 9, and the constant, 5:

$9 \cdot 5 = 45$

(3) We look for factors of 45 that add to 18. Since all coefficients are positive, we need to consider only positive factors.

Pairs of factors	Sums of factors
1, 45	46
3, 15	18
5, 9	14

The numbers we need are 3 and 15.

(4) Rewrite the middle term:

$18x = 3x + 15x$

(5) Factor by grouping:

$$9x^2 + 18x + 5 = 9x^2 + 3x + 15x + 5$$
$$= 3x(3x + 1) + 5(3x + 1)$$
$$= (3x + 1)(3x + 5)$$

We must include the factor of –1 in order to get a factorization of the original trinomial.

$-9x^2 - 18x - 5 = -1(3x + 1)(3x + 5)$, or
$-(3x + 1)(3x + 5)$

55. $10x^2 + 30x - 70$

(1) Factor out the largest common factor, 10:

$10x^2 + 30x - 70 = 10(x^2 + 3x - 7)$

Since $x^2 + 3x - 7$ is prime, this trinomial cannot be factored further.

57. $18x^3+21x^2-9x$

(1) Factor out the largest common factor, $3x$:

$$18x^3+21x^2-9x=3x(6x^2+7x-3)$$

(2) To factor $6x^2+7x-3$ by grouping we first multiply the leading coefficient, 6, and the constant, –3:

$6(-3)=-18$

(3) We look for factors of –18 that add to 7.

Pairs of factors	Sums of factors
–1, 18	17
1, –18	–17
–2, 9	7
2, –9	–7
–3, 6	3
3, –6	–3

The numbers we need are –2 and 9.

(4) Rewrite the middle term:

$7x=-2x+9x$

(5) Factor by grouping:

$$\begin{aligned}6x^2+7x-3&=6x^2-2x+9x-3\\&=2x(3x-1)+3(3x-1)\\&=(3x-1)(2x+3)\end{aligned}$$

The factorization of $6x^2+7x-3$ is $(3x-1)(2x+3)$. We must include the common factor in order to get a factorization of the original trinomial:

$$18x^3+21x^2-9x=3x(3x-1)(2x+3)$$

59. $89x+64+25x^2=25x^2+89x+64$

(1) First note that there is no common factor (other than 1 or –1).

(2) Multiply the leading coefficient, 25, and the constant, 64:

$25\cdot64=1600$

(3) We look for factors of 1600 that add to 89. Since all coefficients are positive, we need to consider only positive factors. The numbers we need are 25 and 64.

(4) Rewrite the middle term:

$89x=25x+64x$

(5) Factor by grouping:

$$\begin{aligned}25x^2+89x+64&=25x^2+25x+64x+64\\&=25x(x+1)+64(x+1)\\&=(x+1)(25x+64)\end{aligned}$$

61. $168x^3+45x^2+3x$

(1) Factor out the largest common factor, $3x$:

$$168x^3+45x^2+3x=3x(56x^2+15x+1)$$

(2) To factor $56x^2+15x+1$ we first multiply the leading coefficient, 56, and the constant, 1:

$56\cdot1=56$

(3) We look for factors of 56 that add to 15. Since all coefficients are positive, we need to consider only positive factors. The numbers we need are 7 and 8.

(4) Rewrite the middle term:

$15x=7x+8x$

(5) Factor by grouping:

$$\begin{aligned}56x^2+15x+1&=56x^2+7x+8x+1\\&=7x(8x+1)+1(8x+1)\\&=(8x+1)(7x+1)\end{aligned}$$

The factorization of $56x^2+15x+1$ is $(8x+1)(7x+1)$. We must include the common factor in order to get a factorization of the original trinomial:

$$168x^3+45x^2+3x=3x(8x+1)(7x+1)$$

63. $-14t^4+19t^3+3t^2$

(1) Factor out $-t^2$. This not only removes the largest common factor, t^2. It also produces a trinomial with a positive leading coefficient.

$$-14t^4+19t^3+3t^2=-t^2(14t^2-19t-3)$$

(2) To factor $14t^2-19t-3$ we first multiply the leading coefficient, 14, and the constant, –3:

$14(-3)=-42$

(3) We look for factors of –42 that add to –19. The numbers we need are –21 and 2.

(4) Rewrite the middle term:

$-19t=-21t+2t$

(5) Factor by grouping:

$$\begin{aligned}14t^2-19t-3&=14t^2-21t+2t-3\\&=7t(2t-3)+1(2t-3)\\&=(2t-3)(7t+1)\end{aligned}$$

The factorization of $14t^2-19t-3$ is $(2t-3)(7t+1)$. We must include the common factor in order to get a factorization of the original trinomial:

$$-14t^4+19t^3+3t^2=-t^2(2t-3)(7t+1)$$

65. $132y+32y^2-54=32y^2+132y-54$

(1) Factor out the largest common factor, 2:

$$32y^2+132y-54=2(16y^2+66y-27)$$

(2) To factor $16y^2+66y-27$ we first multiply the leading coefficient, 16, and the constant, –27:

$16(-27)=-432$

(3) We look for factors of –432 that add to 66. The numbers we need are 72 and –6.

(4) Rewrite the middle term:

$66y=72y-6y$

(5) Factor by grouping:

$$\begin{aligned}16y^2+66y-27&=16y^2+72y-6y-27\\&=8y(2y+9)-3(2y+9)\\&=(2y+9)(8y-3)\end{aligned}$$

The factorization of $16y^2+66y-27$ is $(2y+9)(8y-3)$. We must include the common factor in order to get a factorization of the original trinomial:

$$132y+32y^2-54=2(2y+9)(8y-3)$$

67. $2a^2 - 5ab + 2b^2$

(1) There is no common factor (other than 1 or –1).
(2) Multiply the leading coefficient, 2, and the constant, 2:

$2 \cdot 2 = 4$

(3) We look for factors of 4 that add to –5. The numbers we need are –1 and –4.
(4) Rewrite the middle term:

$-5ab = -ab - 4ab$

(5) Factor by grouping:

$$\begin{aligned}2a^2 - 5ab + 2b^2 &= 2a^2 - ab - 4ab + 2b^2 \\ &= a(2a - b) - 2b(2a - b) \\ &= (2a - b)(a - 2b)\end{aligned}$$

69. $8s^2 + 22st + 14t^2$

(1) Factor out the largest common factor, 2:

$8s^2 + 22st + 14t^2 = 2(4s^2 + 11st + 7t^2)$

(2) Multiply the leading coefficient, 4, and the constant, 7:

$4 \cdot 7 = 28$

(3) We look for factors of 28 that add to 11. The numbers we need are 4 and 7.
(4) Rewrite the middle term:

$11st = 4st + 7st$

(5) Factor by grouping:

$$\begin{aligned}4s^2 + 11st + 7t^2 &= 4s^2 + 4st + 7st + 7t^2 \\ &= 4s(s + t) + 7t(s + t) \\ &= (s + t)(4s + 7t)\end{aligned}$$

The factorization of $4s^2 + 11st + 7t^2$ is $(s + t)(4s + 7t)$. We must include the common factor in order to get a factorization of the original trinomial:

$8s^2 + 22st + 14t^2 = 2(s + t)(4s + 7t)$

71. $27x^2 - 72xy + 48y^2$

(1) Factor out the largest common factor, 3:

$27x^2 - 72xy + 48y^2 = 3(9x^2 - 24xy + 16y^2)$

(2) To factor $9x^2 - 24xy + 16y^2$, we first multiply the leading coefficient, 9, and the constant, 16:

$9 \cdot 16 = 144$

(3) We look for factors of 114 that add to –24. The numbers we need are –12 and –12.
(4) Rewrite the middle term:

$-24xy = -12xy - 12xy$

(5) Factor by grouping:

$$\begin{aligned}9x^2 - 24xy + 16y^2 &= 9x^2 - 12xy - 12xy + 16y^2 \\ &= 3x(3x - 4y) - 4y(3x - 4y) \\ &= (3x - 4y)(3x - 4y)\end{aligned}$$

The factorization of $9x^2 - 24xy + 16y^2$ is $(3x - 4y)(3x - 4y)$. We must include the common factor in order to get a factorization of the original trinomial:

$27x^2 - 72xy + 48y^2 = 3(3x - 4y)(3x - 4y)$ or $3(3x - 4y)^2$

73. $-24a^2 + 34ab - 12b^2$

(1) Factor out –2. This not only removes the largest common factor, 2. It also produces a trinomial with a positive leading coefficient.

$-24a^2 + 34ab - 12b^2 = -2(12a^2 - 17ab + 6b^2)$

(2) To factor $12a^2 - 17ab + 6b^2$, we first multiply the leading coefficient, 12, and the constant, 6:

$12 \cdot 6 = 72$

(3) We look for factors of 72 that add to –17. The numbers we need are –8 and –9.
(4) Rewrite the middle term:

$-17ab = -8ab - 9ab$

(5) Factor by grouping:

$$\begin{aligned}12a^2 - 17ab + 6b^2 &= 12a^2 - 8ab - 9ab + 6b^2 \\ &= 4a(3a - 2b) - 3b(3a - 2b) \\ &= (3a - 2b)(4a - 3b)\end{aligned}$$

The factorization of $12a^2 - 17ab + 6b^2$ is $(3a - 2b)(4a - 3b)$. We must include the common factor in order to get a factorization of the original trinomial:

$-24a^2 + 34ab - 12b^2 = -2(3a - 2b)(4a - 3b)$

75. $19x^3 - 3x^2 + 14x^4 = 14x^4 + 19x^3 - 3x^2$

(1) Factor out the largest common factor, x^2:

$x^2(14x^2 + 19x - 3)$

(2) To factor $14x^2 + 19x - 3$ by grouping we first multiply the leading coefficient, 14, and the constant, –3:

$14(-3) = -42$

(3) We look for factors of –42 that add to 19. The numbers we need are 21 and –2.
(4) Rewrite the middle term:

$19x = 21x - 2x$

(5) Factor by grouping:

$$\begin{aligned}14x^2 + 19x - 3 &= 14x^2 + 21x - 2x - 3 \\ &= 7x(2x + 3) - 1(2x + 3) \\ &= (2x + 3)(7x - 1)\end{aligned}$$

The factorization of $14x^2 + 19x - 3$ is $(2x + 3)(7x - 1)$. We must include the common factor in order to get a factorization of the original trinomial:

$19x^3 - 3x^2 + 14x^4 = x^2(2x + 3)(7x - 1)$

77. $18a^7 + 8a^6 + 9a^8 = 9a^8 + 18a^7 + 8a^6$

(1) Factor out the largest common factor, a^6:

$9a^8 + 18a^7 + 8a^6 = a^6(9a^2 + 18a + 8)$

(2) To factor $9a^2 + 18a + 8$ we first multiply the leading coefficient, 9, and the constant, 8:

$9 \cdot 8 = 72$

(3) Look for factors of 72 that add to 18. The numbers we need are 6 and 12.
(4) Rewrite the middle term:

$18a = 6a + 12a$

(5) Factor by grouping:

$$\begin{aligned} 9a^2+18a+8 &= 9a^2+6a+12a+8 \\ &= 3a(3a+2)+4(3a+2) \\ &= (3a+2)(3a+4) \end{aligned}$$

The factorization of $9a^2+18a+8$ is $(3a+2)(3a+4)$. We must include the common factor in order to get a factorization of the original trinomial:

$$18a^7+8a^6+9a^8 = a^6(3a+2)(3a+4)$$

79. *Writing Exercise.*

81. Graph $2x-5y=10$.

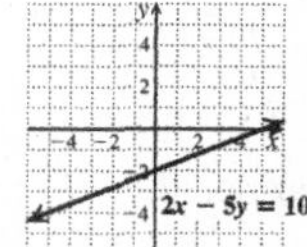

83. Graph $y=\frac{2}{3}x-1$.

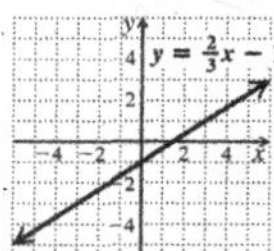

85. Graph $\frac{1}{2}y=1$.

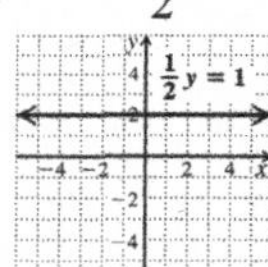

87. For the trinomial ax^2+bx+c to be prime:
(1) Show that there is no common factor (other than 1 or -1)
(2) Multiply the leading coefficient, a, and the constant c.
$a\cdot c = ac$
(3) Show there are <u>no</u> factors of ac that add to b.
(4) The trinomial is prime.

89. $18x^2y^2-3xy-10$
We will factor by grouping.
(1) There is no common factor (other than 1 or -1).
(2) Multiply the leading coefficient, 18, and the constant, -10:
$18(-10)=-180$
(3) We look for factors of -180 that add to -3 The numbers we want are -15 and 12.
(4) Rewrite the middle term:
$-3xy=-15xy+12xy$
(5) Factor by grouping:

$$\begin{aligned} 18x^2y^2-3xy-10 &= 18x^2y^2-15xy+12xy-10 \\ &= 3xy(6xy-5)+2(6xy-5) \\ &= (6xy-5)(3xy+2) \end{aligned}$$

91. $9a^2b^3+25ab^2+16$

We cannot factor the leading term, $9a^2b^3$, in a way that will produce a middle term with variable factors ab^2, so this trinomial is prime.

93. $16t^{10}-8t^5+1=16(t^5)^2-8t^5+1$
(1) There is no common factor (other than 1 or -1).
(2) Because $16t^{10}$ can be factored as $16t^5\cdot t^5$ or $8t^5\cdot 2t^5$ or $4t^5\cdot 4t^5$, we have these possibilities:

$(16t^5+\quad)(t^5+\quad)$ and $(8t^5+\quad)(2t^5+\quad)$ and $(4t^5+\quad)(4t^5+\quad)$

(3) Since the last term is positive and the middle term is negative we need consider only negative factors of 1. The only negative pair of factors is $-1,-1$.
(4) We try some possibilities:

$$\begin{aligned} (16t^5-1)(t^5-1) &= 16t^{10}-17t^5+1 \\ (8t^5-1)(2t^5-1) &= 16t^{10}-10t^5+1 \\ (4t^5-1)(4t^5-1) &= 16t^{10}-8t^5+1 \end{aligned}$$

The factorization is $(4t^5-1)(4t^5-1)$, or $(4t^5-1)^2$.

95. $-15x^{2m}+26x^m-8=-15(x^m)^2+26x^m-8$
(1) Factor out -1 in order to have a trinomial with a positive leading coefficient.

$-15x^{2m}+26x^m-8=-1(15x^{2m}-26x^m+8)$

(2) Because $15x^{2m}$ can be factored as $15x^m\cdot x^m$, or $5x^m\cdot 3x^m$, we have these possibilities:

$(15x^m+\quad)(x^m+\quad)$ and $(5x^m+\quad)(3x^m+\quad)$

(3) Since the last term is positive and the middle term is negative we need consider only negative factors of 8. There are 2 such pairs and they can be listed in two ways:

$-8,-1$ $-4,-2$
and $-1,-8$ $-2,-4$

(4) We try some possibilities:

$$\begin{aligned} (15x^m-8)(x^m-1) &= 15x^{2m}-23x^m+8 \\ (5x^m-8)(3x^m-1) &= 15x^{2m}-29x^m+8 \\ (5x^m-2)(3x^m-4) &= 15x^{2m}-26x^m+8 \end{aligned}$$

The factorization of $15x^{2m}-26x^m+8$ is $(5x^m-2)(3x^m-4)$. We must include the common factor to get a factorization of the original trinomial.

$-15x^{2m}+26x^m-8=-1(5x^m-2)(3x^m-4)$, or $-(5x^m-2)(3x^m-4)$

97. $3a^{6n}-2a^{3n}-1=3(a^{3n})^2-2a^{3n}-1$
(1) There is no common factor (other than 1 or -1).

(2) Because $3a^{6n}$ can be factored as $3a^{3n} \cdot a^{3n}$, we have this possibility:

$(3a^{3n} + \quad)(a^{3n} + \quad)$

(3) The only one pair of factors of -1: 1, –1.

(4) We try some possibilities:

$(3a^{3n} - 1)(a^{3n} + 1) = 3a^{6n} + 2a^{3n} - 1$

$(3a^{3n} + 1)(a^{3n} - 1) = 3a^{6n} - 2a^{3n} - 1$

The factorization of $3a^{6n} - 2a^{3n} - 1$ is

$(3a^{3n} + 1)(a^{3n} - 1)$.

99. $7(t-3)^{2n} + 5(t-3)^n - 2$
$= 7[(t-3)^n]^2 + 5(t-3)^n - 2$

(1) There is no common factor (other than 1 or –1).

(2) Multiply the leading coefficient, 7, and the constant, –2:

$7(-2) = -14$

(3) Look for factors of –14 that add to 5. The numbers we want are 7 and –2.

(4) Rewrite the middle term:

$5(t-3)^n = 7(t-3)^n - 2(t-3)^n$

(5) Factor by grouping:

$7(t-3)^{2n} + 5(t-3)^n - 2$
$= 7(t-3)^{2n} + 7(t-3)^n - 2(t-3)^n - 2$
$= 7(t-3)^n\left[(t-3)^n + 1\right] - 2\left[(t-3)^n + 1\right]$
$= \left[(t-3)^n + 1\right]\left[7(t-3)^n - 2\right]$

The factorization of $7(t-3)^{2n} + 5(t-3)^n - 2$ is $\left[(t-3)^n + 1\right]\left[7(t-3)^n - 2\right]$

101. If x = the length of the section, then the shorter length of the rectangle is $4x + 2$, and the longer side of the rectangle is $7x + 1$. We know area = length × width. So, solve $(7x+1)(4x+2) = 1500$ for x.

$28x^2 + 18x + 2 = 1500$
$28x^2 + 18x - 1498 = 0$
$2(14x^2 + 9x - 749) = 0$
$2(14x + 107)(x - 7) = 0$
$14x + 107 = 0 \quad or \quad x - 7 = 0$
$x = -\frac{107}{14} \quad or \quad x = 7$

Since length must be positive, the length of the section is 7 ft.

Exercise Set 5.4

1. $4x^2 + 49$ is not a trinomial. It is not a difference of squares because the terms do not have different signs. There is no common factor, so $4x^2 + 49$ is a prime polynomial.

3. $t^2 - 100 = t^2 - 10^2$, so $t^2 - 100$ is a difference of squares.

5. $9x^2 + 6x + 1 = (3x)^2 + 2 \cdot 3x \cdot 1 + 1^2$, so this is a perfect-square trinomial.

7. $2t^2 + 10t + 6$ does not contain a term that is a square so it is neither a perfect-square trinomial nor a difference of squares. (We could also say that it is not a difference of squares because it is not a binomial.) There is a common factor, 2, so this is not a prime polynomial. Thus it is none of the given possibilities.

9. $16t^2 - 25 = (4t)^2 - 5^2$, so $16t^2 - 25$ is a difference of squares.

11. $x^2 + 18x + 81$

(1) Two terms, x^2 and 81, are squares.

(2) Neither x^2 nor 81 is being subtracted.

(3) Twice the product of the square roots, $2 \cdot x \cdot 9$, is $18x$, the remaining term.

Thus, $x^2 + 18x + 81$ is a perfect-square trinomial.

13. $x^2 - 10x - 25$

(1) Two terms, x^2 and 25, are squares.

(2) There is a minus sign before 25, so $x^2 - 10x - 25$ is not a perfect-square trinomial.

15. $x^2 - 3x + 9$

(1) Two terms, x^2 and 9, are squares.

(2) There is no minus sign before x^2 or 9.

(3) Twice the product of the square roots, $2 \cdot x \cdot 3$, is $6x$. This is neither the remaining term nor its opposite, so $x^2 - 3x + 9$ is not a perfect-square trinomial.

17. $9x^2 + 25 - 30x$

(1) Two terms $9x^2$, and 25, are squares.

(2) Neither $9x^2$ nor 25 is being subtracted.

(3) Twice the product of the square roots, $2 \cdot 3x \cdot 5$, is $30x$, the opposite of the remaining term, $-30x$.

Thus, $9x^2 + 25 - 30x$ is a perfect-square trinomial.

19. $x^2 + 16x + 64$

$$\begin{array}{l} = x^2 + 2 \cdot x \cdot 8 + 8^2 = (x+8)^2 \\ \quad \uparrow \quad\; \uparrow \;\; \uparrow \;\; \uparrow \;\; \uparrow \\ = A^2 + 2 \quad A \quad B + B^2 = (A+B)^2 \end{array}$$

21. $x^2 - 10x + 25$

$$\begin{array}{l} = x^2 - 2 \cdot x \cdot 5 + 5^2 = (x-5)^2 \\ \quad \uparrow \quad\; \uparrow \;\; \uparrow \;\; \uparrow \;\; \uparrow \\ = A^2 - 2 \quad A \quad B + B^2 = (A-B)^2 \end{array}$$

23. $5p^2 + 20p + 20 = 5(p^2 + 4p + 4)$
$= 5(p^2 + 2 \cdot p \cdot 2 + 2^2)$
$= 5(p+2)^2$

25. $1-2t+t^2=1^2-2\cdot1\cdot t+t^2$
$=(1-t)^2$
We could also factor as follows:
$1-2t+t^2=t^2-2t+1$
$=t^2-2\cdot t\cdot1+1^2$
$=(t-1)^2$

27. $18x^2+12x+2=2(9x^2+6x+1)$
$=2[(3x)^2+2\cdot3x\cdot1+1^2]$
$=2(3x+1)^2$

29. $49-56y+16y^2=16y^2-56y+49$
$=(4y)^2-2\cdot4y\cdot7+7^2$
$=(4y-7)^2$
We could also factor as follows:
$49-56y+16y^2=7^2-2\cdot7\cdot4y+(4y)^2$
$=(7-4y)^2$

31. $-x^5+18x^4-81x^3=-x^3(x^2-18x+81)$
$=-x^3(x^2-2\cdot x\cdot9+9^2)$
$=-x^3(x-9)^2$

33. $2n^3+40n^2+200n=2n(n^2+20n+100)$
$=2n(n^2+2\cdot n\cdot10+10^2)$
$=2n(n+10)^2$

35. $20x^2+100x+125=5(4x^2+20x+25)$
$=5[(2x)^2+2\cdot2x\cdot5+5^2]$
$=5(2x+5)^2$

37. $49-42x+9x^2=7^2-2\cdot7\cdot3x+(3x)^2=(7-3x)^2$, or $(3x-7)^2$

39. $16x^2+24x+9=(4x)^2+2\cdot4x\cdot3+3^2$
$=(4x+3)^2$

41. $2+20x+50x^2=2(1+10x+25x^2)$
$=2[1^2+2\cdot1\cdot5x+(5x)^2]$
$=2(1+5x)^2$, or $2(5x+1)^2$

43. $9p^2+12px+4x^2=(3p)^2+2\cdot3p\cdot2x+(2x)^2$
$=(3p+2x)^2$

45. $a^2-12ab+49b^2$
This is not a perfect square trinomial because $-2\cdot a\cdot7b=-14ab\neq-12ab$. Nor can it be factored using the methods of Sections 5.2 and 5.3. Thus, it is prime.

47. $-64m^2-16mn-n^2=-1(64m^2+16mn+n^2)$
$=-1[(8m)^2+2\cdot8m\cdot n+n^2]$
$=-1(8m+n)^2$, or $-(8m+n)^2$

49. $-32s^2+80st-50t^2=-2(16s^2-40st+25t^2)$
$=-2[(4s)^2-2\cdot4s\cdot5t+(5t)^2]$
$=-2(4s-5t)^2$

51. x^2-100
(1) The first expression is a square: x^2
The second expression is a square: $100=10^2$
(2) The terms have different signs.
Thus, x^2-100 is a difference of squares x^2-10^2.

53. n^4+1
(1) The first expression is a square: $n^4=(n^2)^2$
The second expression is a square: $1=1^2$
(2) The terms do not have different signs.
Thus, n^4+1 is not a difference of squares.

55. $-1+64t^2$ or $64t^2-1$
(1) The first term is a square: $1=1^2$.
The second term is a square: $64t^2=(8t)^2$.
(2) The terms have different signs.
Thus, $-1+64t^2$ is a difference of squares, $(8t)^2-1^2$.

57. $x^2-25=x^2-5^2=(x+5)(x-5)$

59. $p^2-9=p^2-3^2=(p+3)(p-3)$

61. $-49+t^2=t^2-49=t^2-7^2=(t+7)(t-7)$, or $(7+t)(-7+t)$

63. $6a^2-24=6(a^2-4)=6(a^2-2^2)$
$=6(a+2)(a-2)$

65. $49x^2-14x+1=(7x)^2-2\cdot7x\cdot1+1^2=(7x-1)^2$

67. $200-2t^2=2(100-t^2)=2(10^2-t^2)$
$=2(10+t)(10-t)$

69. $-80a^2+45=-5(16a^2-9)=-5[(4a^2)-3^2]$
$=-5(4a+3)(4a-3)$

71. $5t^2-80=5(t^2-16)=5(t^2-4^2)$
$=5(t+4)(t-4)$

73. $8x^2-162=2(4x^2-81)=2[(2x)^2-9^2]$
$=2(2x+9)(2x-9)$

75. $36x-49x^3=x(36-49x^2)=x[6^2-(7x)^2]$
$=x(6+7x)(6-7x)$

77. $49a^4-20$
There is no common factor (other than 1 or –1). Since 20 is not a square, this is not a difference of squares. Thus, the polynomial is prime.

79. $t^4 - 1$
$= (t^2)^2 - 1^2$
$= (t^2 + 1)(t^2 - 1)$
$= (t^2 + 1)(t + 1)(t - 1)$ Factoring further; $t^2 - 1$ is a difference of squares

81. $-3x^3 + 24x^2 - 48x = -3x(x^2 - 8x + 16)$
$= -3x(x^2 - 2 \cdot x \cdot 4 + 4^2)$
$= -3x(x - 4)^2$

83. $75t^3 - 27t = 3t(25t^2 - 9)$
$= 3[(5t)^2 - 3^2]$
$= 3(5t + 3)(5t - 3)$

85. $a^8 - 2a^7 + a^6 = a^6(a^2 - 2a + 1)$
$= a^6(a^2 - 2 \cdot a \cdot 1 + 1^2)$
$= a^6(a - 1)^2$

87. $10a^2 - 10b^2 = 10(a^2 - b^2)$
$= 10(a + b)(a - b)$

89. $16x^4 - y^4 = (4x^2)^2 - (y^2)^2$
$= (4x^2 + y^2)(4x^2 - y^2)$
$= (4x^2 + y^2)(2x + y)(2x - y)$

91. $18t^2 - 8s^2 = 2(9t^2 - 4s^2)$
$= 2[(3t)^2 - (2s)^2]$
$= 2(3t + 2s)(3t - 2s)$

93. *Writing Exercise.*

95. $(2x^3 - x + 3) + (x^2 + x - 5) = 2x^3 + x^2 - 2$

97. $(2x^2 + y)(3x^2 - y) = 6x^4 - 2x^2y + 3x^2y - y^2$
$= 6x^4 + x^2y - y^2$

99. $\dfrac{21x^3 - 3x^2 + 9x}{3x} = \dfrac{21x^3}{3x} - \dfrac{3x^2}{3x} + \dfrac{9x}{3x} = 7x^2 - x + 3$

101. *Writing Exercise.*

103. $x^8 - 2^8 = (x^4 + 2^4)(x^4 - 2^4)$
$= (x^4 + 2^4)(x^2 + 2^2)(x^2 - 2^2)$
$= (x^4 + 2^4)(x^2 + 2^2)(x + 2)(x - 2)$, or
$(x^4 + 16)(x^2 + 4)(x + 2)(x - 2)$

105. $18x^3 - \dfrac{8}{25}x = 2x\left(9x^2 - \dfrac{4}{25}\right)$
$= 2x\left(3x + \dfrac{2}{5}\right)\left(3x - \dfrac{2}{5}\right)$

107. $(y - 5)^4 - z^8$
$= \left[(y - 5)^2 + z^4\right]\left[(y - 5)^2 - z^4\right]$
$= \left[(y - 5)^2 + z^4\right]\left[y - 5 + z^2\right]\left[y - 5 - z^2\right]$
$= (y^2 - 10y + 25 + z^4)(y - 5 + z^2)(y - 5 - z^2)$

109. $-x^4 + 8x^2 + 9 = -1(x^4 - 8x^2 - 9)$
$= -1(x^2 - 9)(x^2 + 1)$
$= -1(x + 3)(x - 3)(x^2 + 1)$

111. $(y + 3)^2 + 2(y + 3) + 1$
$= (y + 3)^2 + 2(y + 3) \cdot 1 + 1^2$
$= \left[(y + 3) + 1\right]^2$
$= (y + 4)^2$

113. $27p^3 - 45p^2 - 75p + 125$
$= 9p^2(3p - 5) - 25(3p - 5)$
$= (3p - 5)(9p^2 - 25)$
$= (3p - 5)(3p + 5)(3p - 5)$, or $(3p - 5)^2(3p + 5)$

115. $81 - b^{4k} = 9^2 - (b^{2k})^2$
$= (9 + b^{2k})(9 - b^{2k})$
$= (9 + b^{2k})[3^2 - (b^k)^2]$
$= (9 + b^{2k})(3 + b^k)(3 - b^k)$

117. $x^2(x + 1)^2 - (x^2 + 1)^2$
$= x^2(x^2 + 2x + 1) - (x^4 + 2x^2 + 1)$
$= x^4 + 2x^3 + x^2 - x^4 - 2x^2 - 1$
$= 2x^3 + x^2 - 2x^2 - 1$
$= 2x^3 - x^2 - 1$

119. $y^2 + 6y + 9 - x^2 - 8x - 16$
$= (y^2 + 6y + 9) - (x^2 + 8x + 16)$
$= (y + 3)^2 - (x + 4)^2$
$= [(y + 3) + (x + 4)][(y + 3) - (x + 4)]$
$= (y + 3 + x + 4)(y + 3 - x - 4)$
$= (y + x + 7)(y - x - 1)$

121. For $c = a^2$, $2 \cdot a \cdot 3 = 24$. Then $a = 4$, so $c = 4^2 = 16$.

123. $(x + 1)^2 - x^2$
$= [(x + 1) + x][(x + 1) - x]$
$= 2x + 1$
$= (x + 1) + x$

Mid-Chapter Review

1. $12x^3y - 8xy^2 + 24x^2y = 4xy(3x^2 - 2y + 6x)$

2. $3a^3 - 3a^2 - 90a = 3a(a^2 - a - 30)$
$= 3a(a - 6)(a + 5)$

3. $6x^5 - 18x^2$ Common factor: $6x^2$
$= 6x^2 \cdot x^3 - 6x^2 \cdot 3$
$= 6x^2(x^3 - 3)$

4. $x^2 + 10x + 16$ No common factor; factor with FOIL.
$= (x + 8)(x + 2)$

5. $2x^2+13x-7$ No common factor; factor with FOIL.
$=(x+7)(2x-1)$

6. x^3+3x^2+2x+6 No common factor; factor with grouping.
$=x^2(x+3)+2(x+3)$
$=(x+3)(x^2+2)$

7. $64n^2-9$ Difference of squares
$=(8n)^2-3^2$
$=(8n+3)(8n-3)$

8. x^2-2x-5 No common factor
The trinomial is prime.

9. $6p^2-6t^2$ Common factor: 6
$=6(p^2-t^2)$ Difference of squares
$=6(p+t)(p-t)$

10. $b^2-14b+49$ Perfect-square trinomial
$=b^2-2\cdot 7\cdot b+7^2$
$=(b-7)^2$

11. $12x^2-x-1$ No common factor; factor with FOIL.
$=(3x-1)(4x+1)$

12. $a-10a^2+25a^3$ Common factor: a
$=a(1-10a+25a^2)$
$=a\left[1^2-2\cdot 5\cdot a+(5a)^2\right]$ Perfect-square trinomial
$=a(1-5a)^2$

13. $10x^4-10$ Common factor: 10
$=10(x^4-1)$ Difference of squares
$=10(x^2+1)(x^2-1)$ Difference of squares
$=10(x^2+1)(x+1)(x-1)$

14. t^2+t-10 No common factor
The trinomial is prime.

15. $15d^2-30d+75$ Common factor: 15
$=15(d^2-2d+5)$ Cannot be factored further.

16. $15p^2+16px+4x^2$ No common factor; factor with FOIL.
$=(3p+2x)(5p+2x)$

17. $-2t^3-10t^2-12t$ Common factor: $-2t$
$=-2t(t^2+5t+6)$ Factor with FOIL.
$=-2t(t+2)(t+3)$

18. $10c^2+20c+10$ Common factor: 10
$=10(c^2+2c+1)$
$=10(c^2+2\cdot c\cdot 1+1^2)$ Perfect-square trinomial
$=10(c+1)^2$

19. $5+3x-2x^2$ Common factor: -1
$=-1(2x^2-3x-5)$ Write in descending order
$=-1(2x-5)(x+1)$ Factor with FOIL.

20. $2m^3n-10m^2n-6mn+30n$ Common factor: $2n$
$=2n(m^3-5m^2-3m+15)$ Factor by grouping.
$=2n\left[m^2(m-5)-3(m-5)\right]$
$=2n(m-5)(m^2-3)$

Exercise Set 5.5

1. $x^3-1=(x)^3-1^3$
This is a difference of two cubes.

3. $9x^4-25=\left(3x^2\right)^2-5^2$
This is a difference of two squares.

5. $1000t^3+1=(10t)^3+1^3$
This is a sum of two cubes.

7. $25x^2+8x$ has a common factor of x so it is not prime, but it does not fall into any of the other categories. It is classified as none of these.

9. $s^{21}-t^{15}=\left(s^7\right)^3-\left(t^5\right)^3$
This is a difference of two cubes.

11. $x^3-64=x^3-4^3$
$=(x-4)\left(x^2+4x+16\right)$
$A^3-B^3=(A-B)\left(A^2+AB+B^2\right)$

13. $z^3+1=z^3+1^3$
$=(z+1)\left(z^2-z+1\right)$
$A^3+B^3=(A+B)\left(A^2-AB+B^2\right)$

15. $t^3-1000=t^3-10^3$
$=(t-10)\left(t^2+10t+100\right)$
$A^3-B^3=(A-B)\left(A^2+AB+B^2\right)$

17. $27x^3+1=(3x)^3+1^3$
$=(3x+1)\left(9x^2-3x+1\right)$
$A^3+B^3=(A+B)\left(A^2-AB+B^2\right)$

19. $64-125x^3=4^3-(5x)^3=(4-5x)\left(16+20x+25x^2\right)$

21. $x^3-y^3=(x-y)\left(x^2+xy+y^2\right)$

23. $a^3+\frac{1}{8}=a^3+\left(\frac{1}{2}\right)^3=\left(a+\frac{1}{2}\right)\left(a^2-\frac{1}{2}a+\frac{1}{4}\right)$

25. $8t^3-8=8\left(t^3-1\right)=8\left(t^3-1^3\right)=8(t-1)\left(t^2+t+1\right)$

27. $54x^3+2=x\left(27x^3+1\right)=2\left[(3x)^3+1^3\right]$
$=2(3x+1)\left(9x^2-3x+1\right)$

29. $rs^4 + 64rs = rs(s^3 + 64)$
$= rs(s^3 + 4^3)$
$= rs(s + 4)(s^2 - 4s + 16)$

31. $5x^3 - 40z^3 = 5(x^3 - 8z^3)$
$= 5[x^3 - (2z)^3]$
$= 5(x - 2z)(x^2 + 2xz + 4z^2)$

33. $y^3 - \frac{1}{1000} = y^3 - \left(\frac{1}{10}\right)^3 = \left(y - \frac{1}{10}\right)\left(y^2 + \frac{1}{10}y + \frac{1}{100}\right)$

35. $x^3 + 0.001 = x^3 + (0.1)^3 = (x + 0.1)(x^2 - 0.1x + 0.01)$

37. $64x^6 - 8t^6 = 8(8x^6 - t^6)$
$= 8\left[(2x^2)^3 - (t^2)^3\right]$
$= 8(2x^2 - t^2)(4x^4 + 2x^2t^2 + t^4)$

39. $54y^4 - 128y = 2y(27y^3 - 64)$
$= 2y\left[(3y)^3 - 4^3\right]$
$= 2y(3y - 4)(9y^2 + 12y + 16)$

41. $z^6 - 1$
$= (z^3)^2 - 1^2$ Writing as a difference of squares
$= (z^3 + 1)(z^3 - 1)$ Factoring a difference of squares
$= (z + 1)(z^2 - z + 1)(z - 1)(z^2 + z + 1)$
Factoring a sum and a difference of cubes

43. $t^6 + 64y^6 = (t^2)^3 + (4y^2)^3$
$= (t^2 + 4y^2)(t^4 - 4t^2y^2 + 16y^4)$

45. $x^{12} - y^3z^{12} = (x^4)^3 - (yz^4)^3$
$= (x^4 - yz^4)(x^8 + x^4yz^4 + y^2z^8)$

47. *Writing Exercise.*

49. ***Translate.***

120 is 5% of what number?

↓ ↓ ↓ ↓ ↓

$120 = 5\% \quad x$

We solve the equation.

$120 = 0.05x \quad (5\% = 0.05)$

$\frac{120}{0.05} = x$

$2400 = x$

The answer is 2400 mg.

51. Let $b =$ the bill without tax.
Rewording and Translating:

The bill plus tax is \$551.20.

↓ ↓ ↓ ↓ ↓

$b \quad + \quad 0.06b \quad = \quad 551.20$

$1.06b = 551.20$

$b = \frac{551.20}{1.06}$

$b = 520$

The bill without tax is \$520.

53. *Writing Exercise.*

55. $x^{6a} - y^{3b} = (x^{2a})^3 - (y^b)^3$
$= (x^{2a} - y^b)(x^{4a} + x^{2a}y^b + y^{2b})$

57. $(x+5)^3 + (x-5)^3$ Sum of cubes
$= [(x+5) + (x-5)][(x+5)^2 - (x+5)(x-5) + (x-5)^2]$
$= 2x[(x^2 + 10x + 25) - (x^2 - 25) + (x^2 - 10x + 25)]$
$= 2x(x^2 + 10x + 25 - x^2 + 25 + x^2 - 10x + 25)$
$= 2x(x^2 + 75)$

59. $5x^3y^6 - \frac{5}{8}$
$= 5\left(x^3y^6 - \frac{1}{8}\right)$
$= 5\left(xy^2 - \frac{1}{2}\right)\left(x^2y^4 + \frac{1}{2}xy^2 + \frac{1}{4}\right)$

61. $x^{6a} - (x^{2a} + 1)^3$
$= [x^{2a} - (x^{2a} + 1)][x^{4a} + x^{2a}(x^{2a} + 1) + (x^{2a} + 1)^2]$
$= (x^{2a} - x^{2a} - 1)(x^{4a} + x^{4a} + x^{2a} + x^{4a} + 2x^{2a} + 1)$
$= -(3x^{4a} + 3x^{2a} + 1)$

63. $t^4 - 8t^3 - t + 8 = t^3(t - 8) - (t - 8)$
$= (t - 8)(t^3 - 1)$
$= (t - 8)(t - 1)(t^2 + t + 1)$

Exercise Set 5.6

1. As a first step when factoring polynomials, always check for a *common factor*.

3. If a polynomial has four terms and no common factor, it may be possible to factor by *grouping*.

5. $5a^2 - 125$
$= 5(a^2 - 25)$ 5 is a common factor.
$= 5(a + 5)(a - 5)$ Factoring the difference of squares

7. $y^2 + 49 - 14y$
$= y^2 - 14y + 49$ Perfect-square trinomial
$= (y - 7)^2$

9. $3t^2 + 16t + 21$
There is no common factor (other than 1 or –1). This trinomial has three terms, but it is not a perfect-square trinomial. Multiply the leading coefficient and the constant, 3 and 21: $3 \cdot 21 = 63$. Try to factor 63 so that the sum of the factors is 16. The numbers we want are 7 and 9: $7 \cdot 9 = 63$ and $7 + 9 = 16$. Split the middle

term and factor by grouping.
$3t^2+16t+21=3t^2+7t+9t+21$
$=t(3t+7)+3(3t+7)$
$=(3t+7)(t+3)$

11. x^3+18^2+81x
$=x(x^2+18x+81)$ x is a common factor.
$=x(x^2+2\cdot x\cdot 9+9^2)$ Perfect-square trinomial
$=x(x+9)^2$

13. $x^3-5x^2-25x+125$
$=x^2(x-5)-25(x-5)$ Factoring by grouping
$=(x-5)(x^2-25)$
$=(x-5)(x+5)(x-5)$ Factoring the difference of squares
$=(x-5)^2(x+5)$

15. $27t^3-3t$
$=3t(9t^2-1)$ $3t$ is a common factor.
$=3t[(3t^2)-1^2]$ Difference of squares
$=3t(3t+1)(3t-1)$

17. $9x^3+12x^2-45x$
$=3x(3x^2+4x-15)$ $3x$ is a common factor.
$=3x(x+3)(3x-5)$ Factoring the trinomial

19. t^2+25
The polynomial has no common factor and is not a difference of squares. It is prime.

21. $6y^2+18y-240$
$=6(y^2+3y-40)$ 6 is a common factor.
$=6(y+8)(y-5)$ Factoring the trinomial

23. $-2a^6+8a^5-8a^4$
$=-2a^4(a^2-4a+4)$ Factoring out $-2a^4$
$=-2a^4(a-2)^2$ Factoring the perfect-square trinomial

25. $5x^5-80x$
$=5x(x^4-16)$ $5x$ is a common factor.
$=5x[(x^2)^2-4^2]$ Difference of squares
$=5x(x^2+4)(x^2-4)$ Difference of squares
$=5x(x^2+4)(x+2)(x-2)$

27. t^4-9 Difference of squares
$=(t^2+3)(t^2-3)$

29. $-x^6+2x^5-7x^4$
$=-x^4(x^2-2x-7)$
The trinomial is prime, so this is the complete factorization.

31. p^2-w^2 Difference of squares
$=(p+w)(p-w)$

33. $ax^2+ay^2=a(x^2+y^2)$

35. $2\pi rh+2\pi r^2$ $2\pi r$ is a common factor
$=2\pi r(r+h)$

37. $(a+b)(5a)+(a+b)(3b)$
$=(a+b)(5a+3b)$ $(a+b)$ is a common factor.

39. $x^2+x+xy+y$
$=x(x+1)+y(x+1)$ Factoring by grouping
$=(x+1)(x+y)$

41. $160a^2m^4-10a^2$
$=10a^2(16m^4-1)$ Factoring out $10a^2$
$=10a^2(4m^2+1)(4m^2-1)$ Difference of squares
$=10a^2(4m^2+1)(2m+1)(2m-1)$ Difference of squares

43. $a^2-2a-ay+2y$
$=a(a-2)-y(a-2)$ Factoring by grouping
$=(a-2)(a-y)$

45. $3x^2+13xy-10y^2=(3x-2y)(x+5y)$

47. $8m^3n-32m^2n^2+24mn$
$=8mn(m^2-4mn+3)$ $8mn$ is a common factor

49. $\frac{9}{16}-y^2$ Difference of squares
$=\left(\frac{3}{4}+y\right)\left(\frac{3}{4}-y\right)$

51. $4b^2+a^2-4ab$
$=4b^2-4ab+a^2$
$=(2b)^2-2\cdot 2b\cdot a+a^2$ Perfect-square trinomial
$=(2b-a)^2$
This result can also be expressed as $(a-2b)^2$.

53. $16x^2+24xy+9y^2$
$=(4x)^2+2\cdot 4x\cdot 3y+(3y)^2$ Perfect-square trinomial
$=(4x+3y)^2$

55. m^2-5m+8
We cannot find a pair of factors whose product is 8 and whose sum is -5, so m^2-5m+8 is prime.

57. $10x^2-11x-6=(2x-3)(5x+2)$

59. a^4b^4-16
$=(a^2b^2)^2-4^2$ Difference of squares
$=(a^2b^2+4)(a^2b^2-4)$ Difference of squares
$=(a^2b^2+4)(ab+2)(ab-2)$

61. $80cd^2 - 36c^2d + 4c^3$
$= 4c(20d^2 - 9cd + c^2)$ 4c is a common factor
$= 4c(4d - c)(5d - c)$ Factoring the trinomial

63. $3b^2 + 17ab - 6a^2 = (3b - a)(b + 6a)$

65. $-12 - x^2y^2 - 8xy$
$= -x^2y^2 - 8xy - 12$
$= -1(x^2y^2 + 8xy + 12)$
$= -1(xy + 2)(xy + 6)$, or $-(xy + 2)(xy + 6)$

67. $14t + 8t^2 - 15$
$= 8t^2 + 14t - 15$
$= (2t + 5)(4t - 3)$

69. $5p^2t^2 + 25pt - 30$
$= 5(p^2t^2 + 5pt - 6)$ 5 is a common factor.
$= 5(pt + 6)(pt - 1)$ Factoring the trinomial

71. $4ab^5 - 32b^4 + a^2b^6$
$= b^4(4ab - 32 + a^2b^2)$ b^4 is a common factor.
$= b^4(a^2b^2 + 4ab - 32)$
$= b^4(ab + 8)(ab - 4)$ Factoring the trinomial

73. $x^6 + x^5y - 2x^4y^2$
$= x^4(x^2 + xy - 2y^2)$ x^4 is a common factor.
$= x^4(x + 2y)(x - y)$ Factoring the trinomial

75. $36a^2 - 15a + \frac{25}{16}$
$= (6a)^2 - 2 \cdot 6a \cdot \frac{5}{4} + \left(\frac{5}{4}\right)^2$ Perfect-square trinomial
$= \left(6a - \frac{5}{4}\right)^2$

77. $\frac{1}{81}x^2 - \frac{8}{27}x + \frac{16}{9}$
$= \left(\frac{1}{9}x\right)^2 - 2 \cdot \frac{1}{9}x \cdot \frac{4}{3} + \left(\frac{4}{3}\right)^2$ Perfect-square trinomial
$= \left(\frac{1}{9}x - \frac{4}{3}\right)^2$
If we had factored out $\frac{1}{9}$ at the outset, the final result would have been $\frac{1}{9}\left(\frac{1}{3}x - 4\right)^2$.

79. $1 - 16x^{12}y^{12}$
$= (1 + 4x^6y^6)(1 - 4x^6y^6)$ Difference of squares
$= (1 + 4x^6y^6)(1 + 2x^3y^3)(1 - 2x^3y^3)$

81. $4a^2b^2 + 12ab + 9$
$= (2ab)^2 + 2 \cdot 2ab \cdot 3 + 3^2$ Perfect-square trinomial
$= (2ab + 3)^2$

83. $z^4 + 6z^3 - 6z^2 - 36z$
$= z(z^3 + 6z^2 - 6z - 36)$ z is a common factor
$= z[z^2(z + 6) - 6(z + 6)]$ Factoring by grouping
$= z(z + 6)(z^2 - 6)$

85. $x^3 + 5x^2 - x - 5$
$= x^2(x + 5) - (x + 5)$ Factoring by grouping
$= (x + 5)(x^2 - 1)$
$= (x + 5)(x + 1)(x - 1)$ Difference of squares

87. *Writing Exercise.*

89. ***Translate.***

1.91 billion	is	42.9%	of	all phone users
↓	↓	↓	↓	↓
1.91	=	0.429	·	x

We solve the equation and convert to percent notation.

$$1.91 = 0.429 \cdot x$$
$$\frac{1.91}{0.429} = x$$
$$4.45 \approx x$$

There were about 4.45 billion cell phone users in 2016.

91. ***Familiarize.*** Let g = the amount spent on Father's Day gifts. Then $g + 8.5$ = the amount spent on Mother's Day gifts.

Translate.

Father's Day gifts	plus	Mother's Day gifts	is	33.9
↓	↓	↓	↓	↓
g	+	$(g + 8.5)$	=	33.9

Carry out. We solve the equation.

$$g + (g + 8.5) = 33.9$$
$$2g + 8.5 = 33.9$$
$$2g = 25.4$$
$$g = 12.7$$

Then $g + 8.5 = 21.2$.

Check. $12.7 + 21.2 = 33.9$. The answer checks.

State. For Mother's Day, \$21.2 billion was spent. For Father's Day, \$12.7 billion was spent.

93. *Writing Exercise.*

95. $36 - 12x + x^2 - a^2$
$= (36 - 12x + x^2) - a^2$
$= (6 - x)^2 - a^2$
$= (6 - x + a)(6 - x - a)$

97. $-(x^5 + 7x^3 - 18x)$
$= -x(x^4 + 7x^2 - 18)$
$= -x(x^2 + 9)(x^2 - 2)$

99. $-x^4+7x^2+18$
$=-1(x^4-7x^2-18)$
$=-1(x^2+2)(x^2-9)$
$=-1(x^2+2)(x+3)(x-3)$, or $-(x^2+2)(x+3)(x-3)$

101. $y^2(y+1)-4y(y+1)-21(y+1)$
$=(y+1)(y^2-4y-21)$
$=(y+1)(y-7)(y+3)$

103. $(y+4)^2+2x(y+4)+x^2$
$=(y+4)^2+2\cdot(y+4)\cdot x+x^2$ Perfect-square trinomial
$=[(y+4)+x]^2$
$=(y+4+x)^2$

105. $2(a+3)^4-(a+3)^3(b-2)-(a+3)^2(b-2)^2$
$=(a+3)^2[2(a+3)^2-(a+3)(b-2)-(b-2)^2]$
$=(a+3)^2[2(a+3)+(b-2)][(a+3)-(b-2)]$
$=(a+3)^2(2a+6+b-2)(a+3-b+2)$
$=(a+3)^2(2a+b+4)(a-b+5)$

107. $49x^4+14x^2+1-25x^6$
$=(7x^2+1)^2-25x^6$ Perfect-square trinomial
$=[(7x^2+1)-5x^3][(7x^2+1+5x^3)]$ Difference of squares
$=(7x^2+1-5x^3)(7x^2+1+5x^3)$

Connecting the Concepts

1. Expression

2. Equation

3. Expression

4. Equation

5. $(2x^3-5x+1)+(x^2-3x-1)$
$=2x^3+x^2+(-5-3)x+(1-1)$
$=2x^3+x^2-8x$

6. $(x^2-x-5)-(3x^2-x+6)$
$=x^2-x-5-3x^2+x-6$
$=(1-3)x^2+(-1+1)x+(-5-6)$
$=-2x^2-11$

7. $t^2-100=0$
$(t+10)(t-10)=0$
$t+10=0$ or $t-10=0$
$t=-10$ or $t=10$
The solutions are –10 and 10.

8. $(3a-2)(2a-5)$
$=6a^2-15a-4a+10$ Using FOIL
$=6a^2-19a+10$

9. $n^2-10n+9=(n-1)(n-9)$ Factor with FOIL

10. $x^2+16=10x$
$x^2-10x+16=0$ Subtracting $10x$ from both sides
$(x-2)(x-8)=0$
$x-2=0$ or $x-8=0$
$x=2$ or $x=8$
The solutions are 2 and 8.

Exercise Set 5.7

1. Equations of the type $ax^2+bx+c=0$, with $a\neq 0$, are quadratic, so choice (c) is correct.

3. The principle of zero products states that $A\cdot B=0$ if and only if $A=0$ or $B=0$, so choice (d) is correct.

5. $(x+2)(x+9)=0$
We use the principle of zero products.
$x+2=0$ *or* $x+9=0$
$x=-2$ *or* $x=-9$
Check:
For –2:

$(x+2)(x+9)=0$	
$(-2+2)(-2+9)$	0
$0\cdot 7$	
$0\overset{?}{=}0$ TRUE	

For –9:

$(x+2)(x+9)=0$	
$(-9+2)(-9+9)$	0
$-7\cdot 0$	
$0\overset{?}{=}0$ TRUE	

The solutions are –2 and –9.

7. $(x+1)(x-8)=0$
$x+1=0$ *or* $x-8=0$
$x=-1$ *or* $x=8$
Check:
For –1:

$(x+1)(x-8)=0$	
$(-1+1)(-1-8)$	0
$0\cdot(-9)$	
$0\overset{?}{=}0$ TRUE	

For 8:

$(x+1)(x-8)=0$	
$(8+1)(8-8)$	0
$9\cdot 0$	
$0\overset{?}{=}0$ TRUE	

The solutions are –1 and 8.

9. $(2t-3)(t+6)=0$
$2t-3=0$ *or* $t+6=0$
$2t=3$ *or* $t=-6$
$t=\frac{3}{2}$ *or* $t=-6$
The solutions are $\frac{3}{2}$ and –6.

11. $4(7x-1)(10x-3)=0$
$(7x-1)(10x-3)=0$ Dividing both sides by 4
$7x-1=0$ *or* $10x-3=0$
$7x=1$ *or* $10x=3$
$x=\frac{1}{7}$ *or* $x=\frac{3}{10}$
The solutions are $\frac{1}{7}$ and $\frac{3}{10}$.

13. $x(x-7)=0$
$x=0$ *or* $x-7=0$
$x=0$ *or* $x=7$
The solutions are 0 and 7.

15. $\left(\frac{2}{3}x-\frac{12}{11}\right)\left(\frac{7}{4}x-\frac{1}{12}\right)=0$
$\frac{2}{3}x-\frac{12}{11}=0$ *or* $\frac{7}{4}x-\frac{1}{12}=0$
$\frac{2}{3}x=\frac{12}{11}$ *or* $\frac{7}{4}x=\frac{1}{12}$
$x=\frac{3}{2}\cdot\frac{12}{11}$ *or* $x=\frac{4}{7}\cdot\frac{1}{12}$
$x=\frac{18}{11}$ *or* $x=\frac{1}{21}$
The solutions are $\frac{18}{11}$ and $\frac{1}{21}$.

17. $6n(3n+8)=0$
$6n=0$ *or* $3n+8=0$
$n=0$ *or* $3n=-8$
$n=0$ *or* $n=-\frac{8}{3}$
The solutions are 0 and $-\frac{8}{3}$.

19. $(20-0.4x)(7-0.1x)=0$
$20-0.4x=0$ *or* $7-0.1x=0$
$-0.4x=-20$ *or* $-0.1x=-7$
$x=50$ *or* $x=70$
The solutions are 50 and 70.

21. $x^2-7x+6=0$
$(x-6)(x-1)=0$ Factoring
$x-6=0$ *or* $x-1=0$
$x=6$ *or* $x=1$
The solutions are 1 and 6.

23. $x^2+4x-21=0$
$(x-3)(x+7)=0$ Factoring
$x-3=0$ *or* $x+7=0$
$x=3$ *or* $x=-7$
The solutions are 3 and –7.

25. $n^2+11n+18=0$
$(n+9)(n+2)=0$
$n+9=0$ *or* $n+2=0$
$n=-9$ *or* $n=-2$
The solutions are –9 and –2.

27. $x^2-10x=0$
$x(x-10)=0$
$x=0$ *or* $x-10=0$
$x=0$ *or* $x=10$
The solutions are 0 and 10.

29. $6t+t^2=0$
$t(6+t)=0$
$t=0$ *or* $6+t=0$
$t=0$ *or* $t=-6$
The solutions are 0 and –6.

31. $x^2-36=0$
$(x+6)(x-6)=0$
$x+6=0$ *or* $x-6=0$
$x=-6$ *or* $x=6$
The solutions are –6 and 6.

33. $4t^2=49$
$4t^2-49=0$
$(2t+7)(2t-7)=0$
$2t+7=0$ *or* $2t-7=0$
$2t=-7$ *or* $2t=7$
$t=\frac{-7}{2}$ *or* $t=\frac{7}{2}$
The solutions are $-\frac{7}{2}$ and $\frac{7}{2}$.

35. $0=25+x^2+10x$
$0=x^2+10x+25$ Writing in descending order
$0=(x+5)(x+5)$
$x+5=0$ *or* $x+5=0$
$x=-5$ *or* $x=-5$
The solution is –5.

37. $64+x^2=16x$
$x^2-16x+64=0$
$(x-8)(x-8)=0$
$x-8=0$ *or* $x-8=0$
$x=8$ *or* $x=8$
The solution is 8.

39. $4t^2=8t$
$4t^2-8t=0$
$4t(t-2)=0$
$t=0$ *or* $t-2=0$
$t=0$ *or* $t=2$
The solutions are 0 and 2.

41. $4y^2=7y+15$
$4y^2-7y-15=0$
$(4y+5)(y-3)=0$
$4y+5=0$ *or* $y-3=0$
$4y=-5$ *or* $y=3$
$y=-\frac{5}{4}$ *or* $y=3$
The solutions are $-\frac{5}{4}$ and 3.

43. $(x-7)(x+1)=-16$
$x^2-6x-7=-16$
$x^2-6x+9=0$
$(x-3)(x-3)=0$
$x-3=0$ *or* $x-3=0$
$x=3$ *or* $x=3$
The solution is 3.

45. $15z^2+7=20z+7$
$15z^2-20z+7=7$
$15z^2-20z=0$
$5z(3z-4)=0$
$5z=0$ *or* $3z-4=0$
$z=0$ *or* $3z=4$
$z=0$ *or* $z=\frac{4}{3}$
The solutions are 0 and $\frac{4}{3}$.

47. $36m^2-9=40$
$36m^2-49=0$
$(6m+7)(6m-7)=0$ *or* $6m-7=0$
$6m+7=0$ *or* $6m-7=0$
$6m=-7$ *or* $6m=0$
$m=-\frac{7}{6}$ *or* $m=\frac{7}{6}$
The solutions are $-\frac{7}{6}$ or $\frac{7}{6}$.

49. $(x+3)(3x+5)=7$
$3x^2+14x+15=7$
$3x^2+14x+8=0$
$(3x+2)(x+4)=0$
$3x+2=0$ *or* $x+4=0$
$3x=-2$ *or* $x=-4$
$x=-\frac{2}{3}$ *or* $x=-4$
The solutions are $-\frac{2}{3}$ and -4.

51. $3x^2-2x=9-8x$
$3x^2+6x-9=0$ Adding $8x$ and -9
$3(x^2+2x-3)=0$
$3(x+3)(x-1)=0$
$x+3=0$ *or* $x-1=0$
$x=-3$ *or* $x=1$
The solutions are –3 and 1.

53. $(6a+1)(a+1)=21$
$6a^2+7a+1=21$
$6a^2+7a-20=0$
$(3a-4)(2a+5)=0$
$3a-4=0$ *or* $2a+5=0$
$3a=4$ *or* $2a=-5$
$a=\frac{4}{3}$ *or* $a=-\frac{5}{2}$
The solutions are $\frac{4}{3}$ and $-\frac{5}{2}$.

55. The solutions of the equation are the first coordinates of the x-intercepts of the graph. From the graph we see that the x-intercepts are (–1, 0) and (4, 0), so the solutions of the equation are –1 and 4.

57. The solutions of the equation are the first coordinates of the x-intercepts of the graph. From the graph we see that the x-intercepts are (–3, 0) and (2, 0), so the solutions of the equation are –3 and 2.

59. We let $y=0$ and solve for x.
$0=x^2-x-6$
$0=(x-3)(x+2)$
The x-intercepts are (3, 0) and (–2, 0).

61. We let $y=0$ and solve for x.
$0=x^2+2x-8$
$0=(x+4)(x-2)$
The x-intercepts are (–4, 0) and (2, 0).

63. We let $y=0$ and solve for x.
$0=2x^2+3x-9$
$0=(2x-3)(x+3)$
$2x-3=0$ *or* $x+3=0$
$2x=3$ *or* $x=-3$
$x=\frac{3}{2}$ *or* $x=-3$
The x-intercepts are $\left(\frac{3}{2}, 0\right)$ and (–3, 0).

65. *Writing Exercise.*

67. The opposite of –65 is 65.

69. $|-1.65|=1.65$

71. $-\frac{2}{3}=-0.666666...=-0.\overline{6}$

73. *Writing Exercise.*

75. $(2x-11)(3x^2+29x+56)=0$
$(2x-11)(3x+8)(x+7)=0$
$2x-11=0$ *or* $3x+8=0$ *or* $x+7=0$
$2x=11$ *or* $3x=-8$ *or* $x=-7$
$x=\frac{11}{2}$ *or* $x=-\frac{8}{3}$ *or* $x=-7$
The solutions are -7, $-\frac{8}{3}$, and $\frac{11}{2}$.

77. **a.** $x=-4$ *or* $x=5$
$x+4=0$ *or* $x-5=0$
$(x+4)(x-5)=0$ Principle of zero products
$x^2-x-20=0$ Multiplying

b. $x=-1$ *or* $x=7$
$x+1=0$ *or* $x-7=0$
$(x+1)(x-7)=0$
$x^2-6x-7=0$

c. $x=\frac{1}{4}$ *or* $x=3$

$x-\frac{1}{4}=0$ *or* $x-3=0$

$\left(x-\frac{1}{4}\right)(x-3)=0$

$x^2-\frac{13}{4}x+\frac{3}{4}=0$

$4\left(x^2-\frac{13}{4}x+\frac{3}{4}\right)=4\cdot 0$ Multiplying both sides by 4

$4x^2-13x+3=0$

d. $x=\frac{1}{2}$ *or* $x=\frac{1}{3}$

$x-\frac{1}{2}=0$ *or* $x-\frac{1}{3}=0$

$\left(x-\frac{1}{2}\right)\left(x-\frac{1}{3}\right)=0$

$x^2-\frac{5}{6}x+\frac{1}{6}=0$

$6x^2-5x+1=0$ Multiplying both sides by 6

e. $x=\frac{2}{3}$ *or* $x=\frac{3}{4}$

$x-\frac{2}{3}=0$ *or* $x-\frac{3}{4}=0$

$\left(x-\frac{2}{3}\right)\left(x-\frac{3}{4}\right)=0$

$x^2-\frac{17}{12}x+\frac{1}{2}=0$

$12x^2-17x+6=0$ Multiplying both sides by 12

f. $x=-1$ *or* $x=2$ *or* $x=3$

$x+1=0$ *or* $x-2=0$ *or* $x-3=0$

$(x+1)(x-2)(x-3)=0$

$(x^2-x-2)(x-3)=0$

$x^3-4x^2+x+6=0$

79. $a(9+a)=4(2a+5)$

$9a+a^2=8a+20$

$a^2+a-20=0$

$(a+5)(a-4)=0$

$a+5=0$ *or* $a-4=0$

$a=-5$ *or* $a=4$

The solutions are –5 and 4.

81. $-x^2+\frac{9}{25}=0$

$x^2-\frac{9}{25}=0$ Multiplying by -1

$\left(x+\frac{3}{5}\right)\left(x-\frac{3}{5}\right)=0$

$x+\frac{3}{5}=0$ *or* $x-\frac{3}{5}=0$

$x=-\frac{3}{5}$ *or* $x=\frac{3}{5}$

The solutions are $-\frac{3}{5}$ and $\frac{3}{5}$.

83. $(t+1)^2=9$

Observe that $t+1$ is a number which yields 9 when it is squared. Thus, we have

$t+1=-3$ *or* $t+1=3$

$t=-4$ *or* $t=2$

The solutions are –4 and 2.

We could also do this exercise as follows:

$(t+1)^2=9$

$t^2+2t+1=9$

$t^2+2t-8=0$

$(t+4)(t-2)=0$

$t+4=0$ *or* $t-2=0$

$t=-4$ *or* $t=2$

Again we see that the solutions are –4 and 2.

85. $x^3-2x^2-x+2=0$

$x^2(x-2)-(x-2)=0$

$(x-2)(x^2-1)=0$

$(x-2)(x+1)(x-1)=0$

$x-2=0$ *or* $x+1=0$ *or* $x-1=0$

$x=2$ *or* $x=-1$ *or* $x=1$

The solutions are –1, 1, and 2.

87. a. $2(x^2+10x-2)=2\cdot 0$ Multiplying (a) by 2

$2x^2+20x-4=0$

(a) and $2x^2+20x-4=0$ are equivalent.

b. $(x-6)(x+3)=x^2-3x-18$ Multiplying

(b) and $x^2-3x-18=0$ are equivalent.

c. $5x^2-5=5(x^2-1)=5(x+1)(x-1)=$

$(x+1)5(x-1)=(x+1)(5x-5)$

(c) and $(x+1)(5x-5)=0$ are equivalent.

d. $2(2x-5)(x+4)=2\cdot 0$ Multiplying (d) by 2

$2(x+4)(2x-5)=0$

$(2x+8)(2x-5)=0$

(d) and $(2x+8)(2x-5)=0$ are equivalent.

e. $4(x^2+2x+9)=4\cdot 0$ Multiplying (e) by 4

$4x^2+8x+36=0$

(e) and $(2x+8)(2x-5)=0$ are equivalent.

f. $3(3x^2-4x+8)=3\cdot 0$ Multiplying (f) by 3

$9x^2-12x+24=0$

(f) and $9x^2-12x+24=0$ are equivalent.

89. *Writing Exercise.*

91. 2.33, 6.77

93. –9.15, –4.59

95. –3.76, 0

97. ***Familiarize***. Let x = the smaller integer. Then $x+1$ = the larger integer.

Translate. We reword the problem. The sum of the two integers is $x+x+1$. Thus the square of the sum of the two integers is $(x+x+1)^2$. The square of the sum of two consecutive integers is 225.

$$(x+x+1)^2 = 225$$

Carry out.

$$\begin{aligned}(x+x+1)^2 &= 225\\ (2x+1)^2 &= 225\\ 4x^2+4x+1 &= 225\\ 4x^2+4x-224 &= 0\\ 4(x^2+x-56) &= 0\\ 4(x+8)(x-7) &= 0\end{aligned}$$

$$x+8=0 \quad or \quad x-7=0$$
$$x=-8 \quad or \quad x=7$$

Check. The solutions of the equation are –8 and 7. When x is –8, then $x+1$ is –7 and $(-8+-7)^2 = 225$. The numbers –8 and –7 are consecutive integers which are solutions of the problem. When x is 7, then $x+1=8$ and $(7+8)^2 = 225$. The numbers 7 and 8 are also consecutive integers which are solutions of the problem.

State. We have two solutions, each of which consists of a pair of numbers: –8 and –7 or 7 and 8.

Exercise Set 5.8

1. If x is an integer, then the next consecutive integer is $x + 1$.

3. A right triangle contains a(n) 90° angle.

5. ***Familiarize.*** Let x = the number.

Translate. We reword the problem.

The square of a number	minus	the number	is	6.
↓	↓	↓	↓	↓
x^2	$-$	x	$=$	6

Carry out. We solve the equation.

$$\begin{aligned}x^2-x &= 6\\ x^2-x-6 &= 0\\ (x+2)(x-3) &= 0\end{aligned}$$

$$x+2=0 \quad or \quad x-3=0$$
$$x=-2 \quad or \quad x=3$$

Check. For 3: The square of 3 is 3^2, or 9, and $9-3=6$. For –2: The square of –2, or 4 and $4-(-2)=4+2=6$. Both numbers check.

State. The numbers are 3 and –2.

7. ***Familiarize.*** The parking spaces are consecutive integers. Let x = the smaller integer. Then $x+1=$ the larger integer.

Translate. We reword the problem.

Smaller integer	times	larger integer	is	132.
↓	↓	↓	↓	↓
x	$\cdot$	$(x+1)$	$=$	132

Carry out. We solve the equation.

$$\begin{aligned}x(x+1) &= 132\\ x^2+x &= 132\\ x^2+x-132 &= 0\\ (x+12)(x-11) &= 0\end{aligned}$$

$$x+12=0 \quad or \quad x-11=0$$
$$x=-12 \quad or \quad x=11$$

Check. The solutions of the equation are –12 and 11. Since a parking space number cannot be negative, we only need to check 11. When $x=11$, then $x+1=12$, and $11\cdot 12=132$. This checks.

State. The parking space numbers are 11 and 12.

9. ***Familiarize.*** Let x = the smaller even integer. Then $x+2$ = the larger even integer.

Translate. We reword the problem.

Smaller even integer	times	larger even integer	is	168.
↓	↓	↓	↓	↓
x	$\cdot$	$(x+2)$	$=$	168

Carry out.

$$\begin{aligned}x(x+2) &= 168\\ x^2+2x &= 168\\ x^2+2x-168 &= 0\\ (x+14)(x-12) &= 0\end{aligned}$$

$$x+14=0 \quad or \quad x-12=0$$
$$x=-14 \quad or \quad x=12$$

Check. The solutions of the equation are –14 and 12. When x is –14, then $x+2$ is –12 and $-14(-12)=168$. The numbers –14 and –12 are consecutive even integers which are solutions of the problem. When x is 12, then $x+2=14$ and $12\cdot 14=168$. The numbers 12 and 14 are also consecutive even integers which are solutions of the problem.

State. We have two solutions, each of which consists of a pair of numbers: –14 and –12 or 12 and 14.

11. ***Familiarize.*** Let w = the width of the porch, in feet. Then $5w=$ the length. Recall that the area of a rectangle is length · width.

Translate.

The area of the rectangle	is	320 ft².
↓	↓	↓
$5w\cdot w$	$=$	320

Carry out. We solve the equation.

$$\begin{aligned}5w\cdot w &= 320\\ 5w^2 &= 320\\ 5w^2-320 &= 0\\ 5(w^2-64) &= 0\\ 5(w+8)(w-8) &= 0\end{aligned}$$

$$w+8=0 \quad or \quad w-8=0$$
$$w=-8 \quad or \quad w=8$$

Check. Since the width must be positive, −8 cannot be a solution. If the width is 8 ft, then the length is $5\cdot 8$ ft, or 40 ft, and the area is $8\text{ ft}\cdot 40\text{ ft} = 320\text{ ft}^2$. Thus, 8 checks.

State. The porch is 40 ft long and 8 ft wide.

13. ***Familiarize***. Let w = the width of the photo, in cm. Then $w+5=$ the length. Recall that the area of a rectangle is length · width.

Translate.

$$\underbrace{\text{The area of the rectangle}}_{w(w+5)} \quad \underset{=}{\text{is}} \quad \underbrace{84\text{ cm}^2}_{84}.$$

Carry out. We solve the equation.

$$w(w+5)=84$$
$$w^2+5w=84$$
$$w^2+5w-84=0$$
$$(w+12)(w-7)=0$$
$$w+12=0 \quad or \quad w-7=0$$
$$w=-12 \quad or \quad w=7$$

Check. Since the width must be positive, −12 cannot be a solution. If the width is 7 cm, then the length is $7+5$ cm, or 12 cm, and the area is

$7\text{ cm}\cdot 12\text{ cm} = 84\text{ cm}^2$. Thus, 7 checks.

State. The photo is 12 cm long and 7 cm wide.

15. ***Familiarize***. Using the labels show on the drawing in the text, we let x = the length of the foot of the sail, in ft, and $x+5=$ the height of the sail, in ft. Recall that the formula for the area of a triangle is $\frac{1}{2}\cdot(\text{base})\cdot(\text{height})$.

Translate.

$$\underset{\frac{1}{2}}{\frac{1}{2}} \quad \underset{\cdot}{\text{times}} \quad \underset{x}{\text{base}} \quad \underset{\cdot}{\text{times}} \quad \underset{(x+5)}{\text{height}} \quad \underset{=}{\text{is}} \quad \underbrace{42\text{ ft}^2}_{42}.$$

Carry out.

$$\frac{1}{2}x(x+5)=42$$
$$x(x+5)=84 \quad \text{Multiplying by 2}$$
$$x^2+5x=84$$
$$x^2+5x-84=0$$
$$(x+12)(x-7)=0$$
$$x+12=0 \quad or \quad x-7=0$$
$$x=-12 \quad or \quad x=7$$

Check. The solutions of the equation are −12 and 7. The length of the base of a triangle cannot be negative, so −12 cannot be a solution. Suppose the length of the foot of the sail is 7 ft. Then the height is $7+5$, or 12 ft, and the area is $\frac{1}{2}\cdot 7\cdot 12$, or 42 ft^2. These numbers check.

State. The length of the foot of the sail is 7 ft, and the height is 12 ft.

17. ***Familiarize***. Let h = the height of triangle, in ft, and $\frac{1}{2}h=$ the width of the base, in ft. Recall that the formula for the area of a triangle is $\frac{1}{2}\cdot(\text{base})\cdot(\text{height})$.

Translate.

$$\underset{\frac{1}{2}}{\frac{1}{2}} \quad \underset{\cdot}{\text{times}} \quad \underset{\frac{h}{2}}{\text{base}} \quad \underset{\cdot}{\text{times}} \quad \underset{h}{\text{height}} \quad \underset{=}{\text{is}} \quad \underbrace{64\text{ ft}^2}_{64}.$$

Carry out.

$$\frac{1}{2}\left(\frac{h}{2}\right)h=64$$
$$h^2=256 \quad \text{Multiplying by 4}$$
$$h=-16 \quad or \quad h=16$$

Check. The solutions of the equation are −16 and 16. The height of the triangle cannot be negative, so −16 cannot be a solution. Suppose the height is 16 ft. Then the base is $\frac{16}{2}$, or 8 ft, and the area is $\frac{1}{2}\cdot 8\cdot 16$, or 64 ft^2. These numbers check.

State. The base is 8 ft, and the height is 16 ft.

19. ***Familiarize***. Let b = the base of triangle, in ft, and $2b-1=$ the height of the triangle, in ft. Recall that the formula for the area of a triangle is $\frac{1}{2}\cdot(\text{base})\cdot(\text{height})$.

Translate.

$$\underset{\frac{1}{2}}{\frac{1}{2}} \quad \underset{\cdot}{\text{times}} \quad \underset{b}{\text{base}} \quad \underset{\cdot}{\text{times}} \quad \underset{(2b-1)}{\text{height}} \quad \underset{=}{\text{is}} \quad \underbrace{60\text{ ft}^2}_{60}.$$

Carry out.

$$\frac{1}{2}b(2b-1)=60$$
$$b(2b-1)=120 \quad \text{Multiplying by 2}$$
$$2b^2-b=120$$
$$2b^2-b-120=0$$
$$(2b+15)(b-8)=0$$
$$2b+15=0 \quad or \quad b-8=0$$
$$b=-\frac{15}{2} \quad or \quad b=8$$

Check. The solutions of the equation are $-\frac{15}{2}$ and 8. The base of the triangle cannot be negative, so $-\frac{15}{2}$ cannot be a solution. Suppose the base is 8 ft. Then the height is $2(8)-1$, or 15 ft, and the area is $\frac{1}{2}\cdot 8\cdot 15$, or 60 ft^2. These numbers check.

State. The base is 8 ft, and the height is 15 ft.

21. ***Familiarize***. We will use the formula $x^2 - x = N$.

Translate. Substitute 56 for N.

$$x^2 - x = 56$$

Carry out.

$$x^2 - x = 56$$
$$x^2 - x - 56 = 0$$
$$(x+7)(x-8) = 0$$
$$x+7=0 \quad or \quad x-8=0$$
$$x=-7 \quad or \quad x=8$$

Check. The solutions of the equation are –7 and 8. Since the number of teams cannot be negative, –7 cannot be a solution. But 8 checks since

$$8^2 - 8 = 64 - 8 = 56.$$

State. There are 8 teams in the league.

23. ***Familiarize***. We will use the formula

$$H = \frac{1}{2}(n^2 - n).$$

Translate. Substitute 12 for n.

$$H = \frac{1}{2}(12^2 - 12)$$

Carry out. We do the computation on the right.

$$H = \frac{1}{2}(12^2 - 12)$$
$$H = \frac{1}{2}(144 - 12)$$
$$H = \frac{1}{2}(132)$$
$$H = 66$$

Check. We can recheck the computation, or we can solve the equation $66 = \frac{1}{2}(n^2 - n)$. The answer checks.

State. 66 handshakes are possible.

25. ***Familiarize***. We will use the formula $H = \frac{1}{2}(n^2 - n)$, since "high fives" can be substituted for handshakes.

Translate. Substitute 66 for H.

$$66 = \frac{1}{2}(n^2 - n)$$

Carry out.

$$66 = \frac{1}{2}(n^2 - n)$$
$$132 = n^2 - n \qquad \text{Multiplying by 2}$$
$$0 = n^2 - n - 132$$
$$0 = (n-12)(n+11)$$
$$n+11=0 \quad or \quad n-12=0$$
$$n=-11 \quad or \quad n=12$$

Check. The solutions of the equation are –11 and 12. Since the number of players cannot be negative, –11 cannot be a solution. However, 12 checks since

$$\frac{1}{2}(12^2 - 12) = \frac{1}{2}(144-12) = \frac{1}{2}(132) = 66.$$

State. 12 players were on the team.

27. ***Familiarize and Translate***. We substitute 150 for A in the formula.

$$A = -50t^2 + 200t$$
$$150 = -50t^2 + 200t$$

Carry out. We solve the equation.

$$150 = -50t^2 + 200t$$
$$0 = -50t^2 + 200t - 150$$
$$0 = -50(t^2 - 4t + 3)$$
$$0 = -50(t-1)(t-3)$$
$$t-1=0 \quad or \quad t-3=0$$
$$t=1 \quad or \quad t=3$$

Check. Since $-50 \cdot 1^2 + 200 \cdot 1 = -50 + 200 = 150,$ the number 1 checks. Since $-50 \cdot 3^2 + 200 \cdot 3$ $= -450 + 600 = 150,$ the number 3 checks also.

State. There will be about 150 micrograms of Albuterol in the bloodstream 1 minute and 3 minutes after an inhalation.

29. ***Familiarize***. We will use the formula

$$H = 0.03x^2 + 0.6x - 6.$$

Translate. Substitute 3 for H.

$$3 = 0.03x^2 + 0.6x - 6$$

Carry out.

$$3 = 0.03x^2 + 0.6x - 6$$
$$300 = 3x^2 + 60x - 600 \qquad \text{Multiplying by 100}$$
$$0 = 3x^2 + 60x - 900$$
$$0 = 3(x^2 + 20x - 300)$$
$$0 = 3(x+30)(x-10)$$
$$x=-30 \quad or \quad x=10$$

Check. The solutions of the equation are –30 and 10. Since the wind speed cannot be negative, –30 cannot be a solution. However, 10 checks since

$$0.03(10)^2 + 0.6(10) - 6 = 3 + 6 - 6 = 3.$$

State. The wind speed would be 10 knots.

31. ***Familiarize***. Let h = the vertical height to which each brace reaches, in feet. We have a right triangle with hypotenuse 15 ft and legs 12 ft and h.

Translate. We use the Pythagorean theorem.

$$a^2 + b^2 = c^2$$
$$12^2 + h^2 = 15^2$$

Carry out. We solve the equation.

$$12^2 + h^2 = 15^2$$
$$144 + h^2 = 225$$
$$h^2 - 81 = 0$$
$$(h+9)(h-9) = 0$$
$$h=-9 \quad or \quad h=9$$

Check. Since the vertical height must be positive, –9 cannot be a solution. If the height is 9 ft, then we have $12^2 + 9^2 = 144 + 81 = 225 = 15^2$. The number 9 checks.

State. Each brace reaches 9 ft vertically.

33. ***Familiarize***. Let l = the length of the leg, in ft. Then $l + 200$ = the length of the hypotenuse in feet.

Translate. We use the Pythagorean theorem.

$$a^2 + b^2 = c^2$$
$$400^2 + l^2 = (l + 200)^2$$

Carry out.

$$400^2 + l^2 = l^2 + 400l + 40,000$$
$$160,000 + l^2 = l^2 + 400l + 40,000$$
$$120,000 = 400l$$
$$300 = l$$

Check. When $l = 300$, then $l + 200 = 500$, and

$400^2 + 300^2 = 160,000 + 90,000 = 250,000 = 500^2$,

so the number 300 checks.

State. The dimensions of the garden are 300 ft by 400 ft by 500 ft.

35. ***Familiarize***. Let x = the width of one side, in ft. Then $x + 50$ = the length of the other side in ft.

Translate. We use the Pythagorean theorem.

$$a^2 + b^2 = c^2$$
$$x^2 + (x + 50)^2 = 250^2$$

Carry out.

$$x^2 + x^2 + 100x + 2500 = 62,500$$
$$2x^2 + 100x - 60,000 = 0$$
$$x^2 + 50x - 30,000 = 0$$
$$(x - 150)(x - 200) = 0$$

$x = 150$ *or* $x = 200$

Check. Since a width of 200 ft would yield a length of 250 ft, the same as the diagonal, which is not possible. When $x = 150$, then $x + 50 = 200$, and

$150^2 + 200^2 = 22,500 + 40,000 = 62,500 = 250^2$, so 150 checks.

State. The dimensions of the parking lot are length 200 ft and width 150 ft.

37. ***Familiarize***. From the drawing, we see the legs are x and $x + 1$, and the hypotenuse is $x + 2$.

Translate. We use the Pythagorean theorem.

$$a^2 + b^2 = c^2$$
$$x^2 + (x + 1)^2 = (x + 2)^2$$

Carry out.

$$x^2 + x^2 + 2x + 1 = x^2 + 4x + 4$$
$$x^2 - 2x - 3 = 0$$
$$(x + 1)(x - 3) = 0$$

$x = -1$ *or* $x = 3$

Check. Since the length must be positive, 0 cannot be a solution. When $x = 3$, then $x + 1 = 4$, and

$x + 2 = 5$, then $3^2 + 4^2 = 9 + 16 = 25 = 5^2$, so the number 3 checks.

State. The lengths of the sides are 3, 4, and 5.

39. ***Familiarize***. Let x = the length of the shortest side, in feet. Then $3x + 1$ = the length of the longest side in feet. The $3x - 1$ = the length of the other side in feet.

Translate. We use the Pythagorean theorem.

$$a^2 + b^2 = c^2$$
$$x^2 + (3x - 1)^2 = (3x + 1)^2$$

Carry out.

$$x^2 + 9x^2 - 6x + 1 = 9x^2 + 6x + 1$$
$$x^2 - 12x = 0$$
$$x(x - 12) = 0$$

$x = 0$ *or* $x = 12$

Check. Since the length must be positive, 0 cannot be a solution. When $x = 12$, then $3x - 1 = 35$, and

$12^2 + 35^2 = 144 + 1225 = 1369 = 37^2$, so the number 12 checks.

State. The lengths of the two sides are 12 ft, 35 ft, and 37 ft.

41. ***Familiarize***. We label the drawing. Let x = the length of a side of the dining room, in ft. Then the dining room has dimensions x by x and the kitchen has dimensions x by 10. The entire rectangular space has dimension x by $x + 10$ Recall that we multiply these dimensions to find the area of the rectangle.

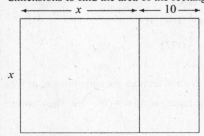

Translate.

The area of the rectangular space	is	264 ft^2.
↓	↓	↓
$x(x + 10)$	$=$	264

Carry out. We solve the equation.

$$x(x + 10) = 264$$
$$x^2 + 10x = 264$$
$$x^2 + 10x - 264 = 0$$
$$(x + 22)(x - 12) = 0$$

$x + 22 = 0$ *or* $x - 12 = 0$

$x = -22$ *or* $x = 12$

Check. Since the length of a side of the dining room must be positive, -22 cannot be a solution. If x is 12 ft, then $x + 10$ is 22 ft, and the area of the space is

$12 \cdot 22$, or 264 ft^2. The number 12 checks.

State. The dining room is 12 ft by 12 ft, and the kitchen is 12 ft by 10 ft.

43. ***Familiarize***. We will use the formula $h = 48t - 16t^2$.

Translate. Substitute $\frac{1}{2}$ for t.

$$h = 48 \cdot \frac{1}{2} - 16\left(\frac{1}{2}\right)^2$$

Carry out. We do the computation on the right.

$$h = 48 \cdot \frac{1}{2} - 16\left(\frac{1}{2}\right)^2$$
$$h = 48 \cdot \frac{1}{2} - 16 \cdot \frac{1}{4}$$
$$h = 24 - 4$$
$$h = 20$$

Check. We can recheck the computation, or we can solve the equation $20 = 48t - 16t^2$. The answer checks.

State. The rocket is 20 ft high $\frac{1}{2}$ sec after it is launched.

45. ***Familiarize***. We will use the formula $h = 48t - 16t^2$.

Translate. Substitute 32 for h.

$$32 = 48t - 16t^2$$

Carry out. We solve the equation.

$$32 = 48t - 16t^2$$
$$0 = -16t^2 + 48t - 32$$
$$0 = -16(t^2 - 3t + 2)$$
$$0 = -16(t-1)(t-2)$$

$t - 1 = 0$ *or* $t - 2 = 0$
$t = 1$ *or* $t = 2$

Check.

When $t = 1$, $h = 48 \cdot 1 - 16 \cdot 1^2 = 48 - 16 = 32$.

When $t = 2$, $h = 48 \cdot 2 - 16 \cdot 2^2 = 96 - 64 = 32$.

Both numbers check.

State. The rocket will be exactly 32 ft above the ground at 1 sec and at 2 sec after it is launched.

47. *Writing Exercise.*

49. $3(x-2) = 5x + 2(3-x)$
$$3x - 6 = 5x + 6 - 2x$$
$$3x - 6 = 3x + 6$$
$$-6 = 6$$

The solution is $\emptyset$. This is a contradiction.

51. $10x + 3 = 5(2x+1) - 2$
$$10x + 3 = 10x + 5 - 2$$
$$10x + 3 = 10x + 3$$
$$0 = 0$$

The solution is $\mathbb{R}$. This is an identity.

53. $\frac{1}{3}x + \frac{1}{2} < \frac{1}{6}$
$$6\left(\frac{1}{3}x + \frac{1}{2}\right) < 6\left(\frac{1}{6}\right)$$
$$2x + 3 < 1$$
$$2x < -2$$
$$x < -1$$

$\{x|x < -1\}$, or $(-\infty, -1)$

55. *Writing Exercise.*

57. ***Familiarize.*** First we find the length of the other leg of the right triangle. Then we find the area of the triangle, and finally we multiply by the cost per square foot of the sailcloth. Let x = the length of the other leg of the right triangle, in feet.

Translate. We use the Pythagorean theorem to find x.

$$a^2 + b^2 = c^2$$
$$x^2 + 24^2 = 26^2 \quad \text{Substituting}$$

Carry out.

$$x^2 + 24^2 = 26^2$$
$$x^2 + 576 = 676$$
$$x^2 - 100 = 0$$
$$(x+10)(x-10) = 0$$

$x + 10 = 0$ *or* $x - 10 = 0$
$x = -10$ *or* $x = 10$

Since the length of the leg must be positive, -10 cannot be a solution. We use the number 10. Find the area of the triangle:

$$\frac{1}{2}bh = \frac{1}{2} \cdot 10 \text{ ft} \cdot 24 \text{ ft} = 120 \text{ ft}^2$$

Finally, we multiply the area, 120 ft^2, by the price per square foot of the sailcloth, \$1.50:

$$120 \cdot (1.50) = 180$$

Check. Recheck the calculations. The answer checks.

State. A new main sail costs \$180.

59. ***Familiarize***. We add labels to the drawing in the text.

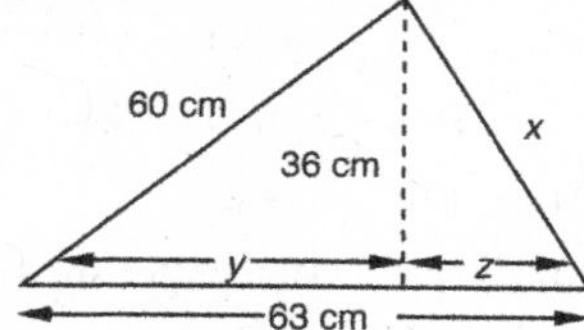

First we will use the Pythagorean theorem to find y. Then we will subtract to find z and, finally, we will use the Pythagorean theorem again to find x.

Translate. We use the Pythagorean theorem to find y.

$$a^2 + b^2 = c^2$$
$$y^2 + 36^2 = 60^2 \quad \text{Substituting}$$

Carry out.

$$y^2 + 36^2 = 60^2$$
$$y^2 + 1296 = 3600$$
$$y^2 - 2304 = 0$$
$$(y+48)(y-48) = 0$$

$y + 48 = 0$ *or* $y - 48 = 0$
$y = -48$ *or* $y = 48$

Since the length y cannot be negative, we use 48 cm. Then $z = 63 - 48 = 15$ cm.

Now we find x. We use the Pythagorean theorem again.

$$15^2+36^2=x^2$$
$$225+1296=x^2$$
$$1521=x^2$$
$$0=x^2-1521$$
$$0=(x+39)(x-39)$$

$x+39=0$ *or* $x-39=0$

$x=-39$ *or* $x=39$

Since the length x cannot be negative, we use 39 cm.

Check. We repeat all of the calculations. The answer checks.

State. The value of x is 39 cm.

61. ***Familiarize***. Let w = the width of the side turned up. The $20-2w=$ the length, in inches of the base. Recall the we multiply these dimensions to find the area of the rectangle.

Translate.

The area of the rectangular cross-section is 48 in^2.

$$w(20-2w)=48$$

Carry out. We solve the equation.

$$w(20-2w)=48$$
$$2w^2-20w+48=0$$
$$2(w^2-10w+24)=0$$
$$2(w-6)(w-4)=0$$

$w-6=0$ *or* $w-4=0$

$w=6$ *or* $w=4$

Check. If $w=6$ in., $20-2(6)=8$ in. and the area is $6\text{ in.}\cdot 8\text{ in.}=48\text{ in}^2$. If $w=4$ in., $20-2(4)=12$ in. and the area is $4\text{ in.}\cdot 12\text{ in.}=48\text{ in}^2$.

State. The possible depths of the gutter are 4 in. or 6 in.

63. ***Familiarize***. First we can use the Pythagorean theorem to find x, in ft. Then the height of the telephone pole is $x+5$.

Translate. We use the Pythagorean theorem.

$$a^2+b^2=c^2$$
$$\left(\frac{1}{2}x+1\right)^2+x^2=34^2$$

Carry out. We solve the equation.

$$\left(\frac{1}{2}x+1\right)^2+x^2=34^2$$
$$\frac{1}{4}x^2+x+1+x^2=1156$$
$$x^2+4x+4+4x^2=4624 \quad \text{Multiplying by 4}$$
$$5x^2+4x+4=4624$$
$$5x^2+4x-4620=0$$
$$(5x+154)(x-30)=0$$

$5x+154=0$ *or* $x-30=0$

$5x=-154$ *or* $x=30$

$x=-30.8$ *or* $x=30$

Check. Since the length x must be positive, -30.8 cannot be a solution. If x is 30 ft, then $\frac{1}{2}x+1$ is $\frac{1}{2}\cdot 30+1$, or 16 ft. Since $16^2+30^2=1156=34^2$, the number 30 checks. When x is 30 ft, then $x+5$ is 35 ft.

State. The height of the telephone pole is 35 ft.

65. ***Familiarize***. Let w = the width of the postcard, in inches. Then $w+\frac{7}{4}=$ the length. Recall that the area of a rectangle is length · width.

Translate.

The area of the postcard is $\frac{51}{2}\text{ in}^2$.

$$w\left(w+\frac{7}{4}\right)=\frac{51}{2}$$

Carry out. We solve the equation.

$$w\left(w+\frac{7}{4}\right)=\frac{51}{2}$$
$$w^2+\frac{7}{4}w=\frac{51}{2}$$
$$w^2+\frac{7}{4}w-\frac{51}{2}=0$$
$$4w^2+7w-102=0 \quad \text{Multiplying by 4}$$
$$(w+6)(4w-17)=0$$

$w+6=0$ *or* $4w-17=0$

$w=-6$ *or* $w=\frac{17}{4}$

Check. Since the width must be positive, -6 cannot be a solution. If the width is $\frac{17}{4}$ in, then the length is $\frac{17}{4}+\frac{7}{4}$ in. or 6 in, and the area is $6\text{ in.}\cdot\frac{17}{4}\text{ in.}=\frac{51}{2}\text{ in}^2$. Thus, $\frac{17}{4}$ checks.

State. The postcard has maximum length of 6 in. and maximum width of $\frac{17}{4}$ in.

67. First substitute 10 for N in the given formula.

$$10=-0.009t(t-12)^3$$

Graph $y_1=10$ and $y_2=-0.009x(x-12)^3$ in the given window and use the TRACE feature to find the first coordinates of the points of intersection of the graphs. We find $x\approx 0.8$ hr and $x\approx 6.4$ hr.

Chapter 5 Review

1. False. The largest common variable factor is the smallest power of the variable in the polynomial.

2. True

3. True

4. False. Only some binomials are the difference of two expressions that are squares.

5. False. Some quadratic equations have two different solutions.

6. True

7. True

8. False. The Pythagorean theorem can be applied to only a right triangle.

9. $20x^3 = (4 \cdot 5)(x \cdot x^2) = (-2x)(-10x^2)$
$= (20x)(x^2)$

10. $-18x^5 = (-9)(2x^5) = (-6x^2)(3x^3)$
$= (2x)(-9x^4)$

11. $12x^4 - 18x^3 = 6x^3 \cdot 2x - 6x^3 \cdot 3 = 6x^3(2x-3)$

12. $100t^2 - 1 = (10t+1)(10t-1)$

13. $x^2 + x - 12 = (x+4)(x-3)$

14. $x^2 + 14x + 49 = (x+7)(x+7) = (x+7)^2$

15. $12x^3 + 12x^2 + 3x = 3x \cdot 4x^2 + 3x \cdot 4x + 3x \cdot 1$
$= 3x(4x^2 + 4x + 1)$
$= 3x(2x+1)(2x+1)$
$= 3x(2x+1)^2$

16. $6x^3 + 9x^2 + 2x + 3$
$= 3x^2 \cdot 2x + 3x^2 \cdot 3 + 2x \cdot 1 + 3 \cdot 1$
$= 3x^2(2x+3) + 1(2x+3)$
$= (2x+3)(3x^2+1)$

17. $6a^2 + a - 5 = (6a-5)(a+1)$

18. $25t^2 + 9 - 30t = 25t^2 - 30t + 9$
$= (5t-3)(5t-3)$
$= (5t-3)^2$

19. $81a^4 - 1 = (9a^2+1)(9a^2-1)$
$= (9a^2+1)(3a+1)(3a-1)$

20. $9x^3 + 12x^2 - 45x = 3x(3x^2 + 4x - 15)$
$= 3x(3x-5)(x+3)$

21. $2x^3 - 250 = 2(x^3 - 125)$
$= 2(x-5)(x^2 + 5x + 25)$

22. $x^4 + 4x^3 - 2x - 8$
$= x^3(x+4) - 2(x+4)$
$= (x+4)(x^3-2)$

23. $a^2b^4 - 64 = (ab^2+8)(ab^2-8)$

24. $-8x^6 + 32x^5 - 4x^4 = -4x^4(2x^2 - 8x + 1)$

25. $75 + 12x^2 - 60x = 12x^2 - 60x + 75$
$= 3(4x^2 - 20x + 25)$
$= 3(2x-5)(2x-5)$
$= 3(2x-5)^2$

26. $y^2 + 9$, prime

27. $-t^3 + t^2 + 42t = -t(t^2 - t - 42)$
$= -t(t-7)(t+6)$

28. $4x^2 - 25 = (2x+5)(2x-5)$

29. $n^2 - 60 - 4n = n^2 - 4n - 60$
$= (n-10)(n+6)$

30. $5z^2 - 30z + 10 = 5(z^2 - 6z + 2)$

31. $8y^3 + 27x^6 = (2y + 3x^2)(4y^2 - 6x^2y + 9x^4)$

32. $2t^2 - 7t - 4 = (2t+1)(t-4)$

33. $7x^3 + 35x^2 + 28x = 7x(x^2 + 5x + 4)$
$= 7x(x+1)(x+4)$

34. $-6x^3 + 150x = -6x(x^2 - 25)$
$= -6x(x+5)(x-5)$

35. $15 - 8x + x^2 = (3-x)(5-x)$
or $(x-3)(x-5)$

36. $3x + x^2 + 5 = x^2 + 3x + 5$, prime

37. $x^2y^2 + 6xy - 16 = (xy+8)(xy-2)$

38. $12a^2 + 84ab + 147b^2 = 3(4a^2 + 28ab + 49b^2)$
$= 3(2a+7b)(2a+7b)$
$= 3(2a+7b)^2$

39. $m^2 + 5m + mt + 5t = m(m+5) + t(m+5)$
$= (m+5)(m+t)$

40. $6r^2 + rs - 15s^2 = (3r+5s)(2r-3s)$

41. $(x-9)(x+11) = 0$
$x - 9 = 0$ *or* $x + 11 = 0$
$x = 9$ *or* $x = -11$

42. $x^2 + 2x - 35 = 0$
$(x+7)(x-5) = 0$
$x + 7 = 0$ *or* $x - 5 = 0$
$x = -7$ *or* $x = 5$

43. $16x^2 = 9$
$16x^2 - 9 = 0$
$(4x+3)(4x-3) = 0$
$4x + 3 = 0$ *or* $4x - 3 = 0$
$x = -\frac{3}{4}$ *or* $x = \frac{3}{4}$

44. $3x^2+2=5x$
$3x^2-5x+2=0$
$(3x-2)(x-1)=0$
$3x-2=0$ *or* $x-1=0$
$x=\frac{2}{3}$ *or* $x=1$

45. $(x+1)(x-2)=4$
$x^2-x-2=4$
$x^2-x-6=0$
$(x+2)(x-3)=0$
$x+2=0$ *or* $x-3=0$
$x=-2$ *or* $x=3$

46. $9t-15t^2=0$
$3t(3-5t)=0$
$3t=0$ *or* $3-5t=0$
$t=0$ *or* $t=\frac{3}{5}$

47. $3x^2+3=6x$
$3x^2-6x+3=0$
$3(x^2-2x+1)=0$
$3(x-1)(x-1)=0$
$x-1=0$
$x=1$

48. ***Familiarize*** Let n = the number.

Translate.

The number squared is 12 more than the number
$n^2 = 12 + n$

Carry out. We solve the equation.
$n^2=12+n$
$n^2-n-12=0$
$(n+3)(n-4)=0$
$n+3=0$ *or* $n-4=0$
$n=-3$ *or* $n=4$

Check.
$4^2=12+4$
$16=16$ True
$(-3)^2=12+(-3)$
$9=9$ True

State. The solutions are –3 and 4.

49. ***Familiarize***. We will use the formula $s=4t+16t^2$.

Translate. Substitute 420 for s.
$420=4t+16t^2$.

Carry out. We solve the equation.
$420=4t+16t^2$
$0=16t^2+4t-420$
$0=4(4t^2+t-105)$
$0=4(4t+21)(t-5)$
$4t+21=0$ *or* $t-5=0$
$t=-\frac{21}{4}$ *or* $t=5$

Check. Since time must be positive, $-\frac{21}{4}$ is not a solution.
$420=4(5)+16(5)^2$
$420=420$ True
The answer checks.

State. The stone will reach the bottom of the cliff in 5 seconds.

50. We let $y=0$ and solve for x
$0=2x^2-3x-5$
$0=(x+1)(2x-5)$
$x+1=0$ *or* $2x-5=0$
$x=-1$ *or* $x=\frac{5}{2}$

The x-intercepts are $\left(\frac{5}{2},0\right)$ and $(-1,\ 0)$.

51. ***Familiarize.*** Let x = height = base of the triangle

Translate. Use the formula $A=\frac{1}{2}bh$
$A=\frac{1}{2}(x)(x)$
$98=\frac{1}{2}(x)(x)$

Carry out. We solve the equation.
$98=\frac{1}{2}x^2$
$196=x^2$
$0=x^2-196$
$0=(x+14)(x-14)$
$x+14=0$ *or* $x-14=0$
$x=14$ *or* $x=14$

Check. Since the height and base must be positive, –14 cannot be a solution.
$98=\frac{1}{2}(14)(14)$
$98=98$ True

State. The height and base are each 14 ft.

52. ***Familiarize.*** Let d = the diagonal of the brace.

Translate. We use the Pythagorean theorem.
$a^2+b^2=c^2$
$8^2+6^2=d^2$

Carry out. We solve the equation.
$8^2+6^2=c^2$
$8^2+6^2=d^2$
$100=d^2$
$0=d^2-100$
$0=(d+10)(d-10)$
$d+10=0$ *or* $d-10=0$
$d=-10$ *or* $d=10$

Check. Since the diagonal must be positive, –10 cannot be a solution.
$8^2+6^2=10^2$
$100=100$ True

State. The diagonal is 10 holes long.

53. *Writing Exercise.*

54. *Writing Exercise.*

55. ***Familiarize*** Let w = the width of the margin. Then $15-2w$ = the width of print in cm, and $20-2w$ = the length of print. Recall that $A = lw$.

Translate.

The area of print is $\frac{1}{2}$ of the area of the page

$$(15-2w)(20-2w) = \frac{1}{2} \cdot (15)(20)$$

Carry out. We solve the equation.

$$(15-2w)(20-2w) = \frac{1}{2}\cdot(15)(20)$$
$$300-70w+4w^2 = 150$$
$$4w^2-70w+150 = 0$$
$$2(2w-5)(w-15) = 0$$

$2w-5=0$ *or* $w-15=0$

$w=2.5$ *or* $w=15$

Check. Since the width of the margin cannot equal the width of the page, 15 cannot be a solution.

$$(15-2(2.5))(20-2(2.5)) = 150$$
$$(10)(15) = 150$$
$$150 = 150 \quad \text{True}$$

State. The width of the margin is 2.5 cm.

56. ***Familiarize*** Let n = the number.

Translate.

The cube of the number is twice the square of the number

$$n^3 = 2\ n^2$$

Carry out. We solve the equation.

$$n^3 = 2n^2$$
$$n^3 - 2n^2 = 0$$
$$n^2(n-2) = 0$$

$n^2=0$ *or* $n-2=0$

$n=0$ *or* $n=2$

Check.

$n^3 = 2n^2$ | $n^3 = 2n^2$

$0^3 = 2(0)^2$ | $3^3 = 2(2)^2$

$0 = 0$ True | $8 = 8$ True

State. The number is 0 or 2.

57. ***Familiarize.***

The length is increased by 20 & Wealth is decreased by 1 the area is 160.

$$(2w+20) \cdot (w-1) = 160$$

Carry out. We solve the equation.

$$(2w+20)\cdot(w-1) = 160$$
$$2w^2+18w-20 = 160$$
$$2w^2+18w-180 = 0$$
$$2(w^2+9w-90) = 0$$
$$2(w+15)(w-6) = 0$$

$w+15=0$ *or* $w-6=0$

$w=-15$ *or* $w=6$

Check. Since the width must be positive, -15 cannot be a solution.

$$(2\cdot 6+20)\cdot(6-1) = 160$$
$$(32)\cdot(5) = 160$$
$$160 = 160 \quad \text{True}$$

State. The width is 6 cm and the length is 12 cm.

58. $(x-2)2x^2 + x(x-2) - (x-2)15 = 0$

$$(x-2)(2x^2+x-15) = 0$$
$$(x-2)(2x-5)(x+3) = 0$$

$x-2=0$ *or* $2x-5=0$ *or* $x+3=0$

$x=2$ *or* $x=\frac{5}{2}$ *or* $x=-3$

59. $x^2+25=0$

No real solution, the sum of two squares cannot be factored.

Chapter 5 Test

1. $x^2-13x+36 = (x-4)(x-9)$

2. $x^2+25-10x = x^2-10x+25$
$= (x-5)(x-5)$
$= (x-5)^2$

3. $6y^2-8y^3+4y^4 = 4y^4-8y^3+6y^2$
$= 2y^2(2y^2-4y+3)$

4. $x^3+x^2+2x+2 = x^2(x+1)+2(x+1)$
$= (x+1)(x^2+2)$

5. $t^7-3t^5 = t^5(t^2-3)$

6. $a^3+3a^2-4a = a(a^2+3a-4)$
$= a(a+4)(a-1)$

7. $28x-48+10x^2 = 10x^2+28x-48$
$= 2(5x^2+14x-24)$
$= 2(5x-6)(x+4)$

8. $4t^2-25 = (2t+5)(2t-5)$

9. $-6m^3-9m^2-3m = -3m(2m^2+3m+1)$
$= -3m(2m+1)(m+1)$

10. $3r^3 - 3 = 3(r^3 - 1)$
$= 3(r-1)(r^2 + r + 1)$

11. $45r^2 + 60r + 20 = 5(9r^2 + 12r + 4)$
$= 5(3r+2)(3r+2)$
$= 5(3r+2)^2$

12. $3x^4 - 48 = 3(x^4 - 16)$
$= 3(x^2+4)(x^2-4)$
$= 3(x^2+4)(x+2)(x-2)$

13. $49t^2 + 36 + 84t = 49t^2 + 84t + 36$
$= (7t+6)(7t+6)$
$= (7t+6)^2$

14. $x^4 + 2x^3 - 3x - 6 = x^3(x+2) - 3(x+2)$
$= (x+2)(x^3 - 3)$

15. $x^2 + 3x + 6$ is prime.

16. $6t^3 + 9t^2 - 15t = 3t(2t^2 + 3t - 5)$
$= 3t(2t+5)(t-1)$

17. $3m^2 - 9mn - 30n^2 = 3(m^2 - 3mn - 10n^2)$
$= 3(m-5n)(m+2n)$

18. $x^2 - 6x + 5 = 0$
$(x-1)(x-5) = 0$
$x - 1 = 0$ *or* $x - 5 = 0$
$x = 1$ *or* $x = 5$

19. $2x^2 - 7x = 15$
$2x^2 - 7x - 15 = 0$
$(2x+3)(x-5) = 0$
$2x + 3 = 0$ *or* $x - 5 = 0$
$2x = -3$ *or* $x = 5$
$x = -\frac{3}{2}$ *or* $x = 5$

The solutions are $-\frac{3}{2}$ and 5.

20. $4t - 10t^2 = 0$
$2t(2-5t) = 0$
$2t = 0$ *or* $2 - 5t = 0$
$t = 0$ *or* $-5t = -2$
$t = 0$ *or* $t = \frac{2}{5}$

The solutions are 0 and $\frac{2}{5}$.

21. $25t^2 = 1$
$25t^2 - 1 = 0$
$(5t+1)(5t-1) = 0$
$5t + 1 = 0$ *or* $5t - 1 = 0$
$5t = -1$ *or* $5t = 1$
$t = -\frac{1}{5}$ *or* $t = \frac{1}{5}$

The solutions are $-\frac{1}{5}$ and $\frac{1}{5}$.

22. $x(x-1) = 20$
$x^2 - x = 20$
$x^2 - x - 20 = 0$
$(x+4)(x-5) = 0$
$x + 4 = 0$ *or* $x - 5 = 0$
$x = -4$ *or* $x = 5$

The solutions are –4 and 5.

23. We let $y = 0$ and solve for x.
$0 = 3x^2 - 5x - 8$
$0 = (x+1)(3x-8)$
$x + 1 = 0$ *or* $3x - 8 = 0$
$x = -1$ *or* $3x = 8$
$x = -1$ *or* $x = \frac{8}{3}$

The intercepts are $(-1, 0)$ and $\left(\frac{8}{3}, 0\right)$.

24. ***Familiarize.*** We make a drawing, Let w = the width, in m, then $w + 6$ = the length, in m.

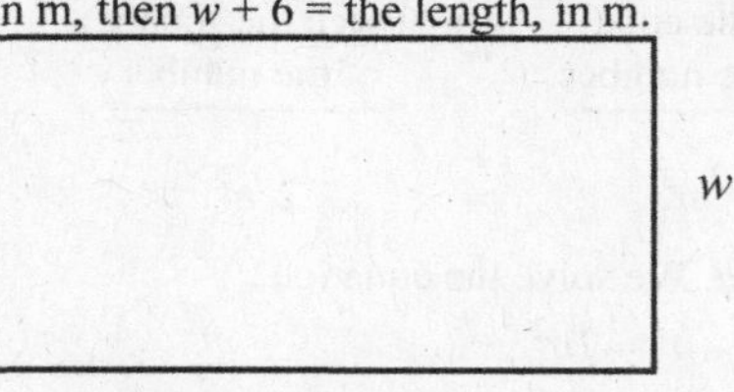

Recall the area of a rectangle is length times width.

Translate. We reword the problem

Length	times	width	is	40 m^2
$(w+6)$	$\cdot$	w	$=$	40

Carry out. We solve the equation.
$(w+6)w = 40$
$w^2 + 6w = 40$
$w^2 + 6w - 40 = 0$
$(w+10)(w-4) = 0$
$w + 10 = 0$ *or* $w - 4 = 0$
$w = -10$ *or* $w = 4$

Check. Since the width must be positive, –10 cannot be a solution. If the width is 4m, the length is 4 + 6, or 10 m, and the area is $10 \cdot 4$, or 40 m^2. Thus 4 checks.

State. The width is 4 m and the length is 10 m.

25. ***Familiarize.*** We will use the formula $H=\frac{1}{2}(n^2-n)$.

Translate. Substitute 45 for H. $45=\frac{1}{2}(n^2-n)$

Carry out. We solve for n.

$$45=\frac{1}{2}(n^2-n)$$
$$90=n^2-n$$
$$0=n^2-n-90$$
$$0=(n+9)(n-10)$$

$n+9=0$ *or* $n-10=0$
$n=-9$ *or* $n=10$

Check. Since the number of people must be positive, -9 cannot be a solution. However, 10 checks since

$$\frac{1}{2}(10^2-10)=\frac{1}{2}(100-10)=\frac{1}{2}(90)=45.$$

State. There were 10 people at the meeting.

26. ***Familiarize.*** From the given drawing x is the distance in feet we are looking for.

Translate. We use the Pythagorean Theorem.

$$3^2+4^2=x^2$$

Carry out. We solve the equation:

$$3^2+4^2=x^2$$
$$9+16=x^2$$
$$25=x^2$$
$$x=5 \text{ or } -5$$

Check. The number -5 is not a solution because distance cannot be negative. If $x=5$,

$3^2+4^2=9+16=5^2$, so the answer checks.

State. The distance is 5 ft.

27. ***Familiarize.*** Let w = the width of the piece of cardboard, in cm. Then $2w$ = the length, in cm. The length and width of the base of the box are $2x-8$ and $x-8$, respectively, and its height is 4.

Recall that the formula for the volume of a rectangular solid is given by length · width · height.

Translate.

$$\underbrace{\text{The volume}}_{(2w-8)(w-8)(4)} \quad \underbrace{\text{is}}_{=} \quad \underbrace{616\text{ cm}^3.}_{616}$$

Carry out. We solve the equation.

$$(2w-8)(w-8)(4)=616$$
$$(2w^2-24w+64)(4)=616$$
$$8w^2-96+256=616$$
$$8w^2-96w-360=0$$
$$8(w^2-12w-45)=0$$
$$w^2-12w-45=0 \quad \text{Dividing by 8}$$
$$(w+3)(w-15)=0$$

$w+3=0$ *or* $w-15=0$
$w=-3$ *or* $w=15$

Check. The width cannot be negative, so we only need to check 15. When $w=15$, then $2w=30$ and the dimensions of the box are $30-8$ by $15-8$ by 4, or 22 by 7 by 4. The volume is $22\cdot 7\cdot 4$, or 616.

State. The original dimension of the cardboard are 15 cm by 30 cm.

28.
$$\begin{aligned}(a+3)^2-2(a+3)-35&=[(a+3)-7][(a+3)+5]\\&=[a+3-7][a+3+5]\\&=(a-4)(a+8)\end{aligned}$$

29.
$$20x(x+2)(x-1)=5x^3-24x-14x^2$$
$$20x(x^2+x-2)=5x^3-14x^2-24x$$
$$20x^3+20x^2-40x=5x^3-14x^2-24x$$
$$15x^3+34x^2-16x=0$$
$$x(15x^2+34x-16)=0$$
$$x(5x-2)(3x+8)=0$$

$x=0$ *or* $5x-2=0$ *or* $3x+8=0$
$x=0$ *or* $5x=2$ *or* $3x=-8$
$x=0$ *or* $x=\frac{2}{5}$ *or* $x=-\frac{8}{3}$

The solutions are $-\frac{8}{3}$, 0 and $\frac{2}{5}$.

Chapter 6

Rational Expressions and Equations

Exercise Set 6.1

1. A rational expression can be written as a *quotient* of two polynomials.

3. A rational expression is simplified when the numerator and the denominator have no *factors* (other than 1) in common.

5. $t+1=0$ when $t=-1$ and $t-4=0$ when $t=4$, so choice (a) is correct.

7. $a^2-a-12=(a-4)(a+3)$; $a-4=0$ when $a=4$ and $a+3=0$ when $a=-3$, so choice (d) is correct.

9. $\frac{18}{-11x}$

We find the real number(s) that make the denominator 0. To do so we set the denominator equal to 0 and solve for x:

$$-11x=0$$
$$x=0$$

The expression is undefined for $x=0$.

11. $\frac{y-3}{y+5}$

Set the denominator equal to 0 and solve for y:

$$y+5=0$$
$$y=-5$$

The expression is undefined for $y=-5$.

13. $\frac{t-5}{3t-15}$

Set the denominator equal to 0 and solve for t:

$$3t-15=0$$
$$3t=15$$
$$t=5$$

The expression is undefined for $t=5$.

15. $\frac{x^2-25}{x^2-3x-28}$

Set the denominator equal to 0 and solve for x:

$$x^2-3x-28=0$$
$$(x+4)(x-7)=0$$
$$x+4=0 \quad or \quad x-7=0$$
$$x=-4 \quad or \quad x=7$$

The expression is undefined for $x=-4$ and $x=7$.

17. $\frac{t^2+t-20}{2t^2+11t-6}$

Set the denominator equal to 0 and solve for t:

$$2t^2+11t-6=0$$
$$(t+6)(2t-1)=0$$
$$t+6=0 \quad or \quad 2t-1=0$$
$$t=-6 \quad or \quad 2t=1$$
$$t=-6 \quad or \quad t=\frac{1}{2}$$

The expression is undefined for $t=-6$ and $t=\frac{1}{2}$.

19. $\frac{50a^2b}{40ab^3}$

$=\frac{5a\cdot 10ab}{4b^2\cdot 10ab}$ Factoring the numerator and denominator. Note the common factor of $10ab$.

$=\frac{5a}{4b^2}\cdot\frac{10ab}{10ab}$ Rewriting as a product of two rational expressions

$=\frac{5a}{4b^2}\cdot 1$ $\frac{10ab}{10ab}=1$

$=\frac{5a}{4b^2}$ Removing the factor 1

21. $\frac{6t+12}{6t-18}=\frac{\cancel{6}(t+2)}{\cancel{6}(t-3)}=\frac{(t+2)}{(t-3)}$

23. $\frac{21t-7}{24t-8}=\frac{7\cancel{(3t-1)}}{8\cancel{(3t-1)}}=\frac{7}{8}$

25.
$$\frac{a^2-9}{a^2+4a+3}=\frac{(a+3)(a-3)}{(a+3)(a+1)}$$
$$=\frac{a+3}{a+3}\cdot\frac{a-3}{a+1}$$
$$=1\cdot\frac{a-3}{a+1}$$
$$=\frac{a-3}{a+1}$$

27.
$$\frac{-36x^8}{54x^5}=\frac{-2x^3\cdot 18x^5}{3\cdot 18x^5}$$
$$=\frac{-2x^3}{3}\cdot\frac{18x^5}{18x^5}$$
$$=\frac{-2x^3}{3}$$

Check: Let $x=1$.

$$\frac{-36x^8}{54x^5}=\frac{-36\cdot 1^8}{54\cdot 1^5}=\frac{-36}{54}=\frac{-2}{3}$$
$$\frac{-2x^3}{3}=\frac{-2\cdot 1^3}{3}=\frac{-2}{3}$$

The answer is probably correct.

29. $\dfrac{-2y+6}{-8y}=\dfrac{-2(y-3)}{-2\cdot 4y}$
$=\dfrac{-2}{-2}\cdot\dfrac{y-3}{4y}$
$=1\cdot\dfrac{y-3}{4y}$
$=\dfrac{y-3}{4y}$

Check: Let $x=2$.
$\dfrac{-2y+6}{-8y}=\dfrac{-2\cdot 2+6}{-8\cdot 2}=\dfrac{2}{-16}=-\dfrac{1}{8}$
$\dfrac{y-3}{4y}=\dfrac{2-3}{4\cdot 2}=\dfrac{-1}{8}=-\dfrac{1}{8}$
The answer is probably correct.

31. $\dfrac{6a^2+3a}{7a^2+7a}=\dfrac{3\not{a}(2a+1)}{7\not{a}(a+1)}=\dfrac{3(2a+1)}{7(a+1)}$

Check: Let $a=1$.
$\dfrac{6a^2+3a}{7a^2+7a}=\dfrac{6\cdot 1^2+3\cdot 1}{7\cdot 1^2+7\cdot 1}=\dfrac{6+3}{7+7}=\dfrac{9}{14}$
$\dfrac{3(2a+1)}{7(a+1)}=\dfrac{3(2\cdot 1+1)}{7(1+1)}=\dfrac{3\cdot 3}{7\cdot 2}=\dfrac{9}{14}$
The answer is probably correct.

33. $\dfrac{t^2-16}{t^2-t-20}=\dfrac{(t-4)(t+4)}{(t-5)(t+4)}$
$=\dfrac{t-4}{t-5}\cdot\dfrac{t+4}{t+4}$
$=\dfrac{t-4}{t-5}\cdot 1$
$=\dfrac{t-4}{t-5}$

Check: Let $t=1$.
$\dfrac{t^2-16}{t^2-t-20}=\dfrac{1^2-16}{1^2-1-20}=\dfrac{-15}{-20}=\dfrac{3}{4}$
$\dfrac{t-4}{t-5}=\dfrac{1-4}{1-5}=\dfrac{3}{4}$
The answer is probably correct.

35. $\dfrac{3a^2+9a-12}{6a^2-30a+24}=\dfrac{3(a^2+3a-4)}{6(a^2-5a+4)}$
$=\dfrac{3(a+4)(a-1)}{3\cdot 2(a-4)(a-1)}$
$=\dfrac{3(a-1)}{3(a-1)}\cdot\dfrac{a+4}{2(a-4)}$
$=1\cdot\dfrac{a+4}{2(a-4)}$
$=\dfrac{a+4}{2(a-4)}$

Check: Let $a=2$.
$\dfrac{3a^2+9a-12}{6a^2-30a+24}=\dfrac{3\cdot 2^2+9\cdot 2-12}{6\cdot 2^2-30\cdot 2+24}=\dfrac{18}{-12}=-\dfrac{3}{2}$
$\dfrac{a+4}{2(a-4)}=\dfrac{2+4}{2(2-4)}=\dfrac{6}{-4}=-\dfrac{3}{2}$
The answer is probably correct.

37. $\dfrac{x^2-8x+16}{x^2-16}=\dfrac{(x-4)(x-4)}{(x+4)(x-4)}$
$=\dfrac{x-4}{x+4}\cdot\dfrac{x-4}{x-4}$
$=\dfrac{x-4}{x+4}\cdot 1$
$=\dfrac{x-4}{x+4}$

Check: Let $x=1$.
$\dfrac{x^2-8x+16}{x^2-16}=\dfrac{1^2-8\cdot 1+16}{1^2-16}=\dfrac{1-8+16}{1-16}=\dfrac{9}{-15}=-\dfrac{3}{5}$
$\dfrac{x-4}{x+4}=\dfrac{1-4}{1+4}=-\dfrac{3}{5}$
The answer is probably correct.

39. $\dfrac{n-2}{n^3-8}=\dfrac{n-2}{(n-2)(n^2+2n+4)}$
$=\dfrac{n-2}{n-2}\cdot\dfrac{1}{n^2+2n+4}=1\cdot\dfrac{1}{n^2+2n+4}$
$=\dfrac{1}{n^2+2n+4}$

Check: Let $n=1$.
$\dfrac{n-2}{n^3-8}=\dfrac{1-2}{1^3-8}=\dfrac{-1}{-7}=\dfrac{1}{7}$
$\dfrac{1}{n^2+2n+4}=\dfrac{1}{1^2+2(1)+4}=\dfrac{1}{7}$
The answer is probably correct.

41. $\dfrac{t^2-1}{t+1}=\dfrac{(t+1)(t-1)}{t+1}$
$=\dfrac{t+1}{t+1}\cdot\dfrac{t-1}{1}$
$=1\cdot\dfrac{t-1}{1}$
$=t-1$

Check: Let $t=2$.
$\dfrac{t^2-1}{t+1}=\dfrac{2^2-1}{2+1}=\dfrac{3}{3}=1$
$t-1=2-1=1$
The answer is probably correct.

43. $\dfrac{y^2+4}{y+2}$ cannot be simplified.

Neither the numerator nor the denominator can be factored.

45. $\dfrac{5x^2+20}{10x^2+40}=\dfrac{5(x^2+4)}{10(x^2+4)}=\dfrac{1\cdot\not{5}\cdot\not{(x^2+4)}}{2\cdot\not{5}\cdot\not{(x^2+4)}}=\dfrac{1}{2}$

Check: Let $x=1$.
$\dfrac{5x^2+20}{10x^2+40}=\dfrac{5\cdot 1^2+20}{10\cdot 1^2+40}=\dfrac{25}{50}=\dfrac{1}{2}$
$\dfrac{1}{2}=\dfrac{1}{2}$
The answer is probably correct.

47. $\frac{y^2+6y}{2y^2+13y+6}=\frac{y(y+6)}{(2y+1)(y+6)}$
$=\frac{y}{2y+1}\cdot\frac{y+6}{y+6}$
$=\frac{y}{2y+1}\cdot 1$
$=\frac{y}{2y+1}$

Check: Let $y=1$.

$\frac{y^2+6y}{2y^2+13y+6}=\frac{1^2+6\cdot 1}{2(1)^2+13\cdot 1+6}=\frac{7}{21}=\frac{1}{3}$

$\frac{y}{2y+1}=\frac{1}{2\cdot 1+1}=\frac{1}{3}$

The answer is probably correct.

49. $\frac{4x^2-12x+9}{10x^2-11x-6}=\frac{(2x-3)(2x-3)}{(2x-3)(5x+2)}$
$=\frac{2x-3}{2x-3}\cdot\frac{2x-3}{5x+2}$
$=1\cdot\frac{2x-3}{5x+2}$
$=\frac{2x-3}{5x+2}$

Check: Let $t=1$.

$\frac{4x^2-12x+9}{10x^2-11x-6}=\frac{4\cdot 1^2-12\cdot 1+9}{10\cdot 1^2-11\cdot 1-6}=\frac{1}{-7}=-\frac{1}{7}$

$\frac{2x-3}{5x+2}=\frac{2\cdot 1-3}{5\cdot 1+2}=\frac{-1}{7}=-\frac{1}{7}$

The answer is probably correct.

51. $\frac{10-x}{x-10}=\frac{-(x-10)}{x-10}=\frac{-1}{1}\cdot\frac{x-10}{x-10}=-1\cdot 1=-1$

Check: Let $x=1$.

$\frac{10-x}{x-10}=\frac{10-1}{1-10}=\frac{9}{-9}=-1$

The answer is probably correct.

53. $\frac{7t-14}{2-t}=\frac{7(t-2)}{-(t-2)}$
$=\frac{7}{-1}\cdot\frac{t-2}{t-2}$
$=\frac{7}{-1}\cdot 1$
$=-7$

Check: Let $t=1$.

$\frac{7t-14}{2-t}=\frac{7\cdot 1-14}{2-1}=\frac{-7}{1}=-7$

The answer is probably correct.

55. $\frac{a-b}{4b-4a}=\frac{a-b}{-4(a-b)}$
$=\frac{1}{-4}\cdot\frac{a-b}{a-b}=-\frac{1}{4}\cdot 1$
$=-\frac{1}{4}$

Check: Let $a=2$ and $b=1$.

$\frac{a-b}{4b-4a}=\frac{2-1}{4\cdot 1-4\cdot 2}=\frac{1}{4-8}=-\frac{1}{4}$

The answer is probably correct.

57. $\frac{3x^2-3y^2}{2y^2-2x^2}=\frac{3(x^2-y^2)}{2(y^2-x^2)}$
$=\frac{3(x^2-y^2)}{2(-1)(x^2-y^2)}$
$=\frac{3}{2(-1)}\cdot\frac{x^2-y^2}{x^2-y^2}=\frac{3}{2(-1)}\cdot 1$
$=-\frac{3}{2}$

Check: Let $x=1$ and $y=2$.

$\frac{3x^2-3y^2}{2y^2-2x^2}=\frac{3\cdot 1^2-3\cdot 2^2}{2\cdot 2^2-2\cdot 1^2}=\frac{-9}{6}=-\frac{3}{2}$

The answer is probably correct.

59. $\frac{7s^2-28t^2}{28t^2-7s^2}$

Note that the numerator and denominator are opposites. Thus, we have an expression divided by its opposite, so the result is –1.

61. *Writing Exercise.*

63. $3x^3+15x^2+x+5=3x^2(x+5)+(x+5)$
$=(x+5)(3x^2+1)$

65. $18y^4-27y^3+3y^2=3y^2(6y^2-9y+1)$

67. $m^3-8m^2+16m=m(m^2-8m+16)$
$=m(m-4)^2$

69. *Writing Exercise.*

71. $\frac{16y^2-x^4}{(x^2+4y^2)(x-2y)}$
$=\frac{(4y^2+x^2)(4y^2-x^2)}{(x^2+4y^2)(x-2y)}$
$=\frac{(4y^2+x^2)(2y+x)(2y-x)}{(x^2+4y^2)(x-2y)}$
$=\frac{(x^2+4y^2)(2y+x)(-1)(x-2y)}{(x^2+4y^2)(x-2y)}$
$=\frac{(x^2+4y^2)(x-2y)}{(x^2+4y^2)(x-2y)}\cdot\frac{(2y+x)(-1)}{1}$
$=-2y-x$, or $-x-2y$ or $-(2y+x)$

73. $\frac{x^5-2x^3+4x^2-8}{x^7+2x^4-4x^3-8}=\frac{x^3(x^2-2)+4(x^2-2)}{x^4(x^3+2)-4(x^3+2)}$
$=\frac{(x^2-2)(x^3+4)}{(x^3+2)(x^4-4)}$
$=\frac{(x^2-2)(x^3+4)}{(x^3+2)(x^2+2)(x^2-2)}$
$=\frac{\cancel{(x^2-2)}(x^3+4)}{(x^3+2)(x^2+2)\cancel{(x^2-2)}}$
$=\frac{x^3+4}{(x^3+2)(x^2+2)}$

75. $\dfrac{(t^4-1)(t^2-9)(t-9)^2}{(t^4-81)(t^2+1)(t+1)^2}$

$=\dfrac{(t^2+1)(t+1)(t-1)(t+3)(t-3)(t-9)(t-9)}{(t^2+9)(t+3)(t-3)(t^2+1)(t+1)(t+1)}$

$=\dfrac{\cancel{(t^2+1)}\,\cancel{(t+1)}(t-1)\cancel{(t+3)}\,\cancel{(t-3)}(t-9)(t-9)}{(t^2+9)\cancel{(t+3)}\,\cancel{(t-3)}\,\cancel{(t^2+1)}\,\cancel{(t+1)}(t+1)}$

$=\dfrac{(t-1)(t-9)(t-9)}{(t^2+9)(t+1)}$, or $\dfrac{(t-1)(t-9)^2}{(t^2+9)(t+1)}$

77. $\dfrac{x^3+6x^2-4x-24}{x^2+4x-12}=\dfrac{x^2(x+6)-4(x+6)}{(x+6)(x-2)}$

$=\dfrac{(x+6)(x^2-4)}{(x+6)(x-2)}$

$=\dfrac{(x+6)(x+2)(x-2)}{(x+6)(x-2)}$

$=\dfrac{\cancel{(x+6)}(x+2)\cancel{(x-2)}}{\cancel{(x+6)}\,\cancel{(x-2)}}$

$=x+2$

79. $\dfrac{(x^2-y^2)(x^2-2xy+y^2)}{(x+y)^2(x^2-4xy-5y^2)}$

$=\dfrac{(x+y)(x-y)(x-y)(x-y)}{(x+y)(x+y)(x-5y)(x+y)}$

$=\dfrac{\cancel{(x+y)}(x-y)(x-y)(x-y)}{\cancel{(x+y)}(x+y)(x-5y)(x+y)}$

$=\dfrac{(x-y)^3}{(x+y)^2(x-5y)}$

81. *Writing Exercise.*

Exercise Set 6.2

1. To simplify a rational expression, *remove a factor equal to 1*. Choice (d) is correct.

3. To find a reciprocal, *interchange the numerator and denominator*. Choice (a) is correct.

5. $\dfrac{x}{2}\cdot\dfrac{5}{y}=\dfrac{5x}{2y}$ Choice (d).

7. $x\cdot\dfrac{5}{y}=\dfrac{5x}{y}$ Choice (a).

9. $x\div\dfrac{5}{y}=\dfrac{x}{1}\cdot\dfrac{y}{5}=\dfrac{xy}{5}$ Choice (b).

11. $\dfrac{3x}{8}\cdot\dfrac{x+2}{5x-1}=\dfrac{3x(x+2)}{8(5x-1)}$

13. $\dfrac{a-4}{a+6}\cdot\dfrac{a+2}{a+6}=\dfrac{(a-4)(a+2)}{(a+6)(a+6)}$, or $\dfrac{(a-4)(a+2)}{(a+6)^2}$

15. $\dfrac{n-4}{n^2+4}\cdot\dfrac{n+4}{n^2-4}=\dfrac{(n-4)(n+4)}{(n^2+4)(n^2-4)}$

17. $\dfrac{8t^3}{5t}\cdot\dfrac{3}{4t}$

$=\dfrac{8t^3\cdot 3}{5t\cdot 4t}$ Multiplying the numerators and the denominators

$=\dfrac{2\cdot 4\cdot t\cdot t\cdot t\cdot 3}{5\cdot t\cdot 4\cdot t}$ Factoring the numerator and the denominator

$=\dfrac{2\cdot\cancel{4}\cdot t\cdot\cancel{t}\cdot\cancel{t}\cdot 3}{5\cdot\cancel{t}\cdot\cancel{4}\cdot\cancel{t}}$ Removing a factor equal to 1

$=\dfrac{6t}{5}$

19. $\dfrac{3c}{d^2}\cdot\dfrac{8d}{6c^3}$

$=\dfrac{3c\cdot 8d}{d^2\cdot 6c^3}$ Multiplying the numerators and the denominators

$=\dfrac{3\cdot c\cdot 2\cdot 4\cdot d}{d\cdot d\cdot 3\cdot 2\cdot c\cdot c\cdot c}$ Factoring the numerator and the denominator

$=\dfrac{\cancel{3}\cdot\cancel{c}\cdot\cancel{2}\cdot 4\cdot\cancel{d}}{\cancel{d}\cdot d\cdot\cancel{3}\cdot\cancel{2}\cdot\cancel{c}\cdot c\cdot c}$

$=\dfrac{4}{c^2d}$

21. $\dfrac{y^2-16}{4y+12}\cdot\dfrac{y+3}{y-4}=\dfrac{(y+4)(y-4)(y+3)}{4(y+3)(y-4)}$

$=\dfrac{(y+4)\cancel{(y-4)}\,\cancel{(y+3)}}{4\cancel{(y+3)}\,\cancel{(y-4)}}$

$=\dfrac{y+4}{4}$

23. $\dfrac{x^2-3x-10}{(x-2)^2}\cdot\dfrac{x-2}{x-5}=\dfrac{(x^2-3x-10)(x-2)}{(x-2)^2(x-5)}$

$=\dfrac{(x-5)(x+2)(x-2)}{(x-2)(x-2)(x-5)}$

$=\dfrac{\cancel{(x-5)}(x+2)\cancel{(x-2)}}{(x-2)\cancel{(x-2)}\,\cancel{(x-5)}}$

$=\dfrac{x+2}{x-2}$

25. $\dfrac{n^2-6n+5}{n+6}\cdot\dfrac{n-6}{n^2+36}=\dfrac{(n^2-6n+5)(n-6)}{(n+6)(n^2+36)}$

$=\dfrac{(n-5)(n-1)(n-6)}{(n+6)(n^2+36)}$

(No simplification is possible.)

27. $\dfrac{a^2-9}{a^2}\cdot\dfrac{7a}{a^2+a-12}=\dfrac{(a+3)(a-3)\cdot 7\cdot a}{a\cdot a(a+4)(a-3)}$

$=\dfrac{(a+3)\cancel{(a-3)}\cdot 7\cdot\cancel{a}}{\cancel{a}\cdot a(a+4)\cancel{(a-3)}}$

$=\dfrac{7(a+3)}{a(a+4)}$

29. $\dfrac{y^2-y}{y^2+5y+4}\cdot(y+4)=\dfrac{y(y-1)(y+4)}{(y+4)(y+1)}$

$=\dfrac{y(y-1)\cancel{(y+4)}}{\cancel{(y+4)}(y+1)}$

$=\dfrac{y(y-1)}{y+1}$

31. $\dfrac{4v-8}{5v}\cdot\dfrac{15v^2}{4v^2-16v+16}=\dfrac{(4v-8)15v^2}{5v(4v^2-16v+16)}$
$=\dfrac{\cancel{4}\,\cancel{(v-2)}\cdot\cancel{5}\cdot 3\cdot v\cdot\cancel{v}}{\cancel{5}\cancel{v}\cdot\cancel{4}\,\cancel{(v-2)}(v-2)}$
$=\dfrac{3v}{v-2}$

33. $\dfrac{t^2+2t-3}{t^2+4t-5}\cdot\dfrac{t^2-3t-10}{t^2+5t+6}=\dfrac{(t^2+2t-3)(t^2-3t-10)}{(t^2+4t-5)(t^2+5t+6)}$
$=\dfrac{(t+3)(t-1)(t-5)(t+2)}{(t+5)(t-1)(t+3)(t+2)}$
$=\dfrac{\cancel{(t+3)}\,\cancel{(t-1)}(t-5)\,\cancel{(t+2)}}{(t+5)\,\cancel{(t-1)}\,\cancel{(t+3)}\,\cancel{(t+2)}}$
$=\dfrac{t-5}{t+5}$

35. $\dfrac{12y+12}{5y+25}\cdot\dfrac{3y^2-75}{8y^2-8}=\dfrac{(12y+12)(3y^2-75)}{(5y+25)(8y^2-8)}$
$=\dfrac{3\cdot\cancel{4}\,\cancel{(y+1)}\,3\,\cancel{(y+5)}(y-5)}{5\,\cancel{(y+5)}\,2\cdot\cancel{4}\,\cancel{(y+1)}(y-1)}$
$=\dfrac{9(y-5)}{10(y-1)}$

37. $\dfrac{x^2+4x+4}{(x-1)^2}\cdot\dfrac{x^2-2x+1}{(x+2)^2}=\dfrac{(x+2)^2(x-1)^2}{(x-1)^2(x+2)^2}=1$

39. $\dfrac{t^2-4t+4}{2t^2-7t+6}\cdot\dfrac{2t^2+7t-15}{t^2-10t+25}$
$=\dfrac{(t^2-4t+4)(2t^2+7t-15)}{(2t^2-7t+6)(t^2-10t+25)}$
$=\dfrac{(t-2)(t-2)(2t-3)(t+5)}{(2t-3)(t-2)(t-5)(t-5)}$
$=\dfrac{\cancel{(t-2)}(t-2)\,\cancel{(2t-3)}(t+5)}{\cancel{(2t-3)}\,\cancel{(t-2)}(t-5)(t-5)}$
$=\dfrac{(t-2)(t+5)}{(t-5)^2}$

41. $(10x^2-x-2)\cdot\dfrac{4x^2-8x+3}{10x^2-11x-6}$
$=\dfrac{(10x^2-x-2)(4x^2-8x+3)}{(10x^2-11x-6)}$
$=\dfrac{\cancel{(5x+2)}(2x-1)(2x-1)\,\cancel{(2x-3)}}{\cancel{(5x+2)}\,\cancel{(2x-3)}}$
$=(2x-1)^2$

43. $\dfrac{49x^2-25}{4x-14}\cdot\dfrac{6x^2-13x-28}{28x-20}$
$=\dfrac{(7x+5)(7x-5)(3x+4)(2x-7)}{2(2x-7)\cdot 4(7x-5)}$
$=\dfrac{(7x+5)\,\cancel{(7x-5)}(3x+4)\,\cancel{(2x-7)}}{2\,\cancel{(2x-7)}\cdot 4\,\cancel{(7x-5)}}$
$=\dfrac{(7x+5)(3x+4)}{8}$

45. $\dfrac{8x^2+14xy-15y^2}{3x^3-x^2y}\cdot\dfrac{3x-y}{4xy-3y^2}$
$=\dfrac{(2x+5y)(4x-3y)(3x-y)}{x^2(3x-y)\cdot y(4x-3y)}$
$=\dfrac{(2x+5y)\,\cancel{(4x-3y)}\,\cancel{(3x-y)}}{x^2\,\cancel{(3x-y)}\cdot y\,\cancel{(4x-3y)}}$
$=\dfrac{2x+5y}{x^2y}$

47. $\dfrac{c^3+8}{c^5-4c^3}\cdot\dfrac{c^6-4c^5+4c^4}{c^2-2c+4}$
$=\dfrac{(c+2)(c^2-2c+4)\cdot c^4(c-2)(c-2)}{c^3(c+2)(c-2)\cdot(c^2-2c+4)}$
$=\dfrac{\cancel{(c+2)}\,\cancel{(c^2-2c+4)}\cdot\cancel{c^3}\cdot c\,\cancel{(c-2)}(c-2)}{\cancel{c^3}\,\cancel{(c+2)}\,\cancel{(c-2)}\,\cancel{(c^2-2c+4)}}$
$=c(c-2)$

49. The reciprocal of $\dfrac{2x}{9}$ is $\dfrac{9}{2x}$ because $\dfrac{2x}{9}\cdot\dfrac{9}{2x}=1$.

51. The reciprocal of a^4+3a is $\dfrac{1}{a^4+3a}$ because

$\dfrac{a^4+3a}{1}\cdot\dfrac{1}{a^4+3a}=1.$

53. $\dfrac{x}{4}\div\dfrac{5}{x}$
$=\dfrac{x}{4}\cdot\dfrac{x}{5}$ Multiplying by the reciprocal of the divisor
$=\dfrac{x\cdot x}{4\cdot 5}$
$=\dfrac{x^2}{20}$

55. $\dfrac{a^5}{b^4}\div\dfrac{a^2}{b}=\dfrac{a^5}{b^4}\cdot\dfrac{b}{a^2}$
$=\dfrac{a^5\cdot b}{b^4\cdot a^2}$
$=\dfrac{a^2\cdot a^3\cdot b}{b\cdot b^3\cdot a^2}$
$=\dfrac{a^2b}{a^2b}\cdot\dfrac{a^3}{b^3}$
$=\dfrac{a^3}{b^3}$

57. $\dfrac{t-3}{6}\div\dfrac{t+1}{8}=\dfrac{t-3}{6}\cdot\dfrac{8}{t+1}$
$=\dfrac{(t-3)(8)}{6\cdot(t+1)}$
$=\dfrac{(t-3)\cdot 4\cdot\cancel{2}}{\cancel{2}\cdot 3(t+1)}$
$=\dfrac{4(t-3)}{3(t+1)}$

59. $\dfrac{4y-8}{y+2} \div \dfrac{y-2}{y^2-4} = \dfrac{4y-8}{y+2} \cdot \dfrac{y^2-4}{y-2}$

$= \dfrac{(4y-8)(y^2-4)}{(y+2)(y-2)}$

$= \dfrac{4\cancel{(y-2)}\cancel{(y+2)}(y-2)}{\cancel{(y+2)}\cancel{(y-2)}(1)}$

$= 4(y-2)$

61. $\dfrac{a}{a-b} \div \dfrac{b}{b-a} = \dfrac{a}{a-b} \cdot \dfrac{b-a}{b}$

$= \dfrac{a(b-a)}{(a-b)(b)}$

$= \dfrac{a(-1)\cancel{(a-b)}}{\cancel{(a-b)}(b)}$

$= \dfrac{-a}{b} = -\dfrac{a}{b}$

63. $(n^2+5n+6) \div \dfrac{n^2-4}{n+3} = \dfrac{(n^2+5n+6)}{1} \cdot \dfrac{(n+3)}{n^2-4}$

$= \dfrac{(n^2+5n+6)(n+3)}{n^2-4}$

$= \dfrac{(n+3)\cancel{(n+2)}(n+3)}{\cancel{(n+2)}(n-2)}$

$= \dfrac{(n+3)^2}{n-2}$

65. $\dfrac{a+2}{a-1} \div \dfrac{3a+6}{a-5} = \dfrac{a+2}{a-1} \cdot \dfrac{a-5}{3a+6}$

$= \dfrac{(a+2)(a-5)}{(a-1)(3a+6)}$

$= \dfrac{(a+2)(a-5)}{(a-1)\cdot 3\cdot (a+2)}$

$= \dfrac{\cancel{(a+2)}(a-5)}{(a-1)\cdot 3\cdot \cancel{(a+2)}}$

$= \dfrac{a-5}{3(a-1)}$

67. $(2x-1) \div \dfrac{2x^2-11x+5}{4x^2-1} = \dfrac{2x-1}{1} \cdot \dfrac{4x^2-1}{2x^2-11x+5}$

$= \dfrac{(2x-1)(4x^2-1)}{1\cdot(2x^2-11x+5)}$

$= \dfrac{(2x-1)(2x+1)(2x-1)}{(2x-1)(x-5)}$

$= \dfrac{\cancel{(2x-1)}(2x+1)(2x-1)}{\cancel{(2x-1)}(x-5)}$

$= \dfrac{(2x-1)(2x+1)}{x-5}$

69. $\dfrac{w^2-14w+49}{2w^2-3w-14} \div \dfrac{3w^2-20w-7}{w^2-6w-16}$

$= \dfrac{w^2-14w+49}{2w^2-3w-14} \cdot \dfrac{w^2-6w-16}{3w^2-20w-7}$

$= \dfrac{(w^2-14w+49)(w^2-6w-16)}{(2w^2-3w-14)(3w^2-20w-7)}$

$= \dfrac{(w-7)\cancel{(w-7)}(w-8)\cancel{(w+2)}}{(2w-7)\cancel{(w+2)}(3w+1)\cancel{(w-7)}}$

$= \dfrac{(w-7)(w-8)}{(2w-7)(3w+1)}$

71. $\dfrac{c^2+10c+21}{c^2-2c-15} \div (5c^2+32c-21)$

$= \dfrac{c^2+10c+21}{c^2-2c-15} \cdot \dfrac{1}{5c^2+32c-21}$

$= \dfrac{(c^2+10c+21)\cdot 1}{(c^2-2c-15)(5c^2+32c-21)}$

$= \dfrac{(c+7)(c+3)}{(c-5)(c+3)(5c-3)(c+7)}$

$= \dfrac{(c+7)(c+3)}{(c+7)(c+3)} \cdot \dfrac{1}{(c-5)(5c-3)}$

$= \dfrac{1}{(c-5)(5c-3)}$

73. $\dfrac{-3+3x}{16} \div \dfrac{x-1}{5} = \dfrac{3x-3}{16} \cdot \dfrac{5}{x-1}$

$= \dfrac{(3x-3)\cdot 5}{16(x-1)}$

$= \dfrac{3(x-1)\cdot 5}{16(x-1)}$

$= \dfrac{3\cancel{(x-1)}\cdot 5}{16\cancel{(x-1)}}$

$= \dfrac{15}{16}$

75. $\dfrac{x-1}{x+2} \div \dfrac{1-x}{4+x^2} = \dfrac{x-1}{x+2} \cdot \dfrac{4+x^2}{1-x}$

$= \dfrac{(x-1)(4+x^2)}{(x+2)(1-x)}$

$= \dfrac{\cancel{(x-1)}(x^2+4)}{-1(x+2)\cancel{(x-1)}}$

$= -\dfrac{x^2+4}{x+2}$ or $\dfrac{-x^2-4}{x+2}$

77. $\dfrac{x-y}{x^2+2xy+y^2} \div \dfrac{x^2-y^2}{x^2-5xy+4y^2}$

$= \dfrac{x-y}{x^2+2xy+y^2} \cdot \dfrac{x^2-5xy+4y^2}{x^2-y^2}$

$= \dfrac{\cancel{(x-y)}(x-y)(x-4y)}{(x+y)(x+y)(x+y)\cancel{(x-y)}}$

$= \dfrac{(x-y)(x-4y)}{(x+y)^3}$

79. $\dfrac{x^3-64}{x^3+64} \div \dfrac{x^2-16}{x^2-4x+16}$
$= \dfrac{x^3-64}{x^3+64} \cdot \dfrac{x^2-4x+16}{x^2-16}$
$= \dfrac{(x-4)(x^2+4x+16)\cdot(x^2-4x+16)}{(x+4)(x^2-4x+16)(x+4)(x-4)}$
$= \dfrac{x^2+4x+16}{(x+4)^2}$

81. $\dfrac{8a^3+b^3}{2a^2+3ab+b^2} \div \dfrac{8a^2-4ab+2b^2}{4a^2+4ab+b^2}$
$= \dfrac{8a^3+b^3}{2a^2+3ab+b^2} \cdot \dfrac{4a^2+4ab+b^2}{8a^2-4ab+2b^2}$
$= \dfrac{(2a+b)(4a^2+2ab+b^2)\cdot(2a+b)(2a+b)}{(a+b)(2a+b)\cdot 2(4a^2+2ab+b^2)}$
$= \dfrac{(2a+b)^2}{2(a+b)}$

83. *Writing Exercise.*

85. Graph $y = \frac{1}{2}x - 5$.

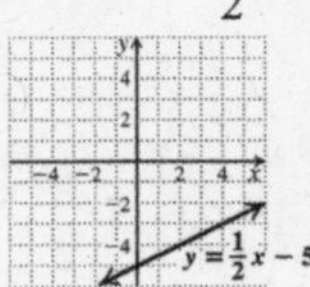

87. Graph $3(x-1) = 4$.

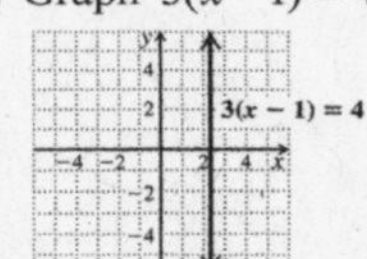

89. Graph $3y = 5x$.

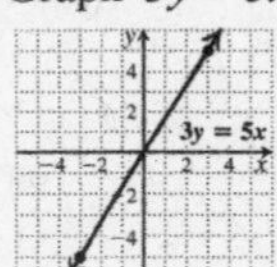

91. *Writing Exercise.*

93. The reciprocal of $2\frac{1}{3}x$ is

$\dfrac{1}{2\frac{1}{3}x} = \dfrac{1}{\frac{7x}{3}} = 1 \div \dfrac{7x}{3} = 1 \cdot \dfrac{3}{7x} = \dfrac{3}{7x}.$

95. $(x-2a) \div \dfrac{a^2x^2-4a^4}{a^2x+2a^3} = \dfrac{x-2a}{1} \cdot \dfrac{a^2x+2a^3}{a^2x^2-4a^4}$
$= \dfrac{(x-2a)(a^2x^2+2a^3)}{(a^2x^2-4a^4)}$
$= \dfrac{(x-2a)\,a^2\,(x+2a)}{a^2\,(x-2a)\,(x+2a)}$
$= 1$

97. $\dfrac{3a^2-5ab-12b^2}{3ab+b^2} \div (3b^2-ab)^2$
$= \dfrac{3a^2-5ab-12b^2}{3ab+4b^2} \cdot \dfrac{1}{(3b^2-ab)^2}$
$= \dfrac{(3a+4b)(a-3b)}{b(3a+4b)\cdot[b(3b-a)]^2}$
$= \dfrac{(3a+4b)(-1)(3b-a)}{b(3a+4b)(b^2)(3b-a)(3b-a)}$
$= \dfrac{(3a+4b)(-1)(3b-a)}{b(3a+4b)(b^2)(3b-a)(3b-a)}$
$= -\dfrac{1}{b^3(3b-a)}$, or $\dfrac{1}{b^3(a-3b)}$

99. $\dfrac{z^2-8z+16}{z^2+8z+16} \div \dfrac{(z-4)^5}{(z+4)^5} \div \dfrac{3z+12}{z^2-16}$
$= \dfrac{(z-4)^2}{(z+4)^2} \cdot \dfrac{(z+4)^5}{(z-4)^5} \cdot \dfrac{(z+4)(z-4)}{3(z+4)}$
$= \dfrac{(z-4)^2(z+4)^2(z+4)^3(z+4)(z-4)}{(z+4)^2(z-4)^2(z-4)(z-4)^2(3)(z+4)}$
$= \dfrac{(z+4)^3}{3(z-4)^2}$

101. $\dfrac{a^4-81b^4}{a^2c-6abc+9b^2c} \cdot \dfrac{a+3b}{a^2+9b^2} \div \dfrac{a^2+6ab+9b^2}{(a-3b)^2}$
$= \dfrac{(a^2+9b^2)(a+3b)(a-3b)}{c(a-3b)^2} \cdot \dfrac{a+3b}{a^2+9b^2} \cdot \dfrac{(a-3b)^2}{(a+3b)^2}$
$= \dfrac{(a^2+9b^2)(a+3b)(a-3b)}{c(a-3b)^2} \cdot \dfrac{a+3b}{a^2+9b^2} \cdot \dfrac{(a-3b)^2}{(a+3b)^2}$
$= \dfrac{a-3b}{c}$

103. $\dfrac{xy-2x+y-2}{xy+4x-y-4} \cdot \dfrac{xy+y+4x+4}{xy-y-2x+2}$
$= \dfrac{x(y-2)+(y-2)}{x(y+4)-1(y+4)} \cdot \dfrac{y(x+1)+4(x+1)}{y(x-1)-2(x-1)}$
$= \dfrac{(y-2)(x+1)(x+1)(y+4)}{(y+4)(x-1)(x-1)(y-2)}$
$= \dfrac{(y-2)(x+1)(x+1)(y+4)}{(y+4)(x-1)(x-1)(y-2)}$
$= \dfrac{(x+1)^2}{(x-1)^2}$

105. $\dfrac{3x^2-12x+bx-4b}{4x^2-16x-bx+4b} \div \dfrac{3bx+b^2+6x+2b}{4bx-b^2+8x-2b}$
$= \dfrac{3x(x-4)+b(x-4)}{4x(x-4)-b(x-4)} \cdot \dfrac{4bx-b^2+8x-2b}{3bx+b^2+6x+2b}$
$= \dfrac{(x-4)(3x+b)}{(x-4)(4x-b)} \cdot \dfrac{b(4x-b)+2(4x-b)}{b(3x+b)+2(3x+b)}$
$= \dfrac{(x-4)(3x+b)}{(x-4)(4x-b)} \cdot \dfrac{(4x-b)(b+2)}{(3x+b)(b+2)}$
$= \dfrac{(3x+b)(4x-b)(b+2)}{(4x-b)(3x+b)(b+2)}$
$= \dfrac{(3x+b)(4x-b)(b+2)}{(4x-b)(3x+b)(b+2)}$
$= 1$

107. $\frac{8n^2-10n+3}{4n^2-4n-3}\cdot\frac{6n^2-5n-6}{6n^2+7n-5}\div\frac{12n^2-17n+6}{6n^2+7n-5}$

$=\frac{8n^2-10n+3}{4n^2-4n-3}\cdot\frac{6n^2-5n-6}{6n^2+7n-5}\cdot\frac{6n^2+7n-5}{12n^2-17n+6}$

$=\frac{8n^2-10n+3}{4n^2-4n-3}\cdot\frac{6n^2-5n-6}{\cancel{6n^2+7n-5}}\cdot\frac{\cancel{6n^2+7n-5}}{12n^2-17n+6}$

$=\frac{(2n-1)(4n-3)(2n-3)(3n+2)}{(2n-3)(2n+1)(4n-3)(3n-2)}$

$=\frac{(2n-1)\cancel{(4n-3)}\cancel{(2n-3)}(3n+2)}{\cancel{(2n-3)}(2n+1)\cancel{(4n-3)}(3n-2)}$

$=\frac{(2n-1)(3n+2)}{(2n+1)(3n-2)}$

109. Enter $y_1=\frac{x-1}{x^2+2x+1}\div\frac{x^2-1}{x^2-5x+4}$ and $y_2=\frac{x^2-5x+4}{(x+1)^3}$, display the values of y_1 and y_2 in a table, and compare the values. (See the Technology Connection in the text.)

Exercise Set 6.3

1. To add two rational expressions when the denominators are the same, add *numerators* and keep the common *denominator*.

3. The least common multiple of two denominators is usually referred to as the *least common denominator* and is abbreviated *LCD*.

5. $\frac{3}{t}+\frac{5}{t}=\frac{8}{t}$ Adding numerators

7. $\frac{x}{12}+\frac{2x+5}{12}=\frac{3x+5}{12}$ Adding numerators

9. $\frac{4}{a+3}+\frac{5}{a+3}=\frac{9}{a+3}$

11. $\frac{11}{4x-7}-\frac{3}{4x-7}=\frac{8}{4x-7}$ Subtracting numerators

13. $\frac{3y+8}{2y}-\frac{y+1}{2y}$

$=\frac{3y+8-(y+1)}{2y}$

$=\frac{3y+8-y-1}{2y}$ Removing parentheses

$=\frac{2y+7}{2y}$

15. $\frac{5x+7}{x+3}+\frac{x+11}{x+3}$

$=\frac{6x+18}{x+3}$ Adding numerators

$=\frac{6(x+3)}{x+3}$ Factoring

$=\frac{6\cancel{(x+3)}}{\cancel{x+3}}$ Removing a factor equal to 1

$=6$

17. $\frac{5x+7}{x+3}-\frac{x+11}{x+3}=\frac{5x+7-(x+11)}{x+3}$

$=\frac{5x+7-x-11}{x+3}$

$=\frac{4x-4}{x+3}$

$=\frac{4(x-1)}{x+3}$

19. $\frac{a^2}{a-4}+\frac{a-20}{a-4}=\frac{a^2+a-20}{a-4}$

$=\frac{(a+5)(a-4)}{a-4}$

$=\frac{(a+5)\cancel{(a-4)}}{\cancel{a-4}}$

$=a+5$

21. $\frac{y^2}{y+2}-\frac{5y+14}{y+2}=\frac{y^2-(5y+14)}{y+2}$

$=\frac{y^2-5y-14}{y+2}$

$=\frac{(y-7)(y+2)}{y+2}$

$=\frac{(y-7)\cancel{(y+2)}}{\cancel{y+2}}$

$=y-7$

23. $\frac{t^2-5t}{t-1}+\frac{5t-t^2}{t-1}$

Note that the numerators are opposites, so their sum is 0. Then we have $\frac{0}{t-1}$, or 0.

25. $\frac{x-6}{x^2+5x+6}+\frac{9}{x^2+5x+6}=\frac{x+3}{x^2+5x+6}$

$=\frac{x+3}{(x+3)(x+2)}$

$=\frac{\cancel{x+3}}{\cancel{(x+3)}(x+2)}$

$=\frac{1}{x+2}$

27. $\frac{3a^2+14}{a^2+5a-6}-\frac{13a}{a^2+5a-6}=\frac{3a^2+14-13a}{a^2+5a-6}$

$=\frac{(3a-7)(a-2)}{(a+6)(a-1)}$

29. $\frac{t^2-5t}{t^2+6t+9}+\frac{4t-12}{t^2+6t+9}=\frac{t^2-t-12}{t^2+6t+9}$

$=\frac{(t-4)(t+3)}{(t+3)^2}$

$=\frac{(t-4)\cancel{(t+3)}}{(t+3)\cancel{(t+3)}}$

$=\frac{t-4}{t+3}$

31. $\dfrac{2y^2+3y}{y^2-7y+12}-\dfrac{y^2+4y+6}{y^2-7y+12}$
$=\dfrac{2y^2+3y-(y^2+4y+6)}{y^2-7y+12}$
$=\dfrac{2y^2+3y-y^2-4y-6}{y^2-7y+12}$
$=\dfrac{y^2-y-6}{y^2-7y+12}$
$=\dfrac{(y-3)(y+2)}{(y-3)(y-4)}$
$=\dfrac{\cancel{(y-3)}(y+2)}{\cancel{(y-3)}(y-4)}$
$=\dfrac{y+2}{y-4}$

33. $\dfrac{3-2x}{x^2-6x+8}+\dfrac{7-3x}{x^2-6x+8}$
$=\dfrac{10-5x}{x^2-6x+8}$
$=\dfrac{5(2-x)}{(x-4)(x-2)}$
$=\dfrac{5(-1)(x-2)}{(x-4)(x-2)}$
$=\dfrac{5(-1)\cancel{(x-2)}}{(x-4)\cancel{(x-2)}}$
$=\dfrac{-5}{x-4}$, or $-\dfrac{5}{x-4}$, or $\dfrac{5}{4-x}$

35. $\dfrac{x-9}{x^2+3x-4}-\dfrac{2x-5}{x^2+3x-4}$
$=\dfrac{x-9-(2x-5)}{x^2+3x-4}$
$=\dfrac{x-9-2x+5}{x^2+3x-4}$
$=\dfrac{-x-4}{x^2+3x-4}$
$=\dfrac{-(x+4)}{(x+4)(x-1)}$
$=\dfrac{-1\cancel{(x+4)}}{\cancel{(x+4)}(x-1)}$
$=\dfrac{-1}{x-1}$, or $-\dfrac{1}{x-1}$, or $\dfrac{1}{1-x}$

37. $15=3\cdot 5$
$36=2\cdot 2\cdot 3\cdot 3$
$\text{LCM}=2\cdot 2\cdot 3\cdot 3\cdot 5=180$

39. $8=2\cdot 2\cdot 2$
$9=3\cdot 3$
$\text{LCM}=2\cdot 2\cdot 2\cdot 3\cdot 3$, or 72

41. $6=2\cdot 3$
$12=2\cdot 2\cdot 3$
$15=3\cdot 5$
$\text{LCM}=2\cdot 2\cdot 3\cdot 5=60$

43. $18t^2=2\cdot 3\cdot 3\cdot t\cdot t$
$6t^5=2\cdot 3\cdot t\cdot t\cdot t\cdot t\cdot t$
$\text{LCM}=2\cdot 3\cdot 3\cdot t\cdot t\cdot t\cdot t\cdot t=18t^5$

45. $15a^4b^7=3\cdot 5\cdot a\cdot a\cdot a\cdot a\cdot b\cdot b\cdot b\cdot b\cdot b\cdot b\cdot b$
$10a^2b^8=2\cdot 5\cdot a\cdot a\cdot b\cdot b\cdot b\cdot b\cdot b\cdot b\cdot b\cdot b$
$\text{LCM}=2\cdot 3\cdot 5\cdot a\cdot a\cdot a\cdot a\cdot b\cdot b\cdot b\cdot b\cdot b\cdot b\cdot b\cdot b,$
$=30a^4b^8$

47. $2(y-3)=2\cdot(y-3)$
$6(y-3)=2\cdot 3\cdot(y-3)$
$\text{LCM}=2\cdot 3\cdot(y-3)=6(y-3)$

49. $x^2-2x-15=(x-5)(x+3)$
$x^2-9=(x-3)(x+3)$
$\text{LCM}=(x-5)(x-3)(x+3)$

51. $t^3+4t^2+4t=t(t^2+4t+4)=t(t+2)(t+2)$
$t^2-4t=t(t-4)$
$\text{LCM}=t(t+2)(t+2)(t-4)=t(t+2)^2(t-4)$

53. $6xz^2=2\cdot 3\cdot x\cdot z\cdot z$
$8x^2y=2\cdot 2\cdot 2\cdot x\cdot x\cdot y$
$15y^3z=3\cdot 5\cdot y\cdot y\cdot y\cdot z$
$\text{LCM}=2\cdot 2\cdot 2\cdot 3\cdot 5\cdot x\cdot x\cdot y\cdot y\cdot y\cdot z\cdot z=120x^2y^3z^2$

55. $a+1=a+1$
$(a-1)^2=(a-1)(a-1)$
$a^2-1=(a+1)(a-1)$
$\text{LCM}=(a+1)(a-1)(a-1)=(a+1)(a-1)^2$

57. $2n^2+n-1=(2n-1)(n+1)$
$2n^2+3n-2=(2n-1)(n+2)$
$\text{LCM}=(2n-1)(n+1)(n+2)$

59. $t-3=t-3$
$t+3=t+3$
$t^2-9=(t+3)(t-3)$
$\text{LCM}=(t+3)(t-3)$

61. $6x^3-24x^2+18x=6x(x^2-4x+3),$
$=2\cdot 3\cdot x(x-1)(x-3)$

$4x^5-24x^4+20x^3=4x^3(x^2-6x+5)$
$=2\cdot 2\cdot x\cdot x\cdot x(x-1)(x-5)$

$\text{LCM}=2\cdot 2\cdot 3\cdot x\cdot x\cdot x(x-1)(x-3)(x-5)$
$=12x^3(x-1)(x-3)(x-5)$

63. $2x^3-2=2(x^3-1)$
$=2\cdot(x-1)(x^2+x+1)$

$x^2-1=(x+1)(x-1)$

$\text{LCM}=2(x+1)(x-1)(x^2+x+1)$

65. $6t^4 = 2\cdot 3\cdot t\cdot t\cdot t\cdot t$

$18t^2 = 2\cdot 3\cdot 3\cdot t\cdot t$

The LCD is $2\cdot 3\cdot 3\cdot t\cdot t\cdot t\cdot t$, or $18t^4$.

$\frac{5}{6t^4}\cdot\frac{3}{3}=\frac{15}{18t^4}$ and

$\frac{s}{18t^2}\cdot\frac{t^2}{t^2}=\frac{st^2}{18t^4}$

67. $3x^4y^2 = 3\cdot x\cdot x\cdot x\cdot x\cdot y\cdot y$

$9xy^3 = 3\cdot 3\cdot x\cdot y\cdot y\cdot y$

The LCD is $3\cdot 3\cdot x\cdot x\cdot x\cdot x\cdot y\cdot y\cdot y$, or $9x^4y^3$.

$\frac{7}{3x^4y^2}\cdot\frac{3y}{3y}=\frac{21y}{9x^4y^3}$ and

$\frac{4}{9xy^3}\cdot\frac{x^3}{x^3}=\frac{4x^3}{9x^4y^3}$

69. $x^2-4=(x+2)(x-2)$

$x^2+5x+6=(x+3)(x+2)$

LCD is $(x+3)(x+2)(x-2)$

$$\frac{2x}{x^2-4}=\frac{2x}{(x+2)(x-2)}\cdot\frac{x+3}{x+3}=\frac{2x(x+3)}{(x+2)(x-2)(x+3)}$$

$$\frac{4x}{x^2+5x+6}=\frac{4x}{(x+3)(x+2)}\cdot\frac{x-2}{x-2}=\frac{4x(x-2)}{(x+3)(x+2)(x-2)}$$

71. *Writing Exercise.*

73. $2x-7=5x+3$

$-7=3x+3$

$-10=3x$

$-\frac{10}{3}=x$

The solution is $-\frac{10}{3}$.

75. $x^2-8x=20$

$x^2-8x-20=0$

$(x+2)(x-10)=0$

$x+2=0$ *or* $x-10=0$

$x=-2$ *or* $x=10$

The solutions are –2 and 10.

77. $2x^2+4x+2=0$

$2(x^2+2x+1)=0$

$2(x+1)^2=0$

$x+1=0$

$x=-1$

The solution is –1.

79. *Writing Exercise.*

81.
$$\frac{6x-1}{x-1}+\frac{3(2x+5)}{x-1}+\frac{3(2x-3)}{x-1}=\frac{6x-1+6x+15+6x-9}{x-1}=\frac{18x+5}{x-1}$$

83.
$$\frac{x^2}{3x^2-5x-2}-\frac{2x}{3x+1}\cdot\frac{1}{x-2}=\frac{x^2}{(3x+1)(x-2)}-\frac{2x}{(3x+1)(x-2)}=\frac{x^2-2x}{(3x+1)(x-2)}=\frac{x\cancel{(x-2)}}{(3x+1)\cancel{(x-2)}}=\frac{x}{3x+1}$$

85. The smallest number of strands that can be used is the LCM of 10 and 3.

$10=2\cdot 5$

$3=3$

$\text{LCM}=2\cdot 5\cdot 3=30$

The smallest number of strands that can be used is 30.

87. If the number of strands must also be a multiple of 4, we find the smallest multiple of 30 that is also a multiple of 4.

$1\cdot 30=30$, not a multiple of 4

$2\cdot 30=60=15\cdot 4$, a multiple of 4

The smallest number of strands that can be used is 60.

89. $4x^2-25=(2x+5)(2x-5)$

$6x^2-7x-20=(3x+4)(2x-5)$

$(9x^2+24x+16)^2=[(3x+4)(3x+4)]^2$

$=(3x+4)(3x+4)(3x+4)(3x+4)$

$\text{LCM}=(2x+5)(2x-5)(3x+4)^4$

91. The first printer prints 24 pages per minute, which is $\frac{24}{60}=\frac{2}{5}$ page per second. The second printer prints 15 pages per minute, which is $\frac{15}{60}$, or $\frac{1}{4}$ page per second. The time it takes until the machines begin printing a page at exactly the same time again is the LCM of their copying rates, 5 and 4.

It takes 20 seconds.

93. The number of minutes after 5:00 A.M. when the shuttles will first leave at the same time again is the LCM of their departure intervals, 15 minutes and 25 minutes.

$15=3\cdot 5$

$25=3\cdot 5$

$\text{LCM}=3\cdot 5\cdot 5=75$

Thus, the shuttles will leave at the same time 75 minutes after 5:00 A.M., or at 6:15 A.M.

95. *Writing Exercise.*

Exercise Set 6.4

1. To add or subtract when denominators are different, first find the *LCD*.

3. Add or subtract the *numerators*, as indicated. Write the sum or difference over the *LCD*.

5. $\frac{3}{x^2}+\frac{5}{x}=\frac{3}{x\cdot x}+\frac{5}{x}$ LCD $= x\cdot x$, or x^2
$$=\frac{3}{x\cdot x}+\frac{5}{x}\cdot\frac{x}{x}$$
$$=\frac{3+5x}{x^2}$$

7. $\left.\begin{array}{l}6r=2\cdot 3\cdot r\\ 8r=2\cdot 2\cdot 2\cdot r\end{array}\right\}$ LCD $=2\cdot 2\cdot 2\cdot 3\cdot r$, or $24r$
$$\frac{1}{6r}-\frac{3}{8r}=\frac{1}{6r}\cdot\frac{4}{4}-\frac{3}{8r}\cdot\frac{3}{3}$$
$$=\frac{4-9}{24r}$$
$$=\frac{-5}{24r},\text{ or }-\frac{5}{24r}$$

9. $\left.\begin{array}{l}uv^2=u\cdot v\cdot v\\ u^3v=u\cdot u\cdot u\cdot v\end{array}\right\}$ LCD $=u\cdot u\cdot u\cdot v\cdot v$, or u^3v^2
$$\frac{3}{uv^2}+\frac{4}{u^3v}=\frac{3}{uv^2}\cdot\frac{u^2}{u^2}+\frac{4}{u^3v}\cdot\frac{v}{v}=\frac{3u^2+4v}{u^3v^2}$$

11. $\left.\begin{array}{l}3xy^2=3\cdot x\cdot y\cdot y\\ x^2y^3=x\cdot x\cdot y\cdot y\cdot y\end{array}\right\}$ LCD $=3\cdot x\cdot x\cdot y\cdot y\cdot y$, or $3x^2y^3$
$$\frac{-2}{3xy^2}-\frac{6}{x^2y^3}=\frac{-2}{3xy^2}\cdot\frac{xy}{xy}-\frac{6}{x^2y^3}\cdot\frac{3}{3}=\frac{-2xy-18}{3x^2y^3}$$
$$=\frac{-2(xy+9)}{3x^2y^3}$$

13. $\left.\begin{array}{l}8=2\cdot 2\cdot 2\\ 6=2\cdot 3\end{array}\right\}$ LCD $=2\cdot 2\cdot 2\cdot 3$, or 24
$$\frac{x+3}{8}+\frac{x-2}{6}=\frac{x+3}{8}\cdot\frac{3}{3}+\frac{x-2}{6}\cdot\frac{4}{4}$$
$$=\frac{3(x+3)+4(x-2)}{24}$$
$$=\frac{3x+9+4x-8}{24}$$
$$=\frac{7x+1}{24}$$

15. $\left.\begin{array}{l}6=2\cdot 3\\ 3=3\end{array}\right\}$ LCD $=2\cdot 3$, or 6
$$\frac{x-2}{6}-\frac{x+1}{3}=\frac{x-2}{6}-\frac{x+1}{3}\cdot\frac{2}{2}$$
$$=\frac{x-2}{6}-\frac{2x+2}{6}$$
$$=\frac{x-2-(2x+2)}{6}$$
$$=\frac{x-2-2x-2}{6}$$
$$=\frac{-x-4}{6},\text{ or }\frac{-(x+4)}{6}$$

17. $\left.\begin{array}{l}15a=3\cdot 5\cdot a\\ 3a^2=3\cdot a\cdot a\end{array}\right\}$ LCD $=5\cdot 3\cdot a\cdot a$, or $15a^2$
$$\frac{a+3}{15a}+\frac{2a-1}{3a^2}=\frac{a+3}{15a}\cdot\frac{a}{a}+\frac{2a-1}{3a^2}\cdot\frac{5}{5}$$
$$=\frac{a^2+3a+10a-5}{15a^2}$$
$$=\frac{a^2+13a-5}{15a^2}$$

19. $\left.\begin{array}{l}3z=3\cdot z\\ 4z=2\cdot 2\cdot z\end{array}\right\}$ LCD $=2\cdot 2\cdot 3\cdot z$, or $12z$
$$\frac{4z-9}{3z}-\frac{3z-8}{4z}=\frac{4z-9}{3z}\cdot\frac{4}{4}-\frac{3z-8}{4z}\cdot\frac{3}{3}$$
$$=\frac{16z-36}{12z}-\frac{9z-24}{12z}$$
$$=\frac{16z-36-(9z-24)}{12z}$$
$$=\frac{16z-36-9z+24}{12z}$$
$$=\frac{7z-12}{12z}$$

21. $\left.\begin{array}{l}cd^2=c\cdot d\cdot d\\ c^2d=c\cdot c\cdot d\end{array}\right\}$ LCD $=c\cdot c\cdot d\cdot d$, or c^2d^2
$$\frac{3c+d}{cd^2}+\frac{c-d}{c^2d}=\frac{3c+d}{cd^2}\cdot\frac{c}{c}+\frac{c-d}{c^2d}\cdot\frac{d}{d}$$
$$=\frac{c(3c+d)+d(c-d)}{c^2d^2}$$
$$=\frac{3c^2+cd+cd-d^2}{c^2d^2}$$
$$=\frac{3c^2+2cd-d^2}{c^2d^2}$$
$$=\frac{(3c-d)(c+d)}{c^2d^2}$$

23. $\left.\begin{array}{l}3xt^2=3\cdot x\cdot t\cdot t\\ x^2t=x\cdot x\cdot t\end{array}\right\}$ LCD $=3\cdot x\cdot x\cdot t\cdot t$, or $3x^2t^2$
$$\frac{4x+2t}{3xt^2}-\frac{5x-3t}{x^2t}=\frac{4x+2t}{3xt^2}\cdot\frac{x}{x}-\frac{5x-3t}{x^2t}\cdot\frac{3t}{3t}$$
$$=\frac{4x^2+2tx}{3x^2t^2}-\frac{15xt-9t^2}{3x^2t^2}$$
$$=\frac{4x^2+2tx-(15xt-9t^2)}{3x^2t^2}$$
$$=\frac{4x^2+2tx-15xt+9t^2}{3x^2t^2}$$
$$=\frac{4x^2-13xt+9t^2}{3x^2t^2}$$

(Although $4x^2-13xt+9t^2$ can be factored, doing so will not enable us to simplify the result further.)

25. The denominators cannot be factored, so the LCD is their product, $(x-2)(x+2)$.
$$\frac{3}{x-2}+\frac{3}{x+2}=\frac{3}{x-2}\cdot\frac{x+2}{x+2}+\frac{3}{x+2}\cdot\frac{x-2}{x-2}$$
$$=\frac{3(x+2)+3(x-2)}{(x-2)(x+2)}$$
$$=\frac{3x+6+3x-6}{(x-2)(x+2)}$$
$$=\frac{6x}{(x-2)(x+2)}$$

27. $\frac{t}{t+3}-\frac{1}{t-1}$ LCD $=(t+3)(t-1)$

$$=\frac{t}{t+3}\cdot\frac{t-1}{t-1}-\frac{1}{t-1}\cdot\frac{t+3}{t+3}$$
$$=\frac{t^2-t}{(t+3)(t-1)}-\frac{t+3}{(t+3)(t-1)}$$
$$=\frac{t^2-t-(t+3)}{(t+3)(t-1)}$$
$$=\frac{t^2-t-t-3}{(t+3)(t-1)}$$
$$=\frac{t^2-2t-3}{(t+3)(t-1)}=\frac{(t-3)(t+1)}{(t+3)(t-1)}$$

(Although t^2-2t-3 can be factored, doing so will not enable us to simplify the result further.)

29. $\left.\begin{array}{l}3x=3\cdot x\\ x+1=x+1\end{array}\right\}$ LCD $=3x(x+1)$

$$\frac{3}{x+1}+\frac{2}{3x}=\frac{3}{x+1}\cdot\frac{3x}{3x}+\frac{2}{3x}\cdot\frac{x+1}{x+1}$$
$$=\frac{9x+2(x+1)}{3x(x+1)}$$
$$=\frac{9x+2x+2}{3x(x+1)}$$
$$=\frac{11x+2}{3x(x+1)}$$

31. $\left.\begin{array}{l}2t^2-2t=2t(t-1)\\ 2t-2=2(t-1)\end{array}\right\}$ LCD $=2t(t-1)$

$$\frac{3}{2t^2-2t}-\frac{5}{2t-2}=\frac{3}{2t(t-1)}-\frac{5}{2(t-1)}$$
$$=\frac{3}{2t(t-1)}-\frac{5}{2(t-1)}\cdot\frac{t}{t}$$
$$=\frac{3-5t}{2t(t-1)}$$

33. LCD $=(a+3)(a-3)$

$$\frac{3a}{a^2-9}+\frac{a}{a+3}$$
$$=\frac{3a}{(a+3)(a-3)}+\frac{a}{a+3}$$
$$=\frac{3a}{(a+3)(a-3)}+\frac{a}{a+3}\cdot\frac{a-3}{a-3}$$
$$=\frac{3a+a(a-3)}{(a+3)(a-3)}$$
$$=\frac{3a+a^2-3a}{(a+3)(a-3)}$$
$$=\frac{a^2}{(a+3)(a-3)}$$

35. LCD $=3(z+4)$

$$\frac{6}{z+4}-\frac{2}{3z+12}=\frac{6}{z+4}-\frac{2}{3(z+4)}$$
$$=\frac{6}{z+4}\cdot\frac{3}{3}-\frac{2}{3(z+4)}$$
$$=\frac{18}{3(z+4)}-\frac{2}{3(z+4)}$$
$$=\frac{16}{3(z+4)}$$

37. $\frac{5}{q-1}+\frac{2}{(q-1)^2}=\frac{5}{q-1}\cdot\frac{q-1}{q-1}+\frac{2}{(q-1)^2}$

$$=\frac{5(q-1)+2}{(q-1)^2}$$
$$=\frac{5q-5+2}{(q-1)^2}$$
$$=\frac{5q-3}{(q-1)^2}$$

39. $\frac{3a}{4a-20}+\frac{9a}{6a-30}$

$$=\frac{3a}{2\cdot 2(a-5)}+\frac{9a}{2\cdot 3(a-5)}\quad \text{LCD}=2\cdot 2\cdot 3(a-5)$$
$$=\frac{3a}{2\cdot 2(a-5)}\cdot\frac{3}{3}+\frac{9a}{2\cdot 3(a-5)}\cdot\frac{2}{2}$$
$$=\frac{9a+18a}{2\cdot 2\cdot 3(a-5)}$$
$$=\frac{27a}{2\cdot 2\cdot 3(a-5)}$$
$$=\frac{\cancel{3}\cdot 9\cdot a}{2\cdot 2\cdot \cancel{3}(a-5)}$$
$$=\frac{9a}{4(a-5)}$$

41. $\frac{y}{y-1}-\frac{y-1}{y}$ LCD $=y(y-1)$

$$=\frac{y}{y-1}\cdot\frac{y}{y}-\frac{y-1}{y}\cdot\frac{y-1}{y-1}$$
$$=\frac{y^2-(y^2-2y+1)}{y(y-1)}$$
$$=\frac{2y-1}{y(y-1)}$$

43. $\frac{6}{a^2+a-2}+\frac{4}{a^2-4a+3}$

$$=\frac{6}{(a+2)(a-1)}+\frac{4}{(a-3)(a-1)}$$

LCD $=(a+2)(a-1)(a-3)$

$$=\frac{6}{(a+2)(a-1)}\cdot\frac{a-3}{a-3}+\frac{4}{(a-3)(a-1)}\cdot\frac{a+2}{a+2}$$
$$=\frac{6(a-3)+4(a+2)}{(a+2)(a-1)(a-3)}$$
$$=\frac{6a-18+4a+8}{(a+2)(a-1)(a-3)}$$
$$=\frac{10a-10}{(a+2)(a-1)(a-3)}$$
$$=\frac{10\cancel{(a-1)}}{(a+2)\cancel{(a-1)}(a-3)}$$
$$=\frac{10}{(a+2)(a-3)}$$

45. $\dfrac{x}{x^2+9x+20}-\dfrac{4}{x^2+7x+12}$

$=\dfrac{x}{(x+4)(x+5)}-\dfrac{4}{(x+3)(x+4)}$

LCD $=(x+3)(x+4)(x+5)$

$=\dfrac{x}{(x+4)(x+5)}\cdot\dfrac{x+3}{x+3}-\dfrac{4}{(x+3)(x+4)}\cdot\dfrac{x+5}{x+5}$

$=\dfrac{x(x+3)-4(x+5)}{(x+3)(x+4)(x+5)}$

$=\dfrac{x^2+3x-4x-20}{(x+3)(x+4)(x+5)}$

$=\dfrac{x^2-x-20}{(x+3)(x+4)(x+5)}$

$=\dfrac{\cancel{(x+4)}(x-5)}{(x+3)\cancel{(x+4)}(x+5)}$

$=\dfrac{x-5}{(x+3)(x+5)}$

47. $\dfrac{3z}{z^2-4z+4}+\dfrac{10}{z^2+z-6}$

$=\dfrac{3z}{(z-2)^2}+\dfrac{10}{(z-2)(z+3)}$

LCD $=(z-2)^2(z+3)$

$=\dfrac{3z}{(z-2)^2}\cdot\dfrac{z+3}{z+3}+\dfrac{10}{(z-2)(z+3)}\cdot\dfrac{z-2}{z-2}$

$=\dfrac{3z(z+3)+10(z-2)}{(z-2)^2(z+3)}$

$=\dfrac{3z^2+9z+10z-20}{(z-2)^2(z+3)}$

$=\dfrac{3z^2+19z-20}{(z-2)^2(z+3)}$

49. $\dfrac{-7}{x^2+25x+24}-\dfrac{0}{x^2+11x+10}$

Note that $\dfrac{0}{x^2+11x+10}=0$, so the difference is

$\dfrac{-7}{x^2+25x+24}$.

51. $3+\dfrac{4}{2x+1}=3\cdot\dfrac{2x+1}{2x+1}+\dfrac{4}{2x+1}$

$=\dfrac{6x+3+4}{2x+1}$

$=\dfrac{6x+7}{2x+1}$

53. $3-\dfrac{2}{4-x}=3\cdot\dfrac{4-x}{4-x}-\dfrac{2}{4-x}$

$=\dfrac{12-3x-2}{4-x}$

$=\dfrac{10-3x}{4-x}$

55. $\dfrac{5x}{4}-\dfrac{x-2}{-4}=\dfrac{5x}{4}-\dfrac{x-2}{-4}\cdot\dfrac{-1}{-1}$

$=\dfrac{5x}{4}-\dfrac{2-x}{4}$

$=\dfrac{5x-(2-x)}{4}$

$=\dfrac{5x-2+x}{4}$

$=\dfrac{6x-2}{4}$

$=\dfrac{2(3x-1)}{2\cdot 2}$

$=\dfrac{\cancel{2}(3x-1)}{\cancel{2}\cdot 2}$

$=\dfrac{3x-1}{2}$

57. $\dfrac{x}{x-5}+\dfrac{x}{5-x}$

Note that the denominators are opposites, so the sum is 0.

59. $\dfrac{y^2}{y-3}+\dfrac{9}{3-y}=\dfrac{y^2}{y-3}+\dfrac{9}{3-y}\cdot\dfrac{-1}{-1}$

$=\dfrac{y^2}{y-3}+\dfrac{-9}{-3+y}$

$=\dfrac{y^2-9}{y-3}$

$=\dfrac{(y+3)\cancel{(y-3)}}{\cancel{y-3}}$

$=y+3$

61. $\dfrac{c-5}{c^2-64}+\dfrac{c-5}{64-c^2}=\dfrac{c-5}{c^2-64}+\dfrac{c-5}{64-c^2}\cdot\dfrac{-1}{-1}$

$=\dfrac{c-5}{c^2-64}+\dfrac{5-c}{c^2-64}$

$=\dfrac{c-5+5-c}{c^2-64}$

$=\dfrac{0}{c^2-64}$

$=0$

63. $\dfrac{t-3}{t^3-1}-\dfrac{2}{1-t^3}=\dfrac{t-3}{t^3-1}+\dfrac{2}{t^3-1}$

$=\dfrac{t-3+2}{t^3-1}$

$=\dfrac{t-1}{(t-1)(t^2+t+1)}$

$=\dfrac{1}{t^2+t+1}$

65. $\dfrac{4-p}{25-p^2}+\dfrac{p+1}{p-5}$

$=\dfrac{4-p}{(5+p)(5-p)}+\dfrac{p+1}{p-5}$

$=\dfrac{4-p}{(5+p)(5-p)}\cdot\dfrac{-1}{-1}+\dfrac{p+1}{p-5}$

$=\dfrac{p-4}{(p+5)(p-5)}+\dfrac{p+1}{p-5}$ LCD $=(p+5)(p-5)$

$=\dfrac{p-4}{(p+5)(p-5)}+\dfrac{p+1}{p-5}\cdot\dfrac{p+5}{p+5}$

$=\dfrac{p-4+p^2+6p+5}{(p+5)(p-5)}$

$=\dfrac{p^2+7p+1}{(p+5)(p-5)}$

67. $\frac{x}{x-4}-\frac{3}{16-x^2}$

$=\frac{x}{x-4}-\frac{3}{(4+x)(4-x)}$

$=\frac{x}{x-4}\cdot\frac{-1}{-1}-\frac{3}{(4+x)(4-x)}$

$=-\frac{x}{4-x}-\frac{3}{(4+x)(4-x)}$ LCD $=(4-x)(4+x)$

$=-\frac{x}{4-x}\cdot\frac{4+x}{4+x}-\frac{3}{(4+x)(4-x)}$

$=\frac{-x(4+x)-3}{(4-x)(4+x)}$

$=\frac{-4x-x^2-3}{(4-x)(4+x)}$

$=\frac{-x^2-4x-3}{(4-x)(4+x)}$

$=\frac{x^2+4x+3}{(x+4)(x-4)}=\frac{(x+3)(x+1)}{(x+4)(x-4)}$

69. $\frac{a}{a^2-1}+\frac{2a}{a-a^2}$

$=\frac{a}{a^2-1}+\frac{2\cdot \cancel{a}}{\cancel{a}(1-a)}$

$=\frac{a}{(a+1)(a-1)}+\frac{2}{1-a}$

$=\frac{a}{(a+1)(a-1)}+\frac{2}{1-a}\cdot\frac{-1}{-1}$

$=\frac{a}{(a+1)(a-1)}+\frac{-2}{a-1}$ LCD $=(a+1)(a-1)$

$=\frac{a}{(a+1)(a-1)}+\frac{-2}{a-1}\cdot\frac{a+1}{a+1}$

$=\frac{a-2a-2}{(a+1)(a-1)}$

$=\frac{-a-2}{(a+1)(a-1)}$, or $\frac{a+2}{(1+a)(1-a)}$

71. $\frac{4x}{x^2-y^2}-\frac{6}{y-x}$

$=\frac{4x}{(x+y)(x-y)}-\frac{6}{y-x}$

$=\frac{4x}{(x+y)(x-y)}-\frac{6}{y-x}\cdot\frac{-1}{-1}$

$=\frac{4x}{(x+y)(x-y)}-\frac{-6}{x-y}$ LCD $=(x+y)(x-y)$

$=\frac{4x}{(x+y)(x-y)}-\frac{-6}{x-y}\cdot\frac{x+y}{x+y}$

$=\frac{4x-(-6)(x+y)}{(x+y)(x-y)}$

$=\frac{4x+6x+6y}{(x+y)(x-y)}$

$=\frac{10x+6y}{(x+y)(x-y)}$

$=\frac{2(5x+3y)}{(x+y)(x-y)}$

73. $\frac{x-3}{2-x}-\frac{x+3}{x+2}+\frac{x+6}{4-x^2}$

$=\frac{x-3}{2-x}-\frac{x+3}{x+2}+\frac{x+6}{(2+x)(2-x)}$ LCD $=(2+x)(2-x)$

$=\frac{x-3}{2-x}\cdot\frac{2+x}{2+x}-\frac{x+3}{x+2}\cdot\frac{2-x}{2-x}+\frac{x+6}{(2+x)(2-x)}$

$=\frac{(x-3)(2+x)-(x+3)(2-x)+(x+6)}{(2+x)(2-x)}$

$=\frac{x^2-x-6-(-x^2-x+6)+x+6}{(2+x)(2-x)}$

$=\frac{x^2-x-6+x^2+x-6+x+6}{(2+x)(2-x)}$

$=\frac{2x^2+x-6}{(2+x)(2-x)}$

$=\frac{(2x-3)\cancel{(x+2)}}{\cancel{(2+x)}(2-x)}$

$=\frac{2x-3}{2-x}$

75. $\frac{2x+5}{x+1}+\frac{x+7}{x+5}-\frac{5x+17}{(x+1)(x+5)}$ LCD $=(x+1)(x+5)$

$=\frac{(2x+5)(x+5)+(x+7)(x+1)-(5x+17)}{(x+1)(x+5)}$

$=\frac{2x^2+15x+25+x^2+8x+7-5x-17}{(x+1)(x+5)}$

$=\frac{3x^2+18x+15}{(x+1)(x+5)}$

$=\frac{3\cancel{(x+1)}\cancel{(x+5)}}{\cancel{(x+1)}\cancel{(x+5)}}$

$=3$

77. $\frac{1}{x+y}+\frac{1}{x-y}-\frac{2x}{x^2-y^2}$

LCD $=(x+y)(x-y)$

$=\frac{1}{x+y}\cdot\frac{x-y}{x-y}+\frac{1}{x-y}\cdot\frac{x+y}{x+y}-\frac{2x}{(x+y)(x-y)}$

$=\frac{(x-y)+(x+y)-2x}{(x+y)(x-y)}$

$=0$

79. $\frac{1}{x^2+7x+12}-\frac{2}{x^2+4x+3}+\frac{3}{x^2+5x+4}$

$=\frac{1}{(x+3)(x+4)}-\frac{2}{(x+1)(x+3)}+\frac{3}{(x+1)(x+4)}$

LCD is $(x+1)(x+3)(x+4)$

$=\frac{1}{(x+3)(x+4)}\cdot\frac{x+1}{x+1}-\frac{2}{(x+1)(x+3)}\cdot\frac{x+4}{x+4}$

$+\frac{3}{(x+1)(x+4)}\cdot\frac{x+3}{x+3}$

$=\frac{x+1-2(x+4)+3(x+3)}{(x+1)(x+3)(x+4)}$

$=\frac{x+1-2x-8+3x+9}{(x+1)(x+3)(x+4)}$

$=\frac{2x+2}{(x+1)(x+3)(x+4)}$

$=\frac{2\cancel{(x+1)}}{\cancel{(x+1)}(x+3)(x+4)}$

$=\frac{2}{(x+3)(x+4)}$

81. *Writing Exercise.*

83. $3-12\div(-4)=3+3=6$

85. $(1.2\times10^8)(2.5\times10^6)=(1.2\times2.5)(10^8\times10^6)$
$=3\times10^{14}$

87. $(3a^{-1}b)^{-2}=3^{-2}a^2b^{-2}=\dfrac{a^2}{9b^2}$

89. *Writing Exercise.*

91. $P=2\left(\dfrac{3}{x+4}\right)+2\left(\dfrac{2}{x-5}\right)$

$=\dfrac{6}{x+4}+\dfrac{4}{x-5}\quad \text{LCD}=(x+4)(x-5)$

$=\dfrac{6}{x+4}\cdot\dfrac{x-5}{x-5}+\dfrac{4}{x-5}\cdot\dfrac{x+4}{x+4}$

$=\dfrac{6x-30+4x+16}{(x+4)(x-5)}$

$=\dfrac{10x-14}{(x+4)(x-5)}=\dfrac{2(5x-7)}{(x+4)(x-5)}$

$A=\left(\dfrac{3}{x+4}\right)\left(\dfrac{2}{x-5}\right)=\dfrac{6}{(x+4)(x-5)}$

93. $\dfrac{x^2}{3x^2-5x-2}-\dfrac{2x}{3x+1}\cdot\dfrac{1}{x-2}$

$=\dfrac{x^2}{(3x+1)(x-2)}-\dfrac{2x}{(3x+1)(x-2)}$

$=\dfrac{x^2-2x}{(3x+1)(x-2)}$

$=\dfrac{x(x-2)}{(3x+1)(x-2)}$

$=\dfrac{x}{3x+1}\cdot\dfrac{x-2}{x-2}$

$=\dfrac{x}{3x+1}$

95. $\dfrac{2x-16}{x^2-x-2}-\dfrac{x+2}{x^2-5x+6}-\dfrac{x+6}{x^2-2x-3}$

$=\dfrac{2x-16}{(x-2)(x+1)}-\dfrac{x+2}{(x-2)(x-3)}-\dfrac{x+6}{(x-3)(x+1)}$

$\text{LCD}=(x+1)(x-2)(x-3)$

$=\dfrac{2x-16}{(x-2)(x+1)}\cdot\dfrac{x-3}{x-3}-\dfrac{x+2}{(x-2)(x-3)}\cdot\dfrac{x+1}{x+1}$
$-\dfrac{x+6}{(x-3)(x+1)}\cdot\dfrac{x-2}{x-2}$

$=\dfrac{2x^2-22x+48-(x^2+3x+2)-(x^2+4x-12)}{(x+1)(x-2)(x-3)}$

$=\dfrac{-29x+58}{(x+1)(x-2)(x-3)}$

$=\dfrac{-29\cancel{(x-2)}}{(x+1)\cancel{(x-2)}(x-3)}$

$=\dfrac{-29}{(x+1)(x-3)}$

97. We recognize that this is the product of the sum and difference of two terms $(A+B)(A-B)=A^2-B^2$.

$\left(\dfrac{x}{x+7}-\dfrac{3}{x+2}\right)\left(\dfrac{x}{x+7}+\dfrac{3}{x+2}\right)$

$=\dfrac{x^2}{(x+7)^2}-\dfrac{9}{(x+2)^2}\quad \text{LCD}=(x+7)^2(x+2)^2$

$=\dfrac{x^2}{(x+7)^2}\cdot\dfrac{(x+2)^2}{(x+2)^2}-\dfrac{9}{(x+2)}\cdot\dfrac{(x+7)^2}{(x+7)^2}$

$=\dfrac{x^2(x+2)^2-9(x+7)^2}{(x+7)^2(x+2)^2}$

$=\dfrac{x^2(x^2+4x+4)-9(x^2+14+49)}{(x+7)^2(x+2)^2}$

$=\dfrac{x^2+4x^3+4x^2-9x^2-126x-441}{(x+7)^2(x+2)^2}$

$=\dfrac{x^4+4x^3-5x^2-126x-441}{(x+7)^2(x+2)^2}$

99. $\left(\dfrac{a}{a-b}+\dfrac{b}{a+b}\right)\left(\dfrac{1}{3a+b}+\dfrac{2a+6b}{9a^2-b^2}\right)$

$=\dfrac{a}{(a-b)(3a+b)}+\dfrac{a(2a+6b)}{(a-b)(9a^2-b^2)}$
$+\dfrac{b}{(a+b)(3a+b)}+\dfrac{b(2a+6b)}{(a+b)(9a^2-b^2)}$

$=\dfrac{a}{(a-b)(3a+b)}+\dfrac{2a^2+6ab}{(a-b)(3a+b)(3a-b)}+$
$+\dfrac{b}{(a+b)(3a+b)}+\dfrac{2ab+6b^2}{(a+b)(3a+b)(3a-b)}$

$\text{LCD}=(a-b)(a+b)(3a+b)(3a-b)$

$=\dfrac{[a(a+b)(3a-b)+(2a^2+6ab)(a+b)}{(a-b)(a+b)(3a+b)(3a-b)}$
$+\dfrac{b(a-b)(3a-b)+(2ab+6b^2)(a-b)]}{(a-b)(a+b)(3a+b)(3a-b)}$

$=\dfrac{(3a^3+2a^2b-ab^2+2a^3+8a^2b+6ab^2}{(a-b)(a+b)(3a+b)(3a-b)}$
$+\dfrac{b^3-4ab^2+3a^2b+4ab^2-6b^3+2a^2b)}{(a-b)(a+b)(3a+b)(3a-b)}$

$=\dfrac{5a^3+15a^2b+5ab^2-5b^3}{(a-b)(a+b)(3a+b)(3a-b)}$

$=\dfrac{5\cancel{(a+b)}(a^2+2ab-b^2)}{(a-b)\cancel{(a+b)}(3a+b)(3a-b)}$

$=\dfrac{5(a^2+2ab-b^2)}{(a-b)(3a+b)(3a-b)}$

101. Answer mays vary. $\dfrac{a}{a-b}+\dfrac{3b}{b-a}$

103. *Writing Exercise.*

Mid-Chapter Review

1. $\frac{a^2}{a-10} \div \frac{a^2+5a}{a^2-100}$
$= \frac{a^2}{a-10} \cdot \frac{a^2-100}{a^2+5a}$
$= \frac{a \cdot a \cdot (a+10) \cdot (a-10)}{(a-10) \cdot a \cdot (a+5)}$
$= \frac{a(a-10)}{a(a-10)} \cdot \frac{a(a+10)}{a+5}$
$= \frac{a(a+10)}{a+5}$

2. $\frac{2}{x} + \frac{1}{x^2+x}$
$= \frac{2}{x} + \frac{1}{x(x+1)}$
$= \frac{2}{x} \cdot \frac{x+1}{x+1} + \frac{1}{x(x+1)}$
$= \frac{2x+2}{x(x+1)} + \frac{1}{x(x+1)}$
$= \frac{2x+3}{x(x+1)}$

3. $\frac{3}{5x} + \frac{2}{x^2} = \frac{3}{5x} \cdot \frac{x}{x} + \frac{2}{x^2} \cdot \frac{5}{5}$ LCD $= 5x^2$
$= \frac{3x+10}{5x^2}$

4. $\frac{3}{5x} \cdot \frac{2}{x^2} = \frac{6}{5x^3}$

5. $\frac{3}{5x} \div \frac{2}{x^2} = \frac{3}{5x} \cdot \frac{x^2}{2} = \frac{3x}{10} \cdot \frac{x}{x} = \frac{3x}{10}$

6. $\frac{3}{5x} - \frac{2}{x^2} = \frac{3}{5x} \cdot \frac{x}{x} - \frac{2}{x^2} \cdot \frac{5}{5}$ LCD $= 5x^2$
$= \frac{3x-10}{5x^2}$

7. $\frac{2x-6}{5x+10} \cdot \frac{x+2}{6x-12} = \frac{\cancel{2}(x-3)\cancel{(x+2)}}{5\cancel{(x+2)} \cdot \cancel{2} \cdot 3(x-2)}$
$= \frac{(x-3)}{15(x-2)}$

8. $\frac{2}{x-5} \div \frac{6}{x-5} = \frac{2}{x-5} \cdot \frac{x-5}{6} = \frac{1}{3} \cdot \frac{2(x-5)}{2(x-5)} = \frac{1}{3}$

9. $\frac{x}{x+2} - \frac{1}{x-1}$
$= \frac{x}{x+2} \cdot \frac{x-1}{x-1} - \frac{1}{x-1} \cdot \frac{x+2}{x+2}$ LCD $= (x+2)(x-1)$
$= \frac{x(x-1)-(x+2)}{(x+2)(x-1)}$
$= \frac{x^2-x-x-2}{(x+2)(x-1)}$
$= \frac{x^2-2x-2}{(x+2)(x-1)}$

10. $\frac{2}{x+3} + \frac{3}{x+4}$
$= \frac{2}{x+3} \cdot \frac{x+4}{x+4} + \frac{3}{x+4} \cdot \frac{x+3}{x+3}$ LCD $= (x+3)(x-4)$
$= \frac{2(x+4)+3(x+3)}{(x+3)(x+4)}$
$= \frac{2x+8+3x+9}{(x+3)(x+4)}$
$= \frac{5x+17}{(x+3)(x+4)}$

11. $\frac{5}{2x-1} + \frac{10}{1-2x}$
$= \frac{5}{2x-1} + \frac{10x}{1-2x} \cdot \frac{-1}{-1}$ LCD $= 2x-1$
$= \frac{5-10x}{2x-1}$
$= \frac{-5(2x-1)}{2x-1}$
$= -5$

12. $\frac{3}{x-4} - \frac{2}{4-x} = \frac{3}{x-4} + \frac{2}{x-4} = \frac{5}{x-4}$

13. $\frac{(x-2)(2x+3)}{(x+1)(x-5)} \div \frac{(x-2)(x+1)}{(x-5)(x+3)}$
$= \frac{(x-2)(2x+3)}{(x+1)(x-5)} \cdot \frac{(x-5)(x+3)}{(x-2)(x+1)}$
$= \frac{(2x+3)(x+3)}{(x+1)(x+1)} \cdot \frac{(x-2)(x-5)}{(x-2)(x-5)}$
$= \frac{(2x+3)(x+3)}{(x+1)^2}$

14. $\frac{a}{6a-9b} - \frac{b}{4a-6b}$
$= \frac{a}{3(2a-3b)} \cdot \frac{2}{2} - \frac{b}{2(2a-3b)} \cdot \frac{3}{3}$ LCD $= 6(2a-3b)$
$= \frac{2a-3b}{6(2a-3b)}$
$= \frac{1}{6}$

15. $\frac{x^2-16}{x^2-x} \cdot \frac{x^2}{x^2-5x+4} = \frac{(x+4)(x-4)x^2}{x(x-1)(x-4)(x-1)}$
$= \frac{x(x+4)}{(x-1)^2}$

16. $\frac{x+1}{x^2-7x+10} + \frac{3}{x^2-x-2}$ LCD $= (x+1)(x-2)(x-5)$
$= \frac{x+1}{(x-5)(x-2)} \cdot \frac{x+1}{x+1} + \frac{3}{(x-2)(x+1)} \cdot \frac{x-5}{x-5}$
$= \frac{x^2+2x+1+3x-15}{(x+1)(x-2)(x-5)}$
$= \frac{x^2+5x-14}{(x+1)(x-2)(x-5)}$
$= \frac{(x+7)\cancel{(x-2)}}{(x+1)\cancel{(x-2)}(x-5)}$
$= \frac{x+7}{(x+1)(x-5)}$

17. $\frac{3u^2-3}{4} \div \frac{4u+4}{3} = \frac{3(u^2-1)}{4} \cdot \frac{3}{4(u+1)}$
$= \frac{9(u+1)(u-1)}{16(u+1)}$
$= \frac{9(u-1)}{16}$

18. $(t^2+t-20) \cdot \frac{t+5}{t-4} = \frac{(t+5)(t-4)(t+5)}{t-4}$
$= (t+5)^2$

19. $= \frac{a^2-2a+1}{a^2-4} \div (a^2-3a+2)$
$= \frac{(a-1)(a-1)}{(a+2)(a-2)} \cdot \frac{1}{(a-2)(a-1)}$
$= \frac{a-1}{(a+2)(a-2)^2}$

20. $\frac{2x-7}{x} - \frac{3x-5}{2} = \frac{2(2x-7)}{2x} - \frac{x(3x-5)}{2x}$ LCD = $2x$
$= \frac{4x-14-3x^2+5x}{2x}$
$= \frac{-3x^2+9x-14}{2x}$

Exercise Set 6.5

1. The expression given above is a *complex* rational expression.

3. The *least common denominator* of the rational expressions within the expression above is $3x$.

5. The LCD is the LCM of x, 2, and 3. It is $6x$.
$$\frac{\frac{1}{x}+\frac{1}{2}}{\frac{1}{3}-\frac{1}{x}} = \frac{\frac{1}{x}+\frac{1}{2}}{\frac{1}{3}-\frac{1}{x}} \cdot \frac{6x}{6x}$$
Choice (b) is correct.

7. We get a single rational expression in the numerator and another in the denominator.
$$\frac{\frac{x-1}{x}}{\frac{x^2}{x^2-1}} = \frac{x-1}{x} \div \frac{x^2}{x^2-1}$$
Choice (a) is correct.

9. $\frac{\frac{1}{2}+\frac{1}{3}}{\frac{1}{4}-\frac{1}{6}}$
$= \frac{\frac{1}{2}\cdot\frac{3}{3}+\frac{1}{3}\cdot\frac{2}{2}}{\frac{1}{4}\cdot\frac{3}{3}-\frac{1}{6}\cdot\frac{2}{2}}$ Getting a common denominator in numerator and in denominator
$= \frac{\frac{3}{6}+\frac{2}{6}}{\frac{3}{12}-\frac{2}{12}}$
$= \frac{\frac{5}{6}}{\frac{1}{12}}$ Adding in the numerator; subtracting in the denominator
$= \frac{5}{6} \cdot \frac{12}{1}$ Multiplying by the reciprocal of the divisor
$= \frac{5 \cdot 6 \cdot 2}{6}$
$= \frac{5 \cdot \not{6} \cdot 2}{\not{6}}$
$= 10$

11. $\frac{1+\frac{1}{4}}{2+\frac{3}{4}}$ LCD is 4.
$= \frac{1+\frac{1}{4}}{2+\frac{3}{4}} \cdot \frac{4}{4}$ Multiplying by $\frac{4}{4}$
$= \frac{\left(1+\frac{1}{4}\right)4}{\left(2+\frac{3}{4}\right)4}$ Multiplying numerator and denominator by 4
$= \frac{1 \cdot 4+\frac{1}{4}\cdot 4}{2 \cdot 4+\frac{3}{4}\cdot 4}$
$= \frac{4+1}{8+3}$
$= \frac{5}{11}$

13. $\frac{\frac{x}{4}+x}{\frac{4}{x}+x} = \frac{\frac{x}{4}+x}{\frac{4}{x}+x} \cdot \frac{4x}{4x}$ LCD is $4x$
$= \frac{\left(\frac{x}{4}+x\right)(4x)}{\left(\frac{4}{x}+x\right)(4x)}$
$= \frac{x^2+4x^2}{16+4x^2}$
$= \frac{5x^2}{16+4x^2}$
$= \frac{5x^2}{4(4+x^2)}$

15. Multiply by the reciprocal of the divisor.
$$\frac{\frac{x+2}{x-1}}{\frac{x+4}{x-3}} = \frac{x+2}{x-1} \cdot \frac{x-3}{x+4} = \frac{(x+2)(x-3)}{(x-1)(x+4)}$$

17. $\frac{\frac{10}{t}}{\frac{2}{t^2}-\frac{5}{t}} = \frac{\frac{10}{t}}{\frac{2}{t^2}-\frac{5}{t}} \cdot \frac{t^2}{t^2}$
$= \frac{\frac{10}{t} \cdot t^2}{\left(\frac{2}{t^2}-\frac{5}{t}\right)t^2}$
$= \frac{10t}{\frac{2}{t^2}\cdot t^2 - \frac{5}{t}\cdot t^2}$
$= \frac{10t}{2-5t}$, or $\frac{-10t}{5t-2}$

19. Multiply by the reciprocal of the divisor.

$$\frac{\frac{2a-5}{3a}}{\frac{a-7}{6a}} = \frac{2a-5}{3a}\cdot\frac{6a}{a-7}$$

$$= \frac{(2a-5)\cdot 2\cdot 3a}{3a\cdot(a-7)}$$

$$= \frac{(2a-5)\cdot 2\cdot \cancel{3a}}{\cancel{3a}\cdot(a-7)}$$

$$= \frac{2(2a-5)}{a-7}$$

21. $\dfrac{\frac{x}{6}-\frac{3}{x}}{\frac{1}{3}+\frac{1}{x}} = \dfrac{\frac{x}{6}-\frac{3}{x}}{\frac{1}{3}+\frac{1}{x}}\cdot\dfrac{6x}{6x}$ LCD is $6x$

$$= \frac{\frac{x}{6}\cdot 6x-\frac{3}{x}\cdot 6x}{\frac{1}{3}\cdot 6x+\frac{1}{x}\cdot 6x}$$

$$= \frac{x^2-18}{2x+6}$$

$$= \frac{x^2-18}{2(x+3)}$$

23. $\dfrac{\frac{1}{s}-\frac{1}{5}}{\frac{s-5}{s}} = \dfrac{\frac{1}{s}-\frac{1}{5}}{\frac{s-5}{s}}\cdot\dfrac{5s}{5s}$ LCD is $5s$

$$= \frac{\frac{1}{s}\cdot 5s-\frac{1}{5}\cdot 5s}{\left(\frac{s-5}{s}\right)(5s)}$$

$$= \frac{5-s}{(s-5)(5)}$$

$$= \frac{-\cancel{(s-5)}}{\cancel{(s-5)}(5)}$$

$$= -\frac{1}{5}$$

25. $\dfrac{\frac{1}{t^2}+1}{\frac{1}{t}-1} = \dfrac{\frac{1}{t^2}+1}{\frac{1}{t}-1}\cdot\dfrac{t^2}{t^2}$ LCD is t^2

$$= \frac{\frac{1}{t^2}\cdot t^2+1\cdot t^2}{\frac{1}{t}\cdot t^2-1\cdot t^2}$$

$$= \frac{1+t^2}{t-t^2}$$

$$= \frac{1+t^2}{t(1-t)}$$

27. Multiply by the reciprocal of the divisor.

$$\frac{\frac{x^2}{x^2-y^2}}{\frac{x}{x+y}} = \frac{x^2}{x^2-y^2}\cdot\frac{x+y}{x}$$

$$= \frac{x^2(x+y)}{(x^2-y^2)(x)}$$

$$= \frac{x\cdot x\cdot(x+y)}{(x+y)(x-y)(x)}$$

$$= \frac{\cancel{x}\cdot x\cdot\cancel{(x+y)}}{\cancel{(x+y)}(x-y)(\cancel{x})}$$

$$= \frac{x}{x-y}$$

29. $\dfrac{\frac{7}{c^2}+\frac{4}{c}}{\frac{6}{c}-\frac{3}{c^3}} = \dfrac{\frac{7}{c^2}+\frac{4}{c}}{\frac{6}{c}-\frac{3}{c^3}}\cdot\dfrac{c^3}{c^3}$ LCD is c^3

$$= \frac{\frac{7}{c^2}\cdot c^3+\frac{4}{c}\cdot c^3}{\frac{6}{c}\cdot c^3-\frac{3}{c^3}\cdot c^3}$$

$$= \frac{7c+4c^2}{6c^2-3}$$

$$= \frac{c(4c+7)}{3(2c^2-1)}$$

31. $\dfrac{\frac{2}{7a^4}-\frac{1}{14a}}{\frac{3}{5a^2}+\frac{2}{15a}} = \dfrac{\frac{2}{7a^4}\cdot\frac{2}{2}-\frac{1}{14a}\cdot\frac{a^3}{a^3}}{\frac{3}{5a^2}\cdot\frac{3}{3}+\frac{2}{15a}\cdot\frac{a}{a}}$

$$= \frac{\frac{4-a^3}{14a^4}}{\frac{9+2a}{15a^2}}$$

$$= \frac{4-a^3}{14a^4}\cdot\frac{15a^2}{9+2a}$$

$$= \frac{15\cdot\cancel{a^2}(4-a^3)}{14a^2\cdot\cancel{a^2}(9+2a)}$$

$$= \frac{15(4-a^3)}{14a^2(9+2a)}$$

33. $\dfrac{\frac{x}{5y^3}+\frac{3}{10y}}{\frac{3}{10y}+\frac{x}{5y^3}}$

Observe that, by the commutative law of addition, the numerator and denominator are equivalent, so the result is 1.

35. $\dfrac{\frac{3}{ab^4}+\frac{4}{a^3b}}{\frac{5}{a^3b}-\frac{3}{ab}} = \dfrac{\frac{3}{ab^4}\cdot\frac{a^2}{a^2}+\frac{4}{a^3b}\cdot\frac{b^3}{b^3}}{\frac{5}{a^3b}-\frac{3}{ab}\cdot\frac{a^2}{a^2}}$

$$= \frac{\frac{3a^2+4b^3}{a^3b^4}}{\frac{5-3a^2}{a^3b}}$$

$$= \frac{3a^2+4b^3}{a^3b^4}\cdot\frac{a^3b}{5-3a^2}$$

$$= \frac{\cancel{a^3b}(3a^2+4b^3)}{\cancel{a^3b}\cdot b^3(5-3a^2)}$$

$$= \frac{3a^2+4b^3}{b^3(5-3a^2)}$$

37. $\dfrac{t-\frac{9}{t}}{t+\frac{4}{t}}=\dfrac{t\cdot\frac{t}{t}-\frac{9}{t}}{t\cdot\frac{t}{t}+\frac{4}{t}}$

$=\dfrac{\frac{t^2-9}{t}}{\frac{t^2+4}{t}}$

$=\dfrac{t^2-9}{t}\cdot\dfrac{t}{t^2+4}$

$=\dfrac{\cancel{t}(t^2-9)}{\cancel{t}(t^2+4)}$

$=\dfrac{t^2-9}{t^2+4}$

$=\dfrac{(t-3)(t+3)}{t^2+4}$

39. $\dfrac{y+y^{-1}}{y-y^{-1}}=\dfrac{y+\frac{1}{y}}{y-\frac{1}{y}}=\dfrac{y\cdot\frac{y}{y}+\frac{1}{y}}{y\cdot\frac{y}{y}-\frac{1}{y}}$

$=\dfrac{\frac{y^2+1}{y}}{\frac{y^2-1}{y}}$

$=\dfrac{y^2+1}{y}\cdot\dfrac{y}{y^2-1}$

$=\dfrac{\cancel{y}(y^2+1)}{\cancel{y}(y^2-1)}$

$=\dfrac{y^2+1}{y^2-1}$

41. $\dfrac{\frac{1}{a-h}-\frac{1}{a}}{h}=\dfrac{\frac{1}{a-h}\cdot\frac{a}{a}-\frac{1}{a}\cdot\frac{a-h}{a-h}}{h}$

$=\dfrac{\frac{a-(a-h)}{a(a-h)}}{h}$

$=\dfrac{\frac{h}{a(a-h)}}{h}$

$=\dfrac{\cancel{h}}{a(a-h)}\cdot\dfrac{1}{\cancel{h}}$

$=\dfrac{1}{a(a-h)}$

43. $\dfrac{\frac{x^{-1}+y^{-1}}{x^2-y^2}}{xy}=\dfrac{\frac{\frac{1}{x}+\frac{1}{y}}{x^2-y^2}}{xy}\cdot\dfrac{xy}{xy}$ LCD = xy

$=\dfrac{y+x}{x^2-y^2}$

$=\dfrac{\cancel{x+y}}{\cancel{(x+y)}(x-y)}$

$=\dfrac{1}{x-y}$

45. $\dfrac{\frac{1}{a}+\frac{1}{b}}{\frac{1}{a^3}+\frac{1}{b^3}}=\dfrac{\frac{1}{a}+\frac{1}{b}}{\frac{1}{a^3}+\frac{1}{b^3}}\cdot\dfrac{a^3b^3}{a^3b^3}$ LCD = a^3b^3

$=\dfrac{a^2b^3+a^3b^2}{b^3+a^3}$

$=\dfrac{a^2b^2(b+a)}{(b+a)(b^2-ab+a^2)}$

$=\dfrac{a^2b^2}{b^2-ab+a^2}$

47. $\dfrac{t+5+\frac{3}{t}}{t+2+\frac{1}{t}}=\dfrac{t+5+\frac{3}{t}}{t+2+\frac{1}{t}}\cdot\dfrac{t}{t}$ LCD is t

$=\dfrac{t\cdot t+5\cdot t+\frac{3}{t}\cdot t}{t\cdot t+2\cdot t+\frac{1}{t}\cdot t}$

$=\dfrac{t^2+5t+3}{t^2+2t+1}$

$=\dfrac{t^2+5t+3}{(t+1)^2}$

49. $\dfrac{x-2-\frac{1}{x}}{x-5-\frac{4}{x}}=\dfrac{x-2-\frac{1}{x}}{x-5-\frac{4}{x}}\cdot\dfrac{x}{x}$

$=\dfrac{x\cdot x-2\cdot x-\frac{1}{x}\cdot x}{x\cdot x-5\cdot x-\frac{4}{x}\cdot x}$

$=\dfrac{x^2-2x-1}{x^2-5x-4}$

51. $\dfrac{\frac{a^2-4}{a^2+3a+2}}{\frac{a^2-5a-6}{a^2-6a-7}}=\dfrac{(a+2)(a-2)}{(a+2)(a+1)}\cdot\dfrac{(a+1)(a-7)}{(a+1)(a-6)}$

$=\dfrac{\cancel{(a+2)}(a-2)\cancel{(a+1)}(a-7)}{\cancel{(a+2)}\cancel{(a+1)}(a+1)(a-6)}$

$=\dfrac{(a-2)(a-7)}{(a+1)(a-6)}$

53. $\dfrac{\frac{x}{x^2+3x-4}-\frac{1}{x^2+3x-4}}{\frac{x}{x^2+6x+8}+\frac{3}{x^2+6x+8}}$

$=\dfrac{\frac{x-1}{x^2+3x-4}}{\frac{x+3}{x^2+6x+8}}$

$=\dfrac{x-1}{(x+4)(x-1)}\cdot\dfrac{(x+4)(x+2)}{x+3}$

$=\dfrac{\cancel{(x-1)}\cancel{(x+4)}(x+2)}{\cancel{(x+4)}\cancel{(x-1)}(x+3)}$

$=\dfrac{x+2}{x+3}$

55. *Writing Exercise.*

57. $6x^3 - 9x^2 - 4x + 6 = 3x^2(2x-3) - 2(2x-3)$
$= (2x-3)(3x^2-2)$

59. $30n^3 - 3n^2 - 9n = 3n(10n^2 - n - 3)$
$= 3n(2n+1)(5n-3)$

61. $n^4 - 1 = (n^2+1)(n^2-1)$
$= (n^2+1)(n+1)(n-1)$

63. *Writing Exercise.*

65. $\dfrac{\frac{x-5}{x-6}}{\frac{x-7}{x-8}}$

This expression is undefined for any value of x that makes a denominator 0. We see that $x-6=0$ when $x=6$, $x-7=0$ when $x=7$, and $x-8=0$ when $x=8$, so the expression is undefined for the x-values 6, 7, and 8.

67. $\dfrac{\frac{2x+3}{5x+4}}{\frac{3}{7}-\frac{x^2}{21}}$

This expression is undefined for any value of x that makes a denominator 0. First we find the value of x for which $5x+4=0$.

$$5x+4=0$$
$$5x=-4$$
$$x=-\frac{4}{5}$$

Then we find the value of x for which $\frac{3}{7}-\frac{x^2}{21}=0$:

$$\frac{3}{7}-\frac{x^2}{21}=0$$
$$21\left(\frac{3}{7}-\frac{x^2}{21}\right)=21\cdot 0$$
$$21\cdot\frac{3}{7}-21\cdot\frac{x^2}{21}=0$$
$$9-x^2=0$$
$$(3-x)(3+x)=0$$
$$x=-3 \text{ or } 3$$

The expression is undefined for the x-values $-\frac{4}{5}$, -3 and 3.

69. For the complex rational expression

$$\frac{\frac{A}{B}}{\frac{C}{D}}=\frac{\frac{A}{B}\cdot BD}{\frac{C}{D}\cdot BD} \quad \text{LCD is } BD.$$
$$=\frac{\frac{ABD}{B}}{\frac{CBD}{D}}=\frac{\frac{A\cancel{B}D}{\cancel{B}}}{\frac{BC\cancel{D}}{\cancel{D}}}$$
$$=\frac{AD}{BC}$$
$$=\frac{A}{B}\cdot\frac{D}{C}$$

71. $$\frac{\frac{x}{x+5}+\frac{3}{x+2}}{\frac{2}{x+2}-\frac{x}{x+5}}=\frac{\frac{x}{x+5}+\frac{3}{x+2}}{\frac{2}{x+2}-\frac{x}{x+5}}\cdot\frac{(x+5)(x+2)}{(x+5)(x+2)}$$
$$=\frac{x(x+2)+3(x+5)}{2(x+5)-x(x+2)}$$
$$=\frac{x^2+2x+3x+15}{2x+10-x^2-2x}$$
$$=\frac{x^2+5x+15}{-x^2+10}$$

73. $$\left[\frac{\frac{x-1}{x-1}-1}{\frac{x+1}{x-1}+1}\right]^5$$

Consider the numerator of the complex rational expression:

$$\frac{x-1}{x-1}-1=1-1=0$$

Since the denominator, $\frac{x+1}{x-1}+1$ is not equal to 0, the simplified form of the original expression is 0.

75. $$\frac{1-\frac{25}{x^2}}{1+\frac{2}{x}-\frac{15}{x^2}}=\frac{1-\frac{25}{x^2}}{1+\frac{2}{x}-\frac{15}{x^2}}\cdot\frac{x^2}{x^2} \quad \text{LCD}=x^2$$
$$=\frac{x^2-25}{x^2+2x-15}$$
$$=\frac{\cancel{(x+5)}(x-5)}{\cancel{(x+5)}(x-3)}$$
$$=\frac{x-5}{x-3}$$

77. $$\frac{1}{\frac{2}{x-1}-\frac{1}{3x-2}}=\frac{1}{\frac{2}{x-1}-\frac{1}{3x-2}}\cdot\frac{(x-1)(3x-2)}{(x-1)(3x-2)}$$
$$=\frac{(x-1)(3x-2)}{\left(\frac{2}{x-1}-\frac{1}{3x-2}\right)(x-1)(3x-2)}$$
$$=\frac{(x-1)(3x-2)}{\frac{2}{x-1}(x-1)(3x-2)-\frac{1}{3x-2}(x-1)(3x-2)}$$
$$=\frac{(x-1)(3x-2)}{2(3x-2)-(x-1)}$$
$$=\frac{(x-1)(3x-2)}{6x-4-x+1}$$
$$=\frac{(x-1)(3x-2)}{5x-3}$$

Connecting the Concepts

1. Expression; $\frac{2}{5n}+\frac{3}{2n-1}$

$$\begin{aligned}&\frac{2}{5n}+\frac{3}{2n-1}\\&=\frac{2(2n-1)}{5n(2n-1)}+\frac{3(5n)}{5n(2n-1)} \quad \text{LCD}=5n(2n-1)\\&=\frac{4n-2+15n}{5n(2n-1)}\\&=\frac{19n-2}{5n(2n-1)}\end{aligned}$$

2. Equation;

$$\begin{aligned}\frac{3}{y}-\frac{1}{4}&=\frac{1}{y} \quad \text{Note } y\neq 0\\4y\left(\frac{3}{y}-\frac{1}{4}\right)&=4y\left(\frac{1}{y}\right) \quad \text{LCD}=4y\\12-y&=4\\-y&=-8\\y&=8\end{aligned}$$

The solution is 8.

3. Equation;

$$\begin{aligned}\frac{5}{x+3}&=\frac{3}{x+2} \quad \text{Note that } x\neq -2,-3\\(x+2)(x+3)\cdot\frac{5}{x+3}&=(x+2)(x+3)\cdot\frac{3}{x+2}\\&\text{LCD is } (x+2)(x+3)\\5(x+2)&=3(x+3)\\5x+10&=3x+9\\5x&=3x-1\\2x&=-1\\x&=-\frac{1}{2}\end{aligned}$$

4. Expression;

$$\begin{aligned}\frac{8t+8}{2t^2+t-1}\cdot\frac{t^2-1}{t^2-2t+1}&=\frac{8(t+1)(t+1)(t-1)}{(2t-1)(t+1)(t-1)(t-1)}\\&=\frac{8(t+1)}{(2t-1)(t-1)}\end{aligned}$$

5. Expression

$$\begin{aligned}\frac{2a}{a+1}-\frac{4a}{1-a^2}&=\frac{2a}{a+1}-\frac{4a}{a^2-1}\\&=\frac{2a(a-1)+4a}{(a+1)(a-1)} \quad \text{LCD}=(a+1)(a-1)\\&=\frac{2a^2-2a+4a}{(a+1)(a-1)}\\&=\frac{2a^2+2a}{(a+1)(a-1)}\\&=\frac{2a(a+1)}{(a+1)(a-1)}\\&=\frac{2a}{a-1}\end{aligned}$$

6. Equation; $\frac{20}{x}=\frac{x}{5}$ Note $x\neq 0$

$$\begin{aligned}5x\cdot\frac{20}{x}&=5x\cdot\frac{x}{5} \quad \text{LCD }=5x\\100&=x^2\\0&=x^2-100\\0&=(x+10)(x-10)\end{aligned}$$

$$\begin{aligned}x+10=0 \quad &or \quad x-10=0\\x=-10 \quad &or \quad x=10\end{aligned}$$

The solutions are -10 and 10.

Exercise Set 6.6

1. The statement is false. See Example 2(c).

3. The statement is true.

5. Because no variable appears in a denominator, no restrictions exist.

$$\begin{aligned}\frac{3}{5}-\frac{2}{3}&=\frac{x}{6}, \text{ LCD}=30\\30\left(\frac{3}{5}-\frac{2}{3}\right)&=30\cdot\frac{x}{6}\\30\cdot\frac{3}{5}-30\cdot\frac{2}{3}&=30\cdot\frac{x}{6}\\18-20&=5x\\-2&=5x\\-\frac{2}{5}&=x\end{aligned}$$

Check:

$$\begin{array}{c|c}\multicolumn{2}{c}{\frac{3}{5}-\frac{2}{3}=\frac{x}{6}}\\\hline \frac{3}{5}-\frac{2}{3} & \frac{-\frac{2}{5}}{6}\\ \frac{18}{30}-\frac{20}{30} & \frac{-2}{5}\cdot\frac{1}{6}\\ -\frac{2}{30} \overset{?}{=} & \frac{-2}{30} \quad \text{TRUE}\end{array}$$

This checks, so the solution is $-\frac{2}{5}$.

7. Note that t cannot be 0.

$$\begin{aligned}\frac{1}{8}+\frac{1}{12}&=\frac{1}{t}, \text{ LCD}=48t\\48t\left(\frac{1}{8}+\frac{1}{12}\right)&=48t\cdot\frac{1}{t}\\48t\cdot\frac{1}{8}+48t\cdot\frac{1}{12}&=48t\cdot\frac{1}{t}\\6t+4t&=48\\10t&=48\\t&=\frac{24}{5}\end{aligned}$$

Check:

$$\begin{array}{c|c}\multicolumn{2}{c}{\frac{1}{8}+\frac{1}{12}=\frac{1}{t}}\\\hline \frac{1}{8}+\frac{1}{12} & \frac{1}{\frac{24}{5}}\\ \frac{3}{24}+\frac{2}{24} & 1\cdot\frac{5}{24}\\ \frac{5}{24} \overset{?}{=} & \frac{5}{24} \quad \text{TRUE}\end{array}$$

This checks, so the solution is $\frac{24}{5}$.

9. Note that x cannot be 0.

$$\frac{x}{6}-\frac{6}{x}=0, \quad \text{LCD}=6x$$
$$6x\left(\frac{x}{6}-\frac{6}{x}\right)=6x\cdot 0$$
$$6x\cdot\frac{x}{6}-6x\cdot\frac{6}{x}=6x\cdot 0$$
$$x^2-36=0$$
$$(x+6)(x-6)=0$$
$$x+6=0 \quad or \quad x-6=0$$
$$x=-6 \quad or \quad x=6$$

Check:

$$\frac{x}{6}-\frac{6}{x}=0$$

$\frac{-6}{6}-\frac{6}{-6}$	0
$-1+1$	

$0 \stackrel{?}{=} 0$ TRUE

$$\frac{x}{6}-\frac{6}{x}=0$$

$\frac{6}{6}-\frac{6}{6}$	0
$1-1$	

$0 \stackrel{?}{=} 0$ TRUE

Both of these check, so the two solutions are –6 and 6.

11. Note that x cannot be 0.

$$\frac{2}{x}=\frac{5}{x}-\frac{1}{4}, \quad \text{LCD}=4x$$
$$4x\cdot\frac{2}{x}=4x\left(\frac{5}{x}-\frac{1}{4}\right)$$
$$4x\cdot\frac{2}{x}=4x\cdot\frac{5}{x}-4x\cdot\frac{1}{4}$$
$$8=20-x$$
$$-12=-x$$
$$12=x$$

Check: $\frac{2}{x}=\frac{5}{x}-\frac{1}{4}$

$\frac{2}{12}$	$\frac{5}{12}-\frac{1}{4}$
	$\frac{5}{12}-\frac{3}{12}$

$\frac{2}{12} \stackrel{?}{=} \frac{2}{12}$ TRUE

This checks, so the solution is 12.

13. Note that t cannot be 0.

$$\frac{5}{3t}+\frac{3}{t}=1, \quad \text{LCD}=3t$$
$$3t\left(\frac{5}{3t}+\frac{3}{t}\right)=3t\cdot 1$$
$$3t\cdot\frac{5}{3t}+3t\cdot\frac{3}{t}=3t\cdot 1$$
$$5+9=3t$$
$$14=3t$$
$$\frac{14}{3}=t$$

Check: $\frac{5}{3t}+\frac{3}{t}=1$

$\frac{5}{3\cdot\frac{14}{3}}+\frac{3}{\frac{14}{3}}$	1
$\frac{5}{14}+\frac{9}{14}$	

$1 \stackrel{?}{=} 1$ TRUE

This checks, so the solution is $\frac{14}{3}$.

15. Note that x cannot be 0.

$$\frac{12}{x}=\frac{x}{3}, \quad \text{LCD}=3x$$
$$3x\cdot\frac{12}{x}=3x\cdot\frac{x}{3}$$
$$36=x^2$$
$$0=x^2-36$$
$$0=(x+6)(x-6)$$
$$x+6=0 \quad or \quad x-6=0$$
$$x=-6 \quad or \quad x=6$$

This checks, so the solutions are –6 and 6.

17. Note that y cannot be 0.

$$y+\frac{4}{y}=-5, \quad \text{LCD is } y.$$
$$y\left(y+\frac{4}{y}\right)=y\cdot(-5)$$
$$y\cdot y+y\cdot\frac{4}{y}=-5y$$
$$y^2+4=-5y$$
$$y^2+5y+4=0$$
$$(y+4)(y+1)=0$$
$$y+4=0 \quad or \quad y+1=0$$
$$y=-4 \quad or \quad y=-1$$

Both numbers check, so the solutions are –4 and –1.

19. To avoid the division by 0, we must have $n-6\neq 0$, or $n\neq 6$.

$$\frac{n+2}{n-6}=\frac{1}{2}, \quad \text{LCD}=2(n-6)$$
$$2(n-6)\cdot\frac{n+2}{n-6}=2(n-6)\cdot\frac{1}{2}$$
$$2(n+2)=n-6$$
$$2n+4=n-6$$
$$n=-10$$

Check:

$$\frac{n+2}{n-6}=\frac{1}{2}$$

$\frac{-10+2}{-10-6}$	$\frac{1}{2}$
$\frac{-8}{-16}$	

$\frac{1}{2} \stackrel{?}{=} \frac{1}{2}$ TRUE

This checks, so the solution is –10.

21. Note that x cannot be 0.

$$x+\frac{12}{x}=-7, \quad \text{LCD is } x$$
$$x\left(x+\frac{12}{x}\right)=x\cdot(-7)$$
$$x\cdot x+x\cdot\frac{12}{x}=-7x$$
$$x^2+12=-7x$$
$$x^2+7x+12=0$$
$$(x+4)(x+3)=0$$
$$x+4=0 \quad or \quad x+3=0$$
$$x=-4 \quad or \quad x=-3$$

Both numbers check, so the solutions are –4 and –3.

23. To avoid division by 0, we must have $x-4\neq 0$ and $x+1\neq 0$, or $x\neq 4$ and $x\neq -1$.

$$\frac{3}{x-4}=\frac{5}{x+1},\quad \text{LCD}=(x-4)(x+1)$$
$$(x-4)(x+1)\cdot\frac{3}{x-4}=(x-4)(x+1)\cdot\frac{5}{x+1}$$
$$3(x+1)=5(x-4)$$
$$3x+3=5x-20$$
$$23=2x$$
$$\frac{23}{2}=x$$

This checks, so the solution is $\frac{23}{2}$.

25. Because no variable appears in a denominator, no restrictions exist.

$$\frac{a}{6}-\frac{a}{10}=\frac{1}{6},\quad \text{LCD}=30$$
$$30\left(\frac{a}{6}-\frac{a}{10}\right)=30\cdot\frac{1}{6}$$
$$30\cdot\frac{a}{6}-30\cdot\frac{a}{10}=30\cdot\frac{1}{6}$$
$$5a-3a=5$$
$$2a=5$$
$$a=\frac{5}{2}$$

This checks, so the solution is $\frac{5}{2}$.

27. Because no variable appears in a denominator, no restrictions exist.

$$\frac{x+1}{3}-1=\frac{x-1}{2},\quad \text{LCD}=6$$
$$6\left(\frac{x+1}{3}-1\right)=6\cdot\frac{x-1}{2}$$
$$6\cdot\frac{x+1}{3}-6\cdot 1=6\cdot\frac{x-1}{2}$$
$$2(x+1)-6=3(x-1)$$
$$2x+2-6=3x-3$$
$$2x-4=3x-3$$
$$-1=x$$

This checks, so the solution is –1.

29. To avoid division by 0, we must have $y-3\neq 0$, or $y\neq 3$.

$$\frac{y+3}{y-3}=\frac{6}{y-3},\quad \text{LCD}=y-3$$
$$(y-3)\cdot\frac{y+3}{y-3}=(y-3)\cdot\frac{6}{y-3}$$
$$y+3=6$$
$$y=3$$

Because of the restriction $y\neq 3$, the number 3 must be rejected as a solution. The equation has no solution.

31. To avoid division by 0, we must have $x+4\neq 0$ and $x\neq 0$, or $x\neq -4$ and $x\neq 0$.

$$\frac{3}{x+4}=\frac{5}{x},\quad \text{LCD}=x(x+4)$$
$$x(x+4)\cdot\frac{3}{x+4}=x(x+4)\cdot\frac{5}{x}$$
$$3x=5(x+4)$$
$$3x=5x+20$$
$$-2x=20$$
$$x=-10$$

This checks, so the solution is –10.

33. To avoid division by 0, we must have $n+2\neq 0$ and $n+1\neq 0$, or $n\neq -2$ and $n\neq -1$.

$$\frac{n+1}{n+2}=\frac{n-3}{n+1},\quad \text{LCD}=(n+2)(n+1)$$
$$(n+2)(n+1)\cdot\frac{n+1}{n+2}=(n+2)(n+1)\cdot\frac{n-3}{n+1}$$
$$(n+1)(n+1)=(n+2)(n-3)$$
$$n^2+2n+1=n^2-n-6$$
$$3n=-7$$
$$n=-\frac{7}{3}$$

This checks, so the solution is $-\frac{7}{3}$.

35. To avoid division by 0, we must have $t-2\neq 0$, or $t\neq 2$.

$$\frac{5}{t-2}+\frac{3t}{t-2}=\frac{4}{t^2-4t+4},\quad \text{LCD is }(t-2)^2$$
$$(t-2)^2\left(\frac{5}{t-2}+\frac{3t}{t-2}\right)=(t-2)^2\cdot\frac{4}{(t-2)^2}$$
$$5(t-2)+3t(t-2)=4$$
$$5t-10+3t^2-6t=4$$
$$3t^2-t-10=4$$
$$3t^2-t-14=0$$
$$(t+2)(3t-7)=0$$
$$t+2=0 \quad or \quad 3t-7=0$$
$$t=-2 \quad or \quad 3t=7$$
$$t=-2 \quad or \quad t=\frac{7}{3}$$

Both numbers check. The solutions are –2 and $\frac{7}{3}$.

37. To avoid division by 0, we must have $x+5\neq 0$ and $x-5\neq 0$, or $x\neq -5$ and $x\neq 5$.

$$\frac{x}{x+5}-\frac{5}{x-5}=\frac{14}{x^2-25},\quad \text{LCD}=(x+5)(x-5)$$
$$(x+5)(x-5)\left(\frac{x}{x+5}-\frac{5}{x-5}\right)$$
$$=\frac{14}{(x+5)(x-5)}(x+5)(x-5)$$
$$x(x-5)-5(x+5)=14$$
$$x^2-5x-5x-25=14$$
$$x^2-10x-39=0$$
$$(x+3)(x-13)=0$$
$$x+3=0 \quad or \quad x-13=0$$
$$x=-3 \quad or \quad x=13$$

Both numbers check. The solutions are –3 and 13.

39. To avoid division by 0, we must have $x-3 \neq 0$ and $x+2 \neq 0$, or $x \neq 3$ and $x \neq -2$.

$$\frac{3}{x-3}+\frac{5}{x+2}=\frac{5x}{x^2-x-6},$$
LCD is $(x+2)(x-3)$

$$(x+2)(x-3)\left(\frac{3}{x-3}+\frac{5}{x+2}\right)=\frac{5x}{(x+2)(x-3)}(x+2)(x-3)$$
$$3(x+2)+5(x-3)=5x$$
$$3x+6+5x-15=5x$$
$$8x-9=5x$$
$$-9=-3x$$
$$3=x$$

Thus, we have $x=3$, but because of the restriction $x \neq 3$, this cannot be a solution. The equation has no solution.

41. To avoid division by 0, we must have $t-3 \neq 0$ and $t+3 \neq 0$, or $t \neq 3$ and $t \neq -3$.

$$\frac{5}{t-3}-\frac{30}{t^2-9}=1, \text{ LCD is } (t-3)(t+3)$$
$$(t-3)(t+3)\cdot\left(\frac{5}{t-3}-\frac{30}{t^2-9}\right)=(t-3)(t+3)\cdot 1$$
$$5(t+3)-30=(t-3)(t+3)$$
$$5t+15-30=t^2-9$$
$$0=t^2-5t+6$$
$$0=(t-3)(t-2)$$
$$t-3=0 \quad or \quad t-2=0$$
$$t=3 \quad or \quad t=2$$

Because of the restriction $t \neq 3$, we must reject the number 3 as a solution. The number 2 checks, so it is the solution.

43. To avoid division by 0, we must have $6-a \neq 0$ (or equivalently $a-6 \neq 0$) or $a \neq 6$.

$$\frac{7}{6-a}=\frac{a+1}{a-6}$$
$$\frac{-1}{-1}\cdot\frac{7}{6-a}=\frac{a+1}{a-6}$$
$$\frac{-7}{a-6}=\frac{a+1}{a-6}, \quad \text{LCD}=a-6$$
$$(a-6)\cdot\frac{-7}{a-6}=(a-6)\cdot\frac{a+1}{a-6}$$
$$-7=a+1$$
$$-8=a$$

This checks, so the solution is –8.

45. $\dfrac{-2}{x+2}=\dfrac{x}{x+2}$

To avoid division by 0, we must have $x+2 \neq 0$, or $x \neq -2$. Now observe that the denominators are the same, so the numerators must be the same. Thus, we have $-2=x$, but because of the restriction $x \neq -2$ this cannot be a solution. The equation has no solution.

47. To avoid division by 0, we must have $3x+3 \neq 0$ and $2x-2 \neq 0$, or $x \neq -1$ and $x \neq 1$.

$$\frac{5}{3x+3}+\frac{1}{2x-2}=\frac{1}{x^2-1}, \quad \text{LCD is } 6(x+1)(x-1)$$
$$6(x+1)(x-1)\left(\frac{5}{3(x+1)}+\frac{1}{2(x-1)}\right)$$
$$=\frac{1}{(x+1)(x-1)}6(x+1)(x-1)$$
$$2(5)(x-1)+3(x+1)=6$$
$$10x-10+3x+3=6$$
$$13x-7=6$$
$$13x=13$$
$$x=1$$

Thus, we have $x=1$, but because of the restriction $x \neq 1$, this cannot be a solution. The equation has no solution.

49. *Writing Exercise.*

51. To find x-intercept, set $y=0$ and solve for x.

$$6x-0=18$$
$$6x=18$$
$$x=3$$

The x-intercept is (3, 0).

To find y-intercept, set $x=0$ and solve for y.

$$6(0)-y=18$$
$$y=-18$$

The y-intercept is (0, –18).

53. $2x+y=5$

$$y=-2x+5$$

From slope-intercept form, slope is –2, and y-intercept is (0, 5).

55. Using slope-intercept form,

$$y=mx+b$$
$$y=\frac{1}{3}x-2$$

57. *Writing Exercise.*

59. To avoid division by 0, we must have $x-3 \neq 0$, or $x \neq 3$.

$$1+\frac{x-1}{x-3}=\frac{2}{x-3}-x, \quad \text{LCD}=x-3$$
$$(x-3)\left(1+\frac{x-1}{x-3}\right)=(x-3)\left(\frac{2}{x-3}-x\right)$$
$$(x-3)\cdot 1+(x-3)\cdot\frac{x-1}{x-3}=(x-3)\cdot\frac{2}{x-3}-(x-3)x$$
$$x-3+x-1=2-x^2+3x$$
$$2x-4=2-x^2+3x$$
$$x^2-x-6=0$$
$$(x+2)(x-3)=0$$
$$x+2=0 \quad or \quad x-3=0$$
$$x=-2 \quad or \quad x=3$$

Because of the restriction $x \neq 3$, we must reject the number 3 as a solution. The number –2 checks, so it is the solution.

61. To avoid division by 0, we must have $x+2 \neq 0$ and $x-2 \neq 0$, or $x \neq -2$ and $x \neq 2$.

$$\frac{12-6x}{x^2-4} = \frac{3x}{x+2} - \frac{3-2x}{2-x}$$
$$\frac{12-6x}{(x+2)(x-2)} = \frac{3x}{x+2} - \frac{3-2x}{2-x} \cdot \frac{-1}{-1}$$
$$\frac{12-6x}{(x+2)(x-2)} = \frac{3x}{x+2} - \frac{2x-3}{x-2}$$

LCD is $(x+2)(x-2)$

$$(x+2)(x-2) \cdot \frac{12-6x}{(x+2)(x-2)} = (x+2)(x-2)\left(\frac{3x}{x+2} - \frac{2x-3}{x-2}\right)$$
$$12-6x = 3x(x-2)-(x+2)(2x-3)$$
$$12-6x = 3x^2-6x-2x^2-x+6$$
$$0 = x^2-x-6$$
$$0 = (x+2)(x-3)$$
$$x+2=0 \quad or \quad x-3=0$$
$$x=-2 \quad or \quad x=3$$

Because of the restriction $x \neq -2$, we must reject the number -2 as a solution. The number 3 checks, so it is the solution.

63. To avoid division by 0, we must have $a+3 \neq 0$, or $a \neq -3$.

$$7 - \frac{a-2}{a+3} = \frac{a^2-4}{a+3} + 5, \quad \text{LCD} = a+3$$
$$(a+3)\left(7 - \frac{a-2}{a+3}\right) = (a+3)\left(\frac{a^2-4}{a+3} + 5\right)$$
$$7(a+3)-(a-2) = a^2-4+5(a+3)$$
$$7a+21-a+2 = a^2-4+5a+15$$
$$6a+23 = a^2+5a+11$$
$$0 = a^2-a-12$$
$$0 = (a+3)(a-4)$$
$$a+3=0 \quad or \quad a-4=0$$
$$a=-3 \quad or \quad a=4$$

Because of the restriction $a \neq -3$, we must reject the number -3 as a solution. The number 4 checks, so it is the solution.

65. To avoid division by 0, we must have $x-1 \neq 0$. or $x \neq 1$.

$$\frac{1}{x-1} + x - 5 = \frac{5x-4}{x-1} - 6, \quad \text{LCD} = x-1$$
$$(x-1)\left(\frac{1}{x-1} + x - 5\right) = (x-1)\left(\frac{5x-4}{x-1} - 6\right)$$
$$1 + x(x-1) - 5(x-1) = 5x-4-6(x-1)$$
$$1 + x^2 - x - 5x + 5 = 5x - 4 - 6x + 6$$
$$x^2-6x+6 = -x+2$$
$$x^2-5x+4 = 0$$
$$(x-1)(x-4) = 0$$
$$x-1=0 \quad or \quad x-4=0$$
$$x=1 \quad or \quad x=4$$

Because of the restriction $x \neq 1$, we must reject the number 1 as a solution. The number 4 checks, so it is the solution.

67. Note that x cannot be 0.

$$\frac{\frac{1}{x}+1}{x} = \frac{\frac{1}{x}}{2}$$
$$\left(\frac{1}{x}+1\right) \cdot \frac{1}{x} = \frac{1}{x} \cdot \frac{1}{2}$$
$$\frac{1}{x^2} + \frac{1}{x} = \frac{1}{2x}, \quad \text{LCD} = 2x^2$$
$$2 + 2x = x$$
$$2 = -x$$
$$-2 = x$$

This checks so the solution is -2.

69.

Exercise Set 6.7

1. False. To find the time that it would take two people working together, we need to solve $\frac{1}{a} + \frac{1}{b} = \frac{1}{t}$ for t, where a and b represent the time needed for each person to complete the work alone.

3. True

5. True

7. Time needed to complete the job: 2 hr

Hourly rate: $\frac{1}{2}$ cake per hour

9. Time needed for Sandy to complete the job: 2 hr

Sandy's hourly rate: $\frac{1}{2}$ cake per hour

Time needed for Eric to complete the job: 3 hr

Eric's hourly rate: $\frac{1}{3}$ cake per hour

Hourly rate working together: $\frac{1}{2} + \frac{1}{3} = \frac{5}{6}$ cake per hour

11. Time needed for Lisa to complete the job: 3 hr

Lisa's hourly rate: $\frac{1}{3}$ lawn per hour

13. ***Familiarize***. The job takes Kelby 10 hours working alone and Natalie 15 hours working alone. Then in 1 hour Kelby does $\frac{1}{10}$ of the job and Natalie does $\frac{1}{15}$ of the job. Working together they can do $\frac{1}{10} + \frac{1}{15}$, or $\frac{1}{6}$ of the job in 1 hour. In 4 hours, Kelby does $4\left(\frac{1}{10}\right)$ of the job and Natalie does $4\left(\frac{1}{15}\right)$ of the job. Working together they can do $4\left(\frac{1}{10}\right) + 4\left(\frac{1}{15}\right)$, or $\frac{2}{3}$ of the job in 4 hours. In 7 hours, Kelby does $7\left(\frac{1}{10}\right)$ of the job and Natalie does $7\left(\frac{1}{15}\right)$ of the job. Working together they can do $7\left(\frac{1}{10}\right)+7\left(\frac{1}{15}\right)$, or $\frac{7}{6}$ of the job which is more of the job than needs to be done. The answer is somewhere between 4 hr and 7 hr.

Translate. If they work together t hours, then Kelby does $t\left(\frac{1}{10}\right)$ of the job and Natalie does $t\left(\frac{1}{15}\right)$ of the job. We want some number t such that

$$\left(\frac{1}{10}+\frac{1}{15}\right)t=1,\text{ or }\frac{1}{6}t=1.$$

Carry out. We solve the equation.

$$\frac{1}{6}t=1$$
$$6\cdot\frac{1}{6}t=6\cdot 1$$
$$t=6$$

Check. We repeat computations. The answer checks. We also expected the result to be between 4 hr. and 7 hr.

State. Working together, it takes Kelby and Natalie 6 hrs.

15. ***Familiarize***. The job takes Bryan 8 hours working alone and Armando 6 hours working alone. Then in 1 hour Bryan does $\frac{1}{8}$ of the job and Armando does $\frac{1}{6}$ of the job. Working together they can do $\frac{1}{8}+\frac{1}{6}$, or $\frac{7}{24}$ of the job in 1 hour. In 2 hours, Bryan does $2\left(\frac{1}{8}\right)$ of the job and Armando does $2\left(\frac{1}{6}\right)$ of the job.

Working together they can do $2\left(\frac{1}{8}\right)+2\left(\frac{1}{6}\right)$, or $\frac{7}{12}$ of the job in 2 hours. In 6 hours, Bryan does $6\left(\frac{1}{8}\right)$ of the job and Armando does $6\left(\frac{1}{6}\right)$ of the job. Working together they can do $6\left(\frac{1}{8}\right)+6\left(\frac{1}{6}\right)$, or $\frac{7}{4}$ of the job which is more of the job than needs to be done. The answer is somewhere between 2 hr and 6 hr.

Translate. If they work together t hours, then Bryan does $t\left(\frac{1}{8}\right)$ of the job and Armando does $t\left(\frac{1}{6}\right)$ of the job. We want some number t such that

$$\left(\frac{1}{8}+\frac{1}{6}\right)t=1,\text{ or }\frac{7}{24}t=1.$$

Carry out. We solve the equation.

$$\frac{7}{24}t=1$$
$$24\cdot\frac{7}{24}t=24\cdot 1$$
$$t=\frac{24}{7},\text{ or }3\frac{3}{7}\text{ hr}$$

Check. We repeat computations. The answer checks. We also expected the result to be between 2 hr. and 6 hr.

State. Working together, it takes Bryan and Armando $3\frac{3}{7}$ hr.

17. ***Familiarize***. The pool can be filled in 12 hours by only the pipe and in 30 hours with only the hose. Then in 1 hour the pipe fills $\frac{1}{12}$ of the pool, and the hose fills $\frac{1}{30}$. Working together, they fill $\frac{1}{12}+\frac{1}{30}$ of the pool in an hour. Let t equal the number of hours it takes them to fill the pool together.

Translate. We want some number t such that

$$t\left(\frac{1}{12}\right)+t\left(\frac{1}{30}\right)=1,\text{ or }\frac{t}{12}+\frac{t}{30}=1.$$

Carry out. We solve the equation. LCD = 60

$$60\left(\frac{t}{12}+\frac{t}{30}\right)=60\cdot 1$$
$$5t+2t=60$$
$$7t=60$$
$$t=\frac{60}{7},\text{ or }8\frac{4}{7}\text{ hr}$$

Check. The pipe fills $\frac{1}{12}\cdot\frac{60}{7}$, or $\frac{5}{7}$ and the hose fills $\frac{1}{30}\cdot\frac{60}{7}$, or $\frac{2}{7}$. Working together, they fill $\frac{5}{7}+\frac{2}{7}=1$, or the entire pool in $\frac{60}{7}$ hr.

State. Working together, the pipe and hose can fill the pool in $\frac{60}{7}$ hr, or $8\frac{4}{7}$ hr.

19. ***Familiarize***. Let t represent the time, in minutes, that it takes the DS-860 to scan the manuscript working alone. Then $2t$ represents the time it takes the DS-6500 to do the job, working alone. In 1 min the DS-860 does $\frac{1}{t}$ of the job and the DS-6500 does $\frac{1}{2t}$ of the job.

Translate. Working together, they can do the entire job in 5 min, so we want to find t such that

$$5\left(\frac{1}{t}\right)+5\left(\frac{1}{2t}\right)=1,\text{ or }\frac{5}{t}+\frac{5}{2t}=1.$$

Carry out. We solve the equation. We multiply both sides by the LCD, $2t$.

$$2t\left(\frac{5}{t}+\frac{5}{2t}\right)=2t\cdot 1$$
$$10+5=2t$$
$$15=2t$$
$$\frac{15}{2}=t,\text{ or }7\frac{1}{2}$$

Check. If the DS-860 can do the job in $\frac{15}{2}$ min, then in 5 min it does $5\cdot\frac{1}{15/2}$, or $\frac{2}{3}$ of the job. If it takes the DS-6500 $2\cdot\frac{15}{2}$, or 15 min, to do the job, then in 5 min it does $5\cdot\frac{1}{15}$, or $\frac{1}{3}$ of the job. Working together, the two machines do $\frac{2}{3}+\frac{1}{3}$, or 1 entire job, in 5 min.

State. Working alone, it takes the DS-860 $7\frac{1}{2}$ min and the DS-6500 15 min to scan the manuscript.

21. ***Familiarize***. Let t represent the number of days it takes Tori to mulch the gardens working alone. Then $t-3$ represents the time it takes Anita to mulch the gardens, working alone. In 1 day, Tori does $\frac{1}{t}$ of the job and Anita does $\frac{1}{t-3}$ of the job.
Translate. Working together, Tori and Anita can mulch the gardens in 2 days.

$$2\left(\frac{1}{t}\right)+2\left(\frac{1}{t-3}\right)=1,\text{ or } \frac{2}{t}+\frac{2}{t-3}=1.$$

Carry out. We solve the equation. Multiply on both sides by the LCD, $t(t-3)$.

$$t(t-3)\left(\frac{2}{t}+\frac{2}{t-3}\right)=t(t-3)\cdot 1$$
$$2(t-3)+2t=t(t-3)$$
$$2t-6+2t=t^2-3t$$
$$0=t^2-7t+6$$
$$0=(t-1)(t-6)$$
$$t=1 \quad or \quad t=6$$

Check. If $t=1$, then $t-3=1-3=-2$. Since negative time has no meaning in this application, 1 cannot be a solution. If $t=6$, then $t-3=6-3=3$. In 2 days Anita does $2\cdot\frac{1}{3}$, or $\frac{2}{3}$ of the job. In 2 days Tori does $2\cdot\frac{1}{6}$, or $\frac{1}{3}$ of the job. Together they do $\frac{2}{3}+\frac{1}{3}$, or 1 entire job. The answer checks.
State. It would take Anita 3 days and Tori 6 days to do the job, working alone.

23. ***Familiarize***. Let t represent the number of months it takes Tristan to program alone. Then $3t$ represents the number of months it takes Sara program alone.
Translate. In 1 month Tristan and Sara will do one entire job, so we have

$$1\left(\frac{1}{t}\right)+1\left(\frac{1}{3t}\right)=1,\text{ or } \frac{1}{t}+\frac{1}{3t}=1$$

Carry out. We solve the equation. Multiply on both sides by the LCD, $3t$.

$$3t\left(\frac{1}{t}+\frac{1}{3t}\right)=3t\cdot 1$$
$$3+1=3t$$
$$4=3t$$
$$\frac{4}{3}=t$$

Check. If Tristan does the job alone in $\frac{4}{3}$ months, then in 1 month he does $\frac{1}{4/3}$, or $\frac{3}{4}$ of the job. If Sara does the job alone in $3\cdot\frac{4}{3}$, or 4 months, then in 1 month she does $\frac{1}{4}$ of the job. Together, they do $\frac{3}{4}+\frac{1}{4}$, or 1 entire job, in 1 month. The result checks.

State. It would take Tristan $\frac{4}{3}$ months and it would take Sara 4 months to program alone.

25. ***Familiarize***. Let t represent the number of minutes it takes Chris to do the job working alone. Then $t+120$ represents the time it takes Kim to do the job working alone.
We will convert hours to minutes:

$$2\text{ hr}=2\cdot 60\text{ min}=120\text{ min}$$
$$2\text{ hr }55\text{ min}=120\text{ min}+55\text{ min}=175\text{ min}$$

Translate. In 175 min Chris and Kim will do one entire job, so we have

$$175\left(\frac{1}{t}\right)+175\left(\frac{1}{t+120}\right)=1,\text{ or } \frac{175}{t}+\frac{175}{t+120}=1$$

Carry out. We solve the equation. Multiply on both sides by the LCD, $t(t+120)$.

$$t(t+120)\left(\frac{175}{t}+\frac{175}{t+120}\right)=t(t+120)\cdot 1$$
$$175(t+120)+175t=t(t+120)$$
$$175t+21{,}000+175t=t^2+120t$$
$$0=t^2-230t-21{,}000$$
$$0=(t-300)(t+70)$$
$$t=300 \quad or \quad t=-70$$

Check. Since negative time has no meaning in this problem -70 is not a solution of the original problem. If Chris does the job alone in 300 min, then in 175 min he does $\frac{175}{300}=\frac{7}{12}$ of the job. If Kim does the job alone in 300 + 120, or 420 min, then in 175 min she does $\frac{175}{420}=\frac{5}{12}$ of the job. Together, they do $\frac{7}{12}+\frac{5}{12}$, or 1 entire job, in 175 min. The result checks.
State. It would take Chris 300 min, or 5 hours to do the job alone.

27. ***Familiarize***. We complete the table shown in the text.

	Distance	Speed	Time
CSX	330	$r-14$	$\frac{330}{r-14}$
AMTRAK	400	r	$\frac{400}{r}$

Translate. Since the time must be the same for both trains, we have the equation

$$\frac{330}{r-14}=\frac{400}{r}.$$

Carry out. We first multiply by the LCD, $r(r-14)$.

$$r(r-14)\cdot\frac{330}{r-14}=r(r-14)\cdot\frac{400}{r}$$
$$330r=400(r-14)$$
$$330r=400r-5600$$
$$-70r=-5600$$
$$r=80$$

If the speed of the AMTRAK train is 80 km/h, then the speed of the CSX train is 80 – 14, or 66 km/h.

Check. The speed of the CSX train is 14 km/h slower than the speed of the AMTRAK train. At 66 km/h the CSX train travels 330 km in 330/66, or 5 hr. At 80 km/h the AMTRAK train travels 400 km in 400/80, or 5 hr. The times are the same, so the answer checks.

State. The speed of the AMTRAK train is 80 km/h, and the speed of the CSX freight train is 66 km/h.

29. ***Familiarize.*** Let $r=$ the speed of Rita's Harley, in mph. Then $r+15=$ the speed of Sean's Camaro. We organize the information in a table using the formula time = distance/rate to fill in the last column.

	Distance	Speed	Time
Harley	120	r	$\frac{120}{r}$
Camaro	156	$r+15$	$\frac{156}{r+15}$

Translate. Since the times must be the same, we have the equation

$$\frac{120}{r}=\frac{156}{r+15}$$

Carry out. We first multiply by the LCD, $r(r+15)$.

$$r(r+15)\cdot\frac{120}{r}=r(r+15)\cdot\frac{156}{r+15}$$
$$120(r+15)=156r$$
$$120r+1800=156r$$
$$1800=36r$$
$$50=r$$

Then $r+15=50+15=65$.

Check. At 50 mph, the Harley travels 120 mi in 120/50, or 2.4 hr. At 65 mph, the Camaro travels 156 mi in 156/65, or 2.4 hr. The times are the same, so the answer checks.

State. The speed of Rita's Harley is 50 mph, and the speed of Sean's Camaro is 65 mph.

31. ***Familiarize.*** We first make a drawing. Let $r=$ the kayak's speed in still water in mph. Then $r-3=$ the speed upstream and $r+3=$ the speed downstream.

Upstream 4 miles $r-3$ mph

10 miles $r+3$ mph Downstream

We organize the information in a table. The time is the same both upstream and downstream so we use t for each time.

	Distance	Speed	Time
Upstream	4	$r-3$	t
Downstream	10	$r+3$	t

Translate. Using the formula Time = Distance/Rate in each row of the table and the fact that the times are the same, we can write an equation.

$$\frac{4}{r-3}=\frac{10}{r+3}$$

Carry out. We solve the equation.

$$\frac{4}{r-3}=\frac{10}{r+3},\quad \text{LCD is } (r-3)(r+3)$$
$$(r-3)(r+3)\cdot\frac{4}{r-3}=(r-3)(r+3)\cdot\frac{10}{r+3}$$
$$4(r+3)=10(r-3)$$
$$4r+12=10r-30$$
$$42=6r$$
$$7=r$$

Check. If $r=7$ mph, then $r-3$ is 4 mph and $r+3$ is 10 mph. The time upstream is $\frac{4}{4}$, or 1 hour. The time downstream is $\frac{10}{10}$, or 1 hour. Since the times are the same, the answer checks.

State. The speed of the kayak in still water is 7 mph.

33. ***Familiarize.*** We first make a drawing. Let $r=$ Roslyn's speed on a nonmoving sidewalk in ft/sec. Then her speed moving forward on the moving sidewalk is $r+1.8$, and her speed in the opposite direction is $r-1.8$.

Forward $r+1.8$ 105 ft

51 ft $r-1.8$ Opposite direction

We organize the information in a table. The time is the same both forward and in the opposite direction, so we use t for each time.

	Distance	Speed	Time
Forward	105	$r+1.8$	t
Opposite direction	51	$r-1.8$	t

Translate. Using the formula Time = Distance/Rate in each row of the table and the fact that the times are the same, we can write an equation.

$$\frac{105}{r+1.8}=\frac{51}{r-1.8}$$

Carry out. We solve the equation.

$$\frac{105}{r+1.8}=\frac{51}{r-1.8}$$
$$\text{LCD is } (r+1.8)(r-1.8)$$
$$(r+1.8)(r-1.8)\frac{105}{r+1.8}=(r+1.8)(r-1.8)\frac{51}{r-1.8}$$
$$105(r-1.8)=51(r+1.8)$$
$$105r-189=51r+91.8$$
$$54r=280.8$$
$$r=5.2$$

Check. If Roslyn's speed on a nonmoving sidewalk is 5.2 ft/sec, then her speed moving forward on the moving sidewalk is 5.2 + 1.8, or 7 ft/sec, and her speed moving in the opposite direction on the sidewalk is 5.2 – 1.8, or 3.4 ft/sec. Moving 105 ft at 7 ft/sec takes $\frac{105}{7}=15$ sec. Moving 51 ft at 3.4 ft/sec takes $\frac{51}{3.4}=15$ sec. Since the times are the same, the answer checks.

State. Roslyn would be walking 5.2 ft/sec on a nonmoving sidewalk.

35. ***Familiarize.*** Let t = the time it takes Caledonia to drive to town and organize the given information in a table.

	Distance	Speed	Time
Caledonia	15	r	t
Manley	20	r	$t+1$

Translate. We can replace each r in the table above using the formula $r = d/t$.

	Distance	Speed	Time
Caledonia	15	$\frac{15}{t}$	t
Manley	20	$\frac{20}{t+1}$	$t+1$

Since the speeds are the same for both riders, we have the equation

$$\frac{15}{t} = \frac{20}{t+1}.$$

Carry out. We multiply by the LCD, $t(t+1)$.

$$t(t+1)\cdot\frac{15}{t} = t(t+1)\cdot\frac{20}{t+1}$$
$$15(t+1) = 20t$$
$$15t+15 = 20t$$
$$15 = 5t$$
$$3 = t$$

If $t = 3$, then $t+1 = 3+1$, or 4.

Check. If Caledonia's time is 3 hr and Manley's time is 4 hr, then Manley's time is 1 hr more than Caledonia's. Caledonia's speed is 15/3, or 5 mph. Manley's speed is 20/4, or 5 mph. Since the speeds are the same, the answer checks.

State. It takes Caledonia 3 hr to drive to town.

37. ***Familiarize.*** Let w = the wind speed, in mph. Then the speed into the wind is $460 - w$, and the speed with the wind is $460 + w$. We organize the information in a table.

	Distance	Speed	Time
Into the wind	525	$460 - w$	t_1
With the wind	525	$460 + w$	t_2

Translate. Using the formula Time = Distance/Rate, we see that $t_1 = \frac{525}{460-w}$ and $t_2 = \frac{525}{460+w}$. The total time into the wind and back is 2.3 hr, so $t_1 + t_2 = 2.3$, or

$$\frac{525}{460-w} + \frac{525}{460+w} = 2.3.$$

Carry out. We solve the equation. Multiply both sides by the LCD, $(460+w)(460-w)$, or $460^2 - w^2$.

$$(460^2 - w^2)\left(\frac{525}{460-w} + \frac{525}{460+w}\right) = (460^2 - w^2)2.3$$
$$525(460-w) + 525(460+w) = 2.3(211{,}600 - w^2)$$
$$241{,}500 - 525w + 241{,}500 + 525w = 486{,}680 - 2.3w^2$$
$$483{,}000 = 486{,}680 - 2.3w^2$$
$$2.3w^2 - 3680 = 0$$
$$2.3(w^2 - 1600) = 0$$
$$2.3(w+40)(w-40) = 0$$
$$w = -40 \text{ or } w = 40$$

Check. We check only 40 since the wind speed cannot be negative. If the wind speed is 40 mph, then the plane's speed into the wind is 460 – 40, or 420 mph, and the speed with the wind is 460 + 40, or 500 mph. Flying 525 mi into the wind takes $\frac{525}{420} = 1.25$ hr.

Flying 525 mi with the wind takes $\frac{525}{500} = 1.05$ hr. The total time is 1.25 + 1.05, or 2.3 hr. The answer checks.

State. The wind speed is 40 mph.

39. ***Familiarize.*** Let r = the speed at which the train actually traveled in mph, and let t = the actual travel time in hours. We organize the given information in a table.

	Distance	Speed	Time
Actual speed	120	r	t
Faster speed	120	$r+10$	$t-2$

Translate. From the first row of the table, we have $120 = rt$, and from the second row we have $120 = (r+10)(t-2)$. Solve the first equation for t, we have $t = \frac{120}{r}$. Substituting for t in the second equation, we have $120 = (r+10)\left(\frac{120}{r} - 2\right)$.

Carry out. We solve the equation.

$$120 = (r+10)\left(\frac{120}{r} - 2\right)$$
$$120 = 120 - 2r + \frac{1200}{r} - 20$$
$$0 = -2r + \frac{1200}{r} - 20$$
$$r\cdot 0 = r\cdot\left(-2r + \frac{1200}{r} - 20\right)$$
$$0 = -2r^2 + 1200 - 20r$$
$$0 = 2r^2 + 20r - 1200$$
$$0 = 2(r^2 + 10r - 600)$$
$$0 = 2(r+30)(r-20)$$
$$r = -30 \text{ or } r = 20$$

Check. Since speed cannot be negative in this problem, –30 cannot be a solution of the original problem. If the speed is 20 mph, it takes $\frac{120}{20}$, or 6 hr to travel 120 mi. If the speed is 10 mph faster, or 30 mph, it takes $\frac{120}{30}$, or 4 hr to travel 120 mi. Since 4 hr is 2 hr less time than 6 hr, the answer checks.

State. The speed was 20 mph.

41. We write a proportion and then solve it.

$$\frac{b}{6} = \frac{7}{4}$$
$$b = \frac{7}{4}\cdot 6$$
$$b = \frac{42}{4}, \text{ or } 10.5$$

(Note that the proportions $\frac{6}{b} = \frac{4}{7}$, $\frac{b}{7} = \frac{6}{4}$, or $\frac{7}{b} = \frac{4}{6}$ could also be used)

43. We write a proportion and then solve it.

$$\frac{4}{f}=\frac{6}{4}$$
$$4f\cdot\frac{4}{f}=4f\cdot\frac{6}{4}$$
$$16=6f$$
$$\frac{8}{3}=f \quad \text{Simplifying}$$

One of the following proportions could also be used:

$$\frac{f}{4}=\frac{4}{6},\ \frac{4}{f}=\frac{9}{6},\ \frac{f}{4}=\frac{6}{9},\ \frac{4}{9}=\frac{f}{6},\ \frac{9}{4}=\frac{6}{f}.$$

45. Consider the two similar right triangles in the drawing. One has legs 5 and 7. The other has legs 9 and r. We use a proportion to find r.

$$\frac{5}{7}=\frac{9}{r}$$
$$7r\cdot\frac{5}{7}=7r\cdot\frac{9}{r}$$
$$5r=63$$
$$r=\frac{63}{5},\text{ or } 12.6$$

47. From the blueprint we see that 9 in. represents 36 ft and that p in. represent 15 ft. We use a proportion to find p.

$$\frac{9}{36}=\frac{p}{15}$$
$$180\cdot\frac{9}{36}=180\cdot\frac{p}{15}$$
$$45=12p$$
$$\frac{15}{4}=p,\text{ or } p=3\frac{3}{4}$$

The length of p is $3\frac{3}{4}$ in.

49. From the blueprint we see that 9 in. represents 36 ft and that 5 in. represents r ft. We use a proportion to find r.

$$\frac{9}{36}=\frac{5}{r}$$
$$36r\cdot\frac{9}{36}=36r\cdot\frac{5}{r}$$
$$9r=180$$
$$r=20$$

The length of r is 20 ft.

51. Consider the two similar right triangles in the drawing. One has legs 4 ft and 6 ft. The other has legs 10 ft and l ft. We use a proportion to find l.

$$\frac{4}{6}=\frac{10}{l}$$
$$6l\cdot\frac{4}{6}=6l\cdot\frac{10}{l}$$
$$4l=60$$
$$l=15\text{ ft}$$

53. Let w = the wing width of the white stork. Then:

$$\frac{24\text{ cm width}}{180\text{ cm span}}=\frac{w\text{ cm width}}{200\text{ cm span}}$$
$$w=\frac{24\cdot 200}{180}\approx 26.7\text{ cm}$$

55. Let x = the amount spent on coffee. Then:

$$\frac{\$17.40}{8\text{ days}}=\frac{x}{30\text{ days}}$$
$$x=\frac{17.40\cdot 30}{8}=\$65.25$$

57. Let p = the number of photos taken. Then:

$$\frac{234\text{ photos}}{14\text{ days}}=\frac{p\text{ photos}}{42\text{ days}}$$
$$p=\frac{234\cdot 42}{14}=702\text{ photos}$$

59. ***Familiarize.*** Let D = the number of defective bulbs in a batch of 1430 bulbs. We can use a proportion to find D.

Translate.

$$\begin{array}{r} \text{defective bulbs} \rightarrow \\ \text{batch size} \rightarrow \end{array} \frac{8}{220}=\frac{D}{1430} \begin{array}{l} \leftarrow \text{defective bulbs} \\ \leftarrow \text{batch size} \end{array}$$

Carry out. We solve the proportion.

$$2860\cdot\frac{8}{220}=2860\cdot\frac{D}{1430}$$
$$104=2D$$
$$52=D$$

Check. $\frac{8}{220}=0.0\overline{36}$, $\frac{52}{1430}=0.0\overline{36}$

The ratios are the same, so the answer checks.

State. In a batch of 1430 bulbs, 52 defective bulbs can be expected.

61. ***Familiarize.*** Let z = the number of ounces of water needed by a Bolognese. We can use a proportion to solve for z.

Translate. We translate to a proportion.

$$\begin{array}{r} \text{dog weight} \rightarrow \\ \text{water} \rightarrow \end{array} \frac{8}{12}=\frac{5}{z} \begin{array}{l} \leftarrow \text{dog weight} \\ \leftarrow \text{water} \end{array}$$

Carry out. We solve the proportion.

$$12z\cdot\frac{8}{12}=12z\cdot\frac{5}{z}$$
$$8z=60$$
$$z=\frac{60}{8}=\frac{15}{2}=7\frac{1}{2}\text{oz}$$

Check. $\frac{8}{12}=0.\overline{6}$, $\frac{5}{7\frac{1}{2}}=0.\overline{6}$

The ratios are the same, so the answer checks.

State. For a 5-lb Bolognese, approximately $7\frac{1}{2}$ oz of water is required per day.

63. Familiarize. Let p = the number of Whale in the pod. We use a proportion to solve for p.

Translate.

$$\begin{array}{r} \text{sighted} \rightarrow \\ \text{pod} \rightarrow \end{array} \frac{12}{27}=\frac{40}{p} \begin{array}{l} \leftarrow \text{sighted} \\ \leftarrow \text{pod} \end{array}$$

Carry out. We solve the proportion.

$$27p\cdot\frac{12}{27}=27p\cdot\frac{40}{p}$$
$$12p=1080$$
$$p=90$$

Check. $\frac{12}{27}=\frac{4}{9}, \frac{40}{90}=\frac{4}{9}$

The ratios are the same, so the answer checks.

State. There are 90 whales in the pod.

65. *Writing Exercise.*

67. $(x^3-3x-7)-(x^2-4x+8)$
$=x^3-3x-7-x^2+4x-8$
$=x^3-x^2+x-15$

69. $(3y^2z-2yz^2+y^2)+(4yz^2+5y^2-6yz)$
$=3y^2z-2yz^2+4yz^2+y^2+5y^2-6yz$
$=3y^2z+2yz^2+6y^2-6yz$

71. $(8n^3+3)(8n^3-3)=64n^6-9$

73. *Writing Exercise.*

75. ***Familiarize***. If the drainage gate is closed, $\frac{1}{9}$ of the bog is filled in 1 hr. If the bog is not being filled, $\frac{1}{11}$ of the bog is drained in 1 hr. If the bog is being filled with the drainage gate left open, $\frac{1}{9}-\frac{1}{11}$ of the bog is filled in 1 hr. Let t = the time it takes to fill the bog with the drainage gate left open.

Translate. We want to find t such that
$t\left(\frac{1}{9}-\frac{1}{11}\right)=1$, or $\frac{t}{9}-\frac{t}{11}=1$.

Carry out. We solve the equation. First we multiply by the LCD, 99.

$$99\left(\frac{t}{9}-\frac{t}{11}\right)=99\cdot 1$$
$$11t-9t=99$$
$$2t=99$$
$$t=\frac{99}{2}, \text{ or } 49\frac{1}{2} \text{ hr}$$

Check. In $\frac{99}{2}$ hr, we have

$$\frac{99}{2}\left(\frac{1}{9}-\frac{1}{11}\right)=\frac{11}{2}-\frac{9}{2}=\frac{2}{2}=1 \text{ full bog}$$

The answer checks.

State. It will take $\frac{99}{2}$, or $49\frac{1}{2}$ hr to fill the bog.

77. ***Familiarize***. Let t = the time it takes Michelle working alone. Then $\frac{t}{2}$ = Sal's time working alone and $t-2$ = Kristen's time working alone. The entire job working together is 1 hr 20 min, or $\frac{4}{3}$ hr. In $\frac{4}{3}$ hr, Michelle does $\frac{4}{3}\cdot\frac{1}{t}$ of the job, Sal does $\frac{4}{3}\cdot\frac{1}{\frac{t}{2}}$ of the job, and Kristen does $\frac{4}{3}\cdot\frac{1}{t-2}$

Translate. We use the information above to write an equation.

$$\frac{4}{3}\cdot\frac{1}{t}+\frac{4}{3}\cdot\frac{1}{\frac{t}{2}}+\frac{4}{3}\cdot\frac{1}{t-2}=1$$

Carry out. We solve the equation.

$$\frac{4}{3}\cdot\frac{1}{t}+\frac{4}{3}\cdot\frac{1}{\frac{t}{2}}+\frac{4}{3}\cdot\frac{1}{t-2}=1, \text{ LCD}=3t(t-2)$$
$$3t(t-2)\left(\frac{4}{3t}+\frac{8}{3t}+\frac{4}{3(t-2)}\right)=3t(t-2)\cdot 1$$
$$4(t-2)+8(t-2)+4t=3t(t-2)$$
$$4t-8+8t-16+4t=3t^2-6t$$
$$0=3t^2-22t+24$$
$$0=(3t-4)(t-6)$$
$$3t-4=0 \quad or \quad t-6=0$$
$$t=\frac{4}{3} \quad or \quad t=6$$

Check. Since Kristen's time, $t-2$, is negative when $t=\frac{4}{3}$, and time cannot be negative in this application, so we check only 6.

$$=\frac{4}{3}\left(\frac{1}{6}+\frac{2}{6}+\frac{1}{4}\right)=1 \text{ complete job}$$

The answer checks.

State. Thus, working alone it would take Michelle 6 hr, Sal 3 hr and Kristen 4 hr to wax the car.

79. Sean's speed downstream is 7 + 3, or 10 mph. Using Time = Distance/Rate, we find that the time it will take Sean to kayak 5 mi downstream is 5/10, or 1/2 hr, or 30 min.

81. ***Familiarize***. Let p = the number of people per hour moved by the 60 cm-wide escalator. Then $2p$ = the number of people per hour moved by the 100 cm-wide escalator. We convert 1575 people per 14 minutes to people per hour:

$$\frac{1575 \text{ people}}{14 \text{ min}}\cdot\frac{60 \text{ min}}{1 \text{ hr}}=6750 \text{ people / hr}$$

Translate. We use the information that together the escalators move 6750 people per hour to write an equation.

$$p+2p=6750$$

Carry out. We solve the equation.

$$p+2p=6750$$
$$3p=6750$$
$$p=2250$$

Check. If the 60 cm-wide escalator moves 2250 people per hour, then the 100 cm-wide escalator moves $2\cdot 2250$, or 4500 people per hour. Together, they move $2250+4500$, or 6750 people per hour. The answer checks.

State. The 60 cm-wide escalator moves 2250 people per hour.

83. ***Familiarize***. Let d = the distance, in miles, the paddleboat can cruise upriver before it is time to turn around. The boat's speed upriver is 12 – 5, or 7 mph, and its speed downriver is 12 + 5, or 17 mph. We organize the information in a table.

	Distance	Speed	Time
Upriver	d	7	t_1
Downriver	d	17	t_2

Translate. Using the formula Time = Distance/Rate we see that $t_1 = \frac{d}{7}$ and $t_2 = \frac{d}{17}$. The time upriver and back is 3 hr, so $t_1 + t_2 = 3$, or $\frac{d}{7} + \frac{d}{17} = 3$.

Carry out. We solve the equation.

$$7\cdot 17\left(\frac{d}{7}+\frac{d}{17}\right) = 7\cdot 17\cdot 3$$
$$17d + 7d = 357$$
$$24d = 357$$
$$d = \frac{119}{8}$$

Check. Traveling $\frac{119}{8}$ mi upriver at a speed of 7 mph takes $\frac{119/8}{7} = \frac{17}{8}$ hr. Traveling $\frac{119}{8}$ mi downriver at a speed of 17 mph takes $\frac{119/8}{17} = \frac{7}{8}$ hr. The total time is $\frac{17}{8} + \frac{7}{8} = \frac{24}{8} = 3$ hr. The answer checks.

State. The pilot can go $\frac{119}{8}$, or $14\frac{7}{8}$ mi upriver before it is time to turn around.

85. ***Familiarize***. Let t = the number of seconds for a net gain of one person. The rate of birth is $\frac{1}{8}$, the rate of death is $-\frac{1}{11}$ and the rate of new migrant is $\frac{1}{27}$.

Translate. We use the information above to write an equation.

$$t\left(\frac{1}{8} - \frac{1}{11} + \frac{1}{27}\right) = 1$$

Carry out. We solve the equation.

$$t\left(\frac{1}{8} - \frac{1}{11} + \frac{1}{27}\right) = 1; \text{ LCD} = 2376$$
$$2376t\cdot\frac{1}{8} - 2376t\cdot\frac{1}{11} + 2376t\cdot\frac{1}{27} = 2376\cdot 1$$
$$297t - 216t + 88t = 2376$$
$$169t = 2376$$
$$t \approx 14.1$$

Check. $\frac{14.1}{8} - \frac{14.1}{11} + \frac{14.1}{27} \approx 1$. The answer checks.

State. It will take approximately 14 sec for a net gain of one person.

87. It helps to first make a drawing.

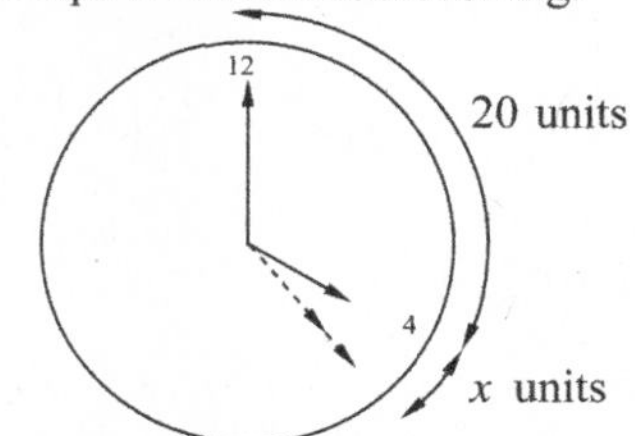

The minute hand moves 60 units per hour while the hour hand moves 5 units per hour, where one unit represents one minute on the face of the clock. When the hands are in the same position the first time, the hour hand will have moved x units and the minute hand will have moved $x + 20$ units. The times are the same.

	Distance	Speed	Time
Minute	$x+20$	60	t
Hour	x	5	t

Solve $\frac{x+20}{60} = \frac{x}{5}$.

$x = 1\frac{9}{11}$, so the hands will first be in the same position at $20 + 1\frac{9}{11}$, or $21\frac{9}{11}$ min, after 4:00.

89. Traveling 100 km at 40 km/h takes $\frac{100}{40}$, or $\frac{5}{2}$ hr.

Traveling 100 km at 60 km/h takes $\frac{100}{60}$, or $\frac{5}{3}$ hr.

The total time of the trip is $\frac{5}{2} + \frac{5}{3}$, or $\frac{25}{6}$ hr. We use the formula

$$\text{Average speed} = \frac{\text{Total distance}}{\text{Total time}} = \frac{200}{25/6} = 48$$

The average speed was 48 km/h.

91. ***Familiarize***. We organize the information in a table. Let r = the speed on the first part of the trip and t = the time driven at that speed.

	Distance	Speed	Time
First part	30	r	t
Second part	20	$r+15$	$1-t$

Translate. From the rows of the table we obtain two equations:

$$30 = rt$$
$$20 = (r+15)(1-t)$$

We solve each equation for t and set the results equal:

Solving $30 = rt$ for t: $t = \frac{30}{r}$

Solving $20 = (r+15)(1-t)$ for t: $t = 1 - \frac{20}{r+15}$

Then $\frac{30}{r} = 1 - \frac{20}{r+15}$.

Carry out. We first multiply the equation by the LCD, $r(r+15)$.

$$r(r+15)\cdot\frac{30}{r}=r(r+15)\left(1-\frac{20}{r+15}\right)$$
$$30(r+15)=r(r+15)-20r$$
$$30r+450=r^2+15r-20r$$
$$0=r^2-35r-450$$
$$0=(r+10)(r-45)$$
$$r+10=0 \quad or \quad r-45=0$$
$$r=-10 \quad or \quad r=45$$

Check. Since the speed cannot be negative, we only check 45. If $r=45$, then the time for the first part is $\frac{30}{45}$, or $\frac{2}{3}$ hr. If $r=45$, then $r+15=60$ and the time for the second part is $\frac{20}{60}$, or $\frac{1}{3}$ hr. The total time is $\frac{2}{3}+\frac{1}{3}$, or 1 hour. The value checks.

State. The speed for the first 30 miles was 45 mph.

93. Equation 1: $\frac{1}{a}\cdot t+\frac{1}{b}\cdot t=1$;

Equation 2: $\left(\frac{1}{a}+\frac{1}{b}\right)t=1$;

Equation 3: $\frac{t}{a}+\frac{t}{b}=1$;

Equation 4: $\frac{1}{a}+\frac{1}{b}=\frac{1}{t}$;

$\frac{1}{a}\cdot t+\frac{1}{b}\cdot t=1$ Equation 1

$t\left(\frac{1}{a}+\frac{1}{b}\right)=1$ Factoring out t = Equation 2

$t\cdot\frac{1}{a}+t\cdot\frac{1}{b}=1$ Using the distributive law

$\frac{t}{a}+\frac{t}{b}=1$ Multiplying; Equation 2=Equation 3

$\frac{1}{t}\cdot\left(\frac{t}{a}+\frac{t}{b}\right)=\frac{1}{t}\cdot 1$ Multiplying both sides by $\frac{1}{t}$

$\frac{1}{t}\cdot\frac{t}{a}+\frac{1}{t}\cdot\frac{t}{b}=\frac{1}{t}\cdot 1$ Using the distributive law

$\frac{1}{a}+\frac{1}{b}=\frac{1}{t}$ Multiplying; Equation 3=Equation 4

95. *Writing Exercise.*

Chapter 6 Review

1. False; some rational expressions like $\frac{x^2+4}{x+2}$ cannot be simplified.

2. True; when $t=2$, the denominator is zero. Thus, the expression is undefined.

3. False; when $t=3$, then $\frac{t-3}{t^2-4}=\frac{3-3}{3^2-4}=\frac{0}{5}=0$.

4. True

5. True

6. False; a common denominator is required to add rational expressions.

7. False; a common denominator is required to subtract rational expressions.

8. False; for example, 0 is the solution to $\frac{x+3}{x+1}=\frac{x+6}{x+2}$.

9. $\frac{17}{-x^2}$

Set the denominator equal to 0 and solve for x.

$$-x^2=0$$
$$x=0$$

The expression is undefined for $x=0$.

10. $\frac{x-5}{x^2-36}$

Set the denominator equal to 0 and solve for x.

$$x^2-36=0$$
$$(x+6)(x-6)=0$$
$$x+6=0 \quad or \quad x-6=0$$
$$x=-6 \quad or \quad x=6$$

The expression is undefined for $x=-6$. and $x=6$.

11. $\frac{x^2+3x+2}{x^2+x-30}$

Set the denominator equal to 0 and solve for x.

$$x^2+x-30=0$$
$$(x+6)(x-5)=0$$
$$x+6=0 \quad or \quad x-5=0$$
$$x=-6 \quad or \quad x=5$$

The expression is undefined for $x=-6$ and $x=5$.

12. $\frac{-6}{(t+2)^2}$

Set the denominator equal to 0 and solve for t.

$$(t+2)^2=0$$
$$t+2=0$$
$$t=-2$$

The expression is undefined for $t=-2$.

13. $\frac{3x^2-9x}{3x^2+15x}=\frac{3x(x-3)}{3x(x+5)}=\frac{x-3}{x+5}$

14. $\frac{14x^2-x-3}{2x^2-7x+3}=\frac{(2x-1)(7x+3)}{(2x-1)(x-3)}=\frac{7x+3}{x-3}$

15. $\frac{6y^2-36y+54}{4y^2-36}=\frac{6(y^2-6y+9)}{4(y^2-9)}$
$=\frac{3(y-3)(y-3)}{2(y+3)(y-3)}=\frac{3(y-3)}{2(y+3)}$

16. $\frac{5x^2-20y^2}{2y-x}=\frac{-5(4y^2-x^2)}{(2y-x)}$
$=\frac{-5(2y+x)(2y-x)}{(2y-x)}=-5(2y+x)$

17. $\frac{a^2-36}{10a}\cdot\frac{2a}{a+6}=\frac{(a+6)(a-6)(2a)}{10a(a+6)}$
$=\frac{a-6}{5}\cdot\frac{2a(a+6)}{2a(a+6)}$
$=\frac{a-6}{5}$

18. $\frac{6y-12}{2y^2+3y-2}\cdot\frac{y^2-4}{8y-8}=\frac{6(y-2)(y-2)(y+2)}{(2y-1)(y+2)(8)(y-1)}$
$=\frac{3(y-2)^2}{4(2y-1)(y-1)}\cdot\frac{2(y+2)}{2(y+2)}$
$=\frac{3(y-2)^2}{4(2y-1)(y-1)}$

19. $\frac{16-8t}{3}\div\frac{t-2}{12t}=\frac{8(2-t)}{3}\cdot\frac{12t}{t-2}$
$=8(-1)(4t)\frac{3(t-2)}{3(t-2)}$
$=-32t$

20. $\frac{4x^4}{x^2-1}\div\frac{2x^3}{x^2-2x+1}=\frac{4x^4}{(x+1)(x-1)}\cdot\frac{(x-1)(x-1)}{2x^3}$
$=\frac{2x(x-1)}{x+1}\cdot\frac{2x^3(x-1)}{2x^3(x-1)}$
$=\frac{2x(x-1)}{x+1}$

21. $\frac{x^2+1}{x-2}\cdot\frac{2x+1}{x+1}=\frac{(x^2+1)(2x+1)}{(x-2)(x+1)}$
No simplification is possible.

22. $(t^2+3t-4)\div\frac{t^2-1}{t+4}=(t+4)(t-1)\cdot\frac{(t+4)}{(t+1)(t-1)}$
$=\frac{(t+4)^2}{t+1}\cdot\frac{t-1}{t-1}$
$=\frac{(t+4)^2}{t+1}$

23. $10a^3b^8=2\cdot5\cdot a\cdot a\cdot a\cdot b\cdot b\cdot b\cdot b\cdot b\cdot b\cdot b\cdot b$
$12a^5b=2\cdot2\cdot3\cdot a\cdot a\cdot a\cdot a\cdot a\cdot b$
$\text{LCM}=2\cdot2\cdot3\cdot5\cdot a\cdot a\cdot a\cdot a\cdot a\cdot b\cdot b\cdot b\cdot b\cdot b\cdot b\cdot b\cdot b$
$=60a^5a^8$

24. $x^2-x=x(x-1)$
$x^5-x^3=x^3(x^2-1)=x\cdot x\cdot x\cdot(x+1)(x-1)$
$x^4=x\cdot x\cdot x\cdot x$
$\text{LCM}=x\cdot x\cdot x\cdot x\cdot(x+1)(x-1)=x^4(x+1)(x-1)$

25. $y^2-y-2=(y-2)(y+1)$
$y^2-4=(y-2)(y+2)$
$\text{LCM}=(y+1)(y+2)(y-2)$

26. $\frac{x+6}{x+3}+\frac{9-4x}{x+3}=\frac{x+6+9-4x}{x+3}=\frac{-3x+15}{x+3}$

27. $\frac{6x-3}{x^2-x-12}-\frac{2x-15}{x^2-x-12}=\frac{6x-3-2x+15}{x^2-x-12}$
$=\frac{4x+12}{(x-4)(x+3)}$
$=\frac{4(x+3)}{x-4(x+3)}$
$=\frac{4}{x-4}$

28. $\frac{3x-1}{2x}-\frac{x-3}{x}=\frac{3x-1}{2x}-\frac{x-3}{x}\cdot\frac{2}{2}$ LCD $=2x$
$=\frac{3x-1-2(x-3)}{2x}$
$=\frac{3x-1-2x+6}{2x}$
$=\frac{x+5}{2x}$

29. $\frac{2a+4b}{5ab^2}-\frac{5a-3b}{a^2b}=\frac{2a+4b}{5ab^2}\cdot\frac{a}{a}-\frac{5a-3b}{a^2b}\cdot\frac{5b}{5b}$
LCM is $5a^2b^2$
$=\frac{2a^2+4ab-25ab+15b^2}{5a^2b^2}$
$=\frac{2a^2-21ab+15b^2}{5a^2b^2}$

30. $\frac{y^2}{y-2}+\frac{6y-8}{2-y}=\frac{y^2}{y-2}+\frac{8-6y}{y-2}$
$=\frac{y^2-6y+8}{y-2}$
$=\frac{(y-4)(y-2)}{y-2}$
$=y-4$

31. $\frac{t}{t+1}+\frac{t}{1-t^2}=\frac{t}{t+1}\cdot\frac{1-t}{1-t}+\frac{t}{(1-t)(1+t)}$
$=\frac{t-t^2+t}{(1-t)(1+t)}$
$=\frac{-t^2+2t}{(1-t)(1+t)}$
$=\frac{-t(t-2)}{(1-t)(1+t)}$ or $\frac{t(t-2)}{(t-1)(t+1)}$

32. $\frac{d^2}{d-2}+\frac{4}{2-d}=\frac{d^2}{d-2}-\frac{4}{d-2}$
$=\frac{d^2-4}{d-2}$
$=\frac{(d+2)(d-2)}{d-2}$
$=d+2$

33. $\frac{1}{x^2-25}-\frac{x-5}{x^2-4x-5}=\frac{1}{(x+5)(x-5)}-\frac{x-5}{(x-5)(x+1)}$
$=\frac{x+1-(x-5)(x+5)}{(x+1)(x+5)(x-5)}$
$=\frac{x+1-x^2+25}{(x+1)(x+5)(x-5)}$
$=\frac{-x^2+x+26}{(x+1)(x+5)(x-5)}$

34. $\frac{3x}{x+2}-\frac{x}{x-2}+\frac{8}{x^2-4}$ LCD $=(x+2)(x-2)$

$=\frac{3x}{x+2}\cdot\frac{x-2}{x-2}-\frac{x}{x-2}\cdot\frac{x+2}{x+2}+\frac{8}{(x+2)(x-2)}$

$=\frac{3x^2-6x-x^2-2x+8}{(x+2)(x-2)}$

$=\frac{2x^2-8x+8}{(x+2)(x-2)}=\frac{2(x^2-4x+4)}{(x+2)(x-2)}$

$=\frac{2(x-2)(x-2)}{(x+2)(x-2)}$

$=\frac{2(x-2)}{x+2}$

35. $\frac{3}{4t}+\frac{3}{3t+2}$ LCD $=4t(3t+2)$

$=\frac{3}{4t}\cdot\frac{3t+2}{3t+2}+\frac{3}{3t+2}\cdot\frac{4t}{4t}$

$=\frac{9t+6+12t}{4t(3t+2)}$

$=\frac{21t+6}{4t(3t+2)}$

$=\frac{3(7t+2)}{4t(3t+2)}$

36. $\frac{\frac{1}{z}+1}{\frac{1}{z^2}-1}=\frac{\frac{1}{z}+1}{\frac{1}{z^2}-1}\cdot\frac{z^2}{z^2}$ LCD $=z^2$

$=\frac{z+z^2}{1-z^2}$

$=\frac{z(1+z)}{(1-z)(1+z)}$

$=\frac{z}{1-z}$

37. $\frac{\frac{5}{2x^2}}{\frac{3}{4x}+\frac{4}{x^3}}=\frac{\frac{5}{2x^2}}{\frac{3}{4x}+\frac{4}{x^3}}\cdot\frac{4x^3}{4x^3}$ LCD $=4x^3$

$=\frac{10x}{3x^2+16}$

38. $\frac{\frac{c}{d}-\frac{d}{c}}{\frac{1}{c}+\frac{1}{d}}=\frac{\frac{c}{d}-\frac{d}{c}}{\frac{1}{c}+\frac{1}{d}}\cdot\frac{cd}{cd}$ LCD $=cd$

$=\frac{c^2-d^2}{d+c}$

$=\frac{(c-d)(c+d)}{c+d}$

$=c-d$

39. $\frac{3}{x}-\frac{1}{4}=\frac{1}{2}$, Note $x\neq 0$

$4x\left(\frac{3}{x}-\frac{1}{4}\right)=4x\left(\frac{1}{2}\right)$, LCD $=4x$

$12-x=2x$

$12=3x$

$4=x$

The solution is 4.

40. $\frac{3}{x+4}=\frac{1}{x-1}$ Note $x\neq 1,-4$

$(x-1)(x+4)\frac{3}{x+4}=(x-1)(x+4)\frac{1}{x-1}$

LCD is $(x-1)(x+4)$

$3(x-1)=x+4$

$3x-3=x+4$

$3x=x+7$

$2x=7$

$x=\frac{7}{2}$

41. $x+\frac{6}{x}=-7$ Note $x\neq 0$

$x\left(x+\frac{6}{x}\right)=-7x$

$x^2+6=-7x$

$x^2+7x+6=0$

$(x+6)(x+1)=0$

$x+6=0$ *or* $x+1=0$

$x=-6$ *or* $x=-1$

The solution are –6 and –1.

42. $1=\frac{2}{x-1}+\frac{2}{x+2}$ Note $x\neq -2$, 1

$(x+2)(x-1)1=(x+2)(x-1)\left(\frac{2}{x-1}+\frac{2}{x+2}\right)$

LCD is $(x+2)(x-1)$

$x^2+x-2=2x+4+2x-2$

$x^2-3x-4=0$

$(x+1)(x-4)=0$

$x+1=0$ *or* $x-4=0$

$x=-1$ *or* $x=4$

43. ***Familiarize***. The job takes Jackson 12 hours working alone and Charis 9 hours working alone. Then in 1 hour Jackson does $\frac{1}{12}$ of the job and Charis does $\frac{1}{9}$ of the job. Working together, they can do $\frac{1}{9}+\frac{1}{12}$, or $\frac{7}{36}$ of the job in 1 hour.

Translate. If they work together t hours, then Jackson does $t\left(\frac{1}{9}\right)$ of the job and Charis does $t\left(\frac{1}{12}\right)$ of the job. We want some number t such that

$\left(\frac{1}{9}+\frac{1}{12}\right)t=1$, or $\frac{7}{36}t=1$.

Carry out. We solve the equation.

$\frac{7}{36}t=1$

$\frac{36}{7}\cdot\frac{7}{36}t=\frac{36}{7}\cdot 1$

$t=\frac{36}{7}$ or $5\frac{1}{7}$

Check. The check can be done by repeating the computation.

State. Working together, it takes them $5\frac{1}{7}$ hrs to complete the job.

44. ***Familiarize***. Let t represent the number of hours it takes the Jon to build one section of trail working alone. Then $t - 15$ represents the time it takes Ben to build one section of trail, working alone. In 1 hour Jon does $\frac{1}{t}$ of the job and Ben does $\frac{1}{t-15}$ of the job.

Translate. Working together, Jon and Ben can build the section of trail in 18 hr to find t such that

$$18\left(\frac{1}{t}\right)+18\left(\frac{1}{t-15}\right)=1, \text{ or } \frac{18}{t}+\frac{18}{t-15}=1.$$

Carry out. We solve the equation. First we multiply both sides by the LCD, $t(t-15)$.

$$\begin{aligned} t(t-15)\left(\frac{18}{t}+\frac{18}{t-15}\right)&=t(t-15)\cdot 1\\ 18(t-15)+18t&=t(t-15)\\ 18t-270+18t&=t^2-15t\\ 0&=t^2-51t+270\\ 0&=(t-6)(t-45)\\ t=6 \text{ or } t&=45 \end{aligned}$$

Check. If $t = 6$, then $t - 15 = 6 - 15 = -9$. Since negative time has no meaning in this application, 6 cannot be a solution. If $t = 45$, then $t - 15 = 45 - 15 = 30$. In 18 hr Jon does $18\cdot\frac{1}{45}$, or $\frac{2}{5}$ of the job. In 18 hr Ben does $18\cdot\frac{1}{30}$, or $\frac{3}{5}$ of the job. Together they do $\frac{2}{5}+\frac{3}{5}$, or 1 entire job. The answer checks.

State. Working alone, Jon can build a section of trail in 45 hr and Ben can build a section of trail in 30 hr.

45. Let r = the speed of the boat in still water in mph.

Solve $\frac{30}{r-6}=\frac{50}{r+6}$.

The solution is 24 mph. This answer checks.

46. ***Familiarize***. Let r = Elizabeth's speed in mph. Then $r + 8$ = Jennifer's speed in mph. We organize the information in a table. The time is the same for both so we use t for each time.

	Distance	Speed	Time
Elizabeth	93	r	t
Jennifer	105	$r+8$	t

Translate. Using the formula Time = Distance/Rate in each row of the table and the fact that the times are the same, we can write an equation.

$$\frac{93}{r}=\frac{105}{r+8}$$

Carry out. We solve the equation.

$$\frac{93}{r}=\frac{105}{r+8}, \quad \text{LCD is } r(r+8)$$

$$\begin{aligned} r(r+8)\cdot\frac{93}{r}&=r(r+8)\cdot\frac{105}{r+8}\\ 93(r+8)&=105r\\ 93r+744&=105r\\ 744&=12r\\ 62&=r \end{aligned}$$

Check. If Elizabeth's speed is 62 mph, then Jennifer's speed is 62 + 8, or 70 mph. Traveling 93 mi at 62 mph takes $\frac{93}{62}=1.5$ hr. Traveling 105 mi at 70 mph takes $\frac{105}{70}=1.5$ hr. Since the times are the same, the answer checks.

State. Elizabeth's speed is 62 mph; Jennifer's speed is 70 mph.

47. ***Familiarize.*** The ratio of seal tagged to the total number of seals in the harbor, T, is $\frac{33}{T}$. Of the 40 seals caught later, 24 were tagged. The ratio of tagged seals to seals caught is $\frac{24}{40}$.

Translate. We translate to a proportion.

$$\begin{array}{rcl} \text{Seals originally tagged} \rightarrow & \dfrac{33}{T}=\dfrac{24}{40} & \leftarrow \text{Tagged seals caught later} \\ \text{Seals in harbor} \rightarrow & & \leftarrow \text{Seals caught later} \end{array}$$

Carry out. We solve the proportion.

$$\begin{aligned} 40T\cdot\frac{33}{T}&=40T\cdot\frac{24}{40}\\ 1320&=24T\\ 55&=T \end{aligned}$$

Check. $\frac{33}{55}=0.6, \quad \frac{24}{40}=0.6$

The ratios are the same, so the answer checks.

State. We estimate that there are 55 seals in the harbor.

48. We write a proportion and then solve it.

$$\frac{x}{8.5}=\frac{2.4}{3.4}$$
$$x=6$$

49. ***Familiarize.*** Let D = the number of defective radios you would expect in a sample of 540 radios. We use a proportion to solve for D.

Translate.

$$\begin{array}{rcl} \text{Defective} \rightarrow & \dfrac{4}{30}=\dfrac{D}{540} & \leftarrow \text{Defective} \\ \text{Radios} \rightarrow & & \leftarrow \text{Radios} \end{array}$$

Carry out. We solve the proportion. We multiply by the LCD, 540.

$$\begin{aligned} 540\cdot\frac{4}{30}&=540\cdot\frac{D}{540}\\ 72&=D \end{aligned}$$

Check. $\frac{4}{30}=\frac{2}{15}, \quad \frac{72}{540}=\frac{2}{15}$

The ratios are the same, so the answer checks.

State. You would expect 72 defective radios in a batch of 540.

50. *Writing Exercise.*

51. *Writing Exercise.*

52. $\frac{2a^2+5a-3}{a^2}\cdot\frac{5a^3+30a^2}{2a^2+7a-4}\div\frac{a^2+6a}{a^2+7a+12}$

$$=\frac{\cancel{(2a-1)}(a+3)}{\cancel{a^2}}\cdot\frac{5\cancel{a^2}\cancel{(a+6)}}{\cancel{(2a-1)}\cancel{(a+4)}}\cdot\frac{(a+3)\cancel{(a+4)}}{a\cancel{(a+6)}}$$

$$=\frac{5(a+3)^2}{a}$$

53. $\frac{12a}{(a-b)(b-c)} - \frac{2a}{(b-a)(c-b)} = \frac{12a-2a}{(a-b)(b-c)}$

Since $(b-a)(c-b) = (a-b)(b-c)$

$= \frac{10a}{(a-b)(b-c)}$

54. $\frac{5(x-y)}{(x-y)(x+2y)} - \frac{5(x-3y)}{(x+2y)(x-3y)}$

$= \frac{5}{x+2y} - \frac{5}{x+2y} = 0$

55. We write a proportion and solve it.

total hits → $\frac{153+x}{395+125} = \frac{4}{10}$ ← total average at-bats
total at-bats →

$\frac{153+x}{520} = \frac{4}{10}$

$520 \cdot \frac{153+x}{520} = 520 \cdot \frac{4}{10}$

$153 + x = 208$

$x = 55$ more hits

new hits → $\frac{55}{125} = 0.44$ or 44%
new at-bats →

He must hit at 44% of the 125 at-bats.

Chapter 6 Test

1. $\frac{2-x}{5x}$

We find the number which makes the denominator 0.

$5x = 0$

$x = 0$

The expression is undefined for $x = 0$.

2. $\frac{x^2+x-30}{x^2-3x+2}$

We find the number which makes the denominator 0.

$x^2 - 3x + 2 = 0$

$(x-2)(x-1) = 0$

$x - 2 = 0$ *or* $x - 1 = 0$

$x = 2$ *or* $x = 1$

The expression is undefined for $x = 1$ and $x = 2$.

3. $\frac{6x^2+17x+7}{2x^2+7x+3} = \frac{(3x+7)(2x+1)}{(x+3)(2x+1)}$

$= \frac{(3x+7)(2x+1)}{(x+3)(2x+1)}$

$= \frac{3x+7}{x+3}$

4. $\frac{t^2-9}{12t} \cdot \frac{8t^2}{t^2-4t+3} = \frac{(t+3)(t-3)}{12t} \cdot \frac{8t^2}{(t-3)(t-1)}$

$= \frac{(t+3)(t-3) \cdot 2 \cdot 2 \cdot 2 \cdot t \cdot t}{2 \cdot 2 \cdot 3 \cdot t(t-3)(t-1)}$

$= \frac{(t+3)(t-3) \cdot 2 \cdot 2 \cdot 2 \cdot t \cdot t}{2 \cdot 2 \cdot 3 \cdot t(t-3)(t-1)}$

$= \frac{2t(t+3)}{3(t-1)}$

5. $\frac{25y^2-1}{9y^2-6y} \div \frac{5y^2+9y-2}{3y^2+y-2}$

$= \frac{25y^2-1}{9y^2-6y} \cdot \frac{3y^2+y-2}{5y^2+9y-2}$

$= \frac{(5y+1)(5y-1)}{3y(3y-2)} \cdot \frac{(3y-2)(y+1)}{(5y-1)(y+2)}$

$= \frac{(5y+1)(5y-1)(3y-2)(y+1)}{3y(3y-2)(5y-1)(y+2)}$

$= \frac{(5y+1)(y+1)}{3y(y+2)}$

6. $\frac{4a^2+1}{4a^2-1} \div \frac{4a^2}{4a^2+4a+1} = \frac{4a^2+1}{4a^2-1} \cdot \frac{4a^2+4a+1}{4a^2}$

$= \frac{4a^2+1}{(2a+1)(2a-1)} \cdot \frac{(2a+1)(2a+1)}{4a^2}$

$= \frac{(4a^2+1)(2a+1)(2a+1)}{4a^2(2a+1)(2a-1)}$

$= \frac{(2a+1)(4a^2+1)}{4a^2(2a-1)}$

7. $(x^2+6x+9) \cdot \frac{(x-3)^2}{x^2-9}$

$= \frac{(x+3)(x+3)}{1} \cdot \frac{(x-3)(x-3)}{(x+3)(x-3)}$

$= \frac{(x+3)(x+3)(x-3)(x-3)}{(x+3)(x-3)}$

$= (x+3)(x-3)$

8. $y^2 - 9 = (y+3)(y-3)$

$y^2 + 10y + 21 = (y+7)(y+3)$

$y^2 + 4y - 21 = (y+7)(y-3)$

$\text{LCM} = (y+3)(y-3)(y+7)$

9. $\frac{2+x}{x^3} + \frac{7-4x}{x^3} = \frac{2+x-4x+7}{x^3} = \frac{-3x+9}{x^3}$

10. $\frac{5-t}{t^2+1} - \frac{t-3}{t^2+1} = \frac{5-t-(t-3)}{t^2+1}$

$= \frac{5-t-t+3}{t^2+1}$

$= \frac{-2t+8}{t^2+1}$

11. $\frac{2x-4}{x-3} + \frac{x-1}{3-x} = \frac{2x-4}{x-3} + \frac{x-1}{-1(3-x)}$

$= \frac{2x-4}{x-3} + \frac{-1(x-1)}{x-3}$

$= \frac{2x-4-x+1}{x-3}$

$= \frac{x-3}{x-3}$

$= 1$

12. $\frac{2x-4}{x-3}-\frac{x-1}{3-x}=\frac{2x-4}{x-3}-\frac{x-1}{-1(x-3)}$
$=\frac{2x-4}{x-3}+\frac{x-1}{x-3}$
$=\frac{2x-4+x-1}{x-3}$
$=\frac{3x-5}{x-3}$

13. $\frac{7}{t-2}+\frac{4}{t}$ LCD is $t(t-2)$
$=\frac{7}{t-2}\cdot\frac{t}{t}+\frac{4}{t}\cdot\frac{t-2}{t-2}$
$=\frac{7t}{t(t-2)}+\frac{4(t-2)}{t(t-2)}$
$=\frac{7t+4t-8}{t(t-2)}=\frac{11t-8}{t(t-2)}$

14. $\frac{y}{y^2+6y+9}+\frac{1}{y^2+2y-3}$
$=\frac{y}{(y+3)(y+3)}+\frac{1}{(y+3)(y-1)}$
LCD is $(y+3)(y+3)(y-1)$
$=\frac{y}{(y+3)(y+3)}\cdot\frac{y-1}{y-1}+\frac{1}{(y+3)(y-1)}\frac{(y+3)}{(y+3)}$
$=\frac{y(y-1)}{(y+3)(y+3)(y-1)}+\frac{(y+3)}{(y+3)(y-1)(y+3)}$
$=\frac{y^2-y}{(y+3)(y+3)(y-1)}+\frac{y+3}{(y+3)(y-1)(y+3)}$
$=\frac{y^2-y+y+3}{(y+3)(y+3)(y-1)}$
$=\frac{y^2+3}{(y-1)(y+3)^2}$

15. $\frac{1}{x-1}+\frac{4}{x^2-1}-\frac{2}{x^2-2x+1}$
$\frac{1}{x-1}+\frac{4}{(x+1)(x-1)}-\frac{2}{(x-1)(x-1)}$
LCD is $(x-1)(x-1)(x+1)$
$=\frac{1}{x-1}\cdot\frac{(x+1)(x-1)}{(x+1)(x-1)}+\frac{4}{(x+1)(x-1)}\cdot\frac{x-1}{x-1}-\frac{2(x+1)}{(x+1)(x-1)^2}$
$=\frac{(x+1)(x-1)}{(x+1)(x-1)^2}+\frac{4(x-1)}{(x+1)(x-1)^2}-\frac{2(x+1)}{(x+1)(x-1)^2}$
$=\frac{(x+1)(x-1)+4(x-1)-2(x+1)}{(x+1)(x-1)^2}$
$=\frac{x^2-1+4x-4-2x-2}{(x+1)(x-1)^2}$
$=\frac{x^2+2x-7}{(x+1)(x-1)^2}$

16. $\frac{9-\frac{1}{y^2}}{3-\frac{1}{y}}$ LCD is y^2
$=\frac{y^2}{y^2}\cdot\left(\frac{9-\frac{1}{y^2}}{3-\frac{1}{y}}\right)=\frac{9y^2-\frac{y^2}{y^2}}{3y^2-\frac{y^2}{y}}$
$=\frac{9y^2-1}{3y^2-y}=\frac{(3y+1)(3y-1)}{y(3y-1)}$
$=\frac{(3y+1)\cancel{(3y-1)}}{y\cancel{(3y-1)}}=\frac{3y+1}{y}$

17. $\frac{\frac{x}{8}-\frac{8}{x}}{\frac{1}{8}+\frac{1}{x}}$ LCD is $8x$
$=\frac{8x}{8x}\cdot\frac{\frac{x}{8}-\frac{8}{x}}{\frac{1}{8}+\frac{1}{x}}=\frac{\frac{8x^2}{8}-\frac{64x}{x}}{\frac{8x}{8}+\frac{8x}{x}}$
$=\frac{x^2-64}{x+8}=\frac{(x+8)(x-8)}{x+8}=\frac{\cancel{(x+8)}(x-8)}{\cancel{x+8}}$
$=x-8$

18. Note that $t\neq 0$.
$\frac{1}{t}+\frac{1}{3t}=\frac{1}{2}$, LCD $=6t$
$6t\left(\frac{1}{t}+\frac{1}{3t}\right)=6t\left(\frac{1}{2}\right)$
$6t\cdot\frac{1}{t}+6t\cdot\frac{1}{3t}=6t\frac{1}{2}$
$6+2=3t$
$8=3t$
$\frac{8}{3}=t$
The solution is $\frac{8}{3}$.

19. To avoid division by 0, we must have $x\neq 0$ and $x-2\neq 0$, or $x\neq 0$ and $x\neq 2$.
$\frac{15}{x}-\frac{15}{x-2}=-2$ LCD $=x(x-2)$
$x(x-2)\left(\frac{15}{x}-\frac{15}{x-2}\right)=x(x-2)(-2)$
$15(x-2)-15x=-2x(x-2)$
$15x-30-15x=-2x^2+4x$
$2x^2-4x-30=0$
$2(x^2-2x-15)=0$
$2(x-5)(x+3)=0$
$x-5=0$ *or* $x+3=0$
$x=5$ *or* $x=-3$
The solutions are –3 and 5.

20. ***Familiarize.*** The job takes the first copier 20 min working alone and the second copier 30 min working alone. In 1 min, the first copier does $\frac{1}{20}$ of the job and the second copier does $\frac{1}{30}$ of the job. Working together they can do $\frac{1}{20}+\frac{1}{30}$, or $\frac{5}{60}$, or $\frac{1}{12}$ of the job in 1 min. In 10 min, the first copier does $10\cdot\frac{1}{30}$ of the job. Working together they can do $10\cdot\frac{1}{20}+10\cdot\frac{1}{30}$, or $\frac{5}{6}$, of the job in 10 min. In 15 min, the first copier does $15\cdot\frac{1}{20}$ of the job and the second copier does $15\cdot\frac{1}{30}$ of the job. Working together they can do $15\cdot\frac{1}{20}+15\cdot\frac{1}{30}$, or $1\frac{1}{4}$ of the job which is more of the job than needs to be done. The answer is somewhere between 10 min and 15 min.

Translate. If they work together t minutes, then the first copier does $t\cdot\frac{1}{20}$ of the job and the second copier does $t\cdot\frac{1}{30}$ of the job. We want a number t such that $(\frac{1}{20}+\frac{1}{30})t=1$, or $\frac{1}{12}\cdot t=1$.

Carry out. We solve the equation.

$$\frac{1}{12}\cdot t=1$$
$$12\frac{1}{12}\cdot t=12\cdot 1$$
$$t=12$$

Check. We repeat the computations. We also expected the result to be between 10 min and 15 min as it is.

State. Working together, it takes the two copiers 12 min to do the job.

21. ***Familiarize***. Let t represent the number of hours it takes Tyler to prepare the meal, working alone. Then $t+6$ represents the time it takes Katie to prepare the meal, working alone. In 1 hour Tyler does $\frac{1}{t}$ of the job and Katie does $\frac{1}{t+6}$ of the job.

Translate. Working together, Tyler and Katie can prepare the meal in $2\frac{6}{7}$ hr.

$$2\frac{6}{7}\left(\frac{1}{t}\right)+2\frac{6}{7}\left(\frac{1}{t+6}\right)=1,\text{ or }\frac{20}{7t}+\frac{20}{7(t+6)}=1.$$

Carry out. We solve the equation. First we multiply both sides by the LCD, $7t(t+6)$.

$$7t(t+6)\left(\frac{20}{7t}+\frac{20}{7(t+6)}\right)=7t(t+6)\cdot 1$$
$$20(t+6)+20t=7t(t+6)$$
$$20t+120+20t=7t^2+42t$$
$$0=7t^2+2t-120$$
$$0=(7t+30)(t-4)$$
$$t=-\frac{30}{7}\quad or\quad t=4$$

Check. Since negative time has no meaning in this application, $-\frac{30}{7}$ cannot be a solution. If $t=4$, then $t+6=4+6=10$. In $2\frac{6}{7}$ hr Tyler does $2\frac{6}{7}\cdot\frac{1}{4}$, or $\frac{5}{7}$ of the job. In $2\frac{6}{7}$ hr Katie does $2\frac{6}{7}\cdot\frac{1}{10}$, or $\frac{2}{7}$ of the job. Together they do $\frac{5}{7}+\frac{2}{7}$, or 1 entire job. The answer checks.

State. Working alone, Tyler can prepare the meal in 4 hr and Katie can prepare the meal in 10 hr.

22. ***Familiarize.*** Burning 320 calories corresponds to walking 4 mi, and we wish to find the number of miles m that corresponds to burning 100 calories. We can use a proportion.

Translate.

$$\begin{matrix}\text{calories burned}\rightarrow\\ \text{miles walked}\rightarrow\end{matrix}\ \frac{320}{4}=\frac{100}{m}\ \begin{matrix}\leftarrow\text{calories burned}\\ \leftarrow\text{miles walked}\end{matrix}$$

Carry out. We solve the proportion.

$$4m\cdot\frac{320}{4}=4m\cdot\frac{100}{m}$$
$$320m=400$$
$$m=\frac{5}{4}=1\frac{1}{4}$$

Check. $\frac{320}{4}=80$, $\frac{100}{5/4}=80$ The ratios are the same so the answer checks.

State. Walking $1\frac{1}{4}$ mi corresponds to burning 100 calories.

23. ***Familiarize.*** Let r = Alicia's speed in km/h. Then $r+20$ = Ryan's speed. We organize the information in a table using the formula time = distance/rate to fill in the last column.

	Distance	Speed	Time
Alicia	225	r	$\frac{225}{r}$
Ryan	325	$r+20$	$\frac{325}{r+20}$

Translate. Since the times must be the same, we have the equation $\frac{225}{r}=\frac{325}{r+20}$.

Carry out. We first multiply by the LCD, $r(r+20)$.

$$r(r+20)\cdot\frac{225}{r}=r(r+20)\cdot\frac{325}{r+20}$$
$$225(r+20)=325r$$
$$225r+4500=325r$$
$$4500=100r$$
$$45=r$$

Then $r+20=45+20=65$.

Check. Then, Ryan's speed is 20 km/h faster than Alicia's speed. At 45 km/h, Alicia travels 225 km in 225/45, or 5 hr. At 65 km/h, Ryan travels 325 km in 325/65, or 5 hr. The times are the same, so the answer checks.

State. Ryan's speed is 65 km/h and Alicia's speed is 45 km/h.

24. $1-\dfrac{1}{1-\dfrac{1}{1-\dfrac{1}{a}}}=1-\dfrac{1}{1-\dfrac{1}{\dfrac{a-1}{a}}}$

$=1-\dfrac{1}{1-\dfrac{a}{a-1}}$

$=1-\dfrac{1}{\dfrac{a-1-a}{a-1}}$

$=1-\dfrac{1}{\dfrac{-1}{a-1}}$

$=1+a-1$

$=a$

25. ***Familiarize.*** Let $x=$ the number. Then $-\frac{1}{x}$ is the opposite of the numbers reciprocal.

Translate. The square of the number, x^2 is equivalent to $-\frac{1}{x}$, so we write an equation.

$$x^2=-\frac{1}{x}$$

Carry out. We solve the equation.

$$x\cdot x^2=x\cdot\left(-\frac{1}{x}\right)$$
$$x^3=-1$$
$$x=\sqrt[3]{-1}=-1$$

Check. $(-1)^2=1$, $-\frac{1}{-1}=1$, so the values are equivalent.

State. The number is -1.

Chapter 7

Functions and Graphs

Exercise Set 7.1

1. A function is a special kind of correspondence between two sets.

3. For any function, the set of all inputs, or first values, is called the domain.

5. When a function is graphed, members of the domain are located on the horizontal axis.

7. The notation $f(3)$ is read "*f* of 3," "*f* at 3," or "the value of *f* at 3."

9. The correspondence is a function, because each member of the domain corresponds to exactly one member of the range.

11. The correspondence is a function, because each member of the domain corresponds to exactly one member of the range.

13. The correspondence is not a function because a member of the domain (Meryl Streep) corresponds to more than one member of the range.

15. The correspondence is a function, because each member of the domain corresponds to exactly one member of the range.

17. The correspondence is a function, because each USB flash drive has only one storage capacity.

19. This correspondence is a function, because each player has only one uniform number.

21. a. The domain is the set of all *x*-values of the set. It is $\{-3, -2, 0, 4\}$.
    b. The range is the set of all *y*-values of the set. It is $\{-10, 3, 5, 9\}$.
    c. The correspondence is a function, because each member of the domain corresponds to exactly one member of the range.

23. a. The domain is the set of all *x*-values of the set. It is $\{1, 2, 3, 4, 5\}$.
    b. The range is the set of all *y*-values of the set. It is $\{1\}$.
    c. The correspondence is a function, because each member of the domain corresponds to exactly one member of the range.

25. a. The domain is the set of all *x*-values of the set. It is $\{-2, 3, 4\}$.
    b. The range is the set of all *y*-values of the set. It is $\{-8, -2, 4, 5\}$.
    c. The correspondence is not a function, because a member of the domain (4) corresponds to more than one member of the range.

27. a. Locate 1 on the horizontal axis and then find the point on the graph for which 1 is the first coordinate. From that point, look to the vertical axis to find the corresponding y-coordinate, –2. Thus, $f(1) = -2$.
    b. To determine which member(s) of the domain are paired with 2, locate 2 on the vertical axis. From there look left and right to the graph to find any points for which 2 is the second coordinate. One such point exists. Its first coordinate is 4. Thus, the *x*-value for which $f(x) = 2$ is 4.

29. a. Locate 1 on the horizontal axis and then find the point on the graph for which 1 is the first coordinate. From that point, look to the vertical axis to find the corresponding *y*-coordinate, –2. Thus, $f(1) = -2$.
    b. To determine which member(s) of the domain are paired with 2, locate 2 on the vertical axis. From there look left and right to the graph to find any points for which 2 is the second coordinate. One such point exists. Its first coordinate is –2. Thus, the *x*-value for which $f(x) = 2$ is –2.

31. a. Locate 1 on the horizontal axis and then find the point on the graph for which 1 is the first coordinate. From that point, look to the vertical axis to find the corresponding *y*-coordinate, 3. Thus, $f(1) = 3$.
    b. To determine which member(s) of the domain are paired with 2, locate 2 on the vertical axis. From there look left and right to the graph to find any points for which 2 is the second coordinate. One such point exists. Its first coordinate is –3. Thus, the *x*-value for which $f(x) = 2$ is –3.

33. a. Locate 1 on the horizontal axis and then find the point on the graph for which 1 is the first coordinate. From that point, look to the vertical axis to find the corresponding *y*-coordinate, 3. Thus, $f(1) = 3$.
    b. To determine which member(s) of the domain are paired with 2, locate 2 on the vertical axis. From there look left and right to the graph to find any points for which 2 is the second coordinate. There are two such points, (–2, 2) and (0, 2). Thus, the *x*-values for which $f(x) = 2$ are –2 and 0.

35. a. Locate 1 on the horizontal axis and then find the point on the graph for which 1 is the first coordinate. From that point, look to the vertical axis to find the corresponding *y*-coordinate, 4. Thus, $f(1) = 4$.
    b. To determine which member(s) of the domain are paired with 2, locate 2 on the vertical axis. From there look left and right to the graph to find any

points for which 2 is the second coordinate. There are two such points, (–1, 2) and (3, 2). Thus, the x-values for which $f(x)=2$ are –1 and 3.

37. We can use the vertical-line test:

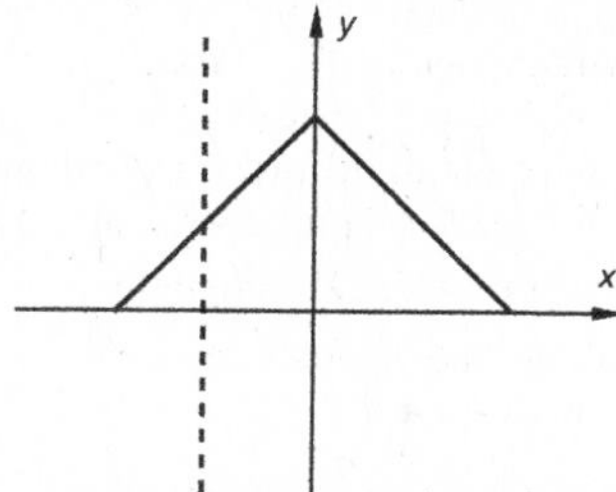

Visualize moving this vertical line across the graph. No vertical line will intersect the graph more than once. Thus, the graph is a graph of a function.

39. We can use the vertical-line test:

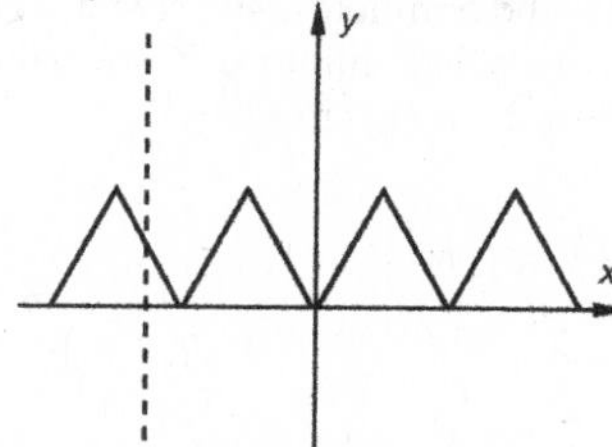

Visualize moving this vertical line across the graph. No vertical line will intersect the graph more than once. Thus, the graph is a graph of a function.

41. We can use the vertical line test.

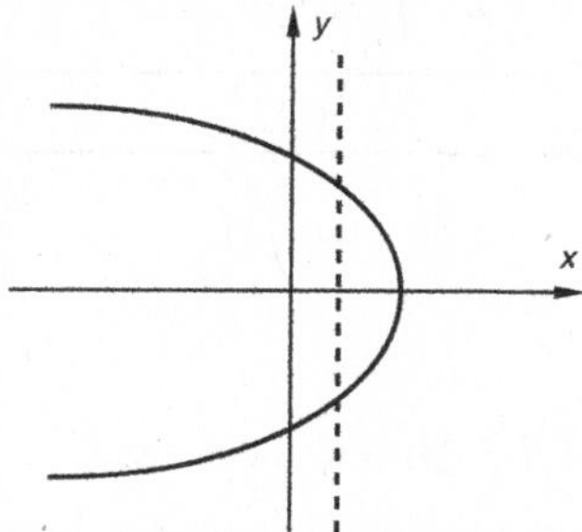

It is possible for a vertical line to intersect the graph more than once. Thus this is not the graph of a function.

43. $g(x)=2x+5$

a. $g(0)=2(0)+5=0+5=5$

b. $g(-4)=2(-4)+5=-8+5=-3$

c. $g(-7)=2(-7)+5=-14+5=-9$

d. $g(8)=2(8)+5=16+5=21$

e. $g(a+2)=2(a+2)+5=2a+4+5=2a+9$

f. $g(a)+2=2(a)+5+2=2a+7$

45. $f(n)=5n^2+4n$

a. $f(0)=5\cdot 0^2+4\cdot 0=0+0=0$

b. $f(-1)=5(-1)^2+4(-1)=5-4=1$

c. $f(3)=5\cdot 3^2+4\cdot 3=45+12=57$

d. $f(t)=5t^2+4t$

e. $f(2a)=5(2a)^2+4\cdot 2a=5\cdot 4a^2+8a=20a^2+8a$

f. $f(3)-9=5\cdot 3^2+4\cdot 3-9=5\cdot 9+4\cdot 3-9$
$=45+12-9=48$

47. $f(x)=\dfrac{x-3}{2x-5}$

a. $f(0)=\dfrac{0-3}{2\cdot 0-5}=\dfrac{-3}{0-5}=\dfrac{-3}{-5}=\dfrac{3}{5}$

b. $f(4)=\dfrac{4-3}{2\cdot 4-5}=\dfrac{1}{8-5}=\dfrac{1}{3}$

c. $f(-1)=\dfrac{-1-3}{2(-1)-5}=\dfrac{-4}{-2-5}=\dfrac{-4}{-7}=\dfrac{4}{7}$

d. $f(3)=\dfrac{3-3}{2\cdot 3-5}=\dfrac{0}{6-5}=\dfrac{0}{1}=0$

e. $f(x+2)=\dfrac{x+2-3}{2(x+2)-5}=\dfrac{x-1}{2x+4-5}=\dfrac{x-1}{2x-1}$

f. $f(a+h)=\dfrac{a+h-3}{2(a+h)-5}=\dfrac{a+h-3}{2a+2h-5}$

49. Given the input, find the output.
$f(x)=2x-5$
$f(8)=2(8)-5=16-5=11$

51. Given the output, find the input.
$f(x)=2x-5$
$-5=2x-5$
$0=2x$
$0=x$

53. Given the output, find the input.
$f(x)=\frac{1}{3}x+4$
$\frac{1}{2}=\frac{1}{3}x+4$
$-\frac{7}{2}=\frac{1}{3}x$
$-\frac{21}{2}=x$

55. Given the input, find the output.
$f(x)=\frac{1}{3}x+4$
$f\left(\frac{1}{2}\right)=\frac{1}{3}\left(\frac{1}{2}\right)+4=\frac{1}{6}+4=\frac{25}{6}$

57. Given the output 7, find the input.
$f(x)=4-x$
$7=4-x$
$3=-x$
$-3=x$

59. Given the output –3, find the input.
$f(x)=0.1x-0.5$
$-3=0.1x-0.5$
$-2.5=0.1x$
$-25=x$

61. $A(s)=s^2\dfrac{\sqrt{3}}{4}$
$A(4)=4^2\dfrac{\sqrt{3}}{4}=4\sqrt{3}\approx 6.93$
The area is $4\sqrt{3}\text{ cm}^2\approx 6.93\text{ cm}^2$.

63. $V(r) = 4\pi r^2$
$V(3) = 4\pi(3)^2 = 36\pi$
The area is $36\pi \text{ in}^2 \approx 113.10 \text{ in}^2$.

65. Locate the point that is directly above 225. Then estimate its second coordinate by moving horizontally from the point to the vertical axis. The rate is about 75 heart attacks per 10,000 men.

67. Locate the point that is directly to the right of 100. Then estimate its first coordinate by moving vertically from the point to the horizontal axis. The blood cholesterol is about 250 milligrams per deciliter.

69. *Writing Exercise.*

71. $$\begin{aligned} 2(x-5)-3 &= 4-(x-1) \\ 2x-10-3 &= 4-x+1 \\ 2x-13 &= 5-x \\ 3x-13 &= 5 \\ 3x &= 18 \\ x &= 6 \end{aligned}$$
The solution is 6.

73. $x^2 = 36$
$x = \pm 6$
The solutions are –6 or 6.

75. $\frac{1}{x} = x$
$1 = x^2$
$\pm 1 = x$
The solutions are –1 or 1.

77. $$\begin{aligned} \frac{1}{3}x + 2 &= \frac{5}{4} + 3x \\ \frac{3}{4} &= \frac{8}{3}x \\ \frac{9}{32} &= x \end{aligned}$$
The solution is $\frac{9}{32}$.

79. *Writing Exercise.*

81. To find $f(g(-4))$, we first find $g(-4)$:
$g(-4) = 2(-4)+5 = -8+5 = -3$.
Then
$f(g(-4)) = f(-3) = 3(-3)^2 - 1 = 3\cdot 9 - 1 = 27 - 1 = 26$.
To find $g(f(-4))$, we first find $f(-4)$:
$f(-4) = 3(-4)^2 - 1 = 3\cdot 16 - 1 = 48 - 1 = 47$.
Then $g(f(-4)) = g(47) = 2\cdot 47 + 5 = 94 + 5 = 99$.

83. $f(\text{tiger}) = \text{dog}$
$f(\text{dog}) = f(f(\text{tiger})) = \text{cat}$
$f(\text{cat}) = f(f(f(\text{tiger}))) = \text{fish}$
$f(\text{fish}) = f(f(f(f(\text{tiger})))) = \text{worm}$

85. a. Locate 1 on the horizontal axis and then find the point on the graph for which 1 is the first coordinate. From that point, look to the vertical axis to find the corresponding *y*-coordinate, 2. Thus, $f(1) = 2$.

b. Locate 2 on the horizontal axis and then find the point on the graph for which 2 is the first coordinate. From that point, look to the vertical axis to find the corresponding *y*-coordinate, 2. Thus, $f(2) = 2$.

c. To determine which member(s) of the domain are paired with 2, locate 2 on the vertical axis. From there look left and right to the graph to find any points for which 2 is the second coordinate. All points in the set $\{x \mid 0 < x \le 2\}$ satisfy this condition. These are the *x*-values for which $f(x) = 2$.

87. Locate the highest point on the graph. Then move horizontally to the vertical axis and read the corresponding pressure. It is about 22 mm.

89. The two largest contractions occurred at about 2 minutes, 50 seconds and 5 minutes, 40 seconds. The difference in these times, is 2 minutes, 50 seconds, so the frequency is about 1 every 3 minutes.

91.

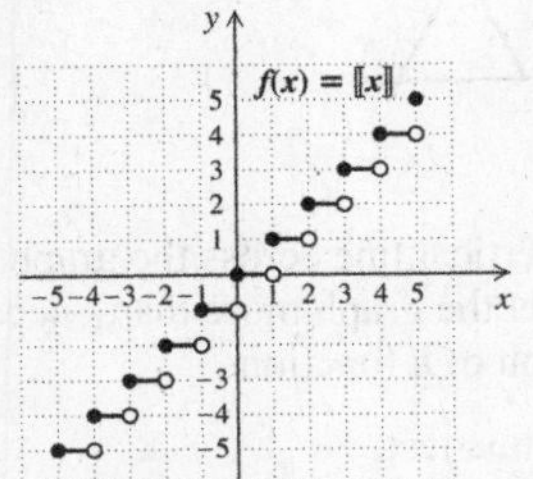

Exercise Set 7.2

1. Domain

3. Domain

5. Domain

7. Since $-1 \le 0 < 10$, function (c) would be used.

9. Since $10 \ge 10$, function (d) would be used.

11. Since $-1 \le -1 < 10$, function (c) would be used.

13. The function of *f* can be written as $\{(-4,-2),\ (-2,-1),\ (0,\ 0),\ (2,\ 1),\ (4,\ 2)\}$.
The domain is the set of all first coordinates, $\{-4, -2,\ 0,\ 2,\ 4\}$ and the range is the set of all second coordinates, $\{-2, -1,\ 0,\ 1,\ 2\}$.

15. The function of *f* can be written as $\{(-5,-1),\ (-3,-1),\ (-1,-1),\ (0,\ 1),\ (2,\ 1),\ (4,\ 1)\}$.
The domain is the set of all first coordinates, $\{-5, -3, -1,\ 0,\ 2,\ 4\}$ and the range is the set of all second coordinates, $\{-1,\ 1\}$.

17. The domain of the function is the set of all x-values that are in the graph, $\{x|-4 \le x \le 3\}$, or $[-4, 3]$.
The range is the set of all y-values that are in the graph, $\{y|-3 \le y \le 4\}$, or $[-3, 4]$.

19. The domain of the function is the set of all x-values that are in the graph, $\{x|-4 \le x \le 5\}$, or $[-4, 5]$.
The range is the set of all y-values that are in the graph, $\{y|-2 \le y \le 4\}$, or $[-2, 4]$.

21. The domain of the function is the set of all x-values that are in the graph, $\{x|-4 \le x \le 4\}$, or $[-4, 4]$.
The range is the set of all y-values that are in the graph, $\{-3, -1, 1\}$.

23. For any x-value and for any y-value there is a point on the graph. Thus,
Domain of $f = \{x|x \text{ is a real number}\}$, or $\mathbb{R}$ and
Range of $f = \{y|y \text{ is a real number}\}$, or $\mathbb{R}$.

25. For any x-value there is a point on the graph. Thus,
Domain of $f = \{x|x \text{ is a real number}\}$, or $\mathbb{R}$.
The only y-value on the graph is 4. Thus,
Range of $f = \{4\}$.

27. For any x-value there is a point on the graph. Thus,
Domain of $f = \{x|x \text{ is a real number}\}$, or $\mathbb{R}$.
The function has no y-values less than 1 and every y-value greater than or equal to 1 corresponds to a member of the domain. Thus,
Range of $f = \{y|y \ge 1\}$, or $[1, \infty)$.

29. The hole in the graph at (–2, –4) indicates that the function is not defined for $x = -2$. For any other x-value there is a point on the graph. Thus,
Domain of $f = \{x|x \text{ is a real number } and\ x \ne -2\}$.
There is no function value at (–2, –4), so –4 is not in the range of the function. For any other y-value there is a point on the graph. Thus,
Range of $f = \{y|y \text{ is a real number } and\ y \ne -4\}$.

31. The function has no x-values less than 0 and every x-value greater than or equal to 0 corresponds to a member of the domain. Thus,
Domain of $f = \{x|x \ge 0\}$, or $[0, \infty)$.
The function has no y-values less than 0 and every y-value greater than or equal to 0 corresponds to a member of the range. Thus,
Range of $f = \{y|y \ge 0\}$, or $[0, \infty)$.

33. $f(x) = \dfrac{5}{x-3}$

Since $\dfrac{5}{x-3}$ cannot be computed when the denominator is 0, we find the x-value that causes $x - 3$ to be 0:

$$x - 3 = 0$$
$$x = 3 \quad \text{Adding 3 to both sides}$$

Thus, 3 is not in the domain of f, while all other real numbers are. The domain of f is
$\{x|x \text{ is a real number } and\ x \ne 3\}$.

35. $f(x) = \dfrac{x}{2x-1}$

Since $\dfrac{x}{2x-1}$ cannot be computed when the denominator is 0, we find the x-value that causes $2x - 1$ to be 0:

$$2x - 1 = 0$$
$$2x = 1$$
$$x = \frac{1}{2}$$

Thus, $\frac{1}{2}$ is not in the domain of f, while all other real numbers are. The domain of f is
$\left\{x \middle| x \text{ is a real number } and\ x \ne \frac{1}{2}\right\}$.

37. $f(x) = 2x + 1$
Since we can compute $2x + 1$ for any real number x, the domain is the set of all real numbers.

39. $g(x) = |5 - x|$
Since we can compute $|5 - x|$ for any real number x, the domain is the set of all real numbers.

41. $f(x) = \dfrac{5}{x^2 - 9}$

The expression $\dfrac{5}{x^2-9}$ is undefined when $x^2 - 9 = 0$.

$$x^2 - 9 = 0$$
$$(x+3)(x-3) = 0$$
$$x + 3 = 0 \quad or \quad x - 3 = 0$$
$$x = -3 \quad or \quad x = 3$$

Thus, Domain of $f = \{x|x \text{ is a real number } and\ x \ne -3\ and\ x \ne 3\}$.

43. $f(x) = x^2 - 9$

Since we can compute $x^2 - 9$ for any real number x, the domain is the set of all real numbers.

45. $f(x) = \dfrac{2x-7}{x^2+8x+7}$

The expression $\dfrac{2x-7}{x^2+8x+7}$ is undefined when $x^2 + 8x + 7 = 0$.

$$x^2 + 8x + 7 = 0$$
$$(x+1)(x+7) = 0$$
$$x + 1 = 0 \quad or \quad x + 7 = 0$$
$$x = -1 \quad or \quad x = -7$$

Thus, Domain of $f = \{x|x \text{ is a real number } and\ x \ne -1\ and\ x \ne -7\}$.

47. $V(t) = 250 - 50t$
Time must be positive, so we have $t > 0$. In addition $V(t)$ must be nonnegative, so we have:

$$250 - 50t \geq 0$$
$$-50t \geq -250$$
$$t \leq 5$$

Then the domain of the function is $\{t | 0 \leq t \leq 5\}$ or [0, 5].

49. $A(p) = -2.5p + 26.5$
The price must be positive, so we have $p > \$0$. In addition $A(p)$ must be nonnegative, so we have:

$$-2.5p + 26.5 \geq 0$$
$$26.5 \geq 2.5p$$
$$10.6 \geq p$$

Then the domain of the function is $\{p | 0 \leq p \leq \$10.60\}$ or $[0, 10.60]$.

51. $P(d) = 0.03d + 1$
The depth must be positive, so we have $d \geq 0$. In addition $P(d)$ must be nonnegative, so we have:

$$0.03d + 1 \geq 0$$
$$0.03d \geq -1$$
$$d \geq -33.\overline{3}$$

Then we have $d \geq 0$ *and* $d \geq -33.\overline{3}$, so the domain of the function is $\{d | d \geq 0\}$ or $[0, \infty)$.

53. $h(t) = -16t^2 + 64t + 80$
The time cannot be negative, so we have $t \geq 0$. The height cannot be negative either, so an upper limit for t will be the positive value of t for which $h(t) = 0$.

$$-16t^2 + 64t + 80 = 0$$
$$-16(t^2 - 4t - 5) = 0$$
$$-16(t-5)(t+1) = 0$$

$$t - 5 = 0 \quad or \quad t + 1 = 0$$
$$t = 5 \quad or \quad t = -1$$

We know that -1 is not in the domain of the function. We also see that 5 is an upper limit for t. Then the domain of the function is $\{t | 0 \leq t \leq 5\}$ or $[0, 5]$.

55. $f(x) = \begin{cases} x, & \text{if } x < 0, \\ 2x + 1, & \text{if } x \geq 0 \end{cases}$

a. Since $-5 < 0$, we use $f(x) = x$. Thus, $f(-5) = -5$

b. Since $0 \geq 0$, we use $f(x) = 2x + 1$. Thus, $f(0) = 2(0) + 1 = 1$

c. Since $10 \geq 0$, we use $f(x) = 2x + 1$. Thus, $f(10) = 2(10) + 1 = 20 + 1 = 21$

57. $G(x) = \begin{cases} x - 5, & \text{if } x \leq -1, \\ x, & \text{if } x > -1 \end{cases}$

a. Since $-10 \leq -1$, we use $G(x) = x - 5$. Thus, $G(-10) = -10 - 5 = -15$

b. Since $0 > -1$, we use $G(x) = x$. Thus, $G(0) = 0$

c. Since $-1 \leq -1$, we use $G(x) = x - 5$. Thus, $G(-1) = -1 - 5 = -6$

59. $f(x) = \begin{cases} x^2 - 10, & \text{if } x < -10, \\ x^2, & \text{if } -10 \leq x \leq 10 \\ x^2 + 10, & \text{if } x > 10 \end{cases}$

a. Since $-10 \leq -10$, we use $f(x) = x^2$. Thus, $f(-10) = (-10)^2 = 100$

b. Since $10 \leq 10$, we use $f(x) = x^2$. Thus, $f(10) = (10)^2 = 100$

c. Since $11 > 10$, we use $f(x) = x^2 + 10$. Thus, $f(11) = (11)^2 + 10 = 131$

61. *Writing Exercise.*

63. $3 - 2(1-4)^2 \div 6 \cdot 2$
$= 3 - 2(-3)^2 \div 6 \cdot 2$
$= 3 - 2(9) \div 6 \cdot 2$
$= 3 - 18 \div 6 \cdot 2$
$= 3 - 3 \cdot 2$
$= 3 - 6$
$= -3$

65. $(2x^6y)^2 = 4x^{12}y^2$

67. *Writing Exercise.*

69. Answers may vary.
Domain: $\mathbb{R}$; range: $\mathbb{R}$

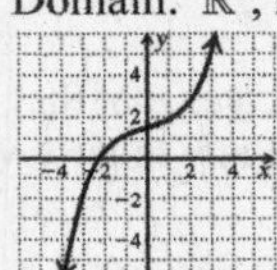

71. Answers may vary.
Domain: $\{x | 1 \leq x \leq 5\}$; range: $\{y | 0 \leq y \leq 2\}$

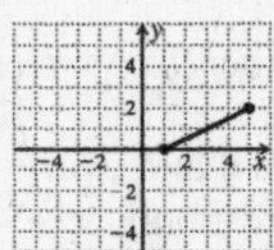

73. The graph indicates that the function is not defined for $x = 0$. For any other x-value there is a point on the graph. Thus,

Domain of $f = \{x | x$ is a real number *and* $x \neq 0\}$.

The graph also indicates that the function is not defined for $y = 0$. For any other y-value there is a point on the graph. Thus,

Range of $f = \{y | y$ is a real number *and* $y \neq 0\}$

75. The function has no x-values for $-2 \le x \le 0$. For any other x-value there is a point on the graph. Thus, the domain of the function is $\{x|x < -2 \text{ or } x > 0\}$.
The function has no y-values for $-2 \le y \le 3$. Every other y-value corresponds to a member of the range. Then the range is $\{y|y < -2 \text{ or } y > 3\}$.

77. From the graph below, we see that the domain of f is $\{x|x \text{ is a real number}\}$ and the range is $\{y|y \ge 0\}$.

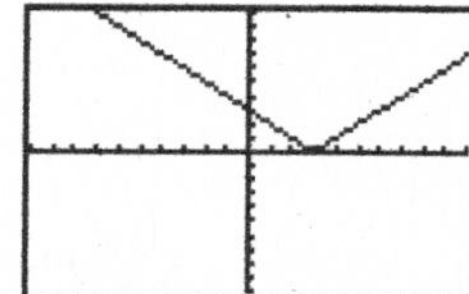

79. From the graph below, we see that the domain of f is $\{x|x \text{ is a real number and } x \ne 2\}$ and the range is $\{y|y \text{ is a real number and } y \ne 0\}$.

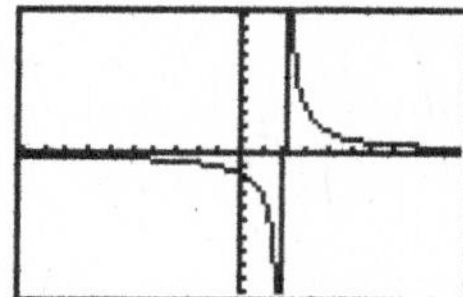

81. We graph the function $h(t) = -16t^2 + 64t + 80$ in the window [0, 5,–5, 150] with Xscl = 1 and Yscl = 15.

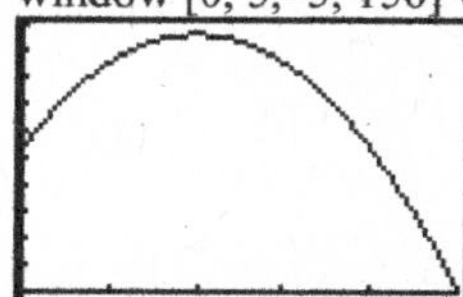

From the graph we estimate that the range of the function is $\{h|0 \le h \le 144\}$, or [0, 144].

83. From Exercise 55: $f(x) = \begin{cases} x, & \text{if } x < 0 \\ 2x+1, & \text{if } x \ge 0 \end{cases}$

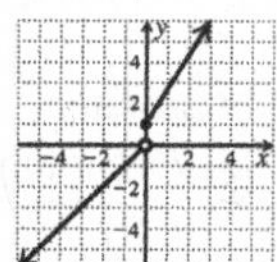

85. From Exercise 57: $G(x) = \begin{cases} x-5, & \text{if } x \le -1, \\ x, & \text{if } x > -1 \end{cases}$

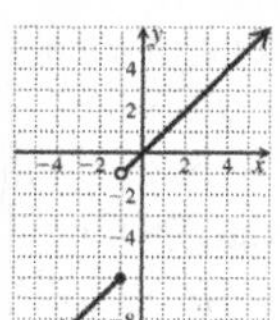

87. *Graphing Calculator Exercise*

Connecting the Concepts

1. Domain of $f: \{-2, 1, 3\}$

2. Range of $f: \{-6, 5\}$

3. The input of 1 corresponds to the output –6.
$f(1) = -6$

4. The output of –6 corresponds to the input 1.
$f(1) = -6$

5. The function is undefined when $x = 0$, so the domain of g is $\{x|x \text{ is a real number and } x \ne 0\}$.

6. $g(1) = \frac{1-1}{2(1)} = 0$

7. $g(4) = \frac{4-1}{2(4)} = \frac{3}{8}$

8. $\frac{1}{4} = \frac{2-1}{2(2)}$ The output of $\frac{1}{4}$ corresponds to the input 2.

Exercise Set 7.3

1. False; a graph of a vertical line cannot pass the vertical-line test.

3. False; unless restricted, the domain of a constant function is the set of all real numbers.

5. True

7. Graph $y = 2x - 1$.
The slope is 2 and the y-intercept is (0, –1). We plot (0,–1). Then, thinking of the slope as $\frac{2}{1}$, we start at (0, –1) and move up 2 units and right 1 unit to the point (1, 1). Alternatively, we can think of the slope as $\frac{-2}{-1}$ and, starting at (0,–1), move down 2 units and left 1 unit to (–1,–3). Using the points found we draw the graph.

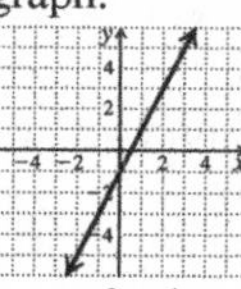

9. Graph $y = -\frac{2}{3}x + 3$.

The slope is $-\frac{2}{3}$ and the y-intercept is (0, 3). We plot (0, 3). Then, thinking of the slope as $\frac{-2}{3}$, we start at (0, 3) and move down 2 units and right 3 units to the point (3, 1). Alternatively, we can think of the slope as $\frac{2}{-3}$ and, starting at (0, 3), move up 2 units and left 3 units to (–3, 5). Using the points found we draw the graph.

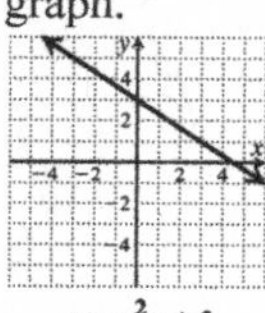

11. Graph $3y = 6 - 4x$.

First we write the equation in slope-intercept form.

$$3y = 6 - 4x$$
$$y = \frac{1}{3}(6 - 4x)$$
$$y = 2 - \frac{4}{3}x$$
$$y = -\frac{4}{3}x + 2$$

The slope is $-\frac{4}{3}$ and the y-intercept is (0, 2). We plot (0, 2). Then, thinking of the slope as $\frac{-4}{3}$, we start at (0, 2) and move down 4 units and right 3 units to the point (3, –2). Alternatively, we can think of the slope as $\frac{4}{-3}$ and, starting at (0, 2), move up 4 units and left 3 units to (–3, 6). Using the points found we draw the graph.

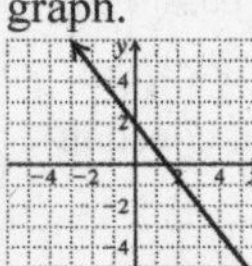

$3y = 6 - 4x$

13. Graph $x - y = 4$.

First we determine the intercepts.

Let $y = 0$ to find the x-intercept.

$$x - 0 = 4$$
$$x = 4$$

The x-intercept is (4, 0).

Let $x = 0$ to find the y-intercept.

$$0 - y = 4$$
$$-y = 4$$
$$y = -4$$

The y-intercept is (0, –4).

We plot (4, 0) and (0,–4) and draw the line.

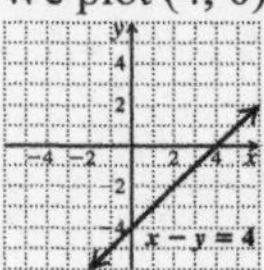

15. Graph $y = -2$.

For every input x, the value of y is –2. The graph is a horizontal line.

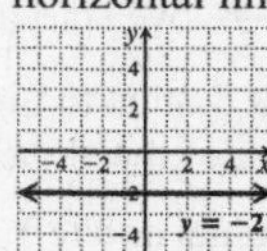

17. Graph $x = 4$.

The value of x is 4 for any value of y. The graph is a vertical line.

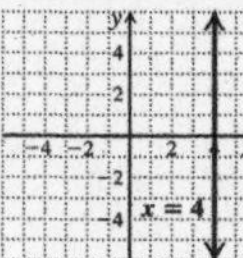

19. Graph $f(x) = x + 3$.

The slope is 1, or $\frac{1}{1}$ and the y-intercept is (0, 3). We plot (0, 3) and move up 1 unit and right 1 unit to the point (1, 4). After we have sketched the line, a third point can be calculated as a check.

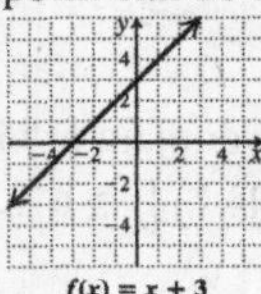

$f(x) = x + 3$

21. Graph $f(x) = \frac{3}{4}x + 1$.

The slope is $\frac{3}{4}$ and the y-intercept is (0, 1). We plot (0, 1) and move up 3 units and right 4 unit to the point (4, 4). After we have sketched the line, a third point can be calculated as a check.

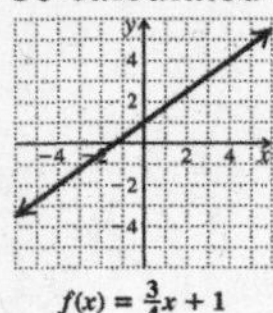

$f(x) = \frac{3}{4}x + 1$

23. Graph $g(x) = 4$.

For every input x, the output is 4. The graph is a horizontal line.

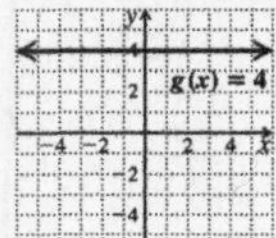

25. ***Familiarize***. After an initial fee of \$30, an additional fee of \$0.75 is charged each mile. After one mile, the total cost is \$30 + \$0.75 = \$30.75. After two miles, the total cost is $30 + 2 \cdot 0.75 = \$31.50$. We can generalize this with a model, letting $C(d)$ represent the total cost, in dollars, for d miles of driving.

Translate. We reword the problem and translate.

$$\underbrace{\text{Total Cost}} \quad \text{is} \quad \underbrace{\text{initial cost}} \quad \text{plus} \quad \underbrace{\text{\$0.75 per mile}}.$$

$$\downarrow \qquad \downarrow \qquad \downarrow \qquad \downarrow \qquad \downarrow$$

$$C(d) \quad = \quad 30 \quad + \quad 0.75d$$

where $d \geq 0$ since we cannot have negative miles driven. So the charge for a one-day rental of a truck driven d miles is $C(d) = 0.75d + 30$.

Carry out. To determine the number of miles required to reach a total cost of \$75, we substitute 75 for $C(d)$ and solve for d.

$C(d) = 0.75d + 30$

$75 = 0.75d + 30$ Substituting

$45 = 0.75d$ Subtracting 30 from both sides

$60 = d$ Dividing both sides by 0.75

Check. We evaluate

$$C(60) = 0.75(60) + 30 = 45 + 30 = 75.$$

State. The function is $C(d) = 0.75d + 30$ and 60 miles were driven for a total cost of \$75.

27. ***Familiarize.*** Lauren's hair starts at a length of 5 in. After one month, the total length is $5+\frac{1}{2}=5\frac{1}{2}$. After two months, the total length is $5+2\cdot\frac{1}{2}=6$. We can generalize this with a model, letting $L(t)$ represent the total length, in inches, after t months.
Translate. We reword the problem and translate.

$$\underbrace{\text{Total length}}\ \text{is}\ \underbrace{\text{initial length}}\ \text{plus}\ \underbrace{\tfrac{1}{2}\text{ in. per month}}.$$
$$\downarrow \quad \downarrow \quad \downarrow \quad \downarrow \quad \downarrow$$
$$L(t) \quad = \quad 5 \quad + \quad \frac{1}{2}t$$

where $t\geq 0$ since we cannot have negative months.
So the length of Lauren's hair t months after the haircut is $L(t)=\frac{1}{2}t+5$.
Carry out. To determine the number of months required to reach a total length of 15 in., we substitute 15 for $L(t)$ and solve for t.

$$L(t)=\frac{1}{2}t+5$$
$$15=\frac{1}{2}t+5 \quad \text{Substituting}$$
$$10=\frac{1}{2}t \quad \text{Subtracting 5 from both sides}$$
$$20=t \quad \text{Multiplying both sides by 2}$$

Check. We evaluate

$$L(20)=\frac{1}{2}(20)+5=10+5=15.$$

State. The function is $L(t)=\frac{1}{2}t+5$ and it will take 20 months after Lauren's haircut for her hair to be 15 in. long.

29. ***Familiarize.*** In 2010, there \$26.7 billion of sales. After one year, the total sales was \$26.7 + \$2.3 = \$29.0 (in billions). After two years, the total sales was $\$26.7+2\cdot\$2.3=\$31.3$ (in billions). We can generalize this with a model, letting $S(t)$ represent the total sales t years after 2010.
Translate. We reword the problem and translate.

$$\underbrace{\text{Total sales}}\ \text{is}\ \underbrace{\$26.7}\ \text{plus}\ \underbrace{\$2.3\text{ per year}}.$$
$$\downarrow \quad \downarrow \quad \downarrow \quad \downarrow \quad \downarrow$$
$$S(t) \quad = \quad 26.7 \quad + \quad 2.3t$$

where $t\geq 0$ since we cannot have negative years.
So the linear function for the number of acres after t years after 2010 is $S(t)=2.3t+26.7$.
Carry out. To determine the number of years required to reach a total sales of \$56.6 billion, we substitute 56.6 for $S(t)$ and solve for t.

$$S(t)=2.3t+26.7$$
$$56.6=2.3t+26.7$$
$$29.9=2.3t$$
$$13=t$$

Check. We evaluate

$$S(13)=2.3(13)+26.7=29.9+26.7=56.6.$$

State. The function is $S(t)=2.3t+26.7$ and 13 years after 2010, or in 2023 is when sales reach \$56.6 billion.

31. a. Let t represent the number of years after 2010, and form the pairs (0, 85) and (3, 87.1). First we find the slope of the function that fits the data:

$$m=\frac{87.1-85}{3-0}=\frac{2.1}{3}=0.7.$$

Use the slope-intercept equation with $m=0.7$ and $(0,\ 85)=(t_1,\ N_1)$.

$$N(t)=0.7t+85$$

b. In 2020, $t=2020-2010=10$

$$N(10)=0.7(10)+85=92\text{ million tons}$$

In 2020, the amount recycled will be approximately 92 million tons.

33. a. Let t represent the number of years since 2000, and form the pairs (0, 79.7) and (10, 81.1). First we find the slope of the function that fits the data:

$$m=\frac{81.1-79.7}{10-0}=\frac{1.4}{10}=0.14.$$

Use the point-slope equation with $m=0.14$ and $(0,\ 79.7)=(t_1,\ E_1)$.

$$E-79.7=0.14(t-0)$$
$$E-79.7=0.14t$$
$$E=0.14t+79.7$$
$$E(t)=0.14t+79.7$$

b. In 2020, $t=2020-2000=20$

$$E(20)=0.14(20)+79.7=82.5$$

The life expectancy of females in 2020 is 82.5 yrs.

35. a. Let t represent the number of years since 1650, and form the pairs (10, 671) and (35, 1137). First we find the slope of the function that fits the data:

$$m=\frac{1137-671}{35-10}=\frac{466}{25}.$$

Use the point-slope equation with $m=\frac{466}{25}$ and $(10,\ 671)=(t_1,\ S_1)$.

$$S-671=\frac{466}{25}(t-10)$$
$$S-671=\frac{466}{25}t-\frac{932}{5}$$
$$S=\frac{466}{25}t+\frac{2423}{5}$$
$$S(t)=\frac{466}{25}t+\frac{2423}{5}$$

b. In 1670, $t=1670-1650=20$

$$S(20)=\frac{466}{25}(20)+\frac{2423}{5}=857.4$$

The average displacement of a ship in 1670 was 857.4 tons.

37. a.

$$m=\frac{7-4}{200-100}=0.03$$
$$P-4=0.03(d-100)$$
$$P(d)=0.03d+1$$

b. $P(690)=0.03(690)+1=21.7$ atm

39. $f(x) = \frac{1}{3}x - 7$

The function is in the form $f(x) = mx + b$, so it is a linear function. We can compute $\frac{1}{3}x - 7$ for any value of x, so the domain is the set of all real numbers.

41. $p(x) = x^2 + x + 1$

The function is in the form $f(x) = ax^2 + bx + c$, $a \neq 0$, so it is a quadratic function. We can compute $x^2 + x + 1$ for any value of x, so the domain is the set of all real numbers.

43. $f(t) = \frac{12}{3t + 4}$

The function is described by a rational equation so it is a rational function. The expression $\frac{12}{3t+4}$ is undefined when $t = -\frac{4}{3}$, so the domain is

$\left\{t \middle| t \text{ is a real number } \textit{and } t \neq -\frac{4}{3}\right\}$

45. $f(x) = 0.02x^4 - 0.1x + 1.7$

The function is described by a polynomial equation that is neither linear nor quadratic, so it is a polynomial function. We can compute $0.02x^4 - 0.1x + 1.7$ for any value of x, so the domain is the set of all real numbers.

47. $f(x) = \frac{x}{2x - 5}$

The function is described by a rational equation, so it is a rational function. The expression $\frac{x}{2x-5}$ is undefined when $x = \frac{5}{2}$, so the domain is

$\left\{x \middle| x \text{ is a real number } \textit{and } x \neq \frac{5}{2}\right\}$.

49. $f(n) = \frac{4n - 7}{n^2 + 3n + 2}$

The function is described by a rational equation so it is a rational function. The expression $\frac{4n-7}{n^2+3n+2}$ is undefined for values of n that make the denominator 0. We find those values:

$$n^2 + 3n + 2 = 0$$
$$(n+1)(n+2) = 0$$
$$n + 1 = 0 \quad or \quad n + 2 = 0$$
$$n = -1 \quad or \quad n = -2$$

Then the domain is $\{n | n \text{ is a real number } \textit{and } n \neq -1 \textit{ and } n \neq -2\}$.

51. $f(n) = 200 - 0.1n$

The function can be written in the form $f(n) = mn + b$, so it is a linear function. We can compute $200 - 0.1n$ for any value of n, so the domain is the set of all real numbers.

53. The function has no y-values less than 0 and every y-value greater than or equal to 0 corresponds to a member of the domain. Thus, the range is $\{y | y \geq 0\}$ or $[0, \infty)$.

55. Every y-value corresponds to a member of the domain, so the range is the set of all real numbers.

57. The function has no y-values greater than 0 and every y-value less than or equal to 0 corresponds to a member of the domain. Thus, the range is $\{y | y \leq 0\}$ or $(-\infty, 0]$.

59. Graph $f(x) = x + 3$.

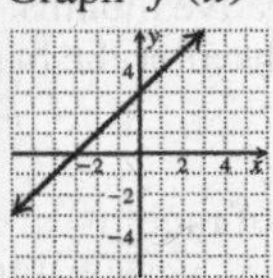

$f(x) = x + 3$

For any x-value and for any y-value there is a point on the graph. Thus,

Domain of $f = \{x | x \text{ is a real number}\}$, or $\mathbb{R}$ and

Range of $f = \{y | y \text{ is a real number}\}$, or $\mathbb{R}$.

61. Graph $f(x) = -1$.

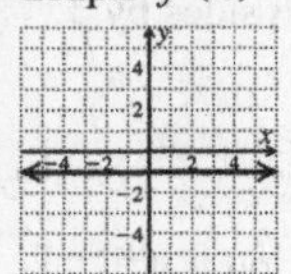

$f(x) = -1$

For any x-value there is a point on the graph, so Domain of $f = \{x | x \text{ is a real number}\}$, or $\mathbb{R}$. The only y-value on the graph is -1, so Range of $f = \{-1\}$.

63. Graph $f(x) = |x| + 1$.

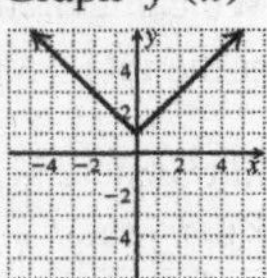

$f(x) = |x| + 1$

For any x-value there is a point on the graph, so Domain of $f = \{x | x \text{ is a real number}\}$, or $\mathbb{R}$. There is no y-value less than 1 and every y-value greater than or equal to 1 corresponds to a member of the domain. Thus,

Range of $f = \{y | y \geq 1\}$ or $[1, \infty)$.

65. Graph $g(x) = x^2$.

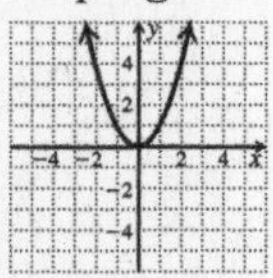

$g(x) = x^2$

For any x-value there is a point on the graph, so Domain of $g = \{x | x \text{ is a real number}\}$, or $\mathbb{R}$.

There is no y-value less than 0 and every y-value greater than or equal to 0 corresponds to a member of the domain. Thus, range of $g = \{y|y \geq 0\}$ or $[0, \infty)$.

67. *Writing Exercise.*

69. $4n^2 - 14n + 49 = (2n-7)^2$

71. $2x^3 - 14x^2 + 12x = 2x(x^2 - 7x + 6)$
$= 2x(x-1)(x-6)$

73. $d^3 - 64d = d(d^2 - 64) = d(d+8)(d-8)$

75. *Writing Exercise.*

77. Let $c = 1$ and $d = 2$. Then
$f(c+d) = f(1+2) = f(3) = 3m + b$, but
$f(c) + f(d) = (m+b) + (2m+b) = 3m + 2b$.
The given statement is false.

79. Let $k = 2$. Then $f(kx) = f(2x) = 2mx + b$, but
$kf(x) = 2(mx+b) = 2mx + 2b$.
The given statement is false.

81. ***Familiarize***. Celsius temperature C corresponding to a Fahrenheit temperature F can be modeled by a line that contains the points (32, 0) and (212, 100).
Translate. We find an equation relating C and F.

$$m = \frac{100-0}{212-32} = \frac{100}{180} = \frac{5}{9}$$
$$C - 0 = \frac{5}{9}(F - 32)$$
$$C = \frac{5}{9}(F-32)$$

Carry out. Using function notation we have $C(F) = \frac{5}{9}(F-32)$. Now we find $C(70)$:

$$C(70) = \frac{5}{9}(70-32) = \frac{5}{9}(38) \approx 21.1.$$

Check. We can repeat the calculations. We could also graph the function and determine that (70, 21.1) is on the graph.
State. A temperature of about 21.1°C corresponds to a temperature of 70°F.

83. ***Familiarize***. The total cost C of the phone, in dollars, after t months, can be modeled by a line that contains the points (5, 410) and (9, 690).
Translate. We find an equation relating C and t.

$$m = \frac{690-410}{9-5} = \frac{280}{4} = 70$$
$$C - 410 = 70(t-5)$$
$$C - 410 = 70t - 350$$
$$C = 70t + 60$$

Carry out. Using function notation we have $C(t) = 70t + 60$. To find the costs already incurred when the service began we find $C(0)$:

$$C(0) = 70 \cdot 0 + 60 = 60$$

Check. We can repeat the calculations. We could also graph the function and determine that (0, 60) is on the graph.
State. Tam had already incurred $60 in costs when the service just began.

85. a. We have two pairs, (3, –5) and (7, –1). Use the point-slope form:

$$m = \frac{-1-(-5)}{7-3} = \frac{-1+5}{4} = \frac{4}{4} = 1$$
$$y - (-5) = 1(x-3)$$
$$y + 5 = x - 3$$
$$y = x - 8$$
$$g(x) = x - 8 \quad \text{Using function notation}$$

b. $g(-2) = -2 - 8 = -10$

c. $g(a) = a - 8$
If $g(a) = 75$, we have
$a - 8 = 75$
$a = 83$.

Mid-Chapter Review

1. $f(x) = 2x^2 - 3x$
$f(-5) = 2(-5)^2 - 3(-5)$
$= 2(25) - (-15)$
$= 50 + 15$
$= 65$

2. $x - 7 = 0$
$x = 7$
Domain $= \{x|x \text{ is a real number } and\ x \neq 7\}$

3. Domain of $f: \{-1, 0, 3, 4\}$

4. Range of $f: \{-2, 0, 6, 8\}$

5. The input of –1 corresponds to the output –2.
$f(-1) = -2$

6. For any x-value there is a point on the graph, so Domain of $g = \{x|x \text{ is a real number}\}$, or $\mathbb{R}$.

7. The function is undefined when $x = 0$, so the domain of h is $\{x|x \text{ is a real number } and\ x \neq 0\}$.

8. $h(x) = \frac{2}{x}$
$h(10) = \frac{2}{10} = \frac{1}{5}$.

9. The function is in the form $f(x) = mx + b$, so it is a linear function.

10. Rational function

11. Range of $g = \{y|y \text{ is a real number}\}$, or $\mathbb{R}$.

12. $x^2 + 6x - 40 = 0$
$(x+10)(x-4) = 0$
$x = -10 \quad x = 4$
Domain $= \{x|x \text{ is a real number } and\ x \neq -10\ and\ x \neq 4\}$

13. Yes; it is a function. It passes the vertical-line test.

14. The input of 2 corresponds to the output –4. $F(2)=-4$

15. The output of 1 corresponds to the input of –1 and $4\frac{1}{2}$.

16. $\{x|x\geq -3\}$, or $[-3,\infty)$.

17. The function has no y-values less than –4 and every y-value greater than or equal to –4 corresponds to a member of the domain. Thus, the range is $\{y|y\geq -4\}$, or $[-4,\infty)$.

18. $G(3)=3+1=4$

19. Since $1=1$, $G(1)=10$

20. $G(-12)=1-(-12)=13$

Exercise Set 7.4

1. If f and g are functions then $(f+g)(x)$ is the *sum* of the functions.

3. One way to compute $(f-g)(2)$ is to simplify $f(x)-g(x)$ and then *evaluate* the result for $x=2$.

5. The domain of f/g is the set of all values common to the domains of f and g, *excluding* any values for which $g(x)$ is 0.

7. Since $f(3)=-2\cdot 3+3=-3$ and $g(3)=3^2-5=4$, we have $f(3)+g(3)=-3+4=1$.

9. Since $f(1)=-2\cdot 1+3=1$ and $g(1)=1^2-5=-4$, we have $f(1)-g(1)=1-(-4)=5$.

11. Since $f(-2)=-2\cdot(-2)+3=7$ and $g(-2)=(-2)^2-5=-1$ we have $f(-2)\cdot g(-2)=7(-1)=-7$.

13. Since $f(-4)=-2\cdot(-4)+3=11$ and $g(-4)=(-4)^2-5=11$, we have $\dfrac{f(-4)}{g(-4)}=\dfrac{11}{11}=1$.

15. Since $g(1)=1^2-5=-4$ and $f(1)=-2\cdot 1+3=1$, we have $g(1)-f(1)=-4-1=-5$.

17. $$\begin{aligned}(f+g)(x)&=f(x)+g(x)=(-2x+3)+(x^2-5)\\&=x^2-2x-2\end{aligned}$$

19. $$\begin{aligned}(g-f)(x)&=g(x)-f(x)=(x^2-5)-(-2x+3)\\&=x^2+2x-8\end{aligned}$$

21. $$\begin{aligned}(F+G)(x)&=F(x)+G(x)\\&=x^2-2+5-x\\&=x^2-x+3\end{aligned}$$

23. $$\begin{aligned}(F-G)(x)&=F(x)-G(x)\\&=x^2-2-(5-x)\\&=x^2-2-5+x\\&=x^2+x-7\end{aligned}$$

Then we have

$$\begin{aligned}(F-G)(3)&=3^2+3-7\\&=9+3-7\\&=5.\end{aligned}$$

25. $$\begin{aligned}(F\cdot G)(a)&=F(a)\cdot G(a)\\&=(a^2-2)(5-a)\\&=5a^2-a^3-10+2a,\text{ or }-a^3+5a^2+2a-10\end{aligned}$$

27. $$\begin{aligned}(F/G)(x)&=F(x)/G(x)\\&=\frac{x^2-2}{5-x},\ x\neq 5\end{aligned}$$

29. $$(G/F)(-2)=\frac{5-(-2)}{(-2)^2-2}=\frac{5+2}{4-2}=\frac{7}{2}$$

31. $$\begin{aligned}(F+F)(x)&=F(x)+F(x)\\&=(x^2-2)+(x^2-2)\\&=2x^2-4\\(F+F)(1)&=2(1)^2-4\\&=-2\end{aligned}$$

33. $$(r\cdot t)(x)=r(x)\cdot t(x)=\frac{5}{x^2}\cdot\frac{3}{2x}=\frac{15}{2x^3}$$

35. $$\begin{aligned}(r-t)(x)&=r(x)-t(x)=\frac{5}{x^2}-\frac{3}{2x}\\&=\frac{5}{x^2}\cdot\frac{2}{2}-\frac{3}{2x}\cdot\frac{x}{x}\\&=\frac{10}{2x^2}-\frac{3x}{2x^2}\\&=\frac{10-3x}{2x^2}\end{aligned}$$

37. $$(t/r)(x)=t(x)\div r(x)=\frac{3}{2x}\div\frac{5}{x^2}=\frac{3}{2x}\cdot\frac{x^2}{5}=\frac{3x}{10}$$

39. $$\begin{aligned}&(g-f)(x)\\&=g(x)-f(x)=\frac{x+2}{x^2-9}-\frac{x-1}{x^2-x-6}\\&=\frac{x+2}{(x+3)(x-3)}-\frac{x-1}{(x+2)(x-3)}\\&=\frac{x+2}{(x+3)(x-3)}\cdot\frac{x+2}{x+2}-\frac{x-1}{(x+2)(x-3)}\cdot\frac{x+3}{x+3}\\&=\frac{x^2+4x+4-(x^2+2x-3)}{(x+3)(x+2)(x-3)}\\&=\frac{2x+7}{(x+3)(x+2)(x-3)}\end{aligned}$$

41. $$\begin{aligned}&(f+g)(x)\\&=f(x)+g(x)=\frac{x-1}{x^2-x-6}+\frac{x+2}{x^2-9}\\&=\frac{x-1}{(x+2)(x-3)}+\frac{x+2}{(x+3)(x-3)}\\&=\frac{x-1}{(x+2)(x-3)}\cdot\frac{x+3}{x+3}+\frac{x+2}{(x+3)(x-3)}\cdot\frac{x+2}{x+2}\\&=\frac{x^2+2x-3+x^2+4x+4}{(x+3)(x+2)(x-3)}\\&=\frac{2x^2+6x+1}{(x+3)(x+2)(x-3)}\end{aligned}$$

43. $(f/g)(x)$
$$= f(x) \div g(x) = \frac{x-1}{x^2-x-6} \div \frac{x+2}{x^2-9}$$
$$= \frac{x-1}{(x+2)(x-3)} \cdot \frac{(x+3)(x-3)}{x+2}$$
$$= \frac{(x-1)(x+3)(x-3)}{(x+2)^2(x-3)}$$
$$= \frac{(x-1)(x+3)}{(x+2)^2}$$

45. $N(2015) = (C+B)(2015) = C(2015) + B(2015)$
$\approx 1.3 + 2.7 = 4.0$ million
We estimate the number of births in 2015 to be 4.0 million.

47. $(B-C)(2015) = B(2015) - C(2015)$
$\approx 2.7 - 1.3 = 1.4$ million
This number estimates how many more non-Caesarean section births than Caesarean section births there were in 2015.

49. $(p+r)('09) = p('09) + r('09)$
$\approx 25 + 60 = 85$ million
This represents the number of tons of municipal solid waste that was composted or recycled in 2009.

51. $F('00) \approx 240$ million
This represents the number of tons of municipal solid waste in 2000.

53. $(F-p)('08) = F('08) - p('08)$
$\approx 240 - 20 = 220$ million
This represents the number of tons of municipal solid waste that was not composted in 2008.

55. The domain of f and of g is all real numbers. Thus,
Domain of $f+g$ = Domain of $f-g$ = Domain of $f \cdot g$
$= \{x \mid x \text{ is a real number}\}$.

57. Because division by 0 is undefined, we have
Domain of $f = \{x \mid x \text{ is a real number } and\ x \neq -5\}$,
and Domain of $g = \{x \mid x \text{ is a real number}\}$.
Thus,
Domain of $f+g$ = Domain of $f-g$ = Domain of $f \cdot g$
$= \{x \mid x \text{ is a real number } and\ x \neq -5\}$.

59. Because division by 0 is undefined, we have
Domain of $f = \{x \mid x \text{ is a real number } and\ x \neq 0\}$,
and Domain of $g = \{x \mid x \text{ is a real number}\}$.
Thus, Domain of
$f+g$ = Domain of $f-g$ = Domain of $f \cdot g$
$= \{x \mid x \text{ is a real number } and\ x \neq 0\}$.

61. Because division by 0 is undefined, we have
Domain of $f = \{x \mid x \text{ is a real number } and\ x \neq 1\}$,
and Domain of $g = \{x \mid x \text{ is a real number}\}$.
Thus,
Domain of $f+g$ = Domain of $f-g$ = Domain of $f \cdot g$
$= \{x \mid x \text{ is a real number } and\ x \neq 1\}$.

63. Because division by 0 is undefined,.we have
Domain of $f = \left\{x \middle| x \text{ is a real number } and\ x \neq -\frac{9}{2}\right\}$,
and Domain of $g = \{x \mid x \text{ is a real number } and\ x \neq 1\}$.
Thus, Domain of
$f+g$ = Domain of $f-g$ = Domain of $f \cdot g$
$= \left\{x \middle| x \text{ is a real number } and\ x \neq -\frac{9}{2} \text{ and } x \neq 1\right\}$.

65. Domain of
f = Domain of $g = \{x \mid x \text{ is a real number}\}$.
Since $g(x) = 0$ when $x - 3 = 0$, we have $g(x) = 0$ when $x = 3$. We conclude that
Domain of $f/g = \{x \mid x \text{ is a real number } and\ x \neq 3\}$.

67. Domain of
f = Domain of $g = \{x \mid x \text{ is a real number}\}$.
Since $g(x) = 0$ when $2x + 8 = 0$, we have $g(x) = 0$ when $x = -4$. We conclude that
Domain of $f/g = \{x \mid x \text{ is a real number } and\ x \neq -4\}$.

69. Domain of $f = \{x \mid x \text{ is a real number and } x \neq 4\}$.
Domain of $g = \{x \mid x \text{ is a real number}\}$.
Since $g(x) = 0$ when $5 - x = 0$, we have $g(x) = 0$ when $x = 5$. We conclude that Domain of
$f/g = \{x \mid x \text{ is a real number and } x \neq 4 \text{ and } x \neq 5\}$.

71. Domain of $f = \{x \mid x \text{ is a real number and } x \neq -1\}$.
Domain of $g = \{x \mid x \text{ is a real number}\}$.
Since $g(x) = 0$ when $2x + 5 = 0$, we have $g(x) = 0$ when $x = -\frac{5}{2}$. We conclude that Domain of f/g
$= \left\{x \middle| x \text{ is a real number and } x \neq -1 \text{ and } x \neq -\frac{5}{2}\right\}$.

73. $(F+G)(5) = F(5) + G(5) = 1 + 3 = 4$
$(F+G)(7) = F(7) + G(7) = -1 + 4 = 3$

75. $(G-F)(7) = G(7) - F(7) = 4 - (-1) = 4 + 1 = 5$
$(G-F)(3) = G(3) - F(3) = 1 - 2 = -1$

77. From the graph we see that Domain of
$F = \{x \mid 0 \le x \le 9\}$ and Domain of $G = \{x \mid 3 \le x \le 10\}$.
Then
Domain of $F+G = \{x \mid 3 \le x \le 9\}$. Since $G(x)$ is never 0,
Domain of $F/G = \{x \mid 3 \le x \le 9\}$.

79. We use $(F+G)(x) = F(x) + G(x)$.

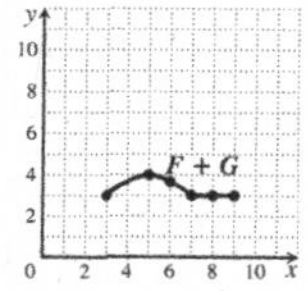

81. *Writing Exercise.*

83. ***Familiarize***. Let x = the second angle of a triangle. Then the first angle is $2x$. The third angle is $3x$.
Translate.

First angle	plus	second angle	plus	third angle	is	180.
↓	↓	↓	↓	↓	↓	↓
$2x$	$+$	x	$+$	$3x$	$=$	180

Carry out. Solve the equation.

$$2x + x + 3x = 180$$
$$6x = 180$$
$$x = 30$$

Then $2x = 2(30^\circ) = 60^\circ$
and $3x = 3(30^\circ) = 90^\circ$
Check. If the angles are 60°, 30° and 90°, the sum of the three angles is 60° + 30° + 90° = 180°. The answer checks.
State. The angles of the triangle are 60°, 30°, and 90°.

85. To find the weight of each molecule, divide the weight of a mole of water by the number of molecules.

$$\frac{18.015 \text{ g}}{6.022 \times 10^{23}} = 2.992 \times 10^{-23} \text{ g}.$$

Therefore, each molecule weighs 2.992×10^{-23} g.

87. *Writing Exercise.*

89. Domain of $F = \{x \mid x \text{ is a real number and } x \neq 4\}$.
Domain of $G = \{x \mid x \text{ is a real number and } x \neq 3\}$.
$G(x) = 0$ when $x^2 - 4 = 0$, or when $x = 2$ or $x = -2$. Then Domain of $F/G = \{x \mid x \text{ is a real number and } x \neq 4 \text{ and } x \neq 3 \text{ and } x \neq 2 \text{ and } x \neq -2\}$.

91. Answers may vary.

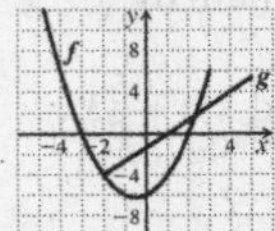

93. The problem states that Domain of $m = \{x \mid -1 < x < 5\}$. Since $n(x) = 0$ when $2x - 3 = 0$, we have $n(x) = 0$ when $x = \frac{3}{2}$. We conclude that Domain of m/n
$= \left\{x \middle| x \text{ is a real number and } -1 < x < 5 \text{ and } x \neq \frac{3}{2}\right\}$.

95. Answers may vary. $f(x) = \dfrac{1}{x+2}$, $g(x) = \dfrac{1}{x-5}$

97. *Graphing Calculator Exercise*

Exercise Set 7.5

1. (d) LCM

3. (e) Product

5. (a) Directly

7. As the number of painters increases, the time required to scrape the house decreases, so we have inverse variation.

9. As the number of laps increases, the time required to swim them increases, so we have direct variation.

11. As the number of volunteers increases, the time required to wrap the toys decreases, so we have inverse variation.

13.
$$f = \frac{L}{d}$$
$$df = L \quad \text{Multiplying by } d$$
$$d = \frac{L}{f} \quad \text{Dividing by } f$$

15.
$$s = \frac{(v_1 + v_2)t}{2}$$
$$2s = (v_1 + v_2)t \quad \text{Multiplying by 2}$$
$$\frac{2s}{t} = v_1 + v_2 \quad \text{Dividing by } t$$
$$\frac{2s}{t} - v_2 = v_1$$

This result can also be expressed as $v_1 = \dfrac{2s - tv_2}{t}$.

17.
$$\frac{t}{a} + \frac{t}{b} = 1$$
$$ab\left(\frac{t}{a} + \frac{t}{b}\right) = ab \cdot 1 \quad \text{Multiplying by the LCD}$$
$$ab \cdot \frac{t}{a} + ab \cdot \frac{t}{b} = ab$$
$$bt + at = ab$$
$$at = ab - bt$$
$$at = b(a - t) \quad \text{Factoring}$$
$$\frac{at}{a - t} = b$$

19.
$$R = \frac{gs}{g + s}$$
$$(g + s) \cdot R = (g + s) \cdot \frac{gs}{g + s} \quad \text{Multiplying by the LCD}$$
$$Rg + Rs = gs$$
$$Rs = gs - Rg$$
$$Rs = g(s - R) \quad \text{Factoring out } g$$
$$\frac{Rs}{s - R} = g \quad \text{Multiplying by } \frac{1}{s - R}$$

21.
$$I = \frac{nE}{R + nr}$$
$$I(R + nr) = \frac{nE}{R + nr} \cdot (R + nr) \quad \text{Multiplying by the LCD}$$
$$IR + Inr = nE$$
$$IR = nE - Inr$$
$$IR = n(E - Ir)$$
$$\frac{IR}{E - Ir} = n$$

23. $\frac{1}{p}+\frac{1}{q}=\frac{1}{f}$

$pqf\left(\frac{1}{p}+\frac{1}{q}\right)=pqf\cdot\frac{1}{f}$ Multiplying by the LCD

$qf+pf=pq$

$pf=pq-qf$

$pf=q(p-f)$

$\frac{pf}{p-f}=q$

25. $S=\frac{H}{m(t_1-t_2)}$

$(t_1-t_2)S=\frac{H}{m}$ Multiplying by t_1-t_2

$t_1-t_2=\frac{H}{Sm}$ Dividing by S

$t_1=\frac{H}{Sm}+t_2$, or $\frac{H+Smt_2}{Sm}$

27. $\frac{E}{e}=\frac{R+r}{r}$

$er\cdot\frac{E}{e}=er\cdot\frac{R+r}{r}$ Multiplying by the LCD

$Er=e(R+r)$

$Er=eR+er$

$Er-er=eR$

$r(E-e)=eR$

$r=\frac{eR}{E-e}$

29. $S=\frac{a}{1-r}$

$(1-r)S=a$ Multiplying by the LCD, $1-r$

$1-r=\frac{a}{S}$ Dividing by S

$1-\frac{a}{S}=r$ Adding r and $-\frac{a}{S}$

This result can also be expressed as $r=\frac{S-a}{S}$.

31. $c=\frac{f}{(a+b)c}$

$\frac{a+b}{c}\cdot c=\frac{a+b}{c}\cdot\frac{f}{(a+b)c}$

$a+b=\frac{f}{c^2}$

33. $P=\frac{A}{1+r}$

$P(1+r)=\frac{A}{1+r}\cdot(1+r)$

$P(1+r)=A$

$1+r=\frac{A}{P}$

$r=\frac{A}{P}-1$, or $\frac{A-P}{P}$

35. $v=\frac{d_2-d_1}{t_2-t_1}$

$(t_2-t_1)v=(t_2-t_1)\cdot\frac{d_2-d_1}{t_2-t_1}$

$(t_2-t_1)v=d_2-d_1$

$t_2-t_1=\frac{d_2-d_1}{v}$

$-t_1=-t_2+\frac{d_2-d_1}{v}$

$t_1=t_2-\frac{d_2-d_1}{v}$, or $\frac{t_2v-d_2+d_1}{v}$

37. $\frac{1}{t}=\frac{1}{a}+\frac{1}{b}$

$tab\cdot\frac{1}{t}=tab\left(\frac{1}{a}+\frac{1}{b}\right)$

$ab=tb+ta$

$ab=t(b+a)$

$\frac{ab}{b+a}=t$

39. $A=\frac{2Tt+Qq}{2T+Q}$

$(2T+Q)\cdot A=(2T+Q)\cdot\frac{2Tt+Qq}{2T+Q}$

$2AT+AQ=2Tt+Qq$

$AQ-Qq=2Tt-2AT$ Adding $-2AT$ and $-Qq$

$Q(A-q)=2Tt-2AT$

$Q=\frac{2Tt-2AT}{A-q}$

41. $p=\frac{-98.42+4.15c-0.082w}{w}$

$pw=-98.42+4.15c-0.082w$

$pw+0.082w=-98.42+4.15c$

$w(p+0.082)=-98.42+4.15c$

$w=\frac{-98.42+4.15c}{p+0.082}$

43. $y=kx$

$30=k\cdot 5$ Substituting

$6=k$

The variation constant is 6.

The equation of variation is $y=6x$.

45. $y=kx$

$3.4=k\cdot 2$ Substituting

$1.7=k$

The variation constant is 1.7.

The equation of variation is $y=1.7x$.

47. $y=kx$

$2=k\cdot\frac{1}{5}$ Substituting

$10=k$ Multiplying by 5

The variation constant is 10.

The equation of variation is $y=10x$.

49. $y=\frac{k}{x}$

$5=\frac{k}{20}$ Substituting

$100=k$

The variation constant is 100.

The equation of variation is $y=\frac{100}{x}$.

51. $y=\frac{k}{x}$

$11=\frac{k}{4}$ Substituting

$44=k$

The variation constant is 44.

The equation of variation is $y=\frac{44}{x}$.

53. $y = \frac{k}{x}$

$27 = \frac{k}{\frac{1}{3}}$ Substituting

$9 = k$

The variation constant is 9.

The equation of variation is $y = \frac{9}{x}$.

55. ***Familiarize***. Because of the phrase "*d* ... varies directly as ... *m*," we express the distance as a function of the mass. Thus we have $d(m) = km$. We know that $d(3) = 20$.

Translate. We find the variation constant and then find the equation of variation.

$d(m) = km$

$d(3) = k \cdot 3$ Replacing *m* with 3

$20 = k \cdot 3$ Replacing $d(3)$ with 20

$\frac{20}{3} = k$ Variation constant

The equation of variation is $d(m) = \frac{20}{3}m$.

Carry out. We compute $d(5)$.

$d(m) = \frac{20}{3}m$

$d(5) = \frac{20}{3} \cdot 5$ Replacing *m* with 5

$= \frac{100}{3}$, or $33\frac{1}{3}$

Check. Reexamine the calculations.

State. The distance is $33\frac{1}{3}$ cm.

57. ***Familiarize***. Because *T* varies inversely as *P*, we write $T(P) = k / P$. We know that $T(7) = 5$.

Translate. We find the variation constant and the equation of variation.

$T(P) = \frac{k}{P}$

$T(7) = \frac{k}{7}$ Replacing *P* with 7

$5 = \frac{k}{7}$ Replacing $T(P)$ with 5

$35 = k$ Variation constant

$T(P) = \frac{35}{P}$ Equation of variation

Carry out. We find $T(10)$.

$T(10) = \frac{35}{10} = 3.5$

Check. Reexamine the calculations.

State. It would take 3.5 hr for 10 volunteers to complete the job.

59. ***Familiarize***. Because cost *C* varies directly as people fed *P*, we write $C(P) = kP$. We know that $C(7) = 15.75$.

Translate. We find the variation constant and the equation of variation.

$C(P) = kP$

$C(7) = k \cdot 7$ Replacing *P* with 7

$15.75 = k \cdot 7$ Replacing $C(7)$ with 15.75

$2.25 = k$ Variation constant

$C(P) = 2.25P$ Equation of variation

Carry out. We find *P* when $C(P) = 27.00$.

$27 = 2.25P$

$12 = P$

Check. Reexamine the calculations.

State. 12 people could be fed with a gift of \$27.00.

61. ***Familiarize***. Because the amount of salt *A*, in tons, varies directly as the number of storms *n*, we write $A(n) = kn$. We know that $A(8) = 1200$.

Translate. We find the variation constant and the equation of variation.

$A(n) = kn$

$W(8) = k \cdot 8$ Replacing *n* with 8

$1200 = k \cdot 8$ Replacing $W(8)$ with 1200

$150 = k$ Variation constant

$A(n) = 150n$ Equation of variation

Carry out. We find $A(4)$.

$A(4) = 150 \cdot 4 = 600$

Check. Reexamine the calculations.

State. For 4 storms, they would need 600 tons of salt.

63. ***Familiarize***. Because the frequency, *f* varies inversely as length *L*, we write $f(L) = k / L$. We know that $f(33) = 260$.

Translate. We find the variation constant and the equation of variation.

$f(L) = \frac{k}{L}$

$f(33) = \frac{k}{33}$ Replacing *L* with 33

$260 = \frac{k}{33}$ Replacing $f(33)$ with 260

$8580 = k$ Variation constant

$f(L) = \frac{8580}{L}$ Equation of variation

Carry out. We find $f(30)$.

$f(30) = \frac{8580}{30} = 286$

Check. Reexamine the calculations.

State. If the string was shortened to 30 cm the new frequency would be 286 Hz.

65. ***Familiarize***. Because of the phrase "*t* varies inversely as ...*u*," we write $t(u) = k / u$. We know that $t(4) = 75$.

Translate. We find the variation constant and then we find the equation of variation.

$t(u) = \frac{k}{u}$

$t(4) = \frac{k}{4}$ Replacing *u* with 4

$75 = \frac{k}{4}$ Replacing $t(4)$ with75

$300 = k$ Variation constant

$t(u) = \frac{300}{u}$ Equation of variation

Carry out. We find $t(14)$.

$$t(14) = \frac{300}{14} \approx 21$$

Check. Reexamine the calculations. Note that, as expected, as the UV rating increases, the time it takes to burn goes down.
State. It will take about 21 min to burn when the UV rating is 14.

67. ***Familiarize***. The CPI c, varies inversely to the stitch length l. We write c as a function of l: $c(l) = \frac{k}{l}$. We know that $c(0.166) = 34.85$.
Translate.

$$c(l) = \frac{k}{l}.$$
$$c(0.166) = \frac{k}{0.166} \quad \text{Replacing } l \text{ with } 0.166$$
$$34.85 = \frac{k}{0.166} \quad \text{Replacing } c(0.166) \text{ with } 34.85$$
$$5.7851 = k \quad \text{Variation constant}$$
$$c(l) = \frac{5.7851}{l} \quad \text{Equation of variation}$$

Carry out. Find $c(0.175)$.

$$c(l) = \frac{5.7851}{l}$$
$$c(0.175) = \frac{5.7851}{0.175}$$
$$\approx 33.06$$

Check. Reexamine the calculations. Answers may vary slightly due to rounding differences.
State. The CPI would be about 33.06 for a stitch length of 0.175 in.

69. $y = kx^2$

$$50 = k(10)^2 \quad \text{Substituting}$$
$$50 = k \cdot 100$$
$$\frac{1}{2} = k \quad \text{Variation constant}$$

The equation of variation is $y = \frac{1}{2}x^2$.

71. $y = \frac{k}{x^2}$

$$50 = \frac{k}{(10)^2} \quad \text{Substituting}$$
$$50 = \frac{k}{100}$$
$$5000 = k \quad \text{Variation constant}$$

The equation of variation is $y = \frac{5000}{x^2}$.

73. $y = kxz$

$$105 = k \cdot 14 \cdot 5 \quad \text{Substituting}$$
$$105 = k \cdot 70$$
$$1.5 = k \quad \text{Variation constant}$$

The equation of variation is $y = 1.5xz$.

75. $y = k \cdot \frac{wx^2}{z}$

$$49 = k \cdot \frac{3 \cdot 7^2}{12} \quad \text{Substituting}$$
$$4 = k \quad \text{Variation constant}$$

The equation of variation is $y = \frac{4wx^2}{z}$.

77. ***Familiarize***. Because the stopping distance d, in feet varies directly as the square of the speed r, in mph, we write $d = kr^2$. We know that $d = 138$ when $r = 60$.
Translate. Find k and the equation of variation.

$$d = kr^2$$
$$138 = k(60)^2$$
$$\frac{23}{600} = k$$
$$d = \frac{23}{600}r^2 \quad \text{Equation of variation}$$

Carry out. We find the value of d when r is 40.

$$d = \frac{23}{600}(40)^2 \approx 61.3 \text{ ft}$$

Check. Reexamine the calculations.
State. It would take a car going 40 mph about 61.3 ft to stop.

79. ***Familiarize***. Because the wind power P, in megawatts varies directly as the cube of the wind speed v, in m/s, we write $P = kv^3$. We know that $P = 400$ when $v = 12$.
Translate. Find k and the equation of variation.

$$P = kv^3$$
$$400 = k(12)^3$$
$$k = \frac{25}{108} \approx 0.231$$
$$P = 0.231v^3 \quad \text{Equation of variation}$$

Carry out. We find the value of P when v is 15.

$$P = 0.231(15)^3 \approx 780 \text{ MW}$$

Check. Reexamine the calculations.
State. At 15 m/s, the generator would create about 780 MW of power.

81. ***Familiarize***. Because I varies inversely as the square of d, we write $I = \frac{k}{d^2}$. We know that $I = 400$ when $d = 1$.
Translate. Find k and the equation of variation.

$$I = \frac{k}{d^2}$$
$$400 = \frac{k}{1^2}$$
$$400 = k$$
$$I = \frac{400}{d^2} \quad \text{Equation of variation}$$

Carry out. Substitute 2.5 for d and find I.

$$I = \frac{400}{(2.5)^2} = 64$$

Check. Reexamine the calculations.
State. The illumination is 64 foot-candles 2.5 ft from the source.

83. ***Familiarize***. The drag W varies jointly as the surface area A and velocity v, so we write $W = kAv$. We know that $W = 222$ when $A = 37.8$ and $v = 40$.
Translate. Find k.

$$W = kAv$$
$$222 = k(37.8)(40)$$
$$\frac{222}{37.8(40)} = k$$
$$\frac{37}{252} = k$$
$$W = \frac{37}{252}Av \quad \text{Equation of variation}$$

Carry out. Substitute 51 for A and 430 for W and solve for v.

$$430 = \frac{37}{252}\cdot 51\cdot v$$
$$57 \text{ mph} \approx v$$

Check. Reexamine the calculations.
State. The car must travel about 57 mph.

85. *Writing Exercise.*

87. $(3x)^2 = 9x^2$

89. $-\frac{24}{35} \div (-3) = -\frac{24}{35}\cdot\left(-\frac{1}{3}\right) = \frac{8}{35}$

91. $-16-(-10) = -16+10 = -6$

93. *Writing Exercise.*

95. Use the result of Example 2.

$$h = \frac{2R^2 g}{V^2} - R$$

We have $V = 6.5$ mi/sec, $R = 3960$ mi, and $g = 32.2$ ft/sec^2. We must convert 32.2 ft/sec^2 to mi/sec^2 so all units of length are the same.

$$32.2\frac{\cancel{ft}}{\text{sec}^2}\cdot\frac{1 \text{ mi}}{5280\ \cancel{ft}} \approx 0.0060984\frac{\text{mi}}{\text{sec}^2}$$

Now we substitute and compute.

$$h = \frac{2(3960)^2(0.0060984)}{(6.5)^2} - 3960$$
$$h \approx 567$$

The satellite is about 567 mi from the surface of Earth.

97. $c = \frac{a}{a+12}\cdot d$

$$c = \frac{2a}{2a+12}\cdot d \quad \text{Doubling } a$$
$$= \frac{2a}{2(a+6)}\cdot d$$
$$= \frac{a}{a+6}\cdot d \quad \text{Simplifying}$$

The ratio of the larger dose to the smaller dose is

$$\frac{\frac{a}{a+6}\cdot d}{\frac{a}{a+12}\cdot d} = \frac{\frac{ad}{a+6}}{\frac{ad}{a+12}}$$
$$= \frac{ad}{a+6}\cdot\frac{a+12}{ad}$$
$$= \frac{\cancel{ad}(a+12)}{(a+6)\cancel{ad}}$$
$$= \frac{a+12}{a+6}$$

The amount by which the dosage increases is

$$\frac{a}{a+6}\cdot d - \frac{a}{a+12}\cdot d$$
$$= \frac{ad}{a+6} - \frac{ad}{a+12}$$
$$= \frac{ad}{a+6}\cdot\frac{a+12}{a+12} - \frac{ad}{a+12}\cdot\frac{a+6}{a+6}$$
$$= \frac{ad(a+12) - ad(a+6)}{(a+6)(a+12)}$$
$$= \frac{a^2d + 12ad - a^2d - 6ad}{(a+6)(a+12)}$$
$$= \frac{6ad}{(a+6)(a+12)}$$

Then the percent by which the dosage increases is

$$\frac{\frac{6ad}{(a+6)(a+12)}}{\frac{a}{a+12}\cdot d} = \frac{\frac{6ad}{(a+6)(a+12)}}{\frac{ad}{a+12}}$$
$$= \frac{6ad}{(a+6)(a+12)}\cdot\frac{a+12}{ad}$$
$$= \frac{6\cdot\cancel{ad}\cdot\cancel{(a+12)}}{(a+6)\cancel{(a+12)}\cdot\cancel{ad}}$$
$$= \frac{6}{a+6}$$

This is a decimal representation for the percent of increase. To give the result in percent notation we multiply by 100 and use a percent symbol. We have $\frac{6}{a+6}\cdot 100\%$, or $\frac{600}{a+6}\%$.

99.
$$a = \frac{\frac{d_4-d_3}{t_4-t_3} - \frac{d_2-d_1}{t_2-t_1}}{t_4-t_2}$$
$$a(t_4-t_2) = \frac{d_4-d_3}{t_4-t_3} - \frac{d_2-d_1}{t_2-t_1} \quad \text{Multiplying by } t_4-t_2$$
$$a(t_4-t_2)(t_4-t_3)(t_2-t_1) = (d_4-d_3)(t_2-t_1) - (d_2-d_1)(t_4-t_3)$$
Multiplying by $(t_4-t_3)(t_2-t_1)$
$$a(t_4-t_2)(t_4-t_3)(t_2-t_1) - (d_4-d_3)(t_2-t_1) = -(d_2-d_1)(t_4-t_3)$$
$$(t_2-t_1)[a(t_4-t_2)(t_4-t_3) - (d_4-d_3)] = -(d_2-d_1)(t_4-t_3)$$
$$t_2 - t_1 = \frac{-(d_2-d_1)(t_4-t_3)}{a(t_4-t_2)(t_4-t_3)-(d_4-d_3)}$$
$$t_2 + \frac{(d_2-d_1)(t_4-t_3)}{a(t_4-t_2)(t_4-t_3)+d_3-d_4} = t_1$$

101. Let w = the wattage of the bulb. Then we have $I = \frac{kw}{d^2}$.

Now substitute $2w$ for w and $2d$ for d.

$$I = \frac{k(2w)}{(2d)^2} = \frac{2kw}{4d^2} = \frac{kw}{2d^2} = \frac{1}{2}\cdot\frac{kw}{d^2}$$

We see that the intensity is halved.

103. ***Familiarize***. We write $T = kml^2f^2$. We know that $T = 100$ when $m = 5$, $l = 2$, and $f = 80$.

Translate. Find *k*.

$$T = kml^2f^2$$
$$100 = k(5)(2)^2(80)^2$$
$$0.00078125 = k$$
$$T = 0.00078125ml^2f^2$$

Carry out. Substitute 72 for *T*, 5 for *m*, and 80 for f and solve for *l*.

$$72 = 0.00078125(5)(l^2)(80)^2$$
$$2.88 = l^2$$
$$1.7 \approx l$$

Check. Recheck the calculations.
State. The string should be about 1 7 m long.

105. ***Familiarize***. Because *d* varies inversely as *s*, we write $d(s) = k/s$. We know that $d(0.56) = 50$.

Translate.

$$d(s) = \frac{k}{s}$$
$$d(0.56) = \frac{k}{0.56} \quad \text{Replacing } s \text{ with } 0.56$$
$$50 = \frac{k}{0.56} \quad \text{Replacing } d(0.56) \text{ with } 50$$
$$28 = k$$
$$d(s) = \frac{28}{s} \quad \text{Equation of variation}$$

Carry out. Find $d(0.40)$.

$$d(0.40) = \frac{28}{0.40} = 70$$

Check. Reexamine the calculations. Also observe that, as expected, when *d* decreases, then *s* increases.

State. The equation of variation is $d(s) = \frac{28}{s}$. The distance is 70 yd.

Chapter 7 Review

1. True
2. True
3. False; (9, 5) and (7, 5) can represent a function and pass the vertical-line test.
4. True
5. False; the *vertical*-line test is a quick way to determine whether a graph represents a function.
6. True
7. True; $(f+g)(x) = f(x) + g(x)$ represents the addition of functions *f* and *g*.
8. True
9. True
10. False

11. a. Locate 2 on the horizontal axis and find the point on the graph for which 2 is the first coordinate. From this point, look to the vertical axis to find the corresponding *y*-coordinate, 3. Thus $f(2) = 3$.

b. The set of all *x*-values in the graph extends from –2 to 4, so the domain is $\{x|-2 \le x \le 4\}$, or [–2, 4].

c. To determine which member(s) of the domain are with 2, locate 2 on the vertical axis. From there, look left and right to the graph to find any points for which 2 is the second coordinate. One such point exists. Its first coordinate is –1. Thus $f(-1) = 2$.

d. The set of all *y*-values in the graph extends from 1 to 5, so the range is $\{y|1 \le y \le 5\}$, or [1, 5].

12. $g(x) = \frac{x}{x+1}$

$$g(-3) = \frac{-3}{-3+1} = \frac{3}{2}$$

13. $f(x) = x^2 + 2x - 3$

$$f(2a) = (2a)^2 + 2(2a) - 3$$
$$= 4a^2 + 4a - 3$$

14. $s(t) = -2.5t + 32$

$$s(9) = -2.5(9) + 32 = 9.5\%$$

15. a. Yes, it is a function.

b. For any *x*-value there is a point on the graph. Thus, Domain of $f = \{x|x \text{ is a real number}\}$ or $\mathbb{R}$.
The function has no *y*-values less than 0 and every *y*-value greater than or equal to 0 corresponds to a member of the domain. Thus,
Range of $f = \{y|y \ge 0\}$ or $[0, \infty)$.

16. a. No, the graph is not a function since it fails the vertical-line test.

17. a. No, the graph is not a function since it fails the vertical-line test.

18. a. Yes, it is a function.

b. For any *x*-value there is a point on the graph. Thus,
Domain of $f = \{x|x \text{ is a real number}\}$, or $\mathbb{R}$.
The only *y*-value on the graph is –2. Thus,
Range of $f = \{-2\}$.

19. $f(x) = 3x^2 - 7$

Since we can compute $3x^2 - 7$ for any real number *x*, the domain is the set of all real numbers.

20. $g(x) = \frac{x^2}{x-1}$

Since *x* – 1 cannot be computer when the denominator is 0, we find *x*-values that cause *x* – 1 to be 0:

$$x - 1 = 0$$
$$x = 1$$

Thus, 1 is not in the domain of *g*, while all other real numbers are. The domain of *g* is $\{x|x \text{ is a real number } \textit{and } x \ne 1\}$.

21. $f(t)=\dfrac{1}{t^2+5t+4}$

The expression $\dfrac{1}{t^2+5t+4}$ is undefined when t^2+5t+4 is zero.

$$t^2+5t+4=0$$
$$(t+4)(t+1)=0$$
$$t+4=0 \quad or \quad t+1=0$$
$$t=-4 \quad or \quad t=-1$$

Thus, the Domain of $f=\{t|t \text{ is a real number } and$ $t\neq-4 \ and \ t\neq-1\}$.

22. $r(t)=900-15t$

The time must be nonnegative, so we have $t\geq 0$. In addition, $r(t)$ must be nonnegative so we have:

$$900-15t\geq 0$$
$$-15t\geq -900$$
$$t\leq 60$$

Then the domain of the function is $\{t|0\leq t\leq 60\}$, or $[0,\ 60]$.

23. $f(x)=\begin{cases} 2-x, & \text{if } x\leq -2 \\ x^2, & \text{if } -2<x\leq 5 \\ x+10, & \text{if } x>5 \end{cases}$

a. Since $-3\leq -2$, we use the equation $f(x)=2-x$.
$f(-3)=2-(-3)=5$

b. Since $-2\leq -2$, we use the equation $f(x)=2-x$.
$f(-2)=2-(-2)=4$

c. Since $-2<4\leq 5$, we use the equation $f(x)=x^2$.
$f(4)=4^2=16$

d. Since $25>5$, we use the equation $f(x)=x+10$.
$f(25)=25+10=35$

24. ***Familiarize***. After an initial fee of \$90, an additional fee of \$30 is charged each month. After one month, the total cost is \$90 + \$30 = \$120. After two months, the total cost is $90+2\cdot 30=\$150$. We can generalize this with a model, letting $C(t)$ represent the total cost, in dollars, for t months.

Translate. We reword the problem and translate.

Total Cost	is	initial cost	plus	\$30 per month.
↓	↓	↓	↓	↓
$C(t)$	=	90	+	$30t$

where $t\geq 0$ since we cannot have negative months.

Carry out. To determine the number of months required to reach a total cost of \$300, we substitute 300 for $C(t)$ and solve for t.

$$C(t)=30t+90$$
$$300=30t+90 \quad \text{Substituting}$$
$$210=30t \quad \text{Subtracting 90 from both sides}$$
$$7=t \quad \text{Dividing both sides by 30}$$

Check. We evaluate

$C(7)=30(7)+90=210+30=300$.

State. The function is $C(t)=30t+90$ and it will take 7 months to reach a total cost of \$300.

25. **a.** $m=\dfrac{19.75-19.19}{3-28}=-\dfrac{0.56}{25}\approx -0.02$

$$R-19.75=-0.02(t-3)$$
$$R-19.75=-0.02t+0.06$$
$$R(t)=-0.02t+19.81$$

b. For 2015, $t=35$.
$R(35)=-0.02(33)+19.81\approx 19.11$ sec
For 2020, $t=40$.
$R(40)=-0.02(40)+19.81\approx 19.01$ sec

26. $f(x)=|3x-7|$
The function is described by an absolute-value equation, so it is an absolute-value function.

27. $g(x)=4x^5-8x^3+7$
The function is described by a polynomial equation that is neither linear nor quadratic, so it is a polynomial function.

28. $p(x)=x^2+x-10$
The function is in the form $f(x)=ax^2+bx+c$, $a\neq 0$, so it is a quadratic function.

29. $h(n)=4n-17$
The function is in the form $f(x)=mx+b$, so it is a linear function.

30. $s(t)=\dfrac{t+1}{t+2}$
The function is described by a rational equation so it is a rational function.

31. Graph $f(x)=3$.

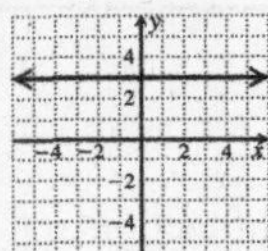

$f(x)=3$

For any x-value there is a point on the graph, so Domain of $f=\{x|x \text{ is a real number}\}$, or $\mathbb{R}$.
The only y-value on the graph is 3, so Range of $f=\{3\}$.

32. Graph $f(x)=2x+1$.

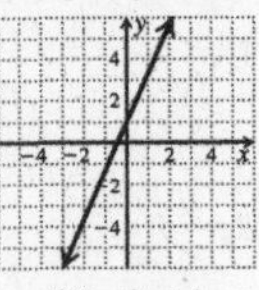

$f(x)=2x+1$

For any x-value and for any y-value there is a point on the graph. Thus,
Domain of $f=\{x|x \text{ is a real number}\}$, or $\mathbb{R}$ and
Range of $f=\{y|y \text{ is a real number}\}$, or $\mathbb{R}$.

33. Graph $g(x) = |x+1|$.

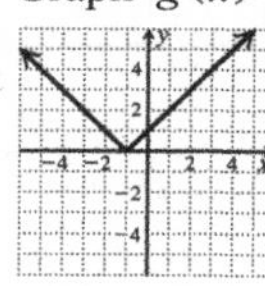

For any x-value there is a point on the graph, so Domain of $g = \{x|x \text{ is a real number}\}$, or $\mathbb{R}$.
There is no y-value less than 0 and every y-value greater than or equal to 0 corresponds to a member of the domain. Thus, Range of $g = \{y|y \geq 0\}$, or $[0, \infty)$.

34. $(g \cdot h)(4) = g(4) \cdot h(4) = [3(4) - 6]\left[(4)^2 + 1\right]$
$= [12 - 6][16 + 1]$
$= 6 \cdot 17 = 102$

35. $(g - h)(-2) = g(-2) - h(-2)$
$= [3(-2) - 6] - \left[(-2)^2 + 1\right]$
$= [-6 - 6] - [4 + 1]$
$= -12 - 5 = -17$

36. $\left(\frac{g}{h}\right)(-1) = \frac{g(-1)}{h(-1)} = \frac{3(-1) - 6}{(-1)^2 + 1} = \frac{-3 - 6}{1 + 1} = -\frac{9}{2}$

37. The domain of g and h is all real numbers. Thus,
Domain of $g + h$ = Domain of $g \cdot h$
$= \{x|x \text{ is a real number}\}$.

38. Domain of g = Domain of h = $\{x|x \text{ is a real number}\}$.
Since $g(x) = 0$ when $3x - 6 = 0$, we have $g(x) = 0$ when $x = 2$. We conclude that
Domain of $h/g = \{x|x \text{ is a real number and } x \neq 2\}$.

39.
$$I = \frac{2V}{R + 2r}$$
$$(R + 2r)I = (R + 2r) \cdot \frac{2V}{R + 2r}$$
$$IR + 2rI = 2V$$
$$2rI = 2V - IR$$
$$r = \frac{2V - IR}{2I}, \text{ or } \frac{V}{I} - \frac{R}{2}$$

40.
$$S = \frac{H}{m(t_1 - t_2)}$$
$$m(t_1 - t_2) \cdot S = m(t_1 - t_2) \cdot \frac{H}{m(t_1 - t_2)}$$
$$mS(t_1 - t_2) = H$$
$$m = \frac{H}{S(t_1 - t_2)}$$

41.
$$\frac{1}{ac} = \frac{2}{ab} - \frac{3}{bc}$$
$$abc \cdot \frac{1}{ac} = abc \cdot \frac{2}{ab} - abc \cdot \frac{3}{bc}$$
$$b = 2c - 3a$$
$$b + 3a = 2c$$
$$\frac{b + 3a}{2} = c$$

42.
$$T = \frac{A}{v(t_2 - t_1)}$$
$$v(t_2 - t_1) \cdot T = v(t_2 - t_1) \cdot \frac{A}{v(t_2 - t_1)}$$
$$vTt_2 - vTt_1 = A$$
$$vTt_2 - A = vTt_1$$
$$\frac{vTt_2 - A}{vT} = t_1, \text{ or } t_1 = t_2 - \frac{A}{vT}$$

43.
$y = kx$
$30 = k \cdot 4$ Substituting
$\frac{15}{2} = k$

The variation constant is $\frac{15}{2}$.

The equation of variation is $y = \frac{15}{2}x$.

44.
$y = \frac{k}{x}$
$3 = \frac{k}{\frac{1}{4}}$ Substituting
$\frac{3}{4} = k$

The variation constant is $\frac{3}{4}$.

The equation of variation is $y = \frac{\frac{3}{4}}{x}$, or $\frac{3}{4x}$.

45.
$y = k \cdot \frac{xw^2}{z}$
$150 = k \cdot \frac{6 \cdot 10^2}{2}$ Substituting
$150 = 300k$
$\frac{1}{2} = k$ Variation constant

The equation of variation is $y = \frac{1}{2} \cdot \frac{xw^2}{z}$, or $\frac{xw^2}{2z}$.

46. ***Familiarize***. Because of the phrase "the base varies inversely as the height," we write $b(h) = \frac{k}{h}$.
Since $b = 8$ when $h = 10$, we know that $b(10) = 8$.
Translate. We find the variation constant and then we find the equation of variation.

$b(h) = \frac{k}{h}$
$b(10) = \frac{k}{10}$ Replacing h with 10
$8 = \frac{k}{10}$ Replacing $b(10)$ with 8
$80 = k$ Variation constant
$b(h) = \frac{80}{h}$ Equation of variation

Carry out. We find $b(4)$.

$b(4) = \frac{80}{4} = 20$

Check. Reexamine the calculations.
State. The base will be 20 cm when the height is 4 cm.

47. ***Familiarize***. Because the time t, in seconds varies inversely as the current I, in amperes, we write $t = \frac{k}{I^2}$.

We know that $t = 3.4$ when $I = 0.089$.
Translate. Find k and the equation of variation.

$$t = \frac{k}{I^2}$$
$$3.4 = \frac{k}{(0.089)^2}$$
$$0.0269314 = k$$
$$t = \frac{0.0269314}{I^2} \quad \text{Equation of variation}$$

Carry out. We find the value of t when I is 0.096.

$$t = \frac{0.0269314}{(0.096)^2} \approx 2.9 \text{ sec}$$

Check. Reexamine the calculations.
State. A 0.096-amp current would be deadly after about 2.9 sec.

48.
$$W(S) = kS$$
$$16.8 = k \cdot 150$$
$$0{,}112 = k \quad \text{Variation constant}$$
$$W(S) = 0.112S \quad \text{Equation of variation}$$
$$W(500) = 0.112(500)$$
$$= 56 \text{ in.}$$

49. *Writing Exercise*. Two functions that have the same domain and range are not necessarily identical. For example, the functions $f: \{(-1, 2), (-3, 2)\}$ and $g: \{(-2, 2), (-3, 1)\}$ have the same domain and range but are different functions.

50. *Writing Exercise*. Jenna is not correct. Any value of the variable that makes a denominator 0 is not in the domain; 0 itself may or may not make a denominator 0.

51. Let x represent the number of packages.
Total cost = package charge + shipping charges
$f(x) = 7.99x + (2.95x + 20)$
$f(x) = 10.94x + 20$

52. The function has no x-values less than –4. The hold in the graph at (2, 3) indicates that the function is not defined for $x = 2$. So for any other x-values greater than or equal to –4 there is a point on the graph. Thus,

Domain of $f = \{x | x \geq -4 \text{ and } x \neq 2\}$.

The function has no y-values less than 0. There is no function value at (2, 3), so 3 is not in the range of the function. For any other y-values greater than or equal to 0, there is a point on the graph. Thus,

Range of $f = \{y | y \geq 0 \text{ and } y \neq 3\}$.

Chapter 7 Test

1. a. $f(-2) = 1$

b. Domain is $\{x | -3 \leq x \leq 4\}$, or $[-3, 4]$.

c. If $f(x) = \frac{1}{2}$, then $x = 3$.

d. Range is $\{y | -1 \leq y \leq 2\}$, or $[-1, 2]$.

2. a. Yes; it passes the vertical-line test.

b. For any x-value and for any y-value there is a point on the graph. Thus,

Domain of $f = \{x | x \text{ is a real number}\}$, or $\mathbb{R}$ and

Range of $f = \{y | y \text{ is a real number}\}$, or $\mathbb{R}$.

3. a. Yes; it passes the vertical-line test.

b. For any x-value there is a point on the graph. Thus,

Domain of $f = \{x | x \text{ is a real number}\}$, or $\mathbb{R}$.

The function has no y-values less than 1 and every y-value greater than or equal to 1 corresponds to a member of the domain. Thus,

Range of $f = \{y | y \geq 1\}$, or $[1, \infty)$.

4. a. No; it fails the vertical-line test.

5. $d(t) = 240 - 60t$

The number of hours must be nonnegative, so $t \geq 0$. In addition $d(t)$ must be nonnegative, so we have:

$$240 - 60t \geq 0$$
$$-60t \geq 240$$
$$t \leq 4$$

Then the domain of the function is $\{t | 0 \leq t \leq 4\}$, or $[0, 4]$

6. $f(x) = \begin{cases} x^2, & \text{if } x < 0 \\ 3x - 5, & \text{if } 0 \leq x \leq 2 \\ x + 7, & \text{if } x > 2 \end{cases}$

a. Since $0 \leq 0 \leq 2$, we use the equation $f(x) = 3x - 5$.

$$f(0) = 3(0) - 5 = -5$$

b. Since $3 > 2$, we use the equation $f(x) = x + 7$.

$$f(3) = 3 + 7 = 10$$

7. $C(t) = 55t + 180$

Solve $55t + 180 = 840$
$$55t = 660$$
$$t = 12 \text{ months}$$

8. a. Let m = the number of miles, $C(m)$ represents the cost. We form the pairs (250, 100) and (300, 115).

$$\text{Slope} = \frac{115 - 100}{300 - 250} = \frac{15}{50} = \frac{3}{10} = 0.3$$

$$y - y_1 = m(x - x_1)$$
$$C - 100 = 0.3(m - 250)$$
$$C = 0.3m + 25$$
$$C(m) = 0.3m + 25$$

b. $C(500) = 0.3(500) + 25 = \$175$

9. $f(x) = \frac{1}{4}x + 7$

The function is in the form $f(x) = mx + b$, so it is a linear function. We can compute $\frac{1}{4}x + 7$ for any value of *n*, so the domain is the set of all real numbers.

10. $g(x) = \frac{3}{x^2 - 16}$

The function is described by a rational equation so it is a rational function. The expression $\frac{3}{x^2 - 16}$ is undefined for values of *x* that make the denominator 0. We find those values:

$$x^2 - 16 = 0$$
$$(x + 4)(x - 4) = 0$$
$$x + 4 = 0 \quad or \quad x - 4 = 0$$
$$x = -4 \quad or \quad x = 4$$

Then the domain is $\{x|x$ is a real number *and* $x \neq -4$ *and* $x \neq 4\}$.

11. $p(x) = 4x^2 + 7$

The function is in the form $f(x) = ax^2 + bx + c$, $a \neq 0$, so it is a quadratic function. We can compute $4x^2 + 7$ for any value of *x*, so the domain is the set of all real numbers.

12. Graph $f(x) = \frac{1}{3}x - 2$.

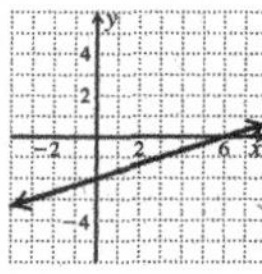

$f(x) = \frac{1}{3}x - 2$

For any *x*-value and for any *y*-value there is a point on the graph. Thus,
Domain of $f = \{x|x$ is a real number$\}$, or $\mathbb{R}$ and
Range of $f = \{y|y$ is a real number$\}$, or $\mathbb{R}$.

13. Graph $g(x) = x^2 - 1$.

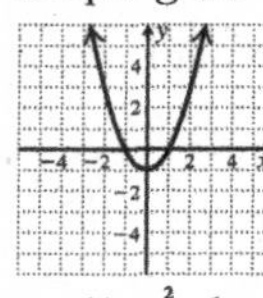

$g(x) = x^2 - 1$

For any *x*-value there is a point on the graph, so
Domain of $g = \{x|x$ is a real number$\}$, or $\mathbb{R}$.
There is no *y*-value less than -1 and every *y*-value greater than or equal to -1 corresponds to a member of the domain. Thus,
Range of $g = \{y|y \geq -1\}$, or $[-1, \infty)$.

14. Graph $h(x) = -\frac{1}{2}$.

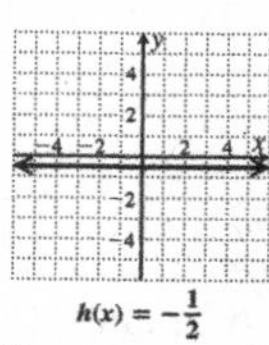

$h(x) = -\frac{1}{2}$

For any *x*-value there is a point on the graph, so
Domain of $h = \{x|x$ is a real number$\}$, or $\mathbb{R}$.
The only *y*-value on the graph is $-\frac{1}{2}$, so
Range of $h = \left\{-\frac{1}{2}\right\}$.

15. $g(-1) = \frac{1}{-1} = -1$

16. $h(x) = 2x + 1$
$h(5a) = 2(5a) + 1 = 10a + 1$

17. $(g + h)(x) = g(x) + h(x) = \frac{1}{x} + 2x + 1$

18. Domain of *g*: $\{x|x$ is a real number and $x \neq 0\}$

19. Domain of *h*: $\{x|x$ is a real number$\}$
Domain of $g + h$: $\{x|x$ is a real number and $x \neq 0\}$

20. $h(x) = 2x + 1 = 0$, if $x = -\frac{1}{2}$
Domain of g/h:
$\left\{x \middle| x \text{ is a real number and } x \neq 0 \text{ and } x \neq -\frac{1}{2}\right\}$

21.
$$R = \frac{gs}{g + s}$$
$$(g + s)R = (g + s) \cdot \frac{gs}{g + s}$$
$$gR + sR = gs$$
$$gr = gs - sR$$
$$gR = s(g - R)$$
$$\frac{gR}{g - R} = s$$

22. $y = kx$
$10 = k \cdot 20$ Substituting
$\frac{1}{2} = k$

The variation constant is $\frac{1}{2}$.

The equation of variation is $y = \frac{1}{2}x$.

23. $n(t)=\frac{k}{t}$

$25=\frac{k}{6}$

$150=k$ Variation constant

$n(t)=\frac{150}{t}$ Equation of variation

$n(5)=\frac{150}{5}$

$=30$ workers

24. ***Familiarize***. Because the surface area A, in square inches varies directly as the square of the radius r, in inches, we write $A=kr^2$. We know that $A=325$ when $r=5$.

Translate. Find k and the equation of variation.

$A=kr^2$

$325=k(5)^2$

$13=k$

$d=13r^2$ Equation of variation

Carry out. We find the value of d when r is 7.

$A=13(7)^2=637\text{ in}^2$

Check. Reexamine the calculations.

State. The area would be 637 in^2 when the radius is 7 in.

25. **a.** 1 hr and 40 min is equal to $1\frac{40}{60}=1\frac{2}{3}=\frac{5}{3}$ hr.

We must find $f\left(\frac{5}{3}\right)$.

$f\left(\frac{5}{3}\right)=5+15\cdot\frac{5}{3}=5+25=30$

The cyclist will be 30 mi from the starting point 1 hr and 40 min after passing the 5-mi marker.

b. From the equation $f(t)=5+15t$, we see that the cyclist is advancing 15 mi for every hour he travels. So the rate is 15 mph.

26. Answers may vary. In order to have the restriction on the domain of $f/g/h$ that $x\neq\frac{3}{4}$ and $x\neq\frac{2}{7}$, we need to find some function $h(x)$ such that $h(x)=0$ when $x=\frac{2}{7}$.

$x=\frac{2}{7}$

$7x=2$

$7x-2=0$

One possible answer is $h(x)=7x-2$.

Chapter 8

Systems of Linear Equations and Problem Solving

Exercise Set 8.1

1. False; see Example 4(b).

3. True

5. True; see Example 4(b).

7. False

9. We use alphabetical order for the variables. We replace x by 2 and y by 3.

$$\begin{array}{c|c} 2x-y=1 & \\ \hline 2\cdot 2-3 & 1 \\ 4-3 & \\ \end{array} \qquad \begin{array}{c|c} 5x-3y=1 & \\ \hline 5\cdot 2-3\cdot 3 & 1 \\ 10-9 & \\ \end{array}$$

$1 \stackrel{?}{=} 1$ TRUE $\qquad 1 \stackrel{?}{=} 1$ TRUE

The pair (2, 3) makes both equations true, so is it a solution of the system.

11. We use alphabetical order for the variables. We replace x by -5 and y by 1.

$$\begin{array}{c|c} x+5y=0 & \\ \hline -5+5\cdot 1 & 0 \\ -5+5 & \\ \end{array} \qquad \begin{array}{c|c} y=2x+9 & \\ \hline 1 & 2(-5)+9 \\ & -10+9 \\ \end{array}$$

$0 \stackrel{?}{=} 0$ TRUE $\qquad 1 \stackrel{?}{=} -1$ FALSE

The pair $(-5, 1)$ is not a solution of $y = 2x + 9$. Therefore, it is not a solution of the system of equations.

13. We replace x by 0 and y by -5.

$$\begin{array}{c|c} x-y=5 & \\ \hline 0-(-5) & 5 \\ 0+5 & \\ \end{array} \qquad \begin{array}{c|c} y=3x-5 & \\ \hline -5 & 3\cdot 0-5 \\ & 0-5 \\ \end{array}$$

$5 \stackrel{?}{=} 5$ TRUE $\qquad -5 \stackrel{?}{=} -5$ TRUE

The pair $(0, -5)$ makes both equations true, so is it a solution of the system.

15. Observe that if we multiply both sides of the first equation by 2, we get the second equation. Thus, if we find that the given points makes the one equation true, we will also know that it makes the other equation true. We replace x by 3 and y by -1 in the first equation.

$$\begin{array}{c|c} 3x-4y=13 & \\ \hline 3\cdot 3-4(-1) & 13 \\ 9+4 & \\ \end{array}$$

$13 \stackrel{?}{=} 13$ TRUE

The pair $(3, -1)$ makes both equations true, so is it a solution of the system.

17. Graph both equations.

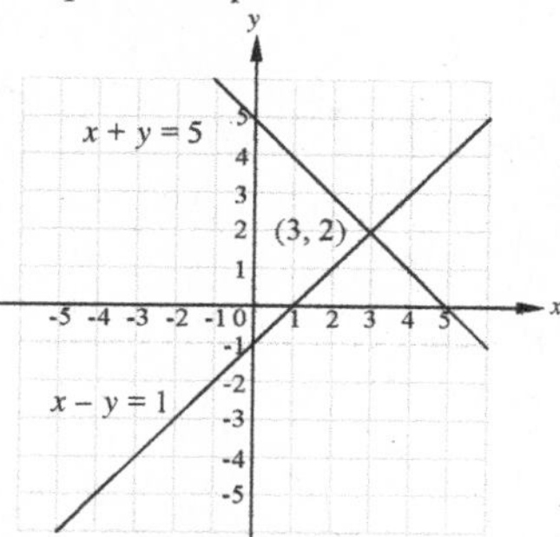

The solution (point of intersection) is apparently (3, 2).
Check:

$$\begin{array}{c|c} x-y=1 & \\ \hline 3-2 & 1 \\ \end{array} \qquad \begin{array}{c|c} x+y=5 & \\ \hline 3+2 & 5 \\ \end{array}$$

$1 \stackrel{?}{=} 1$ TRUE $\qquad 5 \stackrel{?}{=} 5$ TRUE

The solution is (3, 2).

19. Graph the equations.

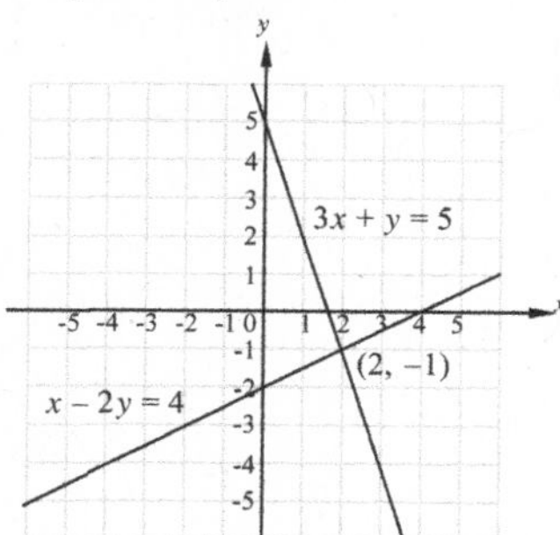

The solution (point of intersection) is apparently $(2, -1)$.
Check:

$$\begin{array}{c|c} 3x+y=5 & \\ \hline 3\cdot 2+(-1) & 5 \\ 6-1 & \\ \end{array} \qquad \begin{array}{c|c} x-2y=4 & \\ \hline 2-2(-1) & 4 \\ 2+2 & \\ \end{array}$$

$5 \stackrel{?}{=} 5$ TRUE $\qquad 4 \stackrel{?}{=} 4$ TRUE

The solution is $(2, -1)$.

21. Graph both equations.

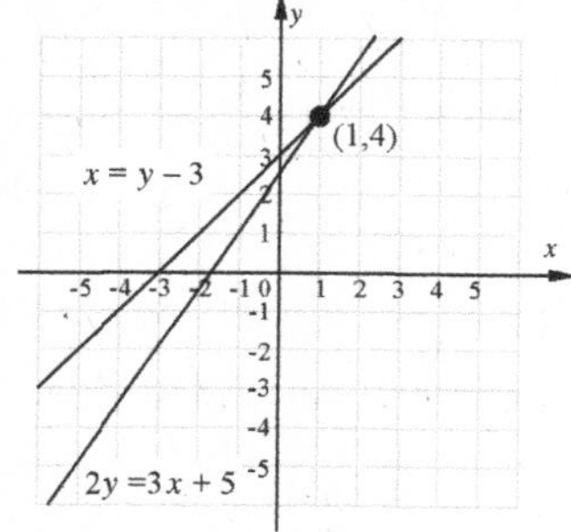

The solution (point of intersection) is apparently (1, 4).
Check:

$$\begin{array}{c|c} 2y = 3x+5 & \\ \hline 2\cdot 4 & 3\cdot 1+5 \\ 8 & 3+5 \\ 8 \overset{?}{=} 8 & \text{TRUE} \end{array} \qquad \begin{array}{c|c} x = y-3 & \\ \hline 1 & 4-3 \\ 1 \overset{?}{=} 1 & \text{TRUE} \end{array}$$

The solution is (1, 4).

23. Graph both equations.

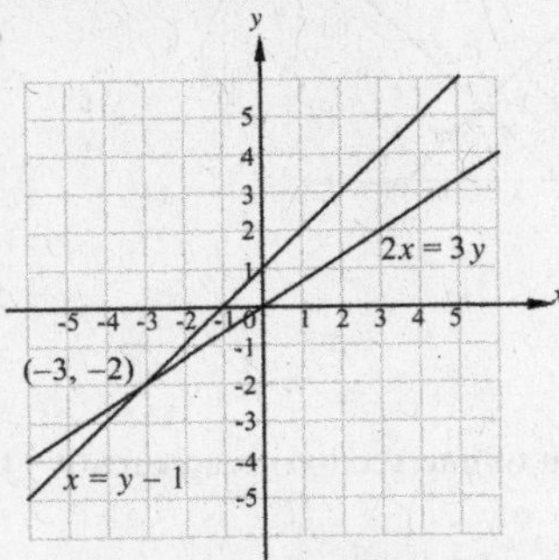

The solution (point of intersection) is apparently (−3, −2).
Check:

$$\begin{array}{c|c} x = y-1 & \\ \hline -3 & -2-1 \\ -3 \overset{?}{=} -3 & \text{TRUE} \end{array} \qquad \begin{array}{c|c} 2x = 3y & \\ \hline 2(-3) & 3(-2) \\ -6 \overset{?}{=} -6 & \text{TRUE} \end{array}$$

The solution is (−3, −2).

25. Graph both equations.

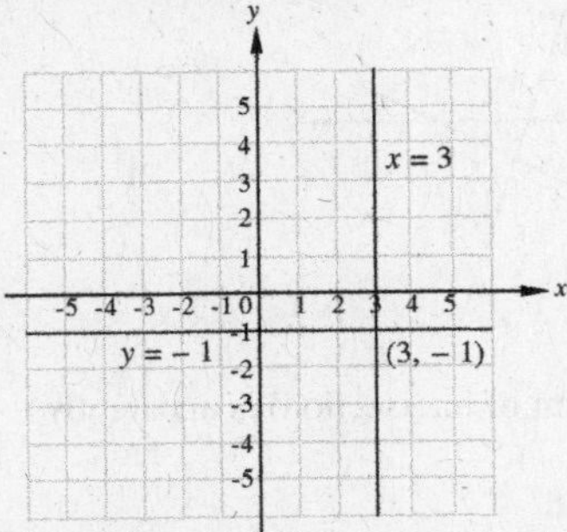

The ordered pairs (3, −1) checks in both equations. It is the solution.

27. Graph both equations.

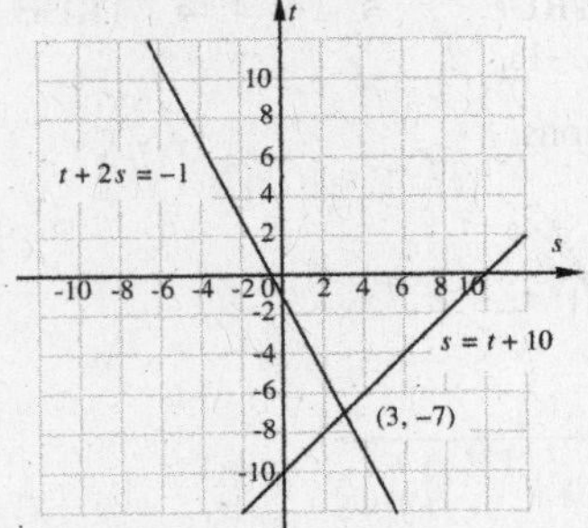

The solution (point of intersection) is apparently (3, −7).
Check:

$$\begin{array}{c|c} t+2s = -1 & \\ \hline -7+2\cdot 3 & -1 \\ -7+6 & \\ -1 \overset{?}{=} -1 & \text{TRUE} \end{array} \qquad \begin{array}{c|c} s = t+10 & \\ \hline 3 & -7+10 \\ 3 \overset{?}{=} 3 & \text{TRUE} \end{array}$$

The solution is (3, −7).

29. Graph both equations.

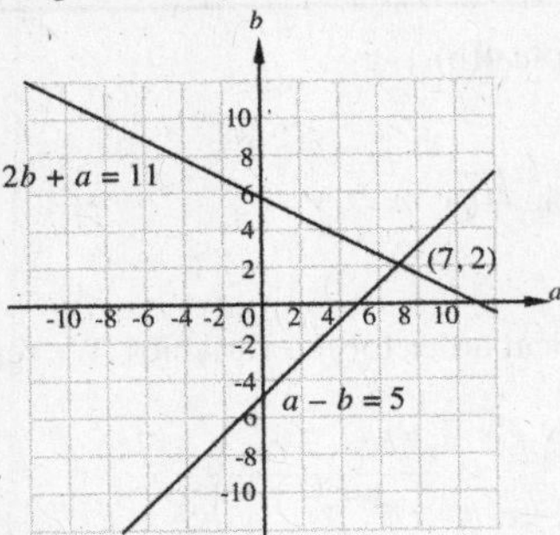

The solution (point of intersection) is apparently (7, 2).
Check:

$$\begin{array}{c|c} 2b+a = 11 & \\ \hline 2\cdot 2+7 & 11 \\ 4+7 & \\ 11 \overset{?}{=} 11 & \text{TRUE} \end{array} \qquad \begin{array}{c|c} a-b = 5 & \\ \hline 7-2 & 5 \\ 5 \overset{?}{=} 5 & \text{TRUE} \end{array}$$

The solution is (7, 2).

31. Graph both equations.

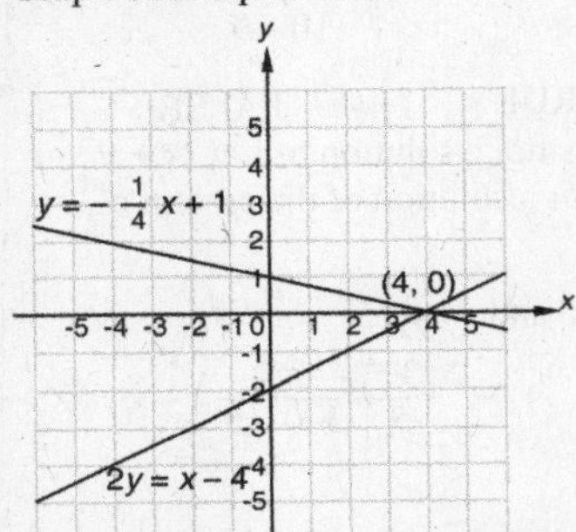

The solution (point of intersection) is apparently (4, 0).
Check:

$$\begin{array}{c|c} y = -\frac{1}{4}x+1 & \\ \hline 0 & -\frac{1}{4}\cdot 4+1 \\ & -1+1 \\ 0 \overset{?}{=} 0 & \text{TRUE} \end{array} \qquad \begin{array}{c|c} 2y = x-4 & \\ \hline 2\cdot 0 & 4-4 \\ 0 \overset{?}{=} 0 & \text{TRUE} \end{array}$$

The solution is (4, 0).

33. Graph both equations.

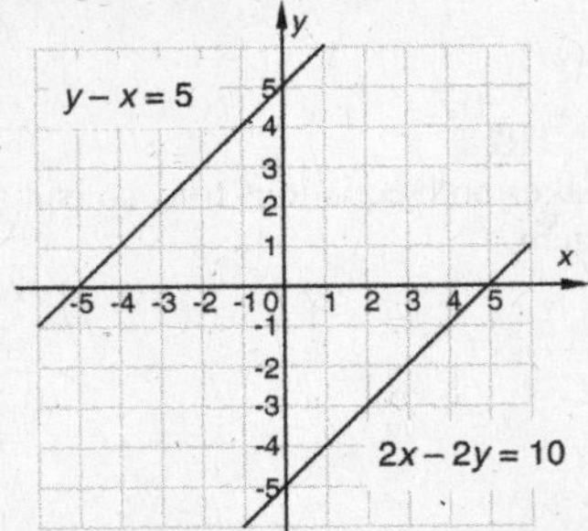

The lines are parallel. The system has no solution.

35. Graph both equations.

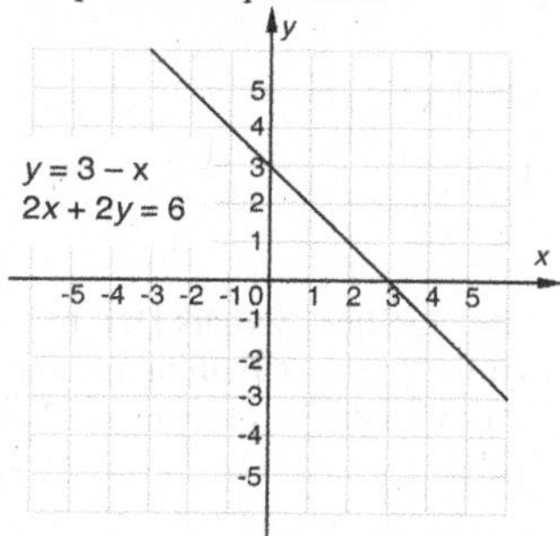

The graphs are the same. Any solution of one equation is a solution of the other. Each equation has infinitely many solutions. The solution set is the set of all pairs (x, y) for which $y = 3 - x$, or $\{(x, y) | y = 3 - x\}$. (In place of $y = 3 - x$ we could have used $2x + 2y = 6$ since the two equations are equivalent.)

37. A system of equations is consistent if it has at least one solution. Of the systems under consideration, only the one in Exercise 33 has no solution. Therefore, all except the system in Exercise 33 are consistent.

39. A system of two equations in two variables is dependent if it has infinitely many solutions. Only the system in Exercise 35 is dependent.

41. ***Familiarize***. Let x = the first number and y = the second number.
Translate.

The sum of the numbers is 10.
$$x + y = 10$$

The first number is $\frac{2}{3}$ of the second number.
$$x = \frac{2}{3} \times y$$

We have a system of equations:
$$x + y = 10,$$
$$x = \frac{2}{3}y.$$

43. ***Familiarize***. Let p = the number of endangered plant species and a = the number of endangered animal species.
Translate.

plants and animals is 1223.
$$p + a = 1223$$

plants are 243 more than animals.
$$p = 243 + a$$

We have a system of equations:
$$p + a = 1223,$$
$$p = a + 243.$$

45. ***Familiarize***. Let x = the measure of one angle and y = the measure of the other angle.
Translate.
Two angles are supplementary.
Rewording: The sum of the measures is 180°.
$$x + y = 180$$

One angle is 3 less than twice the other.
Rewording:

One angle is twice the other angle minus 3°.
$$x = 2y - 3$$

We have a system of equations:
$$x + y = 180,$$
$$x = 2y - 3$$

47. Familiarize. Let x = the number of two-point shots and y = the number of foul shots made.
Translate. We organize the information in a table.

Kind of shot	2-point shot	Foul shot	Total
Number scored	x	y	64
Points per score	2	1	
Points scored	$2x$	x	100

From the "Number scored" row of the table we get one equation:
$$x + y = 64$$
The "Points scored" row gives us another equation:
$$2x + y = 100$$
We have a system of equations:
$$x + y = 64,$$
$$2x + y = 100$$

49. ***Familiarize***. Let w = the number of wrapped strings and u = the number of unwrapped strings.
Translate. We organize the information in a table.

Strings	Wrapped	Unwrapped	Total
Number	w	u	32
Price	\$4.49	\$2.99	
Amount paid	$4.49w$	$2.99u$	107.68

The "Number" row of the table gives us one equation:
$$w + u = 32.$$
The "Amount paid" row gives us a second equation:
$$4.49w + 2.99u = 107.68.$$

51. ***Familiarize***. Let h = the number of hats knitted and s = the number of scarves knitted.
Translate. We organize the information in a table.

Items knitted	Hats	Scarves	Total
Number	h	s	110
Time	8	12	
Time spent	$8h$	$12s$	1072

The "Number" row of the table gives us one equation:
$$h + s = 110.$$
The "Time spent" row gives us a second equation:
$$8h + 12s = 1072.$$

53. ***Familiarize***. The lacrosse field is a rectangular with perimeter 340 yd. Let l = the length, in yards, and w = the width, in yards. Recall that for a rectangle with length l and width w, the perimeter P is given by $P = 2l + 2w$.
Translate. The formula for perimeter gives us one equation:

$$2l + 2w = 340.$$

The statement relating length and width gives us another equation:

$$l = w + 50.$$

We have a system of equations:

$$2l + 2w = 340,$$
$$l = w + 50.$$

55. *Writing Exercise.*

57. $-\frac{1}{2} - \frac{3}{10} = -\frac{5}{10} - \frac{3}{10} = -\frac{8}{10} = -\frac{4}{5}$

59. $-10^{-2} = -\frac{1}{10^2} = -\frac{1}{100}$

61. $(-3)^2 - 2 - 4 \cdot 6 \div 2 \cdot 3$
$= 9 - 2 - 4 \cdot 6 \div 2 \cdot 3$
$= 9 - 2 - 24 \div 2 \cdot 3$
$= 9 - 2 - 12 \cdot 3$
$= 9 - 2 - 36$
$= -29$

63. *Writing Exercise.*

65. a. There are many correct answers. One can be found by expressing the sum and difference of the two numbers:

$$x + y = 6,$$
$$x - y = 4$$

b. There are many correct answers. For example, write an equation in two variables. Then write a second equation by multiplying the left side of the first equation by one nonzero constant and multiplying the right side by another nonzero constant.

$$x + y = 1,$$
$$2x + 2y = 3$$

c. There are many correct answers. One can be found by writing an equation in two variables and then writing a nonzero constant multiple of that equation:

$$x + y = 1,$$
$$2x + 2y = 2$$

67. Substitute 4 for x and –5 for y in the first equation:

$$A(4) - 6(-5) = 13$$
$$4A + 30 = 13$$
$$4A = -17$$
$$A = -\tfrac{17}{4}$$

Substitute 4 for x and –5 for y in the second equation:

$$4 - B(-5) = -8$$
$$4 + 5B = -8$$
$$5B = -12$$
$$B = -\tfrac{12}{5}$$

We have $A = -\frac{17}{4}$, $B = -\frac{12}{5}$.

69. ***Familiarize***. Let x = the number of years Dell has taught and y = the number of years Juanita has taught. Two years ago, Dell and Juanita had taught $x - 2$ and $y - 2$ years, respectively.
Translate.

Together, the number of years of service is 46.
↓ ↓ ↓
$x + y = 46$

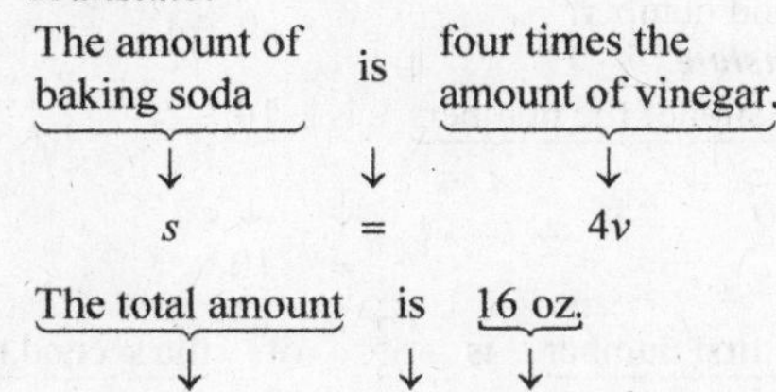

Two years ago
Dell had taught 2.5 times as many years as Juanita.
x − 2 = 2.5(y − 2)

We have a system of equations:

$$x + y = 46,$$
$$x - 2 = 2.5(y - 2)$$

71. ***Familiarize***. Let s = the number of ounces of baking soda and v = the number of ounces of vinegar to be used. The amount of baking soda in the mixture will be four times the amount of vinegar.
Translate.

The amount of baking soda is four times the amount of vinegar.
↓ ↓ ↓
$s = 4v$

The total amount is 16 oz.
↓ ↓ ↓
$s + v = 16$

We have a system of equations.

$$s = 4v,$$
$$s + v = 16$$

73. From Exercise 44, graph both equations:

$$v + m = 16$$
$$m = 2v + 4$$

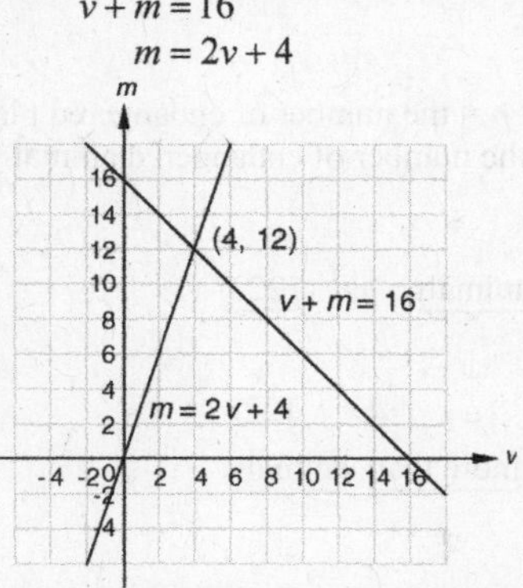

The lines intersect at $v = 4$, $m = 12$. The solution is 4 oz of vinegar and 12 oz of mineral oil.

75. Graph both equations.

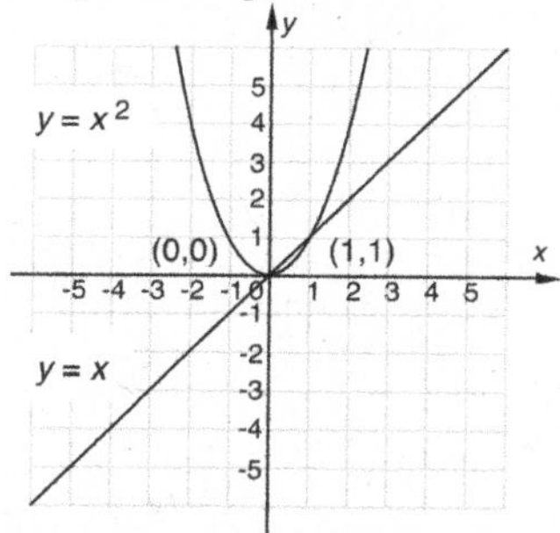

The solutions are apparently (0, 0) and (1, 1). Both pairs check.

77. (0.07, –7.95)

79. (0.00, 1.25)

Connecting the Concepts

1. $x = y,\quad (1)$
$x + y = 2\quad (2)$
Substitute y for x in Equation (2) and solve for y.
$$\begin{aligned} x + y &= 2 \quad (2)\\ y + y &= 2 \quad \text{Substituting}\\ 2y &= 2\\ y &= 1 \end{aligned}$$
Substitute 1 for y in Equation (1) and solve for x.
$$\begin{aligned} x &= y \quad (1)\\ x &= 1 \end{aligned}$$
We obtain (1, 1) as the solution.

2. $$\begin{aligned} x + y &= 10 \quad (1)\\ x - y &= 8 \quad (2)\\ \hline 2x \quad &= 18 \quad \text{Adding}\\ x &= 9 \end{aligned}$$
Substitute 9 for x in Equation (1) and solve for y.
$$\begin{aligned} x + y &= 10 \quad (1)\\ 9 + y &= 10 \quad \text{Substituting}\\ y &= 1 \end{aligned}$$
We obtain (9, 1) as the solution.

3. $y = \frac{1}{2}x + 1,\quad (1)$
$y = 2x - 5\quad (2)$
Substitute $2x - 5$ for y in Equation (1) and solve for x.
$$\begin{aligned} y &= \frac{1}{2}x + 1 \quad (1)\\ 2x - 5 &= \frac{1}{2}x + 1 \quad \text{Substituting}\\ 4x - 10 &= x + 2 \quad \text{Clearing fractions}\\ 3x &= 12\\ x &= 4 \end{aligned}$$
Substitute 4 for x in Equation (2) and solve for y.
$$\begin{aligned} y &= 2x - 5 \quad (2)\\ y &= 2(4) - 5 = 3 \end{aligned}$$
We obtain (4, 3) as the solution.

4. $y = 2x - 3,\quad (1)$
$x + y = 12\quad (2)$
Substitute $2x - 3$ for y in Equation (2) and solve for x.
$$\begin{aligned} x + y &= 12 \quad (2)\\ x + (2x - 3) &= 12 \quad \text{Substituting}\\ x + 2x - 3 &= 12\\ 3x - 3 &= 12\\ 3x &= 15\\ x &= 5 \end{aligned}$$
Substitute 5 for x in Equation (2) and solve for y.
$$\begin{aligned} y &= 2x - 3 \quad (1)\\ y &= 2(5) - 3 = 7 \end{aligned}$$
We obtain (5, 7) as the solution.

5. $$\begin{aligned} 12x - 19y &= 13 \quad (1)\\ 8x + 19y &= 7 \quad (2)\\ \hline 20x \quad &= 20 \quad \text{Adding}\\ x &= 1 \end{aligned}$$
Substitute 1 for x in Equation (2) and solve for y.
$$\begin{aligned} 8x + 19y &= 7 \quad (2)\\ 8(1) + 19y &= 7 \quad \text{Substituting}\\ 19y &= -1\\ y &= -\frac{1}{19} \end{aligned}$$
We obtain $\left(1, -\frac{1}{19}\right)$ as the solution.

6. $2x - 5y = 1,\quad (1)$
$3x + 2y = 11\quad (2)$
We multiply Equation (1) by 2 and Equation (2) by 5.
$$\begin{aligned} 4x - 10y &= 2 \quad \text{Multiplying (1) by 2}\\ 15x + 10y &= 55 \quad \text{Multiplying (2) by 5}\\ \hline 19x \quad &= 57\\ x &= 3 \end{aligned}$$
Substitute 3 for x in Equation (2) and solve for y.
$$\begin{aligned} 3x + 2y &= 11 \quad (2)\\ 3(3) + 2y &= 11\\ 9 + 2y &= 11\\ 2y &= 2\\ y &= 1 \end{aligned}$$
We obtain (3, 1) as the solution.

7. $y = \frac{5}{3}x + 7,\quad (1)$
$y = \frac{5}{3}x - 8\quad (2)$
We see that both lines have the same slope and are parallel. There is no solution.

8. $x = 2 - y,\quad (1)$
$3x + 3y = 6\quad (2)$
Substitute $2 - y$ for x in Equation (2) and solve for y.
$$\begin{aligned} 3x + 3y &= 6 \quad (2)\\ 3(2 - y) + 3y &= 6 \quad \text{Substituting}\\ 6 - 3y + 3y &= 6\\ 6 &= 6 \end{aligned}$$
There are many solutions.
The solution set is $\{(x, y) | x = 2 - y\}$.

Exercise Set 8.2

1. To use the *substitution* method, a variable must be isolated.

3. To eliminate a variable by adding, two terms must be *opposites*.

5. Adding the equations, we get $8x = 7$, so choice (d) is correct.

7. Multiplying the first equation by -5 gives us the system of equations in (a), so choice (a) is correct.

9. Substituting $4x-7$ for y in the second equation gives us $6x+3(4x-7)=19$, so choice (c) is correct.

11. $y = 3-2x,\quad (1)$
$3x + y = 5\quad (2)$
We substitute $3-2x$ for y into the second equation and solve for x.

$$\begin{aligned} 3x+y&=5 && (2)\\ 3x+(3-2x)&=5 && \text{Substituting}\\ x+3&=5\\ x&=2 \end{aligned}$$

Next substitute 2 for x in either equation of the original system and solve for y.

$$\begin{aligned} y&=3-2x && (1)\\ y&=3-2\cdot 2 && \text{Substitute}\\ y&=3-4\\ y&=-1 \end{aligned}$$

We check the ordered pair $(2,-1)$.

$y=3-2x$	
-1	$3-2\cdot 2$
	$3-4$
$-1\overset{?}{=}-1$ TRUE	

$3x+y=5$	
$3\cdot 2+(-1)$	5
$6-1$	
$5\overset{?}{=}5$ TRUE	

Since $(2,-1)$ checks, it is the solution.

13. $3x+5y=3,\quad (1)$
$x=8-4y\quad (2)$
We substitute $8-4y$ for x in the first equation and solve for y.

$$\begin{aligned} 3x+5y&=3 && (1)\\ 3(8-4y)+5y&=3 && \text{Substituting}\\ 24-12y+5y&=3\\ 24-7y&=3\\ -7y&=-21\\ y&=3 \end{aligned}$$

Next we substitute 3 for y in either equation of the original system and solve for x.

$$\begin{aligned} x&=8-4y \quad (2)\\ x&=8-4\cdot 3=8-12=-4 \end{aligned}$$

We check the ordered pair $(-4,3)$.

$x=8-4y$	
-4	$8-4\cdot 3$
	$8-12$
$-4\overset{?}{=}-4$ TRUE	

$3x+5y=3$	
$3(-4)+5\cdot 3$	3
$-12+15$	
$3\overset{?}{=}3$ TRUE	

Since $(-4,3)$ checks, it is the solution.

15. $3s-4t=14,\quad (1)$
$5s+t=8\quad (2)$
We solve the second equation for t.

$$\begin{aligned} 5s+t&=8 && (2)\\ t&=8-5s && (3) \end{aligned}$$

We substitute $8-5s$ for t in the first equation and solve for s.

$$\begin{aligned} 3s-4t&=14 && (1)\\ 3s-4(8-5s)&=14 && \text{Substituting}\\ 3s-32+20s&=14\\ 23s-32&=14\\ 23s&=46\\ s&=2 \end{aligned}$$

Next we substitute 2 for s in Equation (1), (2), or (3). It is easiest to use Equation (3) since it is already solved for t.

$$t=8-5\cdot 2=8-10=-2$$

We check the ordered pair $(2,-2)$.

$3s-4t=14$	
$3\cdot 2-4(-2)$	14
$6+8$	
$14\overset{?}{=}14$ TRUE	

$5s+t=8$	
$5\cdot 2+(-2)$	8
$10-2$	
$8\overset{?}{=}8$ TRUE	

Since $(2,-2)$ checks, it is the solution.

17. $4x-2y=6,\quad (1)$
$2x-3=y\quad (2)$
We substitute $2x-3$ for y in the first equation and solve for x.

$$\begin{aligned} 4x-2y&=6 && (1)\\ 4x-2(2x-3)&=6\\ 4x-4x+6&=6\\ 6&=6 \end{aligned}$$

We have an identity, or an equation that is always true. The equations are dependent and the solution set is infinite: $\{(x,y)\,|\,2x-3=y\}$.

19. $-5s+t=11,\quad (1)$
$4s+12t=4\quad (2)$
We solve the first equation for t.

$$\begin{aligned} -5s+t&=11 && (1)\\ t&=5s+11 && (3) \end{aligned}$$

We substitute $5s+11$ for t in the second equation and solve for s.

$$\begin{aligned} 4s+12t&=4 && (2)\\ 4s+12(5s+11)&=4\\ 4s+60s+132&=4\\ 64s+132&=4\\ 64s&=-128\\ s&=-2 \end{aligned}$$

Next we substitute -2 for s in Equation (3).

$$t=5s+11=5(-2)+11=-10+11=1$$

We check the ordered pair $(-2,1)$.

$-5s+t=11$	
$-5(-2)+1$	11
$10+1$	
$11\overset{?}{=}11$ TRUE	

$4s+12t=4$	
$4(-2)+12\cdot 1$	4
$-8+12$	
$4\overset{?}{=}4$ TRUE	

Since $(-2,1)$ checks, it is the solution.

21. $2x+2y=2,\quad (1)$
$3x-y=1\quad (2)$
We solve the second equation for y.

$$\begin{aligned} 3x-y&=1 && (2)\\ -y&=-3x+1\\ y&=3x-1 && (3) \end{aligned}$$

We substitute $3x-1$ for y in the first equation and solve for x.

$$\begin{aligned} 2x+2y&=2 \quad (1)\\ 2x+2(3x-1)&=2\\ 2x+6x-2&=2\\ 8x-2&=2\\ 8x&=4\\ x&=\tfrac{1}{2} \end{aligned}$$

Next we substitute $\frac{1}{2}$ for x in Equation (3).

$$y=3x-1=3\cdot\frac{1}{2}-1=\frac{3}{2}-1=\frac{1}{2}$$

The ordered pair $\left(\frac{1}{2}, \frac{1}{2}\right)$ checks in both equations. It is the solution.

23. $2a+6b=4,$ (1)
$3a-b=6$ (2)

We solve the second equation for b.

$$\begin{aligned} 3a-b&=6 \quad (2)\\ -b&=-3a+6\\ b&=3a-6 \quad (3) \end{aligned}$$

We substitute $3a-6$ for b in the first equation and solve for a.

$$\begin{aligned} 2a+6b&=4 \quad (1)\\ 2a+6(3a-6)&=4\\ 2a+18a-36&=4\\ 20a-36&=4\\ 20a&=40\\ a&=2 \end{aligned}$$

We substitute 2 for a in Equation 3 and solve for y.

$$\begin{aligned} b&=3a-6 \quad (3)\\ b&=3\cdot2-6\\ b&=6-6\\ b&=0 \end{aligned}$$

The ordered pair (2, 0) checks in both equations. It is the solution.

25. $2x-3=y$ (1)
$y-2x=1,$ (2)

We substitute $2x-3$ for y in the second equation and solve for x.

$$\begin{aligned} y-2x&=1 && (2)\\ 2x-3-2x&=1 && \text{Substituting}\\ -3&=1 && \text{Collecting like terms} \end{aligned}$$

We have a contradiction, or an equation that is always false. Therefore, there is no solution.

27.
$$\begin{aligned} x+3y&=7 && (1)\\ -x+4y&=7 && (2)\\ \hline 0+7y&=14 && \text{Adding}\\ 7y&=14\\ y&=2 \end{aligned}$$

Substitute 2 for y in one of the original equations and solve for x.

$$\begin{aligned} x+3y&=7 && (1)\\ x+3\cdot2&=7 && \text{Substituting}\\ x+6&=7\\ x&=1 \end{aligned}$$

Check:

$$\begin{array}{c|c} x+3y=7 & \\ \hline 1+3\cdot2 & 7\\ 1+6 & \\ 7\overset{?}{=}7 & \text{TRUE} \end{array} \qquad \begin{array}{c|c} -x+4y=7 & \\ \hline -1+4\cdot2 & 7\\ -1+8 & \\ 7\overset{?}{=}7 & \text{TRUE} \end{array}$$

Since (1, 2) checks, it is the solution.

29.
$$\begin{aligned} x-2y&=11 && (1)\\ 3x+2y&=17 && (2)\\ \hline 4x\quad&=28 && \text{Adding}\\ x&=7 \end{aligned}$$

Substitute 7 for x in Equation (1) and solve for y.

$$\begin{aligned} x-2y&=11 && (1)\\ 7-2y&=11 && \text{Substituting}\\ -2y&=4\\ y&=-2 \end{aligned}$$

We obtain (7, –2). This checks, so it is the solution.

31.
$$\begin{aligned} 9x+3y&=-3 && (1)\\ 2x-3y&=-8 && (2)\\ \hline 11x+0&=-11 && \text{Adding}\\ 11x&=-11\\ x&=-1 \end{aligned}$$

Substitute –1 for x in Equation (1) and solve for y.

$$\begin{aligned} 9x+3y&=-3\\ 9(-1)+3y&=-3 && \text{Substituting}\\ -9+3y&=-3\\ 3y&=6\\ y&=2 \end{aligned}$$

We obtain (–1, 2). This checks, so it is the solution.

33. $5x+3y=19,$ (1)
$x-6y=11$ (2)

We multiply Equation (1) by 2 to make two terms become opposites.

$$\begin{aligned} 10x+6y&=38 && \text{Multiplying (1) by 2}\\ x-6y&=11\\ \hline 11x\quad&=49\\ x&=\frac{49}{11} \end{aligned}$$

Substitute $\frac{49}{11}$ for x in Equation (1) and solve for y.

$$\begin{aligned} 5x+3y&=19\\ 5\cdot\frac{49}{11}+3y&=19\\ 3y&=-\frac{36}{11}\\ y&=-\frac{12}{11} \end{aligned}$$

We obtain $\left(\frac{49}{11},-\frac{12}{11}\right)$. This checks, so it is the solution.

35. $5r-3s=24,$ (1)
$3r+5s=28$ (2)

We multiply twice to make two terms become additive inverses.

$$\begin{aligned} \text{From (1): } 25r-15s&=120 && \text{Multiplying by 5}\\ \text{From (2): } 9r+15s&=84 && \text{Multiplying by 3}\\ \hline 34r+0&=204 && \text{Adding}\\ r&=6 \end{aligned}$$

Substitute 6 for r in Equation (2) and solve for s.

$$\begin{aligned} 3r+5s&=28 \\ 3\cdot 6+5s&=28 \quad \text{Substituting} \\ 18+5s&=28 \\ 5s&=10 \\ s&=2 \end{aligned}$$

We obtain (6, 2). This checks, so it is the solution.

37. $6s+9t=12,$ (1)
$4s+6t=5$ (2)

We multiply twice to make two terms become opposites.

From (1): $12s+18t=24$ Multiplying by 2
From (2): $-12s-18t=-15$ Multiplying by -3

$$0=9$$

We get a contradiction, or an equation that is always false. The system has no solution.

39. $\frac{1}{2}x-\frac{1}{6}y=10,$ (1)
$\frac{2}{5}x+\frac{1}{2}y=8$ (2)

We first multiply each equation by the LCM of the denominators to clear fractions.

$3x-y=60,$ (3) Multiplying (1) by 6
$4x+5y=80$ (4) Multiplying (2) by 10

We multiply Equation (3) by 5 and then add.

$15x-5y=300$ Multiplying (3) by 5
$4x+5y=80$ (4)

$$\begin{aligned} 19x &= 380 \\ x&=20 \end{aligned}$$

Substitute 20 for x in one of the equations in which the fractions were cleared and solve for y.

$$\begin{aligned} 3x-y&=60 \quad (3) \\ 3\cdot 20-y&=60 \\ 60-y&=60 \\ -y&=0 \\ y&=0 \end{aligned}$$

We obtain $(20, 0)$. This checks, so it is the solution.

41. $\frac{x}{2}+\frac{y}{3}=\frac{7}{6},$ (1)
$\frac{2x}{3}+\frac{3y}{4}=\frac{5}{4}$ (2)

We first multiply each equation by the LCM of the denominators to clear fractions.

$3x+2y=7$ (3) Multiplying (1) by 6
$8x+9y=15$ (4) Multiplying (2) by 12

We multiply twice to make two terms become opposites.

From (3):
From (4):

$27x+18y=63$ Multiplying by 9
$-16x-18y=-30$ Multiplying by -2
$11x=33$ Adding
$x=3$

Substitute 3 for x in one of the equations in which the fractions were cleared and solve for y.

$$\begin{aligned} 3x+2y&=7 \quad (3) \\ 3\cdot 3+2y&=7 \quad \text{Substituting} \\ 9+2y&=7 \\ 2y&=-2 \\ y&=-1 \end{aligned}$$

We obtain (3, –1). This checks, so it is the solution.

43. $12x-6y=-15,$ (1)
$-4x+2y=5$ (2)

Observe that, if we multiply Equation (1) by $-\frac{1}{3}$, we obtain Equation (2). Thus, any pair that is a solution of Equation (1) is also a solution of Equation (2). The equations are dependent and the solution set is infinite: $\{(x, y)\,|\,-4x+2y=5\}$.

45. $0.3x+0.2y=0.3,$
$0.5x+0.4y=0.4$

We first multiply each equation by 10 to clear decimals.

$3x+2y=3,$ (1)
$5x+4y=4$ (2)

We multiply Equation (1) by –2.

$-6x-4y=-6$ Multiplying (1) by -2
$5x+4y=4$ (2)

$$\begin{aligned} -x &= -2 \\ x&=2 \end{aligned}$$

Substitute 2 for x in Equation (1) and solve for y.

$$\begin{aligned} 3x+2y&=3 \quad (1) \\ 3\cdot 2+2y&=3 \\ 6+2y&=3 \\ 2y&=-3 \\ y&=-\frac{3}{2} \end{aligned}$$

We obtain $\left(2,-\frac{3}{2}\right)$. This checks, so it is the solution.

47. $a-2b=16,$ (1)
$b+3=3a$ (2)

We will use the substitution method. First solve Equation (1) for a.

$$\begin{aligned} a-2b&=16 \\ a&=2b+16 \quad (3) \end{aligned}$$

Now substitute $2b+16$ for a in Equation (2) and solve for b.

$$\begin{aligned} b+3&=3a \quad (2) \\ b+3&=3(2b+16) \quad \text{Substituting} \\ b+3&=6b+48 \\ -45&=5b \\ -9&=b \end{aligned}$$

Substitute –9 for b in Equation (3).

$$a=2(-9)+16=-2$$

We obtain (–2, –9). This checks, so it is the solution.

49. $10x+y=306,$ (1)
$10y+x=90$ (2)

We will use the substitution method. First solve Equation (1) for y.

$10x + y = 306$
$y = -10x + 306 \quad (3)$

Now substitute $-10x + 306$ for y in Equation (2) and solve for y.

$$\begin{aligned} 10y + x &= 90 && (2) \\ 10(-10x + 306) + x &= 90 && \text{Substituting} \\ -100x + 3060 + x &= 90 \\ -99x + 3060 &= 90 \\ -99x &= -2970 \\ x &= 30 \end{aligned}$$

Substitute 30 for x in Equation (3).

$y = -10 \cdot 30 + 306 = 6$

We obtain (30, 6). This checks, so it is the solution.

51. $6x - 3y = 3, \quad (1)$
$4x - 2y = 2 \quad (2)$

Observe that, if we multiply Equation (1) by $\frac{3}{2}$, we obtain Equation (2). Thus, any pair that is a solution of Equation (1) is also a solution of Equation (2). The equations are dependent and the solution set infinite: $\{(x, y) | 4x - 2y = 2\}$.

53. $3s - 7t = 5,$
$7t - 3s = 8$

First we rewrite the second equation with the variables in a different order. Then we use the elimination method.

$$\begin{aligned} 3s - 7t &= 5 && (1) \\ -3s + 7t &= 8 && (2) \\ \hline 0 &= 13 \end{aligned}$$

We get a contradiction, so the system has no solution.

55. $0.05x + 0.25y = 22,$
$0.15x + 0.05y = 24$

We first multiply each equation by 100 to clear decimals.

$5x + 25y = 2200, \quad (1)$
$15x + 5y = 2400 \quad (2)$

We multiply by –5 on both sides of the second equation and add.

$$\begin{aligned} 5x + 25y &= 2200 && (1) \\ -75x - 25y &= -12{,}000 && \text{Multiplying (2) by } -5 \\ \hline -70x \qquad &= -9800 && \text{Adding} \\ x &= \frac{-9800}{-70} \\ x &= 140 \end{aligned}$$

Substitute 140 for x in one of the equations in which the decimals were cleared and solve for y.

$$\begin{aligned} 5x + 25y &= 2200 && (1) \\ 5 \cdot 140 + 25y &= 2200 && \text{Substituting} \\ 700 + 25y &= 2200 \\ 25y &= 1500 \\ y &= 60 \end{aligned}$$

We obtain (140, 60). This checks, so it is the solution.

57. $13a - 7b = 9, \quad (1)$
$2a - 8b = 6 \quad (2)$

We will use the elimination method. First we multiply the equations so that the b-terms can be eliminated.

$$\begin{aligned} &\text{From (1): } & 104a - 56b &= 72 && \text{Multiplying by 8} \\ &\text{From (2): } & -14a + 56b &= -42 && \text{Multiplying by } -7 \\ \hline & & 90a \qquad &= 30 && \text{Adding} \\ & & a &= \frac{1}{3} \end{aligned}$$

Substitute $\frac{1}{3}$ for a in one of the equations and solve for b.

$$\begin{aligned} 2a - 8b &= 6 && (2) \\ 2 \cdot \frac{1}{3} - 8b &= 6 && \text{Substituting} \\ \frac{2}{3} - 8b &= 6 \\ -8b &= \frac{16}{3} \\ b &= -\frac{2}{3} \end{aligned}$$

We obtain $\left(\frac{1}{3}, -\frac{2}{3}\right)$. This checks, so it is the solution.

59. $a - \frac{1}{2}c = 6,$
$c + 2a = 8$

We first multiply first equation by 2 to clear fractions.

$2a - c = 12, \quad (1)$
$2a + c = 8 \quad (2)$

We will use the elimination method.

$$\begin{aligned} 2a - c &= 12 && (1) \\ 2a + c &= 8 && (2) \\ \hline 4a \qquad &= 20 && \text{Adding} \\ a &= 5 \end{aligned}$$

Substitute 5 for a in Equation (2) and solve for c.

$$\begin{aligned} c + 2a &= 8 && (2) \\ c + 2 \cdot 5 &= 8 && \text{Substituting} \\ c + 10 &= 8 \\ c &= -2 \end{aligned}$$

We obtain (5,–2). This checks, so it is the solution.

61. $8x = y - 14,$
$6(y - x) = 63$

We first write the equations in standard form.

$8x - y = -14, \quad (1)$
$-6x + 6y = 63 \quad (2)$

We will use the elimination method. First write each equation in the form $Ax + By = C$.

$$\begin{aligned} 48x - 6y &= -84 && \text{Mutiplying (1) by 6} \\ -6x + 6y &= 63 && (2) \\ \hline 42x \qquad &= -21 && \text{Adding} \\ x &= -\frac{1}{2} \end{aligned}$$

Substitute $-\frac{1}{2}$ for x in one of the equations solve for y.

$$\begin{aligned} -6x + 6y &= 63 && (2) \\ -6\left(-\frac{1}{2}\right) + 6y &= 63 && \text{Substituting} \\ 3 + 6y &= 63 \\ 6y &= 60 \\ y &= 10 \end{aligned}$$

We obtain $\left(-\frac{1}{2}, 10\right)$. This checks, so it is the solution.

63. $2m+6n=4,\quad (1)$
$4m-2n=6\quad (2)$

We will use the elimination method. We multiply Equation (2) by 3.

$$\begin{aligned} 2m+6n&=4 && (1)\\ 12m-6n&=18 && \text{Multiplying (2) by 3}\\ \hline 14m&=22 && \text{Adding}\\ m&=\frac{11}{7} \end{aligned}$$

Substitute $\frac{11}{7}$ for m in Equation (1) and solve for n.

$$\begin{aligned} 2m+6n&=4\\ 2\left(\frac{11}{7}\right)+6n&=4 && \text{Substituting}\\ \frac{22}{7}+6n&=4\\ 6n&=\frac{6}{7}\\ n&=\frac{1}{7} \end{aligned}$$

We obtain $\left(\frac{11}{7}, \frac{1}{7}\right)$. This checks, so it is the solution.

65. $23x-y=5,$
$11x-10=2y$

We will use the elimination method. First write each equation in the form $Ax+By=C$.

$$\begin{aligned} 23x-y&=5, && (1)\\ 11x-2y&=10 && (2) \end{aligned}$$

We multiply by –2 on both sides of the first equation and add.

$$\begin{aligned} -48x+2y&=-10 && \text{Mutiplying (1) by }-2\\ 11x-2y&=10 && (2)\\ \hline -37x\quad&=0 && \text{Adding}\\ x&=0 \end{aligned}$$

Substitute 0 for x in one of the equations solve for y.

$$\begin{aligned} 11x-10&=2y && (2)\\ 11(0)-10&=2y && \text{Substituting}\\ -10&=2y\\ -5&=y \end{aligned}$$

We obtain (0,–5). This checks, so it is the solution.

67. *Writing Exercise.*

69. $(4+m)+n=4+(m+n)$

71. $$\begin{aligned} 8x-3[5x+2(6-9x)]&=8x-3[5x+12-18x]\\ &=8x-15x-36+54x\\ &=47x-36 \end{aligned}$$

73. $30{,}050{,}000=3.005\times10^7$

75. *Writing Exercise.*

77. First write $f(x)=mx+b$ as $y=mx+b$. Then substitute 1 for x and 2 for y to get one equation and also substitute –3 for x and 4 for y to get a second equation:

$$\begin{aligned} 2&=m\cdot1+b\\ 4&=m(-3)+b \end{aligned}$$

Solve the resulting system of equations.

$$\begin{aligned} 2&=m+b\\ 4&=-3m+b \end{aligned}$$

Multiply the second equation by –1 and add.

$$\begin{aligned} 2&=m+b\\ -4&=3m-b\\ \hline -2&=4m\\ -\frac{1}{2}&=m \end{aligned}$$

Substitute $-\frac{1}{2}$ for m in the first equation and solve for b.

$$\begin{aligned} 2&=-\frac{1}{2}+b\\ \frac{5}{2}&=b \end{aligned}$$

Thus, $m=-\frac{1}{2}$ and $b=\frac{5}{2}$.

79. Substitute –4 for x and –3 for y in both equations and solve for a and b.

$$\begin{aligned} -4a-3b&=-26, && (1)\\ -4b+3a&=7 && (2)\\ -12a-9b&=-78 && \text{Multiplying (1) by 3}\\ 12a-16b&=28 && \text{Multiplying (2) by 4}\\ \hline -25b&=-50\\ b&=2 \end{aligned}$$

Substitute 2 for b in Equation (2).

$$\begin{aligned} -4\cdot2+3a&=7\\ 3a&=15\\ a&=5 \end{aligned}$$

Thus, $a=5$ and $b=2$.

81. $\frac{x+y}{2}-\frac{x-y}{5}=1,$
$\frac{x-y}{2}+\frac{x+y}{6}=-2$

After clearing fractions we have:

$$\begin{aligned} 3x+7y&=10, && (1)\\ 4x-2y&=-12 && (2) \end{aligned}$$

$$\begin{aligned} 6x+14y&=20 && \text{Multiplying (1) by 2}\\ 28x-14y&=-84 && \text{Multiplying (2) by 7}\\ \hline 34x\quad&=-64\\ x&=-\frac{32}{17} \end{aligned}$$

Substitute $-\frac{32}{17}$ for x in Equation (1).

$$\begin{aligned} 3\left(-\frac{32}{17}\right)+7y&=10\\ 7y&=\frac{266}{17}\\ y&=\frac{38}{17} \end{aligned}$$

The solution is $\left(-\frac{32}{17}, \frac{38}{17}\right)$.

83. $\frac{2}{x}+\frac{1}{y}=0,$ or $2\cdot\frac{1}{x}+\frac{1}{y}=0,$

$\frac{5}{x}+\frac{2}{y}=-5$ $\quad 5\cdot\frac{1}{x}+2\cdot\frac{1}{y}=-5$

Substitute u for $\frac{1}{x}$ and v for $\frac{1}{y}$.

$2u+v=0,$ (1)
$5u+2v=-5$ (2)

$-4u-2v=0$ Multiplying (1) by -2
$5u+2v=-5$ (2)
$u=-5$

Substitute -5 for u in Equation (1).

$2(-5)+v=0$
$-10+v=0$
$v=10$

If $u=-5$, then $\frac{1}{x}=-5$. Thus $x=-\frac{1}{5}$.

If $v=10$, then $\frac{1}{y}=10$. Thus $y=\frac{1}{10}$.

The solution is $\left(-\frac{1}{5},\frac{1}{10}\right)$.

85. ***Familiarize***. Let w = the number of kilowatt hours of electricity used each month by the toaster oven. Then $4w$ = the number of kilowatt hours used by the convection oven. The sum of these two numbers is 15 kilowatt hours.

Translate.

kilowatt hours for toaster oven	plus	kilowatt hours for convection oven	is	15.
↓	↓	↓	↓	↓
w	$+$	$4w$	$=$	15

Carry out. We solve the equation.

$w+4w=15$
$5w=15$
$w=3$

If $w=3$, then $4w=12$.

Check. We have 3 kWh + 4(3) kWh = 15 kWh. The answer checks.

State. For the month, the toaster oven uses 3 kWh and the convection oven uses 12 kWh.

87. *Writing Exercise.*

Exercise Set 8.3

1. If 10 coffee mugs are sold for \$8 each, the *total value* of the mugs is \$80.

3. To solve a motion problem, we often use the fact that *distance* divided by rate equals time.

5. The Familiarize and Translate steps were done in Exercise 41 of Exercise Set 8.1.

Carry out. We solve the system of equations.

$x+y=10,$ (1)
$x=\frac{2}{3}y$ (2)

where x is the first number and y is the second number.

We use substitution.

$\frac{2}{3}y+y=10$
$\frac{5}{3}y=10$
$y=6$

Now substitute 6 for y in Equation (2).

$x=\frac{2}{3}(6)=4$

Check. The sum of the numbers is 4 + 6, or 10 and $\frac{2}{3}$ times the second number, 6, is the first number 4. The answer checks.

State. The first number is 4, and the second number is 6.

7. The Familiarize and Translate steps were done in Exercise 43 of Exercise Set 8.1.

Carry out. We solve the system of equations.

$p+a=1223,$ (1)
$p=a+243$ (2)

where p is the number of endangered plant species and a is the number of endangered animal species. We use substitution. Substitute $a+243$ for p in Equation (1) and solve for a.

$(a+243)+a=1223$
$2a+243=1223$
$2a=980$
$a=490$

Now substitute 490 for a in Equation (2).

$p=(490)+243=733$

Check. The sum of the endangered plant and animal species is 733 + 490, or 1223. The answer checks.

State. There are 733 endangered plant species and 490 endangered animal species.

9. The Familiarize and Translate steps were done in Exercise 45 of Exercise Set 8.1.

Carry out. We solve the system of equations

$x+y=180,$ (1)
$x=2y-3$ (2)

where x = the measure of one angle and y = the measure of the other angle. We use substitution. Substitute $2y-3$ for x in (1) and solve for y.

$2y-3+y=180$
$3y-3=180$
$3y=183$
$y=61$

Now substitute 61 for y in (2).

$x=2\cdot 61-3=122-3=119$

Check. The sum of the angle measures is 119° + 61°, or 180°, so the angles are supplementary. Also $2\cdot 61°-3°=122°-3°=119°$. The answer checks.

State. The measures of the angles are 119° and 61°.

11. The Familiarize and Translate steps were done in Exercise 47 of Exercise Set 8.1.

Carry out. We solve the system of equations

$x+y=64,$ (1)
$2x+y=100$ (2)

where x = the number of two-point shots and y = the

number of foul shots made. We use elimination.

$$\begin{array}{rl} -x-y=-64 & \text{Multiplying (1) by } -1 \\ \underline{2x+y=100} & \\ x=36 & \end{array}$$

Substitute 36 for x in (1) and solve for y.

$$36+y=64$$
$$y=28$$

Check. The total number of scores was 36 + 28, or 64. The total number of points was $2\cdot 36+28=72+28=100$. The answer checks.

State. Chamberlain made 36 two-point shots and 28 foul shots.

13. The Familiarize and Translate steps were done in Exercise 49 of Exercise Set 8.1.
We can multiply both sides of the second equation by 100 to clear the decimals.

$$w+u=32,$$
$$449w+299u=10,768.$$

Carry out. We solve the system of equations.

$$w+u=32, \quad (1)$$
$$449x+299y=10,768 \quad (2)$$

where w is the number of wrapped strings and u is the number of unwrapped strings. We use elimination. Begin by multiplying Equation (1) by –299.

$$\begin{array}{rl} -299w-299u=-9568 & \text{Multiplying (1) by } -299 \\ \underline{449w+299u=10,768} & (2) \\ 150w \quad =1200 & \\ w=8 & \end{array}$$

Substitute 8 for w in Equation (1) and solve for u.

$$8+u=32$$
$$u=24$$

Check. The number of wrapped and unwrapped strings is 8 + 24, or 32. The amount paid was $\$14.49(8)+\$2.99(24)=\$35.92+\$71.76=\$107.68$. The answer checks.

State. 8 wrapped and 24 unwrapped strings were bought.

15. The Familiarize and Translate steps were done in Exercise 51 of Exercise Set 8.1.

$$h+s=110,$$
$$8h+12s=1072.$$

Carry out. We solve the system of equations.

$$h+s=110, \quad (1)$$
$$8h+12s=1072 \quad (2)$$

where h = the number of hats and s = the number of scarves knitted.. We use elimination.

$$\begin{array}{rl} -8h-8s=-880 & \text{Multiplying (1) by } -8 \\ \underline{8h+12s=1072} & (2) \\ 4s=192 & \\ s=48 & \end{array}$$

Substitute 48 for s in Equation (1) and solve for h.

$$h+48=110$$
$$h=62$$

Check. A total of 48 + 62, or 110 items were knitted. The time spent knitting, in hours, was $8(62)+12(48)=496+576=1072$. The answer checks.

State. 62 hats and 48 scarves were knitted.

17. The Familiarize and Translate steps were done in Exercise 53 of Exercise Set 8.1.

Carry out. We solve the system of equations.

$$2l+2w=340, \quad (1)$$
$$l=w+50 \quad (2)$$

where l = the length, in yards, and w = the width, in yards of the lacrosse field. We use substitution. We substitute w + 50 for l in Equation (1) and solve for w.

$$2(w+50)+2w=340$$
$$2w+100+2w=340$$
$$4w+100=340$$
$$4w=240$$
$$w=60$$

Now substitute 60 for w in Equation (2).

$$l=60+50=110$$

Check. The perimeter is $2\cdot 110+2\cdot 60=220+120=340$. The length, 110 yards, is 50 yards more than the width, 60 yards. The answer checks.

State. The length of the lacrosse field is 110 yards, and the width is 60 yards.

19. ***Familiarize***. Let x = the number of MWH (in thousands) of wind and y = the number of MWH (in thousands) of solar generated.

Translate.

$$\underbrace{\text{The total MWH}}_{x+y} \quad \underbrace{\text{is}}_{=} \quad \underbrace{218.}_{218}$$

$$\underbrace{\text{The wind}}_{x} \quad \underbrace{\text{is}}_{=} \quad \underbrace{2}_{2} \quad \underbrace{\text{more than}}_{+} \quad \underbrace{\text{7 times the solar.}}_{7y}$$

We have translated to a system of equations:

$$x+y=218, \quad (1)$$
$$x=2+7y \quad (2)$$

Carry out. We use the substitution method to solve the system of equations. Substitute 2 + 7y for x in Equation (1) and solve for y.

$$2+7y+y=218$$
$$8y+2=218$$
$$8y=216$$
$$y=27$$

Substitute 27 for y in Equation (1) and solve for x.

$$x+27=218$$
$$x=191$$

Check. A total of 27 + 191, or 218 MWH were generated. The wind was 2 + 7(27) = 191. The answer checks.

State. 191 thousand MWH of wind and 27 thousand MWH of solar were generated.

21. ***Familiarize***. Let x = the number of 3-credit coursess and y = the number of 4-credit courses taken.

Translate. We organize the information in a table.

	3-credit courses	4-credit courses	Total
Number taken	x	y	48
Credits	3	4	
Total credits	$3x$	$4y$	155

We get one equation from the "Number taken" row of the table:

$$x+y=48$$

The "Total credits" row yields a second equation:

$$3x+4y=155$$

We have translated to a system of equations:

$$x+y=48 \quad (1)$$
$$3x+4y=155 \quad (2)$$

Carry out. We use the elimination method to solve the system of equations.

$$-3x-3y=-144 \quad \text{Multiplying (1) by } -3$$
$$3x+4y=155 \quad (2)$$
$$y=11$$

Substitute 11 for y in Equation (1) and solve for x.

$$x+11=48$$
$$x=37$$

Check. A total of 37 + 11, or 48 courses, were taken. The total credits were 3(37) + 4(11) = 111 + 44 = 155. The answer checks.

State. 37 3-credit courses and 11 4-credit courses were taken.

23. ***Familiarize***. Let x = the number of cases of regular paper used and y = the number of cases of recycled paper.

Translate. We organize the information in a table.

	Regular	Recycled paper	Total
Number purchased	x	y	27
Price	\$46.99	\$61.99	
Total cost	$46.99x$	$61.99y$	1433.73

We get one equation from the "Number purchased" row of the table: $x+y=27$

The "Total cost" row yields a second equation. All costs are expressed in dollars.

$$46.99x + 61.99y = 1433.73.$$

We have the problem translated to a system of equations.

$$x+y=27 \quad (1)$$
$$46.99x+61.99y=1433.73 \quad (2)$$

Carry out. We use the elimination method to solve the system of equations.

$$-4699x-4699y=-126{,}873 \quad \text{Multiplying (1) by} -4699$$
$$4699x+6199y=143{,}373 \quad \text{Multiplying (2) by 100}$$
$$1500y=16{,}500$$
$$y=11$$

Substitute 11 for y in Equation (1) and solve for x.

$$x+11=27$$
$$x=16$$

Check. A total of 16 + 11, or 27 cases of paper were used. The total cost was \$46.99(16) + \$61.99(11) = \$751.84 + \$681.89 = \$1433.73. The answer checks.

State. 16 cases of regular paper and 11 cases of recycled paper were purchased.

25. ***Familiarize***. Let x = the number of 8.5-watt bulbs and y = the number of 18-watt bulbs purchased.

Translate. We organize the information in a table.

	8.5-watt bulbs	18-watt bulbs	Total
Number purchased	x	y	200
Price	\$3.97	\$8.97	
Total cost	$3.97x$	$8.97y$	1494

We get one equation from the "Number purchased" row of the table:

$$x+y=200$$

The "Total cost" row yields a second equation:

$$3.97x+8.97y=1494$$

We have translated to a system of equations:

$$x+y=200 \quad (1)$$
$$3.97x+8.97y=1494 \quad (2)$$

Carry out. We use the elimination method to solve the system of equations.

$$-397x-397y=-79{,}400 \quad \text{Multiplying (1) by } -397$$
$$397x+897y=149{,}400 \quad \text{Mulitplying (2) by 100}$$
$$500y=70{,}000$$
$$y=140$$

Substitute 140 for y in Equation (1) and solve for x.

$$x+140=200$$
$$x=60$$

Check. A total of 60 + 140, or 200 bulbs, were purchased. The total cost was

$$\$3.97(60)+\$8.97(140)=\$238.20+\$1255.80=\$1494$$

The answer checks.

State. 60 8.5-watt bulbs and 140 18-watt bulbs were purchased.

27. ***Familiarize***. Let s = the number of Starter customers and a = the number of Already Composting customers.

Translate.

	Starter	Already Composting	Total
Customers	s	a	215
Price	\$160	\$105	
Total revenue	$160s$	$105a$	26,975

We get one equation from the "Customers" row of the table:

$$s+a=215$$

The "Total revenue" row yields a second equation:

$$160s+105a=26{,}975$$

We have translated to a system of equations:

$$s+a=215 \quad (1)$$
$$160s+105a=26{,}975 \quad (2)$$

Carry out. We use the elimination method to solve the system of equations.

$$-105s-105a=-22{,}575 \quad \text{Multiplying (1) by } -105$$
$$160s+105a=26{,}975 \quad (2)$$
$$55s=4400$$
$$s=80$$

Substitute 80 for s in Equation (1) and solve for a.

$s+80=215$
$s=135$

Check. There are a total of 80 + 135, or 215 customers. The total revenue was $\$160(80)+\$105(135)$ $=\$12,800+\$14,175=\$26,975$. The answer checks.

State. There are 135 Starters and 80 Already Composting customers.

29. An immediate solution can be determined from the fact that \$19 is the average of equal parts of \$18 and \$20. So there is 14 lb of Mexican and 14 lb of Peruvian. Or we can solve using the usual method.

Familiarize. Let m = the number of pounds of Mexican coffee and p = the number of pounds of Peruvian coffee to be used in the mixture. The value of the mixture will be \$19(28), or \$432.

Translate. We organize the information in a table.

	Mexican	Peruvian	Mixture
Number of pounds	m	p	28
Price per pound	\$20	\$18	\$19
Value of coffee	$20m$	$18p$	532

The "Number of pounds" row of the table gives us one equation:

$m+p=28$

The "Value of coffee" row yields a second equation:

$20m+18p=532$

We have translated to a system of equations:

$m+p=28$ (1)
$20m+18p=532$ (2)

Carry out. We use the elimination method to solve the system of equations.

$-18m-18p=-504$ Multiplying (1) by -18
$20m+18p=532$ (2)
$2m=28$
$m=14$

Substitute 14 for m in Equation (1) and solve for p.

$14+p=28$
$p=14$

Check. The total mixture contains 14 lb + 14 lb, or 28 lb. Its value is $\$20\cdot 14+\$18\cdot 14$ $=\$280+\$252=\$532$. The answer checks.

State. 14 lb of Mexican coffee and 14 lb of Peruvian coffee should be used.

31. ***Familiarize***. Let x = the number of ounces of custom printed M&Ms and y = the number of ounces of bulk M&Ms. Converting lb to oz: 20 lb = 20(16) = 320 oz. The mixture's value will be \$0.59(320), or \$188.80.

Translate. We organize the information in a table.

	Custom printed	Bulk	Mixture
Number of ounces	x	y	320
Price per ounce	\$1.04	\$0.32	\$0.59
Value of M&Ms	$1.04x$	$0.32y$	188.80

The "Number of ounces" row of the table gives us one equation:

$x+y=320$

The "Value of M&Ms" row yields a second equation:

$1.04x+0.32y=188.80$

After clearing decimals, we have the problem translated to a system of equations:

$x+y=320$ (1)
$104x+32y=18,880$ (2)

Carry out. We use the elimination method to solve the system of equations.

$-32x-32y=-10,240$ Multiplying (1) by -32
$104x+32y=10,240$ (2)
$72x=8640$
$x=120$

Substitute 120 for x in Equation (1) and solve for y.

$120+y=320$
$y=200$

Check. The total mixture contains 120 oz + 200 oz, or 320 oz. Its value is

$\$1.04\cdot 120+\$0.32\cdot 200=\$188.80$. The answer checks.

State. 120 oz of custom-printed M&Ms and 200 oz of bulk M&Ms should be used.

33. ***Familiarize***. Let x = the number of mL of 50% acid solution and y = the number of mL of 80% acid solution. The amount of acid in the mixture is 68%(200 mL), or 0.68(200 mL) =136 mL.

Translate. We organize the information in a table.

	50% acid	80% acid	Mixture
Amount of solution	x	y	200
Percent acid	50%	80%	68%
Amount of acid in solution	$0.50x$	$0.80y$	136

The "Amount of solution" row of the table gives us one equation: $x+y=200$

The last row of the table yields a second equation:

$0.50x+0.80y=136$

After clearing decimals, we have the problem translated to a system of equations:

$x+y=200$ (1)
$5x+8y=1360$ (2)

Carry out. We use the elimination method to solve the system of equations.

$-5x-5y=-1000$ Multiplying (1) by -5
$5x+8y=1360$ (2)
$3y=360$
$y=120$

Substitute 120 for y in Equation (1) and solve for y.

$x+120=200$
$x=80$

Check. The amount of the mixture is 80 mL + 120 mL, or 200 mL. The amount of acid is

$0.50(80)+0.80(120)=40\text{ mL}+96\text{ mL}=136\text{ mL}$. The answer checks.

State. 80 mL of 50% acid solution and 120 mL of 80% acid solution should be used.

35. ***Familiarize***. Let x = the number of pounds 80% bluegrass and y = the number of pounds of 30% bluegrass to be used in the mixture. The blend of the mixture is 60%(50 lb), or $0.60(50 \text{ lb}) = 30 \text{ lb}$.

Translate. We organize the information in a table.

	80%	30%	Mixture
Number of pounds	x	y	50
Percent of blend	80%	30%	60%
Amount of blend	$0.80x$	$0.30y$	30 lb

We get one equation from the "Number of pounds" row of the table:

$$x + y = 50$$

The last row of the table yields a second equation:

$$0.8x + 0.3y = 30$$

After clearing decimals, we have the problem translated to a system of equations:

$$x + y = 50, \quad (1)$$
$$8x + 3y = 300 \quad (2)$$

Carry out. We use the elimination method to solve the system of equations.

$$-3x - 3y = -150 \quad \text{Multiplying (1) by } -3$$
$$8x + 3y = 300$$
$$5x = 150$$
$$x = 30$$

Substitute 30 for x in (1) and solve for y.

$$30 + y = 50$$
$$y = 20$$

Check. The amount of the mixture is 30 lb + 20 lb, or 50 lb. The amount of bluegrass in the mixture is 0.80(30 lb) + 0.30(20 lb) = 24 lb + 6 lb = 30 lb. The answer checks.

State. 30 lb of 80% bluegrass blend and 20 lb of 30% bluegrass blend should be mixed.

37. ***Familiarize***. Let x = the amount of the 3.2% loan and y = the amount of the 4.5% loan. Recall that the formula for simple interest is

Interest = Principal × Rate × Time

Translate. We organize the information in a table.

	3.2% loan	4.5% loan	Total
Principal	x	y	12,000
Interest rate	3.2%	4.5%	
Time	1 yr	1 yr	
Interest	$0.032x$	$0.045y$	442.50

The "Principal" row of the table gives us one equation:

$$x + y = 12{,}000$$

The last row of the table yields a second equation:

$$0.032x + 0.045y = 442.50$$

After clearing decimals, we have the problem translated to a system of equations:

$$x + y = 12{,}000 \quad (1)$$
$$32x + 45y = 442{,}500 \quad (2)$$

Carry out. We use the elimination method to solve the system of equations.

$$-32x - 32y = -384{,}000 \quad \text{Multiplying (1) by } -32$$
$$32x + 45y = 442{,}500 \quad (2)$$
$$13y = 58{,}500$$
$$y = 4500$$

Substitute 4500 for y in Equation (1) and solve for x.

$$x + 4500 = 12{,}000$$
$$x = 7500$$

Check. The loans total \$7500 + \$4500, or \$12,000. The total interest is 0.032(\$7500) 0.045(\$4500) = \$240 + \$202.50 = \$442.50. The answer checks.

State. The 3.2% loan was for \$7500 and the 4.5% loan was for \$4500.

39. ***Familiarize***. Let x = the number of liters of Steady State and y = the number of liters of Even Flow in the mixture. The amount of alcohol in the mixture is $0.15(20 \text{ L}) = 3 \text{ L}$.

Translate. We organize the information in a table.

	18% solution	10% solution	Mixture
Number of liters	x	y	20
Percent of alcohol	18%	10%	15%
Amount of alcohol	$0.18x$	$0.10y$	3

We get one equation from the "Number of liters" row of the table:

$$x + y = 20$$

The last row of the table yields a second equation:

$$0.18x + 0.1y = 3$$

After clearing decimals we have the problem translated to a system of equations:

$$x + y = 20, \quad (1)$$
$$18x + 10y = 300 \quad (2)$$

Carry out. We use the elimination method to solve the system of equations.

$$-10x - 10y = -200 \quad \text{Multiplying (1) by } -10$$
$$18x + 10y = 300 \quad (2)$$
$$8x = 100$$
$$x = 12.5$$

Substitute 12.5 for x in (1) and solve for y.

$$12.5 + y = 20$$
$$y = 7.5$$

Check. The total amount of the mixture is $12.5 \text{ L} + 7.5 \text{ L}$ or 20 L. The amount of alcohol in the mixture is $0.18(12.5 \text{ L}) + 0.1(7.5 \text{ L}) = 2.25 \text{ L} + 0.75 \text{ L} = 3 \text{ L}$. The answer checks.

State. 12.5 L of Steady State and 7.5 L of Even Flow should be used.

41. ***Familiarize***. Let x = the number of gallons of 87-octane gas and y = the number of gallons of 95-octane gas in the mixture. The amount of octane in the mixture can be expressed as 93(10), or 930.

Translate. We organize the information in a table.

	87-octane	95-octane	Mixture
Number of gallons	x	y	10
Octane rating	87	95	93
Total octane	$87x$	$95y$	930

We get one equation from the "Number of gallons" row of the table :

$$x + y = 10$$

The last row of the table yields a second equation:

$87x + 95y = 930$

We have a system of equations:

$$\begin{aligned} x + y &= 10 \quad (1) \\ 87x + 95y &= 930 \quad (2) \end{aligned}$$

Carry out. We use the elimination method to solve the system of equations.

$$\begin{aligned} -87x - 87y &= -870 \quad \text{Multiplying (1) by } -87 \\ 87x + 95y &= 930 \quad (2) \\ \hline 8y &= 60 \\ y &= 7.5 \end{aligned}$$

Substitute 7.5 for y in Equation (1) and solve for x.

$$\begin{aligned} x + 7.5 &= 10 \\ x &= 2.5 \end{aligned}$$

Check. The total amount of the mixture is 2.5 gal + 7.5 gal, or 10 gal. The amount of octane can be expressed as $87(2.5) + 95(7.5) = 217.5 + 712.5 = 930$. The answer checks.

State. 2.5 gal of 87-octane gas and 7.5 gal of 95-octane gas should be used.

43. ***Familiarize***. From the bar graph we see that whole milk is 4% milk fat, milk for cream cheese is 8% milk fat, and cream is 30% milk fat. Let x = the number of pounds of whole milk and y = the number of pounds of cream to be used. The mixture contains 8%(200 lb), or 0.08(200 lb) = 16 lb of milk fat.

Translate. We organize the information in a table.

	Whole milk	Cream	Mixture
Number of pounds	x	y	200
Percent of milk fat	4%	30%	8%
Amount of milk fat	$0.04x$	$0.30y$	16 lb

We get one equation from the "Number of pounds" row of the table:

$x + y = 200$

The last row of the table yields a second equation:

$0.04x + 0.3y = 16$

After clearing decimals, we have the problem translated to a system of equations:

$$\begin{aligned} x + y &= 200, \quad (1) \\ 4x + 30y &= 1600 \quad (2) \end{aligned}$$

Carry out. We use the elimination method to solve the system of equations.

$$\begin{aligned} -4x - 4y &= -800 \quad \text{Multiplying (1) by } -4 \\ 4x + 30y &= 1600 \quad (2) \\ \hline 26y &= 800 \\ y &= \frac{400}{13}, \text{ or } 30\frac{10}{13} \end{aligned}$$

Substitute $\frac{400}{13}$ for y in (1) and solve for x.

$$\begin{aligned} x + \frac{400}{13} &= 200 \\ x &= \frac{2200}{13}, \text{ or } 169\frac{3}{13} \end{aligned}$$

Check. The total amount of the mixture is $\frac{2200}{13}\text{ lb} + \frac{400}{13}\text{ lb} = \frac{2600}{13}\text{ lb} = 200\text{ lb}$. The amount of milk fat in the mixture is $0.04\left(\frac{2200}{13}\text{ lb}\right) + 0.3\left(\frac{400}{13}\text{ lb}\right) = \frac{88}{13}\text{ lb} + \frac{120}{13}\text{ lb} = \frac{208}{13}\text{ lb} = 16\text{ lb}$. The answer checks.

State. $169\frac{3}{13}$ lb of whole milk and $30\frac{10}{13}$ lb of cream should be mixed.

45. ***Familiarize***. We first make a drawing.

Slow train,
d kilometers, 75 km/h $(t+2)$ hr

Fast train,
d kilometers, 125 km/h t hr

From the drawing we see that the distances are the same. Now complete the chart.

$d = r \cdot t$

	Distance	Rate	Time	
Slow train	d	75	$t+2$	$\rightarrow d = 75(t+2)$
Fast train	d	125	t	$\rightarrow d = 125t$

Translate. Using $d = rt$ in each row of the table, we get a system of equations:

$$\begin{aligned} d &= 75(t+2), \\ d &= 125t \end{aligned}$$

Carry out. We solve the system of equations.

$$\begin{aligned} 125t &= 75(t+2) \quad \text{Using substitution} \\ 125t &= 75t + 150 \\ 50t &= 150 \\ t &= 3 \end{aligned}$$

Then $d = 125t = 125 \cdot 3 = 375$

Check. At 125 km/h, in 3 hr the fast train will travel $125 \cdot 3 = 375$ km. At 75 km/h, in 3 + 2, or 5 hr the slow train will travel $75 \cdot 5 = 375$ km. The numbers check.

State. The trains will meet 375 km from the station.

47. ***Familiarize***. We first make a drawing. Let d = the distance and r = the speed of the canoe in still water. Then when the canoe travels downstream, its speed is $r + 6$, and its speed upstream is $r - 6$. From the drawing we see that the distances are the same.

Downstream, 6 mph current
d mi, $r + 6$, 4 hr

Upstream, 6 mph current
d mi, $r - 6$, 10 hr

Organize the information in a table.

	Distance	Rate	Time
Downstream	d	$r+6$	4
Upstream	d	$r-6$	10

Translate. Using $d = rt$ in each row of the table, we get a system of equations:

$$\begin{aligned} d &= 4(r+6), \\ d &= 10(r-t) \end{aligned} \quad \text{or} \quad \begin{aligned} d &= 4r + 24, \\ d &= 10r - 60 \end{aligned}$$

Carry out. Solve the system of equations.

$$\begin{aligned} 4r + 24 &= 10r - 60 \\ 24 &= 6r - 60 \\ 84 &= 6r \\ 14 &= r \end{aligned}$$

Check. When $r = 14$, then $r + 6 = 14 + 6 = 20$, and the

distance traveled in 4 hours is $4 \cdot 20 = 80$ km. Also $r - 6 = 14 - 6 = 8$, and the distance traveled in 10 hours is $10 \cdot 8 = 80$ km. The answer checks.
State. The speed of the canoe in still water is 14 km/h.

49. ***Familiarize***. We make a drawing. Note that the plane's speed traveling toward London is 360 + 50, or 410 mph, and the speed traveling toward New York City is 360 − 50, or 310 mph. Also, when the plane is d mi from New York City, it is 3458 − d mi from London.

New York City London
310 mph t hours t hours 410 mph

3458 mi

d 3458 mi − d

Organize the information in a table.

	Distance	Rate	Time
Toward NYC	d	310	t
Toward London	$3458 - d$	410	t

Translate. Using $d = rt$ in each row of the table, we get a system of equations:

$$d = 310t, \quad (1)$$
$$3458 - d = 410t \quad (2)$$

Carry out. We solve the system of equations.

$$3458 - 310t = 410t \quad \text{Using substitution}$$
$$3458 = 720t$$
$$4.8028 \approx t$$

Substitute 4.8028 for t in (1).

$$d \approx 310(4.8028) \approx 1489$$

Check. If the plane is 1489 mi from New York City, it can return to New York City, flying at 310 mph, in $1489/310 \approx 4.8$ hr. If the plane is $3458 - 1489$, or 1969 mi from London, it can fly to London, traveling at 410 mph, in $1969/410 \approx 4.8$ hr. Since the times are the same, the answer checks.
State. The point of no return is about 1489 mi from New York City.

51. ***Familiarize***. Let l = the length, in feet, and w = the width, in feet. Recall that the formula for the perimeter P of a rectangle with length l and width w is $P = 2l + 2w$.
Translate.

The perimeter is 860 ft.
$$2l + 2w = 860$$

The length is 100 ft. more than the width.
$$l = 100 + w$$

We have translated to a system of equations:

$$2l + 2w = 860, \quad (1)$$
$$l = 100 + w \quad (2)$$

Carry out. We use the substitution method to solve the system of equations.
Substitute 100 + w for l in (1) and solve for w.

$$2(100 + w) + 2w = 860$$
$$200 + 2w + 2w = 860$$
$$200 + 4w = 860$$
$$4w = 660$$
$$w = 165$$

Now substitute 165 for w in (2).

$$l = 100 + 165 = 265$$

Check. The perimeter is $2 \cdot 265 \text{ ft} + 2 \cdot 165 \text{ ft} = 530 \text{ ft} + 330 \text{ ft} = 860 \text{ ft}$. The length, 265 ft, is 100 ft more than the width, 165 ft. The answer checks.
State. The length is 265 ft, and the width is 165 ft.

53. ***Familiarize***. Let x = the number of minutes to a landline and y = the number of minutes to a wireless.
Translate.

\$5.00 and Landline call and Wireless call is Total bill.
$$5.00 + 0.09x + 0.15y = \$59.90$$

Total minutes is 400 minutes.
$$x + y = 400$$

We have a system of equations:

$$5 + 0.09x + 0.15y = 59.90 \quad (1)$$
$$x + y = 400 \quad (2)$$

Carry out. We use the elimination method to solve the system of equations. Subtract 5 from both sides of (1).

$$9x + 15y = 5490 \quad \text{Multiplying (1) by 100}$$
$$-9x - 9y = -3600 \quad \text{Multiplying (2) by } -9$$
$$6y = 1890$$
$$y = 315$$

Now substitute 315 for y in Equation (2).

$$x + 315 = 400$$
$$x = 85$$

Check. The total number of minutes is 85 + 315 = 400. The total bill is \$5 + \$0.09(85) + \$0.15(315) = \$59.90. The answer checks.
State. There were 85 minutes to a landline and 315 minutes to a wireless number.

55. ***Familiarize***. Let x = the number of Basic \$7.99 plans and y = the number of Standard \$7.99 + \$2 plans.
Translate.

Total number of plans is 280.
$$x + y = 280$$

Total value of plans is 2417.20.
$$7.99x + 9.99y = 2417.20$$

After clearing decimals, we have a system of equations:

$$x + y = 280 \quad (1)$$
$$799x + 999y = 241,720 \quad (2)$$

Carry out. We use the elimination method to solve the system of equations.

$$\begin{array}{rl} -799x - 799y = -223{,}720 & \text{Multiplying (1) by } -799 \\ 799x + 999y = 241{,}720 & (2) \\ \hline 200y = 18{,}000 & \\ y = 90 & \end{array}$$

Substitute 90 for y in Equation (1) and solve for x.

$$x + 90 = 280$$
$$x = 190$$

Check. The total number of plans is 190 + 90, or 280. The total value of the plans is $7.99(190) + 9.99(90) = 1518.10 + 899.10 = 2417.20$. The answer checks.

State. 190 of the \$7.99 Basic plans and 90 of the Standard \$7.99 +\$2 plans were purchased.

57. ***Familiarize***. The change from the \$9.25 purchase is \$20 – \$9.25, or \$10.75. Let $x =$ the number of quarters and $y =$ the number of fifty-cent pieces. The total value of the quarters, in dollars, is $0.25x$ and the total value of the fifty-cent pieces, in dollars, is $0.50y$.

Translate.

The total number of coins is 30.

$$x + y = 30$$

The total value of the coins is \$10.75.

$$0.25x + 0.50y = 10.75$$

After clearing decimals we have the following system of equations:

$$x + y = 30, \quad (1)$$
$$25x + 50y = 1075 \quad (2)$$

Carry out. We use the elimination method to solve the system of equations.

$$\begin{array}{rl} -25x - 25y = -750 & \text{Multiplying (1) by } -25 \\ 25x + 50y = 1075 & \\ \hline 25y = 325 & \\ y = 13 & \end{array}$$

Substitute 13 for y in (1) and solve for x.

$$x + 13 = 30$$
$$x = 17$$

Check. The total number of coins is 17 + 13, or 30. The total value of the coins is $\$0.25(17) + \$0.50(13) = \$4.25 + \$6.50 = \$10.75$. The answer checks.

State. There were 17 quarters and 13 fifty-cent pieces.

59. *Writing Exercise.*

61. $h(0) = 0 - 7 = -7$

63. $(h \cdot f)(7) = h(7)f(7) = (7-7)(7^2 + 2) = 0 \cdot 51 = 0$

65. From Exercise 64, $h + f = x^2 + x - 5$

Domain: $\mathbb{R}$

67. *Writing Exercise.*

69. ***Familiarize***. Let $x =$ the number of reams of 0% post-consumer fiber paper purchased and $y =$ the number of reams of 30% post-consumer fiber paper.

Translate. We organize the information in a table.

	0% post-consumer	30% post-consumer	Total
Reams purchased	x	y	60
Percent of post-consumer fiber	0%	30%	20%
Total post-consumer fiber	$0 \cdot x$, or 0	$0.3y$	0.2(60), or 12

We get one equation from the "Reams purchased" row of the table:

$$x + y = 60$$

The last row of the table yields a second equation:

$$0x + 0.3y = 12, \text{ or } 0.3y = 12$$

After clearing the decimal we have the problem translated to a system of equations.

$$x + y = 60, \quad (1)$$
$$3y = 120 \quad (2)$$

Carry out. First we solve (2) for y.

$$3y = 120$$
$$y = 40$$

Now substitute 40 for y in (1) and solve for x.

$$x + 40 = 60$$
$$x = 20$$

Check. The total purchase is 20 + 40, or 60 reams. The post-consumer fiber can be expressed as $0 \cdot 20 + 0.3(40) = 12$. The answer checks.

State. 20 reams of 0% post-consumer fiber paper and 40 reams of 30% post-consumer fiber paper would have to be purchased.

71. ***Familiarize***. Let $x =$ the number of ounces of pure silver.

Translate. We organize the information in a table.

	Coin	Pure silver	New Mixture
Number of ounces	32	x	$32 + x$
Percent silver	90%	100%	92.5%
Amount of silver	0.9(32)	$1 \cdot x$	$0.925(32 + x)$

The last row gives the equation

$$0.9(32) + x = 0.925(32 + x).$$

Carry out. We solve the equation.

$$28.8 + x = 29.6 + 0.925x$$
$$0.075x = 0.8$$
$$x = \frac{32}{3} = 10\tfrac{2}{3}$$

Check. 90% of 32, or 28.8 + 100% of $10\frac{2}{3}$, or $10\frac{2}{3}$ is $39\frac{7}{15}$ and 92.5% of $\left(32 + 10\frac{2}{3}\right)$ is $0.925\left(42\frac{2}{3}\right)$, or $39\frac{7}{15}$. The answer checks.

State. $10\frac{2}{3}$ ounces of pure silver should be added.

73. ***Familiarize***. Let x = the number of 6-cupcake boxes and y = the number of single cupcakes. Note that $6x$ is the number of cupcakes in the gift box.

Translate.

$$\underbrace{\text{Total number of cupcakes}}_{\downarrow\; 6x+y} \quad \underbrace{\text{is}}_{\downarrow\; =} \quad \underbrace{256.}_{\downarrow\; 256}$$

$$\underbrace{\text{Total value sales}}_{\downarrow\; 15.99x+3y} \quad \underbrace{\text{is}}_{\downarrow\; =} \quad \underbrace{701.67.}_{\downarrow\; 701.67}$$

We have a system of equations:

$$6x + y = 256 \quad (1)$$
$$15.99x + 3y = 701.67 \quad (2)$$

Carry out. We use the elimination method to solve the system of equations.

$$\begin{aligned} -18x - 3y &= -768 \quad \text{Multiplying (1) by } -3 \\ 15.99x + 3y &= 701.67 \quad (2) \\ \hline -2.01x \quad &= -66.33 \\ x &= 33 \end{aligned}$$

Although the problem asks for the number of 6-cupcake boxes, we will also find y in order to check. Substitute 33 for x in Equation (1) and solve for y.

$$6(33) + y = 256$$
$$y = 58$$

Check. The total number of cupcakes is $6(33) + 58 = 198 + 58 = 256$. The total sales is $\$15.99(33) + \$3(58) = \$527.67 + \$174 = \$701.67$. The answer checks.

State. 33 6-cupcake boxes were sold.

75. ***Familiarize***. Let x = the number of gallons of pure brown and y = the number of gallons of neutral stain that should be added to the original 0.5 gal. Note that a total of 1 gal of stain needs to be added to bring the amount of stain up to 1.5 gal. The original 0.5 gal of stain contains 20%(0.5 gal), or 0.2(0.5 gal) = 0.1 gal of brown stain. The final solution contains 60%(1.5 gal), or 0.6(1.5 gal) = 0.9 gal of brown stain. This is composed of the original 0.1 gal and the x gal that are added.

Translate.

$$\underbrace{\text{The amount of stain added}}_{\downarrow\; x+y} \quad \underbrace{\text{was}}_{\downarrow\; =} \quad \underbrace{\text{1 gal.}}_{\downarrow\; 1}$$

$$\underbrace{\text{The amount of brown stain in the final solution}}_{\downarrow\; 0.1+x} \quad \underbrace{\text{is}}_{\downarrow\; =} \quad \underbrace{\text{0.9 gal.}}_{\downarrow\; 0.9}$$

We have a system of equations.

$$x + y = 1, \quad (1)$$
$$0.1 + x = 0.9 \quad (2)$$

Carry out. First we solve (2) for x.

$$0.1 + x = 0.9$$
$$x = 0.8$$

Then substitute 0.8 for x in (1) and solve for y.

$$0.8 + y = 1$$
$$y = 0.2$$

Check. Total amount of stain: 0.5 + 0.8 + 0.2 = 1.5 gal
Total amount of brown stain: 0.1 + 0.8 = 0.9 gal
Total amount of neutral stain:
$0.8(0.5) + 0.2 = 0.4 + 0.2 = 0.6 \text{ gal} = 0.4(1.5 \text{ gal})$
The answer checks.

State. 0.8 gal of pure brown and 0.2 gal of neutral stain should be added.

77. ***Familiarize***. Let x and y represent the number of city miles and highway miles that were driven, respectively. Then in city driving, $\frac{x}{18}$ gallons of gasoline are used; in highway driving, $\frac{y}{24}$ gallons are used.

Translate. We organize the information in a table.

Type of driving	City	Highway	Total
Number of miles	x	y	465
Gallons of gasoline used	$\frac{x}{18}$	$\frac{y}{24}$	23

The first row of the table gives us one equation:

$$x + y = 465$$

The second row gives us another equation:

$$\frac{x}{18} + \frac{y}{24} = 23$$

After clearing fractions, we have the following system of equations:

$$x + y = 465, \quad (1)$$
$$24x + 18y = 9936 \quad (2)$$

Carry out. We solve the system of equations using the elimination method.

$$\begin{aligned} -18x - 18y &= -8370 \quad \text{Multiplying (1) by } -18 \\ 24x + 18y &= 9936 \quad (2) \\ \hline 6x \quad &= 1566 \\ x &= 261 \end{aligned}$$

Now substitute 261 for x in Equation (1) and solve for y.

$$261 + y = 465$$
$$y = 204$$

Check. The total mileage is 261 + 204, or 465. In 216 city miles, 261 / 18, or 14.5 gal of gasoline are used; in 204 highway miles, 204 / 24, or 8.5 gal are used. Then a total of 14.5 + 8.5, or 23 gal of gasoline are used. The answer checks.

State. 261 miles were driven in the city, and 204 miles were driven on the highway.

79. The 1.5 gal mixture contains $0.1 + x$ gal of pure brown stain. (See Exercise 75.) Thus, the function $P(x) = \frac{0.1 + x}{1.5}$ gives the percentage of brown in the mixture as a decimal quantity. Using the Intersect feature, we confirm that when $P(x) = 0.6$, or 60%, then $x = 0.8$.

Exercise Set 8.4

1. The equation is equivalent to one in the form $Ax + By + Cz = D$, so the statement is true.

3. False

5. True; see Example 6.

7. Substitute (2, –1, –2) into the three equations, using alphabetical order.

$$\begin{array}{c|c} x+y-2z=5 & \\ \hline 2+(-1)-2(-2) & 5 \\ 2-1+4 & \\ 5 \overset{?}{=} 5 & \text{TRUE} \end{array}$$

$$\begin{array}{c|c} 2x-y-z=7 & \\ \hline 2\cdot 2-(-1)-(-2) & 7 \\ 4+1+2 & \\ 7 \overset{?}{=} 7 & \text{TRUE} \end{array}$$

$$\begin{array}{c|c} -x-2y-3z=6 & \\ \hline -2-2(-1)-3(-2) & 6 \\ -2+2+6 & \\ 6 \overset{?}{=} 6 & \text{TRUE} \end{array}$$

Since the triple (2, –1, –2) is true in all three equations, it is a solution of the system.

9.
$$\begin{aligned} x-y-z&=0, \quad (1)\\ 2x-3y+2z&=7, \quad (2)\\ -x+2y+z&=1 \quad (3) \end{aligned}$$

1., 2. The equations are already in standard form with no fractions or decimals.
3. Use Equations (1) and (2) to eliminate x.

$$\begin{aligned} -2x+2y+2z&=0 \quad \text{Multiplying (1) by } -2\\ 2x-3y+2z&=7 \quad (2)\\ \hline -y+4z&=7 \quad \text{(4) Adding} \end{aligned}$$

4. Use a different pair of equations and eliminate x.

$$\begin{aligned} x-y-z&=0 \quad (1)\\ -x+2y+z&=1 \quad (3)\\ \hline y&=1 \quad \text{Adding} \end{aligned}$$

5. When we used Equation (1) and (3) to eliminate x, we also eliminated z and found $y = 1$. Substitute 1 for y in Equation (4) to find z.

$$\begin{aligned} -1+4z&=7 \quad \text{Substituting 1 for } y \text{ in (4)}\\ 4z&=8\\ z&=2 \end{aligned}$$

6. Substitute in one of the original equations to find x.

$$\begin{aligned} x-1-2&=0\\ x&=3 \end{aligned}$$

We obtain (3, 1, 2). This checks, so it is the solution.

11.
$$\begin{aligned} x-y-z&=1, \quad (1)\\ 2x+y+2z&=4, \quad (2)\\ x+y+3z&=5 \quad (3) \end{aligned}$$

1., 2. The equations are already in standard form with no fractions or decimals.
3. Use Equations (1) and (2) to eliminate y.

$$\begin{aligned} x-y-z&=1 \quad (1)\\ 2x+y+2z&=4 \quad (2)\\ \hline 3x\quad +z&=5 \quad \text{(4) Adding} \end{aligned}$$

4. Use a different pair of equations and eliminate y.

$$\begin{aligned} x-y-z&=1 \quad (1)\\ x+y+3z&=5 \quad (3)\\ \hline 2x\quad +2z&=6 \quad \text{(5) Adding} \end{aligned}$$

5. Now solve the system of Equations (4) and (5).

$$\begin{aligned} 3x+z&=5 \quad (4)\\ 2x+2z&=6 \quad (5) \end{aligned}$$

$$\begin{aligned} -6x-2z&=-10 \quad \text{Multiplying (4) by } -2\\ 2x+2z&=6 \quad (5)\\ \hline -4x\quad &=-4 \quad \text{Adding}\\ x&=1 \end{aligned}$$

$$\begin{aligned} 3\cdot 1+z&=5 \quad \text{Substituting 1 for } x \text{ in (4)}\\ 3+z&=5\\ z&=2 \end{aligned}$$

6. Substitute in one of the original equations to find y.

$$\begin{aligned} 1+y+3\cdot 2&=5 \quad \text{Substituting 1 for } x \text{ and 2 for } z \text{ in (3)}\\ 1+y+6&=5\\ y+7&=5\\ y&=-2 \end{aligned}$$

We obtain (1, –2, 2). This checks, so it is the solution.

13.
$$\begin{aligned} 3x+4y-3z&=4, \quad (1)\\ 5x-y+2z&=3, \quad (2)\\ x+2y-z&=-2 \quad (3) \end{aligned}$$

1., 2. The equations are already in standard form with no fractions or decimals.
3., 4. We eliminate y from two different pairs of equations.

$$\begin{aligned} 3x+4y-3z&=4 \quad (1)\\ 20x-4y+8z&=12 \quad \text{Multiplying (2) by 4}\\ \hline 23x\quad +5z&=16 \quad (4) \end{aligned}$$

$$\begin{aligned} 10x-2y+4z&=6 \quad \text{Multiplying (2) by 2}\\ x+2y-z&=-2 \quad (3)\\ \hline 11x\quad +3z&=4 \quad (5) \end{aligned}$$

5. Now solve the system of Equations (4) and (5).

$$\begin{aligned} 23x+5z&=16 \quad (4)\\ 11x+3z&=4 \quad (5) \end{aligned}$$

$$\begin{aligned} -69x-15z&=-48 \quad \text{Multiplying (4) by } -3\\ 55x+15z&=20 \quad \text{Multiplying (5) by 5}\\ \hline -14x\quad &=-28 \quad \text{Adding}\\ x&=2 \end{aligned}$$

$$\begin{aligned} 11\cdot 2+3z&=4 \quad \text{Substituting 2 for } x \text{ in (5)}\\ 3z&=-18\\ z&=-6 \end{aligned}$$

6. Substitute in one of the original equations to find y.

$$\begin{aligned} 3\cdot 2+4y-3(-6)&=4 \quad \text{Substituting 2 for } x \text{ and } -6 \text{ for } z \text{ in (1)}\\ 6+4y+18&=4\\ 4y&=-20\\ y&=-5 \end{aligned}$$

We obtain (2, –5, –6). This checks, so it is the solution.

15. $x+y+z=0,$ (1)
$2x+3y+2z=-3,$ (2)
$-x-2y-z=1$ (3)

1., 2. The equations are already in standard form with no fractions or decimals.
3., 4. We eliminate x from two different pairs of equations.

$$\begin{aligned} -2x-2y-2z&=0 \quad \text{Multiplying (1) by } -2\\ 2x+3y+2z&=-3 \quad (2)\\ \hline y&=-3 \end{aligned}$$

We eliminated not only x but also z and found that $y=-3$.
5., 6. Substitute –3 for y in two of the original equations to produce a system of two equations in two variables. Then solve this system.

$$\begin{aligned} x-3+z&=0 \quad \text{Substituting in (1)}\\ -x-2(-3)-z&=1 \quad \text{Substituting in (3)} \end{aligned}$$

Simplifying we have

$$\begin{aligned} x+z&=3\\ -x-z&=-5\\ \hline 0&=-2 \end{aligned}$$

We get a false equation, so there is no solution.

17. $2x-3y-z=-9,$ (1)
$2x+5y+z=1,$ (2)
$x-y+z=3$ (3)

1., 2. The equations are already in standard form with no fractions or decimals.
3., 4. We eliminate z from two different pairs of equations.

$$\begin{aligned} 2x-3y-z&=-9 \quad (1)\\ 2x+5y+z&=1 \quad (2)\\ \hline 4x+2y&=-8 \quad (4) \end{aligned}$$

$$\begin{aligned} 2x-3y-z&=-9 \quad (1)\\ x-y+z&=3 \quad (3)\\ \hline 3x-4y&=-6 \quad (5) \end{aligned}$$

5. Now solve the system of Equations (4) and (5).

$$\begin{aligned} 4x+2y&=-8 \quad (4)\\ 3x-4y&=-6 \quad (5) \end{aligned}$$

$$\begin{aligned} 8x+4y&=-16 \quad \text{Multiplying (4) by 2}\\ 3x-4y&=-6 \quad (5)\\ \hline 11x&=-22 \quad \text{Adding}\\ x&=-2 \end{aligned}$$

$$\begin{aligned} 4(-2)+2y&=-8 \quad \text{Substituting } -2 \text{ for } x \text{ in (4)}\\ -8+2y&=-8\\ y&=0 \end{aligned}$$

6. Substitute in one of the original equations to find z.

$$\begin{aligned} 2(-2)+5\cdot 0+z&=1 \quad \text{Substituting } -2 \text{ for } x \text{ and } 0 \text{ for } y \text{ in (2)}\\ -4+z&=1\\ z&=5 \end{aligned}$$

We obtain (–2, 0, 5). This checks, so it is the solution.

19. $a+b+c=5,$ (1)
$2a+3b-c=2,$ (2)
$2a+3b-2c=4$ (3)

1., 2. The equations are already in standard form with no fractions or decimals.
3., 4. We eliminate a from two different pairs of equations.

$$\begin{aligned} -2a-2b-2c&=-10 \quad \text{Multiplying (1) by } -2\\ 2a+3b-c&=2 \quad (2)\\ \hline b-3c&=-8 \quad (4) \end{aligned}$$

$$\begin{aligned} -2a-3b+c&=-2 \quad \text{Multiplying (2) by } -1\\ 2a+3b-2c&=4 \quad (3)\\ \hline -c&=2\\ c&=-2 \end{aligned}$$

We eliminate not only a, but also b and found $c=-2$.
5. Substitute –2 for c in Equation (4) to find b.

$$\begin{aligned} b-3(-2)&=-8 \quad \text{Substituting } -2 \text{ for } c \text{ in (4)}\\ b+6&=-8\\ b&=-14 \end{aligned}$$

6. Substitute in one of the original equations to find a.

$$\begin{aligned} a-14-2&=5 \quad \text{Substituting } -2 \text{ for } c \text{ and } -14 \text{ for } b \text{ in (1)}\\ a-16&=5\\ a&=21 \end{aligned}$$

We obtain (21, –14, –2). This checks, so it is the solution.

21. $-2x+8y+2z=4,$ (1)
$x+6y+3z=4,$ (2)
$3x-2y+z=0$ (3)

1., 2. The equations are already in standard form with no fractions or decimals.
3., 4. We eliminate z from two different pairs of equations.

$$\begin{aligned} -2x+8y+2z&=4 \quad (1)\\ -6x+4y-2z&=0 \quad \text{Multiplying (3) by } -2\\ \hline -8x+12y&=4 \quad (4) \end{aligned}$$

$$\begin{aligned} x+6y+3z&=4 \quad (2)\\ -9x+6y-3z&=0 \quad \text{Multiplying (3) by } -3\\ \hline -8x+12y&=4 \quad (5) \end{aligned}$$

5. Now solve the system of Equations (4) and (5).

$$\begin{aligned} -8x+12y&=4 \quad (4)\\ -8x+12y&=4 \quad (5) \end{aligned}$$

$$\begin{aligned} -8x+12y&=4 \quad (4)\\ 8x-12y&=-4 \quad \text{Multiplying (5) by } -1\\ \hline 0&=0 \quad (6) \end{aligned}$$

Equation (6) indicates that Equations (1), (2), and (3) are dependent. (Note that if Equation (1) is subtracted from Equation (2), the result is Equation (3).) We could also have concluded that the equations are dependent by observing that Equations (4) and (5) are identical.

23. $2u-4v-w=8,$ (1)
$3u+2v+w=6,$ (2)
$5u-2v+3w=2$ (3)

1., 2. The equations are already in standard form with no fractions or decimals.
3., 4. We eliminate w from two different pairs of equations.

$$\begin{array}{ll} 2u-4v-w=8 & (1) \\ 3u+2v+w=6 & (2) \\ \hline 5u-2v\quad=14 & (4) \end{array}$$

$$\begin{array}{ll} 6u-12v-3w=24 & \text{Multiplying (1) by 3} \\ 5u-2v+3w=2 & (3) \\ \hline 11u-14v\quad=26 & (5) \end{array}$$

5. Now solve the system of Equations (4) and (5).

$$\begin{array}{ll} 5u-2v=14 & (4) \\ 11u-14v=26 & (5) \end{array}$$

$$\begin{array}{ll} -35u+14v=-98 & \text{Multiplying (4) by } -7 \\ 11u-14v=26 & (5) \\ \hline -24u\quad=-72 & \\ u=3 & \end{array}$$

$$\begin{array}{ll} 5\cdot 3-2v=14 & \text{Substituting 3 for } u \text{ in (4)} \\ 15-2v=14 & \\ -2v=-1 & \\ v=\frac{1}{2} & \end{array}$$

6. Substitute in one of the original equations to find w.

$$\begin{array}{ll} 2\cdot 3-4\left(\frac{1}{2}\right)-2=8 & \text{Substituting 3 for } u \text{ and } \frac{1}{2} \text{ for } v \text{ in (1)} \\ 6-2-w=8 & \\ w=-4 & \end{array}$$

We obtain $\left(3, \frac{1}{2}, -4\right)$. This checks, so it is the solution.

25. $r+\frac{3}{2}s+6t=2,$
$2r-3s+3t=0.5,$
$r+s+t=1$

1. All equations are already in standard form.
2. Multiply the first equation by 2 to clear the fraction. Also, multiply the second equation by 10 to clear the decimal.

$$\begin{array}{ll} 2r+3s+12t=4, & (1) \\ 20r-30s+30t=5, & (2) \\ r+s+t=1 & (3) \end{array}$$

3., 4. We eliminate s from two different pairs of equations.

$$\begin{array}{ll} 20r+30s+120t=40 & \text{Multiplying (1) by 10} \\ 20r-30s+30t=5 & (2) \\ \hline 40r\quad+150t=45 & \text{(4) Adding} \end{array}$$

$$\begin{array}{ll} 2r+3s+12t=4 & (1) \\ -3r-3s-3t=-3 & \text{Multiplying (3) by } -3 \\ \hline -r\quad+9t=1 & \text{(5) Adding} \end{array}$$

5. Solve the system of Equations (4) and (5).

$$\begin{array}{ll} 40r+150t=45, & (4) \\ -r+9t=1, & (5) \end{array}$$

$$\begin{array}{ll} 40r+150t=45 & (4) \\ -40r+360t=40 & \text{Multiplying (5) by 40} \\ \hline 510t=85 & \\ t=\frac{85}{510} & \\ t=\frac{1}{6} & \end{array}$$

$$\begin{array}{ll} 40r+150\left(\frac{1}{6}\right)=45 & \text{Substituting } \frac{1}{6} \text{ for } t \text{ in (4)} \\ 40r+25=45 & \\ 40r=20 & \\ r=\frac{1}{2} & \end{array}$$

6. Substitute in one of the original equations to find s.

$$\begin{array}{ll} \frac{1}{2}+s+\frac{1}{6}=1 & \text{Substituting } \frac{1}{2} \text{ for } r \text{ and } \frac{1}{6} \text{ for } t \text{ in (3)} \\ s+\frac{2}{3}=1 & \\ s=\frac{1}{3} & \end{array}$$

We obtain $\left(\frac{1}{2}, \frac{1}{3}, \frac{1}{6}\right)$. This checks, so it is the solution.

27. $$\begin{array}{ll} 4a+9b\quad=8, & (1) \\ 8a\quad+6c=-1, & (2) \\ 6b+6c=-1 & (3) \end{array}$$

1., 2. The equations are already in standard form with no fractions or decimals.
3., 4. Note that there is no c in Equation (1). We will use Equations (2) and (3) to obtain another equation with no c-term.

$$\begin{array}{ll} 8a\quad+6c=-1 & (2) \\ -6b-6c=1 & \text{Multiplying (3) by} -1 \\ \hline 8a-6b\quad=0 & \text{(4) Adding} \end{array}$$

5. Now solve the system of Equations (1) and (4).

$$\begin{array}{ll} -8a-18b=-16 & \text{Multiplying (1) by} -2 \\ 8a-6b=0 & \\ \hline -24b=-16 & \\ b=\frac{2}{3} & \end{array}$$

$$\begin{array}{ll} 8a-6\left(\frac{2}{3}\right)=0 & \text{Substituting } \frac{2}{3} \text{ for } b \text{ in (4)} \\ 8a-4=0 & \\ 8a=4 & \\ a=\frac{1}{2} & \end{array}$$

6. Substitute in Equation (2) or (3) to find c.

$$\begin{array}{ll} 8\left(\frac{1}{2}\right)+6c=-1 & \text{Substituting } \frac{1}{2} \text{ for } a \text{ in (2)} \\ 4+6c=-1 & \\ 6c=-5 & \\ c=-\frac{5}{6} & \end{array}$$

We obtain $\left(\frac{1}{2}, \frac{2}{3}, -\frac{5}{6}\right)$. This checks, so it is the solution.

29. $$\begin{array}{ll} x+y+z=57, & (1) \\ -2x+y\quad=3, & (2) \\ x\quad-z=6 & (3) \end{array}$$

1., 2. The equations are already in standard form with no fractions or decimals.
3., 4. Note that there is no z in Equation (2). We will use Equations (1) and (3) to obtain another equation with no z-term.

$$\begin{array}{ll} x+y+z=57 & (1) \\ x\quad-z=6 & (3) \\ \hline 2x+y\quad=63 & (4) \end{array}$$

5. Now solve the system of Equations (2) and (4).

$$\begin{aligned} -2x+y&=3 \quad (2)\\ 2x+y&=63 \quad (4)\\ \hline 2y&=66\\ y&=33 \end{aligned}$$

$2x+33=63$ Substituting 33 for y in (4)

$2x=30$

$x=15$

6. Substitute in Equation (1) or (3) to find z.

$15-z=6$ Substituting 15 for x in (3)

$9=z$

We obtain (15, 33, 9). This checks, so it is the solution.

31. $a \quad -3c=6,$ (1)

$b+2c=2,$ (2)

$7a-3b-5c=14$ (3)

1., 2. The equations are already in standard form with no fractions or decimals.

3., 4. Note that there is no b in Equation (1). We will use Equations (2) and (3) to obtain another equation with no b-term.

$3b+6c=6$ Multiplying (2) by 3

$7a-3b-5c=14$ (3)

$7a \quad +c=20$ (4)

5. Now solve the system of Equations (1) and (4).

$a-3c=6$ (1)

$7a+c=20$ (4)

$a-3c=6$ (1)

$21a+3c=60$ Multiplying (4) by 3

$22a \quad =66$

$a=3$

$3-3c=6$ Substituting in (1)

$-3c=3$

$c=-1$

6. Substitute in Equation (2) or (3) to find b.

$b+2(-1)=2$ Substituting in (2)

$b-2=2$

$b=4$

We obtain (3, 4, –1). This checks, so it is the solution.

33. $x+y+z=83,$ (1)

$y=2x+3,$ (2)

$z=40+x$ (3)

Observe, from Equations (2) and (3), that we can substitute $2+3x$ for y and $40+x$ for z in Equation (1) and solve for x.

$$\begin{aligned} x+y+z&=83\\ x+(2x+3)+(40+x)&=83\\ 4x+43&=83\\ 4x&=40\\ x&=10 \end{aligned}$$

Now substitute 10 for x in Equation (2).

$y=2x+3=2\cdot 10+3=20+3=23$

Finally, substitute 10 for x in Equation (3).

$z=40+x=40+10=50$.

We obtain (10, 23, 50). This checks, so it is the solution.

35. $x \quad +z=0,$ (1)

$x+y+2z=3,$ (2)

$y+z=2$ (3)

1., 2. The equations are already in standard form with no fractions or decimals.

3., 4. Note that there is no y in Equation (1). We use Equations (2) and (3) to obtain another equation with no y-term.

$x+y+2z=3$ (2)

$-y-z=-2$ Multiplying (3) by –1

$x \quad +z=1$ (4) Adding

5. Now solve the system of Equations (1) and (4).

$x+z=0$ (1)

$-x-z=-1$ Multiplying (4) by –1

$0=-1$ Adding

We get a false equation, or contradiction. There is no solution.

37. $x+y+z=1,$ (1)

$-x+2y+z=2,$ (2)

$2x-y \quad =-1$ (3)

1., 2. The equations are already in standard form with no fractions or decimals.

3. Note that there is no z in Equation (3). We will use Equations (1) and (2) to eliminate z:

$x+y+z=1$ (1)

$x-2y-z=-2$ Multiplying (2) by –1

$2x-y \quad =-1$ (4)

Equations (3) and (4) are identical, so Equations (1), (2), and (3) are dependent. (We have seen that if Equation (2) is multiplied by –1 and added to Equation (1), the result is Equation (3).)

39. *Writing Exercise.*

41. $x-3y=7$

$-3y=-x+7$

$y=\frac{1}{3}x-\frac{7}{3}$

The slope is $\frac{1}{3}$.

The y-intercept is $\left(0,-\frac{7}{3}\right)$.

43. To find the y-intercept, let $x=0$ and solve for y.

$0-5y=20$

$y=-4$

The y-intercept is (0,–4).

To find the x-intercept, let $y=0$ and solve for x.

$2x+0=20$

$x=10$

The x-intercept is (10, 0).

45. $3x-y=12$

$-y=-3x+12$

$y=3x-12$

The slope is $m=3$.

$y=3x+7$

The slope is $m=3$.

The lines are parallel.

47. *Writing Exercise.*

49. $\frac{x+2}{3}-\frac{y+4}{2}+\frac{z+1}{6}=0,$

$\frac{x-4}{3}+\frac{y+1}{4}-\frac{z-2}{2}=-1,$

$\frac{x+1}{2}+\frac{y}{2}+\frac{z-1}{4}=\frac{3}{4}$

1., 2. We clear fractions and write each equation in standard form.

To clear fractions, we multiply both sides of each equation by the LCM of its denominators. The LCM's are 6, 12, and 4, respectively.

$$6\left(\frac{x+2}{3}-\frac{y+4}{2}+\frac{z+1}{6}\right)=6\cdot 0$$
$$2(x+2)-3(y+4)+(z+1)=0$$
$$2x+4-3y-12+z+1=0$$
$$2x-3y+z=7$$

$$12\left(\frac{x-4}{3}+\frac{y+1}{4}-\frac{z-2}{2}\right)=12\cdot(-1)$$
$$4(x-4)+3(y+1)-6(z-2)=-12$$
$$4x-16+3y+3-6z+12=-12$$
$$4x+3y-6z=-11$$

$$4\left(\frac{x+1}{2}+\frac{y}{2}+\frac{z-1}{4}\right)=4\cdot\frac{3}{4}$$
$$2(x+1)+2(y)+(z-1)=3$$
$$2x+2+2y+z-1=3$$
$$2x+2y+z=2$$

The resulting system is

$2x-3y+z=7,$ (1)
$4x+3y-6z=-11,$ (2)
$2x+2y+z=2$ (3)

3., 4. We eliminate z from two different pairs of equations.

$12x-18y+6z=42$ Multiplying (1) by 6
$4x+3y-6z=-11$ (2)
$16x-15y=31$ (4) Adding

$2x-3y+z=7$ (1)
$-2x-2y-z=-2$ Multiplying (3) by -1
$-5y=5$ (5) Adding

5. Solve (5) for y: $-5y=5$
$y=-1$

Substitute -1 for y in (4):
$16x-15(-1)=31$
$16x+15=31$
$16x=16$
$x=1$

6. Substitute 1 for x and -1 for y in (1):
$2\cdot 1-3(-1)+z=7$
$5+z=7$
$z=2$

We obtain $(1, -1, 2)$. This checks, so it is the solution.

51. $w+x+y+z=2,$ (1)
$w+2x+2y+4z=1,$ (2)
$w-x+y+z=6,$ (3)
$w-3x-y+z=2$ (4)

The equations are already in standard form with no fractions or decimals.

Start by eliminating w from three different pairs of equations.

$w+x+y+z=2$ (1)
$-w-2x-2y-4z=-1$ Multiplying (2) by -1
$-x-y-3z=1$ (5) Adding

$w+x+y+z=2$ (1)
$-w+x-y-z=-6$ Multiplying (3) by -1
$2x=-4$ (6) Adding

$w+x+y+z=2$ (1)
$-w+3x+y-z=-2$ Multiplying (4) by -1
$4x+2y=0$ (7) Adding

We can solve (6) for x:
$2x=-4$
$x=-2$

Substitute -2 for x in (7):
$4(-2)+2y=0$
$-8+2y=0$
$2y=8$
$y=4$

Substitute -2 for x and 4 for y in (5):
$-(-2)-4-3z=1$
$-2-3z=1$
$-3z=3$
$z=-1$

Substitute -2 for x, 4 for y, and -1 for z in (1):
$w-2+4-1=2$
$w+1=2$
$w=1$

We obtain $(1, -2, 4, -1)$. This checks, so it is the solution.

53. $\frac{2}{x}-\frac{1}{y}-\frac{3}{z}=-1,$

$\frac{2}{x}-\frac{1}{y}+\frac{1}{z}=-9,$

$\frac{1}{x}+\frac{2}{y}-\frac{4}{z}=17$

Let u represent $\frac{1}{x}$, v represent $\frac{1}{y}$, and w represent $\frac{1}{z}$.

Substituting, we have

$2u-v-3w=-1,$ (1)
$2u-v+w=-9,$ (2)
$u+2v-4w=17$ (3)

1., 2. The equations in u, v, and w are in standard form with no fractions or decimals.

3., 4. We eliminate v from two different pairs of equations.

$2u-v-3w=-1$ (1)
$-2u+v-w=9$ Multiplying (2) by -1
$-4w=8$ (4) Adding

$4u-2v-6w=-2$ Multiplying (1) by 2
$u+2v-4w=17$ (3)
$5u-10w=15$ (5) Adding

5. We can solve (4) for w:
$-4w=8$
$w=-2$

Substitute –2 for w in (5):

$$\begin{aligned}5u-10(-2)&=15\\5u+20&=15\\5u&=-5\\u&=-1\end{aligned}$$

6. Substitute –1 for u and –2 for w in (1):

$$\begin{aligned}2(-1)-v-3(-2)&=-1\\-v+4&=-1\\-v&=-5\\v&=5\end{aligned}$$

Solve for x, y, and z. We substitute –1 for u, 5 for v and –2 for w.

$$u=\frac{1}{x} \quad v=\frac{1}{y} \quad w=\frac{1}{z}$$

$$-1=\frac{1}{x} \quad 5=\frac{1}{y} \quad -2=\frac{1}{z}$$

$$x=-1 \quad y=\frac{1}{5} \quad z=-\frac{1}{2}$$

We obtain $\left(-1, \frac{1}{5}, -\frac{1}{2}\right)$. This checks, so it is the solution.

55. $5x-6y+kz=-5$, (1)
$x+3y-2z=2$, (2)
$2x-y+4z=-1$ (3)

Eliminate y from two different pairs of equations.

$$\begin{aligned}5x-6y+kz&=-5 && (1)\\2x+6y-4z&=4 && \text{Multiplying (2) by 2}\\\hline 7x+(k-4)z&=-1 && (4)\end{aligned}$$

$$\begin{aligned}x+3y-2z&=2 && (2)\\6x-3y+12z&=-3 && \text{Multiplying (3) by 3}\\\hline 7x+10z&=-1 && (5)\end{aligned}$$

Solve the system of Equations (4) and (5).

$$\begin{aligned}7x+(k-4)z&=-1 && (4)\\7x+10z&=-1 && (5)\end{aligned}$$

$$\begin{aligned}-7x-(k-4)z&=1 && \text{Multiplying (4) by }-1\\7x+10z&=-1 && (5)\\\hline(-k+14)z&=0 && (6)\end{aligned}$$

The system is dependent for the value of k that makes Equation (6) true. This occurs when $-k+14$ is 0. We solve for k:

$$\begin{aligned}-k+14&=0\\14&=k\end{aligned}$$

57. $z=b-mx-ny$

Three solutions are (1, 1, 2), (3, 2, –6), and $\left(\frac{3}{2}, 1, 1\right)$.

We substitute for x, y, and z and then solve for b, m, and n.

$$\begin{aligned}2&=b-m-n,\\-6&=b-3m-2n,\\1&=b-\tfrac{3}{2}m-n\end{aligned}$$

1., 2. Write the equations in standard form. Also, clear the fraction in the last equation.

$$\begin{aligned}b-m-n&=2, && (1)\\b-3m-2n&=-6, && (2)\\2b-3m-2n&=2 && (3)\end{aligned}$$

3., 4. Eliminate b from two different pairs of equations.

$$\begin{aligned}b-m-n&=2 && (1)\\-b+3m+2n&=6 && \text{Multiplying (2) by }-1\\\hline 2m+n&=8 && \text{(4) Adding}\end{aligned}$$

$$\begin{aligned}-2b+2m+2n&=-4 && \text{Multiplying (1) by }-2\\2b-3m-2n&=2 && (3)\\\hline -m&=-2 && \text{(5) Adding}\end{aligned}$$

5. We solve Equation (5) for m:

$$\begin{aligned}-m&=-2\\m&=2\end{aligned}$$

Substitute in Equation (4) and solve for n.

$$\begin{aligned}2\cdot2+n&=8\\4+n&=8\\n&=4\end{aligned}$$

6. Substitute in one of the original equations to find b.

$$\begin{aligned}b-2-4&=2 && \text{Substituting 2 for } m \text{ and 4 for } n \text{ in (1)}\\b-6&=2\\b&=8\end{aligned}$$

The solution is (8, 2, 4), so the equation is $z=8-2x-4y$.

Mid-Chapter Review

1. $2x-3y=5$, (1)
$y=x-1$ (2)

Substitute $x-1$ for y in Equation (2) and solve for y.

$$\begin{aligned}2x-3(x-1)&=5 && \text{Substituting}\\2x-3x+3&=5\\-x+3&=5\\-x&=2\\x&=-2\end{aligned}$$

Substitute –2 for x in Equation (2) and solve for y.

$$\begin{aligned}y&=x-1 && (2)\\y&=-2-1\\y&=-3\end{aligned}$$

The solution is (–2,–3).

2.
$$\begin{aligned}2x-5y&=1 && (1)\\x+5y&=8 && (2)\\\hline 3x&=9 && \text{Adding}\\x&=3\end{aligned}$$

Substitute 3 for x in Equation (2) and solve for y.

$$\begin{aligned}x+5y&=8 && (1)\\3+5y&=8 && \text{Substituting}\\5y&=5\\y&=1\end{aligned}$$

The solution is (3, 1).

3. $x=y$, (1)
$x+y=2$ (2)

Substitute y for x in Equation (2) and solve for y.

$$\begin{aligned}x+y&=2 && (2)\\y+y&=2 && \text{Substituting}\\2y&=2\\y&=1\end{aligned}$$

Substitute 1 for y in Equation (1) and solve for x.

$$\begin{aligned}x&=y && (1)\\x&=1\end{aligned}$$

We obtain (1, 1) as the solution.

4. $x + y = 10$ (1)
$x - y = 8$ (2)
$2x = 18$ Adding
$x = 9$

Substitute 9 for x in Equation (1) and solve for y.

$x + y = 10$ (1)
$9 + y = 10$ Substituting
$y = 1$

We obtain (9, 1) as the solution.

5. $y = \frac{1}{2}x + 1,$ (1)
$y = 2x - 5$ (2)

Substitute $2x - 5$ for y in Equation (1) and solve for x.

$y = \frac{1}{2}x + 1$ (1)
$2x - 5 = \frac{1}{2}x + 1$ Substituting
$4x - 10 = x + 2$ Clearing fractions
$3x = 12$
$x = 4$

Substitute 4 for x in Equation (2) and solve for y.

$y = 2x - 5$ (2)
$y = 2(4) - 5 = 3$

We obtain $(4, 3)$ as the solution.

6. $y = 2x - 3,$ (1)
$x + y = 12$ (2)

Substitute $2x - 3$ for y in Equation (2) and solve for x.

$x + y = 12$ (2)
$x + (2x - 3) = 12$ Substituting
$x + 2x - 3 = 12$
$3x - 3 = 12$
$3x = 15$
$x = 5$

Substitute 5 for x in Equation (1) and solve for y.

$y = 2x - 3$ (1)
$y = 2(5) - 3 = 7$

We obtain $(5, 7)$ as the solution.

7. $x = 5,$ (1)
$y = 10$ (2)

We obtain $(5, 10)$ as the solution.

8. $3x + 5y = 8$ (1)
$3x - 5y = 4$ (2)
$6x = 12$ Adding
$x = 2$

Substitute 2 for x in Equation (1) and solve for y.

$3x + 5y = 8$ (1)
$3(2) + 5y = 8$ Substituting
$6 + 5y = 8$
$5y = 2$
$y = \frac{2}{5}$

We obtain $\left(2, \frac{2}{5}\right)$ as the solution.

9. $2x - y = 1,$ (1)
$2y - 4x = 3$ (2)

We multiply Equation (1) by 2.

$4x - 2y = 2$ Multiplying (1) by 2
$-4x + 2y = 3$ (2)
$0 = 5$ FALSE

There is no solution.

10. $x = 2 - y,$ (1)
$3x + 3y = 6$ (2)

Substitute $2 - y$ for x in Equation (2) and solve for y.

$3x + 3y = 6$ (2)
$3(2 - y) + 3y = 6$ Substituting
$6 - 3y + 3y = 6$
$6 = 6$

There are many solutions.
The solution set is $\{(x, y)|x = 2 - y\}$.

11. $1.1x - 0.3y = 0.8$ (1)
$2.3x + 0.3y = 2.6$ (2)
$3.4x = 3.4$ Adding
$x = 1$

Substitute 1 for x in Equation (1) and solve for y.

$1.1x - 0.3y = 0.8$ (1)
$1.1(1) - 0.3y = 0.8$ Substituting
$-0.3y = -0.3$
$y = 1$

We obtain (1, 1) as the solution.

12. $\frac{1}{4}x = \frac{1}{3}y,$ (1)
$\frac{1}{2}x - \frac{1}{15}y = 2$ (2)

We solve Equation (1) for x.

$\frac{1}{4}x = \frac{1}{3}y$ (1)
$x = \frac{4}{3}y$

Substitute $\frac{4}{3}y$ for x in Equation (2) and solve for y.

$\frac{1}{2}x - \frac{1}{15}y = 2$ (2)
$\frac{1}{2}\left(\frac{4}{3}y\right) - \frac{1}{15}y = 2$
$\frac{2}{3}y - \frac{1}{15}y = 2$
$10y - y = 30$ Multiplying by 15
$9y = 30$
$y = \frac{10}{3}$

Substitute $\frac{10}{3}$ for y in Equation (1) and solve for x.

$\frac{1}{4}x = \frac{1}{3}y$
$\frac{1}{4}x = \frac{1}{3}\left(\frac{10}{3}\right)$
$\frac{1}{4}x = \frac{10}{9}$
$x = \frac{40}{9}$

We obtain $\left(\frac{40}{9}, \frac{10}{3}\right)$ as the solution.

13. $3x+y-z=-1,$ (1)
$2x-y+4z=2,$ (2)
$x-y+3z=3$ (3)

1., 2. The equations are already in standard form with no fractions or decimals.

3. Use Equations (1) and (2) to eliminate y.

$$\begin{array}{ll} 3x+y-z=-1 & (1) \\ 2x-y+4z=2 & (2) \\ \hline 5x+3z=1 & (4)\text{ Adding} \end{array}$$

4. Use a different pair of equations and eliminate y.

$$\begin{array}{ll} 3x+y-z=-1 & (1) \\ x-y+3z=3 & (3) \\ \hline 4x+2z=2 & (5)\text{ Adding} \end{array}$$

5. Now solve the system of Equations (4) and (5).

$5x+3z=1$ (4)
$4x+2z=2$ (5)

$$\begin{array}{ll} -10x-6z=-2 & \text{Multiplying (4) by } -2 \\ 12x+6z=6 & \text{Multiplying (5) by 3} \\ \hline 2x=4 & \text{Adding} \\ x=2 & \end{array}$$

$5\cdot 2+3z=1$ Substituting 2 for x in (4)
$10+3z=1$
$3z=-9$
$z=-3$

6. Substitute in one of the original equations to find y.

$2-y+3(-3)=3$ Substituting 2 for x and -3 for z in (3)
$2-y-9=3$
$-y=10$
$y=-10$

We obtain (2, –10,–3). This checks, so it is the solution.

14. $2x+y-3z=-4,$ (1)
$4x+y+3z=-1,$ (2)
$2x-y+6z=7$ (3)

1., 2. The equations are already in standard form with no fractions or decimals.

3. Use Equations (1) and (2) to eliminate z.

$$\begin{array}{ll} 2x+y-3z=-4 & (1) \\ 4x+y+3z=-1 & (2) \\ \hline 6x+2y=-5 & (4)\text{ Adding} \end{array}$$

4. Use a different pair of equations and eliminate z.

$$\begin{array}{ll} 4x+2y-6z=-8 & \text{Multiplying (1) by 2} \\ 2x-y+6z=7 & (3) \\ \hline 6x+y=-1 & (5)\text{ Adding} \end{array}$$

5. Now solve the system of Equations (4) and (5).

$6x+2y=-5$ (4)
$6x+y=-1$ (5)

$$\begin{array}{ll} -6x-2y=5 & \text{Multiplying (4) by } -1 \\ 6x+y=-1 & (5) \\ \hline -y=4 & \text{Adding} \\ y=-4 & \end{array}$$

$6x-4=-1$ Substituting -4 for y in (5)
$6x=3$
$x=\frac{1}{2}$

6. Substitute in one of the original equations to find z.

$2\left(\frac{1}{2}\right)-4-3z=-4$ Substituting $\frac{1}{2}$ for x and -4 for y in (1)
$1-4-3z=-4$
$-3z=-1$
$z=\frac{1}{3}$

We obtain $\left(\frac{1}{2},-4,\ \frac{1}{3}\right)$. This checks, so it is the solution.

15. $3x+5y-z=8,$ (1)
$x+6y=4,$ (2)
$x-7y-z=3$ (3)

1., 2. The equations are already in standard form with no fractions or decimals.

3., 4. We eliminate x from two different pairs of equations.

$$\begin{array}{ll} 3x+5y-z=8 & (1) \\ -3x-18y=-12 & \text{Multiplying (2) by } -3 \\ \hline -13y-z=-4 & (4) \end{array}$$

$$\begin{array}{ll} 3x+5y-z=8 & (1) \\ -3x+21y+3z=-9 & \text{Multiplying (3) by } -3 \\ \hline 26y+2z=-1 & (4) \end{array}$$

5. Now solve the system of Equations (4) and (5).

$-13y-z=-4$ (4)
$26y+2z=-1$ (5)

$$\begin{array}{ll} -26y-2z=-8 & \text{Multiplying (4) by 2} \\ 26y+2z=-1 & (5) \\ \hline 0=-9 & \end{array}$$

We get a false equation, so there is no solution.

16. $x-y=4,$ (1)
$2x+y-z=5,$ (2)
$3x-z=9$ (3)

1., 2. The equations are already in standard form with no fractions or decimals.

3. Note that there is no y in Equation (3). We will use Equations (1) and (2) to eliminate y:

$$\begin{array}{ll} x-y=4 & (1) \\ 2x+y-z=5 & (2) \\ \hline 3x-z=9 & \text{Adding} \end{array}$$

Equations (3) and (4) are identical, so Equations (1), (2), and (3) are dependent.

17. ***Familiarize.*** Let x = the amount of text messages received and y = the amount of text messages sent.

Translate.

Total texts is 3853.
↓ ↓ ↓
$x+y = 3853$

texts sent is 191 more than texts received
↓ ↓ ↓ ↓ ↓
$y = 191 + x$

We have a system of equations:

$x+y=3853$ (1)
$y=191+x$ (2)

Carry out. We use the substitution method to solve the system of equations. Substitute $191 + x$ for y in Equation (1).

$$x + 191 + x = 3853$$
$$2x + 191 = 3853$$
$$2x = 3662$$
$$x = 1831$$

Substitute 1831 for x in Equation (1) and solve for x.

$$1831 + y = 3853$$
$$y = 2022$$

Check. The total amount is 1831 + 2022, or 3853 texts. Texts received 191 + 1831 = 2022. The answer checks.
State. There were 2022 text messages sent and 1831 text messages received.

18. Let x = the number of 5-cent bottles or cans and y = the number of 10-cent bottles or cans.
Solve: $x + y = 430,$
$0.05x + 0.10y = 26.20$
The solution is (336, 94). So, 336 5-cent bottles or cans and 94 10-cent bottles or cans were collected.

19. ***Familiarize***. Let x = the number of pounds of Pecan Morning Granola and y = the number of pounds of Oat Dream Granola to be used in the mixture. The amount of nuts and dried fruit in the mixture is 19%(20 lb), or 0.19(20 lb) = 3.8 lb.
Translate. We organize the information in a table.

	Pecan Morning	Oat Dream	Mixture
Number of pounds	x	y	20
Percent of nuts and dried fruit	25%	10%	19%
Amount of nuts and dried fruit	$0.25x$	$0.10y$	3.8 lb

We get one equation from the "Number of pounds" row of the table:
$x + y = 20$
The last row of the table yields a second equation:
$0.25x + 0.1y = 3.8$
After clearing decimals, we have the problem translated to a system of equations:

$$x + y = 20, \quad (1)$$
$$25x + 10y = 380 \quad (2)$$

Carry out. We use the elimination method to solve the system of equations.

$$-10x - 10y = -200 \quad \text{Multiplying (1) by } -10$$
$$25x + 10y = 380$$
$$15x \quad = 180$$
$$x = 12$$

Substitute 12 for x in (1) and solve for y.

$$12 + y = 20$$
$$y = 8$$

Check. The amount of the mixture is 12 lb + 8 lb, or 20 lb. The amount of nuts and dried fruit in the mixture is 0.25(12 lb) + 0.1(8 lb) = 3 lb + 0.8 lb = 3.8 lb. The answer checks.
State. 12 lb of Pecan Morning Granola and 8 lb of Oat Dream Granola should be mixed.

20. Let r = the speed of the boat in still water.

	Distance	Rate	Time
With current	d	$r+6$	1.5
Against current	d	$r-6$	3

Solve: $d = 1.5(r+6)$
$d = 3(r-6)$
We find that $r = 18$ mph.

Exercise Set 8.5

1. a

3. d

5. ***Familiarize***. Let x = the first number, y = the second number, and z = the third number.
Translate.

The sum of the three numbers is 85.
↓ ↓ ↓
$x + y + z \quad = \quad 85$

The second is seven more than the first.
↓ ↓ ↓ ↓ ↓
$y \quad = \quad 7 \quad + \quad x$

The third is two more than four times the second.
↓ ↓ ↓ ↓ ↓ ↓ ↓
$z \quad = \quad 2 \quad + \quad 4 \quad \times \quad y$

We have a system of equations:

$$x + y + z = 85, \quad \text{or} \quad x + y + z = 85$$
$$y = 7 + x \qquad -x + y \quad = 7$$
$$z = 2 + 4y \qquad -4y + z = 2$$

Carry out. Solving the system we get (8, 15, 62).
Check. The sum of the three numbers is 8 + 15 + 62, or 85. The second number 15, is 7 more than the first number 8. The third number, 62 is 2 more than four times the second number, 15. The numbers check.
State. The numbers are 8, 15, and 62.

7. ***Familiarize***. Let x = the first number, y = the second number, and z = the third number.
Translate.

The sum of three numbers is 26.
↓ ↓ ↓
$x + y + z \quad = \quad 26$

Twice the first minus the second is the third less 2
↓ ↓ ↓ ↓ ↓ ↓ ↓
$2x \quad - \quad y \quad = \quad z \quad - \quad 2$

The third is the second minus 3 times the first.
↓ ↓ ↓ ↓ ↓
$z \quad = \quad y \quad - \quad 3x$

We now have a system of equations.

$x+y+z=26,$ or $x+y+z=26,$
$2x-y=z-2,$ $\quad 2x-y-z=-2,$
$z=y-3x$ $\quad 3x-y+z=0$

Carry out. Solving the system we get (8, 21, –3).
Check. The sum of the numbers is $8+21-3$, or 26. Twice the first minus the second is $2\cdot 8-21$, or –5, which is 2 less than the third. The second minus three times the first is $21-3\cdot 8$, or –3, which is the third. The numbers check.
State. The numbers are 8, 21, and –3.

9. ***Familiarize***. We first make a drawing.

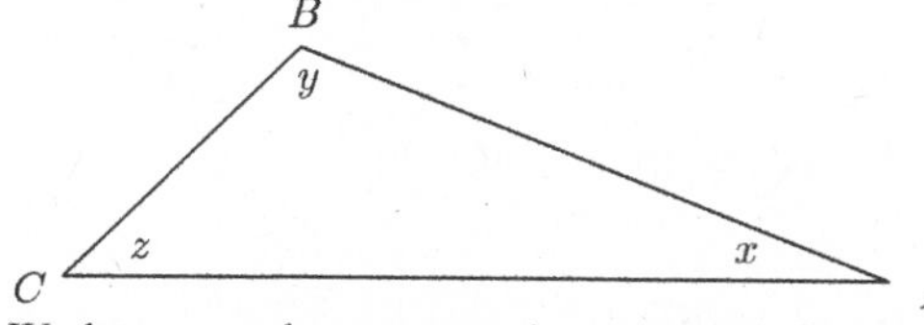

We let *x*, *y*, and *z* represent the measures of angles *A*, *B*, and *C*, respectively. The measures of the angles of a triangle add up to 180°.
Translate.

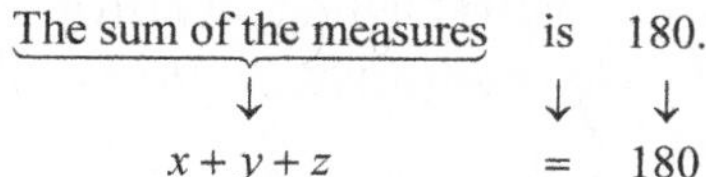

The sum of the measures is 180.
x + y + z = 180

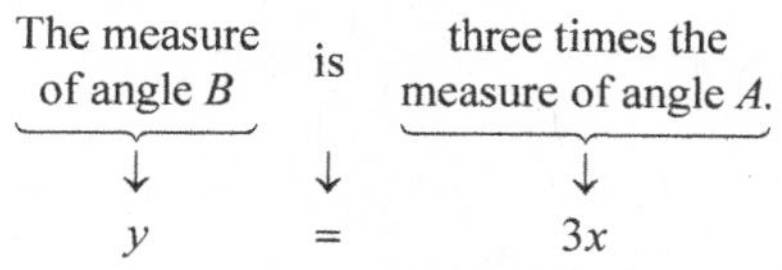

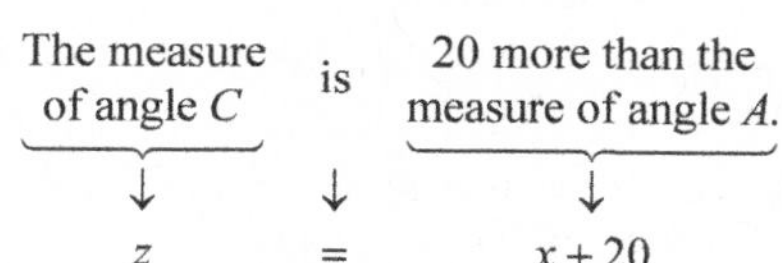

We now have a system of equations.

$$x+y+z=180,$$
$$y=3x,$$
$$z=x+20$$

Carry out. Solving the system we get (32, 96, 52).
Check. The sum of the measures is $32°+96°+52°$, or 180°. Three times the measure of angle *A* is $3\cdot 32°$, or 96°, the measure of angle *B*. 20 more than the measure of angle *A* is $32°+20°$, or 52°, the measure of angle *C*. The numbers check.
State. The measures of angles *A*, *B*, and *C* are 32°, 96°, and 52°, respectively.

11. ***Familiarize***. Let *x*, *y* and *z* represent the GRE critical verbal score, quantitative score, and writing score, respectively.
Translate.

The sum of three scores is 306.3.
↓ ↓ ↓
$x+y+z$ = 306.3

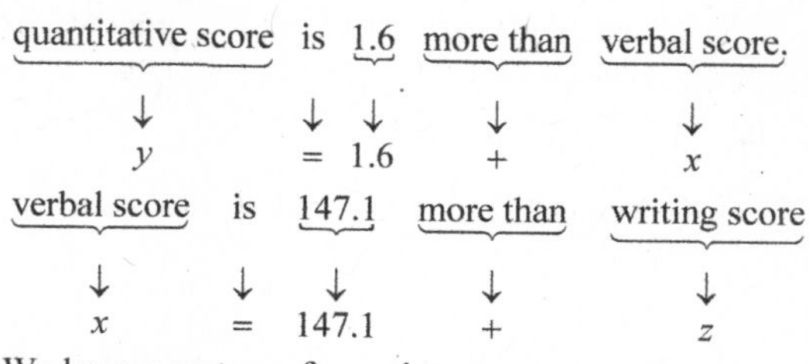

quantitative score is 1.6 more than verbal score.
y = 1.6 + x
verbal score is 147.1 more than writing score
x = 147.1 + z

We have a system of equations:

$x+y+z=306.3,$ or $x+y+z=306.3$
$y=1.6+x$ $\quad -x+y=1.6$
$x=147.1+z$ $\quad x-z=147.1$

Carry out Solving the system we get (150.6, 152.2, 3.5)
Check. The sum of the scores is 150.6 + 152.2 + 3.5, or 306.3. The quantitative score, 152.2, is 1.6 more than the verbal score, 150.6. The verbal score, 150.6 is 147.1 more than the writing score, 3.5. The answer checks.
State. The average score for verbal reasononing was 150.6, for quantitative reasoning 152.2, and for analytical writing 3.5.

13. ***Familiarize***. Let *x*, *y*, and *z* represent the number of grams of fiber in 1 bran muffin, 1 banana, and a 1-cup serving of Wheaties, respectively.
Translate.
Two bran muffins, 1 banana, and a 1-cup serving of Wheaties contain 9 g of fiber, so we have

$$2x+y+z=9.$$

One bran muffin, 2 bananas, and a 1-cup serving of Wheaties contain 10.5 g of fiber, so we have

$$x+2y+z=10.5.$$

Two bran muffins and a 1-cup serving of Wheaties contain 6 g of fiber, so we have

$$2x+z=6.$$

We now have a system of equations.

$$2x+y+z=9,$$
$$x+2y+z=10.5,$$
$$2x+z=6$$

Carry out. Solving the system, we get (1.5, 3, 3).
Check. Two bran muffins, 1 banana, and a 1-cup serving of Wheaties contain $2(1.5)+3+3$, or 9 g of fiber. One bran muffin, 2 bananas, and a 1-cup serving of Wheaties contain $1.5+2\cdot 3+3$, or 10.5 g of fiber. Two bran muffins and a 1-cup serving of Wheaties contain $2(1.5)+3$, or 6 g of fiber. The answer checks.
State. A bran muffin has 1.5 g of fiber, a banana has 3 g, and a 1-cup serving of Wheaties has 3 g.

15. Observe that the basic model plus tow package costs \$24,290 and when a hard top is added the price rises to \$25,285. This tells us that the price of a hard top is \$25,285 – \$24,290, or \$995. Now observe that the base model and hard top costs \$24,890 so the base model costs \$24,890 – \$995, or \$23,895. Finally we know the tow package is \$24,290 – \$23,895, or \$395.

17. ***Familiarize***. Let *x* = the number of 12-oz cups, *y* = the number of 16-oz cups, and *z* = the number of 20-oz cups that Reba filled. Note that six 144-oz brewers contain $6\cdot 144$, or 864 oz of coffee. Also, *x* 12-oz cups

contain a total of $12x$ oz of coffee and bring in $\$1.85x$, y 16-oz cups contain $16y$ oz and bring in $\$2.10y$, and z 20-oz cups contain $20z$ oz and bring in $\$2.45z$.
Translate.

The total number of coffees served was 55.

$$x+y+z = 55$$

The total amount of coffee served was 864 oz.

$$12x+16y+20z = 864$$

The total amount collected was \$115.80.

$$1.85x+2.10y+2.45z = 115.80$$

Now we have a system of equations.

$$\begin{aligned} x+y+z &= 55,\\ 12x+16y+20z &= 864,\\ 1.85x+2.10y+2.45z &= 115.80 \end{aligned}$$

Carry out. Solving the system we get (17, 25, 13).
Check. The total number of coffees served was $17+25+13$, or 55. The total amount of coffee served was $12\cdot 17+16\cdot 25+20\cdot 13=204+400+260=864$ oz. The total amount collected was $\$1.85(17)+\$2.10(25)+\$2.45(13)=\$31.45+\$52.50+\$31.85=\$115.80$. The numbers check.
State. Reba filled 17 12-oz cups, 25 16-oz cups, and 13 20-oz cups.

19. ***Familiarize.*** Let x = the amount of the loan at 7%, y = the amount of the loan at 5%, and z = the amount of the loan at 3.2%.
Translate.

Total of the three loans is \$120,000.

$$x+y+z = 120{,}000$$

Total interest due is \$5040.

$$0.07x+0.05y+0.032z = 5040$$

home-equity interest is \$1190 more than business-equipment loan interest.

$$0.032z = 1190 + 0.07x$$

We have a system of equations:

$$\begin{aligned} x+y+z &= 120{,}000\\ 0.07x+0.05y+0.032z &= 5040\\ -0.07x+0.032z &= 1190 \end{aligned}$$

Carry out. Solving the system we get (15,000, 35,000, 70,000).
Check. The total of the three loans is \$15,000 + \$35,000 + \$70,000 = \$120,000. The total interest due is 0.07(\$15,000) + 0.05(\$35,000) + 0.032(\$70,000) = \$1050 + \$1750 + \$2240 = \$5040. The interest from the home-equity loan, 0.032(\$70,000), or \$2240 is \$1190 more than the business-equipment interest 0.07(\$15,000), or \$1050. The numbers check.
State. The business-equipment loan was \$15,000, the small-business loan was \$35,000, and the home-equity loan was \$70,000.

21. ***Familiarize.*** Let x = the price of 1 g of gold, y = the the price of 1 g of silver, and z = the price of 1 g of copper.
Translate.

Cost of 100 g of red gold is \$4177.15.

$$100(0.75x+0.05y+0.20z) = 4177.15$$

Cost of 100 g of yellow gold is \$4185.25.

$$100(0.75x+0.125y+0.125z) = 4185.25$$

Cost of 100 g of white gold is \$2153.875.

$$100(0.375x+0.625y) = 2153.875$$

We have a system of equations:

$$\begin{aligned} 75x+5y+20z &= 4177.15\\ 750x+125y+125z &= 41{,}852.50\\ 375x+625y &= 21{,}538.75 \end{aligned}$$

Carry out. Solving the system we get (55.62, 1.09, 0.01).
Check. The cost of 100 g of red gold is $100[0.75(\$55.62)+0.05(\$1.09)+0.20(\$0.01)]=\$4171.5+\$5.45+\$0.20=\$4177.15$. The cost of 100 g of yellow gold is $100[0.75(\$55.62)+0.125(\$1.09)+0.125(\$0.01)]=\$4171.50+\$13.625+\$0.125=\$4185.25$. The cost of 100 g of white gold is $100[0.375(\$55.62)+0.625(\$1.09)]=\$2085.75+\$68.125=\$2153.875$. The numbers check.
State. The price of 1 g of gold is \$55.62, of silver is \$1.09 and of copper is \$0.01.

23. ***Familiarize.*** Let r = the number of servings of roast beef, p = the number of baked potatoes, and b = the number of servings of broccoli. Then r servings of roast beef contain $300r$ Calories, $20r$ g of protein, and no vitamin C. In p baked potatoes there are $100p$ Calories, $5p$ g of protein, and $20p$ mg of vitamin C. And b servings of broccoli contain $50b$ Calories, $5b$ g of protein, and $100b$ mg of vitamin C. The patient requires 800 Calories, 55 g of protein, and 220 mg of vitamin C.
Translate. Write equations for the total number of calories, the total amount of protein, and the total amount of vitamin C.

$$\begin{aligned} 300r+100p+50b &= 800 \quad \text{(Calories)}\\ 20r+5p+5b &= 55 \quad \text{(protein)}\\ 20p+100b &= 220 \quad \text{(vitamin C)} \end{aligned}$$

We now have a system of equations.
Carry out. Solving the system we get (2, 1, 2).
Check. Two servings of roast beef provide 600 Calories, 40 g of protein, and no vitamin C. One baked potato provides 100 Calories, 5 g of protein, and 20 mg of vitamin C. And 2 servings of broccoli provide 100 Calories, 10 g of protein, and 200 mg of vitamin C. Together, then, they provide 800 Calories, 55 g of protein, and 220 mg of vitamin C. The values check.
State. The dietician should prepare 2 servings of roast beef, 1 baked potato, and 2 servings of broccoli.

25. ***Familiarize***. Let x, y, and z be the number of tickets sold for the first mezzanine, main floor and second mezzanine, respectively.
Translate.

Total number of tickets	is	40.
$x+y+z$	$=$	40

Number of tickets for first mezzanine and main floor	is	number of tickets for second mezzanine.
$x+y$	$=$	z

Total cost of tickets	is	\$1432.
$52x+38y+28z$	$=$	1432

We have a system of equations:

$$x+y+z=40$$
$$x+y=z$$
$$52x+38y+28z=1432$$

Carry out. Solving the system we get (8, 12, 20).
Check. The total number of tickets is 8 + 12 + 20, or 40. The sum of first mezzanine and main floor is 8 + 12, is the number of second mezzanine, 20. The total cost is $\$52(8)+\$38(12)+\$28(20)$

$=\$416+\$456+\$560=\$1432.$ The numbers check.
State. There were 8 first mezzanine tickets, 12 main floor tickets and 20 second mezzanine tickets sold.

27. ***Familiarize***. Let x, y, and z represent the populations of Asia, Africa, and the rest of the world, respectively, in billions, in 2050.
Translate.

The total world population	will be	9.4 billion.
$x+y+z$	$=$	9.4

Population of Asia	will be	2.9 billion	more than	Population of Africa
x	$=$	2.9	$+$	y

Population of the rest of the world	will be	0.6 billion	more than	one-fourth	of	population of Asia.
z	$=$	0.6	$+$	$\frac{1}{4}$	$\times$	x

We have a system of equations.

$$x+y+z=9.4$$
$$x=2.9+y$$
$$z=0.6+\frac{1}{4}x$$

Carry out. Solving the system we get (5.2, 2.3, 1.9).
Check. The total population will be 5.2 + 2.3 + 1.9, or 9.4 billion. The population of Asia, 5.2 billion, is 2.9 billion more than the population of Africa, 2.3 billion. Also, the rest of the world population, 1.9 billion is 0.6 billion more than $\frac{1}{4}$ of 5.2, or 1.3 billion. The numbers check.
State. In 2050, the population of Asia will be 5.2 billion, the population of Africa will be 2.3 billion and the population of the rest of the world will be 1.9 billion.

29. *Writing Exercise.*

31. Graph $y=4$.

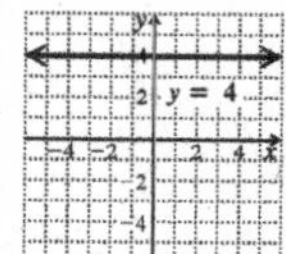

33. Graph $y-3x=3$

$$y=3x+3$$

$y - 3x = 3$

35. Graph $f(x)=2x-1$.

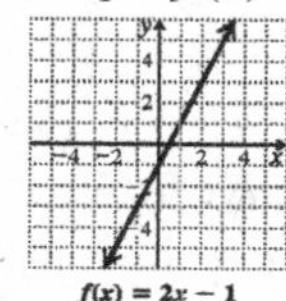

$f(x) = 2x - 1$

37. *Writing Exercise.*

39. ***Familiarize***. Let w, x, y, and z represent the ACT benchmarks for English, reading, mathematics, and science test scores, respectively.
Translate.

science score	is	6 points	more than	English score.
z	$=$	6	$+$	w

reading and math score	is	1 point	more than	English and science score.
$x+y$	$=$	1	$+$	$w+z$

English, math, science score	is	1 point	more than	three times reading score.
$w+y+z$	$=$	1	$+$	$3x$

Sum of all scores	is	85.
$w+x+y+z$	$=$	85

We have a system of equations:

$$-w+z=6 \quad (1)$$
$$-w+x+y-z=1 \quad (2)$$
$$w-3x+y+z=1 \quad (3)$$
$$w+x+y+z=85 \quad (4)$$

Carry out. We solve the system of equations. First add Equations (2) and (4).

$$\begin{array}{rl} -w+x+y-z=1 & (2) \\ w-3x+y+z=1 & (3) \\ \hline -2x+2y \quad =2 & (5) \end{array}$$

Next, add Equations (2) and (4)

$$\begin{array}{rl} -w+x+y-z=1 & (2) \\ w+x+y+z=85 & (4) \\ \hline 2x+2y \quad =86 & (6) \end{array}$$

Then, add Equations (5) and (6).

$$\begin{array}{rl} -2x+2y=2 & (5) \\ 2x+2y=86 & (6) \\ \hline 4y=88 & \\ y=22 & \end{array}$$

Next substitute 22 for y in (6) and solve for x.

$$2x+2(22)=86$$
$$2x=42$$
$$x=21$$

Substitute 21 for x and 22 for y in (4).

$$w+21+22+z=85$$
$$w+z=42 \quad (7)$$

Add Equations (1) and (7).

$$\begin{array}{rl} -w+z=6 & (1) \\ w+z=42 & (7) \\ \hline 2z=48 & \\ z=24 & \end{array}$$

Finally, substitute 24 for z in (1) and solve for w.

$$-w+24=6$$
$$w=18$$

The solution is (18, 21, 22, 24).
Check. The check is left to the student.
State. The benchmarks for English, reading, math, and science scores are 18, 21, 22, and 24, respectively.

41. ***Familiarize***. Let w, x, y, and z represent the ages of Tammy, Carmen, Dennis, and Mark respectively.
Translate.
Tammy's age is the sum of the ages of Carmen and Dennis, so we have

$$w=x+y.$$

Carmen's age is 2 more than the sum of the ages of Dennis and Mark, so we have

$$x=2+y+z.$$

Dennis's age is four times Mark's age, so we have

$$y=4z.$$

The sum of all four ages is 42, so we have

$$w+x+y+z=42.$$

Now we have a system of equations.

$$\begin{array}{rl} w=x+y, & (1) \\ x=2+y+z, & (2) \\ y=4z, & (3) \\ w+x+y+z=42 & (4) \end{array}$$

Carry out. We solve the system of equations. First we will express w, x, and y in terms of z and then solve for z. From (3) we know that $y=4z$. Substitute $4z$ for y in (2):

$$x=2+4z+z=2+5z.$$

Substitute $2+5z$ for x and $4z$ for y in (1):

$$w=2+5z+4z=2+9z.$$

Now substitute $2+9z$ for w, $2+5z$ for x, and $4z$ for y in (4) and solve for z.

$$2+9z+2+5z+4z+z=42$$
$$19z+4=42$$
$$19z=38$$
$$z=2$$

Then we have:

$$w=2+9z=2+9\cdot 2=20,$$
$$x=2+5z=2+5\cdot 2=12, \text{ and}$$
$$y=4z=4\cdot 2=8$$

Although we were asked to find only Tammy's age, we found all of the ages so that we can check the result.
Check. The check is left to the student.
State. Tammy is 20 years old.

43. Let T, G, and H represent the number of tickets Tom, Gary, and Hal begin with, respectively. After Hal gives tickets to Tom and Gary, each has the following number of tickets:

Tom: $T+T$, or $2T$,
Gary: $G+G$, or $2G$,
Hal: $H-T-G$.

After Tom gives tickets to Gary and Hal, each has the following number of tickets:

Gary: $2G+2G$, or $4G$,
Hal: $(H-T-G)+(H-T-G)$, or $2(H-T-G)$,
Tom: $2T-2G-(H-T-G)$, or $3T-H-G$

After Gary gives tickets to Hal and Tom, each has the following number of tickets:

Hal: $2(H-T-G)+2(H-T-G)$, or $4(H-T-G)$
Tom: $(3T-H-G)+(3T-H-G)$, or $2(3T-H-G)$,
Gary: $4G-2(H-T-G)-(3T-H-G)$, or $7G-H-T$.

Since Hal, Tom, and Gary each finish with 40 tickets, we write the following system of equations:

$$4(H-T-G)=40,$$
$$2(3T-H-G)=40,$$
$$7G-H-T=40$$

Solving the system we find that $T=35$, so Tom started with 35 tickets.

Exercise Set 8.6

1. matrix

3. entry

5. rows

7. $x+2y=11,$
$3x-y=5$

Write a matrix using only the constants.

$$\left[\begin{array}{cc|c} 1 & 2 & 11 \\ 3 & -1 & 5 \end{array}\right]$$

Multiply the first row by –3 and add it to the second row.

$\begin{bmatrix} 1 & 2 & | & 11 \\ 0 & -7 & | & -28 \end{bmatrix}$ New Row 2 = -3(Row 1) + Row 2

Reinserting the variables, we have

$x+2y=11,\quad (1)$
$-7y=-28\quad (2)$

Solve Equation (2) for y.

$-7y=-28$
$y=4$

Substitute 4 for y in Equation (1) and solve for x.

$x+2(4)=11$
$x+8=11$
$x=3$

The solution is (3, 4).

9. $3x+y=-1,$
$6x+5y=13$

We first write a matrix using only the constants.

$\begin{bmatrix} 3 & 1 & | & -1 \\ 6 & 5 & | & 13 \end{bmatrix}$

Multiply the first row by –2 and add it to the second row.

$\begin{bmatrix} 3 & 1 & | & -1 \\ 0 & 3 & | & 15 \end{bmatrix}$ New Row 2 = -2(Row 1) + Row 2

Reinserting the variables, we have

$3x+y=-1,\quad (1)$
$3y=15\quad (2)$

Solve Equation (2) for y.

$3y=15$
$y=5$

Substitute 5 for y in Equation (1) and solve for x.

$3x+5=-1$
$3x=-6$
$x=-2$

The solution is (–2, 5).

11. $6x-2y=4,$
$7x+y=13$

Write a matrix using only the constants.

$\begin{bmatrix} 6 & -2 & | & 4 \\ 7 & 1 & | & 13 \end{bmatrix}$

Multiply the second row by 6 to make the first number in row 2 a multiple of 6.

$\begin{bmatrix} 6 & -2 & | & 4 \\ 42 & 6 & | & 78 \end{bmatrix}$ New Row 2 = 6(Row 2)

Now multiply the first row by -7 and add it to the second row.

$\begin{bmatrix} 6 & -2 & | & 4 \\ 0 & 20 & | & 50 \end{bmatrix}$ New Row 2 = -7(Row 1) + Row 2

Reinserting the variables, we have

$6x-2y=4,\quad (1)$
$20y=50.\quad (2)$

Solve Equation (2) for y.

$20y=50$
$y=\frac{5}{2}$

Substitute $\frac{5}{2}$ for y in Equation (1) and solve for x.

$6x-2y=4$
$6x-2\left(\frac{5}{2}\right)=4$
$6x-5=4$
$6x=9$
$x=\frac{3}{2}$

The solution is $\left(\frac{3}{2}, \frac{5}{2}\right)$.

13. $3x+2y+2z=3,$
$x+2y-z=5,$
$2x-4y+z=0$

We first write a matrix using only the constants.

$\begin{bmatrix} 3 & 2 & 2 & | & 3 \\ 1 & 2 & -1 & | & 5 \\ 2 & -4 & 1 & | & 0 \end{bmatrix}$

First interchange rows 1 and 2 so that each number below the first number in the first row is a multiple of that number.

$\begin{bmatrix} 1 & 2 & -1 & | & 5 \\ 3 & 2 & 2 & | & 3 \\ 2 & -4 & 1 & | & 0 \end{bmatrix}$

Multiply row 1 by –3 and add it to row 2.
Multiply row 1 by –2 and add it to row 3.

$\begin{bmatrix} 1 & 2 & -1 & | & 5 \\ 0 & -4 & 5 & | & -12 \\ 0 & -8 & 3 & | & -10 \end{bmatrix}$

Multiply row 2 by –2 and add it to row 3.

$\begin{bmatrix} 1 & 2 & -1 & | & 5 \\ 0 & -4 & 5 & | & -12 \\ 0 & 0 & -7 & | & 14 \end{bmatrix}$

Reinserting the variables, we have

$x+2y-z=5,\quad (1)$
$-4y+5z=-12,\quad (2)$
$-7z=14.\quad (3)$

Solve (3) for z.

$-7z=14$
$z=-2$

Substitute –2 for z in (2) and solve for y.

$-4y+5(-2)=-12$
$-4y-10=-12$
$-4y=-2$
$y=\frac{1}{2}$

Substitute $\frac{1}{2}$ for y and –2 for z in (1) and solve for x.

$x+2\cdot\frac{1}{2}-(-2)=5$
$x+1+2=5$
$x+3=5$
$x=2$

The solution is $\left(2, \frac{1}{2}, -2\right)$.

15. $p-2q-3r=3,$
$2p-q-2r=4,$
$4p+5q+6r=4$

We first write a matrix using only the constants.

$$\left[\begin{array}{ccc|c}1&-2&-3&3\\2&-1&-2&4\\4&5&6&4\end{array}\right]$$

$$\left[\begin{array}{ccc|c}1&-2&-3&3\\0&3&4&-2\\0&13&18&-8\end{array}\right]\begin{array}{l}\\ \text{New Row 2} = -2(\text{Row 1}) + \text{Row 2}\\ \text{New Row 3} = -4(\text{Row 1}) + \text{Row 3}\end{array}$$

$$\left[\begin{array}{ccc|c}1&-2&-3&3\\0&3&4&-2\\0&39&54&-24\end{array}\right]\quad \text{New Row 3} = 3(\text{Row 3})$$

$$\left[\begin{array}{ccc|c}1&-2&-3&3\\0&3&4&-2\\0&0&2&2\end{array}\right]\quad \text{New Row 3} = -13(\text{Row 2}) + \text{Row 3}$$

Reinserting the variables, we have

$p-2q-3r=3,$ (1)
$3q+4r=-2,$ (2)
$2r=2$ (3)

Solve (3) for r.

$2r=2$
$r=1$

Substitute 1 for r in (2) and solve for q.

$3q+4\cdot 1=-2$
$3q+4=-2$
$3q=-6$
$q=-2$

Substitute -2 for q and 1 for r in (1) and solve for p.

$p-2(-2)-3\cdot 1=3$
$p+4-3=3$
$p+1=3$
$p=2$

The solution is (2, –2, 1).

17. $3p+2r=11,$
$q-7r=4,$
$p-6q=1$

We first write a matrix using only the constants.

$$\left[\begin{array}{ccc|c}3&0&2&11\\0&1&-7&4\\1&-6&0&1\end{array}\right]$$

$$\left[\begin{array}{ccc|c}1&-6&0&1\\0&1&-7&4\\3&0&2&11\end{array}\right]\quad \text{Interchange Row 1 and Row 3}$$

$$\left[\begin{array}{ccc|c}1&-6&0&1\\0&1&-7&4\\0&18&2&8\end{array}\right]\quad \text{New Row 3} = -3(\text{Row 1}) + \text{Row 3}$$

$$\left[\begin{array}{ccc|c}1&-6&0&1\\0&1&-7&4\\0&0&128&-64\end{array}\right]\quad \text{New Row3} = -18(\text{Row2}) + \text{Row3}$$

Reinserting the variables, we have

$p-6q=1,$ (1)
$q-7r=4,$ (2)
$128r=-64.$ (3)

Solve (3) for r.

$128r=-64$
$r=-\frac{1}{2}$

Substitute $-\frac{1}{2}$ for r in (2) and solve for q.

$q-7r=4$
$q-7\left(-\frac{1}{2}\right)=4$
$q+\frac{7}{2}=4$
$q=\frac{1}{2}$

Substitute $\frac{1}{2}$ for q in (1) and solve for p.

$p-6\cdot\frac{1}{2}=1$
$p-3=1$
$p=4$

The solution is $\left(4, \frac{1}{2}, -\frac{1}{2}\right)$.

19. We will rewrite the equations with the variables in alphabetical order:

$-2w+2x+2y-2z=-10,$
$w+x+y+z=-5,$
$3w+x-y+4z=-2,$
$w+3x-2y+2z=-6$

Write a matrix using only the constants.

$$\left[\begin{array}{cccc|c}-2&2&2&-2&-10\\1&1&1&1&-5\\3&1&-1&4&-2\\1&3&-2&2&-6\end{array}\right]$$

$$\left[\begin{array}{cccc|c}-1&1&1&-1&-5\\1&1&1&1&-5\\3&1&-1&4&-2\\1&3&-2&2&-6\end{array}\right]\quad \text{New Row 1} = \frac{1}{2}(\text{Row 1})$$

$$\left[\begin{array}{cccc|c}-1&1&1&-1&-5\\0&2&2&0&-10\\0&4&2&1&-17\\0&4&-1&1&-11\end{array}\right]\begin{array}{l}\\ \text{New Row2} = \text{Row1} + \text{Row2}\\ \text{New Row3} = 3(\text{Row1}) + \text{Row3}\\ \text{New Row4} = \text{Row1} + \text{Row4}\end{array}$$

$$\left[\begin{array}{cccc|c}-1&1&1&-1&-5\\0&2&2&0&-10\\0&0&-2&1&3\\0&0&-5&1&9\end{array}\right]\begin{array}{l}\\ \\ \text{New Row3} = -2(\text{Row2}) + \text{Row3}\\ \text{New Row4} = -2(\text{Row2}) + \text{Row4}\end{array}$$

$$\left[\begin{array}{cccc|c}-1&1&1&-1&-5\\0&2&2&0&-10\\0&0&-2&1&3\\0&0&-10&2&18\end{array}\right]\quad \text{New Row 4} = 2(\text{Row 4})$$

$$\left[\begin{array}{cccc|c}-1&1&1&-1&-5\\0&2&2&0&-10\\0&0&-2&1&3\\0&0&0&-3&3\end{array}\right]\quad \text{New Row4} = -5(\text{Row3}) + \text{Row4}$$

Reinserting the variables, we have

$-w+x+y-z=-5,$ (1)
$2x+2y=-10,$ (2)
$-2y+z=3,$ (3)
$-3z=3.$ (4)

Solve (4) for z.

$$-3z = 3$$
$$z = -1$$

Substitute -1 for z in (3) and solve for y.

$$-2y + (-1) = 3$$
$$-2y = 4$$
$$y = -2$$

Substitute -2 for y in (2) and solve for x.

$$2x + 2(-2) = -10$$
$$2x - 4 = -10$$
$$2x = -6$$
$$x = -3$$

Substitute -3 for x, -2 for y, and -1 for z in (1) and solve for w.

$$-w + (-3) + (-2) - (-1) = -5$$
$$-w - 3 - 2 + 1 = -5$$
$$-w - 4 = -5$$
$$-w = -1$$
$$w = 1$$

The solution is $(1, -3, -2, -1)$.

21. ***Familiarize***. Let d = the number of dimes and n = the number of nickels. The value of d dimes is $\$0.10d$, and the value of n nickels is $\$0.05n$.

Translate.

Total number of coins is 42.

$$d + n = 42$$

Total value of coins is \$3.

$$0.10d + 0.05n = 3$$

After clearing decimals, we have this system.

$$d + n = 42,$$
$$10d + 5n = 300$$

Carry out. Solve using matrices.

$$\left[\begin{array}{cc|c} 1 & 1 & 42 \\ 10 & 5 & 300 \end{array}\right]$$

$$\left[\begin{array}{cc|c} 1 & 1 & 42 \\ 0 & -5 & -120 \end{array}\right] \quad \text{New Row 2} = -10(\text{Row 1}) + \text{Row 2}$$

Reinserting the variables, we have

$$d + n = 42, \quad (1)$$
$$-5n = -120 \quad (2)$$

Solve (2) for n.

$$-5n = -120$$
$$n = 24$$
$$d + 24 = 42 \quad \text{Back-substituting}$$
$$d = 18$$

Check. The sum of the two numbers is 42. The total value is $\$0.10(18) + \$0.05(24) = \$1.80 + \$1.20 = \$3$. The numbers check.

State. There are 18 dimes and 24 nickels.

23. ***Familiarize***. Let x = the number of pounds of dried fruit and y = the number of pounds of macadamia nuts. We organize the information in a table.

	Dried fruit	Macadamia nuts	Mixture
Number of pounds	x	y	15
Price per pound	\$5.80	\$14.75	\$9.38
Value of mixture	$5.80x$	$14.75y$	140.70

Translate.

The total number of pounds is 15.

$$x + y = 15$$

The total value of the mixture is \$140.70.

$$5.80x + 14.75y = 140.70$$

After clearing decimals, we have this system:

$$x + y = 15,$$
$$580x + 1475y = 14{,}070$$

Carry out. Solve using matrices.

$$\left[\begin{array}{cc|c} 1 & 1 & 15 \\ 580 & 1475 & 14{,}070 \end{array}\right]$$

Multiply row 1 by -580 and add it to the second row.

$$\left[\begin{array}{cc|c} 1 & 1 & 15 \\ 0 & 895 & 5370 \end{array}\right] \quad \text{New Row 2} = -580(\text{Row 1}) + \text{Row 2}$$

Reinserting the variables, we have

$$x + y = 15$$
$$895y = 5370$$
$$y = 6$$

Back-substitute 6 for y in Equation (1) and solve for x.

$$x + 6 = 15$$
$$x = 9$$

Check. The sum of the numbers, $6 + 9 = 15$. The total value is $\$5.90(9) + \$14.75(6)$, or $\$52.20 + \$88.50$, or \$140.70. The numbers check.

State. 9 pounds of dried fruit and 6 pounds of macadamia nuts should be used.

25. ***Familiarize***. We let x, y, and z represent the amounts invested at 3%, 4%, and 5%, respectively. Recall the formula for simple interest:

Interest = Principal × Rate × Time

Translate. We organize the information in a table.

	First Investment	Second Investment	Third Investment	Total
P	x	y	z	\$2500
R	3%	4%	5%	
T	1 yr	1 yr	1 yr	
I	$0.03x$	$0.04y$	$0.05z$	\$112

The first row gives us one equation:

$$x + y + z = 2500$$

The last row gives a second equation:

$$0.03x + 0.04y + 0.05z = 112$$

Amount invested at 5% is \$1100 more than amount invested at 4%.

$$z = \$1100 + y$$

After clearing decimals, we have this system:

$$x + y + z = 2500,$$
$$3x + 4y + 5z = 11{,}200,$$
$$-y + z = 1100$$

Carry out. Solve using matrices.

$$\left[\begin{array}{ccc|c} 1 & 1 & 1 & 2500 \\ 3 & 4 & 5 & 11,200 \\ 0 & -1 & 1 & 1100 \end{array}\right]$$

$$\left[\begin{array}{ccc|c} 1 & 1 & 1 & 2500 \\ 0 & 1 & 2 & 3700 \\ 0 & -1 & 1 & 1100 \end{array}\right] \quad \text{New Row 2} = -3(\text{Row 1}) + \text{Row 2}$$

$$\left[\begin{array}{ccc|c} 1 & 1 & 1 & 2500 \\ 0 & 1 & 2 & 3700 \\ 0 & 0 & 3 & 4800 \end{array}\right] \quad \text{New Row 3} = \text{Row 2} + \text{Row 3}$$

Reinserting the variables, we have

$$x + y + z = 2500, \quad (1)$$
$$y + 2z = 3700, \quad (2)$$
$$3z = 4800 \quad (3)$$

Solve (3) for z.

$$3z = 4800$$
$$z = 1600$$

Back-substitute 1600 for z in (2) and solve for y.

$$y + 2 \cdot 1600 = 3700$$
$$y + 3200 = 3700$$
$$y = 500$$

Back-substitute 500 for y and 1600 for z in (1) and solve for x.

$$x + 500 + 1600 = 2500$$
$$x + 2100 = 2500$$
$$x = 400$$

Check. The total investment is \$400 + \$500 + \$1600, or \$2500. The total interest is 0.03(\$400) + 0.04(\$500) + 0.05(\$1600) = \$12 + \$20 + \$80 = \$112. The amount invested at 5%, \$1600, is \$1100 more than the amount invested at 4%, \$500. The numbers check.

State. \$400 is invested at 3%, \$500 is invested at 4%, and \$1600 is invested at 5%.

27. *Writing Exercise.*

29.
$$\frac{x^3 - x}{x+2} \cdot \frac{x^2 + 3x + 2}{x-1} = \frac{x(x+1)(x-1)(x+1)(x+2)}{(x+2)(x-1)}$$
$$= \frac{x(x+1)\cancel{(x-1)}(x+1)\cancel{(x+2)}}{\cancel{(x+2)}\cancel{(x-1)}}$$
$$= x(x+1)^2$$

31.
$$\frac{3}{3a^2 + 2a - 1} - \frac{1}{a^2 + 2a + 1}$$
$$= \frac{3}{(3a-1)(a+1)} - \frac{1}{(a+1)(a+1)}$$
$$\text{LCD} = (3a-1)(a+1)^2$$
$$= \frac{3}{(3a-1)(a+1)} \cdot \frac{a+1}{a+1} - \frac{1}{(a+1)(a+1)} \cdot \frac{3a-1}{3a-1}$$
$$= \frac{3(a+1) - (3a-1)}{(3a-1)(a+1)^2}$$
$$= \frac{3a + 3 - 3a + 1}{(3a-1)(a+1)^2}$$
$$= \frac{4}{(3a-1)(a+1)^2}$$

33. *Writing Exercise.*

35. ***Familiarize***. Let w, x, y, and z represent the thousand's, hundred's, ten's, and one's digits, respectively.

Translate.

The sum of the digits is 10.

$$w + x + y + z = 10$$

Twice the sum of the thousand's and ten's digits is the sum of the hundred's and one's digits less one.

$$2(w + y) = x + z - 1$$

The ten's digit is twice the thousand's digit.

$$y = 2 \cdot w$$

The one's digit equals the sum of the thousand's and hundred's digits.

$$z = w + x$$

We have a system of equations which can be written as

$$\begin{aligned} w + x + y + z &= 10, \\ 2w - x + 2y - z &= -1, \\ -2w + y &= 0, \\ w + x - z &= 0. \end{aligned}$$

Carry out. We can use matrices to solve the system. We get (1, 3, 2, 4).

Check. The sum of the digits is 10. Twice the sum of 1 and 2 is 6. This is one less than the sum of 3 and 4. The ten's digit, 2, is twice the thousand's digit, 1. The one's digit, 4, equals 1 + 3. The numbers check.

State. The number is 1324.

Exercise Set 8.7

1. True

3. True

5. False

7. $\begin{vmatrix} 3 & 5 \\ 4 & 8 \end{vmatrix} = 3 \cdot 8 - 4 \cdot 5 = 24 - 20 = 4$

9. $\begin{vmatrix} 10 & 8 \\ -5 & -9 \end{vmatrix} = 10(-9) - 8(-5) = -90 + 40 = -50$

11.
$$\begin{vmatrix} 1 & 4 & 0 \\ 0 & -1 & 2 \\ 3 & -2 & 1 \end{vmatrix}$$
$$= 1\begin{vmatrix} -1 & 2 \\ -2 & 1 \end{vmatrix} - 0\begin{vmatrix} 4 & 0 \\ -2 & 1 \end{vmatrix} + 3\begin{vmatrix} 4 & 0 \\ -1 & 2 \end{vmatrix}$$
$$= 1[-1 \cdot 1 - (-2) \cdot 2] - 0 + 3[4 \cdot 2 - (-1) \cdot 0]$$
$$= 1 \cdot 3 - 0 + 3 \cdot 8$$
$$= 3 - 0 + 24$$
$$= 27$$

13. $\begin{vmatrix} -1 & -2 & -3 \\ 3 & 4 & 2 \\ 0 & 1 & 2 \end{vmatrix}$

$= -1\begin{vmatrix} 4 & 2 \\ 1 & 2 \end{vmatrix} - 3\begin{vmatrix} -2 & -3 \\ 1 & 2 \end{vmatrix} + 0\begin{vmatrix} -2 & -3 \\ 4 & 2 \end{vmatrix}$
$= -1[4\cdot 2 - 1\cdot 2] - 3[-2\cdot 2 - 1(-3)] + 0$
$= -1\cdot 6 - 3\cdot(-1) + 0$
$= -6 + 3 + 0$
$= -3$

15. $\begin{vmatrix} -4 & -2 & 3 \\ -3 & 1 & 2 \\ 3 & 4 & -2 \end{vmatrix}$

$= -4\begin{vmatrix} 1 & 2 \\ 4 & -2 \end{vmatrix} - (-3)\begin{vmatrix} -2 & 3 \\ 4 & -2 \end{vmatrix} + 3\begin{vmatrix} -2 & 3 \\ 1 & 2 \end{vmatrix}$
$= -4[1(-2) - 4\cdot 2] + 3[-2(-2) - 4\cdot 3]$
$\quad + 3[-2\cdot 2 - 1\cdot 3]$
$= -4(-10) + 3(-8) + 3(-7)$
$= 40 - 24 - 21 = -5$

17. $5x + 8y = 1,$
$3x + 7y = 5$

We compute D, D_x, and D_y.

$D = \begin{vmatrix} 5 & 8 \\ 3 & 7 \end{vmatrix} = 35 - 24 = 11$

$D_x = \begin{vmatrix} 1 & 8 \\ 5 & 7 \end{vmatrix} = 7 - 40 = -33$

$D_y = \begin{vmatrix} 5 & 1 \\ 3 & 5 \end{vmatrix} = 25 - 3 = 22$

Then,

$$x = \frac{D_x}{D} = \frac{-33}{11} = -3 \text{ and } y = \frac{D_y}{D} = \frac{22}{11} = 2.$$

The solution is (–3, 2).

19. $5x - 4y = -3,$
$7x + 2y = 6$

We compute D, D_x, and D_y.

$D = \begin{vmatrix} 5 & -4 \\ 7 & 2 \end{vmatrix} = 10 - (-28) = 38$

$D_x = \begin{vmatrix} -3 & -4 \\ 6 & 2 \end{vmatrix} = -6 - (-24) = 18$

$D_y = \begin{vmatrix} 5 & -3 \\ 7 & 6 \end{vmatrix} = 30 - (-21) = 51$

Then,

$$x = \frac{D_x}{D} = \frac{18}{38} = \frac{9}{19}$$

and

$$y = \frac{D_y}{D} = \frac{51}{38}.$$

The solution is $\left(\frac{9}{19}, \frac{51}{38}\right)$.

21. $3x - y + 2z = 1,$
$x - y + 2z = 3,$
$-2x + 3y + z = 1$

We compute D, D_x, D_y and D_z.

$D = \begin{vmatrix} 3 & -1 & 2 \\ 1 & -1 & 2 \\ -2 & 3 & 1 \end{vmatrix}$

$= 3\begin{vmatrix} -1 & 2 \\ 3 & 1 \end{vmatrix} - 1\begin{vmatrix} -1 & 2 \\ 3 & 1 \end{vmatrix} - 2\begin{vmatrix} -1 & 2 \\ -1 & 2 \end{vmatrix}$
$= 3(-7) - 1(-7) - 2(0)$
$= -21 + 7 - 0$
$= -14$

$D_x = \begin{vmatrix} 1 & -1 & 2 \\ 3 & -1 & 2 \\ 1 & 3 & 1 \end{vmatrix}$

$= 1\begin{vmatrix} -1 & 2 \\ 3 & 1 \end{vmatrix} - 3\begin{vmatrix} -1 & 2 \\ 3 & 1 \end{vmatrix} + 1\begin{vmatrix} -1 & 2 \\ -1 & 2 \end{vmatrix}$
$= 1(-7) - 3(-7) + 1(0)$
$= -7 + 21 + 0$
$= 14$

$D_y = \begin{vmatrix} 3 & 1 & 2 \\ 1 & 3 & 2 \\ -2 & 1 & 1 \end{vmatrix}$

$= 3\begin{vmatrix} 3 & 2 \\ 1 & 1 \end{vmatrix} - 1\begin{vmatrix} 1 & 2 \\ 1 & 1 \end{vmatrix} - 2\begin{vmatrix} 1 & 2 \\ 3 & 2 \end{vmatrix}$
$= 3\cdot 1 - 1(-1) - 2(-4)$
$= 3 + 1 + 8$
$= 12$

$D_z = \begin{vmatrix} 3 & -1 & 1 \\ 1 & -1 & 3 \\ -2 & 3 & 1 \end{vmatrix}$

$= 3\begin{vmatrix} -1 & 3 \\ 3 & 1 \end{vmatrix} - 1\begin{vmatrix} -1 & 1 \\ 3 & 1 \end{vmatrix} - 2\begin{vmatrix} -1 & 1 \\ -1 & 3 \end{vmatrix}$
$= 3(-10) - 1(-4) - 2(-2)$
$= -30 + 4 + 4$
$= -22$

Then, $x = \frac{D_x}{D} = \frac{14}{-14} = -1$

and $y = \frac{D_y}{D} = \frac{12}{-14} = -\frac{6}{7}$

and $z = \frac{D_z}{D} = \frac{-22}{-14} = \frac{11}{7}$.

The solution is $\left(-1, -\frac{6}{7}, \frac{11}{7}\right)$.

23. $2x - 3y + 5z = 27,$
$x + 2y - z = -4,$
$5x - y + 4z = 27$

We compute D, D_x, D_y and D_z.

$D = \begin{vmatrix} 2 & -3 & 5 \\ 1 & 2 & -1 \\ 5 & -1 & 4 \end{vmatrix}$

$= 2\begin{vmatrix} 2 & -1 \\ -1 & 4 \end{vmatrix} - 1\begin{vmatrix} -3 & 5 \\ -1 & 4 \end{vmatrix} + 5\begin{vmatrix} -3 & 5 \\ 2 & -1 \end{vmatrix}$

$= 2(7) - 1(-7) + 5(-7)$
$= 14 + 7 - 35 = -14$

$$D_x = \begin{vmatrix} 27 & -3 & 5 \\ -4 & 2 & -1 \\ 27 & -1 & 4 \end{vmatrix}$$
$$= 27\begin{vmatrix} 2 & -1 \\ -1 & 4 \end{vmatrix} - (-4)\begin{vmatrix} -3 & 5 \\ -1 & 4 \end{vmatrix} + 27\begin{vmatrix} -3 & 5 \\ 2 & -1 \end{vmatrix}$$
$= 27(7) + 4(-7) + 27(-7)$
$= 189 - 28 - 189$
$= -28$

$$D_y = \begin{vmatrix} 2 & 27 & 5 \\ 1 & -4 & -1 \\ 5 & 27 & 4 \end{vmatrix}$$
$$= 2\begin{vmatrix} -4 & -1 \\ 27 & 4 \end{vmatrix} - 1\begin{vmatrix} 27 & 5 \\ 27 & 4 \end{vmatrix} + 5\begin{vmatrix} 27 & 5 \\ -4 & -1 \end{vmatrix}$$
$= 2(11) - 1(-27) + 5(-7)$
$= 22 + 27 - 35$
$= 14$

$$D_z = \begin{vmatrix} 2 & -3 & 27 \\ 1 & 2 & -4 \\ 5 & -1 & 27 \end{vmatrix}$$
$$= 2\begin{vmatrix} 2 & -4 \\ -1 & 27 \end{vmatrix} - 1\begin{vmatrix} -3 & 27 \\ -1 & 27 \end{vmatrix} + 5\begin{vmatrix} -3 & 27 \\ 2 & -4 \end{vmatrix}$$
$= 2(50) - 1(-54) + 5(-42)$
$= 100 + 54 - 210$
$= -56$

Then $x = \frac{D_x}{D} = \frac{-28}{-14} = 2$

and $y = \frac{D_y}{D} = \frac{14}{-14} = -1$

and $z = \frac{D_z}{D} = \frac{-56}{-14} = 4$.

The solution is (2, −1, 4).

25. $r - 2s + 3t = 6,$
$2r - s - t = -3,$
$r + s + t = 6$

We compute D, D_r, D_s and D_t.

$$D = \begin{vmatrix} 1 & -2 & 3 \\ 2 & -1 & -1 \\ 1 & 1 & 1 \end{vmatrix}$$
$$= 1\begin{vmatrix} -1 & -1 \\ 1 & 1 \end{vmatrix} - 2\begin{vmatrix} -2 & 3 \\ 1 & 1 \end{vmatrix} + 1\begin{vmatrix} -2 & 3 \\ -1 & -1 \end{vmatrix}$$
$= 1(0) - 2(-5) + 1(5)$
$= 0 + 10 + 5$
$= 15$

$$D_r = \begin{vmatrix} 6 & -2 & 3 \\ -3 & -1 & -1 \\ 6 & 1 & 1 \end{vmatrix}$$
$$= 6\begin{vmatrix} -1 & -1 \\ 1 & 1 \end{vmatrix} - (-3)\begin{vmatrix} -2 & 3 \\ 1 & 1 \end{vmatrix} + 6\begin{vmatrix} -2 & 3 \\ -1 & -1 \end{vmatrix}$$
$= 6(0) + 3(-5) + 6(5)$
$= 0 - 15 + 30$
$= 15$

$$D_s = \begin{vmatrix} 1 & 6 & 3 \\ 2 & -3 & -1 \\ 1 & 6 & 1 \end{vmatrix}$$
$$= 1\begin{vmatrix} -3 & -1 \\ 6 & 1 \end{vmatrix} - 2\begin{vmatrix} 6 & 3 \\ 6 & 1 \end{vmatrix} + 1\begin{vmatrix} 6 & 3 \\ -3 & -1 \end{vmatrix}$$
$= 1(3) - 2(-12) + 1(3)$
$= 3 + 24 + 3$
$= 30$

$$D_t = \begin{vmatrix} 1 & -2 & 6 \\ 2 & -1 & -3 \\ 1 & 1 & 6 \end{vmatrix}$$
$$= 1\begin{vmatrix} -1 & -3 \\ 1 & 6 \end{vmatrix} - 2\begin{vmatrix} -2 & 6 \\ 1 & 6 \end{vmatrix} + 1\begin{vmatrix} -2 & 6 \\ -1 & -3 \end{vmatrix}$$
$= 1(-3) - 2(-18) + 1(12)$
$= -3 + 36 + 12$
$= 45$

Then $r = \frac{D_r}{D} = \frac{15}{15} = 1$

and $s = \frac{D_s}{D} = \frac{30}{15} = 2$

and $t = \frac{D_t}{D} = \frac{45}{15} = 3$.

The solution is (1, 2, 3).

27. *Writing Exercise.*

29. $6x^2 + 23x - 4 = (6x - 1)(x + 4)$

31. $25y^8 - 49z^2 = (5y^4 + 7z)(5y^4 - 7z)$

33. *Writing Exercise.*

35. $\begin{vmatrix} y & -2 \\ 4 & 3 \end{vmatrix} = 44$

$y \cdot 3 - 4(-2) = 44$ Evaluating the determinant
$3y + 8 = 44$
$3y = 36$
$y = 12$

37. $\begin{vmatrix} m+1 & -2 \\ m-2 & 1 \end{vmatrix} = 27$

$(m+1)(1) - (m-2)(-2) = 27$ Evaluating the determinant
$m + 1 + 2m - 4 = 27$
$3m = 30$
$m = 10$

Exercise Set 8.8

1. b

3. h

5. e

7. c

9. $C(x) = 35x + 200{,}000 \quad R(x) = 55x$

a. $P(x) = R(x) - C(x)$
$= 55x - (35x + 200{,}000)$
$= 55x - 35x - 200{,}000$
$= 20x - 200{,}000$

b. Solve the system
$R(x) = 55x,$
$C(x) = 35x + 200{,}000.$

Since both $R(x)$ and $C(x)$ are in dollars and they are equal at the break-even point, we can rewrite the system:
$d = 55x, \quad (1)$
$d = 35x + 200{,}000 \quad (2)$

We solve using substitution.
$55x = 35x + 200{,}000$ Substituting $55x$ for d in (2)
$20x = 200{,}000$
$x = 10{,}000$

Thus, 10,000 units must be produced and sold in order to break even.
The revenue will be
$R(10{,}000) = 55 \cdot 10{,}000 = 550{,}000$.
The break-even point is (10,000 units, $550,000).

11. $C(x) = 15x + 3100 \quad R(x) = 40x$

a. $P(x) = R(x) - C(x)$
$= 40x - (15x + 3100)$
$= 40x - 15x - 3100$
$= 25x - 3100$

b. Solve the system
$R(x) = 40x,$
$C(x) = 15x + 3100.$

Since both $R(x)$ and $C(x)$ are in dollars and they are equal at the break-even point, we can rewrite the system:
$d = 40x, \quad (1)$
$d = 15x + 3100 \quad (2)$

We solve using substitution.
$40x = 15x + 3100$ Substituting $40x$ for d in (2)
$25x = 3100$
$x = 124$

Thus, 124 units must be produced and sold in order to break even.
The revenue will be $R(124) = 40 \cdot 124 = 4960$.
The break-even point is (124 units, $4960).

13. $C(x) = 40x + 22{,}500 \quad R(x) = 85x$

a. $P(x) = R(x) - C(x)$
$= 85x - (40x + 22{,}500)$
$= 85x - 40x - 22{,}500$
$= 45x - 22{,}500$

b. Solve the system
$R(x) = 85x,$
$C(x) = 40x + 22{,}500.$

Since both $R(x)$ and $C(x)$ are in dollars and they are equal at the break-even point, we can rewrite the system:
$d = 85x, \quad (1)$
$d = 40x + 22{,}500 \quad (2)$

We solve using substitution.
$85x = 40x + 22{,}500$ Substituting $85x$ for d in (2)
$45x = 22{,}500$
$x = 500$

Thus, 500 units must be produced and sold in order to break even.
The revenue will be $R(500) = 85 \cdot 500 = 42{,}500$.
The break-even point is (500 units, $42,500).

15. $C(x) = 24x + 50{,}000 \quad R(x) = 40x$

a. $P(x) = R(x) - C(x)$
$= 40x - (24x + 50{,}000)$
$= 40x - 24x - 50{,}000$
$= 16x - 50{,}000$

b. Solve the system
$R(x) = 40x,$
$C(x) = 24x + 50{,}000.$

Since both $R(x)$ and $C(x)$ are in dollars and they are equal at the break-even point, we can rewrite the system:
$d = 40x, \quad (1)$
$d = 24x + 50{,}000 \quad (2)$

We solve using substitution.
$40x = 24x + 50{,}000$ Substituting $40x$ for d in (2)
$16x = 50{,}000$
$x = 3125$

Thus, 3125 units must be produced and sold in order to break even.
The revenue will be $R(3125) = 40 \cdot 3125 = \$125{,}000$.
The break-even point is (3125 units, $125,000).

17. $C(x) = 75x + 100{,}000 \quad R(x) = 125x$

a. $P(x) = R(x) - C(x)$
$= 125x - (75x + 100{,}000)$
$= 125x - 75x - 100{,}000$
$= 50x - 100{,}000$

b. Solve the system
$R(x) = 125x,$
$C(x) = 75x + 100{,}000.$

Since $R(x) = C(x)$ at the break-even point, we can rewrite the system:
$R(x) = 125x, \quad (1)$
$C(x) = 75x + 100{,}000 \quad (2)$

We solve using substitution.
$125x = 75x + 100{,}000$ Substituting $125x$ for $R(x)$ in (2)
$50x = 100{,}000$
$x = 2000$

To break even 2000 units must be produced and sold.
The revenue will be $R(2000) = 125 \cdot 2000 = 250{,}000$.
The break-even point is (2000 units, $250,000).

19. $D(p) = 2000 - 15p,$
$S(p) = 740 + 6p$

Rewrite the system:

$q = 2000 - 15p$, (1)
$q = 740 + 6p$ (2)

Substitute $2000 - 15p$ for q in (2) and solve.

$$2000 - 15p = 740 + 6p$$
$$1260 = 21p$$
$$60 = p$$

The equilibrium price is \$60 per unit.
To find the equilibrium quantity we substitute \$60 into either $D(p)$ or $S(p)$.

$$D(60) = 2000 - 15(60) = 2000 - 900 = 1100$$

The equilibrium quantity is 1100 units.
The equilibrium point is (\$60, 1100).

21. $D(p) = 760 - 13p$,
$S(p) = 430 + 2p$

Rewrite the system:

$q = 760 - 13p$, (1)
$q = 430 + 2p$ (2)

Substitute $760 - 13p$ for q in (2) and solve.

$$760 - 13p = 430 + 2p$$
$$330 = 15p$$
$$22 = p$$

The equilibrium price is \$22 per unit.
To find the equilibrium quantity we substitute \$22 into either $D(p)$ or $S(p)$.

$$S(22) = 430 + 2(22) = 430 + 44 = 474$$

The equilibrium quantity is 474 units.
The equilibrium point is (\$22, 474).

23. $D(p) = 7500 - 25p$,
$S(p) = 6000 + 5p$

Rewrite the system:

$q = 7500 - 25p$, (1)
$q = 6000 + 5p$ (2)

Substitute $7500 - 25p$ for q in (2) and solve.

$$7500 - 25p = 6000 + 5p$$
$$1500 = 30p$$
$$50 = p$$

The equilibrium price is \$50 per unit.
To find the equilibrium quantity we substitute \$50 into either $D(p)$ or $S(p)$.

$$D(50) = 7500 - 25(50) = 7500 - 1250 = 6250$$

The equilibrium quantity is 6250 units.
The equilibrium point is (\$50, 6250).

25. $D(p) = 1600 - 53p$,
$S(p) = 320 + 75p$

Rewrite the system:

$q = 1600 - 53p$, (1)
$q = 320 + 75p$ (2)

Substitute $1600 - 53p$ for q in (2) and solve.

$$1600 - 53p = 320 + 75p$$
$$1280 = 128p$$
$$10 = p$$

The equilibrium price is \$10 per unit.
To find the equilibrium quantity we substitute \$10 into either $D(p)$ or $S(p)$.

$$S(10) = 320 + 75(10) = 320 + 750 = 1070$$

The equilibrium quantity is 1070 units.
The equilibrium point is (\$10, 1070).

27. a. $C(x) = \text{Fixed costs} + \text{Variable costs}$
$C(x) = 45{,}000 + 40x$,
where x is the number of cell phones.

b. Each cell phone sells for \$130. The total revenue is 130 times the number of cell phone sold. We assume that all cell phones produced are sold.

$R(x) = 130x$

c. $P(x) = R(x) - C(x)$

$$P(x) = 130x - (45{,}000 + 40x)$$
$$= 130x - 45{,}000 - 40x$$
$$= 90x - 45{,}000$$

d. $P(3000) = 90(3000) - 45{,}000$

$$= 270{,}000 - 45{,}000$$
$$= \$225{,}000$$

The company will realize a profit of \$225,000 when 3000 cell phones are produced and sold.

$$P(400) = 90(400) - 45{,}000$$
$$= 36{,}000 - 45{,}000$$
$$= -\$9000$$

The company will realize a \$9000 loss when 400 cell phones are produced and sold.

e. Solve the system

$R(x) = 130x$,
$C(x) = 45{,}000 + 40x$.

Since both $R(x)$ and $C(x)$ are in dollars and they are equal at the break-even point, we can rewrite the system:

$d = 130x$, (1)
$d = 45{,}000 + 40x$ (2)

We solve using substitution.

$130x = 45{,}000 + 40x$ Substituting $130x$ for d in (2)

$$90x = 45{,}000$$
$$x = 500$$

The firm will break even if it produces and sells 500 cell phones and takes in a total of $R(500) = 130 \cdot 500 = \$65{,}000$ in revenue. Thus, the break-even point is (500 cell phones, \$65,000).

29. a. $C(x) = \text{Fixed costs} + \text{Variable costs}$
$C(x) = 10{,}000 + 30x$,
where x is the number of pet car seats produced.

b. Each pet car seat sells for \$80. The total revenue is 80 times the number of seats sold. We assume that all seats produced are sold.

$R(x) = 80x$

c. $P(x) = R(x) - C(x)$

$$P(x) = 80x - (10{,}000 + 30x)$$
$$= 80x - 10{,}000 - 30x$$
$$= 50x - 10{,}000$$

d. $P(2000) = 50(2000) - 10{,}000$

$$= 100{,}000 - 10{,}000$$
$$= 90{,}000$$

The company will realize a profit of \$90,000 when 2000 seats are produced and sold.

$$\begin{aligned}P(50) &= 50(50) - 10{,}000\\ &= 2500 - 10{,}000\\ &= -7500\end{aligned}$$

The company will realize a loss of \$7500 when 50 seats are produced and sold.

e. Solve the system

$R(x) = 80x,$
$C(x) = 10{,}000 + 30x.$

Since both $R(x)$ and $C(x)$ are in dollars and they are equal at the break-even point, we can rewrite the system:

$d = 80x, \quad (1)$
$d = 10{,}000 + 30x \quad (2)$

We solve using substitution.

$80x = 10{,}000 + 30x$ Substituting $80x$ for d in (2)
$50x = 10{,}000$
$x = 200$

The firm will break even if it produces and sells 200 seats and takes in a total of $R(200) = 80 \cdot 200 = \$16{,}000$ in revenue. Thus, the break-even point is (200 seats, \$16,000).

31. *Writing Exercise.*

33.

$$\begin{aligned}25x^2 &= 1\\ 25x^2 - 1 &= 0\\ (5x+1)(5x-1) &= 0\end{aligned}$$

$5x + 1 = 0$ *or* $5x - 1 = 0$

$x = -\frac{1}{5}$ *or* $x = \frac{1}{5}$

The solutions are $-\frac{1}{5}$ and $\frac{1}{5}$.

35. Note that x cannot be 0.

$\frac{1}{x} = \frac{2}{5}$, LCD $= 5x$

$5x \cdot \frac{1}{x} = 5x \cdot \frac{2}{5}$
$5 = 2x$
$\frac{5}{2} = x$

This checks, so the solution is $\frac{5}{2}$.

37. *Writing Exercise.*

39. The supply function contains the points (\$2, 100) and (\$8, 500). We find its equation:

$m = \frac{500-100}{8-2} = \frac{400}{6} = \frac{200}{3}$

$y - y_1 = m(x - x_1)$ Point-slope form
$y - 100 = \frac{200}{3}(x-2)$
$y - 100 = \frac{200}{3}x - \frac{400}{3}$
$y = \frac{200}{3}x - \frac{100}{3}$

We can equivalently express supply S as a function of price p:

$S(p) = \frac{200}{3}p - \frac{100}{3}$

The demand function contains the points (\$1, 500) and (\$9, 100). We find its equation:

$m = \frac{100-500}{9-1} = \frac{-400}{8} = -50$

$y - y_1 = m(x - x_1)$
$y - 500 = -50(x-1)$
$y - 500 = -50x + 50$
$y = -50x + 550$

We can equivalently express demand D as a function of price p:

$D(p) = -50p + 550$

We have a system of equations

$S(p) = \frac{200}{3}p - \frac{100}{3},$
$D(p) = -50p + 550.$

Rewrite the system:

$q = \frac{200}{3}p - \frac{100}{3}, \quad (1)$
$q = -50p + 550 \quad (2)$

Substitute $\frac{200}{3}p - \frac{100}{3}$ for q in (2) and solve.

$\frac{200}{3}p - \frac{100}{3} = -50p + 550$
$200p - 100 = -150p + 1650$ Multiplying by 3 to clear fractions
$350p - 100 = 1650$
$350p = 1750$
$p = 5$

The equilibrium price is \$5 per unit.
To find the equilibrium quantity, we substitute \$5 into either $S(p)$ or $D(p)$.

$D(5) = -50(5) + 550 = -250 + 550 = 300$

The equilibrium quantity is 300 yo-yo's.
The equilibrium point is (\$5, 300 yo-yo's).

41. a. Use a graphing calculator to find the first coordinate of the point of intersection of $y_1 = -14.97x + 987.35$ and $y_2 = 98.55x - 5.13$, to the nearest hundredth. It is 8.74, so the price per unit that should be charged is \$8.74.

b. Use a graphing calculator to find the first coordinate of the point of intersection of $y_1 = 87{,}985 + 5.15x$ and $y_2 = 8.74x$. It is about 24,508.4, so 24,509 units must be sold in order to break even.

43. Convert the yearly savings of \$175 to daily savings.

$\frac{\$175}{1 \text{ year}} \cdot \frac{1 \text{ year}}{365 \text{ days}} \approx \$0.48$ per day

Find how many days to reach \$8.14 at \$0.48 per day:

$\frac{\$8.14}{\$0.48 \text{ per day}} \approx 17$ days

It will take about 17 days to break even on the purchase.

Chapter 8 Review

1. substitution; see Section 8.2.
2. elimination, see Section 8.2.
3. graphical; see Sections 8.1 and 8.2.
4. dependent; see Section 8.1.
5. inconsistent; see Section 8.1.
6. contradiction; see Section 8.2.
7. parallel; see Section 8.1.
8. square; see Section 8.7.
9. determinant; see Section 8.7.
10. zero; see Section 8.8.
11. Graph the equations.

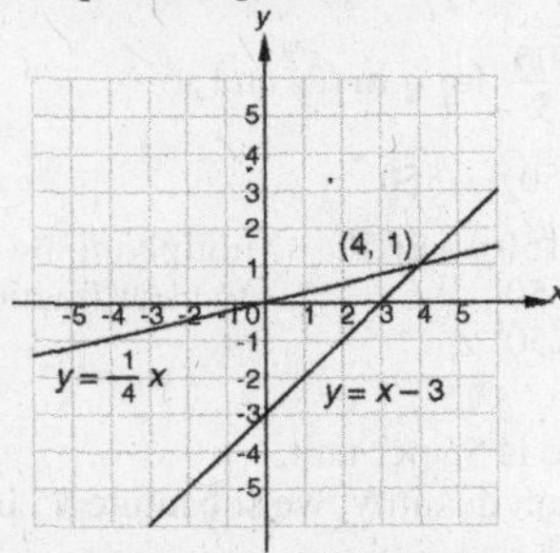

The solution (point of intersection) is (4, 1).

12. Graph the equations:

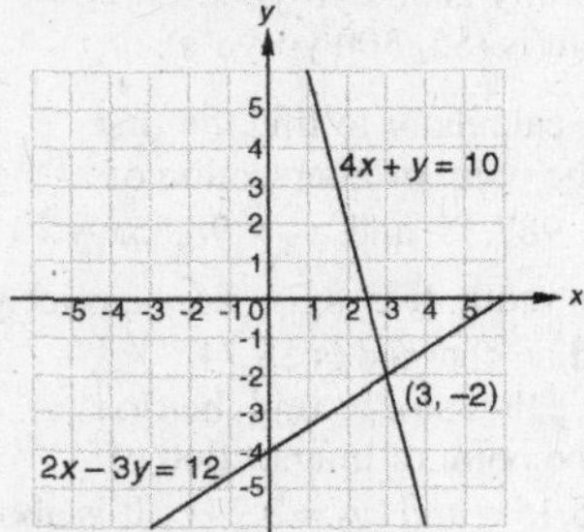

13. $5x-2y=4,$ (1)
$x=y-2$ (2),
We substitute $y-2$ for x in Equation (1) and solve for y.

$$\begin{aligned} 5x-2y&=4 && (1)\\ 5(y-2)-2y&=4 && \text{Substituting}\\ 5y-10-2y&=4\\ 3y-10&=4\\ 3y&=14\\ y&=\frac{14}{3} \end{aligned}$$

Next we substitute $\frac{14}{3}$ for y in either equation of the original system and solve for x.

$$\begin{aligned} x&=y-2 \quad (2)\\ x&=\frac{14}{3}-2=\frac{8}{3} \end{aligned}$$

Since $\left(\frac{8}{3}, \frac{14}{3}\right)$ checks, it is the solution.

14. $y=x+2$ (1)
$y-x=8,$ (2)
We substitute $x+2$ for y in the second equation and solve for x.

$$\begin{aligned} y-x&=8 && (2)\\ x+2-x&=8 && \text{Substituting}\\ 2&=8 \end{aligned}$$

We have a contradiction, or an equation that is always false. Therefore, there is no solution.

15. $2x+5y=8$ (1)
$6x-5y=10$ (2)

$$\begin{aligned} 8x&=18 && \text{Adding}\\ x&=\frac{9}{4} \end{aligned}$$

Substitute $\frac{9}{4}$ for x in Equation (1) and solve for y.

$$\begin{aligned} 2x+5y&=8\\ 2\left(\frac{9}{4}\right)+5y&=8 && \text{Substituting}\\ \frac{9}{2}+5y&=8\\ 5y&=\frac{7}{2}\\ y&=\frac{7}{10} \end{aligned}$$

We obtain $\left(\frac{9}{4}, \frac{7}{10}\right)$. This checks, so it is the solution.

16. $3x-5y=9,$ (1)
$5x-3y=-1$ (2)
We multiply Equation (1) by –3 and Equation (2) by 5.

$$\begin{aligned} -9x+15y&=-27 && \text{Multiplying (1) by } -3\\ 25x-15y&=-5 && \text{Multiplying (2) by 5}\\ 16x&=-32 && \text{Adding}\\ x&=-2 \end{aligned}$$

Substitute –2 for x in Equation (1) and solve for y.

$$\begin{aligned} 3x-5y&=9\\ 3(-2)-5y&=9 && \text{Substituting}\\ -6-5y&=9\\ -5y&=15\\ y&=-3 \end{aligned}$$

We obtain (–2, –3). This checks, so it is the solution.

17. $x-3y=-2,$ (1)
$7y-4x=6$ (2)
We solve Equation (1) for x.

$$\begin{aligned} x-3y&=-2\\ x&=3y-2 \quad (3) \end{aligned}$$

We substitute $3y-2$ for x in the second equation and solve for y.

$$\begin{aligned} 7y-4x&=6 && (2)\\ 7y-4(3y-2)&=6 && \text{Substituting}\\ 7y-12y+8&=6\\ -5y+8&=6\\ -5y&=-2\\ y&=\frac{2}{5} \end{aligned}$$

We substitute $\frac{2}{5}$ for y in Equation (3) and solve for x.

$$x=3y-2 \quad (3)$$
$$x=3\cdot\frac{2}{5}-2=\frac{6}{5}-2=-\frac{4}{5}$$

Since $\left(-\frac{4}{5}, \frac{2}{5}\right)$ checks, it is the solution.

18. $4x-7y=18,\quad (1)$
$9x+14y=40\quad (2)$

We multiply Equation (1) by 2.

$$\begin{aligned} 8x-14y&=36 && \text{Multiplying (1) by 2}\\ 9x+14y&=40 && (2)\\ \hline 17x\quad&=76 && \text{Adding}\\ x&=\frac{76}{17} \end{aligned}$$

Substitute $\frac{76}{17}$ for x in Equation (1) and solve for y.

$$\begin{aligned} 4x-7y&=18\\ 4\left(\frac{76}{17}\right)-7y&=18 && \text{Substituting}\\ \frac{304}{17}-7y&=18\\ -7y&=\frac{2}{17}\\ y&=-\frac{2}{119} \end{aligned}$$

We obtain $\left(\frac{76}{17},-\frac{2}{119}\right)$. This checks, so it is the solution.

19. $1.5x-3=-2y,\quad (1)$
$3x+4y=6\quad (2)$

Rewriting the equations in standard form.

$1.5x+2y=3,\quad (1)$
$3x+4y=6\quad (2)$

Observe that, if we multiply Equation (1) by 2, we obtain Equation (2). Thus, any pair that is a solution of Equation (1) is also a solution of Equation (2). The equations are dependent and the solution set is infinite: $\{(x, y)\,|\,3x+4y=6\}$.

20. $y=2x-5,\quad (1)$
$y=\frac{1}{2}x+1\quad (2)$

We substitute $2x-5$ for y in Equation (2) and solve for x.

$$\begin{aligned} y&=\frac{1}{2}x+1 && (2)\\ 2x-5&=\frac{1}{2}x+1 && \text{Substituting}\\ 2(2x-5)&=2\left(\frac{1}{2}x+1\right)\\ 4x-10&=x+2\\ 3x-10&=2\\ 3x&=12\\ x&=4 \end{aligned}$$

Next we substitute 4 for x in either equation of the original system and solve for y.

$y=2x-5\quad (1)$
$y=2(4)-5=3$

Since (4, 3) checks, it is the solution.

21. ***Familiarize***. Let g = the number of students for group lessons and p = the number of students for private lessons.

Translate. We organize the information in a table.

	Group	Private	Total
Number of students	g	p	12
Price per lesson	\$18	\$25	
Earnings	$18g$	$25p$	\$265

The "Number of students" row of the table gives us one equation:

$g+p=12$

The "Earnings" row yields a second equation:

$18g+25p=265$

We have translated to a system of equations:

$g+p=12\quad (1)$
$18g+25p=265\quad (2)$

Carry out. We use the elimination method to solve the system of equations.

$$\begin{aligned} -18g-18p&=-216 && \text{Multiplying (1) by }-18\\ 18g+25p&=265 && (2)\\ \hline 7p&=49\\ p&=7 \end{aligned}$$

Substitute 7 for p in Equation (1) and solve for g.

$g+7=12$
$g=5$

Check. The total number of students is 5 + 7, or 12. The total earnings is $\$18\cdot 5+\$25\cdot 7=\$90+\$175$ $=\$265$. The answer checks.

State. There were 7 students taking private lessons and 5 students taking group lessons.

22. ***Familiarize***. Let t = the number of hours for the passenger train, and $t+1$ = the number of hours for the freight train.

Now complete the chart.

$d \;=\; r \;\cdot\; t$

	Distance	Rate	Time	
Passenger train	d	55	t	$\rightarrow d=55t$
Freight train	d	44	$t+1$	$\rightarrow d=44(t+1)$

Translate. Using $d=rt$ in each row of the table, we get a system of equations:

$d = 55t,$
$d = 44(t+1)$

Carry out. We solve the system of equations.

$55t = 44(t+1)$ Using substitution
$55t = 44t + 44$
$11t = 44$
$t = 4$

Check. At 55 mph, in 4 hr the passenger train will travel $55 \cdot 4 = 220$ mi. At 44 mph, in 4 + 1, or 5 hr the freight train will travel $44 \cdot 5 = 220$ mi. The numbers check.

State. The passenger train overtakes the freight train after 4 hours.

23. Let x = the number of liters of 15% juice and y = the number of liters of 8% juice.

Solve: $x + y = 14$
$0.15x + 0.08y = 0.10(14)$

The solution is (4, 10). So, 4 liters of 15% juice and 10 liters of 8% juice should be purchased.

24. $x + 4y + 3z = 2,$ (1)
$2x + y + z = 10,$ (2)
$-x + y + 2z = 8$ (3)

1., 2. The equations are already in standard form with no fractions or decimals.

3. Use Equations (1) and (3) to eliminate x.

$x + 4y + 3z = 2$ (1)
$-x + y + 2z = 8$ (3)
$5y + 5z = 10$ (4) Adding

4. Use a different pair of equations and eliminate x.

$2x + y + z = 10$ (2)
$-2x + 2y + 4z = 16$ Multiplying (3) by 2
$3y + 5z = 26$ (5) Adding

5. Now solve the system of Equations (4) and (5).

$5y + 5z = 10$ (4)
$3y + 5z = 26$ (5)

$5y + 5z = 10$ (4)
$-3y - 5z = -26$ Multiplying (5) by -1
$2y = -16$ Adding
$y = -8$

$5(-8) + 5z = 10$ Substituting -8 for y in (4)
$-40 + 5z = 10$
$5z = 50$
$z = 10$

6. Substitute in one of the original equations to find x.

$x + 4(-8) + 3(10) = 2$ Substituting -8 for y and 10 for z in (1)
$x - 2 = 2$
$x = 4$

We obtain (4, −8, 10). This checks, so it is the solution.

25. $4x + 2y - 6z = 34,$ (1)
$2x + y + 3z = 3,$ (2)
$6x + 3y - 3z = 37$ (3)

1., 2. The equations are already in standard form with no fractions or decimals.

3., 4. We eliminate z from two different pairs of equations.

$4x + 2y - 6z = 34$ (1)
$4x + 2y + 6z = 6$ Multiplying (2) by 2
$8x + 4y = 40$ (4) Adding

$2x + y + 3z = 3$ (2)
$6x + 3y - 3z = 37$ (3)
$8x + 4y = 40$ (5) Adding

5. Now solve the system of Equations (4) and (5).

$8x + 4y = 40$ (4)
$8x + 4y = 40$ (5)

$8x + 4y = 40$ (4)
$-8x - 4y = -40$ Multiplying (5) by -1
$0 = 0$ (6)

Equation (6) indicates Equations (1), (2), and (3) are dependent. (Note that if Equation (1) is added to Equation (2), the result is Equation (3).) We could also have concluded that the equations are dependent by observing that Equations (4) and (5) are identical.

26. $2x - 5y - 2z = -4,$ (1)
$7x + 2y - 5z = -6,$ (2)
$-2x + 3y + 2z = 4$ (3)

1., 2. The equations are already in standard form with no fractions or decimals.

3. Use Equations (1) and (2) to eliminate x.

$14x - 35y - 14z = -28$ Multiplying (1) by 7
$-14x - 4y + 10z = 12$ Multiplying (2) by -2
$-39y - 4z = -16$ (4) Adding

4. Use Equations (1) and (3) to eliminate x.

$2x - 5y - 2z = -4$ (1)
$-2x + 3y + 2z = 4$ (3)
$-2y = 0$ (5) Adding
$y = 0$

5. When we used Equation (1) and (3) to eliminate x, we also eliminated z and found $y = 0$. Substitute 0 for y in Equation (4) to find z.

$-39 \cdot 0 - 4z = -16$ Substituting 0 for y in (4)
$-4z = -16$
$z = 4$

6. Substitute in one of the original equations to find x.

$2x - 5 \cdot 0 - 2 \cdot 4 = -4$
$2x = 4$
$x = 2$

We obtain (2, 0, 4). This checks, so it is the solution.

27. $3x + y = 2,$ (1)
$x + 3y + z = 0,$ (2)
$x + z = 2$ (3)

1., 2. The equations are already in standard form with no fractions or decimals.

3., 4. Note that there is no z in Equation (1). We will use Equations (2) and (3) to obtain another equation with no z-term.

$x + 3y + z = 0$ (2)
$-x - z = -2$ Multiplying (3) by -1
$3y = -2$ (4)
$y = -\frac{2}{3}$

5. Now substitute $-\frac{2}{3}$ for y in (1) to solve for x.

$3x - \frac{2}{3} = 2$ Substituting $-\frac{2}{3}$ for y in (1)

$3x = \frac{8}{3}$

$x = \frac{8}{9}$

6. Substitute in Equation (3) to find z.

$\frac{8}{9} + z = 2$ Substituting $\frac{8}{9}$ for x in (3)

$z = \frac{10}{9}$

We obtain $\left(\frac{8}{9}, -\frac{2}{3}, \frac{10}{9}\right)$. This checks, so it is the solution.

28. $2x - 3y + z = 1,$ (1)
$x - y + 2z = 5,$ (2)
$3x - 4y + 3z = -2$ (3)

1., 2. The equations are already in standard form with no fractions or decimals.

3., 4. We eliminate y from two different pairs of equations.

$2x - 3y + z = 1$ (1)
$-3x + 3y - 6z = -15$ Multiplying (2) by -3
$-x - 5z = -14$ (4)

$-4x + 4y - 8z = -20$ Multiplying (2) by -4
$3x - 4y + 3z = -2$ (3)
$-x - 5z = -22$ (4)

5. Now solve the system of Equations (4) and (5).

$-x - 5z = -14$ (4)
$-x - 5z = -22$ (5)

$-x - 5z = -14$ (4)
$x + 5z = 22$ Multiplying (5) by -1
$0 = 8$

We get a false equation, so there is no solution.

29. ***Familiarize***. We let x, y, and z represent the measures of angles A, B, and C, respectively. The measures of the angles of a triangle add up to 180°.

Translate.

The sum of the measures is 180.
$x + y + z = 180$

The measure of angle A is four times the measure of angle C.
$x = 4z$

The measure of angle B is 45° more than the measure of angle C.
$y = z + 45$

We now have a system of equations.

$x + y + z = 180,$
$x = 4z,$
$y = z + 45$

Carry out. Solving the system we get (90, 67.5, 22.5).

Check. The sum of the measures is $90° + 67.5° + 22.5°$, or 180°. Four times the measure of angle C is $4 \cdot 22.5°$, or 90°, the measure of angle A. 45 more than the measure of angle C is $45° + 22.5°$, or 67.5°, the measure of angle B. The numbers check.

State. The measures of angles A, B, and C are 90°, 67.5°, and 22.5°, respectively.

30. ***Familiarize***. Let x, y, and z represent the average number of cries for a man, woman and a one-year-old child, respectively.

Translate.

The sum for each month is 56.7, so we have
$x + y + z = 56.7.$

The number of cries for a woman is 3.9 more than the man, so we have
$y = 3.9 + x.$

The number of cries for the child is 43.3 more than the sum for the man and woman, so we have
$z = 43.3 + x + y.$

We now have a system of equations.

$x + y + z = 56.7,$
$y = 3.9 + x,$
$z = 43.3 + x + y$

Carry out. Solving the system, we get (1.4, 5.3, 50).

Check. The sum of the average number of cry times for a man, a woman, and a one-year-old child is $1.4 + 5.3 + 50$, or 56.7. A woman cries 3.9 + 1.4, or 5.3 times a month. A one-year-old child cries 43.3 + 1.4 + 5.3, or 50 times a month. The answer checks.

State. The monthy average number of cries for a man is 1.4, for a woman is 5.3, and a one-year-old child is 50.

31. $3x + 4y = -13,$
$5x + 6y = 8$

Write a matrix.

$$\left[\begin{array}{cc|c} 3 & 4 & -13 \\ 5 & 6 & 8 \end{array}\right]$$

Multiply the second row by 3

$$\left[\begin{array}{cc|c} 3 & 4 & -13 \\ 15 & 18 & 24 \end{array}\right] \quad \text{New Row 2} = 3(\text{Row 2})$$

Multiply row 1 by –5 and add it to row 2

$$\left[\begin{array}{cc|c} 3 & 4 & -13 \\ 0 & -2 & 89 \end{array}\right] \quad \text{New Row 2} = -5(\text{Row 1}) + \text{Row 2}$$

Reinserting the variables, we have

$3x + 4y = -13,$ (1)
$-2y = 89$ (2)

Solve Equation (2) for y.

$-2y = 89$
$y = -\frac{89}{2}$

Substitute $-\frac{89}{2}$ for y in Equation (1) and solve for x.

$3x + 4\left(-\frac{89}{2}\right) = -13$
$3x - 178 = -13$
$3x = 165$
$x = 55$

The solution is $\left(55, -\frac{89}{2}\right)$.

32. $3x - y + z = -1,$
$2x + 3y + z = 4,$
$5x + 4y + 2z = 5$

We first write a matrix.

$$\left[\begin{array}{ccc|c} 3 & -1 & 1 & -1 \\ 2 & 3 & 1 & 4 \\ 5 & 4 & 2 & 5 \end{array}\right]$$

$$\left[\begin{array}{ccc|c} 3 & -1 & 1 & -1 \\ 0 & 11 & 1 & 14 \\ 0 & 17 & 1 & 20 \end{array}\right] \begin{array}{l} \text{New Row2} = -2(\text{Row1}) + 3(\text{Row2}) \\ \text{New Row3} = -5(\text{Row1}) + 3(\text{Row3}) \end{array}$$

$$\left[\begin{array}{ccc|c} 3 & -1 & 1 & -1 \\ 0 & 11 & 1 & 14 \\ 0 & 187 & 11 & 220 \end{array}\right] \quad \text{New Row 3} = 11(\text{Row 3})$$

$$\left[\begin{array}{ccc|c} 3 & -1 & 1 & -1 \\ 0 & 11 & 1 & 14 \\ 0 & 0 & -6 & -18 \end{array}\right] \quad \text{New Row3} = -17(\text{Row2}) + \text{Row3}$$

Reinserting the variables, we have

$3x - y + z = -1, \quad (1)$
$11y + z = 14, \quad (2)$
$-6z = -18 \quad (3)$

Solve (3) for z.

$-6z = -18$
$z = 3$

Substitute 3 for z in (2) and solve for y.

$11y + 3 = 14$
$11y = 11$
$y = 1$

Substitute 1 for y and 3 for z in (1) and solve for x.

$3x - 1 + 3 = -1$
$3x + 2 = -1$
$3x = -3$
$x = -1$

The solution is $(-1, 1, 3)$.

33. $\begin{vmatrix} -2 & -5 \\ 3 & 10 \end{vmatrix} = -2(10) - (-5)(3) = -20 + 15 = -5$

34. $\begin{vmatrix} 2 & 3 & 0 \\ 1 & 4 & -2 \\ 2 & -1 & 5 \end{vmatrix}$

$= 2\begin{vmatrix} 4 & -2 \\ -1 & 5 \end{vmatrix} - 1\begin{vmatrix} 3 & 0 \\ -1 & 5 \end{vmatrix} + 2\begin{vmatrix} 3 & 0 \\ 4 & -2 \end{vmatrix}$
$= 2[4 \cdot 5 - (-2)(-1)] - 1[3 \cdot 5 - 0(-1)]$
$\quad + 2[3(-2) - 0 \cdot 4]$
$= 2(18) - 1(15) + 2(-6)$
$= 36 - 15 - 12$
$= 9$

35. $2x + 3y = 6,$
$x - 4y = 14$

We compute D, D_x, and D_y.

$D = \begin{vmatrix} 2 & 3 \\ 1 & -4 \end{vmatrix} = -8 - 3 = -11$

$D_x = \begin{vmatrix} 6 & 3 \\ 14 & -4 \end{vmatrix} = -24 - 42 = -66$

$D_y = \begin{vmatrix} 2 & 6 \\ 1 & 14 \end{vmatrix} = 28 - 6 = 22$

Then,

$$x = \frac{D_x}{D} = \frac{-66}{-11} = 6 \text{ and } y = \frac{D_y}{D} = \frac{22}{-11} = -2.$$

The solution is $(6, -2)$.

36. $2x + y + z = -2,$
$2x - y + 3z = 6,$
$3x - 5y + 4z = 7$

We compute D, D_x, D_y and D_z.

$D = \begin{vmatrix} 2 & 1 & 1 \\ 2 & -1 & 3 \\ 3 & -5 & 4 \end{vmatrix}$
$= 2\begin{vmatrix} -1 & 3 \\ -5 & 4 \end{vmatrix} - 2\begin{vmatrix} 1 & 1 \\ -5 & 4 \end{vmatrix} + 3\begin{vmatrix} 1 & 1 \\ -1 & 3 \end{vmatrix}$
$= 2(11) - 2(9) + 3(4)$
$= 22 - 18 + 12 = 16$

$D_x = \begin{vmatrix} -2 & 1 & 1 \\ 6 & -1 & 3 \\ 7 & -5 & 4 \end{vmatrix}$
$= -2\begin{vmatrix} -1 & 3 \\ -5 & 4 \end{vmatrix} - 6\begin{vmatrix} 1 & 1 \\ -5 & 4 \end{vmatrix} + 7\begin{vmatrix} 1 & 1 \\ -1 & 3 \end{vmatrix}$
$= -2(11) - 6(9) + 7(4)$
$= -22 - 54 + 28 = -48$

$D_y = \begin{vmatrix} 2 & -2 & 1 \\ 2 & 6 & 3 \\ 3 & 7 & 4 \end{vmatrix} = 2\begin{vmatrix} 6 & 3 \\ 7 & 4 \end{vmatrix} - 2\begin{vmatrix} -2 & 1 \\ 7 & 4 \end{vmatrix} + 3\begin{vmatrix} -2 & 1 \\ 6 & 3 \end{vmatrix}$
$= 2(3) - 2(-15) + 3(-12)$
$= 6 + 30 - 36 = 0$

$D_z = \begin{vmatrix} 2 & 1 & -2 \\ 2 & -1 & 6 \\ 3 & -5 & 7 \end{vmatrix}$
$= 2\begin{vmatrix} -1 & 6 \\ -5 & 7 \end{vmatrix} - 2\begin{vmatrix} 1 & -2 \\ -5 & 7 \end{vmatrix} + 3\begin{vmatrix} 1 & -2 \\ -1 & 6 \end{vmatrix}$
$= 2(23) - 2(-3) + 3(4)$
$= 46 + 6 + 12 = 64$

Then,

$$x = \frac{D_x}{D} = \frac{-48}{16} = -3$$

$$y = \frac{D_y}{D} = \frac{0}{16} = 0.$$

$$z = \frac{D_z}{D} = \frac{64}{16} = 4.$$

The solution is $(-3, 0, 4)$.

37. $C(x) = 30x + 15{,}800 \quad R(x) = 50x$

a. $P(x) = R(x) - C(x)$
$= 50x - (30x + 15{,}800)$
$= 50x - 30x - 15{,}800$
$= 20x - 15{,}800$

b. Solve the system
$R(x) = 50x,$
$C(x) = 30x + 15{,}800.$

Since both $R(x)$ and $C(x)$ are in dollars and they are equal at the break-even point, we can rewrite the system:

$d = 50x,$ (1)
$d = 30x + 15,800$ (2)

We solve using substitution.

$50x = 30x + 15,800$ Substituting $50x$ for d in (2)

$20x = 15,800$
$x = 790$

Thus, 790 units must be produced and sold in order to break even.

The revenue will be $R(790) = 50 \cdot 790 = 39,500$.

The break-even point is (790 units, \$39,500).

38. $60 + 7p = 120 - 13p$
$20p = 60$
$p = 3$

$S(3) = 60 + 7 \cdot 3 = 81$

The equilibrium point is (\$3, 81).

39. a. $C(x) = 4.75x + 54,000$, where x is the number of pints of honey.

b. Each pint of honey sells for \$9.25. The total revenue is 9.25 times the number of pints of honey. We assume all the honey is sold. $R(x) = 9.25x$

c. $P(x) = R(x) - C(x)$
$= 9.25x - (4.75x + 54,000)$
$= 9.25x - 4.75x - 54,000$
$= 4.5x - 54,000$

d. $P(5000) = 4.5(5000) - 54,000$
$= 22,500 - 54,000$
$= -31,500$

Danae will realize a \$31,500 loss when 5000 pints of honey are produced and sold.

$P(15,000) = 4.5(15,000) - 54,000$
$= 67,500 - 54,000$
$= 13,500$

Danae will realize a profit of \$13,500 when 15,000 pints of honey are produced and sold.

e. Solve the system
$R(x) = 9.25x,$
$C(x) = 4.75x + 54,000.$

Since both $R(x)$ and $C(x)$ are in dollars and they are equal at the break-even point, we can rewrite the system:

$d = 9.25x,$ (1)
$d = 4.75x + 54,000$ (2)

We solve using substitution.

$9.25x = 4.75x + 54,000$ Substituting $9.25x$ for d in (2)

$4.5x = 54,000$
$x = 12,000$

Thus, 12,000 units must be produced and sold in order to break even. The revenue will be $R(12,000) = 9.25 \cdot 12,000 = \$111,000$.

The break-even point is (12,000 pints, \$111,000).

40. *Writing Exercise.* To solve a problem involving four variables, go through the *Familiarize* and *Translate* steps as usual. The resulting system of equations can be solved using the elimination method just as for three variables but likely with more steps.

41. *Writing Exercise.* A system of equations can be both dependent and inconsistent if it is equivalent to a system with fewer equations that has no solution. An example is a system of three equations in three unknowns in which two of the equations represent the same plane, and the third represents a parallel plane.

42. From Exercise 39, we have
$P(x) = 4.5x - 54,000$.

Danae's salary was \$36,000, or
$S(x) = 36,000.$

Since both $P(x)$ and $S(x)$ are in dollars and they are equal, we can rewrite the system:

$d = 4.5x - 54,000$ (1)
$d = 36,000$ (2)

We solve using substitution.

$4.5x - 54,000 = 36,000$
$4.5x = 90,000$
$x = 20,000$

Danae's earnings will be the same if 20,000 pints of honey are produced and sold.

43. Graph both equations.

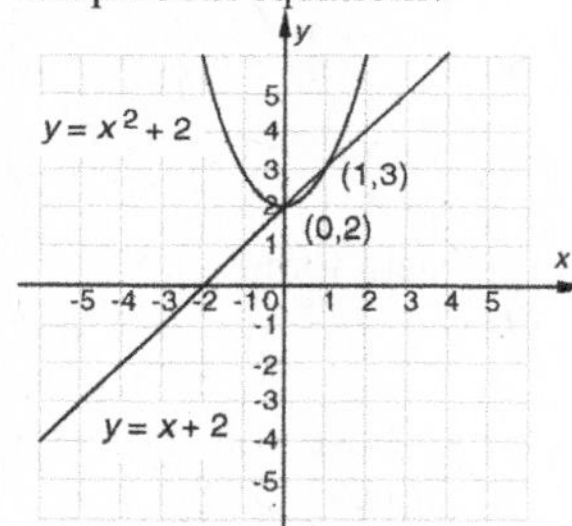

The solutions are apparently (0, 2) and (1, 3). Both pairs check.

Chapter 8 Test

1. Graph the equations.

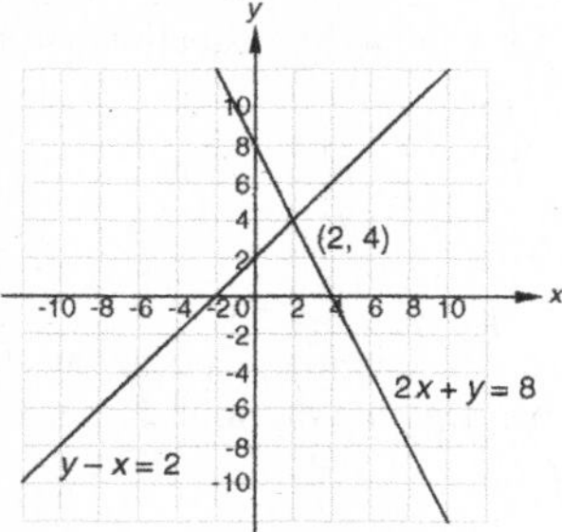

2. $x + 3y = -8,$ (1)
$4x - 3y = 23$ (2)

Solve Equation (1) for x.

$x + 3y = -8$
$x = -3y - 8$

We substitute $-3y - 8$ for x in (2) and solve for y.

$$\begin{aligned} 4x-3y&=23 \quad (2)\\ 4(-3y-8)-3y&=23 \quad \text{Substituting}\\ -12y-32-3y&=23\\ -15y-32&=23\\ -15y&=55\\ y&=-\frac{11}{3}\end{aligned}$$

Next we substitute $-\frac{11}{3}$ for y in either equation of the original system and solve for x.

$$\begin{aligned} x+3y&=-8 \quad (1)\\ x+3\left(-\frac{11}{3}\right)&=-8\\ x-11&=-8\\ x&=3\end{aligned}$$

Since $\left(3,-\frac{11}{3}\right)$ checks, it is the solution.

3. $3x-y=7$ (1)
$x+y=1$ (2)
$4x \quad =8$ Adding
$x=2$

Substitute 2 for x in Equation (2) and solve for y.

$$\begin{aligned} x+y&=1\\ 2+y&=1 \quad \text{Substituting}\\ y&=-1\end{aligned}$$

We obtain (2, –1). This checks, so it is the solution.

4. $4y+2x=18,$
$3x+6y=26$

Rewrite the equations in standard form.

$2x+4y=18,$ (1)
$3x+6y=26$ (2)

We multiply Equation (1) by –3 and Equation (2) by 2.

$-6x-12y=-54$ Multiplying (1) by -3
$6x+12y=52$ Multiplying (2) by 2
$0=-2$ Adding

We have a contradiction, or an equation that is always false. Therefore, there is no solution.

5. $2x-4y=-6$ (1)
$x=2y-3$ (2)

We substitute $2y-3$ for x in the first equation and solve for y.

$$\begin{aligned} 2x-4y&=-6\\ 2(2y-3)-4y&=-6\\ 4y-6-4y&=-6\\ -6&=-6\end{aligned}$$

We have an identity, or an equation that is always true. The equations are dependent and the solution set is infinite: $\{(x, y)|2y-3=x\}$.

6. $4x-6y=3,$ (1)
$6x-4y=-3$ (2)

We multiply Equation (1) by 6 and Equation (2) by –4.

$24x-36y=18$ Multiplying (1) by 6
$-24x+16y=12$ Multiplying (2) by -4
$-20y=30$ Adding
$y=-\frac{3}{2}$

Substitute $-\frac{3}{2}$ for y in Equation (1) and solve for x.

$$\begin{aligned} 4x-6y&=3\\ 4x-6\left(-\frac{3}{2}\right)&=3 \quad \text{Substituting}\\ 4x+9&=3\\ 4x&=-6\\ x&=-\frac{3}{2}\end{aligned}$$

We obtain $\left(-\frac{3}{2},-\frac{3}{2}\right)$. This checks, so it is the solution.

7. ***Familiarize.*** Let w = the width of the basketball court and let l = the length of the basketball court. Recall that the perimeter of a rectangle is given by the formula $P=2w+2l$.

Translate.
The perimeter of the court is 288 ft, so we have
$2w+2l=288.$
The length if 44 longer than the width, so we have
$l=44+w.$
We now have a system of equations.
$2w+2l=288.$ (1)
$l=44+w$ (2)

Carry out. Substitute 44 + w for l in the first equation and solve for w.

$$\begin{aligned} 2w+2l&=288\\ 2w+2(44+w)&=288\\ 2w+88+2w&=288\\ 4w+88&=288\\ 4w&=200\\ w&=50\end{aligned}$$

Finally substitute 50 for w in Equation (2) and solve for l.

$$\begin{aligned} l&=44+w\\ l&=44+50=94\end{aligned}$$

Check. The perimeter of the court is $2(94\text{ ft})$ $+2(50\text{ ft})=188$ ft + 100 ft = 288 ft. The length, 94 ft, is 44 ft more than the width, or 44 + 50 = 94.

State. The basketball court is 94 ft long and 50 ft wide.

8. Let x = the number of grams of Goldfish and y = the number of grams of Pretzels.
Solve: $x+y=620,$
$0.40x+0.09y=93$
The solution is (120, 500). So, the mixture contains 120 g of Goldfish and 500 g of Pretzels.

9. ***Familiarize.*** Let d = the distance and r = the speed of the boat in still water. Then when the boat travels downstream, its speed is $r+5$, and its speed upstream is $r-5$. The distances are the same.
Organize the information in a table.

	Distance	Rate	Time
Downstream	d	$r+5$	3
Upstream	d	$r-5$	5

Translate. Using $d=rt$ in each row of the table, we get a system of equations:

$$d = 3(r+5), \qquad d = 3r + 15,$$
or
$$d = 5(r-5) \qquad d = 5r - 25$$

Carry out. Solve the system of equations.

$$\begin{aligned} 3r + 15 &= 5r - 25 \\ 15 &= 2r - 25 \\ 40 &= 2r \\ 20 &= r \end{aligned}$$

Check. When $r = 20$, then $r + 5 = 20 + 5 = 25$, and the distance traveled in 3 hours is $3 \cdot 25 = 75$ mi. Also $r - 5 = 20 - 5 = 15$, and the distance traveled in 5 hours is $15 \cdot 5 = 75$ mi. The answer checks.

State. The speed of the boat in still water is 20 mph.

10. $-3x + y - 2z = 8$, (1)
$-x + 2y - z = 5$, (2)
$2x + y + z = -3$ (3)

1., 2. The equations are already in standard form with no fractions or decimals.

3., 4. We eliminate x from two different pairs of equations.

$-3x + y - 2z = 8$ (1)
$3x - 6y + 3z = -15$ Multiplying (2) by -3
$-5y + z = -7$ (4)

$-2x + 4y - 2z = 10$ Multiplying (2) by 2
$2x + y + z = -3$ (3)
$5y - z = 7$ (5)

5. Now solve the system of Equations (4) and (5).

$-5y + z = -7$ (4)
$5y - z = 7$ (5)
$0 = 0$ (6)

Equation (6) indicates that Equations (1), (2), and (3) are dependent.

11. $6x + 2y - 4z = 15$, (1)
$-3x - 4y + 2z = -6$, (2)
$4x - 6y + 3z = 8$ (3)

1., 2. The equations are already in standard form with no fractions or decimals.

3., 4. We eliminate x from two different pairs of equations.

$6x + 2y - 4z = 15$ (1)
$-6x - 8y + 4z = -12$ Multiplying (2) by 2
$-6y = 3$ (4)
$y = -\frac{1}{2}$

$-12x - 16y + 8z = -24$ Multiplying (2) by 4
$12x - 18y + 9z = 24$ Multiplying (3) by 3
$-34y + 17z = 0$ (5)

5. Now substitute $-\frac{1}{2}$ for y in (5) to solve for z.

$-34\left(-\frac{1}{2}\right) + 17z = 0$ Substituting $-\frac{1}{2}$ for y in (5)
$17 + 17z = 0$
$z = -1$

6. Substitute in Equation (1) to find x.

$6x + 2\left(-\frac{1}{2}\right) - 4(-1) = 15$ Substituting
$6x - 1 + 4 = 15$
$6x = 12$
$x = 2$

We obtain $\left(2, -\frac{1}{2}, -1\right)$. This checks, so it is the solution.

12. $2x + 2y = 0$, (1)
$4x + 4z = 4$, (2)
$2x + y + z = 2$ (3)

1., 2. The equations are already in standard form with no fractions or decimals.

3., 4. Note that there is no z in Equation (1). We will use Equations (2) and (3) to obtain another equation with no z-term.

$4x + 4z = 4$ (2)
$-8x - 4y - 4z = -8$ Multiplying (3) by -4
$-4x - 4y = -4$ (4)

5. Now solve the system of Equations (1) and (4).

$2x + 2y = 0$ (1)
$-4x - 4y = -4$ (4)

$4x + 4y = 0$ Multiplying (1) by 2
$-4x - 4y = -4$ (4)
$0 = -4$ Adding

We get a false equation, or contradiction. There is no solution.

13. $3x + 3z = 0$, (1)
$2x + 2y = 2$, (2)
$3y + 3z = 3$ (3)

1., 2. The equations are already in standard form with no fractions or decimals.

3., 4. Note that there is no z in Equation (2). We will use Equations (1) and (3) to obtain another equation with no z-term.

$3x + 3z = 0$ (1)
$-3y - 3z = -3$ Multiplying (3) by -1
$3x - 3y = -3$ (4)

5. Now solve the system of Equations (2) and (4).

$2x + 2y = 2$ (2)
$3x - 3y = -3$ (4)

$6x + 6y = 6$ Multiplying (2) by 3
$-6x + 6y = 6$ Multiplying (4) by -2
$12y = 12$
$y = 1$

$2x + 2 \cdot 1 = 2$ Substituting in (2)
$2x = 0$
$x = 0$

6. Substitute in Equation (1) or (3) to find z.

$3 \cdot 0 + 3z = 0$ Substituting in (1)
$3z = 0$
$z = 0$

We obtain (0, 1, 0). This checks, so it is the solution.

14. $\left[\begin{array}{cc|c} 4 & 1 & 12 \\ 3 & 2 & 2 \end{array}\right]$

Multiply row 2 by 4.

$\left[\begin{array}{cc|c} 4 & 1 & 12 \\ 12 & 8 & 8 \end{array}\right]$ New Row 2 = 4(Row 2)

Multiply row 1 by –3 and add it to row 2.

$\left[\begin{array}{cc|c} 4 & 1 & 12 \\ 0 & 5 & -28 \end{array}\right]$ New Row 2 = – 3(Row 1) + Row 2

Then $4x + y = 12,$
$5y = -28$

and $y = -\frac{28}{5}$, $x = \frac{22}{5}$, or $\left(-\frac{28}{5}, \frac{22}{5}\right)$.

15. $x + 3y - 3z = 12,$
$3x - y + 4z = 0,$
$-x + 2y - z = 1$

We first write a matrix.

$\left[\begin{array}{ccc|c} 1 & 3 & -3 & 12 \\ 3 & -1 & 4 & 0 \\ -1 & 2 & -1 & 1 \end{array}\right]$

$\left[\begin{array}{ccc|c} 1 & 3 & -3 & 12 \\ 0 & -10 & 13 & -36 \\ 0 & 5 & -4 & 13 \end{array}\right]$ New Row2 = –3(Row1) + Row2 . New Row3 = (Row1) + Row3

$\left[\begin{array}{ccc|c} 1 & 3 & -3 & 12 \\ 0 & 5 & -4 & 13 \\ 0 & -10 & 13 & -36 \end{array}\right]$ Interchange Row 2 and Row 3

$\left[\begin{array}{ccc|c} 1 & 3 & -3 & 12 \\ 0 & 5 & -4 & 13 \\ 0 & 0 & 5 & -10 \end{array}\right]$ New Row 3 = 2(Row 2) + Row 3

Reinserting the variables, we have

$x + 3y - 3z = 12,$ (1)
$5y - 4z = 13,$ (2)
$5z = -10.$ (3)

Solve (3) for z.

$5z = -10$
$z = -2$

Substitute –2 for z in (2) and solve for y.

$5y - 4(-2) = 13$
$5y = 5$
$y = 1$

Substitute 1 for y and –2 for z in (1) and solve for x.

$x + 3 \cdot 1 - 3(-2) = 12$
$x + 3 + 6 = 12$
$x = 3$

The solution is (3, 1, –2).

16. $\begin{vmatrix} 4 & -2 \\ 3 & -5 \end{vmatrix} = 4(-5) - (-2)(3) = -20 + 6 = -14$

17. $\begin{vmatrix} 3 & 4 & 2 \\ -2 & -5 & 4 \\ 0 & 5 & -3 \end{vmatrix}$

$= 3\begin{vmatrix} -5 & 4 \\ 5 & -3 \end{vmatrix} - (-2)\begin{vmatrix} 4 & 2 \\ 5 & -3 \end{vmatrix} + 0\begin{vmatrix} 4 & 2 \\ -5 & 4 \end{vmatrix}$
$= 3[(-5)(-3) - 5 \cdot 4] + 2[4(-3) - 5 \cdot 2]$
$+ 0[4 \cdot 4 - 2(-5)]$
$= 3(-5) + 2(-22) + 0$
$= -15 - 44 = -59$

18. $3x + 4y = -1,$
$5x - 2y = 4$

We compute D, D_x, and D_y.

$D = \begin{vmatrix} 3 & 4 \\ 5 & -2 \end{vmatrix} = -6 - 20 = -26$

$D_x = \begin{vmatrix} -1 & 4 \\ 4 & -2 \end{vmatrix} = 2 - 16 = -14$

$D_y = \begin{vmatrix} 3 & -1 \\ 5 & 4 \end{vmatrix} = 12 - (-5) = 17$

$x = \frac{D_x}{D} = \frac{-14}{-26} = \frac{7}{13}$ $\quad y = \frac{D_y}{D} = \frac{-17}{26}$.

The solution is $\left(\frac{7}{13}, -\frac{17}{26}\right)$.

19. ***Familiarize***. Let *x*, *y*, and *z* represent the number of hours for the electrician, carpenter and plumber, respectively.

Translate.

The total number of hours worked is 21.5, so we have

$x + y + z = 21.5.$

The total earnings is $673, so we have

$30x + 28.50y + 34z = 673.$

The plumber worked 2 more hours than the carpenter, so we have

$z = 2 + y.$

We have a system of equations:

$x + y + z = 21.5,$
$30x + 28.50y + 34z = 673,$
$z = 2 + y$

Carry out. Solving the system we get (3.5, 8, 10).
Check. The total number of hours is 3.5 + 8 + 10, or 21.5 h. The total amount earned is $\$30(3.5) + \$28.50(8) + \$34(10) = \$105 + \$228 + \$340$, or $673. The plumber worked 10 h, which is 2 h more than the 8 h worked by the carpenter. The numbers check.
State. The electrician worked 3.5 h, the carpenter worked 8 h and the plumber worked 10 h.

20. $79 - 8p = 37 + 6p$
$42 = 14p$
$3 = p$

$D(3) = 79 - 8 \cdot 3 = 55$

The equilibrium point is ($3, 55 units).

21. **a.** $C(x) = 25x + 44{,}000$, where x is the number of hammocks produced.

b. $R(x) = 80x$ We assume all hammocks are sold.

c. $P(x) = R(x) - C(x)$
$= 80x - (25x + 44{,}000)$
$= 80x - 25x - 44{,}000$
$= 55x - 44{,}000$

d. $P(300) = 55(300) - 44{,}000$
$= 16{,}500 - 44{,}000$
$= -27{,}500$

The company will realize a \$27,500 loss when 300 hammocks are produced and sold.

$P(900) = 55(900) - 44{,}000$
$= 49{,}500 - 44{,}000$
$= 5500$

The company will realize a profit of \$5500 when 900 hammocks are produced and sold.

e. Solve the system
$R(x) = 80x,$
$C(x) = 25x + 44{,}000.$

Since both $R(x)$ and $C(x)$ are in dollars and they are equal at the break-even point, we can rewrite the system:

$d = 80x,$ (1)
$d = 25x + 44{,}000$ (2)

We solve using substitution.

$80x = 25x + 44{,}000$ Substituting $80x$ for d in (2)
$55x = 44{,}000$
$x = 800$

Thus, 800 hammocks must be produced and sold in order to break even.

The revenue will be $R(800) = 80 \cdot 800 = \$64{,}000$.

The break-even point is (800 hammocks, \$64,000).

22. For (–1, 3):

$f(-1) = m(-1) + b$
$= -m + b = 3$

For (–2, –4):

$f(-2) = m(-2) + b$
$= -2m + b = -4$

We have a system of equations:

$-m + b = 3$ (1)
$-2m + b = -4$ (2)

Solving we get $m = 7$ and $b = 10$. Thus the function is $f(x) = 7x + 10$.

23. ***Familiarize***. Let x = the number of pounds of Kona coffee

Translate. We organize the information in a table.

	Kona	Mexican	Mixture
Number of pounds	x	40	$x+40$
Percent Kona	100%	0%	30%
Amount of Kona	$1.00x$	0	$0.30(x+40)$

The last row of the table gives us one equation:

$1.00x + 0 = 0.3(x + 40).$

Carry out. After clearing decimals, we solve the equation.

$100x + 0 = 30(x + 40)$
$100x = 30x + 1200$
$70x = 1200$
$x = \frac{120}{7}$

Check. 30% of $\left(\frac{120}{7} + 40\right)$ lb is $0.3\left(\frac{400}{7}\right)$, or $\frac{120}{7}$ lb. of Kona coffee. The answer checks.

State. At least $\frac{120}{7}$ lb of Kona coffee is added to the 40 lb of Mexican coffee to market the mixture as Kona Blend.

Chapter 9

Inequalities and Problem Solving

Exercise Set 9.1

1. An *inequality* is a sentence containing $<, >, \leq, \geq$, or $\neq$.

3. We reverse the direction of the inequality symbol when we multiply both sides of an inequality by a *negative* number.

5. If we add $3x$ to both sides of the equation $5x + 7 = 6 - 3x$, we get the equation $8x + 7 = 6$, so these are equivalent equations.

7. If we add 7 to both sides of the inequality $x - 7 > -2$, we get the inequality $x > 5$ so these are equivalent inequalities.

9. If we multiply the equation $\frac{3}{5}a + \frac{1}{5} = 2$ by 5 on both sides, we get the equation $3a + 1 = 10$, so these are equivalent equations.

11.
$$\begin{aligned} 3x + 1 &< 7 \\ 3x &< 6 \quad \text{Adding } -1 \\ x &< 2 \quad \text{Dividing by 3} \end{aligned}$$

The solution set is $\{x \mid x < 2\}$, or $(-\infty, 2)$.

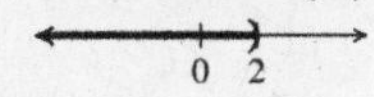

13.
$$\begin{aligned} 3 - x &\geq 12 \\ -x &\geq 9 \quad \text{Adding } -3 \\ x &\leq -9 \quad \text{Dividing by } -1 \text{ and reversing the inequality symbol} \end{aligned}$$

The solution set is $\{x \mid x \leq -9\}$, or $(-\infty, -9]$.

−9 0

15.
$$\begin{aligned} \frac{2x+7}{5} &< -9 \\ 5 \cdot \frac{2x+7}{5} &< 5(-9) \quad \text{Multiplying by 5} \\ 2x + 7 &< -45 \\ 2x &< -52 \quad \text{Adding } -7 \\ x &< -26 \quad \text{Dividing by 2} \end{aligned}$$

The solution set is $\{x \mid x < -26\}$, or $(-\infty, -26)$.

−26 0 26

17.
$$\begin{aligned} \frac{3t-7}{-4} &\leq 5 \\ -4 \cdot \frac{3t-7}{-4} &\geq -4 \cdot 5 \quad \text{Multiplying by } -4 \text{ and reversing the inequality symbol} \\ 3t - 7 &\geq -20 \\ 3t &\geq -13 \quad \text{Adding 7} \\ t &\geq -\frac{13}{3} \quad \text{Dividing by 3} \end{aligned}$$

The solution set is $\left\{t \mid t \geq -\frac{13}{3}\right\}$, or $\left[-\frac{13}{3}, \infty\right)$.

$-\frac{13}{3}$

−4 0 4

19.
$$\begin{aligned} 3 - 8y &\geq 9 - 4y \\ -4y + 3 &\geq 9 \\ -4y &\geq 6 \\ y &\leq -\frac{3}{2} \end{aligned}$$

The solution set is $\left\{y \middle| y \leq -\frac{3}{2}\right\}$, or $\left(-\infty, -\frac{3}{2}\right]$.

21.
$$\begin{aligned} 5(t-3) + 4t &< 2(7+2t) \\ 5t - 15 + 4t &< 14 + 4t \\ 9t - 15 &< 14 + 4t \\ 5t - 15 &< 14 \\ 5t &< 29 \\ t &< \frac{29}{5} \end{aligned}$$

The solution set is $\left\{t \middle| t < \frac{29}{5}\right\}$, or $\left(-\infty, \frac{29}{5}\right)$.

23.
$$\begin{aligned} 5[3m - (m+4)] &> -2(m-4) \\ 5(3m - m - 4) &> -2(m-4) \\ 5(2m-4) &> -2(m-4) \\ 10m - 20 &> -2m + 8 \\ 12m - 20 &> 8 \\ 12m &> 28 \\ m &> \frac{28}{12} \\ m &> \frac{7}{3} \end{aligned}$$

The solution set is $\left\{m \middle| m > \frac{7}{3}\right\}$, or $\left(\frac{7}{3}, \infty\right)$.

25. The graphs intersect at (2, 3). Since the inequality symbol is $\geq$, we include the point in the solution. From the graph, we see that for values to the right of the point (2, 3), $f(x) \geq g(x)$.

The solution set is $\{x \mid x \geq 2\}$, or $[2, \infty)$.

27. The graphs intersect at (0, 3). Since the inequality symbol is <, we do not include the point in the solution. From the graph, we see that for values to the right of the point (0, 3), $f(x) < g(x)$.

The solution set is $\{x \mid x > 0\}$, or $(0, \infty)$.

29. Solve graphically: $x-3<4$.

We graph and solve the system of equations

$y=x-3,$
$y=4$

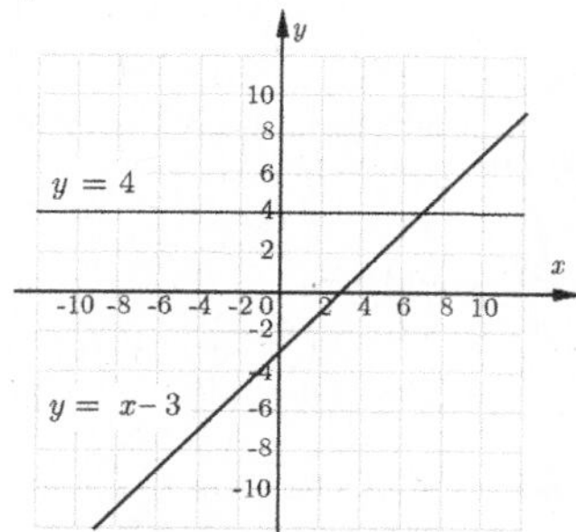

By substitution, we have

$x-3=4$
$x=7$

Then $y=7-3=4$.

Thus, the point of intersection is (7, 4). Since the inequality symbol is <, we do not include this point in the solution.

We see from the graph, $x-3<4$ is to the left of the point of intersection.

Thus, the solution set is $\{x|x<7\}$, or $(-\infty,\ 7)$.

31. Solve graphically: $2x-3\geq 1$.

We graph and solve the system of equations

$y=2x-3,$
$y=1$

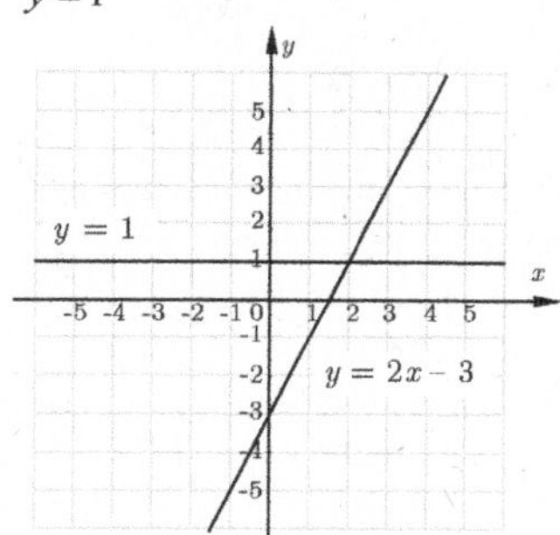

By substitution, we have

$2x-3=1$
$2x=4$
$x=2$

Then $y=2(2)-3=1$.

Thus, the point of intersection is (2, 1). Since the inequality symbol is ≥, we include this point in the solution.

We see from the graph, $2x-3\geq 1$ is to the right of the point of intersection.

Thus, the solution set is $\{x|x\geq 2\}$, or $[2,\ \infty)$.

33. Solve graphically: $x+3>2x-5$.

We graph and solve the system of equations

$y=x+3,$
$y=2x-5$

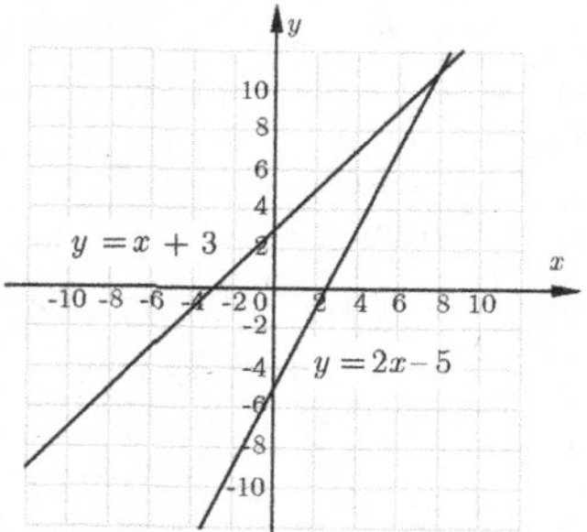

By substitution, we have

$x+3=2x-5$
$3=x-5$
$8=x$

Then $y=8+3=11$.

Thus, the point of intersection is (8, 11). Since the inequality symbol is >, we do not include this point in the solution.

We see from the graph, $x+3>2x-5$ is to the left of the point of intersection.

Thus, the solution set is $\{x|x<8\}$, or $(-\infty,\ 8)$.

35. Solve graphically: $\frac{1}{2}x-2\leq 1-x$.

We graph and solve the system of equations

$y=\frac{1}{2}x-2,$
$y=1-x$

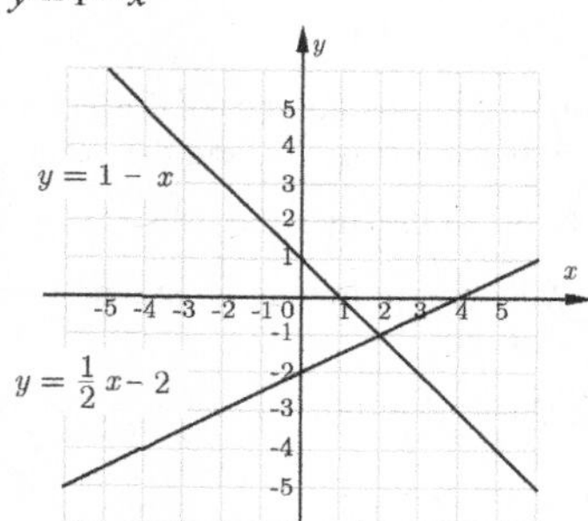

By substitution, we have

$\frac{1}{2}x-2=1-x$
$\frac{3}{2}x-2=1$
$\frac{3}{2}x=3$
$x=2$

Then $y=1-2=-1$.

Thus, the point of intersection is (2, −1). Since the inequality symbol is ≤, we include this point in the solution.

We see from the graph, $\frac{1}{2}x-2\leq 1-x$ is to the left of the point of intersection.

Thus, the solution set is $\{x|x\leq 2\}$, or $(-\infty,\ 2]$.

37. $f(x)=7-3x$, $g(x)=2x-3$

$$f(x)\le g(x)$$
$$7-3x\le 2x-3$$
$$7-5x\le -3 \quad \text{Adding } -2x$$
$$-5x\le -10 \quad \text{Adding } -7$$
$$x\ge 2 \quad \text{Multiplying by } -\tfrac{1}{5} \text{ and reversing the inequality symbol}$$

The solution set is $\{x\,|\,x\ge 2\}$, or $[2,\infty)$.

−2 0 2

39. ***Familiarize***. Let n = the number of hours. Then the total fee using the hourly plan is $120n$.
Translate. We write an inequality stating that the hourly plan costs less than the flat fee.

$$120n<900$$

Carry out.

$$120n<900$$
$$n<\frac{15}{2}, \text{ or } 7\frac{1}{2}$$

Check. We can do a partial check by substituting a value for n less than $\frac{15}{2}$. When n = 7, the hourly plan costs 120(7), or \$840, so the hourly plan is less than the flat fee of \$900. When n = 8, the hourly plan costs 120(8), or \$960, so the hourly plan is more expensive than the flat fee of \$900.
State. The hourly rate is less expensive for lengths of time less than $7\frac{1}{2}$ hours.

41. ***Familiarize***. Let n = the number of correct answers. Then the points earned are $2n$, and the points deducted are $\frac{1}{2}$ of the rest of the questions, $80-n$, or $\frac{1}{2}(80-n)$.
Translate. We write an inequality stating the score is at least 100.

$$2n-\frac{1}{2}(80-n)\ge 100.$$

Carry out.

$$2n-\frac{1}{2}(80-n)\ge 100$$
$$2n-40+\frac{1}{2}n\ge 100$$
$$\frac{5}{2}n-40\ge 100$$
$$\frac{5}{2}n\ge 140$$
$$n\ge 56$$

Check. When n = 56, the score earned is $2(56)-\frac{1}{2}(80-56)$, or $112-\frac{1}{2}(24)$, or $112-12$, or 100. When n = 58, the score earned is $2(58)-\frac{1}{2}(80-58)$, or $116-\frac{1}{2}(22)$, or $116-11$, or 105. Since the score is exactly 100 when 56 questions are answered correctly and more than 100 when 58 questions are correct, we have performed a partial check.
State. At least 56 questions are correct for a score of at least 100.

43. ***Familiarize***. We list the given information in a table.

Plan A: Monthly Income	Plan B: Monthly Income
\$400 salary 8% of sales Total: 400+8% of sales	\$610 5% of sales Total: 610+5% of sales

Suppose Toni had gross sales of \$5000 one month. Then under plan A she would earn

$$\$400+0.08(\$5000), \text{ or } \$800.$$

Under plan B she would earn

$$\$610+0.05(\$5000), \text{ or } \$860.$$

This shows that, for gross sales of \$5000, plan B is better.
If Toni had gross sales of \$10,000 one month, then under plan A she would earn

$$\$400+0.08(\$10{,}000), \text{ or } \$1200.$$

Under plan B she would earn

$$\$610+0.05(\$10{,}000), \text{ or } \$1110.$$

This shows that, for gross sales of \$10,000, plan A is better. To determine all values for which plan A is better we solve an inequality.
Translate.

Income from plan A — is greater than — Income from plan B.

$$400+0.08s \quad > \quad 610+0.05s$$

Carry out.

$$400+0.08s>610+0.05s$$
$$400+0.03s>610$$
$$0.03s>210$$
$$s>7000$$

Check. For $s=\$7000$, the income from plan A is

$$\$400+0.08(\$7000), \text{ or } \$960$$

and the income from plan B is

$$\$610+0.05(\$7000), \text{ or } \$960.$$

This shows that for sales of \$7000 Toni's income is the same from each plan. In the Familiarize step we show that, for a value less than \$7000, plan B is better and, for a value greater than \$7000, plan A is better. Since we cannot check all possible values, we stop here.
State. Toni should select plan A for gross sales greater than \$7000.

45. ***Familiarize***. Let b = the number of bins collected. Then the Purple Plan will cost \$5 + \$3b per month and the Blue Plan will cost \$15 + \$1.75b per month.
Translate. We write an inequality stating that the Blue Plan costs less than the Purple Plan.

$$5+3b>15+1.75b$$

Carry out.

$$5+3b>15+1.75b$$
$$1.25b>10$$
$$b>8$$

Check. We do a partial check by substituting a value for b less than 8 and a value for b greater than 8.
When b = 7, the Purple Plan costs \$5 + \$3(7), or \$26 and the Blue Plan costs \$15 + \$1.75(7), or \$27.25. So the Purple Plan is less expensive.

When $b = 9$, the Purple Plan costs \$5 + \$3(9), or \$32, and the Blue Plan costs \$15 + \$1.75(9), or \$30.75. So the Blue Plan is less expensive.
State. The Blue Plan costs less for more than 8 bins collected per month.

47. ***Familiarize***. Find the values of t for which $c(t) > 3.42$.
Translate. $-0.42t + 11 < 3.42$
Carry out.

$$-0.42t + 11 < 3.42$$
$$-0.42t < -7.58$$
$$t > 18.05$$

Check. $c(18.05) \approx 3.42$.
When $t = 18$, $c(18) = -0.42(18) + 11$, or 3.44.
When $t = 19$, $c(19) = -0.42(19) + 11$, or 3.02.
State. The cost will be less in the United States than the 2011 cost in Germany 19 or more years after 2011, or 2019 and later.

49. a. ***Familiarize***. Find the values of d for which $F(d) > 25$.
Translate.

$$\left(\frac{4.95}{d} - 4.50\right) \times 100 > 25$$

Carry out.

$$\left(\frac{4.95}{d} - 4.50\right) \times 100 > 25$$
$$\frac{495}{d} - 450 > 25$$
$$\frac{495}{d} > 475$$
$$495 > 475d$$
$$\frac{495}{475} > d$$
$$\frac{99}{95} > d, \text{ or } d < 1.04$$

Check. When $d = 1$,

$$F(1) = \left(\frac{4.95}{1} - 4.50\right) \times 100, \text{ or 45 percent.}$$

When $d = 1.05$,

$$F(1.05) = \left(\frac{4.95}{1.05} - 4.50\right) \times 100, \text{ or 21 percent.}$$

State. A man is considered obese for body density less than $\frac{99}{95}$ kg/L, or about 1.04 kg/L.

b. ***Familiarize***. Find the values of d for which $F(d) > 32$.
Translate.

$$\left(\frac{4.95}{d} - 4.50\right) \times 100 > 32$$

Carry out.

$$\left(\frac{4.95}{d} - 4.50\right) \times 100 > 32$$
$$\frac{495}{d} - 450 > 32$$
$$\frac{495}{d} > 482$$
$$495 > 482d$$
$$\frac{495}{482} > d, \text{ or } d < 1.03$$

Check. Our check from part (a) leads to the result that $F(d) > 32$ when $d < 1.03$.
State. A woman is considered obese for body density less than $\frac{495}{482}$ kg/L, or about 1.03 kg/L.

51. ***Familiarize***. Find the values of t for which $p(t) > f(t)$.
Translate.

$$27t + 325 > 16t + 500$$

Carry out.

$$27t + 325 > 16t + 500$$
$$11t > 175$$
$$t > 15.9$$

Check. $p(15.9) = 754.3 \approx r(15.9)$. Calculate $p(t)$ and $f(t)$ for some t greater than 15.9 and for some t less than 15.9.

Suppose $t = 15$:
$p(15) = 27(15) + 325 = 730$ and
$f(15) = 16(15) + 500 = 740$.

In this case $p(t) < f(t)$.

Suppose $t = 16$:
$p(16) = 27(16) + 325 = 757$ and
$f(16) = 16(16) + 500 = 756$.

In this case $p(t) > f(t)$.
Then for $t > 16$, $p(t) > f(t)$.
State. There will be more part-time faculty than full-time faculty in 2011 and later.

53. a. ***Familiarize***. Find the values of x for which $R(x) < C(x)$.
Translate.

$$48x < 90{,}000 + 25x$$

Carry out.

$$23x < 90{,}000$$
$$x < 3913\tfrac{1}{23}$$

Check. $R\left(3913\frac{1}{23}\right) = \$187{,}826.09 = C\left(3913\frac{1}{23}\right)$.

Calculate $R(x)$ and $C(x)$ for some x greater than $3913\frac{1}{23}$ and for some x less than $3913\frac{1}{23}$.

Suppose $x = 4000$:
$R(x) = 48(4000) = 192{,}000$ and
$C(x) = 90{,}000 + 25(4000) = 190{,}000$.

In this case $R(x) > C(x)$.

Suppose $x = 3900$:
$R(x) = 48(3900) = 187{,}200$ and
$C(x) = 90{,}000 + 25(3900) = 187{,}500$.

In this case $R(x) < C(x)$.
Then for $x < 3913\frac{1}{23}$, $R(x) < C(x)$.
State. We will state the result in terms of integers, since the company cannot sell a fraction of a lamp. For 3913 or fewer lamps the company loses money.

b. Our check in part a) shows that for $x > 3913\frac{1}{23}$, $R(x) > C(x)$ and the company makes a profit. Again, we will state the result in terms of an integer. For more than 3913 lamps the company makes money.

55. *Writing Exercise.*

57.
$$x-(9-x)=-3(x+5)$$
$$x-9+x=-3x-15$$
$$2x-9=-3x-15$$
$$5x=-6$$
$$x=-\frac{6}{5}$$

The solution is $-\frac{6}{5}$.

59.
$$2x-3y=5 \quad (1)$$
$$x+3y=-1 \quad (2)$$
$$3x+0=4 \quad \text{Adding}$$
$$3x=4$$
$$x=\frac{4}{3}$$

Substitute $\frac{4}{3}$ for x in Equation (2) and solve for y.

$$x+3y=-1$$
$$\frac{4}{3}+3y=-1 \quad \text{Substituting}$$
$$3y=-\frac{7}{3}$$
$$y=-\frac{7}{9}$$

We obtain $\left(\frac{4}{3},-\frac{7}{9}\right)$. This checks, so it is the solution.

61.
$$ar=b-cr$$
$$ar+cr=b$$
$$r(a+c)=b$$
$$r=\frac{b}{a+c}$$

63. *Writing Exercise.*

65.
$$3ax+2x\ge 5ax-4$$
$$2x-2ax\ge -4$$
$$2x(1-a)\ge -4$$
$$x(1-a)\ge -2$$
$$x\le -\frac{2}{1-a}, \text{ or } \frac{2}{a-1}$$

We reversed the inequality symbol when we divided because when $a>1$, then $1-a<0$.

The solution set is $\left\{x \middle| x\le \frac{2}{a-1}\right\}$.

67.
$$a(by-2)\ge b(2y+5)$$
$$aby-2a\ge 2by+5b$$
$$aby-2by\ge 2a+5b$$
$$y(ab-2b)\ge 2a+5b$$
$$y\ge \frac{2a+5b}{ab-2b}, \text{ or } \frac{2a+5b}{b(a-2)}$$

The inequality symbol remained unchanged when we divided because when $a>2$ and $b>0$, then $ab-2b>0$.

The solution set is $\left\{y \middle| y\ge \frac{2a+5b}{b(a-2)}\right\}$.

69. False. If $a=2$, $b=3$, $c=4$, and $d=5$, then $2<3$ and $4<5$ but $2-4=3-5$.

71. ***Familiarize.*** Let g = the number of gallons used for the business trip. Prepaying for an entire tank of gas would cost \$4.099(14), or \$57.386. Under the wait-and-pay option, it would cost \$7.34g.
Translate. We write an inequality stating that the second option of the wait-and-pay is less than the first option of prepaying.
$$7.34g<57.386$$
Carry out.
$$7.34g<57.386$$
$$g<7.81825613$$
or g is approximately less than 7.8.
Check. We can do a partial check by substituting a value for g less than 7.8 and a value for g greater than 7.8. If the number of gallons used was 7.6, then under the prepayment plan, the cost would be \$57.386, and under the second option, the cost would be \$7.34(7.6), or \$55.784 making the second option the less expensive. However, if the number of gallons used was 7.9, the prepayment plan would still cost \$57.386, and the second option would cost \$7.34(7.9), or \$57.986, making the prepayment plan the less expensive.
State. Abriana will save money under the second plan as long as she uses about 7.8 gal or less.

73. a. Let x = the cost of each dinner.
Solve: $150x>6000$
$$x>40$$
Each dinner should cost more than \$40 in order for the rental fee to be waived.

b. Let x = the cost of each dinner.
Solve: $150x+1500>6000$
$$x>30$$
For costs greater than \$30, the total cost will exceed \$6000, including the rental fee.

c. For a meal over \$30, say \$31, the total cost, including rental fee will be:
$$150(31)+1500=\$6150$$
For a meal under \$40, say \$39, the total cost, including rental fee will be:
$$150(39)+1500=\$7350$$
For meals costing more than \$30 but less than \$40, it would be more economical to choose a \$40 meal.

75. *Writing Exercise.*

Exercise Set 9.2

1. The *intersection* of two sets is the set of all elements that are in both sets.

3. The word “and” corresponds to *intersection*.

5. The solution of a disjunction is the *union* of the solution sets of the individual sentences.

7. h

9. f

11. e

13. b

15. c

17. $\{2, 4, 16\} \cap \{4, 16, 256\}$
The numbers 4 and 16 are common to both sets, so the intersection is $\{4, 16\}$.

19. $\{0, 5, 10, 15\} \cup \{5, 15, 20\}$
The numbers in either or both sets are 0, 5, 10, 15, and 20, so the union is $\{0, 5, 10, 15, 20\}$.

21. $\{a, b, c, d, e, f\} \cap \{b, d, f\}$
The letters *b*, *d*, and *f* are common to both sets, so the intersection is $\{b, d, f\}$.

23. $\{x, y, z\} \cup \{u, v, x, y, z\}$
The letters in either or both sets are *u*, *v*, *x*, *y*, and *z*, so the union is $\{u, v, x, y, z\}$.

25. $\{3, 6, 9, 12\} \cap \{5, 10, 15\}$
There are no numbers common to both sets, so the solution set has no members. It is $\varnothing$.

27. $\{1, 3, 5\} \cup \varnothing$
The numbers in either or both sets are 1, 3, and 5, so the union is $\{1, 3, 5\}$.

29. $1 < x < 3$
This inequality is an abbreviation for the conjunction $1 < x$ *and* $x < 3$. The graph is the intersection of two separate solution sets:
$\{x|1 < x\} \cap \{x|x < 3\} = \{x|1 < x < 3\}$.

0 1 3

Interval notation: $(1, 3)$

31. $-6 \le y \le 0$
This inequality is an abbreviation for the conjunction $-6 \le y$ and $y \le 0$.

−6 0

Interval notation: $[-6, 0]$

33. $x < -1$ *or* $x > 4$
The graph of this disjunction is the union of the graphs of the individual solution sets $\{x \mid x < -1\}$ and $\{x \mid x > 4\}$.

−1 0 4

Interval notation: $(-\infty, -1) \cup (4, \infty)$

35. $x \le -2$ *or* $x > 1$

−2 0 1

Interval notation: $(-\infty, -2] \cup (1, \infty)$

37. $-4 \le -x < 2$
$4 \ge x > -2$ Multiplying by -1 and reversing the inequality symbols
$-2 < x \le 4$ Rewriting

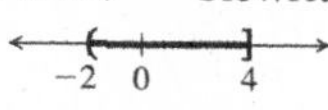

Interval notation: $(-2, 4]$

39. $x > -2$ *and* $x < 4$
This conjunction can be abbreviated as $-2 < x < 4$.

−2 0 4

Interval notation: $(-2, 4)$

41. $5 > a$ *or* $a > 7$

0 5 7

Interval notation: $(-\infty, 5) \cup (7, \infty)$

43. $x \ge 5$ *or* $-x \ge 4$
Multiplying the second inequality by −1 and reversing the inequality symbols, we get $x \ge 5$ or $x \le -4$.

−4 0 5

Interval notation: $(-\infty, -4] \cup [5, \infty)$

45. $7 > y$ *and* $y \ge -3$
This conjunction can be abbreviated as $-3 \le y < 7$.

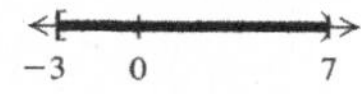

Interval notation: $[-3, 7)$

47. $-x < 7$ *and* $-x \ge 0$
Multiplying the inequalities by −1 and reversing the inequality symbols, we get $x > -7$ *and* $x \le 0$.

−7 0

Interval notation: $(-7, 0]$

49. $t < 2$ *or* $t < 5$
Observe that every number that is less than 2 is also less than 5. Then $t < 2$ *or* $t < 5$ is equivalent to $t < 5$ and the graph of this disjunction is the set $\{t \mid t < 5\}$.

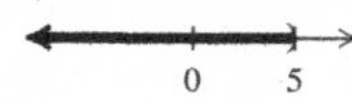

Interval notation: $(-\infty, 5)$

51.
$$-3 \le x + 2 < 9$$
$$-3 - 2 \le x < 9 - 2$$
$$-5 \le x < 7$$
The solution set is $\{x \mid -5 \le x < 7\}$ or $[-5, 7)$.

−5 0 7

53. $0 < t - 4$ *and* $t - 1 \le 7$
$4 < t$ *and* $t \le 8$
We can abbreviate the answer as $4 < t \le 8$.
The solution set is $\{t \mid 4 < t \le 8\}$, or $(4, 8]$.

0 4 8

55. $-7 \le 2a - 3$ *and* $3a + 1 < 7$
$-4 \le 2a$ *and* $3a < 6$
$-2 \le a$ *and* $a < 2$
We can abbreviate the answer as $-2 \le a < 2$. The solution set is $\{a \mid -2 \le a < 2\}$, or $[-2, 2)$.

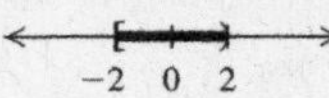

57. $x + 3 \le -1$ *or* $x + 3 > -2$
Observe that any real number is either less than or equal to –1 or greater than or equal to –2. Then the solution set is $\{x \mid x \text{ is a real number}\}$, or $(-\infty, \infty)$.

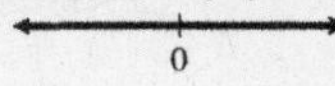

59. $-10 \le 3x - 1 \le 5$
$-9 \le 3x \le 6$
$-3 \le x \le 2$
The solution set is $\{x \mid -3 \le x \le 2\}$, or $[-3, 2]$.

61. $5 > \frac{x-3}{4} > 1$
$20 > x - 3 > 4$ Multiplying by 4
$23 > x > 7$, or
$7 < x < 23$
The solution set is $\{x \mid 7 < x < 23\}$, or $(7, 23)$.

63. $-2 \le \frac{x+2}{-5} \le 6$
$10 \ge x + 2 \ge -30$ Multiplying by -5
$8 \ge x \ge -32$, or
$-32 \le x \le 8$
The solution set is $\{x \mid -32 \le x \le 8\}$, or $[-32, 8]$.

65. $2 \le 3x - 1 \le 8$
$3 \le 3x \le 9$
$1 \le x \le 3$
The solution set is $\{x \mid 1 \le x \le 3\}$, or $[1, 3]$.

67. $-21 \le -2x - 7 < 0$
$-14 \le -2x < 7$
$7 \ge x > -\frac{7}{2}$, or
$-\frac{7}{2} < x \le 7$
The solution set is $\left\{x \middle| -\frac{7}{2} < x \le 7\right\}$, or $\left(-\frac{7}{2}, 7\right]$.

69. $5t + 3 < 3$ *or* $5t + 3 > 8$
$5t < 0$ *or* $5t > 5$
$t < 0$ *or* $t > 1$
The solution set is $\{t \mid t < 0 \text{ or } t > 1\}$, or $(-\infty, 0) \cup (1, \infty)$.

71. $6 > 2a - 1$ *or* $-4 \le -3a + 2$
$7 > 2a$ *or* $-6 \le -3a$
$\frac{7}{2} > a$ *or* $2 \ge a$
The solution set is $\left\{a \middle| \frac{7}{2} > a\right\} \cup \{a \mid 2 \ge a\} = \left\{a \middle| \frac{7}{2} > a\right\}$, or $\left\{a \middle| a < \frac{7}{2}\right\}$, or $\left(-\infty, \frac{7}{2}\right)$.

73. $a + 3 < -2$ *and* $3a - 4 < 8$
$a < -5$ *and* $3a < 12$
$a < -5$ *and* $a < 4$
The solution set is
$\{a \mid a < -5\} \cap \{a \mid a < 4\} = \{a \mid a < -5\}$, or $(-\infty, -5)$.

75. $3x + 2 < 2$ *and* $3 - x < 1$
$3x < 0$ *and* $-x < -2$
$x < 0$ *and* $x > 2$
The solution set is $\{x \mid x < 0\} \cap \{x \mid x > 2\} = \varnothing$.

77. $2t - 7 \le 5$ *or* $5 - 2t > 3$
$2t \le 12$ *or* $-2t > -2$
$t \le 6$ *or* $t < 1$
The solution set is $\{t \mid t \le 6\} \cup \{t \mid t < 1\} = \{t \mid t \le 6\}$, or $(-\infty, 6]$.

79. $f(x) = \frac{9}{x+6}$
$f(x)$ cannot be computed when the denominator is 0. Since $x + 6 = 0$ is equivalent to $x = -6$, we have Domain of $f = \{x \mid x \text{ is a real number } \textit{and } x \ne -6\}$ $= (-\infty, -6) \cup (-6, \infty)$.

81. $f(x) = \frac{1}{x}$
$f(x)$ cannot be computed when the denominator is 0. We have Domain of f $= \{x \mid x \text{ is a real number and } x \ne 0\} = (-\infty, 0) \cup (0, \infty)$.

83. $f(x) = \frac{x+3}{2x-8}$
$f(x)$ cannot be computed when the denominator is 0. Since $2x - 8 = 0$ is equivalent to $x = 4$, we have Domain of $f = \{x \mid x \text{ is a real number } \textit{and } x \ne 4\}$, or $(-\infty, 4) \cup (4, \infty)$.

85. $f(x) = \sqrt{x - 10}$
$x - 10 \ge 0$ $x - 10$ must be nonnegative.
$x \ge 10$ Adding 10
When $x \ge 10$, the expression $x - 10$ is nonnegative. Thus the domain of f is $\{x \mid x \ge 10\}$, or $[10, \infty)$.

87. $f(x)=\sqrt{3-x}$

$3-x\ge 0$ $\quad$ $3-x$ must be nonnegative.
$-x\ge -3$ $\quad$ Adding -3
$x\le 3$ $\quad$ Multiplying by -1 and reversing the inequality symbol

When $x\le 3$, the expression $3-x$ is nonnegative. Thus the domain of f is $\{x\mid x\le 3\}$, or $(-\infty, 3]$.

89. $f(x)=\sqrt{2x+7}$

$2x+7\ge 0$ $\quad$ $2x+7$ must be nonnegative.
$2x\ge -7$ $\quad$ Adding -7
$x\ge -\frac{7}{2}$ $\quad$ Dividing by 2

When $x\ge -\frac{7}{2}$, the expression $2x+7$ is nonnegative.

Thus the domain of f is $\left\{x\middle|x\ge -\frac{7}{2}\right\}$, or $\left[-\frac{7}{2}, \infty\right)$.

91. $f(x)=\sqrt{8-2x}$

$8-2x\ge 0$ $\quad$ $8-2x$ must be nonnegative.
$-2x\ge -8$ $\quad$ Adding -8
$x\le 4$ $\quad$ Dividing by -2 and reversing the inequality symbol

When $x\le 4$, the expression $8-2x$ is nonnegative. Thus the domain of f is $\{x\mid x\le 4\}$, or $(-\infty, 4]$.

93. *Writing Exercise.*

95. $x^2-12x+20=(x-2)(x-10)$

97. $10c^6-10$
$=10(c^6-1)$
$=10(c^3+1)(c^3-1)$
$=10(c+1)(c^2-c+1)(c-1)(c^2+c+1)$

99. *Writing Exercise.*

101. From the graph we observe that the values of x for which $2x-5>-7$ *and* $2x-5<7$ are $\{x\mid -1<x<6\}$, or $(-1, 6)$.

103. Solve $1\le P(d)\le 7$, or $1\le 1+\frac{d}{33}\le 7$.

$$1\le 1+\frac{d}{33}\le 7$$
$$0\le \frac{d}{33}\le 6$$
$$0\le d\le 198$$

Thus, $0\text{ ft}\le d\le 198\text{ ft}$.

105. Solve $25\le F(d)\le 31$.

$$25\le (4.95/d-4.50)\times 100\le 31$$
$$25\le \frac{495}{d}-450\le 31$$
$$475\le \frac{495}{d}\le 481$$
$$\frac{1}{475}\ge \frac{d}{495}\ge \frac{1}{481}$$
$$\frac{495}{475}\ge d\ge \frac{495}{481},$$
or $1.03\le d\le 1.04$

Acceptable body densities are between 1.03 kg/L and 1.04 kg/L.

107. Let c = the number of crossings in six months. Then at the \$6 per crossing rate, the total cost of c crossings is \$6c. A six-month pass costs \$50 and additional \$2 per crossing toll brings the total cost of c crossings to \$50 + \$2c. An unlimited crossing pass costs \$300.

We write an inequality that states that the cost of c crossings using six-month passes is less than the cost using \$6 per crossing toll and is less than the cost of using the unlimited-trip pass. Solve:

$$50+2c<6c \text{ and } 50+2c<300.$$

We get $12.5<c$ *and* $c<125$.
For more than 12 crossings but less than 125 crossings in six months, the reduced-fare pass is more economical.

109. $4m-8>6m+5$ *or* $5m-8<-2$
$-13>2m$ *or* $5m<6$
$-\frac{13}{2}>m$ *or* $m<\frac{6}{5}$

$\left\{m\middle|m<\frac{6}{5}\right\}$, or $\left(-\infty, \frac{6}{5}\right)$

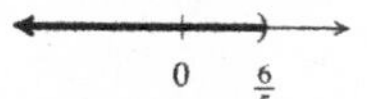

111. $3x<4-5x<5+3x$
$0<4-8x<5$
$-4<-8x<1$
$\frac{1}{2}>x>-\frac{1}{8}$

The solution set is $\left\{x\middle|-\frac{1}{8}<x<\frac{1}{2}\right\}$, or $\left(-\frac{1}{8}, \frac{1}{2}\right)$.

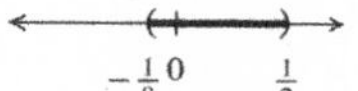

113. Let $a=b=c=2$. Then $a\le c$ and $c\le b$, but $b\not> a$. The given statement is false.

115. If $-a<c$, then $-1(-a)>-1\cdot c$, or $a>-c$. Then if $a>-c$ and $-c>b$, we have $a>-c>b$, so $a>b$ and the given statement is true.

117. $f(x)=\dfrac{\sqrt{3-4x}}{x+7}$

$3-4x\ge 0$ is equivalent to $x\le \frac{3}{4}$ and $x+7=0$ is equivalent to $x=-7$. Then we have Domain of $f=\left\{x\middle|x\le \frac{3}{4} \text{ and } x\ne -7\right\}$, or $(-\infty,-7)\cup\left(-7, \frac{3}{4}\right]$.

119. Observe that the graph of y_2 lies on or above the graph of y_1 and below the graph of y_3 for x in the interval $[-3, 4)$.

121. We substitute 1.2 for s in the formula.
$r=206.835-1.015n-84.6s$
$r=206.835-1.015n-84.6(1.2)$
$r=105.315-1.015n$

5th graders score $90\le r\le 100$. Substitute $105.315-1.015n$ for r in the inequality and solve for n.

$90 \le 105.315 - 1.015n \le 100$
$-15.315 \le -1.015n \le -5.315$
$15.09 \ge n \ge 5.24$
His average sentence should be between 5.24 and 15.09 words per sentence.

123. Let w represent the number of ounces in a bottle; $15.9 \le w \le 16.1$, or [15.9, 16.1]

125. *Graphing Calculator Exercise*

127. *Graphing Calculator Exercise*

Exercise Set 9.3

1. $|x| \ge 0$, so the statement is true.

3. True

5. True

7. False; the solution is all real numbers.

9. g

11. d

13. a

15. $|x| = 10$
$x = -10$ *or* $x = 10$ Using the absolute value principle
The solution set is $\{-10, 10\}$.

17. $|x| = -1$
The absolute value of a number is always nonnegative. Therefore, the solution set is $\varnothing$.

19. $|p| = 0$
The only number whose absolute value is 0 is 0. The solution set is $\{0\}$.

21. $|2x - 3| = 4$
$2x - 3 = -4$ *or* $2x - 3 = 4$ Absolute-value principle
$2x = -1$ *or* $2x = 7$
$x = -\frac{1}{2}$ *or* $x = \frac{7}{2}$
The solution set is $\left\{-\frac{1}{2}, \frac{7}{2}\right\}$.

23. $|3x + 5| = -8$
Absolute value is always nonnegative, so the equation has no solution. The solution set is $\varnothing$.

25. $|x - 2| = 6$
$x - 2 = -6$ *or* $x - 2 = 6$ Absolute-value principle
$x = -4$ *or* $x = 8$
The solution set is $\{-4, 8\}$.

27. $|x - 7| = 1$
$x - 7 = -1$ *or* $x - 7 = 1$
$x = 6$ *or* $x = 8$
The solution set is $\{6, 8\}$.

29. $|t| + 1.1 = 6.6$
$|t| = 5.5$ Adding -1.1
$t = -5.5$ *or* $t = 5.5$
The solution set is $\{-5.5, 5.5\}$.

31. $|5x| - 3 = 37$
$|5x| = 40$ Adding 3
$5x = -40$ *or* $5x = 40$
$x = -8$ *or* $x = 8$
The solution set is $\{-8, 8\}$.

33. $7|q| + 2 = 9$
$7|q| = 7$ Adding -2
$|q| = 1$ Multiplying by $\frac{1}{7}$
$q = -1$ or $q = 1$
The solution set is $\{-1, 1\}$.

35. $\left|\frac{2x-1}{3}\right| = 4$
$\frac{2x-1}{3} = -4$ *or* $\frac{2x-1}{3} = 4$
$2x - 1 = -12$ *or* $2x - 1 = 12$
$2x = -11$ *or* $2x = 13$
$x = -\frac{11}{2}$ *or* $x = \frac{13}{2}$
The solution set is $\left\{-\frac{11}{2}, \frac{13}{2}\right\}$.

37. $|5 - m| + 9 = 16$
$|5 - m| = 7$ Adding -9
$5 - m = -7$ *or* $5 - m = 7$
$-m = -12$ *or* $-m = 2$
$m = 12$ *or* $m = -2$
The solution set is $\{-2, 12\}$.

39. $5 - 2|3x - 4| = -5$
$-2|3x - 4| = -10$
$|3x - 4| = 5$
$3x - 4 = -5$ *or* $3x - 4 = 5$
$3x = -1$ *or* $3x = 9$
$x = -\frac{1}{3}$ *or* $x = 3$
The solution set is $\left\{-\frac{1}{3}, 3\right\}$.

41. $|2x + 6| = 8$
$2x + 6 = -8$ *or* $2x + 6 = 8$
$2x = -14$ *or* $2x = 2$
$x = -7$ *or* $x = 1$
The solution set is $\{-7, 1\}$.

43. $|x| - 3 = 5.7$
$|x| = 8.7$
$x = -8.7$ *or* $x = 8.7$
The solution set is $\{-8.7, 8.7\}$.

45. $\left|\frac{1-2x}{5}\right|=2$

$\frac{1-2x}{5}=-2$ *or* $\frac{1-2x}{5}=2$

$1-2x=-10$ *or* $1-2x=10$

$-2x=-11$ *or* $-2x=9$

$x=\frac{11}{2}$ *or* $x=-\frac{9}{2}$

The solution set is $\left\{-\frac{9}{2}, \frac{11}{2}\right\}$.

47. $|x-7|=|2x+1|$

$x-7=2x+1$ *or* $x-7=-(2x+1)$

$-x=8$ *or* $x-7=-2x-1$

$x=-8$ *or* $3x=6$

$x=2$

The solution set is $\{-8, 2\}$.

49. $|x+4|=|x-3|$

$x+4=x-3$ *or* $x+4=-(x-3)$

$4=-3$ *or* $x+4=-x+3$

False $2x=-1$

$x=-\frac{1}{2}$

The solution set is $\left\{-\frac{1}{2}\right\}$.

51. $|3a-1|=|2a+4|$

$3a-1=2a+4$ *or* $3a-1=-(2a+4)$

$a-1=4$ *or* $3a-1=-2a-4$

$a=5$ *or* $5a-1=-4$

$5a=-3$

$a=-\frac{3}{5}$

The solution set is $\left\{-\frac{3}{5}, 5\right\}$.

53. Since $|n-3|$ and $|3-n|$ are equivalent expressions, the solution set of $|n-3|=|3-n|$ is the set of all real numbers.

55. $|7-4a|=|4a+5|$

$7-4a=4a+5$ *or* $7-4a=-(4a+5)$

$-8a=-2$ *or* $7-4a=-4a-5$

$a=\frac{1}{4}$ *or* $7=-5$

False

The solution set is $\left\{\frac{1}{4}\right\}$.

57. $|a|\le 3$

$-3\le a\le 3$ Part (b)

The solution set is $\{a|-3\le a\le 3\}$, or $[-3, 3]$.

−3 0 3

59. $|t|>0$

$t<0$ or $0<t$ Part (c)

The solution set is $\{t\,|\,t<0 \text{ or } t>0\}$, or $\{t\,|\,t\ne 0\}$, or $(-\infty, 0)\cup(0, \infty)$.

0

61. $|x-1|<4$

$-4<x-1<4$ Part (b)

$-3<x<5$

The solution set is $\{x\,|-3<x<5\}$, or $(-3, 5)$.

−3 0 5

63. $|n+2|\le 6$

$-6\le n+2\le 6$ Part (b)

$-8\le n\le 4$

The solution set is $\{n|-8\le n\le 4\}$, or $[-8, 4]$.

−8 0 4

65. $|x-3|+2>7$

$|x-3|>5$ Adding -2

$x-3<-5$ *or* $5<x-3$ Part(c)

$x<-2$ *or* $8<x$

The solution set is $\{x\,|\,x<-2 \text{ or } x>8\}$, or $(-\infty,-2)\cup(8, \infty)$.

−2 0 8

67. $|2y-9|>-5$

Since absolute value is never negative, any value of $2y-9$, and hence any value of y, will satisfy the inequality. The solution set is the set of all real numbers, or $(-\infty, \infty)$.

0

69. $|3a+4|+2\ge 8$

$|3a+4|\ge 6$ Adding -2

$3a+4\le -6$ *or* $6\le 3a+4$ Part (c)

$3a\le -10$ *or* $2\le 3a$

$a\le -\frac{10}{3}$ *or* $\frac{2}{3}\le a$

The solution set is $\left\{a\middle|a\le -\frac{10}{3} \text{ or } a\ge \frac{2}{3}\right\}$, or

$\left(-\infty, -\frac{10}{3}\right]\cup\left[\frac{2}{3}, \infty\right)$.

$-\frac{10}{3}$ 0 $\frac{2}{3}$

71. $|y-3|<12$

$-12<y-3<12$ Part (b)

$-9<y<15$ Adding 3

The solution set is $\{y\,|-9<y<15\}$, or $(-9, 15)$.

−9 0 15

73. $9-|x+4|\le 5$
$-|x+4|\le -4$
$|x+4|\ge 4$
$x+4\le -4 \quad or \quad 4\le x+4$ Part (c)
$x\le -8 \quad or \quad 0\le x$

The solution set is $\{x|x\le -8 \text{ or } x\ge 0\}$, or $(-\infty,-8]\cup[0,\infty)$.

−8 0

75. $6+|3-2x|>10$
$|3-2x|>4$
$3-2x<-4 \quad or \quad 4<3-2x$
$-2x<-7 \quad or \quad 1<-2x$
$x>\frac{7}{2} \quad or \quad -\frac{1}{2}>x$

The solution set is $\left\{x\middle|x<-\frac{1}{2} \text{ or } x>\frac{7}{2}\right\}$, or $\left(-\infty,-\frac{1}{2}\right)\cup\left(\frac{7}{2},\infty\right)$.

$-\frac{1}{2}$ 0 $\frac{7}{2}$

77. $|5-4x|<-6$
Absolute value is always nonnegative, so the inequality has no solution. The solution set is $\varnothing$.

79. $\left|\frac{1+3x}{5}\right|>\frac{7}{8}$
$\frac{1+3x}{5}<-\frac{7}{8} \quad or \quad \frac{7}{8}<\frac{1+3x}{5}$
$1+3x<-\frac{35}{8} \quad or \quad \frac{35}{8}<1+3x$
$3x<-\frac{43}{8} \quad or \quad \frac{27}{8}<3x$
$x<-\frac{43}{24} \quad or \quad \frac{9}{8}<x$

The solution set is $\left\{x\middle|x<-\frac{43}{24} \text{ or } x>\frac{9}{8}\right\}$, or $\left(-\infty,-\frac{43}{24}\right)\cup\left(\frac{9}{8},\infty\right)$.

$-\frac{43}{24}$ 0 $\frac{9}{8}$

81. $|m+3|+8\le 14$
$|m+3|\le 6$ Adding −8
$-6\le m+3\le 6$
$-9\le m\le 3$
The solution set is $\{m|-9\le m\le 3\}$, or $[-9, 3]$.

−9 0 3

83. $25-2|a+3|>19$
$-2|a+3|>-6$
$|a+3|<3$ Multiplying by $-\frac{1}{2}$
$-3<a+3<3$ Part(b)
$-6<a<0$

The solution set is $\{a|-6<a<0\}$, or $(-6, 0)$.

−6 0

85. $|2x-3|\le 4$
$-4\le 2x-3\le 4$ Part (b)
$-1\le 2x\le 7$ Adding 3
$-\frac{1}{2}\le x\le \frac{7}{2}$ Multiplying by $\frac{1}{2}$

The solution set is $\left\{x\middle|-\frac{1}{2}\le x\le \frac{7}{2}\right\}$, or $\left[-\frac{1}{2}, \frac{7}{2}\right]$.

$-\frac{1}{2}$ 0 $\frac{7}{2}$

87. $5+|3x-4|\ge 16$
$|3x-4|\ge 11$
$3x-4\le -11 \quad or \quad 11\le 3x-4$ Part (c)
$3x\le -7 \quad or \quad 15\le 3x$
$x\le -\frac{7}{3} \quad or \quad 5\le x$

The solution set is $\left\{x\middle|x\le -\frac{7}{3} \text{ or } x\ge 5\right\}$, or $\left(-\infty,-\frac{7}{3}\right]\cup[5,\infty)$.

$-\frac{7}{3}$ 0 5

89. $7+|2x-1|<16$
$|2x-1|<9$
$-9<2x-1<9$ Part (b)
$-8<2x<10$
$-4<x<5$

The solution set is $\{x|-4<x<5\}$, or $(-4, 5)$.

−4 0 5

91. *Writing Exercise.*

93. $y=\frac{1}{3}x-2$

95. $m=\frac{3-(-3)}{-1-(-4)}=\frac{6}{3}=2$
$y-3=2(x+1)$
$y-3=2x+2$
$y=2x+5$

97. *Writing Exercise.*

99. $|3x-5|=x$
$3x-5=-x \quad or \quad 3x-5=x$
$-5=-4x \quad or \quad -5=-2x$
$\frac{5}{4}=x \quad or \quad \frac{5}{2}=x$

The solution set is $\left\{\frac{5}{4}, \frac{5}{2}\right\}$.

101. $2 \le |x-1| \le 5$

$2 \le |x-1|$ *and* $|x-1| \le 5$.

For $2 \le |x-1|$:

$x-1 \le -2$ *or* $2 \le x-1$

$x \le -1$ *or* $3 \le x$

The solution set of $2 \le |x-1|$ is $\{x \mid x \le -1 \text{ or } x \ge 3\}$.

For $|x-1| \le 5$:

$-5 \le x-1 \le 5$

$-4 \le x \le 6$

The solution set of $|x-1| \le 5$ is $\{x \mid -4 \le x \le 6\}$.

The solution set of $2 \le |x-1| \le 5$ is $\{x \mid x \le -1 \text{ or } x \ge 3\} \cap \{x \mid -4 \le x \le 6\}$ $= \{x \mid -4 \le x \le -1 \text{ or } 3 \le x \le 6\}$, or $[-4,-1] \cup [3, 6]$.

103. $t-2 \le |t-3|$

$t-3 \le -(t-2)$ *or* $t-2 \le t-3$

$t-3 \le -t+2$ *or* $-2 \le -3$

$2t-3 \le 2$ False

$2t \le 5$

$t \le \frac{5}{2}$

The solution set is $\left\{t \middle| t \le \frac{5}{2}\right\}$, or $\left(-\infty, \frac{5}{2}\right]$.

105. Using part (b), we find that $-3 < x < 3$ is equivalent to $|x| < 3$.

107. $x \le -6$ or $6 \le x$

$|x| \ge 6$ Using part (c)

109. $x < -8$ *or* $2 < x$

$x+3 < -5$ *or* $5 < x+3$ Adding 3

$|x+3| > 5$ Using part (c)

111. The distance from x to 7 is $|x-7|$ or $|7-x|$, so we have $|x-7| < 2$, or $|7-x| < 2$.

113. The length of the segment from -1 to 7 is $|-1-7| = |-8| = 8$ units. The midpoint of the segment is $\frac{-1+7}{2} = \frac{6}{2} = 3$. Thus, the interval extends 8/2, or 4, units on each side of 3. An inequality for which the closed interval is the solution set is then $|x-3| \le 4$.

115. The length of the segment from -7 to -1 is $|-7-(-1)| = |-6| = 6$ units. The midpoint of the segment is $\frac{-7+(-1)}{2} = \frac{-8}{2} = -4$. Thus, the interval extends 6/2, or 3, units on each side of -4. An inequality for which the open interval is the solution set is $|x-(-4)| < 3$, or $|x+4| < 3$.

117. $|d - 60\text{ft}| \le 10\text{ft}$

$-10\text{ft} \le d - 60\text{ft} \le 10\text{ft}$

$50\text{ft} \le d \le 70\text{ft}$

When the bungee jumper is 50 ft above the river, she is $150-50$, or 100 ft, from the bridge. When she is 70 ft above the river, she is $150-70$, or 80 ft, from the bridge. Thus, at any given time, the bungee jumper is between 80 ft and 100 ft from the bridge.

119. a. Let $x = 4$ (in hundreds of kWh), and solve for y.

$y = 7.2 - |x-5|$

$y = 7.2 - |4-5| = 6.2$

Since y is the number of customers in hundreds, there are 620 customers using 400 kWh per month.

b. Let $y = 5.2$ (in hundreds of customers), and solve for x.

$y = 7.2 - |x-5|$

$5.2 = 7.2 - |x-5|$

$|x-5| = 2$

$x-5 = -2$ *or* $x-5 = 2$

$x = 3$ *or* $x = 7$

Since x is the power used in hundreds of kWh, 520 customers draw 300 kWh and 700 kWh of power each month.

121. *Writing Exercise.*

Mid-Chapter Review

1. $-3 \le x-5 \le 6$

$2 \le x \le 11$

The solution is [2, 11].

2. $|x-1| > 9$

$x-1 < -9$ *or* $9 < x-1$

$x < -8$ *or* $10 < x$

The solution is $(-\infty, -8) \cup (10, \infty)$.

3. $|x| = 15$

$x = -15$ *or* $x = 15$

The solution is $\{-15, 15\}$.

4. $|t| < 10$

$-10 < t < 10$

The solution is $\{t \mid -10 < t < 10\}$, or $(-10, 10)$.

5. $|p| > 15$

$p < -15$ *or* $15 < p$

The solution is $\{p \mid p < -15 \text{ or } p > 15\}$, or $(-\infty, -15) \cup (15, \infty)$.

6. $|2x+1| = 7$

$2x+1 = -7$ *or* $2x+1 = 7$

$x = -4$ *or* $x = 3$

The solution is $\{-4, 3\}$.

7. $-1 < 10-x < 8$

$-11 < -x < -2$

$11 > x > 2$ Reversing the inequality symbol

The solution is (2, 11).

8. $5|t| < 20$
$|t| < 4$
$-4 < t < 4$
The solution is $\{t|-4 < t < 4\}$, or $(-4, 4)$.

9. $x+8<2 \quad or \quad x-4>9$
$x<-6 \quad or \quad x>13$
The solution is $\{x|x<-6 \text{ or } x>13\}$, or $(-\infty,-6)\cup(13, \infty)$.

10. $|x+2| \le 5$
$-5 \le x+2 \le 5$
$-7 \le x \le 3$
The solution is $\{x|-7 \le x \le 3\}$, or $[-7, 3]$.

11. $2+|3x| = 10$
$|3x| = 8$
$3x=-8 \quad or \quad 3x = 8$
$x=-\frac{8}{3} \quad or \quad x=\frac{8}{3}$
The solution is $\left\{-\frac{8}{3}, \frac{8}{3}\right\}$.

12. $2(x-7)-5x > 4-(x+5)$
$2x-14-5x > 4-x-5$
$-3x-14 > -x-1$
$-2x > 13$
$x < -\frac{13}{2}$
The solution is $\left\{x \middle| x < -\frac{13}{2}\right\}$, or $\left(-\infty, -\frac{13}{2}\right)$.

13. $-12 < 2n+6 \quad and \quad 3n-1 \le 7$
$-18 < 2n \quad and \quad 3n \le 8$
$-9 < n \quad and \quad n \le \frac{8}{3}$
The solution is $\left\{n \middle| -9 < n \le \frac{8}{3}\right\}$, or $\left(-9, \frac{8}{3}\right]$.

14. $|2x+5|+1 \ge 13$
$|2x+5| \ge 12$
$2x+5 \le -12 \quad or \quad 12 \le 2x+5$
$2x \le -17 \quad or \quad 7 \le 2x$
$x \le -\frac{17}{2} \quad or \quad \frac{7}{2} \le x$
The solution is $\left\{x \middle| x \le -\frac{17}{2} \text{ or } x \ge \frac{7}{2}\right\}$, or

$\left(-\infty, -\frac{17}{2}\right] \cup \left[\frac{7}{2}, \infty\right)$.

15. $\frac{1}{2}(2x-6) \le \frac{1}{3}(9x+3)$
$x-3 \le 3x+1$
$-2x \le 4$
$x \ge -2$
The solution is $\{x|x \ge -2\}$, or $[-2, \infty)$.

16. $\left|\frac{x+2}{5}\right| = 8$
$\frac{x+2}{5} = -8 \quad or \quad \frac{x+2}{5} = 8$
$x=-42 \quad or \quad x = 38$
The solution is $\{-42, 38\}$.

17. $|8x-11|+6 < 2$
$|8x-11| < -4$
The absolute value of a number is always nonnegative. Therefore, the solution set is $\varnothing$.

18. $8-5|a+6| > 3$
$-5|a+6| > -5$
$|a+6| < 1$
$-1 < a+6 < 1$
$-7 < a < -5$
The solution is $\{a|-7 < a < -5\}$, or $(-7,-5)$.

19. $|5x+7|+9 \ge 4$
$|5x+7| \ge -5$
Since the absolute value of a quantity is always greater than zero, it is always greater than -5.
The solution set is $\mathbb{R}$, or $(-\infty, \infty)$.

20. $3x-7<5 \quad or \quad 2x+1>0$
$3x<12 \quad or \quad 2x>-1$
$x<4 \quad or \quad x>-\frac{1}{2}$
The solution set is $\mathbb{R}$, or $(-\infty, \infty)$.

Connecting the Concepts

1. $x+2=7$
$x=5$

2. $x+2>7$
$x>5$

3. $x+2 \le 7$
$x \le 5$

4. $x+y=2$
$y=-x+2$

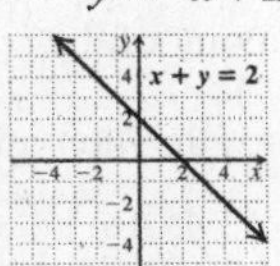

5. $x+y<2$
$y<-x+2$

6. $x+y\geq 2$

$y\geq -x+2$

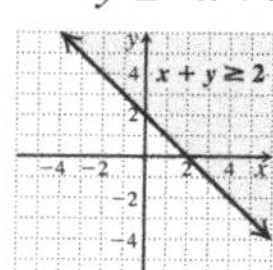

7. $x+2\leq 7$

$x\leq 5$

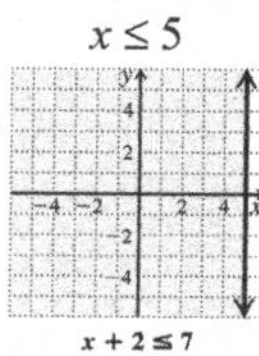

8. $y=x-1,$

$y=-x+1$

9. $y\geq 1-x,$

$y\leq x-3,$

$y\leq 2$

Exercise Set 9.4

1. e

3. d

5. b

7. We replace x with –2 and y with 3.

$$\begin{array}{c|c} \multicolumn{2}{c}{2x-3y>-4} \\ \hline 2(-2)-3 & -4 \end{array}$$

$$-7\overset{?}{>}-4 \quad \text{FALSE}$$

Since –7 > –4 is false, (–2, 3) is not a solution.

9. We replace x with 5 and y with 8.

$$\begin{array}{c|c} \multicolumn{2}{c}{3y-5x\leq 0} \\ \hline 3\cdot 8-5\cdot 5 & 0 \\ 24-25 & \end{array}$$

$$-1\overset{?}{\leq}0 \quad \text{TRUE}$$

Since $-1\leq 0$ is true, (5, 8) is a solution.

11. Graph: $y\geq \frac{1}{2}x$

We first graph the line $y=\frac{1}{2}x$. We draw the line solid since the inequality symbol is $\geq$. To determine which half-plane to shade, test a point not on the line, (0, 1):

$$\begin{array}{c|c} \multicolumn{2}{c}{y\geq \frac{1}{2}x} \\ \hline 1 & \frac{1}{2}\cdot 0 \end{array}$$

$$1\overset{?}{\geq}0 \quad \text{TRUE}$$

Since $1\geq 0$ is true, (0, 1) is a solution as are all of the points in the half-plane containing (0, 1). We shade that half-plane and obtain the graph.

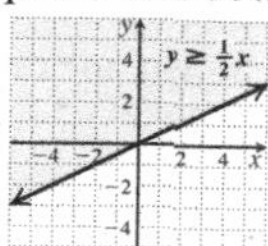

13. Graph: $y>x-3$.

First graph the line $y=x-3$. Draw it dashed since the inequality symbol is >. Test the point (0, 0) to determine if it is a solution.

$$\begin{array}{c|c} \multicolumn{2}{c}{y>x-3} \\ \hline 0 & 0-3 \end{array}$$

$$0\overset{?}{>}-3 \quad \text{TRUE}$$

Since 0 > –3 is true, we shade the half-plane that contains (0, 0) and obtain the graph.

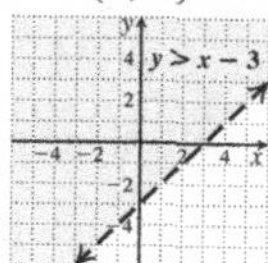

15. Graph: $y\leq x+2$.

First graph the line $y=x+2$. Draw it solid since the inequality symbol is $\leq$. Test the point (0, 0) to determine if it is a solution.

$$\begin{array}{c|c} \multicolumn{2}{c}{y\leq x+2} \\ \hline 0 & 0+2 \end{array}$$

$$0\overset{?}{\leq}2 \quad \text{TRUE}$$

Since $0\leq 2$ is true, we shade the half-plane that contains (0, 0) and obtain the graph.

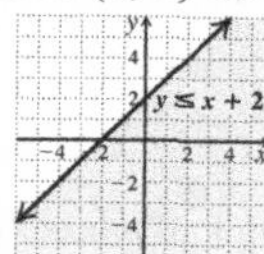

17. Graph: $x-y\leq 4$

First graph the line $x-y=4$. Draw a solid line since the inequality symbol is $\leq$. Test the point (0, 0) to determine if it is a solution.

$$\begin{array}{c|c} \multicolumn{2}{c}{x-y\leq 4} \\ \hline 0-0 & 4 \end{array}$$

$$0\overset{?}{\leq}4 \quad \text{TRUE}$$

Since $0\leq 4$ is true, we shade the half-plane that contains (0, 0) and obtain the graph.

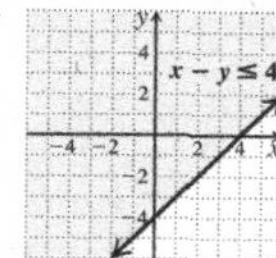

19. Graph: $2x+3y<6$

First graph $2x+3y=6$. Draw the line dashed since the inequality symbol is <. Test the point (0, 0) to determine if it is a solution.

$$\begin{array}{c|c} \multicolumn{2}{c}{2x+3y<6} \\ \hline 2\cdot 0+3\cdot 0 & 6 \end{array}$$

$0 \overset{?}{<} 6$ TRUE

Since $0<6$ is true, we shade the half-plane containing (0, 0) and obtain the graph.

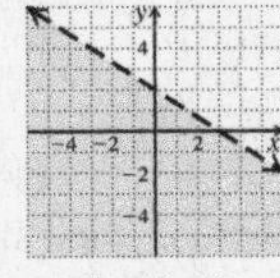

21. Graph: $2y-x\le 4$

We first graph $2y-x=4$. Draw the line solid since the inequality symbol is ≤. Test the point (0, 0) to determine if it is a solution.

$$\begin{array}{c|c} \multicolumn{2}{c}{2y-x\le 4} \\ \hline 2\cdot 0-0 & 4 \end{array}$$

$0 \overset{?}{\le} 4$ TRUE

Since $0\le 4$ is true, we shade the half-plane containing (0, 0) and obtain the graph.

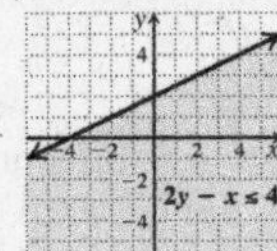

23. Graph: $2x-2y\ge 8+2y$

$2x-4y\ge 8$

First graph $2x-4y=8$. Draw the line solid since the inequality symbol is ≥. Test the point (0, 0) to determine if it is a solution.

$$\begin{array}{c|c} \multicolumn{2}{c}{2x-4y\ge 8} \\ \hline 2\cdot 0-4\cdot 0 & 8 \end{array}$$

$0 \overset{?}{\ge} 8$ FALSE

Since $0\ge 8$ is false, we shade the half-plane that does not contain (0, 0) and obtain the graph.

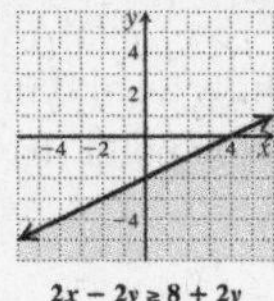

25. Graph: $x>-2$.

We first graph $x=-2$. We draw the line dashed since the inequality symbol is >. Test the point (0, 0) to determine if it is a solution.

$$\begin{array}{c|c} \multicolumn{2}{c}{x>-2} \\ \hline 0 & -2 \end{array}$$

$0 \overset{?}{>} -2$ TRUE

Since $0>-2$ is true, we shade the half-plane that contains (0, 0) and obtain the graph.

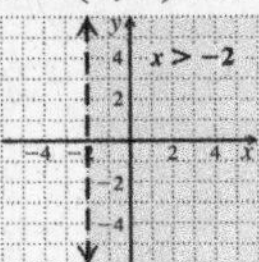

27. Graph: $y\le 6$.

We first graph $y=6$. We draw the line solid since the inequality symbol is ≤. Test the point (0, 0) to determine if it is a solution.

$$\begin{array}{c|c} \multicolumn{2}{c}{y\le 6} \\ \hline 0 & 6 \end{array}$$

$0 \overset{?}{\le} 6$ TRUE

Since $0\le 6$ is true, we shade the half-plane that contains (0, 0) and obtain the graph.

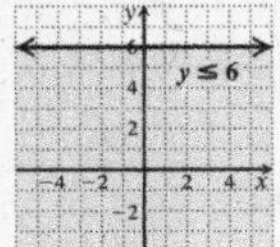

29. Graph: $-2<y<7$

This is a system of inequalities:

$-2<y,$

$y<7$

The graph of $-2<y$ is the half-plane above the line $-2=y$; the graph of $y<7$ is the half-plane below the line $y=7$. We shade the intersection of these graphs.

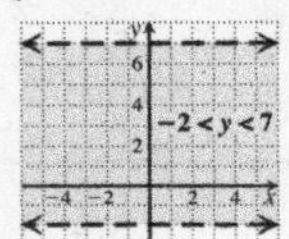

31. Graph: $-5\le x<4$.

This is a system of inequalities:

$-5\le x$

$x<4$

Graph $-5\le x$ and $x<4$. Then shade the intersection of these graphs.

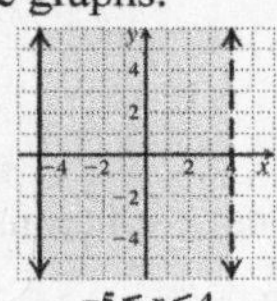

33. Graph: $0\le y\le 3$

This is a system of inequalities:

$0\le y,$

$y\le 3$

Graph $0\le y$ and $y\le 3$.

Then we shade the intersection of these graphs.

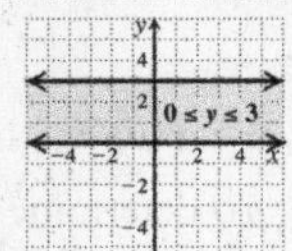

35. Graph: $y > x$,
$y < -x + 3$.

We graph the lines $y = x$ and $y = -x + 3$, using dashed lines. Determine the region for each inequality. Note where the regions overlap and shade the region of solutions.

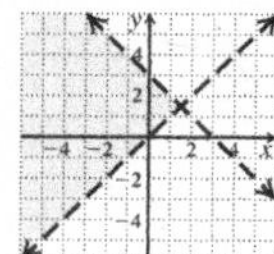

37. Graph: $y \le x$,
$y \le 2x - 5$.

We graph the lines $y = x$ and $y = 2x - 5$, using solid lines. Determine the region for each inequality. Shade the region where they overlap

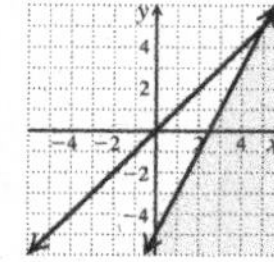

39. Graph: $y \le -3$,
$x \ge -1$

Graph $y = -3$ and $x = -1$ using solid lines. Determine the region for each inequality. Shade the region where they overlap.

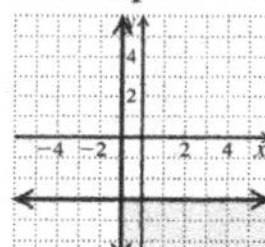

41. Graph: $x > -4$,
$y < -2x + 3$

Graph the lines $x = -4$ and $y = -2x + 3$, using dashed lines. Determine the region for each inequality. Shade the region where they overlap.

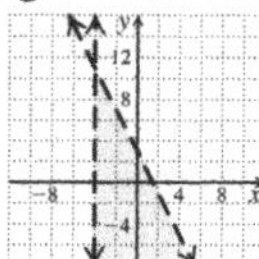

43. Graph: $y \le 5$,
$y \ge -x + 4$

Graph the lines $y = 5$ and $y = -x + 4$, using solid lines. Determine the region for each inequality. Shade the region where they overlap.

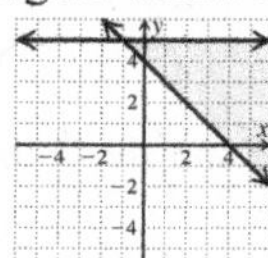

45. Graph: $x + y \le 6$,
$x - y \le 4$

Graph the lines $x + y = 6$ and $x - y = 4$, using solid lines. Determine the region for each inequality. Shade the region where they overlap.

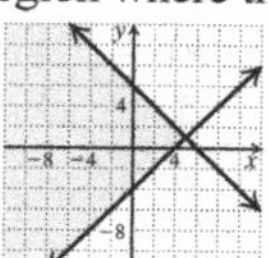

47. Graph: $y + 3x > 0$,
$y + 3x < 2$

Graph the lines $y + 3x = 0$ and $y + 3x = 2$, using dashed lines. Determine the region for each inequality. Shade the region where they overlap.

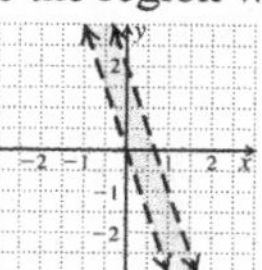

49. Graph: $y \le 2x - 3$, (1)
$y \ge -2x + 1$, (2)
$x \le 5$ (3)

Graph the lines $y = 2x - 3$, $y = -2x + 1$, and $x = 5$ using solid lines. Determine the region for each inequality. Shade the region where they overlap.

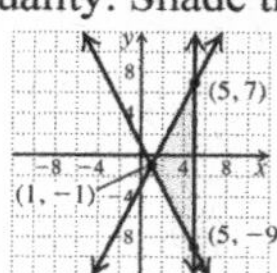

To find the vertex we solve three different systems of related equations.

From (1) and (2) we have $y = 2x - 3$,
$y = -2x + 1$.

Solving, we obtain the vertex $(1, -1)$.

From (1) and (3) we have $y = 2x - 3$,
$x = 5$.

Solving, we obtain the vertex $(5,\ 7)$.

From (2) and (3) we have $y = -2x + 1$,
$x = 5$.

Solving, we obtain the vertex $(5, -9)$.

51. Graph: $x + 2y \le 12$, (1)
$2x + y \le 12$ (2)
$x \ge 0$, (3)
$y \ge 0$ (4)

Graph the lines $x + 2y = 12$, $2x + y = 12$, $x = 0$, and $y = 0$ using solid lines. Determine the region for each inequality. Shade the region where they overlap.

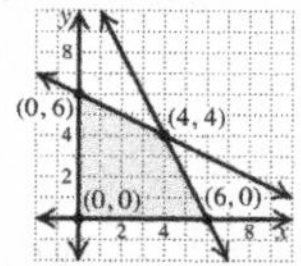

To find the vertices we solve four different systems of equations.

From (1) and (2) we have $x+2y=12$,
$2x+y=12$.
Solving, we obtain the vertex $(4,\ 4)$.
From (1) and (3) we have $x+2y=12$,
$x=0$.
Solving, we obtain the vertex $(0,\ 6)$.
From (2) and (4) we have $2x+y=12$,
$y=0$.
Solving, we obtain the vertex $(6,\ 0)$.
From (3) and (4) we have $x=0$,
$y=0$.
Solving, we obtain the vertex $(0,\ 0)$.

53. Graph: $8x+5y\le 40$, (1)
$x+2y\le 8$ (2)
$x\ge 0$, (3)
$y\ge 0$ (4)

Graph the lines $8x+5y=40$, $x+2y=8$, $x=0$, and $y=0$ using solid lines. Determine the region for each inequality. Shade the region where they overlap.

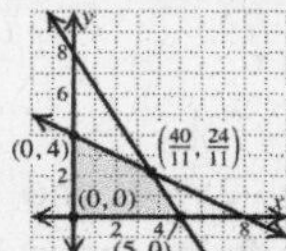

To find the vertices we solve four different systems of equations.
From (1) and (2) we have $8x+5y=40$,
$x+2y=8$.
Solving, we obtain the vertex $\left(\frac{40}{11},\frac{24}{11}\right)$.
From (1) and (4) we have $8x+5y=40$,
$y=0$.
Solving, we obtain the vertex $(5,0)$.
From (2) and (3) we have $x+2y=8$,
$x=0$.
Solving, we obtain the vertex $(0,4)$.
From (3) and (4) we have $x=0$,
$y=0$.
Solving, we obtain the vertex $(0,0)$.

55. Graph: $y-x\ge 2$, (1)
$y-x\le 4$, (2)
$2\le x\le 5$ (3)

Think of (3) as two inequalities:
$2\le x$, (4)
$x\le 5$ (5)

Graph the lines $y-x=2$, $y-x=4$, $x=2$, and $x=5$, using solid lines. Determine the region for each inequality. Shade the region where they overlap.

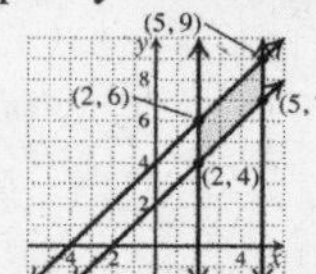

To find the vertices we solve four different systems of equations.

From (1) and (4) we have $y-x=2$,
$x=2$.
Solving, we obtain the vertex $(2,\ 4)$.
From (1) and (5) we have $y-x=2$,
$x=5$.
Solving, we obtain the vertex $(5,\ 7)$.
From (2) and (4) we have $y-x=4$,
$x=2$.
Solving, we obtain the vertex $(2,\ 6)$.
From (2) and (5) we have $y-x=4$,
$x=5$.
Solving, we obtain the vertex $(5,\ 9)$.

57. *Writing Exercise.*

59. $(2p^3-3w^4)^2=4p^6-12p^3w^4+9w^8$

61. $(3t^4+2t^2+1)-(6t^4-t^2+2)$
$=3t^4+2t^2+1-6t^4+t^2-2$
$=-3t^4+3t^2-1$

63. $\dfrac{15x^2-65x-50}{2x^2-x-3}\cdot\dfrac{x^2+2x+1}{10x^2-250}$
$=\dfrac{5(3x^2-13x-10)}{(2x-3)(x+1)}\cdot\dfrac{(x+1)(x+1)}{10(x^2-25)}$
$=\dfrac{5(3x+2)(x-5)(x+1)(x+1)}{(2x-3)(x+1)\cdot 10(x+5)(x-5)}$
$=\dfrac{(3x+2)(x+1)}{2(2x-3)(x+5)}$

65. *Writing Exercise.*

67. Graph: $x+y>8$,
$x+y\le -2$

Graph the line $x+y=8$ using a dashed line and graph $x+y=-2$, using a solid line. Indicate the region for each inequality by arrows. The regions do not overlap (the solution set is $\varnothing$), so we do not shade any portion of the graph.

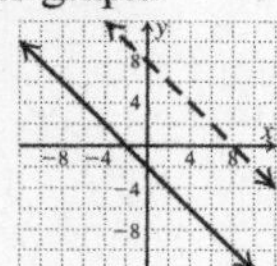

69. Graph: $x-2y\le 0$,
$-2x+y\le 2$,
$x\le 2$,
$y\le 2$,
$x+y\le 4$

Graph the five inequalities above, and shade the region where they overlap.

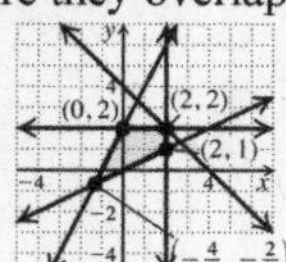

71. Both the width and the height must be positive, so we have

$w > 0,$
$h > 0.$

To be checked as luggage, the sum of the width, height, and length cannot exceed 62 in., so we have

$w + h + 30 \le 62$, or
$w + h \le 32.$

The girth is represented by $2w + 2h$ and the length is 30 in. In order to meet postal regulations the sum of the girth and the length cannot exceed 130 in., so we have:

$2w + 2h + 30 < 130$, or
$2w + 2h \le 100$, or
$w + h \le 50$

Thus, have a system of inequalities:

$w > 0,$
$h > 0$
$w + h \le 32,$
$w + h \le 50$

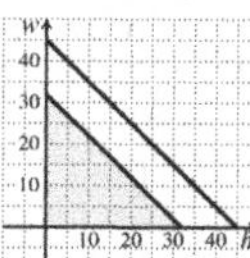

73. We graph the following inequalities:

$q + v \ge 287$
$v \ge 145$
$q \le 170$
$v \le 170$

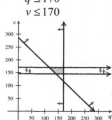

75. Graph: $35c + 75a > 1000,$
$c \ge 0,$
$a \ge 0$

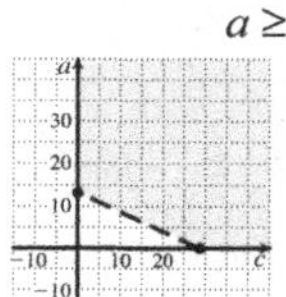

77. $h < 2w$
$w \le 1.5h$
$h \le 3200$
$h \ge 0$
$w \ge 0$

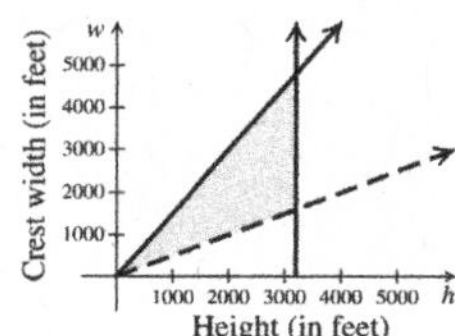

79. a. $3x + 6y > 2$

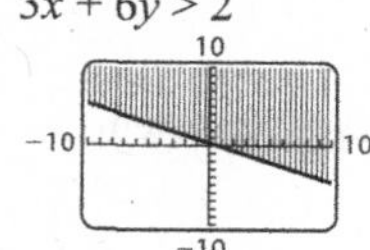

b. $x - 5y \le 10$

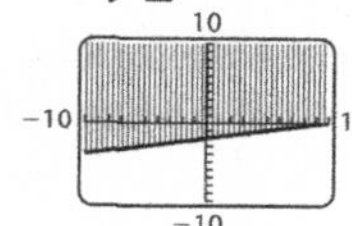

c. $13x - 25y + 10 \le 0$

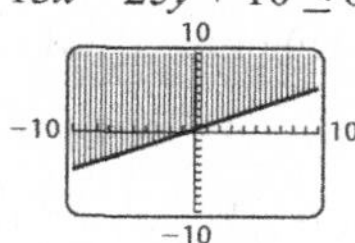

d. $2x + 5y > 0$

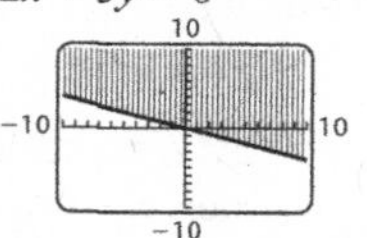

Exercise Set 9.5

1. In linear programming, the quantity we wish to maximize or minimize is represented by the *objective* function.

3. To solve a linear programming problem, we make use of the *corner* principle.

5. In linear programming, the corners of the shaded portion of the graph are referred to as *vertices*.

7. Find the maximum and minimum values of

$F = 2x + 14y,$

subject to

$5x + 3y \le 34, \quad (1)$
$3x + 5y \le 30, \quad (2)$
$x \ge 0, \quad (3)$
$y \ge 0. \quad (4)$

Graph the system of inequalities and find the coordinates of the vertices.

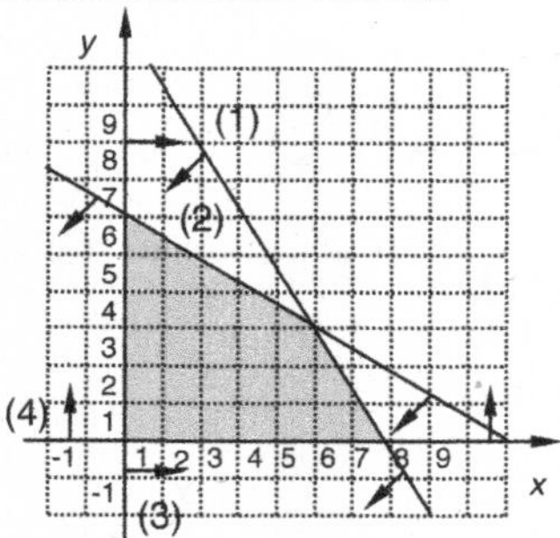

To find one vertex we solve the system

$x = 0,$
$y = 0.$

This vertex is $(0,0)$.

To find a second vertex we solve the system

$5x + 3y = 34,$
$y = 0.$

This vertex is $\left(\frac{34}{5}, 0\right)$.

To find a third vertex we solve the system

$$5x + 3y = 34,$$
$$3x + 5y = 30.$$

This vertex is $(5, 3)$.

To find the fourth vertex we solve the system

$$3x + 5y = 30,$$
$$x = 0.$$

This vertex is $(0, 6)$.

Now find the value of F at each of these points.

Vertex (x, y)	$F = 2x + 14y$
$(0, 0)$	$2\cdot 0 + 14\cdot 0 = 0 + 0 = 0$ ←Minimum
$\left(\frac{34}{5}, 0\right)$	$2\cdot\frac{34}{5} + 14\cdot 0 = \frac{68}{5} + 0 = 13\frac{3}{5}$
$(5, 3)$	$2\cdot 5 + 14\cdot 3 = 10 + 42 = 52$
$(0, 6)$	$2\cdot 0 + 14\cdot 6 = 0 + 84 = 84$ ←Maximum

The maximum value of F is 84 when $x = 0$ and $y = 6$. The minimum value of F is 0 when $x = 0$ and $y = 0$.

9. Find the maximum and minimum values of

$$P = 8x - y + 20,$$

subject to

$$6x + 8y \le 48, \quad (1)$$
$$0 \le y \le 4, \quad (2)$$
$$0 \le x \le 7. \quad (3)$$

Think of (2) as $0 \le y$, (4)
$y \le 4$. (5)

Think of (3) as $0 \le x$, (6)
$x \le 7$. (7)

Graph the system of inequalities.

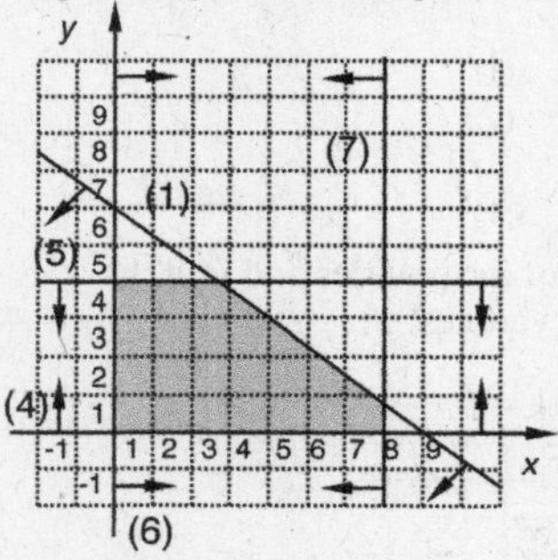

To determine the coordinates of the vertices, we solve the following systems:

$x = 0,$ $y = 0;$ $\quad x = 7,$ $y = 0;$ $\quad 6x + 8y = 48,$ $x = 7;$

$6x + 8y = 48,$ $y = 4;$ $\quad x = 0,$ $y = 4$

The vertices are $(0, 0)$, $(7, 0)$, $\left(7, \frac{3}{4}\right)$, $\left(\frac{8}{3}, 4\right)$, and $(0, 4)$, respectively. Compute the value of P at each of these points.

Vertex (x, y)	$P = 8x - y + 20$	
$(0, 0)$	$8\cdot 0 - 0 + 20$ $= 0 - 0 + 20 = 20$	
$(7, 0)$	$8\cdot 7 - 0 + 20$ $= 56 - 0 + 20 = 76$	← Maximum
$\left(7, \frac{3}{4}\right)$	$8\cdot 7 - \frac{3}{4} + 20$ $= 56 - \frac{3}{4} + 20 = 75\frac{1}{4}$	
$\left(\frac{8}{3}, 4\right)$	$8\cdot\frac{8}{3} - 4 + 20$ $= \frac{64}{3} - 4 + 20 = 37\frac{1}{3}$	
$(0,4)$	$8\cdot 0 - 4 + 20$ $= 0 - 4 + 20 = 16$	← Minimum

The maximum is 76 when $x = 7$ and $y = 0$. The minimum is 16 when $x = 0$ and $y = 4$.

11. Find the maximum and minimum values of

$$F = 2y - 3x,$$

subject to

$$y \le 2x + 1, \quad (1)$$
$$y \ge -2x + 3, \quad (2)$$
$$x \le 3 \quad (3)$$

Graph the system of inequalities and find the coordinates of the vertices.

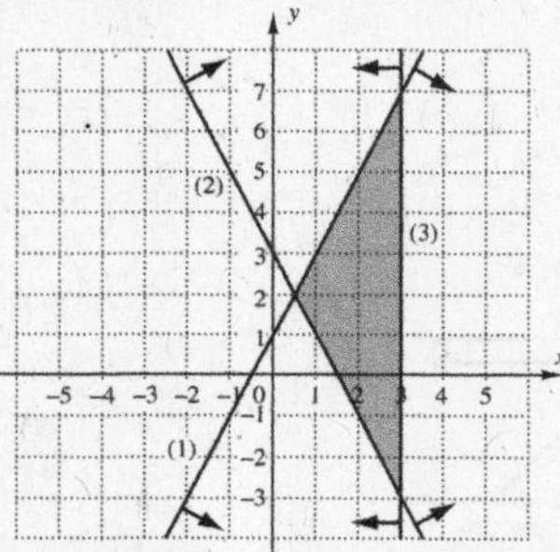

To determine the coordinates of the vertices, we solve the following systems:

$y = 2x + 1,$ $y = -2x + 3;$ $\quad y = 2x + 1,$ $x = 3;$ $\quad y = -2x + 3,$ $x = 3$

The solutions of the systems are $\left(\frac{1}{2}, 2\right)$, $(3, 7)$, and $(3, -3)$, respectively. Now find the value of F at each of these points.

Vertex (x, y)	$F = 2y - 3x$	
$\left(\frac{1}{2}, 2\right)$	$2\cdot 2 - 3\cdot\frac{1}{2} = \frac{5}{2}$	
$(3, 7)$	$2\cdot 7 - 3\cdot 3 = 5$	← Maximum
$(3, -3)$	$2(-3) - 3\cdot 3 = -15$	← Minimum

The maximum value is 5 when $x = 3$ and $y = 7$. The minimum value is -15 when $x = 3$ and $y = -3$.

13. ***Familiarize***. Let x = the number of train rides, and y = the number of bus rides.
Translate. The number of cost C is given by
$C = 5x + 4y$.
We wish to minimize C subject to these constraints.

$x+y\geq 5$
$x+1.5y\leq 6$
$x\geq 0$
$y>0.$

Carry out. We graph the system of inequalities, determine the vertices, and evaluate N at each vertex.

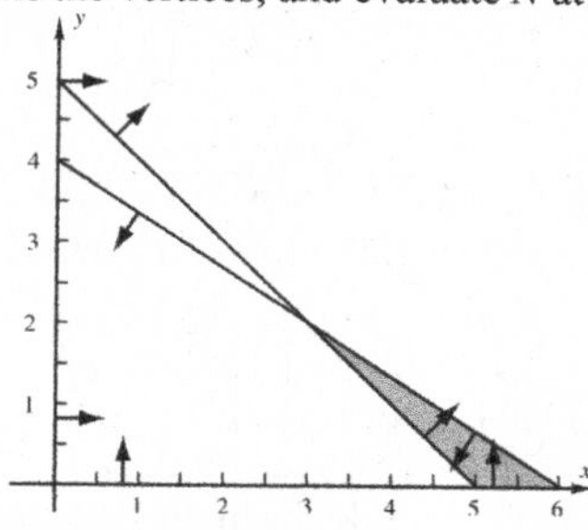

Vertex	$C=5x+4y$
(3, 2)	$5\cdot 3+4\cdot 2=23$
(5, 0)	$5\cdot 5+6\cdot 0=25$
(6, 0)	$5\cdot 6+6\cdot 0=30$

The smallest cost is \$23, obtained when 3 train rides and 2 bus rides are taken.
Check. Go over the algebra and arithmetic.
State. The minimum cost is achieved by taking 3 train rides and 2 bus rides.

15. ***Familiarize***. Let x = the number of 4 photo pages, and y = the number of 6 photo pages.
Translate. The number of photos N is given by
$N=4x+6y$.
We wish to maximize N subject to these constraints.
$x+y\leq 20$
$3x+5y\leq 90$
$x\geq 0$
$y\geq 0.$
Carry out. We graph the system of inequalities, determine the vertices, and evaluate N at each vertex.

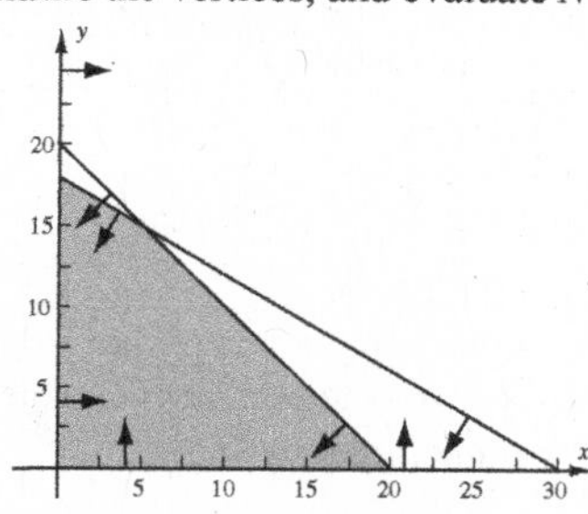

Vertex	$N=4x+6y$
(0, 0)	$4\cdot 0+6\cdot 0=0$
(0, 18)	$4\cdot 0+6\cdot 18=108$
(20, 0)	$4\cdot 20+6\cdot 0=80$
(5, 15)	$4\cdot 5+6\cdot 15=110$

The greatest number of photos is 110, obtained when 5 pages of 4-photos and 15 pages of 6-photos are used.
Check. Go over the algebra and arithmetic.
State. The maximum number of photos is achieved by using 5 pages or 4-photos and 15 pages of 6-photos.

17. In order to earn the most interest Rosa should invest the entire \$40,000. She should also invest as much as possible in the type of investment that has the higher interest rate. Thus, she should invest \$22,000 in corporate bonds and the remaining \$18,000 in municipal bonds. The maximum income is
$0.04(\$22,000)+0.035(\$18,000)=\$1510$.
We can also solve this problem as follows.
Let x = the amount invested in corporate bonds and y = the amount invested in municipal bonds. Find the maximum value of
$I=0.04x+0.035y$
subject to
$x+y\leq \$40,000,$
$\$6000\leq x\leq \$22,000$
$0\leq y\leq \$30,000.$

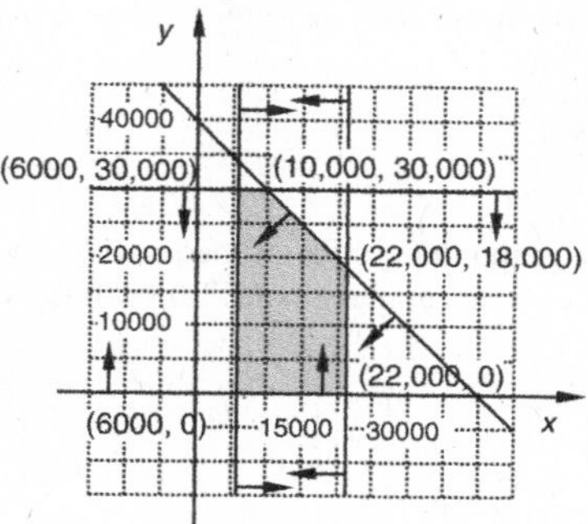

Vertex	$I=0.04x+0.035y$
(\$6000, \$0)	\$240
(\$6000, \$30,000)	\$1290
(\$10,000, \$30,000)	\$1450
(\$22,000, \$18,000)	\$1510
(\$22,000, \$0)	\$880

The maximum income of \$1510 occurs when \$22,000 is invested in corporate bonds and \$18,000 is invested in municipal bonds.

19. ***Familiarize***. Let x = the number of short-answer questions and y = the number of essay questions answered.
Translate. The score S is given by
$S=10x+15y$.
We wish to maximize S subject to these constraints:
$x+y\leq 16$
$3x+6y\leq 60$
$x\geq 0$
$y\geq 0.$
Carry out. We graph the system of inequalities, determine the vertices, and evaluate S at each vertex.

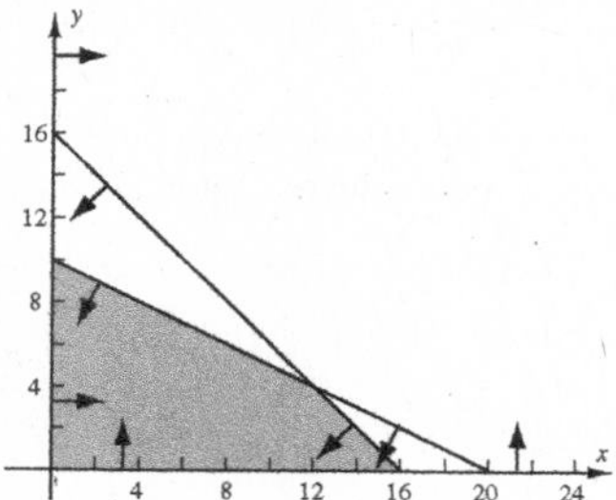

Vertex	$S = 10x + 15y$
(0, 0)	$10 \cdot 0 + 15 \cdot 0 = 0$
(0, 10)	$10 \cdot 0 + 15 \cdot 10 = 150$
(16, 0)	$10 \cdot 16 + 15 \cdot 0 = 160$
(12, 4)	$10 \cdot 12 + 15 \cdot 4 = 180$

The greatest score in the table is 180, obtained when 12 short-answer questions and 4 essay questions are answered.
Check. Go over the algebra and arithmetic.
State. The maximum score is 180 points when 12 short-answer questions and 4 essay questions are answered.

21. ***Familiarize***. Let $x =$ the Merlot acreage and $y =$ the Cabernet acreage.
Translate. The profit P is given by

$P = \$400x + \$300y$.

We wish to maximize P subject to these constraints:

$x + y \le 240,$
$2x + y \le 320,$
$x \ge 0,$
$y \ge 0.$

Carry out. We graph the system of inequalities, determine the vertices, and evaluate P at each vertex.

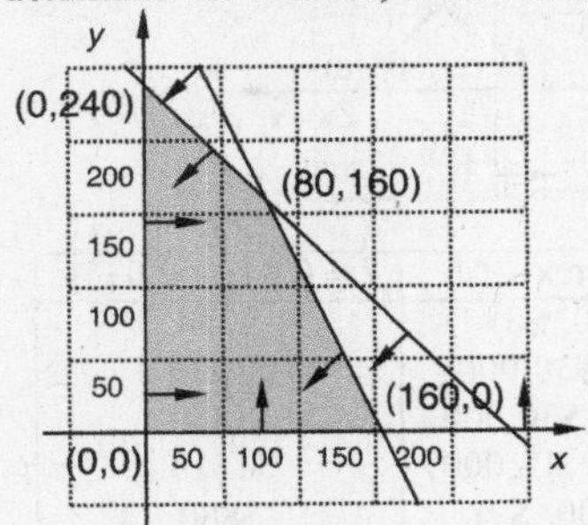

Vertex	$P = \$400x + \$300y$
(0, 0)	\$0
(0, 240)	\$72,000
(80, 160)	\$80,000
(160, 0)	\$64,000

Check. Go over the algebra and arithmetic.
State. The maximum profit occurs by planting 80 acres of Merlot grapes and 160 acres of Cabernet grapes.

23. ***Familiarize***. Let $x =$ the number of servings of goat cheese and $y =$ the number of servings of hazelnuts.
Translate. The total number of calories is given by

$C = 264x + 628y.$

We wish to minimize C subject to these constraints.

$15 \le x + 5y$
$x + 5y \le 45$
$1500 \le 500x + 100y$
$500x + 100y \le 2500$

Carry out. We graph the system of inequalities, determine the vertices and evaluate C at each vertex.

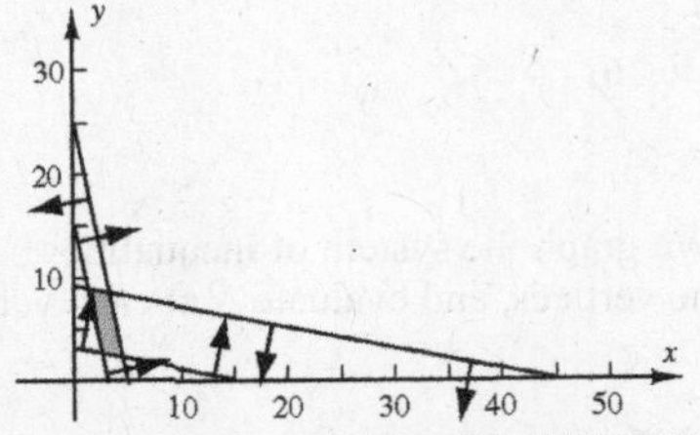

Vertex	$C = 264x + 628y$
(2.5, 2.5)	$264 \cdot 2.5 + 628 \cdot 2.5 = 2230$
(1.25, 8.75)	$264 \cdot 1.25 + 628 \cdot 8.75 = 5825$
$\left(\frac{55}{12}, \frac{25}{12}\right)$	$264 \cdot \frac{55}{12} + 628 \cdot \frac{25}{12} = 2518.3 = \frac{30{,}220}{12}$
$\left(\frac{10}{3}, \frac{25}{3}\right)$	$264 \cdot \frac{10}{3} + 628 \cdot \frac{25}{3} = 6113.3 = \frac{18{,}340}{3}$

The least number of calories in the table is 2230, obtained with 2.5 servings of each.
Check. Go over the algebra and arithmetic.
State. The minimum calories consumed is 2230 with 2.5 servings of each.

25. *Writing Exercise.*

27. $10^{-2} = \frac{1}{100}$

29. $\frac{-6x^2}{3x^{-10}} = -2x^{12}$

31. $\left(\frac{4c^2d}{6cd^4}\right)^{-1} = \frac{6cd^4}{4c^2d} = \frac{3d^3}{2c}$

33. *Writing Exercise.*

35. ***Familiarize***. Let x represent the number of T3 planes and y represent the number of S5 planes. Organize the information in a table.

Plane	Number of planes	Passengers		
		First	Tourist	Economy
T3	x	$40x$	$40x$	$120x$
S5	y	$80y$	$30y$	$40y$

Plane	Cost per mile
T3	$30x$
S5	$25y$

Translate. Suppose C is the total cost per mile. Then $C = 30x + 25y$. We wish to minimize C subject to these facts (constraints) about x and y.

$40x + 80y \ge 2000,$
$40x + 30y \ge 1500,$
$120x + 40y \ge 2400,$
$x \ge 0, \ y \ge 0$

Carry out. Graph the system of inequalities, determine the vertices, and evaluate C at each vertex.

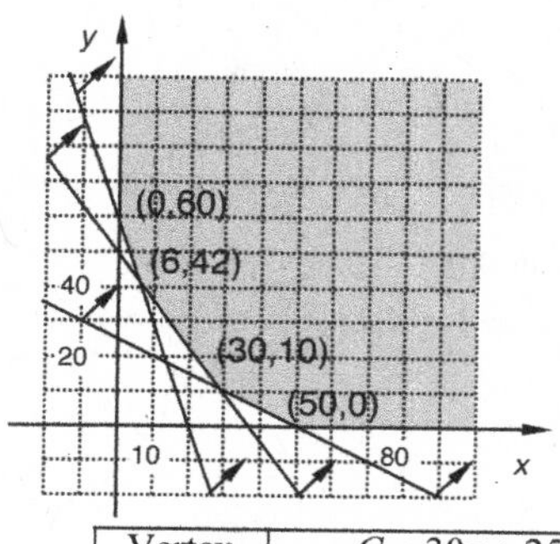

Vertex	$C = 30x + 25y$
(0, 60)	$30(0) + 25(60) = 1500$
(6, 42)	$30(6) + 25(42) = 1230$
(30, 10)	$30(30) + 25(10) = 1150$
(50, 0)	$30(50) + 25(0) = 1500$

Check. Go over the algebra and arithmetic.
State. In order to minimize the operating cost, 30 T3 planes and 10 S5 planes should be used.

Chapter 9 Review

1. True
2. True
3. True
4. True
5. False
6. False
7. $-6x - 5 < 4$
   $-6x < 9$

   Graph: $x > -\frac{3}{2}$

   $-\frac{3}{2}$ 0

   Set builder notation: $\left\{x \middle| x > -\frac{3}{2}\right\}$

   Interval notation: $\left(-\frac{3}{2}, \infty\right)$

8. $$-\frac{1}{2}x - \frac{1}{4} > \frac{1}{2} - \frac{1}{4}x$$
   $$-\frac{1}{2}x - \frac{1}{4} + \frac{1}{4}x > \frac{1}{2} - \frac{1}{4}x + \frac{1}{4}x$$
   $$-\frac{1}{4}x - \frac{1}{4} > \frac{1}{2}$$
   $$-\frac{1}{4}x > \frac{3}{4}$$
   $$x < -3$$
   Graph: $x < -3$.

   −3 0

   Set builder notation: $\{x | x < -3\}$
   Interval notation: $(-\infty, -3)$

9. $-2(x-5) \geq 6(x+7) - 12$
   $-2x + 10 \geq 6x + 42 - 12$
   $-2x + 10 \geq 6x + 30$
   $-20 \geq 8x$
   $-\frac{5}{2} \geq x$

   Graph: $-\frac{5}{2} \geq x$

   $-\frac{5}{2}$ 0

   Set builder notation: $\left\{x \middle| x \leq -\frac{5}{2}\right\}$

   Interval notation: $\left(-\infty, -\frac{5}{2}\right]$

10. Solve graphically: $x - 2 < 3$.
    We graph and solve the system of equations
    $y = x - 2$,
    $y = 3$

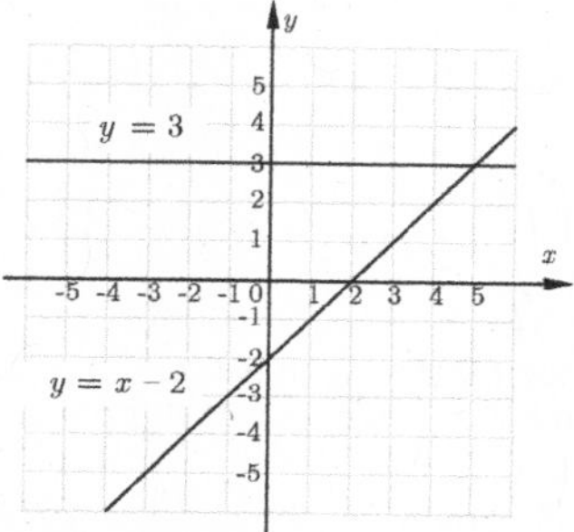

    By substitution, we have
    $x - 2 = 3$
    $x = 5$
    Then $y = 5 - 2 = 3$.
    Thus, the point of intersection is (5, 3). Since the inequality symbol is <, we do not include this point in the solution.
    We see from the graph, $x - 2 < 3$ is to the left of the point of intersection.
    Thus, the solution set is $\{x | x < 5\}$, or $(-\infty, 5)$.

11. Solve graphically: $4 - 3x > 1$.
    We graph and solve the system of equations
    $y = 4 - 3x$,
    $y = 1$

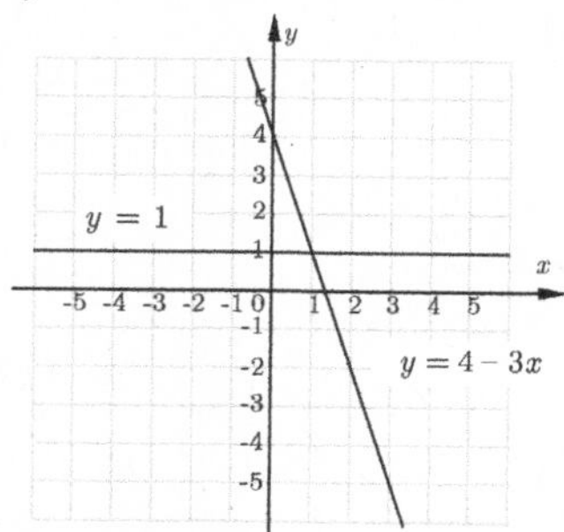

    By substitution, we have
    $4 - 3x = 1$
    $x = 1$

Then $y = 4 - 3(1) = 1$.
Thus, the point of intersection is (1, 1). Since the inequality symbol is >, we do not include this point in the solution.
We see from the graph, $4 - 3x > 1$ is to the left of the point of intersection.
Thus, the solution set is $\{x|x < 1\}$, or $(-\infty, 1)$.

12. Solve graphically: $x - 1 \le 2x + 3$.
We graph and solve the system of equations
$y = x - 1,$
$y = 2x + 3$

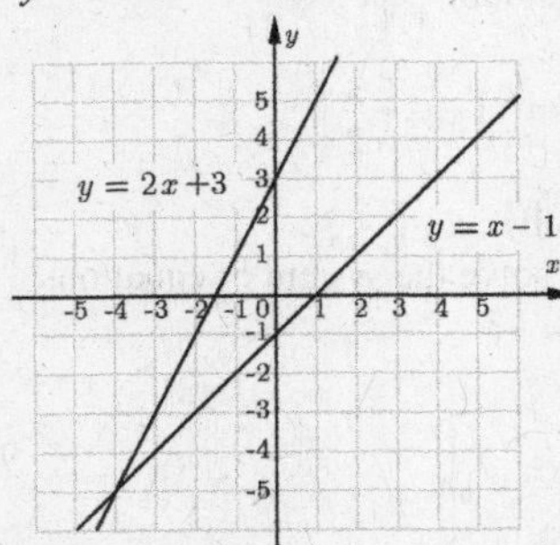

By substitution, we have
$x - 1 = 2x + 3$
$-4 = x$
Then $y = -4 - 1 = -5$.
Thus, the point of intersection is (–4, –5). Since the inequality symbol is ≤, we include this point in the solution.
We see from the graph, $x - 1 \le 2x + 3$ is to the right of the point of intersection.
Thus, the solution set is $\{x|x \ge -4\}$, or $[-4, \infty)$.

13. Solve graphically: $\frac{1}{2}x \ge \frac{1}{3}x + 1$.
We graph and solve the system of equations
$y = \frac{1}{2}x,$
$y = \frac{1}{3}x + 1$

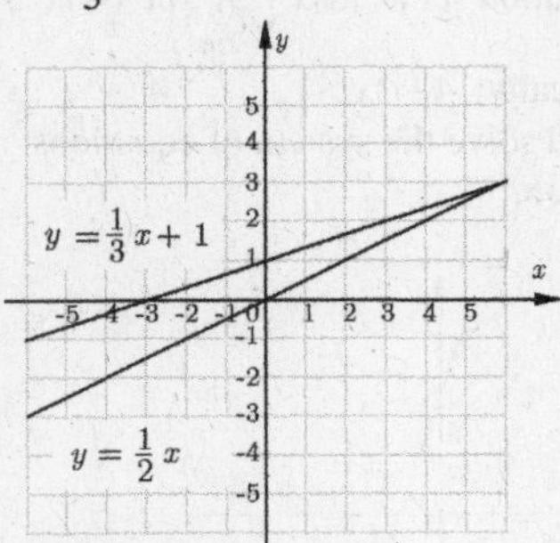

By substitution, we have
$\frac{1}{2}x = \frac{1}{3}x + 1$
$\frac{1}{6}x = 1$
$x = 6$
Then $y = \frac{1}{2}(6) = 3$.

Thus, the point of intersection is (6, 3). Since the inequality symbol is ≥, we include this point in the solution.
We see from the graph, $\frac{1}{2}x \ge \frac{1}{3}x + 1$ is to the right of the point of intersection.
Thus, the solution set is $\{x|x \ge 6\}$, or $[6, \infty)$.

14. $f(x) \le g(x)$
$3x + 2 \le 10 - x$
$4x \le 8$
$x \le 2$
$\{x|x \le 2\}$ or $(-\infty, 2]$

15. Let x = the number of hours worked. Mariah earns \$8.40x at the sandwich shop and in carpentry she earns \$16x, but spends \$950, or \$16x – 950. We solve the inequality.
$8.40x < 16x - 950$
$950 < 7.6x$
$125 < x$
She must work more than 125 hours in carpentry to be more profitable.

16. Let x = the amount invested at 3% and $9000 - x$ = the amount invested at 3.5%. The interest from the first investment is $0.03x$ and the interest from the second investment is $0.035(9000 - x)$. We solve the inequality.
$0.03x + 0.035(9000 - x) \ge 300$
$0.03x + 315 - 0.035x \ge 300$
$-0.005x + 315 \ge 300$
$-0.005x \ge -15$
$x \le 3000$
Clay should invest at most \$3000 at 3%.

17. {a, b, c, d} ∩ {a, c, e, f, g} = {a, c}

18. {a, b, c, d} ∪ {a, c, e, f, g} = {a, b, c, d, e, f, g}

19. Graph: $x \le 2$ *and* $x > -3$
–3 0 2
(–3, 2]

20. Graph: $x \le 3$ *or* $x > -5$
0
$(-\infty, \infty)$

21. $-3 < x + 5 \le 5$
$-8 < x \le 0$
$\{x|-8 < x \le 0\}$
–8 0
(–8, 0]

22. $-15 < -4x - 5 < 0$
$-10 < -4x < 5$
$\frac{5}{2} > x > -\frac{5}{4}$, or $-\frac{5}{4} < x < \frac{5}{2}$

$-\frac{5}{4}$ 0 $\frac{5}{2}$

$\left\{x \middle| -\frac{5}{4} < x < \frac{5}{2}\right\}$ or $\left(-\frac{5}{4}, \frac{5}{2}\right)$

23. $3x < -9$ *or* $-5x < -5$
$x < -3$ *or* $x > 1$

−3 0 1

$\{x | x < -3 \text{ or } x > 1\}$ or $(-\infty, -3) \cup (1, \infty)$

24. $2x + 5 < -17$ *or* $-4x + 10 \le 34$
$2x < -22$ *or* $-4x \le 24$
$x < -11$ *or* $x \ge -6$

−11 −6 0

$\{x | x < -11 \text{ or } x \ge -6\}$ or $(-\infty, -11) \cup [-6, \infty)$

25. $2x + 7 \le -5$ *or* $x + 7 \ge 15$
$2x \le -12$ *or* $x \ge 8$
$x \le -6$

−6 0 8

$\{x | x \le -6 \text{ or } x \ge 8\}$ or $(-\infty, -6] \cup [8, \infty)$

26. $f(x) < -5$ *or* $f(x) > 5$
$3 - 5x < -5$ *or* $3 - 5x > 5$
$-5x < -8$ *or* $-5x > 2$
$x > \frac{8}{5}$ *or* $x < -\frac{2}{5}$

$-\frac{2}{5}$ 0 $\frac{8}{5}$

$\left\{x \middle| x < -\frac{2}{5} \text{ or } x > \frac{8}{5}\right\}$ or $\left(-\infty, -\frac{2}{5}\right) \cup \left(\frac{8}{5}, \infty\right)$

27. $f(x) = \frac{2x}{x+3}$
$x + 3 = 0$
$x = -3$
The domain of f is $(-\infty, -3) \cup (-3, \infty)$.

28. $f(x) = \sqrt{5x - 10}$
$5x - 10 \ge 0$
$5x \ge 10$
$x \ge 2$
The domain of f is $[2, \infty)$.

29. $f(x) = \sqrt{1 - 4x}$
$1 - 4x \ge 0$
$-4x \ge -1$
$x \le \frac{1}{4}$

The domain of f is $\left(-\infty, \frac{1}{4}\right]$.

30. $|x| = 11$
$x = -11$ *or* $x = 11$
$\{-11, 11\}$

31. $|t| \ge 21$
$t \le -21$ *or* $21 \le t$
$\{t | t \le -21 \text{ or } t \ge 21\}$ or $(-\infty, -21] \cup [21, \infty)$

32. $|x - 8| = 3$
$x - 8 = -3$ *or* $x - 8 = 3$
$x = 5$ *or* $x = 11$
$\{5, 11\}$

33. $|4a + 3| < 11$
$-11 < 4a + 3 < 11$
$-14 < 4a < 8$
$-\frac{7}{2} < a < 2$

$\left\{a \middle| -\frac{7}{2} < a < 2\right\}$ or $\left(-\frac{7}{2}, 2\right)$

34. $|3x - 4| \ge 15$
$3x - 4 \le -15$ *or* $15 \le 3x - 4$
$3x \le -11$ *or* $19 \le 3x$
$x \le -\frac{11}{3}$ *or* $\frac{19}{3} \le x$

$\left\{x \middle| x \le -\frac{11}{3} \text{ or } x \ge \frac{19}{3}\right\}$ or $\left(-\infty, -\frac{11}{3}\right] \cup \left[\frac{19}{3}, \infty\right)$

35. $|2x + 5| = |x - 9|$
$2x + 5 = x - 9$ *or* $2x + 5 = -(x - 9)$
$x = -14$ *or* $2x + 5 = -x + 9$
$3x = 4$
$x = \frac{4}{3}$

$\left\{-14, \frac{4}{3}\right\}$

36. $|5n + 6| = -11$
Absolute value is never negative.
The solution is ∅.

37. $\left|\frac{x+4}{6}\right| \le 2$

$-2 \le \frac{x+4}{6} \le 2$
$-12 \le x + 4 \le 12$
$-16 \le x \le 8$
$\{x | -16 \le x \le 8\}$ or $[-16, 8]$

38. $2|x - 5| - 7 > 3$
$2|x - 5| > 10$
$|x - 5| > 5$
$x - 5 < -5$ *or* $5 < x - 5$
$x < 0$ *or* $10 < x$
$\{x | x < 0 \text{ or } x > 10\}$ or $(-\infty, 0) \cup (10, \infty)$

39. $19-3|x+1|\geq 4$
$-3|x+1|\geq -15$
$|x+1|\leq 5$
$-5\leq x+1\leq 5$
$-6\leq x\leq 4$
$\{x|-6\leq x\leq 4\}$ or $[-6, 4]$

40. $|8x-3|<0$
Absolute value is never negative.
The solution is $\varnothing$.

41. Graph $x-2y\geq 6$.

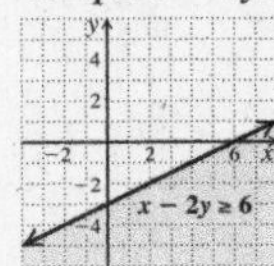

42. Graph $x+3y>-1$,
$x+3y<4$

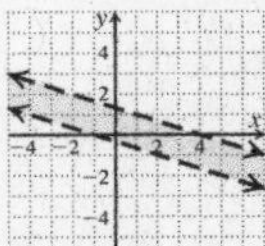

The lines are parallel, there are no vertices.

43. Graph $x-3y\leq 3$,
$x+3y\geq 9$,
$y\leq 6$

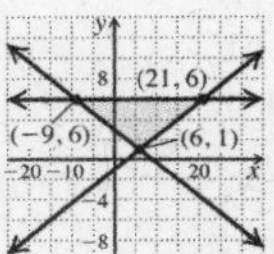

Vertices: (–9, 6), (6, 1) and (21, 6)

44. For $F=3x+y+4$, subject to
$y\leq 2x+1$,
$x\leq 7$,
$y\geq 3$.

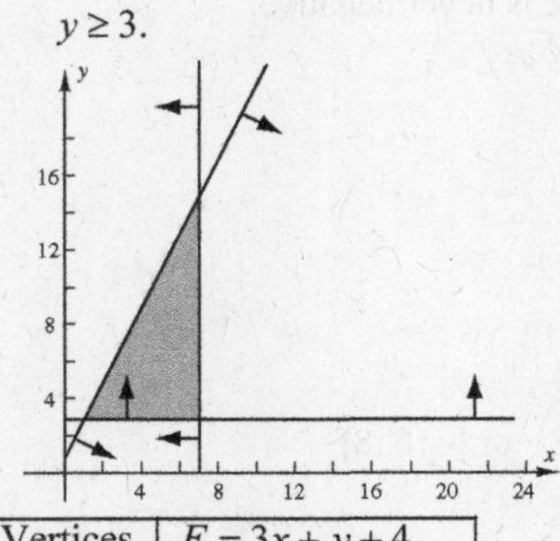

Vertices	$F=3x+y+4$
(7, 3)	$3\cdot 7+3+4=28$
(1, 3)	$3\cdot 1+3+4=10$
(7, 15)	$3\cdot 7+15+4=40$

The maximum value of F is 40 at $x=7, y=15$.
The minimum value of F is 10 at $x=1, y=3$.

45. Let x = the number of books ordered from the East coast supplier and y = the number of books ordered from the West coast supplier. Minimize the time $T=5x+2y$ subject to the constraints.
$2x+4y\leq 320$,
$x+y\geq 100$,
$x\geq 0$
$y\geq 0$

Vertices	$T=5x+2y$
(160, 0)	$5\cdot 160+2\cdot 0=800$
(100, 0)	$5\cdot 100+2\cdot 0=500$
(40, 60)	$5\cdot 40+2\cdot 60=320$

The minimum is 320 when 40 books are ordered from the East coast supplier and 60 books are ordered from the West coast supplier.

46. *Writing Exercise.* The equation $|X|=p$ has two solutions when p is positive because X can be either p or $-p$. The same equation has no solution when p is negative because no number has a negative absolute value.

47. *Writing Exercise.* The solution set of a system of inequalities is all ordered pairs that make *all* the individual inequalities true. This consists of ordered pairs that are common to all the individual solution sets, or the intersection of the graphs.

48. $|2x+5|\leq |x+3|$
$2x+5\leq x+3$ *and* $2x+5\geq -(x+3)$
$x\leq -2$ *and* $2x+5\geq -x-3$
$3x\geq -8$
$x\geq -\frac{8}{3}$

$\left\{x\middle|-\frac{8}{3}\leq x\leq -2\right\}$, or $\left[-\frac{8}{3},-2\right]$

Chapter 9 Test

1. $-4y-3\geq 5$
$-4y\geq 8$
$y\leq -2$ Reversing the inequality symbol
$\{y|y\leq -2\}$, or $(-\infty,-2]$

2. $3(7-x)<2x+5$
$21-3x<2x+5$
$21-5x<5$
$-5x<-16$
$x>\frac{16}{5}$ Reversing the inequality symbol

$\left\{x\middle|x>\frac{16}{5}\right\}$, or $\left(\frac{16}{5},\infty\right)$

3. $-2(3x-1)-5\geq 6x-4(3-x)$
$-6x+2-5\geq 6x-12+4x$
$-6x-3\geq 10x-12$
$-16x-3\geq -12$
$-16x\geq -9$
$x\leq \frac{9}{16}$ Reversing the inequality symbol

$\left\{x\middle|x\leq\frac{9}{16}\right\}$, or $\left(-\infty,\frac{9}{16}\right]$

4. Solve graphically: $3 - x < 2$.
We graph and solve the system of equations
$y = 3 - x,$
$y = 2$

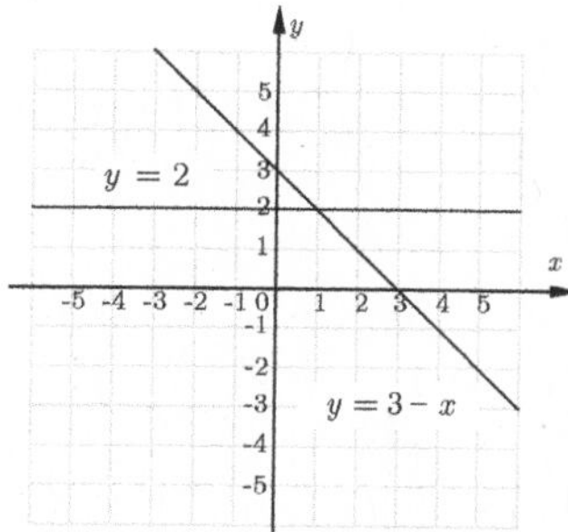

By substitution, we have
$3 - x = 2$
$x = 1$
Then $y = 3 - 1 = 2$.
Thus, the point of intersection is (1, 2). Since the inequality symbol is <, we do not include this point in the solution.
We see from the graph, $3 - x < 2$ is to the right of the point of intersection.
Thus, the solution set is $\{x|x > 1\}$, or $(1, \infty)$.

5. Solve graphically: $2x - 3 \geq x + 1$.
We graph and solve the system of equations
$y = 2x - 3,$
$y = x + 1$

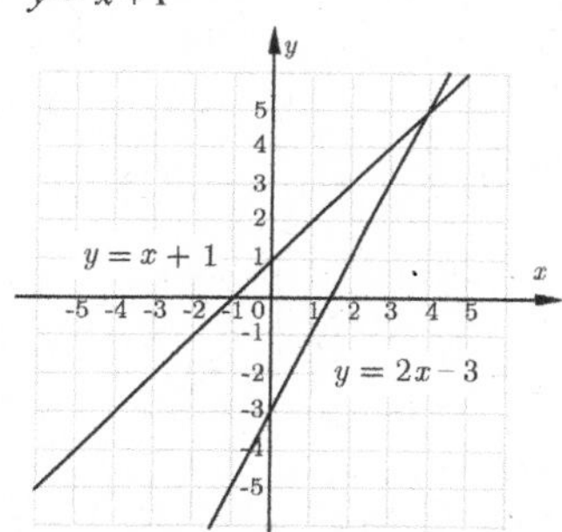

By substitution, we have
$2x - 3 = x + 1$
$x = 4$
Then $y = 4 + 1 = 5$.
Thus, the point of intersection is (4, 5). Since the inequality symbol is ≥, we include this point in the solution.
We see from the graph, $2x - 3 \geq x + 1$ is to the right of the point of intersection.
Thus, the solution set is $\{x|x \geq 4\}$, or $[4, \infty)$.

6. $f(x) > g(x)$
$-5x - 1 > -9x + 3$
$4x - 1 > 3$
$4x > 4$
$x > 1$
$\{x|x > 1\}$ or $(1, \infty)$

7. Let x = the number of miles driven. The cost for unlimited is \$80 and the cost for the other plan is $\$45 + \$0.40x$. We solve the inequality.
$80 < 45 + 0.40x$
$35 < 0.4x$
$87.5 < x$
The unlimited mileage plan is less expensive for more than 87.5 miles.

8. Let x = the number of additional hours. The cost is $\$80 + \$60x$ and \$200 is budgeted. We solve the inequality.
$80 + 60x \leq 200$
$60x \leq 120$
$x \leq 2$
The time of service is $x + \frac{1}{2}$ hr or $2\frac{1}{2}$ hours or less.

9. $\{a, e, i, o, u\} \cap \{a, b, c, d, e\} = \{a, e\}$

10. $\{a, e, i, o, u\} \cup \{a, b, c, d, e\} = \{a, b, c, d, e, i, o, u\}$

11. $f(x) = \sqrt{6 - 3x}$
$6 - 3x \geq 0$
$6 \geq 3x$
$2 \geq x$
The domain of f is $(-\infty, 2]$.

12. $f(x) = \frac{x}{x - 7}$
$x - 7 = 0$
$x = 7$
The domain of f is $(-\infty, 7) \cup (7, \infty)$.

13. $-5 < 4x + 1 \leq 3$
$-6 < 4x \leq 2$
$-\frac{3}{2} < x \leq \frac{1}{2}$
$\left\{x \middle| -\frac{3}{2} < x \leq \frac{1}{2}\right\}$ or $\left(-\frac{3}{2}, \frac{1}{2}\right]$

$-\frac{3}{2}$ 0 $\frac{1}{2}$

14. $3x - 2 < 7$ *or* $x - 2 > 4$
$3x < 9$ *or* $x > 6$
$x < 3$
The solution set is $\{x|x < 3 \text{ or } x > 6\}$ or $(-\infty, 3) \cup (6, \infty)$.

0 3 6

15. $-3x > 12$ *or* $4x \geq -10$
$x < -4$ *or* $x \geq -\frac{5}{2}$
The solution set is $\left\{x \middle| x < -4 \text{ or } x \geq -\frac{5}{2}\right\}$ or
$(-\infty, -4) \cup \left[-\frac{5}{2}, \infty\right)$.

-4 $-\frac{5}{2}$ 0

16. $1 \le 3-2x \le 9$
$-2 \le -2x \le 6$
$1 \ge x \ge -3$
The solution set is $\{x|-3 \le x \le 1\}$ or $[-3, 1]$.

17. $|n| = 15$
$n = -15$ *or* $n = 15$
$\{-15, 15\}$

18. $|a| > 5$
$a < -5$ *or* $5 < a$
$\{a \mid a < -5 \text{ or } a > 5\}$ or $(-\infty, -5) \cup (5, \infty)$

19. $|3x-1| < 7$
$-7 < 3x-1 < 7$
$-6 < 3x < 8$
$-2 < x < \frac{8}{3}$
$\left\{x \middle| -2 < x < \frac{8}{3}\right\}$, or $\left(-2, \frac{8}{3}\right)$

20. $|-5t-3| \ge 10$
$-5t-3 \le -10$ *or* $10 \le -5t-3$
$-5t \le -7$ *or* $13 \le -5t$
$t \ge \frac{7}{5}$ *or* $-\frac{13}{5} \ge t$
$\left\{t \middle| t \le -\frac{13}{5} \text{ or } t \ge \frac{7}{5}\right\}$ or $\left(-\infty, -\frac{13}{5}\right] \cup \left[\frac{7}{5}, \infty\right)$

21. $|2-5x| = -12$
Absolute value is never negative.
The solution is $\varnothing$.

22. $g(x) < -3$ *or* $g(x) > 3$
$4-2x < -3$ *or* $4-2x > 3$
$-2x < -7$ *or* $-2x > -1$
$x > \frac{7}{2}$ *or* $x < \frac{1}{2}$
$\left\{x \middle| x < \frac{1}{2} \text{ or } x > \frac{7}{2}\right\}$ or $\left(-\infty, \frac{1}{2}\right) \cup \left(\frac{7}{2}, \infty\right)$

23. $f(x) = g(x)$
$|2x-1| = |2x+7|$
$2x-1 = 2x+7$ *or* $2x-1 = -(2x+7)$
$-1 = 7$ $\quad 2x-1 = -2x-7$
FALSE $\quad 4x = -6$
$x = -\frac{3}{2}$

$\left\{-\frac{3}{2}\right\}$

24. $y \le 2x+1$

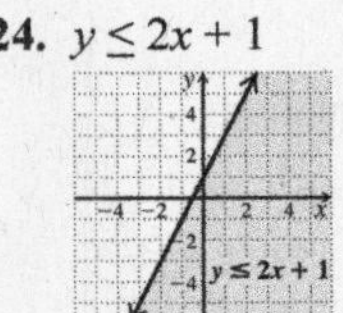

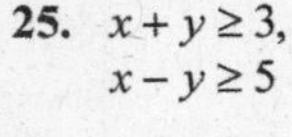

25. $x+y \ge 3,$
$x-y \ge 5$

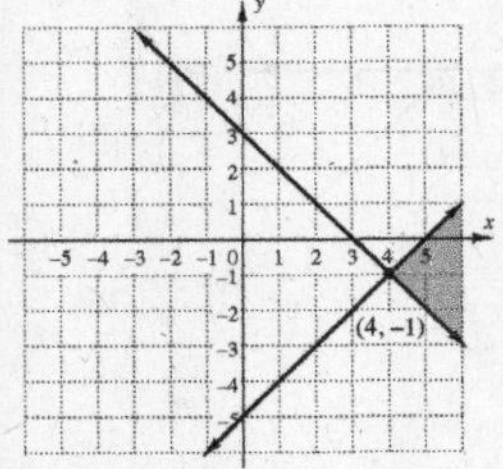

Vertex: $(4, -1)$

26. $2y-x \ge -7,$
$2y+3x \le 15,$
$y \le 0,$
$x \le 0$

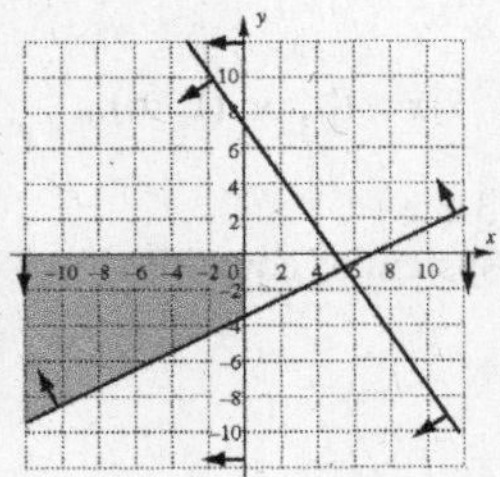

Vertices: $(0, 0)$ and $\left(0, -\frac{7}{2}\right)$

27.

Vertices	$F = 5x+3y$
(1, 0)	$5\cdot1+3\cdot0=5$
(6, 0)	$5\cdot6+3\cdot0=30$
(1, 12)	$5\cdot1+3\cdot12=41$
(3, 12)	$5\cdot3+3\cdot12=51$
(6, 9)	$5\cdot6+3\cdot9=57$

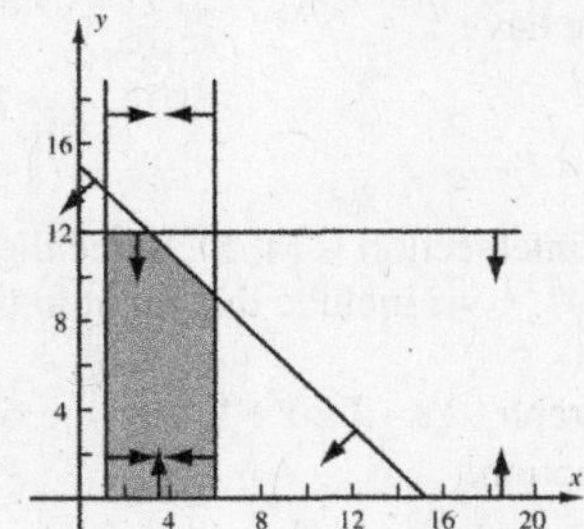

The maximum is 57 when $x = 6, y = 9$.
The minimum is 5 when $x = 1, y = 0$.

28. Let x = the number of manicures and y = the number of haircuts. The profit is $P = 12x + 18y$. We maximize the profit subject to the constraints:

$$30x + 50y \le 5(6)(60)$$
$$x + y \le 50$$
$$x \ge 0$$
$$y \ge 0$$

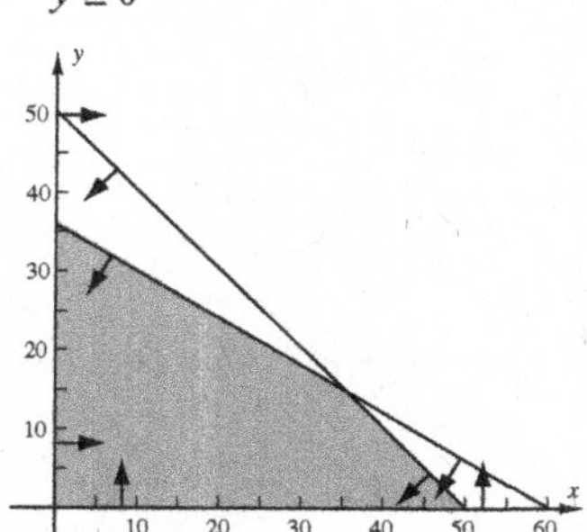

Vertices	$P = 12x + 18y$
(0, 0)	$12 \cdot 0 + 18 \cdot 0 = 0$
(50, 0)	$12 \cdot 50 + 18 \cdot 0 = 600$
(0, 36)	$12 \cdot 0 + 18 \cdot 36 = 648$
(35, 15)	$12 \cdot 35 + 18 \cdot 15 = 690$

The maximum profit is $690 when there are 35 manicures and 15 haircuts.

29. $|2x-5| \le 7$ and $|x-2| \ge 2$

$-7 \le 2x - 5 \le 7$ *and* $x - 2 \le -2$ *or* $2 \le x - 2$

$-2 \le 2x \le 12$ *and* $x \le 0$ *or* $4 \le x$

$-1 \le x \le 6$

$\{x | -1 \le x \le 0 \text{ or } 4 \le x \le 6\}$ or $[-1, 0] \cup [4, 6]$

30. $7x < 8 - 3x < 6 + 7x$

$0 < 8 - 10x < 6$

$-8 < -10x < -2$

$\frac{4}{5} > x > \frac{1}{5}$

$\left\{x \middle| \frac{1}{5} < x < \frac{4}{5}\right\}$ or $\left(\frac{1}{5}, \frac{4}{5}\right)$

31. $\frac{-8+2}{2} = -3$

$\frac{2-(-8)}{2} = 5$

$|x-(-3)| \le 5$ or $|x+3| \le 5$

Chapter 10

Exponents and Radicals

Exercise Set 10.1

1. Every positive number has *two* square roots.

3. For any *positive* number *a*, we have $\sqrt{a^2} = a$.

5. If *a* is a whole number that is not a perfect square, then $\sqrt{a}$ is an *irrational* number.

7. If $\sqrt[4]{x}$ is a real number, then *x* must be *nonnegative*.

9. The square roots of 64 are 8 and –8, because $8^2 = 64$ and $(-8)^2 = 64$.

11. The square roots of 100 are 10 and –10 because $10^2 = 100$ and $(-10)^2 = 100$.

13. The square roots of 400 are 20 and –20 because $20^2 = 400$ and $(-20)^2 = 400$.

15. The square roots of 625 are 25 and –25 because $25^2 = 625$ and $(-25)^2 = 625$.

17. $\sqrt{49} = 7$ Remember, $\sqrt{\ }$ indicates the principle square root.

19. $-\sqrt{16} = -4$ Since, $\sqrt{16} = 4$, $-\sqrt{16} = -4$

21. $\sqrt{\frac{36}{49}} = \frac{6}{7}$

23. $-\sqrt{\frac{16}{81}} = -\frac{4}{9}$ Since, $\sqrt{\frac{16}{81}} = \frac{4}{9}$, $-\sqrt{\frac{16}{81}} = -\frac{4}{9}$

25. $\sqrt{0.04} = 0.2$

27. $\sqrt{0.0081} = 0.09$

29. $f(t) = \sqrt{5t - 10}$
$f(3) = \sqrt{5(3) - 10} = \sqrt{5}$
$f(2) = \sqrt{5(2) - 10} = \sqrt{0} = 0$
$f(1) = \sqrt{5(1) - 10} = \sqrt{-5}$
Since negative numbers do not have real-number square roots, $f(1)$ does not exist.
$f(-1) = \sqrt{5(-1) - 10} = \sqrt{-15}$
Since negative numbers do not have real-number square roots, $f(-1)$ does not exist.

31. $t(x) = -\sqrt{2x^2 - 1}$
$t(5) = -\sqrt{2 \cdot 5^2 - 1} = -\sqrt{49} = -7$
$t(0) = -\sqrt{2 \cdot 0^2 - 1} = \sqrt{-1}$
$t(0)$ does not exist
$t(-1) = -\sqrt{2(-1)^2 - 1} = -\sqrt{1} = -1$
$t\left(-\frac{1}{2}\right) = -\sqrt{2\left(-\frac{1}{2}\right)^2 - 1} = -\sqrt{-\frac{1}{2}}$ does not exist

33. $f(t) = \sqrt{t^2 + 1}$
$f(0) = \sqrt{0^2 + 1} = \sqrt{1} = 1$
$f(-1) = \sqrt{(-1)^2 + 1} = \sqrt{2}$
$f(-10) = \sqrt{(-10)^2 + 1} = \sqrt{101}$

35. $\sqrt{100x^2} = \sqrt{(10x)^2} = |10x| = 10|x|$
Since *x* might be negative, absolute-value notation is necessary.

37. $\sqrt{(-4b)^2} = |-4b| = 4|b|$
Since *b* might be negative, absolute-value notation is necessary.

39. $\sqrt{(8-t)^2} = |8-t|$
Since $8 - t$ might be negative, absolute-value notation is necessary.

41. $\sqrt{y^2 + 16y + 64} = \sqrt{(y+8)^2} = |y+8|$
Since $y + 8$ might be negative, absolute-value notation is necessary.

43. $\sqrt{4x^2 + 28x + 49} = \sqrt{(2x+7)^2} = |2x+7|$
Since $2x + 7$ might be negative, absolute-value notation is necessary.

45. $\sqrt{a^{22}} = |a^{11}|$ Note that $(a^{11})^2 = a^{22}$; *a* could have a negative value.

47. $\sqrt{-25}$ is not a real number, so $\sqrt{-25}$ cannot be simplified.

49. $\sqrt[3]{-1} = -1$ Since $(-1)^3 = -1$

51. $-\sqrt[3]{64} = -4$ $(4^3 = 64)$

53. $-\sqrt[3]{-125y^3} = -(-5y)$ $\left[(-5y^3) = -125y^3\right]$
$= 5y$

55. radicand: $p^2 + 4$; index: 2

57. radicand: $\frac{x}{y+4}$; index: 5

59. $-\sqrt[4]{256}=-4$ Since $4^4=256$

61. $-\sqrt[5]{-\frac{32}{243}}=\frac{2}{3}$ Since $\left(-\frac{2}{3}\right)^5=-\frac{32}{243}$

63. $\sqrt[6]{x^6}=|x|$
The index is even. Use absolute-value notation since x could have a negative value.

65. $\sqrt[9]{t^9}=t$
The index is odd. Absolute-value signs are not necessary.

67. $\sqrt[4]{(6a)^4}=|6a|=6|a|$
The index is even. Use absolute-value notation since a could have a negative value.

69. $\sqrt[10]{(-6)^{10}}=|-6|=6$

71. $\sqrt[414]{(a+b)^{414}}=|a+b|$
The index is even. Use absolute-value notation since $a+b$ could have a negative value.

73. $\sqrt{16x^2}=\sqrt{(4x)^2}=4x$ Assuming x is nonnegative

75. $-\sqrt{(3t)^2}=-3t$ Assuming t is nonnegative

77. $\sqrt{(-5b)^2}=5b$

79. $\sqrt{a^2+2a+1}=\sqrt{(a+1)^2}=a+1$

81. $\sqrt[4]{16x^4}=\sqrt[4]{(2x)^4}=2x$

83. $\sqrt[3]{(x-1)^3}=x-1$

85. $\sqrt{t^{18}}=\sqrt{(t^9)^2}=t^9$

87. $\sqrt{(x-2)^8}=\sqrt{\left[(x-2)^4\right]^2}=(x-2)^4$

89.
$$f(x)=\sqrt[3]{x+1}$$
$$f(7)=\sqrt[3]{7+1}=\sqrt[3]{8}=2$$
$$f(26)=\sqrt[3]{26+1}=\sqrt[3]{27}=3$$
$$f(-9)=\sqrt[3]{-9+1}=\sqrt[3]{-8}=-2$$
$$f(-65)=\sqrt[3]{-65+1}=\sqrt[3]{-64}=-4$$

91.
$$g(t)=\sqrt[4]{t-3}$$
$$g(19)=\sqrt[4]{19-3}=\sqrt[4]{16}=2$$
$$g(-13)=\sqrt[4]{-13-3}=\sqrt[4]{-16}$$
$g(-13)$ does not exist
$$g(1)=\sqrt[4]{1-3}=\sqrt[4]{-2}$$
$g(1)$ does not exist
$$g(84)=\sqrt[4]{84-3}=\sqrt[4]{81}=3$$

93. $f(x)=\sqrt{x-6}$
Since the index is even, the radicand, $x-6$, must be non-negative. We solve the inequality:
$$x-6\ge 0$$
$$x\ge 6$$
Domain of $f=\{x|x\ge 6\}$, or $[6,\infty)$

95. $g(t)=\sqrt[4]{t+8}$
Since the index is even, the radicand, $t+8$, must be non-negative. We solve the inequality:
$$t+8\ge 0$$
$$t\ge -8$$
Domain of $g=\{t|t\ge -8\}$, or $[-8,\infty)$

97. $g(x)=\sqrt[4]{10-2x}$
Since the index is even, the radicand, $10-2x$, must be nonnegative. We solve the inequality:
$$10-2x\ge 0$$
$$-2x\ge -10$$
$$x\le 5$$
Domain of $g=\{x|x\le 5\}$, or $(-\infty,5]$

99. $f(t)=\sqrt[5]{2t+7}$
Since the index is odd, the radicand can be any real number.
Domain of $f=\{t|t \text{ is a real number}\}$, or $(-\infty,\infty)$

101. $h(z)=-\sqrt[6]{5z+2}$
Since the index is even, the radicand, $5z+2$, must be nonnegative. We solve the inequality:
$$5z+2\ge 0$$
$$5z\ge -2$$
$$z\ge -\frac{2}{5}$$
Domain of $h=\left\{z\middle|z\ge -\frac{2}{5}\right\}$, or $\left[-\frac{2}{5},\infty\right)$

103. $f(t)=7+\sqrt[8]{t^8}$
Since we can compute $7+\sqrt[8]{t^8}$ for any real number t, the domain is the set of real numbers, or $\{t|t \text{ is a real number}\}$ or $(-\infty,\infty)$.

105. *Writing Exercise.*

107. $f\left(\frac{1}{3}\right)=3\left(\frac{1}{3}\right)-1=1-1=0$

109. $\{x|x\ne 0\}$, or $(-\infty,0)\cup(0,\infty)$

111. $(fg)(x)=(3x-1)\left(\frac{1}{x}\right)=3-\frac{1}{x}$

113. *Writing Exercise.*

115. $f(p)=118.8\sqrt{p}$
Substitute 50 for p.
$$f(50)=118.8\sqrt{50}$$
$$\approx 840$$
The water flow is about 840 GPM.

117. $S = 88.63\sqrt[4]{A}$
Substitute 63,000 for A.
$S = 88.63\sqrt[4]{63{,}000}$
$S \approx 1404$
There are about 1404 species of plants.

119. $f(x) = \sqrt{x+5}$
Since the index is even, the radicand, $x + 5$, must be non-negative. We solve the inequality:
$x + 5 \ge 0$
$x \ge -5$
Domain of $f = \{x | x \ge -5\}$, or $[-5, \infty)$
Make a table of values, keeping in mind that x must be -5 or greater. Plot these points and draw the graph.

x	$f(x)$
−5	0
−4	1
−1	2
1	2.4
3	2.8
4	3

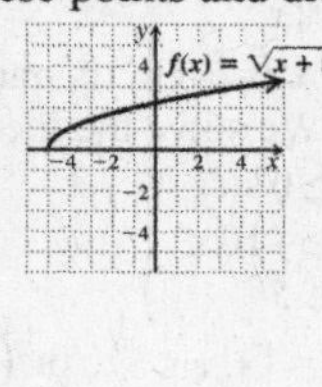

121. $g(x) = \sqrt{x} - 2$
Since the index is even, the radicand, x, must be non-negative, so we have $x \ge 0$.
Domain of $g = \{x | x \ge 0\}$, or $[0, \infty)$
Make a table of values, keeping in mind that x must be 0 or greater. Plot these points and draw the graph.

x	$g(x)$
0	−2
1	−1
4	0
6	0.4
8	0.8

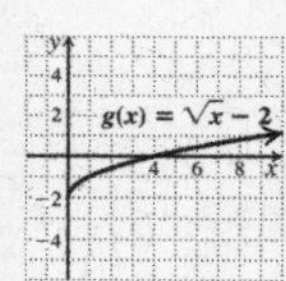

123. $f(x) = \dfrac{\sqrt{x+3}}{\sqrt[4]{2-x}}$
In the numerator we must have $x + 3 \ge 0$, or $x \ge -3$, and in the denominator we must have $2 - x > 0$, or $x < 2$, so Domain of $f = \{x | -3 \le x < 2\}$, or $[-3, 2)$.

125. $F(x) = \dfrac{x}{\sqrt{x^2 - 5x - 6}}$
Since the radical expression in the denominator has an even index, so the radicand, $x^2 - 5x - 6$, must be nonnegative in order for $\sqrt{x^2 - 5x - 6}$ to exist. In addition, the denominator cannot be zero, so the radicand must be positive. We solve the inequality:
$x^2 - 5x - 6 > 0$
$(x+1)(x-6) > 0$
We have $x < -1$ *and* $x > 6$, so
Domain of $F = \{x | x < -1 \text{ or } x > 6\}$, or $(-\infty, -1) \cup (6, \infty)$.

127. $P = 50\sqrt[15]{\dfrac{\text{NYT(Ah + Aw)}}{\text{ENQ(Sc + 5)}} \cdot \text{Md} \cdot \left[\dfrac{\text{Md}}{(\text{Md} + 2)}\right]^{T^2}}$
Substitute 258 for NYT, 29 for Ah, 29 for Aw, 44 for ENQ, 0 for Sc, 120 for Md, and 5 for T.
$P = 50\sqrt[15]{\dfrac{258(29+29)}{44(0+5)} \cdot 120 \cdot \left[\dfrac{120}{(120+2)}\right]^{5^2}}$
$P \approx 88.7$
The probability that the marriage will last 5 years is about 89%.

Exercise Set 10.2

1. The expression $\sqrt{3x}$ is an example of a *radical* expression.

3. The expressions $\sqrt[3]{5mn}$ and $(5mn)^{1/3}$ are *equivalent*.

5. Choice (g) is correct because $a^{m/n} = \sqrt[n]{a^m}$.

7. $x^{-5/2} = \dfrac{1}{x^{5/2}} = \dfrac{1}{(\sqrt{x})^5}$, so choice (e) is correct.

9. $x^{1/5} \cdot x^{2/5} = x^{1/5+2/5} = x^{3/5}$, so choice (a) is correct.

11. Choice (b) is correct because $\sqrt[n]{a^m}$ and $(\sqrt[n]{a})^m$ are equivalent.

13. $y^{1/3} = \sqrt[3]{y}$

15. $36^{1/2} = \sqrt{36} = 6$

17. $32^{1/5} = \sqrt[5]{32} = 2$

19. $64^{1/2} = \sqrt{64} = 8$

21. $(xyz)^{1/2} = \sqrt{xyz}$

23. $(a^2b^2)^{1/5} = \sqrt[5]{a^2b^2}$

25. $t^{5/6} = \sqrt[6]{t^5}$

27. $16^{3/4} = \sqrt[4]{16^3} = (\sqrt[4]{16})^3 = 2^3 = 8$

29. $125^{4/3} = \sqrt[3]{125^4} = (\sqrt[3]{125})^4 = 5^4 = 625$

31. $(81x)^{3/4} = \sqrt[4]{(81x)^3} = \sqrt[4]{81^3x^3}$, or
$\sqrt[4]{81^3} \cdot \sqrt[4]{x^3} = (\sqrt[4]{81})^3 \cdot (\sqrt[4]{x^3}) = 3^3\sqrt[4]{x^3} = 27\sqrt[4]{x^3}$

33. $(25x^4)^{3/2} = \sqrt{(25x^4)^3} = \sqrt{25^3 \cdot x^{12}} = \sqrt{25^3} \cdot \sqrt{x^{12}}$
$= (\sqrt{25})^3 x^6 = 5^3x^6 = 125x^6$

35. $\sqrt[3]{18} = 18^{1/3}$

37. $\sqrt{30} = 30^{1/2}$

39. $\sqrt{x^7} = x^{7/2}$

41. $\sqrt[5]{m^2} = m^{2/5}$

43. $\sqrt[4]{xy} = (xy)^{1/4}$

45. $\sqrt[5]{xy^2z} = \left(xy^2z\right)^{1/5}$

47. $\left(\sqrt{3mn}\right)^3 = (3mn)^{3/2}$

49. $\left(\sqrt[7]{8x^2y}\right)^5 = \left(8x^2y\right)^{5/7}$

51. $\dfrac{2x}{\sqrt[3]{z^2}} = \dfrac{2x}{z^{2/3}}$

53. $8^{-1/3} = \dfrac{1}{8^{1/3}} = \dfrac{1}{(2^3)^{1/3}} = \dfrac{1}{2^{3/3}} = \dfrac{1}{2}$

55. $(2rs)^{-3/4} = \dfrac{1}{(2rs)^{3/4}}$

57. $\left(\dfrac{1}{16}\right)^{-3/4} = \left(\dfrac{16}{1}\right)^{3/4} = \left(2^4\right)^{3/4} = 2^{4(3/4)} = 2^3 = 8$

59. $\dfrac{8c}{a^{-3/5}} = 8a^{3/5}c$

61. $2a^{3/4}b^{-1/2}c^{2/3} = 2 \cdot a^{3/4} \cdot \dfrac{1}{b^{1/2}} \cdot c^{2/3} = \dfrac{2a^{3/4}c^{2/3}}{b^{1/2}}$

63. $3^{-5/2}a^3b^{-7/3} = \dfrac{1}{3^{5/2}} \cdot a^3 \cdot \dfrac{1}{b^{7/3}} = \dfrac{a^3}{3^{5/2}b^{7/3}}$

65. $\left(\dfrac{2ab}{3c}\right)^{-5/6} = \left(\dfrac{3c}{2ab}\right)^{5/6}$ Finding the reciprocal of the base and changing the sign of the exponent

67. $xy^{-1/4} = \dfrac{x}{y^{1/4}}$

69. $11^{1/2} \cdot 11^{1/3} = 11^{1/2+1/3} = 11^{3/6+2/6} = 11^{5/6}$
We added exponents after finding a common denominator.

71. $\dfrac{3^{5/8}}{3^{-1/8}} = 3^{5/8-(-1/8)} = 3^{5/8+1/8} = 3^{6/8} = 3^{3/4}$
We subtracted exponents and simplified.

73. $\dfrac{4.3^{-1/5}}{4.3^{-7/10}} = 4.3^{-1/5-(-7/10)} = 4.3^{-1/5+7/10}$
$= 4.3^{-2/10+7/10} = 4.3^{5/10} = 4.3^{1/2}$
We subtracted exponents after finding a common denominator. Then we simplified.

75. $\left(10^{3/5}\right)^{2/5} = 10^{3/5 \cdot 2/5} = 10^{6/25}$
We multiplied exponents.

77. $a^{2/3} \cdot a^{5/4} = a^{2/3+5/4} = a^{8/12+15/12} = a^{23/12}$
We added exponents after finding a common denominator.

79. $\left(64^{3/4}\right)^{4/3} = 64^{\frac{3}{4}\cdot\frac{4}{3}} = 64^1 = 64$

81. $\left(m^{2/3}n^{-1/4}\right)^{1/2} = m^{2/3 \cdot 1/2}n^{-1/4 \cdot 1/2} = m^{1/3}n^{-1/8}$
$= m^{1/3} \cdot \dfrac{1}{n^{1/8}} = \dfrac{m^{1/3}}{n^{1/8}}$

83. $\sqrt[9]{x^3} = x^{3/9}$ Converting to exponential notation
$= x^{1/3}$ Simplifying the exponent
$= \sqrt[3]{x}$ Returning to radical notation

85. $\sqrt[3]{y^{15}} = y^{15/3}$ Converting to exponential notation
$= y^5$ Simplifying

87. $\sqrt[12]{a^6} = a^{6/12}$ Converting to exponential notation
$= a^{1/2}$ Simplifying the exponent
$= \sqrt{a}$ Returning to radical notation

89. $\left(\sqrt[7]{xy}\right)^{14} = (xy)^{14/7}$ Converting to exponential notation
$= (xy)^2$ Simplifying the exponent
$= x^2y^2$ Using the laws of exponents

91. $\sqrt[4]{(7a)^2} = (7a)^{2/4}$ Converting to exponential notation
$= (7a)^{1/2}$ Simplifying the exponent
$= \sqrt{7a}$ Returning to radical notation

93. $\sqrt[8]{(2x)^6} = (2x)^{6/8}$ Converting to exponential notation
$= (2x)^{3/4}$ Simplifying the exponent
$= \sqrt[4]{(2x)^3}$ Returning to radical notation
$= \sqrt[4]{8x^3}$ Using the laws of exponents

95. $\sqrt{\sqrt[5]{m}} = \sqrt{m^{1/5}}$ Converting to
$= \left(m^{1/5}\right)^{1/2}$ exponential notation
$= m^{1/10}$ Using the laws of exponents
$= \sqrt[10]{m}$ Returning to radical notation

97. $\sqrt[4]{(xy)^{12}} = (xy)^{12/4}$ Converting to exponential notation
$= (xy)^3$ Simplifying the exponent
$= x^3y^3$ Using the laws of exponents

99. $\left(\sqrt[5]{a^2b^4}\right)^{15}$

$= \left(a^2b^4\right)^{15/5}$ Converting to exponential notation

$= \left(a^2b^4\right)^3$ Simplifying the exponent

$= a^6b^{12}$ Using the laws of exponents

101. $\sqrt[3]{\sqrt[4]{xy}} = \sqrt[3]{(xy)^{1/4}}$ Converting to

$= \left[(xy)^{1/4}\right]^{1/3}$ exponential notation

$= (xy)^{1/12}$ Using the laws of exponents

$= \sqrt[12]{xy}$ Returning to radical notation

103. *Writing Exercise.*

105. $2(t+3)-5 = 1-(6-t)$

$2t+6-5 = 1-6+t$

$2t+1 = -5+t$

$t+1 = -5$

$t = -6$

The solution is –6.

109. $\frac{15}{x} - \frac{15}{x+2} = 2$

To ensure that none of the denominators is 0, we note at the outset that $x \neq 0$ and $x \neq -2$. Then we multiply both sides by the LCD, $x(x+2)$.

$$x(x+2)\left(\frac{15}{x} - \frac{15}{x+2}\right) = x(x+2)\cdot 2$$

$$15(x+2) - 15x = 2x(x+2)$$

$$15x + 30 - 15x = 2x^2 + 4x$$

$$30 = 2x^2 + 4x$$

$$0 = 2x^2 + 4x - 30$$

$$0 = 2(x^2 + 2x - 15)$$

$$0 = 2(x+5)(x-3)$$

$x+5=0$ *or* $x-3=0$

$x=-5$ *or* $x=3$

The numbers –5 and 3 check. The solution is –5 or 3.

111. *Writing Exercise.*

113. $\sqrt{x\sqrt[3]{x^2}} = \sqrt{x\cdot x^{2/3}} = \left(x^{5/3}\right)^{1/2} = x^{5/6} = \sqrt[6]{x^5}$

115. $\sqrt[14]{c^2-2cd+d^2} = \sqrt[14]{(c-d)^2} = \left[(c-d)^2\right]^{1/14}$

$= (c-d)^{2/14} = (c-d)^{1/7}$

$= \sqrt[7]{c-d},\ c \geq d$

117. $2^{7/12} \approx 1.498 \approx 1.5$ so the G that is 7 half steps above middle C has a frequency that is about 1.5 times that of middle C.

119. a. $L = \frac{(0.000169)60^{2.27}}{1} \approx 1.8$ m

b. $L = \frac{(0.000169)75^{2.27}}{0.9906} \approx 3.1$ m

c. $L = \frac{(0.000169)80^{2.27}}{2.4} \approx 1.5$ m

d. $L = \frac{(0.000169)100^{2.27}}{1.1} \approx 5.3$ m

121. $T = 0.936d^{1.97}h^{0.85}$

$= 0.936(3)^{1.97}(80)^{0.85}$

≈ 338 cubic feet

123. $BSA = 0.007184w^{0.425}h^{0.725}$

$= 0.007184(29.5)^{0.425}(122)^{0.725}$

$\approx 0.99\ \text{m}^2$

125. *Graphing Calculator Exercise*

Exercise Set 10.3

1. True

3. False; for instance, for $x = 4$, $\sqrt{x^2-9} = \sqrt{4^2-9} = \sqrt{7}$, but $x-3 = 4-3 = 1$.

5. True

7. $\sqrt{3}\sqrt{10} = \sqrt{3\cdot 10} = \sqrt{30}$

9. $\sqrt[3]{7}\,\sqrt[3]{5} = \sqrt[3]{7\cdot 5} = \sqrt[3]{35}$

11. $\sqrt[4]{6}\,\sqrt[4]{9} = \sqrt[4]{6\cdot 9} = \sqrt[4]{54}$

13. $\sqrt{2x}\sqrt{13y} = \sqrt{2x\cdot 13y} = \sqrt{26xy}$

15. $\sqrt[5]{8y^3}\,\sqrt[5]{10y} = \sqrt[5]{8y^3\cdot 10y} = \sqrt[5]{80y^4}$

17. $\sqrt{y-b}\sqrt{y+b} = \sqrt{(y-b)(y+b)} = \sqrt{y^2-b^2}$

19. $\sqrt[3]{0.7y}\,\sqrt[3]{0.3y} = \sqrt[3]{0.7y\cdot 0.3y} = \sqrt[3]{0.21y^2}$

21. $\sqrt[5]{x-2}\,\sqrt[5]{(x-2)^2} = \sqrt[5]{(x-2)(x-2)^2} = \sqrt[5]{(x-2)^3}$

23. $\sqrt{\frac{2}{t}}\sqrt{\frac{3s}{11}} = \sqrt{\frac{2}{t}\cdot\frac{3s}{11}} = \sqrt{\frac{6s}{11t}}$

25. $\sqrt[7]{\frac{x-3}{4}}\,\sqrt[7]{\frac{5}{x+2}} = \sqrt[7]{\frac{x-3}{4}\cdot\frac{5}{x+2}} = \sqrt[7]{\frac{5x-15}{4x+8}}$

27. $\sqrt{12}$

$= \sqrt{4\cdot 3}$ 4 is the largest perfect square factor of 12

$= \sqrt{4}\cdot\sqrt{3}$

$= 2\sqrt{3}$

29. $\sqrt{45}$

$= \sqrt{9\cdot 5}$ 9 is the largest perfect square factor of 45

$= \sqrt{9}\cdot\sqrt{5}$

$= 3\sqrt{5}$

31. $\sqrt{8x^9}$
$= \sqrt{4x^8 \cdot 2x}$ $4x^8$ is a perfect square
$= \sqrt{4x^8} \cdot \sqrt{2x}$ Factoring into two radicals
$= 2x^4\sqrt{2x}$ Taking the square root of $4x^8$

33. $\sqrt{120} = \sqrt{4 \cdot 30} = \sqrt{4} \cdot \sqrt{30} = 2\sqrt{30}$

35. $\sqrt{36a^4b}$
$= \sqrt{36a^4 \cdot b}$ $36a^4$ is a perfect square
$= \sqrt{36a^4} \cdot \sqrt{b}$ Factoring into two radicals
$= 6a^2\sqrt{b}$ Taking the square root of $36a^4$

37. $\sqrt[3]{8x^3y^2}$
$= \sqrt[3]{8x^3 \cdot y^2}$ $8x^3$ is a perfect cube
$= \sqrt[3]{8x^3} \cdot \sqrt[3]{y^2}$ Factoring into two radicals
$= 2x\sqrt[3]{y^2}$ Taking the cube root of $8x^3$

39. $\sqrt[3]{-16x^6}$
$= \sqrt[3]{-8x^6 \cdot 2}$ $-8x^6$ is a perfect cube
$= \sqrt[3]{-8x^6} \cdot \sqrt[3]{2}$ Factoring into two radicals
$= -2x^2\sqrt[3]{2}$ Taking the cube root of $-8x^6$

41. $f(x) = \sqrt[3]{40x^6}$
$= \sqrt[3]{8x^6 \cdot 5}$
$= \sqrt[3]{8x^6} \cdot \sqrt[3]{5}$
$= 2x^2\sqrt[3]{5}$

43. $f(x) = \sqrt{49(x-3)^2}$ $49(x-3)^2$ is a perfect square.
$= |7(x-3)|$, or $7|x-3|$

45. $f(x) = \sqrt{5x^2 - 10x + 5}$
$= \sqrt{5(x^2 - 2x + 1)}$
$= \sqrt{5(x-1)^2}$
$= \sqrt{(x-1)^2} \cdot \sqrt{5}$
$= |x-1|\sqrt{5}$

47. $\sqrt{a^{10}b^{11}}$
$= \sqrt{a^{10} \cdot b^{10} \cdot b}$ Identifying the largest even powers of a and b
$= \sqrt{a^{10}}\sqrt{b^{10}}\sqrt{b}$ Factoring into several radicals
$= a^5b^5\sqrt{b}$

49. $\sqrt[3]{x^5y^6z^{10}}$
$= \sqrt[3]{x^3 \cdot x^2 \cdot y^6 \cdot z^9 \cdot z}$ Identifying the largest perfect-cube powers of x, y and z
$= \sqrt[3]{x^3} \cdot \sqrt[3]{y^6} \cdot \sqrt[3]{z^9} \cdot \sqrt[3]{x^2z}$ Factoring into several radicals
$= xy^2z^3\sqrt[3]{x^2z}$

51. $\sqrt[4]{16x^5y^{11}} = \sqrt[4]{2^4 \cdot x^4 \cdot x \cdot y^8 \cdot y^3}$
$= \sqrt[4]{2^4} \cdot \sqrt[4]{x^4} \cdot \sqrt[4]{y^8} \cdot \sqrt[4]{xy^3}$
$= 2xy^2\sqrt[4]{xy^3}$

53. $\sqrt[5]{x^{13}y^8z^{17}} = \sqrt[5]{x^{10} \cdot x^3 \cdot y^5 \cdot y^3 \cdot z^{15} \cdot z^2}$
$= \sqrt[5]{x^{10}} \cdot \sqrt[5]{y^5} \cdot \sqrt[5]{z^{15}} \cdot \sqrt[5]{x^3y^3z^2}$
$= x^2yz^3\sqrt[5]{x^3y^3z^2}$

55. $\sqrt[3]{-80a^{14}} = \sqrt[3]{-8 \cdot 10 \cdot a^{12} \cdot a^2}$
$= \sqrt[3]{-8} \cdot \sqrt[3]{a^{12}} \cdot \sqrt[3]{10a^2}$
$= -2a^4\sqrt[3]{10a^2}$

57. $\sqrt{5}\sqrt{10} = \sqrt{5 \cdot 10} = \sqrt{50} = \sqrt{25 \cdot 2} = 5\sqrt{2}$

59. $\sqrt{6}\sqrt{33} = \sqrt{6 \cdot 33} = \sqrt{198} = \sqrt{9 \cdot 22} = 3\sqrt{22}$

61. $\sqrt[3]{9}\sqrt[3]{3} = \sqrt[3]{9 \cdot 3} = \sqrt[3]{27} = 3$

63. $\sqrt{24y^5}\sqrt{24y^5} = \sqrt{(24y^5)^2} = 24y^5$

65. $\sqrt[3]{5a^2}\sqrt[3]{2a} = \sqrt[3]{5a^2 \cdot 2a} = \sqrt[3]{10a^3} = \sqrt[3]{a^3 \cdot 10} = a\sqrt[3]{10}$

67. $3\sqrt{2x^5} \cdot 4\sqrt{10x^2} = 12\sqrt{20x^7} = 12\sqrt{4x^6 \cdot 5x} = 24x^3\sqrt{5x}$

69. $\sqrt[3]{s^2t^4}\sqrt[3]{s^4t^6} = \sqrt[3]{s^6t^{10}} = \sqrt[3]{s^6t^9 \cdot t} = s^2t^3\sqrt[3]{t}$

71. $\sqrt[3]{(x-y)^2}\sqrt[3]{(x-y)^{10}} = \sqrt[3]{(x-y)^{12}} = (x-y)^4$

73. $\sqrt[4]{20a^3b^7}\sqrt[4]{4a^2b^5} = \sqrt[4]{80a^5b^{12}} = \sqrt[4]{16a^4b^{12} \cdot 5a}$
$= 2ab^3\sqrt[4]{5a}$

75. $\sqrt[5]{x^3(y+z)^6}\sqrt[5]{x^3(y+z)^4} = \sqrt[5]{x^6(y+z)^{10}}$
$= \sqrt[5]{x^5(y+z)^{10} \cdot x} = x(y+z)^2\sqrt[5]{x}$

77. *Writing Exercise.*

79. $\frac{15a^2x}{8b} \cdot \frac{24b^2x}{5a} = \frac{(\cancel{5a} \cdot 3ax)(\cancel{8b} \cdot 3bx)}{\cancel{8b} \cdot \cancel{5a}} = 9abx^2$

81. $\frac{x-3}{2x-10} - \frac{3x-5}{x^2-25} = \frac{x-3}{2(x-5)} - \frac{3x-5}{(x+5)(x-5)}$
$= \frac{x-3}{2(x-5)} \cdot \frac{x+5}{x+5} - \frac{3x-5}{(x+5)(x-5)} \cdot \frac{2}{2}$
$= \frac{x^2+2x-15-(6x-10)}{2(x+5)(x-5)}$
$= \frac{x^2-4x-5}{2(x+5)(x-5)}$
$= \frac{(x-5)(x+1)}{2(x+5)(x-5)}$
$= \frac{x+1}{2(x+5)}$

83. $\frac{a^{-1}+b^{-1}}{ab} = \frac{\frac{1}{a}+\frac{1}{b}}{ab} = \frac{\frac{1}{a}+\frac{1}{b}}{ab} \cdot \frac{ab}{ab} = \frac{b+a}{a^2b^2}$

85. *Writing Exercise.*

87. $R(x)=\frac{1}{2}\sqrt[4]{\frac{x\cdot 3.0\times 10^6}{\pi^2}}$

$R(5\times 10^4)=\frac{1}{2}\sqrt[4]{\frac{5\times 10^4\cdot 3.0\times 10^6}{\pi^2}}$

$=\frac{1}{2}\sqrt[4]{\frac{15\times 10^{10}}{\pi^2}}$

≈ 175.6 mi

89. a. $T_w=33-\frac{(10.45+10\sqrt{8}-8)(33-7)}{22}$

≈ -3.3 °C

b. $T_w=33-\frac{(10.45+10\sqrt{12}-12)(33-0)}{22}$

≈ -16.6 °C

c. $T_w=33-\frac{(10.45+10\sqrt{14}-14)(33-(-5))}{22}$

≈ -25.5 °C

d. $T_w=33-\frac{(10.45+10\sqrt{15}-15)(33-(-23))}{22}$

≈ -54.0 °C

91. $\left(\sqrt[3]{25x^4}\right)^4=\sqrt[3]{(25x^4)^4}=\sqrt[3]{25^4x^{16}}$

$=\sqrt[3]{25^3\cdot 25\cdot x^{15}\cdot x}=\sqrt[3]{25^3}\sqrt[3]{x^{15}}\sqrt[3]{25x}$

$=25x^5\sqrt[3]{25x}$

93. $\left(\sqrt{a^3b^5}\right)^7=\sqrt{(a^3b^5)^7}=\sqrt{a^{21}b^{35}}$

$=\sqrt{a^{20}\cdot a\cdot b^{34}\cdot b}=\sqrt{a^{20}}\sqrt{b^{34}}\sqrt{ab}=a^{10}b^{17}\sqrt{ab}$

95.

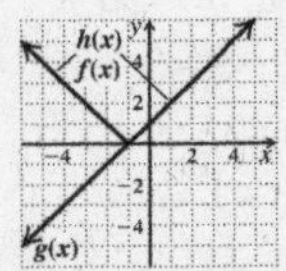

We see that $f(x)=h(x)$ and $f(x)\neq g(x)$.

97. $g(x)=x^2-6x+8$

We must have $x^2-6x+8\geq 0$, or $(x-2)(x-4)\geq 0$.

We graph $y=x^2-6x+8$.

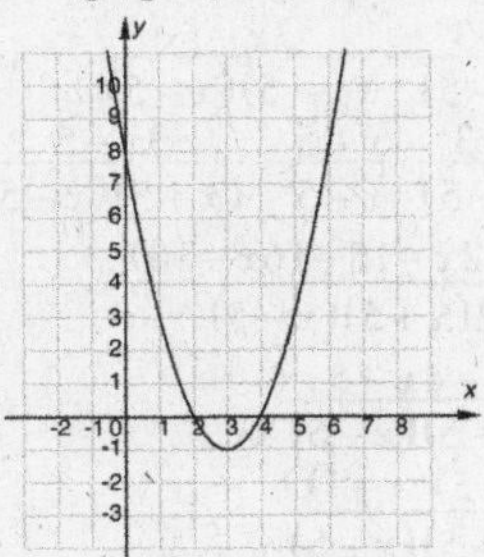

From the graph we see that $y\geq 0$ for $x\leq 2$ or $x\geq 4$, so the domain of g is $\{x\mid x\leq 2 \text{ or } x\geq 4\}$, or $(-\infty, 2]\cup[4, \infty)$.

99. $\sqrt[5]{4a^{3k+2}}\,\sqrt[5]{8a^{6-k}}=2a^4$

$\sqrt[5]{32a^{2k+8}}=2a^4$

$2\sqrt[5]{a^{2k+8}}=2a^4$

$\sqrt[5]{a^{2k+8}}=a^4$

$a^{\frac{2k+8}{5}}=a^4$

Since the base is the same, the exponents must be equal. We have:

$\frac{2k+8}{5}=4$

$2k+8=20$

$2k=12$

$k=6$

101. *Writing Exercise.*

Exercise Set 10.4

1. Quotient rule for radicals

3. Multiplying by 1

5. $\sqrt[4]{\frac{16a^6}{a^2}}=\sqrt[4]{16a^4}=2a$, so choice (f) is correct.

7. $\sqrt[5]{\frac{a^6}{b^4}}=\sqrt[5]{\frac{a^6}{b^4}\cdot\frac{b}{b}}=\sqrt[5]{\frac{a^6b}{b^4\cdot b}}$, so choice (e) is correct.

9. $\frac{\sqrt{5a^4}}{\sqrt{5a^3}}=\sqrt{\frac{5a^4}{5a^3}}=\sqrt{a}$, so choice (c) is correct.

11. $\sqrt{\frac{49}{100}}=\frac{\sqrt{49}}{\sqrt{100}}=\frac{7}{10}$

13. $\sqrt[3]{\frac{125}{8}}=\frac{\sqrt[3]{125}}{\sqrt[3]{8}}=\frac{5}{2}$

15. $\sqrt{\frac{121}{t^2}}=\frac{\sqrt{121}}{\sqrt{t^2}}=\frac{11}{t}$

17. $\sqrt{\frac{36y^3}{x^4}}=\frac{\sqrt{36y^3}}{\sqrt{x^4}}=\frac{\sqrt{36y^2\cdot y}}{\sqrt{x^4}}=\frac{\sqrt{36y^2}\sqrt{y}}{\sqrt{x^4}}=\frac{6y\sqrt{y}}{x^2}$

19. $\sqrt[3]{\frac{27a^4}{8b^3}}=\frac{\sqrt[3]{27a^4}}{\sqrt[3]{8b^3}}=\frac{\sqrt[3]{27a^3\cdot a}}{\sqrt[3]{8b^3}}=\frac{\sqrt[3]{27a^3}\sqrt[3]{a}}{\sqrt[3]{8b^3}}=\frac{3a\sqrt[3]{a}}{2b}$

21. $\sqrt[4]{\frac{32a^4}{2b^4c^8}}=\sqrt[4]{\frac{16a^4}{b^4c^8}}=\frac{\sqrt[4]{16a^4}}{\sqrt[4]{b^4c^8}}=\frac{2a}{bc^2}$

23. $\sqrt[4]{\frac{a^5b^8}{c^{10}}}=\frac{\sqrt[4]{a^5b^8}}{\sqrt[4]{c^{10}}}=\frac{\sqrt[4]{a^4b^8\cdot a}}{\sqrt[4]{c^8\cdot c^2}}=\frac{\sqrt[4]{a^4b^8}\sqrt[4]{a}}{\sqrt[4]{c^8}\sqrt[4]{c^2}}=\frac{ab^2\sqrt[4]{a}}{c^2\sqrt[4]{c^2}}$,

or $\frac{ab^2}{c^2}\sqrt[4]{\frac{a}{c^2}}$

25. $\sqrt[5]{\frac{32x^6}{y^{11}}} = \frac{\sqrt[5]{32x^6}}{\sqrt[5]{y^{11}}} = \frac{\sqrt[5]{32x^5 \cdot x}}{\sqrt[5]{y^{10} \cdot y}}$
$= \frac{\sqrt[5]{32x^5} \cdot \sqrt[5]{x}}{\sqrt[5]{y^{10}} \sqrt[5]{y}} = \frac{2x\sqrt[5]{x}}{y^2\sqrt[5]{y}}$, or $\frac{2x}{y^2}\sqrt[5]{\frac{x}{y}}$

27. $\sqrt[6]{\frac{x^6y^8}{z^{15}}} = \frac{\sqrt[6]{x^6y^8}}{\sqrt[6]{z^{15}}} = \frac{\sqrt[6]{x^6y^6 \cdot y^2}}{\sqrt[6]{z^{12} \cdot z^3}} = \frac{\sqrt[6]{x^6y^6}\sqrt[6]{y^2}}{\sqrt[6]{z^{12}}\sqrt[6]{z^3}}$,
$= \frac{xy\sqrt[6]{y^2}}{z^2\sqrt[6]{z^3}}$, or $\frac{xy}{z^2}\sqrt[6]{\frac{y^2}{z^3}}$

29. $\frac{\sqrt{18y}}{\sqrt{2y}} = \sqrt{\frac{18y}{2y}} = \sqrt{9} = 3$

31. $\frac{\sqrt[3]{26}}{\sqrt[3]{13}} = \sqrt[3]{\frac{26}{13}} = \sqrt[3]{2}$

33. $\frac{\sqrt{40xy^3}}{\sqrt{8x}} = \sqrt{\frac{40xy^3}{8x}} = \sqrt{5y^3} = \sqrt{y^2 \cdot 5y}$
$= \sqrt{y^2}\sqrt{5y} = y\sqrt{5y}$

35. $\frac{\sqrt[3]{96a^4b^2}}{\sqrt[3]{12a^2b}} = \sqrt[3]{\frac{96a^4b^2}{12a^2b}} = \sqrt[3]{8a^2b} = \sqrt[3]{8}\sqrt[3]{a^2b} = 2\sqrt[3]{a^2b}$

37. $\frac{\sqrt{100ab}}{5\sqrt{2}} = \frac{1}{5}\frac{\sqrt{100ab}}{\sqrt{2}} = \frac{1}{5}\sqrt{\frac{100ab}{2}} = \frac{1}{5}\sqrt{50ab}$
$= \frac{1}{5}\sqrt{25 \cdot 2ab} = \frac{1}{5} \cdot 5\sqrt{2ab} = \sqrt{2ab}$

39. $\frac{\sqrt[4]{48x^9y^{13}}}{\sqrt[4]{3xy^{-2}}} = \sqrt[4]{\frac{48x^9y^{13}}{3xy^{-2}}} = \sqrt[4]{16x^8y^{15}} = \sqrt[4]{16x^8y^{12}}\sqrt[4]{y^3}$
$= 2x^2y^3\sqrt[4]{y^3}$

41. $\frac{\sqrt[3]{x^3-y^3}}{\sqrt[3]{x-y}} = \sqrt[3]{\frac{x^3-y^3}{x-y}} = \sqrt[3]{\frac{(x-y)(x^2+xy+y^2)}{x-y}}$
$= \sqrt[3]{x^2+xy+y^2}$

43. $\sqrt{\frac{2}{5}} = \sqrt{\frac{2}{5} \cdot \frac{5}{5}} = \sqrt{\frac{10}{25}} = \frac{\sqrt{10}}{\sqrt{25}} = \frac{\sqrt{10}}{5}$

45. $\frac{2\sqrt{5}}{7\sqrt{3}} = \frac{2\sqrt{5}}{7\sqrt{3}} \cdot \frac{\sqrt{3}}{\sqrt{3}} = \frac{2\sqrt{15}}{21}$

47. $\sqrt[3]{\frac{5}{4}} = \sqrt[3]{\frac{5}{4} \cdot \frac{2}{2}} = \sqrt[3]{\frac{10}{8}} = \frac{\sqrt[3]{10}}{\sqrt[3]{8}} = \frac{\sqrt[3]{10}}{2}$

49. $\frac{\sqrt[3]{3a}}{\sqrt[3]{5c}} = \frac{\sqrt[3]{3a}}{\sqrt[3]{5c}} \cdot \frac{\sqrt[3]{5^2c^2}}{\sqrt[3]{5^2c^2}} = \frac{\sqrt[3]{75ac^2}}{\sqrt[3]{5^3c^3}} = \frac{\sqrt[3]{75ac^2}}{5c}$

51. $\frac{\sqrt[4]{5y^6}}{\sqrt[4]{9x}} = \frac{\sqrt[4]{5y^6}}{\sqrt[4]{9x}} \cdot \frac{\sqrt[4]{9x^3}}{\sqrt[4]{9x^3}} = \frac{\sqrt[4]{5y^6 \cdot 9x^3}}{\sqrt[4]{81x^4}} = \frac{y\sqrt[4]{45x^3y^2}}{3x}$

53. $\sqrt[3]{\frac{2}{x^2y}} = \sqrt[3]{\frac{2}{x^2y} \cdot \frac{xy^2}{xy^2}} = \sqrt[3]{\frac{2xy^2}{x^3y^3}} = \frac{\sqrt[3]{2xy^2}}{\sqrt[3]{x^3y^3}} = \frac{\sqrt[3]{2xy^2}}{xy}$

55. $\sqrt{\frac{7a}{18}} = \sqrt{\frac{7a}{18} \cdot \frac{2}{2}} = \sqrt{\frac{14a}{36}} = \frac{\sqrt{14a}}{\sqrt{36}} = \frac{\sqrt{14a}}{6}$

57. $\sqrt[5]{\frac{9}{32x^5y}} = \sqrt[5]{\frac{9}{32x^5y} \cdot \frac{y^4}{y^4}} = \frac{\sqrt[5]{9y^4}}{\sqrt[5]{32x^5y^5}} = \frac{\sqrt[5]{9y^4}}{2xy}$

59. $\sqrt{\frac{10ab^2}{72a^3b}} = \sqrt{\frac{5b}{36a^2}} = \frac{\sqrt{5b}}{6a}$

61. $\sqrt{\frac{5}{11}} = \sqrt{\frac{5}{11} \cdot \frac{5}{5}} = \sqrt{\frac{25}{55}} = \frac{\sqrt{25}}{\sqrt{55}} = \frac{5}{\sqrt{55}}$

63. $\frac{2\sqrt{6}}{5\sqrt{7}} = \frac{2\sqrt{6}}{5\sqrt{7}} \cdot \frac{\sqrt{6}}{\sqrt{6}} = \frac{2\sqrt{36}}{5\sqrt{42}} = \frac{2 \cdot 6}{5\sqrt{42}} = \frac{12}{5\sqrt{42}}$

65. $\frac{\sqrt{8}}{2\sqrt{3x}} = \frac{\sqrt{8}}{2\sqrt{3x}} \cdot \frac{\sqrt{2}}{\sqrt{2}} = \frac{\sqrt{16}}{2\sqrt{6x}} = \frac{4}{2\sqrt{6x}}$
$= \frac{2}{\sqrt{6x}}$

67. $\frac{\sqrt[3]{7}}{\sqrt[3]{2}} = \frac{\sqrt[3]{7}}{\sqrt[3]{2}} \cdot \frac{\sqrt[3]{7^2}}{\sqrt[3]{7^2}} = \frac{\sqrt[3]{7^3}}{\sqrt[3]{98}} = \frac{7}{\sqrt[3]{98}}$

69. $\sqrt{\frac{7x}{3y}} = \sqrt{\frac{7x}{3y} \cdot \frac{7x}{7x}} = \sqrt{\frac{(7x)^2}{21xy}} = \frac{7x}{\sqrt{21xy}}$

71. $\sqrt[3]{\frac{2a^5}{5b}} = \sqrt[3]{\frac{2a^5}{5b} \cdot \frac{4a}{4a}} = \sqrt[3]{\frac{8a^6}{20ab}} = \frac{2a^2}{\sqrt[3]{20ab}}$

73. $\sqrt{\frac{x^3y}{2}} = \sqrt{\frac{x^3y}{2} \cdot \frac{xy}{xy}} = \sqrt{\frac{x^4y^2}{2xy}} = \frac{\sqrt{x^4y^2}}{\sqrt{2xy}} = \frac{x^2y}{\sqrt{2xy}}$

75. *Writing Exercise.*

77. $-\frac{2}{9} \div \frac{4}{6} = -\frac{2}{9} \cdot \frac{6}{4} = -\frac{12}{36} = -\frac{1}{3}$

79. $12 - 100 \div 5 \cdot (-2)^2 - 3(6-7)$
$= 12 - 100 \div 5 \cdot 4 - 3(-1)$
$= 12 - 20 \cdot 4 + 3$
$= 12 - 80 + 3$
$= -65$

81. $(12x^3 - 6x - 8) \div (x+1)$
$= (12x^3 + 0x^2 - 6x - 8) \div (x+1)$

$$\begin{array}{r|rrrr} -1 & 12 & 0 & -6 & -8 \\ & & -12 & 12 & -6 \\ \hline & 12 & -12 & 6 & -14 \end{array}$$

$(12x^3 - 6x - 8) \div (x+1) = 12x^2 - 12x + 6 + \frac{-14}{x+1}$

83. *Writing Exercise.*

85. a. $T = 2\pi\sqrt{\frac{65}{980}} \approx 1.62$ sec

b. $T = 2\pi\sqrt{\frac{98}{980}} \approx 1.99$ sec

c. $T = 2\pi\sqrt{\frac{120}{980}} \approx 2.20$ sec

87.
$$\begin{aligned}\frac{\left(\sqrt[3]{81mn^2}\right)^2}{\left(\sqrt[3]{mn}\right)^2} &= \frac{\sqrt[3]{(81mn^2)^2}}{\sqrt[3]{(mn)^2}}\\ &= \frac{\sqrt[3]{6561m^2n^4}}{\sqrt[3]{m^2n^2}}\\ &= \sqrt[3]{\frac{6561m^2n^4}{m^2n^2}}\\ &= \sqrt[3]{6561n^2}\\ &= \sqrt[3]{729\cdot 9n^2}\\ &= \sqrt[3]{729}\sqrt[3]{9n^2}\\ &= 9\sqrt[3]{9n^2}\end{aligned}$$

89.
$$\begin{aligned}\sqrt{a^2-3}-\frac{a^2}{\sqrt{a^2-3}} &= \sqrt{a^2-3}-\frac{a^2}{\sqrt{a^2-3}}\cdot\frac{\sqrt{a^2-3}}{\sqrt{a^2-3}}\\ &= \sqrt{a^2-3}-\frac{a^2\sqrt{a^2-3}}{a^2-3}\\ &= \sqrt{a^2-3}\cdot\frac{a^2-3}{a^2-3}-\frac{a^2\sqrt{a^2-3}}{a^2-3}\\ &= \frac{a^2\sqrt{a^2-3}-3\sqrt{a^2-3}-a^2\sqrt{a^2-3}}{a^2-3}\\ &= \frac{-3\sqrt{a^2-3}}{a^2-3}, \text{ or } \frac{-3}{\sqrt{a^2-3}}\end{aligned}$$

91. Step 1: $\sqrt[n]{a} = a^{1/n}$, by definition;

Step 2: $\left(\frac{a}{b}\right)^n = \frac{a^n}{b^n}$, raising a quotient to a power;

Step 3: $a^{1/n} = \sqrt[n]{a}$, by definition

93. $f(x) = \sqrt{18x^3}$, $g(x) = \sqrt{2x}$

$$(f/g)(x) = \frac{f(x)}{g(x)} = \frac{\sqrt{18x^3}}{\sqrt{2x}} = \sqrt{\frac{18x^3}{2x}} = \sqrt{9x^2} = 3x$$

$\sqrt{2x}$ is defined for $2x \geq 0$, or $x \geq 0$. To avoid division by 0, we must exclude 0 from the domain. Thus, the domain of
$f/g = \{x|x \text{ is a real number and } x > 0\}$, or $(0, \infty)$.

95. $f(x) = \sqrt{x^2-9}$, $g(x) = \sqrt{x-3}$

$$\begin{aligned}(f/g)(x) &= \frac{f(x)}{g(x)} = \frac{\sqrt{x^2-9}}{\sqrt{x-3}} = \sqrt{\frac{x^2-9}{x-3}}\\ &= \sqrt{\frac{(x+3)(x-3)}{x-3}} = \sqrt{x+3}\end{aligned}$$

$\sqrt{x-3}$ is defined for $x-3 \geq 0$, or $x \geq 3$. To avoid division by 0, we must exclude 3 from the domain. Thus, the domain of
$f/g = \{x|x \text{ is a real number and } x > 3\}$, or $(3, \infty)$.

Connecting the Concepts

1. $\frac{6}{\sqrt{7}} = \frac{6}{\sqrt{7}}\cdot\frac{\sqrt{7}}{\sqrt{7}} = \frac{6\sqrt{7}}{7}$

2. $\frac{1}{3-\sqrt{2}} = \frac{1}{3-\sqrt{2}}\cdot\frac{3+\sqrt{2}}{3+\sqrt{2}} = \frac{\sqrt{3}+2}{7}$

3. $\frac{2}{\sqrt{xy}} = \frac{2}{\sqrt{xy}}\cdot\frac{\sqrt{xy}}{\sqrt{xy}} = \frac{2\sqrt{xy}}{xy}$

4. $\frac{5}{\sqrt{8}} = \frac{5}{\sqrt{4\cdot 2}} = \frac{5}{2\sqrt{2}}\cdot\frac{\sqrt{2}}{\sqrt{2}} = \frac{5\sqrt{2}}{4}$

5. $\frac{\sqrt{2}}{\sqrt{5}+\sqrt{3}} = \frac{\sqrt{2}}{\sqrt{5}+\sqrt{3}}\cdot\frac{\sqrt{5}-\sqrt{3}}{\sqrt{5}-\sqrt{3}} = \frac{\sqrt{10}-\sqrt{6}}{2}$

6. $\frac{2}{1-\sqrt{5}} = \frac{2}{1-\sqrt{5}}\cdot\frac{1+\sqrt{5}}{1+\sqrt{5}} = \frac{2(1+\sqrt{5})}{-4} = \frac{-1-\sqrt{5}}{2}$

7. $\frac{1}{\sqrt[3]{x^2y}} = \frac{1}{\sqrt[3]{x^2y}}\cdot\frac{\sqrt[3]{xy^2}}{\sqrt[3]{xy^2}} = \frac{\sqrt[3]{xy^2}}{xy}$

8. $\frac{a}{\sqrt[4]{a^3b^2}} = \frac{a}{\sqrt[4]{a^3b^2}}\cdot\frac{\sqrt[4]{ab^2}}{\sqrt[4]{ab^2}} = \frac{a\sqrt[4]{ab^2}}{ab} = \frac{\sqrt[4]{ab^2}}{b}$

Exercise Set 10.5

1. To add radical expressions, both the *radicands* and the *indices* must be the same.

3. To find a product by adding exponents, the *bases* must be the same.

5. To rationalize the *numerator* of $\frac{\sqrt{c}-\sqrt{a}}{5}$, we multiply by a form of 1, using the *conjugate* of $\sqrt{c}-\sqrt{a}$, or $\sqrt{c}+\sqrt{a}$, to write 1.

7. $4\sqrt{3}+7\sqrt{3} = (4+7)\sqrt{3} = 11\sqrt{3}$

9. $7\sqrt[3]{4}-5\sqrt[3]{4} = (7-5)\sqrt[3]{4} = 2\sqrt[3]{4}$

11. $\sqrt[3]{y}+9\sqrt[3]{y} = (1+9)\sqrt[3]{y} = 10\sqrt[3]{y}$

13. $8\sqrt{2}-\sqrt{2}+5\sqrt{2} = (8-1+5)\sqrt{2} = 12\sqrt{2}$

15. $9\sqrt[3]{7}-\sqrt{3}+4\sqrt[3]{7}+2\sqrt{3}$
$= (9+4)\sqrt[3]{7}+(-1+2)\sqrt{3} = 13\sqrt[3]{7}+\sqrt{3}$

17. $4\sqrt{27}-3\sqrt{3}$
$= 4\sqrt{9\cdot 3}-3\sqrt{3}$ Factoring the
$= 4\sqrt{9}\cdot\sqrt{3}-3\sqrt{3}$ first radical
$= 4\cdot 3\sqrt{3}-3\sqrt{3}$ Taking the square root of 9
$= 12\sqrt{3}-3\sqrt{3}$
$= 9\sqrt{3}$ Combining the radicals

19. $3\sqrt{45}-8\sqrt{20}$
$=3\sqrt{9\cdot 5}-8\sqrt{4\cdot 5}$ Factoring the
$=3\sqrt{9}\cdot\sqrt{5}-8\sqrt{4}\cdot\sqrt{5}$ radicals
$=3\cdot 3\sqrt{5}-8\cdot 2\sqrt{5}$ Taking the square roots
$=9\sqrt{5}-16\sqrt{5}$
$=-7\sqrt{5}$ Combining like radicals

21. $3\sqrt[3]{16}+\sqrt[3]{54}=3\sqrt[3]{8\cdot 2}+\sqrt[3]{27\cdot 2}$
$=3\sqrt[3]{8}\cdot\sqrt[3]{2}+\sqrt[3]{27}\cdot\sqrt[3]{2}=3\cdot 2\sqrt[3]{2}+3\sqrt[3]{2}$
$=6\sqrt[3]{2}+3\sqrt[3]{2}=9\sqrt[3]{2}$

23. $\sqrt{a}+3\sqrt{16a^3}=\sqrt{a}+3\sqrt{16a^2\cdot a}=\sqrt{a}+3\sqrt{16a^2}\cdot\sqrt{a}$
$=\sqrt{a}+3\cdot 4a\sqrt{a}=\sqrt{a}+12a\sqrt{a}$
$=(1+12a)\sqrt{a}$

25. $\sqrt[3]{6x^4}-\sqrt[3]{48x}=\sqrt[3]{x^3\cdot 6x}-\sqrt[3]{8\cdot 6x}$
$=\sqrt[3]{x^3}\cdot\sqrt[3]{6x}-\sqrt[3]{8}\cdot\sqrt[3]{6x}=x\sqrt[3]{6x}-2\sqrt[3]{6x}$
$=(x-2)\sqrt[3]{6x}$

27. $\sqrt{4a-4}+\sqrt{a-1}=\sqrt{4(a-1)}+\sqrt{a-1}$
$=\sqrt{4}\sqrt{a-1}+\sqrt{a-1}=2\sqrt{a-1}+\sqrt{a-1}=3\sqrt{a-1}$

29. $\sqrt{x^3-x^2}+\sqrt{9x-9}=\sqrt{x^2(x-1)}+\sqrt{9(x-1)}$
$=\sqrt{x^2}\cdot\sqrt{x-1}+\sqrt{9}\cdot\sqrt{x-1}$
$=x\sqrt{x-1}+3\sqrt{x-1}=(x+3)\sqrt{x-1}$

31. $\sqrt{2}(5+\sqrt{2})=\sqrt{2}\cdot 5+\sqrt{2}\cdot\sqrt{2}=5\sqrt{2}+2$

33. $3\sqrt{5}(\sqrt{6}-\sqrt{7})=3\sqrt{5}\cdot\sqrt{6}-3\sqrt{5}\cdot\sqrt{7}=3\sqrt{30}-3\sqrt{35}$

35. $\sqrt{2}(3\sqrt{10}-\sqrt{8})=\sqrt{2}\cdot 3\sqrt{10}-\sqrt{2}\cdot\sqrt{8}=3\sqrt{20}-\sqrt{16}$
$=3\sqrt{4\cdot 5}-\sqrt{16}=3\cdot 2\sqrt{5}-4$
$=6\sqrt{5}-4$

37. $\sqrt[3]{3}(\sqrt[3]{9}-4\sqrt[3]{21})=\sqrt[3]{3}\cdot\sqrt[3]{9}-\sqrt[3]{3}\cdot 4\sqrt[3]{21}$
$=\sqrt[3]{27}-4\sqrt[3]{63}$
$=3-4\sqrt[3]{63}$

39. $\sqrt[3]{a}(\sqrt[3]{a^2}+\sqrt[3]{24a^2})=\sqrt[3]{a}\cdot\sqrt[3]{a^2}+\sqrt[3]{a}\sqrt[3]{24a^2}$
$=\sqrt[3]{a^3}+\sqrt[3]{24a^3}$
$=\sqrt[3]{a^3}+\sqrt[3]{8a^3\cdot 3}$
$=a+2a\sqrt[3]{3}$

41. $(2+\sqrt{6})(5-\sqrt{6})=2\cdot 5-2\sqrt{6}+5\sqrt{6}-\sqrt{6}\cdot\sqrt{6}$
$=10+3\sqrt{6}-6=4+3\sqrt{6}$

43. $(\sqrt{2}+\sqrt{7})(\sqrt{3}-\sqrt{7})$
$=\sqrt{2}\cdot\sqrt{3}-\sqrt{2}\cdot\sqrt{7}+\sqrt{7}\cdot\sqrt{3}-\sqrt{7}\cdot\sqrt{7}$
$=\sqrt{6}-\sqrt{14}+\sqrt{21}-7$

45. $(2-\sqrt{3})(2+\sqrt{3})=2^2-(\sqrt{3})^2=4-3=1$

47. $(\sqrt{10}-\sqrt{15})(\sqrt{10}+\sqrt{15})=(\sqrt{10})^2-(\sqrt{15})^2$
$=10-15=-5$

49. $(3\sqrt{7}+2\sqrt{5})(2\sqrt{7}-4\sqrt{5})$
$=3\sqrt{7}\cdot 2\sqrt{7}-3\sqrt{7}\cdot 4\sqrt{5}+2\sqrt{5}\cdot 2\sqrt{7}-2\sqrt{5}\cdot 4\sqrt{5}$
$=6\cdot 7-12\sqrt{35}+4\sqrt{35}-8\cdot 5=42-8\sqrt{35}-40$
$=2-8\sqrt{35}$

51. $(4+\sqrt{7})^2=4^2+2\cdot 4\cdot\sqrt{7}+(\sqrt{7})^2=16+8\sqrt{7}+7$
$=23+8\sqrt{7}$

53. $(\sqrt{3}-\sqrt{2})^2=(\sqrt{3})^2-2\cdot\sqrt{3}\cdot\sqrt{2}+(\sqrt{2})^2$
$=3-2\sqrt{6}+2=5-2\sqrt{6}$

55. $(\sqrt{2t}+\sqrt{5})^2=(\sqrt{2t})^2+2\cdot\sqrt{2t}\cdot\sqrt{5}+(\sqrt{5})^2$
$=2t+2\sqrt{10t}+5$

57. $(3-\sqrt{x+5})^2=3^2-2\cdot 3\cdot\sqrt{x+5}+(\sqrt{x+5})^2$
$=9-6\sqrt{x+5}+x+5$
$=14-6\sqrt{x+5}+x$

59. $(2\sqrt[4]{7}-\sqrt[4]{6})(3\sqrt[4]{9}+2\sqrt[4]{5})$
$=2\sqrt[4]{7}\cdot 3\sqrt[4]{9}+2\sqrt[4]{7}\cdot 2\sqrt[4]{5}-\sqrt[4]{6}\cdot 3\sqrt[4]{9}-\sqrt[4]{6}\cdot 2\sqrt[4]{5}$
$=6\sqrt[4]{63}+4\sqrt[4]{35}-3\sqrt[4]{54}-2\sqrt[4]{30}$

61. $\dfrac{6}{3-\sqrt{2}}=\dfrac{6}{3-\sqrt{2}}\cdot\dfrac{3+\sqrt{2}}{3+\sqrt{2}}=\dfrac{6(3+\sqrt{2})}{(3-\sqrt{2})(3+\sqrt{2})}$
$=\dfrac{18+6\sqrt{2}}{3^2-(\sqrt{2})^2}=\dfrac{18+6\sqrt{2}}{9-2}=\dfrac{18+6\sqrt{2}}{7}$

63. $\dfrac{2+\sqrt{5}}{6+\sqrt{3}}=\dfrac{2+\sqrt{5}}{6+\sqrt{3}}\cdot\dfrac{6-\sqrt{3}}{6-\sqrt{3}}$
$=\dfrac{(2+\sqrt{5})(6-\sqrt{3})}{(6+\sqrt{3})(6-\sqrt{3})}=\dfrac{12-2\sqrt{3}+6\sqrt{5}-\sqrt{15}}{36-3}$
$=\dfrac{12-2\sqrt{3}+6\sqrt{5}-\sqrt{15}}{33}$

65. $\dfrac{\sqrt{a}}{\sqrt{a}+\sqrt{b}}=\dfrac{\sqrt{a}}{\sqrt{a}+\sqrt{b}}\cdot\dfrac{\sqrt{a}-\sqrt{b}}{\sqrt{a}-\sqrt{b}}$
$=\dfrac{\sqrt{a}(\sqrt{a}-\sqrt{b})}{(\sqrt{a}+\sqrt{b})(\sqrt{a}-\sqrt{b})}=\dfrac{a-\sqrt{ab}}{a-b}$

67. $\dfrac{\sqrt{7}-\sqrt{3}}{\sqrt{3}-\sqrt{7}}=\dfrac{-1(\sqrt{3}-\sqrt{7})}{\sqrt{3}-\sqrt{7}}=-1\cdot\dfrac{\sqrt{3}-\sqrt{7}}{\sqrt{3}-\sqrt{7}}=-1\cdot 1=-1$

69. $\dfrac{3\sqrt{2}-\sqrt{7}}{4\sqrt{2}+2\sqrt{5}}=\dfrac{3\sqrt{2}-\sqrt{7}}{4\sqrt{2}+2\sqrt{5}}\cdot\dfrac{4\sqrt{2}-2\sqrt{5}}{4\sqrt{2}-2\sqrt{5}}$
$=\dfrac{(3\sqrt{2}-\sqrt{7})(4\sqrt{2}-2\sqrt{5})}{(4\sqrt{2}+2\sqrt{5})(4\sqrt{2}-2\sqrt{5})}$
$=\dfrac{12\cdot 2-6\sqrt{10}-4\sqrt{14}+2\sqrt{35}}{16\cdot 2-4\cdot 5}$
$=\dfrac{24-6\sqrt{10}-4\sqrt{14}+2\sqrt{35}}{32-20}$
$=\dfrac{24-6\sqrt{10}-4\sqrt{14}+2\sqrt{35}}{12}$
$=\dfrac{2(12-3\sqrt{10}-2\sqrt{14}+\sqrt{35})}{2\cdot 6}$
$=\dfrac{12-3\sqrt{10}-2\sqrt{14}+\sqrt{35}}{6}$

71. $\dfrac{\sqrt{5}+1}{4}=\dfrac{\sqrt{5}+1}{4}\cdot\dfrac{\sqrt{5}-1}{\sqrt{5}-1}=\dfrac{(\sqrt{5}+1)(\sqrt{5}-1)}{4(\sqrt{5}-1)}$

$=\dfrac{(\sqrt{5})^2-1}{4(\sqrt{5}-1)}=\dfrac{5-1}{4(\sqrt{5}-1)}=\dfrac{4}{4(\sqrt{5}-1)}$

$=\dfrac{1}{\sqrt{5}-1}$

73. $\dfrac{\sqrt{6}-2}{\sqrt{3}+7}=\dfrac{\sqrt{6}-2}{\sqrt{3}+7}\cdot\dfrac{\sqrt{6}+2}{\sqrt{6}+2}=\dfrac{(\sqrt{6}-2)(\sqrt{6}+2)}{(\sqrt{3}+7)(\sqrt{6}+2)}$

$=\dfrac{6-4}{\sqrt{18}+2\sqrt{3}+7\sqrt{6}+14}=\dfrac{2}{3\sqrt{2}+2\sqrt{3}+7\sqrt{6}+14}$

75. $\dfrac{\sqrt{x}-\sqrt{y}}{\sqrt{x}+\sqrt{y}}=\dfrac{\sqrt{x}-\sqrt{y}}{\sqrt{x}+\sqrt{y}}\cdot\dfrac{\sqrt{x}+\sqrt{y}}{\sqrt{x}+\sqrt{y}}$

$=\dfrac{(\sqrt{x}-\sqrt{y})(\sqrt{x}+\sqrt{y})}{(\sqrt{x}+\sqrt{y})(\sqrt{x}+\sqrt{y})}=\dfrac{x-y}{x+2\sqrt{xy}+y}$

77. $\dfrac{\sqrt{a+h}-\sqrt{a}}{h}=\dfrac{\sqrt{a+h}-\sqrt{a}}{h}\cdot\dfrac{\sqrt{a+h}+\sqrt{a}}{\sqrt{a+h}+\sqrt{a}}$

$=\dfrac{(\sqrt{a+h}-\sqrt{a})(\sqrt{a+h}+\sqrt{a})}{h(\sqrt{a+h}+\sqrt{a})}=\dfrac{a+h-a}{h(\sqrt{a+h}+\sqrt{a})}$

$=\dfrac{h}{h(\sqrt{a+h}+\sqrt{a})}=\dfrac{1}{\sqrt{a+h}+\sqrt{a}}$

79. $\sqrt[3]{a}\,\sqrt[6]{a}$

$=a^{1/3}\cdot a^{1/6}$ Converting to exponential notation

$=a^{1/2}$ Adding exponents

$=\sqrt{a}$ Returning to radical notation

81. $\sqrt{b^3}\,\sqrt[5]{b^4}$

$=b^{3/2}\cdot b^{4/5}$ Converting to exponential notation

$=b^{23/10}$ Adding exponents

$=b^{2+3/10}$ Writing $\frac{23}{10}$ as a mixed number

$=b^2b^{3/10}$ Factoring

$=b^2\sqrt[10]{b^3}$ Returning to radical notation

83. $\sqrt{xy^3}\,\sqrt[3]{x^2y}=(xy^3)^{1/2}(x^2y)^{1/3}$

$=(xy^3)^{3/6}(x^2y)^{2/6}$

$=\left[(xy^3)^3(x^2y)^2\right]^{1/6}$

$=\sqrt[6]{x^3y^9\cdot x^4y^2}$

$=\sqrt[6]{x^7y^{11}}$

$=\sqrt[6]{x^6y^6\cdot xy^5}$

$=xy\sqrt[6]{xy^5}$

85. $\sqrt[4]{9ab^3}\,\sqrt{3a^4b}=(9ab^3)^{1/4}(3a^4b)^{1/2}$

$=(9ab^3)^{1/4}(3a^4b)^{2/4}$

$=\left[(9ab^3)(3a^4b)^2\right]^{1/4}$

$=\sqrt[4]{9ab^3\cdot 9a^8b^2}$

$=\sqrt[4]{81a^9b^5}$

$=\sqrt[4]{81a^8b^4\cdot ab}$

$=3a^2b\sqrt[4]{ab}$

87. $\sqrt{a^4b^3c^4}\,\sqrt[3]{ab^2c}=(a^4b^3c^4)^{1/2}(ab^2c)^{1/3}$

$=(a^4b^3c^4)^{3/6}(ab^2c)^{2/6}$

$=\left[(a^4b^3c^4)^3(ab^2c)^2\right]^{1/6}$

$=\sqrt[6]{a^{12}b^9c^{12}\cdot a^2b^4c^2}$

$=\sqrt[6]{a^{14}b^{13}c^{14}}$

$=\sqrt[6]{a^{12}b^{12}c^{12}\cdot a^2bc^2}$

$=a^2b^2c^2\sqrt[6]{a^2bc^2}$

89. $\dfrac{\sqrt[3]{a^2}}{\sqrt[4]{a}}$

$=\dfrac{a^{2/3}}{a^{1/4}}$ Converting to exponential notation

$=a^{2/3-1/4}$ Subtracting exponents

$=a^{5/12}$ Converting back

$=\sqrt[12]{a^5}$ to radical notation

91. $\dfrac{\sqrt[4]{x^2y^3}}{\sqrt[3]{xy}}$

$=\dfrac{(x^2y^3)^{1/4}}{(xy)^{1/3}}$ Converting to exponential notation

$=\dfrac{x^{2/4}y^{3/4}}{x^{1/3}y^{1/3}}$ Using the power and product rules

$=x^{2/4-1/3}y^{3/4-1/3}$ Subtracting exponents

$=x^{2/12}y^{5/12}$

$=(x^2y^5)^{1/12}$ Converting back to radical notation

$=\sqrt[12]{x^2y^5}$

93. $\dfrac{\sqrt{ab^3}}{\sqrt[5]{a^2b^3}}$

$=\dfrac{(ab^3)^{1/2}}{(a^2b^3)^{1/5}}$ Converting to exponential notation

$=\dfrac{a^{1/2}b^{3/2}}{a^{2/5}b^{3/5}}$

$=a^{1/10}b^{9/10}$ Subtracting exponents

$=(ab^9)^{1/10}$ Converting back to radical notation

$=\sqrt[10]{ab^9}$

95. $\dfrac{\sqrt{(7-y)^3}}{\sqrt[3]{(7-y)^2}}$

$=\dfrac{(7-y)^{3/2}}{(7-y)^{2/3}}$ Converting to exponential notation

$=(7-y)^{3/2-2/3}$ Subtracting exponents

$=(7-y)^{5/6}$

$=\sqrt[6]{(7-y)^5}$ Returning to radical notation

97. $\dfrac{\sqrt[4]{(5+3x)^3}}{\sqrt[3]{(5+3x)^2}}$

$= \dfrac{(5+3x)^{3/4}}{(5+3x)^{2/3}}$ Converting to exponential notation

$= (5+3x)^{3/4-2/3}$ Subtracting exponents

$= (5+3x)^{1/12}$ Converting back to radical notation

$= \sqrt[12]{5+3x}$

99. $\sqrt[3]{x^2y}\left(\sqrt{xy}-\sqrt[5]{xy^3}\right)$

$= (x^2y)^{1/3}\left[(xy)^{1/2}-(xy^3)^{1/5}\right]$

$= x^{2/3}y^{1/3}\left(x^{1/2}y^{1/2}-x^{1/5}y^{3/5}\right)$

$= x^{2/3}y^{1/3}x^{1/2}y^{1/2}-x^{2/3}y^{1/3}x^{1/5}y^{3/5}$

$= x^{2/3+1/2}y^{1/3+1/2}-x^{2/3+1/5}y^{1/3+3/5}$

$= x^{7/6}y^{5/6}-x^{13/15}y^{14/15}$ Writing as a mixed numeral

$= x\cdot x^{1/6}y^{5/6}-x^{13/15}y^{14/15}$

$= x\left(xy^5\right)^{1/6}-\left(x^{13}y^{14}\right)^{1/15}$

$= x\sqrt[6]{xy^5}-\sqrt[15]{x^{13}y^{14}}$

101. $\left(m+\sqrt[3]{n^2}\right)\left(2m+\sqrt[4]{n}\right)$

$= \left(m+n^{2/3}\right)\left(2m+n^{1/4}\right)$ Converting to exponential notation

$= 2m^2+mn^{1/4}+2mn^{2/3}+n^{2/3}n^{1/4}$ Using FOIL

$= 2m^2+mn^{1/4}+2mn^{2/3}+n^{2/3+1/4}$ Adding exponents

$= 2m^2+mn^{1/4}+2mn^{2/3}+n^{11/12}$

$= 2m^2+m\sqrt[4]{n}+2m\sqrt[3]{n^2}+\sqrt[12]{n^{11}}$ Converting back to radical notation

103. $f(x)=\sqrt[4]{x},\ g(x)=2\sqrt{x}-\sqrt[3]{x^2}$

$(f\cdot g)(x)=\sqrt[4]{x}\left(2\sqrt{x}-\sqrt[3]{x^2}\right)$

$= x^{1/4}\cdot 2x^{1/2}-x^{1/4}\cdot x^{2/3}$

$= 2x^{1/4+1/2}-x^{1/4+2/3}$

$= 2x^{1/4+2/4}-x^{3/12+8/12}$

$= 2x^{3/4}-x^{11/12}$

$= 2\sqrt[4]{x^3}-\sqrt[12]{x^{11}}$

105. $f(x)=x+\sqrt{7},\ g(x)=x-\sqrt{7}$

$(f\cdot g)(x)=\left(x+\sqrt{7}\right)\left(x-\sqrt{7}\right)$

$= x^2-\left(\sqrt{7}\right)^2$

$= x^2-7$

107. $f(x)=x^2$

$f\left(3-\sqrt{2}\right)=\left(3-\sqrt{2}\right)^2=3^2-2\cdot 3\cdot\sqrt{2}+\left(\sqrt{2}\right)^2$

$= 9-6\sqrt{2}+2=11-6\sqrt{2}$

109. $f(x)=x^2$

$f\left(\sqrt{6}+\sqrt{21}\right)=\left(\sqrt{6}+\sqrt{21}\right)^2$

$= \left(\sqrt{6}\right)^2+2\cdot\sqrt{6}\cdot\sqrt{21}+\left(\sqrt{21}\right)^2$

$= 6+2\sqrt{126}+21=27+2\sqrt{9\cdot 14}$

$= 27+6\sqrt{14}$

111. *Writing Exercise.*

113. IV

115. Let $y=0$, then $x=10$. Then the x-intercept is $(10, 0)$. Let $x=0$, then $y=-10$. Then the y-intercept is $(0,-10)$.

117. A line perpendicular to $y=\frac{1}{2}x-7$ has slope -2.

The line is $y=-2x+12$.

119. *Writing Exercise.*

121. $f(x)=\sqrt{x^3-x^2}+\sqrt{9x^3-9x^2}-\sqrt{4x^3-4x^2}$

$= \sqrt{x^2(x-1)}+\sqrt{9x^2(x-1)}-\sqrt{4x^2(x-1)}$

$= x\sqrt{x-1}+3x\sqrt{x-1}-2x\sqrt{x-1}$

$= 2x\sqrt{x-1}$

123. $f(x)=\sqrt[4]{x^5-x^4}+3\sqrt[4]{x^9-x^8}$

$= \sqrt[4]{x^4(x-1)}+3\sqrt[4]{x^8(x-1)}$

$= \sqrt[4]{x^4}\cdot\sqrt[4]{x-1}+3\sqrt[4]{x^8}\sqrt[4]{x-1}$

$= x\sqrt[4]{x-1}+3x^2\sqrt[4]{x-1}$

$= \left(x+3x^2\right)\sqrt[4]{x-1}$

125. $7x\sqrt{(x+y)^3}-5xy\sqrt{x+y}-2y\sqrt{(x+y)^3}$

$= 7x\sqrt{(x+y)^2(x+y)}-5xy\sqrt{x+y}$

$\quad -2y\sqrt{(x+y)^2(x+y)}$

$= 7x(x+y)\sqrt{x+y}-5xy\sqrt{x+y}-2y(x+y)\sqrt{x+y}$

$= \left[7x(x+y)-5xy-2y(x+y)\right]\sqrt{x+y}$

$= \left(7x^2+7xy-5xy-2xy-2y^2\right)\sqrt{x+y}$

$= \left(7x^2-2y^2\right)\sqrt{x+y}$

127. $\sqrt{8x(y+z)^5}\sqrt[3]{4x^2(y+z)^2}$

$= \left[8x(y+z)^5\right]^{1/2}\left[4x^2(y+z)^2\right]^{1/3}$

$= \left[8x(y+z)^5\right]^{3/6}\left[4x^2(y+z)^2\right]^{2/6}$

$= \left\{\left[2^3x(y+z)^5\right]^3\left[2^2x^2(y+z)^2\right]^2\right\}^{1/6}$

$= \sqrt[6]{2^9x^3(y+z)^{15}\cdot 2^4x^4(y+z)^4}$

$= \sqrt[6]{2^{13}x^7(y+z)^{19}}$

$= \sqrt[6]{2^{12}x^6(y+z)^{18}\cdot 2x(y+z)}$

$= 2^2x(y+z)^3\sqrt[6]{2x(y+z)}$, or

$4x(y+z)^3\sqrt[6]{2x(y+z)}$

129. $\dfrac{\frac{1}{\sqrt{w}}-\sqrt{w}}{\frac{\sqrt{w}+1}{\sqrt{w}}}=\dfrac{\frac{1}{\sqrt{w}}-\sqrt{w}}{\frac{\sqrt{w}+1}{\sqrt{w}}}\cdot\dfrac{\sqrt{w}}{\sqrt{w}}=\dfrac{1-w}{\sqrt{w}+1}$

$= \dfrac{1-w}{\sqrt{w}+1}\cdot\dfrac{\sqrt{w}-1}{\sqrt{w}-1}=\dfrac{\sqrt{w}-1-w\sqrt{w}+w}{w-1}$

$= \dfrac{(w-1)-\sqrt{w}(w-1)}{w-1}=\dfrac{(w-1)(1-\sqrt{w})}{w-1}$

$= 1-\sqrt{w}$

131. $x-5=\left(\sqrt{x}\right)^2-\left(\sqrt{5}\right)^2=\left(\sqrt{x}+\sqrt{5}\right)\left(\sqrt{x}-\sqrt{5}\right)$

133. $x-a=\left(\sqrt{x}\right)^2-\left(\sqrt{a}\right)^2=\left(\sqrt{x}+\sqrt{a}\right)\left(\sqrt{x}-\sqrt{a}\right)$

135. $\left(\sqrt{x+2}-\sqrt{x-2}\right)^2=x+2-2\sqrt{(x+2)(x-2)}+x-2$
$=x+2-2\sqrt{x^2-4}+x-2$
$=2x-2\sqrt{x^2-4}$

137. $A^2+B^2=(A+B+\sqrt{2AB})(A+B-\sqrt{2AB})$

a. $(A+B+\sqrt{2AB})(A+B-\sqrt{2AB})$
$=(A+B)^2-(\sqrt{2AB})^2$
$=A^2+2AB+B^2-2AB$
$=A^2+B^2$

b. $2AB$ must be a perfect square.

Mid-Chapter Review

1. $\sqrt{6x^9}\cdot\sqrt{2xy}=\sqrt{6x^9\cdot 2xy}$
$=\sqrt{12x^{10}y}$
$=\sqrt{4x^{10}\cdot 3y}$
$=\sqrt{4x^{10}}\cdot\sqrt{3y}$
$=2x^5\sqrt{3y}$

2. $\sqrt{12}-3\sqrt{75}+\sqrt{8}$
$=2\sqrt{3}-3\cdot 5\sqrt{3}+2\sqrt{2}$
$=2\sqrt{3}-15\sqrt{3}+2\sqrt{2}$
$=-13\sqrt{3}+2\sqrt{2}$

3. $\sqrt{81}=9$

4. $-\sqrt{\dfrac{9}{100}}=-\dfrac{3}{10}$

5. $\sqrt{64t^2}=|8t|$, or $8|t|$

6. $\sqrt[5]{x^5}=x$

7. $f(x)=\sqrt[3]{12x-4}$
$f(-5)=\sqrt[3]{12(-5)-4}=\sqrt[3]{-64}=-4$

8. $g(x)=\sqrt[4]{10-x}$
Since the index is even, the radicand, $10-x$, must be nonnegative. We solve the inequality:
$10-x\ge 0$
$-x\ge -10$
$x\le 10$
Domain of $g=\{x|x\le 10\}$, or $(-\infty, 10]$

9. $8^{2/3}=\left(2^3\right)^{2/3}=2^{3\cdot\frac{2}{3}}=2^2=4$

10. $\sqrt[6]{\sqrt{a}}=\sqrt[6]{a^{1/2}}=\left(a^{1/2}\right)^{1/6}=a^{\frac{1}{2}\cdot\frac{1}{6}}=a^{1/12}=\sqrt[12]{a}$

11. $\sqrt[3]{y^{24}}=y^8$

12. $\sqrt{(t+5)^2}=t+5$

13. $\sqrt[3]{-27a^{12}}=\sqrt[3]{\left(-3a^4\right)^3}=-3a^4$

14. $\sqrt{6x}\sqrt{15x}=\sqrt{6x\cdot 15x}=\sqrt{90x^2}=\sqrt{9x^2\cdot 10}=3x\sqrt{10}$

15. $\dfrac{\sqrt{20y}}{\sqrt{45y}}=\sqrt{\dfrac{20y}{45y}}=\sqrt{\dfrac{4}{9}}=\dfrac{\sqrt{4}}{\sqrt{9}}=\dfrac{2}{3}$

16. $\sqrt{6}\left(\sqrt{10}-\sqrt{33}\right)=\sqrt{6}\sqrt{10}-\sqrt{6}\sqrt{33}=\sqrt{6\cdot 10}-\sqrt{6\cdot 33}$
$=\sqrt{60}-\sqrt{198}=\sqrt{4\cdot 15}-\sqrt{9\cdot 22}$
$=2\sqrt{15}-3\sqrt{22}$

17. $\dfrac{\sqrt{t}}{\sqrt[8]{t^3}}=\dfrac{t^{1/2}}{t^{3/8}}=t^{1/2-3/8}=t^{1/8}=\sqrt[8]{t}$

18. $\dfrac{\sqrt[5]{3a^{12}}}{\sqrt[5]{96a^2}}=\sqrt[5]{\dfrac{3a^{12}}{96a^2}}=\sqrt[5]{\dfrac{a^{10}}{32}}=\dfrac{a^2}{2}$

19. $2\sqrt{3}-5\sqrt{12}=2\sqrt{3}-5\sqrt{4\cdot 3}=2\sqrt{3}-5\cdot 2\sqrt{3}$
$=2\sqrt{3}-10\sqrt{3}=-8\sqrt{3}$

20. $\left(\sqrt{5}+3\right)\left(\sqrt{5}-3\right)=\left(\sqrt{5}\right)^2-3^2=5-9=-4$

21. $\left(\sqrt{15}+\sqrt{10}\right)^2=\left(\sqrt{15}\right)^2+2\cdot\sqrt{10}\sqrt{15}+\left(\sqrt{10}\right)^2$
$=15+2\sqrt{25\cdot 6}+10=25+2\cdot 5\sqrt{6}$
$=25+10\sqrt{6}$

22. $\sqrt{25x-25}-\sqrt{9x-9}=\sqrt{25(x-1)}-\sqrt{9(x-1)}$
$=\sqrt{25}\sqrt{x-1}-\sqrt{9}\sqrt{x-1}$
$=5\sqrt{x-1}-3\sqrt{x-1}$
$=2\sqrt{x-1}$

23. $\sqrt{x^3y}\sqrt[5]{xy^4}=x^{3/2}y^{1/2}x^{1/5}y^{4/5}=x^{3/2+1/5}y^{1/2+4/5}$
$=x^{17/10}y^{13/10}=x^{1+7/10}y^{1+3/10}$
$=xy\sqrt[10]{x^7y^3}$

24. $\sqrt[3]{5000}+\sqrt[3]{625}=\sqrt[3]{1000\cdot 5}+\sqrt[3]{125\cdot 5}$
$=\sqrt[3]{1000}\sqrt[3]{5}+\sqrt[3]{125}\sqrt[3]{5}$
$=10\sqrt[3]{5}+5\sqrt[3]{5}=15\sqrt[3]{5}$

25. $\sqrt[3]{12x^2y^5}\sqrt[3]{18x^7y}=\sqrt[3]{216x^9y^6}=\sqrt[3]{6^3x^9y^6}=6x^3y^2$

Exercise Set 10.6

1. When we "square both sides" of an equation, we are using the principle of *powers*.

3. To solve an equation with a radical term, we first *isolate* the radical term on one side of the equation.

5. True by the principle of powers

7. False; if $x^2 = 36$, then $x = 6$, or $x = -6$.

9. $\sqrt{5x+1} = 4$

$(\sqrt{5x+1})^2 = 4^2$ Principle of powers (squaring)

$5x+1 = 16$

$5x = 15$

$x = 3$

Check: $\sqrt{5x+1} = 4$

$\sqrt{5\cdot 3+1} \mid 4$

$\sqrt{16}$

$4 \stackrel{?}{=} 4$ TRUE

The solution is 3.

11. $\sqrt{3x}+1 = 5$

$\sqrt{3x} = 4$ Adding to isolate the radical

$(\sqrt{3x})^2 = 4^2$ Principle of powers (squaring)

$3x = 16$

$x = \frac{16}{3}$

Check: $\sqrt{3x}+1 = 5$

$\sqrt{3\cdot\frac{16}{3}}+1 \mid 5$

$\sqrt{16}+1$

$4+1$

$5 \stackrel{?}{=} 5$ TRUE

The solution is $\frac{16}{3}$.

13. $\sqrt{y+5}-4 = 1$

$\sqrt{y+5} = 5$ Adding to isolate the radical

$(\sqrt{y+5})^2 = 5^2$ Principle of powers (squaring)

$y+5 = 25$

$y = 20$

Check: $\sqrt{y+5}-4 = 1$

$\sqrt{20+5}-4 \mid 1$

$\sqrt{25}-4$

$5-4$

$1 \stackrel{?}{=} 1$ TRUE

The solution is 20.

15. $\sqrt{8-x}+7 = 10$

$\sqrt{8-x} = 3$ Adding to isolate the radical

$(\sqrt{8-x})^2 = 3^2$ Principle of powers (squaring)

$8-x = 9$

$x = -1$

Check: $\sqrt{8-x}+7 = 10$

$\sqrt{8-(-1)}+7 \mid 10$

$\sqrt{9}+7$

$3+7$

$10 \stackrel{?}{=} 10$ TRUE

The solution is -1.

17. $\sqrt[3]{y+3} = 2$

$(\sqrt[3]{y+3})^3 = 2^3$ Principle of powers (cubing)

$y+3 = 8$

$y = 5$

Check: $\sqrt[3]{y+3} = 2$

$\sqrt[3]{5+3} \mid 2$

$\sqrt[3]{8}$

$2 \stackrel{?}{=} 2$ TRUE

The solution is 5.

19. $\sqrt[4]{t-10} = 3$

$(\sqrt[4]{t-10})^4 = 3^4$

$t-10 = 81$

$t = 91$

Check: $\sqrt[4]{t-10} = 3$

$\sqrt[4]{91-10} \mid 3$

$\sqrt[4]{81}$

$3 \stackrel{?}{=} 3$ TRUE

The solution is 91.

21. $6\sqrt{x} = x$

$(6\sqrt{x})^2 = x^2$

$36x = x^2$

$0 = x^2 - 36x$

$0 = x(x-36)$

$x = 0$ *or* $x = 36$

Check:

For $x = 0$: $6\sqrt{x} = x$

$6\sqrt{0} \mid 0$

$0 \stackrel{?}{=} 0$ TRUE

For $x = 36$: $6\sqrt{x} = x$

$6\sqrt{36} \mid 36$

$6\cdot 6$

$36 \stackrel{?}{=} 36$ TRUE

The solutions are 0 and 36.

23. $2y^{1/2} - 13 = 7$

$2\sqrt{y} - 13 = 7$

$2\sqrt{y} = 20$

$\sqrt{y} = 10$

$(\sqrt{y})^2 = 10^2$

$y = 100$

Check: $2y^{1/2} - 13 = 7$

$2\cdot 100^{1/2} - 13 \mid 7$

$2\cdot 10 - 13$

$20 - 13$

$7 \stackrel{?}{=} 7$ TRUE

The solution is 100.

25. $\sqrt[3]{x} = -5$

$(\sqrt[3]{x})^3 = (-5)^3$

$x = -125$

Check: $\sqrt[3]{x} = -5$

$\sqrt[3]{-125} \mid -5$

$\sqrt[3]{(-5)^3}$

$-5 \stackrel{?}{=} -5$ TRUE

The solution is –125.

27. $z^{1/4} + 8 = 10$

$z^{1/4} = 2$

$(z^{1/4})^4 = 2^4$

$z = 16$

Check: $z^{1/4} + 8 = 10$

$16^{1/4} + 8 \mid 10$

$2 + 8$

$10 \stackrel{?}{=} 10$ TRUE

The solution is 16.

29. $\sqrt{n} = -2$

This equation has no solution, since the principal square root is never negative.

31. $\sqrt[4]{3x+1} - 4 = -1$

$\sqrt[4]{3x+1} = 3$

$(\sqrt[4]{3x+1})^4 = 3^4$

$3x + 1 = 81$

$3x = 80$

$x = \frac{80}{3}$

Check: $\sqrt[4]{3x+1} - 4 = -1$

$\sqrt[4]{3 \cdot \frac{80}{3} + 1} - 4 \mid -1$

$\sqrt[4]{81} - 4$

$3 - 4$

$-1 \stackrel{?}{=} -1$ TRUE

The solution is $\frac{80}{3}$.

33. $(21x + 55)^{1/3} = 10$

$\left[(21x + 55)^{1/3}\right]^3 = 10^3$

$21x + 55 = 1000$

$21x = 945$

$x = 45$

Check: $(21x + 55)^{1/3} = 10$

$(21 \cdot 45 + 55)^{1/3} \mid 10$

$(945 + 55)^{1/3}$

$10 \stackrel{?}{=} 10$ TRUE

The solution is 45.

35. $\sqrt[3]{3y+6} + 7 = 8$

$\sqrt[3]{3y+6} = 1$

$(\sqrt[3]{3y+6})^3 = 1^3$

$3y + 6 = 1$

$3y = -5$

$y = -\frac{5}{3}$

Check: $\sqrt[3]{3y+6} + 7 = 8$

$\sqrt[3]{3\left(-\frac{5}{3}\right) + 6} + 7 \mid 8$

$\sqrt[3]{1} + 7$

$1 + 7$

$8 \stackrel{?}{=} 8$ TRUE

The solution is $-\frac{5}{3}$.

37. $3 + \sqrt{5 - x} = x$

$\sqrt{5 - x} = x - 3$

$(\sqrt{5 - x})^2 = (x - 3)^2$

$5 - x = x^2 - 6x + 9$

$0 = x^2 - 5x + 4$

$0 = (x - 1)(x - 4)$

$x - 1 = 0$ *or* $x - 4 = 0$

$x = 1$ *or* $x = 4$

Check:

For 1: $3 + \sqrt{5 - x} = x$

$3 + \sqrt{5 - 1} \mid 1$

$3 + \sqrt{4}$

$3 + 2$

$5 \stackrel{?}{=} 1$ FALSE

For 4: $3 + \sqrt{5 - x} = x$

$3 + \sqrt{5 - 4} \mid 4$

$3 + \sqrt{1}$

$3 + 1$

$4 \stackrel{?}{=} 4$ TRUE

Since 4 checks but 1 does not, the solution is 4.

39. $\sqrt{3t + 4} = \sqrt{4t + 3}$

$(\sqrt{3t + 4})^2 = (\sqrt{4t + 3})^2$

$3t + 4 = 4t + 3$

$4 = t + 3$

$1 = t$

Check:

$\sqrt{3t + 4} = \sqrt{4t + 3}$

$\sqrt{3 \cdot 1 + 4} \mid \sqrt{4 \cdot 1 + 3}$

$\sqrt{7} \stackrel{?}{=} \sqrt{7}$ TRUE

The solution is 1.

41. $3(4-t)^{1/4}=6^{1/4}$

$$\left[3(4-t)^{1/4}\right]^4=\left(6^{1/4}\right)^4$$
$$81(4-t)=6$$
$$324-81t=6$$
$$-81t=-318$$
$$t=\frac{106}{27}$$

The number $\frac{106}{27}$ checks and is the solution.

43. $\sqrt{4x-3}=2+\sqrt{2x-5}$ One radical is already isolated.

$$\left(\sqrt{4x-3}\right)^2=\left(2+\sqrt{2x-5}\right)^2 \quad \text{Squaring both sides}$$
$$4x-3=4+4\sqrt{2x-5}+2x-5$$
$$2x-2=4\sqrt{2x-5}$$
$$x-1=2\sqrt{2x-5}$$
$$x^2-2x+1=8x-20$$
$$x^2-10x+21=0$$
$$(x-7)(x-3)=0$$
$$x-7=0 \quad or \quad x-3=0$$
$$x=7 \quad or \quad x=3$$

Both numbers check. The solutions are 7 and 3.

45. $\sqrt{20-x}+8=\sqrt{9-x}+11$

$$\sqrt{20-x}=\sqrt{9-x}+3 \quad \text{Isolating one radical}$$
$$\left(\sqrt{20-x}\right)^2=\left(\sqrt{9-x}+3\right)^2 \quad \text{Squaring both sides}$$
$$20-x=9-x+6\sqrt{9-x}+9$$
$$2=6\sqrt{9-x} \quad \text{Isolating the remaining radical}$$
$$1=3\sqrt{9-x} \quad \text{Multiplying by } \frac{1}{2}$$
$$1^2=\left(3\sqrt{9-x}\right)^2 \quad \text{Squaring both sides}$$
$$1=9(9-x)$$
$$1=81-9x$$
$$-80=-9x$$
$$\frac{80}{9}=x$$

The number $\frac{80}{9}$ checks and is the solution.

47. $\sqrt{x+2}+\sqrt{3x+4}=2$

$$\sqrt{x+2}=2-\sqrt{3x+4} \quad \text{Isolating one radical}$$
$$\left(\sqrt{x+2}\right)^2=\left(2-\sqrt{3x+4}\right)^2$$
$$x+2=4-4\sqrt{3x+4}+3x+4$$
$$-2x-6=-4\sqrt{3x+4} \quad \text{Isolating the remaining radical}$$
$$x+3=2\sqrt{3x+4} \quad \text{Multiplying by } -\frac{1}{2}$$
$$(x+3)^2=\left(2\sqrt{3x+4}\right)^2$$
$$x^2+6x+9=4(3x+4)$$
$$x^2+6x+9=12x+16$$
$$x^2-6x-7=0$$
$$(x-7)(x+1)=0$$
$$x-7=0 \quad or \quad x+1=0$$
$$x=7 \quad or \quad x=-1$$

Check:

For 7:

$$\frac{\sqrt{x+2}+\sqrt{3x+4}=2}{\sqrt{7+2}+\sqrt{3\cdot 7+4} \;\big|\; 2}$$
$$\sqrt{9}+\sqrt{25}$$
$$8\overset{?}{=}2 \quad \text{FALSE}$$

For -1:

$$\frac{\sqrt{x+2}+\sqrt{3x+4}=2}{\sqrt{-1+2}+\sqrt{3\cdot(-1)+4} \;\big|\;}$$
$$\sqrt{1}+\sqrt{1}$$
$$2\overset{?}{=}2 \quad \text{TRUE}$$

Since -1 checks but 7 does not, the solution is -1.

49. We must have $f(x)=1$, or $\sqrt{x}+\sqrt{x-9}=1$.

$$\sqrt{x}+\sqrt{x-9}=1$$
$$\sqrt{x-9}=1-\sqrt{x} \quad \text{Isolating one radical term}$$
$$\left(\sqrt{x-9}\right)^2=\left(1-\sqrt{x}\right)^2$$
$$x-9=1-2\sqrt{x}+x$$
$$-10=-2\sqrt{x} \quad \text{Isolating the remaining radical term}$$
$$5=\sqrt{x}$$
$$25=x$$

This value does not check. There is no solution, so there is no value of x for which $f(x)=1$.

51. $\sqrt{t-2}-\sqrt{4t+1}=-3$

$$\sqrt{t-2}=\sqrt{4t+1}-3$$
$$\left(\sqrt{t-2}\right)^2=\left(\sqrt{4t+1}-3\right)^2$$
$$t-2=4t+1-6\sqrt{4t+1}+9$$
$$-3t-12=-6\sqrt{4t+1}$$
$$t+4=2\sqrt{4t+1}$$
$$(t+4)^2=\left(2\sqrt{4t+1}\right)^2$$
$$t^2+8t+16=4(4t+1)$$
$$t^2+8t+16=16t+4$$
$$t^2-8t+12=0$$
$$(t-2)(t-6)=0$$
$$t-2=0 \quad or \quad t-6=0$$
$$t=2 \quad or \quad t=6$$

Both numbers check, so we have $f(t)=-3$ when $t=2$ and when $t=6$.

53. We must have $\sqrt{2x-3}=\sqrt{x+7}-2$.

$$\begin{aligned}\sqrt{2x-3}&=\sqrt{x+7}-2\\ \left(\sqrt{2x-3}\right)^2&=\left(\sqrt{x+7}-2\right)^2\\ 2x-3&=x+7-4\sqrt{x+7}+4\\ x-14&=-4\sqrt{x+7}\\ (x-14)^2&=\left(-4\sqrt{x+7}\right)^2\\ x^2-28x+196&=16(x+7)\\ x^2-28x+196&=16x+112\\ x^2-44x+84&=0\\ (x-2)(x-42)&=0\\ x=2 \quad &or \quad x=42\end{aligned}$$

Since 2 checks but 42 does not, we have $f(x)=g(x)$ when $x=2$.

55. We must have $4-\sqrt{t-3}=(t+5)^{1/2}$.

$$\begin{aligned}4-\sqrt{t-3}&=(t+5)^{1/2}\\ \left(4-\sqrt{t-3}\right)^2&=\left[(t+5)^{1/2}\right]^2\\ 16-8\sqrt{t-3}+t-3&=t+5\\ -8\sqrt{t-3}&=-8\\ \sqrt{t-3}&=1\\ \left(\sqrt{t-3}\right)^2&=1^2\\ t-3&=1\\ t&=4\end{aligned}$$

The number 4 checks, so we have $f(t)=g(t)$ when $t=4$.

57. *Writing Exercise.*

59. ***Familiarize.*** Let t = the score of Taylor's last test.

Translate. Write an inequality stating the average score is at least 80.

$$\frac{74+88+76+78+t}{5}\ge 80$$

Carry out. We solve the inequality.

$$\begin{aligned}\frac{74+88+76+78+t}{5}&\ge 80\\ \frac{316+t}{5}&\ge 80\\ 316+t&\ge 400\\ t&\ge 84\end{aligned}$$

Check. If the score is 84%, then the average score is $\frac{74+88+76+78+84}{5}=80$. The answer checks.

State. Taylor needs to score at least 84% on the last test to earn at least a B in the course.

61. ***Familiarize***. Let c = the speed of the current, in mph. Then $10+c$ = the speed downriver and $10-c$ = the speed upriver. We organize the information in a table.

	Distance	Speed	Time
Downriver	7	$10+c$	t_1
Upriver	7	$10-c$	t_2

Translate. Using the formula Time = Distance/Rate we see that $t_1=\frac{7}{10+c}$ and $t_2=\frac{7}{10-c}$. The total time upriver and back is $1\frac{2}{3}$ hr, so $t_1+t_2=1\frac{2}{3}$, or

$$\frac{7}{10+c}+\frac{7}{10-c}=\frac{5}{3}.$$

Carry out. We solve the equation. Multiply both sides by the LCD, $3(10+c)(10-c)$.

$$\begin{aligned}3(10+c)(10-c)\left(\frac{7}{10+c}+\frac{7}{10-c}\right)&=3(10+c)(10-c)\frac{5}{3}\\ 21(10-c)+21(10+c)&=5(100-c^2)\\ 210-21c+210+21c&=500-5c^2\\ 5c^2-80&=0\\ 5(c+4)(c-4)&=0\\ c+4=0 \quad or \quad c-4&=0\\ c=-4 \quad or \quad c&=4\end{aligned}$$

Check. Since speed cannot be negative in this problem, –4 cannot be a solution of the original problem. If the speed of the current is 4 mph, the boat travels upriver at 10 – 4, or 6 mph. At this rate it takes $\frac{7}{6}$, or $1\frac{1}{6}$ hr, to travel 7 mi. The boat travels downriver at 10 + 4, or 14 mph. At this rate it takes $\frac{7}{14}$, or $\frac{1}{2}$ hr, to travel 7 mi. The total travel time is $1\frac{1}{6}+\frac{1}{2}$, or $1\frac{2}{3}$ hr. The answer checks.

State. The speed of the current is 4 mph.

63. *Writing Exercise.*

65. Substitute 100 for $v(p)$ and solve for p.

$$\begin{aligned}v(p)&=12.1\sqrt{p}\\ 100&=12.1\sqrt{p}\\ 8.2645&\approx\sqrt{p}\\ (8.2645)^2&\approx\left(\sqrt{p}\right)^2\\ 68.3013&\approx p\end{aligned}$$

The nozzle pressure is about 68 psi.

67. Let f be the frequency of the string and t be the tension of the string. Substitute 260 for f, 28 for t, and solve for k, the constant of variation.

$$\begin{aligned}f&=k\sqrt{t}\\ 260&=k\sqrt{28}\\ k&=\frac{260}{\sqrt{28}}\approx 49.135\end{aligned}$$

Then substitute 32 for t and solve for f.

$$\begin{aligned}f&=49.135\sqrt{t}\\ f&=49.135\sqrt{32}\approx 277.952\end{aligned}$$

The frequency is about 278 Hz.

69. Substitute 1880 for $S(t)$ and solve for t.

$$1880 = 1087.7\sqrt{\frac{9t+2617}{2457}}$$

$$1.7284 \approx \sqrt{\frac{9t+2617}{2457}} \quad \text{Dividing by 1087.7}$$

$$(1.7284)^2 \approx \left(\sqrt{\frac{9t+2617}{2457}}\right)^2$$

$$2.9874 \approx \frac{9t+2617}{2457}$$

$$7340.0418 \approx 9t + 2617$$

$$4723.0418 \approx 9t$$

$$524.7824 \approx t$$

The temperature is about 524.8°C.

71.

$$S = 1087.7\sqrt{\frac{9t+2617}{2457}}$$

$$\frac{S}{1087.7} = \sqrt{\frac{9t+2617}{2457}}$$

$$\left(\frac{S}{1087.7}\right)^2 = \left(\sqrt{\frac{9t+2617}{2457}}\right)^2$$

$$\frac{S^2}{1087.7^2} = \frac{9t+2617}{2457}$$

$$\frac{2457S^2}{1087.7^2} = 9t + 2617$$

$$\frac{2457S^2}{1087.7^2} - 2617 = 9t$$

$$\frac{1}{9}\left(\frac{2457S^2}{1087.7^2} - 2617\right) = t$$

73.

$$v = \sqrt{2gr}\sqrt{\frac{h}{r+h}}$$

$$v^2 = 2gr \cdot \frac{h}{r+h} \quad \text{Squaring both sides}$$

$$v^2(r+h) = 2grh \quad \text{Multiplying by } r+h$$

$$v^2r + v^2h = 2grh$$

$$v^2h = 2grh - v^2r$$

$$v^2h = r\left(2gh - v^2\right)$$

$$\frac{v^2h}{2gh - v^2} = r$$

75.

$$d(n) = 0.75\sqrt{2.8n}$$

$$\frac{4}{3}d(n) = \sqrt{2.8n} \quad \text{Dividing by 0.75 or 3/4}$$

$$\left[\frac{4}{3}d(n)\right]^2 = \left(\sqrt{2.8n}\right)^2$$

$$\frac{16}{9}[d(n)]^2 = 2.8n$$

$$n = \frac{40}{63}[d(n)]^2$$

77.

$$\frac{x+\sqrt{x+1}}{x-\sqrt{x+1}} = \frac{5}{11}$$

$$11\left(x+\sqrt{x+1}\right) = 5\left(x-\sqrt{x+1}\right)$$

$$11x + 11\sqrt{x+1} = 5x - 5\sqrt{x+1}$$

$$16\sqrt{x+1} = -6x$$

$$8\sqrt{x+1} = -3x$$

$$8\sqrt{x+1} = -3x$$

$$\left(8\sqrt{x+1}\right)^2 = (-3x)^2$$

$$64(x+1) = 9x^2$$

$$64x + 64 = 9x^2$$

$$0 = 9x^2 - 64x - 64$$

$$0 = (9x+8)(x-8)$$

$$9x + 8 = 0 \quad or \quad x - 8 = 0$$

$$9x = -8 \quad or \quad x = 8$$

$$x = -\tfrac{8}{9} \quad or \quad x = 8$$

Since $-\frac{8}{9}$ checks but 8 does not, the solution is $-\frac{8}{9}$.

79.

$$\left(z^2+17\right)^{3/4} = 27$$

$$\left[\left(z^2+17\right)^{3/4}\right]^{4/3} = \left(3^3\right)^{4/3}$$

$$z^2 + 17 = 3^4$$

$$z^2 + 17 = 81$$

$$z^2 - 64 = 0$$

$$(z+8)(z-8) = 0$$

$$z = -8 \quad or \quad z = 8$$

Both –8 and 8 check. They are the solutions.

81.

$$\sqrt{8-b} = b\sqrt{8-b}$$

$$\left(\sqrt{8-b}\right)^2 = \left(b\sqrt{8-b}\right)^2$$

$$(8-b) = b^2(8-b)$$

$$0 = b^2(8-b) - (8-b)$$

$$0 = (8-b)(b^2-1)$$

$$0 = (8-b)(b+1)(b-1)$$

$$8 - b = 0 \quad or \quad b + 1 = 0 \quad or \quad b - 1 = 0$$

$$8 = b \quad or \quad b = -1 \quad or \quad b = 1$$

Since the numbers 8 and 1 check but –1 does not, 8 and 1 are the solutions.

83. We find the values of x for which $g(x) = 0$.

$$6x^{1/2} + 6x^{-1/2} - 37 = 0$$

$$6\sqrt{x} + \frac{6}{\sqrt{x}} = 37$$

$$\left(6\sqrt{x} + \frac{6}{\sqrt{x}}\right)^2 = 37^2$$

$$36x + 72 + \frac{36}{x} = 1369$$

$$36x^2 + 72x + 36 = 1369x \quad \text{Multiplying by } x$$

$$36x^2 - 1297x + 36 = 0$$

$$(36x-1)(x-36) = 0$$

$$36x - 1 = 0 \quad or \quad x - 36 = 0$$

$$36x = 1 \quad or \quad x = 36$$

$$x = \tfrac{1}{36} \quad or \quad x = 36$$

Both numbers check. The x-intercepts are $\left(\frac{1}{36}, 0\right)$ and (36, 0).

85. *Graphing Calculator Exercise*

Exercise Set 10.7

1. The correct choice is (d) Right.

3. The correct choice is (e) Square roots.

5. The correct choice is (f) 30°-60°-90°.

7. $a=5$, $b=3$
Find c.
$c^2=a^2+b^2$ Pythagorean theorem
$c^2=5^2+3^2$ Substituting
$c^2=25+9$
$c^2=34$
$c=\sqrt{34}$ Exact answer
$c\approx 5.831$ Approximation

9. $a=9$, $b=9$
Observe that the legs have the same length, so this is an isosceles right triangle. Then we know that the length of the hypotenuse is the length of a leg times $\sqrt{2}$, or $9\sqrt{2}$, or approximately 12.728.

11. $b=15$, $c=17$
Find a.
$a^2+b^2=c^2$ Pythagorean theorem
$a^2+15^2=17^2$ Substituting
$a^2+225=289$
$a^2=64$
$a=8$

13. $a^2+b^2=c^2$ Pythagorean theorem
$(4\sqrt{3})^2+b^2=8^2$
$16\cdot 3+b^2=64$
$48+b^2=64$
$b^2=16$
$b=4$
The other leg is 4 m long.

15. $a^2+b^2=c^2$ Pythagorean theorem
$1^2+b^2=(\sqrt{20})^2$ Substituting
$1+b^2=20$
$b^2=19$
$b=\sqrt{19}$
$b\approx 4.359$
The length of the other leg is $\sqrt{19}$ in., or about 4.359 in.

17. Observe that the length of the hypotenuse, $\sqrt{2}$, is $\sqrt{2}$ times the length of the given leg, 1 m. Thus, we have an isosceles right triangle and the length of the other leg is also 1 m.

19. We have a right triangle with legs of 150 ft and 200 ft. Let $d=$ the length of the diagonal, in feet. We use the Pythagorean theorem to find d.
$150^2+200^2=d^2$
$22,500+40,000=d^2$
$62,500=d^2$
$250=d$
Clare travels 250 ft across the parking lot.

21. We have a right triangle with legs of 800 ft and 60 ft. Let $d=$ the length of the diagonal, in feet. We use the Pythagorean theorem to find d.
$800^2+60^2=d^2$
$640,000+3600=d^2$
$643,600=d^2$
$d=\sqrt{643,600}=20\sqrt{1609}\approx 802.247$
The zipline was about 802.247 ft.

23. We make a drawing similar to the one in the text.

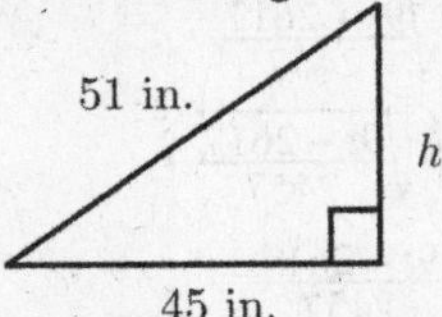

We use the Pythagorean theorem to find h.
$45^2+h^2=51^2$
$2051+h^2=2601$
$h^2=576$
$h=24$
The height of the screen is 24 in.

25. First we will find the diagonal distance, d, in feet, across the room. We make a drawing.

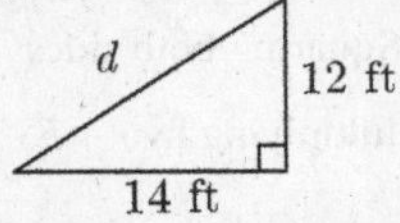

Now we use the Pythagorean theorem.
$12^2+14^2=d^2$
$144+196=d^2$
$340=d^2$
$d=\sqrt{340}=2\sqrt{85}\approx 18.439$
Recall that 4 ft of slack is required on each end. Thus, $\sqrt{340}+2\cdot 4$, or $(\sqrt{340}+8)$ ft, of wire should be purchased. This is about 26.439 ft.

27. The diagonal is the hypotenuse of a right triangle with legs of 70 paces and 40 paces. First we use the Pythagorean theorem to find the length d of the diagonal, in paces.
$70^2+40^2=d^2$
$4900+1600=d^2$
$6500=d^2$
$d=\sqrt{6500}=10\sqrt{65}\approx 80.623$
If Marissa walks along two sides of the quad she takes $70+40$, or 110 paces. Then by using the diagonal she saves $(110-\sqrt{6500})$ paces. This is approximately $110-80.623$, or 29.377 paces.

29. Since one acute angle is 45°, this is an isosceles right triangle with one leg = 5. Then the other leg = 5 also. And the hypotenuse is the length of the a leg times $\sqrt{2}$, or $5\sqrt{2}$.

Exact answer: Leg=5, hypotenuse=$5\sqrt{2}$

Approximation: hypotenuse ≈ 7.071

31. This is a 30-60-90 right triangle with hypotenuse 14. We find the legs:

$2a = 14$, so $a = 7$ and $a\sqrt{3} = 7\sqrt{3}$

Exact answer: shorter leg=7; longer leg=$7\sqrt{3}$

Approximation: longer leg ≈ 12.124

33. This is a 30-60-90 right triangle with one leg = 15. We substitute to find the length of the other leg, a, and the hypotenuse, c.

$$b = a\sqrt{3}$$
$$15 = a\sqrt{3}$$
$$\frac{15}{\sqrt{3}} = a$$
$$\frac{15\sqrt{3}}{3} = a \quad \text{Rationalizing the denominator}$$
$$5\sqrt{3} = a \quad \text{Simplifying}$$
$$c = 2a$$
$$c = 2 \cdot 5\sqrt{3}$$
$$c = 10\sqrt{3}$$

Exact answer: $a = 5\sqrt{3}, c = 10\sqrt{3}$

Approximations: $a \approx 8.660, c \approx 17.321$

35. This is an isosceles right triangle with hypotenuse 13. The two legs have the same length, a.

$$a\sqrt{2} = 13$$
$$a = \frac{13}{\sqrt{2}} = \frac{13\sqrt{2}}{2}$$

Exact answer: $\frac{13\sqrt{2}}{2}$

Approximation: 9.192

37. This is a 30-60-90 triangle with the shorter leg = 14. We find the longer leg and the hypotenuse.

$a\sqrt{3} = 14\sqrt{3}$, and $2a = 2 \cdot 14 = 28$.

Exact answer: longer leg $= 14\sqrt{3}$, hypotenuse = 28

Approximation: longer leg ≈ 24.249

39. h is the longer leg of a 30-60-90 right triangle with shorter leg $= 5$. Then $h = 5\sqrt{3} \approx 8.660$.

41. We make a drawing.

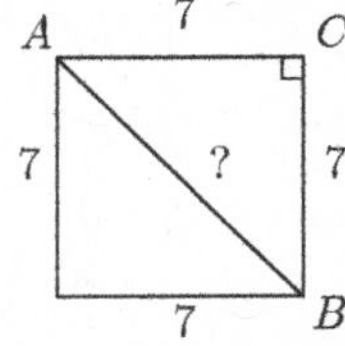

Triangle ABC is an isosceles right triangle with legs of length 7. Then the hypotenuse=$7\sqrt{2} \approx 9.899$.

43. We make a drawing.

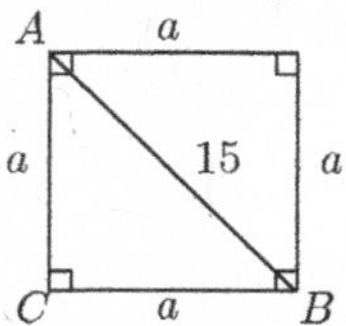

Triangle ABC is an isosceles right triangle with hypotenuse $= 15$. Then $a = \frac{15\sqrt{2}}{2} \approx 10.607$.

45. We will express all distances in feet. Recall that 1 mi = 5280 ft.

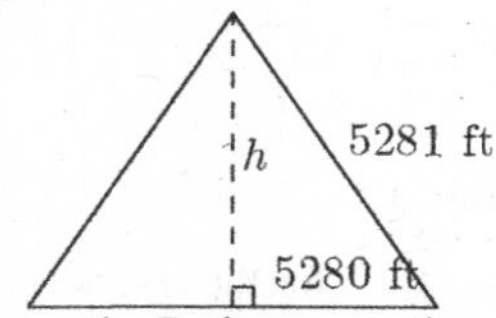

We use the Pythagorean theorem to find h.

$$h^2 + (5280)^2 = (5281)^2$$
$$h^2 + 27{,}878{,}400 = 27{,}888{,}961$$
$$h^2 = 10{,}561$$
$$h = \sqrt{10{,}561}$$
$$h \approx 102.767$$

The height of the bulge is $\sqrt{10{,}561}$ ft, or about 102.767 ft.

47. We make a drawing.

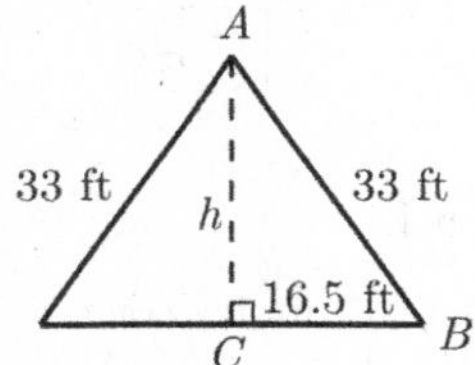

The base of the lodge is an equilateral triangle, so all the angles are 60°. The altitude bisects one angle and one side. Then the triangle ABC is a 30°-60°-90° right triangle with the shorter leg of length $\frac{33}{2}$, or 16.5 ft, and hypotenuse of length 33. Then the height is the length of the shorter leg times $\sqrt{3}$.

Exact answer: $h = \frac{33\sqrt{3}}{2}$ ft

Approximation: $h \approx 28.579$ ft

If the height of triangle ABC is $\frac{33\sqrt{3}}{2}$ and the base is 33 ft, the area is $\frac{1}{2} \cdot 33 \cdot \frac{33\sqrt{3}}{2} = \frac{1089}{4}\sqrt{3}$ ft^2, or about 471.551 ft^2.

49. We make a drawing.

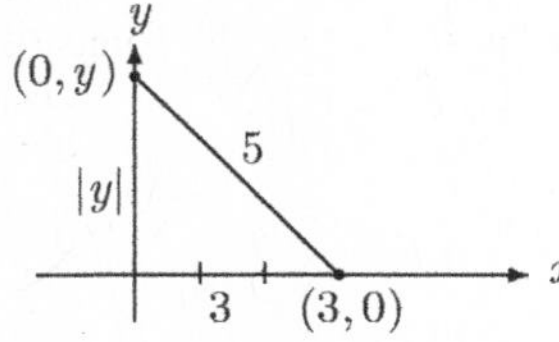

$$|y|^2 + 3^2 = 5^2$$
$$y^2 + 9 = 25$$
$$y^2 = 16$$
$$y = \pm 4$$

The points are $(0, -4)$ and $(0, 4)$.

51. Using the distance formula $d = \sqrt{(x_2 - x_1)^2 + (y_2 - y_1)^2}$ for the points (4, 5) and (7, 1),

$$d = \sqrt{(7-4)^2 + (1-5)^2}$$
$$= \sqrt{3^2 + (-4)^2} = \sqrt{9+16} = \sqrt{25}$$
$$= 5$$

53. Using the distance formula $d = \sqrt{(x_2 - x_1)^2 + (y_2 - y_1)^2}$ for the points (1, –2) and (0, –5),

$$d = \sqrt{(1-0)^2 + (-2-(-5))^2}$$
$$= \sqrt{1^2 + 3^2} = \sqrt{1+9} = \sqrt{10}$$
$$\approx 3.162$$

55. Using the distance formula $d = \sqrt{(x_2 - x_1)^2 + (y_2 - y_1)^2}$ for the points (6, –6) and (–4, 4),

$$d = \sqrt{(6-(-4))^2 + (-6-4)^2}$$
$$= \sqrt{10^2 + (-10)^2} = \sqrt{100+100} = \sqrt{200}$$
$$\approx 14.142$$

57. Using the distance formula $d = \sqrt{(x_2 - x_1)^2 + (y_2 - y_1)^2}$ for the points (–9.2, –3.4) and (8.6, –3.4),

$$d = \sqrt{(-9.2-8.6)^2 + (-3.4-(-3.4))^2}$$
$$= \sqrt{(-17.8)^2 + 0^2} = \sqrt{316.84}$$
$$\approx 17.8$$

59. Using the distance formula $d = \sqrt{(x_2 - x_1)^2 + (y_2 - y_1)^2}$ for the points $\left(\frac{5}{6}, -\frac{1}{6}\right)$ and $\left(\frac{1}{2}, \frac{1}{3}\right)$,

$$d = \sqrt{\left(\frac{5}{6}-\frac{1}{2}\right)^2 + \left(-\frac{1}{6}-\frac{1}{3}\right)^2}$$
$$= \sqrt{\left(\frac{2}{6}\right)^2 + \left(-\frac{3}{6}\right)^2} = \sqrt{\frac{4}{36} + \frac{9}{36}} = \sqrt{\frac{13}{36}} = \frac{\sqrt{13}}{6}$$
$$\approx 0.601$$

61. Using the distance formula $d = \sqrt{(x_2 - x_1)^2 + (y_2 - y_1)^2}$ for the points (0, 0) and $\left(-\sqrt{6}, \sqrt{6}\right)$,

$$d = \sqrt{\left(0-\left(-\sqrt{6}\right)\right)^2 + \left(0-\sqrt{6}\right)^2}$$
$$= \sqrt{\left(\sqrt{6}\right)^2 + \left(-\sqrt{6}\right)^2} = \sqrt{6+6} = \sqrt{12}$$
$$\approx 3.464$$

63. Using the distance formula $d = \sqrt{(x_2 - x_1)^2 + (y_2 - y_1)^2}$ for the points (–2, –40) and (–1, –30),

$$d = \sqrt{(-2-(-1))^2 + (-40-(-30))^2}$$
$$= \sqrt{(-1)^2 + (-10)^2} = \sqrt{1+100} = \sqrt{101}$$
$$\approx 10.050$$

65. Using the midpoint formula $\left(\frac{x_1+x_2}{2}, \frac{y_1+y_2}{2}\right)$ for the points (–2, 5) and (8, 3),

$$\left(\frac{-2+8}{2}, \frac{5+3}{2}\right), \text{ or } \left(\frac{6}{2}, \frac{8}{2}\right), \text{ or } (3, 4)$$

67. Using the midpoint formula $\left(\frac{x_1+x_2}{2}, \frac{y_1+y_2}{2}\right)$ for the points (2, –1) and (5, 8),

$$\left(\frac{2+5}{2}, \frac{-1+8}{2}\right), \text{ or } \left(\frac{7}{2}, \frac{7}{2}\right)$$

69. Using the midpoint formula $\left(\frac{x_1+x_2}{2}, \frac{y_1+y_2}{2}\right)$ for the points (–8, –5) and (6, –1),

$$\left(\frac{-8+6}{2}, \frac{-5+(-1)}{2}\right), \text{ or } \left(-\frac{2}{2}, \frac{-6}{2}\right), \text{ or } (-1, -3)$$

71. Using the midpoint formula $\left(\frac{x_1+x_2}{2}, \frac{y_1+y_2}{2}\right)$ for the points (–3.4, 8.1) and (4.8, –8.1),

$$\left(\frac{-3.4+4.8}{2}, \frac{8.1+(-8.1)}{2}\right), \text{ or } \left(\frac{1.4}{2}, \frac{0}{2}\right), \text{ or } (0.7, 0)$$

73. Using the midpoint formula $\left(\frac{x_1+x_2}{2}, \frac{y_1+y_2}{2}\right)$ for the points $\left(\frac{1}{6}, -\frac{3}{4}\right)$ and $\left(-\frac{1}{3}, \frac{5}{6}\right)$,

$$\left(\frac{\frac{1}{6}+\left(-\frac{1}{3}\right)}{2}, \frac{-\frac{3}{4}+\frac{5}{6}}{2}\right), \text{ or } \left(\frac{-\frac{1}{6}}{2}, \frac{\frac{1}{12}}{2}\right), \text{ or } \left(-\frac{1}{12}, \frac{1}{24}\right)$$

75. Using the midpoint formula $\left(\frac{x_1+x_2}{2}, \frac{y_1+y_2}{2}\right)$ for the points $\left(\sqrt{2}, -1\right)$ and $\left(\sqrt{3}, 4\right)$,

$$\left(\frac{\sqrt{2}+\sqrt{3}}{2}, \frac{-1+4}{2}\right), \text{ or } \left(\frac{\sqrt{2}+\sqrt{3}}{2}, \frac{3}{2}\right)$$

77. *Writing Exercise.*

79. $y = 2x - 3$

Slope is 2; y-intercept is (0, –3).

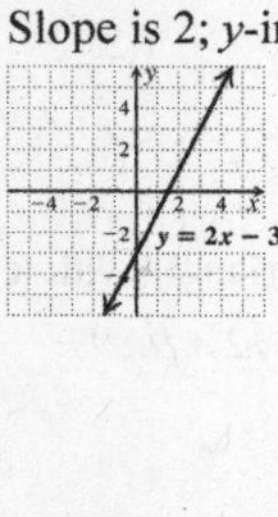

81. $8x - 4y = 8$

To find the y-intercept, let $x = 0$ and solve for y.

$$8 \cdot 0 - 4y = 8$$
$$y = -2$$

The y-intercept is $(0, -2)$.

To find the x-intercept, let $y = 0$ and solve for x.

$$8x - 4 \cdot 0 = 8$$
$$x = 1$$

The x-intercept is $(1, 0)$.

Plot these points and draw the line. A third point could be used as a check.

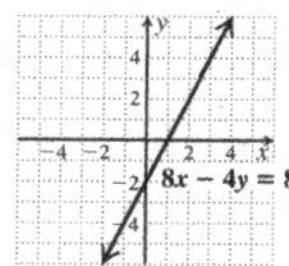

83. $x \geq 1$

Graph the line $x = 1$. Draw the line solid since the inequality symbol is $\geq$. Test the point (0, 0) to determine if it is a solution.

$$\begin{array}{c|c} \multicolumn{2}{c}{x \geq 1} \\ \hline 0 & 1 \end{array}$$

$0 \overset{?}{\geq} 1$ FALSE

Since $0 \geq 1$ is false, we shade the half plane that does not contains (0, 0) and obtain the graph.

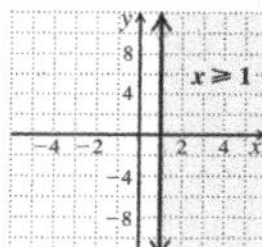

85. *Writing Exercise.*

87. The length of a side of the hexagon is 72/6, or 12 cm. Then the shaded region is a triangle with base 12 cm. To find the height of the triangle, note that it is the longer leg of a 30°-60°-90° right triangle. Thus its length is the length of the length of the shorter leg times $\sqrt{3}$. The length of the shorter leg is half the length of the base, $\frac{1}{2} \cdot 12$ cm, or 6 cm, so the length of the longer leg is $6\sqrt{3}$ cm. Now we find the area of the triangle.

$$A = \frac{1}{2}bh$$
$$= \frac{1}{2}(12 \text{ cm})(6\sqrt{3} \text{ cm})$$
$$= 36\sqrt{3} \text{ cm}^2$$
$$\approx 62.354 \text{ cm}^2$$

89. We make a drawing.

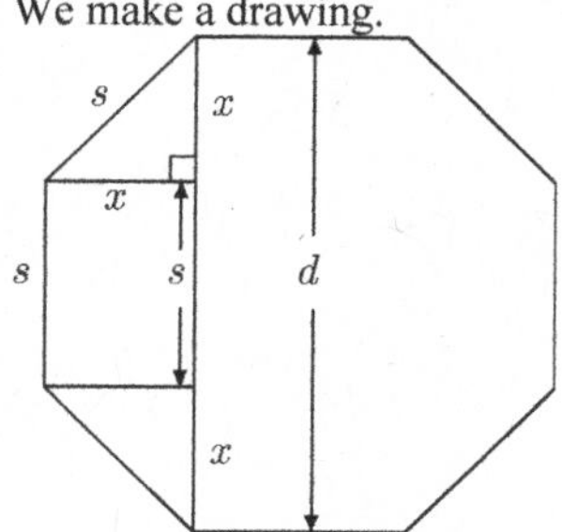

$$d = s + 2x$$

a. Use the Pythagoran theorem to find x.

$$x^2 + x^2 = s^2$$
$$2x^2 = s^2$$
$$x^2 = \frac{s^2}{2}$$
$$x = \frac{s}{\sqrt{2}} = \frac{s}{\sqrt{2}} \cdot \frac{\sqrt{2}}{\sqrt{2}} = \frac{s\sqrt{2}}{2}$$

Then $d = s + 2x = s + 2\left(\frac{s\sqrt{2}}{2}\right) = s + s\sqrt{2}$.

b. From the drawing we see that an octagon is a composite figure consisting of two trapezoids, one on the right and one on the left, with bases s and d, and height x, and a rectangle with sides s and d. Recall that the area of a trapezoid is $\frac{1}{2}h(b_1 + b_2)$. From part a, we know $d = s + s\sqrt{2}$ and $x = \frac{s\sqrt{2}}{2}$.

$$A = 2 \cdot \frac{1}{2}x[s + d] + sd$$
$$A = \frac{s\sqrt{2}}{2}\left[s + \left(s + s\sqrt{2}\right)\right] + s\left(s + s\sqrt{2}\right)$$
$$A = \frac{s\sqrt{2}}{2}\left(2s + s\sqrt{2}\right) + s^2 + s^2\sqrt{2}$$
$$A = s^2\sqrt{2} + s^2 + s^2 + s^2\sqrt{2}$$
$$A = 2s^2 + 2s^2\sqrt{2}, \text{ or } 2s^2(1 + \sqrt{2})$$

91. First we find the radius of a circle with an area of 6160 ft^2. This is the length of the hose.

$$A = \pi r^2$$
$$6160 = \pi r^2$$
$$\frac{6160}{\pi} = r^2$$
$$\sqrt{\frac{6160}{\pi}} = r$$
$$44.28 \approx r$$

Now we make a drawing of the room.

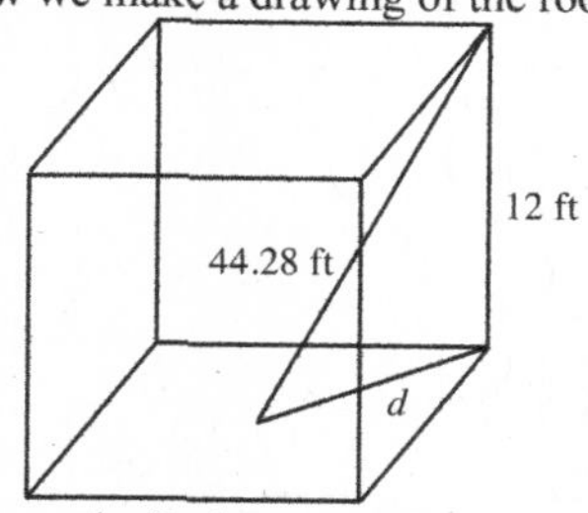

We use the Pythagorean theorem to find d.

$$d^2 + 12^2 = 44.28^2$$
$$d^2 + 144 = 1960.7184$$
$$d^2 = 1816.7184$$
$$d \approx 42.623$$

Now we make a drawing of the floor of the room.

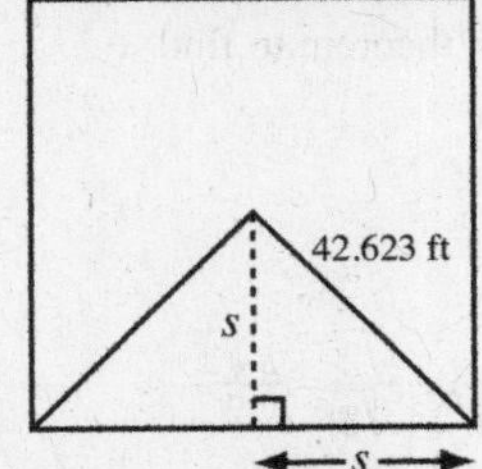

We have an isosceles right triangle with hypotenuse 42.623 ft. We find the length of a side s.

$s\sqrt{2} = 42.623$

$s = \frac{42.623}{\sqrt{2}} \approx 30.14$ ft

Then the length of a side of the room is $2s = 2(30.14 \text{ ft}) = 60.28$ ft; so the dimensions of the largest square room that meets the given conditions are 60.28 ft by 60.28 ft.

93. We make a drawing.

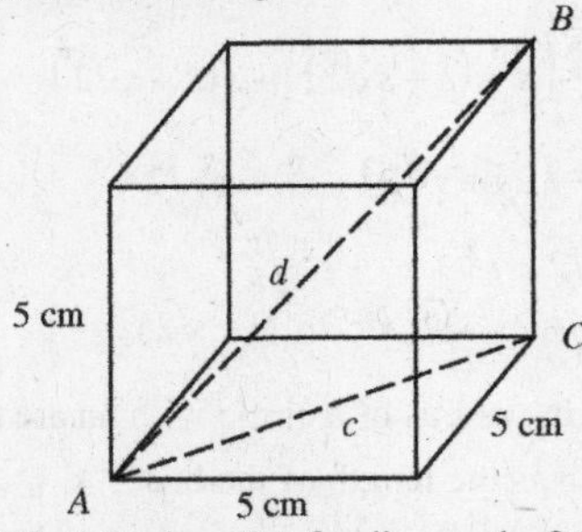

First find the length of a diagonal of the base of the cube. It is the hypotenuse of an isosceles right triangle with legs 5 cm. Then $c = 5\sqrt{2}$ cm.

Triangle ABC is a right triangle with legs of $5\sqrt{2}$ cm and 5 cm and hypotenuse d. Use the Pythagorean theorem to find d, the length of the diagonal that connects two opposite corners of the cube.

$d^2 = (5\sqrt{2})^2 + 5^2$

$d^2 = 25 \cdot 2 + 25$

$d^2 = 50 + 25$

$d^2 = 75$

$d = \sqrt{75}$

Exact answer: $d = \sqrt{75}$ cm

Exercise Set 10.8

1. False

3. True

5. True

7. False

9. $\sqrt{-100} = \sqrt{-1 \cdot 100} = \sqrt{-1} \cdot \sqrt{100} = i \cdot 10 = 10i$

11. $\sqrt{-5} = \sqrt{-1 \cdot 5} = \sqrt{-1} \cdot \sqrt{5} = i \cdot \sqrt{5}$, or $\sqrt{5}i$

13. $\sqrt{-8} = \sqrt{-1} \cdot \sqrt{4} \cdot \sqrt{2} = i \cdot 2 \cdot \sqrt{2} = 2i\sqrt{2}$, or $2\sqrt{2}i$

15. $-\sqrt{-11} = -\sqrt{-1} \cdot \sqrt{11} = -i \cdot \sqrt{11} = -i\sqrt{11}$, or $-\sqrt{11}i$

17. $-\sqrt{-49} = -\sqrt{-1 \cdot 49} = -\sqrt{-1} \cdot \sqrt{49} = -i \cdot 7 = -7i$

19. $-\sqrt{-300} = -\sqrt{-1} \cdot \sqrt{100} \cdot \sqrt{3} = -i \cdot 10 \cdot \sqrt{3} = -10i\sqrt{3}$, or $-10\sqrt{3}i$

21. $6 - \sqrt{-84} = 6 - \sqrt{-1 \cdot 4 \cdot 21} = 6 - i \cdot 2\sqrt{21} = 6 - 2i\sqrt{21}$, or $6 - 2\sqrt{21}i$

23. $-\sqrt{-76} + \sqrt{-125} = -\sqrt{-1 \cdot 4 \cdot 19} + \sqrt{-1 \cdot 25 \cdot 5}$
$= -i \cdot 2\sqrt{19} + i \cdot 5\sqrt{5} = -2i\sqrt{19} + 5i\sqrt{5} = (-2\sqrt{19} + 5\sqrt{5})i$

25. $\sqrt{-18} - \sqrt{-64} = \sqrt{-1 \cdot 9 \cdot 2} - \sqrt{-1 \cdot 64}$
$= i \cdot 3 \cdot \sqrt{2} - i \cdot 8 = 3i\sqrt{2} - 8i$, or $(3\sqrt{2} - 8)i$

27. $(3 + 4i) + (2 - 7i)$
$= (3 + 2) + (4 - 7)i$ Combining the real and the imaginary parts
$= 5 - 3i$

29. $(9 + 5i) - (2 + 3i) = (9 - 2) + (5 - 3)i$
$= 7 + 2i$

31. $(7 - 4i) - (5 - 3i) = (7 - 5) + [-4 - (-3)]i = 2 - i$

33. $(-5 - i) - (7 + 4i) = (-5 - 7) + (-1 - 4)i = -12 - 5i$

35. $5i \cdot 8i = 40 \cdot i^2 = 40(-1) = -40$

37. $(-4i)(-6i) = 24 \cdot i^2 = 24(-1) = -24$

39. $\sqrt{-36}\sqrt{-9} = \sqrt{-1} \cdot \sqrt{36} \cdot \sqrt{-1} \cdot \sqrt{9} = i \cdot 6 \cdot i \cdot 3$
$= i^2 \cdot 18 = -1 \cdot 18 = -18$

41. $\sqrt{-3}\sqrt{-10} = \sqrt{-1} \cdot \sqrt{3} \cdot \sqrt{-1} \cdot \sqrt{10}$
$= i \cdot \sqrt{3} \cdot i \cdot \sqrt{10}$
$= \sqrt{30} \cdot i^2 = \sqrt{30} \cdot (-1)$
$= -\sqrt{30}$

43. $\sqrt{-6}\sqrt{-21} = \sqrt{-1} \cdot \sqrt{6} \cdot \sqrt{-1} \cdot \sqrt{21}$
$= i \cdot \sqrt{6} \cdot i \cdot \sqrt{21} = i^2\sqrt{126} = -1 \cdot \sqrt{9 \cdot 14}$
$= -3\sqrt{14}$

45. $5i(2 + 6i) = 5i \cdot 2 + 5i \cdot 6i = 10i + 30i^2$
$= 10i - 30 = -30 + 10i$

47. $-7i(3 + 4i) = -7i \cdot 3 - 7i \cdot 4i$
$= -21i - 28i^2$
$= -21i + 28 = 28 - 21i$

49. $(1 + i)(3 + 2i) = 3 + 2i + 3i + 2i^2$
$= 3 + 2i + 3i - 2 = 1 + 5i$

51. $(6 - 5i)(3 + 4i) = 18 + 24i - 15i - 20i^2$
$= 18 + 24i - 15i + 20 = 38 + 9i$

53. $(7-2i)(2-6i)=14-42i-4i+12i^2$
$=14-42i-4i-12=2-46i$

55. $(3+8i)(3-8i)=3^2-(8i)^2$ Difference of squares
$=9-64i^2=9-64(-1)$
$=9+64=73$

57. $(-7+i)(-7-i)=(-7)^2-i^2$ Difference of squares
$=49-i^2=49-(-1)$
$=49+1=50$

59. $(4-2i)^2=4^2-2\cdot 4\cdot 2i+(2i)^2=16-16i+4i^2$
$=16-16i-4=12-16i$

61. $(2+3i)^2=2^2+2\cdot 2\cdot 3i+(3i)^2=4+12i+9i^2$
$=4+12i-9=-5+12i$

63. $(-2+3i)^2=(-2)^2+2(-2)(3i)+(3i)^2$
$=4-12i+9i^2=4-12i-9=-5-12i$

65. $\frac{10}{3+i}=\frac{10}{3+i}\cdot\frac{3-i}{3-i}$ Multiplying by 1, using the conjugate
$=\frac{30-10i}{9-i^2}$
$=\frac{30-10i}{9-(-1)}$
$=\frac{30-10i}{10}$
$=3-i$

67. $\frac{2}{3-2i}=\frac{2}{3-2i}\cdot\frac{3+2i}{3+2i}$ Multiplying by 1, using the conjugate
$=\frac{6+4i}{9-4i^2}$
$=\frac{6+4i}{9-4(-1)}$
$=\frac{6+4i}{13}$
$=\frac{6}{13}+\frac{4}{13}i$

69. $\frac{2i}{5+3i}=\frac{2i}{5+3i}\cdot\frac{5-3i}{5-3i}=\frac{10i-6i^2}{25-9i^2}=\frac{10i+6}{25+9}$
$=\frac{10i+6}{34}=\frac{6}{34}+\frac{10}{34}i=\frac{3}{17}+\frac{5}{17}i$

71. $\frac{5}{6i}=\frac{5}{6i}\cdot\frac{i}{i}=\frac{5i}{6i^2}=\frac{5i}{-6}=-\frac{5}{6}i$

73. $\frac{5-3i}{4i}=\frac{5-3i}{4i}\cdot\frac{i}{i}=\frac{5i-3i^2}{4i^2}=\frac{5i+3}{-4}$
$=-\frac{3}{4}-\frac{5}{4}i$

75. $\frac{7i+14}{7i}=\frac{7i}{7i}+\frac{14}{7i}=1+\frac{2}{i}=1+\frac{2}{i}\cdot\frac{i}{i}$
$=1+\frac{2i}{i^2}=1+\frac{2i}{-1}=1-2i$

77. $\frac{4+5i}{3-7i}=\frac{4+5i}{3-7i}\cdot\frac{3+7i}{3+7i}=\frac{12+28i+15i+35i^2}{9-49i^2}$
$=\frac{12+28i+15i-35}{9+49}=\frac{-23+43i}{58}$
$=-\frac{23}{58}+\frac{43}{58}i$

79. $\frac{2+3i}{2+5i}=\frac{2+3i}{2+5i}\cdot\frac{2-5i}{2-5i}=\frac{4-10i+6i-15i^2}{4-25i^2}$
$=\frac{4-10i+6i+15}{4+25}=\frac{19-4i}{29}$
$=\frac{19}{29}-\frac{4}{29}i$

81. $\frac{3-2i}{4+3i}=\frac{3-2i}{4+3i}\cdot\frac{4-3i}{4-3i}=\frac{12-9i-8i+6i^2}{16-9i^2}$
$=\frac{12-9i-8i-6}{16+9}=\frac{6-17i}{25}$
$=\frac{6}{25}-\frac{17}{25}i$

83. $i^{32}=\left(i^2\right)^{16}=(-1)^{16}=1$

85. $i^{15}=i^{14}\cdot i=\left(i^2\right)^7\cdot i=(-1)^7\cdot i=-i$

87. $i^{42}=\left(i^2\right)^{21}=(-1)^{21}=-1$

89. $i^9=\left(i^2\right)^4\cdot i=(-1)^4\cdot i=1\cdot i=i$

91. $(-i)^6=(-1\cdot i)^6=(-1)^6\cdot i^6=1\cdot i^6=(i^2)^3=(-1)^3=-1$

93. $(5i)^3=5^3\cdot i^3=125\cdot i^2\cdot i=125(-1)(i)=-125i$

95. $i^2+i^4=-1+(i^2)^2=-1+(-1)^2=-1+1=0$

97. *Writing Exercise.*

99. $x^2-100=(x+10)(x-10)$

101. $2x-63+x^2=x^2+2x-63$
$=(x+9)(x-7)$

103. $w^3-4w+3w^2-12$
$=w(w^2-4)+3(w^2-4)$
$=(w^2-4)(w+3)$
$=(w+2)(w-2)(w+3)$

105. *Writing Exercise.* Yes; every real number a is a complex number $a+bi$ with $b=0$.

107.

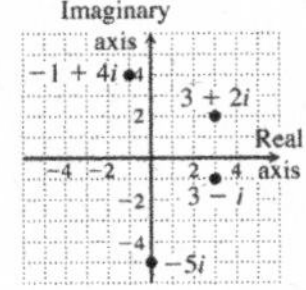

109. $|3+4i|=\sqrt{3^2+4^2}=\sqrt{9+16}=\sqrt{25}=5$

111. $|-1+i|=\sqrt{(-1)^2+1^2}=\sqrt{1+1}=\sqrt{2}$

113. $g(3i) = \frac{(3i)^4 - (3i)^2}{3i - 1} = \frac{81i^4 - 9i^2}{-1+3i} = \frac{81+9}{-1+3i}$
$= \frac{90}{-1+3i} = \frac{90}{-1+3i} \cdot \frac{-1-3i}{-1-3i} = \frac{90(-1-3i)}{1-9i^2}$
$= \frac{90(-1-3i)}{1+9} = \frac{90(-1-3i)}{10} = \frac{9 \cdot 10(-1-3i)}{10}$
$= 9(-1-3i) = -9-27i$

115. First we simplify $g(z)$.

$g(z) = \frac{z^4 - z^2}{z-1} = \frac{z^2(z^2-1)}{z-1} = \frac{z^2(z+1)(z-1)}{z-1}$
$= \frac{z^2(z+1)(z-1)}{z-1} = z^2(z+1)$

Now we substitute.

$g(5i-1) = (5i-1)^2(5i-1+1)$
$= (25i^2 - 10i + 1)(5i)$
$= (-25 - 10i + 1)(5i) = (-24 - 10i)(5i)$
$= -120i - 50i^2 = 50 - 120i$

117. $\frac{1}{w - w^2} = \frac{1}{\frac{1-i}{10} - \left(\frac{1-i}{10}\right)^2}$
$= \frac{1}{\frac{1-i}{10} - \frac{1-2i+i^2}{100}} = \frac{1}{\frac{10-10i-(1-2i-1)}{100}}$
$= \frac{1}{\frac{10-8i}{100}} = \frac{100}{10-8i} = \frac{50}{5-4i}$
$= \frac{50}{5-4i} \cdot \frac{5+4i}{5+4i} = \frac{250+200i}{25+16}$
$= \frac{250+200i}{41} = \frac{250}{41} + \frac{200}{41}i$

119. $(1-i)^3(1+i)^3$
$= (1-i)(1+i) \cdot (1-i)(1+i) \cdot (1-i)(1+i)$
$= (1-i^2)(1-i^2)(1-i^2) = (1+1)(1+1)(1+1)$
$= 2 \cdot 2 \cdot 2 = 8$

121. $\frac{6}{1+\frac{3}{i}} = \frac{6}{\frac{i+3}{i}} = \frac{6i}{i+3} = \frac{6i}{i+3} \cdot \frac{-i+3}{-i+3}$
$= \frac{-6i^2 + 18i}{-i^2 + 9} = \frac{6+18i}{10} = \frac{6}{10} + \frac{18}{10}i$
$= \frac{3}{5} + \frac{9}{5}i$

123. $\frac{i - i^{38}}{1+i} = \frac{i - (i^2)^{19}}{1+i} = \frac{i - (-1)^{19}}{1+i} = \frac{i-(-1)}{1+i} = \frac{i+1}{1+i} = 1$

Chapter 10 Review

1. True
2. False
3. False
4. True
5. True
6. True
7. True
8. False
9. $\sqrt{\frac{100}{121}} = \frac{\sqrt{100}}{\sqrt{121}} = \frac{10}{11}$
10. $-\sqrt{0.36} = -0.6$
11. $f(x) = \sqrt{x+10}$
    $f(15) = \sqrt{15+10} = \sqrt{25} = 5$
12. $f(x) = \sqrt{x+10}$
    Since the index is even, the radicand, $x + 10$, must be non-negative. We solve the inequality:
    $x + 10 \geq 0$
    $x \geq -10$
    Domain of $f = \{x | x \geq -10\}$, or $[-10, \infty)$
13. $\sqrt{64t^2} = \sqrt{(8t)^2} = |8t| = 8|t|$
14. $\sqrt{(c+7)^2} = |c+7|$
15. $\sqrt{4x^2 + 4x + 1} = \sqrt{(2x+1)^2} = |2x+1|$
16. $\sqrt[5]{-32} = \sqrt[5]{(-2)^5} = -2$
17. $\left(\sqrt[3]{5ab}\right)^4 = (5ab)^{4/3}$
18. $(3a^4)^{1/5} = \sqrt[5]{3a^4}$
19. $\sqrt{x^6 y^{10}} = \left(x^6 y^{10}\right)^{1/2} = x^{6/2} y^{10/2} = x^3 y^5$
20. $\sqrt[6]{\left(x^2 y\right)^2} = \left(x^2 y\right)^{2/6} = \left(x^2 y\right)^{1/3} = \sqrt[3]{x^2 y}$
21. $\left(x^{-2/3}\right)^{3/5} = x^{-\frac{2}{3}\cdot\frac{3}{5}} = x^{-2/5} = \frac{1}{x^{2/5}}$
22. $\frac{7^{-1/3}}{7^{-1/2}} = 7^{-1/3+1/2} = 7^{1/6}$
23. $f(x) = \sqrt{25}\sqrt{(x-6)^2} = 5|x-6|$
24. $\sqrt[4]{16x^{20}y^8} = \sqrt[4]{2^4\left(x^5\right)^4\left(y^2\right)^4} = 2x^5y^2$
25. $\sqrt{250x^3y^2} = \sqrt{25x^2y^2 \cdot 10x} = 5xy\sqrt{10x}$
26. $\sqrt{5a}\sqrt{7b} = \sqrt{5a \cdot 7b} = \sqrt{35ab}$
27. $\sqrt[3]{3x^4b}\sqrt[3]{9xb^2} = \sqrt[3]{3x^4b \cdot 9xb^2}$
    $= \sqrt[3]{27x^3b^3 \cdot x^2}$
    $= 3xb\sqrt[3]{x^2}$
28. $\sqrt[3]{-24x^{10}y^8}\sqrt[3]{18x^7y^4} = \sqrt[3]{-24x^{10}y^8 \cdot 18x^7y^4}$
    $= \sqrt[3]{-216x^{15}y^{12} \cdot 2x^2}$
    $= -6x^5y^4\sqrt[3]{2x^2}$

29. $\dfrac{\sqrt[3]{60xy^3}}{\sqrt[3]{10x}} = \sqrt[3]{\dfrac{60xy^3}{10x}} = \sqrt[3]{6y^3} = y\sqrt[3]{6}$

30. $\dfrac{\sqrt{75x}}{2\sqrt{3}} = \dfrac{1}{2}\sqrt{\dfrac{75x}{3}} = \dfrac{1}{2}\sqrt{25x} = \dfrac{5\sqrt{x}}{2}$

31. $\sqrt[4]{\dfrac{48a^{11}}{c^8}} = \dfrac{\sqrt[4]{16a^8\cdot 3a^3}}{\sqrt[4]{c^8}} = \dfrac{2a^2\sqrt[4]{3a^3}}{c^2}$

32. $5\sqrt[3]{4y} + 2\sqrt[3]{4y} = (5+2)\sqrt[3]{4y} = 7\sqrt[3]{4y}$

33.
$$\begin{aligned} 2\sqrt{75} - 9\sqrt{3} &= 2\sqrt{25\cdot 3} - 9\sqrt{3} = 2\cdot 5\sqrt{3} - 9\sqrt{3} \\ &= 10\sqrt{3} - 9\sqrt{3} = (10-9)\sqrt{3} = \sqrt{3} \end{aligned}$$

34.
$$\begin{aligned} &\sqrt{50} + 2\sqrt{18} + \sqrt{32} = \sqrt{25\cdot 2} + 2\sqrt{9\cdot 2} + \sqrt{16\cdot 2} \\ &= 5\sqrt{2} + 2\cdot 3\sqrt{2} + 4\sqrt{2} = 5\sqrt{2} + 6\sqrt{2} + 4\sqrt{2} \\ &= (5+6+4)\sqrt{2} = 15\sqrt{2} \end{aligned}$$

35. $\left(3+\sqrt{10}\right)\left(3-\sqrt{10}\right) = 3^2 - \left(\sqrt{10}\right)^2 = 9 - 10 = -1$

36.
$$\begin{aligned} &\left(\sqrt{3} - 3\sqrt{8}\right)\left(\sqrt{5} + 2\sqrt{8}\right) \\ &= \sqrt{3}\sqrt{5} + \sqrt{3}\cdot 2\sqrt{8} - 3\sqrt{8}\sqrt{5} - 3\sqrt{8}\cdot 2\sqrt{8} \\ &= \sqrt{15} + 2\sqrt{3\cdot 8} - 3\sqrt{8\cdot 5} - 6\sqrt{64} \\ &= \sqrt{15} + 2\sqrt{4\cdot 6} - 3\sqrt{4\cdot 10} - 6\cdot 8 \\ &= \sqrt{15} + 2\cdot 2\sqrt{6} - 3\cdot 2\sqrt{10} - 48 \\ &= \sqrt{15} + 4\sqrt{6} - 6\sqrt{10} - 48 \end{aligned}$$

37. $\sqrt[4]{x}\sqrt{x} = x^{1/4}\cdot x^{1/2} = x^{1/4+2/4} = x^{3/4} = \sqrt[4]{x^3}$

38. $\dfrac{\sqrt[3]{x^2}}{\sqrt[4]{x}} = \dfrac{x^{2/3}}{x^{1/4}} = x^{2/3-1/4} = x^{5/12} = \sqrt[12]{x^5}$

39.
$$\begin{aligned} f\left(2-\sqrt{a}\right) = \left(2-\sqrt{a}\right)^2 &= 2^2 - 2\cdot 2\sqrt{a} + \left(\sqrt{a}\right)^2 \\ &= 4 - 4\sqrt{a} + a \end{aligned}$$

40. $\sqrt{\dfrac{x}{8y}} = \dfrac{\sqrt{x}}{\sqrt{8y}} = \dfrac{\sqrt{x}}{\sqrt{4\cdot 2y}} = \dfrac{\sqrt{x}}{2\sqrt{2y}}\cdot\dfrac{\sqrt{2y}}{\sqrt{2y}} = \dfrac{\sqrt{2xy}}{4y}$

41.
$$\begin{aligned} \frac{4\sqrt{5}}{\sqrt{2}+\sqrt{3}} &= \frac{4\sqrt{5}}{\sqrt{2}+\sqrt{3}}\cdot\frac{\sqrt{2}-\sqrt{3}}{\sqrt{2}-\sqrt{3}} = \frac{4\sqrt{10}-4\sqrt{15}}{2-3} \\ &= -4\sqrt{10} + 4\sqrt{15} \end{aligned}$$

42. $\dfrac{4\sqrt{5}}{\sqrt{2}+\sqrt{3}} = \dfrac{4\sqrt{5}}{\sqrt{2}+\sqrt{3}}\cdot\dfrac{\sqrt{5}}{\sqrt{5}} = \dfrac{4\sqrt{25}}{\sqrt{10}+\sqrt{15}} = \dfrac{20}{\sqrt{10}+\sqrt{15}}$

43.
$$\begin{aligned} \sqrt{y+6} - 2 &= 3 \\ \sqrt{y+6} &= 5 \\ \left(\sqrt{y+6}\right)^2 &= 5^2 \\ y+6 &= 25 \\ y &= 19 \end{aligned}$$

Check:
$$\begin{array}{r|l} \sqrt{y+6} - 2 & = 3 \\ \hline \sqrt{19+6} - 2 & 3 \\ \sqrt{25} - 2 & \\ 5 - 2 & \\ \end{array}$$
$3 \overset{?}{=} 3$ TRUE

The solution is 19.

44.
$$\begin{aligned} (x+1)^{1/3} &= -5 \\ \sqrt[3]{x+1} &= -5 \\ \left(\sqrt[3]{x+1}\right)^3 &= (-5)^3 \\ x+1 &= -125 \\ x &= -126 \end{aligned}$$

Check:
$$\begin{array}{r|l} (x+1)^{1/3} & = -5 \\ \hline (-126+1)^{1/3} & -5 \\ (-125) & \\ \end{array}$$
$-5 \overset{?}{=} -5$ TRUE

The solution is -126.

45.
$$\begin{aligned} 1+\sqrt{x} &= \sqrt{3x-3} \\ \left(1+\sqrt{x}\right)^2 &= \left(\sqrt{3x-3}\right)^2 \\ 1 + 2\sqrt{x} + x &= 3x - 3 \\ 2\sqrt{x} &= 2x - 4 \\ \sqrt{x} &= x - 2 \\ \left(\sqrt{x}\right)^2 &= (x-2)^2 \\ x &= x^2 - 4x + 4 \\ 0 &= x^2 - 5x + 4 \\ 0 &= (x-1)(x-4) \\ x = 1 \ &or\ x = 4 \end{aligned}$$

Check:

For $x = 1$:
$$\begin{array}{r|l} 1+\sqrt{x} & = \sqrt{3x-3} \\ \hline 1+\sqrt{1} & \sqrt{3\cdot 1 - 3} \\ 1+1 & \sqrt{0} \\ \end{array}$$
$2 \overset{?}{=} 0$ FALSE

For $x = 4$:
$$\begin{array}{r|l} 1+\sqrt{x} & = \sqrt{3x-3} \\ \hline 1+\sqrt{4} & \sqrt{3\cdot 4 - 3} \\ 1+2 & \sqrt{9} \\ \end{array}$$
$3 \overset{?}{=} 3$ TRUE

The solution is 4.

46.
$$\begin{aligned} f(a) = \sqrt{a+2} + a &= 4 \\ \sqrt{a+2} + a &= 4 \\ \sqrt{a+2} &= 4 - a \\ \left(\sqrt{a+2}\right)^2 &= (4-a)^2 \\ a + 2 &= 16 - 8a + a^2 \\ 0 &= a^2 - 9a + 14 \\ 0 &= (a-2)(a-7) \\ a = 2 \ &or\ a = 7 \end{aligned}$$

Check:

For $a = 2$: $\sqrt{a+2} + a = 4$

$\sqrt{2+2} + 2 \mid 4$

$2 + 2$

$4 \stackrel{?}{=} 4$ TRUE

For $a = 7$: $\sqrt{a+2} + a = 4$

$\sqrt{7+2} + 7 \mid 4$

$3 + 7$

$10 \stackrel{?}{=} 4$ FALSE

The solution is 2.

47. Let a represent the side of the square. Triangle ABC is an isosceles right triangle with hypotenuse 10 cm. Then the length of a side of the square a, is the length of the legs of the triangle. We have

$$a\sqrt{2} = 10$$
$$a = \frac{10}{\sqrt{2}}, \text{ or}$$
$$a = 5\sqrt{2}$$
$$a \approx 7.071$$

The side is $5\sqrt{2}$ cm or about 7.071 cm.

48. Let b represent the base. We use the Pythagorean theorem to find b.

$$b^2 + 2^2 = 6^2$$
$$b^2 + 4 = 36$$
$$b^2 = 32$$
$$b = \sqrt{32} \approx 5.657$$

The base is $\sqrt{32}$ ft or about 5.657 ft.

49. This is a 30°-60°-90° right triangle with hypotenuse 20. Let $a =$ the shorter leg and $b =$ the longer leg.
$2a = 20$, so $a = 10$, and
$b = a\sqrt{3} = 10\sqrt{3} \approx 17.321$.

50. Using the distance formula
$d = \sqrt{(x_2 - x_1)^2 + (y_2 - y_1)^2}$ for the points (–6, 4) and (–1, 5),

$$d = \sqrt{(-1-(-6))^2 + (5-4)^2}$$
$$= \sqrt{5^2 + 1^2} = \sqrt{25+1} = \sqrt{26}$$
$$\approx 5.099$$

51. Using the midpoint formula $\left(\frac{x_1 + x_2}{2}, \frac{y_1 + y_2}{2}\right)$ for the points (–7, –2) and (3, –1),

$$\left(\frac{-7+3}{2}, \frac{-2+(-1)}{2}\right), \text{ or } \left(\frac{-4}{2}, \frac{-3}{2}\right), \text{ or } \left(-2, -\frac{3}{2}\right)$$

52. $\sqrt{-45} = \sqrt{-1} \cdot \sqrt{9} \cdot \sqrt{5} = 3i\sqrt{5}$ or $3\sqrt{5}\,i$

53. $(-4+3i) + (2-12i) = (-4+2) + (3-12)i = -2-9i$

54. $(9-7i) - (3-8i) = (9-3) + (-7+8)i = 6+i$

55. $(2+5i)(2-5i) = 4 - 25i^2 = 4 + 25 = 29$

56. $i^{34} = (i^2)^{17} = (-1)^{17} = -1$

57. $(6-3i)(2-i) = 12 - 6i - 6i + 3i^2$
$= 12 - 6i - 6i - 3 = 9 - 12i$

58.
$$\frac{7-2i}{3+4i} = \frac{7-2i}{3+4i} \cdot \frac{3-4i}{3-4i} = \frac{21 - 28i - 6i + 8i^2}{9 - 16i^2}$$
$$= \frac{21 - 34i - 8}{9+16} = \frac{13 - 34i}{25} = \frac{13}{25} - \frac{34}{25}i$$

59. *Writing Exercise.* A complex number $a + bi$ is real when $b = 0$. It is imaginary when $b \neq 0$.

60. *Writing Exercise.* An absolute-value sign must be used to simplify $\sqrt[n]{x^n}$ when n is even, since x may be negative. If x is negative while n is even, the radical expression cannot be simplified to x, since $\sqrt[n]{x^n}$ represents the principal, or nonnegative, root. When n is odd, there is only one root, and it will be positive or negative depending on the sign of x. Thus, there is no absolute-value sign when n is odd.

61. Answers may vary. For example, $\frac{2i}{3i}$

62.
$$\sqrt{11x + \sqrt{6+x}} = 6$$
$$\left(\sqrt{11x + \sqrt{6+x}}\right)^2 = (6)^2$$
$$11x + \sqrt{6+x} = 36$$
$$\sqrt{6+x} = 36 - 11x$$
$$\left(\sqrt{6+x}\right)^2 = (36 - 11x)^2$$
$$6 + x = 1296 - 792x + 121x^2$$
$$0 = 121x^2 - 793x + 1290$$
$$0 = (x-3)(121x - 430)$$
$$x = 3 \text{ or } x = \frac{430}{121}$$

Check:

For $x = 3$: $\sqrt{11x + \sqrt{6+x}} = 6$

$\sqrt{11 \cdot 3 + \sqrt{6+3}} \mid 6$

$\sqrt{33+3}$

$6 \stackrel{?}{=} 6$ TRUE

For $x = \frac{430}{121}$: $\sqrt{11x + \sqrt{6+x}} = 6$

$\sqrt{11 \cdot \frac{430}{121} + \sqrt{6 + \frac{430}{121}}} \mid 6$

$\sqrt{\frac{430}{11} + \frac{34}{11}}$

$\sqrt{\frac{464}{11}} \stackrel{?}{=} 6$ FALSE

The solution is 3.

63.
$$\frac{2}{1-3i} - \frac{3}{4+2i} = \frac{2}{1-3i} \cdot \frac{1+3i}{1+3i} + \frac{-3}{4+2i} \cdot \frac{4-2i}{4-2i}$$
$$= \frac{2+6i}{1-9i^2} + \frac{-12+6i}{16-4i^2}$$
$$= \frac{2+6i}{1+9} + \frac{-12+6i}{16+4}$$
$$= \frac{2+6i}{10} + \frac{-12+6i}{20}$$
$$= \frac{1}{5} + \frac{6}{10}i + \frac{-3}{5} + \frac{3}{10}i$$
$$= -\frac{2}{5} + \frac{9}{10}i$$

64. The isosceles right triangle has hypotenuse 6. Let x represent the leg of the triangle. Using the Pythagorean theorem,

$$x^2 + x^2 = 6^2$$
$$2x^2 = 36$$
$$x^2 = 18$$
$$x = \sqrt{18} \text{ ft}$$

The area of the isosceles right triangle is

$$A = \frac{1}{2}x^2 = \frac{1}{2}\left(\sqrt{18}\right)^2 = 9 \text{ ft}^2$$

Then in the 30°-60°-90° triangle, a is the shorter leg and we have

$$6 = 2a$$
$$3 = a$$
$$b = a\sqrt{3}$$
$$b = 3\sqrt{3}$$

The area of the 30-60-90 triangle is

$$A = \frac{1}{2}\left(3\sqrt{3}\right)(3) = \frac{9\sqrt{3}}{2} \text{ ft}^2 \approx 7.794 \text{ ft}^2$$

The area of the isosceles right triangle is larger by about 1.206 ft^2.

65. Using the distance formula $d = \sqrt{(x_2 - x_1)^2 + (y_2 - y_1)^2}$ for the points (0.456, 1.387) and (4.783, 2.865),

$$d = \sqrt{(4.783 - 0.456)^2 + (2.865 - 1.387)^2}$$
$$= \sqrt{(4.327)^2 + (1.478)^2} \approx \sqrt{20.907413} \approx 4.5724$$

The distance is approximately 4.572 miles.

Chapter 10 Test

1. $\sqrt{50} = \sqrt{25 \cdot 2} = \sqrt{25} \cdot \sqrt{2} = 5\sqrt{2}$

2. $\sqrt[3]{-\frac{8}{x^6}} = \frac{\sqrt[3]{-8}}{\sqrt[3]{x^6}} = -\frac{2}{x^2}$

3. $\sqrt{81a^2} = \sqrt{(9a)^2} = |9a| = 9|a|$

4. $\sqrt{x^2 - 8x + 16} = \sqrt{(x-4)^2} = |x - 4|$

5. $\sqrt{7xy} = (7xy)^{1/2}$

6. $\left(4a^3b\right)^{5/6} = \sqrt[6]{\left(4a^3b\right)^5}$

7. $f(x) = \sqrt{2x - 10}$

Since the index is even, the radicand, $2x - 10$, must be non-negative. We solve the inequality:

$$2x - 10 \geq 0$$
$$x \geq 5$$

Domain of $f = \{x | x \geq 5\}$, or $[5, \infty)$

8.
$$f(x) = x^2$$
$$f\left(5 + \sqrt{2}\right) = \left(5 + \sqrt{2}\right)^2 = 5^2 + 2 \cdot 5 \cdot \sqrt{2} + \left(\sqrt{2}\right)^2$$
$$= 25 + 10\sqrt{2} + 2 = 27 + 10\sqrt{2}$$

9.
$$\sqrt[5]{32x^{16}y^{10}} = \sqrt[5]{2^5x^{15}y^{10} \cdot x}$$
$$= \sqrt[5]{2^5x^{15}y^{10}} \cdot \sqrt[5]{x}$$
$$= 2x^3y^2\sqrt[5]{x}$$

10. $\sqrt[3]{4w}\sqrt[3]{4v^2} = \sqrt[3]{4w \cdot 4v^2} = \sqrt[3]{8 \cdot 2v^2w} = 2\sqrt[3]{2v^2w}$

11. $\sqrt{\frac{100a^4}{9b^6}} = \frac{\sqrt{100a^4}}{\sqrt{9b^6}} = \frac{10a^2}{3b^3}$

12. $\frac{\sqrt[5]{48x^6y^{10}}}{\sqrt[5]{16x^2y^9}} = \sqrt[5]{\frac{48x^6y^{10}}{16x^2y^9}} = \sqrt[5]{3x^4y}$

13.
$$\sqrt[4]{x^3}\sqrt{x} = x^{3/4}x^{1/2} = x^{3/4+1/2}$$
$$= x^{5/4} = x^{1+1/4} = x\sqrt[4]{x}$$

14. $\frac{\sqrt{y}}{\sqrt[10]{y}} = \frac{y^{1/2}}{y^{1/10}} = y^{1/2-1/10} = y^{2/5} = \sqrt[5]{y^2}$

15. $8\sqrt{2} - 2\sqrt{2} = (8-2)\sqrt{2} = 6\sqrt{2}$

16.
$$\sqrt{50xy} + \sqrt{72xy} - \sqrt{8xy}$$
$$= \sqrt{25 \cdot 2xy} + \sqrt{36 \cdot 2xy} - \sqrt{4 \cdot 2xy}$$
$$= 5\sqrt{2xy} + 6\sqrt{2xy} - 2\sqrt{2xy}$$
$$= 9\sqrt{2xy}$$

17.
$$\left(7 + \sqrt{x}\right)\left(2 - 3\sqrt{x}\right) = 14 - 7 \cdot 3\sqrt{x} + 2\sqrt{x} - \sqrt{x} \cdot 3\sqrt{x}$$
$$= 14 - 21\sqrt{x} + 2\sqrt{x} - 3\left(\sqrt{x}\right)^2$$
$$= 14 - 19\sqrt{x} - 3x$$

18. $\frac{\sqrt[3]{x}}{\sqrt[3]{4y}} = \frac{\sqrt[3]{x}}{\sqrt[3]{2^2y}} \cdot \frac{\sqrt[3]{2y^2}}{\sqrt[3]{2y^2}} = \frac{\sqrt[3]{2xy^2}}{2y}$

19.
$$6 = \sqrt{x-3} + 5$$
$$1 = \sqrt{x-3}$$
$$1^2 = \left(\sqrt{x-3}\right)^2$$
$$1 = x - 3$$
$$4 = x$$

Check: $6 = \sqrt{x-3} + 5$

6	$\sqrt{4-3} + 5$
	$\sqrt{1} + 5$

$6 \stackrel{?}{=} 6$ TRUE

The solution is 4.

20.
$$x = \sqrt{3x+3} - 1$$
$$x + 1 = \sqrt{3x+3}$$
$$(x+1)^2 = \left(\sqrt{3x+3}\right)^2$$
$$x^2 + 2x + 1 = 3x + 3$$
$$x^2 - x - 2 = 0$$
$$(x+1)(x-2) = 0$$
$$x = -1 \text{ or } x = 2$$

Check:

For $x=-1$: $x=\sqrt{3x+3}-1$
$$\begin{array}{c|l} -1 & \sqrt{3(-1)+3}-1 \\ & \sqrt{0}-1 \\ \end{array}$$
$$-1\overset{?}{=}-1 \quad \text{TRUE}$$

For $x=2$: $x=\sqrt{3x+3}-1$
$$\begin{array}{c|l} 2 & \sqrt{3\cdot 2+3}-1 \\ & \sqrt{9}-1 \\ \end{array}$$
$$2\overset{?}{=}2 \quad \text{TRUE}$$

The solutions are –1 and 2.

21.
$$\begin{aligned} \sqrt{2x} &= \sqrt{x+1}+1 \\ \sqrt{2x}-1 &= \sqrt{x+1} \\ \left(\sqrt{2x}-1\right)^2 &= \left(\sqrt{x+1}\right)^2 \\ 2x-2\sqrt{2x}+1 &= x+1 \\ -2\sqrt{2x} &= -x \\ \left(-2\sqrt{2x}\right)^2 &= (-x)^2 \\ 4(2x) &= x^2 \\ 0 &= x^2-8x \\ 0 &= x(x-8) \end{aligned}$$
$$x=0 \ \text{or} \ x=8$$

Check:

For $x=0$: $\sqrt{2x}=\sqrt{x+1}+1$
$$\begin{array}{c|l} \sqrt{2\cdot 0} & \sqrt{0+1}+1 \\ & 1+1 \end{array}$$
$$0\overset{?}{=}2 \quad \text{FALSE}$$

For $x=8$: $\sqrt{2x}=\sqrt{x+1}+1$
$$\begin{array}{c|l} \sqrt{2\cdot 8} & \sqrt{8+1}+1 \\ \sqrt{16} & 3+1 \end{array}$$
$$4\overset{?}{=}4 \quad \text{TRUE}$$

The solution is 8.

22. We make a drawing.

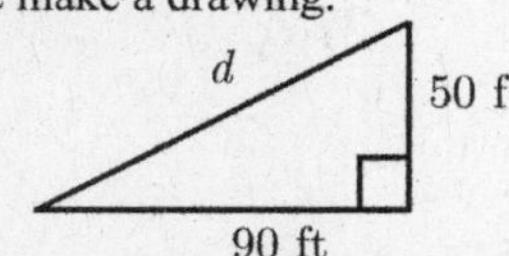

We use the Pythagorean theorem to find *d*.

$d^2=50^2+90^2$

$d^2=2500+8100$

$d^2=10{,}600$

$d=\sqrt{10{,}600}\approx 102.956$

She jogs $\sqrt{10{,}600}$ ft or about 102.956 ft.

23. This is a 30°-60°-90° right triangle with hypotenuse 10. Let $a=$ the shorter leg and $b=$ the longer leg.
$2a=10$, so $a=5$ cm, and
$b=a\sqrt{3}=5\sqrt{3}\approx 8.660$ cm.

24. Using the distance formula
$d=\sqrt{(x_2-x_1)^2+(y_2-y_1)^2}$ for the points (3, 7) and (–1, 8),
$$\begin{aligned} d &= \sqrt{(-1-3)^2+(8-7)^2} \\ &= \sqrt{(-4)^2+(1)^2}=\sqrt{16+1}=\sqrt{17} \\ &\approx 4.123 \end{aligned}$$

25. Using the midpoint formula $\left(\frac{x_1+x_2}{2}, \frac{y_1+y_2}{2}\right)$ for the points (2, –5) and (1, –7),
$\left(\frac{2+1}{2}, \frac{-5+(-7)}{2}\right)$, or $\left(\frac{3}{2}, \frac{-12}{2}\right)$, or $\left(\frac{3}{2}, -6\right)$

26. $\sqrt{-50}=\sqrt{-1}\cdot\sqrt{25}\cdot\sqrt{2}=5i\sqrt{2}$ or $5\sqrt{2}\,i$

27. $(9+8i)-(-3+6i)=(9+3)+(8-6)i=12+2i$

28.
$$\begin{aligned} (4-i)^2 &= 4^2-2\cdot 4\cdot i+(-i)^2=16-8i+i^2 \\ &= 16-8i-1=15-8i \end{aligned}$$

29.
$$\begin{aligned} \frac{-2+i}{3-5i} &= \frac{-2+i}{3-5i}\cdot\frac{3+5i}{3+5i}=\frac{-6-10i+3i+5i^2}{9-25i^2} \\ &= \frac{-6-7i-5}{9+25}=\frac{-11-7i}{34} \\ &= -\frac{11}{34}-\frac{7}{34}i \end{aligned}$$

30. $i^{37}=i^{36}\cdot i=\left(i^2\right)^{18}\cdot i=(-1)^{18}\cdot i=i$

31.
$$\begin{aligned} \sqrt{2x-2}+\sqrt{7x+4} &= \sqrt{13x+10} \\ \left(\sqrt{2x-2}+\sqrt{7x+4}\right)^2 &= \left(\sqrt{13x+10}\right)^2 \\ 2x-2+2\sqrt{2x-2}\sqrt{7x+4}+7x+4 &= 13x+10 \\ 2\sqrt{2x-2}\sqrt{7x+4} &= 4x+8 \\ \left(\sqrt{2x-2}\sqrt{7x+4}\right)^2 &= (2x+4)^2 \\ (2x-2)(7x+4) &= 4x^2+16x+16 \\ 14x^2-6x-8 &= 4x^2+16x+16 \\ 10x^2-22x-24 &= 0 \\ 2\left(5x^2-11x-12\right) &= 0 \\ 2(5x+4)(x-3) &= 0 \end{aligned}$$
$$x=-\frac{4}{5} \ \text{or} \ x=3$$

Check:

For $x=-\frac{4}{5}$:
$$\begin{array}{c|c} \multicolumn{2}{c}{\sqrt{2x-2}+\sqrt{7x+4}=\sqrt{13x+10}} \\ \hline \sqrt{2\cdot\left(-\frac{4}{5}\right)+2}+\sqrt{7\left(-\frac{4}{5}\right)+4} & \sqrt{13\left(-\frac{4}{5}\right)+10} \\ \sqrt{\frac{2}{5}}+\sqrt{-\frac{8}{5}} & \sqrt{-\frac{2}{5}} \end{array}$$

Since the values in the check are not real, $-\frac{4}{5}$ is not a solution.

For $x = 3$:

$$\begin{array}{r|l} \sqrt{2x-2}+\sqrt{7x+4} & =\sqrt{13x+10} \\ \hline \sqrt{2\cdot 3-2}+\sqrt{7\cdot 3+4} & \sqrt{13\cdot 3+10} \\ \sqrt{4}+\sqrt{25} & \sqrt{49} \\ 2+5 & \end{array}$$

$$7 \stackrel{?}{=} 7 \quad \text{TRUE}$$

The solution is 3.

32. $\dfrac{1-4i}{4i(1+4i)^{-1}} = \dfrac{1-4i}{4i} \cdot \dfrac{(1+4i)}{1}$

$= \dfrac{1-16i^2}{4i} = \dfrac{1+16}{4i}$

$= \dfrac{17}{4i} = \dfrac{17}{4i} \cdot \dfrac{i}{i}$

$= \dfrac{17i}{4i^2} = -\dfrac{17}{4}i$

33. Substitute 180 for $D(h)$, and solve for h.

$D(h) = 1.2\sqrt{h}$

$180 = 1.2\sqrt{h}$

$150 = \sqrt{h}$

$(150)^2 = \left(\sqrt{h}\right)^2$

$22{,}500 = h$

The pilot must be above 22,500 ft.

Chapter 11

Quadratic Functions and Equations

Exercise Set 11.1

1. The general form of a *quadratic function* is $f(x) = ax^2 + bx + c$.

3. The quadratic equation $ax^2 + bx + c = 0$ is written in *standard form.*

5. If $x^2 = 7$, we know that $x = \sqrt{7}$ or $x = -\sqrt{7}$ because of the principle of *square roots.*

7. $x^2 = 100$

$x = 10$ *or* $x = -10$ Using the principle of square roots

The solutions are -10 and 10, or ± 10.

9. $p^2 - 50 = 0$

$p^2 = 50$ Isolating p^2

$p = \sqrt{50}$ *or* $p = -\sqrt{50}$ Principle of square roots

$p = 5\sqrt{2}$ *or* $p = -5\sqrt{2}$

The solutions are $5\sqrt{2}$ and $-5\sqrt{2}$ or $\pm 5\sqrt{2}$.

11. $5y^2 = 30$

$y^2 = 6$ Isolating y^2

$y = \sqrt{6}$ *or* $y = -\sqrt{6}$ Principle of square roots

The solutions are $\sqrt{6}$ and $-\sqrt{6}$ or $\pm\sqrt{6}$.

13. $9x^2 - 49 = 0$

$x^2 = \frac{49}{9}$ Isolating x^2

$x = \sqrt{\frac{49}{9}}$ *or* $x = -\sqrt{\frac{49}{9}}$ Principle of square roots

$x = \frac{7}{3}$ *or* $x = -\frac{7}{3}$

The solutions are $\frac{7}{3}$ and $-\frac{7}{3}$ or $\pm\frac{7}{3}$.

15. $6t^2 - 5 = 0$

$t^2 = \frac{5}{6}$

$t = \sqrt{\frac{5}{6}}$ *or* $t = -\sqrt{\frac{5}{6}}$ Principle of square roots

$t = \sqrt{\frac{5}{6}\cdot\frac{6}{6}}$ *or* $t = -\sqrt{\frac{5}{6}\cdot\frac{6}{6}}$ Rationalizing denominators

$t = \frac{\sqrt{30}}{6}$ *or* $t = -\frac{\sqrt{30}}{6}$

The solutions are $\sqrt{\frac{5}{6}}$ and $-\sqrt{\frac{5}{6}}$. This can also be written as $\pm\sqrt{\frac{5}{6}}$ or, if we rationalize the denominator, $\pm\frac{\sqrt{30}}{6}$.

17. $a^2 + 1 = 0$

$a^2 = -1$

$a = \sqrt{-1}$ *or* $a = -\sqrt{-1}$

$a = i$ *or* $a = -i$

The solutions are i and $-i$ or $\pm i$.

19. $4d^2 + 81 = 0$

$d^2 = -\frac{81}{4}$

$d = \sqrt{-\frac{81}{4}}$ *or* $d = -\sqrt{-\frac{81}{4}}$

$d = \frac{9}{2}i$ *or* $d = -\frac{9}{2}i$

The solutions are $\frac{9}{2}i$ and $-\frac{9}{2}i$ or $\pm\frac{9}{2}i$.

21. $(x-3)^2 = 16$

$x - 3 = \sqrt{16}$ *or* $x - 3 = -\sqrt{16}$

$x - 3 = 4$ *or* $x - 3 = -4$

$x = 7$ *or* $x = -1$

The solutions are -1 and 7.

23. $(t+5)^2 = 12$

$t + 5 = \sqrt{12}$ *or* $t + 5 = -\sqrt{12}$

$t + 5 = 2\sqrt{3}$ *or* $t + 5 = -2\sqrt{3}$

$t = -5 + 2\sqrt{3}$ *or* $t = -5 - 2\sqrt{3}$

The solutions are $-5 + 2\sqrt{3}$ and $-5 - 2\sqrt{3}$, or $-5 \pm 2\sqrt{3}$.

25. $(x+1)^2 = -9$

$x + 1 = \sqrt{-9}$ *or* $x + 1 = -\sqrt{-9}$

$x + 1 = 3i$ *or* $x + 1 = -3i$

$x = -1 + 3i$ *or* $x = -1 - 3i$

The solutions are $-1 + 3i$ and $-1 - 3i$, or $-1 \pm 3i$.

27. $\left(y + \frac{3}{4}\right)^2 = \frac{17}{16}$

$y + \frac{3}{4} = \pm\frac{\sqrt{17}}{4}$

$y = -\frac{3}{4} \pm \frac{\sqrt{17}}{4}$, or $\frac{-3 \pm \sqrt{17}}{4}$

The solutions are $-\frac{3}{4} \pm \frac{\sqrt{17}}{4}$, or $\frac{-3 \pm \sqrt{17}}{4}$.

29. $x^2 - 10x + 25 = 64$
$(x-5)^2 = 64$
$x - 5 = \pm 8$
$x = 5 \pm 8$
$x = 13$ or $x = -3$
The solutions are 13 and –3.

31. $f(x) = x^2$
$19 = x^2$ Substituting
$\sqrt{19} = x$ *or* $-\sqrt{19} = x$
The solutions are $\sqrt{19}$ and $-\sqrt{19}$ or $\pm\sqrt{19}$.

33. $f(x) = 16$
$(x-5)^2 = 16$ Substituting
$x - 5 = 4$ *or* $x - 5 = -4$
$x = 9$ *or* $x = 1$
The solutions are 9 and 1.

35. $F(t) = 13$
$(t+4)^2 = 13$ Substituting
$t + 4 = \sqrt{13}$ *or* $t + 4 = -\sqrt{13}$
$t = -4 + \sqrt{13}$ *or* $t = -4 - \sqrt{13}$
The solutions are $-4 + \sqrt{13}$ and $-4 - \sqrt{13}$, or $-4 \pm \sqrt{13}$.

37. $g(x) = x^2 + 14x + 49$
Observe first that $g(0) = 49$. Also observe that when $x = -14$, then
$x^2 + 14x = (-14)^2 - (14)(14) = (14)^2 - (14)^2 = 0$, so $g(-14) = 49$ as well. Thus, we have $x = 0$ or $x = 14$.
We can also do this problem as follows.
$g(x) = 49$
$x^2 + 14x + 49 = 49$ Substituting
$(x+7)^2 = 49$
$x + 7 = 7$ *or* $x + 7 = -7$
$x = 0$ *or* $x = -14$
The solutions are 0 and -14.

39. $x^2 + 16x$
We take half the coefficient of x and square it: Half of 16 is 8, and $8^2 = 64$. We add 64.
$x^2 + 16x + 64 = (x+8)^2$

41. $t^2 - 10t$
We take half the coefficient of t and square it:
Half of –10 is –5, and $(-5)^2 = 25$. We add 25.
$t^2 - 10t + 25 = (t-5)^2$

43. $t^2 - 2t$
We take half the coefficient of t and square it:
$\frac{1}{2}(-2) = -1$, and $(-1)^2 = 1$. We add 1.
$t^2 - 2t + 1 = (t-1)^2$

45. $x^2 + 3x$
We take half the coefficient of t and square it:
$\frac{1}{2}(3) = \frac{3}{2}$, and $\left(\frac{3}{2}\right)^2 = \frac{9}{4}$. We add $\frac{9}{4}$.
$x^2 + 3x + \frac{9}{4} = \left(x + \frac{3}{2}\right)^2$

47. $x^2 + \frac{2}{5}x$
$\frac{1}{2} \cdot \frac{2}{5} = \frac{1}{5}$, and $\left(\frac{1}{5}\right)^2 = \frac{1}{25}$. We add $\frac{1}{25}$.
$x^2 + \frac{2}{5}x + \frac{1}{25} = \left(x + \frac{1}{5}\right)^2$

49. $t^2 - \frac{5}{6}t$
$\frac{1}{2}\left(-\frac{5}{6}\right) = -\frac{5}{12}$, and $\left(-\frac{5}{12}\right)^2 = \frac{25}{144}$. We add $\frac{25}{144}$.
$t^2 - \frac{5}{6}t + \frac{25}{144} = \left(t - \frac{5}{12}\right)^2$

51. $x^2 + 6x = 7$
$x^2 + 6x + 9 = 7 + 9$ Adding 9 to both sides to complete the square
$(x+3)^2 = 16$ Factoring
$x + 3 = \pm 4$ Principle of square roots
$x = -3 \pm 4$
$x = -3 + 4$ *or* $x = -3 - 4$
$x = 1$ *or* $x = -7$
The solutions are 1 and –7.

53. $t^2 - 10t = -23$
$t^2 - 10t + 25 = -23 + 25$ Adding 25 to both sides to complete the square
$(t-5)^2 = 2$ Factoring
$t - 5 = \pm\sqrt{2}$ Principle of square roots
$t = 5 \pm \sqrt{2}$
The solutions are $5 \pm \sqrt{2}$.

55. $x^2 + 12x + 32 = 0$
$x^2 + 12x = -32$
$x^2 + 12x + 36 = -32 + 36$
$(x+6)^2 = 4$
$x + 6 = \pm 2$
$x = -6 \pm 2$
$x = -6 + 2$ *or* $x = -6 - 2$
$x = -4$ *or* $x = -8$
The solutions are –8 and –4.

57. $t^2 + 8t - 3 = 0$
$t^2 + 8t = 3$
$t^2 + 8t + 16 = 3 + 16$
$(t+4)^2 = 19$
$t + 4 = \pm\sqrt{19}$
$t = -4 \pm \sqrt{19}$
The solutions are $-4 \pm \sqrt{19}$.

59. The value of $f(x)$ must be 0 at any x-intercepts.

$$f(x)=0$$
$$x^2+6x+7=0$$
$$x^2+6x=-7$$
$$x^2+6x+9=-7+9$$
$$(x+3)^2=2$$
$$x+3=\pm\sqrt{2}$$
$$x=-3\pm\sqrt{2}$$

The x-intercepts are $(-3-\sqrt{2},\ 0)$ and $(-3+\sqrt{2},\ 0)$.

61. The value of $g(x)$ must be 0 at any x-intercepts.

$$g(x)=0$$
$$x^2+9x-25=0$$
$$x^2+9x=25$$
$$x^2+9x+\frac{81}{4}=25+\frac{81}{4}$$
$$\left(x+\frac{9}{2}\right)^2=\frac{181}{4}$$
$$x+\frac{9}{2}=\pm\frac{\sqrt{181}}{2}$$
$$x=-\frac{9}{2}\pm\frac{\sqrt{181}}{2}$$

The x-intercepts are $\left(-\frac{9}{2}-\frac{\sqrt{181}}{2},\ 0\right)$ and $\left(-\frac{9}{2}+\frac{\sqrt{181}}{2},\ 0\right)$.

63. The value of $f(x)$ must be 0 at any x-intercepts.

$$f(x)=0$$
$$x^2-10x-22=0$$
$$x^2-10x=22$$
$$x^2-10x+25=22+25$$
$$(x-5)^2=47$$
$$x-5=\pm\sqrt{47}$$
$$x=5\pm\sqrt{47}$$

The x-intercepts are $(5-\sqrt{47},\ 0)$ and $(5+\sqrt{47},\ 0)$.

65.

$$9x^2+18x=-8$$
$$x^2+2x=-\frac{8}{9} \quad \text{Dividing both sides by 9}$$
$$x^2+2x+1=-\frac{8}{9}+1$$
$$(x+1)^2=\frac{1}{9}$$
$$x+1=\pm\frac{1}{3}$$
$$x=-1\pm\frac{1}{3}$$
$$x=-1-\frac{1}{3} \quad or \quad x=-1+\frac{1}{3}$$
$$x=-\frac{4}{3} \quad or \quad x=-\frac{2}{3}$$

The solutions are $-\frac{4}{3}$ and $-\frac{2}{3}$.

67.

$$3x^2-5x-2=0$$
$$3x^2-5x=2$$
$$x^2-\frac{5}{3}x=\frac{2}{3} \quad \text{Dividing both sides by 3}$$
$$x^2-\frac{5}{3}x+\frac{25}{36}=\frac{2}{3}+\frac{25}{36}$$
$$\left(x-\frac{5}{6}\right)^2=\frac{49}{36}$$
$$x-\frac{5}{6}=\pm\frac{7}{6}$$
$$x=\frac{5}{6}\pm\frac{7}{6}$$
$$x=\frac{5}{6}-\frac{7}{6} \quad or \quad x=\frac{5}{6}+\frac{7}{6}$$
$$x=-\frac{1}{3} \quad or \quad x=2$$

The solutions are $-\frac{1}{3}$ and 2.

69.

$$5x^2+4x-3=0$$
$$5x^2+4x=3$$
$$x^2+\frac{4}{5}x=\frac{3}{5} \quad \text{Dividing both sides by 5}$$
$$x^2+\frac{4}{5}x+\frac{4}{25}=\frac{3}{5}+\frac{4}{25}$$
$$\left(x+\frac{2}{5}\right)^2=\frac{19}{25}$$
$$x+\frac{2}{5}=\pm\frac{\sqrt{19}}{5}$$
$$x=-\frac{2}{5}\pm\frac{\sqrt{19}}{5}, \text{ or } \frac{-2\pm\sqrt{19}}{5}$$

The solutions are $-\frac{2}{5}\pm\frac{\sqrt{19}}{5}$, or $\frac{-2\pm\sqrt{19}}{5}$.

71. The value of $f(x)$ must be 0 at any x-intercepts.

$$f(x)=0$$
$$4x^2+2x-3=0$$
$$4x^2+2x=3$$
$$x^2+\frac{1}{2}x=\frac{3}{4} \quad \text{Dividing both sides by 4}$$
$$x^2+\frac{1}{2}x+\frac{1}{16}=\frac{3}{4}+\frac{1}{16}$$
$$\left(x+\frac{1}{4}\right)^2=\frac{13}{16}$$
$$x+\frac{1}{4}=\pm\frac{\sqrt{13}}{4}$$
$$x=-\frac{1}{4}\pm\frac{\sqrt{13}}{4}, \text{ or } \frac{-1\pm\sqrt{13}}{4}$$

The x-intercepts are $\left(-\frac{1}{4}-\frac{\sqrt{13}}{4},\ 0\right)$ and $\left(-\frac{1}{4}+\frac{\sqrt{13}}{4},\ 0\right)$, or $\left(\frac{-1-\sqrt{13}}{4},\ 0\right)$ and $\left(\frac{-1+\sqrt{13}}{4},\ 0\right)$.

73. The value of $g(x)$ must be 0 at any x-intercepts.

$$g(x)=0$$
$$2x^2-3x-1=0$$
$$2x^2-3x=1$$
$$x^2-\frac{3}{2}x=\frac{1}{2} \quad \text{Dividing both sides by 2}$$
$$x^2-\frac{3}{2}x+\frac{9}{16}=\frac{1}{2}+\frac{9}{16}$$
$$\left(x-\frac{3}{4}\right)^2=\frac{17}{16}$$
$$x-\frac{3}{4}=\pm\frac{\sqrt{17}}{4}$$
$$x=\frac{3}{4}\pm\frac{\sqrt{17}}{4}, \text{ or } \frac{3\pm\sqrt{17}}{4}$$

The x-intercepts are $\left(\frac{3}{4}-\frac{\sqrt{17}}{4}, 0\right)$ and $\left(\frac{3}{4}+\frac{\sqrt{17}}{4}, 0\right)$, or $\left(\frac{3-\sqrt{17}}{4}, 0\right)$ and $\left(\frac{3+\sqrt{17}}{4}, 0\right)$.

75. ***Familiarize***. We are already familiar with the compound-interest formula.
Translate. We substitute into the formula.

$$A=P(1+r)^t$$
$$2205=2000(1+r)^2$$

Carry out. We solve for r.

$$2205=2000(1+r)^2$$
$$\frac{2205}{2000}=(1+r)^2$$
$$\frac{441}{400}=(1+r)^2$$
$$\pm\sqrt{\frac{441}{400}}=1+r$$
$$\pm\frac{21}{20}=1+r$$
$$-\frac{20}{20}\pm\frac{21}{20}=r$$
$$\frac{1}{20}=r \text{ or } -\frac{41}{20}=r$$

Check. Since the interest rate cannot be negative, we need only check $\frac{1}{20}$, or 5%. If \$2000 were invested at 5% interest, compounded annually, then in 2 years it would grow to $\$2000(1.05)^2$, or \$2205. The number 5% checks.
State. The interest rate is 5%.

77. ***Familiarize***. We are already familiar with the compound-interest formula.
Translate. We substitute into the formula.

$$A=P(1+r)^t$$
$$6760=6250(1+r)^2$$

Carry out. We solve for r.

$$\frac{6760}{6250}=(1+r)^2$$
$$\frac{676}{625}=(1+r)^2$$
$$\pm\frac{26}{25}=1+r$$
$$-\frac{25}{25}\pm\frac{26}{25}=r$$
$$\frac{1}{25}=r \text{ or } -\frac{51}{25}=r$$

Check. Since the interest rate cannot be negative, we need only check $\frac{1}{25}$, or 4%. If \$6250 were invested at 4% interest, compounded annually, then in 2 years it would grow to $\$6250(1.04)^2$, or \$6760. The number 4% checks.
State. The interest rate is 4%.

79. ***Familiarize***. We will use the formula $s=16t^2$.
Translate. We substitute into the formula.

$$s=16t^2$$
$$3593=16t^2$$

Carry out. We solve for t.

$$3593=16t^2$$
$$\frac{3593}{16}=t^2$$
$$\sqrt{\frac{3593}{16}}=t \quad \text{Principle of square roots; rejecting the negative square root}$$
$$15.0\approx t$$

Check. Since $16(15.0)^2=3600\approx 3593$, our answer checks.
State. It would take an object about 15.0 sec to fall freely from the top.

81. ***Familiarize***. We will use the formula $s=16t^2$.
Translate. We substitute into the formula.

$$s=16t^2$$
$$890=16t^2$$

Carry out. We solve for t.

$$890=16t^2$$
$$55.625=t^2$$
$$\sqrt{55.625}=t \quad \text{Principle of square roots; rejecting the negative square root}$$
$$7.5\approx t$$

Check. Since $16(7.5)^2=900\approx 890$, our answer checks.
State. It would take an object about 7.5 sec to fall freely from the bridge to the river.

83. *Writing Exercise.*
1) If the quadratic equation is of the type $x^2=k$, use the principle of square roots.
2) If the quadratic equation is of the type $ax^2+bx+c=0,\ b\neq 0$, use the principle of zero products, if possible.
3) If the quadratic equation is of the type $ax^2+bx+c=0,\ b\neq 0$, and factoring is difficult or impossible, solve by completing the square.

85. $3y^2-300y=3y(y^2-100)$
$=3y(y+10)(y-10)$

87. $6x^2+6x+6=6(x^2+x+1)$

89. $20x^2+7x-6=(4x+3)(5x-2)$

91. *Writing Exercise.*

93. In order for $x^2+bx+81$ to be a square, the following must be true:

$$\left(\frac{b}{2}\right)^2=81$$
$$\frac{b^2}{4}=81$$
$$b^2=324$$
$$b=18 \text{ or } b=-18$$

95. We see that x is a factor of each term, so x is also a factor of $f(x)$. We have $f(x)=x(2x^4-9x^3-66x^2+45x+280)$. Since x^2-5 is a factor of $f(x)$ it is also a factor of $2x^4-9x^3-66x^2+45x+280$. We divide to find another factor.

$$\begin{array}{r} 2x^2-9x-56 \\ x^2-5\,\overline{)\,2x^4-9x^3-66x^2+45x+280} \\ \underline{2x^4\qquad\quad -10x^2\qquad\qquad} \\ -9x^3-56x^2+45x\qquad \\ \underline{-9x^3\qquad\quad +45x\qquad} \\ -56x^2\qquad +280 \\ \underline{-56x^2\qquad +280} \\ 0 \end{array}$$

Then we have $f(x)=x(x^2-5)(2x^2-9x-56)$, or $f(x)=x(x^2-5)(2x+7)(x-8)$. Now we find the values of a for which $f(a)=0$.

$$f(a)=0$$
$$a(a^2-5)(2a+7)(a-8)=0$$

$a=0$ *or* $a^2-5=0$ *or* $2a+7=0$ *or* $a-8=0$
$a=0$ *or* $a^2=5$ *or* $2a=-7$ *or* $a=8$
$a=0$ *or* $a=\pm\sqrt{5}$ *or* $a=-\frac{7}{2}$ *or* $a=8$

The solutions are 0, $\sqrt{5}$, $-\sqrt{5}$, $-\frac{7}{2}$, and 8.

97. ***Familiarize***. It is helpful to list information in a chart and make a drawing. Let r represent the speed of the fishing boat. Then $r-7$ represents the speed of the barge.

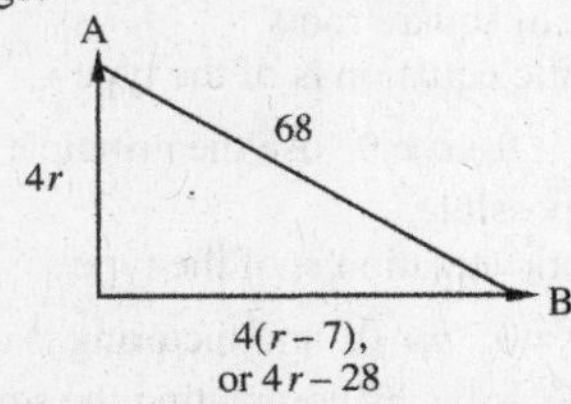

Boat	r	t	d
Fishing	r	4	$4r$
Barge	$r-7$	4	$4(r-7)$

Translate. We use the Pythagorean equation:

$$a^2+b^2=c^2$$
$$(4r-28)^2+(4r)^2=68^2$$

Carry out.

$$(4r-28)^2+(4r)^2=68^2$$
$$16r^2-224r+784+16r^2=4624$$
$$32r^2-224r-3840=0$$
$$r^2-7r-120=0$$
$$(r+8)(r-15)=0$$

$r+8=0$ *or* $r-15=0$
$r=-8$ *or* $r=15$

Check. We check only 15 since the speeds of the boats cannot be negative. If the speed of the fishing boat is 15 km/h, then the speed of the barge is 15 – 7, or 8 km/h, and the distances they travel are $4\cdot 15$ (or 60) and $4\cdot 8$ (or 32). $60^2+32^2=3600+1024=4624$ $=68^2$ The values check.

State. The speed of the fishing boat is 15 km/h, and the speed of the barge is 8 km/h.

99. *Graphing Calculator Exercise*

101. *Writing Exercise*. From a reading of the problem we know that we are interested only in positive values of r and it is safe to assume $r\le 1$. We also know that we want to find the value of r for which $4410=4000(1+r)^2$, so the window must include the y-value 4410. A suitable viewing window might be [0, 1, 4000, 4500], Xscl = 0.1, Yscl = 100.

Connecting the Concepts

1.
$$x^2-3x-10=0$$
$$(x+2)(x-5)=0$$

$x+2=0$ *or* $x-5=0$
$x=-2$ *or* $x=5$

2.
$$x^2=121$$
$$x=\pm 11$$

3.
$$x^2+6x=10$$
$$x^2+6x+9=10+9 \quad \text{Completing the square}$$
$$(x+3)^2=19$$
$$x+3=\pm\sqrt{19}$$
$$x=-3\pm\sqrt{19}$$

4. $x^2+x-3=0$
$a=1$, $b=1$, $c=-3$

$$x=\frac{-1\pm\sqrt{1^2-4\cdot 1\cdot(-3)}}{2\cdot 1}=\frac{-1\pm\sqrt{1+12}}{2}$$
$$=\frac{-1\pm\sqrt{13}}{2}=-\frac{1}{2}\pm\frac{\sqrt{13}}{2}$$

5. $(x+1)^2=2$
$$x+1=\pm\sqrt{2}$$
$$x=-1\pm\sqrt{2}$$

6. $x^2 - 10x + 25 = 0$
$(x-5)(x-5) = 0$
$(x-5)^2 = 0$
$x - 5 = 0$
$x = 5$

7. $x^2 - 2x = 6$
$x^2 - 2x + 1 = 6 + 1$
$(x-1)^2 = 7$
$x - 1 = \pm\sqrt{7}$
$x = 1 \pm \sqrt{7}$

8. $4t^2 = 11$
$t^2 = \frac{11}{4}$
$t = \pm\sqrt{\frac{11}{4}} = \pm\frac{\sqrt{11}}{2}$

Exercise Set 11.2

1. True

3. False; see Example 3.

5. False; the quadratic formula yields at most two solutions.

7. $2x^2 + 3x - 5 = 0$
$(2x+5)(x-1) = 0$ Factoring
$2x + 5 = 0$ *or* $x - 1 = 0$
$x = -\frac{5}{2}$ *or* $x = 1$
The solutions are $-\frac{5}{2}$ and 1.

9. $u^2 + 2u - 4 = 0$
$u^2 + 2u = 4$
$u^2 + 2u + 1 = 4 + 1$ Completing the square
$(u+1)^2 = 5$
$u + 1 = \pm\sqrt{5}$ Principle of square roots
$u = -1 \pm \sqrt{5}$
The solutions are $-1 + \sqrt{5}$ and $-1 - \sqrt{5}$.

11. $t^2 + 3 = 6t$
$t^2 - 6t = -3$
$t^2 - 6t + 9 = -3 + 9$
$(t-3)^2 = 6$
$t - 3 = \pm\sqrt{6}$
$t = 3 \pm \sqrt{6}$
The solutions are $3 + \sqrt{6}$ and $3 - \sqrt{6}$.

13. $x^2 = 3x + 5$
$x^2 - 3x - 5 = 0$
$a = 1,\ b = -3,\ c = -5$
$x = \frac{-b \pm \sqrt{b^2 - 4ac}}{2a}$
$x = \frac{-(-3) \pm \sqrt{(-3)^2 - 4 \cdot 1 \cdot (-5)}}{2 \cdot 1} = \frac{3 \pm \sqrt{9 + 20}}{2}$
$x = \frac{3 \pm \sqrt{29}}{2} = \frac{3}{2} \pm \frac{\sqrt{29}}{2}$
The solutions are $\frac{3}{2} + \frac{\sqrt{29}}{2}$ and $\frac{3}{2} - \frac{\sqrt{29}}{2}$.

15. $3t(t+2) = 1$
$3t^2 + 6t = 1$
$3t^2 + 6t - 1 = 0$
$a = 3,\ b = 6,\ c = -1$
$t = \frac{-b \pm \sqrt{b^2 - 4ac}}{2a}$
$t = \frac{-6 \pm \sqrt{6^2 - 4 \cdot 3 \cdot (-1)}}{2 \cdot 3} = \frac{-6 \pm \sqrt{36 + 12}}{6}$
$t = \frac{-6 \pm \sqrt{48}}{6} = \frac{-6 \pm 4\sqrt{3}}{6}$
$t = -\frac{6}{6} \pm \frac{4\sqrt{3}}{6} = -1 \pm \frac{2\sqrt{3}}{3}$
The solutions are $-1 + \frac{2\sqrt{3}}{3}$ and $-1 - \frac{2\sqrt{3}}{3}$.

17. $\frac{1}{x^2} - 3 = \frac{8}{x}$, LCD is x^2
$x^2\left(\frac{1}{x^2} - 3\right) = x^2 \cdot \frac{8}{x}$
$x^2 \cdot \frac{1}{x^2} - x^2 \cdot 3 = 8x$
$1 - 3x^2 = 8x$
$0 = 3x^2 + 8x - 1$
$a = 3,\ b = 8,\ c = -1$
$x = \frac{-8 \pm \sqrt{8^2 - 4 \cdot 3 \cdot (-1)}}{2 \cdot 3} = \frac{-8 \pm \sqrt{64 + 12}}{6}$
$x = \frac{-8 \pm \sqrt{76}}{6} = \frac{-8 \pm \sqrt{4 \cdot 19}}{6} = \frac{-8 \pm 2\sqrt{19}}{6}$
$x = \frac{-4 \pm \sqrt{19}}{3} = -\frac{4}{3} \pm \frac{\sqrt{19}}{3}$
The solutions are $-\frac{4}{3} - \frac{\sqrt{19}}{3}$ and $-\frac{4}{3} + \frac{\sqrt{19}}{3}$.

19. $t^2 + 10 = 6t$
$t^2 - 6t + 10 = 0$
$a = 1,\ b = -6,\ c = 10$
$t = \frac{-b \pm \sqrt{b^2 - 4ac}}{2a}$
$t = \frac{-(-6) \pm \sqrt{(-6)^2 - 4 \cdot 1 \cdot 10}}{2 \cdot 1} = \frac{6 \pm \sqrt{36 - 40}}{6}$
$t = \frac{6 \pm \sqrt{-4}}{2} = \frac{6 \pm 2i}{2}$
$t = \frac{6}{2} \pm \frac{2i}{2} = 3 \pm i$
The solutions are $3 + i$ and $3 - i$.

21. $p^2 - p + 1 = 0$
$a = 1,\ b = -1,\ c = 1$
$$p = \frac{-b \pm \sqrt{b^2 - 4ac}}{2a}$$
$$p = \frac{-(-1) \pm \sqrt{(-1)^2 - 4 \cdot 1 \cdot 1}}{2 \cdot 1} = \frac{1 \pm \sqrt{1-4}}{2}$$
$$p = \frac{1 \pm \sqrt{-3}}{2} = \frac{1}{2} \pm \frac{\sqrt{3}}{2}i$$
The solutions are $\frac{1}{2} + \frac{\sqrt{3}}{2}i$ and $\frac{1}{2} - \frac{\sqrt{3}}{2}i$.

23. $x^2 + 4x + 6 = 0$
$$x = \frac{-b \pm \sqrt{b^2 - 4ac}}{2a}$$
$$x = \frac{-4 \pm \sqrt{4^2 - 4 \cdot 1 \cdot 6}}{2 \cdot 1} = \frac{-4 \pm \sqrt{16 - 24}}{2}$$
$$x = \frac{-4 \pm \sqrt{-8}}{2} = -\frac{4}{2} \pm \frac{2\sqrt{2}}{2}i$$
$$x = -2 \pm \sqrt{2}i$$
The solutions are $-2 + \sqrt{2}i$ and $-2 - \sqrt{2}i$.

25. $12t^2 + 17t = 40$
$12t^2 + 17t - 40 = 0$
$(3t + 8)(4t - 5) = 0$
$3t + 8 = 0$ *or* $4t - 5 = 0$
$t = -\frac{8}{3}$ *or* $t = \frac{5}{4}$
The solutions are $-\frac{8}{3}$ and $\frac{5}{4}$.

27. $25x^2 - 20x + 4 = 0$
$(5x - 2)(5x - 2) = 0$
$5x - 2 = 0$ *or* $5x - 2 = 0$
$5x = 2$ *or* $5x = 2$
$x = \frac{2}{5}$ *or* $x = \frac{2}{5}$
The solution is $\frac{2}{5}$.

29. $7x(x + 2) + 5 = 3x(x + 1)$
$7x^2 + 14x + 5 = 3x^2 + 3x$
$4x^2 + 11x + 5 = 0$
$a = 4,\ b = 11,\ c = 5$
$$x = \frac{-11 \pm \sqrt{11^2 - 4 \cdot 4 \cdot 5}}{2 \cdot 4} = \frac{-11 \pm \sqrt{121 - 80}}{8}$$
$$x = \frac{-11 \pm \sqrt{41}}{8} = -\frac{11}{8} \pm \frac{\sqrt{41}}{8}$$
The solutions are $-\frac{11}{8} - \frac{\sqrt{41}}{8}$ and $-\frac{11}{8} + \frac{\sqrt{41}}{8}$.

31. $14(x - 4) - (x + 2) = (x + 2)(x - 4)$
$14x - 56 - x - 2 = x^2 - 2x - 8$ Removing parentheses
$13x - 58 = x^2 - 2x - 8$
$0 = x^2 - 15x + 50$
$0 = (x - 10)(x - 5)$
$x - 10 = 0$ *or* $x - 5 = 0$
$x = 10$ *or* $x = 5$
The solutions are 10 and 5.

33. $51p = 2p^2 + 72$
$0 = 2p^2 - 51p + 72$
$0 = (2p - 3)(p - 24)$
$2p - 3 = 0$ *or* $p - 24 = 0$
$p = \frac{3}{2}$ *or* $p = 24$
The solutions are $\frac{3}{2}$ and 24.

35. $x(x - 3) = x - 9$
$x^2 - 3x = x - 9$ Removing parentheses
$x^2 - 4x = -9$
$x^2 - 4x + 4 = -9 + 4$ Completing the square
$(x - 2)^2 = -5$
$x - 2 = \pm\sqrt{-5}$
$x = 2 \pm \sqrt{5}i$
The solutions are $2 + \sqrt{5}i$ and $2 - \sqrt{5}i$.

37. $x^3 - 8 = 0$
$x^3 - 2^3 = 0$
$(x - 2)(x^2 + 2x + 4) = 0$
$x - 2 = 0$ *or* $x^2 + 2x + 4 = 0$
$x = 2$ *or* $x = \frac{-2 \pm \sqrt{2^2 - 4 \cdot 1 \cdot 4}}{2 \cdot 1}$
$x = 2$ *or* $x = \frac{-2 \pm \sqrt{-12}}{2} = \frac{-2 \pm 2i\sqrt{3}}{2}$
$x = 2$ *or* $x = -\frac{2}{2} \pm \frac{2\sqrt{3}}{2}i$
$x = 2$ *or* $x = -1 \pm \sqrt{3}i$
The solutions are 2, $-1 + \sqrt{3}i$, and $-1 - \sqrt{3}i$.

39. $f(x) = 0$
$6x^2 - 7x - 20 = 0$
$(3x + 4)(2x - 5) = 0$
$3x + 4 = 0$ *or* $2x - 5 = 0$
$x = -\frac{4}{3}$ *or* $x = \frac{5}{2}$
$f(x) = 0$ for $x = -\frac{4}{3}$ and $x = \frac{5}{2}$.

41. $f(x) = 1$ Substituting
$$\frac{7}{x} + \frac{7}{x+4} = 1$$
$$x(x+4)\left(\frac{7}{x} + \frac{7}{x+4}\right) = x(x+4) \cdot 1$$
Multiplying by the LCD
$7(x + 4) + 7x = x^2 + 4x$
$7x + 28 + 7x = x^2 + 4x$
$14x + 28 = x^2 + 4x$
$0 = x^2 - 10x - 28$
$a = 1,\ b = -10,\ c = -28$

$$x=\frac{-(-10)\pm\sqrt{(-10)^2-4\cdot1\cdot(-28)}}{2\cdot1}$$
$$x=\frac{10\pm\sqrt{100+112}}{2}=\frac{10\pm\sqrt{212}}{2}$$
$$x=\frac{10\pm\sqrt{4\cdot53}}{2}=\frac{10\pm2\sqrt{53}}{2}$$
$$x=5\pm\sqrt{53}$$
$f(x)=1$ for $x=5+\sqrt{53}$ and $x=5-\sqrt{53}$.

43. $F(x)=G(x)$
$$\frac{3-x}{4}=\frac{1}{4x}$$
$$4x\cdot\frac{3-x}{4}=4x\cdot\frac{1}{4x}$$
$$3x-x^2=1$$
$$0=x^2-3x+1$$
$$x=\frac{-(-3)\pm\sqrt{(-3)^2-4\cdot1\cdot1}}{2\cdot1}=\frac{3\pm\sqrt{5}}{2}$$
$$x=\frac{3}{2}\pm\frac{\sqrt{5}}{2}$$

45. $x^2+6x+4=0$
$$x=\frac{-6\pm\sqrt{6^2-4\cdot1\cdot4}}{2\cdot1}=\frac{-6\pm\sqrt{20}}{2}$$
$$x=\frac{-6+\sqrt{20}}{2}\approx-0.764$$
$$x=\frac{-6-\sqrt{20}}{2}\approx-5.236$$

47. $x^2-6x+4=0$
$a=1$, $b=-6$, $c=4$
$$x=\frac{-(-6)\pm\sqrt{(-6)^2-4\cdot1\cdot4}}{2\cdot1}=\frac{6\pm\sqrt{36-16}}{2}$$
$$x=\frac{6\pm\sqrt{20}}{2}$$
Using a calculator we find that
$\frac{6+\sqrt{20}}{2}\approx5.236$ and $\frac{6-\sqrt{20}}{2}\approx0.764$.
The solutions are approximately 5.236 and 0.764.

49. $2x^2-3x-7=0$
$a=2$, $b=-3$, $c=-7$
$$x=\frac{-(-3)\pm\sqrt{(-3)^2-4\cdot2\cdot(-7)}}{2\cdot2}$$
$$x=\frac{3\pm\sqrt{9+56}}{4}=\frac{3\pm\sqrt{65}}{4}$$
Using a calculator we find that
$\frac{3+\sqrt{65}}{4}\approx2.766$ and $\frac{3-\sqrt{65}}{4}\approx-1.266$.
The solutions are approximately 2.766 and -1.266.

51. *Writing Exercise.*

53. $(-3x^2y^6)^0=1$

55. $x^{1/4}\cdot x^{2/3}=x^{1/4+2/3}=x^{11/12}$

57. $\frac{18a^5bc^{10}}{24a^{-5}bc^3}=\frac{3a^{10}c^7}{4}$

59. *Writing Exercise.*

61. $f(x)=\frac{x^2}{x-2}+1$
To find the x-coordinates of the x-intercepts of the graph of f, we solve $f(x)=0$.
$$\frac{x^2}{x-2}+1=0$$
$$x^2+x-2=0 \quad \text{Multiplying by } x-2$$
$$(x+2)(x-1)=0$$
$$x=-2 \text{ or } x=1$$
The x-intercepts are $(-2, 0)$ and $(1, 0)$.

63. $f(x)=g(x)$
$$\frac{x^2}{x-2}+1=\frac{4x-2}{x-2}+\frac{x+4}{2} \quad \text{Substituting}$$
$$2(x-2)\left(\frac{x^2}{x-2}+1\right)=2(x-2)\left(\frac{4x-2}{x-2}+\frac{x+4}{2}\right)$$
Multiplying by the LCD
$$2x^2+2(x-2)=2(4x-2)+(x-2)(x+4)$$
$$2x^2+2x-4=8x-4+x^2+2x-8$$
$$2x^2+2x-4=x^2+10x-12$$
$$x^2-8x+8=0$$
$a=1$, $b=-8$, $c=8$
$$x=\frac{-(-8)\pm\sqrt{(-8)^2-4\cdot1\cdot8}}{2\cdot1}=\frac{8\pm\sqrt{64-32}}{2}$$
$$x=\frac{8\pm\sqrt{32}}{2}=\frac{8\pm\sqrt{16\cdot2}}{2}=\frac{8\pm4\sqrt{2}}{2}$$
$$x=\frac{8}{2}\pm\frac{4\sqrt{2}}{2}=4\pm2\sqrt{2}$$
The solutions are $4+2\sqrt{2}$ and $4-2\sqrt{2}$.

65. $z^2+0.84z-0.4=0$
$a=1$, $b=0.84$, $c=-0.4$
$$z=\frac{-0.84\pm\sqrt{(0.84)^2-4\cdot1\cdot(-0.4)}}{2\cdot1}$$
$$z=\frac{-0.84\pm\sqrt{2.3056}}{2}$$
$$z=\frac{-0.84+\sqrt{2.3056}}{2}\approx0.339$$
$$z=\frac{-0.84-\sqrt{2.3056}}{2}\approx-1.179$$
The solutions are approximately 0.339 and -1.179.

67. $\sqrt{2}x^2+5x+\sqrt{2}=0$
$$x=\frac{-5\pm\sqrt{5^2-4\cdot\sqrt{2}\cdot\sqrt{2}}}{2\sqrt{2}}=\frac{-5\pm\sqrt{17}}{2\sqrt{2}}, \text{ or}$$
$$x=\frac{-5\pm\sqrt{17}}{2\sqrt{2}}\cdot\frac{\sqrt{2}}{\sqrt{2}}=\frac{-5\sqrt{2}\pm\sqrt{34}}{4}$$
The solutions are $\frac{-5\sqrt{2}\pm\sqrt{34}}{4}$.

69. $kx^2 + 3x - k = 0$

$k(-2)^2 + 3(-2) - k = 0$ Substituting -2 for x

$4k - 6 - k = 0$

$3k = 6$

$k = 2$

$2x^2 + 3x - 2 = 0$ Substituting 2 for k

$(2x-1)(x+2) = 0$

$2x - 1 = 0$ or $x + 2 = 0$

$x = \frac{1}{2}$ or $x = -2$

The other solution is $\frac{1}{2}$.

71. *Graphing Calculator Exercise*

Exercise Set 11.3

1. discriminant

3. two

5. rational

7. $x^2 - 7x + 5 = 0$

$a = 1,\ b = -7,\ c = 5$

We substitute and compute the discriminant.

$$b^2 - 4ac = (-7)^2 - 4 \cdot 1 \cdot 5 = 49 - 20 = 29$$

Since the discriminant is a positive number that is not a perfect square, there are two irrational solutions.

9. $x^2 + 11 = 0$

$a = 1,\ b = 0,\ c = 11$

We substitute and compute the discriminant.

$$b^2 - 4ac = 0^2 - 4 \cdot 1 \cdot 11 = -44$$

Since the discriminant is negative, there are two imaginary-number solutions.

11. $x^2 - 11 = 0$

$a = 1,\ b = 0,\ c = -11$

We substitute and compute the discriminant.

$$b^2 - 4ac = 0^2 - 4 \cdot 1 \cdot (-11) = 44$$

Since the discriminant is a positive number that is not a perfect square, there are two irrational solutions.

13. $4x^2 + 8x - 5 = 0$

$a = 4,\ b = 8,\ c = -5$

We substitute and compute the discriminant.

$$b^2 - 4ac = 8^2 - 4 \cdot 4 \cdot (-5) = 64 + 80 = 144$$

Since the discriminant is a positive number and a perfect square, there are two rational solutions.

15. $x^2 + 4x + 6 = 0$

$a = 1,\ b = 4,\ c = 6$

We substitute and compute the discriminant.

$$b^2 - 4ac = 4^2 - 4 \cdot 1 \cdot 6 = 16 - 24 = -8$$

Since the discriminant is negative, there are two imaginary-number solutions.

17. $9t^2 - 48t + 64 = 0$

$a = 9,\ b = -48,\ c = 64$

We substitute and compute the discriminant.

$$b^2 - 4ac = (-48)^2 - 4 \cdot 9 \cdot 64 = 2304 - 2304 = 0$$

Since the discriminant is 0, there is just one solution and it is a rational number.

19. $9t^2 + 3t = 0$

Observe that we can factor $9t^2 + 3t$. This tells us that there are two rational solutions. We could also do this problem as follows.

$$b^2 - 4ac = 3^2 - 4 \cdot 9 \cdot 0 = 9$$

Since the discriminant is a positive number and a perfect square, there are two rational solutions.

21. $x^2 + 4x = 8$

$x^2 + 4x - 8 = 0$ Standard form

$a = 1,\ b = 4,\ c = -8$

We substitute and compute the discriminant.

$$b^2 - 4ac = 4^2 - 4 \cdot 1 \cdot (-8) = 16 + 32 = 48$$

Since the discriminant is a positive number that is not a perfect square, there are two irrational solutions.

23. $2a^2 - 3a = -5$

$2a^2 - 3a + 5 = 0$ Standard form

$a = 2,\ b = -3,\ c = 5$

We substitute and compute the discriminant.

$$b^2 - 4ac = (-3)^2 - 4 \cdot 2 \cdot 5 = 9 - 40 = -31$$

Since the discriminant is negative, there are two imaginary-number solutions.

25. $7x^2 = 19x$

$7x^2 - 19x = 0$ Standard form

$a = 7,\ b = -19,\ c = 0$

We substitute and compute the discriminant.

$$b^2 - 4ac = (-19)^2 - 4 \cdot 7 \cdot 0 = 361$$

Since the discriminant is a positive number and a perfect square, there are two different rational solutions.

27. $y^2+\frac{9}{4}=4y$

$y^2-4y+\frac{9}{4}=0$ Standard form

$a=1,\ b=-4,\ c=\frac{9}{4}$

We substitute and compute the discriminant.

$$\begin{aligned} b^2-4ac&=(-4)^2-4\cdot 1\cdot \tfrac{9}{4}\\ &=16-9\\ &=7 \end{aligned}$$

The discriminant is a positive number that is not a perfect square. There are two irrational solutions.

29. The solutions are −5 and 4.

$x=-5$ *or* $x=4$

$x+5=0$ *or* $x-4=0$

$(x+5)(x-4)=0$ Principle of zero products

$x^2+x-20=0$ FOIL

31. The only solution is 3. It must be a repeated solution.

$x=3$ *or* $x=3$

$x-3=0$ *or* $x-3=0$

$(x-3)(x-3)=0$ Principle of zero products

$x^2-6x+9=0$ FOIL

33. The solutions are −1 and −3.

$x=-1$ *or* $x=-3$

$x+1=0$ *or* $x+3=0$

$(x+1)(x+3)=0$

$x^2+4x+3=0$

35. The solutions are 5 and $\frac{3}{4}$.

$x=5$ *or* $x=\frac{3}{4}$

$x-5=0$ *or* $x-\frac{3}{4}=0$

$(x-5)\left(x-\frac{3}{4}\right)=0$

$x^2-\frac{3}{4}x-5x+\frac{15}{4}=0$

$x^2-\frac{23}{4}x+\frac{15}{4}=0$

$4x^2-23x+15=0$ Multiplying by 4

37. The solutions are $-\frac{1}{4}$ and $-\frac{1}{2}$.

$x=-\frac{1}{4}$ *or* $x=-\frac{1}{2}$

$x+\frac{1}{4}=0$ *or* $x+\frac{1}{2}=0$

$\left(x+\frac{1}{4}\right)\left(x+\frac{1}{2}\right)=0$

$x^2+\frac{1}{2}x+\frac{1}{4}x+\frac{1}{8}=0$

$x^2+\frac{3}{4}x+\frac{1}{8}=0$

$8x^2+6x+1=0$ Multiplying by 8

39. The solutions are 2.4 and −0.4.

$x=2.4$ *or* $x=-0.4$

$x-2.4=0$ *or* $x+0.4=0$

$(x-2.4)(x+0.4)=0$

$x^2+0.4x-2.4x-0.96=0$

$x^2-2x-0.96=0$

41. The solutions are $-\sqrt{3}$ and $\sqrt{3}$.

$x=-\sqrt{3}$ *or* $x=\sqrt{3}$

$x+\sqrt{3}=0$ *or* $x-\sqrt{3}=0$

$(x+\sqrt{3})(x-\sqrt{3})=0$

$x^2-3=0$

43. The solutions are $2\sqrt{5}$ and $-2\sqrt{5}$.

$x=2\sqrt{5}$ *or* $x=-2\sqrt{5}$

$x-2\sqrt{5}=0$ *or* $x+2\sqrt{5}=0$

$(x-2\sqrt{5})(x+2\sqrt{5})=0$

$x^2-(2\sqrt{5})^2=0$

$x^2-4\cdot 5=0$

$x^2-20=0$

45. The solutions are $4i$ and $-4i$.

$x=4i$ *or* $x=-4i$

$x-4i=0$ *or* $x+4i=0$

$(x-4i)(x+4i)=0$

$x^2-(4i)^2=0$

$x^2+16=0$

47. The solutions are $2-7i$ and $2+7i$.

$x=2-7i$ *or* $x=2+7i$

$x-2+7i=0$ *or* $x-2-7i=0$

$(x-2)+7i=0$ *or* $(x-2)-7i=0$

$[(x-2)+7i][(x-2)-7i]=0$

$(x-2)^2-(7i)^2=0$

$x^2-4x+4-49i^2=0$

$x^2-4x+4+49=0$

$x^2-4x+53=0$

49. The solutions are $3-\sqrt{14}$ and $3+\sqrt{14}$.

$x=3-\sqrt{14}$ *or* $x=3+\sqrt{14}$

$x-3+\sqrt{14}=0$ *or* $x-3-\sqrt{14}=0$

$(x-3)+\sqrt{14}=0$ *or* $(x-3)-\sqrt{14}=0$

$[(x-3)+\sqrt{14}][(x-3)-\sqrt{14}]=0$

$(x-3)^2-(\sqrt{14})^2=0$

$x^2-6x+9-14=0$

$x^2-6x-5=0$

51. The solutions are $1-\frac{\sqrt{21}}{3}$ and $1+\frac{\sqrt{21}}{3}$.

$x=1-\frac{\sqrt{21}}{3}$ *or* $x=1+\frac{\sqrt{21}}{3}$

$x-1+\frac{\sqrt{21}}{3}=0$ *or* $x-1-\frac{\sqrt{21}}{3}=0$

$(x-1)+\frac{\sqrt{21}}{3}=0$ *or* $(x-1)-\frac{\sqrt{21}}{3}=0$

$$\left[(x-1)+\frac{\sqrt{21}}{3}\right]\left[(x-1)-\frac{\sqrt{21}}{3}\right]=0$$
$$(x-1)^2-\left(\frac{\sqrt{21}}{3}\right)^2=0$$
$$x^2-2x+1-\frac{21}{9}=0$$
$$x^2-2x+1-\frac{7}{3}=0$$
$$x^2-2x-\frac{4}{3}=0$$
$$3x^2-6x-4=0 \quad \text{Multiplying by 3}$$

53. The solutions are –2, 1, and 5.
$$x=-2 \quad or \quad x=1 \quad or \quad x=5$$
$$x+2=0 \quad or \quad x-1=0 \quad or \quad x-5=0$$
$$(x+2)(x-1)(x-5)=0$$
$$(x^2+x-2)(x-5)=0$$
$$x^3+x^2-2x-5x^2-5x+10=0$$
$$x^3-4x^2-7x+10=0$$

55. The solutions are –1, 0, and 3.
$$x=-1 \quad or \quad x=0 \quad or \quad x=3$$
$$x+1=0 \quad or \quad x=0 \quad or \quad x-3=0$$
$$(x+1)(x)(x-3)=0$$
$$(x^2+x)(x-3)=0$$
$$x^3-3x^2+x^2-3x=0$$
$$x^3-2x^2-3x=0$$

57. *Writing Exercise.*

59. $\sqrt{270a^7b^{12}}=\sqrt{9a^6b^{12}\cdot 30a}=3a^3b^6\sqrt{30a}$

61. $\sqrt[3]{x}\sqrt{x}=x^{1/3}\cdot x^{1/2}=x^{1/3+1/2}=x^{5/6}=\sqrt[6]{x^5}$

63. $(2-i)(3+i)=6+2i-3i-i^2=6-i+1=7-i$

65. *Writing Exercise.*

67. The graph includes the points (–3, 0), (0, –3), and (1, 0). Substituting in $y=ax^2+bx+c$, we have three equations.
$$\begin{aligned} 0 &= 9a - 3b + c,\\ -3 &= c,\\ 0 &= a + b + c \end{aligned}$$
The solution of this system of equations is $a = 1$, $b = 2$, $c=-3$.

69. a. $kx^2-2x+k=0$; one solution is –3
We first find k by substituting –3 for x.
$$k(-3)^2-2(-3)+k=0$$
$$9k+6+k=0$$
$$10k=-6$$
$$k=-\frac{6}{10}$$
$$k=-\frac{3}{5}$$

b. Now substitute $-\frac{3}{5}$ for k in the original equation.
$$-\frac{3}{5}x^2-2x+\left(-\frac{3}{5}\right)=0$$
$$3x^2+10x+3=0 \quad \text{Multiplying by} -5$$
$$(3x+1)(x+3)=0$$
$$x=-\frac{1}{3} \ or \ x=-3$$
The other solution is $-\frac{1}{3}$.

71. a. $x^2-(6+3i)x+k=0$; one solution is 3.
We first find k by substituting 3 for x.
$$3^2-(6+3i)3+k=0$$
$$9-18-9i+k=0$$
$$-9-9i+k=0$$
$$k=9+9i$$

b. Now we substitute $9+9i$ for k in the original equation.
$$x^2-(6+3i)x+(9+9i)=0$$
$$x^2-(6+3i)x+3(3+3i)=0$$
$$[x-(3+3i)][x-3]=0$$
The other solution is $3+3i$.

73. The solutions of $ax^2+bx+c=0$ are $x=\frac{-b\pm\sqrt{b^2-4ac}}{2a}$. When there is just one solution, $b^2-4ac=0$, so $x=\frac{-b\pm 0}{2a}=-\frac{b}{2a}$.

75. We substitute (–3, 0), $\left(\frac{1}{2},0\right)$, and (0,–12) in $f(x)=ax^2+bx+c$ and get three equations.
$$0=9a-3b+c,$$
$$0=\frac{1}{4}a+\frac{1}{2}b+c,$$
$$-12=c$$
The solution of this system of equations is $a=8$, $b=20$, $c=-12$.

77. If $-\sqrt{2}$ is one solution then $\sqrt{2}$ is another solution. Then
$$x=-\sqrt{2} \ or \ x=\sqrt{2}$$
$$x=\pm\sqrt{2}$$
$$x^2=2 \quad \text{Principle of square roots}$$
$$x^2-2=0$$

79. If $1-\sqrt{5}$ and $3+2i$ are two solutions, then $1+\sqrt{5}$ and $3-2i$ are also solutions. The equation of lowest degree that has these solutions is found as follows.
$$\left[x-(1-\sqrt{5})\right]\left[x-(1+\sqrt{5})\right]\left[x-(3+2i)\right]\left[x-(3-2i)\right]=0$$
$$(x^2-2x-4)(x^2-6x+13)=0$$
$$x^4-8x^3+21x^2-2x-52=0$$

81. *Writing Exercise.*

Exercise Set 11.4

1. c

3. a

5. $A = 4\pi r^2$

$\frac{A}{4\pi} = r^2$ Dividing by 4π

$\frac{1}{2}\sqrt{\frac{A}{\pi}} = r$ Taking the positive square root

7. $A = 2\pi r^2 + 2\pi rh$

$0 = 2\pi r^2 + 2\pi rh - A$ Standard form

$a = 2\pi,\ b = 2\pi h,\ c = -A$

$r = \frac{-2\pi h \pm \sqrt{(2\pi h)^2 - 4 \cdot 2\pi \cdot (-A)}}{2 \cdot 2\pi}$ Using the quadratic formula

$r = \frac{-2\pi h \pm \sqrt{4\pi^2 h^2 + 8\pi A}}{4\pi}$

$r = \frac{-2\pi h \pm 2\sqrt{\pi^2 h^2 + 2\pi A}}{4\pi}$

$r = \frac{-\pi h \pm \sqrt{\pi^2 h^2 + 2\pi A}}{2\pi}$

Since taking the negative square root would result in a negative answer, we take the positive one.

$r = \frac{-\pi h + \sqrt{\pi^2 h^2 + 2\pi A}}{2\pi}$

9. $F = \frac{Gm_1 m_2}{r^2}$

$Fr^2 = Gm_1 m_2$

$r^2 = \frac{Gm_1 m_2}{F}$

$r = \sqrt{\frac{Gm_1 m_2}{F}}$

11. $c = \sqrt{gH}$

$c^2 = gH$ Squaring

$\frac{c^2}{g} = H$

13. $a^2 + b^2 = c^2$

$b^2 = c^2 - a^2$

$b = \sqrt{c^2 - a^2}$

15. $s = v_0 t + \frac{gt^2}{2}$

$0 = \frac{gt^2}{2} + v_0 t - s$ Standard form

$a = \frac{g}{2},\ b = v_0,\ c = -s$

$t = \frac{-v_0 \pm \sqrt{v_0^2 - 4\left(\frac{g}{2}\right)(-s)}}{2\left(\frac{g}{2}\right)}$

$t = \frac{-v_0 \pm \sqrt{v_0^2 + 2gs}}{g}$

Since taking the negative square root would result in a negative answer, we take the positive one.

$t = \frac{-v_0 + \sqrt{v_0^2 + 2gs}}{g}$

17. $N = \frac{1}{2}(n^2 - n)$

$N = \frac{1}{2}n^2 - \frac{1}{2}n$

$0 = \frac{1}{2}n^2 - \frac{1}{2}n - N$

$a = \frac{1}{2},\ b = -\frac{1}{2},\ c = -N$

$n = \frac{-\left(-\frac{1}{2}\right) \pm \sqrt{\left(-\frac{1}{2}\right)^2 - 4 \cdot \frac{1}{2} \cdot (-N)}}{2\left(\frac{1}{2}\right)}$

$n = \frac{1}{2} \pm \sqrt{\frac{1}{4} + 2N}$

$n = \frac{1}{2} \pm \sqrt{\frac{1 + 8N}{4}}$

$n = \frac{1}{2} \pm \frac{1}{2}\sqrt{1 + 8N}$

Since taking the negative square root would result in a negative answer, we take the positive one.

$n = \frac{1}{2} + \frac{1}{2}\sqrt{1 + 8N}$, or $\frac{1 + \sqrt{1 + 8N}}{2}$

19. $T = I\sqrt{\frac{s}{d}}$

$\frac{T}{I} = \sqrt{\frac{s}{d}}$ Multiplying by $\frac{1}{I}$

$\frac{T^2}{I^2} = \frac{s}{d}$ Squaring

$dT^2 = I^2 s$ Multiplying by $I^2 d$

$d = \frac{I^2 s}{T^2}$ Multiplying by $\frac{1}{T^2}$

21. $at^2 + bt + c = 0$

The quadratic formula gives the result.

$t = \frac{-b \pm \sqrt{b^2 - 4ac}}{2a}$

23. a. ***Familiarize and Translate***. From Example 3, we know

$t = \frac{-v_0 + \sqrt{v_0^2 + 19.6s}}{9.8}$.

Carry out. Substituting 500 for s and 0 for v_0, we have

$t = \frac{0 + \sqrt{0^2 + 19.6(500)}}{9.8} \approx 10.1$

Check. Substitute 10.1 for t and 0 for v_0 in the original formula. (See Example 3.)

$s = 4.9t^2 + v_0 t = 4.9(10.1)^2 + 0 \cdot (10.1)^2 \approx 500$

The answer checks.

State. It takes the bolt about 10.1 sec to reach the ground.

b. ***Familiarize and Translate***. From Example 3, we know

$$t=\frac{-v_0+\sqrt{v_0^2+19.6s}}{9.8}$$

Carry out. Substitute 500 for s and 30 for v_0.

$$t=\frac{-30+\sqrt{30^2+19.6(500)}}{9.8}$$
$$t\approx 7.49$$

Check. Substitute 30 for v_0 and 7.49 for t in the original formula. (See Example 3.)

$$s=4.9t^2+v_0t=4.9(7.49)^2+(30)(7.49)\approx 500$$

The answer checks.

State. It takes the ball about 7.49 sec to reach the ground.

c. ***Familiarize and Translate***. We will use the formula in Example 3, $s=4.9t^2+v_0t$.

Carry out. Substitute 5 for t and 30 for v_0.

$$s=4.9(5)^2+30(5)=272.5$$

Check. We can substitute 30 for v_0 and 272.5 for s in the form of the formula we used in part (b).

$$t=\frac{-v_0+\sqrt{v_0^2+19.6s}}{9.8}$$
$$=\frac{-30+\sqrt{(30)^2+19.6(272.5)}}{9.8}=5$$

The answer checks.

State. The object will fall 272.5 m.

25. ***Familiarize***. We will use the formula $4.9t^2=s$.

Translate. Substitute 40 for s.

$$4.9t^2=40$$

Carry out. We solve the equation.

$$4.9t^2=40$$
$$t^2=\frac{40}{4.9}$$
$$t=\sqrt{\frac{40}{4.9}}$$
$$t\approx 2.9$$

Check. Substitute 2.9 for t in the formula.

$$s=4.9(2.9)^2=41.209\approx 40$$

The answer checks.

State. Wyatt will fall for about 2.9 sec before the cord begins to stretch.

27. ***Familiarize***. We will use the formula $V=48T^2$.

Translate. Substitute 36.5 for V.

$$36.5=48T^2$$

Carry out. We solve the equation.

$$36.5=48T^2$$
$$\frac{36.5}{48}=T^2$$
$$T=\sqrt{\frac{36.5}{48}}$$
$$T\approx 0.872$$

Check. Substitute 0.872 for T in the formula.

$$V=48(0.872)^2\approx 36.5$$

The answer checks.

State. His hang time is about 0.872 sec.

29. ***Familiarize and Translate***. We will use the formula in Example 4, $s=4.9t^2+v_0t$.

Carry out. Solve the formula for v_0.

$$s-4.9t^2=v_0t$$
$$\frac{s-4.9t^2}{t}=v_0$$

Now substitute 51.6 for s and 3 for t.

$$\frac{51.6-4.9(3)^2}{3}=v_0$$
$$2.5=v_0$$

Check. Substitute 3 for t and 2.5 for v_0 in the original formula.

$$s=4.9(3)^2+2.5(3)=51.6$$

The solution checks.

State. The initial velocity is 2.5 m/sec.

31. ***Familiarize and Translate***. From Exercise 22 we know that $r=-1+\dfrac{-P_2+\sqrt{P_2^2+4AP_1}}{2P_1}$

where A is the total amount in the account after two years, P_1 is the amount of the original deposit, P_2 is deposited at the beginning of the second year, and r is the annual interest rate.

Carry out. Substitute 3200 for P_1, 1800 for P_2, and 5207 for A.

$$r=-1+\frac{-1800+\sqrt{(1800)^2+4(5207)(3200)}}{2(3200)}$$

Using a calculator we have $r=0.025$.

Check. Substitute in the original formula in Exercise 22.

$$A=P_1(1+r)^2+P_2(1+r)$$
$$A=3200(1.025)^2+1800(1.025)=5207$$

The solution checks.

State. The annual interest rate is 0.025 or 2.5%.

33. ***Familiarize***. We first make a drawing, labeling it with the known and unknown information. We can also organize the information in a table. We let r represent the speed and t the time for the first part of the trip.

r mph t hr $r-10$ mph $4-t$ hr

120 mi 100 mi

Trip	Distance	Speed	Time
1st part	120	r	t
2nd part	100	$r-10$	$4-t$

Translate. Using $r=\dfrac{d}{t}$, we get two equations from the table, $r=\dfrac{120}{t}$ and $r-10=\dfrac{100}{4-t}$.

Carry out. We substitute $\dfrac{120}{t}$ for r in the second equation and solve for t.

$$\frac{120}{t}-10=\frac{100}{4-t}, \quad \text{LCD is } t(4-t)$$
$$t(4-t)\left(\frac{120}{t}-10\right)=t(4-t)\cdot\frac{100}{4-t}$$
$$120(4-t)-10t(4-t)=100t$$
$$480-120t-40t+10t^2=100t$$
$$10t^2-260t+480=0 \quad \text{Standard form}$$
$$t^2-26t+48=0 \quad \text{Multiplying by } \frac{1}{10}$$
$$(t-2)(t-24)=0$$
$t=2$ *or* $t=24$

Check. Since the time cannot be negative (If $t = 24$, $4-t=-20$.), we check only 2 hr. If $t = 2$, then $4-t=2$. The speed of the first part is $\frac{120}{2}$, or 60 mph. The speed of the second part is $\frac{100}{2}$, or 50 mph. The speed of the second part is 10 mph slower than the first part. The value checks.

State. The speed of the first part was 60 mph, and the speed of the second part was 50 mph.

35. ***Familiarize***. We first make a drawing. We also organize the information in a table. We let r = the speed and t = the time of the slower trip.

200 mi r mph t hr

200 mi $r+10$ mph $t-1$ hr

Trip	Distance	Speed	Time
Slower	200	r	t
Faster	200	$r+10$	$t-1$

Translate. Using $t=d/r$, we get two equations from the table:

$$t=\frac{200}{r} \text{ and } t-1=\frac{200}{r+10}.$$

Carry out. We substitute $\frac{200}{r}$ for t in the second equation and solve for r.

$$\frac{200}{r}-1=\frac{200}{r+10}, \quad \text{LCD is } r(r+10)$$
$$r(r+10)\left(\frac{200}{r}-1\right)=r(r+10)\cdot\frac{200}{r+10}$$
$$200(r+10)-r(r+10)=200r$$
$$200r+2000-r^2-10r=200r$$
$$0=r^2+10r-2000$$
$$0=(r+50)(r-40)$$
$r=-50$ *or* $r=40$

Check. Since negative speed has no meaning in this problem, we check only 40. If $r = 40$, then the time for the slower trip is $\frac{200}{40}$, or 5 hours. If $r = 40$, then $r+10=50$ and the time for the faster trip is $\frac{200}{50}$, or 4 hours. This is 1 hour less time than the slower trip took, so we have an answer to the problem.

State. The speed is 40 mph.

37. ***Familiarize***. We make a drawing and then organize the information in a table. We let r = the speed and t = the time of the Cessna.

600 mi r mph t hr

1000 mi $r+50$ mph $t+1$ hr

Plane	Distance	Speed	Time
Cessna	600	r	t
Beechcraft	1000	$r+50$	$t+1$

Translate. Using $t=d/r$, we get two equations from the table:

$$t=\frac{600}{r} \text{ and } t+1=\frac{1000}{r+50}$$

Carry out. We substitute $\frac{600}{r}$ for t in the second equation and solve for r.

$$\frac{600}{r}+1=\frac{1000}{r+50}, \quad \text{LCD is } r(r+50)$$
$$r(r+50)\left(\frac{600}{r}+1\right)=r(r+50)\cdot\frac{1000}{r+50}$$
$$600(r+50)+r(r+50)=1000r$$
$$600r+30{,}000+r^2+50r=1000r$$
$$r^2-350r+30{,}000=0$$
$$(r-150)(r-200)=0$$
$r=150$ *or* $r=200$

Check. If $r = 150$, then the Cessna's time is $\frac{600}{150}$, or 4 hr and the Beechcraft's time is $\frac{1000}{150+50}$, or $\frac{1000}{200}$, or 5 hr. If $r=200$, then the Cessna's time is $\frac{600}{200}$, or 3 hr and the Beechcraft's time is $\frac{1000}{200+50}$, or $\frac{1000}{250}$, or 4 hr. Since the Beechcraft's time is 1 hr longer in each case, both values check. There are two solutions.

State. The speed of the Cessna is 150 mph and the speed of the Beechcraft is 200 mph; or the speed of the Cessna is 200 mph and the speed of the Beechcraft is 250 mph.

39. ***Familiarize***. We make a drawing and then organize the information in a table. We let r represent the speed and t the time of the trip to Hillsboro.

Hillsboro

36 mi r mph t hr

36 mi $r-3$ mph $7-t$ hr

Trip	Distance	Speed	Time
To Hillsboro	36	r	t
Return	36	$r-3$	$7-t$

Translate. Using $t=\frac{d}{r}$, we get two equations from the table,

$$t=\frac{36}{r} \text{ and } 7-t=\frac{36}{r-3}.$$

Carry out. We substitute $\frac{36}{r}$ for t in the second equation and solve for r.

$$7-\frac{36}{r}=\frac{36}{r-3}, \quad \text{LCD is } r(r-3)$$
$$r(r-3)\left(7-\frac{36}{r}\right)=r(r-3)\cdot\frac{36}{r-3}$$
$$7r(r-3)-36(r-3)=36r$$
$$7r^2-21r-36r+108=36r$$
$$7r^2-93r+108=0$$
$$(7r-9)(r-12)=0$$
$$r=\frac{9}{7} \quad or \quad r=12$$

Check. Since negative speed has no meaning in this problem (If $r=\frac{9}{7}$, then $r-3=-\frac{12}{7}$.), we check only 12 mph. If $r=12$, then the time of the trip to Hillsboro is $\frac{36}{12}$, or 3 hr. The speed of the return trip is $12-3$, or 9 mph, and the time is $\frac{36}{9}$, or 4 hr. The total time for the round trip is $3\text{ hr}+4\text{ hr}$, or 7 hr. The value checks.
State. Naoki's speed on the trip to Hillsboro was 12 mph and it was 9 mph on the return trip.

41. ***Familiarize***. We make a drawing and organize the information in a table. Let r represent the speed of the boat in still water, and let t represent the time of the trip upriver.

60 mi $r-3$ mph t hr → Upriver
Downriver ← 60 mi $r+3$ mph $9-t$ hr

Trip	Distance	Speed	Time
Upriver	60	$r-3$	t
Downriver	60	$r+3$	$9-t$

Translate. Using $t=\frac{d}{r}$, we get two equations from the table,

$t=\frac{60}{r-3}$ and $9-t=\frac{60}{r+3}$.

Carry out. We substitute $\frac{60}{r-3}$ for t in the second equation and solve for r.

$$9-\frac{60}{r-3}=\frac{60}{r+3}$$
$$(r-3)(r+3)\left(9-\frac{60}{r-3}\right)=(r-3)(r+3)\cdot\frac{60}{r+3}$$
$$9(r-3)(r+3)-60(r+3)=60(r-3)$$
$$9r^2-81-60r-180=60r-180$$
$$9r^2-120r-81=0$$
$$3r^2-40r-27=0 \quad \text{Dividing by 3}$$

We use the quadratic formula.

$$r=\frac{-(-40)\pm\sqrt{(-40)^2-4\cdot3\cdot(-27)}}{2\cdot3}$$
$$r=\frac{40\pm\sqrt{1924}}{6}$$
$$r\approx14 \quad or \quad r\approx-0.6$$

Check. Since negative speed has no meaning in this problem, we check only 14 mph. If $r\approx14$, then the speed upriver is about $14-3$, or 11 mph, and the time is about $\frac{60}{11}$, or 5.5 hr. The speed downriver is about $14+3$, or 17 mph, and the time is about $\frac{60}{17}$, or 3.5 hr. The total time of the round trip is $5.5+3.5$, or 9 hr. The value checks.
State. The speed of the boat in still water is about 14 mph.

43. ***Familiarize***. Let x represent the time it take the spring to fill the pool. Then $x-8$ represents the time it takes the well to fill the pool. It takes them 3 hr to fill the pool working together, so they can fill $\frac{1}{3}$ of the pool in 1 hr. The spring will fill $\frac{1}{x}$ of the pool in 1 hr, and the well will fill $\frac{1}{x-8}$ of the pool in 1 hr.

Translate. We have an equation.

$$\frac{1}{x}+\frac{1}{x-8}=\frac{1}{3}$$

Carry out. We solve the equation.
We multiply by the LCD, $3x(x-8)$.

$$3x(x-8)\left(\frac{1}{x}+\frac{1}{x-8}\right)=3x(x-8)\cdot\frac{1}{3}$$
$$3(x-8)+3x=x(x-8)$$
$$3x-24+3x=x^2-8x$$
$$0=x^2-14x+24$$
$$0=(x-2)(x-12)$$

Check. Since negative time has no meaning in this problem, 2 is not a solution $(2-8=-6)$. We check only 12 hr. This is the time it would take the spring working alone. Then the well would take $12-8$, or 4 hr working alone. The well would fill $3\left(\frac{1}{4}\right)$, or $\frac{3}{4}$ of the pool in 3 hr, and the spring would fill $3\left(\frac{1}{12}\right)$, or $\frac{1}{4}$ of the pool in 3 hr. Thus, in 3 hr they would fill $\frac{3}{4}+\frac{1}{4}$ of the pool. This is all of it, so the numbers check.
State. It takes the spring, working alone, 12 hr to fill the pool.

45. We make a drawing and then organize the information in a table. We let r represent Kofi's speed in still water. Then $r-2$ is the speed upstream and $r+2$ is the speed downstream. Using $t=\frac{d}{r}$, we let $\frac{1}{r-2}$ represent the time upstream and $\frac{1}{r+2}$ represent the time downstream.

1 mi $r-2$ mph → Upstream
Downstream ← 1 mi $r+2$ mph

Trip	Distance	Speed	Time
Upstream	1	$r-2$	$\frac{1}{r-2}$
Downstream	1	$r+2$	$\frac{1}{r+2}$

Translate. The time for the round trip is 1 hour. We now have an equation.

$$\frac{1}{r-2}+\frac{1}{r+2}=1$$

Carry out. We solve the equation. We multiply by the LCD, $(r-2)(r+2)$.

$$(r-2)(r+2)\left(\frac{1}{r-2}+\frac{1}{r+2}\right)=(r-2)(r+2)\cdot 1$$
$$(r+2)+(r-2)=(r-2)(r+2)$$
$$2r=r^2-4$$
$$0=r^2-2r-4$$

$a=1,\ b=-2,\ c=-4$

$$r=\frac{-(-2)\pm\sqrt{(-2)^2-4\cdot 1(-4)}}{2\cdot 1}$$
$$r=\frac{2\pm\sqrt{4+16}}{2}=\frac{2\pm\sqrt{20}}{2}$$
$$r=\frac{2\pm 2\sqrt{5}}{2}=1\pm\sqrt{5}$$

$1+\sqrt{5}\approx 1+2.236\approx 3.24$

$1-\sqrt{5}\approx 1-2.236\approx -1.24$

Check. Since negative speed has no meaning in this problem, we check only 3.24 mph. If $r\approx 3.24$, then $r-2\approx 1.24$ and $r+2\approx 5.24$. The time it takes to travel upstream is approximately $\frac{1}{1.24}$, or 0.806 hr, and the time it takes to travel downstream is approximately $\frac{1}{5.24}$, or 0.191 hr. The total time is 0.997 which is approximately 1 hour. The value checks.

State. Kofi's speed in still water is approximately 3.24 mph.

47. ***Familiarize***. Let t represent the number of days it takes Katherine to plant new trees, working alone. Then $t+2$ represents the time it takes Julianna to plant new trees, working alone. In 1 day, Katherine does $\frac{1}{t}$ of the job and Julianna does $\frac{1}{t+2}$ of the job.

Translate. Working together, Katherine and Julianna can plant trees in 6 days.

$$6\left(\frac{1}{t}\right)+6\left(\frac{1}{t+2}\right)=1,\text{ or }\frac{6}{t}+\frac{6}{t+2}=1.$$

Carry out. We solve the equation. Multiply on both sides by the LCD, $t(t+2)$.

$$t(t+2)\left(\frac{6}{t}+\frac{6}{t+2}\right)=t(t+2)\cdot 1$$
$$6(t+2)+6t=t(t+2)$$
$$6t+12+6t=t^2+2t$$
$$0=t^2-10t-12$$

$a=1,\ b=-10,\ c=-12$

$$t=\frac{-(-10)\pm\sqrt{(-10)^2-4\cdot 1(-12)}}{2\cdot 1}$$
$$t=\frac{10\pm\sqrt{100+48}}{2}=\frac{10\pm\sqrt{148}}{2}=\frac{10\pm 2\sqrt{37}}{2}$$
$$t=5\pm\sqrt{37}$$

$5+\sqrt{37}\approx 5+6.08\approx 11.08\approx 11$

$5-\sqrt{37}\approx 5-6.08\approx -1.08$

Check. Since negative time has no meaning in this application, we only check 11. If $t=11$, then $t+2=11+2=13$. In 6 days Julianna does $6\cdot\frac{1}{13}$, or $\frac{6}{13}$ of the job. In 6 days the Katherine does $6\cdot\frac{1}{11}$, or $\frac{6}{11}$ of the job. Together they do $\frac{6}{13}+\frac{6}{11}\approx 1$, or 1 entire job. The answer checks.

State. It would take Katherine about 11 days to do the job, working alone.

49. *Writing Exercise.*

51.
$$\frac{1}{2}(x-7)=\frac{1}{3}x+4$$
$$6\cdot\frac{1}{2}(x-7)=6\cdot\left(\frac{1}{3}x+4\right)$$
$$3x-21=2x+24$$
$$x=45$$

53.
$$6|3x+2|=12$$
$$|3x+2|=2$$
$$3x+2=-2 \quad or \quad 3x+2=2$$
$$3x=-4 \quad or \quad 3x=0$$
$$x=-\frac{4}{3} \quad or \quad x=0$$

55.
$$\sqrt{2x-8}=15$$
$$2x-8=225 \qquad \text{Squaring}$$
$$2x=233$$
$$x=\frac{233}{2}$$

57. *Writing Exercise.*

59.
$$A=6.5-\frac{20.4t}{t^2+36}$$
$$(t^2+36)A=(t^2+36)\left(6.5-\frac{20.4t}{t^2+36}\right)$$
$$At^2+36A=(t^2+36)(6.5)-(t^2+36)\left(\frac{20.4t}{t^2+36}\right)$$
$$At^2+36A=6.5t^2+234-20.4t$$
$$At^2-6.5t^2+20.4t+36A-234=0$$
$$(A-6.5)t^2+20.4t+(36A-234)=0$$

$a=A-6.5,\ b=20.4,\ c=36A-234$

$$t=\frac{-20.4+\sqrt{(20.4)^2-4(A-6.5)(36A-234)}}{2(A-6.5)}$$
$$t=\frac{-20.4+\sqrt{416.16-144A^2+1872A-6084}}{2(A-6.5)}$$
$$t=\frac{-20.4+\sqrt{-144A^2+1872A-5667.84}}{2(A-6.5)}$$
$$t=\frac{-20.4+\sqrt{144(-A^2+13A-39.36)}}{2(A-6.5)}$$
$$t=\frac{-20.4+12\sqrt{-A^2+13A-39.36}}{2(A-6.5)}$$
$$t=\frac{2(-10.2+6\sqrt{-A^2+13A-39.36})}{2(A-6.5)}$$
$$t=\frac{-10.2+6\sqrt{-A^2+13A-39.36}}{A-6.5}$$

61. ***Familiarize.*** Let a = the number. Then $a - 1$ is 1 less than a and the reciprocal of that number is $\frac{1}{a-1}$. Also, 1 more than the number is $a + 1$.
Translate.

The reciprocal of 1 less than a number	is	1 more than the number.
↓	↓	↓
$\frac{1}{(a-1)}$	=	$a+1$

Carry out. We solve the equation.

$$\frac{1}{a-1} = a+1, \quad \text{LCD is } a-1$$
$$(a-1)\cdot\frac{1}{a-1} = (a-1)(a+1)$$
$$1 = a^2 - 1$$
$$2 = a^2$$
$$\pm\sqrt{2} = a$$

Check. $\frac{1}{\sqrt{2}-1} \approx 2.4142 \approx \sqrt{2}+1$ and $\frac{1}{-\sqrt{2}-1} \approx -0.4142 \approx -\sqrt{2}+1$. The answers check.

State. The numbers are $\sqrt{2}$ and $-\sqrt{2}$, or $\pm\sqrt{2}$.

63.
$$\frac{w}{l} = \frac{l}{w+l}$$
$$l(w+l)\cdot\frac{w}{l} = l(w+l)\cdot\frac{l}{w+l}$$
$$w(w+l) = l^2$$
$$w^2 + lw = l^2$$
$$0 = l^2 - lw - w^2$$

Use the quadratic formula with $a = 1$, $b = -w$, and $c = -w^2$.

$$l = \frac{-(-w)\pm\sqrt{(-w)^2 - 4\cdot 1(-w^2)}}{2\cdot 1}$$
$$l = \frac{w\pm\sqrt{w^2+4w^2}}{2} = \frac{w\pm\sqrt{5w^2}}{2}$$
$$l = \frac{w\pm w\sqrt{5}}{2}$$

Since $\frac{w - w\sqrt{5}}{2}$ is negative we use the positive square root: $l = \frac{w + w\sqrt{5}}{2}$

65. $mn^4 - r^2pm^3 - r^2n^2 + p = 0$

Let $u = n^2$. Substitute and rearrange.

$$mu^2 - r^2u - r^2pm^3 + p = 0$$
$$a = m,\ b = -r^2,\ c = -r^2pm^3 + p$$
$$u = \frac{-(-r^2)\pm\sqrt{(-r^2)^2 - 4\cdot m(-r^2pm^3 + p)}}{2\cdot m}$$
$$u = \frac{r^2 \pm\sqrt{r^4 + 4m^4r^2p - 4mp}}{2m}$$
$$n^2 = \frac{r^2 \pm\sqrt{r^4 + 4m^4r^2p - 4mp}}{2m}$$
$$n = \pm\sqrt{\frac{r^2 \pm\sqrt{r^4 + 4m^4r^2p - 4mp}}{2m}}$$

67. Let s represent a length of a side of the cube, let S represent the surface area of the cube, and let A represent the surface area of the sphere. Then the diameter of the sphere is s, so the radius r is $s/2$.

From Exercise 5, we know, $A = 4\pi r^2$, so when $r = s/2$ we have $A = 4\pi\left(\frac{s}{2}\right)^2 = 4\pi\cdot\frac{s^2}{4} = \pi s^2$.

From the formula for the surface area of a cube (See Exercise 6.) we know that $S = 6s^2$, so $\frac{S}{6} = s^2$ and then $A = \pi\cdot\frac{S}{6}$, or $A(S) = \frac{\pi S}{6}$.

Exercise Set 11.5

1. True

3. True

5. Since $\left(\sqrt{p}\right)^2 = p$, use $u = \sqrt{p}$.

7. Since $\left(x^2+3\right)^2 = \left(x^2+3\right)^2$, use $u = x^2 + 3$.

9. Since $\left[(1+t)^2\right]^2 = (1+t)^4$, use $u = (1+t)^2$.

11. $x^4 - 13x^2 + 36 = 0$

Let $u = x^2$ and $u^2 = x^4$.

$$u^2 - 13u + 36 = 0 \quad \text{Substituting } u \text{ for } x^2$$
$$(u-4)(u-9) = 0$$
$$u - 4 = 0 \quad or \quad u - 9 = 0$$
$$u = 4 \quad or \quad u = 9$$

Now replace u with x^2 and solve these equations.

$$x^2 = 4 \quad or \quad x^2 = 9$$
$$x = \pm 2 \quad or \quad x = \pm 3$$

The numbers 2, –2, 3, and –3 check. They are the solutions.

13. $t^4 - 7t^2 + 12 = 0$

Let $u = t^2$ and $u^2 = t^4$.

$$u^2 - 7u + 12 = 0 \quad \text{Substituting } u \text{ for } t^2$$
$$(u-3)(u-4) = 0$$
$$u - 3 = 0 \quad or \quad u - 4 = 0$$
$$u = 3 \quad or \quad u = 4$$

Now replace u with t^2 and solve these equations.

$$t^2 = 3 \quad or \quad t^2 = 4$$
$$t = \pm\sqrt{3} \quad or \quad t = \pm 2$$

The numbers $\sqrt{3}, -\sqrt{3}$, 2, and –2 check. They are the solutions.

15. $4x^4 - 9x^2 + 5 = 0$

Let $u = x^2$ and $u^2 = x^4$.

$4u^2 - 9u + 5 = 0$ Substituting u for x^2

$(4u-5)(u-1) = 0$

$4u - 5 = 0$ *or* $u - 1 = 0$

$u = \frac{5}{4}$ *or* $u = 1$

Now replace u with x^2 and solve these equations.

$x^2 = \frac{5}{4}$ *or* $x^2 = 1$

$x = \pm\sqrt{\frac{5}{4}} = \pm\frac{\sqrt{5}}{2}$ *or* $x = \pm 1$

The numbers $\frac{\sqrt{5}}{2}, -\frac{\sqrt{5}}{2}$, 1, and -1 check. They are the solutions.

17. $(x^2-7)^2 - 3(x^2-7) + 2 = 0$

Let $u = x^2 - 7$ and $u^2 = (x^2-7)^2$.

$u^2 - 3u + 2 = 0$ Substituting

$(u-1)(u-2) = 0$

$u = 1$ *or* $u = 2$

$x^2 - 7 = 1$ *or* $x^2 - 7 = 2$ Replacing u with $x^2 - 7$

$x^2 = 8$ *or* $x^2 = 9$

$x = \pm\sqrt{8} = \pm 2\sqrt{2}$ *or* $x = \pm 3$

The numbers $2\sqrt{2}$, $-2\sqrt{3}$, 3, and -3 check. They are the solutions.

19. $x^4 + 5x^2 - 36 = 0$

Let $u = x^2$ and $u^2 = x^4$.

$u^2 + 5u - 36 = 0$

$(u-4)(u+9) = 0$

$u - 4 = 0$ *or* $u + 9 = 0$

$u = 4$ *or* $u = -9$

Now replace u with x^2 and solve these equations.

$x^2 = 4$ *or* $x^2 = -9$

$x = \pm 2$ *or* $x = \pm\sqrt{-9} = \pm 3i$

The numbers 2, -2, $3i$, and $-3i$ check. They are the solutions.

21. $(n^2+6)^2 - 7(n^2+6) + 10 = 0$

Let $u = n^2 + 6$ and $u^2 = (n^2+6)^2$.

$u^2 - 7u + 10 = 0$

$(u-2)(u-5) = 0$

$u - 2 = 0$ *or* $u - 5 = 0$

$u = 2$ *or* $u = 5$

Now replace u with $n^2 + 6$ and solve these equations.

$n^2 + 6 = 2$ *or* $n^2 + 6 = 5$

$n^2 = -4$ *or* $n^2 = -1$

$n = \pm\sqrt{-4} = \pm 2i$ *or* $n = \pm\sqrt{-1} = \pm i$

$2i$, $-2i$, i, and $-i$ check. They are the solutions.

23. $w + 4\sqrt{w} - 12 = 0$

Let $u = \sqrt{w}$ and $u^2 = (\sqrt{w})^2 = w$.

$u^2 + 4u - 12 = 0$

$(u+6)(u-2) = 0$

$u + 6 = 0$ *or* $u - 2 = 0$

$u = -6$ *or* $u = 2$

Now replace u with $\sqrt{w}$ and solve these equations.

$\sqrt{w} = -6$ *or* $\sqrt{w} = 2$

or $w = 4$

Since the principal square root cannot be negative, only 4 checks as a solution.

25. $r - 2\sqrt{r} - 6 = 0$

Let $u = \sqrt{r}$ and $u^2 = r$.

$u^2 - 2u - 6 = 0$

$$u = \frac{-(-2) \pm \sqrt{(-2)^2 - 4\cdot 1\cdot(-6)}}{2\cdot 1}$$

$$u = \frac{2 \pm \sqrt{28}}{2} = \frac{2 + 2\sqrt{7}}{2}$$

$u = 1 \pm \sqrt{7}$

Replace u with $\sqrt{r}$ and solve these equations:

$\sqrt{r} = 1 + \sqrt{7}$ *or* $\sqrt{r} = 1 - \sqrt{7}$

$(\sqrt{r})^2 = (1+\sqrt{7})^2$

$r = 1 + 2\sqrt{7} + 7$

$r = 8 + 2\sqrt{7}$

The number $1 - \sqrt{7}$ is not a solution since it is negative.

The number $8 + 2\sqrt{7}$ checks. It is the solution.

27. $(1+\sqrt{x})^2 + 5(1+\sqrt{x}) + 6 = 0$

Let $u = 1 + \sqrt{x}$ and $u^2 = (1+\sqrt{x})^2$.

$u^2 + 5u + 6 = 0$ Substituting

$(u+3)(u+2) = 0$

$u = -3$ *or* $u = -2$

$1 + \sqrt{x} = -3$ *or* $1 + \sqrt{x} = -2$ Replacing u with $1 + \sqrt{x}$

$\sqrt{x} = -4$ *or* $\sqrt{x} = -3$

Since the principal square root cannot be negative, this equation has no solution.

29. $x^{-2} - x^{-1} - 6 = 0$

Let $u = x^{-1}$ and $u^2 = x^{-2}$.

$u^2 - u - 6 = 0$ Substituting

$(u-3)(u+2) = 0$

$u = 3$ *or* $u = -2$

Now we replace u with x^{-1} and solve these equations:

$x^{-1} = 3$ *or* $x^{-1} = -2$

$\frac{1}{x} = 3$ *or* $\frac{1}{x} = -2$

$\frac{1}{3} = x$ *or* $-\frac{1}{2} = x$

Both $\frac{1}{3}$ and $-\frac{1}{2}$ check. They are the solutions.

31. $4t^{-2} - 3t^{-1} - 1 = 0$

Let $u = t^{-1}$ and $u^2 = t^{-2}$.

$4u^2 - 3u - 1 = 0$ Substituting

$(4u+1)(u-1) = 0$

$4u + 1 = 0$ *or* $u - 1 = 0$

$u = -\frac{1}{4}$ *or* $u = 1$

Now replace u with t^{-1} and solve these equations.

$t^{-1} = -\frac{1}{4}$ *or* $t^{-1} = 1$

$\frac{1}{t} = -\frac{1}{4}$ *or* $\frac{1}{t} = 1$

$-4 = t$ *or* $1 = t$

Both –4 and 1 check. They are the solutions.

33. $t^{2/3} + t^{1/3} - 6 = 0$

Let $u = t^{1/3}$ and $u^2 = t^{2/3}$.

$u^2 + u - 6 = 0$ Substituting

$(u+3)(u-2) = 0$

$u = -3$ *or* $u = 2$

Now we replace u with $t^{1/3}$ and solve these equations:

$t^{1/3} = -3$ *or* $t^{1/3} = 2$

$t = (-3)^3$ *or* $t = 2^3$ Raising to the third power

$t = -27$ *or* $t = 8$

Both –27 and 8 check. They are the solutions.

35. $y^{1/3} - y^{1/6} - 6 = 0$

Let $u = y^{1/6}$ and $u^2 = y^{2/6}$, or $y^{1/3}$.

$u^2 - u - 6 = 0$ Substituting

$(u-3)(u+2) = 0$

$u = 3$ or $u = -2$

Now we replace u with $y^{1/6}$ and solve these equations:

$y^{1/6} = 3$ *or* $y^{1/6} = -2$

$\sqrt[6]{y} = 3$ *or* $\sqrt[6]{y} = -2$

$y = 3^6$

$y = 729$

The equation $\sqrt[6]{y} = -2$ has no solution since principal sixth roots are never negative. The number 729 checks and is the solution.

37. $t^{1/3} + 2t^{1/6} = 3$

$t^{1/3} + 2t^{1/6} - 3 = 0$

Let $u = t^{1/6}$ and $u^2 = t^{2/6} = t^{1/3}$.

$u^2 + 2u - 3 = 0$ Substituting

$(u+3)(u-1) = 0$

$u = -3$ *or* $u = 1$

$t^{1/6} = -3$ *or* $t^{1/6} = 1$ Substituting $t^{1/6}$ for u

$t = 1$

Since principal sixth roots are never negative, the equation $t^{1/6} = -3$ has no solution. The number 1 checks and is the solution.

39. $(3 - \sqrt{x})^2 - 10(3 - \sqrt{x}) + 23 = 0$

Let $u = 3 - \sqrt{x}$ and $u^2 = (3 - \sqrt{x})^2$.

$u^2 - 10u + 23 = 0$ Substituting

$$u = \frac{-(-10) \pm \sqrt{(-10)^2 - 4 \cdot 1 \cdot 23}}{2 \cdot 1}$$

$$u = \frac{10 \pm \sqrt{8}}{2} = \frac{2 \cdot 5 \pm 2\sqrt{2}}{2}$$

$u = 5 \pm \sqrt{2}$

$u = 5 + \sqrt{2}$ *or* $u = 5 - \sqrt{2}$

Now we replace u with $3 - \sqrt{x}$ and solve these equations:

$3 - \sqrt{x} = 5 + \sqrt{2}$ *or* $3 - \sqrt{x} = 5 - \sqrt{2}$

$-\sqrt{x} = 2 + \sqrt{2}$ *or* $-\sqrt{x} = 2 - \sqrt{2}$

$\sqrt{x} = -2 - \sqrt{2}$ *or* $\sqrt{x} = -2 + \sqrt{2}$

Since both $-2 - \sqrt{2}$ and $-2 + \sqrt{2}$ are negative and principal square roots are never negative, the equation has no solution.

41. $16\left(\frac{x-1}{x-8}\right)^2 + 8\left(\frac{x-1}{x-8}\right) + 1 = 0$

Let $u = \frac{x-1}{x-8}$ and $u^2 = \left(\frac{x-1}{x-8}\right)^2$.

$16u^2 + 8u + 1 = 0$ Substituting

$(4u+1)(4u+1) = 0$

$u = -\frac{1}{4}$

Now we replace u with $\frac{x-1}{x-8}$ and solve this equation:

$\frac{x-1}{x-8} = -\frac{1}{4}$

$4x - 4 = -x + 8$ Multiplying by $4(x-8)$

$5x = 12$

$x = \frac{12}{5}$

The number $\frac{12}{5}$ checks and is the solution.

43. The x-intercepts occur where $f(x) = 0$. Thus, we must have $5x + 13\sqrt{x} - 6 = 0$.

Let $u = \sqrt{x}$ and $u^2 = x$.

$5u^2 + 13u - 6 = 0$ Substituting

$(5u-2)(u+3) = 0$

$u = \frac{2}{5}$ *or* $u = -3$

Now replace u with $\sqrt{x}$ and solve these equations:

$\sqrt{x} = \frac{2}{5}$ *or* $\sqrt{x} = -3$ has no solution

$x = \frac{4}{25}$

The number $\frac{4}{25}$ checks. Thus, the x-intercept is $\left(\frac{4}{25}, 0\right)$.

45. The x-intercepts occur where $f(x)=0$. Thus, we must have $(x^2-3x)^2-10(x^2-3x)+24=0$.

Let $u=x^2-3x$ and $u^2=(x^2-3x)^2$.

$u^2-10u+24=0$ Substituting
$(u-6)(u-4)=0$
$u=6$ *or* $u=4$

Now replace u with x^2-3x and solve these equations:

$x^2-3x=6$ *or* $x^2-3x=4$
$x^2-3x-6=0$ *or* $x^2-3x-4=0$

$x=\frac{-(-3)\pm\sqrt{(-3)^2-4(1)(-6)}}{2\cdot 1}$ *or* $(x-4)(x+1)=0$

$x=\frac{3}{2}\pm\frac{\sqrt{33}}{2}$ *or* $x=4$ *or* $x=-1$

All four numbers check. Thus, the x-intercepts are $\left(\frac{3}{2}+\frac{\sqrt{33}}{2}, 0\right)$, $\left(\frac{3}{2}-\frac{\sqrt{33}}{2}, 0\right)$, (4, 0), and (–1, 0).

47. The x-intercepts occur where $f(x)=0$. Thus, we must have $x^{2/5}+x^{1/5}-6=0$.

Let $u=x^{1/5}$ and $u^2=x^{2/5}$.

$u^2+u-6=0$ Substituting
$(u+3)(u-2)=0$
$u=-3$ *or* $u=2$
$x^{1/5}=-3$ *or* $x^{1/5}=2$ Replacing u with $x^{1/5}$
$x=-243$ $x=32$ Raising to the fifth power

Both –243 and 32 check. Thus, the x-intercepts are (–243, 0) and (32, 0).

49. $f(x)=\left(\frac{x^2+2}{x}\right)^4+7\left(\frac{x^2+2}{x}\right)^2+5$

Observe that, for all real numbers x, each term is positive. Thus, there are no real-number values of x for which $f(x)=0$ and hence no x-intercepts.

51. *Writing Exercise.*

53. Graph $2x=-5y$.
We find some ordered pairs, plot points, and draw the graph.

x	y
–5	2
0	0
5	–2

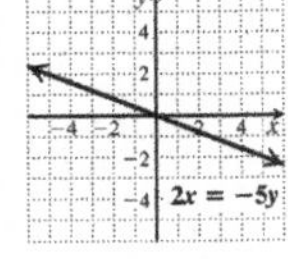

55. Graph $2x-5y=10$.
We find some ordered pairs, plot points, and draw the graph.

x	y
–5	–4
0	–2
5	0

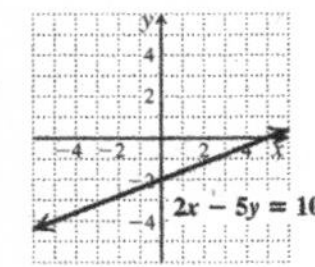

57. Graph $y-2=3(x-4)$.
We find some ordered pairs, plot points, and draw the graph.

x	y
0	–10
2	–4
3	–1
4	2
5	5

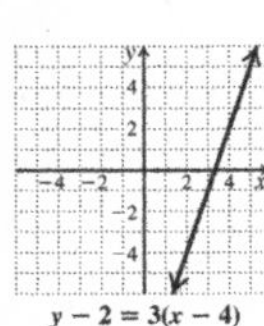

59. *Writing Exercise.*

61. $3x^4+5x^2-1=0$

Let $u=x^2$ and $u^2=x^4$.

$3u^2+5u-1=0$ Substituting

$u=\frac{-5\pm\sqrt{5^2-4\cdot 3\cdot(-1)}}{2\cdot 3}$

$u=\frac{-5\pm\sqrt{37}}{6}$

$x^2=\frac{-5\pm\sqrt{37}}{6}$ Replacing u with x^2

$x=\pm\sqrt{\frac{-5\pm\sqrt{37}}{6}}$

All four numbers check and are the solutions.

63. $\frac{x}{x-1}-6\sqrt{\frac{x}{x-1}}-40=0$

Let $u=\sqrt{\frac{x}{x-1}}$ and $u^2=\frac{x}{x-1}$.

$u^2-6u-40=0$ Substituting
$(u-10)(u+4)=0$
$u=10$ *or* $u=-4$
$\sqrt{\frac{x}{x-1}}=10$ *or* $\sqrt{\frac{x}{x-1}}=-4$ has no solution
$\frac{x}{x-1}=100$
$x=100x-100$ Multiplying by $(x-1)$
$100=99x$
$\frac{100}{99}=x$

The number $\frac{100}{99}$ checks. It is the solution.

65. $a^3-26a^{3/2}-27=0$

Let $u=a^{3/2}$.

$u^2-26u-27=0$ Substituting
$(u-27)(u+1)=0$
$u=27$ *or* $u=-1$

Replace u with $a^{3/2}$.

$a^{3/2} = 27$ *or* $a^{3/2} = -1$ has no solution

$a = 27^{2/3}$

$a = (3^3)^{2/3}$

$a = 9$

The number 9 checks. It is the solution.

67. $x^6 + 7x^3 - 8 = 0$

Let $u = x^3$.

$u^2 + 7u - 8 = 0$

$(u+8)(u-1) = 0$

$u = -8$ *or* $u = 1$

$x^3 = -8$ *or* $x^3 = 1$

$x^3 + 8 = 0$ *or* $x^3 - 1 = 0$

First solve $x^3 + 8 = 0$.

$x^3 + 8 = 0$

$(x+2)(x^2 - 2x + 4) = 0$

$x + 2 = 0$ *or* $x^2 - 2x + 4 = 0$

$x = -2$ *or* $x = 1 \pm \sqrt{3}i$

The solutions of $x^3 - 1 = 0$ are 1 and $-\frac{1}{2} \pm \frac{\sqrt{3}}{2}i$. (See Exercise 66.)

All six numbers check.

69. $-3, -1, 1, 4$

Mid-Chapter Review

1. $x - 7 = \pm\sqrt{5}$

$x = 7 \pm \sqrt{5}$

The solutions are $7 + \sqrt{5}$ and $7 - \sqrt{5}$.

2. $a = 1,\ b = -2,\ c = -1$

$$x = \frac{-(-2) \pm \sqrt{(-2)^2 - 4 \cdot 1 \cdot (-1)}}{2 \cdot 1}$$

$$x = \frac{2 \pm \sqrt{8}}{2}$$

$$x = \frac{2}{2} \pm \frac{2\sqrt{2}}{2}$$

The solutions are $1 + \sqrt{2}$ and $1 - \sqrt{2}$.

3. $x^2 + 4x = 21$

$x^2 + 4x - 21 = 0$

$(x+7)(x-3) = 0$ Factoring

$x + 7 = 0$ *or* $x - 3 = 0$

$x = -7$ *or* $x = 3$

The solutions are -7 and 3.

4. $t^2 - 196 = 0$

$t^2 = 196$

$t = \pm\sqrt{196} = \pm 14$

5. $x^2 = 2x + 5$

$x^2 - 2x + 1 = 5 + 1$

$(x-1)^2 = 6$

$x - 1 = \pm\sqrt{6}$

$x = 1 \pm \sqrt{6}$

The solutions are $1 + \sqrt{6}$ and $1 - \sqrt{6}$.

6. $x^2 = 2x - 5$

$x^2 - 2x + 1 = -5 + 1$

$(x-1)^2 = -4$

$x - 1 = \pm\sqrt{-4}$

$x = 1 \pm 2i$

7. $x^4 = 16$

$x^4 - 16 = 0$

Let $u = x^2$ and $u^2 = x^4$.

$u^2 - 16 = 0$

$(u+4)(u-4) = 0$

$u + 4 = 0$ *or* $u - 4 = 0$

$u = -4$ *or* $u = 4$

Replace u with x^2.

$x^2 = 4$ *or* $x^2 = -4$

$x = \pm 2$ *or* $x = \sqrt{-4} = \pm 2i$

The solutions are -2, 2, $-2i$, and $2i$.

8. $(t+3)^2 = 7$

$t + 3 = \pm\sqrt{7}$

$t = -3 \pm \sqrt{7}$

9. $n(n-3) = 2n(n+1)$

$n^2 - 3n = 2n^2 + 2n$

$0 = n^2 + 5n$

$0 = n(n+5)$

$n = 0$ *or* $n + 5 = 0$

$n = 0$ *or* $n = -5$

The solutions are -5 and 0.

10. $6y^2 - 7y - 10 = 0$

$(6y+5)(y-2) = 0$

$6y + 5 = 0$ *or* $y - 2 = 0$

$y = -\frac{5}{6}$ *or* $y = 2$

11. $16c^2 = 7c$

$16c^2 - 7c = 0$

$c(16c - 7) = 0$

$c = 0$ *or* $16c - 7 = 0$

$c = 0$ *or* $c = \frac{7}{16}$

The solutions are 0 and $\frac{7}{16}$.

12. $3x^2+5x=1$
$3x^2+5x-1=0$
$a=3,\ b=5,\ c=-1$

$$x=\frac{-5\pm\sqrt{(5)^2-4\cdot 3\cdot(-1)}}{2\cdot 3}=\frac{-5\pm\sqrt{25+12}}{6}$$
$$=\frac{-5\pm\sqrt{37}}{6}=-\frac{5}{6}\pm\frac{\sqrt{37}}{6}$$

13. $(t+4)(t-3)=18$
$t^2+t-12=18$
$t^2+t-30=0$
$(t+6)(t-5)=0$
$t+6=0$ *or* $t-5=0$
$t=-6$ *or* $t=5$
The solutions are –6 and 5.

14. $(m^2+3)^2-4(m^2+3)-5=0$

Let $u=m^2+3$ and $u^2=(m^2+3)^2$.
$u^2-4u-5=0$
$(u-5)(u+1)=0$
$u-5=0$ *or* $u+1=0$
$u=5$ *or* $u=-1$

Replace u with m^2+3.
$m^2+3=5$ *or* $m^2+3=-1$
$m^2=2$ *or* $m^2=-4$
$m=\pm\sqrt{2}$ *or* $m=\pm\sqrt{-4}=\pm 2i$

15. $x^2-8x+1=0$
$a=1,\ b=-8,\ c=1$
We substitute and compute the discriminant.
$b^2-4ac=(-8)^2-4\cdot 1\cdot 1=60$
There are two irrational solutions.

16. $b^2-4ac=(-4)^2-4\cdot 3\cdot(-7)=100$
There are two rational solutions.

17. $5x^2-x+6=0$
$a=5,\ b=-1,\ c=6$
We substitute and compute the discriminant.
$b^2-4ac=(-1)^2-4\cdot 5\cdot 6=-119$
There are two imaginary-number solutions.

18. $F=\dfrac{Av^2}{400}$
$400F=Av^2$
$\dfrac{400F}{A}=v^2$
$\sqrt{\dfrac{400F}{A}}=v$
$20\sqrt{\dfrac{F}{A}}=v$, or $v=\dfrac{20\sqrt{FA}}{A}$

19. $D^2-2Dd-2hd=0$
$D^2-2Dd+d^2=d^2+2hd$
$(D-d)^2=d^2+2hd$
$D-d=\sqrt{d^2+2hd}$
$D=d+\sqrt{d^2+2hd}$

20. Let r represent the speed and t the time of the slower trip.

Trip	Distance	Speed	Time
South	225	r	$\frac{225}{r}$
North	225	$r-30$	$\frac{225}{r-30}$

The time for the round trip is 8 hours. We now have an equation

$$\frac{225}{r}+\frac{225}{r-30}=8,$$

Solving for r, we get $r=75$ or $r=11.25$. Only 75 checks in the original problem. The speed South is 75 mph and the speed North is 75 – 30 = 45 mph.

Exercise Set 11.6

1. False. The graph of a quadratic function is a parabola.

3. True

5. $f(x)=x^2$
See Example 1 in the text.

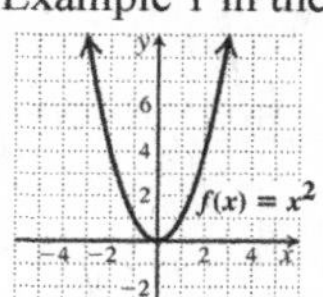

7. $f(x)=-2x^2$
We choose some numbers for x and compute $f(x)$ for each one. Then we plot the ordered pairs $(x, f(x))$ and connect them with a smooth curve.

x	$f(x)=-2x^2$
0	0
1	–2
2	–8
–1	–2
–2	–8

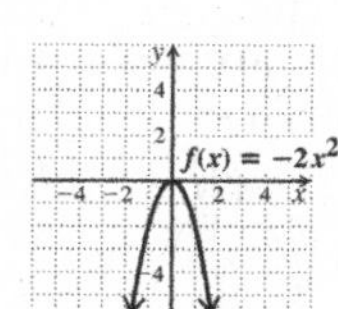

9. $g(x)=\frac{1}{3}x^2$

x	$g(x)=\frac{1}{3}x^2$
0	0
1	$\frac{1}{3}$
2	$\frac{4}{3}$
3	3
–1	$\frac{1}{3}$
–2	$\frac{4}{3}$
–3	3

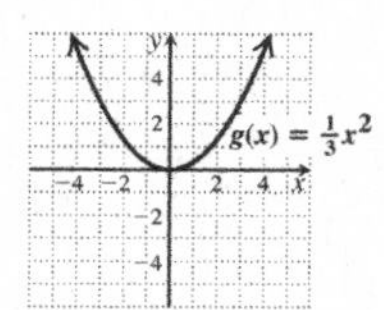

11. $h(x) = -\frac{1}{3}x^2$

Observe that the graph of $h(x) = -\frac{1}{3}x^2$ is the reflection of the graph of $g(x) = \frac{1}{3}x^2$ across the x-axis. We graphed $g(x)$ in Exercise 9, so we can use it to graph $h(x)$. If we did not make this observation we could find some ordered pairs, plot points, and connect them with a smooth curve.

x	$h(x) = -\frac{1}{3}x^2$
0	0
1	$-\frac{1}{3}$
2	$-\frac{4}{3}$
3	-3
-1	$-\frac{1}{3}$
-2	$-\frac{4}{3}$
-3	-3

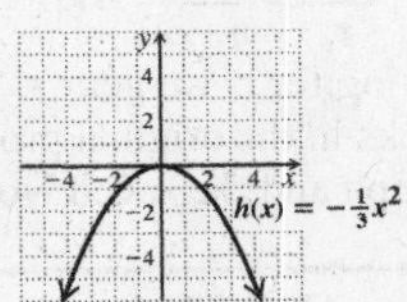

13. $f(x) = \frac{5}{2}x^2$

x	$f(x) = \frac{5}{2}x^2$
0	0
1	$\frac{5}{2}$
2	10
-1	$\frac{5}{2}$
-2	10

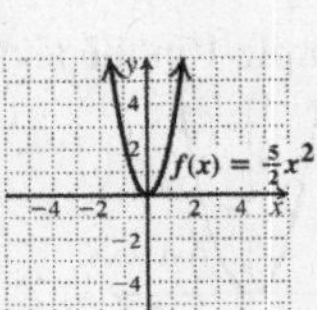

15. $g(x) = (x+1)^2 = [x-(-1)]^2$

We know that the graph of $g(x) = (x+1)^2$ looks like the graph of $f(x) = x^2$ (see Exercise 5) but moved to the left 1 unit.

Vertex: $(-1, 0)$, axis of symmetry: $x = -1$

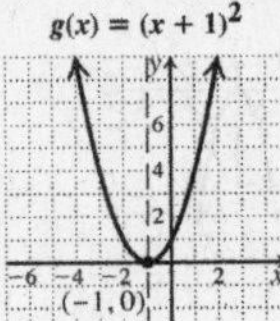

17. $f(x) = (x-2)^2$

The graph of $f(x) = (x-2)^2$ looks like the graph of $f(x) = x^2$ (see Exercise 5) but moved to the right 2 units.

Vertex: $(2, 0)$, axis of symmetry: $x = 2$

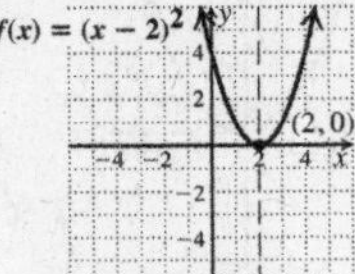

19. $g(x) = -(x+1)^2$

The graph of $g(x) = -(x+1)^2$ looks like the graph of $f(x) = x^2$ (see Exercise 5) but moved to the left 1 unit. It will also open downward because of the negative coefficient, -1.

Vertex: $(-1, 0)$, axis of symmetry: $x = -1$

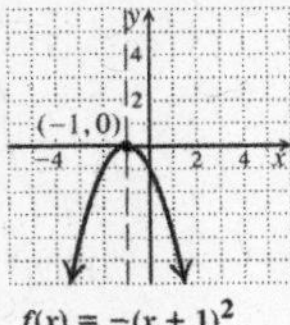

21. $g(x) = -(x-2)^2$

The graph of $g(x) = -(x-2)^2$ looks like the graph of $f(x) = x^2$ (see Exercise 5) but moved to the right 2 units. It will also open downward because of the negative coefficient, -1.

Vertex: $(2, 0)$, axis of symmetry: $x = 2$

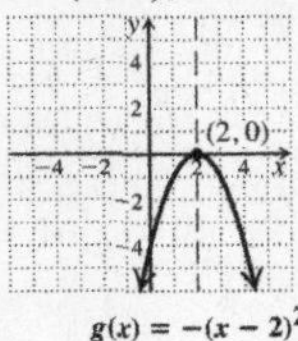

23. $f(x) = 2(x+1)^2$

The graph of $f(x) = 2(x+1)^2$ looks like the graph of $h(x) = 2x^2$ (see graph following Example 1) but moved to the left 1 unit.

Vertex: $(-1, 0)$, axis of symmetry: $x = -1$

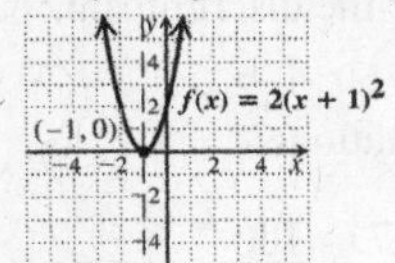

25. $g(x) = 3(x-4)^2$

The graph of $g(x) = 3(x-4)^2$ looks like the graph of $g(x) = 3x^2$ but moved to the right 4 units.

Vertex: $(4, 0)$, axis of symmetry: $x = 4$

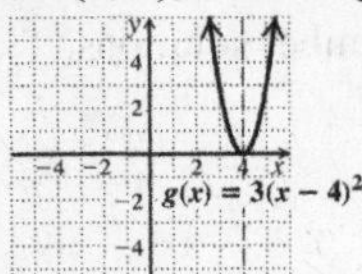

27. $h(x) = -\frac{1}{2}(x-4)^2$

The graph of $h(x) = -\frac{1}{2}(x-4)^2$ looks like the graph of $g(x) = \frac{1}{2}x^2$ (see graph following Example 1) but moved to the right 4 units. It will also open downward because of the negative coefficient, $-\frac{1}{2}$.

Vertex: $(4,0)$, axis of symmetry: $x=4$

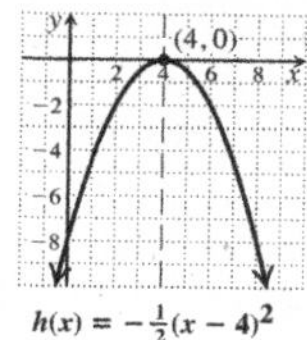

29. $f(x)=\frac{1}{2}(x-1)^2$

The graph of $f(x)=\frac{1}{2}(x-1)^2$ looks like the graph of $g(x)=\frac{1}{2}x^2$ (see graph following Example 1) but moved to the right 1 unit.
Vertex: (1, 0), axis of symmetry: $x = 1$

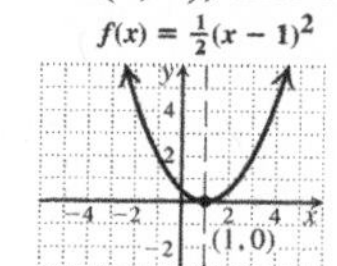

31. $f(x)=-2(x+5)^2=-2[x-(-5)]^2$

The graph of $f(x)=-2(x+5)^2$ looks like the graph of $h(x)=2x^2$ (see the graph following Example 1) but moved to the left 5 units. It will also open downward because of the negative coefficient, –2.
Vertex: (–5, 0), axis of symmetry: $x = -5$

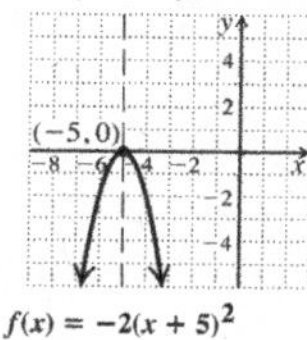

33. $h(x)=-3\left(x-\frac{1}{2}\right)^2$

The graph of $h(x)=-3\left(x-\frac{1}{2}\right)^2$ looks like the graph of $f(x)=-3x^2$ (see Exercise 8) but moved to the right $\frac{1}{2}$ unit.

Vertex: $\left(\frac{1}{2},\ 0\right)$, axis of symmetry: $x=\frac{1}{2}$

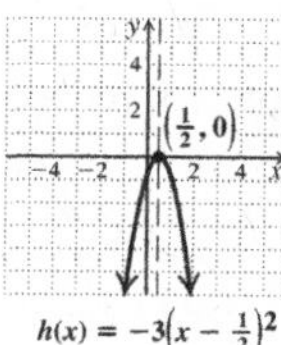

35. $f(x)=(x-5)^2+2$
We know that the graph looks like the graph of $f(x)=x^2$ (see Example 1) but moved to the right 5 units and up 2 units. The vertex is (5, 2), and the axis of symmetry is $x=5$. Since the coefficient of $(x-5)^2$ is positive $(1>0)$, there is a minimum function value, 2.

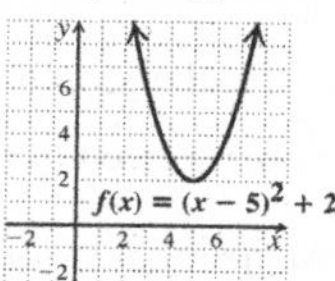

37. $f(x)=(x+1)^2-3$
We know that the graph looks like the graph of $f(x)=x^2$ (see Example 1) but moved to the left 1 unit and down 3 units. The vertex is (–1, –3), and the axis of symmetry is $x=-1$. Since the coefficient of $(x+1)^2$ is positive $(1>0)$, there is a minimum function value, –3.

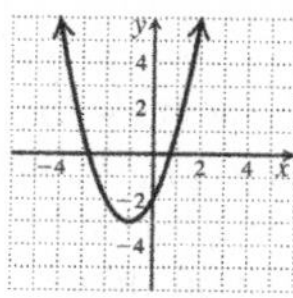

39. $g(x)=\frac{1}{2}(x+4)^2+1$

We know that the graph looks like the graph of $f(x)=\frac{1}{2}x^2$ (see graph following Example 1) but moved to the left 4 units and up 1 unit. The vertex is (–4, 1), and the axis of symmetry is $x = -4$, and the minimum function value is 1.

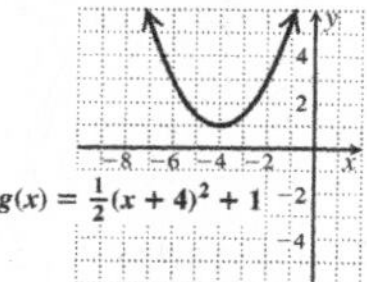

41. $h(x)=-2(x-1)^2-3$
We know that the graph looks like the graph of $h(x)=2x^2$ (see graph following Example 1) but moved to the right 1 unit and down 3 units and turned upside down. The vertex is (1, –3), and the axis of symmetry is $x=1$. The maximum function value is –3.

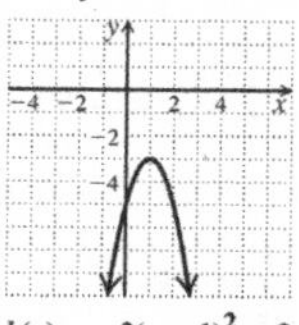

43. $f(x)=2(x+3)^2+1$
We know that the graph looks like the graph of $f(x)=2x^2$ (see graph following Example 1) but moved to the left 3 units and up 1 unit. The vertex is (–3, 1), and the axis of symmetry is $x = -3$. The minimum function value is 1.

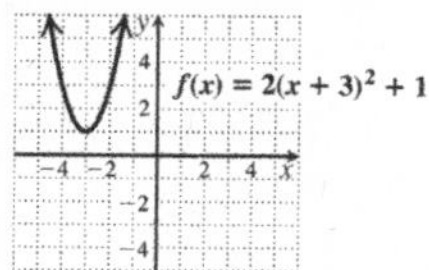

45. $g(x) = -\frac{3}{2}(x-2)^2 + 4$

We know that the graph looks like the graph of $f(x) = \frac{3}{2}x^2$ (see Exercise 14) but moved to the right 2 units and up 4 units and turned upside down. The vertex is (2, 4), and the axis of symmetry is $x = 2$, and the maximum function value is 4.

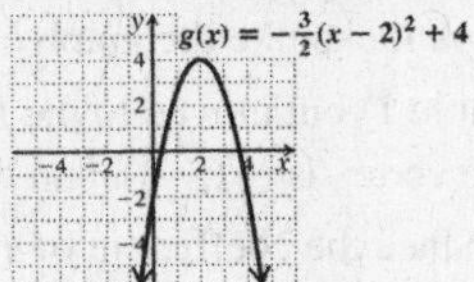

47. $f(x) = 5(x-3)^2 + 9$

The function is of the form $f(x) = a(x-h)^2 + k$ with $a = 5$, $h = 3$, and $k = 9$. The vertex is (h, k), or (3, 9). The axis of symmetry is $x = h$, or $x = 3$. Since $a > 0$, then k, or 9, is the minimum function value.

49. $f(x) = -\frac{3}{7}(x+8)^2 + 2$

The function is of the form $f(x) = a(x-h)^2 + k$ with $a = -\frac{3}{7}$, $h = -8$, and $k = 2$. The vertex is (h, k), or $(-8, 2)$. The axis of symmetry is $x = h$, or $x = -8$. Since $a < 0$, then k, or 2, is the maximum function value.

51. $f(x) = \left(x - \frac{7}{2}\right)^2 - \frac{29}{4}$

The function is of the form $f(x) = a(x-h)^2 + k$ with $a = 1$, $h = \frac{7}{2}$, and $k = -\frac{29}{4}$. The vertex is (h, k), or $\left(\frac{7}{2}, -\frac{29}{4}\right)$. The axis of symmetry is $x = h$, or $x = \frac{7}{2}$. Since $a > 0$, then k, or $-\frac{29}{4}$, is the minimum function value.

53. $f(x) = -\sqrt{2}(x+2.25)^2 - \pi$

The function is of the form $f(x) = a(x-h)^2 + k$ with $a = -\sqrt{2}$, $h = -2.25$, and $k = -\pi$. The vertex is (h, k), or $(-2.25, -\pi)$. The axis of symmetry is $x = h$, or $x = -2.25$. Since $a < 0$, then k, or $-\pi$, is the maximum function value.

55. *Writing Exercise.*

57. $\frac{3}{x} + \frac{x}{x+2} = \frac{3}{x} \cdot \frac{x+2}{x+2} + \frac{x}{x+2} \cdot \frac{x}{x}$

$$= \frac{3x + 6 + x^2}{x(x+2)}$$

$$= \frac{x^2 + 3x + 6}{x(x+2)}$$

59. $\sqrt[3]{8t} - \sqrt[3]{27t} + \sqrt{25t} = 2\sqrt[3]{t} - 3\sqrt[3]{t} + 5\sqrt{t}$

$$= -\sqrt[3]{t} + 5\sqrt{t}$$

61. $\frac{1}{x-1} - \frac{x-2}{x+3}$ LCD is $(x-1)(x+3)$

$$= \frac{1}{x-1} \cdot \frac{x+3}{x+3} - \frac{x-2}{x+3} \cdot \frac{x-1}{x-1}$$

$$= \frac{x+3}{(x-1)(x+3)} - \frac{(x-2)(x-1)}{(x-1)(x+3)}$$

$$= \frac{x+3-x^2+3x-2}{(x-1)(x+3)}$$

$$= \frac{-x^2+4x+1}{(x-1)(x+3)}$$

63. *Writing Exercise.*

65. The equation will be of the form $f(x) = \frac{3}{5}(x-h)^2 + k$ with $h = 1$ and $k = 3$:

$$f(x) = \frac{3}{5}(x-1)^2 + 3$$

67. The equation will be of the form $f(x) = \frac{3}{5}(x-h)^2 + k$ with $h = 4$ and $k = -7$:

$$f(x) = \frac{3}{5}(x-4)^2 - 7$$

69. The equation will be of the form $f(x) = \frac{3}{5}(x-h)^2 + k$ with $h = -2$ and $k = -5$:

$$f(x) = \frac{3}{5}[x-(-2)]^2 + (-5), \text{or}$$

$$f(x) = \frac{3}{5}(x+2)^2 - 5$$

71. Since there is a minimum at (2, 0), the parabola will have the same shape as $f(x) = 2x^2$. It will be of the form $f(x) = 2(x-h)^2 + k$ with $h = 2$ and $k = 0$:

$$f(x) = 2(x-2)^2$$

73. Since there is a maximum at $(0, -5)$, the parabola will have the same shape as $g(x) = -2x^2$. It will be of the form $g(x) = -2(x-h)^2 + k$ with $h = 0$ and $k = -5$: $g(x) = -2(x-0)^2 - 5$, or $g(x) = -2x^2 - 5$.

75. If h is increased, the graph will move to the right.

77. If a is replaced with $-a$, the graph will be reflected across the x-axis.

79. The maximum value of $g(x)$ is 1 and occurs at the point $(5, 1)$, so for $F(x)$ we have $h = 5$ and $k = 1$. $F(x)$ has the same shape as $f(x)$ and has a minimum, so $a = 3$. Thus, $F(x) = 3(x-5)^2 + 1$.

81. The graph of $y = f(x-1)$ looks like the graph of $y = f(x)$ moved 1 unit to the right.

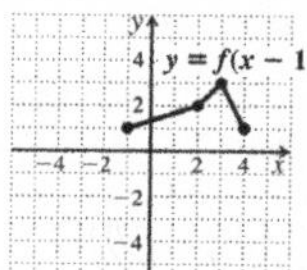

83. The graph of $y = f(x)+2$ looks like the graph of $y = f(x)$ moved up 2 units.

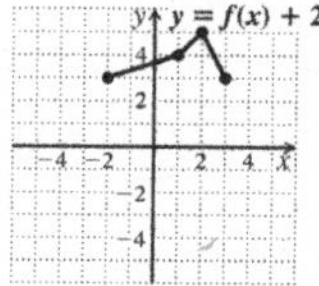

85. The graph of $y = f(x+3)-2$ looks like the graph of $y = f(x)$ moved 3 units to the left and also moved down 2 units.

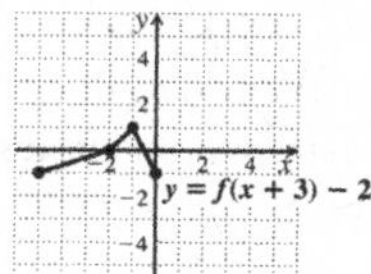

87. *Graphing Calculator Exercise*

89. *Writing Exercise*. The coefficient of x^2 is negative, so the parabola should open down.

Exercise Set 11.7

1. True; since $a = 3 > 0$, the graph opens upward.

3. True

5. False; the axis of symmetry is $x = \frac{3}{2}$.

7. False; the y-intercept is (0, 7).

9. $\frac{1}{2}\cdot(-8) = -4;\ (-4)^2 = 16$

$$\begin{aligned} f(x) &= x^2 - 8x + 2 \\ &= (x^2 - 8x + 16) - 16 + 2 \\ &= (x-4)^2 + (-14) \end{aligned}$$

11. $\frac{1}{2}\cdot 3 = \frac{3}{2};\ \left(\frac{3}{2}\right)^2 = \frac{9}{4}$

$$\begin{aligned} f(x) &= x^2 + 3x - 5 \\ &= \left(x^2 + 3x + \frac{9}{4}\right) - \frac{9}{4} - 5 \\ &= \left[x - \left(-\frac{3}{2}\right)\right]^2 + \left(-\frac{29}{4}\right) \end{aligned}$$

13. $\frac{1}{2}\cdot 2 = 1;\ 1^2 = 1$

$$\begin{aligned} f(x) &= 3x^2 + 6x - 2 \\ &= 3(x^2 + 2x) - 2 \\ &= 3(x^2 + 2x + 1) + 3(-1) - 2 \\ &= 3[x - (-1)]^2 + (-5) \end{aligned}$$

15. $\frac{1}{2}\cdot 4 = 2;\ 2^2 = 4$

$$\begin{aligned} f(x) &= -x^2 - 4x - 7 \\ &= -1(x^2 + 4x) - 7 \\ &= -(x^2 + 4x + 4) + -1(-4) - 7 \\ &= -[x - (-2)]^2 + (-3) \end{aligned}$$

17. $\frac{1}{2}\cdot\left(-\frac{5}{2}\right) = -\frac{5}{4};\ \left(-\frac{5}{4}\right)^2 = \frac{25}{16}$

$$\begin{aligned} f(x) &= 2x^2 - 5x + 10 \\ &= 2\left(x^2 - \frac{5}{2}x\right) + 10 \\ &= 2\left(x^2 - \frac{5}{2}x + \frac{25}{16}\right) + 2\left(-\frac{25}{16}\right) + 10 \\ &= 2\left(x - \frac{5}{4}\right)^2 + \frac{55}{8} \end{aligned}$$

19. a.
$$\begin{aligned} f(x) &= x^2 + 4x + 5 \\ &= (x^2 + 4x + 4 - 4) + 5 \quad \text{Adding } 4-4 \\ &= (x^2 + 4x + 4) - 4 + 5 \quad \text{Regrouping} \\ &= (x+2)^2 + 1 \end{aligned}$$

The vertex is (–2, 1), the axis of symmetry is $x = -2$, and the graph opens upward since the coefficient 1 is positive. We plot a few points as a check and draw the curve.

b.

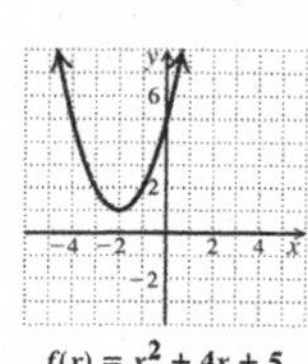

$f(x) = x^2 + 4x + 5$

21. a.
$$\begin{aligned} f(x) &= x^2 + 8x + 20 \\ &= (x^2 + 8x + 16 - 16) + 20 \quad \text{Adding 16–16} \\ &= (x^2 + 8x + 16) - 16 + 20 \quad \text{Regrouping} \\ &= (x+4)^2 + 4 \end{aligned}$$

The vertex is (–4, 4), the axis of symmetry is $x = -4$, and the graph opens upward since the coefficient 1 is positive.

b.

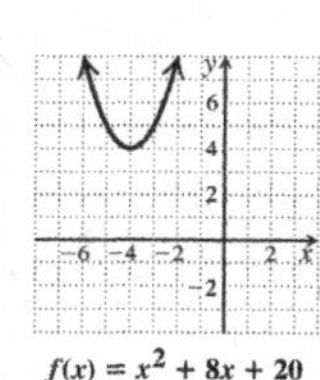

$f(x) = x^2 + 8x + 20$

23. a. $h(x) = 2x^2 - 16x + 25$

$= 2(x^2 - 8x) + 25$ Factoring 2 from the first two terms

$= 2(x^2 - 8x + 16 - 16) + 25$ Adding $16 - 16$ inside the parentheses

$= 2(x^2 - 8x + 16) + 2(-16) + 25$ Distributing to obtain a trinomial square

$= 2(x-4)^2 - 7$

The vertex is (4, –7), the axis of symmetry is $x = 4$, and the graph opens upward since the coefficient 2 is positive.

b.

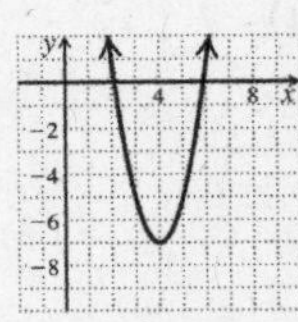

$h(x) = 2x^2 - 16x + 25$

25. a. $f(x) = -x^2 + 2x + 5$

$= -(x^2 - 2x) + 5$ Factoring –1 from the first two terms

$= -(x^2 - 2x + 1 - 1) + 5$ Adding $1 - 1$ inside the parentheses

$= -(x^2 - 2x + 1) - (-1) + 5$

$= -(x-1)^2 + 6$

The vertex is (1, 6), the axis of symmetry is $x = 1$, and the graph opens downward since the coefficient –1 is negative.

b.

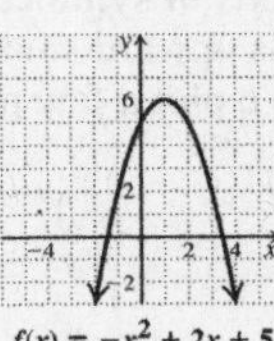

$f(x) = -x^2 + 2x + 5$

27. a. $g(x) = x^2 + 3x - 10$

$= \left(x^2 + 3x + \frac{9}{4} - \frac{9}{4}\right) - 10$

$= \left(x^2 + 3x + \frac{9}{4}\right) - \frac{9}{4} - 10$

$= \left(x + \frac{3}{2}\right)^2 - \frac{49}{4}$

The vertex is $\left(-\frac{3}{2}, -\frac{49}{4}\right)$, the axis of symmetry is $x = -\frac{3}{2}$, and the graph opens upward since the coefficient 1 is positive.

b.

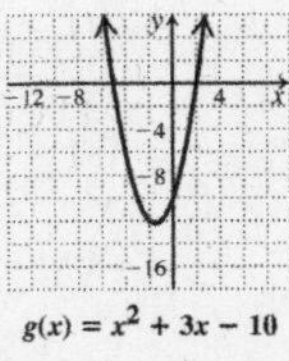

$g(x) = x^2 + 3x - 10$

29. a. $h(x) = x^2 + 7x$

$= \left(x^2 + 7x + \frac{49}{4}\right) - \frac{49}{4}$

$= \left(x + \frac{7}{2}\right)^2 - \frac{49}{4}$

The vertex is $\left(-\frac{7}{2}, -\frac{49}{4}\right)$, the axis of symmetry is $x = -\frac{7}{2}$, and the graph opens upward since the coefficient 1 is positive.

b.

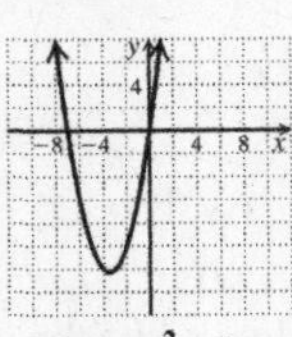

$h(x) = x^2 + 7x$

31. a. $f(x) = -2x^2 - 4x - 6$

$= -2(x^2 + 2x) - 6$ Factoring

$= -2(x^2 + 2x + 1 - 1) - 6$ Adding $1 - 1$ inside the parentheses

$= -2(x^2 + 2x + 1) - 2(-1) - 6$

$= -2(x+1)^2 - 4$

The vertex is (–1, –4), the axis of symmetry is $x = -1$, and the graph opens downward since the coefficient –2 is negative.

b.

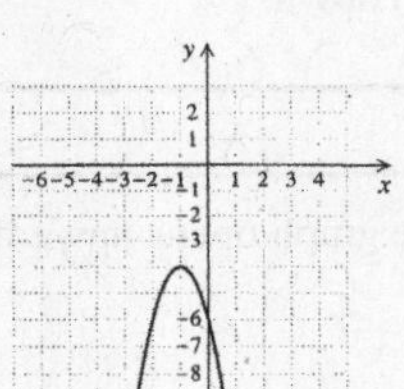

33. a. $g(x) = x^2 - 6x + 13$

$= (x^2 - 6x + 9 - 9) + 13$ Adding $9 - 9$

$= (x^2 - 6x + 9) - 9 + 13$ Regrouping

$= (x-3)^2 + 4$

The vertex is (3, 4), the axis of symmetry is $x = 3$, and the graph opens upward since the coefficient 1 is positive. The minimum is 4.

b.

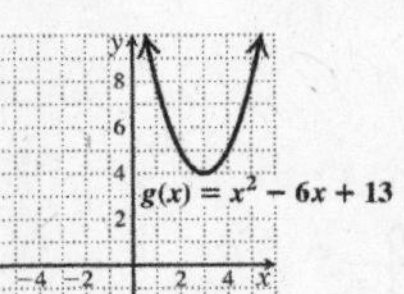

35. a. $g(x) = 2x^2 - 8x + 3$

$= 2(x^2 - 4x) + 3$ Factoring

$= 2(x^2 - 4x + 4 - 4) + 3$ Adding $4 - 4$

$= 2(x^2 - 4x + 4) + 2(-4) + 3$

$= 2(x-2)^2 - 5$

The vertex is (2, –5), the axis of symmetry is

$x = 2$, and the graph opens upward since the coefficient 2 is positive. The minimum is –5.

b.

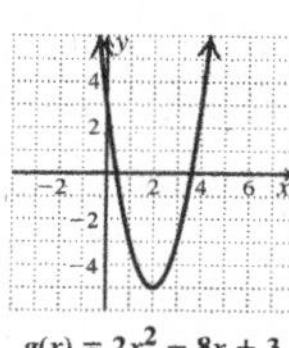

$g(x) = 2x^2 - 8x + 3$

37. a. $f(x) = 3x^2 - 24x + 50$

$= 3(x^2 - 8x) + 50$ Factoring

$= 3(x^2 - 8x + 16 - 16) + 50$ Adding $16 - 16$ inside the parentheses

$= 3(x^2 - 8x + 16) - 3 \cdot 16 + 50$

$= 3(x - 4)^2 + 2$

The vertex is (4, 2), the axis of symmetry is $x = 4$, and the graph opens upward since the coefficient 3 is positive. The minimum is 2.

b.

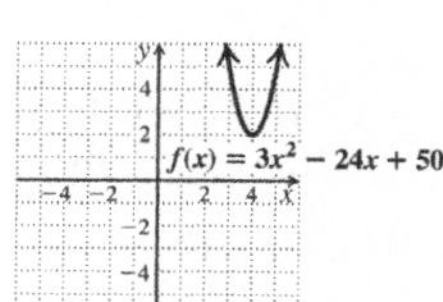

39. a. $f(x) = -3x^2 + 5x - 2$

$= -3\left(x^2 - \frac{5}{3}x\right) - 2$ Factoring

$= -3\left(x^2 - \frac{5}{3}x + \frac{25}{36} - \frac{25}{36}\right) - 2$ Adding $\frac{25}{36} - \frac{25}{36}$ inside the parentheses

$= -3\left(x^2 - \frac{5}{3}x + \frac{25}{36}\right) - 3\left(-\frac{25}{36}\right) - 2$

$= -3\left(x - \frac{5}{6}\right)^2 + \frac{1}{12}$

The vertex is $\left(\frac{5}{6}, \frac{1}{12}\right)$, the axis of symmetry is $x = \frac{5}{6}$, and the graph opens downward since the coefficient –3 is negative. The maximum is $\frac{1}{12}$.

b.

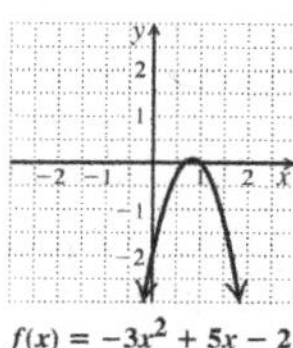

$f(x) = -3x^2 + 5x - 2$

41. a. $h(x) = \frac{1}{2}x^2 + 4x + \frac{19}{3}$

$= \frac{1}{2}(x^2 + 8x) + \frac{19}{3}$ Factoring

$= \frac{1}{2}(x^2 + 8x + 16 - 16) + \frac{19}{3}$ Adding $16 - 16$ inside parentheses

$= \frac{1}{2}(x^2 + 8x + 16) + \frac{1}{2}(-16) + \frac{19}{3}$

$= \frac{1}{2}(x + 4)^2 - \frac{5}{3}$

The vertex is $\left(-4, -\frac{5}{3}\right)$, the axis of symmetry is $x = -4$, and the graph opens upward since the coefficient $\frac{1}{2}$ is positive. The minimum is $-\frac{5}{3}$.

b.

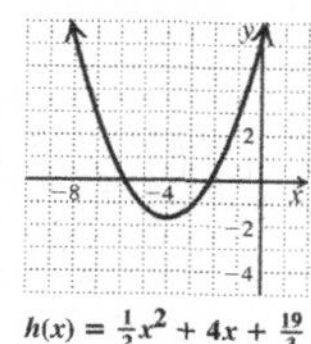

$h(x) = \frac{1}{2}x^2 + 4x + \frac{19}{3}$

43. $f(x) = x^2 - 6x + 3$

To find the x-intercepts, solve the equation $0 = x^2 - 6x + 3$. Use the quadratic formula.

$$x = \frac{-(-6) \pm \sqrt{(-6)^2 - 4 \cdot 1 \cdot 3}}{2 \cdot 1}$$

$$x = \frac{6 \pm \sqrt{24}}{2} = \frac{6 \pm 2\sqrt{6}}{2} = 3 \pm \sqrt{6}$$

The x-intercepts are $(3 - \sqrt{6}, 0)$ and $(3 + \sqrt{6}, 0)$.
The y-intercept is $(0, f(0))$, or (0, 3).

45. $g(x) = -x^2 + 2x + 3$

To find the x-intercepts, solve the equation $0 = -x^2 + 2x + 3$. We factor.

$0 = -x^2 + 2x + 3$

$0 = x^2 - 2x - 3$ Multiplying by -1

$0 = (x - 3)(x + 1)$

$x = 3$ or $x = -1$

The x-intercepts are (–1, 0) and (3, 0).
The y-intercept is $(0, g(0))$, or (0, 3).

47. $f(x) = x^2 - 9x$

To find the x-intercepts, solve the equation $0 = x^2 - 9x$. We factor.

$0 = x^2 - 9x$

$0 = x(x - 9)$

$x = 0$ or $x = 9$

The x-intercepts are (0, 0) and (9, 0).
Since (0, 0) is an x-intercept, we observe that (0, 0) is also the y-intercept.

49. $h(x) = -x^2 + 4x - 4$

To find the x-intercepts, solve the equation $0 = -x^2 + 4x - 4$. We factor.

$0 = -x^2 + 4x - 4$

$0 = x^2 - 4x + 4$ Multiplying by -1

$0 = (x - 2)(x - 2)$

$x = 2$ *or* $x = 2$

The x-intercept is (2, 0).
The y-intercept is $(0, h(0))$, or (0, –4).

51. $g(x) = x^2 + x - 5$
To find the x-intercepts, solve the equation $0 = x^2 + x - 5$. Use the quadratic formula.

$$x = \frac{-1 \pm \sqrt{1^2 - 4\cdot1\cdot(-5)}}{2\cdot1}$$
$$x = \frac{-1 \pm \sqrt{21}}{2} = -\frac{1}{2} \pm \frac{\sqrt{21}}{2}$$

The x-intercepts are $\left(-\frac{1}{2} - \frac{\sqrt{21}}{2},\ 0\right)$ and $\left(-\frac{1}{2} + \frac{\sqrt{21}}{2},\ 0\right)$.
The y-intercept is $(0, g(0))$, or $(0, -5)$.

53. $f(x) = 2x^2 - 4x + 6$
To find the x-intercepts, solve the equation $0 = 2x^2 - 4x + 6$. We use the quadratic formula.

$$x = \frac{-(-4) \pm \sqrt{(-4)^2 - 4\cdot2\cdot6}}{2\cdot2}$$
$$x = \frac{4 \pm \sqrt{-32}}{4} = \frac{4 \pm 4i\sqrt{2}}{2} = 2 \pm 2i\sqrt{2}$$

There are no real-number solutions, so there is no x-intercept.
The y-intercept is $(0, f(0))$, or $(0, 6)$.

55. *Writing Exercise*

57. $(x^2 - 7)(x^2 + 3) = x^4 - 4x^2 - 21$

59.
$$\sqrt[3]{18x^4y}\cdot\sqrt[3]{6x^2y} = \sqrt[3]{18x^4y\cdot6x^2y} = \sqrt[3]{27x^6\cdot4y^2} = 3x^2\sqrt[3]{4y^2}$$

61.
$$\frac{4a^2-b^2}{2ab} \div \frac{2a^2-ab-b^2}{6a^2} = \frac{4a^2-b^2}{2ab}\cdot\frac{6a^2}{2a^2-ab-b^2}$$
$$= \frac{(2a+b)(2a-b)}{2a\cdot b}\cdot\frac{2a\cdot3a}{(2a+b)(a-b)}$$
$$= \frac{\cancel{(2a+b)}(2a-b)}{\cancel{2a}\cdot b}\cdot\frac{\cancel{2a}\cdot3a}{\cancel{(2a+b)}(a-b)}$$
$$= \frac{3a(2a-b)}{b(a-b)}$$

63. *Writing Exercise.*

65. **a.**
$$f(x) = 2.31x^2 - 3.135x - 5.89$$
$$= 2.31(x^2 - 1.357142857x) - 5.89$$
$$= 2.31(x^2 - 1.357142857x + 0.460459183 - 0.460459183) - 5.89$$
$$= 2.31(x^2 - 1.357142857x + 0.460459183) + 2.31(-0.460459183) - 5.89$$
$$= 2.31(x - 0.678571428)^2 - 6.953660714$$

Since the coefficient 2.31 is positive, the function has a minimum value. It is –6.953660714.

b. To find the x-intercepts, solve $0 = 2.31x^2 - 3.135x - 5.89$.

$$x = \frac{-(-3.135) \pm \sqrt{(-3.135)^2 - 4(2.31)(-5.89)}}{2(2.31)}$$
$$x \approx \frac{3.135 \pm 8.015723611}{4.62}$$
$$x \approx -1.056433682 \quad or \quad x \approx 2.413576539$$

The x-intercepts are (–1.056433682, 0) and (2.413576539, 0).
The y-intercept is $(0, f(0))$, or $(0, -5.89)$.

67. $f(x) = x^2 - x - 6$

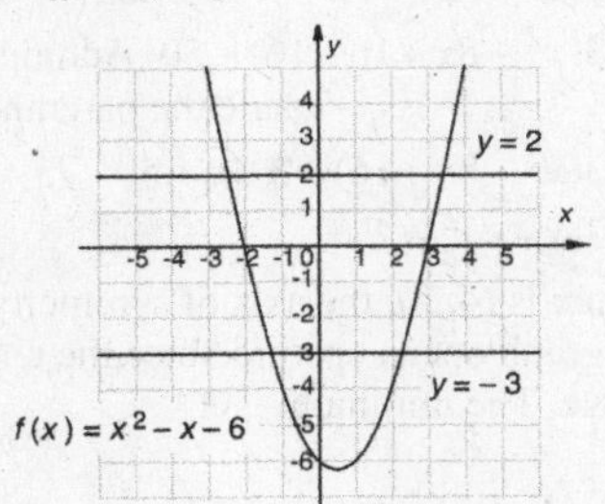

a. The solutions of $x^2 - x - 6 = 2$ are the first coordinates of the points of intersection of the graphs of $f(x) = x^2 - x - 6$ and $y = 2$. From the graph we see that the solutions are approximately –2.4 and 3.4.

b. The solutions of $x^2 - x - 6 = -3$ are the first coordinates of the points of intersection of the graphs of $f(x) = x^2 - x - 6$ and $y = -3$. From the graph we see that the solutions are approximately –1.3 and 2.3.

69.
$$f(x) = mx^2 - nx + p$$
$$= m\left(x^2 - \frac{n}{m}x\right) + p$$
$$= m\left(x^2 - \frac{n}{m}x + \frac{n^2}{4m^2} - \frac{n^2}{4m^2}\right) + p$$
$$= m\left(x - \frac{n}{2m}\right)^2 - \frac{n^2}{4m} + p$$
$$= m\left(x - \frac{n}{2m}\right)^2 + \frac{-n^2 + 4mp}{4m}, \text{or}$$
$$m\left(x - \frac{n}{2m}\right)^2 + \frac{4mp - n^2}{4m}$$

71. The horizontal distance from (–1, 0) to (3, –5) is $|3 - (-1)|$, or 4, so by symmetry the other x-intercept is (3 + 4, 0), or (7, 0). Substituting the three ordered pairs (–1, 0), (3,–5), and (7, 0) in the equation $f(x) = ax^2 + bx + c$ yields a system of equations:

$$0 = a - b + c,$$
$$-5 = 9a + 3b + c,$$
$$0 = 49a + 7b + c$$

The solution of this system of equations is $\left(\frac{5}{16}, -\frac{15}{8}, -\frac{35}{16}\right)$, so $f(x) = \frac{5}{16}x^2 - \frac{15}{8}x - \frac{35}{16}$.
If we complete the square we find that this function can also be expressed as $f(x) = \frac{5}{16}(x-3)^2 - 5$.

73. $f(x) = |x^2 - 1|$

We plot some points and draw the curve. Note that it will lie entirely on or above the x-axis since absolute value is never negative.

x	$f(x)$
−3	8
−2	3
−1	0
0	1
1	0
2	3
3	8

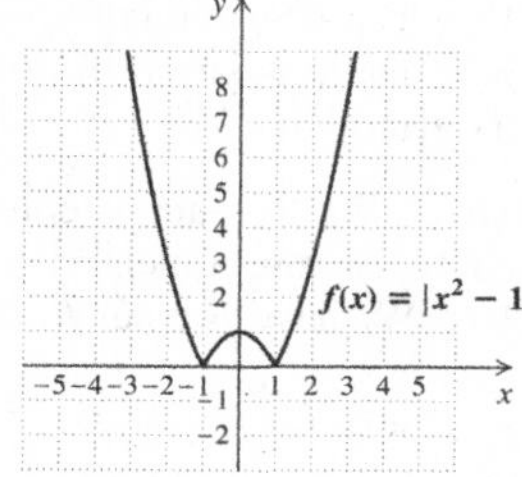

75. $f(x) = \left|2(x-3)^2 - 5\right|$

We plot some points and draw the curve. Note that it will lie entirely on or above the x – axis since absolute value is never negative.

x	$f(x)$
−1	27
0	13
1	3
2	3
3	5
4	3
5	3
6	13

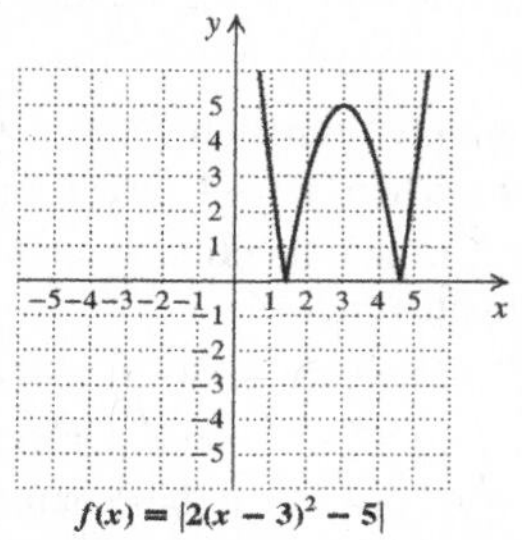

Exercise Set 11.8

1. True

3. True

5. True

7. ***Familiarize and Translate***. We are given the formula $p(x) = -0.2x^2 + 1.3x + 6.2$.

Carry out. To find the value of x for which $p(x)$ is a maximum, we first find $-\frac{b}{2a}$:

$$-\frac{b}{2a} = -\frac{1.3}{2(-0.2)} = 3.25, \text{ or } 3\frac{1}{4}$$

Now we find the maximum value of the function $p(3.25)$:

$$p(3.25) = -0.2(3.25)^2 + 1.3(3.25) + 6.2 = 8.3125$$

The minimum function value of about 8.3 occurs when $x = 3.25$.

Check. We can go over the calculations again. We could also solve the problem again by completing the square. The answer checks.

State. A calf's daily milk consumption is greatest at 3.25 weeks at about 8.3 lb of milk per day.

9. ***Familiarize and Translate***. We want to find the value of x for which $C(x) = 0.1x^2 - 0.7x + 2.425$ is a minimum.

Carry out. We complete the square.

$$C(x) = 0.1(x^2 - 7x + 12.25) + 2.425 - 1.225$$
$$C(x) = 0.1(x-3.5)^2 + 1.2$$

The minimum function value of 1.2 occurs when $x = 3.5$.

Check. Check a function value for x less than 3.5 and for x greater than 3.5.

$$C(3) = 0.1(3)^2 - 0.7(3) + 2.425 = 1.225$$
$$C(4) = 0.1(4)^2 - 0.7(4) + 2.425 = 1.225$$

Since 1.2 is less than these numbers, it looks as though we have a minimum.

State. The minimum average cost is $1.2 hundred, or $120. To achieve the minimum cost, 3.5 hundred, or 350 dulcimers should be built.

11. ***Familiarize***. We make a drawing and label it.

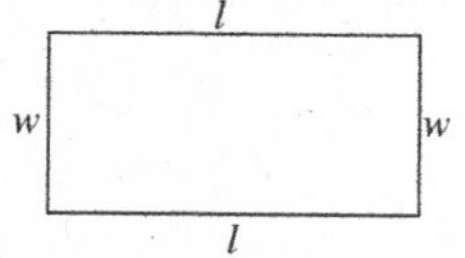

Perimeter: $2l + 2w = 720$ ft

Area: $A = l \cdot w$

Translate. We have a system of equations.

$$2l + 2w = 720,$$
$$A = lw$$

Carry out. Solving the first equation for l, we get $l = 360 - w$. Substituting for l in the second equation we get a quadratic function A:

$$A = (360 - w)w$$
$$A = -w^2 + 360w$$

Completing the square, we get

$$A = -(w-180)^2 + 32{,}400$$

The maximum function value is 32,400. It occurs when w is 180. When $w = 180$, $l = 360 - 180$, or 180.

Check. We check a function value for w less than 180 and for w greater than 180.

$$A(179) = -179^2 + 360 \cdot 179 = 32{,}399$$
$$A(181) = -181^2 + 360 \cdot 181 = 32{,}399$$

Since 32,400 is greater than these numbers, it looks as though we have a maximum.

State. The maximum area occurs when the dimensions are 180 ft by 180 ft.

13. ***Familiarize***. We make a drawing and label it.

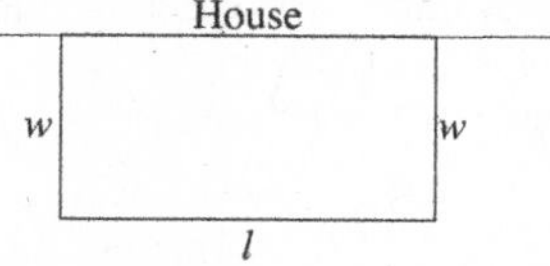

Translate. We have two equations.

$$l + 2w = 60,$$
$$A = lw$$

Carry out. Solve the first equation for l.

$$l = 60 - 2w$$

Substitute for l in the second equation.

$$A = (60 - 2w)w$$
$$A = -2w^2 + 60w$$

Completing the square, we get

$$A = -2(w-15)^2 + 450.$$

The maximum function value of 450 occurs when $w = 15$. When $w = 15$, $l = 60 - 2 \cdot 15 = 30$.
Check. Check a function value for w less than 15 and for w greater than 15.

$$A(14) = -2 \cdot 14^2 + 60 \cdot 14 = 448$$
$$A(16) = -2 \cdot 16^2 + 60 \cdot 16 = 448$$

Since 450 is greater than these numbers, it looks as though we have a maximum.

State. The maximum area of 450 ft^2 will occur when the dimensions are 15 ft by 30 ft.

15. ***Familiarize***. Let x represent the height of the file and y represent the width. We make a drawing.

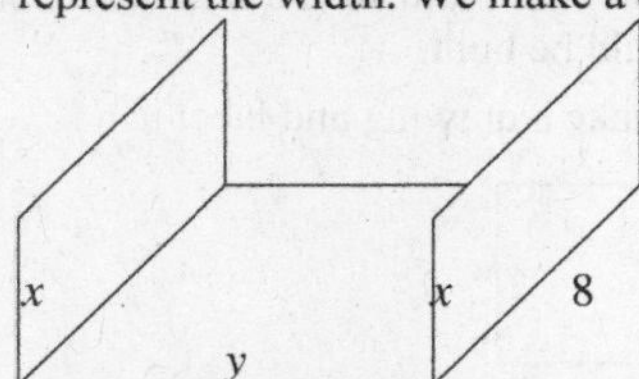

Translate. We have two equations.

$$2x + y = 14$$
$$V = 8xy$$

Carry out. Solve the first equation for y.

$$y = 14 - 2x$$

Substitute for y in the second equation.

$$V = 8x(14 - 2x)$$
$$V = -16x^2 + 112x$$

Completing the square, we get

$$V = -16\left(x - \frac{7}{2}\right)^2 + 196.$$

The maximum function value of 196 occurs when $x = \frac{7}{2}$. When $x = \frac{7}{2}$; $y = 14 - 2 \cdot \frac{7}{2} = 7$.
Check. Check a function value for x less than $\frac{7}{2}$ and for x greater than $\frac{7}{2}$.

$$V(3) = -16 \cdot 3^2 + 112 \cdot 3 = 192$$
$$V(4) = -16 \cdot 4^2 + 112 \cdot 4 = 192$$

Since 196 is greater than these numbers, it looks as though we have a maximum.

State. The file should be $\frac{7}{2}$ in., or 3.5 in. tall.

17. ***Familiarize***. We let x and y represent the numbers, and we let P represent their product.
Translate. We have two equations.

$$x + y = 18,$$
$$P = xy$$

Carry out. Solving the first equation for y, we get $y = 18 - x$. Substituting for y in the second equation we get a quadratic function P:

$$P = x(18 - x)$$
$$P = -x^2 + 18x$$

Completing the square, we get

$$P = -(x - 9)^2 + 81.$$

The maximum function value is 81. It occurs when $x = 9$. When $x = 9$, $y = 18 - 9$, or 9.
Check. We can check a function value for x less than 9 and for x greater than 9.

$$P(10) = -10^2 + 18 \cdot 10 = 80$$
$$P(8) = -8^2 + 18 \cdot 8 = 80$$

Since 81 is greater than these numbers, it looks as though we have a maximum.
State. The maximum product of 81 occurs for the numbers 9 and 9.

19. ***Familiarize***. We let x and y represent the two numbers, and we let P represent their product.
Translate. We have two equations.

$$x - y = 8,$$
$$P = xy$$

Carry out. Solve the first equation for x.

$$x = 8 + y$$

Substitute for x in the second equation.

$$P = (8 + y)y$$
$$P = y^2 + 8y$$

Completing the square, we get

$$P = (y + 4)^2 - 16.$$

The minimum function value is –16. It occurs when $y = -4$. When $y = -4$, $x = 8 + (-4)$, or 4.
Check. Check a function value for y less than –4 and for y greater than –4.

$$P(-5) = (-5)^2 + 8(-5) = -15$$
$$P(-3) = (-3)^2 + 8(-3) = -15$$

Since –16 is less than these numbers, it looks as though we have a minimum.
State. The minimum product of –16 occurs for the numbers 4 and –4.

21. From the results of Exercises 17 and 18, we might observe that the numbers are –5 and –5 and that the maximum product is 25. We could also solve this problem as follows.
Familiarize. We let x and y represent the two numbers, and we let P represent their product.
Translate. We have two equations.

$$x + y = -10,$$
$$P = xy$$

Carry out. Solve the first equation for y.

$$y = -10 - x$$

Substitute for y in the second equation.

$$P = x(-10 - x)$$
$$P = -x^2 - 10x$$

Completing the square, we get

$$P = -(x + 5)^2 + 25$$

The maximum function value is 25. It occurs when $x = -5$. When $x = -5$, $y = -10 - (-5)$, or -5.
Check. Check a function value for x less than –5 and for x greater than –5.

$$P(-6) = -(-6)^2 - 10(-6) = 24$$
$$P(-4) = -(-4)^2 - 10(-4) = 24$$

Since 25 is greater than these numbers, it looks as though we have a maximum.
State. The maximum product of 25 occurs for the numbers –5 and –5.

23. The data points rise and then fall. The graph appears to represent a quadratic function that opens downward. Thus a quadratic function $f(x) = ax^2 + bx + c,\ a < 0,$ might be used to model the data.

25. The data points rise. The graph does not appear to represent a quadratic function in which the data points would rise and then fall or vice versa. Thus a linear function $f(x) = mx + b$ might be used to model the data.

27. The data points do not represent a linear or quadratic pattern. Thus, it does not appear that the data can be modeled with either a quadratic or a linear function.

29. The data points fall and then rise. The graph appears to represent a quadratic function that opens upward. Thus a quadratic function $f(x) = ax^2 + bx + c,\ a > 0,$ might be used to model the data.

31. The data points rise. The graph does not appear to represent a quadratic function in which the data points would rise and then fall or vice versa. Thus a linear function $f(x) = mx + b$ might be used to model the data.

33. The data points fall. The graph does not appear to represent a quadratic function in which the data points would rise and then fall or vice versa. Thus a linear function $f(x) = mx + b$ might be used to model the data.

35. We look for a function of the form $f(x) = ax^2 + bx + c$. Substituting the data points, we get

$$4 = a(1)^2 + b(1) + c,$$
$$-2 = a(-1)^2 + b(-1) + c,$$
$$13 = a(2)^2 + b(2) + c,$$

or

$$4 = a + b + c,$$
$$-2 = a - b + c,$$
$$13 = 4a + 2b + c.$$

Solving this system, we get
$a = 2,\ b = 3,$ and $c = -1.$
Therefore the function we are looking for is

$$f(x) = 2x^2 + 3x - 1.$$

37. We look for a function of the form $f(x) = ax^2 + bx + c$. Substituting the data points, we get

$$0 = a(2)^2 + b(2) + c,$$
$$3 = a(4)^2 + b(4) + c,$$
$$-5 = a(12)^2 + b(12) + c,$$

or

$$0 = 4a + 2b + c,$$
$$3 = 16a + 4b + c,$$
$$-5 = 144a + 12b + c.$$

Solving this system, we get

$$a = -\frac{1}{4},\ b = 3,\ c = -5.$$

Therefore the function we are looking for is

$$f(x) = -\frac{1}{4}x^2 + 3x - 5.$$

39. a. ***Familiarize***. We look for a function of the form $A(s) = as^2 + bs + c$, where $A(s)$ represents the number of nighttime accidents (for every 200 million km) and s represents the travel speed (in km/h).
Translate. We substitute the given values of s and $A(s)$.

$$400 = a(60)^2 + b(60) + c,$$
$$250 = a(80)^2 + b(80) + c,$$
$$250 = a(100)^2 + b(100) + c,$$

or

$$400 = 3600a + 60b + c,$$
$$250 = 6400a + 80b + c,$$
$$250 = 10{,}000a + 100b + c.$$

Carry out. Solving the system of equations, we get
$a = \frac{3}{16},\ b = -\frac{135}{4},\ c = 1750.$
Check. Recheck the calculations.
State. The function
$A(s) = \frac{3}{16}s^2 - \frac{135}{4}s + 1750$ fits the data.

b. Find $A(50)$.

$$A(50) = \frac{3}{16}(50)^2 - \frac{135}{4}(50) + 1750 = 531.25$$

About 531 accidents for every 200 million km driven, occur at 50 km/h.

41. ***Familiarize***. Think of a coordinate system placed on the drawing in the text with the origin at the point where the arrow is released. Then three points on the arrow's parabolic path are (0, 0), (63, 27), and (126, 0). We look for a function of the form $h(d) = ad^2 + bd + c$, where $h(d)$ represents the arrow's height and d represents the distance the arrow has traveled horizontally.
Translate. We substitute the values given above for d and $h(d)$.

$$0 = a \cdot 0^2 + b \cdot 0 + c,$$
$$27 = a \cdot 63^2 + b \cdot 63 + c,$$
$$0 = a \cdot 126^2 + b \cdot 126 + c$$

or

$$0 = c,$$
$$27 = 3969a + 63b + c,$$
$$0 = 15{,}876a + 126b + c$$

Carry out. Solving the system of equations, we get $a \approx -0.0068$, $b \approx 0.8571$, and $c = 0$.
Check. Recheck the calculations.
State. The function $h(d) = -0.0068d^2 + 0.8571d$ expresses the arrow's height as a function of the distance it has traveled horizontally.

43. *Writing Exercise.*

45. $y=-\frac{1}{3}x+16$

47. $m=\frac{0-8}{10-4}=-\frac{8}{6}=-\frac{4}{3}$

$y-0=-\frac{4}{3}(x-10)$

$y=-\frac{4}{3}x+\frac{40}{3}$

49. $2x+y=3$

$y=-2x+3$

Slope of perpendicular line: $\frac{1}{2}$

$y=\frac{1}{2}x-6$

51. *Writing Exercise.*

53. ***Familiarize.*** Position the bridge on a coordinate system as shown with the vertex of the parabola at (0, 30).

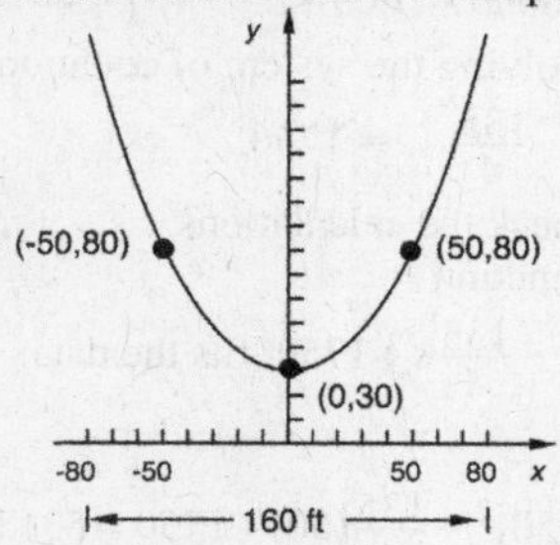

We find a function of the form $y=ax^2+bx+c$ which represents the parabola containing the points (0, 30), (–50, 80), and (50, 80).
Translate. Substitute for x and y.

$$30=a\cdot 0^2+b\cdot 0+c,$$
$$80=a(-50)^2+b(-50)+c,$$
$$80=a(50)^2+b(50)+c,$$

or

$$30=c,$$
$$80=2500a-50b+c,$$
$$80=2500a+50b+c.$$

Carry out. Solving the system of equations, we get
$a=0.02$, $b=0$, $c=30$.

The function $y=0.02x^2+30$ represents the parabola. Because the cable supports are 160 ft apart, the tallest supports are positioned 160/2, or 80 ft, to the left and right of the midpoint. This means that the longest vertical cables occur at $x=-80$ and $x=80$. For $x=\pm 80$,

$$y=0.02(\pm 80)^2+30$$
$$=128+30$$
$$=158 \text{ ft}$$

Check. We go over the calculations.
State. The longest vertical cables are 158 ft long.

55. ***Familiarize.*** Let x represent the number of 25 increases in the admission price. Then $10+0.25x$ represents the admission price, and $80-x$ represents the corresponding average attendance. Let R represent the total revenue.
Translate. Since the total revenue is the product of the cover charge and the number attending a show, we have the following function for the amount of money the owner makes.

$$R(x)=(10+0.25x)(80-x), \text{ or}$$
$$R(x)=-0.25x^2+10x+800$$

Carry out. Completing the square, we get

$$R(x)=-0.25(x-20)^2+900$$

The maximum function value of 900 occurs when $x=20$. The owner should charge \$10 + \$0.25(20), or \$15.
Check. We check a function value for x less than 20 and for x greater than 20.

$$R(19)=-0.25(19)^2+10\cdot 19+800=899.75$$
$$R(21)=-0.25(21)^2+10\cdot 21+800=899.75$$

Since 900 is greater than these numbers, it looks as though we have a maximum.
State. The owner should charge \$15.

57. ***Familiarize.*** We add labels to the drawing in the text.

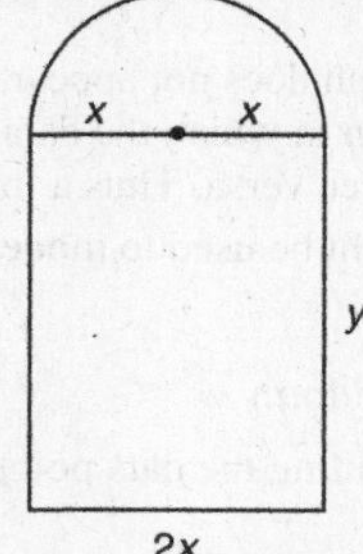

The perimeter of the semicircular portion of the window is $\frac{1}{2}\cdot 2\pi x$, or πx. The perimeter of the rectangular portion is $y+2x+y$, or $2x+2y$. The area of the semicircular portion of the window is $\frac{1}{2}\cdot \pi x^2$, or $\frac{\pi}{2}x^2$. The area of the rectangular portion is $2xy$.
Translate. We have two equations, one giving the perimeter of the window and the other giving the area.

$$\pi x+2x+2y=24,$$
$$A=\frac{\pi}{2}x^2+2xy$$

Carry out. Solve the first equation for y.

$$\pi x+2x+2y=24$$
$$2y=24-\pi x-2x$$
$$y=12-\frac{\pi x}{2}-x$$

Substitute for y in the second equation.

$$A=\frac{\pi}{2}x^2+2x\left(12-\frac{\pi x}{2}-x\right)$$
$$A=\frac{\pi}{2}x^2+24x-\pi x^2-2x^2$$
$$A=-2x^2-\frac{\pi}{2}x^2+24x$$
$$A=-\left(2x+\frac{\pi}{2}\right)x^2+24x$$

Completing the square, we get

$$A=-\left(2+\frac{\pi}{2}\right)\left(x^2+\frac{24}{-\left(2+\frac{\pi}{2}\right)}x\right)$$

$$A=-\left(2+\frac{\pi}{2}\right)\left(x^2-\frac{48}{4+\pi}x\right)$$

$$A=-\left(2+\frac{\pi}{2}\right)\left(x-\frac{24}{4+\pi}\right)^2+\left(\frac{24}{4+\pi}\right)^2$$

The maximum function value occurs when $x=\frac{24}{4+\pi}$. When $x=\frac{24}{4+\pi}$,

$$y=12-\frac{\pi}{2}\left(\frac{24}{4+\pi}\right)-\frac{24}{4+\pi}$$

$$=\frac{48+12\pi}{4+\pi}-\frac{12\pi}{4+\pi}-\frac{24}{4+\pi}=\frac{24}{4+\pi}$$

Check. Recheck the calculations.

State. The radius of the circular portion of the window and the height of the rectangular portion should each be $\frac{24}{4+\pi}$ ft.

59. a. Enter the data and use the quadratic regression operation on a graphing calculator. We get

$a(x)=18.78035714x^2+35.54107143x+165.1107143$,

where x is the number of years after 2006.

b. In 2014, $x = 2014 - 2006 = 8$.

$a(8)\approx 1{,}651{,}382$ subscriptions

Exercise Set 11.9

1. The solutions of $(x-3)(x+2)=0$ are 3 and –2 and for a test value in [–2, 3], say 0, $(x-3)(x+2)$ is negative so the statement is true. (Note that the endpoints must be included in the solution set because the inequality symbol is ≤.)

3. The solutions of $(x-1)(x-6)=0$ are 1 and 6. For a value of x less than 1, say 0, $(x-1)(x-6)$ is positive; for a value of x greater than 6, say 7, $(x-1)(x-6)$ is also positive. Thus, the statement is true. (Note that the endpoints of the intervals are not included because the inequality symbol is >.)

5. Since $x + 2 = 0$ when $x = -2$ and $x - 3 = 0$ when $x = 3$, the statement is false.

7. $p(x)\le 0$ when $-4\le x\le \frac{3}{2}$,

$\left[-4, \frac{3}{2}\right]$ or $\left\{x\middle|-4\le x\le \frac{3}{2}\right\}$

9. $x^4+12x>3x^2+4x^2$ is equivalent to $x^4-3x^2-4x^2+12x>0$, which is the graph in the text.

$p(x)>0$ when $(-\infty,-2)\cup(0, 2)\cup(3, \infty)$ or $\{x|x<-2 \text{ or } 0<x<2 \text{ or } x>3\}$

11. $\frac{x-1}{x+2}<3$ is equivalent to finding the values of x for which the graph $r(x)$ is less than 3, or below $g(x)$.

$\left(-\infty,-\frac{7}{2}\right)\cup(-2, \infty)$ or $\left\{x\middle|x<-\frac{7}{2} \text{ or } x>-2\right\}$

13. $(x-6)(x-5)<0$

The solutions of $(x-6)(x-5)=0$ are 5 and 6. They are not solutions of the inequality, but they divide the real number line in a natural way. The product $(x-6)(x-5)$ is positive or negative, for values other than 5 and 6, depending on the signs of the factors $x - 6$ and $x - 5$.

$x - 6 > 0$ when $x > 6$ and $x - 6 < 0$ when $x < 6$.
$x - 5 > 0$ when $x > 5$ and $x - 5 < 0$ when $x < 5$

We make a diagram.

Sign of $x-6$	–	–	+
Sign of $x-5$	–	+	+
Sign of product	+	–	+
		5	6

For the product $(x-6)(x-5)$ to be negative, one factor must be positive and the other negative. We see from the diagram that numbers satisfying $5 < x < 6$ are solutions. The solution set of the inequality is (5, 6) or $\{x|5<x<6\}$.

15. $(x+7)(x-2)\ge 0$

The solutions of $(x+7)(x-2)=0$ are –7 and 2. They divide the number line into three intervals as shown:

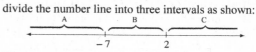

A B C
–7 2

We try test numbers in each interval.

A: Test –8, $f(-8)=(-8+7)(-8-2)=10$

B: Test 0, $f(0)=(0+7)(0-2)=-14$

C: Test 3, $f(3)=(3+7)(3-2)=10$

Since $f(-8)$ and $f(3)$ are positive, the function value will be positive for all numbers in the intervals containing –8 and 3. The inequality symbol is ≤, so we need to include the endpoints. The solution set is $(-\infty,-7]\cup[2, \infty)$, or $\{x\,|\,x\le -7 \text{ or } x\ge 2\}$.

17. $x^2-x-2>0$

$(x+1)(x-2)>0$ Factoring

The solutions of $(x+1)(x-2)=0$ are –1 and 2. They divide the number line into three intervals as shown:

A B C
–1 2

We try test numbers in each interval.

A: Test –2, $f(-2)=(-2+1)(-2-2)=4$

B: Test 0, $f(0)=(0+1)(0-2)=-2$

C: Test 3, $f(3)=(3+1)(3-2)=4$

Since $f(-2)$ and $f(3)$ are positive, the function value will be positive for all numbers in the intervals containing –2 and 3. The solution set is $(-\infty,-1)\cup(2, \infty)$, or $\{x\,|\,x<-1 \text{ or } x>2\}$.

19. $x^2+4x+4<0$

$(x+2)^2<0$

Observe that $(x+2)^2 \geq 0$ for all values of x. Thus, the solution set is $\varnothing$.

21. $x^2-4x \leq 3$

$x^2-4x+4 \leq 3+4$

$(x-2)^2 \leq 7$

$x-2 \leq \pm\sqrt{7}$

$x \leq 2 \pm \sqrt{7}$

The solutions of $x^2-4x-3 \leq 0$ are $2 \pm \sqrt{7}$. They divide the number line into three intervals as shown:

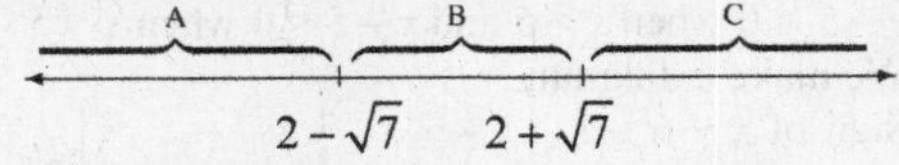

We try test numbers in each interval.

A: Test −1, $f(-1)=(-1)^2-4(-1)-3=2$

B: Test 0, $f(0)=0^2-4(0)-3=-3$

C: Test 5, $f(5)=5^2-4(5)-3=2$

Since $f(0)$ is negative, the function value will be negative for all numbers in the interval containing 0. The solution set is $\left[2-\sqrt{7},\, 2+\sqrt{7}\right]$, or $\left\{x \mid 2-\sqrt{7} \leq x \leq 2+\sqrt{7}\right\}$.

23. $3x(x+2)(x-2)<0$

The solutions of $3x(x+2)(x-2)=0$ are 0, −2, and 2. They divide the real-number line into four intervals as shown:

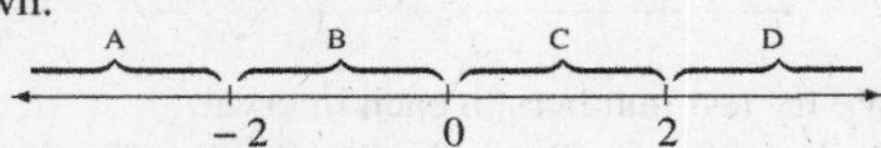

We try test numbers in each interval.

A: Test −3, $f(-3)=3(-3)(-3+2)(-3-2)=-45$

B: Test −1, $f(-1)=3(-1)(-1+2)(-1-2)=9$

C: Test 1, $f(1)=3(1)(1+2)(1-2)=-9$

D: Test 3, $f(3)=3(3)(3+2)(3-2)=45$

Since $f(-3)$ and $f(1)$ are negative, the function value will be negative for all numbers in the intervals containing −3 and 1. The solution set is $(-\infty,-2)\cup(0,\,2)$, or $\{x \mid x<-2 \text{ or } 0<x<2\}$.

25. $(x-1)(x+2)(x-4) \geq 0$

The solutions of $(x-1)(x+2)(x-4)=0$ are 1, −2, and 4. They divide the real-number line in a natural way. The product $(x-1)(x+2)(x-4)$ is positive or negative depending on the signs of $x-1$, $x+2$, and $x-4$.

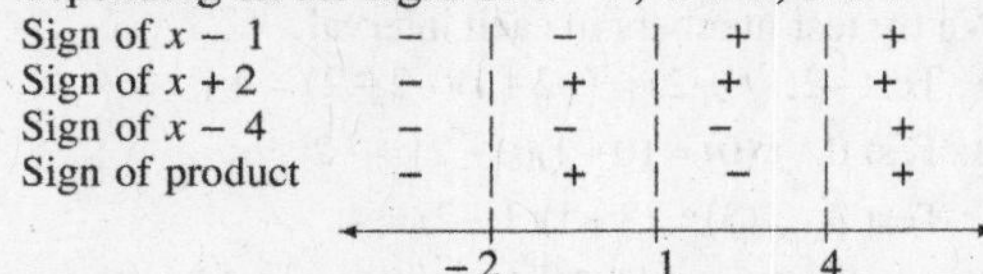

Sign of $x-1$	−	−	+	+
Sign of $x+2$	−	+	+	+
Sign of $x-4$	−	−	−	+
Sign of product	−	+	−	+
		−2	1	4

A product of three numbers is positive when all three factors are positive or when two are negative and one is positive. Since the ≥ symbol allows for equality, the endpoints −2, 1, and 4 are solutions. From the chart we see that the solution set is $[-2,\,1]\cup[4,\,\infty)$, or $\{x \mid -2 \leq x \leq 1 \text{ or } x \geq 4\}$.

27. $f(x) \geq 3$

$7-x^2 \geq 3$

$-x^2+4 \geq 0$

$x^2-4 \leq 0$

$(x-2)(x+2) \leq 0$

The solutions of $(x-2)(x+2)=0$ are 2 and −2. They divide the real-number line as shown below.

Sign of $x-2$	−	−	+
Sign of $x+2$	−	+	+
Sign of product	+	−	+
		−2	2

Because the inequality symbol is ≤, we must include the endpoints in the solution set. From the chart, we see that the solution set is $[-2,\,2]$, or $\{x \mid -2 \leq x \leq 2\}$.

29. $g(x)>0$

$(x-2)(x-3)(x+1)>0$

The solutions of $(x-2)(x-3)(x+1)=0$ are 2, 3, and −1. They divide the real-number line into four intervals as shown below.

A B C D

−1 2 3

We try test numbers in each interval.

A: Test −2, $f(-2)=(-2-2)(-2-3)(-2+1)=-20$

B: Test 0, $f(0)=(0-2)(0-3)(0+1)=6$

C: Test $\frac{5}{2}$, $f\left(\frac{5}{2}\right)=\left(\frac{5}{2}-2\right)\left(\frac{5}{2}-3\right)\left(\frac{5}{2}+1\right)=-\frac{7}{8}$

D: Test 4, $f(4)=(4-2)(4-3)(4+1)=10$

The function value will be positive for all numbers in intervals B and D. The solution set is $(-1,\,2)\cup(3,\,\infty)$, or $\{x \mid -1<x<2 \text{ or } x>3\}$.

31. $F(x) \leq 0$

$x^3-7x^2+10x \leq 0$

$x(x^2-7x+10) \leq 0$

$x(x-2)(x-5) \leq 0$

The solutions of $x(x-2)(x-5)=0$ are 0, 2, and 5. They divide the real-number line as shown below.

Sign of x	−	+	+	+
Sign of $x-2$	−	−	+	+
Sign of $x-5$	−	−	−	+
Sign of product	−	+	−	+
		0	2	5

Because the inequality symbol is ≤ we must include the endpoints in the solution set. From the chart we see that the solution set is $(-\infty,\,0]\cup[2,\,5]$ or $\{x \mid x \leq 0 \text{ or } 2 \leq x \leq 5\}$.

33. $\dfrac{1}{x-5}<0$

We write the related equation by changing the < symbol to =.

$$\frac{1}{x-5}=0$$

We solve the related equation.

$$(x-5)\cdot\frac{1}{x-5}=(x-5)\cdot 0$$
$$1=0$$

The related equation has no solution.
Next we find the values that make the denominator 0 by setting the denominator equation to 0 and solving:

$$x-5=0$$
$$x=5$$

We use 5 to divide the number line into two intervals as shown:

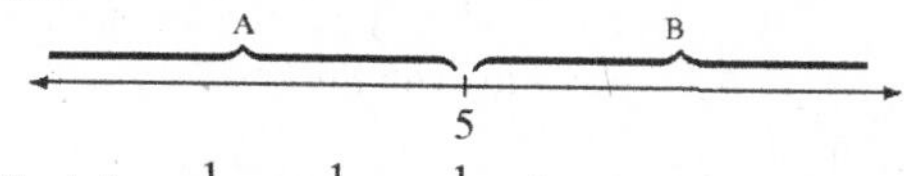

A: Test 0, $\frac{1}{0-5}=\frac{1}{-5}=-\frac{1}{5}<0$

The number 0 is a solution of the inequality, so the interval A is part of the solution set.

B: Test 6, $\frac{1}{6-5}=1\not<0$

The number 6 is not a solution of the inequality, so the interval B is part of the solution set.

The solution set is $(-\infty, 5)$, or $\{x\,|\,x<5\}$.

35. $\frac{x+1}{x-3}\geq 0$

Solve the related equation.

$$\frac{x+1}{x-3}=0$$
$$x+1=0$$
$$x=-1$$

Find the values that make the denominator 0.

$$x-3=0$$
$$x=3$$

Use the numbers -1 and 3 to divide the number line into intervals as shown:

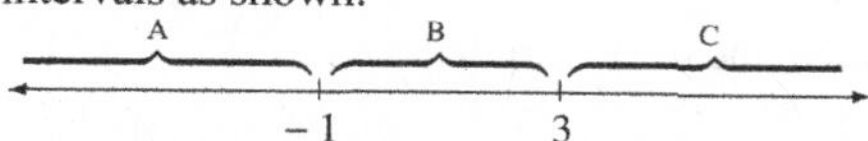

Try test numbers in each interval.

A: Test -2, $\frac{-2+1}{-2-3}=\frac{-1}{-5}=\frac{1}{5}>0$

The number -2 is a solution of the inequality, so the interval A is part of the solution set.

B: Test 0, $\frac{0+1}{0-3}=\frac{1}{-3}=-\frac{1}{3}\not>0$

The number 0 is not a solution of the inequality, so the interval B is not part of the solution set.

C: Test 4, $\frac{4+1}{4-3}=\frac{5}{1}=5>0$

The number 4 is a solution of the inequality, so the interval C is part of the solution set.
The solution set includes intervals A and C. The number -1 is also included since the inequality symbol is $\geq$ and -1 is the solution of the related equation. The number 3 is not included since $\frac{x+1}{x-3}$ is undefined for $x=3$. The solution set is $(-\infty,-1]\cup(3,\infty)$, or $\{x\,|\,x\leq -1 \text{ or } x>3\}$.

37. $\frac{x+1}{x+6}\geq 1$

Solve the related equation $\frac{x+1}{x+6}=1$

$$x+1=x+6$$
$$1=6$$

The related equation has no solution.
Find the values that make the denominator 0.

$$x+6=0$$
$$x=-6$$

Use the number -6 to divide the number line into two intervals.

Try test numbers in each interval.

A: Test -7, $\frac{-7+1}{-7+6}=\frac{-6}{-1}=6>1$.

The number -7 is a solution of the inequality, so the interval A is part of the solution set.

B: Test 0, $\frac{0+1}{0+6}=\frac{1}{6}\not>1$

The number 0 is not a solution of the inequality, so the interval B is not part of the solution set. The number -6 is not included in the solution set since $\frac{x+1}{x+6}$ is undefined for $x=-6$. The solution set is $(-\infty,-6)$, or $\{x\,|\,x<-6\}$.

39. $\frac{(x-2)(x+1)}{x-5}\leq 0$

Solve the related equation.

$$\frac{(x-2)(x+1)}{x-5}=0$$
$$(x-2)(x+1)=0$$
$$x=2 \text{ or } x=-1$$

Find the values that make the denominator 0.

$$x-5=0$$
$$x=5$$

Use the numbers 2, -1, and 5 to divide the number line into intervals as shown:

Try test numbers in each interval.

A: Test -2, $\frac{(-2-2)(-2+1)}{-2-5}=\frac{-4(-1)}{-7}=-\frac{4}{7}\leq 0$

Interval A is part of the solution set.

B: Test 0, $\frac{(0-2)(0+1)}{0-5}=\frac{-2\cdot 1}{-5}=\frac{2}{5}\not\leq 0$

Interval B is not part of the solution set.

C: Test 3, $\frac{(3-2)(3+1)}{3-5}=\frac{1\cdot 4}{-2}=-2\leq 0$

Interval C is part of the solution set.

D: Test 6, $\frac{(6-2)(6+1)}{6-5}=\frac{4\cdot 7}{1}=28\not\leq 0$

Interval D is not part of the solution set.
The solution set includes intervals A and C. The numbers -1 and 2 are also included since the inequality symbol is $\leq$ and -1 and 2 are the solutions of the related equation. The number 5 is not included since $\frac{(x-2)(x+1)}{x-5}$ is undefined for $x=5$. The solution set is $(-\infty,-1]\cup[2,5)$, or $\{x\,|\,x\leq -1 \text{ or } 2\leq x<5\}$.

41. $\dfrac{x}{x+3} \ge 0$

Solve the related equation.

$$\frac{x}{x+3} = 0$$
$$x = 0$$

Find the values that make the denominator 0.

$$x + 3 = 0$$
$$x = -3$$

Use the numbers 0 and –3 to divide the number line into intervals as shown.

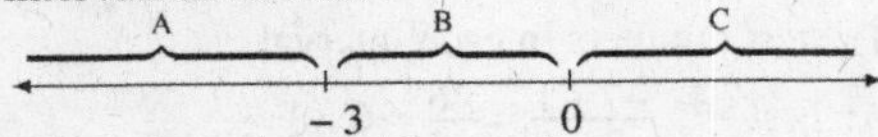

Try test numbers in each interval.

A: Test –4, $\dfrac{-4}{-4+3} = \dfrac{-4}{-1} = 4 \ge 0$

Interval A is part of the solution set.

B: Test –1, $\dfrac{-1}{-1+3} = \dfrac{-1}{2} = -\dfrac{1}{2} \not\ge 0$

Interval B is not part of the solution set.

C: Test 1, $\dfrac{1}{1+3} = \dfrac{1}{4} \ge 0$

The interval C is part of the solution set.

The solution set includes intervals A and C. The number 0 is also included since the inequality symbol is $\ge$ and 0 is the solution of the related equation. The number –3 is not included since $\dfrac{x}{x+3}$ is undefined for $x = -3$. The solution set is $(-\infty, -3) \cup [0, \infty)$, or $\{x \mid x < -3 \text{ or } x \ge 0\}$.

43. $\dfrac{x-5}{x} < 1$

Solve the related equation.

$$\frac{x-5}{x} = 1$$
$$x - 5 = x$$
$$-5 = 0$$

The related equation has no solution.

Find the values that make the denominator 0.

$$x = 0$$

Use the number 0 to divide the number line into two intervals as shown.

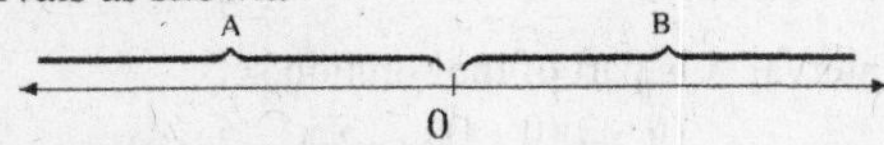

Try test numbers in each interval.

A: Test -1, $\dfrac{-1-5}{-1} = \dfrac{-6}{-1} = 6 \not< 1$

Interval A is not part of the solution set.

B: Test 1, $\dfrac{1-5}{1} = \dfrac{-4}{1} = -4 < 1$

Interval B is part of the solution set.

The solution set is $(0, \infty)$ or $\{x \mid x > 0\}$.

45. $\dfrac{x-1}{(x-3)(x+4)} \le 0$

Solve the related equation.

$$\frac{x-1}{(x-3)(x+4)} = 0$$
$$x - 1 = 0$$
$$x = 1$$

Find the values that make the denominator 0.

$$(x-3)(x+4) = 0$$
$$x = 3 \text{ or } x = -4$$

Use the numbers 1, 3, and –4 to divide the number line into intervals as shown:

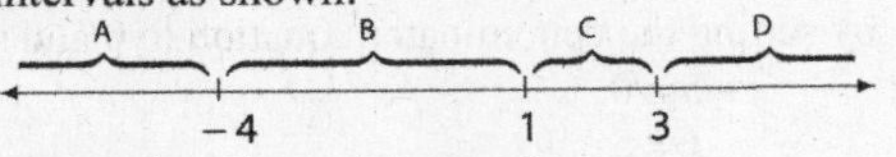

Try test numbers in each interval.

A: Test –5, $\dfrac{-5-1}{(-5-3)(-5+4)} = \dfrac{-6}{-8(-1)} = -\dfrac{3}{4} < 0$

Interval A is part of the solution set.

B: Test 0, $\dfrac{0-1}{(0-3)(0+4)} = \dfrac{-1}{-3 \cdot 4} = \dfrac{1}{12} \not< 0$

Interval B is not part of the solution set.

C: Test 2, $\dfrac{2-1}{(2-3)(2+4)} = \dfrac{1}{-1 \cdot 6} = -\dfrac{1}{6} < 0$

Interval C is part of the solution set.

D: Test 4, $\dfrac{4-1}{(4-3)(4+4)} = \dfrac{3}{1 \cdot 8} = \dfrac{3}{8} \not< 0$

Interval D is not part of the solution set.

The solution set includes intervals A and C. The number 1 is also included since the inequality symbol is $\le$ and 1 is the solution of the related equation. The numbers –4 and 3 are not included since $\dfrac{x-1}{(x-3)(x+4)}$ is undefined for $x = -4$ and for $x = 3$. The solution set is $(-\infty, -4) \cup [1, 3)$, or $\{x \mid x < -4 \text{ or } 1 \le x < 3\}$.

47. $f(x) \ge 0$

$\dfrac{5-2x}{4x+3} \ge 0$

Solve the related equation.

$$\frac{5-2x}{4x+3} = 0$$
$$5 - 2x = 0$$
$$5 = 2x$$
$$\frac{5}{2} = x$$

Find the values that make the denominator 0.

$$4x + 3 = 0$$
$$4x = -3$$
$$x = -\frac{3}{4}$$

Use the numbers $\dfrac{5}{2}$ and $-\dfrac{3}{4}$ to divide the number line as shown:

A B C

$-\dfrac{3}{4}$ $\dfrac{5}{2}$

Try test numbers in each interval.

A: Test –1, $\dfrac{5-2(-1)}{4(-1)+3} = -7 \not\ge 0$

Interval A is not part of the solution set.

B: Test 0, $\dfrac{5-2 \cdot 0}{4 \cdot 0 + 3} = \dfrac{5}{3} > 0$

Interval B is part of the solution set.

C: Test 3, $\frac{5-2\cdot 3}{4\cdot 3+3}=-\frac{1}{15}\not\geq 0$
Interval C is not part of the solution set.

The solution set includes interval B. The number $\frac{5}{2}$ is also included since the inequality symbol is $\geq$ and $\frac{5}{2}$ is the solution of the related equation. The number $-\frac{3}{4}$ is not included since $\frac{5-2x}{4x+3}$ is undefined for $x=-\frac{3}{4}$.

The solution set is $\left(-\frac{3}{4}, \frac{5}{2}\right]$, or $\left\{x \middle| -\frac{3}{4}<x\leq\frac{5}{2}\right\}$.

49. $G(x)\leq 1$

$\frac{1}{x-2}\leq 1$

Solve the related equation.

$$\frac{1}{x-2}=1$$
$$1=x-2$$
$$3=x$$

Find the values of x that make the denominator 0.

$$x-2=0$$
$$x=2$$

Use the numbers 2 and 3 to divide the number line as shown.

A B C
2 3

Try a test number in each interval.

A: Test 0, $\frac{1}{0-2}=-\frac{1}{2}\leq 1$
Interval A is part of the solution set.

B: Test $\frac{5}{2}$, $\frac{1}{\frac{5}{2}-2}=\frac{1}{\frac{1}{2}}=2\not\leq 1$
Interval B is not part of the solution set.

C: Test 4, $\frac{1}{4-2}=\frac{1}{2}\leq 1$
Interval C is part of the solution set.
The solution set includes intervals A and C. The number 3 is also included since the inequality symbol is $\leq$ and 3 is the solution of the related equation. The number 2 is not included since $\frac{1}{x-2}$ is undefined for $x=2$. The solution set is $(-\infty,2)\cup[3,\infty)$, or $\{x\,|\,x<2 \text{ or } x\geq 3\}$.

51. *Writing Exercise.*

53. ***Familiarize.*** Let x represent the number of hours spent on educational activities. Then $x + 0.7$ represents the number of hours spent on leisure activities.
Translate.

$$x+(x+0.7)=7.1$$

Carry out. We solve the equation.

$$x+(x+0.7)=7.1$$
$$2x+0.7=7.1$$
$$2x=6.4$$
$$x=3.2$$

If $x = 3.2$, then $x + 0.7 = 3.2 + 0.7$, or 3.9.

Check. 3.9 is 0.7 more than 3.2. Also, $3.2 + 3.9 = 7.1$ hours. The answer checks.
State. The student spends 3.2 hr on educational activities.

55. ***Familiarize.*** Let n = the number of miles. Then the mileage fee is $0.4n$. There is a flat fee of \$70.
Translate. We write an inequality stating that the truck rental costs less than \$90.

$$0.4n + 70 < 90$$

Carry out.

$$0.4n+70<90$$
$$0.4n<20$$
$$n<50$$

Check. We can do a partial check by substituting a value for n less than 50 When $n = 49$, the truck rental costs $0.4(49) + 70$, or \$89.60, so the mileage is less than the budget of \$90. When $n = 51$, the truck rental costs $0.4(51) + 70$, or \$90.40, so the mileage is more expensive than the budget of \$90.
State. The mileage should be no greater than 50 mi.

57. *Writing Exercise.*

59. $x^2+2x<5$

$x^2+2x-5<0$

Using the quadratic formula, we find that the solutions of the related equation are $x=-1\pm\sqrt{6}$. These numbers divide the real-number line into three intervals as shown:

A B C
$-1-\sqrt{6}$ $-1+\sqrt{6}$

We try test numbers in each interval.

A: Test -4, $f(-4)=(-4)^2+2(-4)-5=3$

B: Test 0, $f(0)=0^2+2\cdot 0-5=-5$

C: Test 2, $f(2)=2^2+2\cdot 2-5=3$

The function value will be negative for all numbers in interval B. The solution set is $(-1-\sqrt{6},-1+\sqrt{6})$, or $\{x|-1-\sqrt{6}<x<-1+\sqrt{6}\}$.

61. $x^4+3x^2\leq 0$

$x^2(x^2+3)\leq 0$

$x^2=0$ for $x=0$, $x^2>0$ for $x\neq 0$, $x^2+3>0$ for all x
The solution set is $\{0\}$.

63. a. $-3x^2+630x-6000>0$

$x^2-210x+2000<0$ Multiplying by $-\frac{1}{3}$

$(x-200)(x-10)<0$

The solutions of $f(x)=(x-200)(x-10)=0$ are 200 and 10. They divide the number line as shown:

A B C
10 200

A: Test 0, $f(0)=0^2-210\cdot 0+2000=2000$

B: Test 20, $f(20)=20^2-210\cdot 20+2000=-1800$

C: Test 300, $f(300) = 300^2 - 210 \cdot 300 + 2000$
$= 29,000$

The company makes a profit for values of x such that $10 < x < 200$, or for values of x in the interval $(10, 200)$.

b. See part (a). Keep in mind that x must be nonnegative since negative numbers have no meaning in this application.
The company loses money for values of x such that $0 \le x < 10$ or $x > 200$, or for values of x in the interval $[0, 10) \cup (200, \infty)$.

65. We find values of n such that $N \ge 66$ *and* $N \le 300$.
For $N \ge 66$:

$$\frac{n(n-1)}{2} \ge 66$$
$$n(n-1) \ge 132$$
$$n^2 - n - 132 \ge 0$$
$$(n-12)(n+11) \ge 0$$

The solutions of $f(n) = (n-12)(n+11) = 0$ are 12 and -11. They divide the number line as shown:

−11 12

However, only positive values of n have meaning in this exercise so we need only consider the intervals shown below:

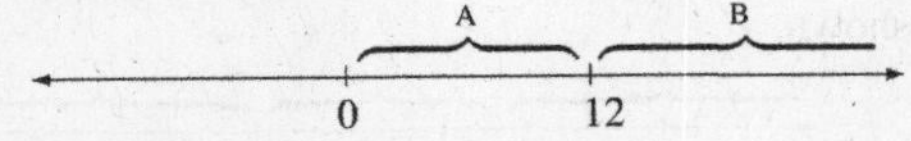

A: Test 1, $f(1) = 1^2 - 1 - 132 = -132$

B: Test 20, $f(20) = 20^2 - 20 - 132 = 248$

Thus, $N \ge 66$ for $\{n \mid n \ge 12\}$.

For $N \le 300$:

$$\frac{n(n-1)}{2} \le 300$$
$$n(n-1) \le 600$$
$$n^2 - n - 600 \le 0$$
$$(n-25)(n+24) \le 0$$

The solutions of $f(n) = (n-25)(n+24) = 0$ are 25 and -24. They divide the number line as shown:

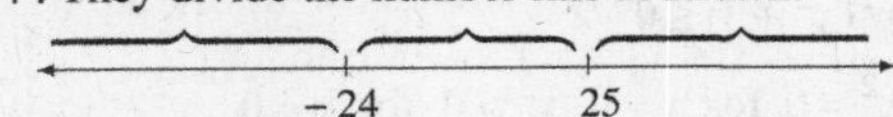

However, only positive values of n have meaning in this exercise so we need only consider the intervals shown below:

A B
0 25

A: Test 1, $f(1) = 1^2 - 1 - 600 = -600$

B: Test 30, $f(30) = 30^2 - 30 - 600 = 270$

Thus, $N \le 300$ (and $n > 0$) for $\{n \mid 0 < n \le 25\}$.

Then $66 \le N \le 300$ for
$\{n \mid n \text{ is an integer } and\ 12 \le n \le 25\}$.

67. From the graph we determine the following:
The solutions of $f(x) = 0$ are -2, 1, and 3.
The solution of $f(x) < 0$ is $(-\infty, -2) \cup (1, 3)$, or $\{x \mid x < -2 \text{ or } 1 < x < 3\}$.
The solution of $f(x) > 0$ is $(-2, 1) \cup (3, \infty)$, or $\{x \mid -2 < x < 1 \text{ or } x > 3\}$.

69. From the graph we determine the following:
$f(x)$ has no zeros.
The solutions of $f(x) < 0$ are $(-\infty, 0)$, or $\{x \mid x < 0\}$;
The solutions of $f(x) > 0$ are $(0, \infty)$, or $\{x \mid x > 0\}$.

71. From the graph we determine the following:
The solutions of $f(x) = 0$ are -1 and 0.
The solution of $f(x) < 0$ is $(-\infty, -3) \cup (-1, 0)$, or $\{x \mid x < -3 \text{ or } -1 < x < 0\}$.
The solution of $f(x) > 0$ is
$(-3, -1) \cup (0, 2) \cup (2, \infty)$, or
$\{x \mid -3 < x < -1 \text{ or } 0 < x < 2 \text{ or } x > 2\}$.

73. For $f(x) = \sqrt{x^2 - 4x - 45}$, we find the domain:

$$x^2 - 4x - 45 \ge 0$$
$$(x+5)(x-9) \ge 0$$

The quadratic is nonnegative when $(-\infty, -5] \cup [9, \infty)$, or $\{x \mid x \le -5 \text{ or } x \ge 9\}$.

75. For $f(x) = \sqrt{x^2 + 8x}$, we find the domain:

$$x^2 + 8x \ge 0$$
$$x(x+8) \ge 0$$

The quadratic is nonnegative when $(-\infty, -8] \cup [0, \infty)$, or $\{x \mid x \le -8 \text{ or } x \ge 0\}$.

77. *Writing Exercise*. Answers may vary.
One example is the rational inequality $\frac{a-x}{x-b} \le 0$.

Chapter 11 Review

1. False; a quadratic equation could have one solution.
2. True
3. True
4. True
5. False; the vertex is $(-3, -4)$.
6. True
7. True; since the coefficient of x^2 is -2, the graph opens down and therefore has no minimum.
8. True

9. False; if the quadratic function has two different imaginary-number zeros, the graph will not have any x-intercepts.

10. True

11. $9x^2 - 2 = 0$
$$9x^2 = 2$$
$$x^2 = \frac{2}{9}$$
$$x = \pm\sqrt{\frac{2}{9}} = \pm\frac{\sqrt{2}}{3}$$
The solutions are $-\frac{\sqrt{2}}{3}$ and $\frac{\sqrt{2}}{3}$.

12. $8x^2 + 6x = 0$
$$2x(4x+3) = 0$$
$$2x = 0 \quad or \quad 4x+3 = 0$$
$$x = 0 \quad or \quad x = -\frac{3}{4}$$

13. $x^2 - 12x + 36 = 9$
$$(x-6)^2 = 9$$
$$x - 6 = \pm 3$$
$$x = 6 \pm 3$$
The solutions are 3 and 9.

14. $x^2 - 4x + 8 = 0$
$$x^2 - 4x + 4 = 4 - 8$$
$$(x-2)^2 = -4$$
$$x - 2 = \pm 2i$$
$$x = 2 \pm 2i$$

15. $x(3x+4) = 4x(x-1) + 15$
$$3x^2 + 4x = 4x^2 - 4x + 15$$
$$0 = x^2 - 8x + 15$$
$$0 = (x-3)(x-5)$$
$$x - 3 = 0 \quad or \quad x - 5 = 0$$
$$x = 3 \quad or \quad x = 5$$
The solutions are 3 and 5.

16. $x^2 + 9x = 1$
$$x^2 + 9x - 1 = 0$$
$$a = 1,\ b = 9,\ c = -1$$
$$x = \frac{-9 \pm \sqrt{9^2 - 4\cdot 1\cdot(-1)}}{2\cdot 1} = \frac{-9 \pm \sqrt{85}}{2}$$
$$x = -\frac{9}{2} \pm \frac{\sqrt{85}}{2}$$

17. $x^2 - 5x - 2 = 0$
$$a = 1,\ b = -5,\ c = -2$$
$$x = \frac{-(-5) \pm \sqrt{(-5)^2 - 4\cdot 1\cdot(-2)}}{2\cdot 1} = \frac{5 \pm \sqrt{25+8}}{2}$$
$$x = \frac{5 \pm \sqrt{33}}{2}$$
$$x \approx -0.372,\ 5.372$$

18. $4x^2 - 3x - 1 = 0$
$$(4x+1)(x-1) = 0$$
$$4x + 1 = 0 \quad or \quad x - 1 = 0$$
$$x = -\frac{1}{4} \quad or \quad x = 1$$

19. $\frac{1}{2}\cdot(-18) = -9;\ (-9)^2 = 81$
$$x^2 - 18x + 81 = (x-9)^2$$

20. $\frac{1}{2}\cdot\frac{3}{5} = \frac{3}{10};\ \left(\frac{3}{10}\right)^2 = \frac{9}{100}$
$$x^2 + \frac{3}{5}x + \frac{9}{100} = \left(x + \frac{3}{10}\right)^2$$

21. $x^2 - 6x + 1 = 0$
$$x^2 - 6x = -1$$
$$x^2 - 6x + 9 = 9 - 1$$
$$(x-3)^2 = 8$$
$$x - 3 = \pm\sqrt{8}$$
$$x = 3 \pm \sqrt{8}$$
$$x = 3 \pm 2\sqrt{2}$$

22. $A = P(1+r)^t$
$$2704 = 2500(1+r)^2$$
$$\frac{2704}{2500} = (1+r)^2$$
$$\pm\frac{26}{25} = 1 + r$$
$$-\frac{25}{25} \pm \frac{26}{25} = r$$
$$\frac{1}{25} = r \quad or \quad -\frac{51}{25} = r$$
Since the interest rate cannot be negative, it is not a solution. $\frac{1}{25} = 0.04$ The interest rate is 4%.

23. $s = 16t^2$
$$443 = 16t^2$$
$$\frac{443}{16} = t^2$$
$$\sqrt{\frac{443}{16}} = t \quad \text{Principle of square roots; rejecting the negative square root.}$$
$$5.3 \approx t$$
It will take an object about 5.3 sec to fall.

24. $x^2 + 3x - 6 = 0$
$$b^2 - 4ac = 3^2 - 4\cdot 1\cdot(-6) = 33$$
There are two irrational real numbers.

25. $x^2 + 2x + 5 = 0$
$$b^2 - 4ac = 2^2 - 4\cdot 1\cdot 5 = -16$$
There are two imaginary numbers.

26. The solutions are $3i$ and $-3i$.

$x = 3i$ *or* $x = -3i$
$x - 3i = 0$ *or* $x + 3i = 0$
$(x - 3i)(x + 3i) = 0$
$x^2 - (3i)^2 = 0$
$x^2 + 9 = 0$

27. The only solution is –5. It must be a repeated solution.

$x = -5$ *or* $x = -5$
$x + 5 = 0$ *or* $x + 5 = 0$
$(x + 5)(x + 5) = 0$
$x^2 + 10x + 25 = 0$

28. Let r represent the plane's speed in still air.

Trip	Distance	Speed	Time
To plant	300	$r - 20$	$\frac{300}{r-20}$
Return	300	$r + 20$	$\frac{300}{r+20}$

Solve the equation $\frac{300}{r-20} + \frac{300}{r+20} = 4.$ We get $r \approx 153$ or $r \approx -2.62$. Only 153 checks in the original problem. The plane's speed in still air is at least 153 mph.

29. ***Familiarize***. Let x represent the time it takes Cheri to reply. Then $x + 6$ represents the time it takes Dani to reply. It takes them 4 hr to reply working together, so they can reply to $\frac{1}{4}$ of the emails in 1 hr. Cheri will reply to $\frac{1}{x}$ of the emails in 1 hr, and Dani will reply to $\frac{1}{x+6}$ of the emails in 1 hr.

Translate. We have an equation.

$$\frac{1}{x} + \frac{1}{x+6} = \frac{1}{4}$$

Carry out. We solve the equation.
We multiply by the LCD, $4x(x+6)$.

$$4x(x+6)\left(\frac{1}{x} + \frac{1}{x+6}\right) = 4x(x+6)\cdot\frac{1}{4}$$
$$4(x+6) + 4x = x(x+6)$$
$$4x + 24 + 4x = x^2 + 6x$$
$$0 = x^2 - 2x - 24$$
$$0 = (x-6)(x+4)$$

Check. Since negative time has no meaning in this problem, –4 is not a solution. We check only 6 hr. This is the time it would take Cheri working alone. Then Dani would take 6 + 6, or 12 hr working alone. Cheri would reply to $4\left(\frac{1}{6}\right)$, or $\frac{2}{3}$ of the emails in 4 hr, and Dani would reply to $4\left(\frac{1}{12}\right)$, or $\frac{1}{3}$ of the emails in 4 hr. Thus, in 4 hr they would reply to $\frac{2}{3} + \frac{1}{3}$ of the emails. This is all of it, so the numbers check.

State. It would take Cheri, working alone, 6 hr to reply to the emails.

30. $x^4 - 13x^2 + 36 = 0$

Let $u = x^2$ and $u^2 = x^4$.

$u^2 - 13u + 36 = 0$
$(u - 4)(u - 9) = 0$
$u = 4$ *or* $u = 9$

Now replace u with x^2 and solve these equations.

$x^2 = 4$ *or* $x^2 = 9$
$x = \pm 2$ *or* $x = \pm 3$

The intercepts are (–3, 0), (–2, 0), (2, 0), and (3, 0).

31. $15x^{-2} - 2x^{-1} - 1 = 0$

Let $u = x^{-1}$ and $u^2 = x^{-2}$.

$15u^2 - 2u - 1 = 0$
$(5u + 1)(3u - 1) = 0$
$5u + 1 = 0$ *or* $3u - 1 = 0$
$u = -\frac{1}{5}$ *or* $u = \frac{1}{3}$

Replace u with x^{-1}.

$x^{-1} = -\frac{1}{5}$ *or* $x^{-1} = \frac{1}{3}$
$x = -5$ *or* $x = 3$

The numbers –5 and 3 check. They are the solutions.

32. $\left(x^2 - 4\right)^2 - \left(x^2 - 4\right) - 6 = 0$

Let $u = x^2 - 4$ and $u^2 = \left(x^2 - 4\right)^2$.

$u^2 - u - 6 = 0$
$(u + 2)(u - 3) = 0$
$u + 2 = 0$ *or* $u - 3 = 0$
$u = -2$ *or* $u = 3$

Replace u with $x^2 - 4$.

$x^2 - 4 = -2$ *or* $x^2 - 4 = 3$
$x^2 = 2$ *or* $x^2 = 7$
$x = \pm\sqrt{2}$ *or* $x = \pm\sqrt{7}$

The numbers $\sqrt{2}, -\sqrt{2}, \sqrt{7}$ and $-\sqrt{7}$ check. They are the solutions.

33. $f(x) = -3(x+2)^2 + 4$

We know that the graph looks like the graph of $h(x) = 3x^2$ but moved to the left 2 units and up 4 units and turned upside down. The vertex is (–2, 4), and the axis of symmetry is $x = -2$. The maximum function value is 4.

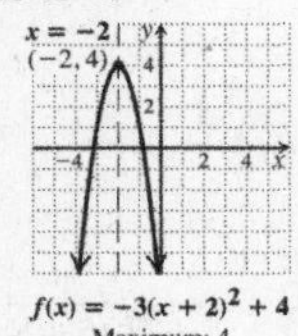

$f(x) = -3(x+2)^2 + 4$
Maximum: 4

34. $f(x) = 2x^2 - 12x + 23$
$= 2(x^2 - 6x) + 23$
$= 2(x^2 - 6x + 9) - 18 + 23$
$= 2(x-3)^2 + 5$

We know that the graph looks like the graph of $h(x) = 2x^2$ but moved to the right 3 units and up 5 units. The vertex is (3, 5), and the axis of symmetry is $x = 3$.

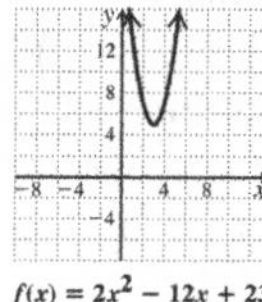

$f(x) = 2x^2 - 12x + 23$

35. $f(x) = x^2 - 9x + 14$

To find the x-intercepts, solve the equation $0 = x^2 - 9x + 14$. We factor.

$$0 = x^2 - 9x + 14$$
$$0 = (x-2)(x-7)$$
$$x = 2 \text{ or } x = 7$$

The x-intercepts are (2, 0) and (7, 0).
The y-intercept is $(0, f(0))$, or (0, 14).

36. $N = 3\pi\sqrt{\frac{1}{p}}$

$$N^2 = 9\pi^2\frac{1}{p}$$
$$p = \frac{9\pi^2}{N^2}$$

37. $2A + T = 3T^2$

$$3T^2 - T - 2A = 0$$
$$a = 3, b = -1, c = -2A$$
$$T = \frac{-(-1) \pm \sqrt{(-1)^2 - 4\cdot 3\cdot(-2A)}}{2\cdot 3}$$
$$T = \frac{1 \pm \sqrt{1+24A}}{6}$$

38. The data points fall. The graph does not appear to represent a quadratic function in which the data points would rise and then fall or vice versa. Thus a linear function $f(x) = mx + b$ might be used to model the data.

39. The data points fall and then rise. The graph appears to represent a quadratic function that opens upward. Thus a quadratic function $f(x) = ax^2 + bx + c,\ a > 0$, might be used to model the data.

40. Since the area is in a corner, only two sides of fencing is needed.

$$l + w = 30$$
$$A = lw$$
$$l = 30 - w$$
$$A = (30 - w)w = -w^2 + 30w = -1(w-15)^2 + 225$$

The maximum function value of 225 occurs when $w = 15$.

When $w = 15$, $l = 30 - 15 = 15$.

Maximum area: 225 ft^2; dimensions: 15 ft by 15 ft

41. a. ***Familiarize***. We look for a function of the form $f(x) = ax^2 + bx + c$, where $f(x)$ represents the national debt, in trillions, and x represents the years after 1990.

Translate. We substitute the given values of x and $f(x)$.

$$3 = a(0)^2 + b(0) + c,$$
$$6 = a(10)^2 + b(10) + c,$$
$$18 = a(25)^2 + b(25) + c,$$

or

$$3 = 0a + 0b + c,$$
$$6 = 100a + 10b + c,$$
$$18 = 625a + 25b + c.$$

Carry out. Solving the system of equations, we get $a = \frac{1}{50},\ b = \frac{1}{10},\ c = 3.$

Check. Recheck the calculations.

State. The function $f(x) = \frac{1}{50}x^2 + \frac{1}{10}x + 3$ fits the data.

b. $2010 - 1990 = 20$

Find $f(20)$.

$$f(20) = \frac{1}{50}(20)^2 + \frac{1}{10}(20) + 3 = 13$$

The national debt in 2010 will be about \$13,000,000,000,000.

42. $x^3 - 3x > 2x^2$

$$x^3 - 2x^2 - 3x > 0$$
$$x(x+1)(x-3) > 0$$

The solutions of $x(x+1)(x-3)$ are -1, 0, and 3. They divide the real-number line into four intervals as shown:

A B C D
−1 0 3

We try test numbers in each interval.

A Test -2, $f(-2) = -2(-2+1)(-2-3) = -10$

B: Test $-\frac{1}{2}$, $f\left(-\frac{1}{2}\right) = -\frac{1}{2}\left(-\frac{1}{2}+1\right)\left(-\frac{1}{2}-3\right) = \frac{7}{8}$

C: Test 1, $f(1) = 1(1+1)(1-3) = -4$

D: Test 4, $f(4) = 4(4+1)(4-3) = 20$

The function value will be positive for all numbers in intervals B and D. The solution set is $(-1, -0) \cup (3, \infty)$, or $\{x \mid -1 < x < 0 \text{ or } x > 3\}$.

43. $\frac{x-5}{x+3} \le 0$

Solve the related equation.

$$\frac{x-5}{x+3} = 0$$
$$x - 5 = 0$$
$$x = 5$$

Find the values that make the denominator 0.

$x+3=0$

$x=-3$

Use the numbers 5 and –3 to divide the number line into intervals as shown:

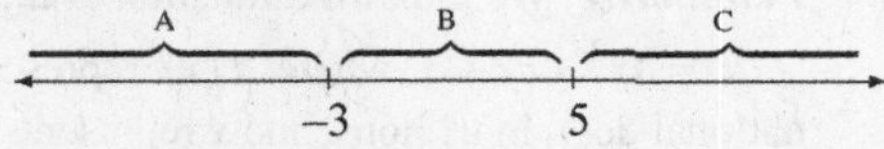

Try test numbers in each interval.

A: Test –4, $\frac{-4-5}{-5+3}=\frac{-9}{-2}=\frac{9}{2}\not<0$

The number –4 is a not solution of the inequality, so the interval A is not part of the solution set.

B: Test 0, $\frac{0-5}{0+3}=\frac{-5}{3}=-\frac{5}{3}<0$

The number 0 is a solution of the inequality, so the interval B is part of the solution set.

C: Test 6, $\frac{6-5}{6+3}=\frac{1}{9}\not<0$

The number 6 is a not solution of the inequality, so the interval C is not part of the solution set.
The solution set includes interval B. The number 5 is also included since the inequality symbol is $\le$ and 5 is the solution of the related equation. The number –3 is not included since $\frac{x-5}{x+3}$ is undefined for $x=-3$. The solution set is $(-3, 5]$, or $\{x\,|\,-3<x\le 5\}$.

44. *Writing Exercise.* The x-coordinate of the maximum or minimum point lies halfway between the x-coordinates of the x-intercepts.

45. *Writing Exercise.* The first coordinate of each x-intercept of f is a solution of $f(x)=0$. Suppose the first coordinates of the x-intercepts are a and b. Then $(x-a)$ and $(x-b)$ are factors of $f(x)$. If the graph of a quadratic function has one x-intercept $(a, 0)$, then $(x-a)$ is a repeated factor of $f(x)$.

46. *Writing Exercise.* Completing the square was used to solve quadratic equations and to graph quadratic functions by rewriting the function in the form $f(x)=a(x-h)^2+k$.

47. Substituting the three ordered pairs (–3, 0), (5, 0), and (0, –7) in the equation $f(x)=ax^2+bx+c$ yields a system of equations:

$$0=9a-3b+c,$$
$$0=25a+5b+c,$$
$$-7=0a+0b+c$$

The solution of this system of equations is $\left(\frac{7}{15},-\frac{14}{5},-7\right)$, so $f(x)=\frac{7}{15}x^2-\frac{14}{5}x-7$.

48. From Section 8.3, we know the sum of the solutions of $ax^2+bx+c=0$ is $-\frac{b}{a}$, and the product is $\frac{c}{a}$.

$3x^2-hx+4k=0$

$a=3,\ b=-h,\ c=4k$

Substituting

For $-\frac{b}{a}$: $-\frac{-h}{3}=20$

$\frac{h}{3}=20$

$h=60$

For $\frac{c}{a}$: $\frac{4k}{3}=80$

$4k=240$

$k=60$

49. Let x and y represent the positive integers. Since one of the numbers is the square root of the other, we let $y=\sqrt{x}$. To find their average, we find their sum and divide by 2.

$$\frac{x+\sqrt{x}}{2}=171$$
$$x+\sqrt{x}=342$$
$$x+\sqrt{x}-342=0$$

Let $u=\sqrt{x}$ and $u^2=x$.

$$u^2+u-342=0$$
$$(u+19)(u-18)=0$$
$$u=-19 \text{ or } u=18$$

Substituting: $\sqrt{x}=-19$ *or* $\sqrt{x}=18$

We use only $\sqrt{x}=18$

$x=324$

The numbers are 18 and 324.

Chapter 11 Test

1. $25x^2-7=0$

$25x^2=7$

$x^2=\frac{7}{25}$

$x=\pm\sqrt{\frac{7}{25}}=\pm\frac{\sqrt{7}}{5}$

The solutions are $-\frac{\sqrt{7}}{5}$ and $\frac{\sqrt{7}}{5}$.

2. $4x(x-2)-3x(x+1)=-18$

$4x^2-8x-3x^2-3x=-18$

$x^2-11x+18=0$

$(x-2)(x-9)=0$

$x-2=0$ *or* $x-9=0$

$x=2$ *or* $x=9$

3. $x^2+2x+3=0$

$a=1,\ b=2,\ c=3$

$$x=\frac{-2\pm\sqrt{2^2-4(1)(3)}}{2(1)}=\frac{-2\pm\sqrt{4-12}}{2}$$
$$=\frac{-2\pm\sqrt{-8}}{2}=\frac{-2\pm 2i\sqrt{2}}{2}=\frac{-2}{2}\pm\frac{2i\sqrt{2}}{2}$$
$$=-1\pm i\sqrt{2} \text{ or } -1\pm\sqrt{2}i$$

The solutions are $-1+\sqrt{2}i$ and $-1-\sqrt{2}i$.

4. $2x+5=x^2$
$5+1=x^2-2x+1$
$6=(x-1)^2$
$\pm\sqrt{6}=x-1$
$1\pm\sqrt{6}=x$

5. $x^{-2}-x^{-1}=\frac{3}{4}$
$x^{-2}-x^{-1}-\frac{3}{4}=0$
$4x^{-2}-4x^{-1}-3=0$ Clearing fractions

Let $u=x^{-1}$ and $u^2=x^{-2}$.
$4u^2-4u-3=0$
$(2u-3)(2u+1)=0$
$2u-3=0$ *or* $2u+1=0$
$u=\frac{3}{2}$ *or* $u=-\frac{1}{2}$

Now we replace u with x^{-1} and solve these equations:
$x^{-1}=\frac{3}{2}$ *or* $x^{-1}=-\frac{1}{2}$
$\frac{1}{x}=\frac{3}{2}$ *or* $\frac{1}{x}=-\frac{1}{2}$
$2=3x$ *or* $2=-x$
$\frac{2}{3}=x$ $\quad -2=x$

The solutions are –2 and $\frac{2}{3}$.

6. $x^2+3x=5$
$x^2+3x-5=0$
$a=1, b=3, c=-5$
$x=\frac{-3\pm\sqrt{3^2-4\cdot1\cdot(-5)}}{2\cdot1}=\frac{-3\pm\sqrt{29}}{2}$
$x=\frac{-3-\sqrt{29}}{2}\approx-4.193$
$x=\frac{-3+\sqrt{29}}{2}\approx1.193$

7. Let $f(x)=0$ and solve for x.
$0=12x^2-19x-21$
$0=(4x+3)(3x-7)$
$x=-\frac{3}{4}$ *or* $x=\frac{7}{3}$

The solutions are $-\frac{3}{4}$ and $\frac{7}{3}$.

8. $\frac{1}{2}\cdot20=10;\ 10^2=100$
$x^2-20x+100=(x-10)^2$

9. $\frac{1}{2}\cdot\frac{2}{7}=\frac{1}{7};\ \left(\frac{1}{7}\right)^2=\frac{1}{49}$
$x^2+\frac{2}{7}x+\frac{1}{49}=\left(x+\frac{1}{7}\right)^2$

10. $x^2+10x+15=0$
$x^2+10x=-15$
$x^2+10x+25=-15+25$
$(x+5)^2=10$
$x+5=\pm\sqrt{10}$
$x=-5\pm\sqrt{10}$

11. $x^2+2x+5=0$
$b^2-4ac=2^2-4(1)(5)=-16$
Two imaginary numbers

12. $x=\sqrt{11}$ *or* $x=-\sqrt{11}$
$x-\sqrt{11}=0$ *or* $x+\sqrt{11}=0$
$(x-\sqrt{11})(x+\sqrt{11})=0$
$x^2-11=0$

13. ***Familiarize***. Let r represent the cruiser's speed in still water. Then $r-4$ is the speed upriver and $r+4$ is the speed downriver. Using $t=\frac{d}{r}$, we let $\frac{60}{r-4}$ represent the time upriver and $\frac{60}{r+4}$ represent the time downriver.

Trip	Distance	Speed	Time
Upriver	60	$r-4$	$\frac{60}{r-4}$
Downriver	60	$r+4$	$\frac{60}{r+4}$

Translate. We have an equation.
$\frac{60}{r-4}+\frac{60}{r+4}=8$
Carry out. We solve the equation.
We multiply by the LCD, $(r-4)(r+4)$.
$(r-4)(r+4).\left(\frac{60}{r-4}+\frac{60}{r+4}\right)=(r-4)(r+4)\cdot8$
$60(r+4)+60(r-4)=8(r-4)(r+4)$
$60r+240+60r-240=8r^2-128$
$0=8r^2-120r-128$
$0=8(r^2-15r-16)$
$0=8(r-16)(r+1)$
$r=16$ *or* $r=-1$
Check. Since negative rate has no meaning in this problem, –1 is not a solution. We check only 16 km/hr. If $r=16$, then the speed upriver is 16 – 4, or 12 km/h, and the time is $\frac{60}{12}$, or 5 hr. The speed downriver is 16 + 4, or 20 km/h, and the time is $\frac{60}{20}$, or 3 hr. The total time of the round trip is 5 + 3, or 8 hr. The value checks.
State. The speed of the cruiser in still water is 16 km/h.

14. Let x represent the time it takes Dal to assemble the swing set. Then $x+4$ represents the time it takes Kim to do the same job. It takes them 1.5 hr to assemble the swing set working together, so they can complete $\frac{2}{3}$ of the job in 1 hr. Dal will assemble $\frac{1}{x}$ of the set in

1 hr, and Kim will assemble $\frac{1}{x+4}$ of the set in 1 hr.

We solve the equation.

$$\frac{1}{x}+\frac{1}{x+4}=\frac{2}{3}$$

The solutions are –3 and 2. Since negative time has no meaning for this application, Dal can assemble the swing set in 2 hr.

15. $f(x)=x^4-15x^2-16$

To find the x-intercepts, solve the equation $0=x^4-15x^2-16$.

Let $u=x^2$ and $u^2=x^4$.

$$0=u^2-15u-16$$
$$0=(u-16)(u+1)$$
$$u=16 \text{ or } u=-1$$

Replace u with x^2.

$$x^2=16 \quad or \quad x^2=-1 \quad \text{Has no real solutions}$$
$$x=\pm 4$$

The x-intercepts are (–4, 0) and (4, 0).

16. $f(x)=4(x-3)^2+5$

We know that the graph looks like the graph of $h(x)=4x^2$ but moved to the right 3 units and up 5 units. The vertex is (3, 5), and the axis of symmetry is $x=3$. The minimum function value is 5.

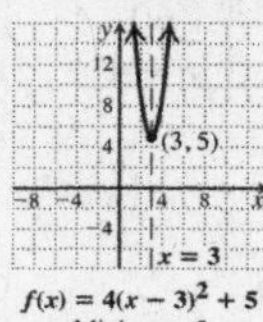

$f(x)=4(x-3)^2+5$
Minimum: 5

17. $f(x)=2x^2+4x-6$

$$=2(x^2+2x)-6$$
$$=2(x^2+2x+1)-6-2$$
$$=2(x+1)^2-8$$

We know that the graph looks like the graph of $h(x)=2x^2$ but moved to the left 1 unit and down 8 units. The vertex is (–1, –8), and the axis of symmetry is $x=-1$.

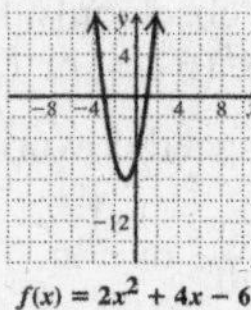
$f(x)=2x^2+4x-6$

18. $f(x)=x^2-x-6$

To find the x-intercepts, solve the equation $0=x^2-x-6$. We factor.

$$0=x^2-x-6$$
$$0=(x+2)(x-3)$$
$$x=-2 \text{ or } x=3$$

The x-intercepts are (–2, 0) and (3, 0).
The y-intercept is $(0, f(0))$, or (0, –6).

19.
$$V=\frac{1}{3}\pi\left(R^2+r^2\right)$$
$$\frac{3V}{\pi}=R^2+r^2$$
$$\frac{3V}{\pi}-R^2=r^2$$
$$\sqrt{\frac{3V}{\pi}-R^2}=r$$

We only consider the positive square root as instructed.

20. The data points rise and then fall. The graph appears to represent a quadratic function that opens downward.

21. $C(x)=0.2x^2-1.3x+3.4025$
$$C(x)=0.2\left(x^2-6.5x\right)+3.4025$$
$$C(x)=0.2(x-6.5x+10.5625)-0.2(10.5625)+3.4025$$
$$C(x)=0.2(x-3.25)^2+1.29$$

3.25 hundred or 325 cabinets should be built to have a minimum at \$1.29 hundred, or \$129 per cabinet.

22. Substituting the three ordered pairs (0, 0), (3, 0), and (5, 2) in the equation $f(x)=ax^2+bx+c$ yields a system of equations:

$$0=0a+0b+c,$$
$$0=9a+3b+c,$$
$$2=25a+5b+c$$

The solution of this system of equations is $\left(\frac{1}{5},-\frac{3}{5},0\right)$, so $f(x)=\frac{1}{5}x^2-\frac{3}{5}x$.

23.
$$x^2+5x<6$$
$$x^2+5x-6<0$$
$$(x+6)(x-1)<0$$

The solutions of $(x+6)(x-1)=0$ are –6 and 1. They divide the number line into three intervals as shown:

A B C
–6 1

We try test numbers in each interval.

A: Test –7, $f(-7)=(-7+6)(-7-1)=8$

B: Test 0, $f(0)=(0+6)(0-1)=-6$

C: Test 2, $f(2)=(2+6)(2-1)=8$

Since $f(0)$ is negative, the function value will be negative for all numbers in the interval containing 0. Because the symbol is $<$, we do not include the endpoints in the solution. The solution set is $(-6, 1)$, or $\{x|-6<x<1\}$.

24. $x-\frac{1}{x}\geq 0$

Solve the related equation.

$$x-\frac{1}{x}=0$$
$$x^2-1=0$$
$$(x+1)(x-1)=0$$
$$x=-1 \text{ or } x=1$$

Find the values of x that make the denominator 0.

$$x=0$$

A: Test −2, $-2-\frac{1}{-2}=-\frac{3}{2}\not\geq 0$

B: Test $-\frac{1}{2}$, $-\frac{1}{2}-\frac{1}{-\frac{1}{2}}=\frac{3}{2}\geq 0$

C: Test $-\frac{1}{2}$, $\frac{1}{2}-\frac{1}{\frac{1}{2}}=-\frac{3}{2}\not\geq 0$

D: Test 4, $4-\frac{1}{4}=\frac{15}{4}\geq 0$

$[-1,\ 0)\cup[1,\ \infty)$, or $\{x|-1\leq x<0 \text{ or } x\geq 1\}$

25.
$$kx^2+3x-k=0$$
$$k(-2)^2+3(-2)-k=0 \quad \text{Substitute } -2 \text{ for } x$$
$$4k-6-k=0$$
$$3k=6$$
$$k=2$$

Then $2x^2+3x-2=0$
$$(x+2)(2x-1)=0$$

The other solution is $\frac{1}{2}$.

26. If $-\sqrt{3}$ and $2i$ are solutions, then $\sqrt{3}$ and $-2i$ must also be solutions. A fourth-degree equations can be found as follows:
$$\left[x-(-\sqrt{3})\right]\left[x-\sqrt{3}\right][x-2i][x+2i]=0$$
$$\left(x^2-3\right)\left(x^2+4\right)=0$$
$$x^4+x^2-12=0$$

Answers may vary.

27. $x^4-4x^2-1=0$

Let $u=x^2$ and $u^2=x^4$.
$$u^2-4u-1=0$$
$$u^2-4u=1$$
$$u^2-4u+4=1+4$$
$$(u-2)^2=5$$
$$u-2=\pm\sqrt{5}$$
$$u=2\pm\sqrt{5}$$

Replace u with x^2.
$$x^2=2+\sqrt{5} \quad \text{or} \quad x^2=2-\sqrt{5}$$
$$x=\pm\sqrt{2+\sqrt{5}} \quad \text{or} \quad x=\pm\sqrt{2-\sqrt{5}}$$

Since $2-\sqrt{5}$ is negative, we can rewrite it as follows:
$$\pm\sqrt{2-\sqrt{5}}=\pm\sqrt{\sqrt{5}-2}i$$

The solutions are $\pm\sqrt{\sqrt{5}+2}$ and $\pm\sqrt{\sqrt{5}-2}i$.

Chapter 12

Exponential Functions and Logarithmic Functions

Exercise Set 12.1

1. True

3. $(g \circ f) = g(f(x)) = x^2 + 3 \neq (x+3)^2$, so the statement is false.

5. False

7. True

9. a. $(f \circ g)(1) = f(g(1)) = f(1-3)$
$= f(-2) = (-2)^2 + 1$
$= 4 + 1 = 5$

b. $(g \circ f)(1) = g(f(1)) = g(1^2 + 1)$
$= g(2) = 2 - 3 = -1$

c. $(f \circ g) = f(g(x)) = f(x-3)$
$= (x-3)^2 + 1 = x^2 - 6x + 9 + 1$
$= x^2 - 6x + 10$

d. $(g \circ f)(x) = g(f(x)) = g(x^2 + 1)$
$= x^2 + 1 - 3 = x^2 - 2$

11. a $(f \circ g)(1) = f(g(1)) = f(2 \cdot 1^2 - 7)$
$= f(-5) = 5(-5) + 1 = -24$

b. $(g \circ f)(1) = g(f(1)) = g(5 \cdot 1 + 1)$
$= g(6) = 2 \cdot 6^2 - 7 = 65$

c. $(f \circ g)(x) = f(g(x)) = f(2x^2 - 7)$
$= 5(2x^2 - 7) + 1 = 10x^2 - 34$

d. $(g \circ f)(x) = g(f(x)) = g(5x+1)$
$= 2(5x+1)^2 - 7 = 2(25x^2 + 10x + 1) - 7$
$= 50x^2 + 20x - 5$

13. a. $(f \circ g)(1) = f(g(1)) = f\left(\frac{1}{1^2}\right)$
$= f(1) = 1 + 7 = 8$

b. $(g \circ f)(1) = g(f(1)) = g(1+7) = g(8) = \frac{1}{8^2} = \frac{1}{64}$

c. $(f \circ g)(x) = f(g(x)) = f\left(\frac{1}{x^2}\right) = \frac{1}{x^2} + 7$

d. $(g \circ f)(x) = g(f(x)) = g(x+7) = \frac{1}{(x+7)^2}$

15. a. $(f \circ g)(1) = f(g(1)) = f(1+3)$
$= f(4) = \sqrt{4} = 2$

b. $(g \circ f)(1) = g(f(1)) = g(\sqrt{1})$
$= g(1) = 1 + 3 = 4$

c. $(f \circ g)(x) = f(g(x)) = f(x+3) = \sqrt{x+3}$

d. $(g \circ f)(x) = g(f(x)) = g(\sqrt{x}) = \sqrt{x} + 3$

17. a. $(f \circ g)(1) = f(g(1)) = f\left(\frac{1}{1}\right) = f(1) = \sqrt{4 \cdot 1} = \sqrt{4} = 2$

b. $(g \circ f)(1) = g(f(1)) = g(\sqrt{4 \cdot 1}) = g(\sqrt{4}) = g(2) = \frac{1}{2}$

c. $(f \circ g)(x) = f(g(x)) = f\left(\frac{1}{x}\right) = \sqrt{4 \cdot \frac{1}{x}} = \sqrt{\frac{4}{x}}$

d. $(g \circ f)(x) = g(f(x)) = g(\sqrt{4x}) = \frac{1}{\sqrt{4x}}$

19. a. $(f \circ g)(1) = f(g(1)) = f(\sqrt{1-1})$
$= f(\sqrt{0}) = f(0) = 0^2 + 4 = 4$

b. $(g \circ f)(1) = g(f(1)) = g(1^2 + 4)$
$= g(5) = \sqrt{5-1} = \sqrt{4} = 2$

c. $(f \circ g)(x) = f(g(x)) = f(\sqrt{x-1})$
$= (\sqrt{x-1})^2 + 4 = x - 1 + 4 = x + 3$

d. $(g \circ f)(x) = g(f(x)) = g(x^2 + 4)$
$= \sqrt{x^2 + 4 - 1} = \sqrt{x^2 + 3}$

21. $h(x) = (3x-5)^4$
This is $3x - 5$ raised to the fourth power, so the two most obvious functions are $f(x) = x^4$ and $g(x) = 3x - 5$.

23. $h(x) = \sqrt{9x+1}$
We have $9x + 1$ and take the square root of the expression, so the two most obvious functions are $f(x) = \sqrt{x}$ and $g(x) = 9x + 1$.

25. $h(x) = \frac{6}{5x-2}$
This is 6 divided by $5x - 2$, so two functions that can be used are $f(x) = \frac{6}{x}$ and $g(x) = 5x - 2$.

27. The graph of $f(x) = -x$ is shown below.

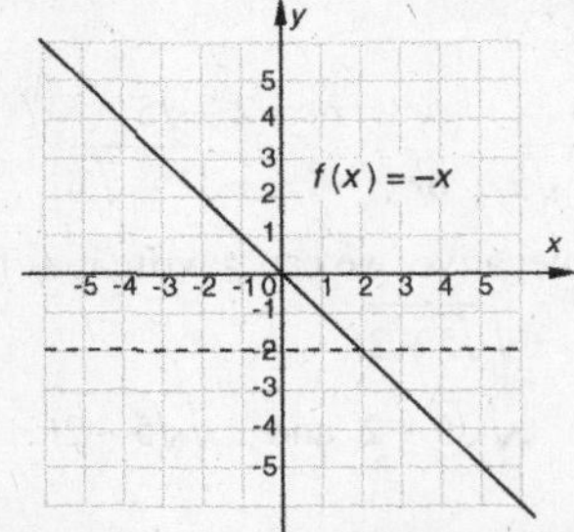

Since there is no horizontal line that crosses the graph more than once, the function is one-to-one.

29. $f(x) = x^2 + 3$

Observe that the graph of this function is a parabola that opens up. Thus, there are many horizontal lines that cross the graph more than once, so the function is not one-to-one. We can also draw the graph as shown below.

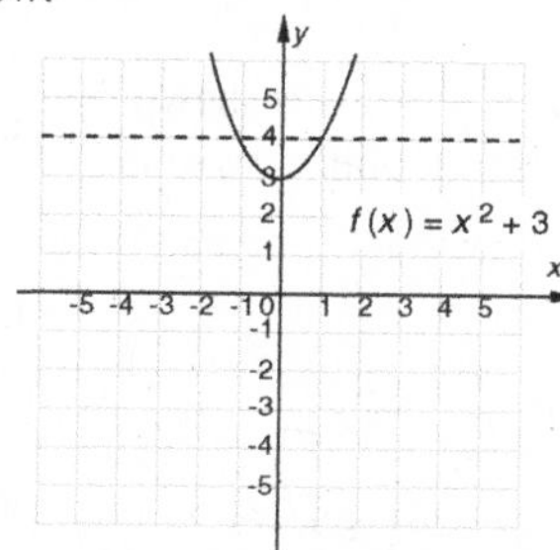

In particular, the line $y = 4$ crosses the graph more than once. The function is not one-to-one.

31. Since there is no horizontal line that crosses the graph more than once, the function is one-to-one.

33. There are many horizontal lines that cross the graph more than once, the function is not one-to-one.

35. a. The function $f(x) = x + 3$ is a linear function that is not constant, so it passes the horizontal-line test. Thus, f is one-to-one.

b. Replace $f(x)$ by y: $y = x + 3$
Interchange x and y: $x = y + 3$
Solve for y: $x - 3 = y$
Replace y by $f^{-1}(x)$: $f^{-1}(x) = x - 3$

37. a. The function $f(x) = 2x$ is a linear function that is not constant, so it passes the horizontal-line test. Thus, f is one-to-one.

b. Replace $f(x)$ by y: $y = 2x$
Interchange x and y: $x = 2y$
Solve for y: $\frac{x}{2} = y$
Replace y by $f^{-1}(x)$: $f^{-1}(x) = \frac{x}{2}$

39. a. The function $g(x) = 3x - 1$ is a linear function that is not constant, so it passes the horizontal-line test. Thus, g is one-to-one.

b. Replace $g(x)$ by y: $y = 3x - 1$
Interchange x and y: $x = 3y - 1$
Solve for y: $x + 1 = 3y$
$\frac{x+1}{3} = y$
Replace y by $g^{-1}(x)$: $g^{-1}(x) = \frac{x+1}{3}$

41. a. The function $f(x) = \frac{1}{2}x + 1$ is a linear function that is not constant, so it passes the horizontal-line test. Thus, f is one-to-one.

b. Replace $f(x)$ by y: $y = \frac{1}{2}x + 1$
Interchange variables: $x = \frac{1}{2}y + 1$
Solve for y: $x - 1 = \frac{1}{2}y$
$2x - 2 = y$
Replace y by $f^{-1}(x)$: $f^{-1}(x) = 2x - 2$

43. a. The graph of $g(x) = x^2 + 5$ is shown below. There are many horizontal lines that cross the graph more than once. For example, the line $y = 8$ crosses the graph more than once. The function is not one-to-one.

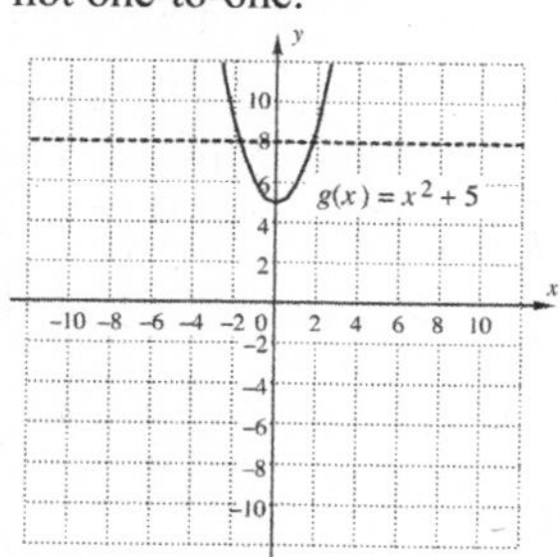

45. a. The function $h(x) = -10 - x$ is a linear function that is not constant, so it passes the horizontal-line test. Thus, h is one-to-one.

b. Replace $h(x)$ by y: $y = -10 - x$
Interchange variables: $x = -10 - y$
Solve for y: $x + 10 = -y$
$-x - 10 = y$
Replace y by $h^{-1}(x)$: $h^{-1}(x) = -x - 10$

47. a. The graph of $f(x) = \frac{1}{x}$ is shown below. It passes the horizontal-line test, so the function is one-to-one.

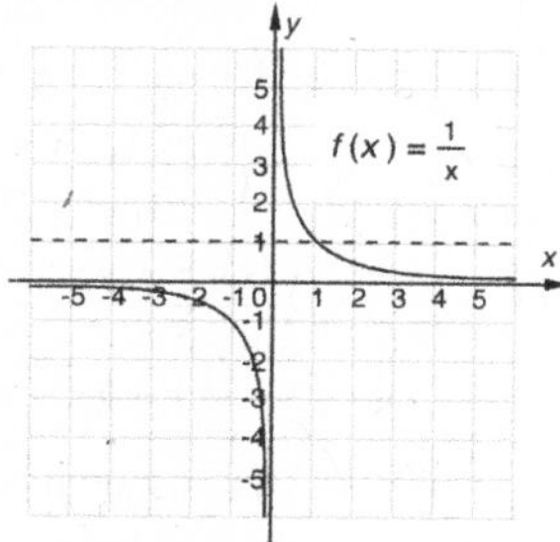

b. Replace $f(x)$ by y: $y = \frac{1}{x}$
Interchange x and y: $x = \frac{1}{y}$
Solve for y: $xy = 1$
$y = \frac{1}{x}$
Replace y by $f^{-1}(x)$: $f^{-1}(x) = \frac{1}{x}$

49. a. The graph of $g(x)=1$ is shown below. The horizontal line $y=1$ crosses the graph more than once, so the function is not one-to-one.

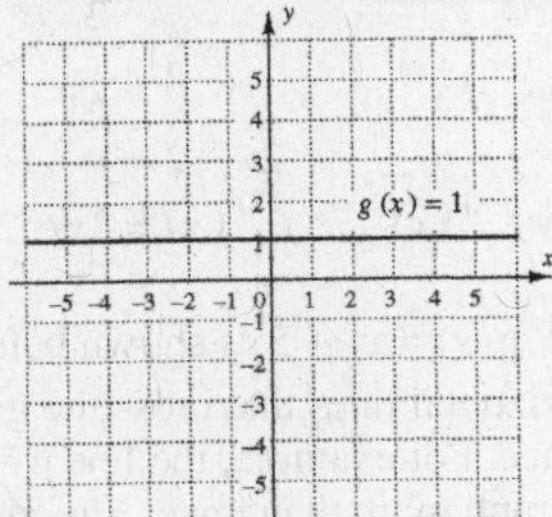

51. a. The function $f(x)=\frac{2x+1}{3}=\frac{2}{3}x+\frac{1}{3}$ is a linear function that is not constant, so it passes the horizontal-line test. Thus, f is one-to-one.

b. Replace $f(x)$ by y: $y=\frac{2x+1}{3}$

Interchange x and y: $x=\frac{2y+1}{3}$

Solve for y:
$$3x=2y+1$$
$$3x-1=2y$$
$$\frac{3x-1}{2}=y$$

Replace y by $f^{-1}(x)$: $f^{-1}(x)=\frac{3x-1}{2}$

53. a. The graph of $f(x)=x^3+5$ is shown below. It passes the horizontal-line test, so the function is one-to-one.

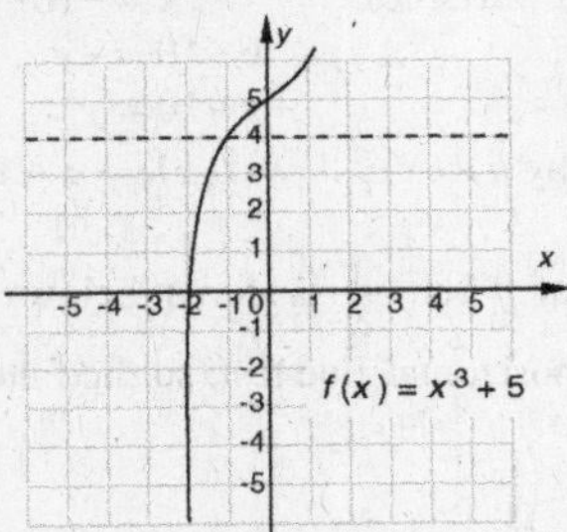

b. Replace $f(x)$ by y: $y=x^3+5$

Interchange x and y: $x=y^3+5$

Solve for y:
$$x-5=y^3$$
$$\sqrt[3]{x-5}=y$$

Replace y by $f^{-1}(x)$: $f^{-1}(x)=\sqrt[3]{x-5}$

55. a. The graph of $g(x)=(x-2)^3$ is shown below. It passes the horizontal-line test, so the function is one-to-one.

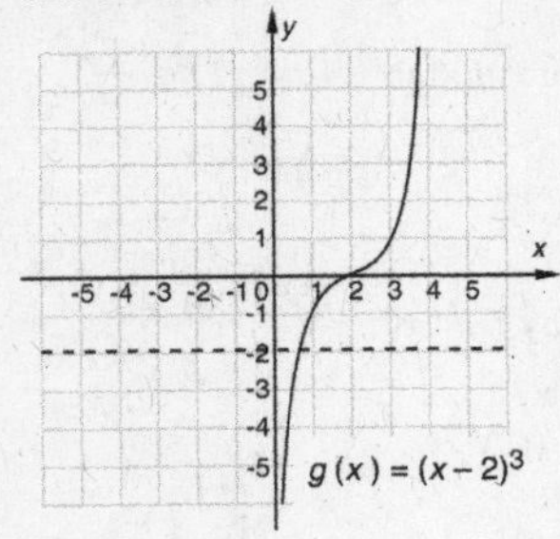

b. Replace $g(x)$ by y: $y=(x-2)^3$

Interchange x and y: $x=(y-2)^3$

Solve for y:
$$\sqrt[3]{x}=y-2$$
$$\sqrt[3]{x}+2=y$$

Replace y by $g^{-1}(x)$: $g^{-1}(x)=\sqrt[3]{x}+2$

57. a. The graph of $f(x)=\sqrt{x}$ is shown below. It passes the horizontal-line test, so the function is one-to-one.

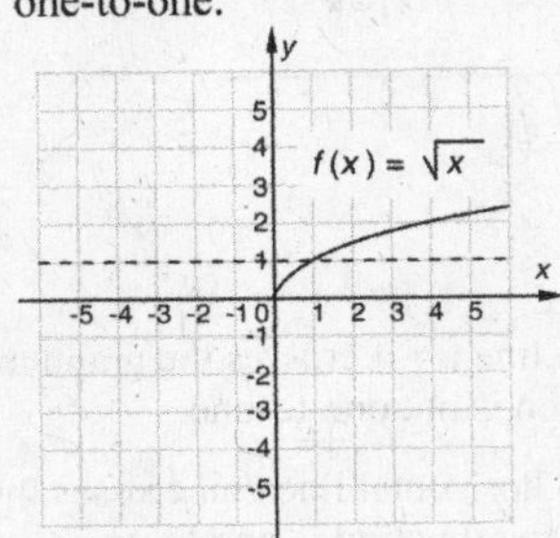

b. Replace $f(x)$ by y: $y=\sqrt{x}$ (Note that $f(x)\ge 0$)

Interchange x and y: $x=\sqrt{y}$

Solve for y: $x^2=y$

Replace y by $f^{-1}(x)$: $f^{-1}(x)=x^2; x\ge 0$

59. a. $f(8)=2(8+12)=2\cdot 20=40$

Size 40 in Italy corresponds to size 8 in the U.S.

$f(10)=2(10+12)=2\cdot 22=44$

Size 44 in Italy corresponds to size 10 in the U.S.

$f(14)=2(14+12)=2\cdot 26=52$

Size 52 in Italy corresponds to size 14 in the U.S.

$f(18)=2(18+12)=2\cdot 30=60$

Size 60 in Italy corresponds to size 18 in the U.S.

b. The function $f(x)=2(x+12)$ is a linear function that is not constant, so it passes the horizontal-line test and has an inverse that is a function.

Replace $f(x)$ by y: $y=2(x+12)$

Interchange x and y: $x=2(y+12)$

Solve for y:
$$x=2y+24$$
$$x-24=2y$$
$$\frac{x-24}{2}=y$$

Replace y by $f^{-1}(x)$: $f^{-1}(x)=\frac{x-24}{2}$ or $\frac{x}{2}-12$

c. $f^{-1}(40)=\frac{40-24}{2}=\frac{16}{2}=8$

Size 8 in the U.S. corresponds to size 40 in Italy.

$f^{-1}(44)=\frac{44-24}{2}=\frac{20}{2}=10$

Size 10 in the U.S. corresponds to size 44 in Italy.

$f^{-1}(52)=\frac{52-24}{2}=\frac{28}{2}=14$

Size 14 in the U.S. corresponds to size 52 in Italy.

$f^{-1}(60)=\frac{60-24}{2}=\frac{36}{2}=18$

Size 18 in the U.S. corresponds to size 60 in Italy.

61. First graph $f(x)=\frac{2}{3}x+4$. Then graph the inverse function by reflecting the graph of $f(x)=\frac{2}{3}x+4$ across the line $y=x$. The graph of the inverse function can also be found by first finding a formula for the inverse, substituting to find function values, and then plotting points.

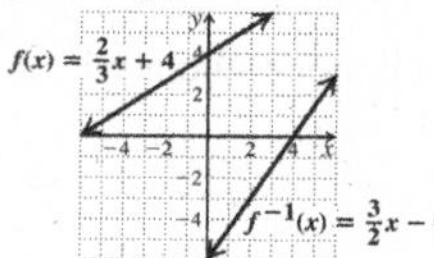

63. Follow the procedure described in Exercise 61 to graph the function and its inverse.

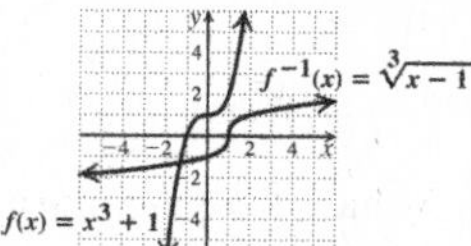

65. Follow the procedure described in Exercise 61 to graph the function and its inverse.

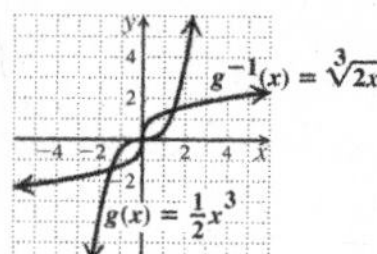

67. Follow the procedure described in Exercise 61 to graph the function and its inverse.

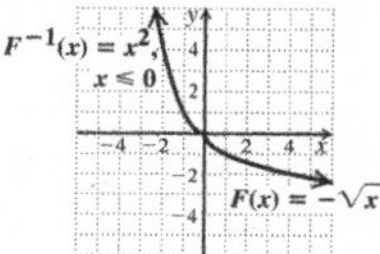

69. Follow the procedure described in Exercise 61 to graph the function and its inverse.

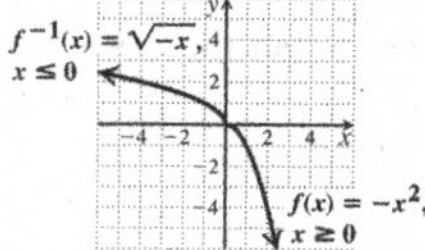

71. We check to see that $\left(f^{-1}\circ f\right)(x)=x$ and $\left(f\circ f^{-1}\right)(x)=x.$

$$(f^{-1}\circ f)(x)=f^{-1}(f(x))=f^{-1}(\sqrt[3]{x-4})$$
$$=\left(\sqrt[3]{x-4}\right)^3+4=x-4+4=x$$
$$(f\circ f^{-1})(x)=f(f^{-1}(x))=f(x^3+4)$$
$$=\sqrt[3]{x^3+4-4}=\sqrt[3]{x^3}=x$$

73. We check to see that $f^{-1}\circ f(x)=x$ and $f\circ f^{-1}(x)=x$.

$$f^{-1}\circ f(x)=f^{-1}(f(x))=f^{-1}\left(\frac{1-x}{x}\right)=\frac{1}{\frac{1-x}{x}+1}$$
$$=\frac{1}{\frac{1-x}{x}+1}\cdot\frac{x}{x}=\frac{x}{1-x+x}=\frac{x}{1}=x$$

$$f\circ f^{-1}(x)=f(f^{-1}(x))=f\left(\frac{1}{x+1}\right)=\frac{1-\frac{1}{x+1}}{\frac{1}{x+1}}$$
$$=\frac{1-\frac{1}{x+1}}{\frac{1}{x+1}}\cdot\frac{x+1}{x+1}=\frac{x+1-1}{1}=\frac{x}{1}=x$$

75. *Writing Exercise.*

77. $t^{1/5}t^{2/3}=t^{1/5+2/3}=t^{13/15}$

79. $(-3x^{-6}y^4)^{-2}=(-3)^{-2}(x^{-6})^{-2}(y^4)^{-2}$
$$=\frac{1}{9}x^{12}y^{-8}$$
$$=\frac{x^{12}}{9y^8}$$

81. $3^3+2^2-(32\div4-16\div8)=27+4-(8-2)$
$$=31-6$$
$$=25$$

83. *Writing Exercise.*

85. Reflect the graph of f across the line $y=x$.

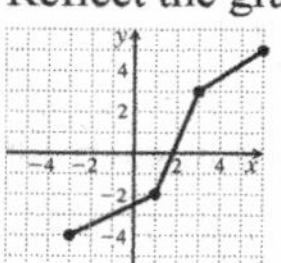

87. From Exercise 59(b), we know that a function that converts dress sizes in Italy to those in the United States is $g(x)=\frac{x-24}{2}$. From Exercise 60, we know that a function that converts dress sizes in the United States to those in France is $f(x)=x+32$. Then a function that converts dress sizes in Italy to those in France is

$$h(x)=(f\circ g)(x)$$
$$h(x)=f\left(\frac{x-24}{2}\right)$$
$$h(x)=\frac{x-24}{2}+32$$
$$h(x)=\frac{x}{2}-12+32$$
$$h(x)=\frac{x}{2}+20.$$

89. *Writing Exercise.*

91. Suppose that $h(x)=(f\circ g)(x)$. First note that for $I(x)=x$, $(f\circ I)(x)=f(I(x))$ for any function f.

i) $((g^{-1}\circ f^{-1})\circ h)(x)=((g^{-1}\circ f^{-1})\circ(f\circ g))(x)$
$$=((g^{-1}\circ(f^{-1}\circ f))\circ g)(x)$$
$$=((g^{-1}\circ I)\circ g)(x)$$
$$=(g^{-1}\circ g)(x)=x$$

ii) $(h\circ(g^{-1}\circ f^{-1}))(x)=((f\circ g)\circ(g^{-1}\circ f^{-1}))(x)$
$$=((f\circ(g\circ g^{-1}))\circ f^{-1})(x)$$
$$=((f\circ I)\circ f^{-1})(x)$$
$$=(f\circ f^{-1})(x)=x$$

Therefore, $(g^{-1}\circ f^{-1})(x)=h^{-1}(x)$.

93. $(f \circ g)(x) = x$ and $(g \circ f)(x) = x$, so the functions are inverses.

95. $(f \circ g)(x) \neq x$, so the functions are not inverses. (It is also true that $(g \circ f)(x) \neq x$.)

97. (1) C; (2) A; (3) B; (4) D

99. We assume f and g are both linear functions.
Find $f(x)$.

$$m = \frac{6.5-6}{7-6} = \frac{1}{2}$$
$$y - 6 = \frac{1}{2}(x-6)$$
$$y = \frac{1}{2}x + 3$$
$$f(x) = \frac{1}{2}x + 3$$

Find $g(x)$.

$$m = \frac{8-6}{7-6} = 2$$
$$y - 6 = 2(x-6)$$
$$y = 2x - 6$$
$$g(x) = 2x - 6$$

Determine whether f and g are inverses of each other.

$$f(g(x)) = f(2x-6) = \frac{1}{2}(2x-6) + 3 = x - 3 + 3$$
$$= x$$
$$g(f(x)) = g\left(\frac{1}{2}x+3\right) = 2\left(\frac{1}{2}x+3\right) - 6 = x + 6 - 6$$
$$= x$$

Yes, f and g are inverses of each other.

101. $(c \circ g)(a)$ makes sense. It represents the cost of sealant for a bamboo floor with area a.

Exercise Set 12.2

1. True

3. True; the graph of $y = f(x-3)$ is a translation of the graph of $y = f(x)$, 3 units to the right.

5. False; the graph of $y = 3^x$ crosses the y-axis at (0, 1).

7. Graph: $y = f(x) = 3^x$
We compute some function values, thinking of y as $f(x)$, and keep the results in a table.

$f(0) = 3^0 = 1$

$f(1) = 3^1 = 3$

$f(2) = 3^2 = 9$

$f(-1) = 3^{-1} = \frac{1}{3^1} = \frac{1}{3}$

$f(-2) = 3^{-2} = \frac{1}{3^2} = \frac{1}{9}$

x	y, or $f(x)$
0	1
1	3
2	9
−1	$\frac{1}{3}$
−2	$\frac{1}{9}$

Next we plot these points and connect them with a smooth curve.

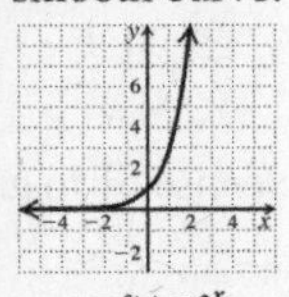

9. Graph: $y = 6^x$
We compute some function values, thinking of y as $f(x)$, and keep the results in a table.

$f(0) = 6^0 = 1$

$f(1) = 6^1 = 6$

$f(2) = 6^2 = 36$

$f(-1) = 6^{-1} = \frac{1}{6^1} = \frac{1}{6}$

$f(-2) = 6^{-2} = \frac{1}{6^2} = \frac{1}{36}$

x	y, or $f(x)$
0	1
1	6
2	36
−1	$\frac{1}{6}$
−2	$\frac{1}{36}$

Next we plot these points and connect them with a smooth curve.

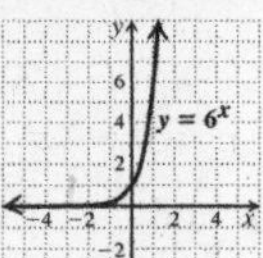

11. Graph: $y = 2^x + 1$
We compute some function values, thinking of y as $f(x)$, and keep the results in a table.

$f(-4) = 2^{-4} + 1 = \frac{1}{2^4} + 1 = \frac{1}{16} + 1 = 1\frac{1}{16}$

$f(-2) = 2^{-2} + 1 = \frac{1}{2^2} + 1 = \frac{1}{4} + 1 = 1\frac{1}{4}$

$f(0) = 2^0 + 1 = 1 + 1 = 2$

$f(1) = 2^1 + 1 = 2 + 1 = 3$

$f(2) = 2^2 + 1 = 4 + 1 = 5$

x	y, or $f(x)$
-4	$1\frac{1}{16}$
-2	$1\frac{1}{4}$
0	2
1	3
2	5

Next we plot these points and connect them with a smooth curve.

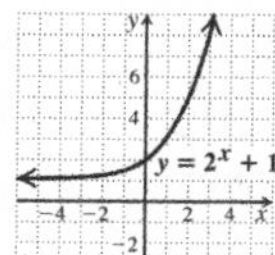

13. Graph: $y = 3^x - 2$

We compute some function values, thinking of y as $f(x)$, and keep the results in a table.

$f(-3) = 3^{-3} - 2 = \frac{1}{3^3} - 2 = \frac{1}{27} - 2 = -\frac{53}{27}$

$f(-1) = 3^{-1} - 2 = \frac{1}{3} - 2 = -\frac{5}{3}$

$f(0) = 3^0 - 2 = 1 - 2 = -1$

$f(1) = 3^1 - 2 = 3 - 2 = 1$

$f(2) = 3^2 - 2 = 9 - 2 = 7$

x	y, or $f(x)$
-3	$-\frac{53}{27}$
-1	$-\frac{5}{3}$
0	-1
1	1
2	7

Next we plot these points and connect them with a smooth curve.

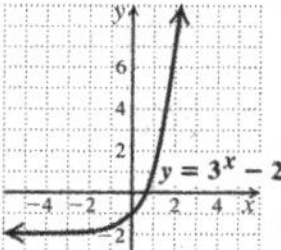

15. Graph: $y = 2^x - 5$

We construct a table of values, thinking of y as $f(x)$. Then we plot the points and connect them with a smooth curve.

$f(0) = 2^0 - 5 = 1 - 5 = -4$

$f(1) = 2^1 - 5 = 2 - 5 = -3$

$f(2) = 2^2 - 5 = 4 - 5 = -1$

$f(3) = 2^3 - 5 = 8 - 5 = 3$

$f(-1) = 2^{-1} - 5 = \frac{1}{2} - 5 = -\frac{9}{2}$

$f(-2) = 2^{-2} - 5 = \frac{1}{4} - 5 = -\frac{19}{4}$

$f(-4) = 2^{-4} - 5 = \frac{1}{16} - 5 = -\frac{79}{16}$

x	y, or $f(x)$
0	-4
1	-3
2	-1
3	3
-1	$-\frac{9}{2}$
-2	$-\frac{19}{4}$
-4	$-\frac{79}{16}$

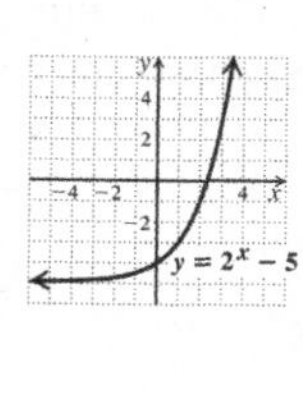

17. Graph: $y = 2^{x-3}$

We construct a table of values, thinking of y as $f(x)$. Then we plot the points and connect them with a smooth curve.

$f(0) = 2^{0-3} = 2^{-3} = \frac{1}{8}$

$f(-1) = 2^{-1-3} = 2^{-4} = \frac{1}{16}$

$f(1) = 2^{1-3} = 2^{-2} = \frac{1}{4}$

$f(2) = 2^{2-3} = 2^{-1} = \frac{1}{2}$

$f(3) = 2^{3-3} = 2^0 = 1$

$f(4) = 2^{4-3} = 2^1 = 2$

$f(5) = 2^{5-3} = 2^2 = 4$

x	y, or $f(x)$
0	$\frac{1}{8}$
-1	$\frac{1}{16}$
1	$\frac{1}{4}$
2	$\frac{1}{2}$
3	1
4	2
5	4

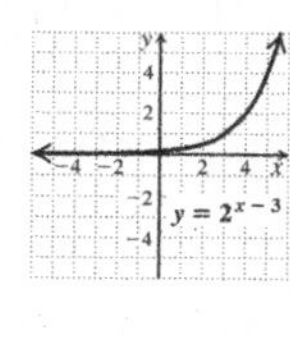

19. Graph: $y = 2^{x+1}$

We construct a table of values, thinking of y as $f(x)$. Then we plot the points and connect them with a smooth curve.

$f(-3) = 2^{-3+1} = 2^{-2} = \frac{1}{4}$

$f(-1) = 2^{-1+1} = 2^0 = 1$

$f(0) = 2^{0+1} = 2^1 = 2$

$f(1) = 2^{1+1} = 2^2 = 4$

x	y, or $f(x)$
-3	$\frac{1}{4}$
-1	1
0	2
1	4

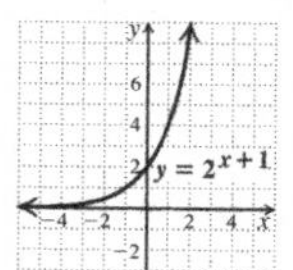

21. Graph: $y=\left(\frac{1}{4}\right)^x$

We construct a table of values, thinking of y as $f(x)$. Then we plot the points and connect them with a smooth curve.

$f(0)=\left(\frac{1}{4}\right)^0=1$

$f(1)=\left(\frac{1}{4}\right)^1=\frac{1}{4}$

$f(2)=\left(\frac{1}{4}\right)^2=\frac{1}{16}$

$f(-1)=\left(\frac{1}{4}\right)^{-1}=\frac{1}{\frac{1}{4}}=4$

$f(-2)=\left(\frac{1}{4}\right)^{-2}=\frac{1}{\frac{1}{16}}=16$

x	y, or $f(x)$
0	1
1	$\frac{1}{4}$
2	$\frac{1}{16}$
−1	4
−2	16

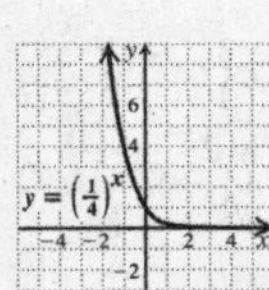

23. Graph: $y=\left(\frac{1}{3}\right)^x$

We construct a table of values, thinking of y as $f(x)$. Then we plot the points and connect them with a smooth curve.

$f(0)=\left(\frac{1}{3}\right)^0=1$

$f(1)=\left(\frac{1}{3}\right)^1=\frac{1}{3}$

$f(2)=\left(\frac{1}{3}\right)^2=\frac{1}{9}$

$f(3)=\left(\frac{1}{3}\right)^3=\frac{1}{27}$

$f(-1)=\left(\frac{1}{3}\right)^{-1}=\frac{1}{\left(\frac{1}{3}\right)^1}=\frac{1}{\frac{1}{3}}=3$

$f(-2)=\left(\frac{1}{3}\right)^{-2}=\frac{1}{\left(\frac{1}{3}\right)^2}=\frac{1}{\frac{1}{9}}=9$

$f(-3)=\left(\frac{1}{3}\right)^{-3}=\frac{1}{\left(\frac{1}{3}\right)^3}=\frac{1}{\frac{1}{27}}=27$

x	y, or $f(x)$
0	1
1	$\frac{1}{3}$
2	$\frac{1}{9}$
3	$\frac{1}{27}$
−1	3
−2	9
−3	27

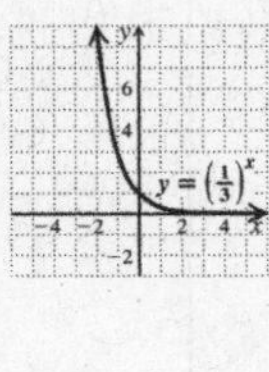

25. Graph: $y=2^{x+1}-3$

We construct a table of values, thinking of y as $f(x)$. Then we plot the points and connect them with a smooth curve.

$f(0)=2^{0+1}-3=2-3=-1$

$f(1)=2^{1+1}-3=4-3=1$

$f(2)=2^{2+1}-3=8-3=5$

$f(-1)=2^{-1+1}-3=1-3=-2$

$f(-2)=2^{-2+1}-3=\frac{1}{2}-3=-\frac{5}{2}$

$f(-3)=2^{-3+1}-3=\frac{1}{4}-3=-\frac{11}{4}$

x	y, or $f(x)$
0	−1
1	1
2	5
−1	−2
−2	$-\frac{5}{2}$
−3	$-\frac{11}{4}$

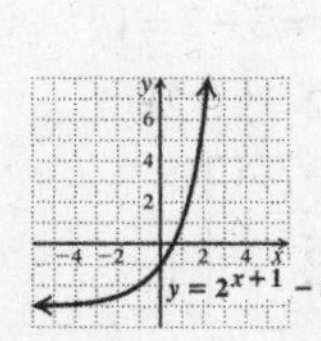

27. Graph: $x=6^y$

We can find ordered pairs by choosing values for y and then computing values for x.

For $y=0$, $x=6^0=1$.

For $y=1$, $x=6^1=6$.

For $y=-1$, $x=6^{-1}=\frac{1}{6^1}=\frac{1}{6}$.

For $y=-2$, $x=6^{-2}=\frac{1}{6^2}=\frac{1}{36}$.

x	y
1	0
6	1
$\frac{1}{6}$	−1
$\frac{1}{36}$	−2

↑ ↑ (1) Choose values for y.

(2) Compute values for x.

We plot the points and connect them with a smooth curve.

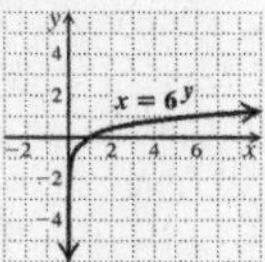

29. Graph: $x = 3^{-y} = \left(\frac{1}{3}\right)^y$

We can find ordered pairs by choosing values for y and then computing values for x. Then we plot these points and connect them with a smooth curve.

For $y = 0$, $x = \left(\frac{1}{3}\right)^0 = 1$.

For $y = 1$, $x = \left(\frac{1}{3}\right)^1 = \frac{1}{3}$.

For $y = 2$, $x = \left(\frac{1}{3}\right)^2 = \frac{1}{9}$.

For $y = -1$, $x = \left(\frac{1}{3}\right)^{-1} = \frac{1}{\frac{1}{3}} = 3$.

For $y = -2$, $x = \left(\frac{1}{3}\right)^{-2} = \frac{1}{\frac{1}{9}} = 9$.

x	y
1	0
$\frac{1}{3}$	1
$\frac{1}{9}$	2
3	−1
9	−2

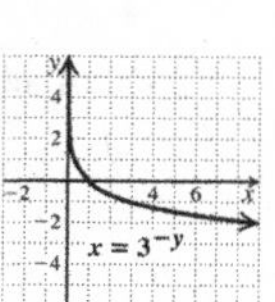

31. Graph: $x = 4^y$

We can find ordered pairs by choosing values for y and then computing values for x. Then we plot these points and connect them with a smooth curve.

For $y = 0$, $x = 4^0 = 1$.

For $y = 1$, $x = 4^1 = 4$.

For $y = 2$, $x = 4^2 = 16$.

For $y = -1$, $x = 4^{-1} = \frac{1}{4}$.

For $y = -2$, $x = 4^{-2} = \frac{1}{16}$.

x	y
1	0
4	1
16	2
$\frac{1}{4}$	−1
$\frac{1}{16}$	−2

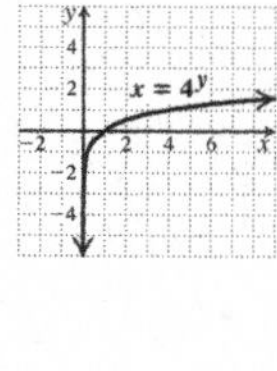

33. Graph: $x = \left(\frac{4}{3}\right)^y$

We can find ordered pairs by choosing values for y and then computing values for x. Then we plot these points and connect them with a smooth curve.

For $y = 0$, $x = \left(\frac{4}{3}\right)^0 = 1$.

For $y = 1$, $x = \left(\frac{4}{3}\right)^1 = \frac{4}{3}$.

For $y = 2$, $x = \left(\frac{4}{3}\right)^2 = \frac{16}{9}$.

For $y = 3$, $x = \left(\frac{4}{3}\right)^3 = \frac{64}{27}$.

For $y = -1$, $x = \left(\frac{4}{3}\right)^{-1} = \frac{3}{4}$.

For $y = -2$, $x = \left(\frac{4}{3}\right)^{-2} = \left(\frac{3}{4}\right)^2 = \frac{9}{16}$.

For $y = -3$, $x = \left(\frac{4}{3}\right)^{-3} = \left(\frac{3}{4}\right)^3 = \frac{27}{64}$.

x	y
1	0
$\frac{4}{3}$	1
$\frac{16}{9}$	2
$\frac{64}{27}$	3
$\frac{3}{4}$	−1
$\frac{9}{16}$	−2
$\frac{27}{64}$	−3

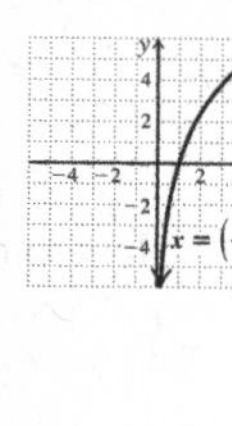

35. Graph $y = 3^x$ (see Exercise 7) and $x = 3^y$ (see Exercise 28) using the same set of axes.

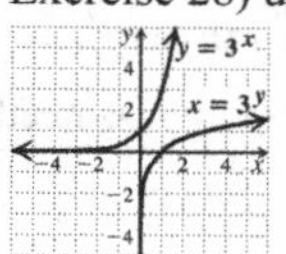

37. Graph $y = \left(\frac{1}{2}\right)^x$ and $x = \left(\frac{1}{2}\right)^y$ (see Exercise 30) using the same set of axes.

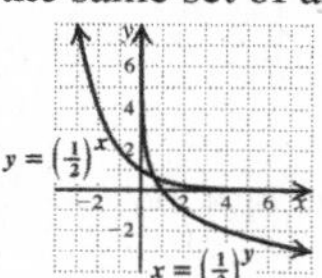

39. a. In 0 days (at birth), $t = 0$

$W(0) = 2000(1.0077)^0 = 2000$ lb

In 30 days, $W(30) = 2000(1.0077)^{30} \approx 2500$ lb

In 60 days, $W(60) = 2000(1.0077)^{60} \approx 3200$ lb

In 90 days, $W(30) = 2000(1.0077)^{90} \approx 4000$ lb

b. Use the function values computed in part (a) and others, if desired, and draw the graph.

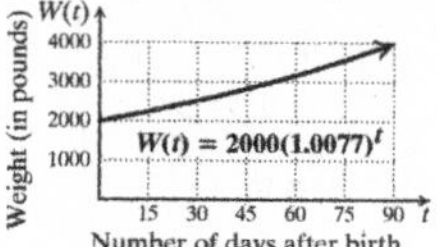

41. a. $P(1) = 21.4(0.914)^1 \approx 19.6\%$

$P(3) = 21.4(0.914)^3 \approx 16.3\%$

1 yr = 12 months; $P(12) = 21.4(0.914)^{12} \approx 7.3\%$

b.

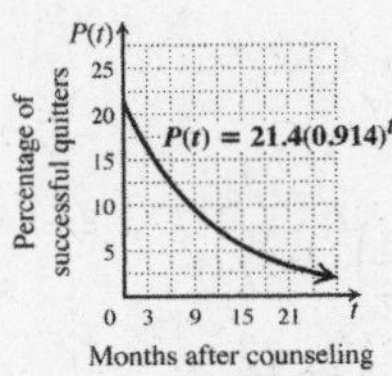

43. a. In 1930, $t = 1930 - 1900 = 30$.

$$P(t) = 150(0.960)^t$$
$$P(30) = 150(0.960)^{30} \approx 44.079$$

In 1930, about 44.079 thousand, or 44,079, humpback whales were alive.
In 1960, $t = 1960 - 1900 = 60$.

$$P(t) = 150(0.960)^t$$
$$P(60) = 150(0.960)^{60} \approx 12.953$$

In 1960, about 12.953 thousand, or 12,953, humpback whales were alive.

b. Plot the points found in part (a), $(30, 44{,}079)$ and $(60, 12{,}953)$ and additional points as needed and graph the function.

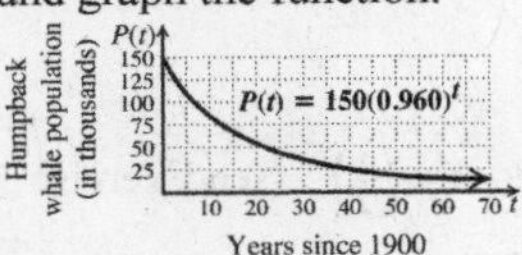

45. a. In 1992, $t = 1992 - 1982 = 10$.

$$P(10) = 5.5(1.047)^{10} \approx 8.706$$

In 1992, about 8.706 thousand, or 8706, humpback whales were alive.
In 2004, $t = 2004 - 1982 = 22$.

$$P(22) = 5.5(1.047)^{22} \approx 15.107$$

In 2004, about 15.107 thousand, or 15,107, humpback whales were alive.

b. Use the function values computed in part (a) and others, if desired, and draw the graph.

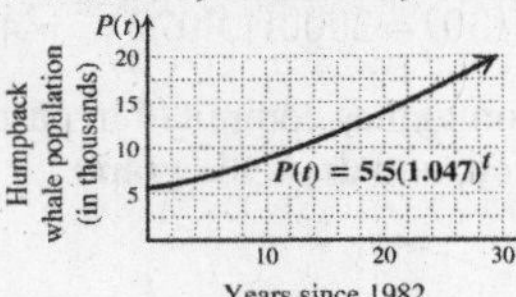

47. a. In 1997, $t = 1997 - 1997 = 0$.

$$M(t) = 50(1.25)^t$$
$$M(0) = 50(1.25)^0 = 50$$

In 1997, about 50 moose were alive.
In 2012, $t = 2012 - 1997 = 15$.

$$M(t) = 50(1.25)^t$$
$$M(15) = 50(1.25)^{15} \approx 1421$$

In 2012, about 1421 moose were alive.
In 2020, $t = 2020 - 1997 = 23$.

$$M(t) = 50(1.25)^t$$
$$M(23) = 50(1.25)^{23} \approx 8470$$

In 2020, about 8470 moose will be alive.

b. Plot the points found in part (a), (0, 50), (15, 1421) and (23, 8470) and additional points as needed and graph the function.

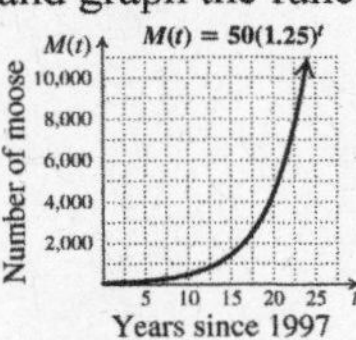

49. *Writing Exercise.*

51. $3x^2 - 48 = 3(x^2 - 16) = 3(x+4)(x-4)$

53. $6x^2 + x - 12 = (2x+3)(3x-4)$

55. $t^2 - y^2 + 2y - 1$

$$= t^2 - (y^2 - 2y + 1)$$
$$= t^2 - (y-1)^2 \qquad \text{Difference of squares}$$
$$= [t - (y-1)][t + (y-1)]$$
$$= (t - y + 1)(t + y - 1)$$

57. *Writing Exercise.*

59. Since the bases are the same, the one with the larger exponent is the larger number. Thus $\pi^{2.4}$ is larger.

61. Graph: $f(x) = 2.5^x$

Use a calculator with a power key to construct a table of values. (We will round values of $f(x)$ to the nearest hundredth.) Then plot these points and connect them with a smooth curve.

x	y
0	1
1	2.5
2	6.25
3	15.63
−1	0.4
−2	0.16

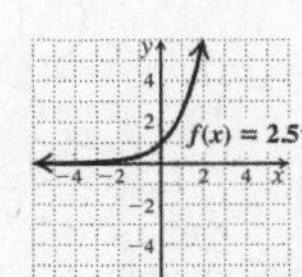

63. Graph: $y = 2^x + 2^{-x}$

Construct a table of values, thinking of y as $f(x)$. Then plot these points and connect them with a curve.

$$f(0) = 2^0 + 2^{-0} = 1 + 1 = 2$$
$$f(1) = 2^1 + 2^{-1} = 2 + \frac{1}{2} = 2\frac{1}{2}$$
$$f(2) = 2^2 + 2^{-2} = 4 + \frac{1}{4} = 4\frac{1}{4}$$
$$f(3) = 2^3 + 2^{-3} = 8 + \frac{1}{8} = 8\frac{1}{8}$$
$$f(-1) = 2^{-1} + 2^{-(-1)} = \frac{1}{2} + 2 = 2\frac{1}{2}$$
$$f(-2) = 2^{-2} + 2^{-(-2)} = \frac{1}{4} + 4 = 4\frac{1}{4}$$

$f(-3) = 2^{-3} + 2^{-(-3)} = \frac{1}{8} + 8 = 8\frac{1}{8}$

x	y, or $f(x)$
0	2
1	$2\frac{1}{2}$
2	$4\frac{1}{4}$
3	$8\frac{1}{8}$
−1	$2\frac{1}{2}$
−2	$4\frac{1}{4}$
−3	$8\frac{1}{8}$

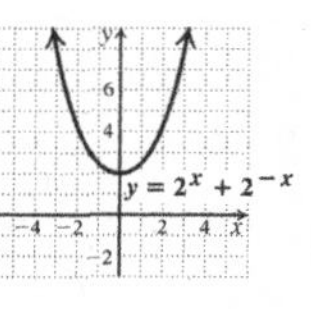

65. Graph: $y = |2^x - 2|$

We construct a table of values, thinking of y as $f(x)$. Then plot these points and connect them with a curve.

$f(0) = |2^0 - 2| = |1 - 2| = |-1| = 1$

$f(1) = |2^1 - 2| = |2 - 2| = |0| = 0$

$f(2) = |2^2 - 2| = |4 - 2| = |2| = 2$

$f(3) = |2^3 - 2| = |8 - 2| = |6| = 6$

$f(-1) = |2^{-1} - 2| = \left|\frac{1}{2} - 2\right| = \left|-\frac{3}{2}\right| = \frac{3}{2}$

$f(-3) = |2^{-3} - 2| = \left|\frac{1}{8} - 2\right| = \left|-\frac{15}{8}\right| = \frac{15}{8}$

$f(-5) = |2^{-5} - 2| = \left|\frac{1}{32} - 2\right| = \left|-\frac{63}{32}\right| = \frac{63}{32}$

x	y, or $f(x)$
0	1
1	0
2	2
3	6
−1	$\frac{3}{2}$
−3	$\frac{15}{8}$
−5	$\frac{63}{32}$

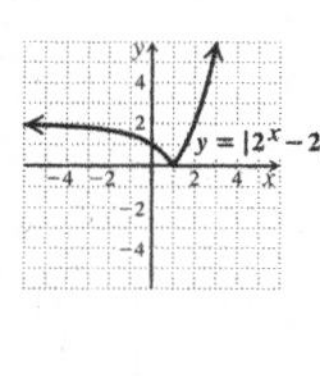

67. Graph: $y = |2^{x^2} - 1|$

We construct a table of values, thinking of y as $f(x)$. Then we plot these points and connect them with a curve.

$f(0) = |2^{0^2} - 1| = |1 - 1| = 0$

$f(1) = |2^{1^2} - 1| = |2 - 1| = 1$

$f(2) = |2^{2^2} - 1| = |16 - 1| = 15$

$f(-1) = |2^{(-1)^2} - 1| = |2 - 1| = 1$

$f(-2) = |2^{(-2)^2} - 1| = |16 - 1| = 15$

x	y, or $f(x)$
0	0
1	1
2	15
−1	1
−2	15

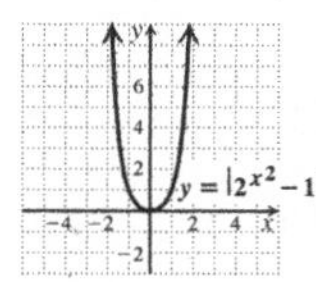

69. $y = 3^{-(x-1)}$ $x = 3^{-(y-1)}$

x	y
0	3
1	1
2	$\frac{1}{3}$
3	$\frac{1}{9}$
−1	9

x	y
3	0
1	1
$\frac{1}{3}$	2
$\frac{1}{9}$	3
9	−1

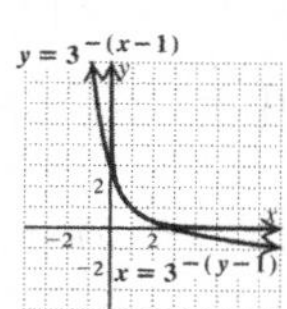

71. Enter the data points (0, 100), (4, 3000) and (8, 14,000) and then use the ExpReg option from the STAT CALC menu of a graphing calculator to find an exponential function that models the data:

$N(t) = 136(1.85)^t$, where $N(t)$ is the number of ruffe, t years after 1984.

In 1990, $t = 1990 - 1984 = 6$

$N(6) \approx 5550$ ruffe.

73. *Writing Exercise.*

75. *Graphing Calculator Exercise*

Exercise Set 12.3

1. The inverse of the function given by $f(x) = 3^x$ is a *logarithmic* function.

3. Logarithm bases are *positive*.

5. $5^2 = 25$, so choice (g) is correct.

7. $5^1 = 5$, so choice (a) is correct.

9. The exponent to which we raise 5 to get 5^x is x, so choice (b) is correct.

11. $5 = 2^x$ is equivalent to $\log_2 5 = x$, so choice (e) is correct.

13. $\log_{10} 1000$ is the exponent to which we raise 10 to get 1000. Since $10^3 = 1000$, $\log_{10} 1000 = 3$.

15. $\log_7 49$ is the exponent to which we raise 7 to get 49. Since $7^2 = 49$, $\log_7 49 = 2$.

17. $\log_5 \frac{1}{25}$ is the exponent to which we raise 5 to get $\frac{1}{25}$. Since $5^{-2} = \frac{1}{25}$, $\log_5 \frac{1}{25} = -2$.

19. Since $8^{-1} = \frac{1}{8}$, $\log_8 \frac{1}{8} = -1$.

21. Since $5^4 = 625$, $\log_5 625 = 4$.

23. Since $7^1 = 7$, $\log_7 7 = 1$.

25. Since $3^0 = 1$, $\log_3 1 = 0$.

27. $\log_6 6^5$ is the exponent to which we raise 6 to get 6^5. Clearly, this power is 5, so $\log_6 6^5 = 5$.

29. Since $10^{-2} = \frac{1}{100} = 0.01$, $\log_{10} 0.01 = -2$.

31. Since $16^{1/2} = 4$, $\log_{16} 4 = \frac{1}{2}$.

33. Since $9 = 3^2$ and $(3^2)^{3/2} = 3^3 = 27$, $\log_9 27 = \frac{3}{2}$.

35. Since $1000 = 10^3$ and $(10^3)^{2/3} = 10^2 = 100$, $\log_{1000} 100 = \frac{2}{3}$.

37. Since $\log_3 29$ is the power to which we raise 3 to get 29, then 3 raised to this power is 29. That is $3^{\log_3 29} = 29$.

39. Graph: $y = \log_{10} x$

The equation $y = \log_{10} x$ is equivalent to $10^y = x$. We can find ordered pairs by choosing values for y and computing the corresponding x-values.

For $y = 0$, $x = 10^0 = 1$.

For $y = 1$, $x = 10^1 = 10$.

For $y = 2$, $x = 10^2 = 100$.

For $y = -1$, $x = 10^{-1} = \frac{1}{10}$.

For $y = -2$, $x = 10^{-2} = \frac{1}{100}$.

x, or 10^y	y
1	0
10	1
100	2
$\frac{1}{10}$	−1
$\frac{1}{100}$	−2

↑ ↑ (1) Select y.
(2) Compute x.

We plot the set of ordered pairs and connect the points with a smooth curve.

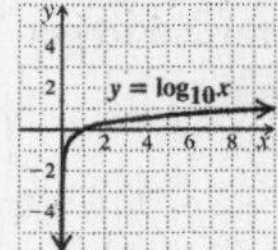

41. Graph: $y = \log_3 x$

The equation $y = \log_3 x$ is equivalent to $3^y = x$. We can find ordered pairs by choosing values for y and computing the corresponding x-values.

For $y = 0$, $x = 3^0 = 1$.

For $y = 1$, $x = 3^1 = 3$.

For $y = 2$, $x = 3^2 = 9$.

For $y = -1$, $x = 3^{-1} = \frac{1}{3}$.

For $y = -2$, $x = 3^{-2} = \frac{1}{9}$.

x, or 3^y	y
1	0
3	1
9	2
$\frac{1}{3}$	−1
$\frac{1}{9}$	−2

We plot the set of ordered pairs and connect the points with a smooth curve.

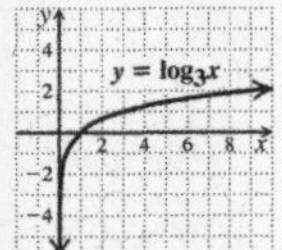

43. Graph: $f(x) = \log_6 x$

Think of $f(x)$ as y. Then $y = \log_6 x$ is equivalent to $6^y = x$. We find ordered pairs by choosing values for y and computing the corresponding x-values. Then we plot the points and connect them with a smooth curve.

For $y = 0$, $x = 6^0 = 1$.

For $y = 1$, $x = 6^1 = 6$.

For $y = 2$, $x = 6^2 = 36$.

For $y = -1$, $x = 6^{-1} = \frac{1}{6}$.

For $y = -2$, $x = 6^{-2} = \frac{1}{36}$.

x, or 6^y	y
1	0
6	1
36	2
$\frac{1}{6}$	−1
$\frac{1}{36}$	−2

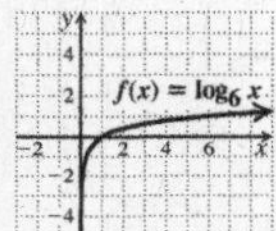

45. Graph: $f(x) = \log_{2.5} x$

Think of $f(x)$ as y. Then $y = \log_{2.5} x$ is equivalent to $2.5^y = x$. We construct a table of values, plot these points and connect them with a smooth curve.

For $y = 0$, $x = 2.5^0 = 1$.

For $y = 1$, $x = 2.5^1 = 2.5$.

For $y = 2$, $x = 2.5^2 = 6.25$.

For $y = 3$, $x = 2.5^3 = 15.625$.

For $y = -1$, $x = 2.5^{-1} = 0.4$.

For $y = -2$, $x = 2.5^{-2} = 0.16$.

x, or 2.5^y	y
1	0
2.5	1
6.25	2
15.625	3
0.4	−1
0.16	−2

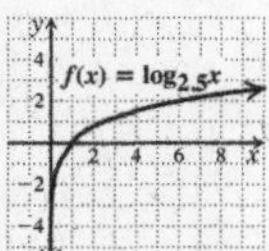

21.
$$M = \log\frac{v}{1.34}$$
$$7.5 = \log\frac{v}{1.34}$$
$$10^{7.5} = \frac{v}{1.34}$$
$$1.34(10^{7.5}) = v$$
$$42{,}400{,}000 \approx v$$
Approximately 42.4 million messages per day are sent by that network.

23. a. Substitute 0.025 for k:
$$P(t) = P_0 e^{0.025t}$$

b. To find the balance after one year, replace P_0 with 5000 and t with 1. We find $P(1)$:
$$P(1) = 5000e^{0.025(1)} = 5000e^{0.025} \approx \$5126.58$$
To find the balance after 2 years, replace P_0 with 5000 and t with 2. We find $P(2)$:
$$P(2) = 5000e^{0.025(2)} = 5000e^{0.05} \approx \$5256.36$$

c. To find the doubling time, replace P_0 with 5000 and $P(t)$ with 10,000 and solve for t.
$$10{,}000 = 5000e^{0.025t}$$
$$2 = e^{0.025t}$$
$$\ln 2 = \ln e^{0.025t}$$
$$\ln 2 = 0.025t$$
$$\frac{\ln 2}{0.025} = t$$
$$27.7 \approx t$$
The investment will double in about 27.7 years.

25. a. $P(t) = 324e^{0.0073t}$, where $P(t)$ is in millions and t is the number of years after 2016.

b. In 2025, $t = 2025 - 2016 = 9$. Find $P(9)$.
$$P(9) = 324e^{0.0073(9)} = 324e^{0.0657} \approx 346$$
The U.S. population will be about 346 million in 2025.

c. Substitute 400 for $P(t)$ and solve for t.
$$400 = 324e^{0.0073t}$$
$$\frac{400}{324} = e^{0.0073t}$$
$$\ln\frac{400}{324} = \ln e^{0.0073t}$$
$$\ln\frac{400}{324} = 0.0073t$$
$$\frac{\ln\frac{400}{324}}{0.0073} = t$$
$$29 \approx t$$
The U.S. population will reach 400 million about 29 yr after 2016, or in 2045.

27. The exponential growth function is $P(t) = P_0e^{3.24t}$. We replace $P(t)$ with $2P_0$ and solve for t.
$$P(t) = P_0e^{0.0324t}$$
$$P2_0 = P_0e^{0.0324t}$$
$$2 = e^{0.0324t}$$
$$\ln 2 = \ln e^{0.0324t}$$
$$\ln 2 = 0.0324t$$
$$\frac{\ln 2}{0.0324} = t$$
$$21.4 \approx t$$
The doubling time is about 21.4 yr.

29. $Y(x) = 88.5\ln\frac{x}{7.4}$

a. $Y(10) = 88.5\ln\frac{10}{7.4} \approx 27$

The world population will reach 10 billion about 27 yr after 2016, or in 2043.

b. $Y(12) = 88.5\ln\frac{12}{7.4} \approx 43$

The world population will reach 12 billion about 43 yr after 2016, or in 2059.

c. Plot the points found in parts (a) and (b) and others as necessary and draw the graph.

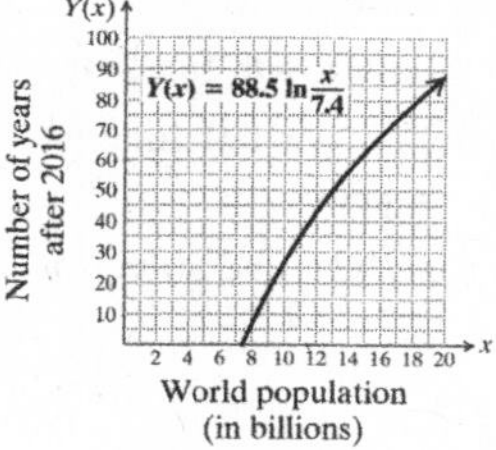

31. $s(t) = 1.1 + 36\ln t$, t is the number of years after 2005.

a. In 2012, $t = 2012 - 2005 = 7$. Find $s(7)$.
$$s(7) = 1.1 + 36\ln 7 \approx 71$$
About 71% of Americans ages 30-49 used social networking sites in 2012.

b. Plot the points found in part (a) and others as necessary and draw the graph.

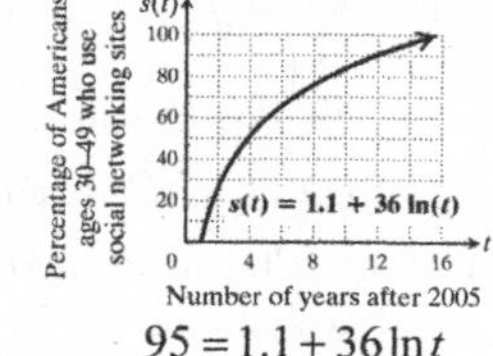

c.
$$95 = 1.1 + 36\ln t$$
$$93.9 = 36\ln t$$
$$\frac{93.9}{36} = \ln t$$
$$e^{93.9/36} = t$$
$$14 \approx t$$
95% of Americans ages 30-49 will use social network sites about 14 yr after 2005, or in 2019.

d. The minimum of the domain is 1. The maximum occurs when 100% is achieved.

$$100 = 1.1 + 36\ln t$$
$$98.9 = 36\ln t$$
$$\frac{98.9}{36} = \ln t$$
$$e^{98.9/36} = t$$
$$15.6 \approx t$$

The domain is [1, 15.6].

33. a. We start with the exponential growth equation. Substituting 12 for P_0, we have

$P(t) = 12e^{kt}$, where t is the number of years after 1967.

To find the exponential growth rate k, observe that the ticket price was \$2670 in 2015, which is 48 years after 1967. We substitute and solve for k.

$$P(t) = 12e^{k \cdot 48}$$
$$2670 = 12e^{48k}$$
$$222.5 = e^{48k}$$
$$\ln 222.5 = \ln e^{48k}$$
$$\ln 222.5 = 48k$$
$$\frac{\ln 222.5}{48} = k$$
$$0.113 \approx k$$

Thus, the exponential growth function is $P(t) = 12e^{0.113t}$, where t is the number of years after 1967.

b. Substitute 5000 for $P(t)$ and solve for t.

$$5000 = 12e^{0.113t}$$
$$\frac{5000}{12} = e^{0.113t}$$
$$\ln \frac{5000}{12} = \ln e^{0.113t}$$
$$\ln \frac{5000}{12} = 0.113t$$
$$\frac{\ln \frac{5000}{12}}{0.113} = t$$
$$53.4 \approx t$$

The ticket price will reach \$5000 about 53 yr after 1967, or in 2020.

35. a. The function $P(t) = P_0e^{-kt}$, $k > 0$, can be used to model decay. We substitute 11 for $P(t)$ and 1 for t. Then we substitute 2 for $P(t)$ and 6 for t. We have two equations and two unknowns:

$$11 = P_0e^{-k(1)} \text{ and } 2 = P_0e^{-k(6)}$$

Solve each equation for P_0.

$$11 = P_0e^{-k(1)} \quad \text{and} \quad 2 = P_0e^{-k(6)}$$
$$11e^k = P_0 \qquad 2e^{6k} = P_0$$

Substitute and solve for k.

$$11e^k = 2e^{6k}$$
$$\ln 11e^k = \ln 2e^{6k}$$
$$\ln 11 + \ln e^k = \ln 2 + \ln e^{6k}$$
$$\ln 11 + k = \ln 2 + 6k$$
$$\ln 11 - \ln 2 = 5k$$
$$\frac{\ln 11 - \ln 2}{5} = k$$
$$0.341 \approx k$$

Solve for P_0.

$$P_0 = 11e^{0.341} \approx 15.5$$

Thus, the exponential decay function is $P(t) = 15.5e^{-0.341t}$, where t is in hours.

b. Substitute 3 for t.

$$P(3) = 15.5e^{-0.341(3)} \approx 5.6$$

After 3 hr, there is about 5.6 mcg/mL concentration.

c. Substitute 4 for $P(t)$ and solve for t.

$$4 = 15.5e^{-0.341t}$$
$$\frac{4}{15.5} = e^{-0.341t}$$
$$\ln \frac{4}{15.5} = \ln e^{-0.341t}$$
$$\ln \frac{4}{15.5} = -0.341t$$
$$\frac{\ln \frac{4}{15.5}}{-0.341} = t$$
$$4 \approx t$$

It will take approximately 4 hr for the dosage to reach 4 mcg/mL.

d. Substitute $\frac{1}{2}(15.5)$, or 7.75 for $P(t)$. Solve for t.

$$7.75 = 15.5e^{-0.341t}$$
$$\frac{1}{2} = e^{-0.341t}$$
$$\ln 0.5 = \ln e^{-0.341t}$$
$$\ln 0.5 = -0.341t$$
$$\frac{\ln 0.5}{-0.341} = t$$
$$2 \approx t$$

The half-life is about 2 hr.

37. We will use the function derived in Example 7:

$$P(t) = P_0e^{-0.00012t}$$

If the seed had lost 21% of its carbon-14 from the initial amount P_0, then $79\%(P_0)$ is the amount present. To find the age t of the seed, we substitute $79\%(P_0)$, or $0.79P_0$ for $P(t)$ in the function above and solve for t.

$$0.79P_0 = P_0e^{-0.00012t}$$
$$0.79 = e^{-0.00012t}$$
$$\ln 0.79 = \ln e^{-0.00012t}$$
$$\ln 0.79 = -0.00012t$$
$$\frac{\ln 0.79}{-0.00012} = t$$
$$1964 \approx t$$

The seed is about 1964 yr old.

39. The function $P(t) = P_0 e^{-kt}$, $k > 0$, can be used to model decay. For iodine-131, $k = 9.6\%$, or 0.096. To find the half-life we substitute 0.096 for k and $\frac{1}{2}$ P_0 for $P(t)$, and solve for t.

$$\frac{1}{2}P_0 = P_0 e^{-0.096t}, \text{ or } \frac{1}{2} = e^{-0.096t}$$
$$\ln\frac{1}{2} = \ln e^{-0.096t} = -0.096t$$
$$t = \frac{\ln 0.5}{-0.096} \approx \frac{-0.6931}{-0.096} \approx 7.2 \text{ days}$$

41. a. The function $P(t) = P_0 e^{-kt}$, $k > 0$, can be used to model decay. We substitute $\frac{1}{2}P_0$ for $P(t)$ and 5 for t and solve for the decay rate k.

$$\frac{1}{2}P_0 = P_0 e^{-k\cdot 5}$$
$$\frac{1}{2} = e^{-5k}$$
$$\ln\frac{1}{2} = \ln e^{-5k}$$
$$-\ln 2 = -5k$$
$$\frac{\ln 2}{5} = k$$
$$0.139 \approx k$$

The decay rate is 0.139, or 13.9% per hour.

b. 95% consumed = 5% remains

$$0.05P_0 = P_0 e^{-0.139t}$$
$$0.05 = e^{-0.139t}$$
$$\ln 0.05 = \ln e^{-0.139t}$$
$$\ln 0.05 = -0.139t$$
$$\frac{\ln 0.05}{-0.139} = t$$
$$21.6 \approx t$$

It will take approximately 21.6 hr for 95% of the caffeine to leave the body.

43. a. We start with the exponential growth equation

$V(t) = V_0 e^{kt}$, where t is the number of years after 1989.

Substituting 20.6 for V_0, we have

$$V(t) = 20.6e^{kt}.$$

To find the exponential growth rate k, observe that the painting sold for \$300 million, or \$300,000,000 in 2015, or 26 years after 1989. We substitute and solve for k.

$$300 = 20.6e^{k\cdot 26}$$
$$\frac{300}{20.6} = e^{26k}$$
$$\ln\frac{300}{20.6} = \ln e^{26k}$$
$$\ln\frac{300}{20.6} = 26k$$
$$\frac{\ln\frac{300}{20.6}}{26} = k$$
$$0.103 \approx k$$

The exponential growth function is $V(t) = 20.6e^{0.103t}$, where $V(t)$ is in millions of dollars and t is the number of years after 1989.

b. In 2025, $t = 2025 - 1989 = 36$.

$$V(36) = 20.6e^{0.103(36)} \approx \$840 \text{ million}$$

c. To find the doubling time, replace $V(t)$ with 41.2 and solve for t.

$$41.2 = 20.6e^{0.103t}$$
$$2 = e^{0.103t}$$
$$\ln 2 = \ln e^{0.103t}$$
$$\ln 2 = 0.103t$$
$$\frac{\ln 2}{0.103} = t$$
$$6.7 \approx t$$

The doubling time is about 6.7 years.

d. \$1 billion = \$1000 million.

$$1000 = 20.6e^{0.103t}$$
$$\frac{1000}{20.6} = e^{0.103t}$$
$$\ln\frac{1000}{20.6} = \ln e^{0.103t}$$
$$\ln\frac{1000}{20.6} = 0.103t$$
$$\frac{\ln\frac{1000}{20.6}}{0.103} = t$$
$$37.7 \text{ yr} \approx t$$

The painting will be worth \$1 billion 37.7 yr after 1989.

45. *Writing Exercise.*

47. $f(x) = 18x + \frac{1}{2}$

49.
$$2x - 3y = 4$$
$$-3y = -2x + 4$$
$$y = \frac{2}{3}x - \frac{4}{3}$$

A line parallel to this line has slope $m = \frac{2}{3}$.

$$y - 7 = \frac{2}{3}(x + 3)$$
$$y - 7 = \frac{2}{3}x + 2$$
$$f(x) = \frac{2}{3}x + 9$$

51. *Writing Exercise.*

53. We will use the exponential growth function $V(t) = V_0 e^{kt}$, where t is the number of years after 2015 and $V(t)$ is in millions of dollars. Substitute 32 for $V(t)$, 0.04 for k, and 8 for t and solve for V_0.

$$V(t) = V_0 e^{kt}$$
$$32 = V_0 e^{0.04(8)}$$
$$\frac{32}{e^{0.32}} = V_0$$
$$23.2 \approx V_0$$

About \$23.2 million would need to be invested.

55. **a.** Substitute 1390 for I and solve for m.

$$m(I) = -(19 + 2.5 \cdot \log I)$$
$$m = -(19 + 2.5 \cdot \log 1390)$$
$$m \approx -26.9$$

The apparent stellar magnitude is about –26.9.

b. Substitute 23 for m and solve for I.

$$m(I) = -(19 + 2.5 \cdot \log I)$$
$$23 = -(19 + 2.5 \cdot \log I)$$
$$-23 = 19 + 2.5 \cdot \log I$$
$$-42 = 2.5 \cdot \log I$$
$$-16.8 = \log I$$
$$10^{-16.8} = I$$
$$1.58 \times 10^{-17} \approx I$$

The intensity is about 1.58×10^{-17} W/m^2.

57. Consider an exponential growth function $P(t) = P_0 e^{kt}$. Suppose that at time T, $P(T) = 2P_0$. Solve for T:

$$2P_0 = P_0 e^{kT}$$
$$2 = e^{kT}$$
$$\ln 2 = \ln e^{kT}$$
$$\ln 2 = kT$$
$$\frac{\ln 2}{k} = T$$

59. Consider an exponential growth function $P(t) = P_0 e^{kt}$. Suppose that at time $T = 5.32$, $P(T) = 2P_0$. Solve for k:

$$2P_0 = P_0 e^{kT}$$
$$2 = e^{k5.32}$$
$$\ln 2 = \ln e^{5.32k}$$
$$\ln 2 = 5.32k$$
$$\frac{\ln 2}{5.32} = k$$
$$k \approx 0.1303$$

The growth rate is about 13.03%.

Chapter 12 Review

1. True

2. True

3. True

4. False; the product rule states $\ln(ab) = \ln a + \ln b$.

5. False; log, which has base 10, is not the same as ln, which has base e. In addition, the power rule states $\log x^a = a \log x$.

6. True; this is the quotient rule for logarithms.

7. False; the domain of $f(x) = 3^x$ is all real numbers.

8. False; the domain of $g(x) = \log_2 x$ is $(0, \infty)$.

9. True; if $F(-2) = F(5)$, then the function has the same output for two different inputs, and therefore is not one-to-one.

10. False; the function g is one-to-one if it passes the horizontal-line test.

11. $(f \circ g)(x) = f(g(x)) = f(2x-3)$

$$= (2x-3)^2 + 1$$
$$= 4x^2 - 12x + 9 + 1$$
$$= 4x^2 - 12x + 10$$

$$(g \circ f)(x) = g(f(x)) = g(x^2 + 1)$$
$$= 2(x^2 + 1) - 3$$
$$= 2x^2 + 2 - 3$$
$$= 2x^2 - 1$$

12. $h(x) = \sqrt{3-x}$

We have $3 - x$ and take the square root of the expression, so the two most obvious functions are $f(x) = \sqrt{x}$ and $g(x) = 3 - x$.

13. $f(x) = 4 - x^2$

The graph of this function is a parabola that opens down. Thus, there are many horizontal lines that cross the graph more than once. In particular, the line $y = -4$ crosses the graph more than once. The function is not one-to-one.

14.

Replace $f(x)$ by y:	$y = x - 10$
Interchange variables:	$x = y - 10$
Solve for y:	$x + 10 = y$
Replace y by $f^{-1}(x)$:	$f^{-1}(x) = x + 10$

15.

$y = \dfrac{3x+1}{2}$	Replace $g(x)$.
$x = \dfrac{3y+1}{2}$	Interchange variables.
$\dfrac{2x-1}{3} = y$	Solve for y.
$g^{-1}(x) = \dfrac{2x-1}{3}$	Replace y.

16.

$y = 27x^3$	Replace $f(x)$.
$x = 27y^3$	Interchange variables.
$\dfrac{\sqrt[3]{x}}{3} = y$	Solve for y.
$f^{-1}(x) = \dfrac{\sqrt[3]{x}}{3}$	Replace y.

17. Graph $f(x) = 3^x + 1$

We compute some function values, thinking of y as $f(x)$, and keep the results in a table.

$$f(-2) = 3^{-2} + 1 = \frac{1}{9} + 1 = \frac{10}{9}$$
$$f(-1) = 3^{-1} + 1 = \frac{1}{3} + 1 = \frac{4}{3}$$
$$f(0) = 3^0 + 1 = 2$$
$$f(1) = 3^1 + 1 = 4$$
$$f(2) = 3^2 + 1 = 10$$

x	y, or $f(x)$
-2	$\frac{10}{9}$
-1	$\frac{4}{3}$
0	2
1	4
2	10

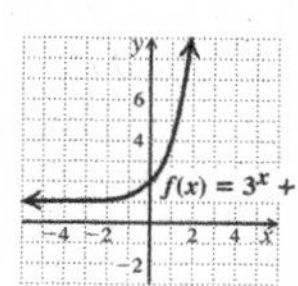

18. Graph $x = \left(\frac{1}{4}\right)^y$.

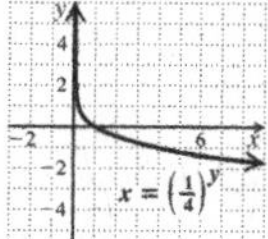

19. Graph: $f(x) = \log_5 x$

Think of $f(x)$ as y. Then $y = \log_5 x$ is equivalent to $5^y = x$. We find ordered pairs by choosing values for y and computing the corresponding x-values. Then we plot the points and connect them with a smooth curve.

For $y = 0$, $x = 5^0 = 1$
For $y = 1$, $x = 5^1 = 5$
For $y = 2$, $x = 5^2 = 25$
For $y = -1$, $x = 5^{-1} = \frac{1}{5}$
For $y = -2$, $x = 5^{-2} = \frac{1}{25}$

x, or 5^y	y
1	0
5	1
25	2
$\frac{1}{5}$	-1
$\frac{1}{25}$	-2

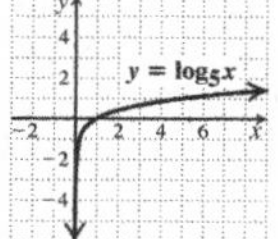

20. $\log_9 81 = \log_9 9^2 = 2$

21. $\log_3 \frac{1}{9} = \log_3 3^{-2} = -2\log_3 3 = -2$

22. $\log_2 2^{11} = 11$

23. $\log_{16} 4 = \log_{16} 16^{1/2} = \frac{1}{2}\log_{16} 16 = \frac{1}{2}$

24. $2^{-3} = \frac{1}{8}$ is equivalent to $-3 = \log_2 \frac{1}{8}$.

25. $25^{1/2} = 5$ is equivalent to $\frac{1}{2} = \log_{25} 5$.

26. $\log_4 16 = x$ is equivalent to $16 = 4^x$.

27. $\log_8 1 = 0$ is equivlant to $1 = 8^0$.

28. $\log_a x^4y^2z^3 = \log_a x^4 + \log_a y^2 + \log_a z^3$
$= 4\log_a x + 2\log_a y + 3\log_a z$

29. $\log_a \frac{x^5}{yz^2} = \log_a x^5 - \log_a y - \log_a z^2$
$= 5\log_a x - \log_a y - 2\log_a z$

30. $\log \sqrt[4]{\frac{z^2}{x^3y}} = \log\left(\frac{z^2}{x^3y}\right)^{1/4} = \frac{1}{4}\log\frac{z^2}{x^3y}$
$= \frac{1}{4}\left(\log z^2 - \log x^3 - \log y\right)$
$= \frac{1}{4}(2\log z - 3\log x - \log y)$

31. $\log_a 5 + \log_a 8 = \log_a (5 \cdot 8) = \log_a 40$

32. $\log_a 48 - \log_a 12 = \log_a \frac{48}{12} = \log_a 4$

33. $\frac{1}{2}\log a - \log b - 2\log c = \log a^{1/2} - \log b - \log c^2$
$= \log \frac{a^{1/2}}{bc^2}$

34. $\frac{1}{3}[\log_a x - 2\log_a y] = \frac{1}{3}\left[\log_a x - \log_a y^2\right]$
$= \frac{1}{3}\left[\log_a \frac{x}{y^2}\right] = \log_a \left(\frac{x}{y^2}\right)^{1/3}$
$= \log_a \sqrt[3]{\frac{x}{y^2}}$

35. $\log_m m = 1$

36. $\log_m 1 = 0$

37. $\log_m m^{17} = 17\log_m m = 17 \cdot 1 = 17$

38. $\log_a 14 = \log_a (2 \cdot 7)$
$= \log_a 2 + \log_a 7$
$= 1.8301 + 5.0999$
$= 6.93$

39. $\log_a \frac{2}{7} = \log_a 2 - \log_a 7$
$= 1.8301 - 5.0999$
$= -3.2698$

40. $\log_a 28 = \log_a (2^2 \cdot 7)$
$= \log_a 2^2 + \log_a 7$
$= 2\log_a 2 + \log_a 7$
$= 2(1.8301) + 5.0999$
$= 8.7601$

41. $\log_a 3.5 = \log_a \frac{7}{2}$
$= \log_a 7 - \log_a 2$
$= 5.0999 - 1.8301$
$= 3.2698$

42. $\log_a \sqrt{7} = \log_a 7^{1/2} = \frac{1}{2}\log_a 7 = \frac{1}{2}(5.0999) = 2.54995$

43. $\log_a \frac{1}{4} = \log_a 1 - \log_a 4$
$= 0 - \log_a 2^2$
$= -2\log_a 2$
$= -2(1.8301)$
$= -3.6602$

44. 1.8751

45. 61.5177

46. −1.2040

47. 0.3753

48. $\log_5 50 = \frac{\ln 50}{\ln 5} \approx 2.4307$

49. We will use common logarithms for the conversion. Let $a = 10$, $b = 6$, and $M = 5$ and substitute in the change-of-base formula.

$$\log_6 5 = \frac{\log 5}{\log 6} \approx \frac{0.6890}{0.7782} \approx 0.8982$$

50. Graph $f(x) = e^x - 1$.

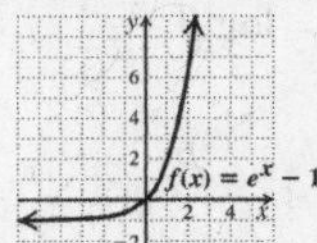

The domain is all real numbers and the range is $(-1, \infty)$.

51. Graph $g(x) = 0.6\ln x$.
We find some function values, plot points, and draw the graph.

x	$0.6\ln x$
0.5	−0.42
1	0
2	0.42
3	0.66

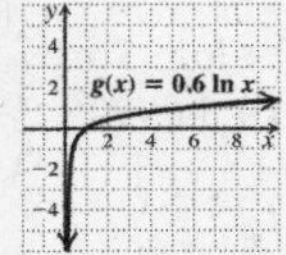

The domain is $(0, \infty)$ and the range is all real numbers.

52. $5^x = 125$
$5^x = 5^3$
$x = 3$

53. $3^{2x} = \frac{1}{9}$
$3^{2x} = 3^{-2}$
$2x = -2$
$x = -1$
The solution is −1.

54. $\log_3 x = -4$
$x = 3^{-4} = \frac{1}{81}$

55. $\log_x 16 = 4$
$x^4 = 16$
$x = \pm\sqrt[4]{16} = \pm 2$
Since x cannot be negative, we use only the positive solution. The solution is 2.

56. $\log x = -3$
$x = 10^{-3} = \frac{1}{1000}$

57. $6\ln x = 18$
$\ln x = 3$
$x = e^3 \approx 20.0855$
The solution is e^3 or approximately 20.0855.

58. $4^{2x-5} = 19$
$2x - 5 = \log_4 19$
$2x = \log_4 19 + 5$
$x = \frac{1}{2}(\log_4 19 + 5)$
$x = \frac{1}{2}\left(\frac{\log 19}{\log 4} + 5\right) \approx 3.5620$

59. $2^x = 12$
$x = \log_2 12 = \frac{\log 12}{\log 2} \approx 3.5850$

60. $e^{-0.1t} = 0.03$
$-0.1t = \ln 0.03$
$t = \frac{\ln 0.03}{-0.1} \approx 35.0656$

61. $2\ln x = -6$
$\ln x = -3$
$x = e^{-3} \approx 0.0498$

62. $\log(2x - 5) = 1$
$2x - 5 = 10^1$
$2x = 15$
$x = \frac{15}{2}$

63. $\log_4 x - \log_4 (x - 15) = 2$
$\log_4 \frac{x}{x-15} = 2$
$\frac{x}{x-15} = 4^2$
$\frac{x}{x-15} = 16$
$x = 16(x - 15)$
$x = 16x - 240$
$240 = 15x$
$16 = x$

64. $\log_3(x-4) = 2 - \log_3(x+4)$
$\log_3(x-4) + \log_3(x+4) = 2$
$\log_3(x^2 - 16) = 2$
$x^2 - 16 = 3^2$
$x^2 = 25$
$x = \pm 5$
The value −5 does not check. The solution is 5.

65. $S(t) = 82 - 18\log(t+1)$

a. $S(0) = 82 - 18\log(0+1) = 82 - 18\log 1 = 82$

b. $S(6) = 82 - 18\log(6+1) = 82 - 18\log 7 \approx 66.8$

c. Substitute 54 for $S(t)$ and solve for t.

$$54 = 82 - 18\log(t+1)$$
$$-28 = -18\log(t+1)$$
$$\frac{14}{9} = \log(t+1)$$
$$t+1 = 10^{14/9}$$
$$t = 10^{14/9} - 1 \approx 35$$

The average score will be 54 after about 35 months.

66. a.

$$900 = 1500(0.8)^t$$
$$0.6 = (0.8)^t$$
$$\ln 0.6 = \ln(0.8)^t$$
$$\ln 0.6 = t\ln 0.8$$
$$\frac{\ln 0.6}{\ln 0.8} = t$$
$$2.3 \text{ yr} \approx t$$

b. $V(0) = 1500(0.8)^0 = 1500,$ and $0.5(1500) = 750$

$$750 = 1500(0.8)^t$$
$$0.5 = (0.8)^t$$
$$\ln 0.5 = \ln(0.8)^t$$
$$\ln 0.5 = t\ln 0.8$$
$$\frac{\ln 0.5}{\ln 0.8} = t$$
$$3.1 \text{ yr} \approx t$$

67. a. We start with the exponential growth equation

$A(t) = A_0e^{kt}$, where t is the number of years after 1936.

Substituting \$400 for A_0, we have

$$A(t) = 400e^{kt}.$$

Substitute 77 for t, \$3,000,000 for $A(t)$ and solve for k.

$$A(77) = 400e^{k(77)}$$
$$3{,}000{,}000 = 400e^{77k}$$
$$\frac{3{,}000{,}000}{400} = e^{77k}$$
$$\ln\frac{3{,}000{,}000}{400} = \ln e^{77k}$$
$$\ln\frac{3{,}000{,}000}{400} = 77k$$
$$\frac{\ln\frac{3{,}000{,}000}{400}}{77} = k$$
$$0.116 \approx k$$

Thus, the exponential growth function is

$A(t) = 400e^{0.116t}$, where t is the number of years after 1936.

b. In 2020, $t = 2020 - 1936 = 84$

$A(84) = 400e^{0.116(84)} \approx 6{,}800{,}000$

The value of the nickel in 2020 will be about \$6,800,000.

c. Substitute \$10,000,000 for $A(t)$ and solve for t.

$$10{,}000{,}000 = 400e^{0.116t}$$
$$\frac{10{,}000{,}000}{400} = e^{0.116t}$$
$$\ln\frac{10{,}000{,}000}{400} = \ln e^{0.116t}$$
$$\ln\frac{10{,}000{,}000}{400} = 0.116t$$
$$\frac{\ln\frac{10{,}000{,}000}{400}}{0.116} = t$$
$$87 \approx t$$

The nickel's value will reach \$10,000,000 about 87 years after 1936, or in 2023.

d. Substitute 2(\$400), or \$800 for $A(t)$ and solve for t.

$$800 = 400e^{0.116t}$$
$$2 = e^{0.116t}$$
$$\ln 2 = \ln e^{0.116t}$$
$$\ln 2 = 0.116t$$
$$\frac{\ln 2}{0.116} = t$$
$$6.0 \approx t$$

The doubling time is about 6.0 years.

68. a. We start with the exponential decay equation

$C(t) = C_0e^{-kt}$, where t is the number of years after 1980.

Substituting 22 for C_0, and 6.1% or 0.061 for k, we have the exponential function

$C(t) = 22e^{-0.061t}$, where t is the number of years after 1980.

b. In 2015, $t = 2015 - 1980 = 35$.

$C(35) = 22e^{-0.061(35)} \approx \$2.60$

c.

$$1 = 22e^{-0.061t}$$
$$\frac{1}{22} = e^{-0.061t}$$
$$\ln\frac{1}{22} = \ln e^{-0.061t}$$
$$\ln\frac{1}{22} = -0.061t$$
$$\frac{\ln\frac{1}{22}}{-0.061} = t$$
$$51 \approx t$$

Installation cost per watt will be \$1 about 51 years after 1980, or in 2031.

69. The doubling time of the initial investment P_0 would be $2P_0$ and $t = 6$ years. Substitute this information into the exponential growth formula and solve for k.

$$2P_0 = P_0e^{k(6)}$$
$$2 = e^{6k}$$
$$\ln 2 = \ln e^{6k}$$
$$\ln 2 = 6k$$
$$\frac{\ln 2}{6} = k$$
$$0.11553 \approx k$$

The rate is 11.553% per year.

70. $2P_0 = 2\cdot 7600,\ k = 4.2\% = 0.042,\ P_0 = 7600$
Substitute into the exponential growth formula and solve for t.

$$\begin{aligned}2\cdot 7600 &= 7600e^{0.042t}\\ 2 &= e^{0.042t}\\ \ln 2 &= \ln e^{0.042t}\\ \ln 2 &= 0.042t\\ \frac{\ln 2}{0.042} &= t\\ 16.5 &\approx t\end{aligned}$$

$7600 will double in about 16.5 years.

71. We will use the function $P(t) = P_0e^{-0.00012t}$. If the skull had lost 34% of its carbon-14 from the initial amount P_0, then $66\%(P_0)$ is the amount present. To find the age t of the skull, we substitute $66\%(P_0)$, or $0.66P_0$ for $P(t)$ in the function above and solve for t.

$$\begin{aligned}0.66P_0 &= P_0e^{-0.00012t}\\ 0.66 &= e^{-0.00012t}\\ \ln 0.66 &= \ln e^{-0.00012t}\\ \ln 0.66 &= -0.00012t\\ \frac{\ln 0.66}{-0.00012} &= t\\ 3463 &\approx t\end{aligned}$$

The skull is about 3463 yr old.

72. $\text{pH} = -\log[H^+] = -\log(7.9\times 10^{-6}) = 5.1$
The pH of coffee is 5.1.

73. The function $P(t) = P_0e^{-kt}$, $k > 0$, can be used to model decay. We substitute $\frac{1}{2}P_0$ for $P(t)$ and 24,360 for t and solve for the decay rate k.

$$\begin{aligned}\frac{1}{2}P_0 &= P_0e^{-k\cdot 24,360}\\ \frac{1}{2} &= e^{-24,360k}\\ \ln\frac{1}{2} &= \ln e^{-24,360k}\\ -\ln 2 &= -24,360k\\ \frac{\ln 2}{24,360} &= k\\ 0.00002845431776 &\approx k\end{aligned}$$

So $P(t) = P_0e^{-0.00002845431776t}$

90% consumed = 10% remains

$$\begin{aligned}0.10P_0 &= P_0e^{-0.00002845431776t}\\ 0.10 &= e^{-0.00002845431776t}\\ \ln 0.10 &= \ln e^{-0.00002845431776t}\\ \ln 0.10 &= -0.00002845431776t\\ \frac{\ln 0.10}{-0.00002845431776} &= t\\ 80,922 &\approx t\end{aligned}$$

A fuel rod of Pu-239 to will lose 90% of its radioactivity in about 80,922 years, or about 80,792 years rounding $k = 0.0000285$.

74. $$\begin{aligned}L &= 10\cdot\log\frac{I}{I_0}\\ &= 10\cdot\log\frac{2.5\times 10^{-1}}{10^{-12}}\\ &= 10\log(2.5\times 10^{11})\\ &\approx 114\end{aligned}$$

The sound is about 114 dB.

75. *Writing Exercise.* Negative numbers do not have logarithms because logarithm bases are positive, and there is no exponent to which a positive number can be raised to yield a negative number.

76. *Writing Exercise.* If $f(x) = e^x$, then to find the inverse function, we let $y = e^x$ and interchange x and y: $x = e^y$. If $x = e^y$, then $\log_e x = y$ by the definition of logarithms. Since $\log_e x = \ln x$, we have $y = \ln x$ or $f^{-1}(x) = \ln x$. Thus, $g(x) = \ln x$ is the inverse of $f(x) = e^x$. Another approach is to find $(f\circ g)(x)$ and $(g\circ f)(x)$:

$$(f\circ g)(x) = e^{\ln x} = x, \text{ and}$$
$$(g\circ f)(x) = \ln e^x = x.$$

Thus, g and f are inverse functions.

77. $$\begin{aligned}\ln(\ln x) &= 3\\ \ln x &= e^3\\ x &= e^{e^3}\end{aligned}$$

78. $2^{x^2+4x} = \frac{1}{8}$ can be written as $2^{x^2+4x} = 2^{-3}$.
So the exponents must be equal.

$$\begin{aligned}x^2 + 4x &= -3\\ x^2 + 4x + 3 &= 0\\ (x+3)(x+1) &= 0\end{aligned}$$
$$x + 3 = 0 \quad \textit{or} \quad x + 1 = 0$$
$$x = -3 \quad \textit{or} \quad x = -1$$

The solutions are –3 and –1.

79. Solve the system:

$$\begin{aligned}5^{x+y} &= 25, \quad (1)\\ 2^{2x-y} &= 64 \quad (2)\end{aligned} \quad \text{or} \quad \begin{aligned}5^{x+y} &= 5^2, \quad (1)\\ 2^{2x-y} &= 2^6 \quad (2)\end{aligned}$$

From the exponents we have:

$$\begin{aligned}x + y &= 2, \quad (3)\\ 2x - y &= 6 \quad (4)\end{aligned}$$

We use elimination.

$$\begin{aligned}x + y &= 2\\ 2x - y &= 6\\ \hline 3x &= 8\\ x &= \frac{8}{3}\end{aligned}$$

From Equation (3) we solve for y.

$$y = 2 - x = 2 - \frac{8}{3} = -\frac{2}{3}$$

The solution is $\left(\frac{8}{3}, -\frac{2}{3}\right)$.

Chapter 12 Test

1. $(f \circ g)(x) = f(g(x)) = f(2x+1)$
$= (2x+1) + (2x+1)^2$
$= 2x + 1 + 4x^2 + 4x + 1$
$= 4x^2 + 6x + 2$

$(g \circ f)(x) = g(f(x)) = g(x + x^2)$
$= 2(x + x^2) + 1$
$= 2x + 2x^2 + 1$
$= 2x^2 + 2x + 1$

2. $h(x) = \frac{1}{2x^2+1}$

This is 1 divided by $2x^2 + 1$, so two functions that can be used are $f(x) = \frac{1}{x}$ and $g(x) = 2x^2 + 1$.

3. $f(x) = x^2 + 3$

Observe that the graph of this function is a parabola that opens up. Thus, there are many horizontal lines that cross the graph more than once. In particular, the line $y = 4$ crosses the graph more than once. The function is not one-to-one.

4.

$y = 3x + 4$	Replace $f(x)$.
$x = 3y + 4$	Interchange variables.
$\frac{x-4}{3} = y$	Solve for y.
$f^{-1}(x) = \frac{x-4}{3}$	Replace y.

5.

$y = (x+1)^3$	Replace $g(x)$.
$x = (y+1)^3$	Interchange variables.
$\sqrt[3]{x} - 1 = y$	Solve for y.
$g^{-1}(x) = \sqrt[3]{x} - 1$	Replace y.

6. Graph $f(x) = 2^x - 3$.

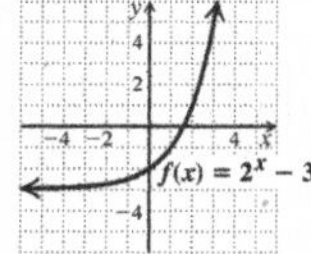

7. Graph: $f(x) = \log_7 x$

Think of $f(x)$ as y. Then $y = \log_7 x$ is equivalent to $7^y = x$. We find ordered pairs by choosing values for y and computing the corresponding x-values. Then we plot the points and connect them with a smooth curve.

For $y = 0$, $x = 7^0 = 1$
For $y = 1$, $x = 7^1 = 7$
For $y = 2$, $x = 7^2 = 49$
For $y = -1$, $x = 7^{-1} = \frac{1}{7}$
For $y = -2$, $x = 7^{-2} = \frac{1}{49}$

x, or 7^y	y
1	0
7	1
49	2
$\frac{1}{7}$	-1
$\frac{1}{49}$	-2

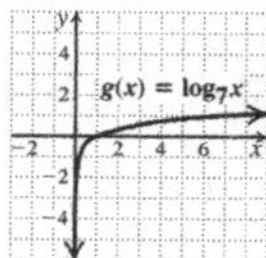

8. $\log_5 125 = \log_5 5^3 = 3\log_5 5 = 3$

9. $\log_{100} 10 = \log_{100} 100^{1/2} = \frac{1}{2}\log_{100} 100 = \frac{1}{2} \cdot 1 = \frac{1}{2}$

10. $\log_n n = 1$

11. $\log_c 1 = 0$

12. $5^{-4} = \frac{1}{625}$ is equivalent to $-4 = \log_5 \frac{1}{625}$.

13. $m = \log_2 \frac{1}{2}$ is equivalent to $2^m = \frac{1}{2}$.

14. $\log \frac{a^3 b^{1/2}}{c^2} = \log a^3 + \log b^{1/2} - \log c^2$
$= 3\log a + \frac{1}{2}\log b - 2\log c$

15. $\frac{1}{3}\log_a x + 2\log_a z = \log_a x^{1/3} + \log_a z^2$
$= \log_a \sqrt[3]{x} + \log_a z^2$
$= \log_a (z^2 \sqrt[3]{x})$

16. $\log_a 14 = \log_a (2 \cdot 7)$
$= \log_a 2 + \log_a 7$
$= 0.301 + 0.845$
$= 1.146$

17. $\log_a 3 = \log_a \frac{6}{2}$
$= \log_a 6 - \log_a 2$
$= 0.778 - 0.301$
$= 0.477$

18. $\log_a 16 = \log_a 2^4 = 4\log_a 2 = 4(0.301) = 1.204$

19. 1.3979

20. 0.1585

21. –0.9163

22. 121.5104

23. We will use common logarithms for the conversion. Let $a = 10$, $b = 3$, and $M = 14$ and substitute in the change-of-base formula.

$$\log_3 14 = \frac{\log 14}{\log 3} \approx \frac{1.1461}{0.4771} \approx 2.4022$$

24. Graph $f(x) = e^x + 3$.

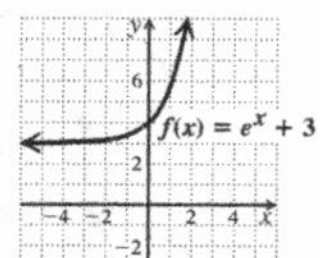

The domain is all real numbers and the range is $(3, \infty)$.

25. Graph $g(x)=\ln(x-4)$.
We find some function values, plot points, and draw the graph.

x	$\ln(x-4)$
4.5	−0.69
5	0
6	0.69
7	1.10

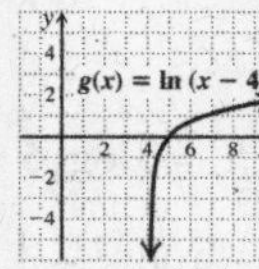

The domain is $(4, \infty)$ and the range is all real numbers.

26. $2^x=\frac{1}{32}$
$2^x=2^{-5}$
$x=-5$

27. $\log_4 x=\frac{1}{2}$
$x=4^{1/2}$
$x=2$

28. $\log x=-2$
$x=10^{-2}=\frac{1}{100}$

29. $7^x=1.2$
$x=\log_7 1.2=\frac{\log 1.2}{\log 7}\approx 0.0937$

30. $\log(x-3)+\log(x+1)=\log 5$
$\log(x-3)(x+1)=\log 5$
$\log(x^2-2x-3)=\log 5$
$x^2-2x-3=5$
$x^2-2x-8=0$
$(x-4)(x+2)=0$
$x=4 \text{ or } x=-2$
The value −2 does not check. The solution is 4.

31. $R=0.37\ln P+0.05$

a. Substitute 384 for *P* and solve for *R*.
$R=0.37\ln 384+0.05\approx 2.3$
The average walking speed is about 2.3 ft/sec.

b. Substitute 3.0 for *R* and solve for *P*.
$3.0=0.37\ln P+0.05$
$2.95=0.37\ln P$
$\frac{2.95}{0.37}=\ln P$
$P=e^{2.95/0.37}\approx 2900$
The population is approximately 2,900,000.

32. a. $P(t)=186e^{0.026t}$, where $P(t)$ is in millions and *t* is the number of years after 2016.

b. In 2020, $t=2020-2016=4$. Find $P(4)$.
$P(4)=186e^{0.026(4)}=186e^{0.104}\approx 206$ million
In 2050, $t=2050-2016=34$. Find $P(34)$.
$P(34)=186e^{0.026(34)}=186e^{0.884}\approx 450$ million

c. Substitute 500 for $P(t)$ and solve for *t*.
$500=186e^{0.026t}$
$\frac{500}{186}=e^{0.026t}$
$\ln\frac{500}{186}=\ln e^{0.026t}$
$\ln\frac{500}{186}=0.026t$
$\frac{\ln\frac{500}{186}}{0.026}=t$
$38\approx t$
The population will reach 500 million about 38 yr after 2016, or in 2054.

d. $P(0)=186e^{0.026(0)}=186$, and $2(186)=372$
$372=186e^{0.026t}$
$2=e^{0.026t}$
$\ln 2=0.026t$
$\frac{\ln 2}{0.026}=t$
$26.7 \text{ yr}\approx t$

33. a. We start with the exponential growth equation
$C(t)=C_0e^{kt}$, where *t* is the number of years after 2005-2006.
Substituting \$35,106 for C_0, we have
$C(t)=35{,}106e^{kt}$.
Substitute 10 for *t*, \$43,921 for $C(t)$ and solve for *k*.
$43{,}921=35{,}106e^{k(10)}$
$\frac{43{,}921}{35{,}106}=e^{10k}$
$\ln\frac{43{,}921}{35{,}106}=\ln e^{10k}$
$\ln\frac{43{,}921}{35{,}106}=10k$
$\frac{\ln\frac{43{,}921}{35{,}106}}{10}=k$
$0.022\approx k$
Thus, the exponential growth function is $C(t)=35{,}106e^{0.022t}$, where *t* is the number of years after 2005-2006.

b. In 2019-2020, $t=2019-2005=14$
$C(14)=35{,}106e^{0.022(14)}\approx 47{,}769$
The cost in 2019-2020 will be about \$47,769.

c. Substitute \$60,000 for $C(t)$ and solve for t.

$$\begin{aligned} 60{,}000 &= 35{,}106e^{0.022t} \\ \frac{60{,}000}{35{,}106} &= e^{0.022t} \\ \ln\frac{60{,}000}{35{,}106} &= \ln e^{0.022t} \\ \ln\frac{60{,}000}{35{,}106} &= 0.022t \\ \frac{\ln\frac{60{,}000}{35{,}106}}{0.022} &= t \\ 24 &\approx t \end{aligned}$$

Cost for college will be \$60,000 about 24 years after 2005-2006, or in 2029-2030.

34. The doubling time of the initial investment P_0 would be $2P_0$ and $t = 16$ years. Substitute this information into the exponential growth formula and solve for k.

$$\begin{aligned} 2P_0 &= P_0e^{k(16)} \\ 2 &= e^{16k} \\ \ln 2 &= \ln e^{16k} \\ \ln 2 &= 16k \\ \frac{\ln 2}{16} &= k \\ 0.043 &\approx k \end{aligned}$$

The rate is 4.3% per year.

35.
$$\begin{aligned} \text{pH} &= -\log\left[H^+\right] \\ &= -\log\left(1.0\times10^{-7}\right) \\ &= 7.0 \end{aligned}$$

The pH of water is 7.0.

36.
$$\begin{aligned} \log_5|2x-7| &= 4 \\ |2x-7| &= 5^4 \\ |2x-7| &= 625 \end{aligned}$$

$$\begin{aligned} 2x-7 &= -625 \quad \textit{or} \quad 2x-7 = 625 \\ x &= -309 \quad \textit{or} \quad x = 316 \end{aligned}$$

37.
$$\begin{aligned} &\log_a \frac{\sqrt[3]{x^2z}}{\sqrt[3]{y^2z^{-1}}} \\ &= \log_a \sqrt[3]{\frac{x^2z}{y^2z^{-1}}} \\ &= \log_a \sqrt[3]{\frac{x^2z^2}{y^2}} \\ &= \log_a \left(\frac{x^2z^2}{y^2}\right)^{1/3} \\ &= \frac{1}{3}\log_a \frac{x^2z^2}{y^2} \\ &= \frac{1}{3}\left(\log_a x^2 + \log_a z^2 - \log_a y^2\right) \\ &= \frac{1}{3}(2\log_a x + 2\log_a z - 2\log_a y) \\ &= \frac{1}{3}(2\cdot 2 + 2\cdot 4 - 2\cdot 3) \\ &= 2 \end{aligned}$$

Chapter 13

Conic Sections

Exercise Set 13.1

1. Parabolas and circles are examples of *conic sections.*

3. A parabola with a *horizontal* axis of symmetry opens to the right or to the left.

5. In the equation of a circle, the point (*h*, *k*) represents the *center* of the circle.

7. $(x-2)^2+(y+5)^2=9$, or $(x-2)^2+[y-(-5)]^2=3^2$, is the equation of a circle with center $(2,-5)$ and radius 3, so choice (f) is correct.

9. $y=(x-2)^2-5$ is the equation of a parabola with vertex $(2,-5)$ that opens upward, so choice (c) is correct.

11. $x=(y-2)^2-5$ is the equation of a parabola with vertex $(-5, 2)$ that opens to the right, so choice (d) is correct.

13. $y=-x^2$

This is equivalent to $y=-(x-0)^2+0$. The vertex is (0, 0). We choose some *x*-values on both sides of the vertex and compute the corresponding values of *y*. The graph opens down, because the coefficient of x^2, –1, is negative.

x	y
0	0
1	–1
2	–4
–1	–1
–2	–4

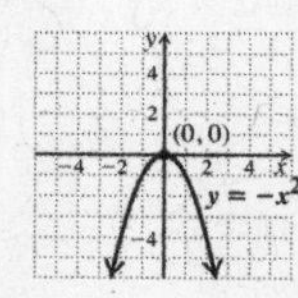

15. $y=-x^2+4x-5$

We can find the vertex by computing the first coordinate, $x=-b/2a$, and then substituting to find the second coordinate:

$$x=-\frac{b}{2a}=-\frac{4}{2(-1)}=2$$

$$y=-x^2+4x-5=-(2)^2+4(2)-5=-1$$

The vertex is $(2,-1)$.

We choose some *x*-values and compute the corresponding values for *y*. The graph opens downward because the coefficient of x^2, –1, is negative.

x	y
2	–1
3	–2
4	–5
1	–2
0	–5

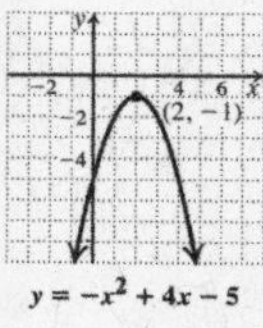

17. $x=y^2-4y+2$

We find the vertex by completing the square.

$$x=(y^2-4y+4)+2-4$$

$$x=(y-2)^2-2$$

The vertex is $(-2,2)$.

To find ordered pairs, we choose values for *y* and compute the corresponding values of *x*. The graph opens to the right, because the coefficient of y^2, 1, is positive.

x	y
7	–1
2	0
–1	1
–2	2
–1	3

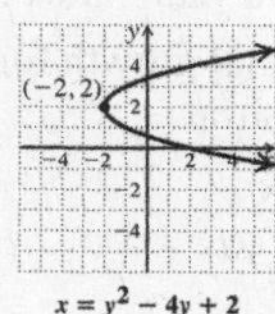

19. $x=y^2+3$

$$x=(y-0)^2+3$$

The vertex is (3, 0).

To find the ordered pairs, we choose *y*-values and compute the corresponding values for *x*. The graph opens to the right, because the coefficient of y^2, 1, is positive.

x	y
3	0
4	1
7	2
4	–1
7	–2

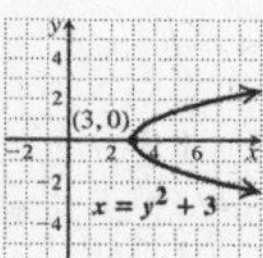

21. $x=2y^2$

$$x=2(y-0)^2+0$$

The vertex is (0, 0).

We choose *y*-values and compute the corresponding values for *x*. The graph opens to the right, because the coefficient of y^2, 2, is positive.

x	y
0	0
2	1
2	–1
8	2
8	–2

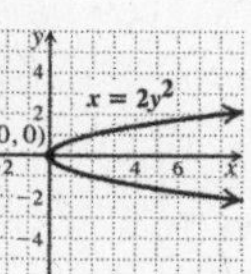

23. $x=-y^2-4y$

We find the vertex by computing the second coordinate, $y=-b/2a$, and then substituting to find the first coordinate:

$$y=-\frac{b}{2a}=-\frac{-4}{2(-1)}=-2$$
$$x=-y^2-4y=-(-2)^2-4(-2)=4$$

The vertex is $(4,-2)$.

We choose y-values and compute the corresponding values for x. The graph opens to the left, because the coefficient of y^2, -1, is negative.

x	y
4	−2
−5	1
0	0
3	−1
3	−3

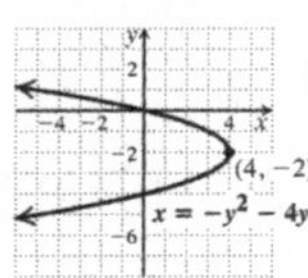

25. $y=x^2-2x+1$

$y=(x-1)^2+0$

The vertex is (1, 0).

We choose x-values and compute the corresponding values for y. The graph opens upward, because the coefficient of x^2, 1, is positive.

x	y
1	0
0	1
−1	4
2	1
3	4

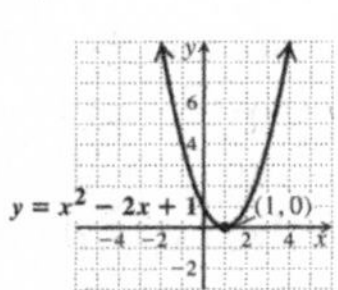

27. $x=-\frac{1}{2}y^2$

$x=-\frac{1}{2}(y-0)^2+0$

The vertex is (0, 0).

We choose y-values and compute the corresponding values for x. The graph opens to the left, because the coefficient of y^2, $-\frac{1}{2}$, is negative.

x	y
0	0
−2	2
−8	4
−2	−2
−8	−4

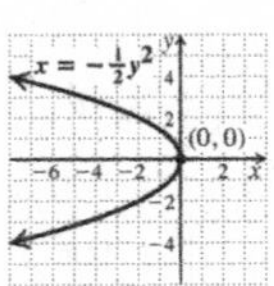

29. $x=-y^2+2y-1$

We find the vertex by computing the second coordinate, $y=-b/2a$, and then substituting to find the first coordinate.

$$y=-\frac{b}{2a}=-\frac{2}{2(-1)}=1$$
$$x=-y^2+2y-1=-(1)^2+2(1)-1=0$$

The vertex is (0, 1).

We choose y-values and compute the corresponding values for x. The graph opens to the left, because the coefficient of y^2, -1, is negative.

x	y
−4	3
−1	2
0	1
−1	0
−4	−1

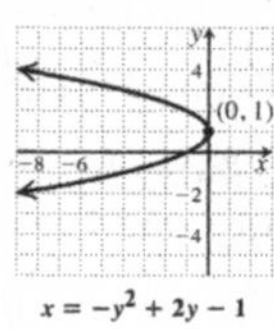

31. $x=-2y^2-4y+1$

We find the vertex by completing the square.

$$x=-2(y^2+2y)+1$$
$$x=-2(y^2+2y+1)+1+2$$
$$x=-2(y+1)^2+3$$

The vertex is $(3,-1)$.

We choose y-values and compute the corresponding values for x. The graph opens to the left, because the coefficient of y^2, -2, is negative.

x	y
3	−1
1	−2
−5	−3
1	0
−5	1

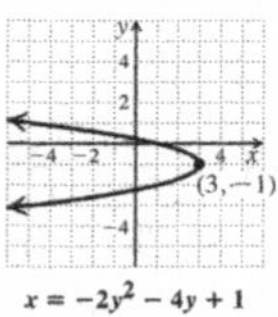

33. $(x-h)^2+(y-k)^2=r^2$ Standard form

$(x-0)^2+(y-0)^2=8^2$ Substituting

$x^2+y^2=64$ Simplifying

35. $(x-h)^2+(y-k)^2=r^2$ Standard form

$(x-7)^2+(y-3)^2=(\sqrt{6})^2$ Substituting

$(x-7)^2+(y-3)^2=6$

37. $(x-h)^2+(y-k)^2=r^2$

$[x-(-4)]^2+(y-3)^2=(3\sqrt{2})^2$

$(x+4)^2+(y-3)^2=18$

39. $(x-h)^2+(y-k)^2=r^2$

$[x-(-5)]^2+[y-(-8)]^2=(10\sqrt{3})^2$

$(x+5)^2+(y+8)^2=300$

41. Since the center is (0, 0), we have

$(x-0)^2+(y-0)^2=r^2$ or $x^2+y^2=r^2$

The circle passes through (−3, 4). We find r^2 by substituting −3 for x and 4 for y.

$$(-3)^2+4^2=r^2$$
$$9+16=r^2$$
$$25=r^2$$

Then $x^2+y^2=25$ is an equation of the circle.

43. Since the center is (–4, 1), we have

$$[x-(-4)]^2+(y-1)^2=r^2, \text{ or}$$
$$(x+4)^2+(y-1)^2=r^2.$$

The circle passes through (–2, 5). We find r^2 by substituting –2 for x and 5 for y.

$$(-2+4)^2+(5-1)^2=r^2$$
$$4+16=r^2$$
$$20=r^2$$

Then $(x+4)^2+(y-1)^2=20$ is an equation of the circle.

45. We write standard form.

$$(x-0)^2+(y-0)^2=1^2$$

The center is (0, 0), and the radius is 1.

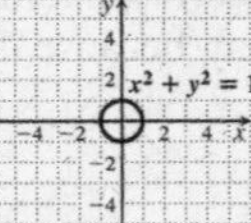

47.
$$(x+1)^2+(y+3)^2=49$$
$$[x-(-1)]^2+[y-(-3)]^2=7^2 \quad \text{Standard form}$$

The center is (–1, –3), and the radius is 7.

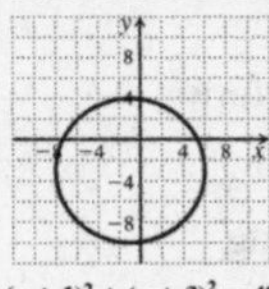
$(x+1)^2+(y+3)^2=49$

49.
$$(x-4)^2+(y+3)^2=10$$
$$(x-4)^2+[y-(-3)]^2=\left(\sqrt{10}\right)^2$$

The center is (4, –3), and the radius is $\sqrt{10}$.

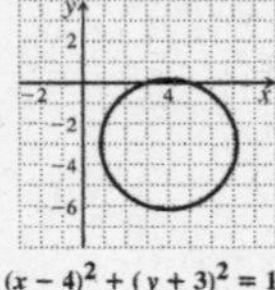
$(x-4)^2+(y+3)^2=10$

51.
$$x^2+y^2=8$$
$$(x-0)^2+(y-0)^2=\left(\sqrt{8}\right)^2 \quad \text{Standard form}$$

The center is (0, 0), and the radius is $\sqrt{8}$, or $2\sqrt{2}$.

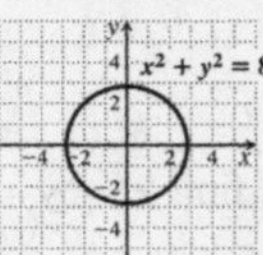

53.
$$(x-5)^2+y^2=\frac{1}{4}$$
$$(x-5)^2+(y-0)^2=\left(\frac{1}{2}\right)^2 \quad \text{Standard form}$$

The center is (5, 0), and the radius is $\frac{1}{2}$.

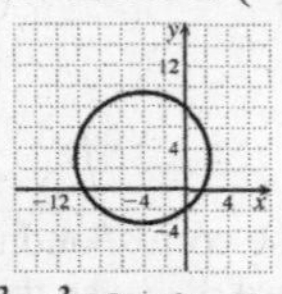

55.
$$x^2+y^2+8x-6y-15=0$$
$$x^2+8x+y^2-6y=15$$
$$(x^2+8x+16)+(y^2-6y+9)=15+16+9 \quad \text{Completing the square twice}$$
$$(x+4)^2+(y-3)^2=40$$
$$[x-(-4)]^2+(y-3)^2=\left(\sqrt{40}\right)^2 \quad \text{Standard form}$$

The center is (–4, 3), and the radius is $\sqrt{40}$, or $2\sqrt{10}$.

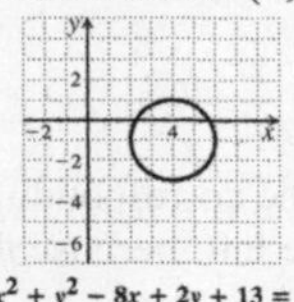
$x^2+y^2+8x-6y-15=0$

57.
$$x^2+y^2-8x+2y+13=0$$
$$x^2-8x+y^2+2y=-13$$
$$(x^2-8x+16)+(y^2+2y+1)=-13+16+1 \quad \text{Completing the square twice}$$
$$(x-4)^2+(y+1)^2=4$$
$$(x-4)^2+[y-(-1)]^2=2^2 \quad \text{Standard form}$$

The center is (4, –1), and the radius is 2.

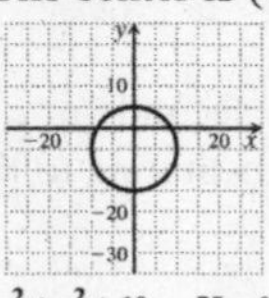
$x^2+y^2-8x+2y+13=0$

59.
$$x^2+y^2+10y-75=0$$
$$x^2+y^2+10y=75$$
$$x^2+(y^2+10y+25)=75+25$$
$$(x-0)^2+(y+5)^2=100$$
$$(x-0)^2+[y-(-5)]^2=10^2$$

The center is (0, –5), and the radius is 10.

$x^2+y^2+10y-75=0$

61.
$$x^2+y^2+7x-3y-10=0$$
$$x^2+7x+y^2-3y=10$$
$$\left(x^2+7x+\frac{49}{4}\right)+\left(y^2-3y+\frac{9}{4}\right)=10+\frac{49}{4}+\frac{9}{4}$$
$$\left(x+\frac{7}{2}\right)^2+\left(y-\frac{3}{2}\right)^2=\frac{98}{4}$$
$$\left[x-\left(-\frac{7}{2}\right)\right]^2+\left(y-\frac{3}{2}\right)^2=\left(\sqrt{\frac{98}{4}}\right)^2$$

The center is $\left(-\frac{7}{2}, \frac{3}{2}\right)$, and the radius is $\sqrt{\frac{98}{4}}$, or $\frac{\sqrt{98}}{2}$, or $\frac{7\sqrt{2}}{2}$.

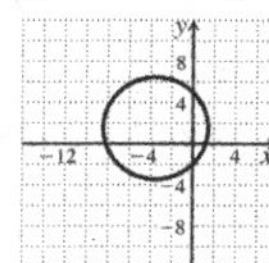

$x^2 + y^2 + 7x - 3y - 10 = 0$

63. $36x^2 + 36y^2 = 1$

$x^2 + y^2 = \frac{1}{36}$ Multiplying by $\frac{1}{36}$ on both sides

$(x-0)^2 + (y-0)^2 = \left(\frac{1}{6}\right)^2$

The center is (0, 0), and the radius is $\frac{1}{6}$.

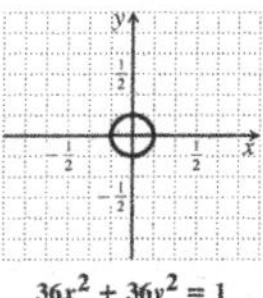

$36x^2 + 36y^2 = 1$

65. *Writing Exercise.*

67. $\sqrt[4]{48x^7y^{12}} = \sqrt[4]{16x^4y^{12} \cdot 3x^3} = 2xy^3\sqrt[4]{3x^3}$

69. $\frac{\sqrt{200x^4w^2}}{\sqrt{2w}} = \sqrt{\frac{200x^4w^2}{2w}} = \sqrt{100x^4w} = 10x^2\sqrt{w}$

71. $\sqrt{8} - 2\sqrt{2} + \sqrt{12} = \sqrt{4 \cdot 2} - 2\sqrt{2} + \sqrt{4 \cdot 3}$
$= 2\sqrt{2} - 2\sqrt{2} + 2\sqrt{3}$
$= 2\sqrt{3}$

73. *Writing Exercise.*

75. We make a drawing of the circle with center (3, –5) and tangent to the *y*-axis.

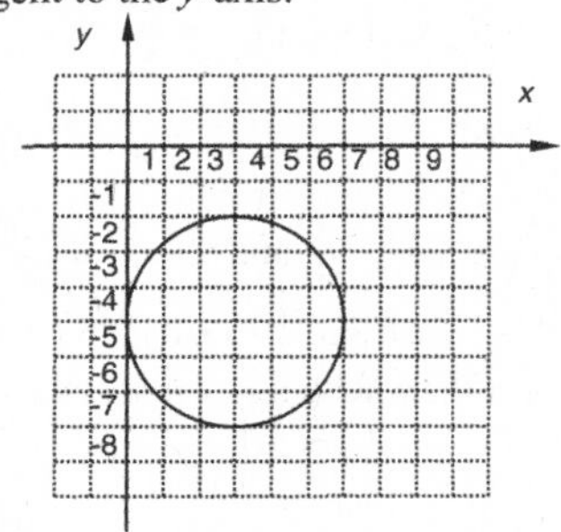

We see that the circle touches the *y*-axis at (0, –5). Hence the radius is the distance between (0, –5) and (3, –5), or $\sqrt{(3-0)^2 + [-5-(-5)]^2}$, or 3. Now we write the equation of the circle.

$(x-h)^2 + (y-k)^2 = r^2$
$(x-3)^2 + [y-(-5)]^2 = 3^2$
$(x-3)^2 + (y+5)^2 = 9$

77. First we use the midpoint formula to find the center:
$\left(\frac{7+(-1)}{2}, \frac{3+(-3)}{2}\right)$, or $\left(\frac{6}{2}, \frac{0}{2}\right)$, or (3, 0)

The length of the radius is the distance between the center (3, 0) and either endpoint of a diameter. We will use endpoint (7, 3) in the distance formula:

$r = \sqrt{(7-3)^2 + (3-0)^2} = \sqrt{25} = 5$

Now we write the equation of the circle:

$(x-h)^2 + (y-k)^2 = r^2$
$(x-3)^2 + (y-0)^2 = 5^2$
$(x-3)^2 + y^2 = 25$

79. For the outer circle, $r^2 = \frac{81}{4}$. For the inner circle, $r^2 = 16$. The area of the red zone is the difference between the areas of the outer and inner circles. Recall that the area A of a circle with radius r is given by the formula $A = \pi r^2$.

$\pi \cdot \frac{81}{4} - \pi \cdot 16 = \frac{81}{4}\pi - \frac{64}{4}\pi = \frac{17}{4}\pi$

The area of the red zone is $\frac{17}{4}\pi$ m^2, or about 13.4 m^2.

81. Superimposing a coordinate system on the snowboard as in Exercise 80, and observing that 1160/2 = 580, we know that three points on the circle are (–580, 0), (0, 23.5) and (580, 0). Let (0, *k*) represent the center of the circle. Use the fact that (0, *k*) is equidistant from (–580, 0) and (0, 23.5).

$\sqrt{(-580-0)^2 + (0-k)^2} = \sqrt{(0-0)^2 + (23.5-k)^2}$
$\sqrt{336{,}400 + k^2} = \sqrt{552.25 - 47k + k^2}$
$336{,}400 + k^2 = 552.25 - 47k + k^2$
$335{,}847.75 = -47k$
$-7145.7 \approx k$

Then to find the radius, we find the distance from the center (0, –7145.7) to any one of the three known points on the circle. We use (0, 23.5).

$r = \sqrt{(0-0)^2 + (-7145.7 - 23.5)^2} \approx 7169$ mm

83. a. When the circle is positioned on a coordinate system as shown in the text, the center lies on the *y*-axis. To find the center, we will find the point on the *y*-axis that is equidistant from (–4, 0) and (0, 2). Let (0, *y*) be this point.

$\sqrt{[0-(-4)]^2 + (y-0)^2} = \sqrt{(0-0)^2 + (y-2)^2}$
$4^2 + y^2 = 0^2 + (y-2)^2$ Squaring both sides
$16 + y^2 = y^2 - 4y + 4$
$12 = -4y$
$-3 = y$

The center of the circle is (0, –3).

b. We find the radius of the circle.

$$\begin{aligned} (x-0)^2+[y-(-3)]^2 &= r^2 && \text{Standard form} \\ x^2+(y+3)^2 &= r^2 \\ (-4)^2+(0+3)^2 &= r^2 && \text{Substituting} \\ 16+9 &= r^2 && (-4,\ 0) \text{ for } (x,\ y) \\ 25 &= r^2 \\ 5 &= r \end{aligned}$$

The radius is 5 ft.

85. We write the equation of a circle with center (0, 30.6) and radius 24.3: $x^2+(y-30.6)^2=590.49$

87. Substitute 6 for N.

$$H=\frac{D^2N}{2.5}=\frac{D^2\cdot 6}{2.5}=2.4D^2$$

Find some ordered pairs for $2.5 \le D \le 8$ and draw the graph.

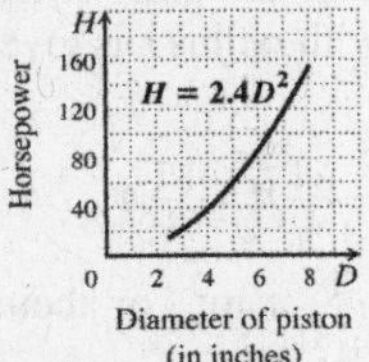

Using the graph, a horse power of 120, on the vertical axis, relates to a diameter of 7 in., on the horizontal axis.

89. *Writing Exercise.*

Exercise Set 13.2

1. Ellipse A

3. Center B

5. True

7. True

9. $\frac{x^2}{1}+\frac{y^2}{4}=1$

$\frac{x^2}{1^2}+\frac{y^2}{2^2}=1$

The x-intercepts are (1, 0) and (–1, 0), and the y-intercepts are (0, 2) and (0, –2). We plot these points and connect them with an oval-shaped curve.

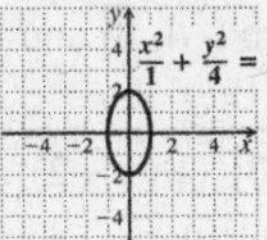

11. $\frac{x^2}{25}+\frac{y^2}{9}=1$

$\frac{x^2}{5^2}+\frac{y^2}{3^2}=1$

The x-intercepts are (5, 0) and (–5, 0), and the y-intercepts are (0, 3) and (0, –3). We plot these points and connect them with an oval-shaped curve.

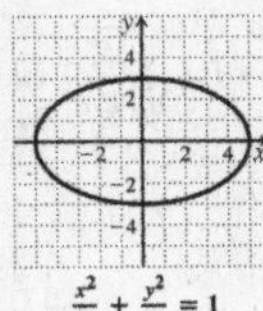

13.

$$\begin{aligned} 4x^2+9y^2 &= 36 \\ \frac{1}{36}(4x^2+9y^2) &= \frac{1}{36}(36) && \text{Multiplying by } \frac{1}{36} \\ \frac{x^2}{9}+\frac{y^2}{4} &= 1 \\ \frac{x^2}{3^2}+\frac{y^2}{2^2} &= 1 \end{aligned}$$

The x-intercepts are (–3, 0) and (3, 0), and the y-intercepts are (0, –2) and (0, 2). We plot these points and connect them with an oval-shaped curve.

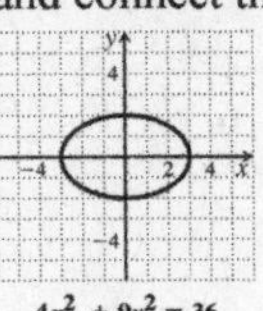

15. $16x^2+9y^2=144$

$\frac{x^2}{9}+\frac{y^2}{16}=1$ Multiplying by $\frac{1}{144}$

$\frac{x^2}{3^2}+\frac{y^2}{4^2}=1$

The x-intercepts are (3, 0) and (–3, 0), and the y-intercepts are (0, 4) and (0, –4). We plot these points and connect them with an oval-shaped curve.

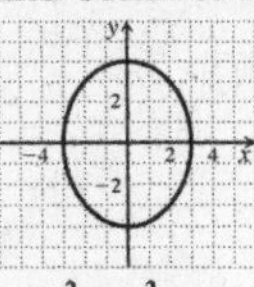

17. $2x^2+3y^2=6$

$\frac{x^2}{3}+\frac{y^2}{2}=1$ Multiplying by $\frac{1}{6}$

$\frac{x^2}{(\sqrt{3})^2}+\frac{y^2}{(\sqrt{2})^2}=1$

The x-intercepts are $(\sqrt{3},\ 0)$ and $(-\sqrt{3},\ 0)$, and the y-intercepts are $(0,\ \sqrt{2})$ and $(0, -\sqrt{2})$. We plot these points and connect them with an oval-shaped curve.

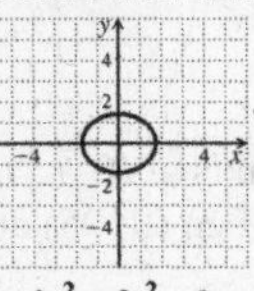

19. $5x^2+5y^2=125$

Observe that the x^2- and y^2-terms have the same coefficient. We divide both sides of the equation by 5 to obtain $x^2+y^2=25$. This is the equation of a circle with center (0, 0) and radius 5.

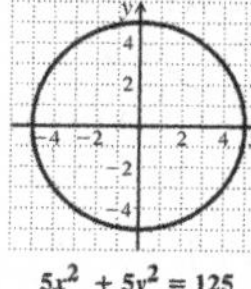

21. $3x^2 + 7y^2 - 63 = 0$

$$3x^2 + 7y^2 = 63$$

$$\frac{x^2}{21} + \frac{y^2}{9} = 1 \quad \text{Multiplying by } \frac{1}{63}$$

$$\frac{x^2}{(\sqrt{21})^2} + \frac{y^2}{3^2} = 1$$

The x-intercepts are $(\sqrt{21}, 0)$ and $(-\sqrt{21}, 0)$, or about $(4.583, 0)$ and $(-4.583, 0)$. The y-intercepts are (0, 3) and (0, –3). We plot these points and connect them with an oval-shaped curve.

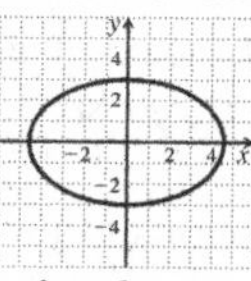

23. $16x^2 = 16 - y^2$

$$16x^2 + y^2 = 16$$

$$\frac{x^2}{1} + \frac{y^2}{16} = 1$$

The x-intercepts are (1, 0) and (–1, 0), and the y-intercepts are (0, 4) and (0, –4). We plot these points and connect them with an oval-shaped curve.

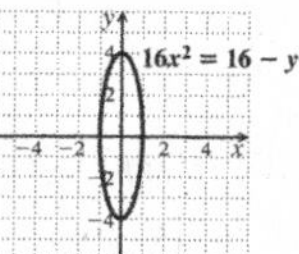

25. $16x^2 + 25y^2 = 1$

Note that $16 = \frac{1}{\frac{1}{16}}$ and $25 = \frac{1}{\frac{1}{25}}$. Thus, we can rewrite the equation:

$$\frac{x^2}{\frac{1}{16}} + \frac{y^2}{\frac{1}{25}} = 1$$

$$\frac{x^2}{\left(\frac{1}{4}\right)^2} + \frac{y^2}{\left(\frac{1}{5}\right)^2} = 1$$

The x-intercepts are $\left(\frac{1}{4}, 0\right)$ and $\left(-\frac{1}{4}, 0\right)$, and the y-intercepts are $\left(0, \frac{1}{5}\right)$ and $\left(0, -\frac{1}{5}\right)$. We plot these points and connect them with an oval-shaped curve.

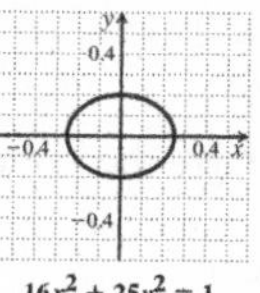

27. $\frac{(x-3)^2}{9} + \frac{(y-2)^2}{25} = 1$

$$\frac{(x-3)^2}{3^2} + \frac{(y-2)^2}{5^2} = 1$$

The center of the ellipse is (3, 2). Note that $a = 3$ and $b = 5$. We locate the center and then plot the points $(3+3, 2)$ $(3-3, 2)$, $(3, 2+5)$, and $(3, 2-5)$, or (6, 2), (0, 2), (3, 7), and (3, –3). Connect these points with an oval-shaped curve.

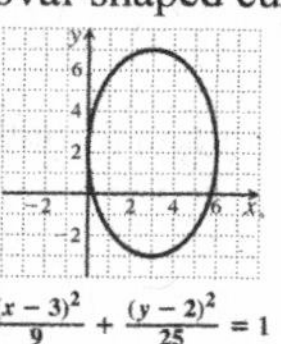

29. $\frac{(x+4)^2}{16} + \frac{(y-3)^2}{49} = 1$

$$\frac{(x-(-4))^2}{4^2} + \frac{(y-3)^2}{7^2} = 1$$

The center of the ellipse is (–4, 3). Note that $a = 4$ and $b = 7$. We locate the center and then plot the points $(-4+4, 3)$, $(-4-4, 3)$, $(-4, 3+7)$, and $(-4, 3-7)$, or (0, 3), (–8, 3), (–4, 10), and (–4, –4). Connect these points with an oval-shaped curve.

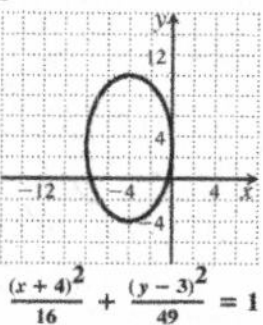

31. $12(x-1)^2 + 3(y+4)^2 = 48$

$$\frac{(x-1)^2}{4} + \frac{(y+4)^2}{16} = 1$$

$$\frac{(x-1)^2}{2^2} + \frac{(y-(-4))^2}{4^2} = 1$$

The center of the ellipse is (1, –4). Note that $a = 2$ and $b = 4$. We locate the center and then plot the points $(1+2, -4)$, $(1-2, -4)$, $(1, -4+4)$, and $(1, -4-4)$, or (3, –4), (–1, –4), (1, 0), and (1, –8). Connect these points with an oval-shaped curve.

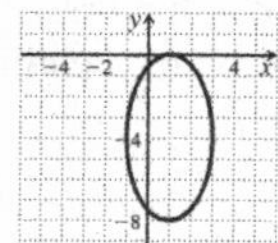

33. $4(x+3)^2+4(y+1)^2-10=90$
$$4(x+3)^2+4(y+1)^2=100$$

Observe that the x^2 - and y^2 -terms have the same coefficient. Dividing both sides by 4, we have
$$(x+3)^2+(y+1)^2=25.$$
This is the equation of a circle with center (–3, –1) and radius 5.

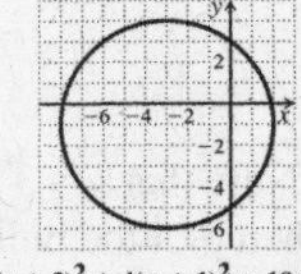

35. *Writing Exercise.*

37. $x^2-5x+3=0$
$$x=\frac{5\pm\sqrt{25-4\cdot1\cdot3}}{2\cdot1} \quad \text{Using the quadratic formula}$$
$$=\frac{5\pm\sqrt{13}}{2}$$

39. $\frac{4}{x+2}+\frac{3}{2x-1}=2$ Note $x\neq-2$ and $x\neq\frac{1}{2}$
$$4(2x-1)+3(x+2)=2(x+2)(2x-1)$$
$$8x-4+3x+6=4x^2+6x-4$$
$$4x^2-5x-6=0$$
$$(4x+3)(x-2)=0$$
$$4x+3=0 \quad \text{or} \quad x-2=0$$
$$x=-\tfrac{3}{4} \quad \text{or} \quad x=2$$

41. $x^2=11$
$$x=\pm\sqrt{11}$$

43. *Writing Exercise.*

45. Plot the given points.

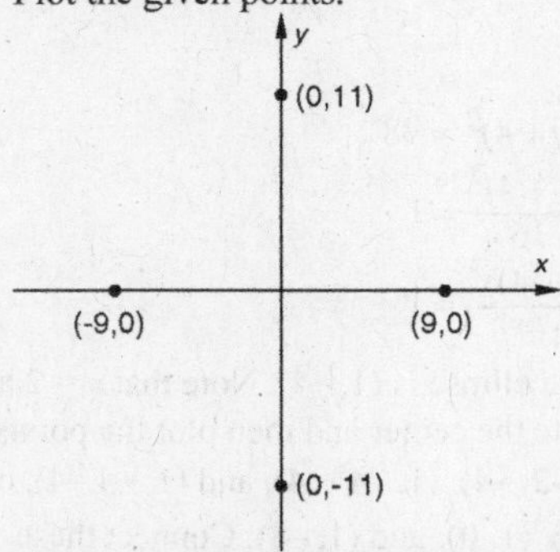

From the location of these points, we see that the ellipse that contains them is centered at the origin with $a=9$ and $b=11$. We write the equation of the ellipse:
$$\frac{x^2}{9^2}+\frac{y^2}{11^2}=1$$
$$\frac{x^2}{81}+\frac{y^2}{121}=1$$

47. Plot the given points.

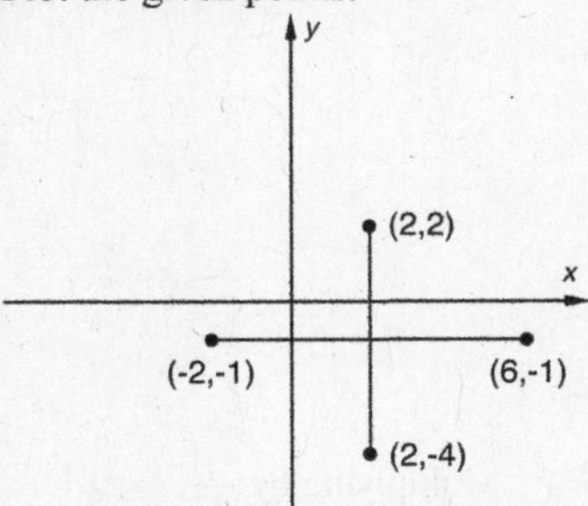

The midpoint of the segment from (–2, –1) to (6, –1) is $\left(\frac{-2+6}{2}, \frac{-1-1}{2}\right)$, or (2, –1). The midpoint of the segment from (2, –4) to (2, 2) is $\left(\frac{2+2}{2}, \frac{-4+2}{2}\right)$, or $(2,-1)$. Thus, we can conclude that (2, –1) is the center of the ellipse. The distance from (–2, –1) to (2, –1) is $\sqrt{[2-(-2)]^2+[-1-(-1)]^2}=\sqrt{16}=4$, so $a=4$. The distance from (2, 2) to (2, –1) is $\sqrt{(2-2)^2+(-1-2)^2}$ $=\sqrt{9}=3$, so $b=3$. We write the equation of the ellipse.
$$\frac{(x-2)^2}{4^2}+\frac{(y-(-1))^2}{3^2}=1$$
$$\frac{(x-2)^2}{16}+\frac{(y+1)^2}{9}=1$$

49. We have a vertical ellipse centered at the origin with $a=6/2$, or 3, and $b=10/2$, or 5. Then the equation is $\frac{x^2}{3^2}+\frac{y^2}{5^2}=1$, or $\frac{x^2}{9}+\frac{y^2}{25}=1$.

51. a. Let $F_1=(-c,\ 0)$ and $F_2=(c,\ 0)$. Then the sum of the distances from the foci to P is $2a$. By the distance formula,
$$\sqrt{(x+c)^2+y^2}+\sqrt{(x-c)^2+y^2}=2a, \text{ or}$$
$$\sqrt{(x+c)^2+y^2}=2a-\sqrt{(x-c)^2+y^2}.$$
Squaring, we get
$$(x+c)^2+y^2=4a^2-4a\sqrt{(x-c)^2+y^2}+(x-c)^2+y^2,$$
or $x^2+2cx+c^2+y^2$
$$=4a^2-4a\sqrt{(x-c)^2+y^2}+x^2-2cx+c^2+y^2.$$
Thus
$$-4a^2+4cx=-4a\sqrt{(x-c)^2+y^2}$$
$$a^2-cx=a\sqrt{(x-c)^2+y^2}.$$
Squaring again, we get
$$a^4-2a^2cx+c^2x^2=a^2(x^2-2cx+c^2+y^2)$$
$$a^4-2a^2cx+c^2x^2=a^2x^2-2a^2cx+a^2c^2+a^2y^2,$$
or $x^2(a^2-c^2)+a^2y^2=a^2(a^2-c^2)$
$$\frac{x^2}{a^2}+\frac{y^2}{a^2-c^2}=1.$$

b. When P is at $(0, b)$, it follows that $b^2 = a^2 - c^2$.

Substituting, we have $\frac{x^2}{a^2} + \frac{y^2}{b^2} = 1$.

53. For the given ellipse, $a = 6/2$, or 3, and $b = 2/2$, or 1. The patient's mouth should be at a distance 2c from the light source, where the coordinates of the foci of the ellipse are $(-c, 0)$ and $(c, 0)$. From Exercise 51(b), we know $b^2 = a^2 - c^2$. We use this to find c.

$$b^2 = a^2 - c^2$$
$$1^2 = 3^2 - c^2 \quad \text{Substituting}$$
$$c^2 = 8$$
$$c = \sqrt{8}$$

Then $2c = 2\sqrt{8} \approx 5.66$. The patient's mouth should be about 5.66 ft from the light source.

55.
$$x^2 - 4x + 4y^2 + 8y - 8 = 0$$
$$x^2 - 4x + 4y^2 + 8y = 8$$
$$x^2 - 4x + 4(y^2 + 2y) = 8$$
$$(x^2 - 4x + 4 - 4) + 4(y^2 + 2y + 1 - 1) = 8$$
$$(x^2 - 4x + 4) + 4(y^2 + 2y + 1) = 8 + 4 + 4 \cdot 1$$
$$(x-2)^2 + 4(y+1)^2 = 16$$
$$\frac{(x-2)^2}{16} + \frac{(y+1)^2}{4} = 1$$
$$\frac{(x-2)^2}{4^2} + \frac{(y-(-1))^2}{2^2} = 1$$

The center of the ellipse is $(2, -1)$. Note that $a = 4$ and $b = 2$. We locate the center and then plot the points $(2+4, -1)$, $(2-4, -1)$, $(2, -1+2)$, $(2, -1-2)$, or $(6, -1)$, $(-2, -1)$, $(2, 1)$, and $(2, -3)$. Connect these points with an oval-shaped curve.

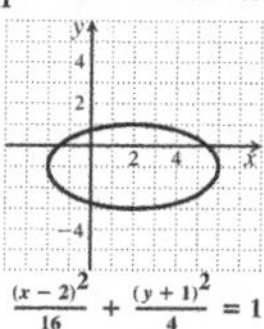

$\frac{(x-2)^2}{16} + \frac{(y+1)^2}{4} = 1$

57. The maximum distance between Earth at the vertex $(-a, 0)$ and the sun at the focus $(c, 0)$ is the same as the distance between vertex $(a, 0)$ and the other focus $(-c, 0)$, or $a + c$. Then

$$a + c = 149.7 + 2.4 = 152.1 \text{ million km}.$$

Connecting the Concepts

1. $y = 3(x-4)^2 + 1$ parabola
Vertex: $(4, 1)$
Axis of symmetry: $x = 4$

2. $x = y^2 + 2y + 3$ parabola
$x = y^2 + 2y + 1 + 2$
$x = (y+1)^2 + 2$
Vertex: $(2, -1)$
Axis of symmetry: $y = -1$

3. $(x-3)^2 + (y-2)^2 = 5$ circle
Center: $(3, 2)$

4.
$$x^2 + 6x + y^2 + 10y = 12 \quad \text{circle}$$
$$x^2 + 6x + 9 + y^2 + 10y + 25 = 12 + 9 + 25$$
$$(x+3)^2 + (y+5)^2 = 46$$
Center: $(-3, -5)$

5. $\frac{x^2}{144} + \frac{y^2}{81} = 1$ ellipse
x-intercepts: $(-12, 0)$ and $(12, 0)$
y-intercepts: $(0, 9)$ and $(0, -9)$

6. $\frac{x^2}{9} - \frac{y^2}{121} = 1$ hyperbola
Vertices: $(-3, 0)$ and $(3, 0)$

7. $4y^2 - x^2 = 4$ hyperbola
$\frac{y^2}{1} - \frac{x^2}{4} = 1$
Vertices: $(0, -1)$ and $(0, 1)$

8. $\frac{y^2}{9} - \frac{x^2}{4} = 1$ hyperbola
$a = 2, b = 3$
Asymptotes: $y = \frac{3}{2}x$ and $y = -\frac{3}{2}x$

Exercise Set 13.3

1. Asymptote B

3. Branch A

5. Hyperbola F

7. $\frac{y^2}{16} - \frac{x^2}{16} = 1$
$\frac{y^2}{4^2} - \frac{x^2}{4^2} = 1$

$a = 4$ and $b = 4$, so the asymptotes are $y = \frac{4}{4}x$ and $y = -\frac{4}{4}x$, or $y = x$ and $y = -x$. We sketch them.

Replacing x with 0 and solving for y, we get $y = \pm 4$, so the intercepts are $(0, 4)$ and $(0, -4)$.
We plot the intercepts and draw smooth curves through them that approach the asymptotes.

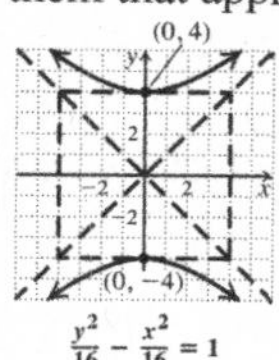

$\frac{y^2}{16} - \frac{x^2}{16} = 1$

9. $\frac{x^2}{4}-\frac{y^2}{25}=1$

$\frac{x^2}{2^2}-\frac{y^2}{5^2}=1$

$a=2$ and $b=5$, so the asymptotes are $y=\frac{5}{2}x$ and $y=-\frac{5}{2}x$. We sketch them.

Replacing y with 0 and solving for x, we get $x=\pm 2$, so the intercepts are (2, 0) and (−2, 0).
We plot the intercepts and draw smooth curves through them that approach the asymptotes.

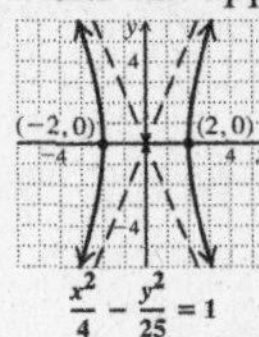

11. $\frac{y^2}{36}-\frac{x^2}{9}=1$

$\frac{y^2}{6^2}-\frac{x^2}{3^2}=1$

$a=3$ and $b=6$, so the asymptotes are $y=\frac{6}{3}x$ and $y=-\frac{6}{3}x$, or $y=2x$ and $y=-2x$. We sketch them.

Replacing x with 0 and solving for y, we get $y=\pm 6$, so the intercepts are (0, 6) and (0, −6).
We plot the intercepts and draw smooth curves through them that approach the asymptotes.

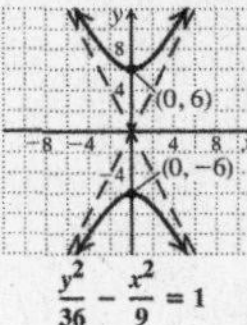

13. $y^2-x^2=25$

$\frac{y^2}{25}-\frac{x^2}{25}=1$

$\frac{y^2}{5^2}-\frac{x^2}{5^2}=1$

$a=5$ and $b=5$, so the asymptotes are $y=\frac{5}{5}x$ and $y=-\frac{5}{5}x$, or $y=x$ and $y=-x$. We sketch them.

Replacing x with 0 and solving for y, we get $y=\pm 5$, so the intercepts are (0, 5) and (0, −5).
We plot the intercepts and draw smooth curves through them that approach the asymptotes.

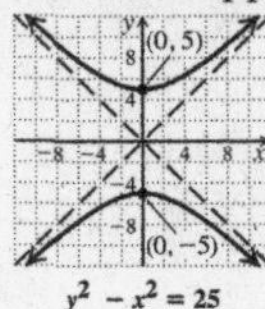

15. $25x^2-16y^2=400$

$\frac{x^2}{16}-\frac{y^2}{25}=1$ Multiplying by $\frac{1}{400}$

$\frac{x^2}{4^2}-\frac{y^2}{5^2}=1$

$a=4$ and $b=5$, so the asymptotes are $y=\frac{5}{4}x$ and $y=-\frac{5}{4}x$. We sketch them.

Replacing y with 0 and solving for x, we get $x=\pm 4$, so the intercepts are (4, 0) and (−4, 0).
We plot the intercepts and draw smooth curves through them that approach the asymptotes.

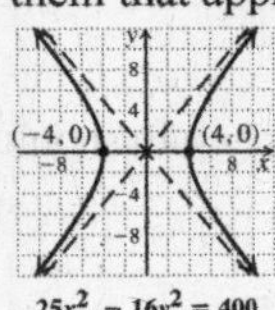

17. $xy=-6$

$y=-\frac{6}{x}$ Solving for y

We find some solutions, keeping the results in a table.

x	y
$\frac{1}{6}$	36
1	−6
6	−1
12	$-\frac{1}{2}$
$-\frac{1}{6}$	36
−1	6
−6	1
−12	$\frac{1}{2}$

Note that we cannot use 0 for x. The x-axis and the y-axis are the asymptotes.

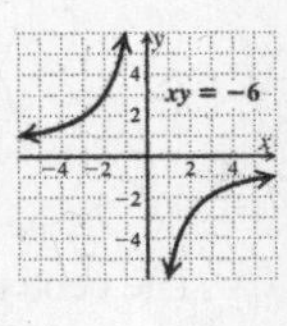

19. $xy=4$

$y=\frac{4}{x}$ Solving for y

We find some solutions, keeping the results in a table.

x	y
$\frac{1}{2}$	8
1	4
4	1
8	$\frac{1}{2}$
$-\frac{1}{2}$	−8
−1	−4
−2	−2
−4	−1

Note that we cannot use 0 for x. The x-axis and the y-axis are the asymptotes.

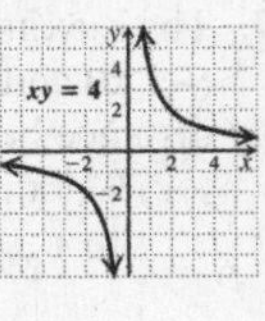

21. $xy = -2$

$y = -\frac{2}{x}$ Solving for y

x	y
$\frac{1}{2}$	-4
1	-2
2	-1
4	$-\frac{1}{2}$
$-\frac{1}{2}$	4
-1	2
-2	1
-4	$\frac{1}{2}$

Note that we cannot use 0 for x. The x-axis and the y-axis are the asymptotes.

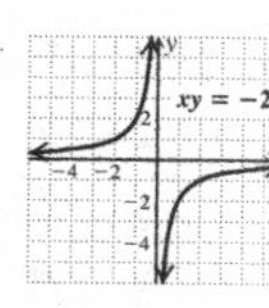

23. $xy = 1$

$y = \frac{1}{x}$ Solving for y

x	y
$\frac{1}{4}$	4
$\frac{1}{2}$	2
1	1
2	$\frac{1}{2}$
4	$\frac{1}{4}$
$-\frac{1}{4}$	-4
$-\frac{1}{2}$	-2
-1	-1
-2	$-\frac{1}{2}$
-4	$-\frac{1}{4}$

Note that we cannot use 0 for x. The x-axis and the y-axis are the asymptotes.

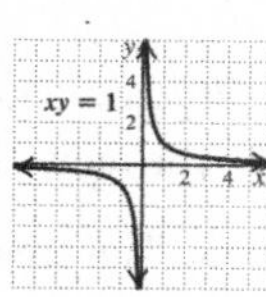

25. $x^2 + y^2 - 6x + 10y - 40 = 0$

Completing the square twice, we obtain an equivalent equation:

$$(x^2 - 6x) + (y^2 + 10y) = 40$$
$$(x^2 - 6x + 9) + (y^2 + 10y + 25) = 40 + 9 + 25$$
$$(x-3)^2 + (y+5)^2 = 74$$

The graph is a circle.

27. $9x^2 + 4y^2 - 36 = 0$

$$9x^2 + 4y^2 = 36$$
$$\frac{x^2}{4} + \frac{y^2}{9} = 1$$

The graph is an ellipse.

29. $4x^2 - 9y^2 - 72 = 0$

$$4x^2 - 9y^2 = 72$$
$$\frac{x^2}{18} - \frac{y^2}{8} = 1$$

The graph is a hyperbola.

31. $y^2 = 20 - x^2$

$$x^2 + y^2 = 20$$

The graph is a circle.

33. $x - 10 = y^2 - 6y$

$$x - 10 + 9 = y^2 - 6y + 9$$
$$x - 1 = (y-3)^2$$

The graph is a parabola.

35. $x - \frac{3}{y} = 0$

$$x = \frac{3}{y}$$
$$xy = 3$$

The graph is a hyperbola.

37. $y + 6x = x^2 + 5$

$$y = x^2 - 6x + 5$$

The graph is a parabola

39. $25y^2 = 100 + 4x^2$

$$25y^2 - 4x^2 = 100$$
$$\frac{y^2}{4} - \frac{x^2}{25} = 1$$

The graph is a hyperbola.

41. $3x^2 + y^2 - x = 2x^2 - 9x + 10y + 40$

$$x^2 + y^2 + 8x - 10y = 40$$

Both variables are squared, so the graph is not a parabola. The plus sign between x^2 and y^2 indicates that we have either a circle or an ellipse. Since the coefficients of x^2 and y^2 are the same, the graph is a circle.

43. $16x^2 + 5y^2 - 12x^2 + 8y^2 - 3x + 4y = 568$

$$4x^2 + 13y^2 - 3x + 4y = 568$$

Both variables are squared, so the graph is not a parabola. The plus sign between x^2 and y^2 indicates that we have either a circle or an ellipse. Since the coefficients of x^2 and y^2 are different, the graph is an ellipse.

45. *Writing Exercise.*

47. $(16 - y^4) = (4 + y^2)(4 - y^2)$

$$= (4 + y^2)(2 + y)(2 - y)$$

49. $10c^3 - 80c^2 + 150c = 10c(c^2 - 8c + 15)$

$$= 10c(c-3)(c-5)$$

51. $8t^4 - 8t = 8t(t^3 - 1) = 8t(t-1)(t^2 + t + 1)$

53. *Writing Exercise.*

55. Since the intercepts are (0, 6) and (0, –6), we know that the hyperbola is of the form $\frac{y^2}{b^2}-\frac{x^2}{a^2}=1$ and that $b = 6$. The equations of the asymptotes tell us that $b/a=3$, so

$$\frac{6}{a}=3$$
$$a=2.$$

The equation is $\frac{y^2}{6^2}-\frac{x^2}{2^2}=1$, or $\frac{y^2}{36}-\frac{x^2}{4}=1$.

57. $\frac{(x-5)^2}{36}-\frac{(y-2)^2}{25}=1$

$\frac{(x-5)^2}{6^2}-\frac{(y-2)^2}{5^2}=1$

$h = 5$, $k = 2$, $a = 6$, $b = 5$
Center: (5, 2)
Vertices: (5 – 6, 2) and (5 + 6, 2), or (–1, 2) and (11, 2)
Asymptotes: $y-2=\frac{5}{6}(x-5)$ and $y-2=-\frac{5}{6}(x-5)$

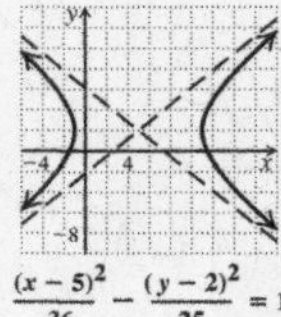

$\frac{(x-5)^2}{36}-\frac{(y-2)^2}{25}=1$

59. $8(y+3)^2-2(x-4)^2=32$

$\frac{(y+3)^2}{4}-\frac{(x-4)^2}{16}=1$

$\frac{(y-(-3))^2}{2^2}-\frac{(x-4)^2}{4^2}=1$

$h = 4$, $k = -3$, $a = 4$, $b = 2$
Center: (4, –3)
Vertices: (4, –3 + 2) and (4, – 3 – 2), or (4, –1) and (4,–5)
Asymptotes: $y-(-3)=\frac{2}{4}(x-4)$ and

$y-(-3)=-\frac{2}{4}(x-4)$, or $y+3=\frac{1}{2}(x-4)$ and

$y+3=-\frac{1}{2}(x-4)$

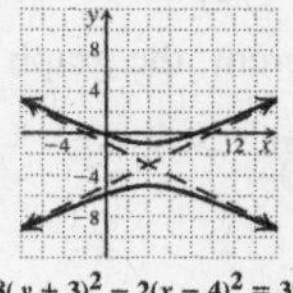

$8(y+3)^2-2(x-4)^2=32$

61.
$$4x^2-y^2+24x+4y+28=0$$
$$4(x^2+6x)-(y^2-4y)=-28$$
$$4(x^2+6x+9-9)-(y^2-4y+4-4)=-28$$
$$4(x^2+6x+9)-(y^2-4y+4)=-28+4\cdot 9-4$$
$$4(x+3)^2-(y-2)^2=4$$
$$\frac{(x+3)^2}{1}-\frac{(y-2)^2}{4}=1$$
$$\frac{(x-(-3))^2}{1^2}-\frac{(y-2)^2}{2^2}=1$$

$h = -3$, $k = 2$, $a = 1$, $b = 2$
Center: (–3, 2)
Vertices: (–3 – 1, 2), and (–3 + 1, 2), or (–4, 2) and (–2, 2)
Asymptotes: $y-2=\frac{2}{1}(x-(-3))$ and

$y-2=-\frac{2}{1}(x-(-3))$, or $y-2=2(x+3)$ and

$y-2=-2(x+3)$

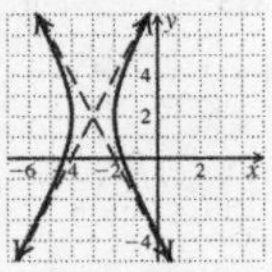

$4x^2-y^2+24x+4y+28=0$

63. *Graphing Calculator Exercise*

Mid-Chapter Review

1.
$$(x^2-4x)+(y^2+2y)=6$$
$$(x^2-4x+4)+(y^2+2y+1)=6+4+1$$
$$(x-2)^2+(y+1)^2=11$$

The center of the circle is (2,–1).
The radius is $\sqrt{11}$.

2. **a.** Is there both an x^2-term and a y^2-term? Yes.

b. Do both the x^2-term and the y^2-term have the same sign? No.

c. The graph of the equation is a hyperbola.

3. $[x-(-4)]^2+(y-9)^2=(2\sqrt{5})^2$

$(x+4)^2+(y-9)^2=20$

4.
$$x^2-10x+y^2+2y=10$$
$$x^2-10x+25+y^2+2y+1=10+25+1$$
$$(x-5)^2+(y+1)^2=36$$

Center: (5, –1); radius: 6

5. $x^2+y^2=36$ is a circle.

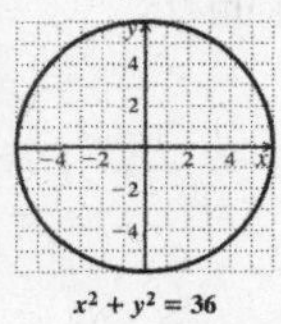

$x^2+y^2=36$

6. $y=x^2-5$ is a parabola.

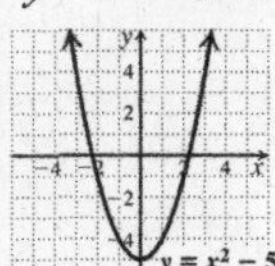

7. $\frac{x^2}{25}+\frac{y^2}{49}=1$ is an ellipse.

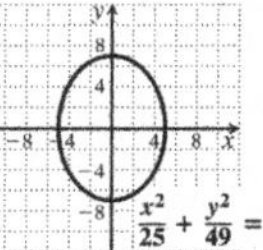

8. $\frac{x^2}{25}-\frac{y^2}{49}=1$ is a hyperbola.

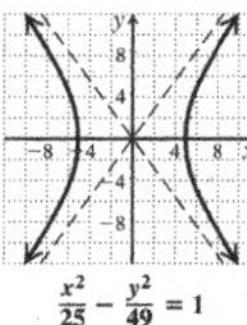

9. $x=(y+3)^2+2$ is a parabola.

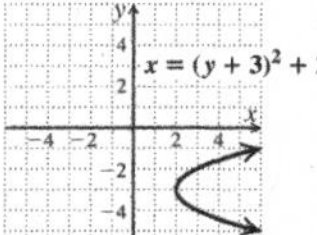

10. $4x^2+9y^2=36$

$\frac{x^2}{9}+\frac{y^2}{4}=1$ is an ellipse.

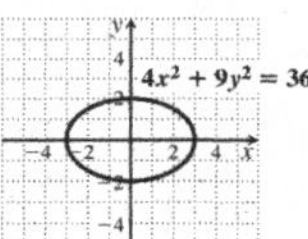

11. $xy=-4$ is a hyperbola.

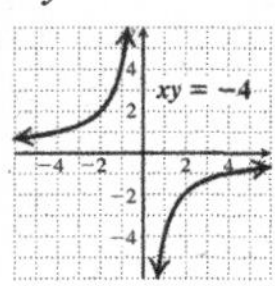

12. $(x+2)^2+(y-3)^2=1$ is a circle.

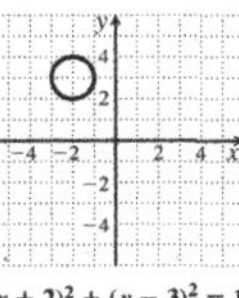

13. $x^2+y^2-8y-20=0$

$x^2+y^2-8y+16=20+16$

$x^2+(y-4)^2=36$ is a circle.

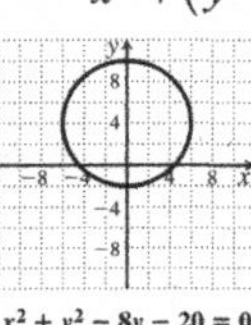

14. $x=y^2+2y$

$x=y^2+2y+1-1$

$x=(y+1)^2-1$ is a parabola.

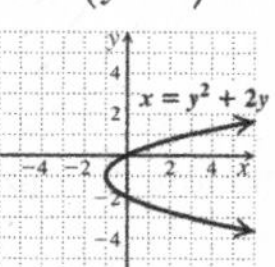

15. $16y^2-x^2=16$

$\frac{y^2}{1}-\frac{x^2}{16}=1$ is a hyperbola.

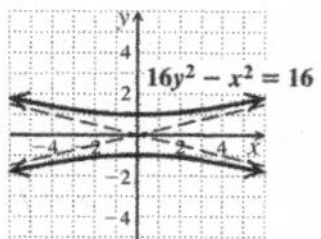

16. $x=\frac{9}{y}$

$xy=9$ is a hyperbola.

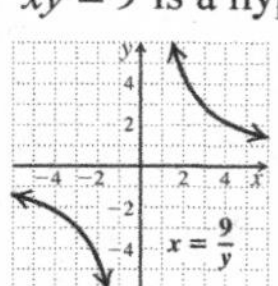

Exercise Set 13.4

1. True

3. False

5. True

7. $x^2+y^2=41,$ (1)

$y-x=1$ (2)

First solve Equation (2) for y.

$y=x+1$ (3)

Then substitute $x+1$ for y in Equation (1) and solve for x.

$$x^2+y^2=41$$
$$x^2+(x+1)^2=41$$
$$x^2+x^2+2x+1=41$$
$$2x^2+2x-40=0$$
$$x^2+x-20=0 \quad \text{Multiplying by } \tfrac{1}{2}$$
$$(x+5)(x-4)=0$$

$x+5=0$ *or* $x-4=0$ Principle of zero products

$x=-5$ *or* $x=4$

Now substitute these numbers in Equation (3) and solve for y.

For $x=-5$, $y=-5+1=-4$

For $x=4$, $y=4+1=5$

The pairs (–5, –4) and (4, 5) check, so they are the solutions.

9. $4x^2+9y^2=36,\quad (1)$
$3y+2x=6\quad (2)$
First solve Equation (2) for y.
$$3y=-2x+6$$
$$y=-\tfrac{2}{3}x+2\quad (3)$$
Then substitute $-\frac{2}{3}x+2$ for y in Equation (1) and solve for x.
$$4x^2+9y^2=36$$
$$4x^2+9\left(-\tfrac{2}{3}x+2\right)^2=36$$
$$4x^2+9\left(\tfrac{4}{9}x^2-\tfrac{8}{3}x+4\right)=36$$
$$4x^2+4x^2-24x+36=36$$
$$8x^2-24x=0$$
$$x^2-3x=0$$
$$x(x-3)=0$$
$$x=0 \text{ or } x=3$$
Now substitute these numbers in Equation (3) and solve for y.

For $x=0$, $y=-\frac{2}{3}\cdot 0+2=2$

For $x=3$, $y=-\frac{2}{3}\cdot 3+2=0$

The pairs (0, 2) and (3, 0) check, so they are the solutions.

11. $y^2=x+3,\quad (1)$
$2y=x+4\quad (2)$
First solve Equation (2) for x.
$$2y-4=x\quad (3)$$
Then substitute $2y-4$ for x in Equation (1) and solve for y.
$$y^2=x+3$$
$$y^2=(2y-4)+3$$
$$y^2=2y-1$$
$$y^2-2y+1=0$$
$$(y-1)(y-1)=0$$
$$y-1=0 \text{ or } y-1=0$$
$$y=1 \text{ or } y=1$$
Now substitute 1 for y in Equation (3) and solve for x.
$$2\cdot 1-4=x$$
$$-2=x$$
The pair (–2, 1) checks. It is the solution.

13. $x^2-xy+3y^2=27,\quad (1)$
$x-y=2\quad (2)$
First solve Equation (2) for y.
$$x-2=y\quad (3)$$
Then substitute $x-2$ for y in Equation (1) and solve for x.
$$x^2-xy+3y^2=27$$
$$x^2-x(x-2)+3(x-2)^2=27$$
$$x^2-x^2+2x+3x^2-12x+12=27$$
$$3x^2-10x-15=0$$
$$x=\frac{-(-10)\pm\sqrt{(-10)^2-4(3)(-15)}}{2\cdot 3}$$
$$x=\frac{10\pm\sqrt{100+180}}{6}=\frac{10\pm\sqrt{280}}{6}$$
$$x=\frac{10\pm 2\sqrt{70}}{6}=\frac{5\pm\sqrt{70}}{3}$$
Now substitute these numbers in Equation (3) and solve for y.

For $x=\frac{5+\sqrt{70}}{3}$, $y=\frac{5+\sqrt{70}}{3}-2=\frac{-1+\sqrt{70}}{3}$

For $x=\frac{5-\sqrt{70}}{3}$, $y=\frac{5-\sqrt{70}}{3}-2=\frac{-1-\sqrt{70}}{3}$

The pairs $\left(\frac{5+\sqrt{70}}{3}, \frac{-1+\sqrt{70}}{3}\right)$ and $\left(\frac{5-\sqrt{70}}{3}, \frac{-1-\sqrt{70}}{3}\right)$ check, so they are the solutions.

15. $x^2+4y^2=25,\quad (1)$
$x+2y=7\quad (2)$
First solve Equation (2) for x.
$$x=-2y+7\quad (3)$$
Then substitute $-2y+7$ for x in Equation (1) and solve for y.
$$x^2+4y^2=25$$
$$(-2y+7)^2+4y^2=25$$
$$4y^2-28y+49+4y^2=25$$
$$8y^2-28y+24=0$$
$$2y^2-7y+6=0$$
$$(2y-3)(y-2)=0$$
$$y=\frac{3}{2} \text{ or } y=2$$
Now substitute these numbers in Equation (3) and solve for x.

For $y=\frac{3}{2}$, $x=-2\cdot\frac{3}{2}+7=4$

For $y=2$, $x=-2\cdot 2+7=3$

The pairs $\left(4, \frac{3}{2}\right)$ and (3, 2) check, so they are the solutions.

17. $x^2-xy+3y^2=5,\quad (1)$
$x-y=2\quad (2)$
First solve Equation (2) for y.
$$x-2=y\quad (3)$$
Then substitute $x-2$ for y in Equation (1) and solve for x.
$$x^2-xy+3y^2=5$$
$$x^2-x(x-2)+3(x-2)^2=5$$
$$x^2-x^2+2x+3x^2-12x+12=5$$
$$3x^2-10x+7=0$$
$$(3x-7)(x-1)=0$$
$$x=\frac{7}{3} \text{ or } x=1$$
Now substitute these numbers in Equation (3) and solve for y.

For $x=\frac{7}{3}$, $y=\frac{7}{3}-2=\frac{1}{3}$
For $x=1$, $y=1-2=-1$

The pairs $\left(\frac{7}{3}, \frac{1}{3}\right)$ and $(1, -1)$ check, so they are the solutions.

19. $3x+y=7$, (1)
$4x^2+5y=24$ (2)

First solve Equation (1) for y.
$y=7-3x$ (3)

Then substitute $7-3x$ for y in Equation (2) and solve for x.

$$4x^2+5y=24$$
$$4x^2+5(7-3x)=24$$
$$4x^2+35-15x=24$$
$$4x^2-15x+11=0$$
$$(4x-11)(x-1)=0$$
$$x=\frac{11}{4} \text{ or } x=1$$

Now substitute these numbers into Equation (3) and solve for y.

For $x=\frac{11}{4}$, $y=7-3\cdot\frac{11}{4}=-\frac{5}{4}$
For $x=1$, $y=7-3\cdot1=4$

The pairs $\left(\frac{11}{4}, -\frac{5}{4}\right)$ and $(1, 4)$ check, so they are the solutions.

21. $a+b=6$, (1)
$ab=8$ (2)

First solve Equation (1) for a.
$a=-b+6$ (3)

Then substitute $-b+6$ for a in Equation (2) and solve for b.

$$(-b+6)b=8$$
$$-b^2+6b=8$$
$$0=b^2-6b+8$$
$$0=(b-2)(b-4)$$
$$b-2=0 \text{ or } b-4=0$$
$$b=2 \text{ or } b=4$$

Now substitute these numbers in Equation (3) and solve for a.

For $b=2$, $a=-2+6=4$
For $b=4$, $a=-4+6=2$

The pairs $(4, 2)$ and $(2, 4)$ check, so they are the solutions.

23. $2a+b=1$, (1)
$b=4-a^2$ (2)

Equation (2) is already solved for b. Substitute $4-a^2$ for b in Equation (1) and solve for a.

$$2a+4-a^2=1$$
$$0=a^2-2a-3$$
$$0=(a-3)(a+1)$$
$$a=3 \text{ or } a=-1$$

Substitute these numbers in Equation (2) and solve for b.

For $a=3$, $b=4-3^2=-5$
For $a=-1$, $b=4-(-1)^2=3$

The pairs $(3, -5)$ and $(-1, 3)$ check, so they are the solutions.

25. $a^2+b^2=89$, (1)
$a-b=3$ (2)

First solve Equation (2) for a.
$a=b+3$ (3)

Then substitute $b+3$ for a in Equation (1) and solve for b.

$$(b+3)^2+b^2=89$$
$$b^2+6b+9+b^2=89$$
$$2b^2+6b-80=0$$
$$b^2+3b-40=0$$
$$(b+8)(b-5)=0$$
$$b=-8 \text{ or } b=5$$

Substitute these numbers in Equation (3) and solve for a.

For $b=-8$, $a=-8+3=-5$
For $b=5$, $a=5+3=8$

The pairs $(-5, -8)$ and $(8, 5)$ check, so they are the solutions.

27. $y=x^2$, (1)
$x=y^2$ (2)

Equation (1) is already solved for y. Substitute x^2 for y in Equation (2) and solve for x.

$$x=y^2$$
$$x=\left(x^2\right)^2$$
$$x=x^4$$
$$0=x^4-x$$
$$0=x(x^3-1)$$
$$0=x(x-1)(x^2+x+1)$$
$$x=0 \text{ or } x=1 \text{ or } x=\frac{-1\pm\sqrt{1^2-4\cdot1\cdot1}}{2}$$
$$x=0 \text{ or } x=1 \text{ or } x=-\frac{1}{2}\pm\frac{\sqrt{3}}{2}i$$

Substitute these numbers in Equation (1) and solve for y.

For $x=0$, $y=0^2=0$
For $x=1$, $y=1^2=1$
For $x=-\frac{1}{2}+\frac{\sqrt{3}}{2}i$, $y=\left(-\frac{1}{2}+\frac{\sqrt{3}}{2}i\right)^2=-\frac{1}{2}-\frac{\sqrt{3}}{2}i$
For $x=-\frac{1}{2}-\frac{\sqrt{3}}{2}i$, $y=\left(-\frac{1}{2}-\frac{\sqrt{3}}{2}i\right)^2=-\frac{1}{2}+\frac{\sqrt{3}}{2}i$

The pairs $(0, 0)$, $(1, 1)$, $\left(-\frac{1}{2}+\frac{\sqrt{3}}{2}i, -\frac{1}{2}-\frac{\sqrt{3}}{2}i\right)$, and $\left(-\frac{1}{2}-\frac{\sqrt{3}}{2}i, -\frac{1}{2}+\frac{\sqrt{3}}{2}i\right)$ check, so they are the solutions.

29. $x^2 + y^2 = 16$, (1)
$x^2 - y^2 = 16$ (2)

Here we use the elimination method.

$$\begin{aligned} x^2 + y^2 &= 16 \quad (1) \\ x^2 - y^2 &= 16 \quad (2) \\ \hline 2x^2 \quad &= 32 \quad \text{Adding} \\ x^2 &= 16 \\ x &= \pm 4 \end{aligned}$$

If $x = 4$, $x^2 = 16$, and if $x = -4$, $x^2 = 16$, so substituting 4 or –4 in Equation (2) gives us

$$\begin{aligned} x^2 + y^2 &= 16 \\ 16 + y^2 &= 16 \\ y^2 &= 0 \\ y &= 0 \end{aligned}$$

The pairs (4, 0) and (–4, 0) check. They are the solutions.

31. $x^2 + y^2 = 25$, (1)
$xy = 12$ (2)

First we solve Equation (2) for y.

$$\begin{aligned} xy &= 12 \\ y &= \frac{12}{x} \end{aligned}$$

Then we substitute $\frac{12}{x}$ for y in Equation (1) and solve for x.

$$\begin{aligned} x^2 + y^2 &= 25 \\ x^2 + \left(\frac{12}{x}\right)^2 &= 25 \\ x^2 + \frac{144}{x^2} &= 25 \\ x^4 + 144 &= 25x^2 \quad \text{Multiplying by } x^2 \\ x^4 - 25x^2 + 144 &= 0 \\ u^2 - 25u + 144 &= 0 \quad \text{Letting } u = x^2 \\ (u-9)(u-16) &= 0 \\ u = 9 \quad &or \quad u = 16 \end{aligned}$$

We now substitute x^2 for u and solve for x.

$$\begin{aligned} x^2 = 9 \quad &or \quad x^2 = 16 \\ x = \pm 3 \quad &or \quad x = \pm 4 \end{aligned}$$

Since $y = 12/x$, if $x = 3$, $y = 4$; if $x = -3$, $y = -4$; if $x = 4$, $y = 3$; and if $x = -4$, $y = -3$. The pairs (3, 4), $(-3,-4)$, (4, 3), and $(-4,-3)$ check. They are the solutions.

33. $x^2 + y^2 = 9$, (1)
$25x^2 + 16y^2 = 400$ (2)

$$\begin{aligned} -16x^2 - 16y^2 &= -144 \quad \text{Multiplying (1) by } -16 \\ 25x^2 + 16y^2 &= 400 \\ \hline 9x^2 \quad &= 256 \quad \text{Adding} \\ x &= \pm \frac{16}{3} \end{aligned}$$

$$\begin{aligned} \frac{256}{9} + y^2 &= 9 \quad \text{Substituting in (1)} \\ y^2 &= 9 - \frac{256}{9} \\ y^2 &= -\frac{175}{9} \\ y &= \pm\sqrt{-\frac{175}{9}} = \pm\frac{5\sqrt{7}}{3}i \end{aligned}$$

The pairs $\left(\frac{16}{3}, \frac{5\sqrt{7}}{3}i\right)$, $\left(\frac{16}{3}, -\frac{5\sqrt{7}}{3}i\right)$, $\left(-\frac{16}{3}, \frac{5\sqrt{7}}{3}i\right)$, and $\left(-\frac{16}{3}, -\frac{5\sqrt{7}}{3}i\right)$ check. They are the solutions.

35. $x^2 + y^2 = 14$, (1)

$$\begin{aligned} x^2 - y^2 &= 4 \quad (2) \\ \hline 2x^2 \quad &= 18 \quad \text{Adding} \\ x^2 &= 9 \\ x &= \pm 3 \end{aligned}$$

$$\begin{aligned} 9 + y^2 &= 14 \quad \text{Substituting in Eq. (1)} \\ y^2 &= 5 \\ y &= \pm\sqrt{5} \end{aligned}$$

The pairs $(-3,-\sqrt{5})$, $(-3, \sqrt{5})$, $(3,-\sqrt{5})$, and $(3, \sqrt{5})$ check. They are the solutions.

37. $x^2 + y^2 = 10$, (1)
$xy = 3$ (2)

First we solve Equation (2) for y.

$$\begin{aligned} xy &= 3 \\ y &= \frac{3}{x} \end{aligned}$$

Then we substitute $\frac{3}{x}$ for y in Equation (1) and solve for x.

$$\begin{aligned} x^2 + y^2 &= 10 \\ x^2 + \left(\frac{3}{x}\right)^2 &= 10 \\ x^2 + \frac{9}{x^2} &= 10 \\ x^4 + 9 &= 10x^2 \quad \text{Multiplying by } x^2 \\ x^4 - 10x^2 + 9 &= 0 \\ u^2 - 10u + 9 &= 0 \quad \text{Letting } u = x^2 \\ (u-1)(u-9) &= 0 \\ u = 1 \quad &or \quad u = 9 \\ x^2 = 1 \quad &or \quad x^2 = 9 \quad \text{Substitute } x^2 \text{ for } u \\ x = \pm 1 \quad &or \quad x = \pm 3 \quad \text{and solve for } x. \end{aligned}$$

$y = 3/x$, so if $x = 1$, $y = 3$; if $x = -1$, $y = -3$; if $x = 3$, $y = 1$; if $x = -3$, $y = -1$. The pairs (1, 3), (–1, –3), (3, 1), and (–3, –1) check. They are the solutions.

39. $x^2 + 4y^2 = 20$, (1)
$xy = 4$ (2)

First we solve Equation (2) for y.

$$y = \frac{4}{x}$$

Then we substitute $\frac{4}{x}$ for y in Equation (1) and solve for x.

$$x^2+4\left(\frac{4}{x}\right)^2=20$$
$$x^2+\frac{64}{x^2}=20$$
$$x^4+64=20x^2$$
$$x^4-20x^2+64=0$$
$$u^2-20u+64=0 \quad \text{Letting } u=x^2$$
$$(u-16)(u-4)=0$$
$$u=16 \quad or \quad u=4$$
$$x^2=16 \quad or \quad x^2=4$$
$$x=\pm 4 \quad or \quad x=\pm 2$$

$y=4/x$, so if $x=4$, $y=1$; if $x=-4$, $y=-1$; if $x=2$, $y=2$; and if $x=-2$, $y=-2$. The pairs (4, 1), $(-4,-1)$, (2, 2), and $(-2,-2)$ check. They are the solutions.

41. $2xy+3y^2=7$, (1)
$3xy-2y^2=4$ (2)

$$6xy+9y^2=21 \quad \text{Multiplying (1) by 3}$$
$$-6xy+4y^2=-8 \quad \text{Multiplying (2) by } -2$$
$$13y^2=13$$
$$y^2=1$$
$$y=\pm 1$$

Substitute for y in Equation (1) and solve for x.

When $y=1$: $2\cdot x\cdot 1+3\cdot 1^2=7$
$$2x=4$$
$$x=2$$

When $y=-1$: $2\cdot x\cdot(-1)+3(-1)^2=7$
$$-2x=4$$
$$x=-2$$

The pairs (2, 1) and $(-2,-1)$ check. They are the solutions.

43. $4a^2-25b^2=0$, (1)
$2a^2-10b^2=3b+4$ (2)

$$4a^2-25b^2=0$$
$$-4a^2+20b^2=-6b-8 \quad \text{Multiplying (2) by } -2$$
$$-5b^2=-6b-8$$

$$0=5b^2-6b-8$$
$$0=(5b+4)(b-2)$$
$$b=-\frac{4}{5} \quad or \quad b=2$$

Substitute for b in Equation (1) and solve for a.

When $b=-\frac{4}{5}$: $4a^2-25\left(-\frac{4}{5}\right)^2=0$
$$4a^2=16$$
$$a^2=4$$
$$a=\pm 2$$

When $b=2$: $4a^2-25(2)^2=0$
$$4a^2=100$$
$$a^2=25$$
$$a=\pm 5$$

The pairs $\left(2,-\frac{4}{5}\right)$, $\left(-2,-\frac{4}{5}\right)$, (5, 2) and $(-5, 2)$ check. They are the solutions.

45. $ab-b^2=-4$, (1)
$ab-2b^2=-6$ (2)

$$ab-b^2=-4$$
$$-ab+2b^2=6 \quad \text{Multiplying (2) by } -1$$
$$b^2=2$$
$$b=\pm\sqrt{2}$$

Substitute for b in Equation (1) and solve for a.

When $b=\sqrt{2}$: $a(\sqrt{2})-(\sqrt{2})^2=-4$
$$a\sqrt{2}=-2$$
$$a=-\frac{2}{\sqrt{2}}=-\sqrt{2}$$

When $b=-\sqrt{2}$: $a(-\sqrt{2})-(-\sqrt{2})^2=-4$
$$-a\sqrt{2}=-2$$
$$a=\frac{-2}{-\sqrt{2}}=\sqrt{2}$$

The pairs $(-\sqrt{2}, \sqrt{2})$ and $(\sqrt{2}, -\sqrt{2})$ check. They are the solutions.

47. ***Familiarize***. We first make a drawing. We let l and w represent the length and width, respectively.

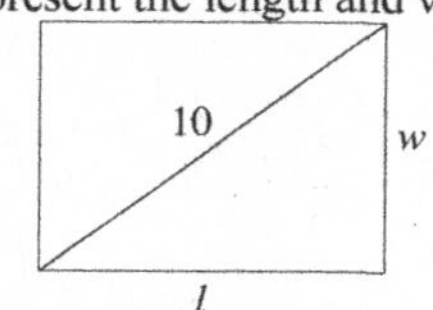

Translate. The perimeter is 28 cm.

$2l+2w=28$, or $l+w=14$

Using the Pythagorean theorem we have another equation.

$l^2+w^2=10^2$, or $l^2+w^2=100$

Carry out. We solve the system:

$l+w=14$, (1)
$l^2+w^2=100$ (2)

First solve Equation (1) for w.

$w=14-l$ (3)

Then substitute $14-l$ for w in Equation (2) and solve for l.

$$l^2+w^2=100$$
$$l^2+(14-l)^2=100$$
$$l^2+196-28l+l^2=100$$
$$2l^2-28l+96=0$$
$$l^2-14l+48=0$$
$$(l-8)(l-6)=0$$
$$l=8 \text{ or } l=6$$

If $l=8$, then $w=14-8$, or 6. If $l=6$, then

$w = 14 - 6$, or 8. Since the length is usually considered to be longer than the width, we have the solution $l = 8$ and $w = 6$, or $(8, 6)$.
Check. If $l = 8$ and $w = 6$, then the perimeter is $2 \cdot 8 + 2 \cdot 6$, or 28. The length of a diagonal is $\sqrt{8^2 + 6^2}$, or $\sqrt{100}$, or 10. The numbers check.
State. The length is 8 cm, and the width is 6 cm.

49. ***Familiarize.*** Let l = the length and w = the width of the rectangle.
Translate. The perimeter is 6 in., so we have one equation:

$$2l + 2w = 6, \text{ or } l + w = 3$$

Using the Pythagorean theorem we have another equation.

$$l^2 + w^2 = (\sqrt{5})^2, \text{ or } l^2 + w^2 = 5$$

Carry out. We solve the system of equations:

$$l + w = 3, \quad (1)$$
$$l^2 + w^2 = 5 \quad (2)$$

Solve Equation (1) for l: $l = 3 - w$. Substitute $3 - w$ for l in Equation (2) and solve for w.

$$l^2 + w^2 = 5$$
$$(3 - w)^2 + w^2 = 5$$
$$9 - 6w + w^2 + w^2 = 5$$
$$2w^2 - 6w + 4 = 0$$
$$2(w^2 - 3w + 2) = 0$$
$$2(w - 2)(w - 1) = 0$$
$$w = 2 \text{ or } w = 1$$

For $w = 2$, $l = 3 - 2 = 1$.
For $w = 1$, $l = 3 - 1 = 2$.
Check. The solutions are (1, 2) and (2, 1). We choose the larger number for the length. $2 + 1 = 3$, and $2^2 + 1^2 = 5$. The solution checks.
State. The length is 2 in. and the width is 1 in.

51. ***Familiarize.*** We first make a drawing. Let l = the length and w = the width of the cargo area, in feet.

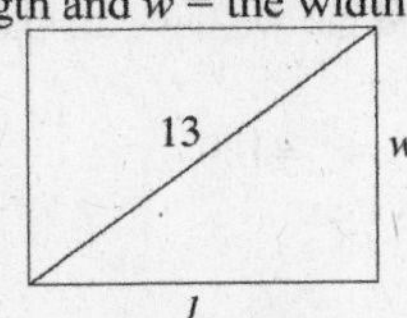

Translate. The cargo area must be 60 ft^2, so we have one equation:

$$lw = 60$$

The Pythagorean equation gives us another equation:

$$l^2 + w^2 = 13^2, \text{ or } l^2 + w^2 = 169$$

Carry out. We solve the system of equations.

$$lw = 60, \quad (1)$$
$$l^2 + w^2 = 169 \quad (2)$$

First solve Equation (1) for w:

$$lw = 60$$
$$w = \frac{60}{l} \quad (3)$$

Then substitute $60/l$ for w in Equation (2) and solve for l.

$$l^2 + w^2 = 169$$
$$l^2 + \left(\frac{60}{l}\right)^2 = 169$$
$$l^2 + \frac{3600}{l^2} = 169$$
$$l^4 + 3600 = 169l^2$$
$$l^4 - 169l^2 + 3600 = 0$$

Let $u = l^2$ and $u^2 = l^4$ and substitute.

$$u^2 - 169u + 3600 = 0$$
$$(u - 144)(u - 25) = 0$$
$$u = 144 \quad \text{or} \quad u = 25$$
$$l^2 = 144 \quad \text{or} \quad l^2 = 25 \quad \text{Replacing } u \text{ with } l^2$$
$$l = \pm 12 \quad \text{or} \quad l = \pm 5$$

Since the length cannot be negative, we consider only 12 and 5. We substitute in Equation (3) to find w. When $l = 12$, $w = 60/12 = 5$; when $l = 5$, $w = 60/5 = 12$. Since we usually consider length to be longer than width, we check the pair (12, 5).
Check. If the length is 12 ft and the width is 5 ft, then the area is $12 \cdot 5$, or 60 ft^2. Also $12^2 + 5^2 = 144 + 25 = 169 = 13^2$. The answer checks.
State. The length is 12 ft and the width is 5 ft.

53. ***Familiarize.*** Let x and y represent the numbers.
Translate. The product of the numbers is 90, so we have

$$xy = 90 \quad (1)$$

The sum of the squares of the numbers is 261, so we have

$$x^2 + y^2 = 261 \quad (2)$$

Carry out. We solve the system of equations.

$$xy = 90, \quad (1)$$
$$x^2 + y^2 = 261 \quad (2)$$

First solve Equation (1) for y:

$$xy = 90$$
$$y = \frac{90}{x} \quad (3)$$

Then substitute $90/x$ for y in Equation (2) and solve for x.

$$x^2 + y^2 = 261$$
$$x^2 + \left(\frac{90}{x}\right)^2 = 261$$
$$x^2 + \frac{8100}{x^2} = 261$$
$$x^4 + 8100 = 261x^2$$
$$x^4 - 261x^2 + 8100 = 0$$

Let $u = x^2$ and $u^2 = x^4$ and substitute.

$$u^2 - 261u + 8100 = 0$$
$$(u - 36)(u - 225) = 0$$
$$u = 36 \quad \textit{or} \quad u = 225$$
$$x^2 = 36 \quad \textit{or} \quad x^2 = 225 \quad \text{Replacing } u \text{ with } x^2$$
$$x = \pm 6 \quad \textit{or} \quad x = \pm 15$$

We use Equation (3) to find y.
When $x = 6$, $y = 90/6 = 15$;

when $x = -6, y = 90/(-6) = -15$;
when $x = 15, y = 90/15 = 6$;
when $x = -15, y = 90/(-15) = -6$. We see that the numbers can be 6 and 15 or –6 and –15.

Check. $6 \cdot 15 = 90$ and $6^2 + 15^2 = 261$; also $-6(-15) - 90$ and $(-6)^2 + (-15)^2 = 261$. The solutions check.

State. The numbers are 6 and 15 or –6 and –15.

55. ***Familiarize.*** Let $l =$ the length and $w =$ the width of the screen, in inches.

Translate. The area must be 90 in^2, so we have one equation: $lw = 90$

The Pythagorean equation gives us another equation:

$$l^2 + w^2 = \sqrt{200.25}^2, \text{ or } l^2 + w^2 = 200.25$$

Carry out. We solve the system of equations.

$$lw = 90, \quad (1)$$
$$l^2 + w^2 = 200.25 \quad (2)$$

First solve Equation (1) for l:

$$l = \frac{90}{w} \quad (3)$$

Then substitute $90/w$ for l in Equation (2) and solve for w.

$$l^2 + w^2 = 200.25$$
$$\left(\frac{90}{w}\right)^2 + w^2 = 200.25$$
$$\frac{8100}{w^2} + w^2 = 200.25$$
$$8100 + w^4 = 200.25w^2$$
$$w^4 - 200.25w^2 + 8100 = 0$$

Let $u = w^2$ and $u^2 = w^4$ and substitute.

$$u^2 - 200.25u + 8100 = 0$$
$$(u - 56.25)(u - 144) = 0$$
$$u = 56.25 \quad or \quad u = 144$$
$$w^2 = 56.25 \quad or \quad w^2 = 144 \quad \text{Replacing } u \text{ with } w^2$$
$$w = \pm 7.5 \quad or \quad w = \pm 12$$

Since the length cannot be negative, we consider only 7.5 and 12. We substitute in Equation (3) to find l. When $w = 7.5$, $l = 90/7.5 = 12$; when $w = 12$, $l = 90/12 = 7.5$. Since we usually consider length to be longer than width, we check the pair (12, 7.5).

Check. If the length is 12 in and the width is 7.5 in, then the area is $12 \cdot 7.5$, or 90 in^2. Also $12^2 + 7.5^2 = 144 + 56.25 = 200.25 = \sqrt{200.25}^2$. The answer checks.

State. The length is 12 in and the width is 7.5 in.

57. ***Familiarize.*** Let $l =$ the length and $w =$ the width of the rectangle area, in meters.

Translate. The area must be $\sqrt{3} \text{ m}^2$, so we have one equation: $lw = \sqrt{3}$

The Pythagorean equation gives us another equation:

$$l^2 + w^2 = 2^2, \text{ or } l^2 + w^2 = 4$$

Carry out. We solve the system of equations.

$$lw = \sqrt{3}, \quad (1)$$
$$l^2 + w^2 = 4 \quad (2)$$

First solve Equation (1) for l:

$$l = \frac{\sqrt{3}}{w} \quad (3)$$

Then substitute $\sqrt{3}/w$ for l in Equation (2) and solve for w.

$$l^2 + w^2 = 4$$
$$\left(\frac{\sqrt{3}}{w}\right)^2 + w^2 = 4$$
$$\frac{3}{w^2} + w^2 = 4$$
$$3 + w^4 = 4w^2$$
$$w^4 - 4w^2 + 3 = 0$$

Let $u = w^2$ and $u^2 = w^4$ and substitute.

$$u^2 - 4u + 3 = 0$$
$$(u - 1)(u - 3) = 0$$
$$u = 1 \quad \text{or} \quad u = 3$$
$$w^2 = 1 \quad \text{or} \quad w^2 = 3 \quad \text{Replacing } u \text{ with } w^2$$
$$w = \pm 1 \quad \text{or} \quad w = \pm\sqrt{3}$$

Since the length cannot be negative, we consider only 1 and $\sqrt{3}$. We substitute in Equation (3) to find l. When $w = \sqrt{3}$, $l = \sqrt{3}/\sqrt{3} = 1$; when $w = 1$, $l = \sqrt{3}/1 = \sqrt{3}$. Since we usually consider length to be longer than width, we check the pair $(\sqrt{3}, 1)$.

Check. If the length is $\sqrt{3}$ m and the width is 1 m, then the area is $\sqrt{3} \cdot 1$, or $\sqrt{3} \text{ m}^2$. Also $\sqrt{3}^2 + 1^2 = 3 + 1 = 4 = 2^2$. The answer checks.

State. The length is $\sqrt{3}$ m and the width is 1 m.

59. *Writing Exercise.*

61.
$$(3a^{-4})^2(2a^{-5})^{-1} = 3^2(a^{-4})^2(2)^{-1}(a^{-5})^{-1}$$
$$= 9a^{-8}2^{-1}a^5$$
$$= \frac{9}{2a^3}$$

63. $\log 10{,}000 = \log 10^4 = 4 \log 10 = 4$

65.
$$-10^2 \div 2 \cdot 5 - 3 = -100 \div 2 \cdot 5 - 3$$
$$= -50 \cdot 5 - 3$$
$$= -250 - 3$$
$$= -253$$

67. *Writing Exercise.*

69. $p^2+q^2=13$, (1)

$\frac{1}{pq}=-\frac{1}{6}$ (2)

Solve Equation (2) for p.

$$\frac{1}{q}=-\frac{p}{6}$$
$$-\frac{6}{q}=p$$

Substitute $-6/q$ for p in Equation (1) and solve for q.

$$\left(-\frac{6}{q}\right)^2+q^2=13$$
$$\frac{36}{q^2}+q^2=13$$
$$36+q^4=13q^2$$
$$q^4-13q^2+36=0$$
$$u^2-13u+36=0 \quad \text{Letting } u=q^2$$
$$(u-9)(u-4)=0$$
$$u=9 \quad \text{or} \quad u=4$$
$$x^2=9 \quad \text{or} \quad x^2=4$$
$$x=\pm 3 \quad \text{or} \quad x=\pm 2$$

Since $p=-6/q$, if $q=3$, $p=-2$; if $q=-3$, $p=2$; if $q=2$, $p=-3$; and if $q=-2$, $p=3$. The pairs $(-2, 3)$, $(2,-3)$, $(-3, 2)$, and $(3,-2)$ check. They are the solutions.

71. ***Familiarize***. Let l = the length of the rectangle, in feet, and let w = the width.
Translate. 100 ft of fencing is used, so we have

$l+w=100$. (1)

The area is 2475 ft^2, so we have

$lw=2475$. (2)

Carry out. Solving the system of equations, we get (55, 45) and (45, 55). Since length is usually considered to be longer than width, we have l = 55 and w = 45.
Check. If the length is 55 ft and the width is 45 ft, then $55+45$, or 100 ft, of fencing is used. The area is $55\cdot 45$, or 2475 ft^2. The answer checks.
State. The length of the rectangle is 55 ft, and the width is 45 ft.

73. ***Familiarize***. We let x and y represent the length and width of the base of the box, in inches, respectively. Make a drawing.

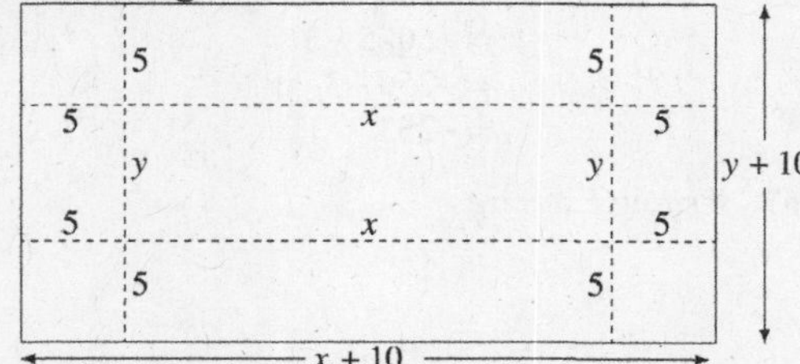

The dimensions of the metal sheet are x + 10 and y + 10.

Translate. The area of the sheet of metal is 340 in^2, so we have

$(x+10)(y+10)=340$. (1)

The volume of the box is 350 in^3, so we have

$x\cdot y\cdot 5=350$. (2)

Carry out. Solving the system of equations, we get (10, 7) and (7, 10). Since length is usually considered to be longer than width, we have l = 10 and w = 7.
Check. The dimensions of the metal sheet are 10 + 10, or 20, and 7 + 10, or 17, so the area is $20\cdot 17$, or 340 in^2. The volume of the box is $7\cdot 10\cdot 5$, or 350 in^3. The answer checks.
State. The dimensions of the box are 10 in. by 7 in. by 5 in.

Chapter 13 Review

1. True
2. False
3. False
4. True
5. True
6. True
7. False
8. True

9.
$$(x+3)^2+(y-2)^2=16$$
$$[x-(-3)]^2+(y-2)^2=4^2 \quad \text{Standard form}$$

The center is (–3, 2) and the radius is 4.

10. $(x-5)^2+y^2=11$

Center: (5, 0); radius: $\sqrt{11}$

11.
$$x^2+y^2-6x-2y+1=0$$
$$x^2-6x+y^2-2y=-1$$
$$(x^2-6x+9)+(y^2-2y+1)=-1+9+1$$
$$(x-3)^2+(y-1)^2=9$$
$$(x-3)^2+(y-1)^2=3^2$$

The center is (3, 1) and the radius is 3.

12.
$$x^2+y^2+8x-6y=20$$
$$x^2+8x+y^2-6y=20$$
$$(x^2+8x+16)+(y^2-6y+9)=20+16+9$$
$$(x+4)^2+(y-3)^2=45$$

Center: (–4, 3); radius: $\sqrt{45}$ or $3\sqrt{5}$

13.
$$(x-h)^2+(y-k)^2=r^2$$
$$[x-(-4)]^2+(y-3)^2=4^2$$
$$(x+4)^2+(y-3)^2=16$$

14.
$$(x-7)^2+[y-(-2)]^2=(2\sqrt{5})^2$$
$$(x-7)^2+(y+2)^2=20$$

15. Circle

$$5x^2+5y^2=80$$
$$x^2+y^2=16$$

Center: (0, 0); radius: 4

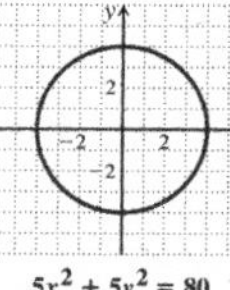

$5x^2+5y^2=80$

16. Ellipse

$$9x^2+2y^2=18$$
$$\frac{x^2}{2}+\frac{y^2}{9}=1$$

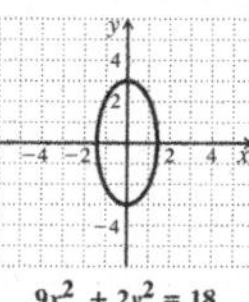

$9x^2+2y^2=18$

17. Parabola

$$y=-x^2+2x-3$$
$$x=-\frac{b}{2a}=-\frac{2}{2(-1)}=1$$
$$y=-x^2+2x-3=-1^2+2\cdot 1-3=-2$$

The vertex is (1, –2). The graph opens downward because the coefficient of x^2, –1 is negative.

x	y
−1	−6
0	−3
1	−2
2	−3
3	−6

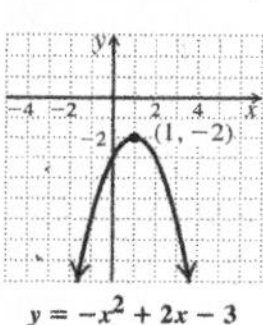

$y=-x^2+2x-3$

18. Hyperbola

$$\frac{y^2}{9}-\frac{x^2}{4}=1$$

$\frac{y^2}{9}-\frac{x^2}{4}=1$

19. Hyperbola

$$xy=9$$
$$y=\frac{9}{x}\quad \text{Solving for } y$$

x	y
1	9
3	3
9	1
$\frac{1}{2}$	18
−1	−9
−3	−3
−9	−1
$-\frac{1}{2}$	−18

Note that we cannot use 0 for x. The x-axis and the y-axis are the asymptotes.

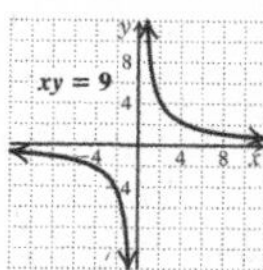

20. Parabola

$$x=y^2+2y-2$$

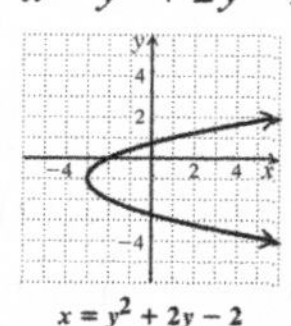

$x=y^2+2y-2$

21. Ellipse

$$\frac{(x+1)^2}{3}+(y-3)^2=1\quad 3>1 \text{ ellipse is horizontal}$$

The center is (–1, 3). Note $a=\sqrt{3}$ and $b=1$.

Vertices: (–1, 2), (–1, 4), $\left(-1-\sqrt{3},\ 3\right)$, $\left(-1+\sqrt{3},\ 3\right)$

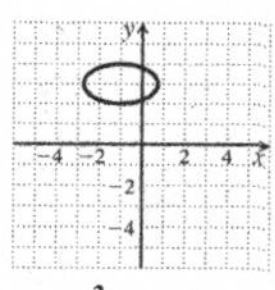

$\frac{(x+1)^2}{3}+(y-3)^2=1$

22. Circle

$$x^2+y^2+6x-8y-39=0$$
$$x^2+6x+y^2-8y=39$$
$$\left(x^2+6x+9\right)+\left(y^2-8y+16\right)=39+9+16$$
$$(x+3)^2+(y-4)^2=64$$

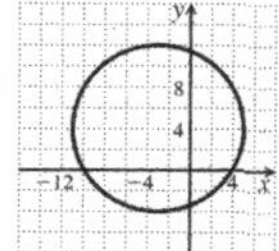

$x^2+y^2+6x-8y-39=0$

23. $x^2-y^2=21,\quad (1)$

$x+y=3\quad (2)$

First we solve Equation (2) for y.

$y=-x+3\quad (3)$

Then we substitute $-x+3$ for y in Equation (1) and solve for x.

$$x^2-(-x+3)^2=21$$
$$x^2-x^2+6x-9=21$$
$$6x=30$$
$$x=5$$

Substitute 5 for x in Equation (3).

$y=-5+3=-2$

The solution is (5, –2).

24. $x^2-2x+2y^2=8,$ (1)

$2x+y=6$ (2)

First we solve Equation (2) for y.

$y=-2x+6$ (3)

Then we substitute $-2x+6$ for y in Equation (1) and solve for x.

$$x^2-2x+2(-2x+6)^2=8$$
$$x^2-2x+8x^2-48x+72=8$$
$$9x^2-50x+64=0$$
$$(x-2)(9x-32)=0$$
$$x=2 \text{ or } x=\frac{32}{9}$$

Now we substitute these numbers for x in Equation (3).

For $x=2,\ y=-2\cdot 2+6=2$

For $x=\frac{32}{9},\ y=-2\left(\frac{32}{9}\right)+6=-\frac{10}{9}$

The solutions are (2, 2) and $\left(\frac{32}{9},-\frac{10}{9}\right)$.

25. $x^2-y=5,$ (1)

$2x-y=5$ (2)

First we solve Equation (2) for y.

$y=2x-5$ (3)

Here we multiply Equation (2) by –1 and then add.

$$x^2-y=5$$
$$-2x+y=-5 \quad \text{Multiplying by } -1$$
$$x^2-2x=0 \quad \text{Adding}$$
$$x(x-2)=0$$
$$x=0 \text{ or } x=2$$

Now we substitute these numbers for x in Equation (3).

If $x=0,\ y=2\cdot 0-5=-5$

If $x=2,\ y=2\cdot 2-5=-1$

The solutions are (0, –5) and (2, –1).

26. $x^2+y^2=25$ (1)

$x^2-y^2=7$ (2)

$$2x^2=32$$
$$x^2=16$$
$$x=\pm 4$$

Simultaneously substitute 4 and –4 for x in Equation (1).

$$(\pm 4)^2+y^2=25$$
$$16+y^2=25$$
$$y^2=9 \text{ or } y=\pm 3$$

The solutions are (4, –3), (4, 3), (–4, –3), and (–4, 3).

27. $x^2-y^2=3,$ (1)

$y=x^2-3$ (2)

First we solve Equation (2) for x.

$$y+3=x^2$$
$$\pm\sqrt{y+3}=x \quad (3)$$

Adding Equations (1) and (2).

$$x^2-y^2=3 \quad (1)$$
$$-x^2+y=-3 \quad (2)$$
$$-y^2+y=0 \quad \text{Adding}$$
$$-y(y-1)=0$$
$$y=0 \text{ or } y=1$$

Now we substitute these numbers for y in Equation (3).

For $y=0,\ x=\pm\sqrt{0+3}=\pm\sqrt{3}$

For $y=1,\ x=\pm\sqrt{1+3}=\pm\sqrt{4}=\pm 2$

The solutions are $(\sqrt{3},0)$, $(-\sqrt{3},0)$, (–2, 1) and (2, 1).

28. $x^2+y^2=18,$ (1)

$2x+y=3$ (2)

First we solve Equation (2) for y.

$y=-2x+3$ (3)

Then we substitute $-2x+3$ for y in Equation (1) and solve for x.

$$x^2+(-2x+3)^2=18$$
$$x^2+4x^2-12x+9=18$$
$$5x^2-12x-9=0$$
$$(x-3)(5x+3)=0$$
$$x=3 \text{ or } x=-\frac{3}{5}$$

Now we substitute these numbers for x in Equation (3).

For $x=3,\ y=-2\cdot 3+3=-3$

For $x=-\frac{3}{5},\ y=-2\left(-\frac{3}{5}\right)+3=\frac{21}{5}$

The solutions are (3, –3) and $\left(-\frac{3}{5},\frac{21}{5}\right)$.

29. $x^2+y^2=100,$ (1)

$2x^2-3y^2=-120$ (2)

We use the elimination method.

$$-2x^2-2y^2=-200 \quad \text{Multiplying (1) by } -2$$
$$2x^2-3y^2=-120 \quad (2)$$
$$-5y^2=-320 \quad \text{Adding}$$
$$y^2=64$$
$$y=\pm 8$$

Solve Equation (1) for x.

$$x^2+y^2=100$$
$$x^2=100-y^2$$
$$x=\pm\sqrt{100-y^2} \quad (3)$$

Since x is solved in terms of y^2, we need only substitute once in Equation (3).

For $y^2=64,\ x=\pm\sqrt{100-64}=\pm\sqrt{36}=\pm 6$

The solutions are (6, 8), (6, –8), (–6, 8) and (–6, –8).

30. $x^2 + 2y^2 = 12,\quad (1)$
$xy = 4\quad (2)$
First we solve Equation (2) for y.
$$y = \frac{4}{x}$$
Then we substitute $\frac{4}{x}$ for y in Equation (1) and solve for x.
$$x^2 + 2\left(\frac{4}{x}\right)^2 = 12$$
$$x^2 + \frac{32}{x^2} = 12$$
$$x^4 + 32 = 12x^2$$
$$x^4 - 12x^2 + 32 = 0$$
$$u^2 - 12u + 32 = 0 \quad \text{Letting } u = x^2$$
$$(u-8)(u-4) = 0$$
$$u = 8 \quad or \quad u = 4$$
$$x^2 = 8 \quad or \quad x^2 = 4$$
$$x = \pm 2\sqrt{2} \quad or \quad x = \pm 2$$
$y = 4/x$, so if $x = 2\sqrt{2}$, $y = \sqrt{2}$; if $x = -2\sqrt{2}$, $y = -\sqrt{2}$; if $x = 2$, $y = 2$; and if $x = -2$, $y = -2$. The pairs $(2\sqrt{2}, \sqrt{2})$, $(-2\sqrt{2}, -\sqrt{2})$, (2, 2), and (–2,–2) check. They are the solutions.

31. ***Familiarize***. Let l = the length and w = the width of the bandstand.
Translate.
Perimeter: $2l + 2w = 38$, or $l + w = 19$
Area: $lw = 84$
Carry out. We solve the system:
Solve the first equation for l: $l = 19 - w$.
Substitute $19 - w$ for l in the second equation and solve for w.
$$(19 - w)w = 84$$
$$19w - w^2 = 84$$
$$0 = w^2 - 19w + 84$$
$$0 = (w-7)(w-12)$$
$$w = 7 \quad or \quad w = 12$$
If $w = 7$, then $l = 19 - 7$, or 12. If $w = 12$, then $l = 19 - 12$, or 7. Since length is usually considered to be longer than width, we have the solution $l = 12$ and $w = 7$, or (12, 7)
Check. If $l = 12$ and $w = 7$, the area is $12 \cdot 7$, or 84. The perimeter is $2 \cdot 12 + 2 \cdot 7$, or 38. The numbers check.
State. The length is 12 m and the width is 7 m.

32. Let l and w represent the length and width, respectively. Solve the system:
$$l^2 + w^2 = 15^2,$$
$$lw = 108$$
The solutions are (12, 9), (–12, –9), (9, 12), and $(-9, -12)$. Only (12, 9) and (9, 12) have meaning in this problem. Choosing the larger number as the length, we have the solution. The length is 12 in., and the width is 9 in.

33. ***Familiarize***. Let x represent the length of a side of the mounting board and y represent the length of a side of the square mirror.
Translate.
$$4x = 4y + 12\text{, or } x = y + 3$$
$$x^2 = y^2 + 39$$
Carry out. We solve the system of equations.
$$x = y + 3 \quad (1)$$
$$x^2 = y^2 + 39 \quad (2)$$
We substitute $y + 3$ for x into Equation (2).
$$(y+3)^2 = y^2 + 39$$
$$y^2 + 6y + 9 = y^2 + 39$$
$$6y + 9 = 39$$
$$6y = 30$$
$$y = 5$$
Then $x = 5 + 3 = 8$.
Check. If $x = 8$ and $y = 5$, then $4 \cdot 5 + 12 = 32 = 4 \cdot 8$, and $5^2 + 39 = 64 = 8^2$. The numbers check. The perimeter of the mounting board is $4 \cdot 8$, or 32, and the perimeter of the mirror is $4 \cdot 5$, or 20.
State. The perimeter of the mounting board is 32 cm and 20 cm, respectively.

34. Let x and y represent the radii of the two circles. Solve the system:
$$\pi x^2 + \pi y^2 = 130\pi \quad \text{Sum of the areas}$$
$$2\pi x - 2\pi y = 16\pi \quad \text{Difference in circumferences}$$
The system can be simplified to:
$$x^2 + y^2 = 130$$
$$x - y = 8$$
The solution (–11, –3) has no meaning for this application, so the only solution is (3, 11).
The radius of one circle is 3 ft, and the other is 11 ft.

35. *Writing Exercise.*

36. *Writing Exercise.*

37. $4x^2 - x - 3y^2 = 9,\quad (1)$
$-x^2 + x + y^2 = 2\quad (2)$
We use the elimination method.
$$4x^2 - x - 3y^2 = 9 \quad (1)$$
$$-3x^2 + 3x + 3y^2 = 6 \quad \text{Multiplying (2) by 3}$$
$$x^2 + 2x = 15 \quad \text{Adding}$$
$$x^2 + 2x - 15 = 0$$
$$(x+5)(x-3) = 0$$
$$x = -5 \quad or \quad x = 3$$
Solving Equation (2) for y:
$$-x^2 + x + y^2 = 2$$
$$y^2 = x^2 - x + 2$$
$$y = \pm\sqrt{x^2 - x + 2}$$
For $x = -5$, $y = \pm\sqrt{(-5)^2 - (-5) + 2} = \pm\sqrt{32} = \pm 4\sqrt{2}$

For $x = 3$, $y = \pm\sqrt{3^2 - 3 + 2} = \pm\sqrt{8} = \pm 2\sqrt{2}$

The solutions are $(-5, -4\sqrt{2})$, $(-5, 4\sqrt{2})$, $(3, -2\sqrt{2})$, $(3, 2\sqrt{2})$.

38. The three points are equidistant from the center of the circle, (h, k). Using each of the three points in the equation of a circle, we get three different equations.

For $(-2, -4)$,
$$[x-(-2)]^2 + [y-(-4)]^2 = r^2$$
$$(x+2)^2 + (y+4)^2 = r^2$$
$$x^2 + 4x + 4 + y^2 + 8y + 16 = r^2$$
$$x^2 + 4x + y^2 + 8y + 20 = r^2 \quad (1)$$

For $(5, -5)$,
$$(x-5)^2 + [y-(-5)]^2 = r^2$$
$$(x-5)^2 + (y+5)^2 = r^2$$
$$x^2 - 10x + 25 + y^2 + 10y + 25 = r^2$$
$$x^2 - 10x + y^2 + 10y + 50 = r^2 \quad (2)$$

For $(6, 2)$,
$$(x-6)^2 + (y-2)^2 = r^2$$
$$x^2 - 12x + 36 + y^2 - 4y + 4 = r^2$$
$$x^2 - 12x + y^2 - 4y + 40 = r^2 \quad (3)$$

Since the radius is equal, we can set Equation (1) equal to Equation (2) and simplify.
$$x^2 + 4x + y^2 + 8y + 20 = x^2 - 10x + y^2 + 10y + 50$$
$$14x - 2y = 30$$
$$7x - y = 15 \quad (4)$$

Next, we set Equation (1) equal to Equation (3).
$$x^2 + 4x + y^2 + 8y + 20 = x^2 - 12x + y^2 - 4y + 40$$
$$16x + 12y = 20$$
$$4x + 3y = 5 \quad (5)$$

We solve the system of Equations (4) and (5) using the elimination method.
$$\begin{aligned} 21x - 3y &= 45 && \text{Multiplying (4) by 3} \\ 4x + 3y &= 5 && (5) \\ \hline 25x \quad &= 50 && \text{Adding} \\ x &= 2 \\ y &= -1 \end{aligned}$$

The center of the circle is $(2, -1)$. Thus the equation is
$$(x-2)^2 + (y+1)^2 = r^2.$$

We may choose any of the three points on the circle to determine r^2.
$$(6-2)^2 + (2+1)^2 = r^2$$
$$4^2 + 3^2 = r^2$$
$$25 = r^2$$

The equation of the circle is $(x-2)^2 + (y+1)^2 = 25$.

39. From the x-intercepts, $(-10, 0)$ and $(10, 0)$, we know $a = 10$; from the y-intercepts, $(0, -1)$ and $(0, 1)$, we know

$b = 1$. So, we have the equation:
$$\frac{x^2}{10^2} + \frac{y^2}{1^2} = 1$$
$$\frac{x^2}{100} + \frac{y^2}{1} = 1$$

Chapter 13 Test

1. For circle with center $(3, -4)$ and radius $2\sqrt{3}$,
$$(x-3)^2 + [y-(-4)]^2 = (2\sqrt{3})^2$$
$$(x-3)^2 + (y+4)^2 = 12$$

2.
$$(x-4)^2 + (y+1)^2 = 5$$
$$(x-4)^2 + [y-(-1)]^2 = (\sqrt{5})^2$$

Center: $(4, -1)$; radius: $\sqrt{5}$

3.
$$x^2 + y^2 + 4x - 6y + 4 = 0$$
$$x^2 + 4x + y^2 - 6y = -4$$
$$(x^2 + 4x + 4) + (y^2 - 6y + 9) = -4 + 4 + 9$$
$$(x+2)^2 + (y-3)^2 = 3^2$$

The center is $(-2, 3)$ and the radius is 3.

4. Parabola

$y = x^2 - 4x - 1$

$y = x^2 - 4x - 1$

5. Circle
$$x^2 + y^2 + 2x + 6y + 6 = 0$$
$$x^2 + 2x + y^2 + 6y = -6$$
$$(x^2 + 2x + 1) + (y^2 + 6y + 9) = -6 + 1 + 9$$
$$(x+1)^2 + (y+3)^2 = 4$$

The center is $(-1, -3)$ and the radius is 2.

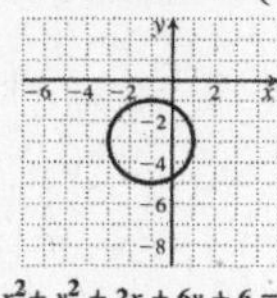

$x^2 + y^2 + 2x + 6y + 6 = 0$

6. Hyperbola

$$\frac{x^2}{16} - \frac{y^2}{9} = 1$$

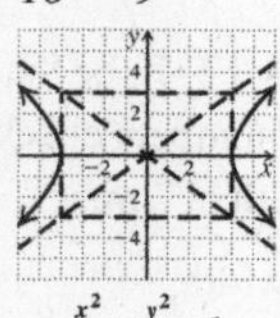

$\frac{x^2}{16} - \frac{y^2}{9} = 1$

7. Ellipse

$16x^2 + 4y^2 = 64 \quad 4<16$ ellipse is vertical
$$\frac{x^2}{4} + \frac{y^2}{16} = 1$$

The center is $(0, 0)$. Note $a = 2$ and $b = 4$.

Vertices: (0, 2), (0, –2), (4, 0), and (–4, 0)

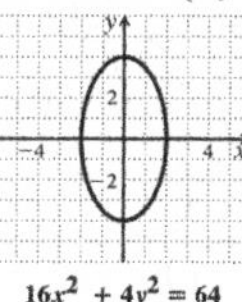

8. Hyperbola

$xy=-5$

$y=\frac{-5}{x}$ Solving for y

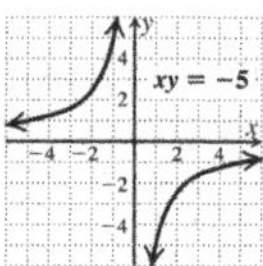

9. Parabola

$x=-y^2+4y$

$y=-\frac{b}{2a}=-\frac{4}{2(-1)}=2$

$x=-y^2+4y=-2^2+4\cdot 2=4$

The vertex is (4, 2). The graph opens to the left because the coefficient of y^2, –1 is negative.

x	y
0	0
3	1
4	2
3	3
0	4

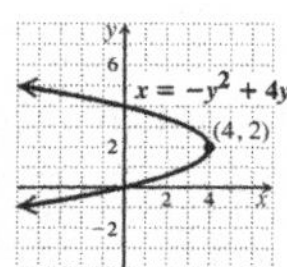

10. $x^2+y^2=36$, (1)

$3x+4y=24$ (2)

First solve Equation (2) for y.

$4y=-3x+24$

$y=-\frac{3}{4}x+6$ (3)

Then substitute $-\frac{3}{4}x+6$ for y in Equation (1) and solve for x.

$$x^2+y^2=36$$
$$x^2+\left(-\tfrac{3}{4}x+6\right)^2=36$$
$$x^2+\tfrac{9}{16}x^2-9x+36=36$$
$$\tfrac{25}{16}x^2-9x=0$$
$$25x^2-144x=0$$
$$x(25x-144)=0$$
$$x=0 \text{ or } x=\frac{144}{25}$$

Now substitute these numbers in Equation (3) and solve for y.

For $x=0$, $y=-\frac{3}{4}\cdot 0+6=6$

For $x=\frac{144}{25}$, $y=-\frac{3}{4}\cdot\frac{144}{25}+6=\frac{42}{25}$

The pairs (0, 6) and $\left(\frac{144}{25},\frac{42}{25}\right)$ check, so they are the solutions.

11. $x^2-y=3$, (1)

$2x+y=5$ (2)

Use the elimination method.

$$x^2-y=3 \quad (1)$$
$$2x+y=5 \quad (2)$$
$$x^2+2x=8 \quad \text{Adding}$$
$$x^2+2x-8=0$$
$$(x+4)(x-2)=0$$
$$x=-4 \text{ or } x=2$$

Substitute for x in Equation (2).

$y=5-2x$ Solving Eq. (2) for y

For $x=-4$, $y=5-2(-4)=13$

For $x=2$, $y=5-2(2)=1$

The pairs (–4, 13) and (2, 1) check. They are the solutions.

12. $x^2-2y^2=1$, (1)

$xy=6$ (2)

First we solve Equation (2) for y.

$xy=6$

$y=\frac{6}{x}$

Then we substitute $\frac{6}{x}$ for y in Equation (1) and solve for x.

$$x^2-2y^2=1$$
$$x^2-2\left(\frac{6}{x}\right)^2=1$$
$$x^2-\frac{72}{x^2}=1$$
$$x^4-72=x^2 \quad \text{Multiplying by } x^2$$
$$x^4-x^2-72=0$$
$$u^2-u-72=0 \quad \text{Letting } u=x^2$$
$$(u-9)(u+8)=0$$
$$u=9 \text{ or } u=-8$$

We now substitute x^2 for u and solve for x.

$$x^2=9 \text{ or } x^2=-8$$
$$x=\pm 3 \text{ or } x=\pm 2\sqrt{2}i$$

Since $y=6/x$, if $x=3$, $y=2$; if $x=-3$, $y=-2$; if $x=2\sqrt{2}i$, $y=-\frac{3\sqrt{2}}{2}i$; and if $x=-2\sqrt{2}i$, $y=\frac{3\sqrt{2}}{2}i$.

The pairs (3, 2), (–3, –2), $\left(-2\sqrt{2}i, \frac{3\sqrt{2}}{2}i\right)$ and $\left(2\sqrt{2}i, -\frac{3\sqrt{2}}{2}i\right)$ check. They are the solutions.

13. $x^2+y^2=10$, (1)

$x^2=y^2+2$ (2)

Substitute y^2+2 for x^2 in Equation (1) and solve for y.

$$x^2+y^2=10$$
$$y^2+2+y^2=10$$
$$2y^2-8=0$$
$$2(y^2-4)=0$$
$$2(y+2)(y-2)=0$$
$$y=-2 \text{ or } y=2$$

Now substitute these numbers in Equation (2) and solve for x.

For $y=-2$, $x^2=(-2)^2+2=6$, so $x=\pm\sqrt{6}$
For $y=2$, $x^2=2^2+2=6$, so $x=\pm\sqrt{6}$

The pairs $(-\sqrt{6},-2)$, $(-\sqrt{6},2)$, $(\sqrt{6},-2)$, $(\sqrt{6},2)$ check, so they are the solutions.

14. Let l = the length and w = the width of the bookmark. We solve the system:

$$lw=22,$$
$$l^2+w^2=(5\sqrt{5})^2$$

The negative solutions (–2, –11) and (–11, –2) are not included since they have no meaning in this application. The remaining solutions are (11, 2) and (2, 11). We choose the larger number to be the length, so the length is 11 and the width is 2.

15. ***Familiarize.*** We let x = the length of a side of one square, in meters, and y = the length of a side of the other square. Make a drawing.

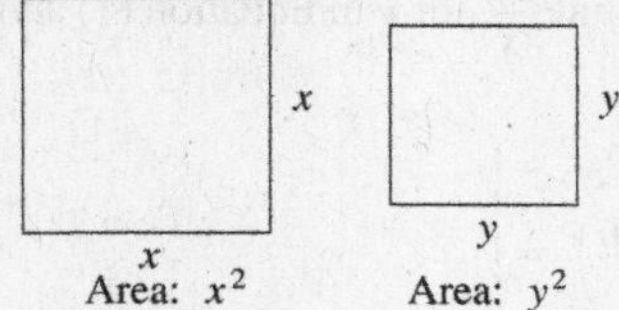

Translate. The sum of the areas is 8 m^2, so we have

$x^2+y^2=8$. (1)

The difference of the areas is 2 m^2, so we have

$x^2-y^2=2$. (2)

Carry out. We solve the system of equations.

$$x^2+y^2=8 \quad (1)$$
$$x^2-y^2=2 \quad (2)$$
$$2x^2=10 \quad \text{Adding}$$
$$x^2=5$$
$$x=\pm\sqrt{5}$$

Since the length cannot be negative we consider only $\sqrt{5}$. We substitute $\sqrt{5}$ for x in Equation (1) and solve for y.

$$(\sqrt{5})^2+y^2=8$$
$$5+y^2=8$$
$$y^2=3$$
$$y=\pm\sqrt{3}$$

Again we consider only the positive number.

Check. If the lengths of the sides of the squares are $\sqrt{5}$ m and $\sqrt{3}$ m, the areas of the squares are $(\sqrt{5})^2$, or 5 m^2, and $(\sqrt{3})^2$, or 3 ft^2, respectively. Then $5\text{ m}^2+3\text{ m}^2=8\text{ m}^2$, and $5\text{ m}^2-3\text{ m}^2=2\text{ m}^2$, so the answer checks.

State. The lengths of the sides of the squares are $\sqrt{5}$ m and $\sqrt{3}$ m.

16. Let l = the length and w = the width of the floor. We solve the system:

$$l^2+w^2=(40)^2$$
$$2l+2w=112$$

The solutions are (32, 24) and (24, 32). We choose the larger number to be the length, so the length is 32 ft and the width is 24 ft.

17. ***Familiarize.*** Let p = the principal and r = the interest rate. We recall the formula $I=prt$. Since the time is one year, $t=1$, and the formula simplifies to $I=pr$.

Translate.

Brett invested p dollars at interest rate r with \$72 in interest.

Erin invested $p+240$ dollars at $\frac{5}{6}r$ interest rate with \$72 in interest. We have two equations.

$$72=pr \quad (1)$$
$$72=(p+240)\frac{5}{6}r \quad (2)$$

Carry out. We solve the system of equations using substitution.

We solve Equation (1) for r: $r=\frac{72}{p}$.

We substitute $\frac{72}{p}$ for r in Equation (2) and solve for p.

$$72=(p+240)\frac{5}{6}\left(\frac{72}{p}\right)$$
$$72=(p+240)\frac{60}{p}$$
$$72p=60p+14{,}400$$
$$12p=14{,}400$$
$$p=1200$$
$$r=\frac{72}{1200}=0.06 \text{ or } 6\%$$

Check. $1200\cdot 0.06=72$ and $(1200+240)\frac{5}{6}\cdot 0.06=72$, so the numbers check.

State. The principal was \$1200 and the interest rate was 6%.

18. Plot the given points. From the location of these points, we see that the ellipse which contains them is centered at (6, 3). The distance from (1, 3) to (6, 3) is 5, so $a=5$. The distance from (6, 6) to (6, 3) is 3, so $b=3$. We write the equation of the ellipse.

$$\frac{(x-6)^2}{5^2}+\frac{(y-3)^2}{3^2}=1$$
$$\frac{(x-6)^2}{25}+\frac{(y-3)^2}{9}=1$$

19. Let x and y represent the numbers. We solve the system.

$$x + y = 36 \quad (1)$$
$$xy = 4 \quad (2)$$

The numbers are $18 - 8\sqrt{5}$ and $18 + 8\sqrt{5}$.
The sum of the reciprocals is

$$\frac{1}{18+8\sqrt{5}} + \frac{1}{18-8\sqrt{5}} = 9.$$

20. Let the actor be in the center at (0, 0). Using the information, we have (–4, 0) and (4, 0) and (0, –7) and (0, 7). Thus, $a = 4$ and $b = 7$. We write the equation of the ellipse.

$$\frac{x^2}{4^2} + \frac{y^2}{7^2} = 1$$
$$\frac{x^2}{16} + \frac{y^2}{49} = 1$$

Chapter 14

Sequences, Series, and the Binomial Theorem

Exercise Set 14.1

1. B

3. B

5. A

7. f

9. d

11. c

13. $a_n = 5n+3$
$a_8 = 5\cdot 8+3 = 40+3 = 43$

15. $a_n = (3n+1)(2n-5)$
$a_9 = (3\cdot 9+1)(2\cdot 9-5) = 28\cdot 13 = 364$

17. $a_n = (-1)^{n-1}(3.4n-17.3)$
$a_{12} = (-1)^{12-1}[3.4(12)-17.3] = -23.5$

19. $a_n = 3n^2(9n-100)$
$a_{11} = 3\cdot 11^2(9\cdot 11-100) = 3\cdot 121(-1) = -363$

21. $a_n = \left(1+\frac{1}{n}\right)^2$
$a_{20} = \left(1+\frac{1}{20}\right)^2 = \left(\frac{21}{20}\right)^2 = \frac{441}{400}$

23. $a_n = 3n-1$
$a_1 = 3\cdot 1-1 = 2$
$a_2 = 3\cdot 2-1 = 5$
$a_3 = 3\cdot 3-1 = 8$
$a_4 = 3\cdot 4-1 = 11$
$a_{10} = 3\cdot 10-1 = 29$
$a_{15} = 3\cdot 15-1 = 44$

25. $a_n = n^2+2$
$a_1 = 1^2+2 = 3$
$a_2 = 2^2+2 = 6$
$a_3 = 3^2+2 = 11$
$a_4 = 4^2+2 = 18$
$a_{10} = 10^2+2 = 102$
$a_{15} = 15^2+2 = 227$

27. $a_n = \frac{n}{n+1}$
$a_1 = \frac{1}{1+1} = \frac{1}{2}$
$a_2 = \frac{2}{2+1} = \frac{2}{3}$
$a_3 = \frac{3}{3+1} = \frac{3}{4}$
$a_4 = \frac{4}{4+1} = \frac{4}{5}$
$a_{10} = \frac{10}{10+1} = \frac{10}{11}$
$a_{15} = \frac{15}{15+1} = \frac{15}{16}$

29. $a_n = \left(-\frac{1}{2}\right)^{n-1}$
$a_1 = \left(-\frac{1}{2}\right)^{1-1} = 1$
$a_2 = \left(-\frac{1}{2}\right)^{2-1} = -\frac{1}{2}$
$a_3 = \left(-\frac{1}{2}\right)^{3-1} = \frac{1}{4}$
$a_4 = \left(-\frac{1}{2}\right)^{4-1} = -\frac{1}{8}$
$a_{10} = \left(-\frac{1}{2}\right)^{10-1} = -\frac{1}{512}$
$a_{15} = \left(-\frac{1}{2}\right)^{15-1} = \frac{1}{16,384}$

31. $a_n = \frac{(-1)^n}{n}$
$a_1 = \frac{(-1)^1}{1} = -1$
$a_2 = \frac{(-1)^2}{2} = \frac{1}{2}$
$a_3 = \frac{(-1)^3}{3} = -\frac{1}{3}$
$a_4 = \frac{(-1)^4}{4} = \frac{1}{4}$
$a_{10} = \frac{(-1)^{10}}{10} = \frac{1}{10}$
$a_{15} = \frac{(-1)^{15}}{15} = -\frac{1}{15}$

33. $a_n = (-1)^n(n^3 - 1)$
$a_1 = (-1)^1(1^3 - 1) = 0$
$a_2 = (-1)^2(2^3 - 1) = 7$
$a_3 = (-1)^3(3^3 - 1) = -26$
$a_4 = (-1)^4(4^3 - 1) = 63$
$a_{10} = (-1)^{10}(10^3 - 1) = 999$
$a_{15} = (-1)^{15}(15^3 - 1) = -3374$

35. 2, 4, 6, 8, 10,...
These are even integers beginning with 2, so the general term could be $2n$.

37. $-1,\ 1, -1,\ 1, \ldots$
-1 and 1 alternate, beginning with -1, so the general term could be $(-1)^n$.

39. $1, -2,\ 3, -4, \ldots$
These are the first four natural numbers, but with alternating signs, beginning with a positive number. The general term could be $(-1)^{n+1} \cdot n$.

41. 3, 5, 7, 9,...
These are odd integers beginning with 3, so the general term could be $2n + 1$.

43. 0, 3, 8, 15, 24,...
We can see a pattern if we write the sequence $1^2 - 1,\ 2^2 - 1,\ 3^2 - 1,\ 4^2 - 1,\ 5^2 - 1, \ldots$ The general term could be $n^2 - 1$, or $(n+1)(n-1)$.

45. $\frac{1}{2}, \frac{2}{3}, \frac{3}{4}, \frac{4}{5}, \frac{5}{6}, \ldots$
These are fractions in which the denominator is 1 greater than the numerator. Also, each numerator is 1 greater than the preceding numerator. The general term could be $\frac{n}{n+1}$.

47. 0.1, 0.01, 0.001, 0.0001,...
This is negative powers of 10, or positive powers of 0.1, so the general term is 10^{-n}, or $(0.1)^n$.

49. $-1,\ 4,\ -9,\ 16, \ldots$
This is the squares of the first four natural numbers, but with alternating signs, beginning with a negative number. The general terms could be $(-1)^n \cdot n^2$.

51. $-1,\ 2, -3,\ 4, -5,\ 6, \ldots$
$S_{10} = -1 + 2 - 3 + 4 - 5 + 6 - 7 + 8 - 9 + 10 = 5$

53. $1, \frac{1}{10}, \frac{1}{100}, \frac{1}{1000}, \ldots$

$$S_6 = 1 + \frac{1}{10} + \frac{1}{100} + \frac{1}{1000} + \frac{1}{10{,}000} + \frac{1}{100{,}000}$$
$$= 1.11111, \text{ or } 1\frac{11{,}111}{100{,}000}$$

55.
$$\sum_{k=1}^{5} \frac{1}{2k} = \frac{1}{2\cdot 1} + \frac{1}{2\cdot 2} + \frac{1}{2\cdot 3} + \frac{1}{2\cdot 4} + \frac{1}{2\cdot 5}$$
$$= \frac{1}{2} + \frac{1}{4} + \frac{1}{6} + \frac{1}{8} + \frac{1}{10}$$
$$= \frac{60}{120} + \frac{30}{120} + \frac{20}{120} + \frac{15}{120} + \frac{12}{120}$$
$$= \frac{137}{120}$$

57.
$$\sum_{k=0}^{4} 10^k = 10^0 + 10^1 + 10^2 + 10^3 + 10^4$$
$$= 1 + 10 + 100 + 1000 + 10{,}000$$
$$= 11{,}111$$

59.
$$\sum_{k=2}^{8} \frac{k}{k-1}$$
$$= \frac{2}{2-1} + \frac{3}{3-1} + \frac{4}{4-1} + \frac{5}{5-1} + \frac{6}{6-1} + \frac{7}{7-1} + \frac{8}{8-1}$$
$$= \frac{2}{1} + \frac{3}{2} + \frac{4}{3} + \frac{5}{4} + \frac{6}{5} + \frac{7}{6} + \frac{8}{7}$$
$$= \frac{1343}{140}$$

61.
$$\sum_{k=1}^{8} (-1)^{k+1} 2^k$$
$$= (-1)^{1+1} 2^1 + (-1)^{2+1} 2^2 + (-1)^{3+1} 2^3 + (-1)^{4+1} 2^4$$
$$+ (-1)^{5+1} 2^5 + (-1)^{6+1} 2^6 + (-1)^{7+1} 2^7 + (-1)^{8+1} 2^8$$
$$= 2 - 4 + 8 - 16 + 32 - 64 + 128 - 256$$
$$= -170$$

63.
$$\sum_{k=0}^{5} (k^2 - 2k + 3)$$
$$= (0^2 - 2\cdot 0 + 3) + (1^2 - 2\cdot 1 + 3) + (2^2 - 2\cdot 2 + 3)$$
$$+ (3^2 - 2\cdot 3 + 3) + (4^2 - 2\cdot 4 + 3) + (5^2 - 2\cdot 5 + 3)$$
$$= 3 + 2 + 3 + 6 + 11 + 18$$
$$= 43$$

65.
$$\sum_{k=3}^{5} \frac{(-1)^k}{k(k+1)} = \frac{(-1)^3}{3(3+1)} + \frac{(-1)^4}{4(4+1)} + \frac{(-1)^5}{5(5+1)}$$
$$= \frac{-1}{3\cdot 4} + \frac{1}{4\cdot 5} + \frac{-1}{5\cdot 6}$$
$$= -\frac{1}{12} + \frac{1}{20} - \frac{1}{30}$$
$$= -\frac{4}{60} = -\frac{1}{15}$$

67. $\frac{2}{3} + \frac{3}{4} + \frac{4}{5} + \frac{5}{6} + \frac{6}{7}$
This is a sum of fractions in which the denominator is one greater than the numerator. Also, each numerator is 1 greater than the preceding numerator. Sigma notation is

$$\sum_{k=1}^{5} \frac{k+1}{k+2}.$$

69. $1 + 4 + 9 + 16 + 25 + 36$
This is the sum of the squares of the first six natural

numbers. Sigma notation is

$$\sum_{k=1}^{6} k^2.$$

71. $4-9+16-25+...+(-1)^n n^2$

This is a sum of terms of the form $(-1)^k k^2$, beginning with $k=2$ and continuing through $k=n$. Sigma notation is

$$\sum_{k=2}^{n} (-1)^k k^2.$$

73. $6+12+18+24+...$

This is the sum of all the positive multiples of 6. It is an infinite series. Sigma notation is

$$\sum_{k=1}^{\infty} 6k.$$

75. $\frac{1}{1\cdot 2}+\frac{1}{2\cdot 3}+\frac{1}{3\cdot 4}+\frac{1}{4\cdot 5}+...$

This is a sum of fractions in which the numerator is 1 and the denominator is a product of two consecutive integers. The larger integer in each product is the smaller integer in the succeeding product. It is an infinite series. Sigma notation is

$$\sum_{k=1}^{\infty} \frac{1}{k(k+1)}.$$

77. *Writing Exercise.*

79. $\frac{t^3+1}{t+1}=\frac{(t+1)(t^2-t+1)}{t+1}=t^2-t+1$

81.
$$\begin{aligned}\frac{3}{a^2+a}+\frac{4}{2a^2-2}&=\frac{3}{a(a+1)}+\frac{2\cdot 2}{2(a^2-1)}\\&=\frac{3}{a(a+1)}+\frac{2}{(a+1)(a-1)}\\&=\frac{3}{a(a+1)}\cdot\frac{a-1}{a-1}+\frac{2}{(a+1)(a-1)}\cdot\frac{a}{a}\\&=\frac{3a-3+2a}{a(a+1)(a-1)}\\&=\frac{5a-3}{a(a+1)(a-1)}\end{aligned}$$

83.
$$\begin{aligned}\frac{x^2-6x+8}{4x+12}\cdot\frac{x+3}{x^2-4}&=\frac{(x-2)(x-4)}{4(x+3)}\cdot\frac{x+3}{(x+2)(x-2)}\\&=\frac{x-4}{4(x+2)}\end{aligned}$$

85. *Writing Exercise.*

87. $a_1=1,\ a_{n+1}=5a_n-2$

$a_1=1$

$a_2=5\cdot 1-2=3$

$a_3=5\cdot 3-2=13$

$a_4=5\cdot 13-2=63$

$a_5=5\cdot 63-2=313$

$a_6=5\cdot 313-2=1563$

89. Find each term by multiplying the preceding term by 0.80: \$2500, \$2000, \$1600, \$1280, \$1024, \$819.20, \$655.36, \$524.29, \$419.43, \$335.54.

91. $a_n=(-1)^n$

This sequence is of the form −1, 1, −1, 1, ... Each pair of terms adds to 0. S_{100} has 50 such pairs, so $S_{100}=0$. S_{101} consists of the 50 pairs in S_{100} that add to 0 as well as a_{101}, or −1, so $S_{101}=-1$.

93. $a_n=i^n$

$a_1=i^1=i$

$a_2=i^2=-1$

$a_3=i^3=i^2\cdot i=-1\cdot i=-i$

$a_4=i^4=(i^2)^2=(-1)^2=1$

$a_5=i^5=(i^2)^2\cdot i=(-1)^2\cdot i=1\cdot i=i$

$S_5=i-1-i+1+i=i$

95. Enter $y_1=x^5-14x^4+6x^3+416x^2-655x-1050.$ Then scroll through a table of values. We see that $y_1=6144$ when $x=11$, so the 11th term of the sequence is 6144.

Exercise Set 14.2

1. 5 + 7 + 9 + 11 is an example of an *arithmetic series*.

3. In 5, 7, 9, 11, the *first term* is 5.

5. 8, 13, 18, 23, ...

$a_1=8$

$d=5$ (13 − 8 = 5, 18 − 13 = 5, 23 − 18 = 5)

7. 7, 3, −1, −5, ...

$a_1=7$

$d=-4$ (3 − 7 = −4, −1 − 3 = −4, −5 − (−1) = −4)

9. $\frac{3}{2}, \frac{9}{4}, 3, \frac{15}{4},...$

$a_1=\frac{3}{2}$

$d=\frac{3}{4}$ $\left(\frac{9}{4}-\frac{3}{2}=\frac{3}{4},\ 3-\frac{9}{4}=\frac{3}{4}\right)$

11. \$8.16, \$8.46, \$8.76, \$9.06, ...

$a_1=\$8.16$

$d=\$0.30$ (\$8.46 − \$8.16 = \$0.30, \$8.76 − \$8.46 = \$0.30, \$9.06 − \$8.76 = \$0.30)

13. 10, 18, 26, ...

$a_1=10,\ d=8$, and $n=19$

$a_n=a_1+(n-1)d$

$a_{19}=10+(19-1)8=10+18\cdot 8=10+144=154$

15. 8, 2, −4, ...

$a_1=8,\ d=-6$, and $n=18$

$a_n=a_1+(n-1)d$

$a_{18}=8+(18-1)(-6)=8+17(-6)=8-102=-94$

17. \$1200, \$964.32, \$728.64, …

$a_1 = \$1200,\ d = \$964.32 - \$1200 = -\$235.68$ and $n = 13$

$$a_n = a_1 + (n-1)d$$

$$\begin{aligned} a_{13} &= \$1200 + (13-1)(-\$235.68) \\ &= \$1200 + 12(-\$235.68) = \$1200 - \$2828.16 \\ &= -\$1628.16 \end{aligned}$$

19. $a_1 = 10,\ d = 8$

$a_n = a_1 + (n-1)d$

Let $a_n = 210$, and solve for n.

$$\begin{aligned} 210 &= 10 + (n-1)8 \\ 210 &= 10 + 8n - 8 \\ 210 &= 2 + 8n \\ 208 &= 8n \\ 26 &= n \end{aligned}$$

The 26th term is 210.

21. $a_1 = 8,\ d = -6$

$a_n = a_1 + (n-1)d$

$$\begin{aligned} -328 &= 8 + (n-1)(-6) \\ -328 &= 8 - 6n + 6 \\ -328 &= 14 - 6n \\ -342 &= -6n \\ 57 &= n \end{aligned}$$

The 57th term is –328.

23. $a_n = a_1 + (n-1)d$

$a_{18} = 8 + (18-1)10$ Substituting 18 for n, 8 for a_1, and 10 for d

$$\begin{aligned} &= 8 + 17 \cdot 10 \\ &= 8 + 170 \\ &= 178 \end{aligned}$$

25. $a_n = a_1 + (n-1)d$

$33 = a_1 + (8-1)4$ Substituting 33 for a_8, 8 for n, and 4 for d

$$\begin{aligned} 33 &= a_1 + 28 \\ 5 &= a_1 \end{aligned}$$

27. $a_n = a_1 + (n-1)d$

$-76 = 5 + (n-1)(-3)$ Substituting -76 for a_n, 5 for a_1, and -3 for d

$$\begin{aligned} -76 &= 5 - 3n + 3 \\ -76 &= 8 - 3n \\ -84 &= -3n \\ 28 &= n \end{aligned}$$

29. We know that $a_{17} = -40$ and $a_{28} = -73$. We would have to add d eleven times to get from a_{17} to a_{28}. That is,

$$\begin{aligned} -40 + 11d &= -73 \\ 11d &= -33 \\ d &= -3. \end{aligned}$$

Since $a_{17} = -40$, we subtract d sixteen times to get a_1.

$$a_1 = -40 - 16(-3) = -40 + 48 = 8$$

We write the first five terms of the sequence:
8, 5, 2, –1, –4.

31. $a_{13} = 13$ and $a_{54} = 54$

Observe that for this to be true, $a_1 = 1$ and $d = 1$.

33. $1 + 5 + 9 + 13 + \ldots$

Note that $a_1 = 1$, $d = 4$, and $n = 20$. Before using the formula for S_n, we find a_{20}:

$a_{20} = 1 + (20-1)4$ Substituting into the formula for a_n

$$\begin{aligned} &= 1 + 19 \cdot 4 \\ &= 77 \end{aligned}$$

Then using the formula for S_n,

$$S_{20} = \frac{20}{2}(1 + 77) = 10(78) = 780.$$

35. The sum is $1 + 2 + 3 + \ldots + 249 + 250$. This is the sum of the arithmetic sequence for which $a_1 = 1$, $a_n = 250$, and $n = 250$. We use the formula for S_n.

$$S_n = \frac{n}{2}(a_1 + a_n)$$

$$S_{250} = \frac{250}{2}(1 + 250) = 125(251) = 31{,}375$$

37. The sum is $2 + 4 + 6 + \ldots + 98 + 100$. This is the sum of the arithmetic sequence for which $a_1 = 2$, $a_n = 100$, and $n = 50$. We use the formula for S_n.

$$S_n = \frac{n}{2}(a_1 + a_n)$$

$$S_{50} = \frac{50}{2}(2 + 100) = 25(102) = 2550$$

39. The sum is $6 + 12 + 18 + \ldots + 96 + 102$. This is the sum of the arithmetic sequence for which $a_1 = 6$, $a_n = 102$, and $n = 17$. We use the formula for S_n.

$$S_n = \frac{n}{2}(a_1 + a_n)$$

$$S_{17} = \frac{17}{2}(6 + 102) = \frac{17}{2}(108) = 918$$

41. Before using the formula for S_n, we find a_{20}:

$a_{20} = 4 + (20-1)5$ Substituting into the formula for a_n

$$\begin{aligned} &= 4 + 19 \cdot 5 \\ &= 99 \end{aligned}$$

Then using the formula for S_n,

$$S_{20} = \frac{20}{2}(4 + 99) = 10(103) = 1030.$$

43. ***Familiarize***. We want to find the fifteenth term and the sum of an arithmetic sequence with $a_1 = 7$, $d = 2$, and $n = 15$. We will first use the formula for a_n to find a_{15}. This result is the number of musicians in the last row. Then we will use the formula for S_n to find S_{15}. This is the total number of musicians.

Translate. Substituting into the formula for a_n, we have

$$a_{15} = 7 + (15-1)2.$$

Carry out. We first find a_{15}.

$$a_{15} = 7 + 14 \cdot 2 = 35$$

Then use the formula for S_n to find S_{15}.

$$S_{15} = \frac{15}{2}(7+35) = \frac{15}{2}(42) = 315$$

Check. We can do the calculations again. We can also do the entire addition.

$$7 + 9 + 11 + \ldots + 35.$$

State. There are 35 musicians in the last row, and there are 315 musicians altogether.

45. ***Familiarize.*** We want to find the sum of the arithmetic sequence $36 + 32 + \ldots + 4$. Note that $a_1 = 36$, and $d = -4$. We will first use the formula for a_n to find n. Then we will use the formula for S_n.

Translate. Substituting into the formula for a_n, we have

$$4 = 36 + (n-1)(-4).$$

Carry out. We solve for n.

$$\begin{aligned} 4 &= 36 + (n-1)(-4) \\ 4 &= 36 - 4n + 4 \\ 4 &= 40 - 4n \\ -36 &= -4n \\ 9 &= n \end{aligned}$$

Now we find S_9.

$$S_9 = \frac{9}{2}(36+4) = \frac{9}{2}(40) = 180$$

Check. We can do the calculations again. We can also do the entire addition.

$$36 + 32 + \ldots + 4.$$

State. There are 180 stones in the pyramid.

47. ***Familiarize.*** We want to find the sum of the arithmetic sequence with $a_1 = 10¢$, $d = 10¢$, and $n = 31$. First we will find a_{31} and then we will find S_{31}.

Translate. Substituting in the formula for a_n, we have

$$a_{31} = 10 + (31-1)(10).$$

Carry out. First we find a_{31}.

$$a_{31} = 10 + 30 \cdot 10 = 10 + 300 = 310$$

Then we use the formula for S_n to find S_{31}.

$$S_{31} = \frac{31}{2}(10+310) = \frac{31}{2}(320) = 4960$$

Check. We can do the calculations again.

State. The amount saved is 4960¢, or \$49.60.

49. *Writing Exercise.*

$$\begin{aligned} &1 + 2 + 3 + \ldots + 100 \\ &= (1+100) + (2+99) + (3+98) + \ldots + (50+51) \\ &= \underbrace{101 + 101 + 101 + \ldots + 101}_{\text{50 addends of 101}} \\ &= 50 \cdot 101 \\ &= 5050 \end{aligned}$$

51. Using the slope-intercept form, where $m = \frac{1}{3}$ and $b = 10$, we have $y = \frac{1}{3}x + 10$.

53. Rewrite the equation.

$$\begin{aligned} 2x + y &= 8 \\ y &= -2x + 8 \end{aligned}$$

The slope of the parallel line is –2.
Use point-slope form.

$$\begin{aligned} y - y_1 &= m(x - x_1) \\ y - 0 &= -2(x-5) \\ y &= -2x + 10 \end{aligned}$$

55. A circle with center (0, 0) and radius 4 is $x^2 + y^2 = 16$.

57. *Writing Exercise.*

59. Find the sequence with $a_1 = 150$. When $n = 21$, $a_{21} = 135$, find d. Note that we could use any of the other points from the table.

$$\begin{aligned} a_n &= a_1 + (n-1)d \\ 135 &= 150 + (21-1)d \\ -15 &= 20d \\ -0.75 &= d \end{aligned}$$

Substitute to find the general term of the sequence.

$$\begin{aligned} a_n &= a_1 + (n-1)d \\ a_n &= 150 + (n-1)(-0.75) \\ a_n &= 150 - 0.75n + 0.75 \\ a_n &= -0.75n + 150.75 \end{aligned}$$

61. Let d = the common difference. Since p, m, and q form an arithmetic sequence, $m = p + d$, and $q = p + 2d$. Then

$$\frac{p+q}{2} = \frac{p + (p+2d)}{2} = p + d = m.$$

63. Each integer from 501 through 750 is 500 more than the corresponding integer from 1 through 250. There are 250 integers from 501 through 750, so their sum is the sum of the integers from 1 to 250 plus $250 \cdot 500$. From Exercise 39, we know that the sum of the integers from 1 through 250 is 31,375. Thus, we have

$$31{,}375 + 250 \cdot 500, \text{ or } 156{,}375.$$

Connecting the Concepts

1. $d = 112 - 115 = -3$, or
$d = 109 - 112 = -3$, or
$d = 106 - 109 = -3$

2. $\frac{1}{3}, -\frac{1}{6}, \frac{1}{12}, -\frac{1}{24}, \ldots$

$$r = \frac{-\frac{1}{6}}{\frac{1}{3}} = -\frac{1}{6} \cdot \frac{3}{1} = -\frac{1}{2}$$

3. 10, 15, 20, 25, …
Note that $a_1 = 10$, $d = 5$, and $n = 21$.

$$a_{21} = 10 + (21-1)5 = 10 + 100 = 110$$

4. 5, 10, 20, 40, ...

$a_1 = 5$, and $r = \frac{10}{5} = 2$

$a_n = a_1 r^{n-1}$

$a_8 = 5(2)^{8-1} = 5(2)^7 = 5(128) = 640$

5. $2+12+22+32+\ldots$

$a_1 = 1$, $d = 10$, and $n = 30$.

$a_{30} = 2+(30-1)10 = 292$

$S_{30} = \frac{30}{2}(2+292) = 4410$

6. $\$100 + \$100(1.03) + \$100(1.03)^2 + \ldots$

$a_1 = \$100$, $n = 10$, and $r = \frac{\$100(1.03)}{\$100} = 1.03$

$S_n = \frac{a_1(1-r^n)}{1-r}$

$S_{10} = \frac{\$100\left[1-(1.03)^{10}\right]}{1-1.03} \approx \frac{\$100(1-1.3439164)}{-0.03}$
$\approx \$1146.39$

7. $0.9+0.09+0.009+\ldots$

$|r| = \left|\frac{0.09}{0.9}\right| = |0.1| = 0.1 < 1$, so the series has a limit.

$S_\infty = \frac{0.9}{1-0.1} = \frac{0.9}{0.9} = 1$

Thus, $0.9+0.09+0.009+\ldots = 1$.

8. $0.9+9+90+\ldots$

$|r| = \left|\frac{9}{0.9}\right| = |10| = 10 \not< 1$, so the series does not have a limit.

Exercise Set 14.3

1. The list 16, 8, 4, 2, 1, ... is an *infinite geometric sequence*.

3. For $16+8+4+2+1+\cdots$, the common *ratio* is *less* than 1, so the limit *does* exist.

5. $\frac{a_{n+1}}{a_n} = 2$, so this is a geometric sequence.

7. $\frac{a_{n+1}}{a_n} = 5$, so this is a geometric series.

9. $\frac{a_{n+1}}{a_n} = -\frac{1}{2}$, so this is a geometric series.

11. 10, 20, 40, 80, ...

$\frac{20}{10} = 2$, $\frac{40}{20} = 2$, $\frac{80}{40} = 2$

$r = 2$

13. 6, –0.6, 0.06, –0.006, ...

$-\frac{0.6}{6} = -0.1$, $\frac{0.06}{-0.6} = -0.1$, $\frac{-0.006}{0.06} = -0.1$

$r = -0.1$

15. $\frac{1}{2}, -\frac{1}{4}, \frac{1}{8}, -\frac{1}{16}, \ldots$

$\frac{-\frac{1}{4}}{\frac{1}{2}} = -\frac{1}{4}\cdot\frac{2}{1} = -\frac{2}{4} = -\frac{1}{2}$, $\frac{\frac{1}{8}}{-\frac{1}{4}} = \frac{1}{8}\cdot\left(-\frac{4}{1}\right) = -\frac{4}{8} = -\frac{1}{2}$,

$\frac{-\frac{1}{16}}{\frac{1}{8}} = -\frac{1}{16}\cdot\frac{8}{1} = -\frac{8}{16} = -\frac{1}{2}$

$r = -\frac{1}{2}$

17. 75, 15, 3, $\frac{3}{5}$, ...

$\frac{15}{75} = \frac{1}{5}$, $\frac{3}{15} = \frac{1}{5}$, $\frac{\frac{3}{5}}{3} = \frac{3}{5}\cdot\frac{1}{3} = \frac{1}{5}$

$r = \frac{1}{5}$

19. $\frac{1}{m}, \frac{6}{m^2}, \frac{36}{m^3}, \frac{216}{m^4}, \ldots$

$\frac{\frac{6}{m^2}}{\frac{1}{m}} = \frac{6}{m^2}\cdot\frac{m}{1} = \frac{6}{m}$, $\frac{\frac{36}{m^3}}{\frac{6}{m^2}} = \frac{36}{m^3}\cdot\frac{m^2}{6} = \frac{6}{m}$

$\frac{\frac{216}{m^4}}{\frac{36}{m^3}} = \frac{216}{m^4}\cdot\frac{m^3}{36} = \frac{6}{m}$

$r = \frac{6}{m}$

21. 2, 6, 18, ...

$a_1 = 2$, $n = 7$, and $r = \frac{6}{2} = 3$

We use the formula $a_n = a_1 r^{n-1}$.

$a_7 = 2\cdot 3^{7-1} = 2\cdot 3^6 = 2\cdot 729 = 1458$

23. $\sqrt{3}, 3, 3\sqrt{3}, \ldots$

$a_1 = \sqrt{3}$, $n = 10$, and $r = \frac{3\sqrt{3}}{3} = \sqrt{3}$

$a_n = a_1 r^{n-1}$

$a_{10} = \sqrt{3}\left(\sqrt{3}\right)^{10-1} = \sqrt{3}\left(\sqrt{3}\right)^9 = \left(\sqrt{3}\right)^{10} = 243$

25. $-\frac{8}{243}, \frac{8}{81}, -\frac{8}{27}, \ldots$

$a_1 = -\frac{8}{243}$, $n = 14$, and $r = \frac{\frac{8}{81}}{-\frac{8}{243}} = \frac{8}{81}\left(-\frac{243}{8}\right) = -3$

$a_n = a_1 r^{n-1}$

$a_{14} = -\frac{8}{243}(-3)^{14-1} = -\frac{8}{243}(-3)^{13}$
$= -\frac{8}{243}(-1{,}594{,}323) = 52{,}488$

27. \$1000, \$1040, \$1081.60, …

$a_1 = \$1000$, $n = 10$, and $r = \frac{1040}{1000} = 1.04$

$a_n = a_1 r^{n-1}$

$a_{10} = \$1000(1.04)^{10-1} \approx \$1000(1.423311812)$
$\approx \$1423.31$

29. 1, 5, 25, 125,...

$a_1 = 1$, and $r = \frac{5}{1} = 5$

$a_n = a_1 r^{n-1}$

$a_n = 1 \cdot 5^{n-1} = 5^{n-1}$

31. 1, −1, 1, −1,...

$a_1 = 1$, and $r = \frac{-1}{1} = -1$

$a_n = a_1 r^{n-1}$

$a_n = 1(-1)^{n-1} = (-1)^{n-1}$, or $(-1)^{n+1}$

33. $\frac{1}{x}, \frac{1}{x^2}, \frac{1}{x^3}, \ldots$

$a_1 = \frac{1}{x}$, and $r = \frac{\frac{1}{x^2}}{\frac{1}{x}} = \frac{1}{x^2} \cdot \frac{x}{1} = \frac{1}{x}$

$a_n = a_1 r^{n-1}$

$a_n = \frac{1}{x}\left(\frac{1}{x}\right)^{n-1} = \frac{1}{x} \cdot \frac{1}{x^{n-1}} = \frac{1}{x^{1+n-1}} = \frac{1}{x^n}$, or x^{-n}

35. $6 + 12 + 24 + \ldots$

$a_1 = 6$, $n = 9$, and $r = \frac{12}{6} = 2$

$S_n = \frac{a_1(1 - r^n)}{1 - r}$

$S_9 = \frac{6(1 - 2^9)}{1 - 2} = \frac{6(1 - 512)}{-1} = \frac{6(-511)}{-1} = 3066$

37. $\frac{1}{18} - \frac{1}{6} + \frac{1}{2} - \ldots$

$a_1 = \frac{1}{18}$, $n = 7$, and $r = \frac{-\frac{1}{6}}{\frac{1}{18}} = -\frac{1}{6} \cdot \frac{18}{1} = -3$

$S_n = \frac{a_1(1 - r^n)}{1 - r}$

$S_7 = \frac{\frac{1}{18}\left[1 - (-3)^7\right]}{1 - (-3)} = \frac{\frac{1}{18}(1 + 2187)}{4} = \frac{\frac{1}{18}(2188)}{4}$
$= \frac{1}{18}(2188)\left(\frac{1}{4}\right) = \frac{547}{18}$

39. $1 + x + x^2 + x^3 + \ldots$

$a_1 = 1$, $n = 8$, and $r = \frac{x}{1}$, or x

$S_n = \frac{a_1(1 - r^n)}{1 - r}$

$S_8 = \frac{1(1 - x^8)}{1 - x} = \frac{(1 + x^4)(1 - x^4)}{1 - x}$
$= \frac{(1 + x^4)(1 + x^2)(1 - x^2)}{1 - x}$
$= \frac{(1 + x^4)(1 + x^2)(1 + x)(1 - x)}{1 - x}$
$= (1 + x^4)(1 + x^2)(1 + x)$

41. $\$200 + \$200(1.06) + \$200(1.06)^2 + \ldots$

$a_1 = \$200$, $n = 16$, and $r = \frac{\$200(1.06)}{\$200} = 1.06$

$S_n = \frac{a_1(1 - r^n)}{1 - r}$

$S_{16} = \frac{\$200\left[1 - (1.06)^{16}\right]}{1 - 1.06} \approx \frac{\$200(1 - 2.540351685)}{-0.06}$
$\approx \$5134.51$

43. $18 + 6 + 2 + \ldots$

$|r| = \left|\frac{6}{18}\right| = \left|\frac{1}{3}\right| = \frac{1}{3}$, and since $|r| < 1$, the series does have a limit.

$S_\infty = \frac{a_1}{1 - r} = \frac{18}{1 - \frac{1}{3}} = \frac{18}{\frac{2}{3}} = 18 \cdot \frac{3}{2} = 27$

45. $7 + 3 + \frac{9}{7} + \ldots$

$|r| = \left|\frac{3}{7}\right| = \frac{3}{7}$, and since $|r| < 1$, the series does have a limit.

$S_\infty = \frac{a_1}{1 - r} = \frac{7}{1 - \frac{3}{7}} = \frac{7}{\frac{4}{7}} = 7 \cdot \frac{7}{4} = \frac{49}{4}$

47. $3 + 15 + 75 + \ldots$

$|r| = \left|\frac{15}{3}\right| = |5| = 5$, and since $|r| \not< 1$, the series does not have a limit.

49. $4 - 6 + 9 - \frac{27}{2} + \ldots$

$|r| = \left|\frac{-6}{4}\right| = \left|-\frac{3}{2}\right| = \frac{3}{2}$, and since $|r| \not< 1$, the series does not have a limit.

51. $0.43 + 0.0043 + 0.000043 + \ldots$

$|r| = \left|\frac{0.0043}{0.43}\right| = |0.01| = 0.01$, and since $|r| < 1$, the series does have a limit.

$S_\infty = \frac{a_1}{1 - r} = \frac{0.43}{1 - 0.01} = \frac{0.43}{0.99} = \frac{43}{99}$

53. $\$500(1.02)^{-1}+\$500(1.02)^{-2}+\$500(1.02)^{-3}+...$

$|r|=\left|\dfrac{\$500(1.02)^{-2}}{\$500(1.02)^{-1}}\right|=|(1.02)^{-1}|=(1.02)^{-1}$, or $\dfrac{1}{1.02}$,

and since $|r|<1$, the series does have a limit.

$$S_\infty=\frac{a_1}{1-r}=\frac{\$500(1.02)^{-1}}{1-\left(\frac{1}{1.02}\right)}=\frac{\frac{\$500}{1.02}}{\frac{0.02}{1.02}}=\frac{\$500}{1.02}\cdot\frac{1.02}{0.02}$$
$$=\$25,000$$

55. $0.5555...=0.5+0.05+0.005+0.0005+...$

This is an infinite geometric series with $a_1=0.5$.

$|r|=\left|\dfrac{0.05}{0.5}\right|=|0.1|=0.1<1$, so the series has a limit.

$$S_\infty=\frac{a_1}{1-r}=\frac{0.5}{1-0.1}=\frac{0.5}{0.9}=\frac{5}{9}$$

Fractional notation for 0.5555... is $\dfrac{5}{9}$.

57. $3.4646...=3+0.4646...$

$0.464646...=0.46+0.0046+0.000046+...$

$|r|=\left|\dfrac{0.0046}{0.46}\right|=|0.01|=0.01<1$, so the series has a limit.

$$S_\infty=\frac{a_1}{1-r}=\frac{0.46}{1-0.01}=\frac{0.46}{0.99}=\frac{46}{99}$$

Fractional notation for 0.4646... is $\dfrac{46}{99}$.

Fractional notation for 3.4646... is $3+\dfrac{46}{99}=\dfrac{343}{99}$.

59. $0.15151515...=0.15+0.0015+0.000015+...$

This is an infinite geometric series with $a_1=0.15$.

$|r|=\left|\dfrac{0.0015}{0.15}\right|=|0.01|=0.01<1$, so the series has a limit.

$$S_\infty=\frac{a_1}{1-r}=\frac{0.15}{1-0.01}=\frac{0.15}{0.99}=\frac{15}{99}=\frac{5}{33}$$

Fractional notation for 0.15151515... is $\dfrac{5}{33}$.

61. ***Familiarize***. The rebound distances form a geometric sequence:

$$\frac{1}{4}\times 20,\ \left(\frac{1}{4}\right)^2\times 20,\ \left(\frac{1}{4}\right)^3\times 20,...,$$

$$\text{or } 5,\ \frac{1}{4}\times 5,\ \left(\frac{1}{4}\right)^2\times 5,...$$

The height of the 6th rebound is the 6th term of the sequence.

Translate. We will use the formula $a_n=a_1r^{n-1}$, with $a_1=5$, $r=\dfrac{1}{4}$, and $n=6$:

$$a_6=5\left(\frac{1}{4}\right)^{6-1}$$

Carry out. We calculate to obtain $a_6=\dfrac{5}{1024}$.

Check. We can do the calculation again.

State. It rebounds $\dfrac{5}{1024}$ ft the 6th time.

63. ***Familiarize***. In one year, the population will be $100,000+0.03(100,000)$, or $(1.03)100,000$. In two years, the population will be $(1.03)100,000+0.03(1.03)100,000$, or $(1.03)^2100,000$. Thus, the populations form a geometric sequence:

$100,000,\ (1.03)100,000,\ (1.03)^2100,000,...$

The population in 15 years will be the 16th term of the sequence.

Translate. We will use the formula $a_n=a_1r^{n-1}$ with $a_1=100,000$, $r=1.03$, and $n=16$:

$$a_{16}=100,000(1.03)^{16-1}$$

Carry out. We calculate to obtain $a_{16}\approx 155,797$.

Check. We can do the calculation again.

State. In 15 years the population will be about 155,797.

65. ***Familiarize***. At the end of each minute the population is 96% of the previous population.

We have a geometric sequence:

$$5000,\ 5000(0.96),\ 5000(0.96)^2,...$$

The number of fruit flies remaining alive after 15 minutes is given by the 16th term of the sequence.

Translate. We use the formula $a_n=a_1r^{n-1}$ with $a_1=5000$, $r=0.96$, and $n=16$:

$$a_{16}=5000(0.96)^{16-1}$$

Carry out. We calculate to obtain $a_{16}\approx 2710$.

Check. We can do the calculation again.

State. About 2710 flies will be alive after 15 min.

67. ***Familiarize***. The lengths of the falls form a geometric sequence:

$$556,\ 556\left(\frac{3}{4}\right),\ 556\left(\frac{3}{4}\right)^2,\ 556\left(\frac{3}{4}\right)^3,...$$

The total length of the first 6 falls is the sum of the first six terms of this sequence. The heights of the rebounds also form a geometric sequence:

$$556\left(\frac{3}{4}\right),\ 556\left(\frac{3}{4}\right)^2,\ 556\left(\frac{3}{4}\right)^3,...\text{ or}$$

$$417,\ 417\left(\frac{3}{4}\right),\ 417\left(\frac{3}{4}\right)^2,...$$

When the ball hits the ground for the 6th time, it will have rebounded 5 times. Thus the total length of the rebounds is the sum of the first five terms of this sequence.

Translate. We use the formula $S_n=\dfrac{a_1(1-r^n)}{1-r}$ twice, once with $a_1=556$, $r=\dfrac{3}{4}$, and $n=6$ and a second time with $a_1 = 417$, $r = \frac{3}{4}$, and $n = 5$.

D = Length of falls + length of rebounds

$$= \frac{556\left[1-\left(\frac{3}{4}\right)^6\right]}{1-\frac{3}{4}} + \frac{417\left[1-\left(\frac{3}{4}\right)^5\right]}{1-\frac{3}{4}}$$

Carry out. We use a calculator to obtain $D \approx 3100.35$.
Check. We can do the calculations again.
State. The ball will have traveled about 3100.35 ft.

69. ***Familiarize***. The heights of the stack form a geometric sequence:

$$0.02,\ 0.02(2),\ 0.02(2)^2,\ \ldots$$

The height of the stack after it is doubled 10 times is given by the 11th term of this sequence.
Translate. We have a geometric sequence with $a_1 = 0.02$, $r = 2$, and $n = 11$. We use the formula

$a_n = a_1 r^{n-1}$.

Carry out. We substitute and calculate.

$$a_{11} = 0.02\left(2^{11-1}\right)$$
$$a_{11} = 0.02(1024) = 20.48$$

Check. We can do the calculations again.
State. The final stack will be 20.48 in. high.

71. *Writing Exercise.*

73. $|x-3| = 11$

$x-3=-11$ *or* $x-3=11$
$x=-8$ *or* $x=14$

The solution set is $\{-8, 14\}$.

75. $|3x-7| \ge 1$

$3x-7 \le -1$ *or* $1 \le 3x-7$
$3x \le 6$ *or* $8 \le 3x$
$x \le 2$ *or* $\frac{8}{3} \le x$

The solution set is $\left\{x \middle| x \le 2 \text{ or } x \ge \frac{8}{3}\right\}$, or

$(-\infty, -2] \cup \left[\frac{8}{3}, \infty\right)$.

77. $x^2 - 5x - 14 < 0$

$(x+2)(x-7) < 0$

The solutions of $(x+2)(x-7) = 0$ are -2 and 7. They divide the number line into three intervals as shown:

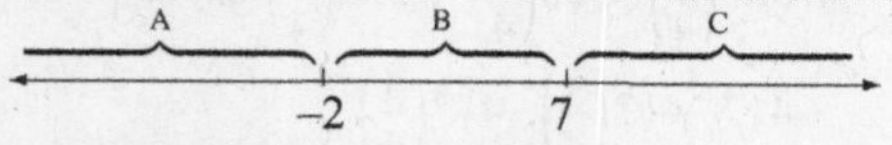

We try test numbers in each interval.
A: Test -3, $f(-3) = (-3+2)(-3-7) = 10$
B: Test 0, $f(0) = (0+2)(0-7) = -14$
C: Test 8, $f(8) = (8+2)(8-7) = 10$
Since $f(-3)$ and $f(8)$ are positive, the function value will be negative for all numbers in the interval containing -2 and 7. The inequality symbol is <, so we do not include the endpoints. The solution set is $(-2, 7)$, or $\{x \mid -2 < x < 7\}$.

79. *Writing Exercise.*

81. $\sum_{k=1}^{\infty} 6(0.9)^k = 6(0.9) + 6(0.9)^2 + 6(0.9)^3 + \ldots$

$|r| = \left|\frac{6(0.9)^2}{6(0.9)}\right| = |0.9| = 0.9 < 1$, so the series has a limit.

$$S_\infty = \frac{a_1}{1-r} = \frac{6(0.9)}{1-0.9} = \frac{5.4}{0.1} = 54$$

83. $x^2 - x^3 + x^4 - x^5 + \ldots$

This is a geometric series with $a_1 = x^2$ and $r = -x$.

$$S_n = \frac{a_1(1-r^n)}{1-r} = \frac{x^2\left[1-(-x)^n\right]}{1-(-x)} = \frac{x^2\left[1-(-x)^n\right]}{1+x}$$

85. The length of a side of the first square is 16 cm. The length of a side of the next square is the length of the hypotenuse of a right triangle with legs 8 cm and 8 cm, or $8\sqrt{2}$ cm. The length of a side of the next square is the length of the hypotenuse of a right triangle with legs $4\sqrt{2}$ cm and $4\sqrt{2}$ cm, or 8 cm. The areas of the squares form a sequence:

$(16)^2,\ (8\sqrt{2})^2,\ (8)^2,\ \ldots$, or
$256,\ 128,\ 64,\ \ldots$

This is a geometric series with $a_1 = 256$ and $r = \frac{1}{2}$.

We find the sum of the infinite geometric series $256 + 128 + 64 + \ldots$

$$S_\infty = \frac{a_1}{1-r} = \frac{256}{1-\frac{1}{2}} = \frac{256}{\frac{1}{2}} = 512 \text{ cm}^2$$

87. *Writing Exercise.*

Mid-Chapter Review

1. $a_n = a_1 + (n-1)d$

$n = 14,\ a_1 = -6,\ d = 5$
$a_{14} = -6 + (14-1)5$
$a_{14} = 59$

2. $a_n = a_1 r^{n-1}$

$n = 7,\ a_1 = \frac{1}{9},\ r = -3$
$a_7 = \frac{1}{9} \cdot (-3)^{7-1}$
$a_7 = 81$

3. $a_n = n^2 - 5n$

$a_{20} = 20^2 - 5 \cdot 20 = 300$

4. $\frac{1}{2}, \frac{1}{3}, \frac{1}{4}, \frac{1}{5}, \ldots$

The general term is $\frac{1}{n+1}$.

5. 1, 2, 3, 4, …
Note that $a_1 = 1$, $d = 1$, and $n = 12$. Before using the formula to find S_{12}, we find a_{12}.

$$a_{12} = 1 + (12-1)1 = 12$$

Then using the formula for S_n,

$$S_{12} = \frac{12}{2}(1+12) = 6(13) = 78.$$

6. $\sum_{k=2}^{5} k^2 = 2^2 + 3^2 + 4^2 + 5^2 = 4 + 9 + 16 + 25 = 54$

7. $1 - 2 + 3 - 4 + 5 - 6 = \sum_{k=1}^{6} (-1)^{k+1} \cdot k$

8.
$$\begin{aligned} a_n &= a_1 + (n-1)d \\ 22 &= 10 + (n-1)0.2 \\ 22 &= 10 + 0.2n - 0.2 \\ 12.2 &= 0.2n \\ 61 &= n \end{aligned}$$
22 is the 61st term.

9. $a_n = a_1 + (n-1)d$
$a_{25} = 9 + (25-1)(-2) = 9 + (24)(-2) = -39$

10. $a_n = a_1 r^{n-1}$
$n = 12,\ a_1 = 1000,\ r = \frac{1}{10}$

$$a_{12} = 1000 \cdot \left(\frac{1}{10}\right)^{12-1}$$
$$a_{12} = \frac{1}{10^8}$$

11. 2, −2, 2, −2, …
$a_1 = 2$, and $r = \frac{-2}{2} = -1$
$a_n = 2(-1)^{n+1}$

12. $|r| = \left|\frac{-20}{100}\right| = \frac{1}{5} < 1$, so the series has a limit.

$$S_\infty = \frac{100}{1-\left(-\frac{1}{5}\right)} = \frac{100}{\frac{6}{5}} = \frac{250}{3} = 83\frac{1}{3}$$

13. $\$1 + \$2 + \$3 + \$4 + \ldots$
This is an arithmetic sequence $a_1 = 1$, $d = 1$, and $n = 30$. Before using the formula to find S_{30}, we find a_{30}.

$$a_{30} = 1 + (30-1)1 = 30$$

Then using the formula for S_n,

$$S_{30} = \frac{30}{2}(1+30) = 15(31) = 465.$$

She earns \$465.

14. $\$1 + \$2 + \$4 + \$8 + \ldots$
This is a geometric sequence. $a_1 = 1, n = 30,\ r = \frac{2}{1} = 2$
Then using the formula for S_n,

$$S_n = \frac{a_1(1-r^n)}{1-r}$$
$$S_{30} = \frac{1(1-2^{30})}{1-2} = \frac{1-1{,}073{,}741{,}824}{-1} = 1{,}073{,}741{,}823$$

He earns \$1,073,741,823.

Exercise Set 14.4

1. The expression $(x+y)^2$ is a *binomial* squared.

3. The *first* number in every row of Pascal's triangle is 1.

5. 8! is an example of *factorial* notation.

7. The last term in the expansion of $(x+2)^5$ is 2^5*, or 32.*

9. In the expansion of $(a+b)^9$, the sum of the exponents in each term is *9.*

11. $4! = 4 \cdot 3 \cdot 2 \cdot 1 = 24$

13. $10! = 10 \cdot 9 \cdot 8 \cdot 7 \cdot 6 \cdot 5 \cdot 4 \cdot 3 \cdot 2 \cdot 1 = 3{,}628{,}800$

15. $\frac{10!}{8!} = \frac{10 \cdot 9 \cdot 8!}{8!} = 10 \cdot 9 = 90$

17. $\frac{9!}{4!5!} = \frac{9 \cdot 8 \cdot 7 \cdot 6 \cdot 5!}{4!5!} = \frac{9 \cdot 8 \cdot 7 \cdot 6}{4 \cdot 3 \cdot 2 \cdot 1} = 3 \cdot 7 \cdot 6 = 126$

19. $\binom{10}{4} = \frac{10!}{6!4!} = \frac{10 \cdot 9 \cdot 8 \cdot 7 \cdot 6!}{6!4!} = \frac{10 \cdot 9 \cdot 8 \cdot 7}{4 \cdot 3 \cdot 2 \cdot 1} = 10 \cdot 3 \cdot 7 = 210$

21. $\binom{9}{9} = \frac{9!}{0!9!} = \frac{9!}{1 \cdot 9!} = \frac{9!}{9!} = 1$

23. $\binom{30}{2} = \frac{30!}{28!2!} = \frac{30 \cdot 29 \cdot 28!}{28!2!} = \frac{30 \cdot 29}{2 \cdot 1} = 15 \cdot 29 = 435$

25. $\binom{40}{38} = \frac{40!}{2!38!} = \frac{40 \cdot 39 \cdot 38!}{2!38!} = \frac{40 \cdot 39}{2 \cdot 1} = 20 \cdot 39 = 780$

27. Expand $(a-b)^4$.
We have $a = a$, $b = -b$, and $n = 4$.
Form 1: We use the fifth row of Pascal's triangle:
1 4 6 4 1

$$\begin{aligned}(a-b)^4 &= 1 \cdot a^4 + 4a^3(-b) + 6a^2(-b)^2 + 4a(-b)^3 + 1 \cdot (-b)^4 \\ &= a^4 - 4a^3b + 6a^2b^2 - 4ab^3 + b^4\end{aligned}$$

Form 2:

$$\begin{aligned}(a-b)^4 &= \binom{4}{0}a^4 + \binom{4}{1}a^3(-b) + \binom{4}{2}a^2(-b)^2 \\ &\quad + \binom{4}{3}a(-b)^3 + \binom{4}{4}(-b)^4 \\ &= \frac{4!}{4!0!}a^4 + \frac{4!}{3!1!}a^3(-b) + \frac{4!}{2!2!}a^2(-b)^2 \\ &\quad + \frac{4!}{1!3!}a(-b)^3 + \frac{4!}{0!4!}(-b)^4 \\ &= a^4 - 4a^3b + 6a^2b^2 - 4ab^3 + b^4\end{aligned}$$

29. Expand $(p+w)^7$.
We have $a = p$, $b = w$, and $n = 7$.
Form 1: We use the 8th row of Pascal's triangle:
1 7 21 35 35 21 7 1

$$(p+w)^7 = p^7 + 7p^6w^1 + 21p^5w^2 + 35p^4w^3 + 35p^3w^4 + 21p^2w^5 + 7pw^6 + w^7$$

Form 2:

$$\begin{aligned}&(p+q)^7\\&=\binom{7}{0}p^7+\binom{7}{1}p^6w^1+\binom{7}{2}p^5w^2+\binom{7}{3}p^4w^3\\&\quad+\binom{7}{4}p^3w^4+\binom{7}{5}p^2w^5+\binom{7}{6}pw^6+\binom{7}{7}w^7\\&=\frac{7!}{7!0!}p^7+\frac{7!}{6!1!}p^6w^1+\frac{7!}{5!2!}p^5w^2+\frac{7!}{4!3!}p^4w^3\\&\quad+\frac{7!}{3!4!}p^3w^4+\frac{7!}{2!5!}p^2w^5+\frac{7!}{1!6!}pw^6+\frac{7!}{0!7!}w^7\\&=p^7+7p^6w^1+21p^5w^2+35p^4w^3\\&\quad+35p^3w^4+21p^2w^5+7pw^6+w^7\end{aligned}$$

31. Expand $(3c-d)^7$.
We have $a = 3c$, $b = -d$, and $n = 7$.
Form 1: We use the 8th row of Pascal's triangle:
1 7 21 35 35 21 7 1

$$\begin{aligned}&(3c-d)^7\\&=(3c)^7+7(3c)^6(-d)^1+21(3c)^5(-d)^2+35(3c)^4(-d)^3\\&\quad+35(3c)^3(-d)^4+21(3c)^2(-d)^5+7(3c)(-d)^6+(-d)^7\\&=2187c^7-5103c^6d+5103c^5d^2-2835c^4d^3\\&\quad+945c^3d^4-189c^2d^5+21cd^6-d^7\end{aligned}$$

Form 2:

$$\begin{aligned}&(3c-d)^7\\&=\binom{7}{0}(3c)^7+\binom{7}{1}(3c)^6(-d)^1+\binom{7}{2}(3c)^5(-d)^2\\&\quad+\binom{7}{3}(3c)^4(-d)^3+\binom{7}{4}(3c)^3(-d)^4+\binom{7}{5}(3c)^2(-d)^5\\&\quad+\binom{7}{6}(3c)(-d)^6+\binom{7}{7}(-d)^7\\&=\frac{7!}{7!0!}(3c)^7+\frac{7!}{6!1!}(3c)^6(-d)^1+\frac{7!}{5!2!}(3c)^5(-d)^2\\&\quad+\frac{7!}{4!3!}(3c)^4(-d)^3+\frac{7!}{3!4!}(3c)^3(-d)^4+\frac{7!}{2!5!}(3c)^2(-d)^5\\&\quad+\frac{7!}{1!6!}(3c)(-d)^6+\frac{7!}{0!7!}(-d)^7\\&=2187c^7-5103c^6d+5103c^5d^2-2835c^4d^3\\&\quad+945c^3d^4-189c^2d^5+21cd^6-d^7\end{aligned}$$

33. Expand $(t^{-2}+2)^6$.
We have $a = t^{-2}$, $b = 2$, and $n = 6$.
Form 1: We use the 7th row of Pascal's triangle:
1 6 15 20 15 6 1

$$\begin{aligned}&(t^{-2}+2)^6\\&=1\cdot(t^{-2})^6+6(t^{-2})^5(2)^1+15(t^{-2})^4(2)^2+20(t^{-2})^3(2)^3\\&\quad+15(t^{-2})^2(2)^4+6(t^{-2})^1(2)^5+1\cdot(2)^6\\&=t^{-12}+12t^{-10}+60t^{-8}+160t^{-6}+240t^{-4}+192t^{-2}+64\end{aligned}$$

Form 2:

$$\begin{aligned}&(t^{-2}+2)^6\\&=\binom{6}{0}(t^{-2})^6+\binom{6}{1}(t^{-2})^5(2)^1+\binom{6}{2}(t^{-2})^4(2)^2\\&\quad+\binom{6}{3}(t^{-2})^3(2)^3+\binom{6}{4}(t^{-2})^2(2)^4\\&\quad+\binom{6}{5}(t^{-2})^1(2)^5+\binom{6}{6}(2)^6\\&=\frac{6!}{6!0!}(t^{-2})^6+\frac{6!}{5!1!}(t^{-2})^5(2)^1+\frac{6!}{4!2!}(t^{-2})^4(2)^2\\&\quad+\frac{6!}{3!3!}(t^{-2})^3(2)^3+\frac{6!}{2!4!}(t^{-2})^2(2)^4\\&\quad+\frac{6!}{1!5!}(t^{-2})^1(2)^5+\frac{6!}{0!6!}(2)^6\\&=t^{-12}+12t^{-10}+60t^{-8}+160t^{-6}+240t^{-4}+192t^{-2}+64\end{aligned}$$

35. Expand $\left(3s+\frac{1}{t}\right)^9$.

We have $a = 3s$, $b = \frac{1}{t}$, and $n = 9$.

Form 1: We use the tenth row of Pascal's triangle:
1 9 36 84 126 126 84 36 9 1

$$\begin{aligned}&\left(3s+\frac{1}{t}\right)^9\\&=1\cdot(3s)^9+9(3s)^8\left(\frac{1}{t}\right)^1+36(3s)^7\left(\frac{1}{t}\right)^2+84(3s)^6\left(\frac{1}{t}\right)^3\\&\quad+126(3s)^5\left(\frac{1}{t}\right)^4+126(3s)^4\left(\frac{1}{t}\right)^5+84(3s)^3\left(\frac{1}{t}\right)^6\\&\quad+36(3s)^2\left(\frac{1}{t}\right)^7+9(3s)^1\left(\frac{1}{t}\right)^8+1\cdot\left(\frac{1}{t}\right)^9\\&=19{,}683s^9+\frac{59{,}049s^8}{t}+\frac{78{,}732s^7}{t^2}+\frac{61{,}236s^6}{t^3}\\&\quad+\frac{30{,}618s^5}{t^4}+\frac{10{,}206s^4}{t^5}+\frac{2268s^3}{t^6}+\frac{324s^2}{t^7}+\frac{27s}{t^8}+\frac{1}{t^9}\end{aligned}$$

Form 2:

$\left(3s+\frac{1}{t}\right)^9$

$=\binom{9}{0}(3s)^9+\binom{9}{1}(3s)^8\left(\frac{1}{t}\right)^1+\binom{9}{2}(3s)^7\left(\frac{1}{t}\right)^2$

$+\binom{9}{3}(3s)^6\left(\frac{1}{t}\right)^3+\binom{9}{4}(3s)^5\left(\frac{1}{t}\right)^4+\binom{9}{5}(3s)^4\left(\frac{1}{t}\right)^5$

$+\binom{9}{6}(3s)^3\left(\frac{1}{t}\right)^6+\binom{9}{7}(3s)^2\left(\frac{1}{t}\right)^7+\binom{9}{8}(3s)^1\left(\frac{1}{t}\right)^8$

$+\binom{9}{9}\left(\frac{1}{t}\right)^9$

$=\frac{9!}{9!0!}(3s)^9+\frac{9!}{8!1!}(3s)^8\left(\frac{1}{t}\right)^1+\frac{9!}{7!2!}(3s)^7\left(\frac{1}{t}\right)^2$

$+\frac{9!}{6!3!}(3s)^6\left(\frac{1}{t}\right)^3+\frac{9!}{5!4!}(3s)^5\left(\frac{1}{t}\right)^4+\frac{9!}{4!5!}(3s)^4\left(\frac{1}{t}\right)^5$

$+\frac{9!}{3!6!}(3s)^3\left(\frac{1}{t}\right)^6+\frac{9!}{2!7!}(3s)^2\left(\frac{1}{t}\right)^7+\frac{9!}{1!8!}(3s)^1\left(\frac{1}{t}\right)^8$

$+\frac{9!}{0!9!}\left(\frac{1}{t}\right)^9$

$=19{,}683s^9+\frac{59{,}049s^8}{t}+\frac{78{,}732s^7}{t^2}$

$+\frac{61{,}236s^6}{t^3}+\frac{30{,}618s^5}{t^4}+\frac{10{,}206s^4}{t^5}$

$+\frac{2268s^3}{t^6}+\frac{324s^2}{t^7}+\frac{27s}{t^8}+\frac{1}{t^9}$

37. Expand $\left(x^3-2y\right)^5$.

We have $a=x^3$, $b=-2y$, and $n=5$.

Form 1: We use the 6th row of Pascal's triangle:

1 5 10 10 5 1

$\left(x^3-2y\right)^5$

$=1\cdot\left(x^3\right)^5+5\left(x^3\right)^4(-2y)^1+10\left(x^3\right)^3(-2y)^2$

$+10\left(x^3\right)^2(-2y)^3+5\left(x^3\right)^1(-2y)^4+1\cdot(-2y)^5$

$=x^{15}-10x^{12}y+40x^9y^2-80x^6y^3+80x^3y^4-32y^5$

Form 2:

$\left(x^3-2y\right)^5$

$=\binom{5}{0}\left(x^3\right)^5+\binom{5}{1}\left(x^3\right)^4(-2y)+\binom{5}{2}\left(x^3\right)^3(-2y)^2$

$+\binom{5}{3}\left(x^3\right)^2(-2y)^3+\binom{5}{4}\left(x^3\right)(-2y)^4+\binom{5}{5}(-2y)^5$

$=\frac{5!}{5!0!}\left(x^3\right)^5+\frac{5!}{4!1!}\left(x^3\right)^4(-2y)+\frac{5!}{3!2!}\left(x^3\right)^3(-2y)^2$

$+\frac{5!}{2!3!}\left(x^3\right)^2(-2y)^3+\frac{5!}{1!4!}\left(x^3\right)(-2y)^4+\frac{5!}{0!5!}(-2y)^5$

$=x^{15}-10x^{12}y+40x^9y^2-80x^6y^3+80x^3y^4-32y^5$

39. Expand $\left(\sqrt{5}+t\right)^6$.

We have $a=\sqrt{5}$, $b=t$, and $n=6$.

Form 1: We use the 7th row of Pascal's triangle:

1 6 15 20 15 6 1

$\left(\sqrt{5}+t\right)^6$

$=1\cdot\left(\sqrt{5}\right)^6+6\left(\sqrt{5}\right)^5t^1+15\left(\sqrt{5}\right)^4t^2+20\left(\sqrt{5}\right)^3t^3$

$+15\left(\sqrt{5}\right)^2t^4+6\left(\sqrt{5}\right)^1t^5+1\cdot t^6$

$=125+150\sqrt{5}t+375t^2+100\sqrt{5}t^3+75t^4+6\sqrt{5}t^5+t^6$

Form 2:

$\left(\sqrt{5}+t\right)^6$

$=\binom{6}{0}\left(\sqrt{5}\right)^6+\binom{6}{1}\left(\sqrt{5}\right)^5t^1+\binom{6}{2}\left(\sqrt{5}\right)^4t^2+\binom{6}{3}\left(\sqrt{5}\right)^3t^3$

$+\binom{6}{4}\left(\sqrt{5}\right)^2t^4+\binom{6}{5}\left(\sqrt{5}\right)^1t^5+\binom{6}{6}t^6$

$=\frac{6!}{6!0!}\left(\sqrt{5}\right)^6+\frac{6!}{5!1!}\left(\sqrt{5}\right)^5t^1+\frac{6!}{4!2!}\left(\sqrt{5}\right)^4t^2+\frac{6!}{3!3!}\left(\sqrt{5}\right)^3t^3$

$+\frac{6!}{2!4!}\left(\sqrt{5}\right)^2t^4+\frac{6!}{1!5!}\left(\sqrt{5}\right)^1t^5+\frac{6!}{0!6!}t^6$

$=125+150\sqrt{5}t+375t^2+100\sqrt{5}t^3+75t^4+6\sqrt{5}t^5+t^6$

41. Expand $\left(\frac{1}{\sqrt{x}}-\sqrt{x}\right)^6$.

We have $a=\frac{1}{\sqrt{x}}$, $b=-\sqrt{x}$, and $n=6$.

Form 1: We use the 7th row of Pascal's triangle:

1 6 15 20 15 6 1

$\left(\frac{1}{\sqrt{x}}-\sqrt{x}\right)^6$

$=1\cdot\left(\frac{1}{\sqrt{x}}\right)^6+6\left(\frac{1}{\sqrt{x}}\right)^5\left(-\sqrt{x}\right)^1+15\left(\frac{1}{\sqrt{x}}\right)^4\left(-\sqrt{x}\right)^2$

$+20\left(\frac{1}{\sqrt{x}}\right)^3\left(-\sqrt{x}\right)^3+15\left(\frac{1}{\sqrt{x}}\right)^2\left(-\sqrt{x}\right)^4$

$+6\left(\frac{1}{\sqrt{x}}\right)^1\left(-\sqrt{x}\right)^5+1\cdot\left(-\sqrt{x}\right)^6$

$=x^{-3}-6x^{-2}+15x^{-1}-20+15x-6x^2+x^3$

Form 2:

$\left(\frac{1}{\sqrt{x}}-\sqrt{x}\right)^6$

$=\binom{6}{0}\left(\frac{1}{\sqrt{x}}\right)^6+\binom{6}{1}\left(\frac{1}{\sqrt{x}}\right)^5\left(-\sqrt{x}\right)^1+\binom{6}{2}\left(\frac{1}{\sqrt{x}}\right)^4\left(-\sqrt{x}\right)^2$

$+\binom{6}{3}\left(\frac{1}{\sqrt{x}}\right)^3\left(-\sqrt{x}\right)^3+\binom{6}{4}\left(\frac{1}{\sqrt{x}}\right)^2\left(-\sqrt{x}\right)^4$

$+\binom{6}{5}\left(\frac{1}{\sqrt{x}}\right)^1\left(-\sqrt{x}\right)^5+\binom{6}{6}\left(-\sqrt{x}\right)^6$

$=\frac{6!}{6!0!}\left(\frac{1}{\sqrt{x}}\right)^6+\frac{6!}{5!1!}\left(\frac{1}{\sqrt{x}}\right)^5\left(-\sqrt{x}\right)^1+\frac{6!}{4!2!}\left(\frac{1}{\sqrt{x}}\right)^4\left(-\sqrt{x}\right)^2$

$+\frac{6!}{3!3!}\left(\frac{1}{\sqrt{x}}\right)^3\left(-\sqrt{x}\right)^3+\frac{6!}{2!4!}\left(\frac{1}{\sqrt{x}}\right)^2\left(-\sqrt{x}\right)^4$

$+\frac{6!}{1!5!}\left(\frac{1}{\sqrt{x}}\right)^1\left(-\sqrt{x}\right)^5+\frac{6!}{0!6!}\left(-\sqrt{x}\right)^6$

$=x^{-3}-6x^{-2}+15x^{-1}-20+15x-6x^2+x^3$

43. Find the 3rd term of $(a+b)^6$.

First we note that $3=2+1$, $a=a$, $b=b$, and $n=6$.

Then the 3rd term of the expansion of $(a+b)^6$ is

$\binom{6}{2}a^{6-2}b^2$, or $\frac{6!}{4!2!}a^4b^2$, or $15a^4b^2$.

45. Find the 12th term of $(a-3)^{14}$.
First we note that $12 = 11+1$, $a = a$, $b = -3$, and $n = 14$.
Then the 12th term of the expansion of $(a-3)^{14}$ is

$$\binom{14}{11}a^{14-11}\cdot(-3)^{11} = \frac{14!}{3!11!}a^3(-177,147)$$
$$= 364a^3(-177,147)$$
$$= -64,481,508a^3$$

47. Find the 5th term of $\left(2x^3+\sqrt{y}\right)^8$.
First we note that $5 = 4+1$, $a = 2x^3$, $b = \sqrt{y}$, and $n = 8$.
Then the 5th term of the expansion of $\left(2x^3+\sqrt{y}\right)^8$ is

$$\binom{8}{4}(2x^3)^{8-4}\left(\sqrt{y}\right)^4 = \frac{8!}{4!4!}(2x^3)^4\left(\sqrt{y}\right)^4$$
$$= 70\left(16x^{12}\right)\left(y^2\right)$$
$$= 1120x^{12}y^2$$

49. The expansion of $\left(2u+3v^2\right)^{10}$ has 11 terms so the 6th term is the middle term. Note that $6 = 5+1$, $a = 2u$, $b = 3v^2$, and $n = 10$. Then the 6th term of the expansion of $\left(2u+3v^2\right)^{10}$ is

$$\binom{10}{5}(2u)^{10-5}\left(3v^2\right)^5 = \frac{10!}{5!5!}(2u)^5\left(3v^2\right)^5$$
$$= 252\left(32u^5\right)\left(243v^{10}\right)$$
$$= 1,959,552u^5v^{10}$$

51. The 9th term of $(x-y)^8$ is the last term, y^8.

53. *Writing Exercise.*

55. Graph $y = x^2 - 5$.
This is a parabola with vertex (0, –5).

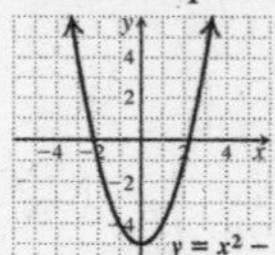

57. Graph $y \ge x - 5$.
Use a solid line to form the line $y = x - 5$. Since the test point (0, 0) is a solution, shade this side of the line.

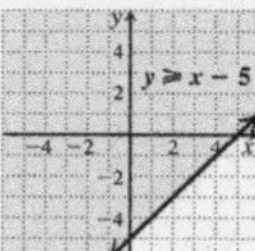

59. Graph $f(x) = \log_5 x$.

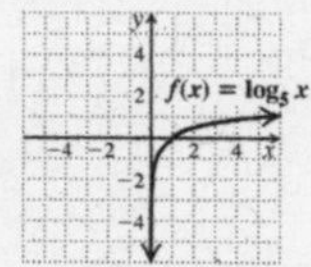

61. *Writing Exercise.*

63. Consider the set of 5 elements $\{a, b, c, d, e\}$. List all the subsets of size 3:
$\{a, b, c\}$, $\{a, b, d\}$, $\{a, b, e\}$, $\{a, c, d\}$, $\{a, c, e\}$, $\{a, d, e\}$, $\{b, c, d\}$, $\{b, c, e\}$, $\{b, d, e\}$, $\{c, d, e\}$.
There are exactly 10 subsets of size 3 and $\binom{5}{3} = 10$, so there are exactly $\binom{5}{3}$ ways of forming a subset of size 3 from a set of 5 elements.

65. Find the sixth term of $(0.15+0.85)^8$.

$$\binom{8}{5}(0.15)^{8-5}(0.85)^5 = \frac{8!}{3!5!}(0.15)^3(0.85)^5 \approx 0.084$$

67. Find and add the 7th through 9th terms of $(0.15+0.85)^9$.

$$\binom{8}{6}(0.15)^2(0.85)^6 + \binom{8}{7}(0.15)(0.85)^7 + \binom{8}{8}(0.85)^8$$
$$\approx 0.89$$

69. $$\binom{n}{n-r} = \frac{n!}{[n-(n-r)]!(n-r)!} = \frac{n!}{r!(n-r)!} = \binom{n}{r}$$

71. The expansion of $\left(x^2-6y^{3/2}\right)^6$ has 7 terms, so the 4th term is the middle term.

$$\binom{6}{3}\left(x^2\right)^3\left(-6y^{3/2}\right)^3 = \frac{6!}{3!3!}\left(x^6\right)\left(-216y^{9/2}\right)$$
$$= -4320x^6y^{9/2}$$

73. The $(r+1)$st term of $\left(\sqrt[3]{x}-\frac{1}{\sqrt{x}}\right)^7$ is

$\binom{7}{r}\left(\sqrt[3]{x}\right)^{7-r}\left(-\frac{1}{\sqrt{x}}\right)^r$. The term containing $\frac{1}{x^{1/6}}$ is the term in which the sum of the exponents is $-1/6$. That is,

$$\left(\frac{1}{3}\right)(7-r)+\left(-\frac{1}{2}\right)(r) = -\frac{1}{6}$$
$$\frac{7}{3}-\frac{r}{3}-\frac{r}{2} = -\frac{1}{6}$$
$$-\frac{5r}{6} = -\frac{15}{6}$$
$$r = 3$$

Find the $(3+1)$st, or 4th term.

$$\binom{7}{3}\left(\sqrt[3]{x}\right)^4\left(-\frac{1}{\sqrt{x}}\right)^3 = \frac{7!}{4!3!}\left(x^{4/3}\right)\left(-x^{-3/2}\right) = -35x^{-1/6}$$

or $-\frac{35}{x^{1/6}}$.

75. The degree of $\left(x^3+3\right)^4$ is the degree of $\left(x^3\right)^4 = x^{12}$, or 12.

Chapter 14 Review

1. False; the next term of the arithmetic sequence 10, 15, 20, ... is 20 + 5, or 25.

2. True; the next term of the geometric sequence 2, 6, 18, 54, ... is $54\cdot 3$, or 162.

3. True.

4. False; for $a_n = 3n-1$, $a_{17} = 3(17)-1 = 50$.

5. False; a geometric sequence has a common *ratio*.

6. True; $|r| = \left|\frac{-5}{10}\right| = \left|-\frac{1}{2}\right| = \frac{1}{2} < 1$, so the series has a limit.

7. False; $n! = n\cdot(n-1)\cdot(n-2)\cdot\ldots\cdot 3\cdot 2\cdot 1$.

8. False; $(x+y)^{17}$ has 18 terms.

9. $$\begin{aligned} a_n &= 10n-9 \\ a_1 &= 10\cdot 1-9 = 1 \\ a_2 &= 10\cdot 2-9 = 11 \\ a_3 &= 10\cdot 3-9 = 21 \\ a_4 &= 10\cdot 4-9 = 31 \\ a_8 &= 10\cdot 8-9 = 71 \\ a_{12} &= 10\cdot 12-9 = 111 \end{aligned}$$

10. $$\begin{aligned} a_n &= \frac{n-1}{n^2+1} \\ a_1 &= \frac{1-1}{1^2+1} = 0 \\ a_2 &= \frac{2-1}{2^2+1} = \frac{1}{5} \\ a_3 &= \frac{3-1}{3^2+1} = \frac{2}{10} = \frac{1}{5} \\ a_4 &= \frac{4-1}{4^2+1} = \frac{3}{17} \\ a_8 &= \frac{8-1}{8^2+1} = \frac{7}{65} \\ a_{12} &= \frac{12-1}{12^2+1} = \frac{11}{145} \end{aligned}$$

11. $-5, -10, -15, -20, \ldots$
These are negative multiples of 5 beginning with –5, so the general term could be $-5n$.

12. $-1, 3, -5, 7, -9, \ldots$
These are odd integers beginning with 1, but with alternating signs, beginning with a negative number.
The general terms could be $(-1)^n(2n-1)$.

13. $$\begin{aligned} \sum_{k=1}^{5}(-2)^k &= (-2)^1+(-2)^2+(-2)^3+(-2)^4+(-2)^5 \\ &= -2+4+(-8)+16+(-32) = -22 \end{aligned}$$

14. $$\begin{aligned} \sum_{k=2}^{7}(1-2k) &= (1-2\cdot 2)+(1-2\cdot 3)+(1-2\cdot 4)+(1-2\cdot 5) \\ &\quad +(1-2\cdot 6)+(1-2\cdot 7) \\ &= -3+(-5)+(-7)+(-9)+(-11)+(-13) \\ &= -48 \end{aligned}$$

15. $7+14+21+28+35+42$
This is the sum of the first six positive multiples of 7. It is an finite series. Sigma notation is
$$\sum_{k=1}^{6} 7k.$$

16. $\frac{-1}{2}+\frac{1}{4}+\frac{-1}{8}+\frac{1}{16}+\frac{-1}{32}$
This is a sum of fractions in which the numerator is 1, but with alternating signs, and the denominator is the first five powers of two. Sigma notation is
$$\sum_{k=1}^{5}\frac{(-1)^k}{2^k} \text{ or } \sum_{k=1}^{5}\frac{1}{(-2)^k}.$$

17. $-3, -7, -11, \ldots$
$$\begin{aligned} &a_1 = -3,\ d = -4, \text{ and } n = 14 \\ &a_n = a_1 + (n-1)d \\ &a_{14} = -3+(14-1)(-4) = -3+13(-4) = -55 \end{aligned}$$

18. $$\begin{aligned} a_n &= a_1 + (n-1)d \\ 14 &= 11+(16-1)d \quad \text{Substituting 14 for } a_{16}, \text{ 16 for } n, \text{ and 11 for } a_1 \\ 14 &= 11+15d \\ 3 &= 15d \\ \frac{1}{5} &= d \end{aligned}$$

19. We know that $a_8 = 20$ and $a_{24} = 100$. We would have to add d sixteen times to get from a_8 to a_{24}. That is,
$$\begin{aligned} 20+16d &= 100 \\ 16d &= 80 \\ d &= 5. \end{aligned}$$
Since $a_8 = 20$, we subtract d seven times to get a_1.
$$a_1 = 20-7(5) = 20-35 = -15$$

20. Before using the formula for S_n, we find a_{17}:
$$\begin{aligned} a_{17} &= -8+(17-1)(-3) \quad \text{Substituting into the formula for } a_n \\ &= -8+16(-3) \\ &= -56 \end{aligned}$$
Then using the formula for S_n,
$$S_{17} = \frac{17}{2}(-8+-56) = \frac{17}{2}(-64) = -544.$$

21. The sum is $5+10+15+\ldots+495+500$. This is the sum of the arithmetic sequence for which $a_1 = 5$, $a_n = 500$, and $n = 100$. We use the formula for S_n.
$$\begin{aligned} S_n &= \frac{n}{2}(a_1+a_n) \\ S_{100} &= \frac{100}{2}(5+500) = 50(505) = 25{,}250 \end{aligned}$$

22. $2,\ 2\sqrt{2},\ 4,\ldots$

$a_1 = 2,\ n = 20$, and $r = \frac{2\sqrt{2}}{2} = \sqrt{2}$

$a_n = a_1 r^{n-1}$

$a_{20} = 2(\sqrt{2})^{20-1} = 2(\sqrt{2})^{19} = 2(512\sqrt{2}) = 1024\sqrt{2}$

23. $r = \frac{30}{40} = \frac{3}{4}$

24. $-2,\ 2,\ -2,\ldots$

$a_1 = -2$, and $r = \frac{2}{-2} = -1$

$a_n = a_1 r^{n-1}$

$a_n = -2(-1)^{n-1} = 2(-1)^n$

25. $3,\ \frac{3}{4}x,\ \frac{3}{16}x^2,\ldots$

$a_1 = 3$, and $r = \frac{\frac{3}{4}x}{3} = \frac{3x}{4}\cdot\frac{1}{3} = \frac{x}{4}$

$a_n = a_1 r^{n-1}$

$a_n = 3\left(\frac{x}{4}\right)^{n-1}$

26. $3 + 15 + 75 + \ldots$

$a_1 = 3,\ n = 6$, and $r = \frac{15}{3} = 5$

$S_n = \frac{a_1(1-r^n)}{1-r}$

$S_6 = \frac{3(1-5^6)}{1-5} = \frac{3(1-15{,}625)}{-4} = \frac{3(-15{,}624)}{-4} = 11{,}718$

27. $3x - 6x + 12x - \ldots$

$a_1 = 3x,\ n = 12$, and $r = \frac{-6x}{3x} = -2$

$S_n = \frac{a_1(1-r^n)}{1-r}$

$S_{12} = \frac{3x(1-(-2)^{12})}{1-(-2)} = \frac{3x(1-4096)}{3} = x(-4095)$
$= -4095x$

28. $6 + 3 + 1.5 + 0.75 + \ldots$

$|r| = \left|\frac{3}{6}\right| = \left|\frac{1}{2}\right| = \frac{1}{2} < 1$, the series has a limit.

$S_\infty = \frac{a_1}{1-r} = \frac{6}{1-\frac{1}{2}} = \frac{6}{\frac{1}{2}} = 6\cdot\frac{2}{1} = 12$

29. $7 - 4 + \frac{16}{7} - \ldots$

$|r| = \left|\frac{-4}{7}\right| = \frac{4}{7} < 1$, the series has a limit.

$S_\infty = \frac{a_1}{1-r} = \frac{7}{1-\left(-\frac{4}{7}\right)} = \frac{7}{\frac{11}{7}} = 7\cdot\frac{7}{11} = \frac{49}{11}$

30. $-\frac{1}{2} + \frac{1}{2} + \left(-\frac{1}{2}\right) + \frac{1}{2} + \ldots$

$|r| = \left|\frac{\frac{1}{2}}{-\frac{1}{2}}\right| = |-1| = 1$, and since $|r| \not< 1$, the series does not have a limit.

31. $0.04 + 0.08 + 0.16 + 0.32 + \ldots$

$|r| = \left|\frac{0.08}{0.04}\right| = |2| = 2$, and since $|r| \not< 1$, the series does not have a limit.

32. $\$2000 + \$1900 + \$1805 + \$1714.75\ldots$

$|r| = \left|\frac{\$1900}{\$2000}\right| = \left|\frac{19}{20}\right| = \frac{19}{20} < 1$, the series has a limit.

$S_\infty = \frac{a_1}{1-r} = \frac{\$2000}{1-\frac{19}{20}} = \frac{\$2000}{\frac{1}{20}} = \frac{\$2000}{1}\cdot\frac{20}{1} = \$40{,}000$

33. $0.5555\ldots = 0.5 + 0.05 + 0.005 + 0.0005 + \ldots$
This is an infinite geometric series with $a_1 = 0.5$.

$|r| = \left|\frac{0.05}{0.5}\right| = |0.1| = 0.1 < 1$, so the series has a limit.

$S_\infty = \frac{a_1}{1-r} = \frac{0.5}{1-0.1} = \frac{0.5}{0.9} = \frac{5}{9}$

Fractional notation for $0.5555\ldots$ is $\frac{5}{9}$.

34. $1.454545\ldots = 1 + 0.45 + 0.0045 + 0.000045 + \ldots$

$|r| = \left|\frac{0.0045}{0.45}\right| = |0.01| = 0.01 < 1$, so the series has a limit.

$S_\infty = \frac{a_1}{1-r} = \frac{0.45}{1-0.01} = \frac{0.45}{0.99} = \frac{45}{99} = \frac{5}{11}$

Fractional notation for $1.4545\ldots$ is $1 + \frac{5}{11} = \frac{16}{11}$.

35. ***Familiarize***. A \$0.40 raise every 3 months for 8 years, is 32 raises. We want to find the 33th term of an arithmetic sequence with $a_1 = \$11.50$, $d = \$0.40$, and $n = 33$. We will use the formula for a_n to find a_{33}. This result is the hourly wage after 8 years.
Translate. Substituting into the formula for a_n, we have

$$a_{33} = \$11.50 + (33-1)(\$0.40).$$

Carry out. We find a_{33}.

$$a_{33} = \$11.50 + 32(\$0.40) = \$24.30$$

Check. We can do the calculations again.
State. After 8 years, Jaykob earns \$24.30 an hour.

36. We go from 42 poles in the bottom layer to one pole in the top layer, where each layer has 1 pole less than the layer below it, so there must be 42 layers.
Thus, we want to find the sum of the first 42 terms of an arithmetic sequence with $a_1 = 42$ and $a_{42} = 1$.

$$S_{42} = \frac{42}{2}(42+1) = 21(43) = 903 \text{ poles}$$

37. ***Familiarize***. At the end of each year, the interest at 4% will be added to the previous year's amount.
We have a geometric sequence:

$\$12,000,\ \$12,000(1.04),\ \$12,000(1.04)^2,...$

We are looking for the amount after 7 years.

Translate. We use the formula $a_n = a_1 r^{n-1}$ with $a_1 = \$12,000,\ r = 1.04$, and $n = 8$:

$$a_8 = \$12,000(1.04)^{8-1}$$

Carry out. We calculate to obtain $a_8 \approx \$15,791.18$.
Check. We can do the calculation again.
State. After 7 years, the amount of the loan will be about \$15,791.18.

38. $S_\infty = \dfrac{12}{1-\frac{1}{3}} = \dfrac{12}{\frac{2}{3}} = \dfrac{12}{1}\cdot\dfrac{3}{2} = 18$

The total distance is 18 m, the fall distance is 12 m, so the rebound distance is 18 – 12, or 6 m.
The total rebound distance is 6 m.

39. $7! = 7\cdot6\cdot5\cdot4\cdot3\cdot2\cdot1 = 5040$

40. $\dbinom{10}{3} = \dfrac{10!}{7!3!} = \dfrac{10\cdot9\cdot8\cdot7!}{7!3!} = \dfrac{10\cdot9\cdot8}{3\cdot2\cdot1} = 10\cdot3\cdot4 = 120$

41. $\dbinom{20}{2}a^{20-2}b^2 = 190a^{18}b^2$

42. Expand $(x-2y)^4$.
We have $a = x$, $b = -2y$, and $n = 4$.
Form 1: We use the fifth row of Pascal's triangle:
1 4 6 4 1

$$(x-2y)^4$$
$$=1\cdot x^4 + 4x^3(-2y) + 6x^2(-2y)^2 + 4x(-2y)^3 + 1\cdot(-2y)^4$$
$$= x^4 - 8x^3y + 24x^2y^2 - 32xy^3 + 16y^4$$

Form 2:

$$(x-2y)^4 = \binom{4}{0}x^4 + \binom{4}{1}x^3(-2y) + \binom{4}{2}x^2(-2y)^2 + \binom{4}{3}x(-2y)^3 + \binom{4}{4}(-2y)^4$$
$$= \frac{4!}{4!0!}x^4 + \frac{4!}{3!1!}x^3(-2y) + \frac{4!}{2!2!}x^2(-2y)^2 + \frac{4!}{1!3!}x(-2y)^3 + \frac{4!}{0!4!}(-2y)^4$$
$$= x^4 - 8x^3y + 24x^2y^2 - 32xy^3 + 16y^4$$

43. *Writing Exercise.* For a geometric sequence with $|r| < 1$, as *n* gets larger, the absolute value of the terms gets smaller, since $|r^n|$ gets smaller.

44. *Writing Exercise.* The first form of the binomial theorem draws the coefficients from Pascal's triangle; the second form uses factorial notation. The second form avoids the need to compute all preceding rows of Pascal's triangle, and is generally easier to use when only one term of an expression is needed. When several terms of an expansion are needed and *n* is not large (say, $n \le 8$), it is often easier to use Pascal's triangle.

45. $1 - x + x^2 - x^3 + ...$

$a_1 = 1,\ n = n$, and $r = \dfrac{-x}{1} = -x$

$$S_n = \frac{1(1-(-x)^n)}{1-(-x)} = \frac{1-(-x)^n}{1+x}$$

46. Expand $(x^{-3} + x^3)^5$.
We have $a = x^{-3}$, $b = x^3$, and $n = 5$.
Form 1: We use the 6th row of Pascal's triangle:
1 5 10 10 5 1

$$(x^{-3}+x^3)^5 = 1\cdot(x^{-3})^5 + 5(x^{-3})^4(x^3)^1 + 10(x^{-3})^3(x^3)^2 + 10(x^{-3})^2(x^3)^3 + 5(x^{-3})^1(x^3)^4 + 1\cdot(x^3)^5$$
$$= x^{-15} + 5x^{-9} + 10x^{-3} + 10x^3 + 5x^9 + x^{15}$$

Form 2:

$$(x^{-3}+x^3)^5 = \binom{5}{0}(x^{-3})^5 + \binom{5}{1}(x^{-3})^4(x^3) + \binom{5}{2}(x^{-3})^3(x^3)^2 + \binom{5}{3}(x^{-3})^2(x^3)^3 + \binom{5}{4}(x^{-3})(x^3)^4 + \binom{5}{5}(x^3)^5$$
$$= \frac{5!}{5!0!}(x^{-3})^5 + \frac{5!}{4!1!}(x^{-3})^4(x^3) + \frac{5!}{3!2!}(x^{-3})^3(x^3)^2 + \frac{5!}{2!3!}(x^{-3})^2(x^3)^3 + \frac{5!}{1!4!}(x^{-3})(x^3)^4 + \frac{5!}{0!5!}(x^3)^5$$
$$= x^{-15} + 5x^{-9} + 10x^{-3} + 10x^3 + 5x^9 + x^{15}$$

Chapter 14 Test

1. $a_n = \dfrac{1}{n^2+1}$

$a_1 = \dfrac{1}{1^2+1} = \dfrac{1}{2}$

$a_2 = \dfrac{1}{2^2+1} = \dfrac{1}{5}$

$a_3 = \dfrac{1}{3^2+1} = \dfrac{1}{10}$

$a_4 = \dfrac{1}{4^2+1} = \dfrac{1}{17}$

$a_5 = \dfrac{1}{5^2+1} = \dfrac{1}{26}$

$a_{12} = \dfrac{1}{12^2+1} = \dfrac{1}{145}$

2. $\dfrac{4}{3}, \dfrac{4}{9}, \dfrac{4}{27},...$

These are positive fractions in which the numerator is 4 and the denominator is a power of 3. The general terms could be $4\left(\dfrac{1}{3}\right)^n$.

3. $\displaystyle\sum_{k=2}^{5}(1-2^k) = (1-2^2)+(1-2^3)+(1-2^4)+(1-2^5)$

$$= -3 + (-7) + (-15) + (-31) = -56$$

4. $1+(-8)+27+(-64)+125$
This is the sum of the cubes of the first five natural numbers, but with alternating signs. Sigma notation is
$$\sum_{k=1}^{5}(-1)^{k+1}k^3.$$

5. $\frac{1}{2}, 1, \frac{3}{2}, 2, \ldots$
$a_1=\frac{1}{2}, d=\frac{1}{2}$, and $n=13$
$a_n=a_1+(n-1)d$
$a_{13}=\frac{1}{2}+(13-1)\left(\frac{1}{2}\right)=\frac{1}{2}+12\left(\frac{1}{2}\right)=\frac{13}{2}$

6. We know that $a_5=16$ and $a_{10}=-3$. We would have to add d five times to get from a_5 to a_{10}. That is,
$16+5d=-3$
$5d=-19$
$d=-3.8.$
Since $a_5=16$, we subtract d four times to get a_1.
$a_1=16-4(-3.8)=16+15.2=31.2$

7. The sum is $24+36+48+\ldots+228+240$. This is the sum of the arithmetic sequence for which $a_1=24$, $a_n=240$, and $n=19$. We use the formula for S_n.
$S_n=\frac{n}{2}(a_1+a_n)$
$S_{19}=\frac{19}{2}(24+240)=\frac{19}{2}(264)=2508$

8. $-3, 6, -12,\ldots$
$a_1=-3$, $n=10$, and $r=\frac{6}{-3}=-2$
$a_n=a_1r^{n-1}$
$a_{10}=-3(-2)^{10-1}=-3(-2)^9=-3(-512)=1536$

9. $r=\dfrac{15}{22\frac{1}{2}}=\frac{15}{1}\cdot\frac{2}{45}=\frac{2}{3}$

10. $3, 9, 27,\ldots$
$a_1=3$, and $r=\frac{9}{3}=3$
$a_n=a_1r^{n-1}$
$a_n=3(3)^{n-1}=3^n$

11. $11+22+44+\ldots$
$a_1=11$, $n=9$, and $r=\frac{22}{11}=2$
$S_n=\dfrac{a_1(1-r^n)}{1-r}$
$S_9=\dfrac{11(1-2^9)}{1-2}=\dfrac{11(1-512)}{-1}=\dfrac{11(-511)}{-1}=5621$

12. $0.5+0.25+0.125+\ldots$
$|r|=\left|\frac{0.25}{0.5}\right|=|0.5|=0.5<1$, the series has a limit.
$S_\infty=\dfrac{a_1}{1-r}=\dfrac{0.5}{1-0.5}=\dfrac{0.5}{0.5}=1$

13. $0.5+1+2+4+\ldots$
$|r|=\left|\frac{1}{0.5}\right|=|2|=2$, and since $|r|\not<1$, the series does not have a limit.

14. $\$1000+\$80+\$6.40+\ldots$
$|r|=\left|\frac{\$80}{\$1000}\right|=|0.08|=0.08<1$, the series has a limit.
$S_\infty=\dfrac{a_1}{1-r}=\dfrac{\$1000}{1-0.08}=\dfrac{\$1000}{0.92}=\dfrac{\$25{,}000}{23}\approx\$1086.96$

15. $0.85858585\ldots=0.85+0.0085+0.000085+\ldots$
This is an infinite geometric series with $a_1=0.85$.
$|r|=\left|\frac{0.0085}{0.85}\right|=|0.01|=0.01<1$, so the series has a limit.
$S_\infty=\dfrac{a_1}{1-r}=\dfrac{0.85}{1-0.01}=\dfrac{0.85}{0.99}=\dfrac{85}{99}$
Fractional notation for $0.85858585\ldots$ is $\frac{85}{99}$.

16. ***Familiarize***. We want to find the seventeenth term of an arithmetic sequence with $a_1=31$, $d=2$, and $n=17$. We will use the formula for a_n to find a_{17}. This result is the number of seats in the 17th row.
Translate. Substituting into the formula for a_n, we have
$a_{17}=31+(17-1)2.$
Carry out. We find a_{17}.
$a_{17}=31+16\cdot 2=63$
Check. We can do the calculations again.
State. There are 63 seats in the 17th row.

17. The sum is $\$100+\$200+\$300+\ldots+\$1800$. This is the sum of the arithmetic sequence for which $a_1=\$100$, $a_n=\$1800$, and $n=18$. We use the formula for S_n.
$S_n=\frac{n}{2}(a_1+a_n)$
$S_{18}=\frac{18}{2}(\$100+\$1800)=9(\$1900)=\$17{,}100$
Her uncle gave her a total of $17,100.

18. ***Familiarize***. At the end of each week, the price will be 95% of the previous week's price.
We have a geometric sequence:

$\$10{,}000,\ \$10{,}000(0.95),\ \$10{,}000(0.95)^2,\ldots$

We are looking for the price after 10 weeks.

Translate. We use the formula $a_n = a_1 r^{n-1}$ with $a_1 = \$10{,}000,\ r = 0.95$, and $n = 11$:

$$a_{10} = \$10{,}000(0.95)^{11-1}$$

Carry out. We calculate to obtain $a_{11} \approx \$5987.37$.

Check. We can do the calculation again.

State. After 10 weeks, the price of the boat will be about \$5987.37.

19. $S_\infty = \dfrac{18}{1-\frac{2}{3}} = \dfrac{18}{\frac{1}{3}} = \dfrac{18}{1}\cdot\dfrac{3}{1} = 54$

The total distance is 54 m, the fall distance is 18 m, so the rebound distance is 54 – 18, or 36 m.
The total rebound distance is 36 m.

20. $\dbinom{12}{9} = \dfrac{12!}{3!9!} = \dfrac{12\cdot 11\cdot 10\cdot 9!}{3!9!} = \dfrac{12\cdot 11\cdot 10}{3\cdot 2\cdot 1} = 4\cdot 11\cdot 5 = 220$

21. Expand $(x-3y)^5$.

We have $a = x,\ b = -3y,$ and $n = 5$.

Form 1: We use the 6th row of Pascal's triangle:

1 5 10 10 5 1

$$\begin{aligned}
&(x-3y)^5\\
&= 1\cdot x^5 + 5x^4(-3y)^1 + 10x^3(-3y)^2\\
&\quad + 10x^2(-3y)^3 + 5x^1(-3y)^4 + 1\cdot(-3y)^5\\
&= x^5 - 15x^4y + 90x^3y^2 - 270x^2y^3 + 405xy^4 - 243y^5
\end{aligned}$$

Form 2:

$$\begin{aligned}
&(x-3y)^5\\
&= \binom{5}{0}x^5 + \binom{5}{1}x^4(-3y) + \binom{5}{2}x^3(-3y)^2\\
&\quad + \binom{5}{3}x^2(-3y)^3 + \binom{5}{4}x(-3y)^4 + \binom{5}{5}(-3y)^5\\
&= \frac{5!}{5!0!}x^5 + \frac{5!}{4!1!}x^4(-3y) + \frac{5!}{3!2!}x^3(9y^2)\\
&\quad + \frac{5!}{2!3!}x^2(-27y^3) + \frac{5!}{1!4!}x(81y^4) + \frac{5!}{0!5!}(-243y^5)\\
&= x^5 - 15x^4y + 90x^3y^2 - 270x^2y^3 + 405xy^4 - 243y^5
\end{aligned}$$

22. $\dbinom{12}{3}a^{12-3}x^3 = 220a^9x^3$

23. The sum is $2+4+6+\ldots+2n$. This is the sum of the arithmetic sequence for which $a_1 = 2,\ a_n = 2n,$ and $n = n$. We use the formula for S_n.

$$S_n = \frac{n}{2}(a_1 + a_n)$$

$$S_n = \frac{n}{2}(2+2n) = \frac{n}{2}[2(1+n)] = n(n+1)$$

24. $1 + \dfrac{1}{x} + \dfrac{1}{x^2} + \dfrac{1}{x^3} + \ldots$

$a_1 = 1,\ n = n,$ and $r = \dfrac{\frac{1}{x}}{1} = \dfrac{1}{x}$

$$S_n = \frac{1\left(1-\left(\frac{1}{x}\right)^n\right)}{1-\frac{1}{x}} = \frac{1-\left(\frac{1}{x}\right)^n}{1-\frac{1}{x}},\ \text{or}\ \frac{x^n-1}{x^{n-1}(x-1)}$$